AGRICULTURAL ENGINEERING
VOLUME 1

C I
G R

PROCEEDINGS OF THE ELEVENTH INTERNATIONAL CONGRESS ON AGRICULTURAL ENGINEERING / DUBLIN / 4-8 SEPTEMBER 1989

Agricultural Engineering

Edited by

VINCENT A.DODD & PATRICK M.GRACE

Agricultural and Food Engineering Department, University College Dublin

VOLUME 1

Land and water use

A.A.BALKEMA / ROTTERDAM / BROOKFIELD / 1989

The texts of the various papers in this volume were set individually by typists under the supervision of each of the authors concerned.

Published by

A.A.Balkema, P.O.Box 1675, 3000 BR Rotterdam, Netherlands

A.A.Balkema Publishers, Old Post Road, Brookfield, VT 05036, USA

For the complete set of four volumes ISBN 90 6191 980 0

For volume 1: ISBN 90 6191 976 2

For volume 2: ISBN 90 6191 977 0

For volume 3: ISBN 90 6191 978 9

For volume 4: ISBN 90 6191 979 7

Printed in the Netherlands

Land and Water Use, Dodd & Grace (eds), © 1989 Balkema, Rotterdam. ISBN 90 6191 980 0

Contents

1.3 *Environmental aspects of soil water*

1.4 *Soil/water engineering*

Land and Water Use, Dodd & Grace (eds), © 1989 Balkema, Rotterdam. ISBN 90 6191 980 0

Preface

This book and its companion volumes contain the papers presented at the Eleventh International Congress on Agricultural Engineering held in Dublin, Ireland, in September 1989.

The congress was organised under the auspices of the International Commission of Agricultural Engineering (CIGR); CIGR is the co-ordinating body for agricultural engineering in over thirty countries and co-ordinates the international congress every five years. The Eleventh Congress was organised by the Irish National Association of Agricultural Engineers, Institution of Engineers of Ireland.

The book presents a broad coverage of basic and applied research projects dealing with the application of engineering principles to both food production and processing and illustrates the considerable contribution made by agricultural engineers in the development of one of man's oldest industries, agriculture. Approximately 450 papers from over 50 countries world wide, were presented at the congress and are included in these proceedings.

The papers have been classified into a series of four volumes as follows:

Volume 1 – Land and water use: Land use of peatlands, including peat production, drainage and hydrology; traffic and transportation systems in agriculture including the design, construction and maintenance of rural and farm roads, on and off-farm transportation systems; environmental aspects of land use including the land spreading of animal manures and sewage waters, the processing of animal manures; drainage, irrigation, soil erosion and water management.

Volume 2 – Agricultural buildings: Environmental control in buildings for animal production including heating, ventilation, dust and other contaminants, odour emissions and control; Current design practice for the layout of buildings for animal and crop production and storage; the design of farm buildings in relation to the landscape; the use of new materials and construction methods; low cost farm buildings including greenhouses; computerised controlled feeding and milking of livestock and their influence on the layout of livestock buildings and the consequences for the management and welfare of livestock.

Volume 3 – Agricultural mechanisation: Optimising soil cultivation and seed bed preparation; problems with soil compaction, soil management systems; separation, sorting and grading during harvesting and handling of agricultural products at farm level including grain, potatoes, fruits; physical properties; application and control of distribution of plant nutrients and pesticides; application techniques, quality of the spray, sprayers and spreaders.

Volume 4 – Power, processing and systems: Controlling energy operated processes in agriculture with emphasis on post harvest technology including grain drying; fundamental and applied aspects of processing agricultural products including rice, coffee, carrots and milk; recent developments in tractors and fuels. Management and decision making systems for agricultural enterprises including farm machine costs and ergonomic standards in agricultural machinery. The final section of this volume is devoted to papers dealing with the application of new techniques in education, research and development.

Vincent A.Dodd

Patrick M.Grace

1 Land and water use

1.1 Use of peat land

Land and Water Use, Dodd & Grace (eds), © 1989 Balkema, Rotterdam. ISBN 90 6191 980 0

Keynote paper: A strategic analysis of milled peat production

R.Kavanagh
Bord na Mona, Newbridge, Co. Kildare, Ireland

ABSTRACT: The question of the long-term viability of Bord na Mona arising from market, price and debt considerations led to the undertaking of a strategic review of milled peat production in the Company. The various parameters impacting on the review are described and the analytic approach which used computer simulation and mathematical techniques, is outlined. The results on the analysis indicate the boundary conditions for milled peat production technology on both the short- and long-term horizons.

1. INTRODUCTION

The three major fuel peat producing countries in the world are Finland, Ireland and the USSR. The production capacities of Finland and Ireland are similar (about 6 million tonnes of peat annually) while the capacity of the USSR is about 200 million tonnes annually. Two parameters, both of which are determined by events outside its control, have a major influence on the industry - world energy prices and weather conditions. As a result the industry cost structure must be responsive to weather conditions and to market price levels. The very nature of peat itself is such that climatic conditions in peat-producing countries tend to be poor in terms of annual rainfall. Because almost all peat produced today involves substantial air-drying the industry must ensure, through the technology and systems it uses, the magnitude of the area under production, as well as stocking policies, that it can isolate demand for its products from the effects of extreme variations in weather conditions. In addition it must be responsive to market parameters.

2. BORD NA MONA

Bord na Mona is a state-owned organization created by the Irish Government in 1946 to develop the peat resources of Ireland. In the last financial year (1987/88) turnover was IR£116 million. Sales of milled peat for electricity production amounted to 3.3 million tonnes valued at IR£62 million, while sales of peat briquettes and sod peat used for heating in residences amounted to 0.6 million tonnes, valued at IR£32 million. Sales of moss peat (1.1 million cubic meters), mainly for export markets, were valued at £22 million. Average employment, mostly located in the relatively underdeveloped Midlands region of the country, was 4800 for the year, rising to a peak of 6000 people during the summer production season.

3. STATEMENT OF THE PROBLEM

Like many industries both in Ireland and elsewhere the peat industry was slow to react to the enormous decline in energy prices in the early 1980's and has been caught in the classic cost-price squeeze in which prices fall but unit costs are excessively high thereby limiting the scope for the price reductions demanded by the market place.

In Bord na Mona labour, a large part of it fixed and independent of production and sales values, account for a very substantial part of unit costs (60% of turnover). Also the capital structure of the company is such that net assets (IR£196m) are funded by loans (IR£165m) and from reserves (IR£31m). This puts severe strains on the profitability of the company especially in bad years. Debt servicing amounts to £20m annually.

Thus at a time of declining energy prices its high cost structure has severely limited Bord na Mona's room for manoevre.

The situation has been worsened by the low levels of production in 1985 and 1986, arising from poor weather conditions. As a result reserve stocks have been severely depleted.

The bad years in 1985 and 1986 have also highlighted that high-cost labour on the permanent pay-roll is inconsistent with the weather dependent nature of the peat industry. It was clear that a cost-base which fluctuated in line with annual production was more appropriate, especially at a time of falling prices.

In the last number of years a database has been developed by Bord na Mona to form the basis of a strategic analysis which would address these fundamental issues facing the industry and generate options which would assure the future viability of the industry in Ireland. In parallel a structural review of the company as well as a technical development programme were undertaken so that the structures, systems and the technology would be consistent with the results of the strategic review.

4. MILLED PEAT PRODUCTION SYSTEMS

Some understanding of the technology of milled peat production is required if the process and output of the strategic analysis are to be understood. The air-drying of a wet milled peat layer accounts for all milled peat produced in the world today. Extensive research and development throughout the world has not, to date, produced a more cost-effective alternative.

The end product is milled peat at a moisture content of 50-55% H_2 (wb) which is used for generating electricity or for producing peat briquettes (12% H_2O) which are used for heating purposes in houses, buildings and industry. Two types of production systems are used to produce milled peat, referred to as Peco and Haku systems. Both systems are widely used in the USSR; Haku is the main system in Finland and the Peco system is, at present, the only system used in Ireland for milled peat production.

4.1 Peco production system

Figure 1 shows the typical bog lay-out for a Peco production system. The main distinctive features to note are:

- Long production fields (up to 4km long):

- Stockpile fields with 5 production fields on either side:

- Temporary rail-line on stockpile field:

- Main-line rail around the bog area.

4.2 Haku production system

Figure 2 shows a typical lay-out for a Haku production system. The distinctive features to note are:

- Short production fields (below 1km long):

- Large central stockpiles:

- Rail access.

5. THE PRODUCTION PROCESS

Virgin bog consists mainly of water (98% water, 2% solids).After drainage and when ready for production the water content will have fallen to 85-90% H_2O. The production of milled peat, whether using the Peco or the Haku system, involves drainage and development of the bog from the virgin state. This operation can take up to 7 years to complete. During this stage the drain lay-out, the field-lengths and stockpile locations are determined to suit either the Peco or the Haku system.

The production cycle starts in May-June each year depending on weather and bog conditions. The first three steps of the operation are common to both Haku and Peco systems. These include milling a layer (10mm) of raw peat using a tractor-drawn miller (Figure 3). To facilitate air-drying the layer of peat is subsequently turned at various intervals using a tractor-drawn harrow unit (Figure 4). The drying cycle depends on weather conditions and varies from 1 day in excellent drying conditions during an anti-cyclonic spell to 5-6 days in poor drying conditions. In normal weather conditions in Ireland 12-15 such cycles will be achieved per annum.

When the optimum moisture content (45% H_2O) is achieved the peat is ridged using a "double ridger" unit (Figure 5) and is then harvested. This final operation is undertaken at 45% H_2O because some re-wetting will take place over the period until final sale which may be many months later.

The harvesting operation differs for the Peco and Haku systems. In the former case self-propelled (Figure 6) harvesting units pick up the ridge and place it on top of the ridge in the adjacent field (see Figure 1). This double ridge is then harvested onto the third field. This process continues until peat on the five fields is finally placed on the stockpile field. The stockpiles are loaded for sale into railway wagons which travel on a temporary track laid beside the stockpile.

In the Haku system harvesters load the peat into tractor-drawn trailers (Figure 7) on the adjacent field. The trailer units then proceed to the stockpile area (Figure 2), empty their load and return to the harvesting unit. The stockpiles are loaded directly into trucks for transport to point of sale, as required.

6 RESEARCH AND DEVELOPMENT CONSIDERATIONS

In recent years Bord na Mona has undertaken extensive research and development activities to establish a data-base for strategic as well as operational analysis of the milled peat production system. In addition a series of machine and field tests have been undertaken to establish the boundary conditions for technical development, given the available level of technology. The data-base enabled the development of simulation and mathematical models which could be used as a basis for analysing the impact of change on operational and technical parameters. On the technical side efforts were focussed on the milling/ harvesting cycle in terms of optimizing the efficiency of drying. In addition a variant of the Haku system, which involved combining elements of both Peco and Haku, was tested (Ring, 1988) in 1988 at Oweninny Works in Co. Mayo. The variant was based on discussions with experts in Finland and on work undertaken in the USSR Alekseyev 1986, Klinov 1987, Kuznetsov 1987 and 1988, Malkov 1987 and 1988 and Sibiryev 1988).

7. STRATEGIC ANALYSIS

Based on the considerations discussed in previous sections the strategic analysis focussed on three areas:

- A weather risk assessment:
- The impact of technology and systems:
- The impact of alternative operating regimes.

7.1 Risk assessment

The objective of the risk assessment (Murphy et al. 1988 and 1989) was to analyse past weather, production, stock and sales patterns and to produce a risk-profile for the next five years given sales forecasts, specific production areas and the Peco production system. The model consisted of a random number generator based on past weather conditions. Many iterations were undertaken to eliminate year-to-year fluctuations. This was then used to estimate production, stocks and sales levels and to provide a stock-out risk probability function and the size, if any, of the stock-out (i.e. a mismatch between supply and demand) in each year.

Having generated a risk profile for the Peco system the process was repeated with the Haku system variant to be introduced on a gradual basis, and accounting for 100% of production in 1993. Typical results for the Peco analysis are shown in Table 1.

Thus for example, in 1989 there is a 10% chance that Bord na Mona will be unable to meet the projected demand. The figures in the right hand columns of the Table show the probabilities of different sizes of shortages, e.g. in 1989 there is a 5.6% chance of a shortage of between 1.0 and 1.5 million tonnes.

Table 1. Risk Assessment - Peco System

Year	Over-all risk*	Stockout (tonnes x 10^6)			
		0.5	0.5-1.0	1.0-1.5	1.5
		Individual Risk			
	%	%	%	%	%
1989	10	4.7	0.1	5.2	0
1990	16	7.5	3.6	4.9	0
1991	14	6.9	3.1	3.2	0.8
1992	13	6.9	3.3	1.9	0.9
1993	11	5.7	2.5	1.7	1.1

* Of not meeting demand in that year.

The Table shows that there is an unacceptably high risk level of a major stockout with the Peco system over the next 5 years. With the introduction of the Haku variant (Table 2) the level of risk as well as the magnitude of the expected stock-out declines to acceptable levels over the longer term. The main risk is for 1989 which arises because the Haku variant will account for only 10% of total production and the benefits of Haku are too small to significantly impact on the overall picture. In addition stock-levels going into 1989 are at a historically very low level. The impact of introducing additional area into production was investigated but its effect was only marginal and the investment was not considered to be viable.

Table 2. Risk assessment - Haku system

Year	Over-all risk*	Stockout (tonnes x 10^6) 0.5	0.5-1.0	1.0-1.5	1.5
		Individual Risk			
	%	%	%	%	%
1989	5	0	5.0	0	0
1990	5	1.9	3.1	0	0
1991	2	1.2	0	0.8	0
1992	1	0.6	0.4	0	0
1993	0	0	0	0	0

* Of not meeting demand in that year.

The main conclusion drawn from the analysis was that full implementation of the Haku variant over a 5-year period was the only cost-effective means, given the available technology, of reducing the stock-out risk to acceptable levels. This would enable Bord na Mona to break the direct link between production and sales, especially in poor weather conditions.

A separate analysis, undertaken on the basis of the Oweninny tests with Haku in 1988 and already referred to, confirmed the results of the risk assessment.

7.2 The impact of technology

A milled peat production simulation model (SIMPEAT) has been developed (Kavanagh et al, 1985 and 1988) for use in analysing the impact of change in Bord na Mona's production systems and operations. The model was also used to determine the boundary conditions set by the available technology. The overall system impact of their introduction on a widespread basis was calculated using SIMPEAT. It was found that in the medium-term (up to 5 years) boundary conditions were set by the following technical parameters:

- Introduce the Haku variant as a substitute for Peco:

- Develop reliable and simple machinery in terms of low break-down frequencies and durations:

- Improve cycle efficiency.

The latter refers to the requirement, given weather and bog parameters, to mill the optimal quantity of raw milled peat and to ensure maximum efficiency for other activities in the production cycle. Thus the existing focus on machine speeds and hourly throughput would change to a greater emphasis on the efficiency of the drying cycle itself and each element in it. In the longer term (over 5 years) the boundary conditions are set solely by the cycle efficiency. Considering both the short- and long-term perspectives together provides an indicator of the potential of technology change.

7.3 The impact of alternative operating regimes

Reference has already been made to the necessity in Bord na Mona's operations to ensure that costs are related to production and weather parameters. At present the major cost element is labour and is incurred independently of weather conditions during production. Thus in a good year unit costs are low while in a poor year the cost per tonne reaches very high levels. The only viable solution to this dilemma would appear to be the use of separate and independent enterprises to undertake production. As a result Bord na Mona would release many of its employees who would then provide production services on an enterprise basis and share the risks and benefits of peat production.

One of the significant features of this new system of operation is the high level of motivation and incentive to make maximum use of the drying cycle. SIMPEAT was used to assess the impact of this change in conjunction with the introduction of the Haku variant. It was found that a further significant improvement in productivity could be achieved.

7.4 The impact of change

The impact of technical change and the new enterprise-based operating regimes, as discussed in Sections 7.2 and 7.3, on productivity is shown in Table 3. It is apparent that immediate and significant gains can be made by the introduction of technical change, in the form of the Haku variant. This arises from the increased production area (Peco stockpile fields represent 10% of total Peco area) and greater harvesting speeds (114 vs 100 for Peco). Greater cycle efficiency in terms of increased harvester pick-up efficiency and lower stockpile losses increase output to 127 compared to normal Peco. The combined impact of Haku would be to increase output from 100 (normal Peco) to 145.

Table 3. Impact of technical and operational changes

	Index*
Peco - Normal operating conditions	100
Haku - Normal operating conditions	114
- Up-tonnage alone	127
- Enterprise system alone	125
- Additional technology alone	160
- Combined Impact	290

* Normalized to 100 - Peco normal operating conditions

In addition the introduction of an enterprise system would enhance output from 100 to 125. The combined near-term impact would bring the output index to 181 (1.25 x 1.45). The impact of further technical development, viz. improvements in cycle efficiency and each of its component parts, will take time to achieve but could increase output from a base of 100 to 160, leading to a combined impact of all changes of 290 compared to the baseline of existing operational regimes and technology of 100.

A similar analysis for the Peco system showed that the output could be increased in the short-term but not at all to the extent shown for Haku. The main impact for Peco would arise from further long-term (over 5 years) technical development.

The results of the analyses are consistent with work (Kennedy and McNulty 1988) undertaken independently on optimal milling depth (i.e. milling depth optimized for weather and bog parameters) and the optimal use of drying conditions (optimal milling and cycle efficiency).

8. CONCLUSIONS

The main features to emerge from the strategic analysis of milled peat production were:

- The introduction of the Haku variant would significantly reduce the exposure of Bord na Mona to poor weather conditions, as compared to the Peco system.

- The introduction of the Haku variant would have a significant impact in terms of higher sales of peat per unit area.

- The introduction of an enterprise system, in conjunction with the Haku system, would further increase output.

- Further improvements can be gained by focussing research and development activities on the overall efficiency of the milling/ harvesting cycle and each of its component parts.

- The improvements indicated can be gained as increased output, lower unit costs or some combination of the two.

- The use of the SIMPEAT milled peat production simulation model as well as the risk assessment model have provided valuable insights into the strategic aspects of milled peat production, operations and research and development.

- The strategic analysis has highlighted how Bord na Mona can reduce its exposure to adverse weather conditions, how it can ensure costs can be linked to production and how technology can be implemented to ensure greater cost-effectiveness in its operations.

REFERENCES

Alekseyev, V.G. 1986. Structural reliability of the technological scheme using separate harvesting. Torfyanaya Promeshlennost 8, 14-16.

Kavanagh, R., Coffey, P.J. and Murphy, C. 1985. Towards a model of the milled peat production process. Procs. Peat and Environment '85, Jonkoping, Sweden. International Peat Society, September 1985.

Kavanagh, R., Barry, C. and Murphy, C. 1988. A simulation model of the milled peat production system. In VII International peat Congress, Leningrad, USSR.

Kennedy, D. 1988. The instrumented miller and depth control. In "The Futureworld of Peat Technology, Bord na Mona and Institution of Engineers of Ireland Conference, University College Dublin.

Klinov, V.S. 1987. The MTT -17 Peat Loader. Torfyanaya Promeshlennost 3, 1987, p8-9.

Kuznetsov, N.V. 1988. Peat stockpiling and storage in the context of separate harvesting. Malkov, L.M., Torfyanaya Promeshlennost 1, p6-8

Kuznetsov, N.V., Malkov, L.M. 1987 Ridging and milling according to the separate harvesting method of peat production. Torfyanaya Promeshlennost 6, p4-7

McNulty, M. 1988. Control of milled peat using modern weather forecasting techniques. In "The Futureworld of Peat Technology", Bord na Mona and Institution of Engineers of Ireland Conference, University College Dublin.

Malkov, L.M., Kuznetsov, N.V., Sheyde, V.P., Isaeva, L.S., Galkin, A.I. and Kuznetsov, S.P. 1987. Pilot-scale trials of the technological process of milled peat harvesting by the separate method. Torfyanaya Promeshlennost 4, 1987, p14-16.

Malkov, L.M. 1987. Loading peat from ridges by the separate harvesting method Torfyanaya Promeshlennost, 9, p9-10.

Malkov, L.M., Kuznetsov, N.V., Sheyde, V.P., Galkin, A.I., Kalinin, I.A., Yurov, V.M. 1987. The transport of milled peat harvested by separate collection. Torfyanaya Promeshlennost 12, p5-6

Malkov, L.M., Kuznetsov, N.V. Sheyde, V.P., Siayeva, L.S., Galkin, A.I. and Balabolin, V.G. 1988. Peat stockpiling and storage in the context of separate harvesting. Torfyanaya Promeshlennost 1, p6-8

Murphy, C., Cole, M. and Moran, A. 1988 and 1989. Risk assessment simulation Statistics Department, University College Cork (preliminary report).

Ring, P. 1988. The Haku Production System. In "The Futureworld of Peat Technology", Bord na Mona and the Institution of Engineers of Ireland Conference, University College Dublin.

Sibiryev, V.P. and Salyayev, G.I. 1988. The MTP-24V-1 Tracked Tippling Trailer. Torfyanaya Promeshlennost 3, 1988 p18-19.

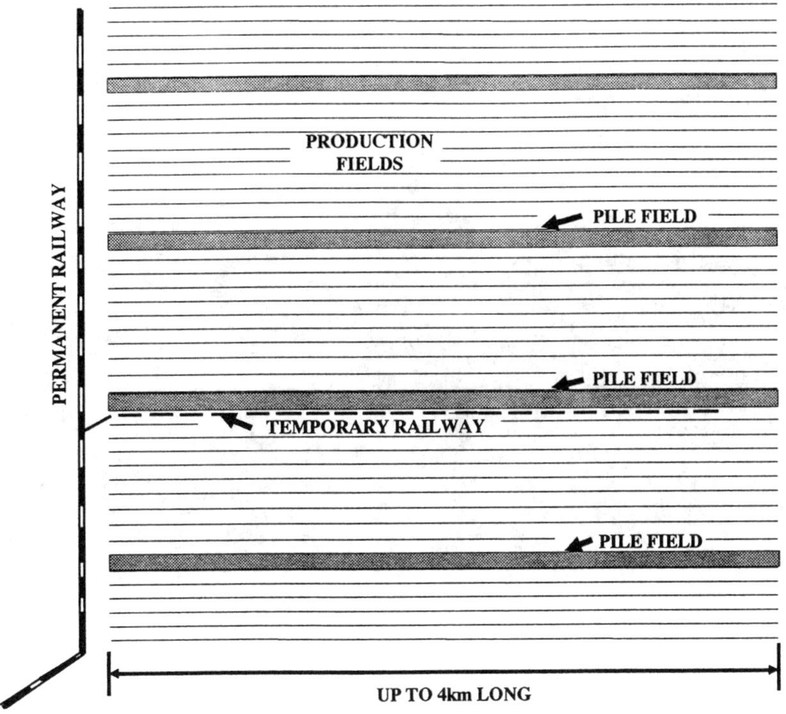

Figure 1. Layout of Peco production system

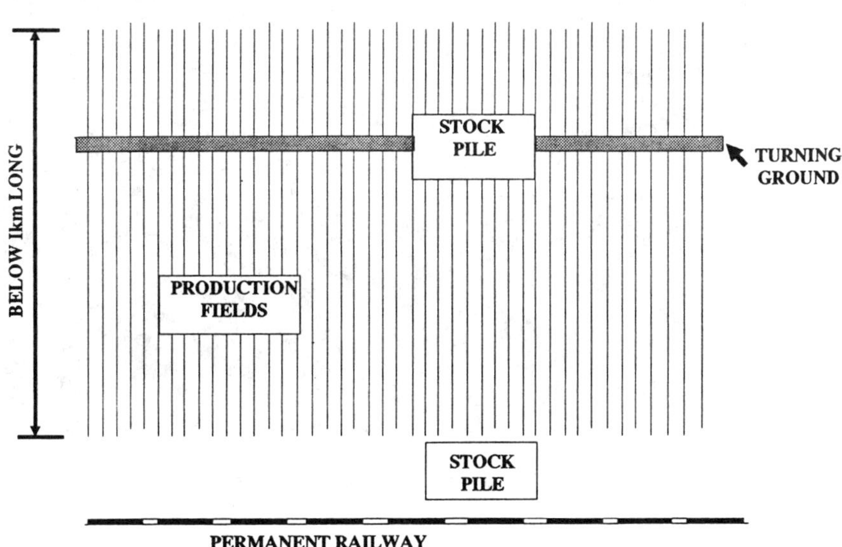

Figure 2. Layout of Haku production system

Figure 3. Milled peat miller

Figure 4 Milled peat harrow

Figure 5. Milled peat ridger

Figure 6. Milled peat harvester

Figure 7. Milled peat tractor - trailed unit

Land and Water Use, Dodd & Grace (eds), © 1989 Balkema, Rotterdam. ISBN 90 6191 980 0

The simulation of open sun peat drying and bog phenomena on a laboratory scale

T.T.Järvinen
Technical Research Centre of Finland, Domestic Fuel Laboratory, Jyväskylä, Finland

ABSTRACT: Peat production is based on the solar drying of peat on the open fields of bogs. The empirical simulation of the open field drying of peat on a laboratory scale provides a simple and relatively cheap way to study and model drying. It is also possible to describe drying using heat and mass transfer equations and solve them for instance numerically. The difficulty is that the loosened peat layer does not form a continuous, homogeneous material layer. Empirical simulation describes the actual drying caused by solar radiation and convective heat flux quite well. The weakness has been, how to adapt a natural peat field to laboratory conditions. However this problem can be solved quite easily.

RÉSUMÉ: La production de tourbe repose sur l'utilisation de l'énergie solaire lors du séchage sur le lieu d'extraction. La simulation en laboratoire des conditions de séchage sur place est un moyen relativement simple et bon marché d'étudier et de modéliser la dessication. Celle-ci peut aussi très représentée au moyen d'équations différentielles partielles qui rendent compte du transfert de chaleur et de matière et qui sont résolvables, par exemple numériquement. La difficulté de la modelisation mathématique provient du fait que la couche de tourbe extraite sur place ne constitue pas une couche de matière continue et homogène. Le recours à la simulation en laboratoire donne une assez bonne image de la dessication résultant du rayonnement solaire et de la perte de chaleur par convection. Le problème est de faire correspondre les conditions d'emploi du simulateur aux conditions réelles d'exploitation. Il est cependant assez facile de trouver une solution satisfaisante.

ABSTRACT: Die Torfproduktion beruht auf der Trocknung des Torfs durch Sonnenenergie an der Oberfläche der Produktionsfelder. Die experimentelle Simulierung der Feldtrocknung im Labor ist ein relativ einfaches und billiges Mittel zur Untersuchung und Normung des Trocknungsvorgangs. Die Trocknung kann auch durch eine Teildifferentialgleichung von Wärme- und Stoffwechsel dargestellt werden, die beispielsweise numerisch gelöst werden kann. Als Schwierigkeit ergibt sich, dass die vom Feld losgelöste Torfschicht keine dauerhaft homogene Materialschicht bildet. Die experimentelle Simulation beschreibt die durch Sonnenstrahlung und konvektiven Wärmefluss verursachte Trocknung recht gut. Ein Schwachpunkt war bisher die Angleichung der "Feldbedingungen" des Simulators an die tatsächlichen Gegebenheiten des Produktionsfeldes. Dieses Problem ist jedoch ziemlich leicht zu lösen.

1 INTRODUCTION

Peat production is based on the solar drying of peat on the open fields of bogs. Solar peat drying can be described using the general partial differential equations for the heat and mass transfer. The difficulty is the inhomogeneous character of peat especially when it is loosened from a bog to be dried on a peat field. The presupposition for a good application of equations is, that the material is continuous and homogeneous. They are difficult to solve analytically in general form. Also numerical solutions need a great deal of precise information about the drying phenomena and a reasonably good computer calculation capacity.

The reliable, simple procedure for a better understanding, development and control of solar peat drying is to simulate natural conditions, both weather and field (bog) conditions empirically on a laboratory scale. It is quite clear, that the accurate simulation and modelling of natural phenomena in a bog on a laboratory scale is impossible. The only way to solve this problem is to simplify the drying situation:

- the net solar radiation is assumed to be converted into heat immediately on the surface of peat layer to be dried. That is why it is not necessary to simulate the total spectral effect of the sun.
- the heat effect of irradiance (solar radiation balance or total radiation) is same as under natural conditions
- the convective heat flux into the peat layer is small under good conditions (during high irradiance)
- the "peat bog" in the simulation chamber is big enough to describe the total soil water potential (raising potential).

The pre-condition for simulation is the possibility of controlling all the weather and peat field conditions: solar radiation (balance), temperature, humidity of air, wind and precipitation (rain), peat field (in as natural a form as possible), ground water level and temperature. It must have the capacity to control these circumstances according to the field measurements following natural cycles: a programmable simulation chamber.

2 THE PHYSICAL AND MATHEMATICAL DESCRIPTION AND ANALYSIS OF THE OPEN FIELD DRYING PHENOMENA

The energy budget can be described according the application of energy conservation equations to a thin homogeneous surface. This application allows the combination of the various modes of energy exchange between the layer and soil/atmosphere (Fig 1.) /Seguin 1986/.

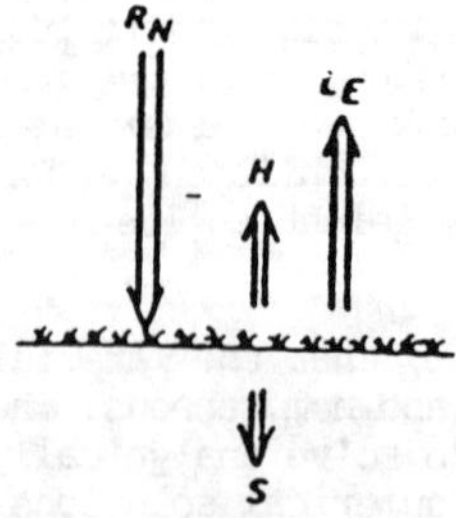

Figure 1. Surface-atmosphere interface and energy echange

This leads to the energy budget equation (1).

$$(1)\quad R_n + S + H + LE \pm (\rho C_p \frac{\partial T}{\partial t}) = 0$$

where R_n is the net radiation
H is the convective atmospheric heat flux
LE is heat flux to the ambient due to evaporation
The last term is the heat storage term, which can be neglected by choosing an adequate time scale.

The temperature distribution inside the product is characterized by the general heat transfer and conduction equation (2). The general mass transfer equation (3) describes the distribution of water in the layer. Those equations will be normally simplified for solution. They can be solved both analytically and numerically. Both type of solutions require a great deal of information and empirical data and finally the calculations must be proven some how either empirically and/or by field experiments and measurements /Antonov 1981/.

$$(2)\quad \frac{\partial T}{\partial t} = \frac{\partial}{\partial x}\left(a_T \frac{\partial T}{\partial x}\right) + \frac{\varepsilon L}{c}\frac{\partial W}{\partial t}$$

$$(3)\quad \frac{\partial W}{\partial t} = \frac{\partial}{\partial x}\left(a_m \frac{\partial W}{\partial x} + a_m \delta \frac{\partial T}{\partial x}\right)$$

where T is temperature in each point
t is time
α_T is thermal diffusivity ($k/\rho C$)
ε is void fraction, volume fraction of non-solids
L is latent heat of evaporation of water
c is heat capacity per mass unit
W is water content in each point
α_m is (water)potential diffusivity
δ is thermal gradient coefficient

3 EMPIRICAL SIMULATION ON A LABORATORY SCALE

In order to simulate the field drying of peat, a solar simulation chamber for conventional sod and milled peat production has been erected in the Domestic Fuel Laboratory of Technical Research Centre of Finland (fig. 2) /Järvinen 1987/.

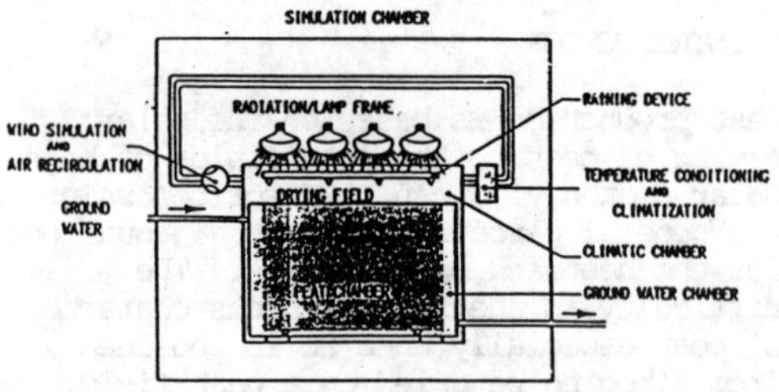

Figure 2. Diagram of the simulation chamber

The chamber consists of a peat block with a certain adjustable ground water level, and temperature and climatic conditions for drying. The radiation is as similar to the natural sunlight spectrum as is possible within reasonable cost limits and its maximum heat output is approx. 2.000 W/m^2. The radiation, the air temperature, humidity and wind are controlled and programmed. The radiation can also be programmed to follow field net radiation measurements. In addition the chamber includes a raining device. The climatic conditions can be kept very stable if needed. Also the drying (curves) of peat samples on the peat ground (block) have been repeatable.

4 THE RESULTS OF EMPIRICAL SIMULATION

The loosened peat layer to be dried has a good heat and mass transfer connection with the atmosphere and ground (peat field). This phenomena is also decisive for drying and both connections have nearly equal importance. The following list includes most important factors affecting drying:

- the effect of total (global) solar radiation (the irradiance)
- the reflectance of the global radiation
- the heat flux loss to the ambient due to convection
- the heat flux conducted into the surface of ground (peat field).

The net thermal radiation seems to cause a quite minimal share of losses because of the low emissivity and temperatures of the peat layer surface. Also a small amount of the incoming effect is conducted into the deeper peat layers of a bog, but a larger share goes into the immediate surface of the peat "ground" and evaporates water from the field. Depending on the field conditions sooner or later the evaporated water will be replaced by new water. The total soil water potential raises "new" water to the field surface. Also continuous excavation of peat bog keeps the field surface wet, because it brings the ground water table closer to the peat field surface and exposes active capillary connections.

In fig. 3 there is an example of drying curves, which have appeared in the chamber and under the same conditions on a field. The two curves, which are very similar, are the chamber tests and upper curve is measured on the field. In this case the water tables were at a low level in the field and the peat field surface was rather dry.

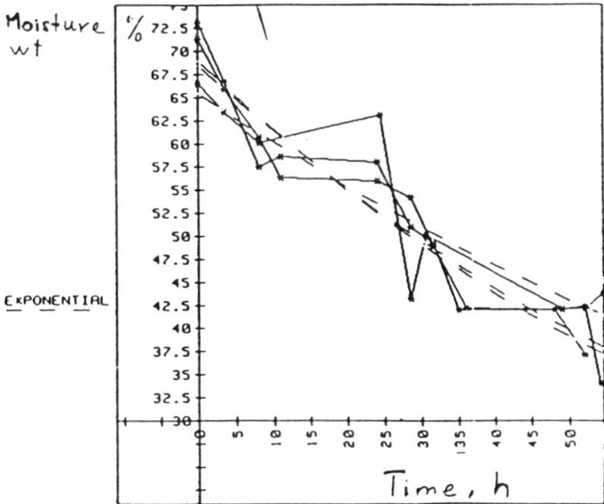

Figure 3. Drying curves on the field and under the same conditions in the simulation chamber.

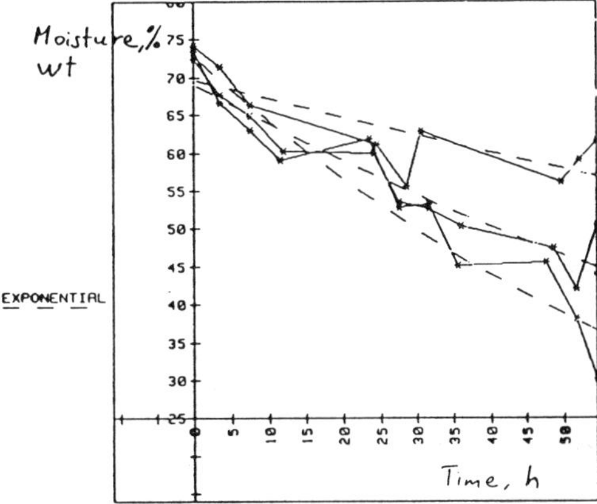

Figure 4. The drying curves on a field and in the chamber.

In fig. 4 there is another example of drying curves on a field and in the chamber under the "same" conditions. In this case, the chamber drying differs considerably from the field drying. The water table was very close to the field surface and the surface layer had a higher moisture content than in the chamber. In most experiments the dryness of peat block surface has been a bit higher than on the field. Also the temperatures have been higher than under "natural" conditions.

Fig. 5 shows an experiment in the simulator, in which on one side of the simulator "field" the milled peat layer in normal contact with the field has been dried and on the other side the milled peat layer being mass (not thermal insulation) isolated from the "field" has been dried. The drying curves follow each other quite well up to 40 ... 50 % moisture content, after which they separate clearly.

This phenomena, and the simulation of field drying show that at the beginning of drying until close to the target moisture content (40 %), the heat transfer and mass transfer as a vapor dominate and after that too mass transfer as a liquid, because of the concentration gradient, has a significant importance.

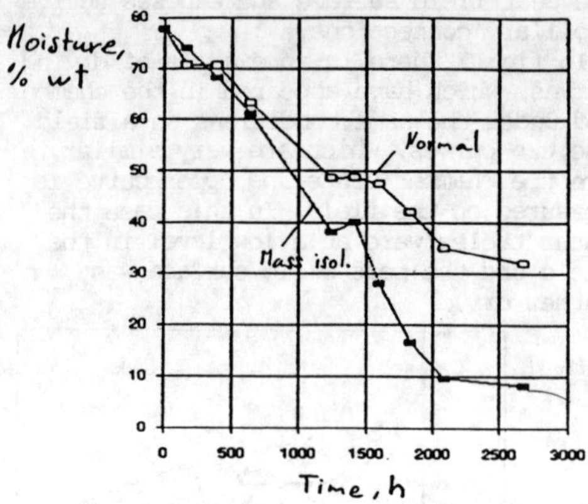

Figure 5. Drying curves of milled peat layers in the simulation chamber in normal and mass isolated contact to the "field".

5 DISCUSSION

The empirical simulation of the open field drying of peat on a laboratory scale provides simple and relatively cheap way to study and model drying. The weakness of empirical simulation has been the "correct" creation of peat field (bog) conditions in the laboratory. The actual drying (caused by solar radiation and convective heat flux) can be simulated quite well.

It is also possible to describe the drying using heat and mass transfer equations and solve them for instance numerically. The difficulty is, that the loosened peat layer on the peat field does not form continuous, homogeneous material. This means, that it is very difficult to obtain reliable information and data on material properties. On the other hand nowadays computer capacity creates new possibilities of handling the information and solving partial differential equations. Perhaps in the future mathematical modelling (computer science) and empirical simulation can provide the best results concerning this kind of problem.

REFERENCES

Antonov, V.J., Malkov, L.M. and Gamajunov, N.I., Technology of the field drying of pea (Tehnologija polevoi suski torfa), second ed., Moscow, Nedra 1981, 238 p.

Järvinen, T., Solar drying of peat, Advances in Solar Energy Technology, Volume 2, ISES Solar World Congress 1987, Hamburg, BRD. International Solar Energy Society, Pergamon Press, New York, p. 1483-1487.

Seguin, B., Solar radiation and wind in agronomy and hydrology, Physical Climatology for Solar and Wind Energy Edited by Guzzi, R. and Justus, C.G., ICTP Course 1986, Miramare-Trieste, Italy, World Scientific, Singapore, p. 951-971.

Land and Water Use, Dodd & Grace (eds), © 1989 Balkema, Rotterdam. ISBN 90 6191 980 0

Economic consequences of reallotment in a peat area with high natural and scenic values

J.A.Kester
The Winand Staring Centre for Integrated Land, Soil and Water Research, Wageningen, Netherlands

ABSTRACT: According to the Relationship Memorandum substantial parts of peatlands in the Netherlands have been designated as reserve areas. Farmers can conclude management agreements for the period before purchase by the government. The consequences of reallotment on labour proceeds and needed hours of labour with alternative presuppositions concerning designation of reserve areas, entering into management agreements and (European Community) milk production restrictions have been studied in a peat area.

Used methods are: a computerised interactive reallotment method, a 'Monte Carlo' method to forecast future land use and AGREVAL, a model-based method for agricultural evaluation of land layout.

The profit of reallotment depends on the location of the farmbuildings (inside or outside the villages). The direct income consequences of management agreements appeared small. The main reason is the extensive use of the peat-meadows (compared with Dutch standards). When production restrictions are in force the direct income decrease as a consequence of management restrictions is mostly negligible.

RESUME: Selon la 'Note de relations' du gouvernement Néerlandais des parts substantiels des terrains tourbeux ont été désignés comme réserves. Pendant la période avant l'achat des terrains désignés par le gouvernement les paysans peuvent passer un contrat concernant une exploitation agricole adaptée. Les conséquences d'un remembrement pour le rendement des fermes et pour le besoin de travail ont été recherchés pour une région tourbeuse avec des qualitées naturelles supérieures. La recherche comprend des alternatives diverses, concernant l'exploitation agricole (exploitation regulière versus exploitation adaptée) et le niveau de production de lait (production libre contre production restrictée par la Commission Europeènne).

Les méthodes de recherche: une méthode pour tracer un plan de remembrement à l'aide d'un ordinateur, une méthode (dit 'Monte Carlo') pour éstimer la structure agricole future d'une région et une méthode d'évaluation économique.

L'effet de remembrement dépend de la localisation des fermes (dans le village ou hors du village). Les effets des contrats de l'exploitation adaptée pour le rendement agricole sont faibles. En cas des restrictions de production sont en vigueur ces effets sont négligables.

KURZFASSUNG: Entsprechend der 'Zusammenhang-Note' werden in den Niederlanden beträchtliche Teile der Niederungsmoorfläche für Naturschutzgebiete bestimmt. Während der Zeit, vorangehend an dem Grundankauf von den staatlichen Behörden können die Bauer einen Landbewirtschaftungsvertrag eingehen. Die Folge von Flurzusammenlegungsmassnahmen auf bauerliches Betriebseinkommen und Arbeitsbedarf wurden in einem Niederungsmoorgebiet studiert mit alternativen Voraussetzungen in Bezug auf Naturschutzgebietsbestimmungen, Landbewirtschaftungsvertragseingehen und (Europäische Gemeinschaft)-initierte Milchproduktionseinschränkungen.

Verwendete Methoden sind: eine komputerisierte, interaktive Flurzusammenlegungsmethode, eine 'Monte Carlo'-Methode um die zukünftige Bodenbenutzung zu voraussagen und AGREVAL, eine modelmässige Methode zur landwirtschaftlichen Bewertung der Flurlage.

Die Erfolge der Zusammenlegungsmassnahmen sind abhängig von der Hoflage (im Dorf oder im Polder). Die mittelbaren Einkommenskonsequenzen von Landbewirtschaftungsverträgen zeigen sich niedrig. Der wichtigste Grund ist die extensive Nutzung der Moorwiesen (Niederländischen Verhältnissen nach). Sind Produktionseinschränkungen durchgeführt worden, dann ist die mittelbare Einkommenssenkung im Folge Landbewirtschaftungsbedingungen fast zu ignorieren.

1 INTRODUCTION

The so-called Relationship Memorandum (i.e. a memorandum on environmentally sensitive areas) introduces a set of policy instruments to protect natural values. These instruments include among other things the designation of certain areas as reserves or environmentally sensitive areas. Land situated in designated reserves will be acquired by the Bureau of Land Management, an agency of the Ministry of Agriculture and Fisheries. As the land has to be acquired by voluntary cession the acquisition may take a long time. In these areas farmers can conclude management agreements for the period between designation and purchase by the government. The Bureau for Land Management has been charged with the implementation of these agreements. For that purpose a management plan is drawn up. It specifies which type of nature and landscape management is most suitable for conserving the values of the concerned area. Mostly several packages of management conditions are determined, that vary in harshness. To every package fits a specific management compensation.

In peatland regions the main goal of management agreements is mostly the preservation or creation of good breeding circumstances for waders. The agreement conditions have to safeguard opportunities for the young birds to be reared before mowing. An extra goal is often the conservation of existing floristic values. A package of medium harshness contains conditions as:

- no use of chain harrow or roller after 15 March;
- no mowing before 15 June;
- mowing once a year at least, grazing the aftermath with no more than four grazing units per ha;
- no use of pesticides;
- no use of inorganic fertilizers and the use of manure must be limited to 20 tons per year;
- no ploughing, leveling or reseeding the land;
- no change in groundwater table.

The main effects are a decrease of the nett grasland production and the impossibility to use the land as meadows for milkcattle. Also the gained fodder is not suitable for high-performing milkcows.

In several peatland regions a substantial part of the area used by agriculture has been designated as reserve areas according to the Relationship Memorandum. In the designated areas farmers fear a deterioration of their possibilities to get a sufficient living standard in the next decades. To find satisfying solutions both for the farmers and for nature conservancy a research program has been drawn up.

The study described in this paper forms a part of this program. It deals with the question what the prospects are for agriculture at several scenarios concerning land acquisition, layout and management. The study has been located in the municipality of Haren in the province of Groningen (Section 2). To examine the possibilities for ameliorating parcellation (with and without reserve areas) reallotment plans are designed first for the whole area of the municipality. These plans are drawn for two situations, namely for the land use situation in 1981 (actual) and for an land use situation in 1998 (estimated) (Section 3).

The effects of the so obtained four land layout situations on required hours of labour and on labour proceeds for dairy farms are evaluated with a model called AGREVAL. The economic evaluation has been carried out for a study area in the eastern part of the municipality. For each land layout situation several alternatives have been made concerning farm management. As consequence of recent EC-production restriction policy all alternatives are calculated for three levels of imposed production restrictions (Section 4).

Based on the calculated results some conclusions can be formulated about the prospects of farmers under the different assumptions (Section 5). A full description of the framework of the study and the acquired results is given by Kester, Kik & Reinds (1989).

2 STUDY AREA

In the municipality of Haren (the province of Groningen in the northern part of the Netherlands) a high ridge of sandy-loam soils is flanked by low wet peatlands (Fig. 1). The drainage of peatlands occurs by means of open drains in a polder system; layout and accessibility are very poor. A substantial part of these peatlands in the eastern polders has been designated as reserve areas. As the research problem is directly connected with the designation of these reserve areas the study area for the examination of the management effects has been confined to the eastern polders (1490 ha, see Fig. 1).

The confined study area consists mostly of open permanent grasslands, used as meadows for milk-cattle. The external production circumstances of the peatlands in the polders are poor: a high water table, insufficient accessibility by a system of unmetalled rural roads and very scattered, narrow and deep parcels. Compared to Dutch standards the land use has been extensive up to now. As a result of local peat digging in the past several peat holes are present in the landscape. These peat holes are in different stages of growing solid by new peat formation.

The variety in natural circumstances, the high water table and the extensive land use have created a diversified landscape with high natural values, especially a good breading area for waders. In the

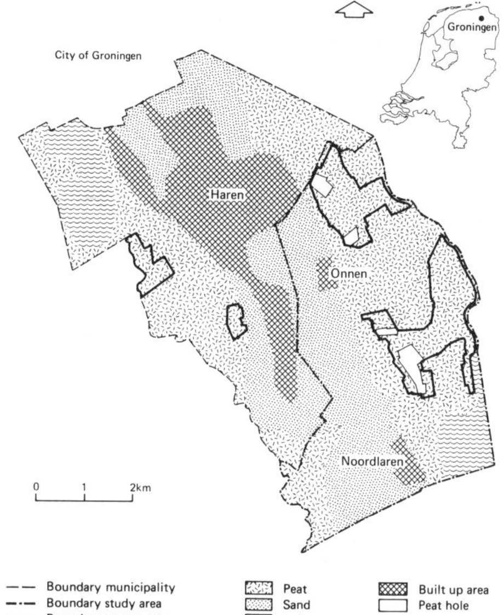

Fig.1 Research areas in the municipality of Haren (province of Groningen)

study area 85% of the grasslands (i.e 1276 ha) are used by 45 dairy farms. So the average size of these farms is 28.4 ha. Only a minority have their farmbuildings inside the polder (i.e. on the eastern side of the road that runs just on the edge of the sandy soils). Most farmbuildings are situated within the villages Onnen and Noordlaren (see Fig. 1).

3 REALLOTMENT RESEARCH

3.1 Method

Reallotment plans were designed for all agricultural lands in the municipality of Haren (2716 ha). To this end a computerised method has been used, which was developed especially for reallotment research in the preparatory phase of a land consolidation project (Kik, 1989). The method takes advantage of an algorithm, based on operational research techniques. The result is a minimizing of the mean lot distance with a limitation of the calculated number of lots.

For the calculation geographic information is needed about existing land-use. When not all land is mutually exchangeable (because of differences in soil suitability or otherwise) exchange classes must be determined. Reallotment calculations are done for each exchange class separately. When the results of the reallotment calculations have been processed a new reallotment plan can be designed by means of an interactive computer-aided procedure.

3.2 Reallotment plan based on actual parcellation in 1981

Basic information on land-use was available from an survey, carried out in 1981 by the Government Service for Land and Water Use, a service charged with preparation and implementation of land development plans. In that scope a user map was produced. The accompanying list of land-users contains also several data about registrated farm size, type of land-use, etc. For the study handled with in this paper, the user map must be digitized first. After digitizing computerised calculations have been produced about actual holding sizes, lot sizes and lot distances. The results have been presented on maps and in overall tables (Kester, Kik & Reinds 1988).

Based on these data a new reallotment plan has been designed. In this designing procedure the lots inside the proposed boundaries of the designated reserve areas has been handled as a separate exchange-class. The analysis of effects of a new allocation has been directed especially at those holdings where the main occupation of the farmer is milkcattle husbandry (later on shortly indicated as dairy farms). A distinction has been made between holdings situated within the villages (Onnen and Noordlaren) and holdings situated in the polder.

For both groups the actual parcellation is unfavourable:

- a great number of lots with considerable mutual transport distances;
- long average transport distances between the lots and the farm-buildings;
- transport mostly over unmetalled roads or across land;
- small working parcels.

Also most farms have only a small fraction of the land near the farm-buildings (the average for the farms in the villages is 6% and for the farms in the polder 18%). For modern milking practices this fraction must be approximately 60% and preferably 80%.

After implementation of a reallotment plan parcellation characteristics for both groups of holdings improve. The fact that the lots within the boundaries of the reserve areas have to be handled as a separate exchange-class, diminishes this improvement. Also it is obvious that the positive effects are much higher for the 'polder'-holdings than for the 'village'-holdings. For the first group the fraction of land near the farm-buildings becomes 68% against only 16% for the latter group.

3.3 Reallotment plan based on estimated parcellation in 1998

In the period preceding the actual formation of reserve areas the government will be actively present on the ground market on behalf of the acquisition of the needed area of land. In case of an intended land consolidation project also land can be bought outside the boundaries of the future reserves and be exchanged afterwards with parcels within the reserve boundaries. These land purchasing activities of the government can effect the development of the agricultural structure in the next decades. So the reallotment opportunities on the mid long term can alter. To examine the possible occurrence and the concequences of such an effect, the results of a study with a Monte Carlo model for forecasting changes in agricultural structure of the study area Haren were used (Bouma 1989). A general description of the method has been given by Bouma (1988).

A prognosis of the agricultural structure in 1998 was made under the assumption that the government buys 50% of all farmland coming on the market as a result of termination or diminition of farms. A part of the other 50% can be bought by continuing farmers and a part goes to non-agricultural uses. As the prognosis has been made on an individual basis it is possible to estimate which lots have got an other user (owner or tenant). So it was possible to produce an estimated parcellation for 1998.

According to the assumptions of the prognosis many farms will reduce in size. As the percentage of free coming land that will be purchased by the government under those assumptions, is very high the average farm size of the continuing dairy farm will be less in 1998 than it was in 1981. On the other hand the amount of land, bought by the government in 1998 equals nearly the size of the designated reserve areas, but this acquired land is scattered all over the territory of the municipality.

The calculated parcellation characteristics for the estimated parcellation 1998 differ very few from the figures, that characterize the actual parcellation of 1981. The unfavourable situation, existing in 1981, must also be expected in 1998 in the case of the estimated 'autonomous' development.

For this estimated 1998 parcellation also a reallotment plan was designed. In this plan all land acquired by the government has been exchanged with land situated inside the boundaries of the reserve areas. So most farmers can get nearly all their lands outside those boundaries. The calculated characteristics for the reallotted area in 1998 show only few differences with the situation in 1981 after reallotment. The average transport distances have been shortened a bit more but the fraction land near the farmbuildings has been improved a bit less. The average of this fraction becomes 56% for the group dairy farms with farmbuildings in the polders and 14% for the group with farmbuildings inside the villages, i.e. still lower than it was in the 1981 situation after reallotment.

4 ECONOMIC EVALUATION

4.1 Method

For the economic evaluation a model-based method for agricultural evaluation of land layout (Righolt & Reinds 1980) was used. The input of the program AGREVAL consists of a number of parcellation characteristics and a set of time elements and characteristics regarding farm labour, machinery and traction. The data concerning parcellation and layout have been derived from the reallotment study, described in Section 3. Data concerning time elements, farm labour, machinery and traction were taken from a task-times data service (Personal communication, Institute of Agricultural Engineering ,IMAG, Wageningen). General data conerning dairy husbandry were derived from Pelzer (1984). Specific study areas determined premises concerning farm plans, practised working methods and production levels have been chosen in consultation with local experts of the Agricultural Extension Service.

The program gives among other things a set of physical output data of crop production and working capacity required. Transformation of this physical output into farm income is performed by introducing market prices for the relevant factors.

4.2 Alternatives

Two main alternatives have been distinguished:

- an alternative in which the farmer has not (yet) entered a management agreement;
- an alternative in which the farmer has entered (already) a management agreement.

A management agreement contains a package of conditions, which ensure the protection of the existing natural and scenic values until the government can purchase the land to realise the reserve status. As for the designated reserves in the study area the first goal of management agreements is the conservation of breeding opportunities for waders, the economic evaluation was carried out under the assumption that in case of management agreements the package of management conditions described in Section 2 of this paper will be in force. With regard to agricultural exploitation such a package results in a decrease of the nett grasland

Table 1. Influence of parcellation, management conditions and milk production restrictions on needed hours of labour and labour proceeds in Dutch guilders (Dfl); average values for 1998 for all dairy farms within the study area. Distinguished are: calculated average values concerning needed hours of labour, total labour proceeds, decrease of labour proceeds per 'ha with management conditions' and labour proceeds per hour

	Autonomous development				After reallotment			
	'village' farms		'polder' farms		'village' farms		'polder' farms'	
Group of farm sizes	<30 ha	>30 ha	<30 ha	>30 ha	<30 ha	>30 ha	<30 ha	>30 ha
Number of dairy farms	15	6	7	5	15	6	7	5
Average area of grassland (ha)	22.17	39.70	22.80	43.60	22.17	39.80	22.80	44.00
Average area within nature reserves	5.98	11.56	1.46	8.18	2.29	2.50	0.77	1.48
NO MILK PRODUCTION RESTRICTIONS								
Average hours of labour/farm								
without management conditions (h)	3801	5178	4003	5422	3759	4962	3225	4992
with management conditions (h)	3602	4903	3982	5290	3683	4943	3215	4976
Decrease per ha with conditions ($h.ha^{-1}$)	33	24	14	16	33	8	13	11
Average labour proceeds/farm								
without management conditions (Dfl)	32,302	70,086	39,899	93,515	41,166	91,519	55,035	116,870
with management conditions (Dfl)	26,862	59,883	39,579	89,118	38,950	90,991	54,852	116,515
Decrease per ha with conditions ($Dfl.ha^{-1}$)	910	883	219	538	968	211	238	240
Average labour proceeds/hour								
without management conditions ($Dfl.h^{-1}$)	8.50	13.54	9.97	17.25	10.95	18.44	17.07	23.41
with management conditions ($Dfl.h^{-1}$)	7.46	12.21	9.94	16.85	10.58	18.41	17.06	23.42
9% MILK PRODUCTION RESTRICTIONS								
Average hours of labour/farm								
without management conditions (h)	3666	5013	3853	5236	3607	4778	3124	4842
with management conditions (h)	3520	4707	3832	5147	3564	4760	3113	4826
Decrease per ha with conditions ($h.ha^{-1}$)	24	26	14	11	19	7	14	11
Average labour proceeds/farm								
without management conditions (Dfl)	26,785	59,612	33,990	81,620	35,422	80,353	48,299	103,374
with management conditions (Dfl)	24,167	57,991	33,670	80,049	34,430	79,954	48,115	103,028
Decrease per ha with conditions ($Dfl.ha^{-1}$)	438	140	219	192	433	160	239	234
Average labour proceeds/hour								
without management conditions ($Dfl.h^{-1}$)	7.31	11.89	8.82	15.59	9.82	16.82	15.46	21.35
with management conditions ($Dfl.h^{-1}$)	6.86	12.32	8.79	15.55	9.66	16.80	15.46	21.35
18% MILK PRODUCTION RESTRICTIONS								
Average hours of labour/farm								
without management conditions (h)	3529	4848	3705	5051	3455	4595	3033	4694
with management conditions (h)	3424	4707	3685	4963	3425	4576	3022	4677
Decrease per ha with conditions ($h.ha^{-1}$)	18	12	14	11	13	8	14	11
Average labour proceeds/farm								
without management conditions (Dfl)	20,554	47,863	27,378	68,331	28,826	67,740	41,256	89,294
with management conditions (Dfl)	19,556	47,405	27,272	67,693	28,523	67,507	41,069	89,009
Decrease per ha with conditions ($Dfl.ha^{-1}$)	167	40	73	78	132	93	243	196
Average labour proceeds/hour								
without management conditions ($Dfl.h^{-1}$)	5.82	9.87	7.39	13.53	8.34	14.74	13.60	19.02
with management conditions ($Dfl.h^{-1}$)	5.71	10.07	7.40	13.64	8.32	14.75	13.59	19.03

production and in the impossibility to use the land as meadows for milkcattle. Also the gained fodder is not suitable for high-performing milkcows.

The allocation study results in four layout alternatives:

- existing parcellation 1981;
- parcellation after reallotment 1981;
- estimated 'autonomous development' 1998;
- parcellation after reallotment 1998.

In 1984 on EC level a system of milk production restrictions was implemented. As this milk quota system may influence the effects of management conditions on directly generated farm income all above mentioned alternatives were calculated for three levels of imposed production restrictions:

a. no production restrictions (production is only limited by the grass production capacity of the farm);

b. the milk production has to be reduced with 9% (with respect to a);

c. the milk production capacity has to be reduced with 18 % (with respect to a).

4.3 Results

Model calculations have been made for several groups of farms differing in farm size and in situation of the farmbuildings (within the villages or in the polder). Finally average values have been calculated for the groups with a farm size <30 ha and for the groups with a farm size >30 ha. This critical level of 30 ha corresponds with a change in mechanisation level. So the calculations have finally resulted in average values for the four layout alternatives.

The calculated hours required farm labour for the studied layouts have been compared with the analogue value for optimal farm layout. The extra need for labour appears to be about 30% for a parcellation before reallotment and about 20% for a parcellation after reallotment. Comparison of the corresponding figures for 1981 and 1998 learns that no significant differences occur. For the (remaining) farms the results for 1998 are slightly better than those for 1981, especially after reallotment. The main reason is that the land acquired by the government has been reallocated inside the boundaries of the designed reserve areas. So in 1998, after the implementation of a reallotment plan the area of farm land within the boundaries of the reserve areas is less than in the 1981 situation. Table 1 shows the values for the two 1998 alternatives. Table 1 shows for all farm layout situations the notable difference in the ratio 'hours of labour'/'labour proceeds' between the farms <30 ha and the farms >30 ha. However, this is a well-known 'scale' phenomenon. More specific is the difference in the location of the farmbuildings. The 'village' farms have already a disadvantage in the situation before reallotment. They also have less benefit of a reallotment, so the disadvantage in respect to the 'polder' farms increases. Meaningful in that regard is the calculated 'labour proceeds per hour'. After reallotment the effect of management agreements is small for most farms. When milk production restrictions are in force the effect of management agreements decreases. At a milk production restriction level of 18% the effect of management agreements on direct labour proceeds and on the calculated labour proceeds per hour is almost neglectible.

5 CONCLUSIONS

The results of the described reallotment research and the model calculations lead to the following conclusions for the study area:

- The present poor parcellation requires relatively many hours of labour per ha in spite of the low intensity of actual grassland use compared with Dutch standards. This means low average labour proceeds per hour. The high labour input per unity of product threatens the sustainability of many dairy farms on the long term. Without reallotment the earning of a sufficient income under reasonable labour input conditions will remain problematic for almost all considered groups of farms.

- The introduction of management agreements for land within the boundaries of designated reserve areas will give a substantial decrease of direct farm income (before possible agreed compensation) for the farms for which the amount of this land with restrictions forms more than 25% of the total farm lands.

- After reallotment the farms <30 ha with farmbuildings within the villages improve very little and stay very poor. The group of 'village' farms >30 ha shows a substantial improvement, but also for this group the direct farm income stays far beyond that of the comparable group of 'polder' farms. The main reason is a lasting lack of sufficient land near the farmbuildings.

- Dairy farms of sufficient size with an improved farm layout and enough land near the farmbuildings (60% or more) may expect a reasonable farm income for the next decades. However, their number is only small. For the majority of the farms a reallotment can give less or more improvement but their sustainability on mid long terms remains uncertain.

- Most farms in the study area use their grasslands extensively compared with Dutch standards. In these circumstances the quota-based EC milk production policy shows a substantial negative effect on farm income in a situation without management agreements. If these restrictions will still be valid in 1998 there will be the same negative effect in case the agricultural structure in the study area will be as prognosticated.

- When the quota-based EG milk production policy will continue in the next decade special management conditions on behalf of nature conservation will hardly effect direct labour proceeds for most dairy farms in the study area. Some farms with a high percentage of 'restricted' land will have to cope with a more substantial decrease in farm income per ha 'restricted' land. This counts the more for the total farm income. So a fixed compensation per ha with management conditions giving full compensation for farms in average circumstances, will undercompensate farms with a high percentage of 'restricted' land.

- Reallotment can ease the process of land acquisition by the government for founding nature reserves and can diminish the negative effects of such a policy for the farmers.

REFERENCES

Bouma, F. 1988. A Monte Carlo model for forecasting changes in agricultural structure. Tijdschr. voor Sociaalwetensch. Onderzoek van de Landbouw 3:5-24.

Bouma, F. 1989. Mogelijkheden en gevolgen van onttrekking van cultuurgrond in Haren. Interne nota 373. Landbouw-Economisch Instituut, The Hague, The Netherlands. 50 pp.

Kester, J.A., R. Kik & G.H. Reinds 1988. Landinrichtingsaspecten met betrekking tot de relatienotaproblematiek in het gebied Haren (Gr.). Rapport 36. Institute for Land and Water Management Research (ICW), Wageningen, The Netherlands (in press).

Kik, R. 1989. Reallotment research in land improvement projects. Agricultural systems. Submitted for publication.

Pelzer, L. (ed.) 1984. Handboek voor de Rundveehouderij, 4e druk. Research Station for Cattle, Sheep and Horse Husbandry (PR), Lelystad, The Netherlands. 370 pp.

Righolt, J.W. and G.H. Reinds 1980. Agricultural evaluation of land layout. In E.W. Schierbeek (ed.), Research Digest 1980. Institute for Land and Water Management Research (ICW), Wageningen, The Netherlands:215-222.

Land and Water Use, Dodd & Grace (eds), © 1989 Balkema, Rotterdam. ISBN 90 6191 980 0

A milled peat production simulation model (SIMPEAT)

R.Kavanagh
Bord na Mona, Newbridge, Co. Kildare, Ireland

C.Barry
University College Galway, Ireland

C.Murphy
University College Cork, Ireland

ABSTRACT: A simulation model, refered to as SIMPEAT, has been developed for use as a planning tool in milled peat production. The model, which is particularly user-friendly is described in detail. The various bog and production parameters, how they were developed and are used in the model are also discussed. The model was applied in actual bog operating circumstances and was validated against actual data. Following validation the model was used for scenario analysis which attempted to assess the impact of change in milled peat production.

1 INTRODUCTION

Bord na Mona is an Irish Semi-State organisation that is charged with the production of peat from state-owned boglands. There are three principal products harvested from Irish bogs. These are; milled peat, sod peat and moss peat.

Milled peat is a very important element of the operations of the Bord. It is the raw material for turf, briquettes and is the fuel used for five of the country's electricity generating stations. It accounts for over seventy five percent of the revenues of the company. This paper concentrates on the development of a production planning system for the milled peat operation.

In Ireland the production of peat is difficult in that the climatic conditions are not favourable to continuous production. Typically, the summer weather is characterised by short lengths of dry weather suitable for production, broken by periods of rain. The uncertain weather patterns underly the importance of making a range of production decisions before the harvesting season commences. Some of these decision include:

- examining alternative production systems
- fixing the location of stockpiles on the bog
- determining the appropriate mix of machines to be used in a particular bog area
- examining the effect of increased machine reliability
- examining the effect of changes in work practices.

It is not possible to use any of the mathematical optimisation techniques to assist in solving these problems. Many of the production variables are not independent or do not exhibit linear relationships, and where many of the decisions are made on the basis of experience, it is difficult to describe the behaviour of the system in terms of equations. Nor is it practical to carry out direct experimentation so that each alternative of a decision can be examined. Undoubtedly the results of such experimentation would yield accurate and authentic information, however, the time and cost involved in exploring various scenarios, from problem formulation to making a decision, are too great. The issue is further complicated by the seasonal nature of production - much of the planning for the coming year's production takes place in the off-season period (from October to April). The effect of decisions made during the planning phase cannot be assessed until production is well under way when it can be difficult and expensive to change or fine-tune a particular production policy.

It was decided to develop a simulation model of the production process to be used as a planning tool. This paper describes a large-scale simulation model of a peat harvesting system. The model uses the discrete event simulation concept (Pidd 1984, Tocher 1963) to evaluation changes in assumptions about the system quickly and economically. The benefits of simulation have been covered in a number of articles

(Emshoff and Sisson 1970, Fishman 1973, Naylor 1966, Neelamkavil 1987, Smith 1968). Essentially, it allows the decision maker to assess rapidly the implications of changes in resources or in assumptions about the constraints. Computer-based simulation can be used in a 'trial and error' mode that allows a manager to consider various scenarios and the consequences of implementing these. It can help to increase understanding of the system being studied and to solve problems. The refinement of the simulation model (called SIMPEAT) to allow Bord na Mona managers to compare the results of alternative decisions, was an important consideration in the design process.

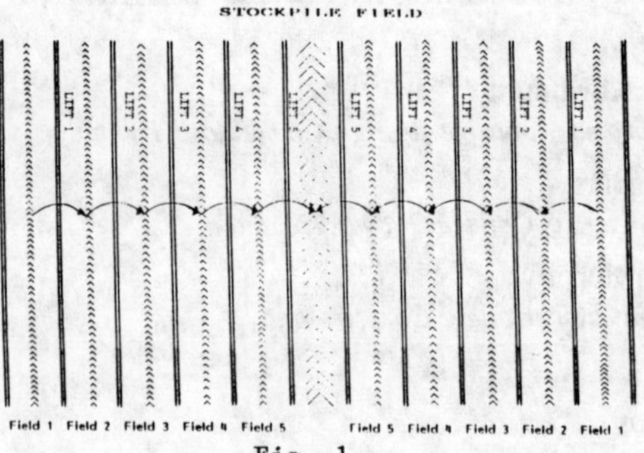

Fig. 1

2 DESCRIPTION OF THE MILLED PEAT PRODUCTION SYSTEMS

2.1 The PECO production system

The normal production system used by Bord na Mona is called the PECO production system (Barry, Kavanagh & Murphy 1988, Coffey, Kavanagh & Murphy 1985). The system is basically circular in nature. The processes involved are:

- milling
- harrowing
- ridging
- harvesting

The main unit of harvesting is a railway which contains a central field for stockpiling, surrounded on both sides by typically five fields. The peat is first milled. Milling involves the peat surface being granulated to a depth of up to 20mm. It is then harrowed. This is done to invert the peat that has been milled, thus disturbing the surface crust and exposing the underlying peat to the atmosphere for drying. The number of harrowings that the peat receives is dependent on the quality of the peat and the weather conditions. The peat is then ridged. This involves pushing the peat that has been milled into the centre of its field so as to form a ridge of peat down the length of the field. Finally the peat is harvested. The harvesting is a complex process. A harvester goes down the first field, i.e. the one furthest from the stockpile and transfers the peat onto the ridge of peat on the second field (this is known as a first lift). The process is repeated until all of the peat is finally transferred from the first field onto the stockpile field (See Fig. 1).

2.2 The HAKU production system

Bord na Mona has recently been evaluating a production system (HAKU), used in Finland, Russia and Sweden, as an alternative to that currently in use (PECO).

The HAKU production system differs from PECO only in the harvesting operation and the location of stockpiles. Under the PECO system the harvester handles the peat, on average, three times in transferring the peat to the stockpile field. Peat is lost each time the harvester moves it. It is either left behind or blown away. Peat that remains on the field surface makes subsequent milling more difficult. Under the HAKU system the harvester handles the peat only once and therefore the amount of losses would be reduced considerably. Peat that is ready for harvesting is loaded directly into trailers which transport it to a large central stockpile convenient to a road or railway. Four tractor trailer units are required to service each harvester. This system releases the stockpile at the centre of each railway (under the PECO system) for production and reduces the exposed surface area of stockpiled peat. A further advantage of the HAKU system is the improved average speed for harvesters. Because harvesters operate at 'first-lift' speeds, harvesting is quicker and wear and tear of machines is reduced.

3 OVERVIEW OF THE SIMULATION SYSTEM

All of the assumptions and parameters of the model are maintained in data files. The separation of assumptions from the simulation programs increases the flexibility and 'user-friendliness' of the model. This was greatly assisted by using the Microsoft QuickBASIC compiler on an IBM PS/2 (Model 60) system. The user sets up the assumptions and parameter through the menu-driven interface. The overall structure of the simulation system can be seen in the Hierarchy Chart in Fig. 2. The user first of all selects a bog area for analysis. Since Bord na Mona bog areas are given a numeric code, it is convenient to use this code as a unique identifier for all files that must be established and maintained for that bog area. This task is accomplished by the program, which concatenates the appropriate file name and the numeric code. Those files maintained for each bog area include; bog area data, model assumptions, machine mix, stockpile number and location and the number of tractor trailer units (for the HAKU system). The program will not allow the user to enter the simulation mode until all this data is established. The attributes of all machine types and models are held on a separate file where for example, a record contains all relevant data on a Type III harvester. This file is only modified when some machine attribute is to be altered, such as a decrease in breakdown duration for a Type III harvester. The file is not related to any particular bog area.

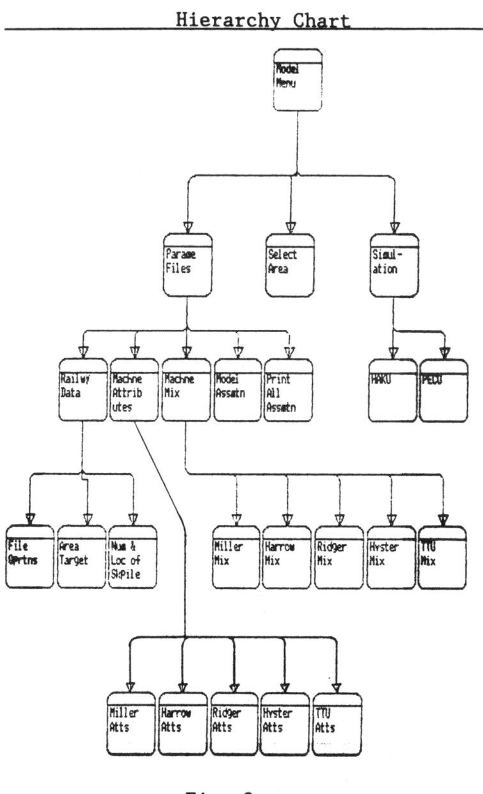

Fig. 2

The user may at any stage, view on screen or print out the assumptions and attributes of the model. Then, when the user is satisfied with the data, the simulation mode may be selected, at which point the desired production system (PECO or HAKU) is chosen.

4 MODEL DESCRIPTION

The simulation methodology used was a variant of the three phase approach first illustrated by Tocher 1963. A very good description of the approach is to be found in Pidd 1984, in which activities are classified as bound activities or conditional. A bound activity is one which must be executed when its scheduled time is reached. A conditional activity is one whose execution is dependent on the cooperation of different entities or when various conditions within the simulation are satisfied.

The entities are the elements of the system which are being simulated. In the model examples of entities are the machines, i.e. millers, harrows, ridgers, harvesters and the railways. In the simulation the railways are processed by the machines; the machine would be regarded as active entities and the railway as passive ones.

The entities of the system move from state to state because of the activities in which they engage - the activities transform the state of the entities. For example, the activity "begin milling" transforms a railway from a state of "waiting for a miller" to the state of "being milled".

It is the interaction between the entities which produce the distinctive behaviour of the model and it is that behaviour which is monitored and measured by the simulation model. Each entity has associated with it a number of attributes. In the model the railway entities have the following attributes:

- time cell (scheduled time of next activity
- next activity (code for type of next activity)
- last activity (code for type of last activity)
- railway number (ID of railway)
- number of fields (number of fields in the railway)
- railway length (average length of field in railway)
- harvests completed (no. completed to date)
- machine link (ID of machine working on railway)

The machine entities have the following attributes:

- time cell (scheduled time of next activity

- next activity (code for type of next activity)
- last activity (code for type of last activity)
- machine type (i.e. miller or harvester etc.)
- machine model (which type of miller etc)
- next breakdown time
- breakdown frequency type (type of distribution for breakdowns)
- railway link (ID of railway)
- mean or range of breakdown frequency distribution
- standard deviation of breakdown duration distribution
- breakdown duration type (distribution for duration of breakdowns)
- mean or range of breakdown duration distribution
- standard deviation of breakdown duration distribution
- speed type (distribution of speeds for the machine)
- mean or range for speed distribution
- standard deviation of speed distribution.

All of this information on the entities in the system is held in an N x M array where N is the number of entities and M is the number of attributes that an entity can have. The state of the system at any point in time is reflected in the values of the attributes.

The simulation process can be represented by the flowchart (Fig. 3).

An important parameter the user must determine is the run time. This is the length of time that the user wishes to have simulated, it can be as short or as long as is required. It is better to give a run time which is sufficiently long to enable a number of harvests to be simulated. Typically, in running the model a year's production is simulated. Experience with the model has shown that the first two harvests in a season represent the transition state and it is only from the third harvest onwards that the system reaches a steady state situation.

Once the simulation has started the model scans each entity to check if it is ready to be activated for that entity against the simulation clock. If this is less than or equal to the simulation clock and if the activity is of the B (Bound Activity) type then this activity will be executed. On the other hand if the activity is of the C (Conditional Activity) type then the other condition(s) necessary to carry out that activity are tested. If all of these are satisfied then the activity is carried out, otherwise the entity is queued for that activity (see Fig. 4). An example of this would be a railway which has been ridged and is waiting for a harvester. The harvester should be available as soon as the railway has been ridged. However, if all of the harvesters are already working on other rail ways then the railway would have to wait for one to be available. The model stores all occurances of queueing by railway as well as the idle time of machines.

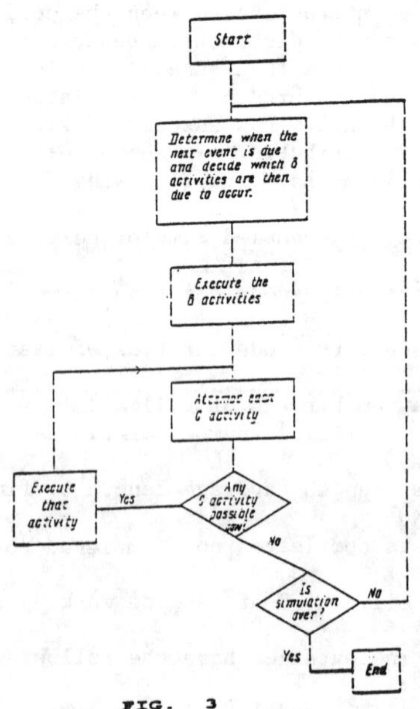

FIG. 3

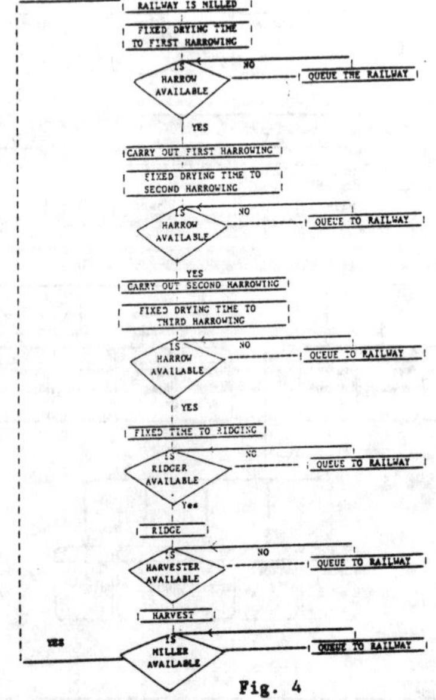

Fig. 4

To illustrate the logic of the model it is helpful to consider the milling operation. The model first checks if any railway has been scheduled to be milled before or at the current clock time. If one has, the model checks on miller availability. If a miller is available the model calculates the time, if any, that the railway was waiting for a miller and records this. It also determines the length of time that the miller had been idle. The duration of the milling activity is then determined. This is done by sampling the speed of the miller from a normal distribution (see Section 5). The milling time is obtained by dividing the length of the field by the speed and allowing for turning times. Turning times are the time it takes the milling machine to turn at the end of each field in the railway. If the miller is going to breakdown during the milling operation (this is obtained from the sampling of breakdown frequencies) the following rule is used. If the breakdown is less than two hours in duration (this is obtained from sampling the distribution of breakdown duration for millers) then the completion of the milling operation and the miller will be delayed by this amount of time. If the duration of the breakdown is greater than two hours then another miller will be assigned to the railway as soon as possible after the first one breaks down and the replacement one will start from the position where the first broke down. These rules reflect the decision that would be taken by the bog supervisor. If a meal break occurs during the milling operation then the milling time is increased by the duration of the break. Similar modules are used for the harrowing, ridging, and harvesting operations.

The harvesting module (under PECO) is more complex than the others because of the requirements to use five different speed distributions for the five lifts involved in harvesting. If a railway is waiting to be harvested, the model first checks that the number of complete harvests for the whole bog area. This ensures that all railways are harvested in this cycle before any are harvested in the next cycle. This does not mean, of course, that the other activities such as milling cannot proceed for railways that have been harvested in the current cycle. The model then checks if any harvesters are available. Commonly, up to three harvesters will be used for the PECO system (assuming they are available). For the HAKU system at least one tractor-trailer unit must be ready to serve a harvester. The routine to calculate the completion time for harvesting is complicated (particularly for the PECO system). It involves holding the 'mix' of harvesters and the unique speed attributes for each in an array and matching the railway parameters with them. For example, certain harvester types are used for heavier - third, fourth and fifth - lifts. However, not all railways have five lifts either side of the central stockpile field, and the machines must be allocated to the fields accordingly. The speeds of the harvesters are obtained for the appropriate lifts, giving the harvesting time for each field and the total harvesting time for that railway. Under the HAKU system the harvester loads the peat directly into tractor-trailer units (see Section 2.2). The trailers are filled after about two hundred and fifty metres (most fields exceed this in length), and the change-over time to another tractor-trailer unit takes about fifteen seconds. The return time of the units are calculated by finding the nearest stockpile. This involves tracking the (X, Y) coordinates of the tractor-trailer units and determining the distance to the (X, Y) coordinate of each stockpile. If these units cannot keep up with the harvesting speed of the harvester then the process is delayed.

5 DATA COLLECTION

The model is dependent on a number of key operational assumptions and parameters (Wyman 1977). In particular the speeds, breakdown duration and breakdown frequencies of the various types and models of machines are critical. The data for these parameters were collected from the Works Study Section within Bord na Mona. The data were tested to see if they could be represented by standared statistical distributions. (Kleijnen 1974, Poole & Szymankiewicz 1977). In most cases the goodness of fit were significant. Speeds of machines followed normal or uniform distributions. Breakdown frequencies and durations followed exponential distributions. A list of these details are given in the Appendix.

Bord na Mona has a number of bog works throughout Ireland. Each works has a number of bog areas. Each bog area is a separate production unit. The size of these bog areas can vary but a typical size would be one thousanc acres (400 Hectares). Specific information for each bog area was obtained from the appropriate works. This information included:

- size of bog area
- number of railways
- number of fields in each railway
- average length of field in each railway
- machine allocation by type and model

- daily hours of operation
- number of harrowings per harvest
- average time between harrowings
- distribution of lunch and tea breaks.

The number of fields in each railway was important for determining the speeds of harvesters under the PECO system. The distribution of harvester speeds for each type of lift had been obtained previously and knowing the number of fields in a railway allows the model to determine the time taken to lift the peat from that railway.

6 OUTPUT FROM THE MODEL

All simulation assumptions and parameters are printed out. Some of these include:

- simulation duration
- tea break and lunch break times
- pick-up tonnage
- machine configuration
- bog area details
- stockpile locations (HAKU only)

Results of the simulation include:

- net kilometer available for harvesting
- target tonnes
- harvested tonnes
- % target harvested
- average number of crops
- average trip to stockpile distance (HAKU only)

In addition, machine usage and performance statistics are provided:

- mean idle time for each machine type
- mean down time for each machine type
- mean utilization for each machine type.

7 VALIDATION OF THE MODEL

Preliminary testing of the PECO production model was carried out during development of the model with satisfactory results. The model was further refined following discussions with senior management personnel and a rigorous programme for validating the model was proposed. Representative areas (16 and 20) selected were in the Blackwater Works located in County Galway in the West of Ireland. The objective of the experiment was to validate the design of the model as a tool to predict the outcome of any machine/bog mix, and the result of any alteration of machine allocations, through the use of historic information as test data. To this end the experiment was successful. The principal measure of production performance is target tonnes - this is established for each bog area before the harvesting season commences. The simulation outputs matched the measures of performance of the real life system. Tonnes harvested achieved in the simulation experiment were within two percent of the actual results for the 1987 harvesting season. Other outputs of the model suggested that the millers and harvesters had the greatest down times, that machines in Area 16 were utilized more heavily than in Area 20, that harvesters are rarely idle and in most cases were the bottleneck of the system.

SIMPEAT was modified in August 1988 to facilitate modelling of the HAKU production system. The changes included identifying good locations for stockpiles and the number of tractor-trailer units to deliver the peat to them, and adjusting harvester speeds (see Section 2). Since all other parameters of the model were the same as the PECO system, similar test data was used for validation. Results were compared with experiments carried out by the Research and Development group at Bord na Mona and were found to be accurate.

Weather parameters such as rainfall, hours of sunshine, wind speed have not been explicitly excluded in the model. However, activity durations which would ordinarily be influenced by the weather have been determined in a number of ways.

(a) the average number of harrowings, which is heavily influenced by the weather, was based on the historical data from the particular bog area that is being simulated.

(b) the drying times between harrowing activities are built into the model based on figures obtained from the Torraun/Derries Works.

Because the activities above are critically affected by the prevailing weather conditions, any change in weather could result in changes in these activities. An effort to establish the relationship between the weather and these weather-dependent activities will involve extensive research. If weather conditions were to be randomly sampled, the variations between different simulation runs could be large. Consequently, the task of isolating the effect of changing a decision such as the number of harvesters, would be very difficult. While the present means of dealing with weather may not be completely satisfactory, it does offer the advantage of being able to examine the effect of a single decision. This is adequate for planning purposes since it is impossible to predict the weather a year in advance in Ireland. However, work is ongoing to allow the user to select weather conditions of good, fair or poor without having to change the source code for the model. At the moment, the weather conditions built into the model are average weather conditions for Ireland.

8 APPLICATION OF THE MODEL

8.1 Scenario Analysis of HAKU Vs. PECO Systems.

The scenarios examined are outlined in the paragraphs below.

Bord na Mona management are introducing the HAKU production system on a phased basis in all works. The use of this system in other countries has suggested that substantial gains in output are possible and that reduced moisture content in peat stock will result. An analysis of a changeover from PECO to HAKU was required.

In the past, Bord na Mona has employed all personnel involved in harvesting as full time staff and have purchased (or developed) and maintained all machinery used in the production process. A decision to contract out some harvesting operations for the 1989 season was taken. The impact of contracting was examined with the simulation model.

Bord na Mona management wished to examine the impact of further technical changes in higher machine speeds and greater reliability. They asked that this be done on the basis of 1987 production parameters and weather patterns so as to be able to assess the potential impact of different strategies.

SIMPEAT was used to analyse the above scenarios. A plan for analysis was determined and carried out by the authors. A Base Case (Case 1) was established and various scenarios (given below) were compared with it (see Table 1).

Table 1: Scenarios for Analysis

CASE 1:	Output based on assumptions for the production year 1987 for the PECO system and projected outputs for the HAKU system for that year. This Case simply simulates the 1987 production season.
CASE 2:	Same as CASE 1 except that the working week is reduced from 7 to 6.25 days, and that lunch and tea breaks increase by 15 minutes each. This Case would represent a 'worst case' situation.
CASE 3:	Same as CASE 1 except that the duration of breakdowns is reduced by 50%. This represents some improvement in operational performance, viz that contractors will not tolerate any break-downs.
CASE 4:	Same as CASE 1 except that the duration of breakdowns is reduced by 50%, and all tea and lunch breaks are eliminated. This CASE represents a full improvement in operational performance through contracting but excludes pick-up efficiency.
CASE 5:	Same as CASE 4 except that the frequency interval for harvesters down-time is doubled, and forward speeds for harvesters are increased by 50%. This represents technological improvement, i.e. higher speed and improved reliability.

A further five cases identical to the HAKU production system replacing the PECO system were carried out. The results are shown in Table 2. Management requested that the increase in pick-up efficiency under HAKU be excluded until further field work confirms theri initial findings on these losses. To include the effect of fewer losses, requires only a change in one input parameter to the model.

Table 2 Summarised Results of Scenarios Analysis

SYSTEM: PECO
AREA: 20

	Tonnes	% Target	No. Crops
Actual	72058	116.50	14.50
CASE 1	72936	117.90	14.26
CASE 2	58895	95.18	11.52
CASE 3	78219	126.42	15.30
CASE 4	95569	154.51	18.69
CASE 5	103783	167.73	20.30

SYSTEM: HAKU
AREA: 20

	Tonnes	% Target	No. Crops
CASE 1	83148	134.39	14.71
CASE 2	67905	109.75	12.01
CASE 3	84600	136.73	14.97
CASE 4	104168	168.34	18.34
CASE 5	107273	173.37	18.97

A two-pronged strategy resulted from the analysis:

(a) that the HAKU system and contracting be introduced, yielding as much as a ninety per cent increase in output.

(b) that the focus should be on the milling/harvesting cycle and pick-up efficiency and not on machine speeds and machine output, resulting in a sixty percent increase in output.

8.2 Determination of stockpile location with the HAKU system

As outlined above, Bord na Mona plans to introduce the HAKU production system on a phased basis for each bog works in Ireland. Research and Development staff had been asked to determine the appropriate number and locations of stockpiles. While the simulation cannot be used as an optimization model, alternative scenarios can be examined to identify the sensitivity of the production system to changes in the number and location of stockpiles or in the number of tractor trailer units. The simulation outputs helped support the decision making of Research and Development staff and enabled them to fix the location of stockpiles in various bog areas.

9 CONCLUSIONS

The simulation model has been developed as a tool to study the milled peat production system. It reproduces the complexities of the milled peat production process. It supports managers in their planning function and gives them the facility to assess a greater degree of alternatives than were possible heretofore. The model has been extensively validated and has proven to be capable of dealing with different types of problems, production systems and bog areas. It is the first known simulation model of the entire milled peat production system in the world.

REFERENCES

Barry, C.J., Kavanagh, R. & Murphy, C. 1988. A simulation model of the milled peat production system. Leningrad. Proceedings of the VIII International Peat Congress.

Coffey, P.J., Kavanagh, R., & Murphy, C. 1985 Towards a model of the milled peat production process. Jonkoping. Proceedings of the Peat and Environment '85 International Peat Society.

Emshoff, J.R., & Sisson, R.L. 1970. The design and use of computer simulation models. New York: Macmillan.

Fishman, G.S. 1973. Concepts and methods of discrete event digital simulation. New York: Wiley.

Kleignen, J.P.C. 1974. Statistical techniques in simulation. New York: Marcel Dekker.

Naylor, T.H. 1966. Computer simulation techniques. Wiley.

Neelamkavil, F. 1987. Computer simulation and modelling. New York: Wiley.

Pidd, M. 1984. Computer simulation in management science. Wiley

Poole, T.G., Szymankiewicz, J. 1977. Using simulation to solve problems. London: Mcgraw-Hill.

Smith, J. 1968. Computer Simulation Models. London: Charles Griffin.

Tocher, K.D. 1963. The art of simulation. London: English Universities Press.

Wyman, F.P. 1979. Simulation of the sands mining operations. Interfaces. Vol. 8. No. 1.

APPENDIX

The following are the key distributions which are used in the model.

Millers

- speed (normal distribution mean = 8.33 k/hour, std = .89)
- breakdown frequency (exponential distribution mean = 33.9 hours, std = 35.7)
- breakdown duration (exponential distribution mean = 7.9 hours, std = 8.1).

Harrows

- speed (normal distribution mean = 18k/hour, std = 3)
- breakdown frequency (exponential distribution mean = 195 hours, std. = 196)
- breakdown duration (exponential distribution mean = 3.5 hours, std = 3.75).

Ridgers

- speed (uniform distribution mean = 10.9k/hour, std = 16.2)
- breakdown frequency (exponential distribution mean = 22.3 hours, std = 21)
- breakdown duration (exponential distribution mean = 14.7 hours, std = 5.1)

Harvesters

- speed for lift 1 (normal distribution mean = 4.8k/hour, std = .9)
- speed for lift 2 (uniform distribution range (3-4k/hour))
- speed for lift 3 (uniform distribution range (2-3.8k/hour))
- speed for lift 4 (uniform distribution range (1.5-3k/hour))
- speed for lift 5 (normal distribution mean = 2k/hour, std = .53)
- breakdown frequency (exponential distribution mean = 28.9 hours, std = 26.7)
- breakdown duration (exponential distribution mean = 12.1 hours, std = 13.3).

Land and Water Use, Dodd & Grace (eds), © 1989 Balkema, Rotterdam. ISBN 90 6191 980 0

Milled peat production control

M.McNulty
Bord na Mona, Newbridge, Ireland

L.Campbell
Meteorological Service, Dublin, Ireland

ABSTRACT: In volume terms, milled peat operations constitute the main production system for fuel peat worldwide. Production whether by Haku or Peco harvesting systems is totally weather dependant. However, the difficulties caused by bad weather and changes in weather conditions can be reduced using specialised weather forecasts. Research has been carried out in Bord na Mona to test whether such forecasts can be used to control production, and ensure greater use of available drying. In 1987 and 1988 field trials were undertaken and two thirds of the variation in 4-day evaporation was accounted for. The sales yield in 1987 was increased by the order of 7% and in 1988 by the order of 100%.

1 INTRODUCTION

Most fuel peat throughout the world is produced as a friable crumb called milled peat. Drying of the raw milled peat is achieved by the utilisation of solar energy.

2 THE PRODUCTION CYCLE

The term milling refers to the cutting and shredding of the top layer of the bog surface. The greater the depth of this layer, the greater the yield of peat. The moisture content of the peat is reduced through solar radiation from approximately 75 - 85% after milling, to 45 - 50% at harvesting. This process generally takes 3 to 4 days and is known as the harvest cycle. The machinery allocated to each production area is based on this cycle.

The harrowing operation undertaken after milling, aids drying by breaking the "crust" on the surface of the milled peat layer and by exposing the bottom layer to the elements, thereby increasing evaporation efficiency. There are generally 3 harrowing operations for each harvest cycle.

The peat, when dry, is ridged and then harvested into the stockpile.

3 WEATHER FORECASTING AND PEAT PRODUCTION CONTROL

Peat production can be controlled using specialised weather forecasts for specified areas. These forecasts are used to:

- Vary the yield of peat milled according to the available drying.
- Minimise the milling and harrowing operations during spells of unsettled weather.
- Maximise harvesting in unsettled weather conditions.

It is therefore possible to relate peat production more directly to weather and bog conditions, thereby increasing stocks and reducing production costs.

4.0 DRYING CHARACTERISTICS OF PEAT

The air-drying of peat consists of 3 phases

Phase 1 - Saturated surface phase (90-70% moisture on a wet basis)

During this phase there is a constant rate of drying, as the water is evaporated in the same manner as from an open water surface. The controlling weather parameters are temperature, humidity, solar radiation and wind.

Phase 2 - Unsaturated surface phase
(70-50% moisture)

The rate of drying decreases linearly during this phase, and is influenced by air velocity and humidity. In this phase the properties of the drying layer play a role in determining the rate of drying.

Phase 3 - Internal diffusion phase
(50% - to equilibrium)

This is characterised by a further decrease in the rate of drying. The limiting drying factor is the resistance to the movement of liquid or vapour through the peat rather than from the surface of the peat. Hence the thickness of the sample (yield), the sample shape and temperature, grain size distribution, and equilibrium moisture content are the governing factors for drying.

5.0 FACTORS AFFECTING THE DRYING RATE OF PEAT

The three main factors which affect milled peat drying include:

- Atmospheric variables:
- Field variables:
- Milling depth.

5.1 Atmospheric Variables

The drying rate/evaporation for a given harvest cycle is governed by solar radiation, air velocity, humidity and precipitation.

Solar radiation

The bulk of the energy required for the drying of peat comes from solar energy. Therefore all other things being equal, the drying time is inversely proportional to the solar radiation intensity. Solar radiation is at its greatest intensity at noon (approximately 2 or 3 times greater than at 0600 h and 1700 h) and thus maximum drying is achieved around midday. The intensity of solar radiation is a function of the time of the year reaching its peak in June/July in Irish conditions.

Air velocity

The drying mechanism is the result of a velocity gradient formed in the proximity of the drying surface, coupled with solar radiation. The laminar boundary layer formed near the surface of the gradient resists heat and moisture transfer from the peat.

As the velocity of the wind decreases, the thickness of the laminar boundary layer decreases, which results in an increase in moisture transfer. However, as the wind moves over the surface of the peat, removing the evaporated moisture, it also removes some of the heat received by the peat, thus reducing the effectiveness of the drying process.

Humidity

The rate of evaporation is directly proportional to the difference in partial pressure of the vapour of the surrounding air and that at the surface of the drying material.

Precipitation

Small amounts of rain ($<$1mm/day) will not affect the drying time in good drying conditions.

5.2 Field Variables

The two important field variables are the depth of the ground water table and peat quality.

Depth to ground water table

The moisture content at the surface of the bog varies according to the depth of the water table from the surface. It may also affect the relative humidity of the nearby air, thus affecting the drying rate.

Degree of decomposition of the peat

The degree of decomposition becomes a major factor in poor weather conditions, when the effect of rain showers is more dominant. The lower the density of peat, the greater the absorption capacity and the longer the drying cycle

5.3 Milling depth

The thinner the layer of peat milled, the shorter the harvest cycle time, for a given cumulative cycle evaporation. Generally, as cumulative cycle evaporation increases, for a given harvest cycle time, the thickness of the layer of peat milled can also be increased - taking into account peat quality and bog conditions.

6.0 VARIABLE MILLING DEPTH CONTROL

There are many factors which must be taken into account when planning production including available drying,peat type, bog conditions and the harvest cycle time. In order to optimize milling operations, the layer of peat milled should vary as cumulative cycle evaporation varies.

The more decomposed the peat, the greater the volume reduction and drying effectiveness. The quality of peat varies widely from area to area within Bord na Mona, with bulk density values ranging from 150 grams per litre (g/l) to 300 g/l. It is found in practice that as quality (i.e. density) decreases, so too does the maximum allowable peat layer thickness.

In practice there are two options available to increase milling output:

- An increase in the depth of the layer of peat milled during spells of anti-cyclonic weather. This makes optimal use of available drying and the peat is harvested within the 3 to 4 day harvest cycle.

- A decrease in the layer of peat milled. This allows the peat to dry to 45-50% moisture in a shorter, say 2 day, harvest cycle time. The major disadvantage of this method is the high cost incurred in purchasing extra machinery, in order to allow a shorter harvest cycle to be achieved.

The essential features of a variable milling depth control system include:

- Accurate weather forecasts:

- A milling depth control system. This would mill to a set depth, taking into account bog conditions, as well as weather parameters.

7.0 PRODUCTION CONTROL USING SPECIALISED WEATHER FORECASTS

In the 1988 production season, field trials using specialised weather projections were undertaken at Cloncreen bog, in the Derrygreenagh group. There were two critical parts of the field work:

- Weather and evaporation forecasts:
- Production control.

7.1 Weather and evaporation forecasts

Meteorological aspects

The Irish Meteorological Service, in conjunction with the European Centre for Medium Range Weather Forecasting (ECMWF)* supply 4-day predictions of the following weather parameters:

- Surface temperature
- Dew point
- Wind
- Rainfall amount, type, likelihood and time of arrival
- Cloudiness
- Surface solar radiation
- Evaporation

These parameters are translated into predicted cycle evaporation using a computer model, which is based on the use of the Penman equations (McNulty et al 1988). The forecasts are drawn up at 0800hrs. and updated at 1500hrs. every day.

Field weather station

An automatic meteorological station was set up at Cloncreen bog. The objectives were to:

- Correlate predicted and actual weather data

- Check the accuracy of the forecast as received.

- Check the drying rate by examining the humidity readings and evaporation rate.

The parameters recorded were:

- Air temperature and relative humidity at 2m height, using an aspirated sensor (Psychrometer H301)

- Wind speed at 3m height using a wind Porton anemometer.

- Solar radiation at 2m height using a Kipp and Zonen type solarimeter.

*For further details on ECMWF weather forecasting techniques, see McNulty et al (1988)

- The evaporation rate measured on a continuous basis using a standard class "A" evaporation pan mounted on a weighing transducer, from which the weight of water is recorded at regular intervals.

- Precipitation measured using a simple rain guage. The time and intensity is read from a rain recorder chart.

The data were recorded in a semi-automatic manner, onto a data logger. The data logger was manually debriefed every 2 weeks.

The predicted and actual weather data have been examined to determine accuracy and predictability. The nature of any error is important. Thus if there is a systematic bias, i.e. if the forecasts are consistently higher or lower than observed values, it will be possible to apply corrections to the forecasts in 1989.

Operation of the weather station and difficulties encountered.

A number of practical difficulties were encountered:

- There was no direct power supply at the field station, so power was provided by two heavy duty lead acid batteries connected in parallel. A wind charger was available to supply a back-up service. As a result, it was possible to leave the meteorological station for up to 7 days without changing the batteries.

- Dust on the bog permeates the muslin surrounding the wet bulb, thus obstructing the flow of water. Therefore it was necessary to change the wet bulb wick frequently (once a day). This proved to be impractical and it is proposed to use a solid state sensor for future work. This sends signals to the data loger which are proportional to the relative humidity. This type of sensor has a long-term drift problem. However, it is possible to build in a correction factor to counteract this.

The dome of the solarimeter must also be cleaned daily if readings are to be regarded as accurate.

7.2 Production control

On the basis of accurate weather forecasting and the drying and bog parameters discussed earlier, it is possible to control and plan the work to be carried out on the bog. The algorithim used for this purpose is discussed below and the production control process is described.

Algorithm for deep milling operation

<u>Day 1</u> - Start daily weather prediction using meteorological office data - rainfall, solar radiation, wind speed, cloud cover, air temperature, dew point and evaporation for 4-day cycle.

<u>Day 2, Day 3 and Day 4</u> - Obtain actual evaporation in Derrygreenagh.

<u>Day 5</u> - Start production operation/weather algorithm. Get 4-day cumulative evaporation prediction from days 2-5 = Y1.
Get 4-day cumulative actual evaporation from days 2-5 = X1.
Get rainfall prediction for days 5-8 = R1.
P1 = predicted 4-day cumulative evaporation for days 5-8.

Harrowing and Harvesting operations are determined:

If $R1 > 4.0$mm - cancel deep milling operation
If $R1 \leq 4.0$mm - proceed " " "
If $Y1/X1 > 1.88$ - cancel " " "
If $Y1/X1 < 1.88$ - proceed " " "
If $P1/1.88 < 12$ - cancel " " "
If $P1/1.88 > 12$ - Deep milling operation to be carried out.

Day 6 - start procedure as for day 5, except take cumulative 4-day evaporation from days 2-6. Get rainfall prediction for days 6-9, R2 and carry on as before.

Method 1.

When rain is predicted, work such as milling and harrowing can be stopped and the effort concentrated on the collection of dry or almost dry peat.

This technique was applied in Cloncreen during the 1988 production season. As a result crops were harvested before the rain arrived and the number of (unnecessary) harrowing operations was reduced.

If rain is predicted, a shift or the arrival of contractors can be cancelled. This generates large cost savings, especially for weekend work. Servicing of machinery may be timed to coincide with lengthy periods of rain.

With the prediction of fine weather, it is important to know when it will arrive, so that all available drying energy can be fully used.

In unsettled weather conditions, where the drying rate is not good, or there are frequent spells of rainfall, milling operations should be adjusted to suit. Thus the milling yield should be reduced - to say 30 tonnes per kilometer (T/Km). This could, despite poor weather conditions be harvested in the 3-4 day harvest cycle.

Method 2

When there is a predicted period of anti-cyclonic weather cycle efficiency can be increased by increasing the milling thickness (yield).

The decision to mill deeply is based on ECMWF predictions of 4-day cycle rainfall and evaporation. If cumulative predicted rainfall is less than 4mm (Ref. Algorithim), and the predicted cumulative evaporation divided by 1.88 is greater than 12mm, - the deep milling operation is performed. This is achieved by dropping the setting of the miller and reducing the forward speed. Otherwise the millers are allowed to operate as normal.

This method of deep milling worked satisfactorily in practice. However, there was a slowing down in the milling operation as a result of the reduction in the miller forward speed. This is a prime mover behind the idea of a miller with automatic depth control (Kennedy 1988).

Analysis of the field and sales results for the 1988 production season demonstrated that output can be improved through control of milled peat production. This is achieved by:

- Increased milling output by increasing the thickness of the layer in line with weather conditions:

- Correct timing of work phases:

- Taking into account the demands placed on production machinery by weather conditions, and machine availability.

The drying supervisor, who determines when the crop is fit to be ridged and harvested found no time delays in the harvest cycle due to unfit crops. It was also found that low quality peat (100-200g/l) could not be deep milled as it was unable to reach 45-50% moisture content within the 3-4 day harvest cycle.

It is of paramount importance, therefore that deep milling is performed on a selective basis depending on field conditions and peat quality.

8.0 HARROWING

The timing of harrowing is very important in the deep milling operation (Ward et al 1988). Harrowing ensures that all the peat is turned and exposed to the drying elements.

The standard Bord na Mona harrow spoon (4.1kg) is not capable of turning the heavy crops in deep milling. To counteract this weights were added above the spoon (total weight 5.5kg). This harrow was found to be more effective in turning the peat (Flood et al, 1972)

The speed of the harrow was reduced to allow small ridges to be formed across the field, exposing more peat to the atmosphere.

9.0 THE OPTIMUM MILLING DEPTH/YIELD

This question is extremely difficult to answer, as the optimum depth will depend on the effectiveness of drainage (field preparation) and the quality of the peat. Thus deep milling must be on a selective basis. The milling depth should be as great as possible during spells of anti-cyclonic weather, but should be limited to a point such that milling of the subsequent harvest can be undertaken immediately after harvesting the previous one.

10.0 PRODUCTION CONTROL - THE FUTURE

Accurate weather forecasting and automatic depth control of the miller are essential to the further development of milled peat production control.

The future meteorological station will be an automatic "on-line" station, which will allow collection and transfer of data on a near real-time basis. The station will be able to store up to 10 days of hourly data. The data logger, which is linked to a local microcomputer, will collect data at set times e.g. daily or hourly.

The Microcomputer can be used to receive and process forecast information, and to collect and send field data to a central data base. Bord na Mona is developing this type of weather station for use in the 1990 production season. The computer will also be used to run a mathematical model, which will correct the weather prediction according to the field data. A second mathematical model will be developed to determine optimum milling depth according to the predicted weather parameters, bog conditions, and the quality of the peat. Thus a computerised production control algorithm can be developed.

11.0 RESULTS OF CLONCREEN PROJECT

11.1 Weather results

Rainfall data

The comparison of actual rainfall at Cloncreen with predicted meteorological values, (cumulative over 4 days) and based on rainfall greater than or less than 4mm showed that the:

Number of "correct"predictions = 79%
Numberof "incorrect" predictions = 21%

Based on the rainfall predictions deep milling would have taken place 29 days in 112 -26%. However 16% of the time deep milling would have been cancelled incorrectly.

Similar results were found by a comparison of actual rainfall at Derrygreenagh with predicted meteorological values cumulative over 4 days.

Rainfall Correct % Prediction by Month

Month	% Derrygreenagh	% Cloncreen
May	79	76
June	80	80
July	75	83
August	81	78

Evaporation data

Values are cumulative over 4 days. The "Penman" values are divided by a factor of 1.88. This corection factor was determined in the 1987 production season (McNulty 1987)

Comparison of actual evaporation levels at Derrygreenagh with Penman and Meteorological Office predicted values showed that based on greater than or less than 12mm of evaporation:

Predicted evaporation "PENMAN"

Number of "correct" predictions = 76%
Number of "incorrect" predictions = 24%

Coefficient of determination R^2 = 0.66

Predicted evaporation Meteorological Office.

Number of "correct" predictions = 64%
Number of "incorrect" predictions = 36%

Coefficient of determination R^2 = 0.23

There is a difference in evaporation correlation figures predicted by the Met Office model and that based on the use of the Penman equations. This unexplained difference is at present being investigated by the Meteorological Service.

11.2 Sales results

A total of 7 stockpiles have been sold from Cloncreen to date (Appendix 1).

Stockpiles	% M.C.	T/Km	% Gain
1988	46.1	46.7	+100%

On examining historical figures for the area, there is an average ouput in Cloncreen of 18% above the Bord na Mona standard of 23.4 T/Km

Thus a net increase of 82% over and above the standard Cloncreen tonnage has been achieved.

However, the 100% gain is over an average of 8 harvests and the Bord na Mona standard is 12 harvests for a season. If 12 harvests had been achieved the increase would have been of the order of 133%.

It should be noted that the average moisture content is 46.1% with a range of 43.4% to 53% which is within the acceptable moisture content range. In other words the quality is up to standard.

It must be stressed however, that the increased yield is dependent on the weather characteristics of the year in question.

The variation in evaporation patterns can be seen in Table 1. The percentage of total annual 4-day cycles with cumulative evaporation greater than 12mm for the 1987 and 1988 production seasons is 31%.

APPENDIX 1

SALES RESULTS 1988 PRODUCTION SEASON

CLONCREEN BOG
DERRYGREENAGH GROUP

Week Ending (1988)	Normal Tonnes 55% M.C.	Actual Tonnes 55% M.C	Sales Moisture Content (M.C.)	Gain Tonnes
21/9	2234	4104	53.0	1870
12/10	2578	5231	45.6	2653
19/10	2639	4907	46.0	2268
2/11	738	1539	44.7	801
2/11	2482	4586	44.7	2104
7/12	1048	2032	45.0	984
21/12	911	2156	43.4	1245
		Average	46.06	

REFERENCES

Anon (1985), VTT Symposium 59, Control of peat production, Technical Research Centre of Finland, Domestic Fuel Laboratory.

Bogatov, B.A., Zheltko, V.E., Kuptel, G.A. 1985. Milling depth Control on a peat bog. Torfyanaya Promeshlennost, 12 p7-11.

Dalton, J. and Mooney, P.C. 1956. Comparative drying rate tests and some recommendations on deep milling during intensive drying periods. Scientific Office, Bord na Mona.

Derov, A.F. 1985 Predicting the technological process for estimating the weather potential for harvesting milled peat. Torfyanaya Promeshlennost, 5 p11-13.

Derov, A.F., Fomina, G.M. 1984. Some fundamentals of utilizing weather conditions in milled peat production. Torfyanaya Promeshlennost, 12, p4-7.

Flood, H.C. and Coffey, P.J. 1972. Estimation of the effects of 3 different harrows on the rate of drying of milled peat. Scientific Office, Bord na Mona.

Lazaryev, A.V. et al 1983. Operational analysis of weather utilisation for milled peat harvesting by enterprises. Torfyanaya Promeshlennost, 6, p7-11.

Lazaryev, A.V. 1986. Operating pre-selected milling depth. Torfyanaya Promeshlennost, 7, 14-16.

O'Kane, J.P., Kuhnel, V. and Sanders, G. 1987. Mathematical simulations of drying peat, Report and computer model developed for Research and Development Department, Bord na Mona, Ireland.

Shier, C. 1984. Fuel peat in developing countries. Report for the World Bank Produced by Bord na Mona, Ireland

Ward, S., O'Kane, J.P., Turner, J., McNulty, J. 1988. Optimising milled peat harrowing operations. In the "Futureworld of Peat Technology" Bord na Mona and the Institute of Engineers of Ireland Conference, University College Dublin.

McNulty, M., Campbell, L. 1988, Production using modern weather forecasting techniques. In the "Futureworld of Peat Technology" Bord na Mona and the Institute of Engineers of Ireland Conference, University College Dublin

TABLE 1

Percentage of total annual 4-day cycles with cumulative evaporation greater than 16mm

(May to August, inclusive , at Derrygreenagh)

1976	1977	1978	1979	1980	1981	1982	1983	1984	1985	1986	1987	1988
35	38	31	13	17	17	55	29	44	50	0	4	12

Mean = 26.5% Max. = 55% Min. = 0.

Land and Water Use, Dodd & Grace (eds), © 1989 Balkema, Rotterdam. ISBN 90 6191 980 0

Developments in milled peat production systems

P.Ring
Bord na Mona, Ireland

ABSTRACT: The new Haku production system has been used in areas originally set out for Peco production. The Haku-Peco system utilises a standard Peco harvester to load a single ridge of peat into a trailer towed by a tractor. The tractor/trailer unit transports the peat to a central stockpile where a bulldozer is used to level and compact the peat. In this way stockpiles with up to 10,000 tonnes of peat are formed. Before production the areas are modified to suit Haku production. This involves the preparation of stockpile sites and the construction of turning grounds to minimise tractor travel time. The stockpiles are built on prepared gravel bases. Initially the peat in the stockpile is formed into a wedge shape such that the peat can be carried onto the stockpile by the tractor/trailer units. A dozer forms the final shape of the stockpile. The Haku system offers the advantage of increasing the quantity of saleable stock as well as improving the quality of peat at sale. In addition 10% more production area is available due to the absence of Peco stockpiles.

1. INTRODUCTION

The Haku system of milled peat production was used for the first time by Bord na Mona at Oweninny Works in Co. Mayo during the 1988 production season. The system was tested in a production area of 125 hectares which had been in production under the Peco system for over 20 years. The system was tested with the aim of reducing the weather risk associated with milled peat production under the Peco system and establish whether peat output and quality would improve.

2. SYSTEM DESCRIPTION

The milling, harrowing and ridging operations are the same as in the Peco system. In the Peco system a harvester is used to transfer the peat ridges from field to field until the peat on the five fields at each side of the stockpile field is harvested into the stockpile. The harvester actually handles the peat three times in transferring it to the stockpile. In the Haku system a harvester is used as a loader to load a single ridge of peat into a tractor trailer unit (TTU) of $18m^3$ capacity towed by a tractor, on the field next to the harvester. In this case the harvester handles the peat once. The T.T.U. transports the peat to a central stockpile area. The peat is tipped along the centre line of the stockpile and the dozer levels and compacts this first layer such that the T.T.U.'s can drive on the compacted peat to unload. This process continues and the stockpile takes on a wedge shape in side elevation with the T.T.U.'s driving up the shallow ramp and down the steep ramp. This reduces the required dozer time as the T.T.U.'s carry the peat onto the stockpile, while the dozer levels and compacts the loose peat. Eventually the stage is reached where the angle of the approached ramp is too steep for the T.T.U.'s. This was reached at just over 8°. A ramp can now be made across the lower end of the stockpile such that more peat can be carried onto the stockpile or the final shape of the stockpile can be formed by the dozer pushing peat from the base to the apex. In the test production area a layer of peat was placed at the base of the stockpiles on which the dozer travelled while shaping the stockpile. This was done to avoid contamination of the peat with sand and gravel being brought up in the tracks of the dozer. At the end of the production season this layer of which was then wet was pushed onto the stockpile to form a protective cover to minimise wind losses.

In this way stockpiles with up to 10,000 tonnes of peat were formed. The final dimensions of the largest Haku stockpile were 46m wide at the base, 9m high and 100m long. A stockpile of these dimensions contains up to 100 tonnes of peat per metre run, compared to 3 to 5 tonnes per metre in Peco stockpiles. Another difference between the Haku and Peco methods of stockpiling is that the peat is very well compacted in a Haku stockpile whereas the peat is quite loose in a Peco stockpile. Compaction factors of 1.5 - 1.6 were achieved in high density peat. This was achieved through the intensive trafficking on the stockpile of the dozer and in particular the T.T.U.s.

3. PREPARATION OF PRODUCTION AREA

The area in which the Haku system was tested had been in production under the Peco system for over 20 years. Therefore the challenge was to successfully operate the new production system in an area which was set out for Peco production. The first and most vital step was to prepare the stockpile areas. As the production area contained a number of gravel ridges it was decided that these would provide an ideal base on which to build the stockpiles. Three gravel ridges were levelled and a perimeter drain was excavated around each site. The surface was graded towards the perimeter drain such that the rain water would flow away from the stockpile. Turning grounds were then formed at various locations to minimize the travel time of the tractor/trailers units. The turning grounds were formed by piping the field drains and backfilling with dry peat.

The trailers were developed and constructed at Oweninny works. The criteria on which the trailer design was based included the bearing capacity of the bog, the heaped density of the peat and the desired operating characteristics of the trailer. The trailer proved very successful during the production season.

4. PLANT

The major concern prior to the production season was whether the bog surface would be damaged by the intense trafficking of the T.T.U.'s. With this in mind a number of different combinations of tractor and trailer were tested. Both tracked and wheeled trailers were constructed. In addition tractors with flotation tyres, twin wheels and wrap-around track and with 100 HP (74kW) and 140 HP (104kW) engines were used to tow the two types of trailer. The combination which proved most satisfactory when considering both operating performance and cost was the tracked trailer towed by a 100 HP tractor with twin wheels.

The Type III Peco harvester proved effective as a loader in that its output in tonnes per hour was up by 23% compared to its output in the Peco system. The physical operation of loading a ridge of peat into a trailer was carried out without major losses of peat and the coordination of movement of the harvester loader should be developed to replace these harvesters which have high mechanical downtime, are slow during turning and during idle travel and are costly to overhaul.

5. ADVANTAGES OF CENTRAL STOCKPILES

The main advantages of the Haku system lie in the method of storing the peat. The large Haku stockpiles are superior to small Peco stockiles for a number of reasons:-

5.1 Lower moisture content

Even though the peat is harvested at about 45% m.c.(% H_2O) in both systems, small Peco stockpiles are susceptible to an increase in moisture content as the exposed surface area is large relative to the volume of peat in the stockpile. In addition as Peco stockpiles are formed on peat fields, moisture can enter through the base of the stockpile. This problem of an increase in moisture content is exacerbated in poor production seasons as a smaller quantity of peat is harvested, leaving stockpiles of less than 2 tonnes per metre run. Large Haku stockpiles of 50 to 100 tonnes per metre run can be formed even in poor production seasons, as all the peat is transported to central locations. Generally each stockpile will have a catchement area of not less than 40 ha. These Haku stockpiles have a low surface/volume ratio and are located on well drained bases.

The two factors lead to greater stability of the moisture content during the storage period.

5.2 Lower rail laying costs

The cost of laying temporary track along the stockpile at sale is greatly reduced with the Haku system. If a Haku stockpile is adjacent to a permanent railway line 300m of temporary track should be sufficient to load a stockpile with up to 10,000 tonnes of peat. An average Peco stockpile with 10,000 tonnes would require the laying of 2,800m of temporary track.

5.3 Increased saleability of stock

In years of poor production much of the stock is stored in very small Peco stockpiles which are uneconomical to sell as the cost of laying temporary track is inversely proportional to the size of stockpile. It is possible to have hundreds of thousands of tonnes of stock in very small Peco stockpiles and yet be unable to meet the sales demand. The Haku method of stockpiling should eliminate this and every tonne of peat harvested becomes available for sale as the peat is transported to central locations.

5.4 Blending

Haku stockpiles act as blending bunkers, where peat of varying quality can be hauled to one location. The central stockpile system ensures that an even quality is achieved. This should prove most beneficial in the further processing of the peat in the power stations and briquette factories.

5.5 Minimal wind losses

The results of the test production season suggest that well compacted and well-shaped central stockpiles will lead to a reduction in the loss of peat during storage due to storms. A storm with wind gusts of up to 142 kms/hr was experienced at the end of the production season. This led to a considerable loss of stock from the Peco stockpiles yet the losses from the Haku stockpiles were neglegible.

6. ADVANTAGES OF HAKU SYSTEM OPERATION

6.1 10% increase in production area

One obvious advantage of the system is that there is a 10% increase in production area due to the absence of Peco stockpiles. This leads to an increase in the quantity of peat produced.

6.2 Double ridges

Another advantage of the system is that it allows the double ridge method to be utilised. Instead of loading a single ridge, it is left on the field and the field is milled and harrowed at each side of the existing ridge. The ridger now forms what is in effect a double ridge and the peat is loaded in the normal manner. This method reduces the time taken by the harvesting (loading) operation as two harvests are loaded in one operation. Clearly the output of the harvester in terms of tonnes per hour is greatly increased.

6.3 Flexible system

The Haku system is a flexible means of producing peat. It can be used in small production areas unsuited to Peco Production. It can also be used effectively in production areas which have become fragmented through timber infestation or the presence of gravel ridges.

6.4 Increased yields

The initial indications are that the system will lead to increased yields in terms of tonnes per hectare through a reduction in losses at harvesting and loading of the peat at sale (Ref. Table 1).

Harvester losses in the Peco system arise as the peat is transferred from field to field. The indications are that these losses will be reduced as the peat is loaded directly. In addition the losses while loading at sale should be reduced as the stockpiles are located on well drained bases.

TABLE 1

Details of Haku Stockpiles

Stockpile No.	Stockpile Shape	Bulk Density	Standard Output @ 55% m.c.	Average % m.c. at Sale	Output at Sale @ 55% m.c.	Output Increase %
1	Peaked	375 g/1	7390	51.5	9223	25
2	Flat	345 g/1	3380	54.4	4579	35
3	Flat Top	310 g/1	5970	49.6	7920	32

7. SALES RESULTS

Table 1 shows the sales details of the three Haku stockpiles. The peat in general was high density except for stockpile no. 3 which had a mixture of high density and low density peat.

Each of the stockpiles was constructed to a different shape at the end of the production season, such that the moisture increase due to rainfally could be assessed. It seems that the high average moisture content of stockpile No. 2 was due to its flat shape and low rainfall runoff.

The sales yield of the Haku production area was up by 30% on average compared to the standard yield of the Peco system.

8. DISADVANTAGES OF HAKU SYSTEM

The high capital cost is the main disadvantage of the Haku system.
The system requires a large investment by Bord na Mona in tractors and trailers. In addition the cost of modifying existing production areas for Haku production will be considerable.

Production costs will also increase in line with increased manning levels. Four tractor/trailer units per harvester are envisaged, as well as an extra dozer driver for every two harvesters.
In overall terms however the increase in revenue due to quantity and quality improvements more than offsets the greater capital and operating costs.

9. CONCLUSION

The Haku Production System should improve the capacity of Bord na Mona to fully meet the demand for milled peat with a homogenous product of stable moisture content.

This will be achieved through increased production area, increased yields and a greater saleability of stocks. In addition the quality of milled peat supplied to the power stations and briquette factories should be improved through stability of moisture content and the blending of peat in the stockpiles.

References

Ring, P. The Haku production system. In the "Futureworld of Peat Technology", Bord na Mona and the Institute of Engineers of Ireland Conference, University College Dublin, November 1988.

Kavanagh, R. Barry, C. & Murphy, C. A simulation model of the milled peat production system in VII International Peat Congress, Leningrad, USSR, August, 1988.

Land and Water Use, Dodd & Grace (eds), © 1989 Balkema, Rotterdam. ISBN 90 6191 980 0

Hydrologic effects of peat mining in eastern North Carolina

R.W.Skaggs
North Carolina State University, Raleigh, N.C., USA

R.G.Broadhead
Department of Scientific and Industrial Research, Christchurch, New Zealand

ABSTRACT: A hydrologic simulation model was used to evaluate the effects of peat mining on total runoff and peak outflow rates for a 6,000 ha site in eastern, North Carolina. Land uses were fescue pasture before mining and corn-wheat-soybeans after mining. Predicted average monthly outflows after mining were higher than prior to mining for the winter months but lower during the growing season. Peak outflow rates were increased by mining when a conventional open ditch drainage system was used. The use of subsurface drain tubes reduced peak outflow rates below premined predictions, but increased total outflow by about 10%.

RESUME: Un modèle de simulation hydrolique été utilisé afin d'évaluer les effects de l'exploitation de la tourbe sur les taux d'écoulement à la region Est dela Caroline du Nord. Avant l'exploitation du terrain, il y avait des pâturages et après l'exploitation il y avait du maïs-blé-soja. Les écoulements moyens prévus mensuellement après l'exploitation ont été plus élevés qu'avant la mise en exploitation pendant les mois d'hiver, maximum ont augmentés lors de l'exploitation, lorsqu'un système de fossé d'assechement conventionnel était utilisé. L'utilisation de tubes d'assechement subtéranéens a réduit les taux d'écoulement maximum, bien en dessous des prédictions faites avant l'exploitation, mais augmentèrent l'écoulement total d'environ 10%.

ZUSAMMENFASSUNG: Ein hydrologisches Simulationsmodel wurde verwendet, um die Auswirkungen der Torfentfernung auf den Gesamtabflusses sowie auf Höchstgrad des Abflusses von einem Grundstuck von 6.000 Hektaren im östlichen Teil des amerikanischen Bundesstaates North Carolina zu bewerten. Der Boden war vor der Torfentfernung mit Schwingelgras und danach mit Kikuruz, Weizen, und Sojabohnen bepflanzt. Der vorausgesagte monatliche Durchschnittsabfluss war für die Wintermonate hoher nach der Torfentfernung als vorher, niedriger allerdings in der Zuchtzeit der Pflanzen. Der Höchstgrad des Wasserabflusses steigerte sich nach der Torfentfernung bei Benützung von gewöhnlichen offenen Graben. Die Benützung unterirdischer Wasserableitungsröhren verminderte den Höchstgrad des Abflusses unter den vorausgesatgen Grad vor der Torfentfernung, vermehrte aber den Gesamtabluss um zirka 10%.

1 INTRODUCTION

The mining of eastern North Carolina peat on a scale previously unknown in the United States, was proposed in the early 1980's. While the large mining projects are currently on hold, acreage permitted for mining exceeded 10,000 ha at one time and mining plans in the area are still active. There has been much concern about potential impacts of such large scale mining on sensitive coastal waters located nearby.

A three-year study was conducted to determine the hydrologic effects of peat mining. The study concentrated on a 6000 ha tract owned by First Colony Farms (FCF) (Fig. 1). The research involved an intensive field monitoring program and a computer modeling effort. The purpose of this paper is to report the results of a computer simulation study to analyze the effects of peat mining on the hydrology of these flat, high water table organic soils. The computer model described elsewhere in this issue (Broadhead and Skaggs, 1989) was applied to simulate the hydrology of the subject tract before and after mining. At the time of the field study the dominant land cover on the site was pasture. The analysis presented herein considers the effects of mining the peat and using the underlying mineral soil for row crop agriculture.

Fires on part of the tract, and on adjacent lands, changed the natural cover before it could be mined. The hydrologic effects of mining land covered by native pocosin vegetation, that had been burned to various degrees, were analyzed by Konyha et al. (1988). They considered the postmining land use to be forestry, rather than the agricultural uses considered here.

Table 1. Summary of soil property and site parameter inputs to DRAINMOD for postmining conditions.

INPUT		VALUE
1. Soil properties and crop inputs		
Depth to restricting layer:		1.60 m
Hydraulic conductivity, K		
		Depth
K = 3.60 m/day		0 - .25 m
K = .48 m/day		.25 - .70 m
K = .106 m/day		.70 - 1.60 m
Water content at lower limit available to plants:		0.31 mm^3/mm^3
Saturated water content in root zone:		0.60 mm^3/mm^3
Green-Ampt infiltration parameters;		
Water table depth (m)	A (mm^2/hr)	B (mm/hr)
1.00	330	10
2.00	920	10
Planting dates:	April 1, corn	
	Oct. 1 wheat	
Maximum rooting depth:	30 cm	
2a. Drainage system parameters (open ditches)		
Drain depth		1.0 m
Surface depressional storage		15.0 mm
Drain spacing		100 m
2b. Drainage system parameters (drain tubes)		
Drain depth		1.00 m
Drain diameter		100 mm
Effective drain radius		5.1 mm
Drain coefficient (as limited by hydraulics)		25.0 mm/day
Drain spacing		20 m

2 SITE DESCRIPTION

The 6000 ha site is comprised of forty-five, approximately 130 ha blocks (1600 x 800 m^2) (Fig. 1). Most of the blocks are drained by sixteen field ditches, spaced 100 m apart, which drain into a 1600 m long collector canal. The collector canals, running east-west on the site, drain into main canals which all flow north to south.

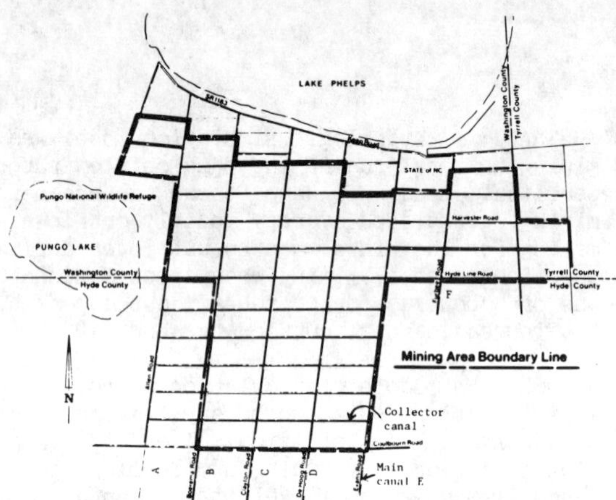

Figure 1. Peat mining area on First Colony Farms.

Soil properties and site conditions prior to mining were assumed to be the same as given for the experimental site in the previous paper (Broadhead and Skaggs, 1989). The soil is deep colloidal muck with an average depth of harvestable peat of approximately 1.9 m. The surface cover was 5-year old fescue pasture with the root depth limited to approximately 0.12 m by acid subsoil conditions. Land surface elevation is greater than 5 m above sea level, so gravity drainage outlets would still be available after the peat is removed.

Mining would remove peat down to the transition layer, which grades from about 90% organic matter (O.M.) to mineral layers with less than 10% O.M. With liming, fertilization, cultivation and incorporation of plant materials, the new surface layers are expected to resemble those of high organic mineral soils in the area. Mixing and tillage of the new surface should decrease its bulk density and thus increase the hydraulic conductivity (K). A two-year corn-wheat-soybean rotation was assumed for the postmined land use. The hydrologic analyses reported herein concentrates on the corn-wheat years of the rotation. The drainage system analyzed for both premining and postmining conditions consisted of conventional open ditches spaced 100 m apart with the land surface sloped toward the ditches to give good surface drainage, the customary drainage system for this region. A subsurface drainage system to increase crop yields and alter the drainage water distribution was also considered as an alternative postmining option.

The canal system after mining was assumed to remain similar, relative to the ground level, to the premining situation. This assumption anticipated canal deepening as the peat was removed to maintain gravity flow in the canals.

3 SIMULATIONS

The combination DRAINMOD-Flood Routing model described in the previous paper was used to simulate the hydrology of the 6000 ha block for soil and site conditions before and after mining. Model inputs for conditions prior to mining are given by Broadhead and Skaggs (1989, Table 1). Inputs for conditions after mining are given in Table 1 herein. Simulations were conducted for a 20-year period of climatological data from Elizabeth City, NC, about 70 km from the site. The modified DRAINMOD predicted water table depths, ET, surface runoff, and subsurface drainage from individual fields on a continuous basis. Outflows for selected storms were routed through the canal network to determine the effect of mining on outflow hydrographs.

4 EFFECTS OF MINING ON THE HYDROLOGY

4.1 Water Table Depth

Predicted water table depths midway between drainage ditches for premined and postmined conditions are plotted in Fig. 2 for 1960, a year of average rainfall (1275 mm). The mean water table after mining was 23 m deeper than that prior to mining. The widely spaced ditches provide poor subsurface drainage in both cases, and have little effect on the midpoint water table depth. The water table rises and falls primarily as a function of infiltration and evapotranspiration (ET). The differences observed in Fig. 2 are a result of different soil properties, profile depths and crop rooting depths. Deeper water tables during the summer months for the postmining situation are a result of deeper effective rooting depths and increased upward flux from the water table causing increased ET.

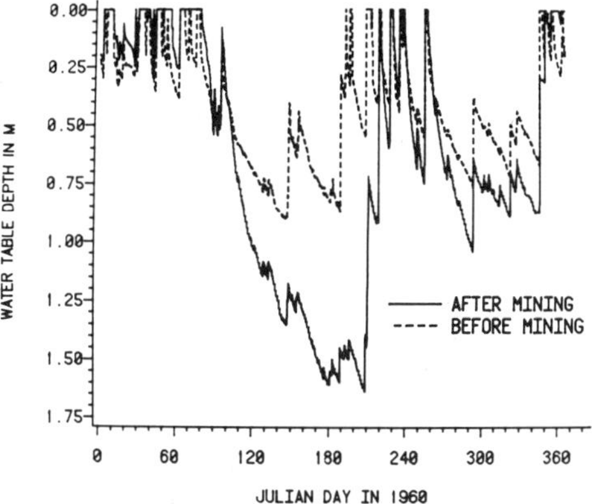

Figure 2. Effect of mining on predicted water table depths.

4.2. Cumulative Drainage

The effects of mining on the total amount of water drained (surface runoff plus subsurface drainage) from the site are illustrated in Fig. 3. Lower average monthly outflows were predicted for the premined as compared to the postmined condition for the winter months when the fields were fallow. However predicted monthly outflows for the peak growing season (June-August) were much higher for conditions prior to mining than for the reclaimed soils after mining. ET was frequently limited due to the lack of available soil water during the hot summer months. Shallower root depths for the pasture that existed prior to mining caused ET to be less than predicted for postmined conditions, which would have improved agricultural soils and deeper rooted crops. Increased ET provides more storage for infiltrating rainfall and reduces runoff.

While the monthly averages give a good indication of trends, changes in the distribution of runoff from individual storms may be more informative. Flow

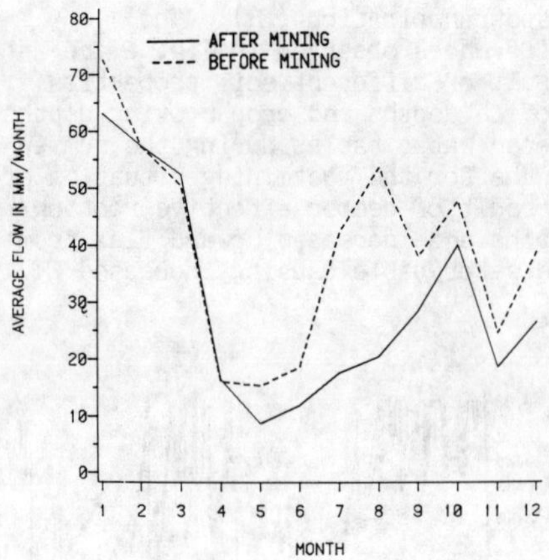

Figure 3. Predicted flow volumes for before and after mining (averages from 20 years of simulation).

duration curves for a collector canal draining a 130 ha block, are given in Fig. 4 for conditions before and after mining. A separate analysis showed that the flow duration curve obtained by routing flows down the collector canal to its outlet was nearly identical to that obtained by plotting the average daily flows from the area (Gregory et al., 1984). Results in Fig. 4 are based on average daily flows predicted for the 20 year simulation period. Differences between pre- and postmining are small for high flow rates where flows are limited by the capacity of the channel network, rather than by field controlled mechanisms. The flow duration curve for the premined condition exhibits a plateau around the 5 to 10% mark. This corresponds to conditions when the water table is at or near the soil surface. The very high K of the surface layer allows high subsurface flow rates (primarily in the cured layer of peat within 0.25 m of the surface) and delays or eliminates surface runoff for some relatively large rainfall events. The plateau is absent for the postmining case which doesn't have the high K layer. However, greater rooting depths for this case cause deeper antecedent water tables, reducing the number of storms having surface runoff, as well as the subsurface drainage rates. The slightly higher flows from the premining case at flow rates less than about 10 l/s are caused by greater release of bank storage from the peat soils.

Because of the wide ditch spacing, the flow rates from the collector canals for both cases are less than 10 l/s for over 80% of the time.

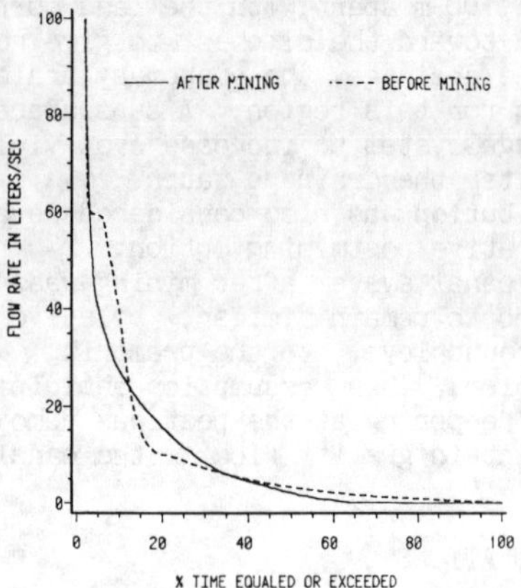

Figure 4. Flow duration curves for a 130 ha block before.

4.3. Storm Events

Storm events are of major concern for flooding, nutrient and sediment losses and their effects on estuaries downstream. The effects of mining and reclamation on individual storm hydrographs were examined. A main canal (canal D, Fig. 1) was selected as a typical outlet for the permitted area and storm hydrographs at the boundary of the permitted area were simulated. Hydrographs for several storms and sequences of rainfall events were compared. Outflows from the field ditches were routed down the collector canals to the main canal D; these flows were in turn routed down canal D to the outlet of the permitted area. Eleven blocks (a total of 1425 ha) contribute to the flow in canal D at the outlet of the tract.

Detailed flood routing analyses were conducted for events that had daily flow rates with 50, 75, 100 and 200 day return periods. These events only cover a narrow range of the flow duration curve, but include the range of flows likely to cause problems. It is very important to examine effects of the flow capacity of the canals in this analysis. Often very dramatic differences in runoff at the field edge will be significantly reduced by time delays involved in a canal reach. Predicted outflow hydrographs for an event with a 75 day return period flow are presented in Fig. 5. The abrupt change on the rising limb of the hydrographs at about 950 l/s corresponds to the onset of overbank flow. The bulge in the falling limb of the 'before mining' case corresponds to the switch from surface runoff to flow through the high K surface layer.

Results similar to those given in Fig. 5 were obtained for the other return periods. Peak outflow rates predicted for after mining conditions were greater than those for premining conditions by an average of 16% for the 200, 100 and 75 day return period storms and by about 107% for the 50 day return period storm. For flows in excess of the 100 day return period storm all canal banks were breached and the hydrographs were very similar for the two conditions.

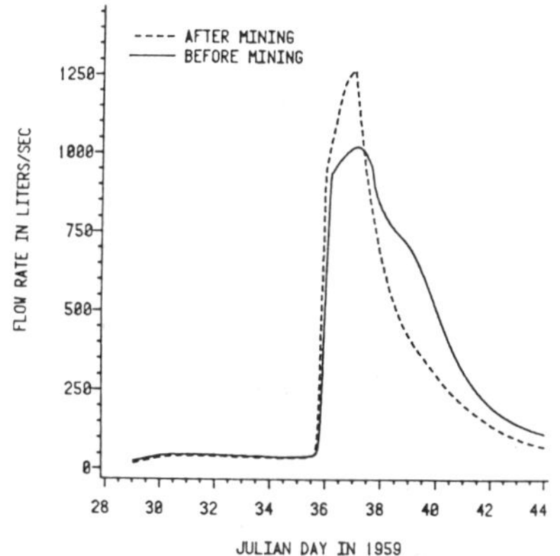

Figure 5. Hydrographs predicted for a 75 day return period.

The differences in peak flow rates and the shapes of the hydrographs are primarily dependent on antecedent soil water conditions, the soil surface conditions and the hydraulic conductivity of the soil surface layers. Except when the water table was in the top 0.25 m surface layer of the peat prior to mining, the 100 m drain spacing provided poor subsurface drainage for both cases shown in Fig. 5. For all four storms analyzed, the water table prior to rainfall was deeper for the premining simulation. All of these storms occurred during October to February, in a period of fallow conditions and shallow rooting depths for the postmining condition. In this period the water table would be expected to be deeper for the premining situation which had a permanent pasture cover. Thus more storage was available in the soil profile and the peak flow rates predicted for premining conditions were lower than those predicted for postmining.

To show the effects of deeper water tables predicted for postmining during the growing season, a simulation was conducted using data for the entire month of July, 1959. The summer months usually contain many convective storms, often occurring on several consecutive days. This makes it difficult to select individual events for analysis. Fig. 6 shows the simulated hydrographs at the D canal outlet for July, 1959. A total of 254 mm of rain occurred during this month with rain falling on 18 of the 31 days. The initial water table depth on July 1st was deeper for the postmining case by 40 mm but, more importantly, the predicted depth of soil which had dried to the wilting point at the soil surface was 300 mm for postmining but only 120 mm for premining. These depths correspond to the effective rooting depths for the respective crops and are proportional to the volumes of water storage available in the soil profile. Lack of rainfall during the preceding

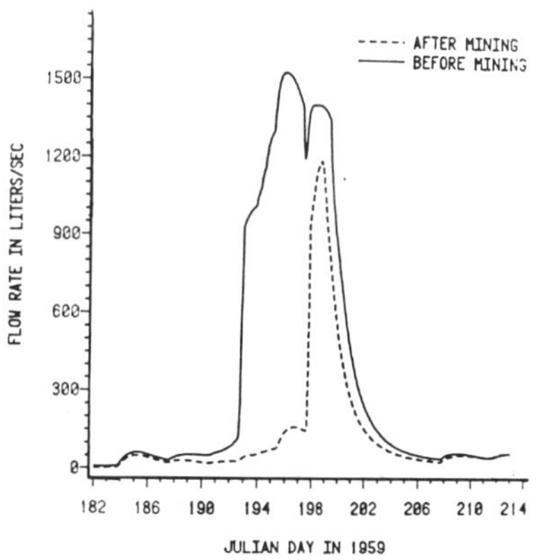

Figure 6. Predicted hydrographs for a midsummer storm.

month had allowed the root zone to completely dry out. Acid subsoils prevented deeper rooting depths in the peat soil prior to mining. Thus rains in July filled the soil profile more quickly for the premining case, producing surface runoff on Julian day 192, five days before the runoff was predicted for the postmining condition. As a result, the premined condition produced considerably more total runoff during the month.

5 EFFECTS OF IMPROVED SUBSURFACE DRAINAGE

Results presented in the previous analyses assumed conventional open ditch drainage for both pre- and postmined conditions. This system provides mostly surface drainage with relatively poor subsurface drainage for most soils. An alternative, for the mineral soils that will remain after mining, is the use of drain tubes to improve subsurface drainage. Methods described by Hardjoamidjojo and Skaggs (1982) were used to determine the drain spacing required to maximize long-term average corn yields on the reclaimed soils after mining. The optimum design required drain tubes to be placed 1.0 m deep and 20 m apart. An analysis was conducted to determine the hydrologic effects of using this improved subsurface drainage. Results were compared to those obtained for postmining conditions with the conventional open ditch drainage system.

Good subsurface drainage lowers water table depths as expected. The average water table depth predicted for 1960 was 868 mm which was 219 mm greater than obtained for the conventional drainage case plotted in Fig. 2. The average monthly flow depths are plotted in Fig. 7. Good subsurface drainage increased total outflow by about 10%, or 35 mm/year, for this soil. The same root depth relationship was assumed for both soils. In fact better subsurface drainage might permit a somewhat greater root depth which would tend to decrease outflows.

The high flow end of the flow duration curves for good subsurface drainage and for the conventional open ditch system (poor surface drainage) are plotted in Fig. 8. Good subsurface drainage increases the low flow rates, but reduced flow rates in the range above 60 l/s. For example, the daily flow rate which is expected to be equaled or exceeded 1% of the time equals 285 l/s for poor subsurface drainage, compared to 180 l/s for good subsurface drainage (a decrease of 37%). It is these higher flow rates that are important from both design and environmental perspectives.

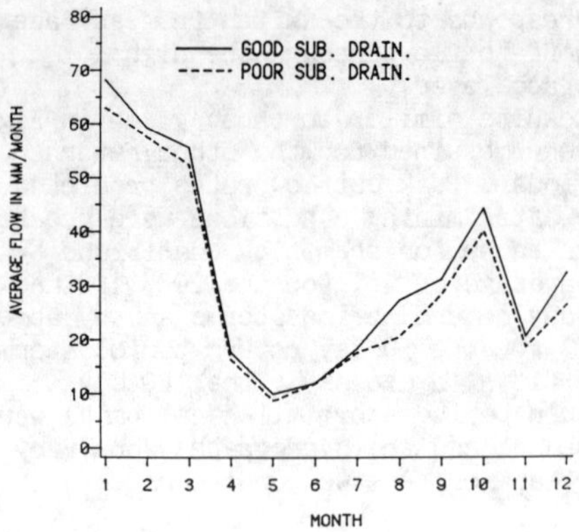

Figure 7. Predicted monthly flow depths for good and poor subsurface drainage, postmining.

Good subsurface drainage lowers the water table and increases the average outflow. This increases storage capacity for subsequent rainfall and leads to reduced outflow rates as illustrated in Fig. 8. In a field with good subsurface drainage the proportion of subsurface flow and surface runoff will be significantly altered from that which will occur for poor subsurface drainage. Good subsurface drainage for postmined conditions reduced predicted surface runoff by 57%, from 93 to 40 mm/yr, compared to conventional drainage on this site.

The effects of improving the subsurface drainage on individual hydrographs was investigated for peak flows of varying magnitudes. There was little difference between drainage treatments for the extremely high flows. If the rainfall is sufficient to bring both water tables to the surface, and flooding occurs (i.e. the canal banks are breached), then the predicted hydrographs for both cases will be similar. Major differences occur for storms with peak flows which fall between 60 and 1000 l/s. These effects are illustrated in Fig. 9 for a 50 day return period event on a 1425 ha block

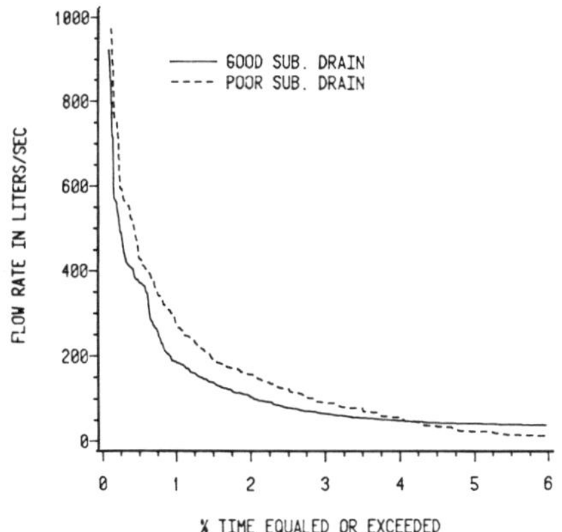

Figure 8. Flow duration curves for good and poor subsurface drainage on a 1425 ha block of reclaimed soils.

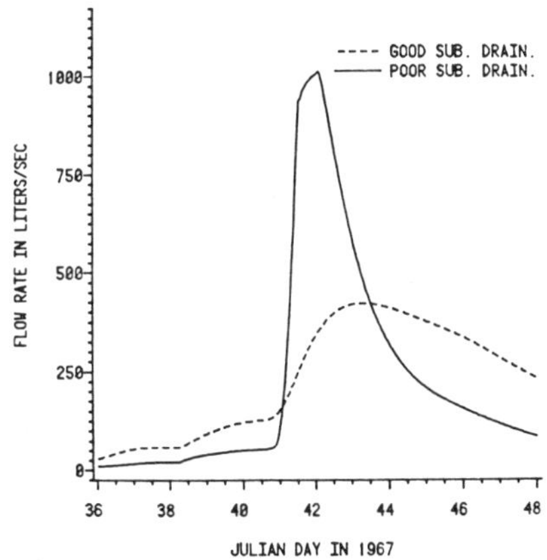

Figure 9. Predicted hydrographs for a 50 day period on a 1425 ha area.

of reclaimed soil. Drainage prior to the rainfall event lowered the water table such that surface runoff did not occur in the soil with good subsurface drainage, thus keeping the peak outflow significantly less than for the poor subsurface drainage case. Note, however, that the flow rates before and after the rainfall event were higher for the soil with good subsurface drainage than for poor subsurface drainage.

6 SUMMARY AND CONCLUSIONS

The computer simulation model, DRAINMOD, was used in combination with a numerical flood routing model to analyze the effects of peat mining on the hydrology of a 6,000 ha tract in eastern North Carolina. Land use was unrenovated pasture prior to mining and a corn-wheat-soybean rotation after mining. Twenty year simulations were conducted to determine the effects of changes in soil properties, land use and drainage system design on the hydrology of the site.

Results showed that mining and subsequent row crop agriculture would result in deeper water table depths but reduced total drainage from the site. Monthly drainage volumes would be significantly reduced during the growing season because of increased ET. Analysis of flow events showed that peak flow rates during the winter months tended to be somewhat higher for postmined conditions. However, the same conditions produced substantially lower flow rates during the growing season.

Improved subsurface drainage on the reclaimed soil will further lower water tables, but will increase average drainage outflow by about 10%. Good subsurface drainage will have a significant effect on peak flow rates from the larger storm events, however. Flow duration relationships showed that subsurface drainage lowered the flow rates from a 1425 ha block which were exceeded less than 4% of the time. In one case the predicted peak flow rate was reduced by a factor of 2.5 compared to that obtained for conventional drainage.

7 REFERENCES

Amein, Michael. 1972. Report on the numerical simulation of unsteady flows in rivers and reservoirs, to Office of Hydrology, National Weather Service, NOAA, U.S. Dept. of Commerce. Contract No. 0-3528, 73 pp.

Broadhead, R. G. and R. W. Skaggs. 1989. A Hydrologic model for artificially drained North Carolina peatlands. This issue.

Gregory, J. D., R. W. Skaggs, R. G. Broadhead, R. H. Culbreath, J. R. Bailey and T.L. Foutz. 1984. Hydrologic and water quality impacts of peat mining in North Carolina. Report No. 214 of the Wat. Resour. Res. Instit. of The Univ. of N. C., Raleigh.

Hardjoamidjojo, S. and R. W. Skaggs. 1982. Predicting the effects of drainage systems on corn yields. Agric. Water Management, 5:127-144.

Konyha, K. D., K. D. Robbins and R. W. Skaggs. 1988. Evaluating peat mining hydrology using DRAINMOD. J. Irrigation and Drainage, Vol. 114(3):490-504.

Purisinsit, Pitsamai. 1982. Evaluation of two hydrologic models for the North Carolina Blacklands. Ph.D. Thesis, Dept. of Biological and Agricultural Engineering, N. C. State Univ., Raleigh.

Skaggs, R.W. 1978. A water management model for shallow water table soils. Report No. 134, Wat. Resour. Res. Instit. of The Univ. of N.C., Raleigh. 178 pp.

Skaggs, R.W., J.W. Gilliam, T.J. Sheets and J.S. Barnes. 1980. Effect of agricultural land development on drainage waters in the North Carolina tidewater region. Rep. No. 159 of the Wat. Resour. Res. Instit. of The Univ. of N. C., Raleigh.

Land and Water Use, Dodd & Grace (eds), © 1989 Balkema, Rotterdam. ISBN 90 6191 980 0

Origin of organic soils in the middle reach of the Noteć River valley and design improvement of flood outflow

B.Przedwojski
Department of Hydraulic Engineering, Agricultural University of Poznań, Poland

ABSTRACT: The middle reach of the Noteć River valley, located between two rightbank tributaries: the Gwda River and the Bydgoski Channel is a flat peat-bog about 5 km wide and 65 km long. The maximum thickness of the organic soils is approximately 15 m. The level of the valley surface has lowered by about 0,89÷1,18 m in the last 93 years. During spring floods about 30%÷70% of the valley is covered with water layer of 0.5 m ÷ 1.2 m for 1 to 2 months. The paper will provide a brief description of investigations and designed works carried out to find the best way of lowering the water surface during floods and to accelerate its outflow from the valley.

RÉSUMÉ: Le secteur central de la vallée de la revière Noteć situé entre la revière Gwda et le canal Bydgoski est une tourbière de 5 km largeur environ et 65 km longueur. L'épaisseur maximum s'approche de 15 m. La surface de la vallée a abaissé a' 0,89 + 1,18 m pendant 93 ans dernier. Pendant la crue printanière de 30 % a' 70% de la surface de la valée est inondée par l'eau de 0,5 a'1,2 m de la profondeur, pendant de 1 a' 2 mois. Dans ce travail on propose une brève description des études déjà faites dans le but d'attendre une solution valable pour abaisser les niveaux des eaux surtout durant la période d'innondation et aussi pour accélerer l'écoulement emanent de la valée.

ZUSAMMENFASSUNG: Die Mittelstromstrecke von Noteć, die zwischen zwei rechtsseitigen Nebenflüssen wie Gwda und Bydgoski-Kanal liegt, bildet ein Niedermoor, dessen Breite 5 km und Lange 65 km beträgt. Die Maximalschichtdicke von organischen Böden reicht bis zu 15 m. Die Talfläche hat sich um 0,89 bis 1,18 m innerhalb von letzten 93 Jahre gesenkt. Zur Zeit des Frühlingshochwassers ist von etwa 30% bis 70% der Talfläche mit Wasser bedect. Die Wasserschicht beträgt von 0,5 bis 1,2 m und hält sich einen bis zu zwei Monaten. In der Arbeit stellte man kurze Beschreibung der gemachten Untersuchungen und Projektarbeiten dar. Das Ziel dieser Bestrebungen war die Bestimmung der bestmöglichen Lösung, die zur Senkung des Wasserpiegels z.z. des Hochwassers führen wird und seinen Abfluß aus dem Tal beschleunigt.

1 INTRODUCTION

The Noteć River reach from km 0,0 (the mouth to the Warta River) to km 187,2 is a waterway of IIB class with the mean depth of the navigable channel 1.80 m. Together with the lower reach of the Warta River in the west and the Bydgoski Channal in the east it establishes a navigation connection between the Odra River and the Vistula River (Fig.1) Period of navigation begins on March 15 (after the end of ice season) and finishes on December 15. The regulation of the water level by weirs - to achieve a sufficient depth for navigation - begins between May 1 and May 30 ie. after the passing of the flood wave. The project of canalization of the river with the aid of weirs and sluices was made in 1907. The weirs and sluices were built up by 1914. The Noteć River from 0,0 to 187,2 km, flowing through a flat valley of 2÷6 km wide as it was shown in Figs 2 and 3. This valley is used for agricultural production such as meadows and pastures. Together with the canalization of the river the drainage-irrigation system was built in order to improve the agriculture condition of using the valley. This drainage-irrigation system and intensification of agriculture production gave rise to an advancing process of mineralization and self-consolidation of peat on the reach 120 ÷ 187,2 km (Fig.2). During 93 years the mean value of the self-consolidation of the surface of the peat-bog has the mean value from 0,89 to 1,18 m ie. from 0,96 cm/year up to 1,27 cm/year. The above mentioned process caused distinct deterioration of underground water level and agricultural conditions on the middle reach of the Noteć River valley. Each year the spring highwater level floods about 30% ÷ 70% of the valley surface particularly on the section between 120 km (gauging station "Ujście") and 157,2 (sluice No. 12). The flooded area depends on the volume of flood wave in the given year. It was estimated in 1988 that this caused a loss of about 1 mln dollars each year. On the other hand the navigable water levels - in dry season - are the same as they were designed in 1907. Considering the self-consolidation of the peat-bog it makes impossible the free gravitational outflow from the drainage system and to establish the optimum underground water levels in the valley. This caused distinct deterioration of meadows particularly in lowered parts of the valley. Besides the losses caused by the spring high-water levels also during low discharges an explicit collision occurred between agricultural and navigation needs. In the years 1983÷ 1988 hydrological and hydraulic analyses and bed and bed-load sediment investigations were carried out (Lupa 1983, Przedwojski 1988). Based on these investigations a choice of optimum design solutions was made - considering the actual financial possibilities.

The aims of the solution are:

1. To protect the valley from floods caused by discharges occurring once per 10 years.

2. To accelerate the outflow of the water during floods with bigger discharges.

3. An improvement of agriculture and navigable conditions in the middle reach of the Noteć River valley.

2 HYDROLOGICAL DATA

Characteristic discharges in the Gwda River at the mouth Q_G (cross-section in 0,0 km), in the Noteć River at 120,3 km Q_{NL} (gauging station "Ujście"-upstream of the Gwda) and in the Noteć River at 119,7 Q_{NB} (downstream of the Gwda) (Fig.3) are presented in Table 1. Measurements made in those cross-sections in the years 1983,1985,1988 are shown in Table 1 too (Lupa 1983, Przedwojski 1988). These measurements are made in the same time. In Table 1: A-catchment area, SNQ-mean of the lowest recorded value of water discharge, SSQ - mean recorded value of water discharge, SWQ -mean of the annual highest recorded value of water discharge, WQ - the highest annual value of water discharge. The discharges ratio Q_G/Q_{NL} in Table 1 indicates that the discharges in the Gwda River are 1,09÷ 5,33 times greater than the ones in the Noteć River (up to gauging station "Ujście"). With the increasing discharge in the Noteć River the ratio Q_G/Q_{NL} decreases.

The greater discharges in the Gwda River are caused by :

1. Higher rainfalls in the basin of the Gwda River (P ≃ 700 mm/year) than the ones in the Noteć River (P ≃ 550 mm/year).

2. During flood flows ($Q \geq 70$ m^3/s) the middle reach of the Noteć valley between 120 ÷ 187,2 km creates a temporary storage. In time of flood in 1979 the maximum storage volume of the above mentioned valley was about 100 mln m^3.

Table 1. Characteristic discharges (m^3/s)

Discharges	Gwda River km 0,0 (Mouth) A=4942,8km² Q_G	Noteć River km 120,30 upstream of the Gwda A=6345,0 km² Q_{NL}	Noteć River km 119,70 downstream of the Gwda A=11 287 km² Q_{NB}	$\frac{Q_G}{Q_{NL}}$
SNQ	13,5	7,77	20,6	1,74
SSQ	27,4	18,8	45,4	1,45
SWQ	57,0	44,0	86,5	1,30
WQ	88,26(1979)	97,0 (1980)	133,3 (1979)	
15.08.1983	18,41	3,45	21,86	5,33
27.03.1985	41,35	24,45	65,80	1,69
1.08.1985	23,81	13,46	35,27	1,77
2.08.1985	20,69	15,43	36,12	1,34
29.03.1988	74,7	68,5	143,2	1,09
5.04.1988	77,6	71,2	148,8	1,09

With the removal of this temporary storage (eg. by means of an embankment) - during floods lasted 1 month - the increment of discharge between levees in the cross-section located upstream of the Gwda is :

$$\frac{100\ 000\ 000\ m^3}{86\ 400\ s * 30\ days} \simeq 39\ m^3/s$$

Characterictic of 4 highest floods - in the last 18 years - is presented in Table 2. The spring and summer high-water levels in the middle reach of the Noteć valley lasted for 30÷65 days.

Table 2. Maximum floods in the years 1970 ÷ 1988.

Occurrence period	Duration in days	Water level at km 120,3 in m
2.04 ÷ 5.06.1970	63	49,61
15.03 ÷29.04.1979	45	49,66
5.07 ÷ 7.09.1980	64	49,30
19.03 ÷18.04.1988	30	49,12

3 BED SEDIMENT

17 bed sediment samples were taken in 29.03.1988 and 05.04.1988. To do this a special device was used. The samples were taken from the top layer of the channel bottom in the section of the Gwda River between 0,15 km ÷ 2,5 km and the Noteć River between 115,1 km ÷ 130,1 km. The sediment characteristic diameters as a result of sieve analysis are given in Table 3. The analysis of samples coming from the channel bottom of the Noteć River upstream of the cross-section at 125,9 km proved that the top layer is made up of sedimentary peat [8]. In the same sections the structure drilling was made to investigate the soil beneath the top layer of

Table 3. Grain size of bed sediment

River	Km	Diameters in mm d_{50}	Diameters in mm d_{90}
Gwda	0,15	0,36	1,20
	0,30	0,37	1,20
	0,80	0,54	1,15
	1,80	0,60	1,95
	2,00	0,39	1,40
	2,50	0,40	0,90
Noteć	130,1	sedimentary peat	
	128,1	"	
	125,9	"	
	125,1	0,38	1,12
	122,1	0,94	38,0
	121,1	0,50	16,0
	119,7	12,5	78,0
	119,1	15,0	74,0
	118,1	0,85	2,10
	117,1	0,65	4,00
	116,1	0,55	1,80
	115,1	0,78	4,00

the channel bottom. The holes were up to 9 m down from the surface of the bottom. The sieve analysis indicates that the basement soil is sand in the Gwda River and the Noteć River downstream of the cross-section at 125,1 km. Upstream of this cross-section the basement soil is sedimentary peat (Przedwojski 1988) (Figs 5 and 6). The investigations of bed sediment proved also that the channel bottom of the Noteć River between 126,9 ÷ 123,0 km is covered with coarser material than the underlying ones. Particulary in the reach located between km 119,7 and 118,1 the top layer is made up of gravel and locally even of boulders in diameter >70 mm. This top layer protect the underlying ones and is referred to as armoured bed. As a result of the armouring process bottom thresholds occur in the channel of the Noteć River (Fig.5).

4. MAIN FEATURES OF RIVER CHANNELS

4.1 The Noteć River

As is shown in the longitudinal profile of the Noteć River (Fig.6) three bottom thresholds occur downstream of the Gwda River mouth (120 km). Field measurements of cross-sections from 1970 to 1988 proved that these armoured thresholds has not been eroded and changed. Their heigths are 1,0÷1,5 m. These thresholds distinctly influence the flow of the river and cause variations in the water surface slope during an open river flow. These variations increase with decrease in the discharge. The field measurements made on March 27, 1985 during open river flow indicate that the water surface slope over above mentioned thresholds is 56,3 cm per km (Fig.5). At lower flows this slope achieve 65,3 cm per km. During floods the mean water surface slope between sluice No 1 and the mouth of the Gwda River varies from 29,5 to 30,7 cm per km. Upstream of the Gwda river mouth the water surface slope changes rapidly. In time of floods its values vary from 1,72 ÷ 3,65 cm per km. For this reason the reach of the Noteć River between km 120 ÷ 187,2 is called the Slow Noteć River and the reach downstream the Gwda River mouth is called the Rapid Noteć River. The investigations of the sediment trasport indicate that the washload sediment only occur in the water flowing through the channel of the Noteć River reach upstream of the Gwda River mouth. Together with the material washed out from the banks of this channel by the going ships it results in the rise of the bottom level. This is shown in the cross-section at km 125,6 (Fig.7) where the bottom level rose about 0,5 m from 1975 to 1988 and in the cross-section at sluice No 12 (Fig.5) where the bottom level rose about 0,70 m above the level of the lock sill. In time of dry season the navigable water surface slope is sufficiently flat so that velocities are not excessive in the channel of the Noteć River. The bedload transported by the Gwda River is accumulated in the reach between km 117÷120 and the bed level rose. In an open river flow during floods the sediment transport capacity in this reach rises rapidly. Calculated from Samov formulae (Samov 1954) the sediment transport capacity in the reach of the Noteć River is twice bigger than the one in the Gwda River during high flows. Then in time of floods the sediment carried by the Gwda River and settled during low flow in the reach of the Rapid Noteć River are transported downstream through the weir sill of the sluice No. 11. This process caused local changes of the bottom level of the Rapid Noteć River as it is shown in the cross-section at km 116.3 (Fig. 7).

4.2 The Gwda River

In time of high flows the water surface slope in the lower reach attains the value 33 cm per km as is shown in Fig. 6 (Lupa 1983, Przedwojski 1988).
In dry season the slope decreases rapidaly at the mouth and increases upstream achieving the value 21,6 cm per km in 2,0÷2,6 km and 33 cm per km in the cross-section at 3,5 km. Thus the bedload sediment transport occurs throughout the year. The grain size of bedload is 5 times smaller than the one in the top layer of the bed. Because the banks of the channel are not prote-

cted the concave bank is being scoured and part of the material is deposited at the convex bank. This process shifting the bends in both: lateral and longitudinal directions (see Fig. 4). The apexes of bends entirely change position in a period of 20 years.

5 ORIGIN OF ORGANIC SOILS

During the last icing the Vistula River was flowing through the Noteć River valley from east to west. When the continental glacier drew off the Vistula River bed changed its direction to north and a new channel formed by the Gwda River began to influence the Noteć River valley. The original Vistula River valley had probably a uniform slope in the reach of the actual Noteć River valley. This slope was smaller than the present one downstream of the Gwda River mouth and steeper than the present one upstream. At the beginning of its life the Gwda River probably carried a large volume of sediment. Because of the flat surface slope of the Noteć River valley the transport capacity of the lower reach of Gwda River -crossing the Noteć River valley - was very small and all the sediment was deposited. This alluvial deposits formed an alluvial fan. In the course of time the alluvial fan had enlarged its size in both: lateral and longitudinal directions. Displacement of the alluvial fan caused the closure of the valley by a cross-bar consisting of noncohesive soils and shifted the Noteć River channel towards the left high bank of the valley. Up till now the Noteć River bed is located alongside the left high bank as is shown in Fig. 3. As a result of lateral erosion of this high bank is a natural armoured bed. The bedload sediment transported by the Gwda River was accumuleted downstream and caused a rise of the bottom of the Noteć River valley. Upstream of the cross-bar a reservoir with a huge storage was formed (65 km long, 2÷5 km wide and to 15 m deep). Peat in shallower parts and sedimentary peat in deeper ones of the reservoir was generated from suspended sediment and decayed animal and plant remains settled on the bottom. This organic soil is now forming the valley surface. The Noteć River valley downstream of the cross-bar consists of noncohesive soils. Only near the periphery of the valley shallow peat-bogs were generated of a thickness of strata about 3 ÷ 4 m (Przedwojski 1988).

6 DESIGNED FLOOD OUTFLOW IMPROVEMENT

The aims of the Noteć River training are to lower the water surface level and to accelerate the outflow from the valley during floods. To do this the following is designed:

1. Dredging the channel of the Noteć River in the reach between sluices No 11 and 12. The designed bed level in longitudinal profile and cross-section are shown on Figs 5 and 7 (dotted line). The designed longitudinal bed slope in the reach between sluice No. 11 and the Gwda River mouth is 15,6 cm per km and in the reach between the Gwda River mouth and sluice No. 12 is 5.03 cm per km.

2. To prevent the channel from scouring an artificial armoured layer is designed downstream of the Gwda River.

3. Levees will be erected alongside the channel to protect the valley from flooding.

4. An additional lowering of the water level is designed by widening and lowering of the weir sills.

5. Transverse sills, being 0,60÷1,10 m high, to prevent the Gwda River bed from scouring are designed. The eroded channel banks will be protected by means of groynes and revetments.

7 CONCLUSIONS

The Gwda River has played a distinct role in forming different kinds of soils in the Noteć River valley. Influx of a large volume of bedload sediment from the Gwda River to the Noteć River valley generated a natural cross-bar. Downstream of the cross-bar the valley is formed by alluvial deposits coming from the Gwda River. This alluvial sediment has caused an increase of the valley surface. This process had lasted until a dynamic equilibrium betwenn the Gwda River and the lower reach of the Noteć River was attained. These mobile rivers can be considered in some form of equilibrium. On the other hand upstream of the cross-bar a reservoir about 65 km long, 2÷5 km wide and up to 15 m deep was formed. As a result of settlement in the reservoir a suspended sediment and decayed animal and plant remains the now existing peat-bog was generated. Because of irrigation and drainage made at the begining of this centuary and intensification of agricultural production the surface of the peat-bog has been lowered by about 0,89 ÷ 1,18 m during the last 93 years. At present, in order to make improvement in agricultural production about 79 km of the Noteć River canalizing system has to be rebuilt and 22 km reach of the Gwda River regulated.

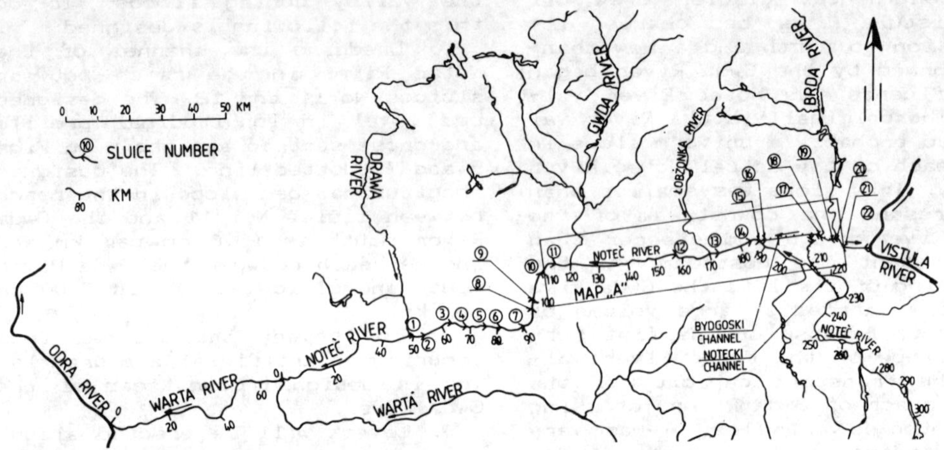

Fig. 1 Schematic plan view of the Noteć River

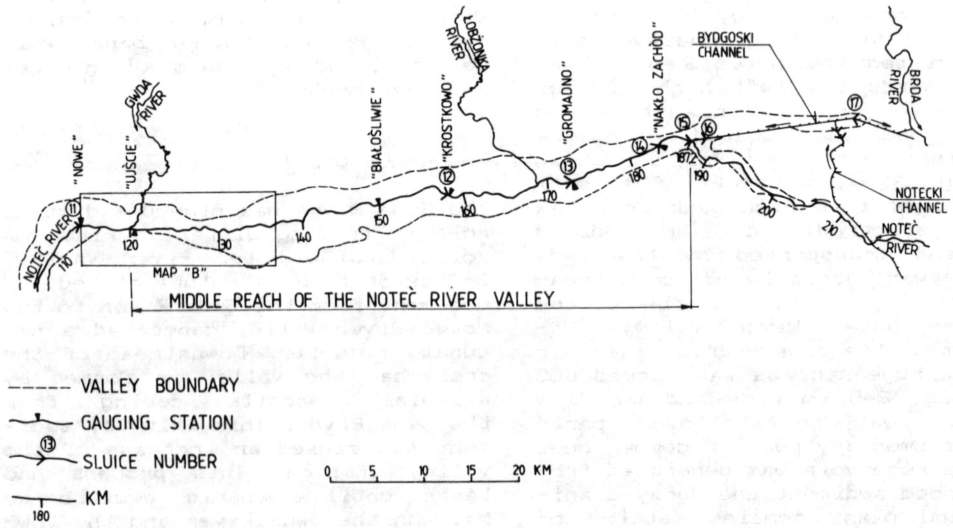

Fig.2 Map A - Middle reach of the Noteć River valley

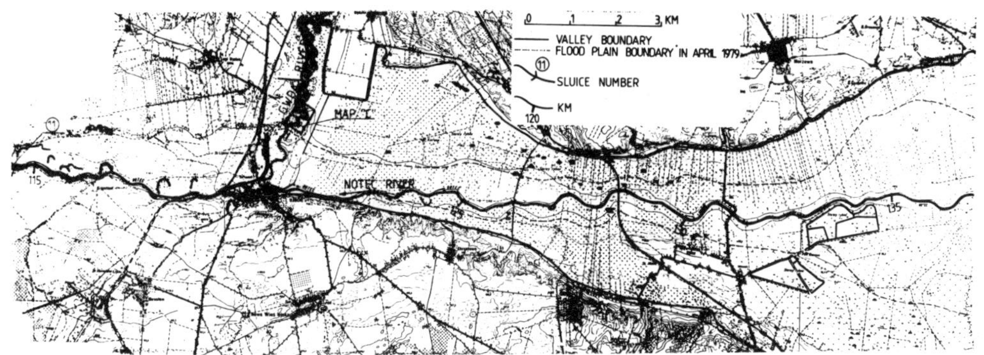

Fig.3 Map B - The Noteć River with the lower section of the Gwda River

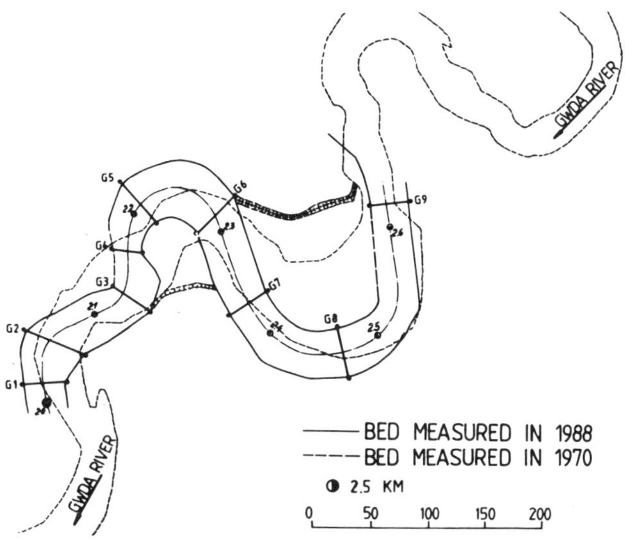

Fig.4 Map C - Section of the Gwda River

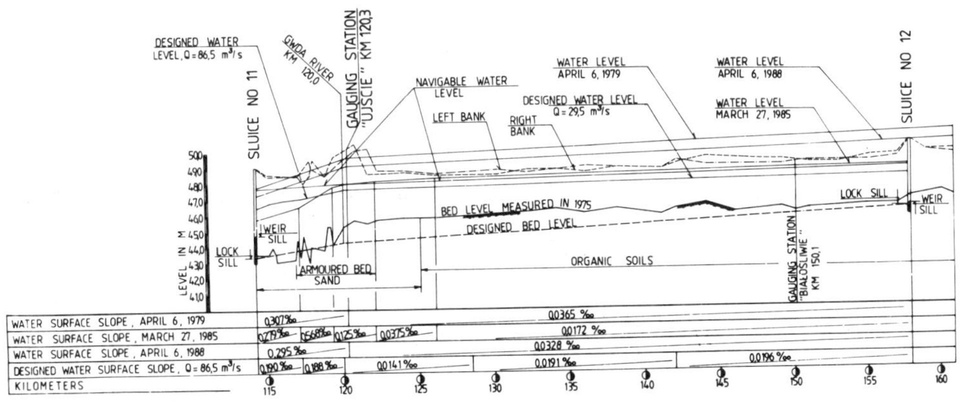

Fig.5 Longitudinal profile of the Noteć River

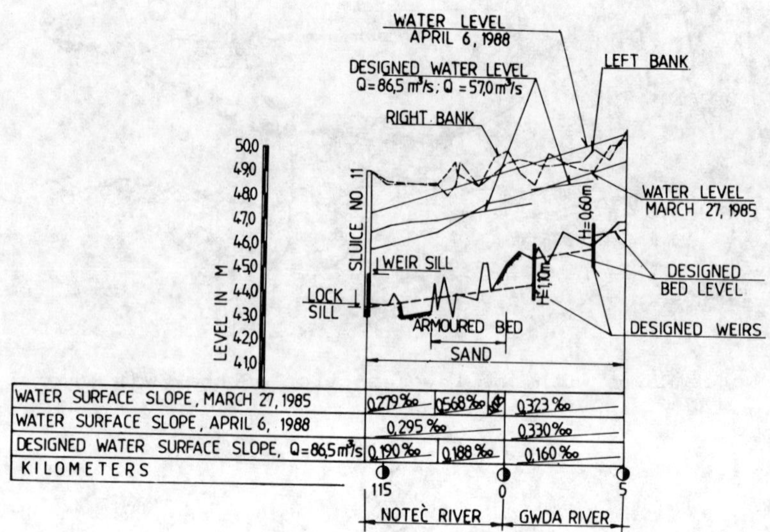

Fig.6 Longitudinal profile of the Noteć River and the Gwda River

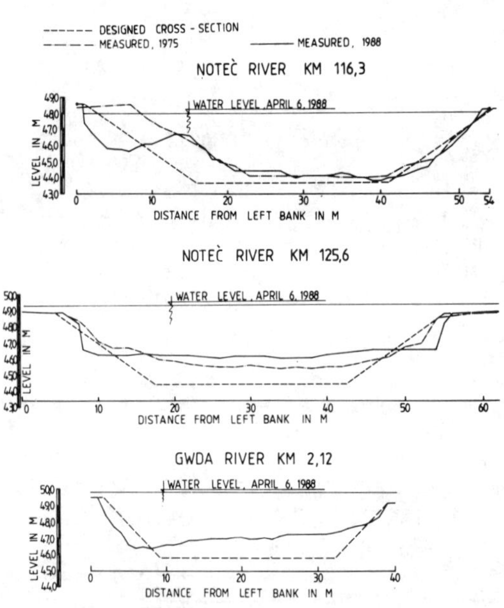

Fig.7 Cross sections

8 REFERENCES

Gajewski, B. 1988. The concept of increase of the channel capacity of the Noteć River at Ujście. Design Bureau Hydroprojekt Poznań. (Polish)

Graf, W.H. 1971. Hydraulics of sediment transport. McGraw-Hill Book Company. New York.

Kowalczak, P. 1984. Hydrology of the middle and lower reaches of the Noteć River. Inst. of Meteorology and Water Manag. Poznań (Polish)

Lupa, H. et al. 1983. Hydraulic research of the Noteć River and the Gwda River at Ujście confluence. Tech. Univ. Gdańska (Polish)

Entwurf zur Kanalisierung der Netze von Stan I (sluice No. 11) bis Stan IV (sluice No. 1). 1907.

Przedwojski, B. 1988. Preliminary analysis of sediment transport and the concept of regulation of the Noteć and the Gwda River. Group of experts. Poznań. (Polish)

Przedwojski, B. et al. 1988. The investigation of bedload sediment motion and geological analysis of the basement soil in channels of the Noteć River and the Gwda River. Poznań (polish)

Samov, G. 1954. Sediment in rivers. Gidroment. Leningrad. (Russian)

Land and Water Use, Dodd & Grace (eds), © 1989 Balkema, Rotterdam. ISBN 90 6191 980 0

A hydrologic model for the artificially drained North Carolina peatlands

R.G.Broadhead
Department of Scientific and Industrial Research, Christchurch, New Zealand

R.W. Skaggs
North Carolina State University, Raleigh, N.C., USA

ABSTRACT: The water management simulation model, DRAINMOD, was modified for use in peatlands. The field scale simulation model was linked with a numerical flood routing model to predict flow rate at the drainage canal outlet. The validity of the model was tested using measured data from an experimental site in eastern North Carolina. Simulated water table depths, monthly runoff values, and flow duration curves were in close agreement with observed values for the test period. While some problems still exist, good agreement was also obtained between simulated and measured hydrographs at the outlet of the 130 ha site.

RÉSUMÉ: Le modèle de simulation d'exploitation de l'eau, qui s'appelle DRAINMOD, a été modifié afin d'être utilisé sur des sols tourbeux. L'echelle du terrain du modèle de simulation était relié à un modèle de contrôle numérique des terrains inondables afin de prédire le taux d'écoulement a la sortie du canal d'assechement. La validite du modèle a été soumise à des essais en utilisant des données mesurées sur un terrain expérimental dans la région Est de la Caroline du Nord. Des profondeurs simulées de nappes phréatiques des valeurs finales mensuelles et les courbes de durée d'écoulement, correspondaient aux valeurs observées pendant la période d'analyse. Bien que certains problèmes existent encore, des résultats similaires ont été obtenus entre des hydrographies simulées et mesurées à l'embouchure d'un terrain d'une superficie égale à 130 hectares.

ZUSAMMENFASSUNG: Das Wasserabführungssimulationsmodell DRAINMOD wurde zur Anwendung im Moorland modifiziert. Das Simulationsmodel im Feldmassstab wurde mit einem bezifferten Hochwasserabfuhrungs-model verbunden, um den Grad des Abflusses am Auslass des Entwasserungskanals vorauszusagen. Die Gultigkeit des Models wurde anhand von Messungen getestet, die auf einer Versuchsstation im ostlichen Teil des amerikanischen Bundesstaates North Carolina gesammelt wurden. Simulierte Wasserspiegeltiefen, monatliche Abflusswerte, und Abflussdauerkurven befanden sich in naher Übereinstimmung mit beobachteten Werten fu die Testperiode. Obwohl einige Probleme noch ubrigbleiben, wurde gute Ubereinstimmung auch zwischen simulierten und ausgefuhrten hydrographischen Messungen am Auslass des Grundstucks von 130 Hektaren erreicht.

1 INTRODUCTION

Eastern North Carolina has over 260,000 ha of peat deposits ranging in thickness from a few centimeters to over five meters. Ingram and Otte (1982) estimated that there are over 75,000 ha comprising nearly 200 million metric tons of peat, on the Albemarle - Pamlico Peninsula alone, that could be mined for fuel or other uses. Most of the deep peat deposits are colloidal in texture, have a very low pH (3 to 4) and contain large amounts of buried wood. As a result, neither agriculture nor timber production is economical on these lands. There have been a number of proposals to mine the peat and reclaim the underlying mineral soil for agricultural or forestry production. While these projects are now on hold due to economic or regulatory restrictions, peat mining is still being planned in the region. The lands are characterized by flat topography and high water tables. They are drained by parallel drainage canals spaced 1600 m apart intersecting collector canals which divide the land into 1600 by 800 m blocks. The hydrologic impacts of peat mining on this scale are not known and the closeness of the proposed mines to environmentally sensitive areas led to much apprehension over their potential effects.

In order to study the effects of peat mining on watershed hydrology, methods were required to simulate the combined hydrologic effects of different conditions within a watershed. This was obtained by linking together the field scale water management model, DRAINMOD (Skaggs, 1978), and a numerical flood routing model. Both models are deterministic and have proven reliability over a wide range of conditions. With the combined model it was possible to predict the effects of alternative land uses, and drainage treatments, on the rate and quantity of outflow from the watershed.

The objectives of this paper are to describe briefly the model, with emphasis on the modifications made for this application, and to present results of tests to determine the reliability of the model for an experimental site in eastern North Carolina.

2 MODEL DEVELOPMENT

2.1. Field Hydrology Model

The computer model, DRAINMOD, was developed and validated for shallow water table mineral soils as a tool to be used in the design of water management systems (Skaggs, 1978).

DRAINMOD is composed of a number of separate components, incorporated as subroutines, which quantify the various mechanisms of water movement and storage in the soil profile. The component descriptions given below focus on the modifications developed for this project. A complete description of DRAINMOD is given by Skaggs (1978). Details of field tests of the model are given by Skaggs (1982), Skaggs et al. (1981), Gayle et al. (1984) and Chang et al. (1983). A new version, DRAINMOD 4.0 was recently released and is available from the second author.

Precipitation: Hourly rainfall data are used as input to the model. These are readily available for many North Carolina stations (Wiser, 1975).

Infiltration: Infiltration is determined by the Green and Ampt (1911) equation with the model input parameters dependent on soil properties, initial water content and surface conditions.

Surface Runoff: Surface runoff is predicted when the average surface depressional storage has been filled with water. It is assumed that the depressional storage is evenly distributed over the field. This parameter is estimated from site observation.

Subsurface Drainage: The rate of subsurface drainage is estimated from the Hooghoudt equation when the water table is below the surface and Kirkham's (1957) equation when the field is flooded.

Bank Storage: The theory leading to the Hooghoudt equation predicts an elliptical water table profile between drains. This assumption often fails for very poorly drained soils such as the peats considered in this study. In this situation the water lost from the banks to the ditches prior to the development of an elliptical profile, becomes significant (Figure 1).

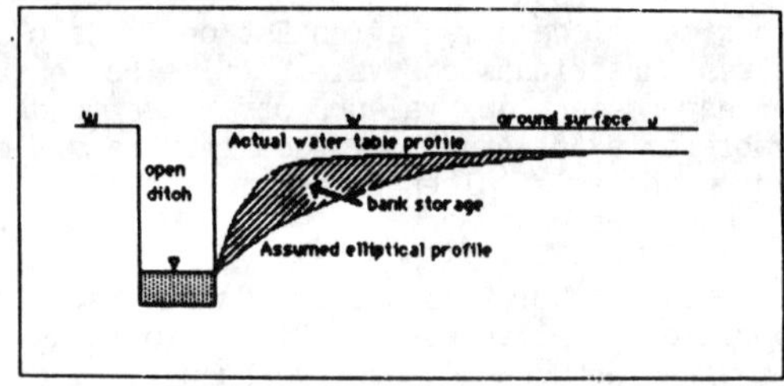

Figure 1. Much of the subsurface water draining to ditches in peat soils comes from bank storage.

The effect of bank storage on the amount and rate of subsurface drainage was determined by solving the Boussinesq equation for transient flow in the profile. This approach depends on the Dupuit-Forchheimer assumptions and neglects unsaturated flow above the water table. Numerical solutions to the non-linear Boussinesq equation (Moody, 1967; Skaggs, 1975) were used to predict the position and shape of the water table profile and the rate of drainage to the ditches as a function of time. Solutions for typical ditch spacings of about 100 m on the tight N.C. peat soils showed that drainage to the open ditches would not affect the midpoint water table depth even over the longest period of dry weather on record for that area.

Solutions were obtained for a range of initial water table depths, rainfall events and ET rates and the drainage rate at the ditch was plotted as a function of average water table depth. Nonlinear regression methods were used to determine coefficients in an exponential equation relating the drainage flux to the water table depth for a typical 800 m * 100 m field on the test site. This equation was included in the subsurface drainage subroutine. Including this equation caused the water balance to be computed for a cross-section of the profile extending from ditch to ditch in the horizontal direction and from the surface to a restricting layer in the vertical direction. Consequently the water table depth predicted is average water table depth for the field. It is somewhat greater than the midpoint water table depth. Ideally, bank outflow should be related to infiltration or ET rates and time since the last rainfall, as well as to water table depth (McCarthy and Skaggs, 1987).

Evapotranspiration: Daily potential evapotranspiration (PET) may be calculated in DRAINMOD by empirical methods presented by Thornthwaite (1948) or read in directly from a file. The Thornthwaite method is used because required input data (daily maximum and minimum temperatures) are available for many locations and because it was found to be reliable over the growing season (Mohammad, 1978). However, comparisons with pan measurements showed that the Thornthwaite method underpredicts PET for the winter months.

Several methods were considered to provide a more accurate prediction of PET. The Christianson R_s (Jansen, 1973) method was selected as being the most suitable following an evaluation of available climatological data and comparisons with Thornthwaite, Penman, Jensen-Haize and pan methods. Required input data for the Christianson method were available for an 18 month period from 6/1/77 to 12/31/78 for the test site. The daily Christianson PET values for this period were calculated independently and read into the model directly.

The lack of wind speed and relative humidity data made it impossible to use the Christianson method for long term simulations. Correction factors for the Thornthwaite method were developed from published data to provide reliable estimates of the PET on a monthly if not daily basis. This should allow reliable predictions of antecedent conditions prior to storm events.

Hourly Flow Rates: DRAINMOD determines the water lost from the soil profile by subsurface and surface drainage and outputs the results as a daily water loss depth. In order to use the DRAINMOD output as lateral flow input to the flood routing model it was necessary to convert this unit area water loss to a flow rate. A typical drainage canal network in the area being investigated consists of 1600*800 m^2 (1*0.5 mi^2) blocks. Each has 16 parallel field ditches spaced 100 m apart which empty into a deep 1600 m long collector canal (Figure 2). For the purpose of these analyses each block was divided into 16 fields, each field being 800 m long and 100 m wide with a field ditch running through the middle. All water loss values from DRAINMOD were converted into flow rates (in l/s) for input to the flood routing model.

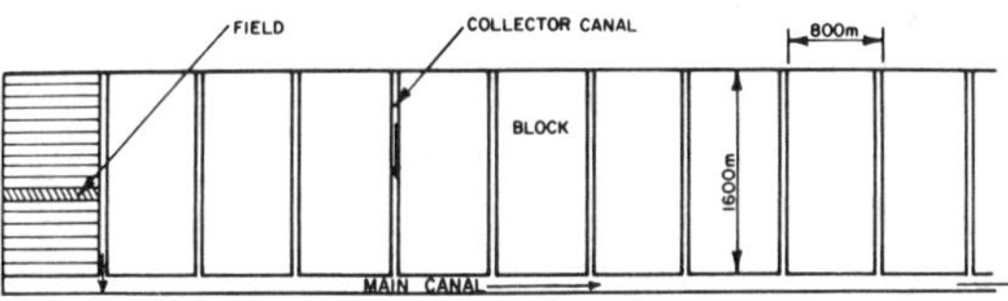

Figure 2. Layout of a typical stretch of main canal on the peat soil test site.

2.2. Flood Routing Model

A large portion of the peatlands in eastern North Carolina, have been cut over for timber, and thousands of ha have been developed for agriculture. As a result a system of regularly spaced drainage channels has been established. Thus, drainage from large areas can be characterized by routing the water lost from the individual fields down through the network of canals, continually accounting for new input along the way. In this way an assessment can be made of how stage and flow rate vary with time throughout the network, as well as, at the outfall of a watershed. Because the area is very flat (surface slopes less than 0.02 %) and because of the intensive drainage canal network, each watershed is clearly defined.

Amein (1968, 1972) presented methods for solving the Saint-Venant equations for unsteady flow in open channels. He wrote the governing equations in implicit finite difference form and developed a rapidly converging method of solving the large system of simultaneous equations that resulted.

Amein's model was selected because of its proven ability to handle rapidly varying lateral inflows, which occur following heavy rainfall on peatland sites. It has been tested for a wide range of applications in natural and artificial channels, reservoirs and estuaries (Amein and Fang, 1969, Amein, 1972, and Wardak, 1977). Observed and computed flows have been in excellent agreement for all cases.

Several modifications were made to the basic methods presented by Amein (1972). Details are given in Gregory et al. (1984). A monthly simulation required 3500 Kilobytes of memory and 1.5 minutes to run on an IBM 3081.

Channel Description: The outflow predicted by DRAINMOD, for a 800*100 m^2 field, was taken directly as lateral inflow to the flood routing model at 16 regularly spaced stations along the 1600 m long collector canal. The outflow from the collector canals could, in turn, be routed down the main canal to the watershed outlet. Average cross-sectional areas were used for the regular shaped collector and main canals.

Overbank Flow: There was no provision in Amein's original model to deal with flooding. The channel depth was allowed to increase indefinitely during storms to remove all water. This resulted in very high and unrealistic flow rates for our application. This was overcome by specifying a maximum channel depth and switching to a composite channel section once the water level had breached the banks and initiated overbank flow (Figure 3). This was achieved by using Manning's equation and rewriting the equation for the friction slope to include effects of overbank flow.

The Manning's roughness coefficient for the overbanks was much higher than for the channel. The overbank width was set equal to half the distance to the next canal but assumed that the water continued to flow in the same direction as the water in the canal. In reality, once the canal levels began to drop, the flow direction would be back towards the canal. However this could not be represented in the model. While this modification is an improvement over the original version, it is still a one-dimensional approximation of a two-dimensional problem.

Boundary Conditions: The original purpose of Amein's model was to route flow from its source to the outlet through a channel reach. The lateral inflow component was primarily used to cope with seepage losses from the channel. In the application for this project, all of the channel flow originated as lateral inflow from each of the field drain outlets. Also the channels all terminated at their farthest upstream points. In order to obtain a solution to the St. Venant equations, the flow in the farthest upstream field drain outlet was taken as the upstream inflow. This introduced a small error in the momentum equation at the upstream end but it was found to be negligible in trial simulations. The downstream boundary condition was fixed by assuming normal flow and using Manning's equation to relate flow rate to depth.

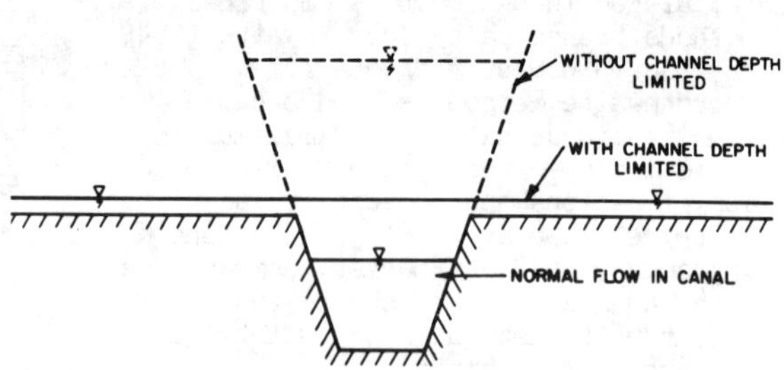

Figure 3. Effect of limiting channel depth.

3 MODEL TESTING

Rainfall, water table depth and runoff measurements were made on three experimental sites on First Colony Farms for a three year period from 1976 to 1979 (Skaggs et al. 1980). The data from a 130 ha field site, (Site 103), which was located within a 6000 ha tract permitted for mining, was used to test the model. The soil is a deep colloidal organic (Typic medisaprist) extending to a depth greater than 1.5 m. The soil has a bulk density of 0.2 to 0.4 g/cc, an organic matter content greater than 88% (Badr, 1978) and large amounts of buried wood in the profile. It is typical of the peats considered for mining in this area. The experimental site is bounded on each end by main drainage canals spaced 1600 m (one mile) apart and on the sides by collector canals 800 m apart. It had been drained and planted to fescue pasture with field ditches spaced 100 m apart emptying into the collector canal (Figure 2).

The model was tested in three ways: (1) water table depth predictions were compared with the observed for the experimental record; (2) predicted monthly outflow volumes were compared to the measured values; and (3) predicted hydrographs at the collector canal outlet were compared to the observed on a storm by storm basis.

3.1. Input Data

A summary of soil property and other input parameters for the simulation models is given in Table 1. The hydraulic conductivity (K) of the surface layer of the soil is very dependent on surface cover and the length of time since development. Badr (1978) measured saturated hydraulic conductivity on vertical cores and found K values of 2.4 to 4.8 m/day for this layer when a fescue sod had been established. However, field observations and water table measurements indicated a higher lateral conductivity of 17 m/day. The high K value in the surface layer is attributed to the irreversible drying and curing of the surface peat layer resulting in soil aggregation as well as the porous root mat close to the surface. Hydraulic conductivity values for the deeper layers were taken from data presented by Badr (1978), Purisinsit (1982) and Foutz (1983). Other data on soil properties, site parameters and canal specifications are summarized in Table 1.

Daily evapotranspiration data were calculated by the Christianson R_s method and read into DRAINMOD for the period available (5/77 - 12/78). The corrected Thornthwaite method was used for the remainder of the test period (1/77 - 4/77 and 1/79 - 9/79).

Table 1. Summary of soil property, drainage system and crop parameter inputs to DRAINMOD for deep organic soil (Site 103).

Input		Value
1. Soil properties		
Depth to restricting layer		2.34 m
		depth range
Saturated hydraulic conductivity, K	K=17 m/day	0-.25 m
	K=.005 m/day	.25-1.40 m
	K=.10 m/day	1.40-2.34 m
Water content at lower limit available to plants		0.45 mm^3/mm^3
Saturated water content in root zone		0.74 mm^3/mm^3

Water Table Depth (m)	Drainage Volume (mm)	Upward Flux (mm/day)
0	0	>20
0.2	3.3	0.60
0.4	8.3	0.30
0.6	17	0.23
0.8	27	0.17
1.0	38	0.12
1.2	50	0.08
1.4	63	0.06
1.6	76	0.03

2. Drainage system parameters (open ditches).	
Drain depth	1.40 m
Effective drain radius	190 mm
Surface depressional storage	12.5 mm
Drain spacing	100 m
3. Crop parameters (grass)	
Effective root depth	120 m
Length of growing season	365 days
4. Canal specifications	
Channel dimensions	
Depth	2.30 m
Bottom width	0.76 m
Side slope (horizontal: vertical)	1:1
Channel slope	0.02%
Manning's roughness coefficients	
Channel	0.20
Overbanks	1.50

3.2. Results and Discussion

3.2.1. Water Table Depth Comparisons:

Predicted and measured water table depths are compared in Fig. 4. Only the periods for which unbiased, independently obtained evaporation data were available were used in the comparisons. These are periods for which

Christianson PET were available and for summer months when Thornthwaite was considered reliable (a total of 25 months between May 1977 and Sept. 1979). The average daily absolute difference between observed and simulated water table depths was 0.10 m for the test periods.

Discrepancies occur close to the surface. The observed water table depths rarely rose completely to the surface as predicted. We believe this to be a combination of errors in the well calibration and of the surface conductivity effects discussed earlier. The water table at the midpoint between drains is essentially controlled by ET and infiltration. Differences between observed and simulated water table depths are primarily a result of differences in the actual and predicted values of these parameters. These differences, however, appear minor as shown by the high degree of correlation exhibited in Fig. 4 and the low average absolute differences.

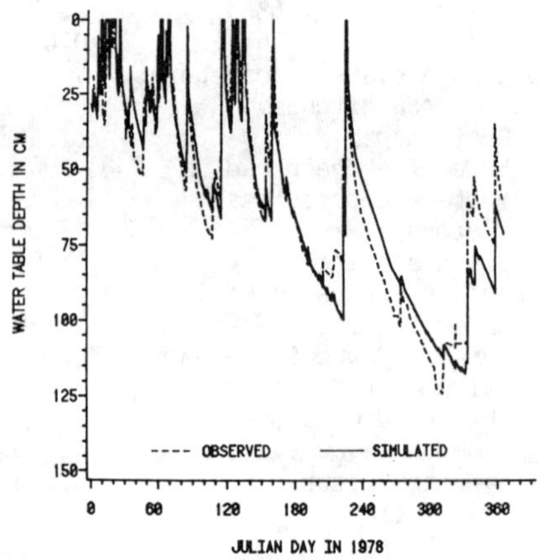

Figure 4. Observed and simulated water table depths on the experimental site for 1978.

3.2.2. Monthly Flow Volumes:

The observed and simulated monthly, unit-area flow volumes are plotted in Fig. 5. Outflow rates were determined in the field study by continuously recording the water level upstream from a triangular sharp crested weir (Skaggs et al. 1980). The weir was subjected to repeated submergence during some high rainfall events. The flow volumes for these periods were approximated by computing a water balance for the period of submergence (Purisinsit, 1982). Actual hydrograph shape and magnitude of the flow peak, however, could not be determined for those storms during which the weirs were submerged.

For the analysis, only the periods 5/77 to 12/78 and 5/79 to 9/79, in which data for calculating ET were available, were considered. On average the observed flows were greater than the simulated flows by 3.5 mm/month or 13.6%. This could be partly due to inaccurate corrections for the periods of submergence.

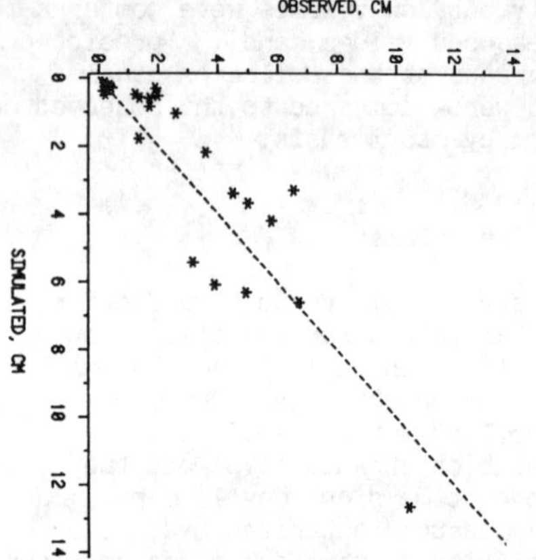

Figure 5. Observed versus simulated monthly flow volumes per unit area for 24 months.

3.2.3. Hydrograph Comparison:

The hourly water losses predicted by DRAINMOD were routed down the collector canal to the main canal for a number of storms during the period of observed record. The results were encouraging but clearly indicated areas for future work. The flows in liters/sec at the collector canal outlet are compared with observed flows in Figs. 6-9.

Fig. 6 illustrates one of the best agreements obtained. The slight displacement would appear to be caused by chart timing error as the initial observed flow rate increase began several hours before rainfall was recorded. In this comparison the observed and simulated water table depths matched very closely, rising to just below the soil surface at each peak. No surface runoff was simulated and most likely none occurred in reality. Thus, when water tables were predicted accurately the DRAINMOD - flood routing combination did a reasonable job in predicting collector canal outflows.

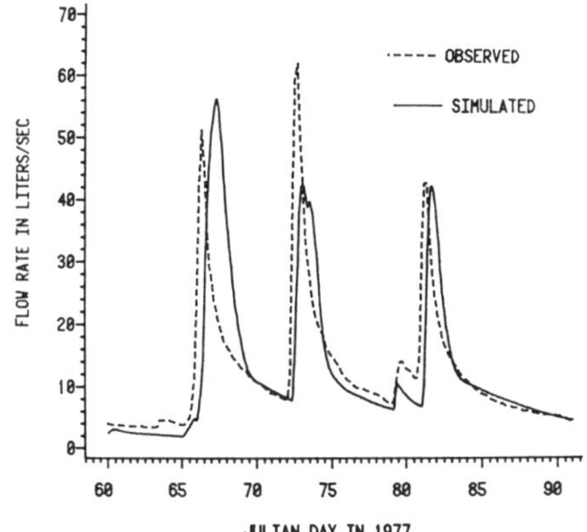

Figure 6. Observed and simulated hydrographs from site 103, when no surface runoff occurred.

The model tends to overpredict the flow rate and volume once surface runoff occurs. This is illustrated in Fig. 7. The flood routing model assumes an unlimited outlet for all water. If for some reason the downstream outlet is restricted, water backs up reducing the outflow rate. While the flood routing model has the capability to model such occurrences, there was no way of knowing exactly when they occurred for the experimental site. We suspect that the channel banks were breached more often than predicted because of restrictions downstream.

The simulation model accounts for all water movement within the system. Not only does the unit area, soil water budget balance, but the water balance for the whole block is maintained. During peak flows with overland runoff occurring it is likely that water is being lost directly from the fields into the main canal, and possibly in other directions, bypassing the collector canal weir. Thus the rigorous accounting procedure of the models would be expected to overpredict the observed flows under these conditions.

When the water table was deep (greater than 0.5 m below the surface) and a moderate rainfall occurred, the observed runoff often exceeded the simulated by a significant amount. Fig. 8 is a example

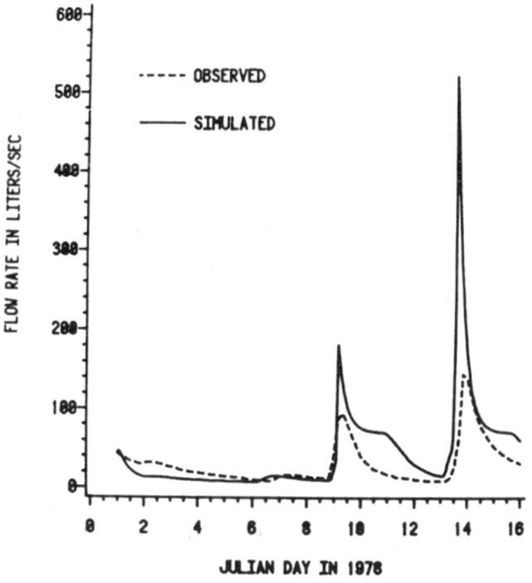

Figure 7. Observed and simulated hydrographs from site 103 for a large event with surface runoff.

of this. The observed flow, while higher than the simulated, is unlikely to be from overland runoff because of the low rainfall intensity. It seems more likely that the rainfall infiltrated the surface layer of dry, cured peat but then was perched on top of the tighter peat layer about 0.3 m below the surface. Since the ground slope has a gradient of about 0.3% toward the ditches, lateral flow through the surface layer to the field ditches could have been significant. The high conductivity surface layer has no effect on the model predictions unless the water table rises up into that layer.

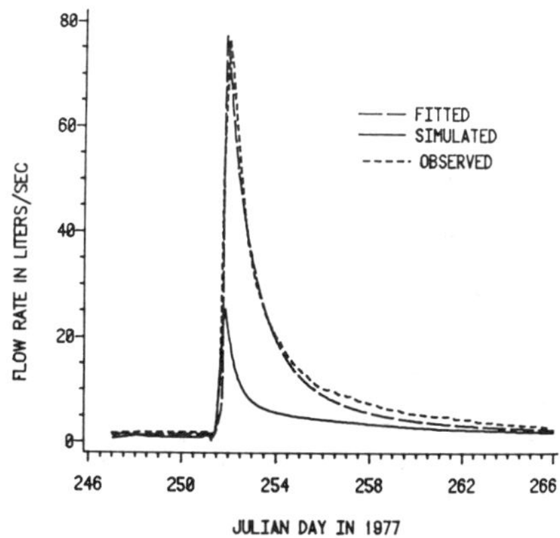

Figure 8. Uncalibrated simulations were in poor agreement with observed for this event.

The observed hydrograph in Fig. 8 and other such events could be fitted by manipulating parameters to control infiltration or antecedent conditions. This is demonstrated in Fig. 8 by the plot labeled "fitted" which was matched to the observed hydrograph. That is, the model parameters and antecedent conditions were adjusted so that the predicted hydrograph was in close agreement with the observed. Hydrologic models of a less deterministic nature than the one used here are often "calibrated" in this manner. However, unless long periods (years) of data are available for such calibration, results such as given in Fig. 8 may be extremely misleading. The fitted hydrographs usually make the model appear more precise than it really is. Altering parameters to fit the hydrographs, while fixing one simulation, could adversely affect others and the results would be misleading. When calibration is used, it is extremely important that predictions of the "calibrated model" be compared with observed data which have not been "fitted" to test the validity of the result. In our case, model parameters were determined independently and calibration was not used.

Model results indicate that changes in surface properties, affecting infiltration capacity of the peat, occur with wetting and drying in a manner that is not completely described by methods presently used in the model. More work is needed in this area if complete understanding and modeling of the hydrology of these organic soils is to be accomplished.

Fig. 9 illustrates a low flow condition. The models match the observed base drainage rate to within $\pm$ 0.5 liters/sec. The hydrograph peaks in Fig. 10 are a result of rain falling directly into the ditches and canals. The percentage of channel surface area for a block was found to be approximately 2% of the total block area. Under most situations the water falling directly into a channel will find its way to the outlet. Thus 2% of the rain falling on a field was considered to be direct runoff in the models. Because of the very low drainage rates from the peat soils this 2% rainfall is significant during low flow situations as evidenced by Fig. 9.

For the period 1/1/77 to 9/31/79, periods containing rainfall events were simulated to produce collector canal outflow. The average length of these periods was 28 days. This was a little over half the total period. Data collected from rainfall events during the remainder of the period were not used because of weir submergence, missing periods, equipment failures, or simply incorrect flow readings. There were also periods of no rainfall or runoff which were excluded from the analysis. In addition to testing the model for individual hydrograph comparisons, the flow rates for the 19 periods simulated were collected

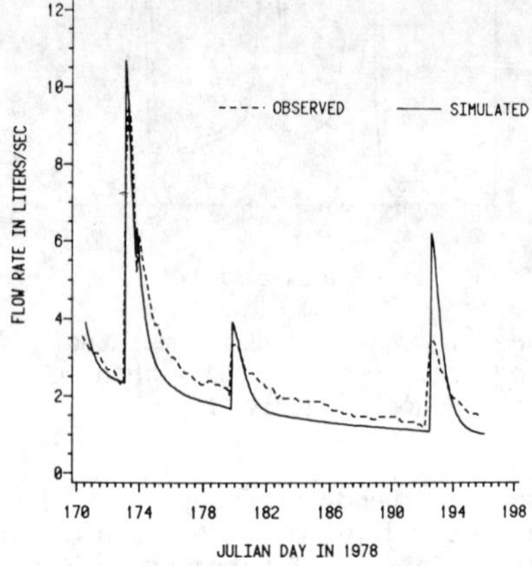

Figure 9. Observed and simulated results for low flow rates.

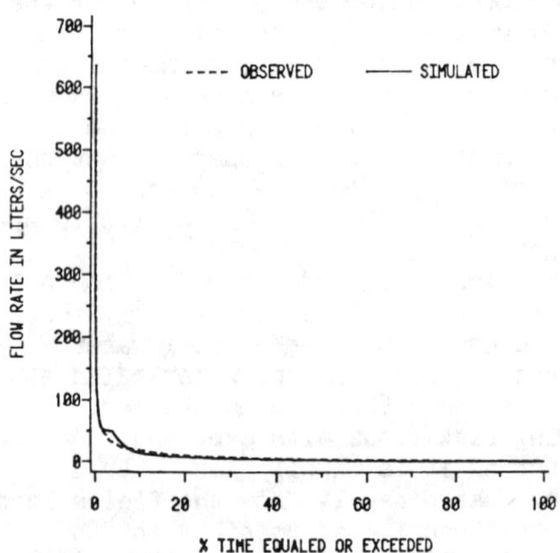

Figure 10. Observed and simulated flow duration curves for experimental site 103.

together and compared on a flow duration curve (Fig. 10). This curve shows the probability of occurrence of a flow of any given magnitude. The simulated and observed flow duration curves are in good agreement illustrating the capability of the model to predict outflow from the site accurately over the observed range in flow rates.

4 SUMMARY AND CONCLUSIONS

This paper describes the modification and linkage of two existing models to produce a comprehensive hydrologic watershed model. A field scale model for predicting subsurface and surface runoff was linked with a numerical flood routing model to predict the watershed outflow hydrograph. The models are deterministic and are based on generally available data.

The validity of both models has been previously tested separately with excellent results. The field scale model was applied separately, and then in conjunction with the flood routing model, to a well channelized experimental site of deep peat in eastern North Carolina. Approximately three years of experimental data were available for testing the models from this site. The observed and simulated water table depths for the peat soil were in close agreement for the test period. While problems still exist, good agreement was observed between most of the observed and simulated collector canal hydrographs. Also the flow duration curve comparison between observed and simulated collector canal flow rates illustrated close agreement. The model presented herein can be used for predicting the hydrologic effects of land use change and alternative water management practices.

5 REFERENCES

Amein, Michael. 1968. An implicit method for numerical flood routing. Water Resources Research, 4(4):719-728

Amein, Michael and C. S. Fang. 1969. Streamflow routing with applications to North Carolina rivers. Report No. 17, Water Resour. Res. Inst. of The Univ. of N.C. Raleigh.

Amein, Michael. 1972. Report on the numerical simulation of unsteady flows in rivers and reservoirs, to Office of Hydrology, National Weather Service, NOAA, U.S. Dept. of Commerce. Contract No. 0-3528, 73 pp.

Badr, A. W. 1978. Physical properties of some North Carolina organic soils and the effect of land development on these properties. M.S. Thesis, Dept. of Biological and Agricultural Engr., N. C. State University, Raleigh.

Chang, A. C., R. W. Skaggs, L. F. Hermsmeier and W. R. Johnson. 1983. Evaluation of a water management model for irrigated agriculture. TRANSACTIONS of the ASAE 26(2):412-418.

Foutz, T. L. 1983. Effects of peat mining on ground water processes. M.S. Thesis, Dept. of Biological and Agricultural Engr., N. C. State Univ., Raleigh.

Gayle, G. A., R. W. Skaggs and C. E. Carter. 1984. Evaluation of a water management model for a Louisiana sugar cane field. Journal of the American Society of Sugar Cane Technologists. In press.

Green, W. H. and G. Ampt. 1911. Studies of soil physics, part I - the flow of air and water through soils. J. of Agricultural Sci., 4:1-24.

Gregory, J. D., R. W. Skaggs, R. G. Broadhead, R. H. Culbreath, J. R. Bailey and T. L. Foutz. 1984. Hydrologic and water quality impacts of peat mining in the coastal zone of North Carolina. Report No.214 Wat. Resour. Res. Inst. of The Univ. of N. C. Raleigh.

Ingram, R. L. and L. J. Otte. 1982. Peat deposits of Pamlimarle Peninsula - Dare, Hyde, Tyrrell, and Washington Counties, North Carolina. Report to the U.S. Dept. of Energy and the N. C. Energy Institute.

Jensen, M. E. (ed.) 1973. Consumptive use of water and irrigation water requirements. ASCE, NY. 215 pp.

Kirkham, D. 1957. Theory of land drainage. In J. Luthin (ed.), Drainage of Agricultural Lands, Am. Soc. of Agronomy, Madison, WI.

McCarthy, E. J. and R. W. Skaggs. 1988. A simplified model for predicting drainage rates for changing boundary conditions. ASAE Paper No. 88-2058, ASAE, St. Joseph, MI.

Mohammad, F. S. 1978. Evaluation of methods for predicting potential evapotranspiration in humid regions. M.S. Thesis, Dept. of Biological and Agricultural Engr., N. C. State Univ., Raleigh.

Moody, W. T. 1966. Nonlinear differential equation of drain spacing. J. of the Irrigation and Drainage Div., ASCE, 92(IR2):1-9.

NOAA. 1982. Evaporation atlas for the 48 contiguous United States. NOAA Tech. Rep. NWS 33, Office of

Hydrology, National Weather Service, Washington, D.C.
Purisinsit, Pitsamai. 1982. Evaluation of two hydrologic models for the North Carolina Blacklands. Ph.D. Thesis, Dept. of Biological and Agricultural Engr., N. C. State Univ., Raleigh.
Skaggs, R. W. 1975. Drawdown solutions for simultaneous drainage and ET. Journal of Irrigation and Drainage Division, ASCE, Vol. 101(IR4):279-291.
Skaggs, R. W. 1978. A water management model for shallow water table soils. Report No. 134, Wat. Res. Res. Inst. of The Univ. of N. C., Raleigh. 178 pp.
Skaggs, R. W., J. W. Gilliam, T. J. Sheets and J. S. Barnes. 1980. Effect of agricultural land development on drainage waters in the North Carolina tidewater region. Rep. No. 159 of the Wat. Res. Res. Inst. of The Univ. of N. C., Raleigh.
Skaggs, R. W., N. R. Fausey, and B. H. Nolte. 1981. Water management model evaluation for north central Ohio. TRANSACTIONS of the ASAE, Vol. 24(4):922-928
Skaggs, R. W. and G. W. Gilliam. 1981. Effect of drainage system design and operation an nitrate transport. TRANSACTIONS of the ASAE, Vol. 24(4):929-934.
Skaggs, R. W. 1982. Field evaluation of a water management simulation model. TRANSACTIONS of the ASAE, Vol. 25(3):666-674.
Thornthwaite, C. W. 1948. An approach toward a rational classification of climate. Geog. Rev., 38:55-94.
Wardak, S. G. 1977. Implicit numerical solution of unsteady flows in open channels and shallow river basins. Ph.D. Thesis, Dept. of Civil Engr., N. C. State Univ., Raleigh.
Wiser, E. H. 1975. HISARS - Hydrologic information and retrieval system, Reference Manual, North Carolina Agricultural Experiment Station Tech. Bull. No. 215, 218pp.

Land and Water Use, Dodd & Grace (eds), © 1989 Balkema, Rotterdam. ISBN 90 6191 980 0

Water management of lowland peat soils

A.C.Armstrong, A.A.Thorburn, C.W.Dennis* & D.A.Castle
ADAS Field Drainage Experimental Unit, Trumpington, Cambridge, UK

**Now: ADAS, Thorpe Road, Norwich, UK*

ABSTRACT: Lowland peats are important agriculturally, but are also often located in areas of environmental concern. In the past, techniques have been developed for the drainage of these soils, and many of the associated problems resolved. The physical and chemical properties of lowland peat soils are reviewed, and the problems associated with the installation and maintenance of effective drainage discussed. Problems encountered include peat wastage, ochre deposition, and slurry formation at installation. The use of pipeless drainage for peat soils is also discussed. These techniques are also relevant to the problems of managing soil water regimes in peat soils through sub-irrigation, either for increased crop production, or for the maintenance of wetland status.

RESUME: Les tourbières des terres basse ont de l'importance agricole, mais elles se situet souvent aussi aux étendues de l'importance en ce qui concerne l'environment. Dans le passé on a developpé des technqiues pour le drainage de ces sols et beaucoup de problèmes associés ont été resolus. Les propriétés chimiques et physiques des tourbières sont examinées et les problèmes associés à l'installation et à l'entretien du drainage effectif sont discutés. Des problèmes qui se presentent comprennent la déperdition de tourbe, le depôt d'ocre et la formation du coulis de sol à l'installation. L'utilisation du drainage par charrue-taupe pour les tourbières est discutée également. Ces techniques sont relevenates aussi aux problèmes du maniement des régimes de la nappe phréatique dans les tourbières au moyen de l'irrigation souterraine, ou bien pour l'augmentation de la recolte, ou bien pour l'entretien des qualites de les tourbières.

ZUSAMMENFASSUNG: Boden des Niedemoores sind landwirtschaftlich wichtig, befinden sich aber oft in Gegenden, die im Hinblick auf die Umwelt bedeuten sind. Fruher hat man Methoden fur die Dränierung dieser Böden entwickelt und viele der damit verbunden Probleme gelöst. Die physischen und chemischen Eigenschaften der Böden des Niedermoors werden erneut überprüft und die Probleme im Zusammenhang mit der Installation und Instandhaltung einer wirksamen Dränung diskutiert. Die Probleme, auf die wir stossen, beziehen sich auf die Moorböden-Sackung, Dränverockerung and Schlammbilding bei die Installation. Der Gebrauch einer rohrlosen Dranung fur Moorboden wird ebenfalls diskutiert. Diese Methoden helfen ebenfalls mit dem Problem des Grundwasserstandes in Moorböden durch Tiefen-Irrigation entweder fur einen erhohten Ernteertag oder fur die Aufrechterhaltung des Marschlandes.

1 INTRODUCTION

Lowland peat soils are located in a number of agriculturally important areas of England. Most prominent is the Fenland area of Eastern England, but other significant areas include the Somerset levels, and the Lancashire Moss peats. An inventory of lowland peat in England and Wales (Burton & Hodgson 1987) identified 103,000 ha of this soil, and also presents details of its distribution and classification. Initially formed under saturated conditions, these areas can be highly productive when drained, often being used for high value horticultural crops. They are however, also in environmentally sensitive areas, where there are pressures to maintain their "wet" status. The techniques for draining these soils, however, can be equally applied to the management of the soil water for conservation purposes. This paper reviews experience with the water management of these soils. It suggests that the knowledge so gained is capable of use for comprehensive water management strategies for the maintenance of agricultural productivity compatable with the conservation requirements of environmentally sensitive areas and land set-aside from agriculture.

2 PROPERTIES OF LOWLAND PEAT

Lowland peat has formed under saturated conditions, when vegetation accumulation rates have been high. They thus typically occupy basin situations, and have complex depositional histories related to variations in sea level and changes in river courses. The Fen peats for example have been much studied, and consist in outline of a lower peat, overlain by a marine clay of neolithic age, and an upper peat layer laid down since 2000 B.C., over which more recent alluvium has locally been deposited (see, e.g. Darby 1983). The Somerset peats have a simpler structure, being laid down in a single phase in a basin whose depth increased from the coast, and which has been overlain only at the seaward edge by a later marine clay. Locally recent alluvial deposits overly the peat, with the consequence that the soil is in detail variable (Avery 1955, Burton & Hodgson 1987). The Moss peats of Lancashire are simpler still, and in general represent single accumulations of peat in hollows (Hall & Folland 1967), although variations in depositional conditions differentiate the individual deposits.

2.1 Physical Properties

Like any other soil, peat is highly variable in both the horizontal and vertical directions. Much of the variation is due to the depositional history. In the Fenland peats for example, old river courses are clearly visible on aerial photographs, and also on the ground as ridges (known locally as "rodhams"), where the mineral content of the soil is appreciably higher (Seale 1975a & b). Further, many standard soil physical measurements are of doubtful applicability because of the very high organic matter content. Neverthless, it is possible to make some generalisations about the behaviour of peat soils.

Saturated hydraulic conductivity of most peat soils is in general quite high, but broadly related to the degree of humification of the organic material (Eggelsmann 1972). Some of the values recorded by ADAS are given in table 1. Even where there are substantial clay layers in the profile, these are frequently penetrated by dead root channels so that these layers are rarely limiting in the hydraulic sense.

Peat soils also however, suffer from problems of drying, during which they become hydrophobic. It has been frequently observed that a dry surface layer can become quite impermeable, so that water will remain perched above an unsaturated layer with a high conductivity. Because of this problem, much peat soil is managed in such a way as to keep it moist at all times. A moist regime also helps to reduce peat wastage by oxidation and erosion by wind.

2.2 Chemical properties

A frequent problem is the inherently acid nature of many organic soils, and the presence of iron pyrite, which on oxidation generates extreme acidity and leads to problems of iron ochre formation. Particularly problematic is the formation of ochre in drain pipes - as in the case reported in detail by Dennis (1978), which lead to complete failure of the drainage system. The initial source of the iron was thought to be pyrite formed when the peat was deposited. It is often particularly prevalent in the strata of higher mineral content at the edge of the peat deposit or interleaved with the peat, reflecting intermittent brackish or estuarine conditions. The iron is mobilised by the introduction of aerobic conditions following drainage, and subsequently deposited in the drains via the interaction with biological processes, notably the agency of the bacteria such as thiobacillus ferro-oxidans (Trafford et.al. 1973, Thorburn 1976, Ap Dewi et.al. 1987a & b). Experimental work has established techniques for predicting the ochre potential of a small soil sample, by determing the sulphide content. However, the problem of adequately sampling this highly variable soil property, and the further very variable relationship between pyrite content and rate of ochre production, renders such methods unreliable for predictive field scale design applications (FDEU - unpublished report).

The acid nature of the peat may often lead to problems. Although well humified peat is only slightly acid, "raw" peat can frequently have pH values around 4, and values as low as pH 2 have been reported. Apart from the agricultural problems of acidity, which can be partly controlled by conventional techniques such as liming, the aggressive soil water that develops can endanger drainage installations. Particularly problematic are the acid sulphate soils (Dent 1986). Not only is the highly acidic nature of the soil damaging to concrete and metal installations, but it may also release highly acid drainage waters, with subsequent harmful effects on the receiving aquatic environment (Gosling & Baker 1980).

3 DRAINAGE OF THE PEAT

Because lowland peats are usually found in basins close to sea level, they require thorough arterial drainage. This may include elements of flood protection from inundation by the sea, and the transmission of "foreign" water which enters the basins from the tributary uplands. Where such protection fails, the extent of inundation may be catastophic, as happened in the Fens in 1947 (Darby 1983). Typically there will therefore be a need for the embanking of rivers that cross peat areas, to carry water entering the basin from adjacent uplands, in addition to the channels for internal drainage. However, even after the neccessary arterial control of water, field drainage is normally neccessary for arable cultivation (Armstrong 1982). This need is in part the result of the fact that these soils are frequently used for the cultivation of high value horticultural crops, which are both sensitive to short periods of waterlogging and are harvested at unfavourable times of the year. Consequently under-drainage is needed to prevent even short periods of waterlogging at critical times of the year. The process is, however, often technically difficult (Castle et al 1984). The difficulties are associated with four main problems: design standards, topography, installation and maintenace.

3.1 Design.

Before subsurface drainage can be attempted, it is essential that the arterial drainage of the area be established. This is normally followed by the excavation of field boundary ditches and the establishment of a suitable level of regional groundwater control by these ditches and associated engineering works. It may be neccessary to use surface drainage for a number of years (typically 2 - 5 years) as an initial de-watering phase. Only then is it possible to consider the installation of pipe drains.

Criteria for the design of pipe drainage systems are often difficult to determine, since they relate to short periods of waterlogging. Conventional drainage design criteria (eg Castle et.al. 1984) rely on steady state rainfall (perhaps as low as 8 mm/day in the dry areas of eastern England) and a 5 day removal period. Under this sort of assumption, consideration of the hydraulic conductivity data might be expected to suggest that a system of arterial ditches at spacings in excess of 150 m might well afford adequate control of the watertable. The situation is complicated by the fact that the hydraulic conductivity of the peat decreases as it is humified (Egglesman 1972), and allowance has to be made for this decrease.

The intimate relationship between ditch and in-field water levels has been demonstrated by Armstrong (1980) at Rickwood EHF. Data showed that land adjacent to pumped ditches responded rapidly to changes in the pump regime, to the extent that diurnal fluctuations in the ditch water level due to the pumping regime, were transmitted back into the field. From this experiment, and from theoretical analysis of the same data, it was concluded that ditch drainage can control the water levels in fields up to 100 m away. Data from a nearby site (Dennis 1986) has demonstrated that deep ditching and associated pumping can also provide adequate in-field control of the watertables. However, the pumping requirements can be onerous, and so the more conventional solution is to install field drains. These have the advantage of providing short term watertable control, by the rapid disposal of storm events (see the drainflow hydrographs in Armstrong 1980).

3.2 Topography.

Frequent ditch maintenance and spreading of excavated material on field boundaries has resulted in many fields having a "bowl shaped" topography. In addition, the topography within the field may also be complex, and can amount to a variation of as much as 1500 mm over a 5 ha block, without there being any overall gradient on the field boundaries. This presents the twin problems of grading pipes through the higher areas, and also ensuring adequate provision for low areas. Frequently drainage is installed specifically as a response to problems of waterlogging in the low "wet spots". However, freeboard available is frequently limited. Consequently, it may well be the case that the constraints on freeboard mean that drain depths in the centre of the field are less than ideal. The problem is not aided by the fact that peat wastage effectively reduces pipe cover with time.

3.3 Installation.

Because in many fields, pipes are installed at zero or very low gradients, a very high standard of installation is demanded. It is however frequently difficult to maintain such standards where heavy machinery may well sink into weak soil surfaces, making the practical issue of maintaining grade difficult. Subsequent sinking of dense pipes (particularly clayware and concrete pipes) or differential peat shrinkage can also cause misalignment and backfalls in grading, which are especially critical when design gradients are very low.

The installation of drains may be hampered by the fact that "raw" peat soils are frequently both structureless and waterlogged at drain installation depth. Consequently, drains are often laid under slurry conditions, and subequently block very rapidly. This seems to have been a particular problem in the Somerset peats, where problems of drain failure have been proportionately more common than on mineral soils. The mechanical act of disturbing the soil with a trenching machine, with the material in a saturated state below the watertable, may create a slurried mix of material which may well be thixotropic, structureless, and impermeable. The low mechanical strength of the peat may also lead to problems of access by drain laying equipment. Additionally because of its low strength, peat may move under water pressure as the confining overpressure is removed. Consequently it has been observed that trenches may infill from below as the unconfined peat rises.

The use of pipes with larger than usual slot sizes has been investigated, both in lysimeters and in the field, as a possible remedy to the problems of water entry into drain pipes (Dennis et al 1985). However, the use of such measures did not noticably improve the performance of the drainage systems, and in some instances plugs of peat material blocked the slots rather then forming bridges over them, with a consequent deterioration in drain performance. Consequently, the use of wider slot sizes has not been recommended in England and Wales. Specifically designed "skeletal" pipes have been tested with varying degrees of success for combatting sealing by both peat and by iron ochre, but they are as yet not widely available.

More successful however, has been the use of pipe-less drainage. The mole plough, which has been found to be highly effective for the drainage of mineral soils of low conductivity, has also been demonstrated to be successful on the peats. Because of the soft nature of the material, techniques that create large soil voids, such as tunnel ploughing are also possible. Godfrey & Tucker (1984) investigated the use of the tunnel plough on the Somerset peats of the Altcar series. They found that this technique, particularly where graded, gave significant benefits in terms of both watertable control and ground conditions. The large channels that are left in the peat may well last for substantial periods of time, and were certainly effective after three years. Dennis (1986) also demonstrated the value of such pipe-less drainage in the Fens, and extended the range of options to include graded mole drains, which also provided an effective and long lasting technique. It was noted that such techniques offered suitably cheap methods for the initial drainage of ochreous sites, which avoided the problem of rapid sealing of pipes and be could renewed as neccessary. These systems retained their effectiveness despite periods of outfall submergence.

3.4 Maintenance.

The problem of ochre deposition in pipe drains has already be mentioned. It is recommended that drains be installed in such a way as to permit frequent cleaning out by jetting machines. Methods for the chemical control of ochre deposition have not as a rule been successful or feasible for routine situations. Jetting of sectional drains is also problematic, as in some peat soils the pressure of the water from the jetter will cause the pipe to move within the soil, with subsequent disruption or misalignment of the pipe.

Many problems are associated with the continued wastage of the peat (Richardson & Smith 1977), a process that includes both de-watering and oxidation. After the initial dewatering associated with drainage, continued wastage is mainly through oxidation (Caldwell & Richardson 1975), particularly when the soils are used for arable agriculture. As a result of this wasting, drains in effect become shallower. Continued wastage of the peat also causes problems associated with the subsidence of structures, and the resultant need to continually improve both levels of arterial protection and the capacity of the pumps (as detailed for example by Darby 1940). The problem is not aided by incidents of wind erosion, which are common in the Fens. Consequently it is frequently neccessary to redrain peat fields, at intervals as close as every 25 years, and in addition to lower the ditch levels through the installation of more efficient pump drainage systems. A detailed record of the process of wastage over 130 years at the Holme Fen Post is given by Hutchinson (1980), who indicates the over-riding importance of the control of the pumping regime in determining the rate of wastage.

Maintenance of ditch efficiency is also a problem, particularly due to vegetation growth. However, little "hard" data is available to relate ditch cutting regimes to drainage efficiency.

4 SUB-IRRIGATION

Because they occupy low-lying situations, surrounded by water-carrying ditches, and are themselves moderately conductive, lowland peat soils would seem to provide an ideal situation for sub-irrigation. This process involves holding ditch water at such a level that it flows into drain, and from there into the soil.

The experimental experience at Rickwood EHF (Dennis 1978) demonstrated that this technique was indeed feasible. The major problem, however, arose that in general it was possible only to create a flat watertable within the field. In the presence of a substantial microtopography, of the order of 50 cm height difference, (associated with depositional variation in the soil), it was not possible to successfully implement this technique without the prior grading of the soil surface. Such planing of the soil surface is both expensive, difficult, and has the consequence that quite raw subsoil, which is often unfavourable for crop growth, can be exposed on the higher areas. In the event, the experiment itself was discontinued because of the severe problems associated with the deposition of ochre in and around the drains. Unless the risk of this process can be assessed, and the relevant areas avoided, this makes the investment in sub-irrigation control equipment a high risk decision.

It should be noted that the solution to the re-drainage of the experimental field was to use pipe-less drainage of various types. It is possible that such pipe-less drainage channels could also be used for sub-irrigation. The feasibilty of such a proposal has been demonstrated in one commercially farmed field, but has not yet been widely adopted. In principle, it should provide excellent water table control, and supply, because of the close spacing of the channels.

5 DISCUSSION.

Clearly, then, techniques exist for the maintenance of any desired water regime in peat soils. The situation is probably better than in many mineral soils, because of the relatively uniform high transmission properties. Drainage of the peat is however an expensive operation, both because of the high initial costs, and the continued maintenance neccessary. The investment must therefore be considered carefully. The same methodology that has been used to drain the material can also be used to protect the wetland status of chosen areas should this be thought desirable.

In the Fens, Hutchinson (1980) has observed that during periods when the watertables were not lowered and the peat not subject to arable cultivation, then wastage rates can be reduced to very near zero. This can be illustrated by reference to Wicken Sedge Fen, which has been managed as a wetland nature reserve since the turn of the century (Bingley & Walters 1966, Darby 1983, Rowell 1986). By importing water to maintain the wet status of the fen, peat wastage has been minimised, so that the fen now stands some 2 m higher than the neighbouring land. The experience of managing this fen also indicates the need for positive management of the vegetation in order to achieve desirable conservations aims.

6 CONCLUSIONS

Technically, we have the ability to drain our peat wetlands and to turn them into productive agricultural land. Equally, the same technology can be used to maintain the wetland status of those same areas. The choice between the two rests not with the engineer, but with the public policy makers.

7 REFERENCES

Ap Dewi, I., W.I.Kelso, & D.B.Johnson 1987a A study of field drain ochre deposits. 1. Abiotic chemical precipitation. Journal of Agricultural Science, Cambridge 108: 341-345

Ap Dewi,I., D.B.Johnson, & W.I.Kelso 1987b A study of field drain ochre deposits. 2. The distribution of micro-organisms. Journal of Agricultural Science, Cambridge, 108, 347-351

Armstrong,A.C. 1980 Rickwood EHF, - Little Rough Lots, - Cambridgeshire. In S.Le Grice (ed.) A Progress report on FDEU sites for 1978 FDEU Technical Report 79/3, p.28-44, Trumpington, Cambridge: MAFF

Armstrong,A.C. 1982 The design of in-field drainage systems in the English Fenlands. In J.C.Van Dam (ed.) Polders of the World Vol.1 p. 433-442, Wageningen: ILRI

Avery,B.W. 1955 The soils of the Glastonbury district of Somerset. Memoirs of the Soil Survey of Great Britain, London: HMSO

Bingley,F.J. & S.M.Walters, 1966 Wicken Sedge Fen: a topographical and botanical guide. London: The National Trust.

Burton,R.G.O. & J.M.Hodgson 1987 Lowland peat in England and Wales. Soil Survey Special Survey No. 15. Harpenden: Soil Survey of England & Wales.

Caldwell,T. & S.Richardson 1975 Field behaviour of lowland peats and organic soils. MAFF Technical Bulletin 29, p. 94-111, London: HMSO

Castle, D.A., J.McCunnall & I.M.iring (1984) Field drainage: principles and practices. London: Batsford

Darby,H.C. 1940 The draining of the fens. Cambridge, CUP

Darby,H.C. 1983 The chamging fenland. Canbridge: CUP

Dennis,C.W. 1978 The failure of a pipe drainage system in an organic soil and subsequent remedial measures. FDEU Technical Report 78/3. Trumpington, Cambridge: MAFF

Dennis,C.W. 1986 Re-drainage of the Rickwood EHF Experimental site. FDEU Report RD/FE/35, Trumpington, Cambridge: MAFF

Dennis,C.W., J.Davies, G.L.Godfrey & M.Tucker 1985 Investigations into the effect of pipe slot size on drain flow in peat soils. FDEU LAWS Report RD/FE/24, Trumpington, Cambridge: MAFF

Dent,D. 1986 Acid sulphate soils: a baseline for research and development. Wageningen: ILRI

Eggelsmann,R. 1972 Dranbemussung im Moor nach Tiefe, Abstant und Art. TELMA 2: 91-108

Godfrey,G.L. & M.Tucker 1984 An investigation to determine the effect of tunnel ploughing as a method of dewatering peat soil. FDEU Report RD/FE/22. Trumpington, Cambridge: MAFF

Gosling,L.M. & S.J.Baker 1980 Acidity fluctuations at a broadland site in Norfolk. Journal of Applied Ecology 17: 479 - 490.

Hutchinson,J.N. 1980 The record of peat wastage in the East Anglian Fenlands at Holme Post, 1848-1978 A.D. Journal of Ecology 68: 229-249

Kellet,A.J. 1975 Some values for the hydraulic conductivity of British Soils FDEU Technical Bulletin 75/4, Trumpington, Cambridge: MAFF

Richardson,S.J. & J.Smith 1977 Peat wastage in the East Anglian Fens. Journal of Soil Science 38: 485-489

Rowell,T.A. 1986 The history of drainage at Wicken Fen, Cambridgeshire, England, and its relevance to conservation. Biological Conservation 35: 111-142

Seale,R. 1975a Soils of the Ely district. Memoirs of the Soil Survey of England and Wales. Harpenden: Soil Survey of England and Wales.

Seale,R. 1975b Soils of the Chatteris district of Cambridgeshire. Soil Survey Special Survey No.9. Harpenden: Soil Survey of England and Wales

Trafford,B.D., C.Bloomfield, W.I.Kelso & G.Pruden 1973 Ochre formation in field drains in pyritic soils. Journal of Soil Science 24: 453-460.

Thorburn, A.A. 1976 Iron ochre in drains: a summary of present knowledge. FDEU Technical Bulletin 76/1, Trumpington, Cambridge: MAFF

Table 1 Some recorded hydraulic conductivity values for peat soils.

Site	Soil Series	Depth (m)	K (m/day)	Source
Rickwood EHF	Adventurers	0.6	5.5	1
Rickwood EHF	Adventurers	1.0	8.5	2
Rickwood EHF	Adventurers	1.5	3.0	2
Rickwood EHF	Prickwillow		4.5	3
Methwold	Adventurers	0.8	0.6	5
W.Sedgemoor	Sedgemoor	0.4	0.02	1
W.Sedgemore			0.42	4
Cayton	Altcar	0.8	2.75	5
Silloth	-	0.8	1.6	5

Sources
1 Kellet (1975)
2 Armstrong (1980)
3 Dennis (1978)
4 Dennis et.al. (1985)
5 Unpublished ADAS data

Land and Water Use, Dodd & Grace (eds), © 1989 Balkema, Rotterdam. ISBN 90 6191 980 0

Drainage and subirrigation of peatlands in the Upper Noteć region of Poland

T.Brandyk, E.Kaca & K.Skąpski
Warsaw Agricultural University, Warsaw, Poland

ABSTRACT: The principal aims of the water management realized in the organic soils in the Upper Noteć region are the reduction of the mineralization process and the fulfilment of the plant water requirements. To achieve these two aims the drainage-subirrigation systems consist of the network of the open ditches and checks were constructed. Using existing Stara Noteć I system as an example, the parameters characterizing operation and maintenance of the drainage-subirrigation system are presented. Also the results of the soil moisture content measurements and the measurements of groundwater level are shown and applied for soil water management estimation.

RESUME: Pour un système hydraulique sur les sols organiques, dans la région de Górna Noteć le plus importante pour tenir un régime d'humidité du limite un procès de minéralisation permettant l'utilisation agricole des sols. Pour cela on utilise un système gravitationel composé des fossés parallèles. Sur l' exemple de système Stara Noteć I on a presenté des paramètres d'état technique du fonctionnement du système drainage-irrigation, autant que les résultats des mésures d'humidité de sol et niveau de l'eau souterraine.

ZUSAMMENFASSUNG: Die Hauptziele der Boden- Wasserwirtschaft im Bereich Obere Noteć Tall ist die Feuchtigkeit in Wurzelzone zo bilden, dass die Mineralisierung organischen Böden begrenzt wird und die rationelle landwirtschaftliche Nutzung diesen Boden ermöglichen. Für diesen Zwecken werden die Gravitations - Ent. - u. Bewässerungs Systeme genuzt. Sie bestehen aus einem Netz parallerer Gräben mit Stauwehren. Die Parametern welche den technischen Stand und leistungsfehigkeit der Ent. - u. Bewässerungs Systeme charakterisieren wurden an Beispiel der vorhandenen Systeme auf dem Obiekt Stara Noteć I dargestellt. Auch die Bemässungen den Bodenfeuchte und Grundwasserstände, welche die Wasserwirtschaft des Bodens bewerten, wurden dargestellt.

1 INTRODUCTION

Drainage of peatlands in the Noteć valley began in the 17th century. However the implementation of the planned land reclamation works was possible after the construction of the Bydgoszcz Channel and after sewering the Noteć river. At the begining of the 19th century land reclamation companies were established and then drainage works were done in the whole Noteć river valley and also in the valleys of its affluents. Agricultural utilization of the peat soils in the Noteć river valley has been lasting for 100-150 years. As the result of the drainage works the swamping process was interrupted and then the process of mineralization was initiated. This process caused the changes of the physical and water soil properties and also the lowering of the soil surface position of about 0,5-1,0 m for fen peats and 1,5 - 2,5 m for bog peats as reported by Roguski (1973). An average annual lowering of the soil surface position of 1 cm on reclaimed meadows and 3-4 cm on arable lands were observed in recent years. The lowering soil surface process due to the settlement is continuous and is finally finished when the peat - muck soil is transformed into organic - mineral soil. This causes the appearence of the large differences in the soil surface level. Beside of the local areas left without drainage facilities or swamped again, the dry areas with the deep groundwater table position and poor retention abilities have appeared. The meadows in these dry areas are very often changed for arable fields. The disappearance of the meadows being natural retention reservoirs, which evaporate large amounts of water and reduce deficit of moisture content in the air during summer periods, has the negative influence on the whole natural environment in the Noteć river valley. For the rational water management, i,e, taking into account both the requirements of the natural environment protection and certain agricultural production effects, the pilot Upper Noteć water management system was established in 1976. It occupies the area of about 490 thousend hectares in the Noteć basin from the river sources to the town of Nakło. In this area, in the river valleys about 47 thousand hectares of meadows are located, from which about 7 thousand hectares are equiped with the systematic network of ditches and drains and also with checks. Using these drainage-subirrigation systems it is possible to regulate the groundwater position which has an influence on the environmental conditions as well as on the agricultural productivity of meadows. The most important problems of the operation and maintenance of the drainage-subirrigation systems in the Upper Noteć region are presented in this paper. These problems are illustrated for example in the drainage-subirrigation system Stara Noteć I which is typical for the Upper Noteć region.

2 CHARACTERISTICS OF THE STARA NOTEC I DRAINAGE-SUBIRRIGATION SYSTEM

The Stara Noteć system is located in the Bydgoszcz district in Poland. It occupies the area of about 900 hectares of meadows in the Noteć river valley within the pilot Upper Noteć water management system. The area of the system is characterized by climatic conditions typical for the region of the great plains in Poland. The average annual precipitation rate for this area is about 550 mm. In the distribution of the annual precipitation over the particular year seasons the spring-summer rainfalls prevail, the highest amounts of rainfalls ocurring usually in July. The average annual air temperature is 7. 8°C, the average temperature of July, which is the warmest month, is 18.2°C.

During the period 1972-1978 the drainage-subirrigation system was constructed. This system consists od 7 watercourses, systematic ditch network, checks and communication structures. The spacing between the ditches varies from 60 m to 120 m, the depth from 0.6 m to 1.0 m, bottom width is equal to 0.4 m and side slopes are 1:1.

The scheme presented in Fig.1 shows the fragment of the Stara Noteć I system of the area about 125 ha, which is a typical example of the solutions used in the designing of the drainage-subirrigation systems. The watercourse B plays the principal role in the water distribution over this area.

In the early spring and after the intensive rainfalls the system performs the soil drainage process, whereas during the relatively dry summer seasons the checks are closed and the subirrigation takes place. The Noteć river is used as the main drainage water receiver and also as the main irrigation water source in this area. Irrigation water is supplied through the turnout located near the barrage on the Noteć river (Fig.1) and then distributed over the area by watercourses B, B-1, B-2 and B-3 which supply with water the systematic ditch network. Peat-muck soils overlaid by sand occur in this area. For the performance of the rational water management all the turnout water structures in the Stara Noteć I system were calibrated and the observational wells and also the observational well with the water level gauges were installed in the characteristic points within the system. Since the year 1986 the following problems have been investigated for the description of the system operation efficiency:

1. Watercourse flow capacity.
2. Velocity of the water level changes in the ditches and in the middle of the ditch spacing during the performance of the drainage and subirrigation processes.
3. Allowable groundwater levels in the middle of the ditch spacing.
4. Allowable soil moisture contents in the root zone.

These research problems are investigated using the data observed and recorded by the measuring devices as presented in Fig.1

3 WATERCOURSE FLOW CAPACITY

The flow capacity of the watercourse depends on several parameters. One of them is roughness coefficient. The Manning roughnes coefficient n was developed empirically as a coefficient which remained a constant for a given boundary condition, regardless slope of channel, size of channel, or depth of flow. The value of n can be calculated from the following equation:

$$n = \frac{R^{2/3} . I^{1/2} . A}{Q} \tag{1}$$

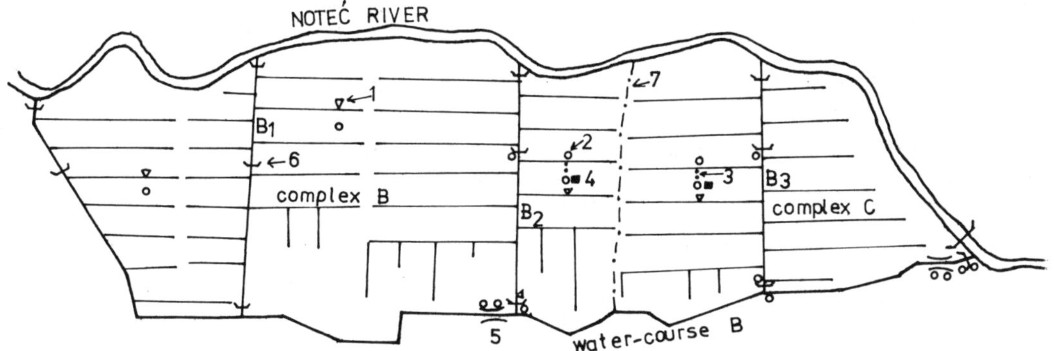

Fig. 1 Typical fragment of the Stara Noteć I drainage-subirrigation system, where: 1 - ditch water - level gauge, 2 - observation well with water-level recorder, 3 - observation wells, 4 - soil tensiometers, 5 - Parshall flumes, 6 - checks, 7 - soil - moisture complex boundary.

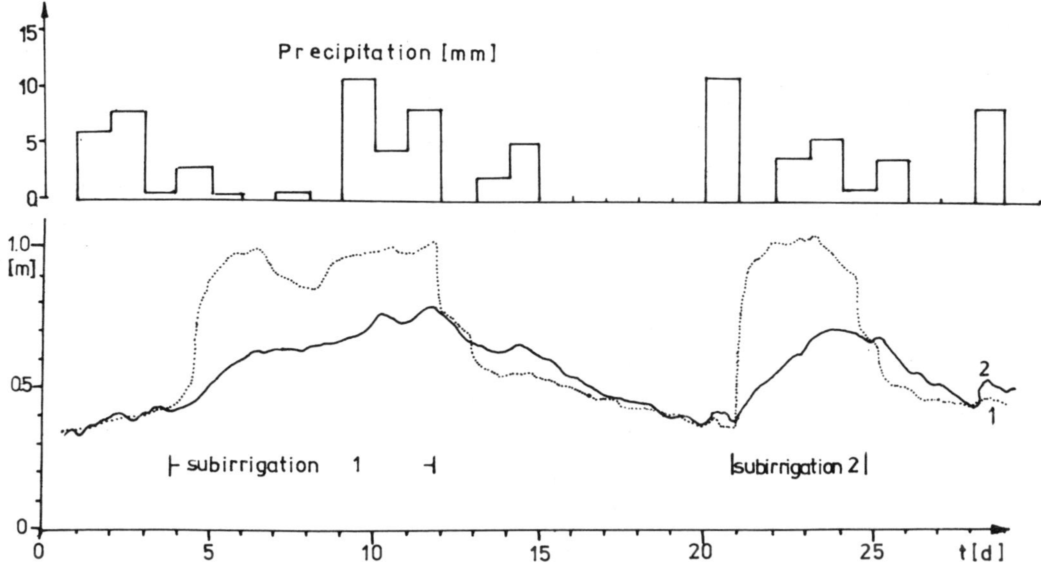

Fig. 2 Changes of the water - levels during the subirrigations in 1988, where: 1 - ditch open water levels and 2 - groundwater levels in the middle of the ditch spacing.

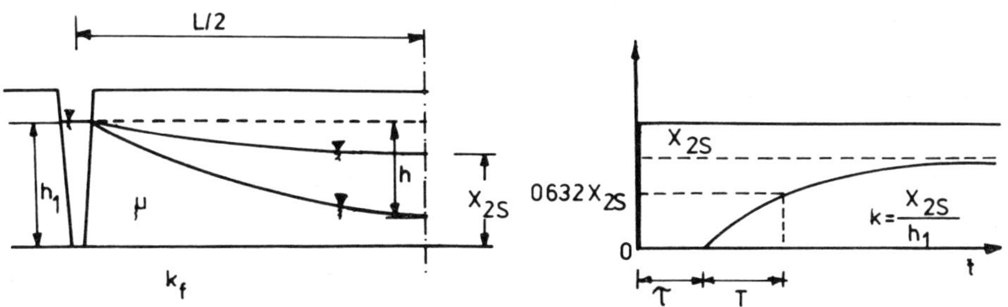

Fig. 3 Graphical interpretation of the parameters k, T, τ .

where:

R - hydraulic radius (m)

I - the slope of channel (-)

A - the area of water cross section (m^2)

Q - discharge ($m^3 s^{-1}$)

The measurements of n were performed for the reprezentative section of the watercourse B. Surveying measurements were used for the determination of the area of water cross-section and the slope of the watercourse. The measurements of the plant height and plant cover on the side slopes were also made. The Parshall flume was used for discharge measurements. The area of the water cross-section and wetted perimeter, required for the determination of the hydraulic radius were calculated as the weighted mean of the measurements at the begining, at the end and in the middle of the investigated section of the watercourse B. The weight value of 0,5 was used for the measuerements in the middle and the weight value of 0,25 was used for the measurements at the both ends of the investigated section. During the time of the measurements the cross section of the watercourse B was occupied in about 30% by plants, what can be accepted as its typical state during the vegetation period. The value of the Manning roughnes coefficient n=0,05 was determined for these conditions.

4 THE EXPERIMENTAL SUBIRRIGATION RESULTS

Two experimental subirrigations were performed in the Stara Noteć I open ditch system during the vegetation period in 1988 year. The open water levels in the ditches and the groundwater levels in the middle of the ditch spacing were measured during these subittigations. The changes of these water levels during the subirrigation and the drainage after the subirrigation are presented in Fig.2. The twenty-four hours sums of the precipitation are also shown in this figure.

The changes of the groundwater levels during the subirrigation can be described using the formula of Ostromęcki (1969), which was modyfied by Kaca (1987) to the following form:

$$t = \tau + \frac{l^2}{a} \ln \left[\frac{ah_1 + b_k x_{2s} l^2}{ah + b_k x_{2s} l^2} \right] \quad \text{for } t > \tau \qquad (2)$$

where:

h - groundwater level in the middle of the ditch spacing (Fig.3)

h_1 - open water level in the ditch (Fig.3)

x_{2s} - stabilized groundwater level in the middle of the ditch spacing (Fig.3)

l - half of the ditch (or drain) spacing

τ - subirrigation time delay

t - time required for the rise of the groundwater level from h_1 to h (Fig.3)

a, b_k - environmental parameters

The environmental parameters a, b_k can be determined from the following relationships:

$$a = \frac{l^2}{T} \qquad (3a)$$

$$b_k = \frac{k - 1}{k \cdot T} \qquad (3b)$$

where k and T are the parameters characterizing the subirrigation process.

The values of these parameters can be determined graphically or analytically from the measurements of the changes of the open ditch water levels and the changes of the groundwater levels in the middle of the ditch spacing. An example of the determination of k, T, τ , parameters is presented Fig.3.

Stabilized groundwater level in the middle of ditch spacing x_{2s} can be determined from the following formula:

$$x_{2s} = k \cdot h_1 \cdot \frac{a}{a - b_k l^2} \qquad (4)$$

The values of the parameters k, T, τ obtained for the Stara Noteć I system are presented in Tab.1.

Table 1. The values of the parameters k, T, τ obtained for the Stara Noteć I system

Experimental subirrigation No	Duration time	k [-]	T [h]	τ [h]
1	8-16 of July 1988	0.74	48	0
2	25-29 of July 1988	0.70	35	0

Calculated values of the parameters k, T, τ, show that the Stara Noteć I drainage - subirrigation ditch system assures the efficient regulation of the groundwater level. Such efficient operation of the ditch network fulfils the requirements of the farms in this area.

5 GROUNDWATER LEVEL MEASUREMENTS RESULTS

Hydrogenic soils in Poland are joined in groups, the so-called soil-moisture complexes according to the similarity of the soil physical properties.

The following soil-moisture complexes

have been specified (Okruszko 1986): A - wet, AB - periodically wet, B - moist, BC - periodically dry, C - dry, CD - periodically semi-arid, D-semi-arid. Recomended in Poland for subirrigation are complexes B, BC, C and CD. The required groundwater levels for these complexes are in Table 2.

Table 2. Required groundwater levels below the soil surface for the soil-moisture complexes according to Szuniewicz (1979)

Groundwater levels corresponding to the following air contents in the root zone in (cm)	Soil-moisture complexes			
	B	BC	C	CD
6% - z_1	35	30	25	20
8% - z_2	45	35	30	30
10% - z_{opt}	70	50	35	35
Maximum admissible groundwater level - z_3	95	85	60	50

As the example of the groundwater level changes the data recorded on the water-level gauge (Fig.1) during the vegetation period in 1988 are presented in Fig.4.

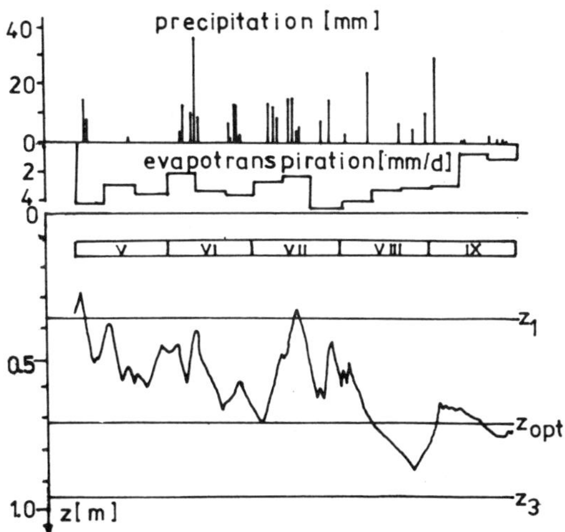

Fig. 4 Groundwater levels recorded in the soil-moisture complex B during the vegetation period in 1988.

The values of the precipitation and the evapotranspiration for the vegetation period in 1988 as well as the required groundwater levels for the soil-moisture complex B are also shown in this figure. From the groundwater changes shown in Fig.4 it follows that during the nearly all considered period of time the groundwater level was the allowable range, i.e. between minimum (z_1) and maximum (z_3) admissible levels. Only during the short time periods in May and July the groundwater level exceeded lower allowable limit (z_1).

6 SOIL MOISTURE CONTENT MEASUREMENTS RESULTS

The efficient operation of the drainage - subirrigation system should keep the soil moisture content in the root zone within the allowable range. The upper limit of this range is the soil moisture content corresponding to the minimum air content. For the grass root zone the minimum value of the air content 6% to 10% is used, what is in agreement with the data presented by Wesseling (1974). The minimum value of soil moisture content for the soil-water management purposes should exceed that of the permanent wilting point by safety margin. For grasslands on the organic soils, a pF value of 2,7 was proposed by: Szuniewicz (1979), Brandyk and Wesseling (1983), Okruszko (1986) as the minimum allowable value of the soil moisture content for the soil-water management purposes. The assumption of such minimum allowable moisture content is also required for the protection of the peat-muck soils against the mineralization process. The results of the soil moisture pressure head measurements in the upper 30 cm soil layer are shown in Fig.5. These measurements were performed in the soil-moisture complex B, using the soil tensiometers (Fig.1). The straight lines representing the soil moisture pressure heads corresponding to the air contents of 6% and 10% are also shown in Fig.5. It can be followed from the data presented in Fig.5 that the soil moisture pressure heads in the root zone were within the allowable range during almost all considered period of time. The maximum allowed moisture content was exceeded only for the short periods of time during the intensive rainfalls.

7 CONCLUSIONS

Drainage of the peatlands in the Noteć river valley caused serious changes in the natural environment. Presently the rational water management of the organic soils should take into account not only the fulfilment of the plant water requirements but also the protection of the peat-muck soils against the mineralization process.

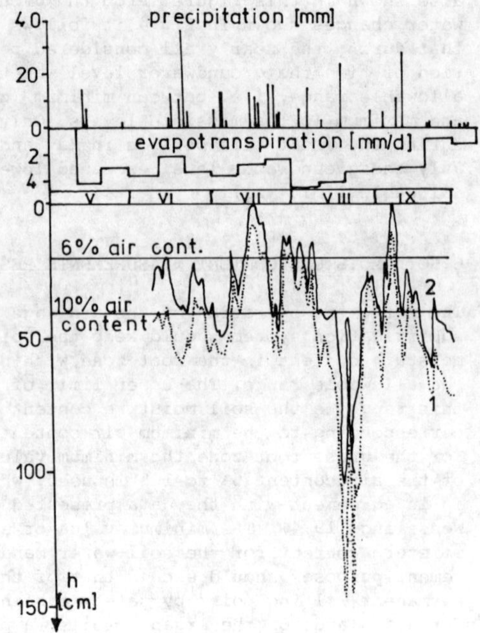

Fig. 5 Changes of the soil moisture pressure heads in the root zone during the vegetation period in 1988, where: 1 - depth 10 cm and 2 - depth 20 cm.

These requirements can be fulfiled by the efficient drainage - subirrigation system. For the proper operation and maintenance of the such systems the basic parameters characterizing water flow should be estimated. Also the systematic measurements of the groundwater levels and the soil moisture pressure heads are necessary for the proper scheduling of the subirrigation and drainage processes under the changing weather conditions.

REFERENCES

Brandyk, T. and Wesseling, J.G. 1983. Estimation of the optimal depth of groundwater for a layered peat-muck profile using a steady state moisture flow solution. Nota 1517, ICW Wageningen.

Kaca, E. 1987. Modyfiakcja metody J.Ostromęckiego eksperymentalnego opisu dynamiki zwierciadła wody gruntowej w nawodnieniu podsiąkowym (in Polish). Roczniki Nauk Rolniczych seria F (in press).

Okruszko, H. 1986. Przyrodniczo-techniczne podstawy melioracji terenów dolinowych. W Podstawy melioracji rolnych pod redakcją P.Prochala (in Polish).

Ostromęcki, J. 1969. Obliczanie nawodnień podsiąkowych. Biblioteczka Wiadomości IMUZ Nr 30 (in Polish).

Roguski, W. 1973. Proces grądowienia i różnicowanie się łąk w dorzeczu Noteci. IMUZ Falenty (in Polish).

Szuniewicz, J. 1979. Charakterystyka kompleksów wilgotnościowo-glebowych pod kątem parametrów systemu melioracyjnego. Biblioteczka Wiadomości IMUZ Nr 58: 29-50 (in Polish).

Wesseling, J. 1974. Crop growth and wet soils. Chapter 2 in Drainage for agriculture. American Society of Agronomy, Madison, Wisconsin.

Land and Water Use, Dodd & Grace (eds), © 1989 Balkema, Rotterdam. ISBN 90 6191 980 0

Methods to improve efficiency of drainage on clay and peat soils in Finland

J.Saavalainen & S.Virtanen
Finnish Field Drainage Centre, Finland

ABSTRACT: Soil compaction and its consequence the deterioration of drainage causes problems for soil cultivation in Finland. Most of the fields in southern and south-western Finland are composed of clay soils while peat fields are more predominant in the north. Clay fields are primarily used for the cultivation of cereals and sugarbeats while peat fields are primarily for producing silage. The mechanisation of agriculture has occured very rapidly over the past few decades. Today heavy machinery is commonly used in farming, even in very wet conditions. Under such circumstances, clay and peat soils are more sensitive to compaction than other field types with the result that the drainage system often fails. One solution to the problem is to improve the efficiency of the drainage system.

Methods to improve the efficiency of drainage are determined by soil type although some of the methods are applicable to both clay and peat fields. On clay soils, experiments have been made with slit trench drainage, blind inlets, insertion of new drains and with mole drainage. At the same time, a bedding system with a bed width of 40 to 50 metres is being tested on peat soils. New drains and blind inlets are also filled with other materials in addition to gravel, for example with wood chips. The techniques, cost and effects on overall drainage performance will be discussed at the conference.

1 THE NEED FOR IMPROVED EFFICIENCY OF DRAINAGE IN FINLAND

1.1 Demands for more efficient drainage

Modern mechanized agriculture is placing ever greater demands on subsurface drainage in regard to the drying of fields. Big and heavy machines are now used in agriculture and drainage must be efficient so that the trafficability of fields is sufficient to allow the machines to work. Finland's short summer sets demands of its own. It is important to get into fields as early as possible in spring so that the best possible usage can be made of the short growing periods of plants. The hydraulic conductivity of the field is the most important factor in drainage, especially in subsurface drainage. The heavy machinery and the one-sided choice of plants in modern farming causes deterioration in the drainage of fields by reducing the amount of macro pores in the soil and further lowering the hydraulic conductivity of the soil. The right way to improve drainage would be use lighter machinery and to cultivate, for example hay. If that is not possible, the efficiency of drainage can be improved so cultivation can be made in dry conditions and no more compaction would occur.

1.2 Soil types

Naturally compact clay and decomposed peat soils have low hydraulic conductivity. In clay soils the hydraulic conductivity is higher in the plow layer than in the underlying layers. Because of the low hydraulic conductivity, water flows toward pipes very slowly in the deep layers. In practice the only way that the water flows into the pipe is along the plow layer to the trench and from there into the pipe. In clay soil it has been found that after the drain-age construction, a subsurface drainage trench which is structured of loose and well permeable soil, becomes compacted with time and resembles the surrounding soil. In that case it is obvious that drainage fails although the system has been correctly carried out. It has been verified that the compaction of the drain trench interferes with drainage already 4-5 years after construction (Vakkilainen 1980). In peat soils the deterioration of the structure and the compaction of the soil takes place primarily in the topmost layer (Fig. 1.)

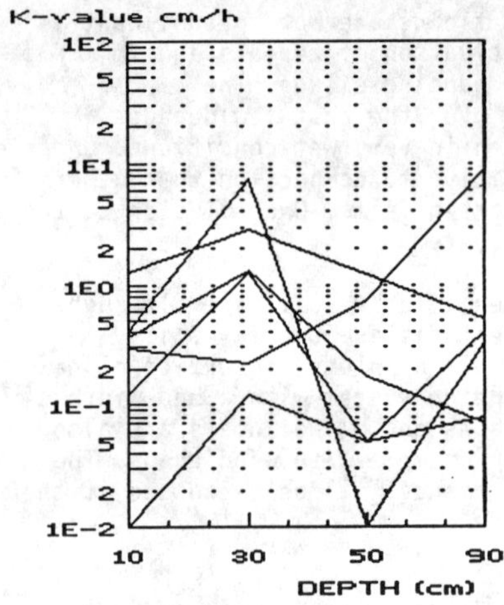

Figure 1. Variation of hydraulic conductivity with depth in some peat field in Finland (Kasurinen 1987).

In Finland, clay soils sensitive to compaction are found particularly in the western and southern areas. The proportion of peat soil to arable land on the other hand increases rapidly when moving from south to north. The maps in Fig. 2 depict the proportional shares of clay and peat soils to arable land in Finland.

2 METHODS TO IMPROVE DRAINAGE

Methods to improve drainage have developed from experimental work in Finland. A part of experimental fields have been established in the fields owned by research stations but the majority have been established in the fields of

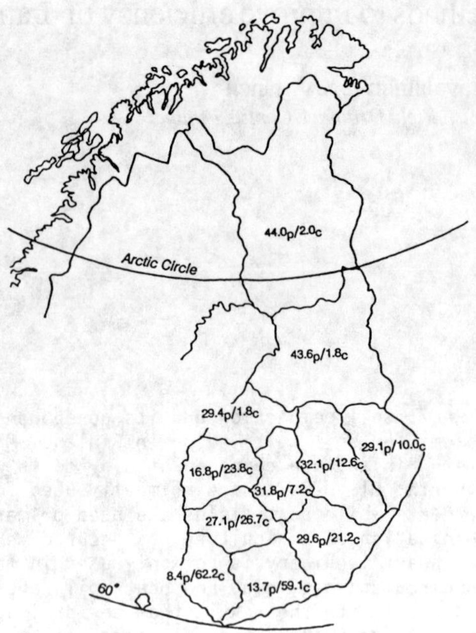

Figure 2. The proportion of soil types in the arable land by regions of Finland (p = peat soil, c =clay soil).

farmers. In the fields owned by research stations the effects of improvement methods are followed up by measuring, among other things the moisture content of the soil, the compaction of the soil and the yield. On the farmers fields the efficiency of drainage methods have been followed by observing the differences in the trafficability of the fields from the wheel tracks of tractors and by interviewing the farmers.

2.1 Insertion of new drains

One method to improve the subsurface drainage is insert to new drains. In the experimental fields new subdrains have been made between every lateral drain. By close drain spacing, and by gravelling ditches right to the surface, drainage has been improved. But due to the great consumption of gravel and the number of ditches, expenditures have become great. In the experimental areas the placing of supplementary drains into every second absorbing drain has been investigated but the results have been that the drainage of the field has been uneven (Puustinen et al. 1987).

Research has also been directed at the ways of filling subsurface trenches. In clay soils research is directed at among other things how well subsurface drainage trenches remain high hydraulic conductivity when the dug clay soil is left to dry before being put back into the trench. Results have been promising (Aura 1988). There have also been experiments in both peat and clay soils to fill the subsurface drainage trenches entirely with gravel or chips of wood which allow good permeability.

2.2 Slit trench drainage

Slit trench drainage experiments have also been carried out to improve drainage in Finland. In slit trench drainage narrow trenches are dug into the field and then filled with a material of high hydraulic conductivity. Slit trench drains are made in the field so that they lead the surface water accumulated on the subsoil either into the field boundary ditches or into subsurface drains or into both (Fig. 3). The slit trench drains are made in a way that they are perpendicular to the subsurface drainage thus crossing the subsurface drains as often as possible. In the field the slit trench drains and the subsurface drains form a net of squares and at the points where the trenches meet the subsurface drains the water flows into the subsurface drains. A prerequisite to this however, is that maps of the old subsurface drain system exist, as is most often the case in Finland.

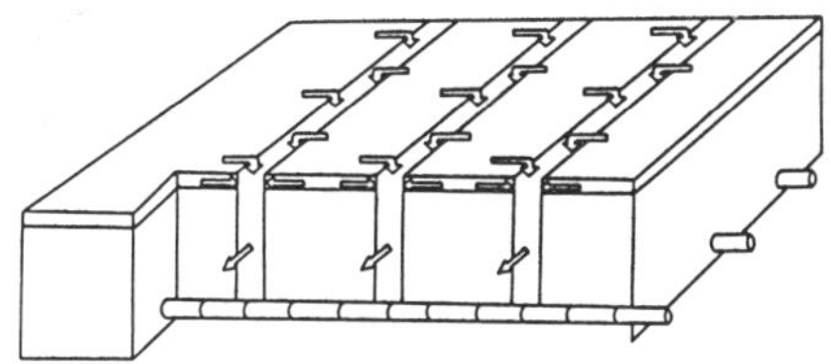

Figure 3. The principle of slit trench drainage.

In Finland slit trench drains have been made by a chain digger or a digger which can be pulled by the farm's own tractor, for example a small chain digger or trencher. Depending on the digger, the width of trench varies from 9 to 20 cm. Slit trenches are dug to a depth of 60-70 cm in such a way that the trenches reach down almost to the gravel layer on the subsurface drains.

The filling of slit trench drains in experimental fields has been carried out with chips of wood, gravel or bark, and in addition, in some trenches there has been installed a tiny pipe with a diameter of 25 mm. The impact of various material on the improvement of drainage is being compared because organic material is usually lighter than gravel and thus drainage construction would cause no more compaction of the soil. An advantage of woodchips is their price if they are made of a farm's own raw materials. Bark also has a reasonable price in the neighbourhood of paper mills. An advantage of gravel is its good hydraulic conductivity and its preservation in the soil. In the experimental fields, one of the factors to be followed up is the decomposition of organic material and the life span of slit trench drains.

The Finnish Field Drainage Centre established the first slit trench drainage experiments in clay fields in southern Finland. During the summer of 1989 slit trench drainage experiments will also be established on peat fields. A double drainage system will also be tried on the peat fields. In Finland mole draining was experimented with in clay soils in the 1950's. Mole draining was not successful because the length of its lifetime, 3-5 years, was considered too short. Now mole draining is going to be tested in peat soils.

The results from the slit trench drainage experiments have been promising. Cultivation of the field has begun at the normal time and the surface water has not accumulated as in previous years. Also, in Estonia the method has proven to be effective (Juske and Tomberg 1986).

2.3 Blind inlets

In case of inadequate drainage, blind inlets can be made in the wettest places of the field. Blind inlets made by big excavators improve drainage more than small, drilled blind inlets. However, because the consumption of the filling material is lower with drilled blind inlets, it is possible to do more of them and achieve more even drainage.

2.4 Bedding

In peat soil where the surface water causes problems there have been attempts to make drainage more effective by bedding. The Norwegians have developed the traditional open ditch system by increasing the breadth of bed to 40-50 m (Lindberg 1987). In Finland, the bedding system has been coupled to the subsurface drainage. In the experimental fields, the combined subdrainage-bedding is compared with a simple bedding system and with a simple subdrainage. In the combined subdrainage-bedding system the drain trenches are filled with highly permeable materials and the bedding is carried out so that furrows are on drain trenches. The slope of the bed is 3 % and breadth 30 m (fig 4). There are also experiments with the mixing of a thin peat layer on the mineral soil below it. This experimental work is aimed at increasing the hydraulic conductivity of the top most layer.

Figure 4. Combined bedding-subdrainage system.

CONCLUSIONS

The need of the improved efficiency of drainage is increasing in Finland. Promising experiences have been occured with many methods. Methods will be further developed and new experimental fields will be established.

REFERENCES

Aura, E. 1988. Salaojakaivannon peittäminen.Käytännön Maamies 1/88.

Juske, A., Tomberg, H. 1986. Intensiivinen teknologia maanparannuksessa. Sotsialistlik Pollumajandus n:o 2 1986.

Juusela, T. 1954. Myyräsalaojituksesta erinomaisia tuloksia sateisenakin kesänä.Maaseudun Tulevaisuus 21.8.1954.

Kasurinen, O. 1987. Kalottialueen salaojitusongelmat. Loppuraportti vuodelta 1987. Salaojakeskuksen moniste.

Lindberg, K. 1987. Profilering av myr for bedre ovarflareavrenning. Aktuellt fra SFFL nr. 1:121-128.

Peltomaa, R. and Kasurinen, O. 1987. Practical experiences of peat land drainage in North-Calotte. Proceedings Third International Workshop on Land Drainage. Columbus.

Puustinen, M. and Pehkonen, A. 1986. Salaojien toimintahäiriöt. Helsingin yliopiston maatalousteknologian laitos. Tutkimustiedote 48.

Puustinen, M., Varis, R. and Luoma, T. 1987. Salaojien toimintahäiriöiden korjaus savi- ja turvemailla. Helsingin yliopiston maatalousteknologian laitos. Tutkimustiedote 51.

Vakkilainen, P. 1980. Tutkimus pellon tiivistymisen vaikutuksesta salaojien toimintaan. Vesitalous 2:23-28.

RESUME

Par conséquence de l'épaissement du sol l'assèchement des champs en même temps que la possibilité de les cultiver sont devenus plus difficiles dans plusieurs endroits en Finlande. Les terrains argileux sont dominants dans le Sud et dans l'Ouest de la Finlande tandis que le nombre des terrains tourbeux augmente remarquablement vers les parties nord du pays. La culture des céréales et des betteraves est étendue dans les terrains argileux. Dans les terrains tourbeux la plante dominante est le gazon. Dans les dernières décennies la mécanisation de l'agriculture a été très forte et actuellement les grandes machines sont utilisées dans l'agriculture. L'usage des machines lourdes dans les conditions mouillées cause l'épaissement nuisible du sol surtout dans les terrains argileux et tourbeux. Par conséquence de cet épaissement l'assèchement du sol devient souvent moins efficace. L'intensification de l'assèchement peut alors donner une solution.

Les méthodes utilisées pour l'assèchement varient selon le sol mais quelques-unes sont communes aussi bien pour les terrains argileux que pour les terrains tourbeux. Dans les terrains argileux nous avons des expériences sur les fossés infiltrants, sur les yeux de gravier, sur les fossés supplémentaires et sur les fossés souterrains. Dans les terrains tourbeux nous avons essayé des fossés à grands intervalles, l'intervalle étant de 40 à 50 mètres et en plus des fossés supplémentaires et des yeux de gravier. Pour remplir les fossés supplémentaires et les yeux de gravier nous avons utilisé aussi autres matériaux en plus de gravier, par exemple du bois coupé en petits morceaux. Dans la conférence seront traités les différentes méthodes tout en prenant en considération la réalisation technique, les frais et leurs effets pour l'assèchement.

RESUMEE

Durch die Verdichtung des Bodens ist die Entwässerung schwächer geworden, und das hat Schwierigkeiten bei der Bebauung der Äcker an vielen Stellen in Finnland verursacht. Ton ist die vorherrschende Bodenart in Süd- und Westfinnland, Torf dagegen zieht sich immer mehr von Süden nach Norden. Auf Tonboden wird im allgemeinen Getreide und Zuckerrübe angebaut, während auf Torfboden Rasen am gewöhnlichsten ist. Die Landwirtschaft ist während der letzten Jahrzehnten sehr schnell mechanisiert worden, und heutzutage werden in der Landwirtschaft schwere Maschinen verwendet. Der Gebrauch schwerer Maschinen auf nassem Boden verursacht schädliche Verdichtung des Bodens, besonders auf Ton- bzw. Torfböden. Durch diese Verdichtung wird die Entwässerung oft schwächer. Dieses Problem kann zum Beispiel durch die Intensivierung der Entwässerung gelöst werden.

Die zur Intensivierung der Entwässerung verwendeten Verfahrensweisen unterscheiden sich nach Bodenart, aber einige Entwässerungsmethoden sind sowohl für Ton-, als auch für Torfböden geeignet. Auf Tonböden wurden Experimente mit Stichgräben, Schluckern, Zusatz- und Maulwurfdränung durchgeführt. Auf Torfböden wurden die Möglichkeiten der Ackerbeetenentwässerung bei 40-50 Meter breiten Beeten untersucht, zur gleichen Zeit wurde aber auch mit Zusatzdränung und Schluckern experimentiert. Die Zusatzdräne und die Schlucker wurden mit Kies und anderen Stoffen, zB. Hackspäne gefüllt. Die technische Verwirklichung, die Kosten und die Entwässerungseffekte der verschiedenen Verfahrensweisen werden an der Konferenz behandelt.

Land and Water Use, Dodd & Grace (eds), © 1989 Balkema, Rotterdam. ISBN 90 6191 980 0

Development of methods for purifying runoff water from peat production areas

R.Ihme
Technical Research Centre of Finland, Finland

E.Lakso
University of Oulu, Finland

A.Lehtinen
Finnish National Board of Waters and the Environment, Finland

ABSTRACT: A description is given of a project on "The Development of Water Purification Technologies for use in Peat Production" begun in 1987, and some preliminary results are presented. The aim was to evolve new methods for reducing the loading on rivers and lakes caused by peat production and to improve existing purification methods. The methods tested to date are surface drainage or filtering as means of purifying runoff water, retention of the load in open ditches, and improvement of the efficiency and purification capacity of sedimentation basins. Preliminary results suggest that surface drainage succeeds in removing both suspended solids and nutrients from the runoff water.

Entwicklung der Reinigungsverfahren der von Torfgewinnungsgebieten abfließenden Sickerwässer

RESÜMEE: In diesem Bericht werden Inhalt und vorläufige Forschungsergebnisse des 1987 angelaufenen Projekts "Entwicklung der Wasserschutztechnologie der Torfgewinnung" dargestellt. Zweck der Untersuchungen ist es, neue Verfahren zur Verringerung der Gewässerbelastung bei der Torfgewinnung zu entwickeln und bestehende Verfahren zu verbessern. Begonnene Teiluntersuchungen: Verwendung der Oberflächenrieselung zur Reinigung der Sickerwässer, Verwendung der Filterung zur Reinigung der Sickerwässer, Zurückhaltung der Belastungen in den Entwässerungsgräben sowie Verbesserung der Funktionsfähigkeit und des Reinigungsvermögens der Absetzbecken. Aufgrund der vorläufigen Forschungsergebnisse können u.a. mit Hilfe der Oberflächenrieselung aus dem Ablaufwasser neben Schwebestoffen auch Nährstoffe entfernt werden.

Développement des méthodes d'épuration des eaux de ruissellement provenant des zones d'extraction de la tourbe.

RÉSUMÉ: Nous présentons ci-après un projet lancé en 1987, "Développement des technologies de protection des cours d'eau dans le cadre de la production de la tourbe" ainsi que les premiers résultats de nos recherches. L'objectif de la recherche est de développer de nouvelles méthodes et d'améliorer les procédés existants afin de réduire la charge des cours d'eau résultant de la production de la tourbe. Voici les recherches partielles commencées : Utilisation de l'écoulement de surface dans l'épuration des eaux de ruissellement, utilisation d'un filtre de tourbe dans l'épuration des eaux de ruissellement, rétention de la charge dans des fossés intérieurs de drainage ainsi que l'amélioration de l'efficacité des bassins de sédimentation et de leur aptitude à se nettoyer. D'après les premiers résultats de nos recherches, l'écoulement de surface permet d'éliminer des eaux de ruissellement outre les matériaux solides, les nutriments.

1 INTRODUCTION

The use of peat in the heat and electricity generating industry in Finland has increased greatly in recent years, and can be expected to increase still further in the future with the completion of a fairly large peat-fired power station in Central Finland. In addition, Kemira Ltd. in Oulu is planning to use peat as a source of energy in its ammonia plant.

Approx. 43 000 ha of mires in Finland were being used for peat production in 1986, with a total production of over 20 mill. m^3. A considerable proportion of the peat reserves are located in the province of Oulu and the southern part of Lapland, the area devoted to peat mining in the former province alone being approx. 10 000 ha (Fig. 1). The mire area used for peat production can be expected in the light of existing plans for expanding the consumption of peat to double in the next few years.

The purpose of the water purification arrangements made in peat production areas is to reduce loading of the river systems during the various stages of mire preparation and peat production as effectively as possible. As peat production becomes more widespread, more attention than ever will have to be paid to this loading factor, which comprises mainly suspended solids and nutrients. The present method employed for doing this is by means of sedimentation basins.

Hydrological research has been used to monitor the amounts of runoff water originating from peat production areas, the quality of this water and areal fluctuations in these measures (Sallantaus 1983), and the producers have cooperated with the water authorities to investigate the possibilities of using sedimentation basins to reduce loading (Selin & Koskinen 1985). Some research has also been carried out into the effects of runoff water from peat production areas on water quality in rivers below its discharge point (Marja-aho & Koskinen 1988).

The increased industrial use of peat led to the initiation in spring 1987 of a research project entitled "The Development of Water Purification Technologies for Peat Production", the general design of which is presented here, together with some preliminary results.

2 PURPOSE AND ORGANIZATION OF THE PROJECT

The purpose of the project, which commenced in March 1987, is to develop new methods for reducing the loading created during the various stages of mire preparation and peat production as effectively as possible and to improve existing methods. The intention is to carry out fieldwork from the beginning of the spring flooding season until early November in three consecutive years, and to make surveys during the winter time as well at some sites. The final report will be produced in 1990. The project, costing about 2 mill. Fmk, is carried out by the Ministry of Trade and Industry, the peat producers, the water conservation authorities and the Technical Research Centre of Finland. In addition, the actual experimental construction work, costing a further 1.5 mill. Fmk is being financed chiefly by the Ministry of Labour.

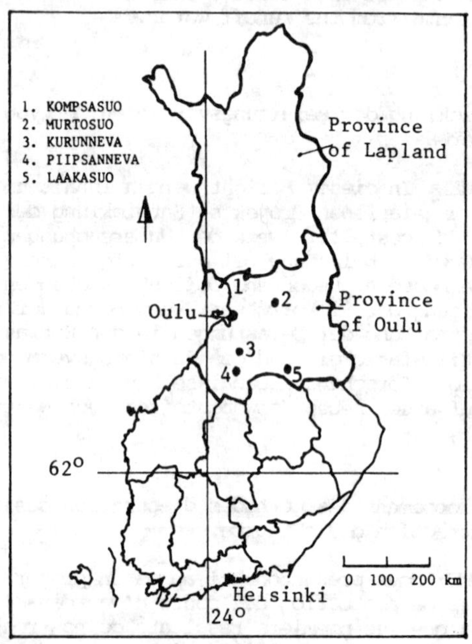

Fig. 1 Location of the sites studied.

The following aspects of the work were commenced in summer 1987:

- use of surface drainage for the purification of runoff water,
- use of filtering for purification of runoff water
- retention of the runoff loading in open ditches

- improvement of the efficiency and purification capacity of existing sedimentation basins.

The sites concerned are all situated in the province of Oulu (Fig. 1). Some mires are at the draining stage and some are already being used to produce milled peat.

3 THE SURFACE DRAINAGE METHOD

3.1 Background

"Surface drainage" refers here to a procedure by which the water from a peat production area is distributed over a certain tract of mire or forest and allow to run off from it (Fig. 2). The method has been used previously mainly for the treatment of waste water. The purification process that takes place has been shown to be largely biological in nature. The water is aerated as it flows, favouring the development of a vigorous fauna of decay micro-organisms which adhere to the plants and soil particles, extract nutrients from the water as it flows past and decompose the organic matter contained in it. In this way the nutrients are made available for the use of the plants.

The purpose of this aspect of the research is to determine the usefulness of the surface drainage method for the processing of runoff water from peat production areas. A study is to be made of details connected with the construction and management of surface drainage areas, the purification capacity achievable with this procedure, figures for use in the planning and dimensioning of such areas and their establishment costs.

The research was carried out in 1987 and 1988 on two mires in the course of being drained for peat production purposes, Kompsasuo and Murtosuo, and one already being used for production, Laakasuo (Fig. 1).

3.2 Research at Kompsasuo

Kompsasuo is a mire area that had remained in a natural state up to 1986, when its preparation for peat production commenced. A small amount of sod peat was produced in June 1988, and it is intended in future to use the area mainly for milled peat, although with some sod peat continuing to be produced. The mire contains predominantly various types of Sphagnum peat with a degree of humification of 4.0. Peat depths in the production area are in the range 1.5 - 5.0 m.

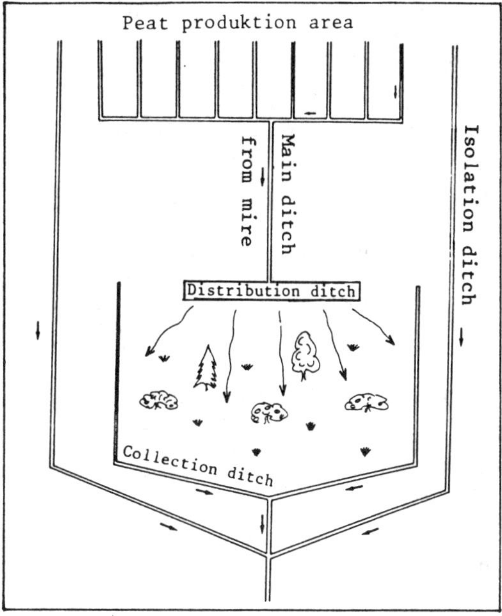

Fig. 2 Principle of the surface drainage method.

The total extent of the production area is 180 ha, the water from which is conducted into the river system below it via a surface drainage area constructed to the south of the production area and a combination of a sedimentation basin and surface drainage area to the north (Fig. 3). The characteristics of the surface drainage areas are listed in Table 1.

Table 1. Properties of the surface drainage area.

Property	Area	
	Northern	Southern
Size (ha)	2.2	1.0
% of catchment area	4.5	0.8*
Gradient %	10.2	4.5

* Note that the surface drainage area also receives a certain amount of runoff water from outside the peat production site.

The northern surface drainage area serves a catchment area of 50 ha and the southern drainage area a catchment area of 130 ha, in addition to which the latter receives a certain amount of water from outside the peat production area.

The efficiency of the surface drainage areas was monitored by means of water samples taken from above, within and below them. In the case of the northern area, samples were also taken from the inflow channel to the sedimentation basin and from below the sedimentation basin (Fig. 3). The samples, taken once a week from April to November, were used from the determination of suspended solids, electrical conductivity, pH, colour, COD, total nitrogen, nitrate nitrogen, ammonium nitrogen, total phosphorus, phosphate phosphorus and iron content according to accepted standards. The quantities of water entering the surface drainage areas were measured by means of triangular measuring weirs at the same time as the samples were taken.

The vegetation of the surface drainage areas was monitored each year from quadrats of 2 x 2 m, which were photographed and the standing water surface and proportions of the dominant plant species estimated.

The amount of sludge remaining on the surface was measured by installing sludge collector plates of size 0.5 x 0.5 m. When the plates became full, they were emptied and the amount of sludge contained on them determined.

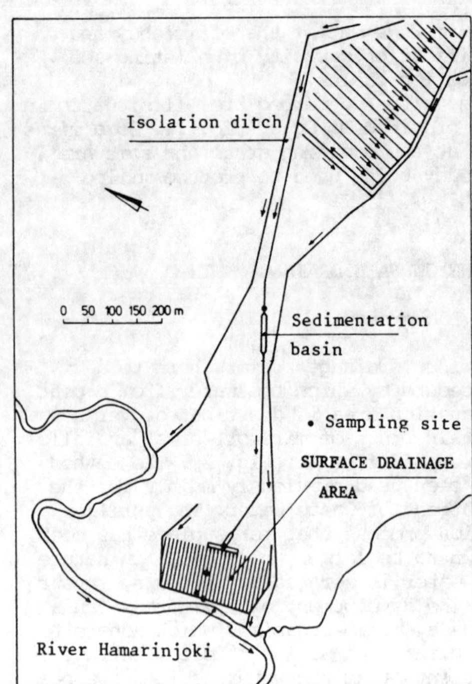

Fig. 3 Northern surface drainage area at Kompsasuo.

3.3 Results

The water flowing from the peat production area was fairly clean, having a suspended solids content of 4 - 8 mg l^{-1}, a mean COD of 40 mg l^{-1}, a total nitrogen content of 0.5 - 2.9 mg l^{-1} and a total phosphorus content of 0.01 - 0.09 mg l^{-1} (Figs. 4 and 5).

The best purification results at Kompsasuo were obtained in the northern surface drainage area, where the mean reductions in concentrations were as follows: suspended solids 70 - 80 %, COD 20%, total nitrogen 40 - 60 % and total phosphorus 60 - 70 %. The suspended solids and total phosphorus results for the northern area at Kompsasuo in 1988 are shown in Figures 4 and 5. The sedimentation basin located above the surface drainage area had relatively little effect.

The purification results for the southern drainage area were considerably poorer, and the concentrations even increased in many cases as the water flowed across the area.

Experiences acquired so far suggest that a surface drainage area should be 2 - 4 % of the size of its catchment area, and that the distribution of the water over the area (lengthwise and breadthwise) should be planned carefully in order to avoid short cuts. Old ditches remaining in the area, even when blocked up, can cause short cuts of this kind. It would also seem advantageous for the ground surface to consist of a peat horizon, through which at least some of the water will be filtered. Research will be needed to determine how long the same area can be used for surface drainage purposes, i.e. whether the soil is capable of retaining only a certain quantity of nutrients, and what effects differences in soil and vegetation may have on the purification result.

4 FILTRATION

4.1 Background

Peat-based slurry filters have be tried out earlier for the drying of excess sludge from waste water purification plants, septic tanks and fish breeding

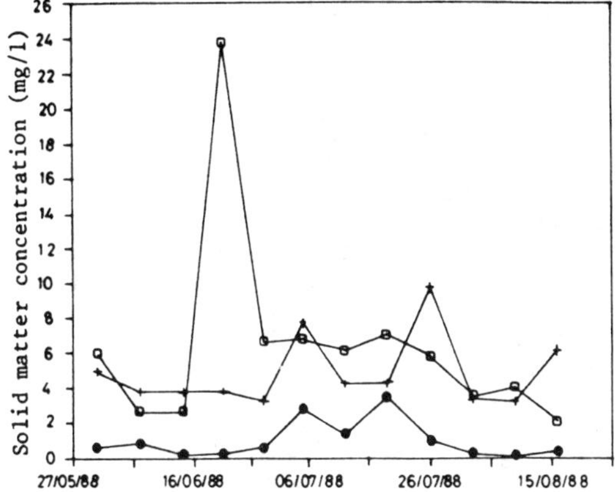

Fig. 4 Effects of the sedimentation basin and surface drainage area in the northern part of Kompsasuo on suspended solids in the runoff water from the peat production area in 1988.

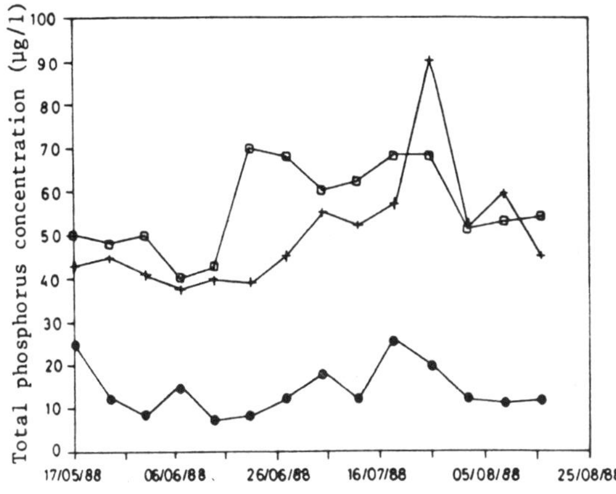

Fig. 5 Effects of the sedimentation basin and surface drainage area in the northern part of Kompsasuo on total phosphorus in the runoff water from the peat production area in 1988.

stations, the peat serving as a mechanical and chemical filter. Preliminary experiments have also been carried out on the purification of water from peat production areas by means of peat filters.

Waste water purification results show that slurry filters have an excellent retention capacity with respect to phosphorus, nitrogen, BOD and suspended solids, i.e. over 90 %. Preliminary tests suggested that it would be worthwhile to conduct further research into the possibilities for using peat filters for the processing of water from peat production areas.

4.2 Research programme

A model of a peat filter was constructed at the Hydraulic Engineering Laboratory of the University of Oulu in 1988 for the purposes of assessing under laboratory conditions certain factors connected with the building of an actual full-scale filter:

- determination of the thickness of the filtration layer and surface loading
- choice of grain size for the draining sand
- duration of use of the filter (point at which it becomes blocked)
- effect of drying out of the peat layer on the results
- practical arrangements involved in constructing a full-scale filter system, e.g. distribution of water to the filter.

The general experimental set-up for the model is depicted in Fig. 6.

Three parallel experimental filters with a combined surface area of 273 m^2 were constructed beside the main outflow ditch of the peat production area of Piipsanneva in 1988 (Fig. 7). The filters are isolated from the natural hydrology of the mire, i.e. the water is pumped into them from the main outflow ditch. The construction work was completed in November 1988. Raw peat is employed as the filtering material, as in the laboratory model. Field experiments will begin in summer 1989, as soon as the filters have thawed out. Particular attention will be paid to the filtration technique, purification being achieved in principle in the same manner as with a slow sand filter.

The results obtained with the laboratory model suggest that this peat filtration method efficiently reduces suspended solids, by approx. 90 % and also those substances bound to them, whereas substances dissolved in the water passed through the filter practically unchanged. A loading of at least 1.0 $m^3m^{-2}h^{-1}$ can apparently be achieved.

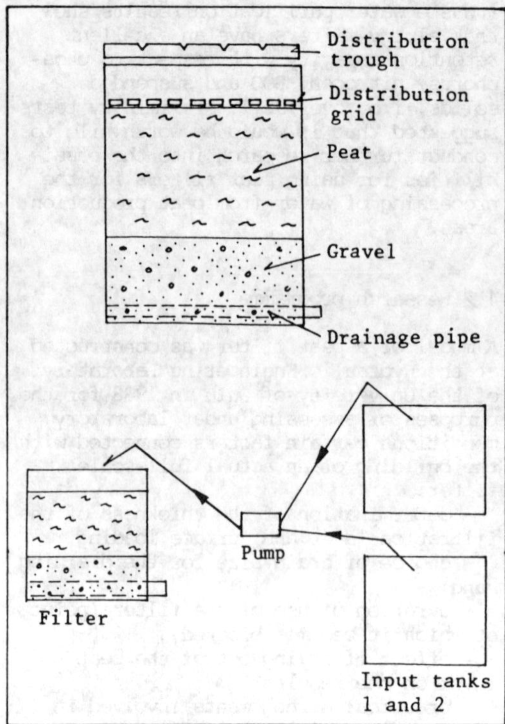

Fig. 6 Cross-section of the peat filtration model and principle of the experiment.

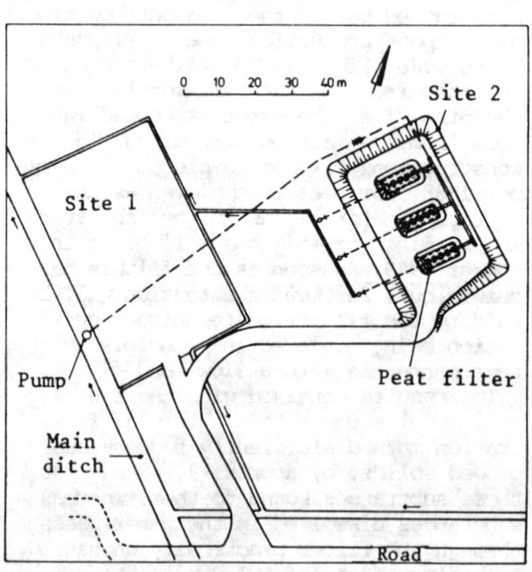

Fig. 7 Site used for studying peat filtration at Piipsanneva.

5 RETENTION DRAINPIPES

5.1 Background

It is necessary to dig a dense network of open ditches in the peat production areas (500 m ha^{-1}) to drain these areas. It has been shown that such ditches can also promote sedimentation. The catchment area of each ditch in this network will be small, and thus the maximum amounts of water will be low and flow conditions remain stable (Selin 1985). Drainpipes are being installed at the lower end of each open ditch to serve as bridges enabling machines to drive from one strip of peat to the next. Various retention devices can then be fitted at the ends of the pipes nearest to the ditches in order to keep the pipes open and improve the slurry retention capacity of the ditches themselves. Small sedimentation pools have also been dug at the lower ends of the ditches, according to dimensions determined experimentally (Fig. 8). Earlier research suggests that the use of retention drainpipes can reduce the suspended solids content of the water

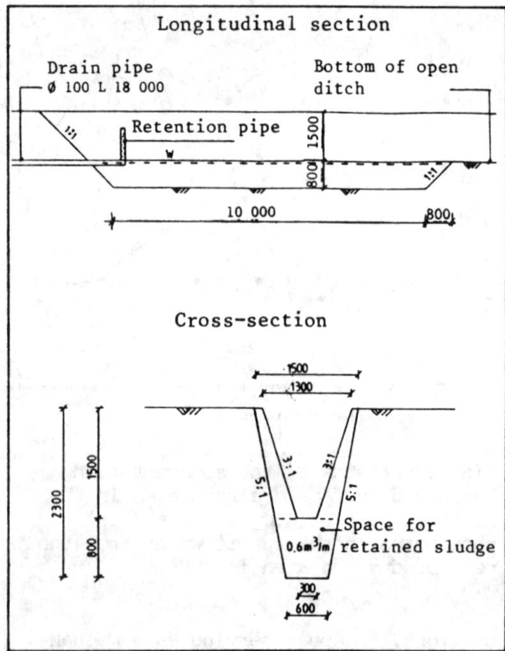

Fig. 8 Proposed watercourse protection measures for the lower parts of the open ditches: drain pipe, retention pipe and ditch basin.

flowing into the river system, whereas they have no appreciable effect on dissolved substances (Selin 1985).

The purpose of the present research is to obtain information on the efficiency of various alternative retention devices. Research was commenced at Piipsanneva near Haapavesi in 1987 (Fig. 1).

5.2 Research programme

The retention devices selected for testing were various plastic and metal pipes, wooden slats and perforated metal sheets (Fig. 9). Their efficiency was monitored by means of water samples, discharge measurements and practical observations. Water samples were taken from above the ditch basins and below the drainpipes, and the discharge through the drainpipes was assessed with a measuring vessel.

The preliminary research results show that the retention devices affect the suspended solids content of the water, but not the nutrient content.

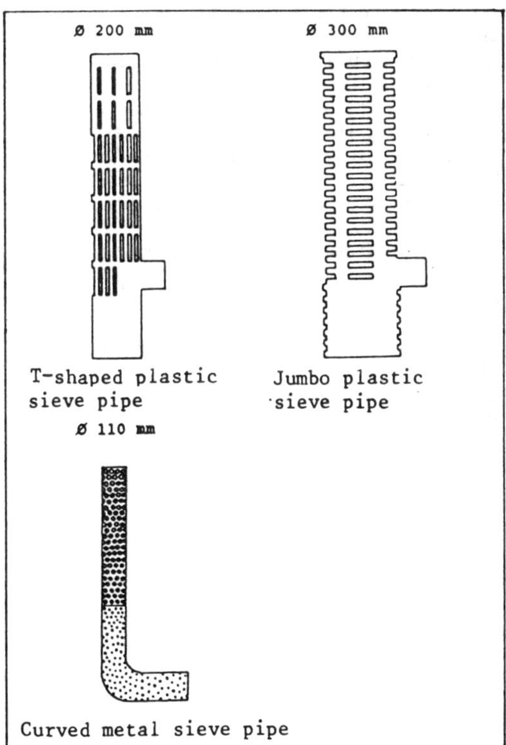

Fig. 9 Examples of different types of retention pipe.

6 SEDIMENTATION BASINS

6.1 Background

The most common means of purifying water from a peat production area employs sedimentation basins, the further development and testing of which has been recommended in the light of earlier findings (Selin 1985). Addition research is also needed to define appropriate dredging frequencies and methods. Research was commenced at Kurunneva and Murtosuo in 1987.

6.2 Research programme

The use of a box weir as an outlet device is being evaluated. This construction allows regulation of the discharge through the basin. When the discharge is small, the water will pass through the holes in the weir, but once it increases beyond 100 l s^{-1} km^{-2} the planks of the set weir restrain the water. The uppermost set plank corresponds to a calculated discharge of 300 l s^{-1} km^{-2}, above which value the water will flow out over the top of the set planks. It should be remembered, however, that the cross-section of the basin also increases as

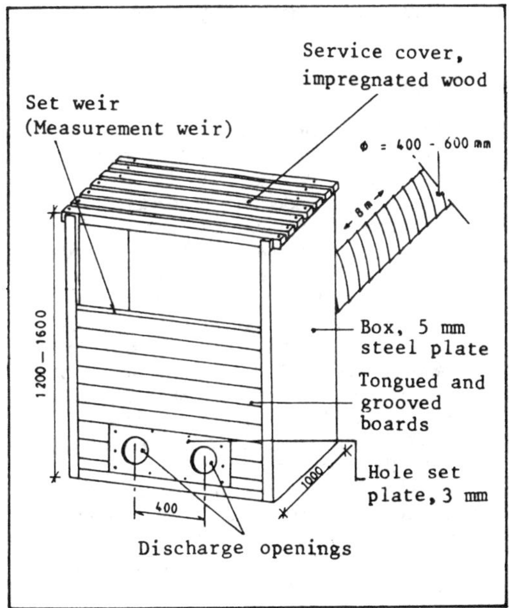

Fig. 10 'Box weir' outlet device for a sedimentation basin.

the water level rises, so that the flow rate in the basin remains more or less constant (Fig. 10).
Water samples were taken from above and below the sedimentation basin for determination of suspended solids, residue of suspended solids on ignition and COD according to accepted standards. Discharge rates above the sedimentation basin were measured by means of a triangular measuring weir.

The water from the Kurunneva peat production area is purified in two sedimentation basins, and this site is also being used to assess the suitability of an excavator-mounted suction scoop dredger for removing the sludge accumulating in the basin. Work is similarly in progress to determine the effect of the frequency of dredging of the basin on the quality of the outgoing water. One of the basins was dredged once in autumn 1988, following the normal procedure, and the other three times during the summer. Water samples were taken from above and below the basins for determination of suspended solids, residue of suspended solids on ignition, COD, electrical conductivity, pH, colour, total nitrogen, ammonium nitrogen, nitrate nitrogen, total phosphorus, phosphate phosphorus and iron content according to accepted standards. Discharge rates below the sedimentation basins were determined using triangular measuring weirs.

7 SUMMARY

The aim of this project on "The Development of Water Purification Technologies for use in Peat Production" is to evolve new methods for reducing watercourse loading from mires used in peat production and to improve existing methods. The work, which is being carried out jointly by the Ministry of Trade and Industry, the peat producers, the water conservation authorities and the Technical Research Centre of Finland at a total cost of approx. 3.5 mill. Fmk, was begun in 1987 and will continue for three years. The final report is to be issued in 1990. The sections of the research commenced to date concern the use of surface drainage to purify the runoff water, retention of suspended solids by means of open ditches, purification of runoff water by filtration and improvement of the efficiency and purification capacity of the sedimentation basins already in use. The experimental sites are all located in the province of Oulu, Finland.

Results to date suggest that a surface drainage area should correspond in size to 2 - 4 % of the peat production area. The method successfully removes both suspended solids and nutrients from the runoff water, leading to reductions of 70 - 80 % in suspended solids, 40 - 60 % in total nitrogen and 60 - 70 % in total phosphorus. Too small surface drainage area can even lead to increased concentrations in the runoff water, however. Future research will be devoted to assessing the effects of the soil and vegetation on the purification result achieved and the duration of time for which the same area of land can be used for this purpose.

Experiments with a small-scale model of a peat filter show that this method allows suspended solids, at least, to be removed very efficiently from the runoff water from peat production areas. Experiments will be carried out with a full-scale peat filtration system in 1989.

Retention drainpipes installed at the lower ends of the open ditches coming from the peat production area will reduce suspended solids, but have no effect on the concentrations of dissolved nutrients. Further experiments will be conducted in 1989 using different retention devices and testing the effects of sedimentation basins located on the open ditches in relation to drainpipes alone.

Work was also carried out to ascertain the possibilities of using a box weir installed at the outflow from the sedimentation basin. One effect of this should be to even out the discharge in the basin, in addition to which it should prevent erosion of the slopes of the basin at the outflow end. The frequency of removing the sludge from the basin will also be investigated with respect to its effect on the quality of the outgoing water. Tests have been carried out with a suction scoop dredger fixed to an excavator.

REFERENCES

Marja-aho, J. & Koskinen, K. 1988. The environmental effects of the peat production (in Finnish). Reports of National Board of Waters and the Environment. In press.

Sallantaus, T. 1983. The load to watercourses from peat mining (in Finnish with English summary). Helsinki. Ministry of Trade and Industry, serie D:29. 122 pp.

Selin, P. & Koskinen, K. 1985. The effects of the sedimentation ponds on the load coming from the peat production areas to the watercourses (in Finnish with English summary). Helsinki. Reports of National Board of Waters, number 262. 112 pp.

Land and Water Use, Dodd & Grace (eds), © 1989 Balkema, Rotterdam. ISBN 90 6191 980 0

A practical electric instrument for in situ measurement of peat properties

M.V.Toikka & M.Hallikainen
Laboratory of Space Technology, Helsinki University of Technology, Finland

ABSTRACT: A new instrument for measuring peat properties has been developed. It consists of an electronics box and a probe. The instrument measures the dielectric properties of peat in the 100 to 200 MHz range. The relation between the dielectric and physical properties of peat is used to calculate the moisture content or dry density. The moisture content measurement range is from 80% to 100%. The instrument is portable and battery operated.

RESUME: Un nouveau appareil pour mesurer les proprietes de la tourbe a ete developpe. L'appareil consiste d'une sonde d'humidité á ondes radioélectriques et une boite électronique. Avec la sonde, il est possible de mesurer les propriétés diélectriques de la tourbe de 100 á 200 MHz. La teneur d'humidité, la teneur énergétique et la densité de la tourbe sont calculées des propriétés diélectriques . L'instrument est portable et il fonctionne à la batterie.

ZUSAMMENFASSUNG: Ein neues Instrument für Torfmessungen ist dargestellt. Dieses Instrument besteht aus einer Sonde und aus einem elektronischen Einheit. Mit dieser Sonde ist es möglich die dielektrischen Eigenschaften von Torf im Frequenzbereich zwischen 100 und 200 MHz zu messen. Wassergehalt, Trockengewicht und Wärmeleistung des Torfes können durch der Elektrischen Werten kalkuliert werden. Das Messbereich bei Wassergehalt messungen ist von 80% bis 100%. Das Instrument ist tragbar und funktioniert mit Batterie.

1 INTRODUCTION

In surveying the peat resources, the main parameters are the dry matter content (bulk density) and the energy content of peat. The in situ bulk density (kg/m^3) and the calorific value (J/kg) can be determined by sampling the peat and analyzing the samples in laboratory. This is rather labourious and time consuming. Consequently, it may take several years, even decades to collect enough peat geological data in order to make plans for peat winning. Therefore, development of rapid methods for peat resource inventory is essential.

A promising new instrument is a radio wave probe developed for in situ measurements of the properties of peat. The radio wave probe is based on the fact that the dielectric properties of a wet material are determined by its water content. On the other hand, there is a good correlation between the in situ water content of peat and the bulk density of peat (Korpijaakko et al. 1981, Laine and Päivänen 1982).

The measurement principle and the first prototype of a radio wave probe was introduced in 1982 (Tiuri and Toikka 1982). The probe has been developed further to make it more practical. Its reliability has been confirmed in extensive field tests (Klemetti and Mäki-Kuutti 1986, Lappalainen and Hänninen 1986).

2 DIELECTRIC PROPERTIES OF PEAT

The dielectric properties of a material can be described by its relative dielectric constant, ϵ_r, also called relative permittivity. The value of the dielectric constant depends on the molecular structure of the material. If the material is lossy, its dielectric constant is complex, $\epsilon_r = \epsilon_r' - \epsilon_r''$, and the imaginary part ϵ_r'' is related to the losses.

The dielectric constant of a mixture depends on the dielectric properties and the volumetric fractions of the materials.

Wet peat can be assumed to be a mixture of three substances: gas, dry matter, and water. The relative dielectric constant of gas is $\epsilon_r = 1$ and it is usually lossless. The amount of gas in normal peat is very small, thus it has a minor effect on the bulk dielectric constant. The relative dielectric constant of dry matter is approximately $\epsilon_r = 3$ and it has low losses. In virgin peatlands the dry matter content is usually less than 20%, thus it has a small effect on the bulk dielectric constant. The relative dielectric constant of water is much higher and it has also high losses. The real part is $\epsilon_r' = 80$ (at 20°C, below the frequency of 10 GHz) and the imaginary part, ϵ_r'', may vary from about four up to several hundreds depending on frequency, temperature and the electrical conductivity of water. The behavior of the dielectric constant of water as a function of temperature and salinity is well known and its value can be calculated using empirical formulas (Klein and Swift 1977).

The dielectric constant of the mixture of water and dry material is approximately given by the Brown formula (Tinga et al. 1973)

$$\epsilon_r = k\epsilon_w + (1-k)\epsilon_d \qquad (1)$$

where ϵ_r is the relative dielectric constant of the mixture, k is the volume fraction of water, ϵ_w is the relative dielectric constant of water and ϵ_d is that for dry matter. If the only interest is in measuring the water content, the imaginary part can be omitted and the real part is used. If we assume $\epsilon_w = 80$, $\epsilon_r = 3$, and the water content is 85 % we get $\epsilon_r = 68$. If the water content is 95 % we get $\epsilon_r = 76$.

In reality, the behavior of the dielectric constant is more complicated than that given by equation (1), due to the absorption (bonding) of water by dry material, and the shape of inclusions. In order to find out the correct dependence of the relative dielectric constant on the water content , natural peat samples have been measured in laboratory (Toikka et al. 1985). Figure 1 shows the water content of peat as a function of the relative dielectric constant. A high degree of correlation, $R^2 = 0.88$, was obtained. It was concluded that the water content, the in situ bulk density or

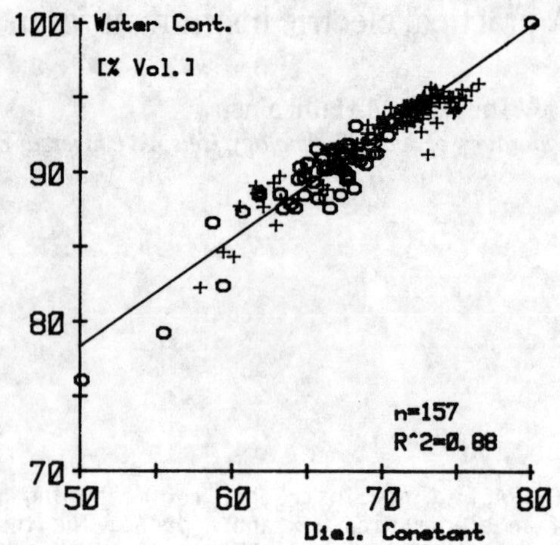

Fig.1 Water content versus relative dielectric constant of peat samples measured in laboratory. 110 Finnish peat samples, marked with +, and 46 tropical peat samples (Jamaica), marked with o (Toikka et al. 1985).

the energy content per volume can be determined by measuring the in situ dielectric constant of peat (Tiuri et al. 1984).

3 RADIO WAVE PROBE FOR IN SITU MEASUREMENTS

3.1 Measurement principle

The wavelength of a radio wave in a material depends on the dielectric constant of this material. The wavelenght, λ can be calculated by using the relative dielectric constant, ϵ_r and the free space wavelenght, λ_o:

$$\lambda = \frac{\lambda_o}{\sqrt{\epsilon_r}} \qquad (2).$$

Thus the relative dielectric constant can be calculated if the wavelength in free space and in the medium under study is known. The wavelength can be determined by measuring the resonant wavelength or frequency of a resonator filled with the medium. The wavelength and frequency are related by Equation (3),

$$\lambda = \frac{c}{f\sqrt{\epsilon_r}} \qquad (3)$$

where λ is the wavelength in the medium, c is the radiowave propagation velocity in free space (speed of light $3*10^8$ m/s), f is the frequency and ϵ_r is the relative dielectric constant of the medium.

A mechanically suitable resonator for peat measurements is a slot at the end of a pipe, Figure 2. The first resonance occurs when the length of the slot is half wavelength. If the length of the slot is 15 cm and the probe is pushed into water, the measured resonant frequency is 112 MHz indicating that the relative dielectric constant of water is 80.

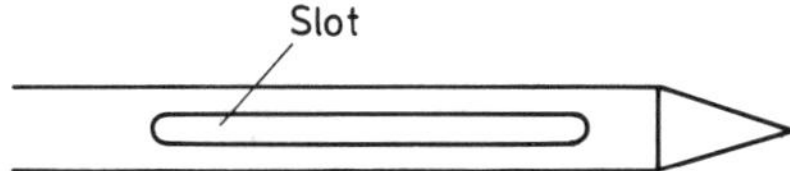

Fig.2 The slot resonator at the end of a pipe

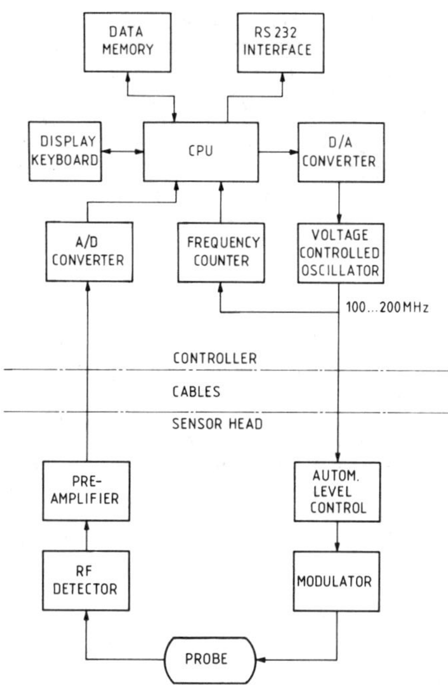

Fig.3 The block diagram of the instrument.

3.2 Description of instrument

The resonant frequency of a radio wave resonator can be measured by feeding a signal through the resonator and determining the signal attenuation. The signal goes through the resonator with low attenuation at resonance whereas outside the resonance the attenuation is high. The resonant frequency can be searched by sweeping the frequency over the possible resonant range. Figure 3 shows the block diagram of the completed instrument. For practical reasons, the instrument is divided in two parts: controller and sensor head. The sensor head contains an automatic level control and a preamplifier. This allows the use of very long cables (>10 m) between the sensor head and the control unit. The modulator is for better sensitivity and to avoid temperature dependency. The radio frequency detector is needed to convert the signal to a lower frequency for the controller. The sensor head is built inside a steel pipe, 26 mm in diameter. The probe is a 20 cm long slot at the end of this pipe.

The controller consists of a microprocessor and appropriate units. The microprocessor searches the resonant frequency by sweeping the frequency of a voltage controlled oscillator. The signal is fed through the probe and then returned back to the controller through an analog to digital converter. The microprocessor finds the signal maximum and then the frequency counter measures the corresponding frequency. The resonant frequency is used to calculate the dielectric constant and the properties of peat.

The user communicates with the system through a hand-held keyboard-display. The instrument has also a solid state data memory. The measurement results can be stored with additional information, for example the date and a code number. The data can be printed out or transmitted to a computer for further analysis.

The whole system is light-weight and portable even under heavy conditions on mires. The instrument is designed to operate in extreme conditions ranging from a rainy summer day (temperature +30^oC) to a cold winter day (temperature -20^oC). The instrument is battery operated and works at least one day without recharge.

The measured data can be stored in the memory. The memory capacity is 32 kilobytes and it is usually enough for three days' operation. The data memory block can easily be changed. It does not need any battery back up and the stored data cannot be destroyed by keyboard operations.

The measured data can be readily checked in the field using the file retrieval program of the probe. The daily data can be listed in a field camp with a printer or a portable microcomputer. After completing the measurements, all the data can be checked and preprocessed with a microcomputer into a suitable form to fit a larger data handling system.

4 RESULTS

In order to determine the feasibility and reliability of the radio wave probe, extensive field tests have been carried out. The new method has been compared with conventional sampling method. On a test site the peat layer has been first measured with the probe and, after that, samples have been taken to determine the water content, bulk density and the energy content of peat. Figure 4 shows a comparison of measured bulk density values. The probe measurements were carried out three times from top to bottom in 10 cm intervals. The sampling was made only once from top to bottom and the sample length was 20 cm. In order to compare a sampling result with the probing results, an average of six individual measurements has to be made. The bulk densities show a reasonably good linear correlation (R^2 = 0.67). The correlation coefficient is not as high as in the laboratory measurements. This is due to the inhomogeneity of peat strata, which makes it impossible to obtain completely corresponding sample-measurement pairs (Tolonen and Ijäs 1982). The inhomogeneity could be seen often as variations in redundant probe measurements. If the amount of peat per area is calculated, the problem of corresponding sample-measurement pairs disappears and the correlation is higher, R^2 = 0.92, Figure 5.

A Finnish peat harvesting company, the Vapo Ltd. has carried out a large peat resource evaluation project using the radiowave probe (Klemetti and Mäki-Kuutti 1986). A comparative sampling was carried out at 50 sites. The results are practically identical to our results. The Geological Survey of Finland has also carried out similar tests for the radio wave probe using over 1000 samples (Lappalainen and Hänninen 1986). The results show that the differences between the sampling method and the probe could be reduced by making a new equation for every mire with one calibration measurement-sampling set. In this way the difference in the moisture content given by the radio wave probe and by the

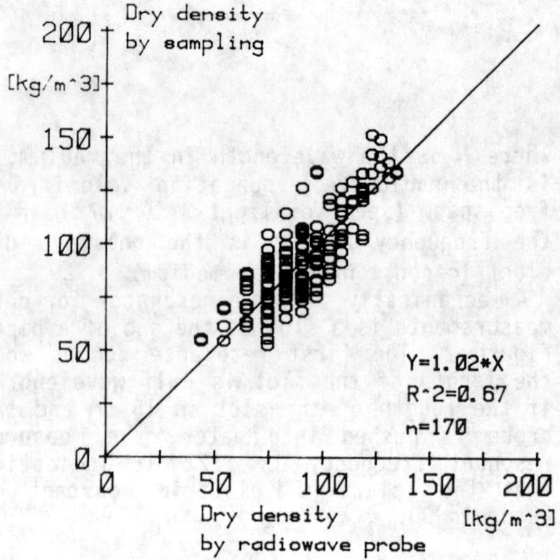

Fig.4 Dry density of peat samples versus in situ measured dry density. Also shown is the liner regression equation, the correlation coefficient and the number of samples.

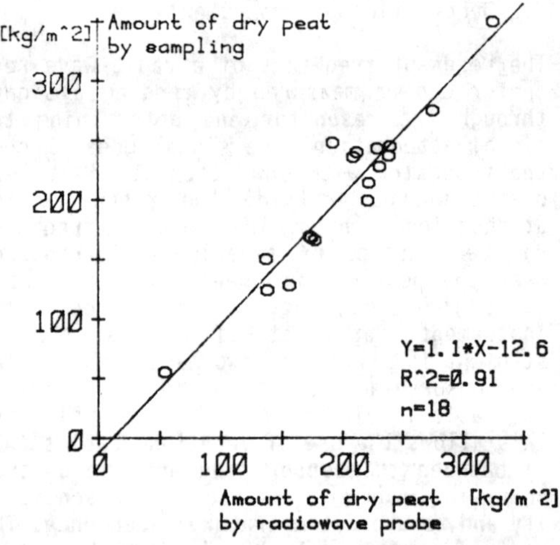

Fig.5 The amount of dry peat per area measured by sampling and by radiowave probe.

sampling method was ±0.5%. Without any specific calibration the difference was about ±1%.

5 CONCLUSIONS

The stratigraphical sampling of peat is labourious and time consuming. Two persons are usually needed in a sampling crew. Their daily accomplishment is around five sites. In an average bog this means about 30 samples, each 50 cm long, which must be transported for further analysis. The drying time is about three days. The determination of the calorific value of 30 samples takes about one week, if only one calorimeter is available, and if proper accuracy is maintained. Thus, what one sampling crew achieves in a summer's work employs one laboratory worker for a full year (Klemetti and Mäki-Kuutti 1986).

The measurement with a radio wave probe is faster and easier than conventional peat sampling. One measurement site can be measured twice from the surface to the base of the peat layer in fifteen minutes. Two persons are usually operating the system. One is pushing the probe into the peat and the other is operating the keyboard. Additional sampling and laboratory analysis of the peat is not necessary.

The instrument described in this paper has several remarkable advantages compared with conventional sampling methods:

1. It is a light-weight and easily portable instrument. There is no need to transport heavy samples.
2. It is fast. The measurement of a profile several meters deep takes only a few minutes and there is no need for additional laboratory measurements; thus, redundant measurements can be made.
3. It uses automated data handling methods and there is no need for paper and pencil.
4. It is an all-year instrument. The measurement is almost as easy in winter as in summer, depending on the frost-ice thickness.

REFERENCES

Korpijaakko,M., Häikiö,J. and Leino,J. 1981. Effect of water content and degree of humification on dry density of peat (in Finnish with English summary). Suo 31 (2):39-43.

Klein,L.A. and Swift,C.T. 1977. An improved model for the dielectric constant of sea water at microwave frequencies. IEEE J. Oceanic Engineering 2 (1):104-111.

Klemetti,V. and Mäki-Kuutti,T. 1986. A practical application experiment using a new radio-wave probe in a large-scale peat resource evaluation project. Proc. Advances in Peatlands Engineering. Ottawa, Canada:7-15.

Laine,J. and Päivänen,J. 1982. Water content and bulk density of peat. Proc. Int. Symposium of the IPS Commissions IV and II. Minsk, USSR:422-430.

Lappalainen,E. and Hänninen,P. 1986. New electric methods of fast inventory of peat and peatland resources. Proc. Advances in Peatlands Engineering. Ottawa, Canada:71-77.

Tinga,W., Voss,A. and Blossey,D. 1973. Generalized approach to multiphase dielectric mixture theory. J. Applied Physics 44 (9).

Tiuri,M. and Toikka,M. 1982. Radio wave probe for in situ water content measurements of peat. Suo 33 (3):65-70.

Tiuri,M., Toikka,M., Tolonen,K. and Rummukainen,A. 1984. Capability of new radiowave moisture probe in peat resource inventory. Proc. 7th International Peat Congress. Dublin,Ireland (1):157-525.

Toikka,M., Tiuri,M. and Tolonen,K. 1985. Capability of new radiowave moisture probe in tropical peat resource inventory. Proc. IPS Symposium on Tropical Peats. Kingston, Jamaica:215-223.

Tolonen,K. and Ijäs,L. 1982. Comparison of two peat samplers used in estimating of dry peat yield in field inventories. Suo 33 (2):33-42.

Land and Water Use, Dodd & Grace (eds), © 1989 Balkema, Rotterdam. ISBN 90 6191 980 0

Comportement des sols tourbeux drainés

M.Soutter & A.Musy
Institut de Génie Rural, Ecole Polytechnique Fédérale, Lausanne, Suisse

RÉSUMÉ: Le tassement des sols tourbeux se traduit notamment par une diminution progressive de l'efficacité des réseaux d'assainissement. La prévision de cette évolution est rendue difficile par la nécessité de prendre en compte les modification des caractéristiques hydrodynamiques des sols. La simulation permet une analyse plus détaillée de cet aspect du phénomène et de ses effets. Une brève présentation d'une région de Suisse Romande illustre finalement la problématique actuelle du drainage des tourbes.

ZUSAMMENFASSUNG: Die Sackung der Moorböden entspricht vor allem einer progressiven Abnahme der Effizienz des Entwässerungsystems. Die Vorhersage deren weiteren Entwicklung wird deshalb erschwert, weil die Veränderungen der hydrodynamischen Charakteristika der Böden mitberücksichtigt werden müssen. Die Simulation erlaubt bezüglich jenes Aspektes des Phänomens und dessen Auswirkungen genauere Analysen durchgeführen. Die Fallstudie einer Westschweizerischer Region veranschaulicht die heutige Drainageproblematik der Torfböden.

ABSTRACT: The compaction of peat soils involves among others a progressive decrease of the drainage networks efficiency. The forecast of this evolution is made difficult by the need to take into account the modifications of the soil hydrodynamic characteristics. Simulation allows a more detailed analysis of this aspect of the phenomenon and its effects. To conclude, a short case study of an area in western Switzerland illustrates the present problems of peat soils drainage.

1 INTRODUCTION

Les sols tourbeux constituent un cas particulier de sols hydromorphes, dont la spécificité tient à leur nature organique et au caractère permanent des nappes au sein desquelles ils se forment. Sans entrer dans le détail des écosystèmes qui président à la formation des tourbières, on peut dire de manière schématique que la constitution de ces sols résulte d'une accumulation de débris végétaux dont la décomposition est très lente et incomplète, suite à l'inhibition presque totale du cycle biologique des éléments nutritifs, caractéristique de conditions anaérobies.

Cette genèse confère à ce type de sols un potentiel d'évolution dynamique très important qui constitue une autre de leurs particularités. Celle-ci se manifeste de manière très active sous l'effet d'agents naturels et surtout artificiels, tel que le drainage, opération préalable très souvent nécessaire à leur mise en valeur agricole. Il suffit en effet de quelques années de drainage pour modifier fondamentalement leurs propriétés physiques, chimiques et bactériologiques.

L'un des aspects les plus spectaculaires de cette évolution consiste en un abaissement prononcé de la surface du sol qui résulte à la fois d'une diminution de la porosité et d'une perte de matière. Ce phénomène se poursuivant constamment dans le temps, bien qu'avec une intensité décroissante, le niveau du sol se rapproche inexorablement de celui des drains, de sorte que les réseaux d'assainissement ne répondent souvent plus aux besoins pour lesquels ils avaient été conçus et réalisés.

Nous nous attacherons à présenter dans cette étude une brève synthèse des divers processus contribuant à l'abaissement ou tassement des sols organiques drainés, des principaux facteurs qui les influencent et des caractéristiques de l'évolution spatio-temporelle de ce phénomène. Nous dégagerons ensuite ses conséquences, en particulier sur le comportement hydraulique de ces sols, avant d'exposer la problématique globale qui en résulte, dans le contexte de l'exploitation agricole des sols d'une plaine tourbeuse de Suisse romande.

2 LES MÉCANISMES DU TASSEMENT DES TOURBES

L'abaissement des sols organiques résulte de l'action combinée de divers processus relevant aussi bien de déformations mécaniques que de réactions chimiques et microbiologiques. Ces processus, soit l'affaissement, le rétrécissement, la minéralisation, le lessivage et l'érosion, font l'objet d'une description succincte ci-dessous. Leurs interactions respectives, peu connues, ne seront pas traitées dans le cadre de cet article.

Par ailleurs pour que cette énumération soit complète il faut encore mentionner les prélèvements d'origine végétale ou anthropique que nous négligerons par la suite.

2.1 L'affaissement

L'affaissement est un processus physique d'altération structurelle du sol, qui n'affecte pas spécifiquement les sols organiques mais également, quoique dans une moindre mesure, les sols hydromorphes minéraux, consécutivement à une opération de drainage.

A la suite de l'élimination de l'eau contenue dans l'espace poral du sol, le squelette formé par ses éléments constitutifs, n'étant plus soutenu par un fluide faiblement compressible, se déforme sous l'effet des contraintes résultant de son poids propre. Cette déformation fait intervenir les propriétés mécaniques caractéristiques du contact et du frottement entre les éléments structuraux du sol et celles de ces éléments eux-mêmes.

La grande sensibilité à l'affaissement que manifestent les sols tourbeux découle dès lors de leur faible compacité et de leur forte porosité. Elle varie toutefois selon le type de sol et son degré de décomposition. En effet les débris végétaux peu décomposés exercent de par leur relative élasticité une action protectrice face à la déstructuration.

Par ailleurs, lorsque le sol est soumis à la circulation d'engins agricoles lourds, la contrainte qu'il subit s'en trouve augmentée et le processus d'affaissement renforcé. Toutefois, les importants progrès réalisés récemment dans le domaine de la distribution dans le sol des charges dues à ces engins tendent à minimiser l'importance de cet aspect du phénomène.

Les principaux facteurs qui influencent l'affaissement, outre l'importance des contraintes et les propriétés mécaniques du sol, sont la profondeur de drainage et les fluctuations saisonnières du niveau de la nappe. Ces dernières accentuent en effet le processus d'affaissement, de par la succession de contraintes et relâchements qu'elles engendrent au niveau du squelette.

2.2 Le rétrécissement

Le rétrécissement constitue également un processus physique, qui affecte les fibres végétales de la tourbe. Lorsque le sol se dessature, la succion croissante que l'eau exerce sur ces fibres entraîne leur contraction. Cette déformation étant inélastique, la faculté qu'elles ont de reprendre leur forme initiale est d'autant plus faible que la succion subie a été forte.

Globalement, ce processus tend à une pulvérisation des débris végétaux. Il se manifeste essentiellement dans les zones où de fortes succions peuvent apparaître, notamment l'horizon superficiel et les zones racinaires des plantes.

Dans un premier temps, lorsque les débris végétaux sont peu décomposés, le rétrécissement induit une augmentation de la porosité, la structuration des éléments du sol n'étant que peu affectée. Par la suite, la perte de volume due à ce processus est d'autant plus importante que la tourbe est homogène, composée de fibres courtes et à un stade de décomposition avancé.

2.3 La minéralisation

La décomposition de la matière organique, qui relève d'un ensemble de réactions d'oxydation biotiques, abiotiques et photo-chimiques, peut être schématiquement décrite en trois phases. Les molécules organiques complexes subissent tout d'abord une décomposition microbienne qui les réduit en molécules plus simples. La minéralisation primaire transforme ensuite une partie de ces dernières en composés minéraux solubles ou gazeux. L'autre partie sert de base à l'édification de molécules nouvelles, à nouveau plus complexes, par humification. Celles-ci, qui de par leur nature colloïdale jouent un rôle important dans la structuration du sol, se décomposent à leur tour en un processus plus lent, la minéralisation secondaire.

Initialement, en raison du manque d'oxygène propre aux milieux anaérobies, l'activité biologique est très faible et, par conséquent, la minéralisation primaire fortement ralentie et l'humification incomplète.

L'aération du sol consécutive à son drainage accélère puissamment cette dynamique, entraînant la libération sous forme soluble ou gazeuse de nombreux éléments. En outre, la minéralisation secondaire exerce alors peu à peu une action déstructurante.

L'intensité de la décomposition, c'est à dire la vitesse des réactions d'oxydation, dépend essentiellement des proportions d'air et d'eau dans le sol et de sa température. Ainsi, les principaux facteurs qui influencent ce processus sont les paramètres climatiques (précipitations, température) et le régime hydrique du sol. L'humidité des horizons supérieurs désaturés est en effet conditionnée par la position du niveau de la nappe.

L'aération du sol quant à elle est liée notamment à la fréquence des labours. Celle-ci est extrêmement réduite dans le cas d'une culture en prairie permanente, limitant ainsi sérieusement la minéralisation du sol.

Finalement, le bilan global de la décomposition dépend de la profondeur de drainage, qui détermine le volume de sol aéré et soumis à la minéralisation.

2.4 Le lessivage

Les composés minéraux libérés par la décomposition sont solubilisés par l'eau avant d'être entraînés vers les zones plus profondes voire même directement évacués avec les eaux de drainage, occasionnant une perte de matière.

Le lessivage est par conséquent un processus étroitement lié au régime des précipitations.

D'autre part, les fluctuations du niveau de la nappe entraînent la formation d'un courant d'infiltration de direction et d'intensité variable, qui constitue également une source de lessivage que l'on désigne parfois sous le terme d'érosion interne.

2.5 L'érosion

Le dernier facteur qui intervient dans l'abaissement des sols organiques est l'érosion, et plus particulièrement l'érosion éolienne.

En effet, comme cela a été mentionné précédemment, le phénomène du rétrécissement des fibres végétales suite à l'alternance dessiccation-humectation est particulièrement marqué à la surface du sol. La tendance à la pulvérisation qui en découle rend ces sols très sensibles à l'érosion éolienne, puisque le taux de déflation dépend, outre de la vitesse et de la turbulence du vent, de la taille et de la densité des particules.

De plus, la déflation est approximativement inversement proportionnelle au carré de l'humidité du sol en surface, endroit qui s'assèche précisément le plus rapidement.

Les sols organiques sont également sensibles à l'érosion hydrique. Toutefois, celle-ci ne prend de l'importance que dans certains cas peu répandus, en fonction du relief.

3 ÉVOLUTION SPATIO-TEMPORELLE DU TASSEMENT DES SOLS TOURBEUX DRAINÉS

L'analyse de l'intensité du phénomène du tassement des sols organiques drainés conduit à mettre en évidence, au delà des variations saisonnières et interannuelles, une évolution temporelle caractéristique et générale. Celle-ci, bien que suivant toujours le même canevas, connaît une forte variabilité spatiale d'amplitude et de rapidité, tant à l'échelle géographique, de par l'influence des facteurs de station, qu'à l'échelle locale, de par l'hétérogénéité des conditions.

Ce canevas se caractérise globalement par une phase de tassement prononcé peu après le drainage, suivie d'une phase au cours de laquelle l'intensité du phénomène diminue progressivement, jusqu'à se stabiliser. On distingue ainsi l'abaissement primaire, essentiellement dû à l'affaissement des couches de tourbe drainées, de l'abaissement secondaire qui lui succède, au cours

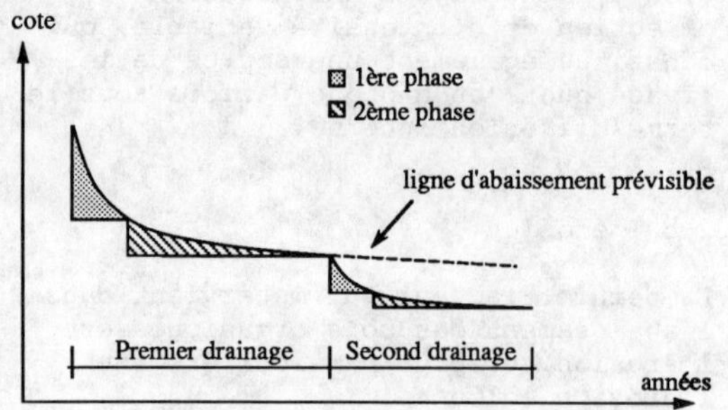

figure 1. Les phases du tassement

duquel la décomposition et l'érosion éolienne constituent les processus dominants. La même évolution se manifeste à nouveau à la suite d'un second drainage, mais avec une amplitude quelque peu atténuée. La figure 1 illustre les deux phases du tassement dans le cas de deux drainages successifs.

Cette évolution est suivie plus ou moins régulièrement selon les variations climatiques interannuelles de la station. En effet, l'intensité du tassement augmente globalement d'autant plus que la hauteur annuelle des précipitations est inférieure à la moyenne.

La répartition saisonnière de l'abaissement annuel dépend des fluctuations du niveau de la nappe. C'est pendant les périodes de nappe basse, c'est à dire la période de drainage intense au printemps et surtout la période sèche en été, que la plus grande partie du tassement se produit. Elle est ensuite très partiellement compensée par la remontée de la nappe en hiver qui provoque un léger regonflement des couches de tourbe.

La vitesse de l'abaissement varie selon le régime climatique de la station. L'intensité du phénomène augmente à mesure que l'on passe de régions froides et humides à des régions chaudes. Elle dépend également des caractéristiques du sol, notamment de son type et de son épaisseur. Lorsque cette dernière est importante (>1.0m), les variations interannuelles se font particulièrement sentir, du fait de la dessication périodique des couches profondes du sol.

A l'échelle d'une station, le tassement constitue un phénomène différentiel. Cette variabilité spatiale résulte à la foi. du rabattement de la nappe, qui voit sa cote et par conséquent le volume de sol drainé varier d'un endroit à l'autre, et de l'hétérogénéité du sol.

L'intensité du phénomène varie en effet selon la composition physicochimique de la tourbe. Du fait des variations locales d'épaisseur des diverses couches composant le sol, le montant total de l'abaissement, somme de leurs contributions respectives, varie d'un profil à l'autre.

La figure 2 (Presler & Sticher 1987) illustre les modifications du microrelief enregistrées au domaine de Bellechasse entre les années 1865 et 1985 (la trame unitaire est de 0.25ha; les surfaces situées hors de l'exploitation sont à la cote 430.5m, alors que, pour 1985, les autres altitudes sont comprises entre 430.54 et 433.07m).

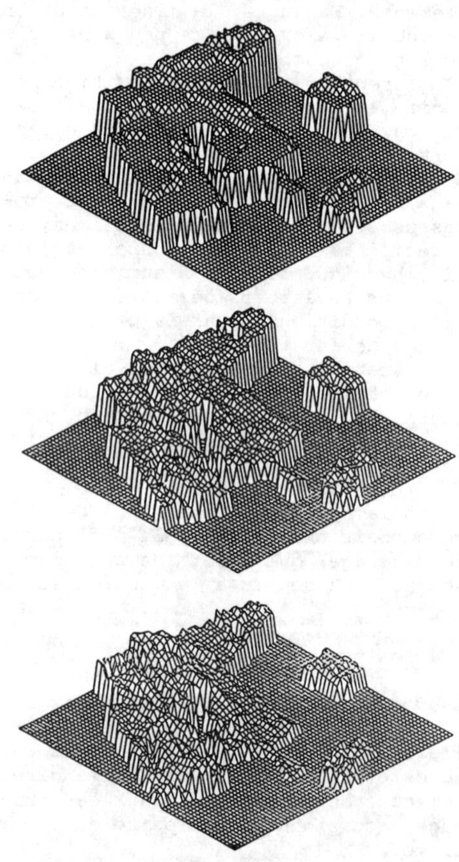

figure 2. Modifications du micro-relief de l'exploitation de Bellechasse. (Presler & Sticher 1987)

4 EFFICACITÉ DU DRAINAGE DES TOURBES

De manière générale, l'évolution des sols organiques à la suite de leur drainage conduit à une diminution progressive de l'efficacité du système d'assainissement. Cete perte d'efficacité résulte conjointement du tassement des sols et du colmatage des drains, autre conséquence majeure du drainage des tourbes.

4.1 Influence du tassement

Le tassement des tourbes engendre une modification de leurs propriétés physiques et chimiques. Le changement de ces dernières, suite à la décomposition et au lessivage, influence essentiellement les qualités agronomiques du sol (excès d'azote notamment). Cet aspect de la question ne sera toutefois pas abordé dans le cadre de cette analyse.

Sur le plan physique, c'est bien entendu la structure du sol qui est affectée par le tassement. Sous l'effet surtout de l'affaissement et du rétrécissement, mais également de la décomposition, la porosité diminue. Les deux premiers processus se traduisent surtout par une réduction de la macroporosité alors que la décomposition entraîne notamment une augmentation du nombre de pores de très petites dimensions. C'est donc l'ensemble de la répartition de la porosimétrie qui se trouve modifiée. Simultanément, sous l'effet des contraintes verticales, les éléments structuraux du sol se réarrangent préférentiellement selon une orientation horizontale.

La diminution de volume de l'espace poral et l'altération de ses caractéristiques dimensionnelles se répercutent sur les caractéristiques hydrodynamiques du sol. La conductivité hydraulique globale diminue, de même que la perméabilité à l'air. Cette diminution de la conductivité est particulièrement sensible dans la gamme des teneurs en eau élevées, qui concernent au premier chef les opérations de drainage; de plus elle est davantage marquée dans la direction verticale du fait de la réorientation des éléments constitutifs du sol.

Suite à la multiplication des pores de faible dimensions, liée au processus de décomposition, la succion de l'eau augmente plus rapidement pour une teneur en eau décroissante. Ainsi, plus le sol est décomposé et plus la valeur de la capacité de rétention maximale est élevée, réduisant de ce fait le volume d'eau susceptible d'être éliminé par drainage.

L'incidence des modifications des propriétés physiques du sol sur son comportement hydraulique se traduit donc progressivement par une diminution des vitesses d'écoulement et par l'augmentation des forces de rétention.

Une autre limitation à l'efficacité du drainage découle de la perte de profondeur des drains consécutive à l'abaissement de la surface du sol. Dans un premier temps cela a pour conséquence de favoriser l'engorgement périodique des zones racinaires des plantes, la nappe ne pouvant plus être suffisamment rabattue. A terme, les drains se trouvent situés dans l'horizon de labour et menacés de destruction, ce qui rend leur remplacement nécessaire. De plus, l'apparition de dépressions locales limitées suite au caractère différentiel du tassement favorise, surtout lorsque la tourbe repose sur un substrat peu perméable, la formation de "mouilles", zones qui restent presque constamment gorgées d'eau.

Le décompactage du sol par diverses techniques telle que le sous-solage profond, permet de restructurer le sol et ainsi d'en freiner l'évolution. Toutefois, cette forme de réhabilitation ne reste praticable que pour autant que la profondeur des drains soit suffisante et que l'incorporation éventuelle d'horizons minéraux ne menace pas sa fertilité. Elle est donc d'un emploi limité dans le temps et ne constitue pas une panacée.

4.2 Le colmatage

Il s'agit surtout, dans le cas des sols tourbeux, du colmatage par gel d'hydroxydes de fer. En effet, des ions ferreux Fe^{2+} solubles se trouvent dans les eaux mal aérées des milieux confinés, d'autres étant libérés et solubilisés au cours du processus de décomposition. Ces ions s'associent avec des composés organiques pour former des complexes mobiles (chélates) qui, lors du drainage, migrent en direction des drains. Lorsqu'ils atteignent la région du filtre du drain, qui en raison de son aération constitue un milieu oxydant, le complexant organique est détruit et les ions ferreux insolubilisés sous forme d'hydroxydes ferriques, soit par voie biologique (ferro-bactéries), soit par voie chimique.

Le colmatage revêt trois formes. Tout

d'abord il se produit un colmatage externe, pouvant aller jusqu'à l'imperméabilisation totale, qui affecte l'entourage des drains. Ensuite, un colmatage interne, par dépôt d'hydroxydes dans le drain, provoque la modifications de ses caractéristiques hydrauliques (coeff. de Strickler, section) et réduit ainsi l'écoulement. Finalement, la formation d'une pellicule solide à l'intérieur du drain, en période de sécheresse, provoque l'obstruction de ses orifices, ce qui empêche tout ou partie de la pénétration de l'eau.

Le colmatage par dépôt d'hydroxydes de fer, en réduisant la capacité d'évacuation du système d'assainissement exerce donc une influence directe sur le comportement hydraulique du sol.

5 LA PRÉVISION DE L'ÉVOLUTION DES TOURBES DRAINÉES

Au vu de l'importance du tassement et de ses conséquences, il est nécessaire que le praticien soit en mesure de tenir compte de ce phénomène lors de l'élaboration d'un projet d'assainissement, tant pour déterminer les caractéristiques géométriques du réseau (profondeurs de pose des collecteurs et des drains, écartement) que pour évaluer les modalités de sa gestion et son efficacité futures. L'estimation du tassement prévisible et des modifications de comportement hydraulique qui en découlent permet ainsi de prendre des mesures préventives.

5.1 La prévision du tassement

Il n'existe pas à ce jour et à notre connaissance de méthode de calcul exacte du tassement des sols tourbeux drainés. La plupart des formules que l'on trouve dans la littérature (Hallakorpi, Segeberg, Paniadi & Kostiakow, Ostromecki, Haefeli & Schaad, Folkens) reposent sur des constantes empiriques. Leur principal désavantage, outre le fait qu'elles ne tiennent compte que du processus de l'affaissement (premier drainage), est l'omission du facteur temps. Elles ne permettent donc pas l'estimation de l'évolution du phénomène et ne fournissent de ce fait que des valeurs indicatives.

La détermination la plus fiable de l'abaissement prévisible du sol repose sur le prolongement de la ligne d'abaissement. Cette méthode n'est toutefois pas applicable au stade d'un projet.

Ilnicki propose une méthode d'estimation de l'évolution de l'abaissement qui tient compte de tous les processus à l'oeuvre (Ilnicki 1977). Selon les données disponibles, on peut ainsi recourir à une régression linéaire multiple ou à diverses lignes d'abaissement au cours du temps. Dans le cas de la plaine de l'Orbe, qui sera présenté par la suite, la confrontation des abaissements calculés et des valeurs effectivement mesurées a montré une divergence de l'ordre de 25 % dans le cas de la régression linéaire et de 10 % dans celui des lignes d'abaissement.

Tableau 1. Prévision du tassement

Ilnicki (1977)		
$S = -60.5 + 19\,x_1 + 3.67\,x_2 + 0.15\,x_3 + 2.21\,x_4$		
S = Abaissement du niveau [cm/an]		r=0.932
x_1 = épaisseur avant l'abaissement [cm]		
x_2 = 1 – porosité [%]		
x_3 = abaissement de la nappe en été [cm]		
x_4 = durée après le drainage [an]		
$S = 14.3 \cdot x_4^{\,0.412}$	tourbe meuble	r=0.983
$S = 7.35 \cdot x_4^{\,0.476}$	tourbe assez meuble	r=0.979
$S = 5.14 \cdot x_4^{\,0.485}$	tourbe dense	r=0.978

Cette méthode constitue donc un outil de travail intéressant qui peut fournir une bonne appréciation de l'ampleur prévisible du tassement.

Toutefois, à l'instar des autres méthodes de calcul de l'abaissement, elle ne permet pas d'intégrer les modifications des caractéristiques hydrodynamiques du sol à l'analyse de l'évolution de l'efficacité du système d'assainissement, ce qui constitue une limitation importante.

5.2 Le comportement hydraulique

Il convient donc d'adjoindre, à la prévision du tassement, une méthode d'appréciation de l'évolution du comportement hydraulique du sol. La difficulté majeure d'une telle approche consiste à établir une relation quantitative entre le tassement du sol et les modifications de ses caractéristiques hydrodynamiques. La détermination de cette relation constitue une tâche des plus délicates dans la mesure où elle fait intervenir des grandeurs peu accessibles.

Dans un premier temps, une évaluation de la diminution de la conductivité hydraulique à saturation, par des mesures sur un sol similaire déjà drainé, peut fournir des indications intéressantes dans le cadre du dimensionnement des ouvrages.

Il semble toutefois souhaitable de disposer d'une analyse plus approfondie et détaillée des effets potentiels de ce facteur. La détermination des modifications des courbes caractéristiques d'une tourbe drainée au cours de son évolution pose des problèmes d'une grande complexité, dont la précision n'est pas le moindre. Dès lors, la simulation du comportement hydraulique d'un tel sol, à l'aide d'un modèle adéquat et dans une gamme de caractéristiques hydrodynamiques déterminée empiriquement (par analogie avec ce que l'on sait des sols minéraux), semble fournir une bonne méthode d'investigation. Ainsi, le recours à un modèle de simulation des écoulements tridimensionnels variablement saturés, basé sur l'équation de Richards :

$$\mathrm{div}\left(K\ \overrightarrow{\mathrm{grad}H} \right) = -\ c(h)\ \frac{\partial h}{\partial t}$$

permet une appréciation plus précise des effets potentiels du tassement sur les écoulements et l'efficacité du drainage. L'avantage d'une discrétisation par éléments finis est ici de permettre la reproduction d'une stratification du sol et de pouvoir ainsi tenir compte d'une évolution différenciée de ses caractéristiques hydrodynamiques selon la profondeur.

Les recherches que nous menons actuellement s'orientent dans cette direction.

6 LE CAS DE LA PLAINE DE L'ORBE

La plaine tourbeuse de l'Orbe, d'une superficie de 5000 ha, est un ancien fond lacustre creusé par les glaciers qui s'est peu à peu, par comblement, transformé en marais.

L'abaissement du niveau moyen du lac sur lequel elle débouche, par deux opérations au milieu des XIXème et XXème siècle, a déclenché un tassement des sols. Des corrections de cours d'eau dans les années 60, qui ont supprimé les inondations et les apports de matériaux qu'elles représentaient, et le drainage souterrain, dès 1865 mais surtout pendant les dernières décennies, ont amplifié le phénomène, dont l'amplitude moyenne atteint 1.2 m.

Actuellement le niveau moyen des sols de la plaine est à peine supérieur à celui du lac, les zones les plus basses étant fréquemment inondées. En outre, un grand nombre de réseaux d'assainissemnt ne fonctionnent plus de manière satisfaisante, s'ils n'ont pas étés détruits.

Face à cette situation, il est possible d'envisager le retour des zones les plus menacées (1500-2000 ha) à leur état antérieur de marécage ou encore le renouvellement des installations de drainage et la limitation de l'agriculture à la prairie permanente. Ces deux solutions permettraient d'assurer une meilleure conservation des sols.

Toutefois, l'ensemble de la plaine s'intègre à un contexte agroéconomique régional dans lequel ces surfaces sont indispensables à l'équilibre des exploitations agricoles. Celles-ci, qui manquent de terres cultivables, se sont équipées, au prix de lourds investissements, pour la culture intensive. Le retour à la prairie constituerait donc une solution défavorable.

La poursuite de l'exploitation intensive soulève cependant deux difficultés majeures. D'une part, les déclivités n'étant plus suffisantes pour assurer un drainage gravitaire, le renouvellement des réseaux implique le recours au pompage, ce qui élève considérablement les coûts. D'autre part, à plus ou moins long terme selon l'épaisseur du sol (de 30 cm à 10 m) se posera le problème inéluctable du substrat minéral qui, étant composé essentiellement dans ce cas de craies lacustres, ne présente pas à l'heure actuelle de bonnes perspectives de fertilité.

Dans cette optique, une gestion appropriée du niveau de la nappe devrait permettre de minimiser le phénomène de tassement et de retarder ainsi dans la mesure du possible la dégradation de ces sols.

Sur le plan du dimensionnement, les collecteurs devraient être situés à une profondeur telle que le renouvellement ultérieur du système puisse se limiter au remplacement des drains. En outre, une attention particulière devrait être portée à la lutte contre le colmatage des drains, favorisé par la nature minérale du substrat. La méthode la plus favorable semblant être l'utilisation de filtres spéciaux, riches en tanins, enrobant les drains.

Il ressort clairement de cette brève présentation que les intérêts de la conservation des sols et les intérêts politico-économiques ne convergent pas toujours.

BIBLIOGRAPHIE SUCCINTE

Duchaufour,Ph. 1984. Pédologie. Paris:Masson

Guérif,J. 1988. Mécanismes du compactage. Comptes rendus de l'académie d'agriculture de France. 74.1:9-28

Ilnicki,P. 1977. Sackung in wiederholt entwässerten Hochmooren des Nordwestdeutsches Flachland.3.Bericht: Vorausberechnung der Moorsackung. Z.f. Kulturtech.u.Flurberein. 18:153-165.

Jaton,J.F. 1977. Etude de quelques propriétés des sols tourbeux et leur influence sur le drainage. Publication IGR/EPFL

Musy,A. 1972. Les bases scientifiques de l'assainissement des sols, les méthodes d'assainissement et leur efficacité. Paris & Lausanne: CIGR-Ire section

Musy,A.,M.Soutter & P.Perrochet 1988. Evaluation of the hydrological response of a drainage network. Nato Advanced Research Workshop, Arles, France.

Presler,J. & H.Sticher 1987. Die Veränderung des Mikroreliefs und des Bodeninventars als Folge der durch Moorsackung und Torfverzehr verursachten Höhenverluste in einem meliorierten Niedermoorgebiet. Mitteilgn. Dtsch. bodenkndl. Ges.

Presler,J. 1987. Les sols organiques, situation actuelle et évolution probable: cas de la basse plaine de l'Orbe. Publication IGR/EPFL.

Regamey,P. 1984. La mise en valeur de la plaine de l'Orbe. Publication IGR/EPFL.

Land and Water Use, Dodd & Grace (eds), © 1989 Balkema, Rotterdam. ISBN 90 6191 980 0

The effect of incorporation of calcareous mineral subsoil on the growth of Salix aquatica gigantea

F.S.MacNaeidhe
Teagasc, Johnstown Castle Research Centre, Wexford, Ireland

ABSTRACT: The effect of depth of incorporation of calcareous mineral subsoil on the growth of Salix aquatica gigantea and on the pH and mineral content of peat was investigated in sod-cutover raised bog. The investigations showed that incorporation of calcareous mineral subsoil at a depth greater than 110 cm was necessary for maximum growth. The main benefit of application was an increase in the pH of the peat. Subsoil additions of 3% by volume (20% by weight) increased the pH to greater than 6.0 thus creating a suitable environment for root growth. The incorporation of excessive quantities of this material reduced plant growth. Growth was influenced more by depth of incorporation than by peat depth. The indications from those investigations are that the minimum depth of incorporation required for maximum yields of Salix aquatica gigantea in acid peat is one meter.

RESUME : L'effet de la profondeur d'incorporation de sous-sol minéral calcaire sur la croisance de Salix aquatica gigantea et sur le pH et les composants minéraux de la tourbe a été examiné sur une tourbière haute exploitée. Les examens ont montré que l'incorporation profonde de sous-sol minéral calcaire était nécessaire pour une croissance maximum dans la tourbe acide. Le principal avantage de l'application se traduit par une augmentation de pH de la tourbe. Un apport de sous-sol de 3% en volume et de 20% en poids porte le pH a plus de 6.0 créant ainsi un environnement qui convient au développement au niveau des racines. L'incorporation de quantités excessives de cette matière réduit la croissance de la plante. La profondeur d'incorporation a plus d'impact sur la croissance que la profondeur de la tourbe. Ces examens indiquent que la profondeur minimum d'incorporation nécessaire a un rendement maximum de Salix aquatica gigantea dans la tourbe acide est d'un mètre.

ZUSAMMENFASSUNG : Der EinfluB der Einarbeitungstiefe von kalkhaltigem Mineralunterboden auf das Wachstum von Salix aquatica gigantea auf den pH-und Mineralgehalt von Torf wurde auf einer abgeatochenen Hochmoorflache untersucht. Die Untersuchungen zeigten, daB fur optimales Wachstum in saurem Torf eine tiefe Einarbeitung von Kalkhaltigem Mineralunterboden erforderlich ist. Der Hauptgewinn der Anwendung bestand im Anstieg des pH-Werts im Torf. Eine Hinzufugung von 3% nach Volumen (20% nach Gewicht) erhohte den pH-Wert auf mehr als 6.0 und bewirkte damit eine gunstige Umgebung fur das Wurzelwachstum. Das Hinzufugen von zu hohen Mengen dieses Materials reduzierte das Pflanzenwachstum. Das Wachstum wurde mehr vonder Einarbeitungstiefe als von der Torftiefe beeinfluBt. Die durchgecuhrten Untersuchungen deuten darauf hin, daB die Minimum-Einarbeitungstiefe, die fur optimale Ertrage von Salix aquatica gigantea in saurem Torf erforderlich ist, bei einem Meter liegt.

1 INTRODUCTION

The area of peatland in Ireland is 1.35 million ha. A small area of Irish peatland (37,424 ha) is harvested commercially for fuel and it is on this peatland that the investigations were carried out. It is predicted that harvesting will be completed early in the next century and alternative uses for this cutover peatland are being investigated. The production of short rotation forestry for fuel is of particular interest because the productive life of the peatlands as a producer of energy could be considerably prolonged. Preliminary trials on short rotation forestry were planted on cutover raised peat in the midlands in 1978. Salix aquatica gigantea was the most productive species in preliminary trials in 1978 (Neenan and Lyons 1980). Commercial plantations of 200 ha were planted on sod-cutover raised peat in 1980 and in 1981 for the purpose of producing fuel for electricity generation in power plants. Growth over a three year period was poor to moderately good but the target yield of 12 tonnes/ha was not achieved. The purpose of this investigation was to determine the soil-related factors responsible for the poor growth and to examine the possibility of improving the overall growth of future plantations on peatland.

2 SITE DESCRIPTION AND MEASUREMENTS

The experimental site was at Clonsast, Co. Offaly.

The peat type was typical of Clonsast complex as described by Finch et al. (1983). The sequence of operations which give rise to this peat complex is described in detail by Callinan (1972). A layer of redistributed acid Sphagnum peat (pH 3.9 - 4.7) of variable depth (70-150 cm) covered the surface with fen peat beneath. Before cultivation a layer of alkaline mineral subsoil of variable depth and width was overlying the Sphagnum peat at the edges of the main drains. The main drains were spaced at 250 m centres. Investigations were carried out in three transects taken across different plantations at this site. At Transect 1 the land was ploughed to a depth of 90 cm. During this operation there was considerable disturbance of the profile by earth-moving and cultivating machinery and a relatively uniform mixture of peat and mineral material was formed to a depth of 30-50 cm. At Transect 2 the mineral overburden was incorporated to a depth of 50-90 cm near the edge of the main drain. This mixture of peat and mineral material was relatively coarse. Ground mineral limestone was incorporated to a depth of 20-30 cm in the remaining 120 m wide portion. At Transect 3 some mixing of the peat and mineral material had occurred but the peat and mineral phases were clearly visible within these mixtures in most cases. Ground mineral limestone was disked to a depth of 20 cm in the 200 m wide portion of the area which was not covered by mineral material. Salix aquatica gigantea was planted at all three sites in spring 1980.

All measurements were carried out between spring 1983 and spring 1984 at 10 m intervals along each transect and at five random points about each plant using a gouge auger of 2.5 cm diameter and dutch augers of from 2.5 cm to 10.0 cm diameter. The soil pH was measured at 10 cm intervals within the soil profile to a depth of 50 cm by digital pH meter using a 2:1 water/soil ratio. The peat depth and composition were assessed using a gouge auger. The moisture content was determined by drying in an oven for 16 hours at 105°C. The mineral (ash) content was determined by loss on ignition at 450°C for 16 hours.

Rooting depth was determined by visual inspection of 10 cm x 10 cm dissected cores taken at 10 cm intervals in the profile using a dutch auger. The bulk density of peat and mineral soil samples was calculated using 2.0 cm x 5.0 cm metal cores. The samples were oven dried at 105°C for 16 hours. Assessments of total neutralizing value (TNV) of mineral subsoils were made by extraction with N HCl + phenolphthalein and titration with 0.5 N NaOH (Byrne, 1979).

A correlation materix was calculated for plant growth, depth of mineral subsoil incorporation, peat depth, rooting depth, mineral content of the peat and volume/volume ratios of peat and mineral material within the samples. The large correlation coefficients were selected for further analysis. Regression lines were calculated for those with a correlation coefficient which was significant at the one per cent level.

3 RESULTS

The mean bulk density of mineral subsoils from Transects 1, 2 and 3 were 1.30, 1.92

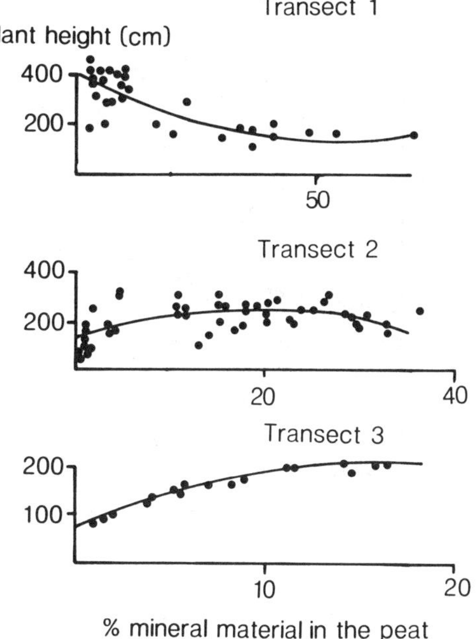

Figure 1. The effect of mineral content of peat on the height of Salix aquatica gigantea.

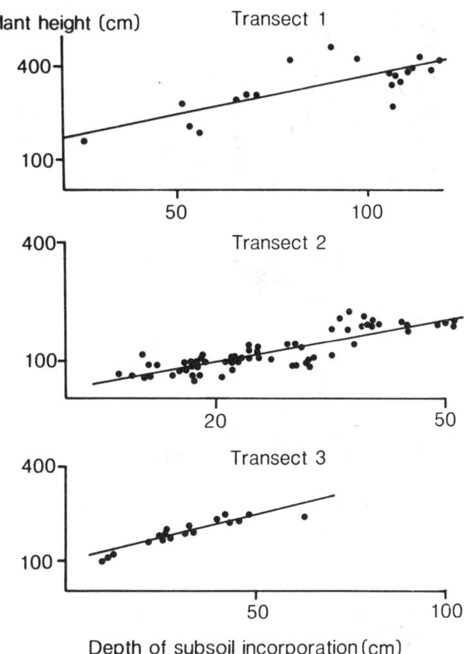

Figure 2. The effect of depth of subsoil incorporation on the height of Salix aquatica gigantea.

and 1.47 g/cc respectively. The mean bulk density of peat from sod-cutover raised peatland which was uncontaminated by additions of ground mineral limestone, fertiliser or other mineral material was 0.135 g/cc. The TNV of mineral subsoils with undeveloped profiles in Transects 1, 2 and 3 were 44.6, 46.8 and 45.6 respectively compared with a TNV of 98.5% for the commercial grade ground mineral limestone which was used in the experiment.

Bulk density measurements of the peat and of mineral subsoils were used to convert weight/weight ratios of both phases in the mixtures into volume/volume ratios. Regressions of plant growth on the volume ratios of peat and mineral soil in the mixtures are given in Figure 1. Regression analysis of the data showed that plant height decreased as the mineral content increased in Transect 1. The Transects 2 and 3 the plant height increased until the mineral content exceeded 25% by volume and decreased thereafter.
Plant height increased with an increasing depth of incorporation of calcareous mineral subsoil and ground mineral limestone in all three transects (Figure 2).

Plant height increased as the depth of peat with a pH of greater than 6.0 increased (Figure 3) but was reduced in areas where acid peat occurred near the surface.

At Transect 1 plant height increased with an increasing peat depth of up to 100 cm and decreased thereafter (Figure 4). At Transect 2 there was an increase in plant growth with an increasing peat depth of up to 150 cm but in Transect 3 there was no relationship between plant height and peat depth.

An increase in plant height was accompanied by an increase in rooting depth in all cases (Figure 5). An increased depth of incorporation of calcareous mineral subsoil also gave an increase in rooting depth (Figure 6). Rooting depth increased with peat depth at Transects 1 and 2. There was no

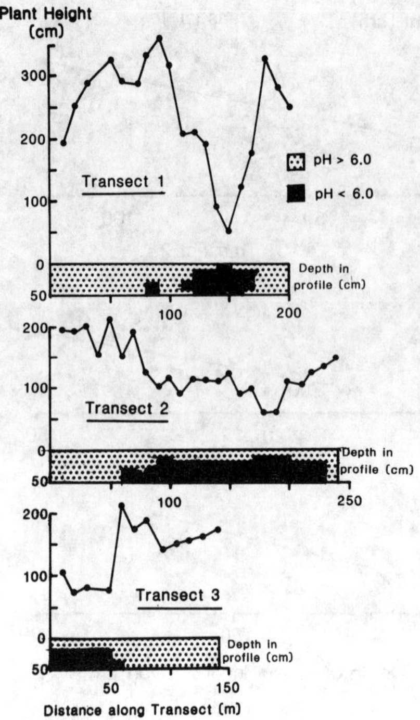

Figure 3. The effect of soil pH on the height of Salix aquatica gigantea.

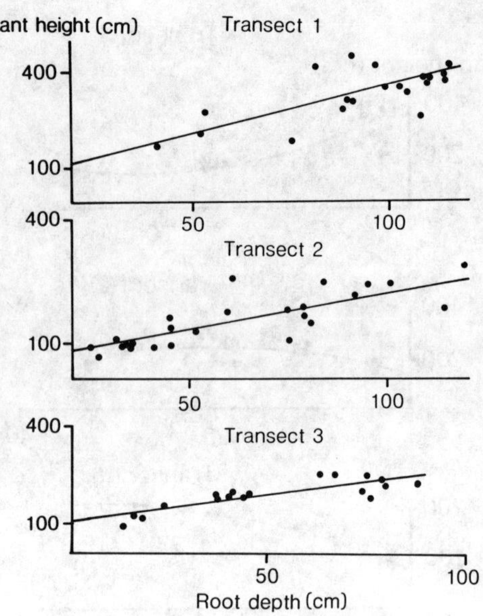

Figure 5. The effect of rooting depth on the height of Salix aquatica gigantea.

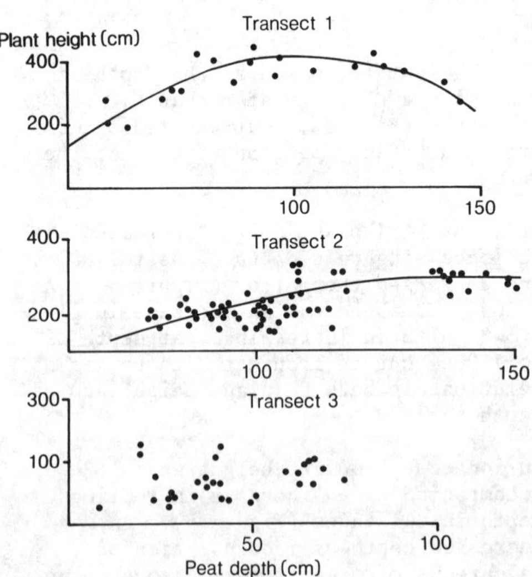

Figure 4. The effect of peat depth on the height of Salix aquatica gigantea.

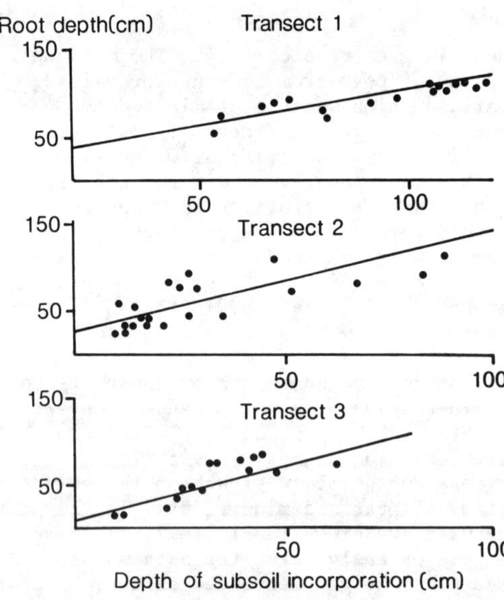

Figure 6. The effect of depth of subsoil incorporation on the root depth of Salix aquatica gigantea.

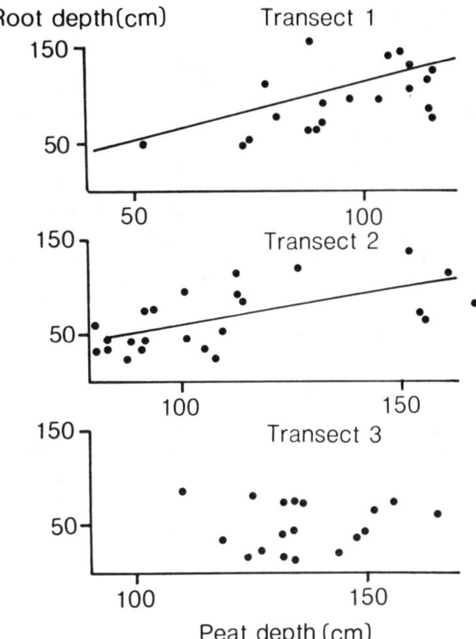

Figure 7. The effect of peat depth on the root depth of Salix aquatica gigantea.

correlation between peat depth and rooting depth at Transect 3 (Figure 7).

4 DISCUSSION

On sod-cutover raised bog calcarious mineral subsoil is excavated from deep drains spaced at 250 m centres, deposited on the drain edges and spread by bulldozer in a band 50-130 m wide and covering from 20 to 50% of the area between the drains. The depth of subsoil is greatest at the drain edge (20-45 cm) and decreases gradually from the drain edge outwards. The area between the drains is used as a spread ground for harvested sod peat (Callinan 1972). This system is ideal for peat harvesting but further reclamation is necessary for the successful production of short rotation forestry. The subsoil deposits have a poor structure are poorly drained and are rich in calcium carbonate. The peat underlying these deposits is extremely acid. Mixing of both phases in the correct proportions improves the structure and drainage (Galvin 1979) and adjusts the soil pH to a value which gives the maximum plant growth. These investigations have shown that it is necessary to spread the mineral material uniformly over the entire area to obtain a suitable blend of peat and mineral material. Skoropanov (1961) made a similar recommendation. In cases where the peat and mineral material are so well mixed that macroscopic examination does not show the presence of any mineral aggregates a mineral content in excess of 3% by volume (20% by weight) will decrease the growth of the Salix and a mixture of 3% or less should be made. A 3% mixture is equivalent to the incorporation of a 3 cm layer of mineral subsoil to a depth of 100 cm. When the mixture is less thorough and mineral aggregates are clearly visible in the peat a mineral content of up to 10% by volume gives an increase in plant growth. With a coarse mixture the plant is tolerant to a mineral content of 25% by volume but growth decrease as the percentage is increased above this value. This is equivalent to the incorporation of a 25 cm layer of mineral material to a depth of 100 cm. Skoropanov (1961) recommends the incorporation of a subsoil layer with a maximum depth of 10 cm, a mixture equivalent to the optimum intermediate blends in this experiment.
In trials in the Netherlands, Wind and Pot (1976) reported that coarse mixtures of peat and subsoil obtained with ploughs and rippers gave higher yields of potatoes, sugar beet, spring wheat and oats than fine mixtures obtained with rotary cultivators. These findings indicate that medium to coarse mixtures are more beneficial than fine mixtures. In the case of perennial crops such as Salix a minimum disturbance of the profile takes place after planting and coarse mixtures are easy to sustain. A volume of 12-30 m^3 of subsoil per linear meter is deposited on the drain edge over time. This material when spread in a uniform layer is equivalent to a 5-13 cm depth of subsoil or a maximum mineral content of 13% when incorporated to 100 cm. This amount of material can be safely incorporated as a medium to coarse mixture with no adverse effect on growth. Subsequent cultivations should be reduced to a minimum to avoid excessive mixing. A single pass of the disk is sufficient to provide a suitable mixture for good plant growth.
The main benefit obtained from the incorporation of subsoil is an increase in the pH of the peat. The pH of sod-cutover raised peat before subsoil or lime incorporation is 3.9-4.7. A pH of 6.0 is necessary for optimum plant growth (Lucas 1983). It is necessary to increase the pH to this value throughout the full depth of

the profile. Increasing the pH to values greater than 6.0 in the surface 10-20 cm is only of marginal benefit (Lilly, 1981). Growth continues to increase as the depth of incorporation increases up to 120 cm. The effects of incorporation to depths greater than 120 cm on the growth of Salix have not been considered in these experiments and deserve further investigation. The amount of calcareous subsoil required to obtain an adequate increase in pH is dependent on the total neutralizing value (TNV) of the subsoil. These investigations show that subsoils from raised bog floor concavities have a TNV of approximately 45% and are suitable for use as a lime substitute. The incorporation of a 3 cm layer of subsoil with a TNV of 45% to a depth of 100 cm will increase the pH of peat soil from 4.4 to 6.3. The TNV of the subsoil should be assessed before it is used as a liming material. In cases where the TNV is low large quantities of subsoil are required to obtained pH and this may reduce the biomass yield.

Deep incorporation increases the volume of soil supporting root growth underneath the crop. As the volume of soil supporting root growth increases plant height also increases. The depth of root penetration exceeds the depth of incorporation of subsoil or lime in all cases showing that roots will invade an unfavorable environment in search of plant nutrients and moisture even when the volume of soil which is suitable for root growth is relatively large. Poor soil physical conditions inhibits root growth. In places where the underlying peat is compacted and dry, contains poorly structured impermeable Eriophorum peat, or has a high percentage of mineral material the roots will penetrate a distance of only 3-10 cm below the zone of incorporation. Where layers of extremely acid (pH 3.6-3.9) undecomposed Sphagnum peat are found few roots of Salix are present. When such layers occur near the peat surface plant growth is the least vigorous. These acid layers are usually made up of poorly decomposed Sphagnum peat of low density, loose structure and with a large pore size. This peat type yields moisture to the surrounding soil but cannot regain it due to the large pore size (Lilly, 1981). The peat becomes dry and compact after a few years of cropping and forms a barrier near the soil surface which prevents the movement of moisture and root penetration (Burke, 1978; Egglesman, 1972). This dry barrier must be broken by cultivation.

5 CONCLUSION

The above discussion outlines the peat type, the peat depth and the sequence of reclamation operations which are necessary to obtain the maximum growth of Salix aquatica gigantea on sod-cutover peatland. A peat depth of greater than 100 cm is necessary for maximum growth but peat depth is of no advantage unless the method of reclamation used is the correct one. A soil pH of greater than 6.0 is necessary for the production of Salix. When the pH is less than 6.0 mineral subsoil with a minimum TNV of 45% or ground mineral limestone must be incorporated. In order to increase the pH the mineral subsoil from the main drains is spread uniformity over the peat surface to a maximum depth of 13 cm and ploughed in to a depth of 110 cm. The ploughed surface is then disked to provide a uniform but relatively coarse mixture. There is evidence that deeper incorporation than 110 cm will provide added growth and this should be investigated.

REFERENCES

Burke, W. (1978). Longterm effects of drainage and land use on some physical properties of blanket peat. Ir. J. agric. Res, 17: 315-322.

Byrne, E. (1979). 'Chemical analysis of agricultural materials'. An Foras Taluntais, Dublin, 197 pp.

Callinan, P.F. (1972). Sod peat production in Ireland. Proc. 4th Int. Peat Congr. Helsinki, 2 : 27-38.

Egglesman, R. (1972). Physical effects of drainage in peat soils of the temperate zone and their forecasting. Int. Symp. on the Hydrology of Marsh-Ridden Areas, Minsk, 1972.

Galvin, L.F. (1979). Land drainage and reclamation. The Institute of Engineers of Ireland Transactions, 103 : 55-63.

Lilly, J.P. (1981). The black soils of North Carolina, their characteristics and management for agriculture. Tech. Bull. No. 270. North Carolina Agric. Res. Service, North Carolina Sta. Univ. Raleigh 70 pp.

Lucas, R.E. (1983). Organic soils (Histosols), Formation, Distribution Physical and Chemical properties and Management for Crop Production. Res. Rep. 435 Farm Science 77 pp.

Neenan, M. and Lyons, G. (1980). (Eds), Production of Energy from Short Rotation Forestry. An Foras Taluntais, Dublin 128 pp.

Skoropanov, S.G. (1961). Osvoenie i ispol 'zovanie torfyano-bolotnykh pochv. Izdatel 'stov Akademii Sel 'skokhozyaistvennykh Nauk BSSR, Minsk, 223pp (Reclamation and cultivar of peat-bog soils). Translated from Russian by Israel Program for Scientific Translations, Jerusalem, 1968). Ch. 4. pp 79-104.

Wind, G.P and Pot, R.A. (1976). Bodenverbesserung in den Hollandischen Veenkolonien. Verspreide overdrukken 191, Z.F. Kulturtechni und Flurbereinigung, 17, 193-205. Instituut voor cultuurtechniek en Waterhuishouding, Wageningen, Netherlands.

Land and Water Use, Dodd & Grace (eds), © 1989 Balkema, Rotterdam. ISBN 90 6191 980 0

To test five methods of incorporation of ground limestone at three rates into strongly acidic Sphagnum derived peat materials

R.F.Hammond
Teagasc, Peatland Experimental Station, Lullymore, Rathangan, Co. Kildare, Ireland

D.Harrington
Teagasc, Statistics Department, Dublin, Ireland

ABSTRACT: Lime incorporation is essential to attain maximum dry matter production from grassland reclaimed from strongly acid Sphagnum peat materials. Five incorporation techniques were tested using a rotovator, plough and specialist rotary spading machine. The rotary spading treatment showed significant yield increases under adverse drought conditions. However, for average climatic conditions in the Irish midlands rotovation creates a topsoil capable of maintaining high levels of dry matter production.

ABRÉGÉ: L'incorporation de la chaux est essential pour obtenir le maximum matière sèche d'un pré récupérée de tourbe Sphagnum fortemente acide. Cette incorporation a été faite comprenant cinq techniques différentes utilisant le motoculteur, la charue et un motoculteur spécial des Pays Bas. Le traitement utilisant le motoculteur spécial a donné un rendement plus élevé dans des conditions de sécheresse sévère. Toutefois avec des conditions climatiques moyennes trouvées au centre Irelande la technique du motoculteur crée une terre arable capable de maintenir un haut niveau de production de matière sèche.

ABSCHNITT: Die Kalkmischung ist wichtig um maximale Trockenmasseproduktion aus kultiviertem Gruenland - welches aus ebenem stark saurem Hochmoor Sphagnumtorfboden gewonnen werden - zu erreichen. Fuenf Vermischungstechniken wurden angewendet: Bodenfräse, Pflug und Kreiselspatenegge. Bei der Anwendung der Kreiselspatenegge wurde ein wesentlich hoeherer Ertrag bei widrigen Trockenheitsverhaeltnissen erzeilt. Allerdings schafft die Anwendung der Bodenfrase bei durchschnittlich klimatisch Bedingungen im mittleren von Irland eine Ackerkrumenkapazitaet, aus der sich ein hoher Grad an Trockenmassenproduktion ergibt.

1. INTRODUCTION

Experience of land preparation at the Peatland Experimental Station, County Kildare has shown that lime additions to acid peat materials only affects soil pH to the depth of incorporation. Research on grass production from peat soils developed from strongly acid peat materials of Sphagnum origin has indicated that yields were influenced by the amount of lime added and depth of incorporation to create an adequate thickness of top-soil (Hammond et al., in preparation). These effects have also been observed by Dampney (1985). The thickness of the rooting zone is an essential element in the supply of water and adequate nutrients to a growing crop (Pearson, 1971).

The experiment reported here was designed to test the efficacy of five methods of incorporating three rates of ground limestone to four depths into strongly acid peat materials. The methods employed utilised a range of easily available farm machinery and a specialist "Rotary Spading Machine" manufactured by Imants/Reusel NV, Holland.

2. EXPERIMENTAL METHODOLOGY

2.1 Site preparation

The experimental site situated in County Kildare, Grid Ref. 28052341 was typical of the range of ground conditions encountered in the drier aspects of the Turbary Complex (Conry et al., 1970). The well-drained site was covered by a dense growth of 15-20 year old heather which camouflaged a very uneven micro-topography resulting from turf cutting by hand.

The experimental area was prepared by burning the heather, levelling with a Caperpillar D6 bulldozer, heavy discing to break up the more resistant top sod material and to facilitate refined levelling using a tractor pulled Eversman automatic leveller. The experiment, of five main treatments with three rates of lime applied as sub-plot treatments, was laid out to a split plot design replicated five times with each block containing fifteen plots. Plot size for the main treatments was 2 x 30 m with a sub-plot size of 2 x 10 m.

Each sub-plot was sampled to 150 mm depth and the initial pH measured. The pH values varied between 3.20-4.65 with a mean value of 3.44±0.08. Lime rates to give pH values of 4.5, 5.0 and 5.5 were calculated using the rule that "for every 2.5 t ha^{-1} of ground limestone incorporated to 150 mm, the pH(water) is raised by 0.2 of a pH unit". The amounts of lime applied to raise pH levels to the required value varied between 12 t ha^{-1} incorporated to 150 mm at the low pH value to 76 t ha^{-1} incorproated to 450 mm at the high pH value. Copper in the form of copper sulphate applied at the rate of 45 kg ha^{-1} is an essential trace element for organic soils. It also requires incorporation for efficient utilisation. The lime and copper sulphate were spread by hand individually on each plot and incorporated to depth by the five main treatment methods listed as follows:

1. Rotovating ground limestone to 150 mm.
2. Rotary spading ground limestone to 450 mm.
3. Rotary spading ground limestone to 300 mm.
4. Half the ground limestone applied to surface and ploughed in to 200 mm, the remainder applied and rotovating in to 200 mm.
5. Rotary spading to 450 mm, lime applied and rotovating to 150 mm.

The plots were then rolled to firm the surface and a seeds mixture of mid- to late season Perennial Ryegrass cv. Talbot and New Zealand White Clover cv. Huia sown at rates of 11 kg and 1 kg ha^{-1} respectively. A compound fertiliser ratio 10:10:20 was applied to the surface at the rate of 494 kg ha^{-1} and both seeds and fertiliser were rolled in. Checks on the efficiency and accuracy of the different techniques for whole plot treatments, especially the rotary spading machine showed them to be incorporating lime to the required depth. The depth for each whole plot treatment as of 1983 is taken as the reference level to which subsequent measurements for subsidence are compared.

2.2 Soil acidity and top soil thickness

Samples for pH measurement were taken from all the plots in July 1983, February 1984 and July 1986. In each plot samples were collected at three depths 0-150, 150-300 and 300-450 mm. After air-drying the samples were ground to pass through a 2 mm screen prior to reading pH values (Byrne, 1981) in suspensions of 3 g peat in 25 ml of water.

The decrease in top soil depth due to subsidence through drainage and the effects of biological oxidation were measured annually over the duration of the experiment. Each plot was divided into ten 1 x 2 m rectangles and using a Dutch gouge auger ten depth values were recorded per plot.

3. RESULTS

3.1 Dry matter yields

The differences stated in yield data in the results that follow are significant at the 5% level or higher. Dry matter yields kg ha^{-1} for each cut and seasonal means (1983-1986) for whole plot treatments are presented in Table 1.

Annual total dry matter yields from the whole plot treatments show that the rotary spading treatments significantly outyielded all other treatments for the years 1983-1984. In 1985 only the rotary spading treatment to 450 mm was significantly higher in dry matter production. In 1986 no significant differences were recorded between whole plot treatments of the experiment. The yields of dry matter recorded from

Table 1. Dry matter yields kg ha-1 for each cut and the seasonal means for whole plot treatments (1983-1986)

Whole plot	Roto 150 mm	Spading 450 mm	Spading 300 mm	Ploughroto 200 mm	Spading 450 mm Roto 150 mm	SED
Cut 1	5129	5956	5437	4784	4842	378.8
1983 Cut 2	4453	4624	5070	4625	3965	295.0
Cut 3	2559	3436	3137	2773	2524	207.8
Season mean	12141	14061	13644	12182	11331	208.6
Cut 1	3204	3507	3249	3761	3301	280.4
1984 Cut 2	2338	4035	3426	2706	2529	350.8
Cut 3	1303	2448	2324	1632	1127	134.6
Season mean	6845	9990	8999	8099	6957	181.6
Cut 1	3311	3582	3236	2983	2684	266.9
1985 Cut 2	1505	2097	1594	1655	1394	187.0
Cut 3	2730	2590	2496	2781	2920	196.0
Season mean	7546	8269	7326	7419	6998	112.4
Cut 1	5278	5083	5289	5097	5021	262.3
1986 Cut 2	4059	4070	4040	4039	3824	226.9
Cut 3	1877	1294	1386	1386	1575	282.0
Season mean	11214	10447	10715	10552	10420	108.6

rotary spading plus rotovation for 1985 were significantly lower than all other treatments. In 1984 this treatment and rotovation were significantly lower than the other three treatments.

The dry matter yields for the three cuts per season from the whole plot treatments 1983-1986 show treatment differences between cuts. In 1983, Cut 1 from the rotary spading to 450 mm treatment and Cut 3 from both rotary spading treatments gave significantly higher dry matter production than all other treatments. Yields from the rotary spading treatments in 1984 for Cuts 2 and 3 were again signficantly higher than the remaining treatments. This was also the case for rotary spading to 450 mm in the the second cut in 1985. In 1984 for Cut 3 the shallower incorporation, rotovation and rotary spading plus rotovation were significantly lower than the ploughing rotovation treatment. In 1986 there were no significant differences between treatments for the three cuts. The differences in dry matter production between the cuts and treatment effects

Table 2. Dry matter yields (kg ha-1) per lime sub-plot treatments from 1983-1986

	Sub-plot	Roto 150 mm	Spading 450 mm	Spading 300 mm	Ploughroto 200 mm	Spading 450 mm Roto 150 mm	SED
	H	13035	13911	13584	12315	11790	851.4*
1983	M	12186	13794	14493	11964	11523	707.1^{+}
	L	11595	13722	12855	11259	10587	
	H	6993	10224	8730	7506	6609	739.2*
1984	M	7506	10001	9327	7800	7086	611.7^{+}
	L	6138	9777	8976	7986	6882	
	H	7443	8658	7707	7419	6777	519.0*
1985	M	7386	7704	7329	7260	6930	483.0^{+}
	L	7809	8445	7329	7578	7284	
	H	11151	10605	10794	10479	10761	641.1*
1986	M	10812	10665	10725	10170	10242	676.2^{+}
	L	10479	10071	10671	10914	10659	

*SED when comparing means with different whole plot/sub-plot
*SED when comparing means with the same level of whole plot

in combination with soil and weather factors will be discussed in more detail in the conclusions.

The effect of sub-plot lime treatments on dry matter yields are given in Table 2. Throughout the years 1983-1986 no significant differences were recorded in dry matter production between the sub-plot lime treatments.

3.2 Lime incorporation - pH values and top-soil thickness

The mean initial pH values for each block, mean lime rate applied and attained pH are given in Table 3.

The pH values (Table 4) measured at the three depths for the lime treatments illustrate how effectively the lime was incorporated into these strongly acid peat materials.

The pH values of the three different depths for the whole plot treatments reflect the method of incorporation employed. Throughout the course of the experiment irrespective of method of incorporation the pH was above 5.27 in the surface layer. At depths greater than 150 mm the treatment effect becomes noticeable. The spaded treatments at 150-300 mm were still greater than 5.0 with the rotovation and combinations of rotovating, spading and ploughing dropping below 5.0 and in the case of rotovation alone in 1983 and 1984 values below 4.0 were recorded. With all treatments except spading to 450 mm the pH was below 5.0 and for treatments involving rotovation the pH was less than 4.0. In all cases the lowest pH value recorded from field sampling was above the initial value determined before lime was applied. This will be treated in greater detail in the discussion.

3.3 Thickness of cultivated horizon (Oap) and rooting depths

The thickness of the cultivated horizon and rooting depth influence herbage dry matter production. measurements were made of these two parameters as well as the degree of overall shrinkage relative

Table 3: Mean initial pH, calculated lime rate (t ha^{-1}) applied and attained pH for surface 150 mm

Blocks	Initial pH	Reqd. pH 4.5 Lime rate	Field pH	Initial pH	Reqd. pH 5.0 Lime rate	Field pH	Initial pH	Reqd. pH 5.5 Lime rate	Field pH
1	3.39	13.8	5.55	3.49	18.5	6.29	3.51	24.7	6.13
2	3.46	13.6	5.42	3.50	18.5	6.03	3.46	25.2	6.03
3	3.46	12.9	5.48	3.55	17.9	5.84	3.48	25.0	6.49
4	3.37	14.1	4.93	3.35	21.1	6.21	3.33	26.8	6.24
5	3.42	13.3	5.27	3.39	20.9	5.50	3.44	26.1	6.55
Mean	3.42	13.5	5.33	3.46	19.4	5.98	3.44	25.6	6.29

Table 4. pH data of the whole plot treatments with depth (mm)

Depth (mm)	Year	Roto 150 mm	Spaded 450 mm	Spaded 300 mm	Ploughroto 200 mm	Spaderoto 450/150 mm	Treatment SED
	1983	5.67	6.17	5.76	5.97	5.40	0.12
0-150	1984	5.42	5.88	5.38	5.38	5.27	0.12
	1986	5.83	5.96	5.84	5.87	5.64	0.07
	1983	3.98	6.38	5.85	5.40	4.17	0.16
150-300	1984	3.89	6.12	5.46	4.12	4.09	0.11
	1986	4.36	6.31	5.99	4.62	4.25	0.15
	1983	3.78	5.56	4.38	3.86	3.85	0.14
300-450	1984	3.78	5.00	4.30	3.84	3.74	0.14
	1986	3.81	5.20	4.65	3.94	3.76	0.14

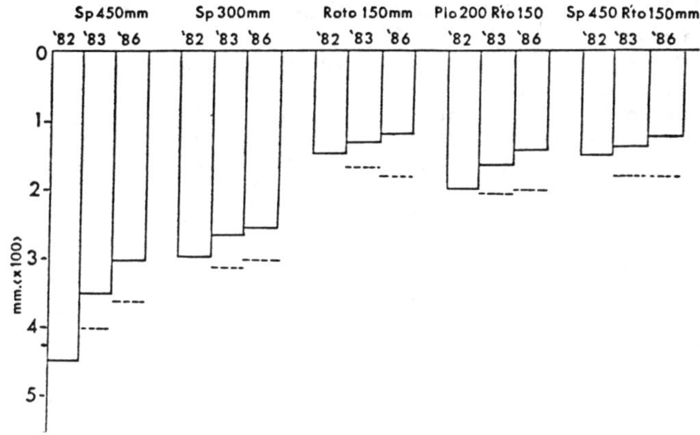

Fig. 1: Measured thicknesses of the cultivated horizon (bar lines) and rooting depths (pecked lines) for the duration of the experiment

to a fixed ordnance datum height. The data collected for thickness of Oap and rooting depth are presented in histogram form in Fig. 1 illustrating the changes that occurred over the duration of the experiment.

In this time shrinkage of the Oap horizon as a percentage of the initial depth varies between 35% and 15% for the deeper spaded treatments and from 30% to 19% for the rotovation and ploughing treatments. Measurement of surface levels across several transects within the plot areas indicated the overall effect of shrinkage. One such transect is presented in Fig. 2.

Shrinkage data plotted against time is asymptotic in form. Shrinkage/subsidence is influenced by several components. The initial large increment was a combination of settlement after land development and the effect of the exceptionally dry summer of 1984. Subsequently there has been a very gradual incremental rate of subsidence. The water table in the area is uncontrolled and shrinkage rates will vary according to the seasonal weather patterns. It is known that controlled water tables decrease the rate of shrinkage of peat soils (Snyder, 1978).

3.4 Soil Moisture

The depth of top soil has a strong influence over available soil water to the grass crop growing on land developed from strongly acid peat residues. Even in prolonged dry periods crop roots will

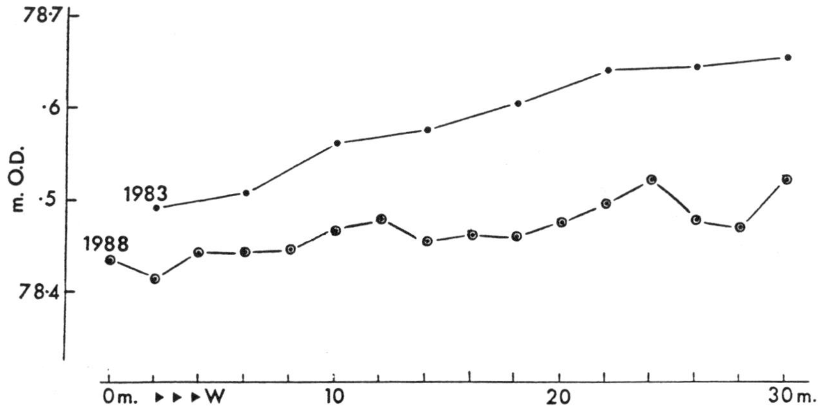

Fig. 2: Surface levels measured in 1983 and 1988 for Section F. Plot G4 illustrate the subsidence that has occurred within the experimental area

not penetrate into underlying acidic materials although they contain abundant supplies of moisture giving rise to "induced drought condition". Soil moisture deficit is an important factor in influencing grass yields (Brereton and Keane, 1982). It is estimated that in five years out of ten in the south east of Ireland there can be a 15% reduction in grass yield with reductions being as high as 40% in two out of ten years (Anon., 1986). Water uptake by the crop may not be restricted where the accumulated soil moisture deficit is less than 30 mm (Aysling, 1965) for up to this point the actual evapotranspiration is equal to the potential evapotranspiration. At accumulated soil moisture deficit levels greater than this a restriction on the supply of water to the crop can be expected. Perle (1954) gives root constant values for different crops in mm of water below which the crop is severely restricted. For grass the values range from 75-125 mm of water.

At field capacity (5 kPa), water occupies approximately 70% (70 mm/100 mm) (Galvin, 1976) of the soil volume in peat materials of the type under investigation. Available Water Capacity (AWC) for the soils in the experimental area was calculated as 50 mm/100 mm of the soil volume. In the topsoils created in this experiment in the first growing season the maximum AWC possible ranged from 75 mm to 225 mm. As pointed out above the shrinkage that has occurred over the years has caused a reduction in the top soil thickness and therefore a diminution in the range of AWC available to the grass crop. As of 1986 the maximum possible range of AWC was 60-150 mm. No actual data on the accumulated soil moisture deficit which is an important factor which determines productivity throughout the growing season are available for the experimental site.

Over the duration of the experiment 1983-1986 the accumulated soil moisture deficits (Anon., 1983-86) of 30 mm and greater recorded at Kinsealy, Co. Dublin are tabulated below. These are considered to be the best approximation available of conditions prevailing at the experimental site.

Year	1983	1984	1985	1986
Month				
March	-	-	-	-
April	-	30	-	-
May	-	58	-	-
First cut circa May 28				
June	33	88	-	-
July	89	58	-	-
Second cut circa July 30				
August	70	90	-	-
Sept.	-	40	-	56
Third cut circa Sept. 20				
Oct.	-	15	-	-

An accumulated soil moisture deficit for the three months June-August in 1983 had no significant effect on dry matter yields in the second cut but gave rise to significant differences in the third cut between the non-specialist incorporation techniques and the specialist spading technique. In the 1984 season an accumulated soil moisture deficit of 30 mm had accrued by the end of March. This pattern continued throughout the growing season attaining 90 mm by the end of August. Yields of dry matter were reduced in all treatments but it is noteworthy that the rotary spaded treatments were significnatly higher than the other treatments for the second and third cuts. In 1985 accumulated soil moisture deficits did not exceed 30 mm until the end of September and should have had no effect on dry matter yields.

Throughout the major part of the growing season in 1986 no accumulated soil moisture deficit was likely until September and no significant differences were recorded throughout that season between main incorporation treatments.

There was a reduction in dry matter production for the first cut of 1985 in comparison to the first cut of 1983 though there was no accumulated soil moisture deficit in either case. The explanation for this was resolved in 1985 when sulphur deficiency was diagnosed in the second cut in all treatments but especially in those with shallow incorporation. The deeper rotary spading had two effects - one to increase moisture supply and also to give the roots a larger soil volume for sulphur nutrition. Sulphur applied as gypsum for the second cut increased dry matter yields significantly bringing them into line with the first year. Thereafter, in 1986 the yields were much better for the second and third cuts restoring total dry matter production to 10-11 tonnes per hectare. Under the total cutting regime of the experiment the reserves of available soil sulphur to supply the plant were adequate for the season of 1983 but were unable to do so in subsequent seasons with eventual exhaustion in 1985.

With no accumulated soil moisture deficit in 1986 the prophylactic application of sulphur has solved the poor dry matter production under a total cutting system. In 1987 the experiment was redesigned to test efficacy of split applications of sulphur in the form of sulphate of ammonia. This will be reported on at a later date.

4. DISCUSSION AND CONCLUSIONS

Throughout the country there are 172,00 hectares of Turbary Complex peatlands (Hammond, 1979). It is calculated that within this landscape unit there are 21,000 hectares of varying sized areas of relatively dry peatland which does not require under drainage. This constitutes a pool of low cost land resource capable of increasing farm sizes. When reclaimed for grassland such soils can efficiently graze dry stock (Rogers, Poole and Cole, 1985) thus allowing a better utilisation of grassland on mineral soils for silage production. It is calculated that a further 69,000 hectares have a potential for development if some under drainage is installed and strict grazing management is followed throughout the grazing season.

The research reported here was designed to assess the merits and efficiency of five techniques of incorporating ground limestone and to determine the effect of three rates of ground limestone on herbage dry matter production on the drier soil type within this landscape unit.

Lime incorporation was carried out using easily available farm equipment capable of incorporating ground limestone to a depth of 200 mm, plus a specialist rotary spading machine to incorporate liming materials to a depth of 450 mm. Incorporation techniques used and their cost effectiveness can be considered an important element in the reclamation of Turbary areas. All machines performed to capability incorporating lime to the required depth and creating well mixed topsoils.

The rates of lime applied which were calculated on the basis of 2.5 t ground limestone ha^{-1} incorporated in the top 150 mm would raise the pH by 0.2 units gave pH levels in the surface 150 mm of 5.33, 5.73 and 6.03 compared with targeted values of 4.5, 5.0 and 5.5 respectively pH values of this order were maintained to the depth of incorporation. The attained values were in the order of 14%-19% above those expected indicating that under the experimental conditions 1.4-1.8 t ha^{-1} 150 mm^{-1} of ground limestone would be sufficient to raise the pH by 0.2 units for 150 mm depth. Statistical analysis of herbage dry matter yields showed no significant differences between the three lime treatments.

At the optimum rates determined in this experiment the cost per hectare to neutralise acidity in organic soils is greater than for mineral soils. The rates as quoted are in line with those established elsewhere (Hamilton and Bernier, 1973; Dampney, 1985). Whilst higher rates are required for organic soils they are nonetheless strongly resistant to leaching of lime. A profile sampled in an area of a similar nature, reclaimed and limed some 16 years ago still maintains a pH(water) of 5.4 in the top 150 mm (Hammond, 1988). In comparison to mineral soils there is a higher initial lime cost per hectare but in the long term the lime cost may be no different to that for mineral soils where five year intervals have to be considered.

Though lime rates applied had no significant effect on dry matter yields the depth of incorporation was significant and therefore top soil thickness can be critical. This is an important element in crop production in particular from soils derived from very acid parent materials. The thickness of the topsoil has a marked effect on root proliferation and availability of water and plant nutrients. The boundary between topsoil and sub-surface horizon in these soils demarcates the zone of rooting with very little or no penetration below this boundary. Where living roots do occur 50 mm below the cultivated interface (Fig. 1) this suggests that some amelioration through leaching of bases could be taking place. This is a long term process but has been shown to occur through time by increasing values with depth for the exchangeable calcium/magnesium ratio in soils initially reclaimed some generations previously (Hammond, 1984).

The dry matter yields recorded in the initial year of the experiment were in excess of 11 t ha^{-1} for all treatments with the deep rotary spading showing a significant increase in dry matter production over all other treatments. These differences can be related to the accumulated soil moisture deficit in the growing season. Whilst the deficit for the three months June-August in 1983 had no significant effect on dry matter yields in the second cut but gave rise to significant differences in the third cut between the non-specialist incorporation techniques and the specialist spading technique. This was repeated in 1984 and 1985.

It is clear that the techniques tested were efficient in incorporating ground limestone throughout the cultivated surface horizon. However, when considering the economics of developing such land for grass production the deep rotary spading treatment can only be considered justified in years when growing conditions are similar to 1984. It is extremely important to ensure an even spread of ground limestone when reclaiming this type of peat soil to agriculture. Without exception such soils are recommended for grassland farming. For these situations under the average climatic conditions which prevail rotovation is a satisfactory incorporation method and with good grassland management such soils can sustain high levels of dry matter production.

5. ACKNOWLEDGEMENTS

Mr. Ted Brennan's technical assistance and the help of staff members at An Foras Taluntais Peatland Experimental Station throughout the experiment is gratefully acknowledged. Thanks are also due to Dr. G. Smillie and Dr. J. F. Collins for helpful discussion and correcting the draft manuscript.

6. REFERENCES

Anon. 1983-86. Agrometeorological Bulletin. Meteorological Service, Dublin.

Anon. 1986. Weather and agriculture. Agricultural Credit Corporation, Dublin.

Aslyng, H. G. 1965. Evaporation, evapotranspiration and water balance investigations at Copenhagen 1955-64. Acta Agriculturae Scandinavica 15: 284-300.

Brereton, A. J. and Keane, T. 1982. The effect of water on grassland productivity in Ireland. Irish Journal of Agricultural Research 21: 227-248.

Byrne, E. 1979. "The analysis of agricultural materials". An Foras Taluntais, Dublin.

Conry, M. J., Hammond, R. F,. and O'Shea, A. 1970. The Soils of County Kildare. Soil Survey Bulletin No. 22, An Foras Taluntais, Dublin.

Dampney, P. M. R. 1985. A trial to determine the lime requirement for reseeded grassland on a peaty soil. Soil Use and Management 1: 95-100.

Galvin, L. F. 1976. The physical nature of Irish peats. Irish Journal of Agricultural Research 15: 207-221.

Hamilton, H. and Bernier, R. 1973. Effects of lime on some chemical characteristics, nutrient availability and crop response of a newly broken organic soil. Canadian Journal of Soil Science 53: 1-8.

Hammond, R. F. 1979. Peatlands of Ireland. Soil Survey Bulletin No. 35, An Foras Taluntais, Dublin.

Hammond, R. F. 1984.Field and laboratory studies on the distribution and classification of peat formations in Ireland, and the genesis of their derived and associated soils. Ph.D. thesis, University College Dublin.

Hammond, R. F. 1988. The characteristics of Irish Midland peatlands which can influence future land use programmes. Workshop on the Utilisation of Irish Midland Peatlands. Scientific Proceedings of the Royal Dublin Society (in press).

Hammond, R. F., Murphy, W. E. and O'Keeffe, M. Grass production from midland cut-over peats - responses to nitrogen (in preparation).

Pearson, R. W. 1971. Significance of rooting pattern and some problems of root reseach. In: "The Plant Root and its Environment". (Editor: E. W. Carson). University Press of Virginia, Charlottesville, Chapter 10, pages 247-270.

Perle, R. T. 1954. British Ministry of Agriculture and Fisheries Technical Bulletin, HMSO, London.

Rogers, P. A. M., Poole, D. B. R. and Cole, A. J. 1985. Copper supplementation of beef cattle on pasture reclaimed from farmer cut-over Sphagnum peat. Soils and Grassland Research Report, An Foras Taluntais, Dublin.

Synder, G. H. 1978. Water table management for organic soil conservation and crop production in the Florida Everglades. Florida Agricultural Experimental Station Bulletin 801.

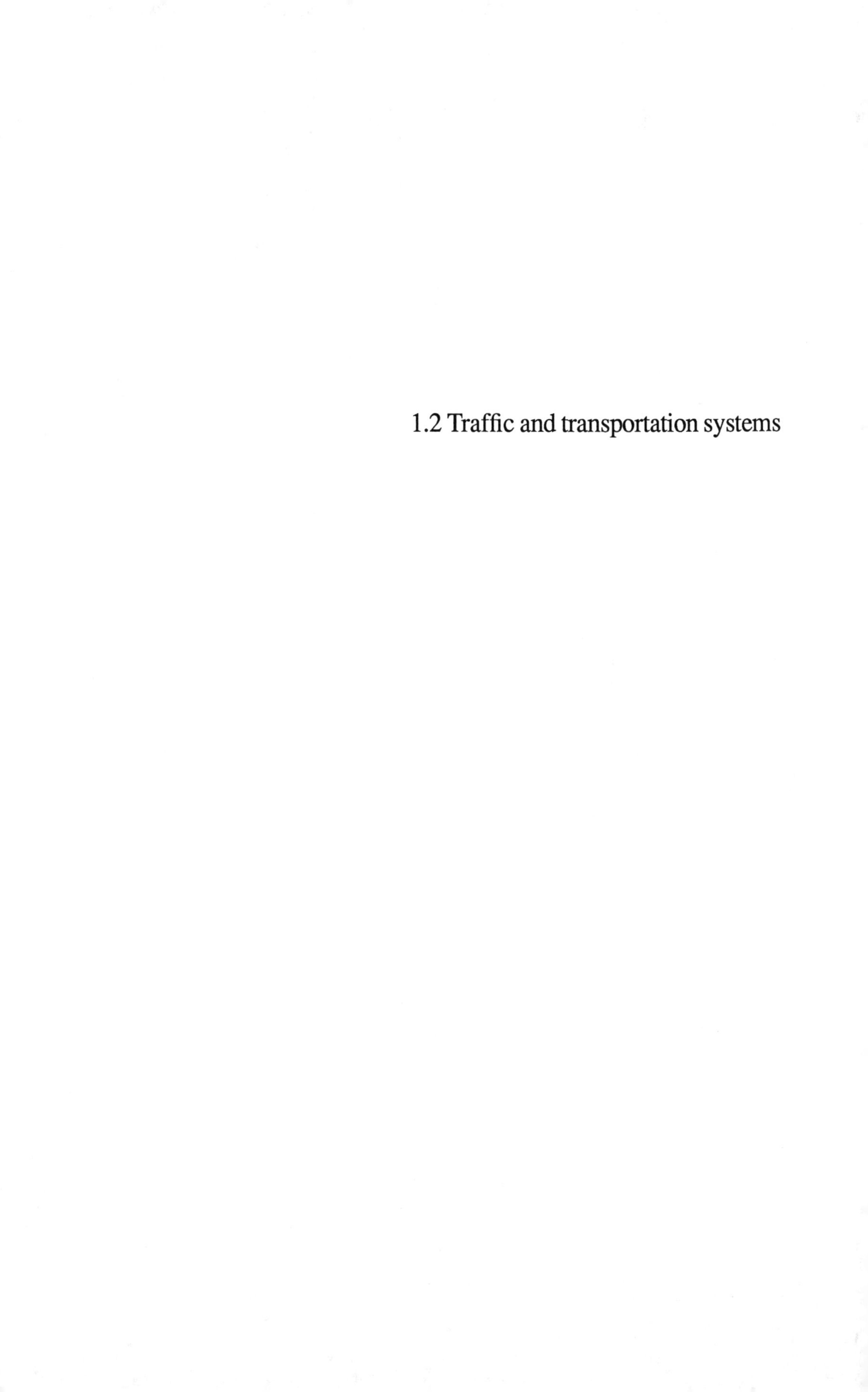

1.2 Traffic and transportation systems

Land and Water Use, Dodd & Grace (eds), © *1989 Balkema, Rotterdam. ISBN 90 6191 980 0*

Keynote paper: Traffic and road systems in rural areas

Th.Michels
Winand Staring Centre for Integrated Land, Soil and Water Research, Wageningen, Netherlands

ABSTRACT: Minor rural road networks in industrialized regions are no longer purely for giving access to agricultural or forest land. Additional urban and outdoor recreation activities and traffic developments put higher demands upon the road system. A functional road classification should precede successive decisions on road layout and construction, the overall process being subject to long term network cost minimization.

Trading off plan options from several viewpoints is important for plan optimization and public enquiry. For this purpose the development of adequate information systems and methods for impact assessment and evaluation should continue. Further research activities should primarily focus on traffic forecasts, low-cost and low-maintenance constructions, ecology, landscaping, road safety and energy consumption. Much existing know-how from the developed regions is worth transferring to developing regions at this stage already.

RESUME: Les réseaux routiers inférieurs en rase campagne dans les régions développées n'offrent plus seulement l'accès aux terrains agricoles ou forestiers. Les activités additionelles de l'urbanisation et des loisirs en plein air et les développements du trafic ont posé de nouvelles demandes au système de transport. Une classification des fonctions routiers doit précéder les décisions successives sur l'aménagement et la construction de la route, le procédé entier étant sujet à la minimisation des frais du réseau à longue échéance.

Evaluer les alternatives selon des critères divers est important pour optimiser le projet et pour la consultation publique. A cette fin il faut continuer le développement des systèmes d'information et des méthodes d'impact. D'autres efforts de recherche devraient se concentrer sur le prognostic de trafic, les constructions à bas frais et faible entretien, l'écologie, le paysage, la sécurité routière et la consommation énergétique. Beaucoup des compétences actuelles dans les régions développées peuvent déjà être transferées aux régions en développement.

KURZFASSUNG: Die untergeordneten ländlichen Strassennetze in den industrialisierten Regionen dienen nicht mehr nur die Erschliessung landwirtschaftlich oder forstlich benutzter Grundstücke. Die Verstädterung und die Freilufterhohlung wie auch verkehrliche Entwicklungen haben neue Ansprüche an das Strassensystem gemacht. Eine Klassifizierung der Strassenfunktionen soll vorangehen an aufeinanderfolgenden Bestimmungen der Strassengestaltung und der Konstruktion, wobei das ganze Transportsystem einer langfristigen Kostenminimalisierung zu unterwerfen ist.

Bewertung von Planalternativen an Hand verschiedener Kriterien ist wichtig für optimale Projektauswahl und öffentliche Beratungen. Dazu sollten Informationssysteme und Wirkungsbeschreibungsmethoden weiterentwickelt werden. Weitere Forschungen sollten sich richten auf Verkehrsvoraussagen, auf billige und unterhaltsfreundliche Konstruktionen, und auf Aspekte der Oekologie, der Landschaft, der Strassensicherheit und Energiebedarf. Vieles der heutigen Sachkunde in entwickelten Regionen eignet sich nach Entwicklungsregionen zu übertragen.

1 INTRODUCTION

From the very first moment of human settling, man has created transportation systems, both for meeting his demands and for satisfying his curiosity. This is not likely to change in future. The rural areas of our today's industrialized world are to a large extent given access by a vaste network of major and minor rural roads, most of these being minor with relatively low traffic flows. Minor rural roads, covering 75 to 95% of the total road network, are of vital significance, both for economic development and for social communication. In fact they constitute the capillaries of the transportation system by giving local access to dwellings, economic activities and land, by collecting or distributing local traffic to or from major roads and by connecting villages and towns. Network densities may vary widely depending on land-use intensity, but the impact of roads upon economic output and quality of life is always considerable. On the other hand, both increasing environmental concern of society and budgetary restrictions will justify certain constraints to road development. It is clear that the cost-effective development and management of rural road systems will create common problems throughout the world. From this point of view it was a good initiative of the Congress organization to include a seperate session on the theme of transportation into the programme.

For a proper optimization of rural transport systems in a broad sense its planning and design should be regarded as an integrated part of physical planning on one hand and in relation to agricultural engineering on the other. When designing or ameliorating the system, road function, road layout and road construction, each representing successive steps in the design process, must be related one to another in order to arrive at optimal solutions.

For each of the three features I will indicate main points of interest in the next three sections respectively, followed by a section on plan evaluation. As a result, future research challenges will be reviewed as well as the possibilities of technology transfer to developing regions.

2 PLANNING ROAD FUNCTIONS

For the determination of the appropriate road layout and construction it is necessary to have a proper idea of the traffic function a road should fulfil as part of the whole network. Planning road functions should form part of an overall planning policy of which the objectives may be defined at either the national or the regional level. These objectives concern the socio-economic management of the network on the long term. The basic aim is to meet total transport needs at the least socio-economic cost, in line with economic management principles like reducing unit investment costs, optimizing the allocation of highway funds, enhancing the quality of the existing road network and extending its lifetime (OECD 1986). Therefore basic information should result from a functional analysis of the network for the determination of medium- and long term needs, with due regard to land-use and traffic developments.

For centuries agriculture and forestry have been the dominating land-use in most rural areas. This is most likely to continue in future. But considerable changes within the agricultural sector are going on and affect transport demands. Economies of scale have brought us bulk carriers and large sized lorries for transporting raw materials and agricultural products. These heavier vehicles are putting increasingly higher axle loads upon the minor road network which is largely not designed for this. On the other hand, continuing mechanization has increased the dimensions of self-driving agricultural machines, putting higher demands upon carriageway widths.

Moreover, in urbanizing rural regions, also residential, industrial and outdoor recreation land-use has developed and generated additional pressures upon the road system. Especially in densely populated areas like in the Netherlands, the United Kingdom, the Federal Republic of Germany or

Switzerland the resulting traffic flows have caused in many cases serious problems with road capacity, especially on major roads. Furthermore, in areas with dense road networks, offering a lot of through-routes, road authorities are facing the problem of rat-run traffic: large flows of passenger cars, overloading low level roads to avoid congestions and delays on major roads. In many of these cases the local agricultural traffic is highly dominated by the passenger car traffic, resulting into traffic unsafety problems.

At last but not to the least, the increased social emphasis upon environmental quality, both in scenic and in ecological terms, has lead to more restrictive policy attitudes towards road network extension and improvement.

The demands for all traffic functions as well as the environmental and budgetary constraints make it necessary to be quite selective in ameliorating particular routes in the network. This requires a clear policy vision on how to differentiate between road functions (i.e. connectors, collectors, local access roads) in order to reach a hierarchical structure inside a road network.

A complication when implementing such a functional hierarchy in a network is that in most rural areas there is a variety of local, regional and even national authorities responsible for specific parts of the road network. This often makes it difficult to make a long term planning of coordinated technical and investment programmes. Therefore it is worth pointing at the possibility of integrated projects for land development and rural reconstruction, as we know these in several countries, supported by national government authorities. These projects offer great opportunities for implementing structural changes in land layout including road infrastructure. Recent developments in several European countries show the public valuation of these amelioration programmes (Boss and Flury 1988; Gevaert 1988; Kollmer 1988; Van Putten 1988; Schatt 1988).

But adjusting road functions is not always necessarily a matter of infrastructural investment; in dense road networks many problems can be solved by an adequate traffic management through traffic regulations, signing and, occasionally, closing off roads for specific traffic categories. Especially in areas with seasonal outdoor recreation pressures there are successful examples of such a regulatory planning (Buck 1988; Hawkins and Hatt 1988).

The overall conclusion from the above is that the need of basic traffic data collection is evident in order to decide upon road functions. Not only total flow but also the breakdown into traffic purposes and vehicle categories must be analyzed and forecasted for defining required changes in functional road classifications, road layout improvements or management actions. In complex road networks it may even be useful to apply traffic models for this purpose (Jaarsma 1988).

In developing regions transportation demands differ considerably from the type of problems discussed above. Here it is much more a matter of giving basic access to rural resources for cultivating land, promoting economic productivity and connecting rural areas to markets and main centres. Nevertheless, here too it is important for long-term cost-effectiveness to set up an integral planning concept for road functions and implement an efficient hierarchical structure in an early stage.

3 LAYOUT DESIGN

Given the traffic function the road is planned for, the appropriate design of its layout (alignment and cross section c.a.) implies supplying a reliable and safe provision, meeting demands of traffic flow, vehicle dimensions, speeds, driving comfort and safety. Especially in case of low traffic densities, high speeds and traffic safety are often controversial. Impacts of road layout features upon road user behaviour and accident rates make up a theme of major concern for highway engineers (Gambard 1988; Stewart 1988). It is generally found that accident rates on minor rural roads are many times higher than those on major roads. An adequate traffic behaviour control according to the road function requires more knowledge and understanding of the traffic process on minor rural roads than we have today.

It is not unusual that the various road authorities in rural areas all follow their own traditions and methods for road layout design. Consequently a great variety of road outfits can often be found in these areas. This is clearly not contributing to a predictable and safe road user behaviour. Therefore it is important to work on standardization of layout design, at least country-wide but preferably internationally. OECD (1986) has made a good contribution to the latter with the publication of such standards, synthesized from various national standards, and an overall design methodology for low-traffic roads.

Quite an essential problem related to layout design is the impact of rural roads upon scenic and ecologic values. This is in many countries a point of increasing concern, often even deciding for the layout choice (Schatt 1988). And this surely deserves attention; road structures can indeed seriously intrude both into landscapes at the large scale and into the micro-habitat of small animals and vegetation. On the other hand roads offer opportunities, mainly through layout design, to strengthen scenic qualities and, through shoulder vegetations and plantations, to contribute to an ecological infrastructure facilitating wildlife of higher quality (Harms & Knaapen 1988). In this respect it is also interesting to notice a certain revival of wheel track pavements, a.o. in the Federal Republic of Germany. In earlier days these wheel tracks have mainly been applied for economic reasons; at present the construction costs are at least equal to those of full width pavements (Unger 1986), but its 'transparancy', both scenic and ecological, makes it an agreeable compromise between accessibility and environmental demands, especially for very low volume roads with mainly agricultural or recreational functions.

In developing regions road layout is often a second stage concern, following after the first stage of giving basic access to land. Here, above all the road has to meet and keep in tune with developing demands in terms of flow and loads. Like anywhere else, when adapting layout to this, the control of traffic speed and accidents deserves main attention; user comfort and environmental demands may necessarely take second place due to economic constraints.

4 CONSTRUCTION DESIGN

After having decided upon road function and layout, the determination of the optimal road construction is basically a minimization of the sum of construction costs and capitalized maintenance costs over its lifetime. So, where the estimation of axle loads will always lead to a technical minimum design, the maintenance cost criterion may require an optimum at higher construction costs. For this purpose it is above all important to have good forecasts of axle load patterns. This question appears in most situations difficult to answer and deserves more research efforts to develop simple axle load forecasting models using input variables on land layout and productivity.

The next question in this design process, after determining the technical demands, is how to arrive at appropriate low-cost and low-maintenance constructions, meeting the overall cost minimization objective. There are several key issues in this framework like the applicability of local materials, both traditional and alternative, the reuse of old materials and, especially in areas with soils of a poor bearing capacity, the use of light-weight constructions. As to the latter, experiments with polystyrene and foamed concrete on Dutch peatland seem quite encouraging (Leewis 1988).

In developing regions it is in many cases often more difficult to make axle load forecasts. In these situations it may be wise to opt for low cost construction standards and make necessary adjustments according to later developments of traffic.

5 EVALUATION OF PLAN OPTIONS

The socio-economic appraisal of public investment in infrastructure is a generally accepted tool for selecting optimal options from a variety of alternatives for a particular project and also for priority ranking of different projects. Moreover, for public enquiry it is a useful means of presenting plan options to the judgement of parties involved.

In the industrialized regions, where the total amount of passenger trips and amount of goods to be transported is only marginally affected by the road system as such, plan optimization means basically an overall minimization of social costs of the system; here the breakdown into cost components (construction, user costs, management costs, environmental losses) is dependent on the quality of the supplied network. In developing regions, where infrastructural investments can be highly decisive for economic growth, it is much more a matter of maximizing the cost-effectiveness of measures, maintaining an optimal balance between road infrastructure investments and rural development measures in general.

Traditions in several countries differ widely in the extent to which the application of plan evaluation methods is an obligatory part of planning procedures. Some renounce it entirely, some concentrate on specific trade off issues like environmental impacts versus agricultural benefits (Schatt 1988), and some go quite far by making a detailed cost-effectiveness analysis of all land development plans, including road ameliorations (Michels 1988).

Apart from this there is a general tendency of taking ever more aspects into account when trading off plan options. For this reason, multi-criteria evaluation methods, presenting impacts in their own different units, are gradually taking the place of the classical cost-benefit analysis in strict monetary terms. This is quite obviously the case for minor rural roads where so many criteria are relevant for plan comparison and can never be expressed in one common unit.

Particular impact categories which demand more than the actual knowledge for an adequate plan evaluation are beyond doubt road safety, ecological quality and scenic value, as discussed in the previous chapters. But also energy consumption should be much more explicitly considered as an evaluation criterion, given the general limitation of our energy resources. This can be done both for the fuel consumption inside the rural transportation system as such, and for the overall rural land layout and land-use, in terms of an aggregated input/output analysis of the area's energy balance. Some first steps into research of this kind have recently been put in the Netherlands (Jaarsma 1987).

A final point deserving emphasis is the further development of information systems. A rapid and flexible data availability is a basic condition for an effective selection and evaluation of plan options. Indirectly this applies not only to plan evaluation but to road design as a whole. Quite clearly there are many activities around information systems going on in many countries. Professionals in research and practice could make more progress and avoid double work if they would put more efforts into exchanging their experiences in this field.

6 RESEARCH CHALLENGES

Reviewing the above considerations in terms of actual research demands it may be concluded that there are indeed specific priorities to emphasize. The following list, however, does not give any priority ranking. Such a ranking is highly dependent on national circumstances and the stage of development of a transportation system. Nevertheless this listing may serve as a contribution to discussions about international exchange and cooperation in research on rural traffic and road systems.

1. Forecasts of traffic flows and loads should be based upon systematic data collection on one hand and on simple forecasting models on the other. Either of these should be elaborated, specifically for minor rural road systems.

2. Further development of low-cost and low-maintenance road constructions will certainly be cost-effective, especially in the case of developing regions. Such research should include the application of alternative materials, reuse of old materials and development of light-weight constructions on poorly bearing soils.

3. Environmental impact assessment should concentrate on scenic impact perception of road plans on one hand and on ecological effects of roads on the other, the latter related to ecological infrastructure in general.

4. The effects of road layout on road user behaviour, road accident rates and on total road user costs should be further elaborated for the purpose of economic appraisal.

5. The impacts of a rural transportation system on energy consumption deserves a more comprehensive study in order to evaluate road plan options from this viewpoint.

6. Methods for socio-economic plan evaluation taking into account all relevant criteria should be further elaborated.

7. Information systems as a tool for both designing and evaluating plan options are generally in development but deserve more efforts for the exchange of know-how.

7 TECHNOLOGY TRANSFER TO DEVELOPING REGIONS

It is beyond doubt that much of the expertise on rural transportation, already existing in many countries and brought together on congresses like this, is potentially worth transferring to developing regions. In many cases this will demand some modification before being really applicable, but the required additional research is often only marginal related to the basic efforts already made. It is very important that researchers remain keen on such opportunities for increasing the returns of their work.

A second basic condition for the transfer of know-how is clearly that technicians keep on presenting their practical experiences and research results through international congresses and literature in order to make these accessible for potential users. The spin-off of this can hardly be underestimated. Therefore it is worth encouraging all the colleagues in the field of rural transportation to make use of these facilities as much as possible.

REFERENCES

Boss, Ch. & U.Flury 1988. Integral planning and design of minor rural roads in Switzerland. In Th.Michels & C.F.Jaarsma (eds.), Minor rural roads, p.67-84. Wageningen, PUDOC.

Buck, P.E.M. 1988. Planning and design of minor roads in some rural areas in the Netherlands. In Th.Michels & C.F.Jaarsma (eds.), Minor rural roads, p.97-106. Wageningen, PUDOC.

Gambard, J.M. 1988. Design of low-traffic roads in France. In Th. Michels & C.F.Jaarsma (eds.), Minor rural roads, p.129-134. Wageningen, PUDOC.

Gevaert, F.J. 1988. The planning of rural roads in land development in Belgium. In Th.Michels & C.F. Jaarsma (eds.), Minor rural roads, p.47-54. Wageningen, PUDOC.

Harms, W.B. & J.P.Knaapen 1988. Landscape planning and ecological infrastructure: the Randstad study. In K.F.Schreiber (ed.), Connectivity in Landscape Ecology, p.163-167. Proceedings of the 2nd International Seminar of the 'International Association for Landscape Ecology'. Münsterse Geographische Arbeiten 29. Münster.

Hawkins, M.R. & M.J. Hatt 1988. Managing a rural road network. In Th.Michels & C.F.Jaarsma (eds.), Minor rural roads, p.55-66. Wageningen, PUDOC.

Jaarsma, C.F. 1987. Energiehuishouding van landelijke gebieden. Landbouwkundig Tijdschrift 99,11: 17-18.

Jaarsma, C.F. 1988. Traffic research in Dutch rural areas. In Th.Michels & C.F.Jaarsma (eds.), Minor rural roads, p.85-96. Wageningen, PUDOC.

Kollmer, K. 1988. Planning, design and realisation of minor rural road networks in German land consolidation projects. In Th.Michels & C.F.Jaarsma (eds.), Minor rural roads, p.25-38. Wageningen, PUDOC.

Leewis, M. 1988. Roads and dams in the peatlands. In Th.Michels & C.F.Jaarsma (eds.), Minor rural roads, p.183-185. Wageningen, PUDOC.

Michels, Th. 1988. Impact assessment and evaluation of road improvement in Dutch land development planning. In Th.Michels & C.F.Jaarsma (eds.), Minor rural roads, p.135-142. Wageningen, PUDOC.

Michels, Th. & C.F. Jaarsma (eds.) 1988. Minor rural roads: planning, design and evaluation. Proceedings Workshop Minor rural roads, Wageningen, October 28-30, 1987. Wageningen, PUDOC. 198 pp.

OECD. 1986. Economic design of low-traffic roads. Report prepared by an OECD scientific expert group. OECD, Paris. 128 pp.

Putten, Th.H. van 1988. Transport and road planning in Dutch land development projects. In Th.Michels & C.F.Jaarsma (eds.), Minor rural roads, p.15-24. Wageningen, PUDOC.

Schatt, H. 1988. Planning and evaluation of rural road construction - illustrated by Bavarian land consolidation schemes. In Th.Michels & C.F.Jaarsma (eds.), Minor rural roads, p.39-46. Wageningen, PUDOC.

Stewart, D. 1988. Prevention of accidents at bends on rural roads. In Th.Michels & C.F.Jaarsma (eds.), Minor rural roads, p.151-158. Wageningen, PUDOC.

Unger, W. 1986. Landschaftsgerechter Wegebau. Bauen für die Landwirtschaft 23,1:17-18.

Land and Water Use, Dodd & Grace (eds), © 1989 Balkema, Rotterdam. ISBN 90 6191 980 0

Influence of agricultural traffic (slow moving vehicles) over traffic flow on rural roads

U.Brannolte
University of Karlsruhe, Institute for Transport Studies, FR Germany

ABSTRACT: An overview is given on recent research projects dealing with empirical data from traffic measurements, traffic flow simulation and road guidelines. It focuses on special types of high-standard rural road providing better facilities for easier and safer overtakings. Therefore these roads are also more suitable for agricultural traffic.

1 INTRODUCTION

Traffic flow on two-lane rural roads can be characterised by the different kinds of vehicle and by the different travel speeds of these.

Speeds of motorised vehicles normally range from travel speeds of cars to those of the so-called slow moving vehicles - and most of these are agricultural vehicles. Cars often have speeds of about 100 km/h and even more, whereas most of the slow moving vehicles (e.g. farm tractors) move at speeds of about 25 km/h.

In order to quantify the conditions of traffic flow for road planning and road design purposes, concepts have been developed which can be found in road capacity manuals (e.g. HCM (1985) in the U.S. and RAS-Q (1982) in W.Germany). In these concepts the achievable "levels of service" depend on speeds cars can maintain under specific road and traffic conditions.

It is obvious that car speeds will be influenced significantly by slow moving vehicles, because each of these on the road has to be overtaken by succeeding cars if the drivers want to maintain their travel speeds. But not each overtaking manoeuvre can be fulfilled immediately because of oncoming traffic and other restrictions. In these situations the drivers have to decide whether to overtake or not. They decide in accordance with their estimations of distances and speeds of oncoming vehicles and of the performance of their own car - and also according to their own risk-taking behaviour.

Numerous investigations on traffic flow on rural roads have been performed and published, but only very few of them refer especially to questions in connection with the slow moving vehicles. There are several different approaches to deal with these problems in order to improve flow conditions and traffic safety.

From the scientific and practical point of view, first of all it seems to be necessary to gather declaring field data on flows and on accidents. These data may be used for further treatment, e.g. for modelling and/or in decision-making processes being due to road planning.

In the following a short description is given on some German research projects concerning slow moving vehicles on rural roads.

2 APPROACHES

2.1 Empirical measurements

In a recent research project by Leutzbach et al. (1987) a series of measurement trips on a typical German rural road with two lanes, each of about 3.75 m width, had been made. This road corresponds to the most generous layout of a cross-section (called: type b2) of a typical two-lane rural road - other types have narrower lane widths. A floating car was used and this was a farm tractor with trailer having a constant speed of about 25 km/h.

This tractor carried a measurement equipment which mainly consisted of a radar meter, three video cameras and recording devices. The radar meter was used to measure speeds of overtaking and oncoming vehicles. The video cameras were used to

record vehicle movements behind, beside and in front of the floating tractor.
When observing the overtakings two different kinds of manoeuvre had been elaborated.

- the flying overtaking (fl.)
- the accelerated overtaking (acc.)

If the overtaker's speed is not significantly influenced by the slow moving vehicle, it is considered to be a flying overtaking. If an immediate overtaking is not possible, e.g. because of the oncoming traffic (or sight-distance restrictions), the overtaker has to slow down and has to follow the slow moving vehicle for a certain time before proceeding to an accelerated overtaking.

One of the essential data are the gaps (time headways) between oncoming vehicles which are either accepted or rejected by the drivers for overtaking. The distribution functions of accepted and rejected gaps are shown in figure 1.

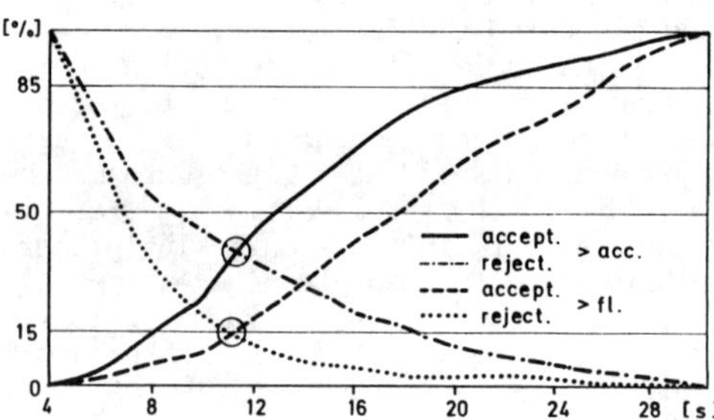

Fig.1 Distribution functions of all accepted and rejected gaps in seconds - overtaking vehicles are cars

Besides gaps further parameters of overtaking manoeuvres had been analysed. For this purpose sections of overtaking had to be fixed in order to observe the tasks which are necessary for overtaking. By determining these tasks quantitively it is possible to describe safety aspects.

It has been found out that the overtaking time lasts on average between 5.4 seconds (flying, cars) and 8.9 seconds (flying, trucks). The medians of overtaking distances range from 80 m to 130 m. The safety margins to the oncoming vehicles were shorter in cases of accelerated overtakings in comparison to the flying ones. In 50 % of accelerated overtakings by cars these were less than 4 seconds or 120 m. In 15 % of all cases, there were no or even negative safety margins, respectively clear distances to the next oncoming vehicles, meaning collisions could only be avoided by lateral movements or decelerations of oncoming vehicles, see figure 2.

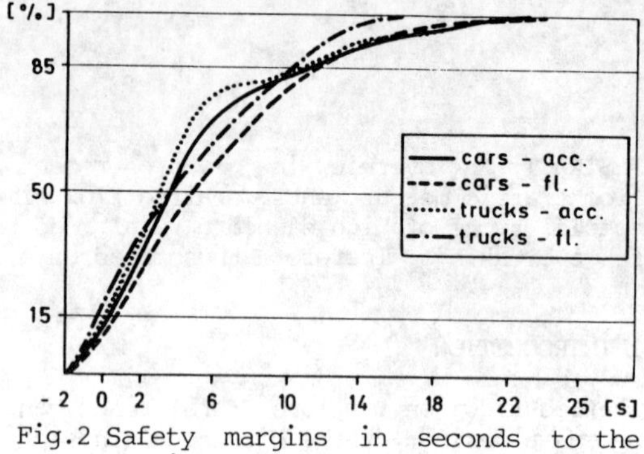

Fig.2 Safety margins in seconds to the next oncoming vehicle, overtaking vehicles are cars resp. trucks.

Mean speeds were lower in cases of accelerated overtakings; here the trucks averaged 47 km/h and the cars averaged 53 km/h. In case of flying overtaking trucks and cars had nearly the same speed of about 68 km/h. The mean speed of directly influenced oncoming vehicles was about 68 km/h.

By comparing these results with mean speeds of all recorded vehicles on the cross-section one can see a difference of about minus 10 km/h. The probability of an overtaking seems to be higher, when the next oncoming vehicle is estimated to be a comparatively slow one. Another reason may be, that oncoming drivers feel impeded by overtakers and apply brakes because of safety reasons.

These observations allow a detailed analysis of overtaking manoeuvres on a microscopic level in case the overtaken vehicle is a slow moving one. The results can complete knowledge available from other research projects, in case the overtaken vehicle is a car or a truck. Such an inventory is given by Brannolte et al.(1989), it includes a more detailed description of the measurements mentioned and a comparison between German and Dutch data.

An interpretation of some other empirical data also presented in the quoted research work gives more hints concerning slow moving vehicles. Concerning relationships on a macroscopic level between speeds, traffic volumes and lane widths it could be perceived that the different existing types of rural roads could be grouped into those having "overwide" widths of more than 3.75 m, those having "medium" widths between 3.25 m and 3.75 m and those having "narrow" widths of about 3.00 m and less. Inside these appropriate groups the relationships are quite simliar. In distinction from this the overtaking behaviour seems to be more dissimliar, yet smaller increments of lane width lead to more overtakings and vice versa. Not very visible influences of the occurence of slow moving vehicles on macroscopic parameters of traffic flow could be constituted in that observed practice. One reason may be that there was a recorded maximum of less than ten slow moving vehicles per hour - even in regions and in periods where one might expect some more. But this is an observation which is closely related with parameters of the infrastructure.

2.2 Simulation

Traffic flow consists of a lot of interactions between the single elements (vehicles-drivers) which are influenced by external parameters (e.g. road design, traffic control). Because there are stochastic influences and complex relations and because it is very time-consuming, expensive and sometimes impossible to get all-out empirical data, there is a call for applying simulation techniques.

A simulation software has been developed which can duplicate traffic flow on two-lane rural roads on a computer. This simulation works on a microscopic level, i.e. the different performances of vehicles and the different behavioural patterns of drivers are modelled explicitly.

This model, called SIMLA-2, had been presented in detail by Brannolte and Holz (1983) and it includes all attributes of traffic flow on two-lane rural roads, comprising the horizontal and vertical alignment, different traffic load and different traffic mix. The user of the model has to define the provided traffic and roadside conditions in an input data set and as an output he gets quantitative terms on traffic flow operations of the simulated stretch of the road. This simulation technique has been proven to be a mighty tool for the traffic engineer to examine traffic flows.

In the already mentioned project by Leutzbach et al. (1987) this model has been enhanced by including explicitly the slow moving vehicles. The model SIMLA-2 has been applied in a computing series to questions in connection with slow moving vehicles.

Some results of these computer runs are to be represented here.

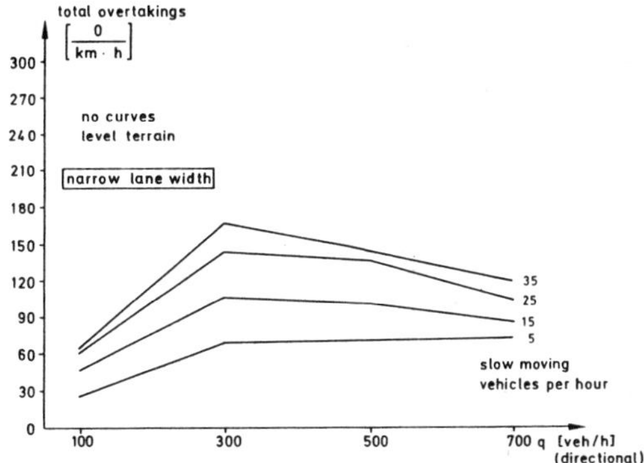

Fig.3 Simulations of overtakings as a function of traffic volume and different quantities of slow moving vehicles - road width "narrow"

Figure 3 shows the overtaking rate as a function of traffic volume and as a function of different quantities of slow moving vehicles on the road. The traffic is characterised by 20 % of heavy vehicles, the simulated road section has a medium degree of curvature. The road width is defined "narrow", i.e. all overtaking manoeuvres can only be fulfilled in a very restricted manner taking into account oncoming traffic and horizontal alignment and according to safety requirements.

Figure 4 shows results in a manner analogous to those in Figure 3 but with the only difference that the road width is defined as "medium" - overtakings are much more easier. The application of the simulation model allows to elaborate these aspects quantitatively.

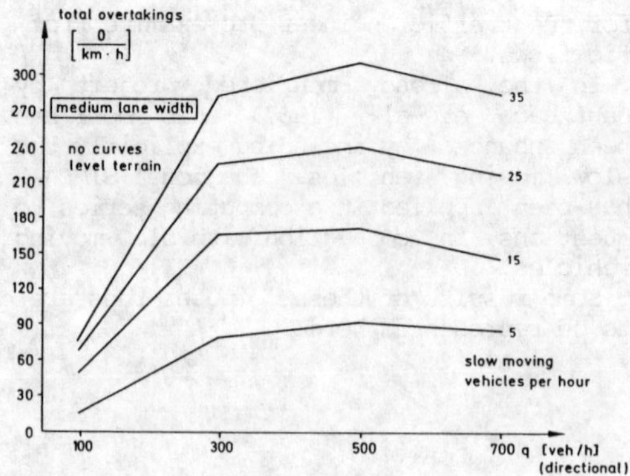

Fig.4 Simulation of overtakings as a function of traffic volume and different quantities of slow moving vehicles - road width "medium"

2.3 Road construction

In the last years various measures have been discussed and performed to make overtakings on rural roads easier; possible solutions are to enlarge sight distances and corner-radii. In this connection it is intended to consider, what types of road or special cross-sections might be provided for better overtaking opportunities. Two existing types of rural road in W. Germany have enlarged cross-sections they are part of the so-called intermediate types of cross-sections. One of these types is already included in the current guidelines (RAS-Q) 1982; see Figure 5.

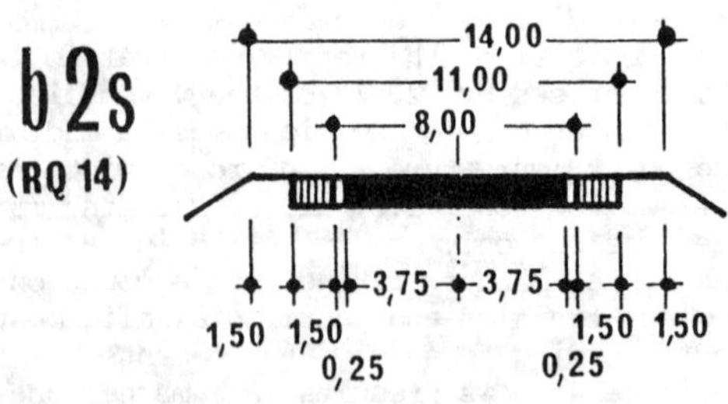

Fig.5 Cross-section of the type of road b2s

The type b2s is a rural road with lane widths of 3.75 m, lane making of 0.25 m and additional multiple- purpose lanes of 1.50 m on both sides. These multiple-purpose lanes are used temporarily by slower vehicles in order to enable faster vehicles to overtake more easily and safely.

Another type is an overwide type of rural road. This type is only different from the mentioned b2s by the fact that there is no lane marking separating the multiple-purpose lane from the traffic lane. This type is already existing quite often and is called b2ü.

On the b2s the slower vehicles, i.e. trucks, farm tractors and even some cars usually move aside to the multiple-purpose lane to allow to be overtaken by the faster ones; this, however, is not an obligation. Here it has to be mentioned that the normal overtaking manoeuvre is in some way reversed. The slower driver has to enable an unimpeded path to the faster one by giving way by changing lane.

On the b2ü the slower vehicle normally is driving close to the outside edge of the road; most of the overtaking manoeuvres can be carried out on the one overwide lane.

2.4 Road safety aspects

As results from measurements it has to be mentioned that on the intermediate cross-sections b2s and b2ü driven speeds are at least about 10 km/h higher than on typical rural roads. But the higher risks, resulting from higher speeds, seem to be overcompensated by safer overtakings. On rural roads overtaking accidents normally have the highest portion comparing all types of accidents. It can be concluded, that accident rates on intermediate cross-sections are substantially lower than on typical rural roads.

Accidents with involvements of slow moving vehicles very seldom occur on b2s and b2ü; this is one result of a present-day survey, Brannolte (1989). For instance only 9 accidents could be found with involvement of slow moving vehicles in an preliminary investigated sample of about 2150 accidents over a period of 3 years.

2.5 Guidelines

In guidelines or capacity manuals the conditions of traffic flow are described by the relationship between travel speeds (v) of cars and actual traffic load (q) for the appropriate types of road.

These v-q-relations can be presented by mathematical curves (using regression methods). The shapes of the curves are depending on several parameters.

In order to include the influence of slow moving vehicles on v-q-relations the current German guidelines RAS-Q (1982) work according to the following idea:

Figure 6 shows such a curve in principle: If a travel speed of V_B is to be kept as lowest level as an index of "level of service", the traffic load must not exceed Q - in the absence of slow moving vehicles.

In case of slow moving vehicles the maximum limit for traffic load has to be decreased appropriately Q_{zul} by using a specific calculation factor which is derived from former estimations and which is presented explicitly in the guideline.

It can be noticed - under practical conditions found in Germany - that the influences of slow moving vehicles on macroscopic traffic flow parameters are not very visible in general because the quantities of these vehicles in traffic are often negligible. But it has to be recognised that in connection with the overtaking behaviour serious safety problems may occur as a matter of principle.

The circumstances of the diversity of travel speeds on two-lane rural roads affect a high amount of potentially overtaking vehicles. In case of high traffic volumes, most of the overtaking needs can not be satisfied, see Figure 7.

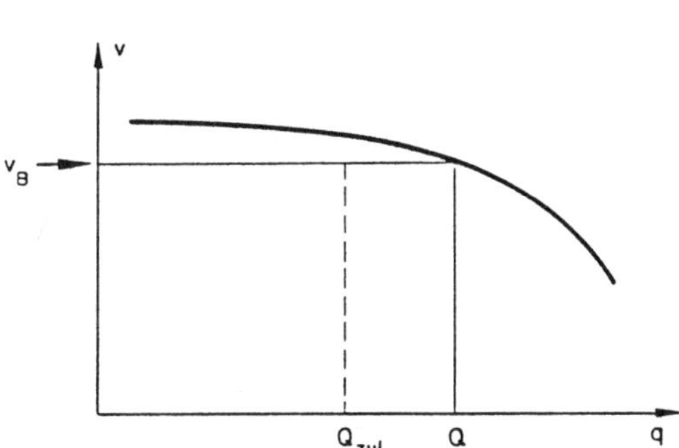

Fig. 6 Broad outline of a procedure considering slow moving vehicles in guideline (RAS-Q)

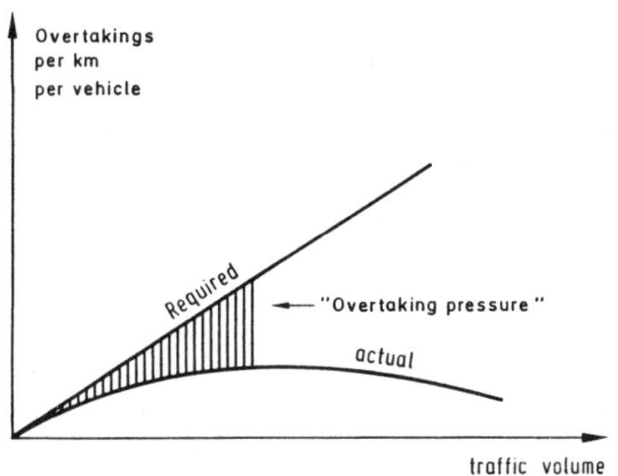

Fig.7 Diagrammatic view on "overtaking pressure" as the discrepancy of required and actual overtakings.

It becomes obvious that this procedure depends on the occurance of a decreasing v with an increasing q in the v-q-relationship. But from more recent measurements it can often be derived that v is nearly independent from q in an extended range of traffic load.

3 CONCLUSIONS

It is obvious that the mix of very different types of vehicle in one single traffic flow requires some special treatments under the aspects of traffic safety and road capacity. One possible solution would be segregation, i.e. in this connection the provision of separate tracks for slow moving vehicles. These tracks are often to be used simultaneously by the non-motorised traffic. Such solutions are not in the focus of this presentation.

In addition the prevailing topographic situations often do not provide the overtaking sight distances which are needed independently from actual traffic volumes. Those overtaking needs which cannot be satisfied can be defined as the so-called "overtaking-pressure" and may play an important role in evaluating new safety and design standards for different types of road which could also include the requirements in connection with slow moving vehicles. A hypothesis on these aspects is given by Leutzbach and Brannolte (1989).

Under these difficult conditions (traffic mix and alignment) the cross-section types b2s and b2ü still offer almost sufficient overtaking opportunities: Experiences show that the risk of having an accident on these roads is statistically lower than that on typical two-lane rural roads. Likewise the types b2s and b2ü belong to the group classified as intermediate cross-sections; the maximum acceptable traffic volumes are higher than on corresponding typical two-lane rural roads. Therefore often the main purpose for installing these types is to accommodate higher traffic volumes regardless of slow moving vehicles. This could not be achieved with the solution using a typical cross-section plus a separate track as mentioned above.

In general there are different opinions whether to install intermediate cross-sections, and, if so, to use either a b2s- or a b2ü-type. The discussions also include considerations how to provide good facilities for slow moving vehicles. All these items are being discussed widely in order to realise best adjustments of safety, traffic and environmental requirements to the selection of types of road. This includes an updating of current guidelines, too.

The research work mentioned here could only be given as an overview. The results ought to be kept in mind in case agricultural traffic is concerned.

REFERENCES

Brannolte, U. and Holz, S.1983. Simulation des Verkehrsablaufs auf Landstraßen - Modellerweiterung. Schriftenreihe Straßenbau und Straßenverkehrstechnik, Heft 402

Brannolte, U. 1989. Unfallauswertung für ausgewählte Streckenabschnitte mit Zwischenquerschnitten. Forschungsprojekt 8527/6 der Bundesanstalt für Straßenwesen

Brannolte, U.; Botma, H. and Schmidt, M. 1989. Überholungen gegenüber langsamen Fahrzeugen auf Landstraßen. Vorläufiger Bericht Nr. 26 des Instituts für Verkehrswesen, Universität (TH) Karlsruhe

HCM 1985. Highway Capacity Manual. Highway Research Board, Washington D.C., U.S.A.

Leutzbach, W.; Brannolte, U.; Schmidt, M. and Baumann, N. 1987. Untersuchungen des Verkehrsablaufs auf einbahnigen Straßen unter besonderer Berücksichtigung langsamer Fahrzeuge. Forschungsprojekt 2.090 des Bundesministers für Verkehr

Leutzbach, W. and Brannolte, U. 1989. Der Überholdruck - Zum Zusammenhang zwischen Überholparametern und Unfallraten auf Landstraßen. Straße und Autobahn, 2/1989

RAS-Q 1982. Richtlinien für die Anlage von Straßen, Teil: Querschnitte. Forschungsgesellschaft für das Straßen- und Verkehrswesen

Land and Water Use, Dodd & Grace (eds), © 1989 Balkema, Rotterdam. ISBN 90 6191 980 0

Rural roads: Results of research carried out in some areas of Lombardy

P.Bonfanti
Institute of Crop Production, University of Udine, Italy

A.Toccolini
Institute of Agricultural Engineering, University of Milan, Italy

ABSTRACTS: Rural roads should meet the following needs: transportation within a farm (field-farm and vice versa); transportation between field and service or collection centers and between the farm and these centers; transportation outside the farm, that is, long-distance transport to link production and consumption centers. This subject is constantly evolving as a result of the increasing connection between agriculture and markets. The result is that agriculture is being called on to innovate its design criteria and policies. For this purpose, it is necessary to outline actual requirements in terms of farm roads and transportation and the state (quality and quantity) of the existing road network. In this context, research has been carried out on some rural areas located in Lombard section of the Po Valley. These areas are representative of a dynamic agriculural situation: a series of parameters have been defined and analyzed with the aid of photo interpretation in order to evaluate various construction characteristics (carriageway width, design speed, bending radius, visibility and pavement) and land features (disaggregated density according to farm type, degree of integration, shape and size of fields, degree of mechanization).

RESUME': Les fonctions que la viabilitè agricole et rurale est appelèe à remplir concernent: les mouvements intèrieur-intèrieur (transferts du champ à l'exploitation et vice versa); les mouvements intèrieur-extèrieur (liaison entre le champ et les centres de service ou de ramassage et entre l'exploitation et ces sevices); les mouvements extèrieur-extèrieur (transports à longue distance pour la liaison entre centres de production et centres de consommation). Cette thèmatique, qui èvolue constamment en raison de l'ouverture aux marchès de l'agriculture, doit nècessairement rènover les modalitès avec lesquelles on affronte les projects en tenant compte des problèmes territoriaux. Il faut pour cela connaître les vèritables exigences des exploitants agricoles en fait de viabilitè et de transports ainsi que la situation (qualitative et quantitative) du rèseau routier existant. On a donc procèdè à une ètude sur certains territoires ruraux de la plaine lombarde, reprèsentatifs d'une rèalitè agricole dynamique; on a relevè et èlaborè - en se servant aussi des mèthodes de photo-interprétation - une série de paramétres permettant d'évaluer les caractéristiques de la construction des routes (largeur de la chaussée, vitesse de base, tortuosité, visibilité, pavage) et celles du territoire (densité divisée par type d'exploitation, degré de remembrement, forme et dimension des champs, niveau de mécanisation).

INHALT: Die landwirtschaftliche und ländliche Fahrbarkeit hat folgende Funktionen zu erfüllen: intern-intern Bewegungen (Verschiebungen Feld-Betriebszentrum und umgekehrt); intern-extern Bewegungen (Verbindungen zwischen Feld und Dienstoder Erntezentren und zwischen Betriebszentrum und Zentren selbst); extern-extern Bewegungen (Langstreckentransporte für die Verbindung zwischen Produktions - und Konsumzentren). Diese sich ständig entwickelnde Themastellung muss sich durch die allmähliche Offnung der landwirtschaftlichen Märkte notwendigerweise mit innovativen Aspekten des Territoriums und der Planung beschäftigen. Zu diesem Zweck muss man die tatsächlichen Anforderungen zum Thema Fahrbarkeit und Transport im Rahmen der landwirtschaftlichen Produktion und die qualitative und quantitative Lage des bestehenden Wegenetzes kennen.

Es ist daher eine Untersuchung in einigen ländlichen, landwirtschaftlich dynamischen Gebieten der lombardischen Ebene durchgeführt worden. Es werden dann, auch mit Hilfe von Photointerpretation, eine Reihe von Parametern zur Beurteilung der Baumerkmale (Fahrwegbreite, Basisgeschwindigkeit, Gewundenheit, Sicht - und

Baumerkmale (Fahrwegbreite, Basisgeschwindigkeit, Gewundenheit, Sicht - und Bodenverhältnisse) und der Gebietsmerkmale (getrennte Dichte pro Betriebstypologie, Stand der Zusammenschlüsse, Form und Ausmasse der Felder, Mechanisierungsstand) ermittelt und verarbeitet.

1. Outline of the Problem

1.1 When dealing with the subject of "roads in agriculture", the most important thing to do is to clearly define the nature of the problem. Therefore, examination of the various definitions (both Italian and foreign) of minor roads is extremely important. This consideration of meanings should be followed by an evaluation of the most pressing aspects that are determinant in the development of agriculture and the "food system" connected with it, as well as the rural situation in general. Indeed, it is clear that roads are a fundamental infrastructure used for the transfer of goods and services needed to carry out productive activity and for meeting the social needs of the rural population.

The various definitions and classifications do indeed turn out to be different. During a recent seminar (October, 1987) on this subject held in Wageningen, Holland, the amount of traffic (defined as 1,500 vehicles/day) was accepted as a means of distinguishing between rural and non-rural roads. This value (with which everyone was not in agreement) would be quite high in many areas, and in fact would be unacceptable in many developing countries. Consequently, various authors talk about "rural roads" and "minor rural roads", where the latter term refers to rural roads whose geometric characteristics, materials and maintenance are inferior.

In order to have some idea of the extent of the problem in several OECD countries, Table 1 provides some statistical data on the length of roads, the population and the surface area. Similar data is not available for Italy; however, it is estimated that there are approximately 142,000 km of road classifiable as "minor" (mainly local rural roads), out of a total of 298,000 km (including highways, trunk, regional and local rural roads). In terms of what may be called "minimum rural roads" (cooperative and farm roads), the total should be on the order of 300,000-400,000 km. This would mean that the density of low-traffic roads (minor plus minimum) in Italy is approximately 1.5 km/km2, which is similar to the situation in Holland and France.

Consequently, the problem is clearly posed in quantitative terms. If we assume a construction cost, at current values, ranging from 50 to 300 Mlire/km (the former value refers to simple farm roads produced by scarifying the land and then spreading and crushing gravel; the latter refers to local rural roads on flat land with carriageway widths of 3-4 meters and bituminous macadam pavement), that would imply a total of 40-50,000 billion lire of minor rural roads (142,000 km) and approximately 30-40,000 billion lire of minimum rural roads (300,000-400,000 km), for a grand total of 70-90,000 billion lire. It may be observed that the OECD proposes an average construction cost (1984 values, excluding unpaved roads) on the order of 80-90,0000 dollars per kilometer (approximately 100,000,000 lire), which is consistent with the estimates provided above.

As far as maintenance is concerned, the same OECD report estimates an annual cost between 1 and 3,000,000 lire/km (1984 values). In our opinion, those figures can be raised to 2-7,000,000 lire/km at current values, depending on the road's characteristics and the climate. Consequently, the total annual maintenance cost can be estimated at 1,500-2,000 billion lire. As we can see, these figures are significant and represent (in terms of reconstruction costs) approximately one-tenth of GNP, which in 1987 would have been approximately 850,000 billion lire.

1.2 With this as our frame of reference, it is important to analyze the various definitions of minor rural roads. Sermonti classifies the following roads as rural and distinguishes between them on the basis of ownership and maintenance expenses:

- <u>reclamation roads</u>: these are the principal district roads, and they are overseen by the government; depending on their characteristics, they are classified as trunk, regional or local roads;
- <u>service roads</u>: which are needed for the maintenance of public works (such as irrigation and drainage) or the

Table 1 - Road length, population and surface area in some OECD countries (source: OECD)

Country	LTR network (km)	Total road network (km)	Population (10^6)	Population/ road (inhabitant /km)	Area (10^6 km^2)	Population/ Area (inhabitant/ km^2)
Australia	752 162	791 750	15.5	19.6	7.64	2.0
Belgium	94 350	108 266	10.0	92.4	0.03	333.3
Canada	729 700	959 200	24.6	25.6	9.99	2.5
Denmark	35 000	69 420	5.0	72.0	0.04	116.2
Finland	66 277	75 000	4.9	64.0	0.34	14.3
France	1260 000	1500 000	55.0	36.0	0.55	100.0
Netherlands	57 500	109 000	14.2	130.2	0.04	344.8
Norway	42 000(2)	84 500(2)	4.4	52.3	0.32	13.6
Spain	324 345	351 672	37.7	107.2	0.50	75.4
Sweden	81 463(1)	97 719(1)	8.3	84.9	0.45	18.4
Switzerland	85 000	125 000	6.3	50.3	0.04	152.6
Turkey	51 189	59 112	44.7	756.8	0.78	57.3
U.S.A.	4202 467	6243 672	266.5	36.3	9.36	24.1

(1) State roads only
(2) Norwegian data cover private roads like forestry roads, etc.

accomplishment of operations concerning hill or mountainside reclamation; for these reasons, these roads are considered to be public works which are the government's responsibility;

- interfarm roads: (also called local/dirt roads) whose purpose is to ensure connections between farms and with reclamation roads or other important public roads; their maintenance is handled by users or frontagers;
- farm roads: within farms, to connect various farm facilities; their maintenance is handled by private individuals and their construction is considered to be part of land improvement.

A functional classification (Kollmer), on the other hand, provides the following breakdown:

- interconnecting roads, between rural villages and for access to product supply and delivery;
- main farm roads, which provide access to farms and to inhabited areas outside rural villages, as well as to large farms with over 250 ha;
- minor farm roads, used by farmers to reach their plots;
- village roads, within the urbanized area of the community;
- farm roads in fields and areas with specialized crops;
- other minor rural roads, also used for recreational purposes and for short connections between villages (school routes); in addition, they include special roads to separate the slow traffic of farm vehicles from faster automobile traffic, often found on one or both sides of the main road.

Apart from the definitions supplied above, which tend to emphasize either ownership, maintenance or function, the problem of rural roads has specific connotations of both a territorial and construction-related nature.

Aspects of the territorial problem include:

- redesign of the existing road network on the basis of the current trend in favor of an increase in the surface area of individual plots, the well-known need for mechanization (increase in plot length) and optimization of the irrigated network;
- the possibility of orienting fields according to the direction of the prevailing winds during land redevelopment processes in order to increase crop yields (up to 10%), as proposed by some authors (Gostovic and Peic);
- the relationship between roads and landscape, which has recently become important with the revival of the scenery issue, at various planning levels; this requires the careful study of environmental resources in the area undergoing road-related construction, as well as a precise evaluation of actual traffic requirements in terms of design speed, section, bending radii, visibility, etc;
- the function of the road

infrastructure, viewed as a decisive action for the recovery of marginal areas and the protection of areas which act as territorial protection (woods, summer pastures).

The following requirements are related to construction and management problems:

- improvement in the geometric characteristics of road lay-out (carriageway width, visibility, solution of junction points, etc.) on the basis of the changed characteristics of traffic and mechanical farm vehicles, and to prevent and/or decrease possible accidents; up-to-date data on the specific subject of minor rural roads do not exist in Italy. Mention should also be made of the phenomenon defined abroad as "rat run traffic", that is, traffic that uses minor roads as a "shortcut" in order to reach work or a recreation area more quickly during rush hour. This phenomenon, which is also widespread in certain rural sections of Italy that belong to greater metropolitan areas, is on the rise and requires a choice (which may vary from one case to another) between forceful restraint of the phenomenon or improvement of the roads so that they can handle their new function;
- improvement in characteristics concerning the load-bearing capacity of the superstructure on the basis of the new and more demanding features of farm vehicles noted above; for example, a tire on a trailer or a sewage-spreading wagon today supports a load of over 2500 daN with an inflation pressure of over 500 kPa. This means that the pressure measured on hard surfaces under the tire's tread can be up to twice as high;
- use of materials that are poor and/or produced by industrial processes and energy production (fly-ash) for the stabilization of the ground, both to decrease costs and find a use for materials which cannot be disposed of.

All of this should be viewed within the framework of the high cost of construction and maintenance.

This demonstrates that a significant increase in the quality of minimum farm roads (from farm to fields) is very difficult, if not impossible. Consequently, because of the high costs noted above, these roads are left in a state of repair that is barely adequate for the passage of farm vehicles (to some degree, this is also the case abroad). On the other hand, there is a serious and real problem concerning the possible and necessary improvement (conceivably at the same time as hoped-for land redevelopment) of rural roads starting from the interfarm level and working our way up.

2. Suggestions Made by the 1986 OECD Report

Some of the most important results of the report "Conception économiques des routes à faible trafic", which was written on behalf of the OECD by a work group established for this purpose (unfortunately, Italy was not included), are interesting within the context of the present paper.

The report's objective was to define a general methodology for the design of low-traffic roads. With reference to the problem of the definition of a low-traffic road, the following criteria were used: the amount of traffic, the function, the administrative classification and the management and financing procedure. The traffic intensity criterion (rural roads with traffic less than or equal to 1500 vehicles/hour) was accepted, though the need to integrate it with functional inspections was also noted.

The design themes considered were the following:

- design speed, defined as the conventional speed that makes it possible to establish some limited geometric characteristics;
- cross-section of the road, showing the carriageway width, the shoulders and the strip which is the subject of the intervention;
- horizontal bending radius, whose determination requires consideration of the design speed, the inclines and the coefficient of transverse friction. The bending radius naturally influences fuel consumption and vehicle safety;
- longitudinal profile, which is determined first by the slope of the lay-out; clearly, the longitudinal slope influences vehicle speed and energy consumption, as well as user safety in terms of braking visibility.

In addition, it should be noted that current design standards are based on laws that govern vehicle behavior on the one hand and that of the driver on the other. These laws define minimum values for the geometric characteristics of the lay-out that guarantee a basic level of safety and comfort. Moreover, experience shows that the non-critical application of design standards does not necessarily lead to satisfactory lay-out from a management cost, energy and environmental standpoint. Indeed, it is a well-known fact that so-called "good design" is the result of the synthesis of constructive, economic and environmental requirements (Table 2).

The report contains numerous tables summarizing the design parameters adopted by various OECD countries. Here it is useful to present a summary of the proposed improvements with special reference to the minimal recommended values for design parameters (Tables 3 and 4).

3. Initial Results of an Investigation Carried out on the Lombard Countryside

In order to better define certain territorial, construction and administrative aspects of rural roads, an investigation was carried out on some parts of the Lombard countryside.

The aim was to analyze the correlations between rural roads and parameters connected with agricultural production; in particular, the study considered land development, farm size, crop lay-out and mechanization.

Table 2 - Impact of geometric design (Source: OECD)

Design Feature	Safety	Comfort	Trip time	Fuel cons.	Const. cost	Env. cost
Carriageway width ↗	(↗)	↗	↘	↘	↗	↗
Carr. width + hard shoulder + shoulder ↗↗	↗	↗	↘	↘	↗	↗
Gradient ↗	(↘)	↘	↗	↗	↘	↘
Horizontal radius of curvature ↗	↗	↗	↘	↘	↗	↗
Vertical radius of curvature ↗	↗	↗	↘	↘	↗	↗
Bendiness ↗	(↘)	↘	↗	↗	↘	↘
Hilliness ↗	↘	↘	↗	↗	↘	↘

( ) possible impact

Table 3 - Recommended minimum carriageway widths in relation to conditions of opposing traffic (Source: OECD)

Carriageway width (m)	Meeting conditions for opposing traffic
6,0	Almost all vehicles can meet with normal speeds
5,5	Lorries/buses can meet with reduced speeds. Lorries can meet passenger cars with reduced speeds. Passenger cars can meet with normal speeds.
5,0	Lorries/buses cannot meet without special widenings. Lorries can meet passenger cars with reduced speeds. Passenger cars can meet with almost normal speeds.
4,5	Special widenings are required for meeting of lorries/ buses and of lorries and passenger cars. Passenger cars can meet with reduced speeds.
4,0	Special widenings are required for meeting of lorries/ buses and of lorries and passenger cars. Passenger cars can meet with considerably reduced speeds.
3,5	Special widenings are required for meeting.

NOTE: Minimum shoulder width: 0.50 m (1m would be desirable)

Table 4 - Minimum design parameters as a function of type of road

Road type (Design speed: DS)	Minimum horizontal radius (m)	Maximum super-elevation (%)	Maximum gradient (%)	Minimum crest radius (m)	Stopping sight distance (m)
Very low traffic and single-lane roads	23-30	8	11-12	Depends on visibility conditions > 200 m	20-30
DS 40/50	40-80	8	10-11	250-800	40-60
DS 60/70	90-130	7-8	9-10	800-2000	70-100
DS 80/90	230-300	7-8	8	2000-5000	120-160

The research methodology adopted (Fig.1) provided for the acquisition of information through photo interpretation of aerial color pictures (average scale 1:20,000), in addition to special tests carried out on the ground, regional cartography and questionnaires for the survey of farm data. Thus, the rural roads in each local area were studied and classified in administrative, construction and functional terms. Later, the use of the soil and the development of the land was reconstructed on regional, technical maps in order to determine the size and shape of the fields.

The following information was obtained at farm level: the agricultural surface area in use (SAU), the number of fields, the average distance and the machinery located on the farm.

During the first phase of data processing, the indexes relating to scattering of farm surface area (ratio between scattered SAU and total SAU), shape (irregular SAU/total SAU) and crop lay-out (sowable land/total SAU) were determined. Then, these indexes were correlated with farm roads in terms of unit of surface area.

In the first area under investigation, rural roads occupied 57.5 km of a total surface area of 1,340 ha of SAU, which corresponded to a length per unit of surface area of 42.9 m/ha. The classification made it possible to determine (Fig. 2): transportation outside the farm, that is, on minor roads or local roads paved with bituminous macadam having a width of at least 5 m and corresponding to 26% of all rural roads and a value per ha of 11 m; transportation between field or farm and service centers, that is, on main farm roads without pavement but with a stabilized subgrade, with a width ranging from 3 to 5 m (33% equal to 14.2 m/ha), and transportation within the farm, that is, on minimum farm roads with a width of less than 3 m and with a stabilized subgrade (18%; 7.9 m/ha), or without stabilization and often grassed over (23%; 9.7 m/ha).

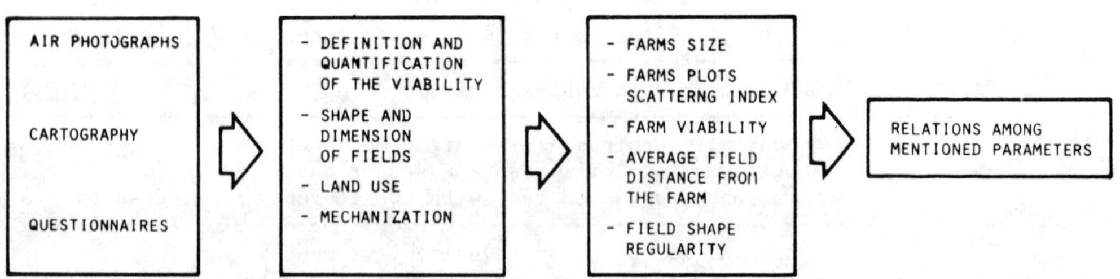

Fig. 1 - Block diagram of methodology used for data collection.

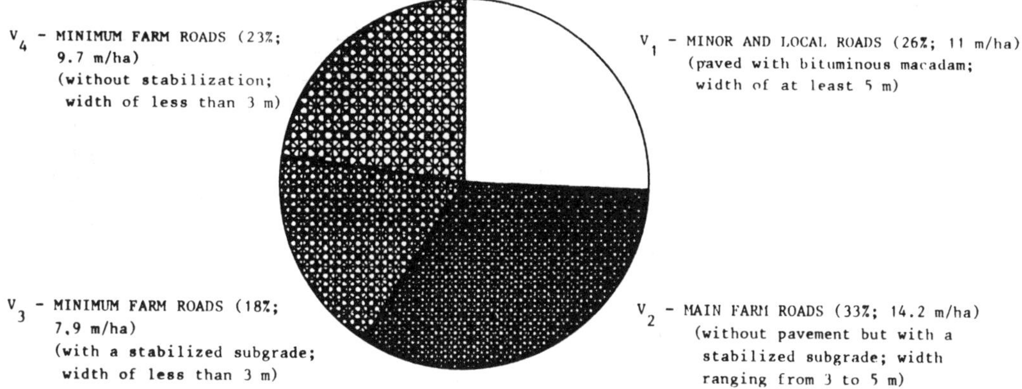

Fig. 2 - Percentage breakdown of various types of rural roads.

In terms of farm utilization of each category of road as a function of the scattering index, the investigation showed a tendency towards use of the first two road types by farms with scattered surface area, while integral farms tended to use the various road types equally, with a slight preference for the last two types (Fig. 3). Analysis of the size and shape of the fields in the area under consideration showed a tendency towards the creation of large fields (average surface area: 2.32 ha) with an irregular shape. Indeed, fields over 2 ha turned out to be 43% of the total number of fields and 74% of the total surface area. In terms of shape, irregular fields represented 38% of the total number and 50% of the surface area (Fig. 4). By correlating the value of rural roads per hectare of cultivated surface area with the average surface area covered by plots, it becomes clear that as the latter increases the amount of road decreases significantly from values of 200-350 m/ha with surfaces of approximately 1 ha to 50 m/ha in the case of surfaces over 3 ha (Fig. 5). Comparison with the shape index (regular surface/SAU), however, does not provide enough information to form a hypothesis, since the values were variously distributed over the graph (Fig. 6).

Figure 7 shows the correlation between farm roads and the scattering index (scattered surface area/SAU); the distribution of the points confirms the fact that large, integral farms have an advantage in terms of roadway over small, scattered farms.

Data relating to the crop lay-out of farms, expressed by the ratio of sowable surface area to SAU, did not provide significant results when compared with road data, and this excluded the influence of soil use on farm roads in the case under consideration.

4. Conclusions

Examination of available data and the results of initial field investigations brings to light some key problems and hence some specific research topics to be pursued in this area. In summary, it is necessary to:

- expand our quantitative and qualitative knowledge (presently scanty) about minor and minimum rural roads; one of the techniques selected for this purpose should be remote sensing;
- define a series of standards (similar to those proposed by the OECD) relating to access from minimum rural roads to higher level roads and improvement of geometric and construction features (including maintenance) of minor rural roads, by evaluating the possibility of creating separate lanes for slow vehicles in certain situations;
- define construction characteristics for farm roads, with special reference to current mechanization requirements (increase in axle loads) and the organization of work (transport sites);

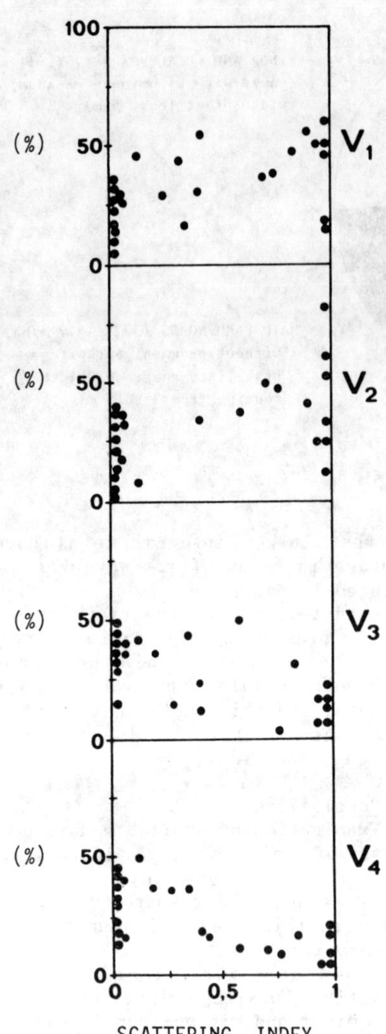

Fig. 3 - Utilization of different types of rural roads (ref.Fig.2) by farms as a function of the scattering index (scattered surface area/SAU).

- define the key elements in the road-countryside relationship, with the dual purpose of focusing attention on concrete themes and providing guidelines on an extremely important subject, especially in the case of areas which rely on agriculture and tourism and/or forestry.

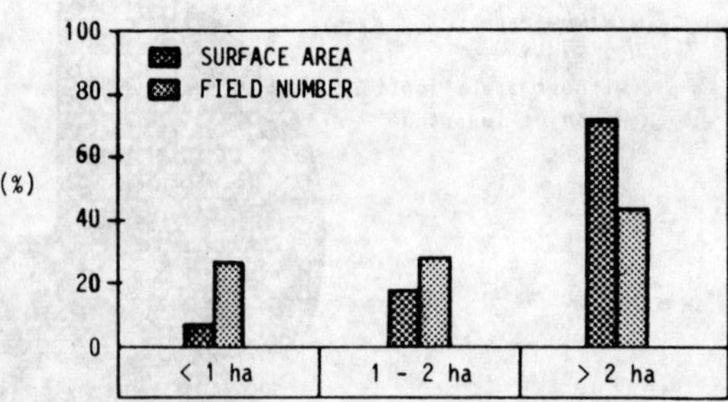

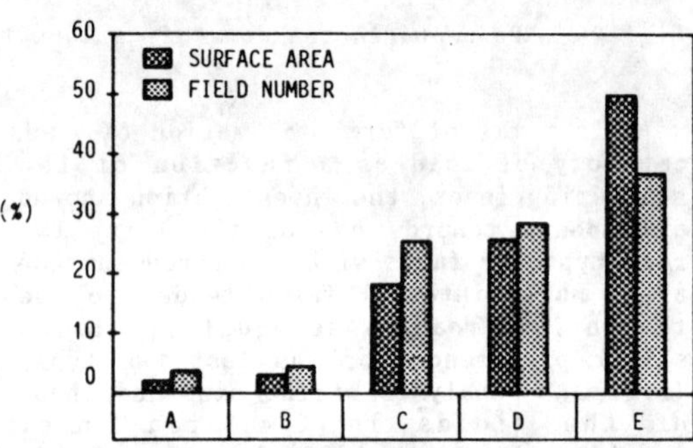

Fig. 4 - Percentage values of surface area and field number according to farm size (above) and shape (below: A = triangular, B = square, C = rectangular, D = trapezoidal, E = irregular).

REFERENCES

Michels Th. & Jaarsma C.F. (eds.) 1988. Minor Rural Roads. Planning, design and evaluation. Wageningen: Pudoc

OECD (ed.) 1986. Economic design of low-traffic roads. Paris

Ministère de la Région Walloune (ed.) 1985. Chemins Ruraux. Bruxelles: Service de remembrement rural.

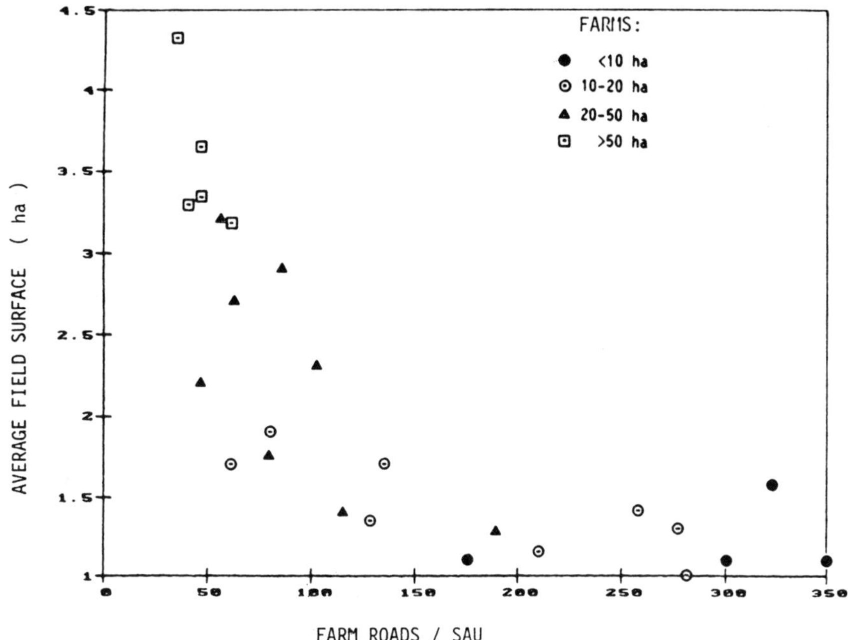

Fig. 5 - Comparison of farm road density and average field surface area.

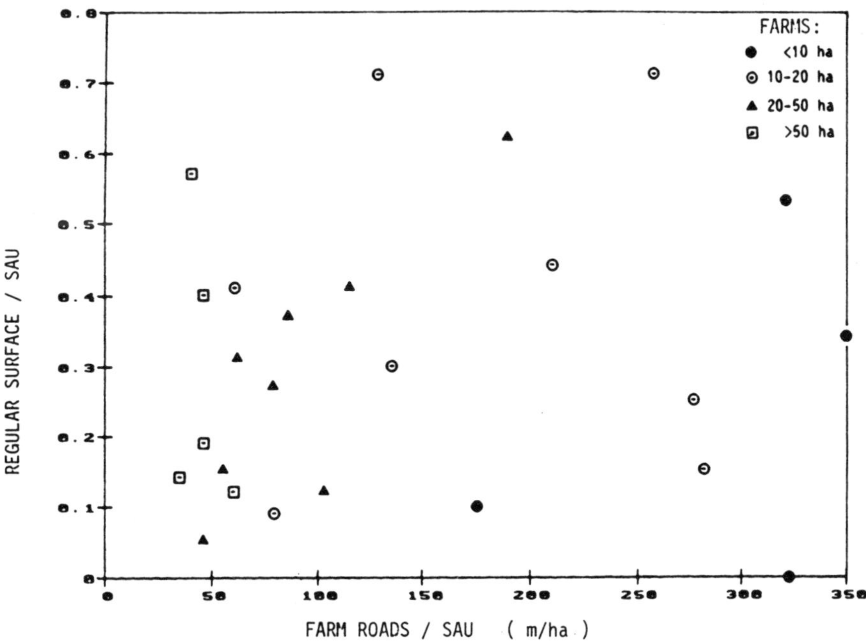

Fig. 6 - Comparison of farm road density and field shape index.

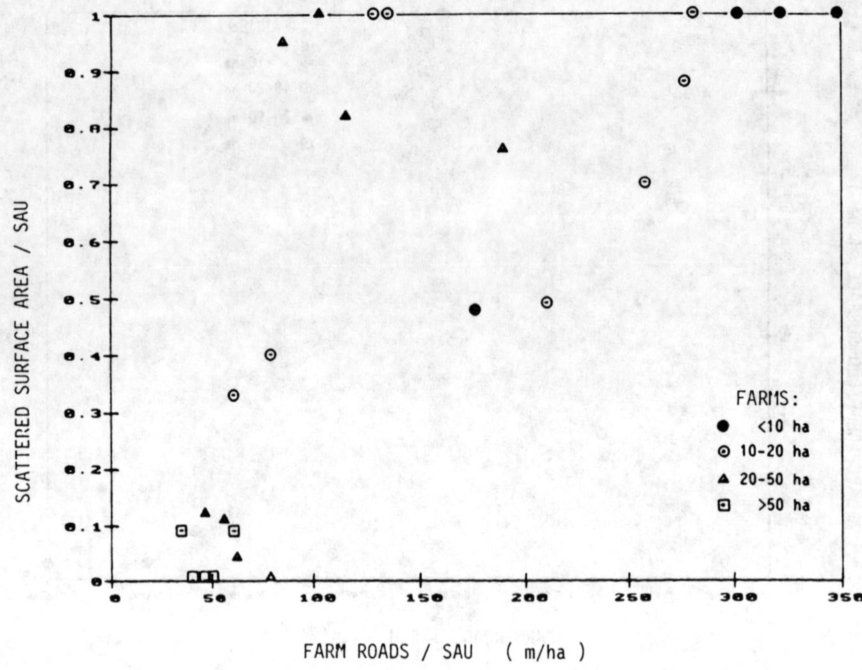

Fig. 7 - Comparison of farm road density and field scattering index.

Land and Water Use, Dodd & Grace (eds), © 1989 Balkema, Rotterdam. ISBN 90 6191 980 0

Reinforced asphalt for rural roads in the province Noord-Holland, Netherlands

D.R.Didden
Government Service for Land and Water Use, Civil Engineering Department Noord-Holland, Haarlem, Netherlands

ABSTRACT: In the province of Noord-Holland there are many rural roads constructed on very poor conditions in relation to the subsoil.This subsoil mainly consists of layers of peat 2-5 meters thick and even more on layers of clay.

Due to this subsoil in relation to the increase of motorized traffic the maintenance of these roads became a tremendous problem.Most of the time it was necessary to resurface the roads within 4 years or less.

Therefore it was decided to apply reinforced asphalt resurfacing in order to prevent longitudial cracking to achieve a longer interval between resurfacing and thus a cheaper way of maintenance was possible. The results of this application of reinforced asphalt are only monitored by visual inspection of these roads.

A test was also done to find out if reinforcement could reduce the calculated asphalt layer thickness.This was done on a selected rural road by a falling- weight-deflectometer.

In 1984,the Government Service for Land and Water Use started an experiment to insert a reinforcement of steel grids in a road to be resurfaced with a new asphalt layer.Afterwards when the expirement seemed to be successfull, reinforcement of other materials were applied.

This contribution to the workshop will give a review of the experience obtained up till now.

The financial aspects of the application of reinforcement in asphalt courses will not be taken into consideration.

Résumé

Beaucoup de chemins ruraux dans la province de la Hollande septentrionale ont été construits sur un très mauvais sous-sol.

Celui-ci se compose notamment d'une couche de tourbe d'une épaisseur de 2 à 5 mètres ou davantage reposant sur une couche d'argile.

Par suite de l'intensification de la circulation motorisée, l'entretien de ces chemins est devenu un énorme problème. En général, le resurfaçage s'impose au moins tous les 4 ans.

Face à cette situation, il a été décidé d'utiliser l'asphalte armé afin d'eviter la fissuration longitudinale et de réaliser un plus grand intervalle entre les opérations de resurfaçage, permettant une diminution des coûts d'entretien. L'effet de l'utilisation dudit matériau est uniquement apprécié par inspection visuelle des chemins concernés. Un test a été effectué pour vérifier si le renforcement permettait également de réduire l'épaisseur calculée des couches d'asphalte.

Ce test a été effectué à l'aide d'un fleximètre à masse tombante et ne concernait qu'un seul chemin.

En 1984, le Service national de l'aménagement rural a entrepris un essai consistant à armer un chemin à asphalter d'un grillage métallique. A la suite du succès de cet essai, le renforcement à l'aide d'autres matériaux a éte appliqué.

Le présent exposé passera en revue l'expérience acquise jusqu'a présent sans prendre en compte les aspects financiers du revêtement à l'aide d'asphalte armé.

Zusammenfassung
Viele Landstrassen in der Provinz Nordholland sind auf sehr slechtem Untergrund gebaut: hauptsächlich eine Torfschicht bis 2 bis 5 Meter Tiefe, manchmal sogar dicker, darunter eine Tonschicht.
Bei diesem Untergrund hat die Intensivierung des Autoverkehrs den Unterhalt der Strassen zu einem gigantischen Problem werden lassen. Meistens musste die Fahrbahndecke schon innerhalb von vier Jahren erneuert werden.
Deshalb wurde beschlossen, als Vorbeugemassnahme gegen Längsrisse bei Überbauung Asphaltarmierungen zu verwenden, um so die Erneuerungsfrequenz und damit die Instandhaltungskosten zu senken. Der Effekt dieser Massnahme wird lediglich durch Inspektion der behandelten Strassen nach dem Augenschein gemessen.
Erprobt wurde auch, ob eine armierte Asphaltschicht eine geringere Dicke haben könnte. Dieses Experiment wurde auf einer Versuchstrasse mit Hilfe eines Fallgewicht-Deflektometers durchgeführt.

Im Jahre 1984 begann der Landentwicklungsdienst ein Experiment mit Stahlrosteinlagen in einer Strasse, die neu asphaltiert werden sollte. Als dieser versuch zu positieven Ergebnissen geführt hatte, wurden auch Armierungen aus anderen Materialen getestet.
In diesem Seminarbeitrag werden die bisher gemachten Erfahrungen besprochen.
Nicht berücksichtigt werden dabei die finanziellen Aspekte der Anwendung von Armierungen in Asphaltschichten.

INTRODUCTION

The past

The province of Noord-Holland consists for the greater part of peat,clay and water.This is caused by the fact that the the dewatering behind the dunes was very poor until about 5 centuries ago so it was possible that an area was created with layers of peat up to several meters thick.
During and also after this period the area was flooded by the sea so that layers of clay were deposited in this area.

For the people who settled here it was a hard way of living,they had to defend themselves against flooding and mostly the higher spots were used to build their farms and also the connections between these farms and later the small villages, were built on the most suitable locations,that is, also on the higher spots.

At that time the roads were improved by all kind of materials such as sand,rubble and so on.These materials however were not available in large quantities,so the roads remained narrow as were the shoulders of these roads. The stability of these roads was poor but sufficient for the use at that time.

Then the motorized traffic became common and the roads had to be paved,first with bricks,later on by all kinds of asphalt.
Sometimes,the asphalt was just put on the existing materials,however this was not the most suitable way of road building.The traffic,specially for its use in agriculture, grew in quantity and in weight.The road construction was not in proportion to this traffic and the roads had to be resurfaced or reconstructed frequently.

One of the main reasons underlying longitudial cracking of these roads was the horizontal forces due to this traffic.This occured particularly where the shoulders were small resulting in hardly any resistance to these forces.

There was only one adequate way of reconstruction, namely excavating the subsoil until a layer of sufficient bearing capacity was reached and filling in the excavation with sand. As stated before the layers of peat were several meters thick so excavation and refilling could be very expensive.This way of working is only applíed by the construction of some motorhighways.
A variant on this construction method was to dump the sand on the natural ground and this sand will sink.This progress will be repeated until a certain balance is reached and the construction of the pavement of the road can be made.
This way of roadbuilding is also expensive and is only used for main rural roads.

New construction methods

As many roads will be reconstructed by the Government Service of Land and Water Use a cheaper way of reconstruction,without resurfacing again within a short period, had to be found.

One possible way to strengthen the road so that cracking was prevented was reinforcing the asphalt concrete.
The reinforcement grids which were used are:

-galvanized steel wire Mesh Track S
this is a twisted wire netting reinforced by three wire transverse strands.Wire diameter is 3 mm.,strands 3*3.5mm.

-polymer grid Tensar AR 1
this is polymer sheet,punched and stretched so that the longitudinal rib and transverse bars are provided with a high tensile strength,
the melting point is ca 148 oC.

-woven polyester HaTelit

this is a polyester fibre with a special treatment to give a good adhensiveniss to the bitumen.

ROAD CONSTRUCTION WITH REINFORCEMENTS

Reinforcing by steel grids

In 1984 the first reinforced resurfaced road was constructed by the Government Service for Land and Water Use.
This concerned a road located about 50 km. north of Amsterdam.
The subsoil shows ca 1.00 m of clay mixed with peat and under this layer there is a layer of peat over 6.00 m.thick.
In cross section the road showed a pavement of 3.00 m. and the shoulders are < 1.00 m.The slopes of the ditches are ca 1:1,the water-level in the ditches is approx 1.25 m. below the road.
The pavement consists of a asphalt concrete layer ca 150 mm.thick on a layer of sand and rubble ca 200 mm.thick. There were longitudial cracks in the pavement,some with a length of several meters and 50 mm or more wide.In addition there was rutting till 20-30 mm.deep.
First a test section was selected with a length of 100 m.The sequence of the construction activities were as follows:
-after a tack coat was applied the rutting was filled with asphalt concrete.
-after this the steel grid was installed and tensioned by a roller and then the meshes were fixed to the underlaying asphalt pavement by nails (ca 4 nails/m2.).
In longitudial direction,the grids overlapped each other by ca 0.50 m.In the bends the grids were cut and the grids also overlapped each other by ca 0.50 m.
-Two layers of asphalt were laid over the grids.The first layer with a thickness of 60 mm. and the second layer with a thickness of 50 mm.

For several weeks,after the test section was completed,heavy transport of trucks was present on this section because of other construction works.The section was monitored before and after this period of heavy dynamic stresses. The section did not show any rutting or cracking.
Given this result the whole road,with a length of 1700 m.was constructed in this way.
An extension of this road continued with the same subsoil conditions and also the same traffic conditions.This road however had already been resurfaced previously so a new asphalt resurfing was not necessary at that time.
After ca 4 years it turned out that the road with the grid reinforcement did not show any cracking or rutting.The road without the grid reinforcement will have to be reconstructed within a short time.
The conclusion after visual inspection of this experiment is that the steelwire reinforced road shows a surprisingly good surface after more then 4 years and the impression is that steelwire reinforcement gives a positive addition to the strength of the road.

Reinforcement by polymergrids

In order to get experience with other grids a road was resurfed and reinforced with a polymer grid in 1985.
This concerned a road on a reservoir dike The water level on one side was approx 1.00 m. below the roadsurface.On the other side there was a polder ca 3.00 m. below the roadsurface.
The dike was built with sand and clay.
The cross section showed on the reservoir side a shoulder of 2.50 m.,a pavement of 3.50 m.and a shoulder on the polder side of ca 1.00 m.The slope of the dike to the polder was 1:3 and this slope ended in the drainage ditch.
First,the road was widened on the reservoir side by constructing a bed next to the exsisting sub-base.This bed was made of blast-furnace slag, 0.40 m.thick. Over this bed,a layer of asphalt was made, ca 40 mm.thick.
Because of stability reasons so no cracks would appear between the new bed and the existing road,it was decided the asphalt course on both the road and the bed should be constucted at least a year later.
To give more horizontal strenght also a polymer grid was applied.
The grids are installed on the surface of the road and fastened to the underlaying surface on one end by a strip which is fixed by nails (3 nails/ml) to this surface. The grid is then brought under tension.Per 200 m. length and for a width of 3.80 m.the tension must be uniformly distributed at 3-5 kN. If the tensioning is carried out correctly the grid must,after tightening, have an extension of 250 mm. per 50 m. of grid.
After the grid is brought under tension then the other end is also fixed to the underlying surface by a strip and nails.
When this is done the whole grid is sprayed with a cationic instable emulsion (ca 1.2 kg/m2) and spread with ca 10 kg/m2 chippings. The chippings are rolled with a rubber-tyred roller. The surplus then is removed and the

asphalt pavement is applied. On this road a base course and a wearing course of asphalt concrete,total,120 mm.thick is used.
The road has now been in use for ca 4 years and cracks have not shown up in this period.

The conclusion after this period is that polymer grids contribute well against horizontal forces of the road.

After the work on this road was finished, an other road was resurfaced in the same way. However the soil- and dewatering conditions were very bad. The shoulders of this road had a width of 0.50 m upto 1.00 m. The pavement was ca 3.50 m. wide and the water level in the side ditches was approx 0.25m below the pavement. The soil conditions were also very bad.The road was constructed on a layer of peat ca 5.00 m to 10.00 m thick.

The existing sub base was a mixture of all kind of materials such as rubble ,slag and bricks and this sub base was covered by an asphalt concrete layer 150-250 mm.thick. This asphalt showed longitidual cracks and cracking.This road had to be surfaced every two to three years.Besides it was also not possible to construct the road too heavy otherwise it would sunk deeper and deeper due to the subsoil.

After the rutting was filled and the cracking removed and refilled with asphalt,the Tensar grid was placed and a layer of asphalt, 120 mm thick,was laid.
In the ca 3 years the road has been in use no cracks and cracking have appeared,so it seems the grid is an usefull addition to the horizontal strength of this road,also under very poor subsoil conditions.

Woven polyester

After the succesfull application with steel and polymer grids,a test was done with a woven polyester.To make a comparision with polymer grids and to find a possible relation to reduce the thickness of the calculated asphalt concrete layers when grids are used, it was decided a comprehensive test should be made.

For this test an average rural road was selected.
The cross section of this road showed a pavement of 3.50m. and on one side a shoulder of ca. 0.50 m. and a dewatering ditch with a water level of approx 0.40 m.below the pavement. Between this shoulder and the dewatering ditch was a concrete revetment.
On the other side of the pavement was an other shoulder of 2.00-3.00 m.There was no dewatering ditch but the land was ca 0.30 m. lower than the pavement. The dewatering of the road was partly through this land.The sub base of the road was a mixture of sand, rubble and slag,up to 0.40 m.thick,covered by a layer of asphalt concrete,100-200 mm.thick. The subsoil was a layer of peat mixed with clay.

For the test,four locations in the road were selected.
100 m. Hatelit (woven polyester)
300 m. Tensar (polymer grid)
200 m. Hatelit
200 m. Resurfacing by asphalt concrete layers (This location was selected as reference section for the reinforced locations.)

The test locations were not concatinated but the most suitable locations were selected i.e.in relation to the subsoil and thickness of the asphalt concrete.

To determine the value of the polymer grid and woven polyester in relation to thickness of the asphalt layers the road was,after the filling up the ruts,tested by a falling-weight-deflectometer.The reinforcement then was applied and a first layer of asphalt, ca.50 mm.thick laid.
Then a second test by the falling-weight-deflectometer was done. After this the wearing course was applied and the last test by the falling-weight-deflectometer was done.
The falling weight of the deflectometer was 50 kN and the deflections were measured by 5 geophones connected to a recorder.
After this the results were computed by a computer program and the HEC values (equivalent construction thickness) determined.
The elastic modulus of the subsoil was also calculated.
The tests of the deflectometer were,all three times,carried out on the exact (measured) same location.
The distance between the test points in a test location was 20 m.and 3 tests were done per cross section.

The test was done in close co-operation with the contractor K.W.S. and his consultant R&E Consult,Utrecht,who also did the deflection tests,calculations and computer runs.

The results of the determination of the increase of the HEC values are shown in table 1

Conclusions from this test

The increase of the HEC values after the second resurfing is bigger in the locations with the Hatelit and the Tensar than in the reference location. This increase can be attributed to the Hatelit c.q.Tensar.

The difference for Hatelit is
87*(2.7-1.9) =69.6 mm HEC.,this is equivalent to ca 20 mm asphalt concrete.

The difference for Tensar is
101*(2.8-1.9)=90.9 mm HEC.,which is equivalent to ca 25-30 mm asphalt concrete.

The results can be influenced by:

-temperature.
With a higher temperature the relation between the elastic moduli of the reinforcement and the asphalt will increase. So by a high temperature the contribution of reinforcement will be higher.

-Thickness of asphalt layer.
When the thickness of the asphalt layer increases the HEC value per mm. asphalt will decrease. This can be seen in the table .
-Location of the tension.
When the thickness of the asphalt layer increases the tension will be increasingly assimilated by the asphalt layer.

REMARKS AND FINAL CONCLUSIONS

Asphalt in relation to reinforcement

The Tensar and Hatelit mats are fixed to the road surface by emulsion and the Mesh Track is fixed by nails.
When the asphalt concrete is applied, some problems can occur with the Tensar and Hatelit and to a less extent with Mesh Track because of the good fixing by the nails.
It should be noted however that Mesh Track is rather inflexible in laying. Some remarks and experiences in this field are mentioned below.

First there is the curling of the Tensar and Hatelit, during the application of the asphalt concrete. However when the Tensar is under sufficient tension it is a minor problem for the Tensar grid. In bends, special attention must be paid to the connection between the grids and to fixing to the road because they are cut in to follow the curve of the road.
Also regulation of the rutting must be done carefully so that the grids can be attached to a flat pavement. Hatelit must be rolled out after the emulsion is broken and this must not be done over a long distance. If possible the Hatelit must be rolled just in front of the asphalt truck. When this truck is empty, the role is flipped over at 45 degrees to the shoulder, the next truck can come and the role is relaid from this position.
To avoid curling under the tracks of the paver it is helpfull to bring some asphalt under these tracks.
Besides this the asphalt trucks should not be pushed by the paver because this will also result in to much force from the tracks to the grids and will curl the grids.
The thickness of the first asphalt course on the grids must be at least 50 mm., the total asphalt courses on Tensar must be at least 80 mm. and on Hatelit at least 50 mm.

Asphalt concrete with a higher percentage of bitumen gives a better result than with a lower bitumen percentage.

table 1

Testloc.	HEC before resurf.	HEC first resurf.	HEC incr.	resurf. thickness in mm	increase HEC per mm asphalt
Hatelit	565	750	185	59	3.1
Tensar	498	679	181	58	3.1
Reference	504	694	190	59	3.2
		second resurf.		thickness cumulative	
Hatelit	565	796	231	87	2.7
Tensar	498	784	286	101	2.8
Reference	504	700	196	106	1.9

Of course rolling of asphalt with a high bitumen percentage can give troubles in relation to fatting up.
An asphalt concrete with a high percentage of gravel does not give many problems in relation to the grids.

Final conclusions

After visual inspection it turns out that reinforcing asphalt concrete by steel grids, polymer grids and woven polyester has been successfull even under poor subsoil conditions and also gives an addition to the horizontal strength of the road.

The test shows that it is possible that the calculated thickness of asphalt courses can be reduced due to the reinforcement, however, further tests should be done to confirm this conclusion.

Land and Water Use, Dodd & Grace (eds), © 1989 Balkema, Rotterdam. ISBN 90 6191 980 0

The application of light weight materials in road constructions

Th.H.van Putten
Government Service for Land and Water Use, Utrecht, Netherlands

ABSTRACT: In areas with thick layers of peat under a thin cover of clay in the subsoil, the task of managing and maintaining a road network is a heavy and expensive one. Traditional measures to ameliorate a road construction bring about an increase in the weight on the subsoil which reinitiate the consolidation proces of the layers of peat.
The continuing and local varying deformation of the subsoil will sooner or later exceed the maximum permitted tensions and strains in the road construction. The result will be cracking of the road construction.
Instead of a total reconstruction of the road by replacing the road base by strong and often heavy materials an experiment has been started to lighten the weights on the subsoil caused by a new construction of the road.
The aim of this sort of design of a road construction is not to disturb the existing balance of the total weights of road construction and subsoil. In this situation the reconstruction starts with the excavation of the existing road construction including a part of the subsoil. The construction of the new road starts with light materials straight on to the subsoil. On these materials the road base and the asphalt concrete layers will be built up.
The feature of this construction is that the light weight materials compensate the heavier weights of the asphalt concrete layer in such a way that the consolidation does not take place.
An experiment was started in 1987 with different types of light weight materials. The constructions, materials and experience to date will be described.

RÉSUMÉ: Dans les zones où le sol est formé d'une mince couche d'argile reposant sur une épaisse couche de tourbe, la gestion et l'entretien du réseau routier est une tâche à la fois difficile et coûteuse.
Les mesures traditionnelles de restauration des chaussées sont de nature à augmenter la charge exercée sur le sous-sol, ce qui déclenche de nouveau le processus de consolidation des couches de tourbe.
Par suite de la déformation continue et, par endroits, variable du sous-sol, la tension maximale dans le corps de la chaussée sera tôt ou tard dépassée, ce qui se manifestera par la fissuration de la surface de roulement.
Au lieu de procéder à la reconstruction totale par la mise en place d'une nouvelle couche de base à l'aide de matériaux solides et souvent lourds, on a entrepris un essai pour alléger la charge que la chaussée exerce sur le sol.
La chaussée de type léger a été concue pour ajuster le poids de l'installation à la portance du sol, de facon à ne pas perturber l'équilibre qui s'est créé.
Après l'excavation de la chaussée existante et l'enlèvement de la couche superficielle du sous-sol, les traveaux commencent par mise en place de matériaux légers sur le sous-sol. Sur ces matériaux seront superposées la couche de base et la couche de surface (en béton ou en asphalte).
Ce type de chaussée présente l'avantage de compenser, grâce à l'utilisation de matériaux légers, le poids assez lourd des couches d'asphalte ou de béton de facon á empêcher la consolidation.
En 1987 un essai a été lancé, mettant en oeuvre différents matériaux légers. Le

présent exposé comporte une description de chaussées de ce type, des caractéristiques de matériaux utilisés ainsi que de l'expérience acquise jusqu'á présent.

KURZFASSUNG: In gebieten, wo der Untergrund aus dicken Torfschichten unter einer dünnen Tonschicht besteht, ist die Verwaltung und Instandhaltung des Strassennetzes eine schwere und kostenaufwendige Aufgabe. Herkömmliche Massnahmen zur eine Verbesserung de Strassenkörpers ergeben eine stärkere Belastung des Untergrunds, wodurch in der Torfschicht der Bodenverdichtungsprozess wieder in Gang gesetzt wird.
Nach kürzerer oder längerer Zeit wird die unablässige und örtlich verschiedene Verformung des Untergrunds dazu führen, dass im Strassenkörper die Spannungsgrenzwerte überschritten werden und Dehnung auftritt. Die Folge: Risse in der Fahrbahndecke.
Statt Totalerneuerung durch Einbau einer neuen Tragschicht aus starkem und oft schwerem Material wird seit einiger Zeit mit Konstruktionen experimentiert, die die Belastung des Bodens durch die neue Anlage verringern sollen.
Ziel dieses strassenbaulichen Konzepts ist eine Konstruktion, bei der das Neugewicht der Gesamtanlage die Tragfähigkeit des Untergrunds nicht überstrapaziert und das entstandene Gleichgewicht nicht zerstört. Bei der Erneuerung werden daher zuerst die Altanlage und eine Schicht Untergrund ausgeschachtet. Der Bau der neuen Anlage fängt damit an, dass leichtes Material auf den Untergrund geschüttet wird. Darauf kommen dann die Tragschicht und die Asphalt- oder Betonschicht.
Der Vorteil dieser Konstruktion besteht darin, dass die leichten Materialien das Gewicht der Asphalt- oder Betonschichten soweit ausgleichen, dass keine Verdichtung der Torfschicht auftritt.
Ein Experiment mit verschiedenen leichten Materialien ist im Jahre 1987 angelaufen. Beschrieben werden die Konstruktionen, die verwendeten Materialien und die bisherigen Erfahrungen.

INTRODUCTION

In The Netherlands there are about 97,000km of paved roads available for all kinds of traffic. Out of these about 43,000 km belong to urban areas. In the rural areas about 4,400 km of motorways and other national highways exist and over 7,000 km belong to the provincial road network. There are about 42,000 km of roads available with a function mainly or largely for the rural areas. Moreover there are about 15,000 km of unpaved roads with an important interest in the rural areas. Table 1 gives a complete presentation of the total length of roads in The Netherlands.

Table 1. Length of the paved and unpaved road network (km) in The Netherlands, 1-1-1985 (after CBS,1986).

	Paved	Unpaved
Total	97,189	14,703
Built-up areas	43,110	174
Rural areas	54,078	14,529
State motorways and dual carriage ways	2354	
Provincial secundary roads	3114	
Provincial tertiary roads	4023	
Minor rural roads	44,587	14,529

In The Netherlands exists a numerous variation of soil types. This is a feature of our country: being the result of what the big rivers Rhine and Maas have transported and sedimented from other parts of Europe, what the sea has partly brought and partly taken away, what nature has eroded, allowed to grow and to die and mankind has built, moved and changed.
From the beginning of human use of the place known nowadays as "The Netherlands" there has always been a need for transport. In the past there were the natural possibilities of rivers, brooks, lakes and ***** to use as waterways. Man starts to link these natural ways by digging canals and so on. As an example, history shows that, as a result of Roman influence in our countries, the Corbulo canal was dug in the Western part of The Netherlands. Although in the beginning, the human activities were adapted to the natural circumstances of the areas and directed to the technical possibilities and constraints of man. As his technical knowledge grows, man starts to adapt the natural circumstances to his activities and his needs. At the moment the techni-

cal development has led to the wish to fulfill almost every social and economic demand of transport. This leads for national and international purposes to a network of motorways, dual carriageways, and provincial highways. For the countryside, this leads to the transportation specification that not only the enclosure of rural areas in the quantitative sense but also in qualitative sense must be taken into account. All kinds of roads and paths that have been used to give access to the origins and destinations in the country must be suitable for all kinds of traffic. Traffic that can serve both the social and the economic activities required by society in the country. These specifications mean a lot for many areas in The Netherlands, both for areas with a rural, used road network in a technically bad condition as for areas with a subsoil which has a bad bearing capacity.
In these situations the specification for the need to transport caused an increasing need for road reconstruction not only to allow clearance of outstanding maintenance work but also to shape the conditions for the changing function for which the road will be used in future.

The philosophy of road design

There are important differences between motorways and rural roads. By their function for national and international transport the motorways have a higher traffic volume than rural roads that are used mainly for origin and destination traffic. Although both types of roads are very different where traffic volume is concerned, there is no difference between the weights and sizes of the vehicles that make use of both. The admission of vehicles on the public road network is legally arranged and bounded by the sizes of vehicles and the axleloads. This means that the specifications for the design of the pavement and the basement of roads do not vary between different types of roads. The only variation in the specification for the design is the adaptation of roads to the traffic volume.
The investments for construction and reconstruction of motorways is higher than for rural roads. On the contrary the investments related to the use of the roads gives a totally different impression: the investments for the rural roads per vehicle-km driven are much higher than the investments for the motorways per vehicle-km driven.
These differences have led to a design philosophy that is common in the construction and reconstruction of rural roads in The Netherlands. From the early fifties it is common sense to make as much use as possible of the materials that are present in the subsoil. As an example of the construction of a stable basement or subbase of blast-furnace slag or stone, in most cases it can be built up on present layers of clay. This can also occur in the case of a subsoil of peat which is covered with a layer of clay of a sufficient thickness. The design of rural roads has always consisted of a basement which had to guarantee the stability of the construction and a surface to cover it and to protect it against the influence of the entrance of water.
The combination of the use of existing materials and engineering judgement has until now succeeded in many cases in the construction and reconstruction of rural roads.

The situation in Driebruggen

In the western part of The Netherlands a land development project in Driebruggen is in planning. This rural district is situated between the towns and villages of Gouda, Bodegraven, Woerden and Oudewater. The subsoil of this district consists of thick layers of peat which are covered with a thin layer of clay. Hundreds of years ago the human activities started with the reclamation of this district by dewatering and lowering the groundwater level to make the soil fit for agricultural purposes. These activities also caused a continuing lowering of the soil surface. This process is still going on, so that a permanent agricultural use can only be assured by a lowering of the groundwater level from time to time. The agricultural activities are largely dairy farming. The land is used exclusively for pasture. The road system is used for the supply in bulk of concentrates and fertilizers and for the collection of milk and the transportation and distribution of dung. The admittance of heavy vehicles is legally restricted to the total weights of 16 ton to prevent as little damage as possible to the roads. The opening-up of the district is assured by a network of very narrow roads. Because of the absence of shoulders along the roads and the presence of wide ditches and water courses there have been few occasions in the past to ameliorate the road width. Up till now the bad baering capacity of the subsoil have caused the road managing authorities much work and high expenses for the management and maintenance of the road network. Every measure to ameliorate the condition of the road by covering the old construction with an asphalt layer brings an increase in the weight of the road construction. This restarts the process of consolidation. This process of consolidation varies from place to place, which causes at the time an exaggeration of maximum permitted tensions and strains. The deformations lead to a very short life span of the measure to maintain the road because the measure itself causes, by the increase of weights on the subsoil, the demolishing of it.

The experiment

To prevent the consolidation process it is important to maintain an existing equilibrium of weights, while reconstruction of the road occurs. In 1987 an experiment was started in the district of Driebruggen to gain experience of materials which could be used and how the construction had to be built up to prevent or to diminish the consolidation. The experiment took place on a part of the Oukoopsedyk. As a part of the land development plan this whole road had to be reconstructed. The reconstruction consisted of the building of 2.5m width shoulders, the adaptation of the pavement width from 3.0m to 3.5m and the replacement of the basement and the pavement construction. Therefore the construction of road basement and shoulder had to be built in the ditch on one side of the road. This ditch was very wide so that it was not necessary to replace the ditch in the direction of the pasture.
The experiment consisted of three parts:
- the use of light weight materials to fill up the ditch in order to broaden the road basement;
- the use and application of different types of light and voluminous materials to diminish the weights on the subsoil;
- the use of different types of basement and pavement constructions;

The filling up

For the adaptation of the road basement to the necessary width, the ditch was partly filled with different types of light weight materials. Two types of materials were used in two sections of 140m each.
The materials were *eifelith-lava* and foam slag. The first material has a natural origin and consists of the light fraction lava from vulcanic activities. Foam slag are produced in combination with rough iron from erts. The liquid stonefraction of the erts is tapped from a blast-furnace and cooled with a lot of water. During this proces the liquid phase changes to the cristalline phase and a voluminous material exists. Both materials were imported from Western-Germany.
To prevent the disappearing of material in the ditch and in its peat subsoil or the blending of material with the existing mud a geotextile was used as a separation.
The total height of the filling in the ditch was 2.5m. The consolidation of the subsoil in the ditch was not taken into account. This subsoil consists of a 8m thick combination of layers of peat and weak clay. The expected consolidation of the peat subsoil of the ditch as a result of the filling up was 0.40m after a year. The realized values of the consolidation gives a good comparison in the expected worths.
After the consolidation of the filling was completed, the new formed shoulder was covered with ground.

The design of the light weight road constructions

The design of the constructions was based on the principle not to exaggerate the weights of the existing road construction. Therefore it was important to remove as much as necessary from the existing road construction and to replace it by a combination of light and heavy material. The function of the light material was to fill up the volume of the original heavier materials and to compensate the weight of the heavy one. The weight of this combination of materials was not to exaggerate the weight of the total construction of the existing situation. In this the height of the construction related to the shoulder of the road and the exits and entrances of farms, houses and pastures along the road had also to be taken into account. The materials to apply in this constructions were foam slag, sand, polystyreen foam, expanded clay grains and a hydraulic mixture of rubble.

The second part of the experiment was the construction of the basement of the new road construction. Three sections were constructed which had a length of 85m each. One of the sections was built up with the normal used materials in order to have a reference for the other two sections where the light materials were applied. The used material in the reference section was blast-furnace slag. The existing layers of asphalt and the old materials of the subbase were spread out over the total width of the new basement. A layer of 500mm of blast-furnace slag was placed on the subsoil. On this a layer of 100mm asphalt concrete was placed. To prevent the entrance of water the construction will be covered with a surface dressing. The weight of this construction is 1250 kg/m2.
For the other two sections the basement was constructed after the old construction of asphalt and the subbase were removed.
The second section was built up with a layer of 300mm of polystyreen foam which was placed on the subsoil. On the foam a layer of 250mm of a hydraulic mixture of rubble was placed. On this a layer of 100mm asphalt concrete was placed. The weight of this construction is 1125 kg/m2.
The third section consisted of a subbase of 500mm expanded clay grains covered with a layer of 250mm of a hydraulic mixture of rubble. Also on this construction a layer of 100mm asphaltconcrete was placed.

CONCLUSIONS

Until now the different constructions have shown few differences in the behaviour concerning the deformation and deviation of the crossection.
It is possible to construct the road without deformation during the execution period.
To draw further conclusions about the resistance of the road constructions to heavy traffic it is necessary to wait until the road has functioned for some time. In the meantime the behaviour of the constructions and the development will be followed and measured.

Land and Water Use, Dodd & Grace (eds), © 1989 Balkema, Rotterdam. ISBN 90 6191 980 0

Restauration, amélioration et développement progressif des voies en terre agricoles en Hongrie

L.Gáspár
Institut Scientifique des Transports (KTI), Budapest, Hongrie

ABSTRACT: Significant damage can be observed as a consequence of the n negligence of agricultural earth roads. A relative small amount of central credit will be available for the construction of new agricultural roads. First the neglected earth roads are improved The earth roads with higher traffic volume and/or those used continuously are to be improved so that they should be passable all year around. This can be obtained preferably by adopting mechanical stabilization consisting of granular industrial by-products - rock and gravel waste, slags and demolition materials. If no suitable granular material is available in the vicinity a strong base course can be constructed by the lime or chemical treatment, as well as by cement or preferably fly ash stabilization of cohesive soils. The earth roads on sandy areas are made passable by mixing slightly cohesive soils to it or by building cohesive soil "stripe road". These restoration and repair works should be incorporated into the melioration plan of agriculture. In case of roads with higher traffic size the stabilization with sufficient strength can be developed using it as a base course when gradually overlayed by bituminous surface treatment or thin asphalt layer. To promote these tasks professional courses are organized for agricultural engineers. Besides in each agricultural factory special sections are formed for the regular and professional implementation of these contruction and maintenance activities, excluding asphalt works.

ZUSAMMENFASSUNG: Die Vernachlässigung der Erdwegen verursacht grosse Verluste. Für den Bau der neuen landwirtschaftlichen Wegen werden nur sehr wenige Zentralkredite künftig zur Verfügung stehen. In der Zukunft werden nur solche ökonomische Methode verwendet, die meistens mit dem Verwerten von industriellen Nebenprodukten in Regie ausführbar sind. Die Befestigungen der Wegen mit grosserem Verkehr werden nur stufenweise ausgebaut. Der Verfasser gibt ein Überblick von der ungarischen Praxis des Fachgebietes.

RÉSUMÉ: La négligence des voies en terre agricoles engendre des dégats importants.Pour la construction des routes neuves agricoles on ne seront que peu de crédits centrals désormais. On n'applique à l'avenir que des méthodes économiques qui peuvent être réalisées en majorité en régie, avec l'utilisation des sous-produits industriels; les corps de chaussée des routes avec un trafic plus important ne seront développées pourtant que par paliers. L'auteur donne un aperçu sur la pratique hongroise dans ce domaine.

1 LA RESTAURATION DES VOIES EN TERRE

La première intervention indispensable est un assainissement efficace. Les voies en bon état doivent être entretenues systématiquement. Pour la suppression de l'orniérage et des autres déformations nuisibles on utilise la niveleuse (grader) très efficacement.

Les routes en mauvais état seront restaurées par un travail plus important - de préférence dans le cadre du plan d'amélioration agricole de la région. On surélève les voies tracées sur le terrain selon du profil en traverse de la figure 1. Les tronçons courts de site profonde - qui deviennent intraversables dans les périodes pluvieuses - doivent être suélevés par des remblais avec ponçeau tubulaire.

2 L'AMÉLIORATION DES ROUTES EN TERRE D'UN TRAFIC PLUS IMPORTANT ET/OU UTILISÉES CONTINUELLEMENT

Pour que ces routes soient viables pendant toute l'année on les muni d'une couche d'amélioration qui peut être plus tard la couche de base du corps de chaussée définitive

2.1 La stabilisation mécanique

On construit économiquement cette couche en régie par l'utilisation de sous-produits indrustriels ou d'autres déchets granulaires bien compactables et résistentes de la région. C'est la stabilisation mécanique ou le "béton minéral" (non traité) dont l'éxécution n'exige pas de materiels importants et de connaissances spéciales. Les dimmensions de la section transversale des routes améliorées peuvent voire sur la figure 2.

Nous utilisons les matériaux granualaires suivants avec une bonne efficacité:

- les déchets de carrières et de ballastières,
- les laitiers de haut fourneau et d'acierie,
- les schistes houillers - rouges calcinés,
- les scories d'incinération des ordures ménagères,
- les matériaux de démolition.

Nous avons particulierement beaucoup d'expériences très favorables et économiques avec l'utilisation des déchets de carrières et de ballastières locales dont le prix de revient est très modéré. Dans beaucoup de communes on a amélioré les rues non-viables de cette manière en régie par la participation des habitants interessés.

Les ballastes de la région occidentale du pays contiennent oxide de fer et pour cela possedent une prise hydraulique lente. Elles constituent souvent l'assise des chaussées des routes départementales et des routes nationales aussi (Gáspár 1982).

On constate une prise lente analogue dans les scories d'incinération des ordures ménagères à cause de leurs contenu de cendres volantes hydrauliques (Gáspár 1988).

L'utilisation des laitiers a une tradition ancienne et favolable chez nous. En cas des voies où les sols cohésifs sout très humids (il y a le péril d'enfoncement des graves) on interpose des géotextiles non tissés avec succès (Gáspár 1986).

Dans les régions à sable fin (à lapin) on construit dans la bande centrale - selon la section transversale de la figure 3. - une couche de sol sodé ou d'argile qui possède une cohésion importante en périodes sèches: ainsi devient viable. Dans les intervalles pluvieux les véhicules roulent sur les bandes laterales, dont le sable humid à une cohésion virtuelle. Cette solution exige - naturellement - une discipline de trafic.

2.2 Le traitement des sols cohésifs

La solution est beaucoup plus difficile dans les régions dépourvues en graves. Les argiles acidiques peuvent être traités au chaux. A cause d'économie on utilise souvent des déchets de chaux (hydraté ou de carbide, ect.).

Une autre possibilité consiste dans le traitement chimique des sols non-basiques. Nous avons contruit environ 20 tronçous expérimentaux - avec RRP (Reynolds Road Packer) ou CBV (Chemische Bodenverbesserung) - avec succès variable (Gáspár 1981).

Nous stabilisons les sols pulvérulables (limons,loesses) au ciment,

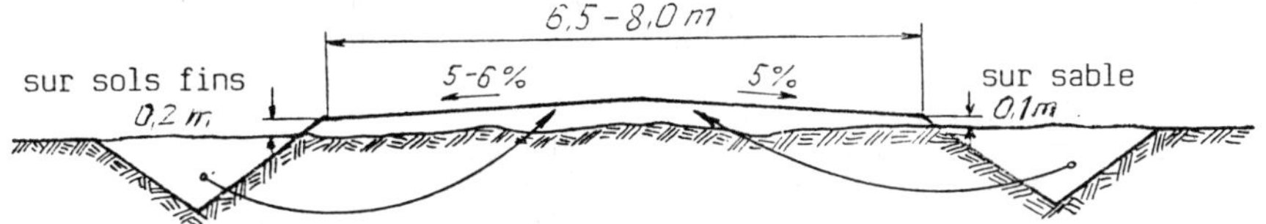

Fig. 1. La surélévation des voies en terre tracées sur le terrain avec niveleuse

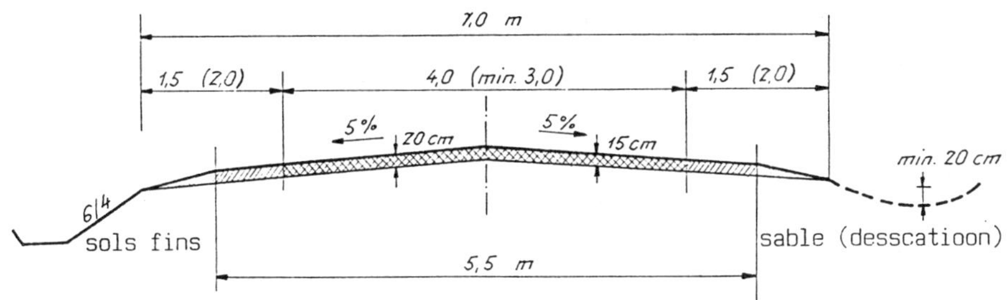

Fig. 2. La section transversale des routes en terre améliorées de stabilisation mécanique

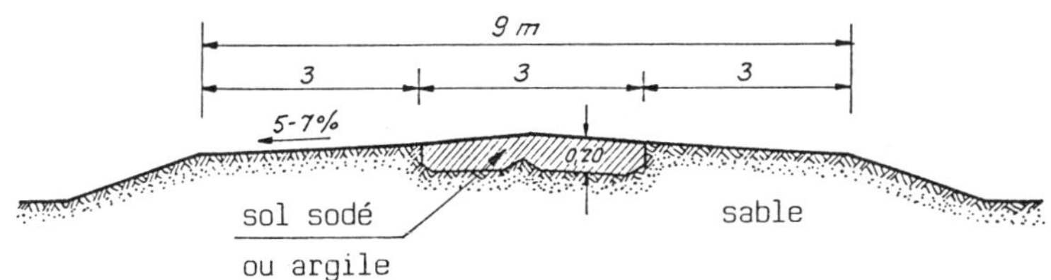

Fig. 3. La section transversale d'une voie en sable avec une bande centrale à cohésion

récemment aux cendres volantes avec la méthode "mixed in place". Sur la surface de la couche compactée doit être répendu un film de bitumes dans la forme de "cut backs" ou émulsions. Cette stabilisation peut supporter un trafic léger pendant quelques années (Gáspár 1982).

3 LE DÉVELOPPEMENT DÉFINITIF DES CHAUSSÉES

Les couches d'amélioration construites selon les méthodes précédentes après quelques ans peuvent être la couche de base des chaussées. Le trafic a vérifié la portance nécessaire. Au début un enduit superficel mono-ou bicouche peut être convenable. La solution définitive consiste dans la construction d'une couche d'usure bitumineuse - souvent tapis mince. Une possibilité plus simple est le macadam pénétré mince. Les travaux bitumineux susdits sont éxécutés par les entreprises routiers de la région le plus souvent.

Dans le résau routier agricole hongrois il y a beaucoup de tronçons

- souvent routes d'accès - dont la couche de base est une stabilisation au ciment (mixed in place) qui fonctionnent très bien depuis plusieurs dizaines d'années, sauf les couches d'usures bitumineuses devaient être renouvelées.

4 AUTRES DEVOIRES

A l'initiative de l'auteur on a commencé la formation routière des ingénieurs agricoles en Hongrie qui dirigent l'entretien et l'amélioration des voies et des routes agricoles dans ses usines agricoles. Pour ce but sont suffisantes les machines agricoles (grader, scraper, doser, pulvimixer, pelles ect.) et les véhicules de transport qui peuvent être exploitées quand les travaux agricoles ne les utilisent pas.

C'est indispensable l'organisation d'un entretien systématique et adéquant du résau routier agricole - de préférence en régie, à l'exception de travaux bitumineux dont l'entretien préventif exige un matériel plus spécial et de techniques adéquantes.

5 CONCLUSIONS

L' amélioration des routes en terres agricoles plus fréquantées avec l'utilisation des sous-produits industriels s'est avéré un solution trés efficace pour diminuer les dégats causés par la négligation du réseau routier. La couche d'amélioration peut être la couche de base du corps de chaussée définitive.

RÉFÉRENCES

Gáspár, L. 1982. Utilisation des sous-produits industriels dans la construction routière en Hongrie, Revue Générale des Routes et des Aérodromes 590.

Gáspár, L. 1988. Prescriptions techniques hongroises de l'utilisation des scories d'incinération des ordures en technique routière, VI ème Conférence Routière de Budapest

Gáspár, L.; Oroszlán, M.; Schneider A. 1986. Ungarische Erfahrungen mit der Verwendung von Geotextilien im Strassenbau, III. Internationaler Geotextil-Kongress, Wien

Gáspár, L. 1981. Essais du traitement chimique de sol en Hongrie, Tenth. Intern. Confer. on Soil Mechanics and Foundation Engineering, Stockholm

Land and Water Use, Dodd & Grace (eds), © 1989 Balkema, Rotterdam. ISBN 90 6191 980 0

Agricultural traffic and road safety in the Netherlands

C.F.Jaarsma
Agricultural University Wageningen, Wageningen, Netherlands

Th. Michels
Winand Staring Centre for Integrated Land, Soil and Water Research, Wageningen, Netherlands

ABSTRACT: Although land consolidation measures have basically reduced agricultural transport, still a considerable number of agricultural vehicles make use of public rural roads. Accidents involving agricultural vehicles show relatively high injury and mortality rates. Therefore this category is quite relevant for road safety considerations.

In a recent case study on a major rural road (9400 vehicles per day) the impact upon the level-of-service was determined through simulation models, calculating the delay and number of overtaking manoeuvres of passenger cars. For estimating the crossing opportunities for agricultural vehicles, a formula developed for bicycle crossings proved to be underestimating and demands further research.

Two types of remedial measures for improving rural road safety are recommended respectively. First, land layout improvement (e.g. reallotment) should take away agricultural vehicles from major roads. If this is impossible, the second option is to adapt road layout in order to seperate agricultural vehicles from other traffic.

RESUME: Aussi bien que le remembrement rural a fondamentalement réduit les transports agricoles, encore un nombre considérable de véhicules agricoles se sert des routes ruraux publiques. Les accidents avec le trafic agricole montrent un taux de blessure et de mortalité relativement haut. En conséquence cette catégorie de trafic est bien importante dans les considérations de la sécurité routière.

Dans une étude exemplaire récente d'une route supérieure en rase campagne (9400 véhicules par jour), l'influence des véhicules agricoles sur la circulation a été déterminée par des modèles de simulation, calculant le nombre et les pertes de temps des manoeuvres de dépassement par des véhicules légères. Pour calculer le nombre d'opportunités de traverser la route par les véhicules agricoles, une formule développée pour les bicyclettes donnait des estimations trop basses et demande plus de recherche.

Pour améliorer la sécurité routière la première proposition est à prendre des mesures de remembrement rural afin de réduire le trafic agricole sur les routes principales . Dès que cette option est impossible, la deuxième proposition est à adapter l'aménagement de la route afin de séparer les véhicules agricoles de la circulation normale.

KURZFASSUNG: Wenn auch Flurbereinigungsmassnahmen die landwirtschaftlichen Transporte schon grundsätzlich reduziert haben, werden die öffentlichen ländlichen Strassen immer noch von einer beträchtlichen Anzahl von Landwirtschaftsfahrzeugen benutzt. Verkehrsunfälle mit dieser Fahrzeugkategorie zeigen relativ hohe Verletzungs- und Mortalitätsraten. Deswegen ist diese Kategorie wichtig in Verkehrssicherheitsbetrachtungen.

In einer neulichen Beispielforschung an einer übergeordneten Landesstrasse (9400 Fahrzeuge pro Tag) ist die Auswirkung von Landwirtschaftsverkehr untersucht worden an Hand von Simulationsmodellen zur Bestimmung von Anzahlen und Verzögerungen von Fahrzeugüberhohlungen von

PKW's. Ein zweites Modell, entwickelt zur Bestimmung von Überquerungsmöglichkeiten für Radfahrer, zeigte sich unterschätzend heraus und fordert weitere Forschung.

Zwei Typen von Verkehrssicherheitsmassnahmen werden empfohlen. An erster Stelle sollte durch Flurbereinigung der Landwirtschaftsverkehr von den übergeordneten Strassen weggenommen werden. Falls dies unmöglich ist, wird eine derartige Anpassung der Strassengestaltung empfohlen dass der landwirtschaftliche Verkehr vom übrigen Verkehr getrennt werde.

1 INTRODUCTION

It is well known that land consolidation projects have as a basic objective to improve agricultural transport efficiency through concentrating the land per holding and minimizing mean lot distance. This has clearly reduced the presence of agricultural traffic on public roads in ameliorated areas. Nevertheless this traffic still plays a substantial role in rural road unsafety.

Agricultural traffic refers to all categories of agricultural vehicles like tractors, either with or without trailers, machine carriers and automobile field machines, as far as these are allowed to drive on public roads; heavy goods vehicles for transporting agricultural products are left out of scope. The agricultural vehicles considered are subject to a legal speed limit of 16 km/hour; on the other hand they may have extraordinary dimensions, e.g. widths of loads up to 3.5 m. Owing to these two features - low speeds and large sizes - agricultural traffic is likely to conflict with other traffic categories, especially on narrow rural roads with relatively high traffic volumes. Conflicts may be bilateral and twofold: both agricultural traffic and other road users will suffer from delays and annoyance on one hand, and from road unsafety risks on the other.

The role of agricultural traffic in rural road accidents is analyzed in the following section. Next, a case study of agricultural vehicle's impacts upon other traffic and vice versa on a major rural road in the northern part of the Netherlands shows how complicated the problem turns out in trying to quantify it in more detail. In conclusion we recommend countermeasures to improve rural road safety.

2 AGRICULTURAL VEHICLE ACCIDENTS

2.1. Available information

Recent information on road accidents involving agricultural vehicles in the Netherlands is available in several studies, which are all using data files from the State Service for Road Accident Registration ('Dienst Verkeers Ongevallen Registratie', VOR). These files are based upon a country-wide registration standard, carried out by national and local authority police officers.

First of all, the Netherlands Central Bureau of Statistics publishes a statistical yearbook on injury accidents on public roads (e.g. CBS, 1988). It presents global information on accidents, casualties and vehicles involved, both for urban and rural roads respectively, not going into details of specific road categories. Data from this source show that the share of accidents involving agricultural traffic varies between 2.5% and 2.9% of all rural road accidents, with no particular trend over time for the period 1982-1987 (CBS, 1988).

A specific selection of injury accident data for major vs. minor rural roads was made by Nieuwenhof and Michels (1983); out of this, data on agricultural vehicle accidents were combined with files of a vehicle insurance company by De Bruijn and Meindertsma (1983).

Still further detailed information has recently been sampled for the preparation of a Governmental policy note on minor rural roads on one hand and for the determination of accident rates related to road geometry and traffic characteristics by road category on the other. This was done through a 6%-sample analysis of road and traffic data at the level of rural road links and intersections, in combination with VOR data on injury and material damage accidents (Michels and Meijer, 1988).

These studies mentioned will all be used in sections 2.2 and 2.3 for describing agricultural vehicle accidents on public rural roads.

2.2 Road link accidents

The average share of accidents involving agricultural traffic, lying between 2.5% and 2.9% of all rural road accidents (CBS, 1988) clearly varies between road categories. Because slow agricultural vehicles have no access to motorways, their share on lower level roads must be higher; this is obvious from the sample inventory (Michels and Meijer, 1988), of which Table 1 summarizes some key figures. Major rural roads (excluded motorways) show a share of 2.4%, but minor rural roads mount up to 3.8%. Yet this figure is not remarkable when compared with the average shares of agricultural vehicles in traffic volumes.

Considering the accident density (acc. per 100 km.year) in Table 1, which is for all accidents on major roads much higher than the one on minor roads, we find the agricultural vehicle accident densities less proportional, due to

Table 1. Rural road link unsafety figures from a sample survey, Netherlands, 1983-1986 (source: Michels and Meijer, 1988). AV = agricultural vehicles

	Major rural roads (non-motorways)	Minor rural roads
AV accidents/all accidents x 100%	2.4	3.8
Accidents per 100 km. year		
all accidents	191.6	42.9
AV accidents	4.7	1.6
Accidents per 10^9 motorvehicles.km		
all accidents	1.224	1.992
AV accidents	.030	.077
Injury rate (%)		
all accidents	21.7	19.2
AV accidents	22.1	11.0

the higher share on minor roads.

The accident rates (acc. per 10^9 motor vehicles.km) show differences in the opposite direction; the minor road rate for all accidents is 1.5 times the major road one, but for agricultural vehicles this is 2.5 times the major road rate.

On the other hand, agricultural vehicle accidents on minor roads end up less seriously than the average, as can be seen in the bottom of Table 1. So, agricultural traffic on major rural roads appears the most serious problem to tackle. According to De Bruijn and Meindertsma (1983), also fatality rates of agricultural vehicle accidents on major rural roads are considerably above the average.

Looking at conflicting vehicle categories and manoeuvres, we find that the major part of these accidents on both road types comprises collisions with motor vehicles (mainly passenger cars and lorries), mostly driving in the same direction (with material damage only) or in the opposite direction (with relatively more injuries). Out of the injury conflicts, bicycles and mopeds have a share of 25% on major roads and 35% on minor roads. Of the agricultural vehicle manoeuvres involved, roughly 50% are left turns (mainly into farm yards and field parcels) and almost 50% are straight-on movements; right turn movements figure out very little in accident courses (Michels and Meijer, 1988). These results may again emphasize the general problem of large sizes and low speeds of agricultural vehicles on one hand, and of poor visibility of flashing signals on tractors and, in many cases, insufficient rear view possibilities or inadequate rear view mirrors on the other.

2.3 Accidents at intersections

Intersection accidents are in general a poorly investigated area. From the sample survey, intersections with three (T-junctions), four and more branches (crossings) respectively can be compared, both for all and for agricultural vehicle accidents. Table 2 presents key figures similar to those given

Table 2. Rural intersection unsafety figures from a sample survey, Netherlands, 1983-1986 (source: Michels and Meijer, 1988). AV = agricultural vehicles

	Three branches	Four branches	More branches
AV accidents/all accidents x 100%	2.8	1.5	3.4
Accidents per 100 intersections per year			
all accidents	22.1	89.9	131.0
AV accidents	.6	1.4	4.5
Accidents per 10^6 motor vehicles passed			
all accidents	.345	.799	1.080
AV accidents	.010	.012	.037
Injury rate (%)			
all accidents	21.9	22.8	19.4
AV accidents	20.5	17.2	8.3

for road links in the foregoing section.

As may be well known, T-junctions in general turn out to be much safer than crossings; this is also shown by the accident densities and accident rates in Table 2. Agricultural vehicle accidents show a similar tendency, however, with less difference between three- and four-branch intersections, on the other hand the more so stronger between four- and more branches.

A remarkable distinction of agricultural vehicle accidents, however, is that on T-junctions these are the most serious ones; the four- and more branches intersections show lower injury rates. The majority of the injury accidents on three- and four-branch intersections is a collision with a bicycle or moped.

The manoeuvres with these accidents are largely similar to those on road links. In material damage accidents more than 50% of the agricultural vehicles make left turn movements, whereas the conflicting passenger cars move straight-on; for the rest, the agricultural vehicles make straight-on movements. In injury accidents the majority of conflicts involves straigt-on driving tractors and left or right turning two-wheelers.

3 CASE STUDY GRONINGEN

3.1 Background and objectives

The case study was performed on the S16 in the province of Groningen in the northern part of The Netherlands (Runsink et al., 1988). The S16 is the former state highway connection between the towns of Leeuwarden and Groningen. The present route between these towns is more to the south via Drachten, partly using motorway A7 (Figure 1).

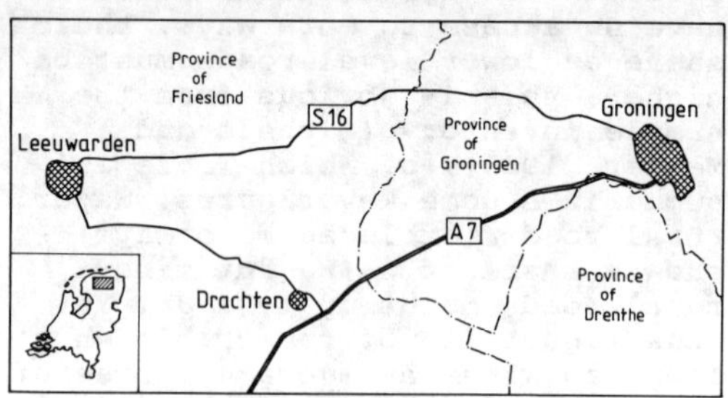

Fig.1 Geographical situation S16 in the province of Groningen

The objective of the case study was to quantify:

- traffic unsafety caused by agricultural vehicles;
- volume, trip lengths, routes and number of crossings of agricultural vehicles;
- impacts of agricultural vehicles on the quality of traffic flow (measured by the so-called level-of-service);
- possibilities for major road crossings by agricultural vehicles.

3.2 Traffic unsafety involving agricultural vehicles

The problem of agricultural vehicles and traffic safety was discussed more in general in Section 2 of this paper. In the case study the number of accidents on the S16 was established: about 60 per year in the period 1984-1987. Agricultural vehicles were involved in 4.1% of these accidents on average. This is much higher than the Dutch average rate between 2.5 and 2.9%, which was mentioned in Section 2. The provincial average for the road category of the S16 is only 1.2% of the accidents (Runsink, 1987).

From this it can be concluded that the S16 is a dangerous road in terms of agricultural vehicle accidents. This leads to the hypotheses that the level-of-service is seriously disturbed and/or that the crossing opportunities may be insufficient. Both phenomena have been analysed in the case study and are described below. First a description is given of counted and calculated volumes c.a. of agricultural vehicles.

3.3 Traffic volumes: the counting approach

Mechanical and visual countings give insight into volumes of agricultural vehicles. On major rural roads in the Netherlands, on average only about 0.16% of the 24-hour volumes consist of agricultural vehicles. Since agricultural trip lengths are rather small, the percentage of the kilometrage is still smaller: about 0.10% on average.

Of course volumes vary from hour

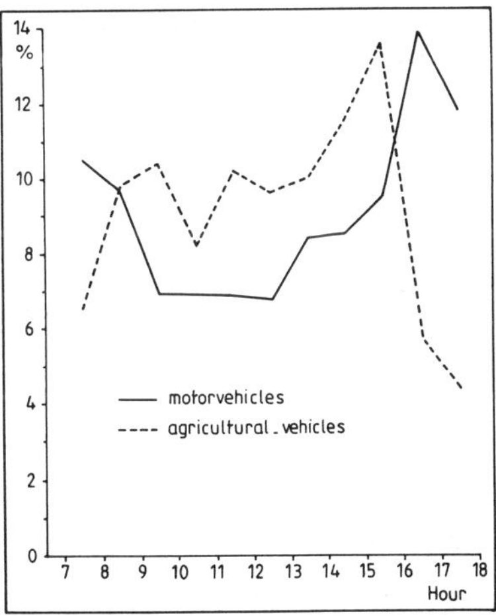

Fig.2 Mean hourly traffic volume for agricultural vehicles and for motor vehicles on weekdays in 1986 for major rural roads in the province of Groningen (7-18 hour = 100%)

to hour. The peak hours for agricultural traffic are between 11-12 hour and 14-16 hour. These hours are off-peak hours for the other motor vehicles (Figure 2). Mechanical and visual countings give insufficient information, however, since data on trip lengths and routes cannot be gathered. This is done by means of a model.

3.4 Traffic volumes: the model approach

In the case study a model was applied named AGREVAL (AGRicultural EVALuation land development). This model, developed by Righolt and Reinds (1980), calculates the number of agricultural vehicle trips between farm building and parcels. The model has been applied to all parcels along the S16 of which the user was known. Parcels further away from the S16, but only accessible via this road, were not included in the model. The result of the model is the number of vehicle trips per year per hectare for each parcel. Assuming the number of weekdays per year, the average daily traffic for agricultural vehicles on weekdays in the period of agricultural field work (the months May-October) can be calculated. Formulas applied in the model take account of:

- type of farm, mechanisation and parcellation (arable farming, dairy farming);
- milking at the farm or in the pasture and the fraction of the land located at the farm building (dairy farming);
- cultivated crops (arable farming).

For application of the model and for calculation of the routes the following information per holding is needed:

- farm size (in hectares);
- number, size, location and distance of the parcels, related to the road and to the farm building;
- the fraction of the land located at the farm building (dairy farming).

These data can be gathered from a so-called land-users map. On that map all parcels of a land-user are indicated. A land-users map is drafted when a land development project is prepared. In case this map is not available, data can be gathered from waterboards, agricultural organisations ('Landbouwschap'), farmers and perhaps the Government Service for Land and Water Use. Cadastral data are not useful, since only the owners of the parcels, and not the users, are recorded in these data.

The application of AGREVAL results in an estimation of the agricultural vehicle volumes of 0.11% of the total 24-hour volumes on the S16 in the period considered. With that the percentage of agricultural vehicles on the S16 is significantly lower than the provincial average for this type of roads: 0.29%. This is caused by the relatively high volume on the S16: 9380 motor vehicles per day, where the provincial average is only 4640 motor vehicles per day.

It was concluded that AGREVAL is a useful model for calculating routes and volumes of agricultural traffic. A disadvantage of this method is that it requires a lot of work, especially if the land-users map is not available. The output of AGREVAL can be used in traffic safety analyses as presented below.

3.5 Agricultural vehicles and traffic flow

Traffic flow is measured by the so-called level-of-service. The level-of-service is not influenced by agricultural traffic, because of low volumes. However, this does not mean that other motor vehicles are never hampered by the agricultural traffic.

At the Technical University of Delft two models were developed for calculation the effect of a slowly driving vehicle on traffic flow (Botma, 1981, 1987). The first model is based on the assumption that the slow vehicle cannot be overtaken. This assumption is a little pessimistic. Therefore in a second model overtaking is possible, where a free visibility length is assumed. This leads to a too optimistic result. Both models are calculating the effect on traffic flow (for one lane) in terms of:
- the number of overtakers;
- duration and volume of the disturbance (number of motor vehicles following the agricultural vehicle);
- the number of vehicles meeting with hindrance;
- the total amount of delay time for these vehicles.

Application of the models requires data on:
- trip length of the agricultural vehicle (determined from the land-users map);
- speed of the agricultural vehicle;
- volume per lane of the other motor vehicles.

As an indication some results for the heaviest link are given in Table 3. Using the (pessimistic) model 1, an agricultural vehicle trip of 1 km with a 25 km/hour speed results in 41 vehicles meeting with hindrance in the morning peak, only 13 vehicles in the off-peak and 51 in the afternoon peak. With the (optimistic) model 2 the numbers are 30, 13 and 45 respectively. The (optimistic) model 2 with overtaking produces mean delay times which are more than 50% lower than the values of the (pessimistic) model 1 without overtaking. More details are given by Runsink (1987).

It was concluded that the models are applicable for an impression of the disturbance caused by agricultural vehicles on major rural roads. Without a thorough verification of the models the results should not be interpreted as absolute quantities.

Table 3. Disturbance of the main flow, forced by an agricultural vehicle (trip length 1 km, speed 25 km/hour), on the heaviest link in the heaviest direction using the (pessimistic) model 1 and the (optimistic) model 2

	Duration of disturbance (minutes)			Vehicles with hindrance			Total delay (minutes)		
	(1)	(2)	(3)	(1)	(2)	(3)	(1)	(2)	(3)
Model 1	4.2	3.5	4.6	41	13	51	39	13	50
Model 2	3.6	2.9	4.3	30	13	45	7	3	22

(1) morning peak (776 vehicles/hour)
(2) off peak (300 vehicles/hour)
(3) afternoon peak (869 vehicles/hour)

3.6 Crossing opportunities for agricultural vehicles

In 3.5 the influence of agricultural vehicles on other motor vehicles was dealt with. The reverse situation, namely when agricultural vehicles have to cross a major road will be discussed in this section. Moreover, three movements can be distinguished: crossing, inserting to the left and inserting to the right. Type of movement, road width of the major road and length of the agricultural vehicle determine the representative crossing length. The number of crossing opportunities per hour depends on volume and speed on the major road on the one side, and on representative crossing length and crossing speed of the agricultural vehicle on the other side.

In the province of Groningen a formula was developed for bicycle crossings. It was expected that this formula could be applied for agricultural traffic too, in the next form (Beleidsnota Personenvervoer, 1984):

$$N = c \times (1000 \times V/B) \times e^{-I \times B/1000V}$$

where N is minimum number of crossing or inserting opportunities per hour, I is peak hour volume (motor vehicles/hour), V is crossing speed (km/hour), B is representative crossing length (m) and c is coefficient depending on main flow speed, for speeds over 75 km/hour, c is 0.5.

The formula is only applicable for:

- uninterrupted traffic flows (no traffic lights);
- independent movements of cars (overtaking is always possible).

Table 4 shows the number of crossing opportunities for agricultural traffic, calculated for the peak hour for four links of the S16. Three movements were distinguished: crossing, inserting to the left and inserting to the right. Calculations were made for single tractors and for tractors with trailers. It was established that crossing gives more problems for the agricultural traffic than inserting to the left or to the right. Especially for crossing the heaviest link extreme low values for N were found, even less than 1.

Table 4. Number of crossing and inserting opportunities (N) by movement for agricultural traffic between 16 and 17 hours on four links of the S16

Link volume (motor vehicles/hour)	Number of opportunities (N)					
	crossing		inserting to the left		right	
	(1)	(2)	(1)	(2)	(1)	(2)
1325	1	0.0	11	0.2	90	8
1155	3	0.0	17	0.4	112	12
645	19	1	66	6	218	47
515	33	2	94	12	258	66

(1) single tractor; (2) tractor with trailor

Nevertheless, agricultural vehicles do cross the road in reality.

Possible explanations for the too low results are:

- the conditions mentioned are not fulfilled (in reality there will be 'gaps' in the flow, which allow the agricultural vehicle to cross the road);
- the assumed crossing speed (1.4 m/s for an unloaded vehicle) is too low.

It was concluded that the formula underestimates the opportunities for crossing the S16. Further research is recommended.

3.7 Conclusions of the case study

The first conclusion from the S16 case study was that this road is a dangerous one in terms of agricultural vehicle accidents. The numbers of agricultural vehicles on the road is relatively small: a calculation with the AGREVAL-model results in 0.11% of the total 24-hour volume on the S16.

For an impression of the disturbance of the main flow caused by agricultural vehicles, two models were applied. The results are applicable for relative comparisons only.

For calculating the number of crossing opportunities for agricultural vehicles a formula developed for bicycle crossings was applied. The model proved to be underestimating and needs further research.

4 REMEDIAL MEASURES

If problems involving agricultural vehicles demand remedial measures, two approaches are possible: one from an agricultural and one from a traffic point of view.

The agricultural approach implies the creation of a land layout in which the major rural road considered is no longer needed for giving access to the agricultural area. This new layout can be created through land development, reallotment and so on. It should be realised that land use planning requires much preparation time. Therefore in practice this is only applicable for areas where the Government Service for Land and Water Use is preparing a project.

Whenever it is necessary to use a major rural road for agricultural access too, measures from a traffic point of view can be considered. This implies an adaptation of road layout. In order to separate agricultural vehicles from other traffic, parallel roads (on one or two sides) are appropriate. Problems with crossing agricultural vehicles can be solved by building small tunnels, viaducts or installation of traffic lights, which will improve the crossing opportunities.

From case to case a choice should be made. In the final decision agricultural vehicles will not be the only factor. Aspects like traffic safety, bicycle traffic, motor vehicle traffic and public transport are a matter of concern too. Furthermore, other effects of the improvements, like on the landscape, nature and environment, should be taken into account in order to made a final choice.

REFERENCES

Beleidsnota personenvervoer. 1984. Provinciale Waterstaat van Groningen, Groningen, The Netherlands. 80 pp.

Botma, H. 1981. Landbouwverkeer op tweestrookswegen. Het effect van een langzaam rijdend voertuig op de verkeersafwikkeling, beschreven met een macroscopisch model zonder inhalen. Meded. 14. Studiecentrum Verkeerstechniek, Driebergen, The Netherlands. 48 pp.

Botma, H. 1987. Het effect van langzame voertuigen op de verkeersafwikkeling op tweestrookswegen. CROW-publ. 4. Ede, The Netherlands. 31 pp.

De Bruijn, C.P.M. and Meindertsma, A. 1983. Onveiligheid door het landbouwverkeer. Vakgroep Landbouwtechniek, Landbouwhogeschool Wageningen/Interpolis, Tilburg, The Netherlands. 51 pp.

CBS. 1988. Statistiek van de verkeersongevallen op de openbare weg 1987. Centraal Bureau voor de Statistiek, 's-Gravenhage, The Netherlands. 64 pp.

Michels, Th. and Meijer, E. 1988. Onderzoek probleemsituaties 80 km/u-wegen, fase 2. 1e Tussenrapportage: kwantitatieve analyse. Nota 1922. ICW, Wageningen, The Netherlands.

Michels, Th. and Meijer, E. 1988. Onderzoek probleemsituaties 80 km/u-wegen, fase 2. 2e Tussenrapportage: kwalitatieve analyse. Nota 1927. ICW, Wageningen, The Netherlands.

Nieuwenhof, H. and Michels, Th. 1983. Verkeersonveiligheid op plattelandswegen. Verkeerskunde 34,9: 433-436. Cultuurtechnisch Tijdschrift 23,2: 69-80.

Righolt, J.W. and Reinds, G.H. 1980. Agricultural evaluation of land layout. In: E.W. Schierbeek (ed.). Research Digest 1980. Techn. Bull. 117. ICW, Wageningen, The Netherlands: 215-222.

Runsink, S.C. 1987. De problematiek van het landbouwverkeer op de S16. Een onderzoek in het kader van de overdracht van de S16 (Rijksweg 855) aan de provincie Groningen. Provinciale Waterstaat, Groningen, The Netherlands. 68 pp.

Runsink, S.C., Jaarsma C.F. and Ten Raa, J.G. 1988. Landbouwverkeer op hoofdwegen. Verkeerskunde 40: 426-429.

Land and Water Use, Dodd & Grace (eds), © 1989 Balkema, Rotterdam. ISBN 90 6191 980 0

Untersuchungen des Abfahrens von dünnen Strassenbefestigungen an der Teststrecke von Makk-puszta

M.Kosztka
Universität für Forstwirtschaft und Holzindustrie, Sopron, Ungarn

ABSTRACT: Flexible road pavement structures have been designed in Hungary since 197o by a method based on an adaption of the AASHO experiments. This relatively simple method is accepted for designing agricultural and forestry roads, too. Investigations have shown, however, that due to certain characteristics of the traffic volume the method results in overdimensioning. In order to clarify this problem three test roads were constructed in the mid-seventies in a Hungarian-GDR cooperation. These roads were built on different kinds of subsoil and they were loaded by artificial traffic. The Hungarian test road is at Makk-puszta on quick ground /sand drift/ without a danger of freezing, it is a pavement being 6 m wide with 3 x 45 m sections, enabling the investigation of 72 different types of road pavement structures. The paper describes the experiments and summarizes results and proposals to improve dimensioning of agricultural and forestry roads.
EXTRAIT: Les structures des tabliers de routes à épaisseur fléxible sont dimensionnées en Hongrie depuis 197o avec la méthode basée sur les résultats des expériences AASHO. Cette méthode relativement simple est répandue aussi pour le dimensionnement des structures des routes en agriculture et en sylviculture. En consideration des recherches differentes toutefois on avait l' impression que en consequant de spécificités de traffic certaines réserves ont produit un surdimensionnement. Pour l'éclaircissement de ce probléme on a costruit trois routes expérimentales chargées de traffic artificel, en cadre d'une collaboration entre Hongrie a la RDA, sur sous-sols differents. La route expérimentale a été costruite à Makk-Puszta, sur sable mouvant, sur sol sans danger de gel, munie un revêtement de 6 m de largeur, en sections de 3 x 45 m., rendant possible les etudes sur 72 types différentes de structures de tablier . L'article explique le déroulement de l' expérience et précise les resultats et les suggestions a base desquelles on peut ameliorer la realisation des structures de routes en agriculture et en sylviculture.
AUSZUG: Elastische Strassenbefestigungen werden in Ungarn seit 197o mit einer Methode dimensioniert, die die Ergebnisse der AASHO-Experimente adaptiert. Diese verhältnissmässig einfache Methode hat sich auch bei der Dimensionierung der Befestigungen von Güter- und Waldstrassen verbreitet. Verschiedene Untersuchungen haben doch gezeigt, dass infolge der Eigenschaften des Verkehrs gewisse Reserven entstehen, die dann als Überdimensionierung erscheinen. Um diese Frage klären zu können, wurden in der Mitte der 7o-er Jahren drei mit Testverkehr gelasteten Strassen in einer ungarisch-DDR Kooperation auf verschiedenen Böden gebaut. Die ungarische Teststrecke hat man in Makk-puszta auf Triebsand, einem forstsicheren Boden mit einer 6 m breiten Befestigung in 3 x 45 m langen Abschnitten gebaut. Sie ermöglicht die Untersuchung von 72 verschieden aufgebauten Befestigungen. Die Studie beschreibt den Verlauf des Experimentes und fasst die Beobachtungen und Vorschläge zusammen, die eine Verfeinerung der Methode der Dimensionierung von Güter- und Waldtstrassen ermöglichen.

1. ZWECK DES EXPERIMENTES

Strassenbefestigungen können mit semiempirischen oder mit mechanisch begründeten Methoden dimensioniert werden. Die semiempirische Methode beruht auf Ergebnissen der AASHO-Grossmodellexperimente, und liefert die äquivalente Dicke /äcm/ der Befestigung in Abhängigkeit der Bodentragfähigkeit /CBR %/ und des durch die Durchfahrt von Normachslasten von loo kN ausgedrückten Verkehrs /F_{100}, Stück/.

Die Grundlage der Methode zur Dimensionierung der elastischen Befestigungen der ungarischen Verkehrsstrassen lieferte auch dieses Experiment. Im Falle von Befestigungen, die durch einen gleichmässigen Verkehr belastet sind - wie es bei der Grossmodelluntersuchungen auch modelliert wurde - können somit Befestigungen geplant werden, die aufgrund der Erfahrungen während der Lebensdauer bei einer regelmässigen Instandhaltung nicht abgefahren werden.

Es schien zweckmässig zu sein, diese verhältnissmässig verbreitete Dimensionierungsmethode wegen der günstigen Erfahrungen und ihrer Einfachheit auch bei Güter- und Waldstrassen zu benutzen. Ihre Verwendung ist auch dadurch unterstützt, dass die Befestigung dieser Strassen ähnlich zu den Befestigungen in den AASHO-Experimenten sind, deshalb beziehen sich die dort gewonnenen Erfahrungen auch auf diese Strassen.

Bei Güter- und Waldstrassen erscheint jedoch die Belastung in Form von Verkehrspitzen, zwischen denen der Verkehr praktisch vernachlässigt werden kann, d.h. der Verkehr unterscheidet sich von dem im AASHO-Experiment untersuchten Fall. Es tauchte daher die Frage auf, ob in solchen Fällen die erwähnte Methode zur Dimensionierung benutzt werden könne, ob durch die Ruhepausen eine Verbesserung der Belastbarkeit auftrete, d. h., der gleiche Verkehr mit einer dünneren Befestigung abgewickelt werden kann, wobei aus der Ersparung der Baukosten weitere Strassen gebaut werden können.

Diese Frage ist einerseits dadurch berechtigt, dass Befestigungen von Güter- und Waldstrassen, die mit Hilfe des semiempirischen Zusammenhanges für einen kleinen Verkehr richtig dimensioniert sind, einen grösseren Verkehr ertragen, als der zur Grundlage der Dimensionierung angenommene Verkehr war, andererseits hat man bei der Untersuchung der Materialkonstante, die bei der mechanischen Dimensionierungsmethode angewendet werden, festgestellt, dass Ruhepausen während Ermüdungsversuche eine Regeneration der Probekörper ergaben und ihre Starrheit den Anfangswert der Starrheit annäherte.

Teils um diese Annahmen zu bestätigen, teils um die Genauigkeit der Dimensionierung und das Abfahren der dünnen Befestigungen zu untersuchen, wurde eine DDR-ungarische Forschungsgemeinschaft geschaffen. Im Rahmen dieses Projektes wurde als ein Teil der Güterstrasse zwischen Szőke-puszta und Makk-puszta im Gebiet von Nagyigmánd eine 1,7 km lange Teststrecke gebaut, deren Pläne am Lehrstuhl für Forstliches Transportwesen der Universität für Forstwirtschaft und Holzindustrie Sopron entwickelt worden sind, und die Grossmodelluntersuchungen ermöglichte /Fig.1./

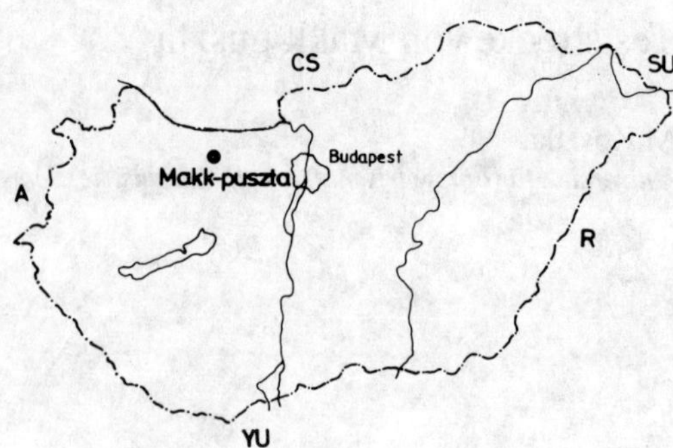

Fig.1. Ort der Teststrecke in Ungarn

2. DIE TESTSTRECKE UND IHR VERKEHR

Die Teststrecke wurde auf Treibsand gebaut, sie ist völlig frostsicher, d.h. Frost- und Tauschäden können nicht auftreten. Die 6,oo m breite Befestigung des Weges wurde in Abschnitten von 3 x 45 m gebaut, wobei 2 x 36= 72 verschiedene Arten von Befestigungen untersucht werden konnten.

Die Befestigung wurde durch einen Testverkehr belastet, d.h. die Strasse wurde von Lastwagen mit bekannten Achslasten befahren, wobei die Anzahl der Durchfahrte mit Zählung bestimmt wurde. In der ersten Belastungsperiode wurden Lastwagen des Typs IFA W 5o benutzt, dann wurden in den folgenden Belastungsperioden Lastwagen des Typs SKODA eingesetzt, da sonst die Abnutzung der Befestigungen zu lang gedauert hätte. Während des Experimentes fuhren 155o3 IFA und 8627 SKODA Lastwagen durch. Das entspricht der Durchfahrt von 11417 Normachslasten.

3. MESSUNGEN AUF DER TESTSTRECKE

Während des Experimentes wurden die von den Testverkehr unabhängigen Daten fortdauernd, die von der Belastung abhängigen Daten dagegen in Messreihen komprimiert gemessen. Die Nullmessreihe, gemessen vom Anfang der Belastungsperioden registrierte den Anfangszustand der Befestigung, dann wurden die Veränderungen nach der Durchfahrt von etwa 25oo Normachslasten festgestellt. Als Ergebnis erhielten wir drei Nullmessreihen und 5 Messreihen nach Belastung.

Die Messungen während des Experimentes umfassten die folgenden:

- fortlaufende Messung der meteorologischen Daten
- Messung der Tragfähigkeit durch Ermittlung der Verformung in jeder Messreihe

- Charakterisierung der Befahrbarkeit und Benutzbarkeit in jeder Messreihe.

Die meteorologischen Daten wurden im Laufe der Jahre kontinuierlich gemessen, wobei das Minimum und das Maximum der Lufttemperatur, die Niederschlagsmenge und die Bodentemperatur an jedem Tag aufgezeichnet wurden. Der Einfluss des Frostes wurde im Winter auf die Befestigung beobachtet, wobei die Eindringung des Frostes unter die Befestigung und die Höhe des Frosthebens gemessen worden sind. In jeder Messreihe wurde auch der Wassergehalt unter der Befestigung festgestellt.

Die Veränderungen der Tragfähigkeit der Befestigung wurden mittels Messung der Verformungen infolge der Belastung festgestellt. Die elastische Durchbiegung wurde an je 6 Stellen des inneren und des äusseren Spures bei jedem Abschnitt gemessen, gleichzeitig wurde auch der Krümmungsradius der Durchbiegungsmulde beobachtet.

Um Befestigungen auch aus dem Gesichtspunkt des Verbrauchers bewerten zu können, wurdenneben der Tragfähigkeit auch die Befahrbarkeit und Benutzbarkeit in jeder Messreihe untersucht. Dabei wurden Längswellen im inneren und äusseren Spur, sowie die Ebenheit und Verformung der Querprofile an zwei Stellen je Abschnitt aufgenommen. Stelle und Grösse der Risse, der Schlaglöcher und anderer Oberflächenfehler wurden aufgezeichnet. Die Befahrbarkeit wurde mit einem Benutzbarkeitsindex charakterisiert. Dieser Index wurde subjektiv von einer zehnköpfigen Gruppe von Experten festgestellt. Note 1 charakterisierte man die Abschnitte, die fehlerlos gefunden wurden, Note 5 die Abschnitte, die mit einer Geschwindigkeit von 4o km/h wegen Unfallgefahr nicht befahren werden können.

4. AUSWERTUNG DER ERGEBNISSE

Die Auswertung der Messergebnisse ist durch den Umstand begrenzt, dass die Durchfahrt von etwa 115oo Normachslasten nur 4 Abschnitte und sogar diese eindeutig wegen Baufehler verfuhr, deshalb kann man ohne Kenntnis der Lebensdauer einen zur Dimensionierung geeigneten numerischen Zusammenhang nicht angeben. Aus der Tendenz der Daten kann man dagegen gewisse Folgerungen ziehen, die bei einer Verfeinerung der Methode der Dimensionierung gebraucht, bzw. bei weiteren Untersuchungen als Grundlage benutzt werden können.

Der Anfang des Abfahrens der Befestigung wird zuerst von den Verbrauchern beobachtet, da Oberflächenschäden, Längswellen und Spurrinen erscheinen, die den Komfort des Befahrens beeinträchtigen. Eine Untersuchung der Einwirkung der Durchfahrt von 115oo Normachslasten gab aus diesem Gesichtspunkt folgende Ergebnisse. In der ersten Belastungsperiode erschienen Oberflächenschäden, die eindeutig auf Baufehler zurückführbar waren. Weitere Schäden traten nicht auf, was durch den subjektiven Benutzbarkeitsindex unterstützt wird. Der Index verändert sich zwischen 1,oo und 1,9o, bei dickeren Schichten schwankt er bei 1,oo mit kleiner Streuung, bei dünneren um 1,5o mit einer grösseren Streuung, aber ohne Anzeichen von einem einsetzenden Abfahren. Ähnliches wird bei den Längswellen erfahren, wobei im äusseren Spur, bzw. bei Abschnitten mit niedrigerer Belastbarkeit die Wellen grösser sind, als im inneren Spur, bzw. bei stärkeren Befestigungen, aber die Längswellen selbst veränderten sich noch nicht unter Einwirkung des Verkehrs während des Experimentes. Eine Deformation der Querprofile, d.h. die Ausbildung von Spurrinnen erschien nur bei dünnen Befestigungen, aber bleibt überall unter einem Wert, der den Abfluss des Wassers von der Oberfläche der Befestigung verhindern würde.

Die Veränderung der Tragfähigkeit der Befestigung wurde durch die unter der Einwirkung der Belastung auftretenden elastischen Verformung, durch die Durchbiegung und durch den Krümmungsradius der Durchbiegungsmulde charakterisiert.

Die Ergebnisse der Messung des Krümmungsradius sind mit einer grossen Streuung belastet. Die Ursache liegt in der Unvollkommenheit der Messmethode, deshalb konnte man die Veränderung dieses Parameters zur Charakterisierung nicht einbeziehen.

Bei der Auswertung der Durchbiegung konnte man gewisse Ergebnisse bestätigen, die auch auf einem Weg der Logik abgeleitet werden konnten, bzw. die aus früheren Versuchen schon bekannt waren. Sie umfassen die folgenden:

- die tragfähigen Befestigungen haben kleinere Verformungen
- die Verformung im äusseren Spur ist grösser, als im inneren Spur
- der Unterschied der Durchbiegungen im äusseren und im inneren Spur vermindert sich bei einer Abnahme der Tragfähigkeit der Befestigungen.

Neben diesen Untersuchungen wurde auch die Tendenz der Abnutzung der Befestigung beobachtet; dabei fand man, dass die Tragfähigkeit der Befestigungen - wahrscheinlich infolge der eingeschalteten Pausen bis zur Durchfahrt von etwa 7ooo Normachslasten sich nicht vermindert hat, danach beginnt langsam der Prozess der Abnutzung, aber Anzeichen eines endgültigen Abfahrens konten sogar bei den schwächsten Abschnitten nicht gefunden werden /Fig.2./

Diese Folgerungen sind auch durch die

schon früher behandelten Änderungen der Oberflächenunebenheit und des Indexes der Benutzbarkeit bestätigt.

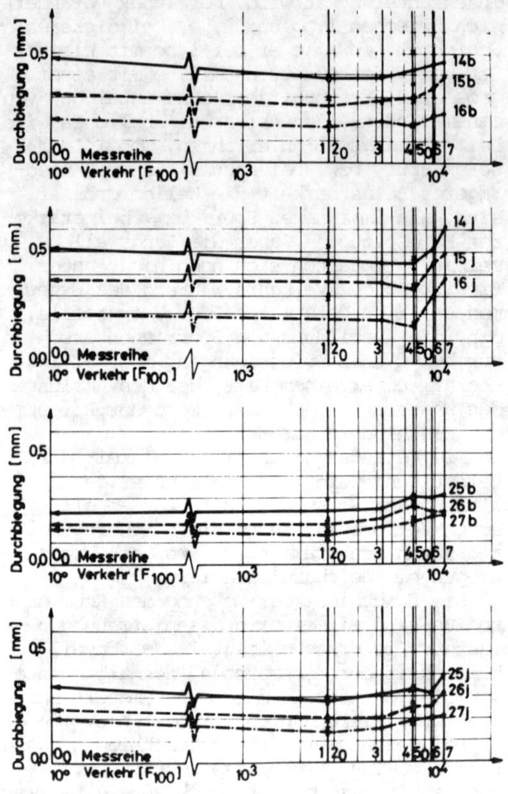

Fig.2. Die Tendenz der Tragfähigkeitsänderungen

Ein für die Dimensionierung interessantes Ergebnis erhält man, wenn man den tatsächlich abgewickelten Verkehr mit dem vergleicht, der aufgrund der Tragfähigkeit der Befestigung erlaubt ist, bzw. der entsprechend der Durchbiegung berechnet wird. Aufgrund der Tragfähigkeit der Befestigungen sollten nach der Durchfahrt von 115oo Normachslasten alle Befestigungen zugrundegehen, die dünner als 17,5 äcm sind, da augfrund des Dimensionierungsdiagrammes eine solche Befestigung die Durchfahrt von 11ooo Normachslasten erlaubt. Die Erfahrung zeigte dagegen, dass sogar bei der dünnsten d.h. bei einer 9,o äcm dicken Befestigung nur auf eine anfängliche Tragfähigkeitsverminderung zu schliessen war /14 j auf Fig. 2/, obgleich die Tragfähigkeit dieser Befestigung nur 1ooo Normachslasten wäre. Aufgrund der Durchbiegungen wäre die zu erwartende Tragfähigkeit bei solchen Befestigungen die Durchfahrt von etwa weiteren 4ooooo Normachslasten. Dieser Wert ist kaum als Grenze der Tragfähigkeit anzunehmen, da es vermutet wird, dass das Abfahren im weiteren rasch vor sich geht, dennoch folgert man aus diesem Ergebnis, dass noch weiterer Verkehr ertragen würde. Der gleiche Wert würde mit einer Durchbiegung von o,74 mm bei einer Dicke von 17,5 äcm eine Tragfähigkeit von weiteren 9ooooo Normachslasten bedeuten.

Wenn man das Verhältnis der erlaubten grössten Durchbiegung der Befestigung zu der nach der Durchfahrt von 115oo Normachslasten gemessenen Durchbiegung untersucht, findet man, dass die Verformungen unter dem erlaubten Wert bleiben, d.h. die Tragfähigkeit ist noch geeignet /Fig.3./

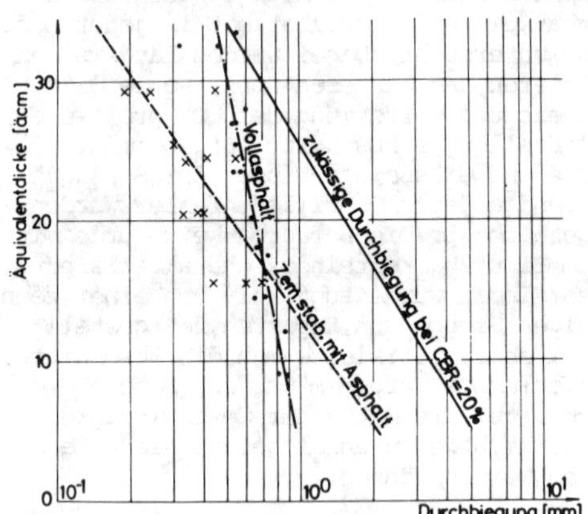

Fig.3. Die zulässigen und die in letzter Messreihe gemessten Durchbiegungen bei verschiedenen Befestigungen

In diesem System unterscheiden sich recht deutlich die nur Asphaltschichten beinhaltenden Befestigungen /Vollasphalt/ von denen, die auch eine zementstabilisierende Schicht beinhalten.

5. FOLGERUNGEN

Obgleich numerische Ergebnisse wegen der zu frühen Beendigung des Experimentes nicht festzustellen sind, gibt es einige Beobachtungen, die bei der Dimensionierung der Befestigungen von Strassen mit einer kleinen

Verkehrsdichte benutzt werden können, wobei es betont werden soll, dass die jetzige Methode der Dimensionierung bei dem vorhandenen Verkehr und den entsprechenden Verkehrverhältnissen gut geeignet ist.

Das Experiment hat uns gezeigt, dass der Zusammenhang der Dimensionierung auch im Bereich der kleineren Verkehrsdichten gültig ist, wie es schon früher bei der Dimensionierung von Waldstrassen angenommen werden musste, wobei aber eine gewisse Überdimensionierung der Befestigungen in Kauf genommen werden sollte.

Es wurde weiters bestätigt, dass die in der Dimensionierungsvorschrift angegebene minimale Befestigungsdicke von 2o äcm übertrieben ist, dieser Wert darf bei Güter- und Waldstrassen mit einer geringen Verkehrsdichte bis lo äcm vermindert werden.

Es kann aufgrund der experimentellen Ergebnissen angenommen werden, dass das aus dem im Versuchsgelände vorkommenden Treibsand gebaute Planum eine Tragfähigkeit von über 2o CBR % besitzt, daher wäre es zweckmässig, die bisher erlaubte maximale Bodentragfähigkeit von maximal 2o CBR % auf 3o % zu erhöhen, die Verwendung dieses Wertes sollte aber nur nach bodenmechanischen Untersuchungen erlaubt sein.

Bei Güter- und Waldstrassen mit einer geringen Verkehrsdichte können auch von der klassischen Auffassung abweichende Befestigungen benutzt werden, die technisch geeignet sind und die ein Ersparnis bedeuten. Es muss aber betont werden, dass diese Methode der Dimensionierung bei Befestigungen, die eine zementstabilisierende Schicht beinhalten, nur nach Verfeinerungen verwendet werden kann.

Die Messungen auf der Teststrecke brachten neue Erfahrungen in der Bewertung des Zustandes von dünnen Befestigungen.

Es hat sich herausgestellt, dass die Veränderung des Krümmungsradius der Durchbiegungsmulde zur Ermittlung der zu erwartenden Lebensdauer gut angewendet werden kann, die Messmethode soll aber dazu noch verfeinert und die theoretische Grundlage geklärt werden.

Als ein weiteres Ergebnis zu betrachten ist, dass man ein Bild über die Ausdehnung der Durchbiegunsmulde von dünnen Befestigungen gewonnen konnte, was zu einer Modifikation der in Ungarn gebrauchten Durchbiegungsmessinstrumentes führte, wobei das Tragwerk des Instrumentes weiter von der Belastung entfernt, ausserhalb der Durchbiegungsmulde gebracht wurde. /Spätere vergleichende Untersuchungen bestätigten die Richtigkeit unserer Annahmen/.

Schliesslich können auch die Erfahrungen als Ergebnis betrachtet werden, die bei der Vorbereitung, Organisation und Verwirklichung der Experimente gewonnen worden sind. Diese Erfahrungen stehen allen zur Verfügung, die solche Aufgaben lösen wollen.

Land and Water Use, Dodd & Grace (eds), © 1989 Balkema, Rotterdam. ISBN 90 6191 980 0

An evaluation method for road reconstruction

W.Hauptmeijer
Government Service for Land and Water Use, Netherlands

ABSTRACT: Within the scope of Dutch land redevelopment projects a working group of the Ministry of Agriculture and Fisheries has developed a new method to carry out cost-effectiveness analysis of road reconstruction plans and alternatives for minor rural roads. This paper will give a short description of the method and its impacts, as well as examples of case studies and an analysis of the significance of the impacts.

Méthode d'évaluation sur l'aménagement et la construction des routiers inférieures.

RÉSUMÉ: Au cadre des projets d'aménagement parcellaire, au Pays Bas une équipe de travail du Ministère de l'Agriculture et de la Pêche a développer une méthode nouvelle pour déterminer les proportions des frais et profits au plans du reconstruction des routiers inférieurs et des alternatives.
Dans cette note il y a une description bref du méthode et la production des effets. Ce faire met d'examples des projets pratique et les effets du chaque aspect fait examiner.

Eine Evaluation methode zu der Renovierung ländlicher Strassen.

KURZFASSUNG: Im Rahmen der Holländischen landeinteilung hat eine Arbeitsgruppe des Landwirtschaftsministeriums eine neue Methode zum Kosten- und Nutzverhältnis von Plänen und Alternativen zur Rekonstruktion ländlicher Strassen entwickelt.
In dieser Notiz wird eine Kurze Bescheibung dieser Methode und der Nutzeffekte gegeben. Beispiele aus der Praxis werden vorgezeigt und die Effekte von jedem Aspekt werden analysiert.

1 INTRODUCTION

Of each land redevelopment project a costeffectiveness analysis is a standard part of plan preparation. Therefore the so-called HELP-method (Staatsuitgeverij, 1983) is used. This method offers the possibility to define the economic impacts, social effects and the effects on nature and landscape of land redevelopment projects.

A gap in the method is the lack of possibility to define the effects of the provisions made for the road network. The existing method only consists of calculating the modification of maintenance costs of the minor rural roads. The provisions on the roads require a governmental subsidy of about Dfl 60 millions a year. Thanks to a new evaluation method (Werkgroep Evaluatie Wegen, 1986; Michels, 1988) it is possible to have a better insight in the impacts of governmental investments. At the same time this method offers the possibility to set priorities at the road section level within the plan design stage.

The new method has been tested within five land development projects. This contribution contains description of the method and an analysis of the results from practical testing.

2 DESCRIPTION OF THE IMPACTS IN THE NEW METHOD

2.1 General features

Within the developed method difference has made between two types of impacts; on one hand a quantitative part calculating economic benefits and on the other hand a qualitative part describing some social impacts.

In the quantitative part of the developed method the considered impacts are expressed in money or in other units (hours, accidents, etc.). These quantitative benefits of the plan are defined per year as the difference between the autonomous situation and the plan situation. The benefits resulting from the plan are assumed to flow constantly during 20 years, being a feasible lifetime for road constructions. Consequently all impacts have to be estimated as an average over this plan period.

2.2 Impacts on road maintenance costs

Road maintenance costs are regarded as having three basic components: structural restore and reconstruction, regular pavement maintenance and shoulders maintenance.

2.2.1 Structural restore and reconstruction

Structural restore and reconstruction is supposed to be especially necessary in the autonomous situation, the so called do-nothing option. In this option a road is not ameliorated at the right time to an adequate extent. As a consequence of this, the road owner, given his limited maintenance budget, will be forced to a sub-optimal strategy of successive small improvement measures in order to keep up a minimum standard of road serviceability. This strategy over time demands higher expenses than the considered plan situation would have done. In the plan option the extra costs of the suboptimal maintenance are assumed to be absent. Consequently these extra costs are regarded as benefits resulting from the plan.

2.2.2 Regular pavement maintenance

The second cost component results from comparing regular pavement maintenance with and without a plan. In the do-nothing option, maintenance is supposed to be more costly; how much more depends on the actual condition of the road and is to be estimated by means of expert's judgement. For this purpose a regular visual inspection of road pavement will be of help.

2.2.3 Shoulders maintenance

The third maintenance cost category refers to shoulders: leveling, mowing and mainenance of plantings. Given the moderate pavement widths of minor rural roads, these costs relate highly to the volume/capacity ratio and, of course to the bearing capacity of shoulder soils (Regterschot and Mooy, 1981).

2.3 Impacts on road user costs

Road user costs consist of two components: travel time and vehicle operating costs.

2.3.1 Travel time

Road traffic time savings depend on journey speeds along a road section. For this reason, journey speeds of passengers cars and heavy goods vehicles are derived from road features like visibility length, lateral clearance, road pavement width and the traffic volume (Regterschot and Oenema, 1981). Journey time is then calculated by dividing road length by journey speed. The plan impact results from comparing total traffic journey time with and without the plan.
Because time savings appear to be a major part of total benefit in many plans, a more detailed method for travel time calculation is elaborated incorporating time delay caused by meeting or overtaking other vehicles as a function of road width, vehicle category and traffic density (Boerma, 1987).

A particular aspect of the new development method is that not the entire time savings has been expressed in money. Based upon information about trip purposes on minor rural roads 16% of time savings has been estimated as non-business traffic. This part of the savings has been regarded as a social impact. The remaining of 84% of passenger car hours and heavy good vehicle hours has been expressed in money (COBA,

1981 and Ministerie van Verkeer en Waterstaat, 1984).

2.3.2 Vehicle operating costs

Vehicle operating costs are assumed to relate to road pavement quality. This can be expressed through the Pavement Serviceability Index (PSI). According to American research (Foltz, 1980), vehicle operating costs increase exponentially with decreasing PSI. Starting from basic vehicle operating costs (Centrale Adviesdienst, 1985) at the optimal PSI, the incremental costs are derived from PSI forecasts with and without plan. In order to avoid costly PSI measurements some further research should be undertaken to find a relationship between PSI and visual road inspection features.

2.4 Impacts on road safety

Road safety impact assesment at the level of road section improvement is still poorly developed. The present advice of the working group is to estimate the potential number of total accidents, injury accidents, injuries and mortalities which can be avoided by the plan measures. The impact forecast has to be done by expert's judgement based upon accident registration of past years, considering specific unsafety features of the road section.

Another way of forecasting road safety impact with and without plan is based upon standard accident rates for a minor rural road categories. In that case, given a change of road category, accident numbers with and without a plan could be derived from traffic flow forecasts and category accidents rates.

The valuation for money of road safety impacts seems to be a general trend (SWOV, 1984) but is nevertheless subject to some discussion. For that reason the working group has advised both to make a money valuation of accidents and victims safed, regarded as economic impacts, and to make an additional presentation of the numbers of accidents and victims considered as social impacts.

2.5 Impacts on abolishing axle load restrictions

After a road reconstruction, as a result of the land development project axle load restrictions should be abolished. This is resulting in transport cost savings because undertakers may transport their products in a more efficient way (Boorsma, 1988).

3 RESULTS AND ANALYSIS FROM PRACTICAL TESTING

3.1 Selected road improvement projects

The new method has been tested in five land development projects. Table 1 gives a brief characterization of the five projects (Vrolijk, 1988; Boorsma, 1988; Rademaker, 1987; Beerda, 1988; Weiland, 1988).

Table 1. Brief characterization of the five road improvement projects.

Area	Soil type	Surface (ha)	Minor rural road improvement length (m)	Investment of which benefits are calculated (Dfl)
De Hilver	sand	8,600	11,910	3,094,750
Driebruggen	peat	5,900	25,810	20,075,000
Eemland	clay/peat	7,910	39,720	15,785,000
Wymbritseradeel	clay	8,440	34,030	14,519,580
Walcheren	clay	14,550	62,800	11,500,000

For all areas the new method of road evaluation has been tested, based on the six aspects as described in Section 2, with the exception of Driebruggen en Wymbritseradeel. In these two projects also calculations have been made to determine the effect of abolishing axle load restrictions. Concerning the investments a restriction is made to that part of which benefits can be calculated.

3.2 Results

Table 2 shows the comparison of the results from the existing method and the new method for the five road improvement projects characterized in Table 1. (For the projects De Hilver and Walcheren information is not complete, because of the planning stage of the objects). Under the existing method only savings of maintenance costs for the road owner have been calculated. Under the new method the calculation of the benefits from road reconstruction is according to the Werkgroep Evaluatie Wegen (1986), as described in Section 2. For each project the investments, the benefits and the internal rate of return (IRR) are given. The benefits of the investments are assumed to flow constantly during 20 years, being the average lifetime for road reconstructions. Difference has been made between investments in road reconstructions with a mainly agricultural objective and road reconstructions with a more general community interest. These difference has been made to fit the new method of road evaluation into the evaluation of the total land development project according to the HELP-method.

Table 2. Comparison of the results from the existing and new method of road evaluation.

Area	Investments of which benefits are calculated (Dfl)	Benefits calculated according to existing method (Dfl)	new method (Dfl)	IRR according to new method of road evaluation
De Hilver	3,094,750	*	262,032	6%
Driebruggen	16,200,000	- 3,981	1,125,962	3%
Eemland	14,050,000	199,375	785,331	1%
Wymbritseradeel	13,571,313	- 5,000	1,334,088	7%
Walcheren	11,500,000	*	1,274,776	9%

* Concerning the projects De Hilver and walcheren all information is not complete because of the planning stage.

Table 2 shows that benefits are increasing with the new method. The increase especially appears to roads with a more general interest. This occurs because, apart from costs of maintenance, the new method also calculates traffic costs. Consequently high volume roads contribute to higher benefits. For roads with a more agricultural interest impacts of the new method are less because farmer benefits (within the HELP-evaluation of the total plan) are included in the calculation of private-economic benefits from reallotment and lot distance reduction. This part of the HELP-method is left out of scope.

In spite of the substantial increase of benefits, the calculated IRR in the projects Driebruggen and Eemland is rather low. Concerning Driebruggen there was evidence that the method of calculating "travel time" especially on narrow roads in the peatlands is inadequate. To get a more realistic method of calculating travel time when roads are widened in the peatlands the whole process of meeting and overtaking other vehicles has been calculated in detail. Calculating in this way benefits are increasing by 130%.

In Eemland a great number of farmplants are moved inside the area. This has been great influence on the benefits calculated by the new method. These benefits have a more private-economic character and are part of the benefits calculated according to the HELP-method. Nevertheless the private-economic impacts are only possible thanks to a proper road network for general interest. According to this integrated way of calculating benefits, connected with moving farmplants, the benefits of the new road evaluation method are only 1%.

3.3 Analysis

Table 3 shows the distribution of benefits about the impacts of the new road evaluation method.

Table 3. Distribution of benefits about the impacts of the new road evaluation method.

Area	Structural restore (%)	Pavement maintenance (%)	Shoulders maintenance (%)	Travel time (%)	Vehicle operating costs (%)	Road safety (%)	TOTAL BENEFITS (100%) (%)	Abolishing axlo load restrictions (%)
De Hilver	21	13	3	46	17	0	262,032	*
Driebruggen	25	0	2	37	19	17	783,049	54
Eemland	56	12	2	14	9	7	785,331	*
Wymbritseradeel	29	3	3	51	7	7	1,152,418	16
Walcheren	53	7	1	4	7	28	1,274,776	*

* Not calculated.

In general the aspects "Structural restore" and "Travel time" appear to supply a main contribution to benefits. Table 3 shows that differences between the areas are more or less related to the extent to which investments are made in road construction (impacts on structural restore, road maintenance and vehicle operating costs) or in widening the pavement (impacts on shoulders maintenance, travel time and road safety). The first case especially appears to Eemland and Walcheren; the latter to Driebruggen and Wymbritseradeel. De Hilver shows a balance between the benefits from traffic and constructive impacts. Furthermore table 3 shows that the benefits calculated by abolishing axle load restrictions are substantial. In relation to impacts on road safety, differences still are difficult to explain. More research to this item is demanded.

3.4 Non-economic impacts

Next to the economic approach of impacts of the road reconstruction plan the new evaluation method obviously has produced some ways of quantifying social impacts. As described in paragraph 2.3.1 16% of time savings has been estimated as non-business traffic. According to the HELP-method this time has been expressed in hours spare time. Within the new road evaluation this spare time has been considered as a social effect. As described in paragraph 2.4 prevention of road accidents not only has been expressed in money. There also has been payed attention to the saving of harm and sorrow for the victims, expressed in its own specific units, and regarded as a social effect of road reconstruction.

CONCLUSIONS

a) the new road evaluation method turned out to be a useful instrument. Both economic and social benefits have been made visible and the IRR of the road reconstruction plan has been improved;

b) with the new method significant higher benefits can be calculated when comparing the benefit levels with the existing method's output;

c) there is a better insight into the characteristic elements concerning the road reconstruction plan;

d) with the new method there is more information to make a balanced judgement of comprehensive plans;

e) the new method seems to be an instrument for road section priority ranking in the plan design stage;

f) the new method is an excellent instrument for a design by evaluation. During the plan design stage road evaluation can gradually be adjusted;

g) there is a basis for impact assessment when abolishing axle load restrictions;

h) in area's with moving a lot of framplants the new method shows insufficient benefits. The main part of agricultural benefits is resulting from new parcellation and is calculated in an other part of the HELP-method;

i) some items like "travel time" and "road safety" demand more research;

j) finally computerization of the new method is an urgent request.

REFERENCES

Beerda, K.J. 1988. Wegenevaluatie Wymbritseradeel. Government Service for Land and Water Use, Utrecht.

Boerma, J.S. 1987. Vertragingen door ontmoeten en inhalen van verkeer op plattelandswegen. Nota 1776. Institute for Land and Water Management Research, Wageningen.

Boorsma, S. 1988. Evaluatie Wegenplan Driebruggen. Nota 1857. Institute for Land and Water Management Research, Wageningen.

Centrale Adviesdienst 1985. Publikatie autokosten, Deventer.

COBA 1981. Cost/benefit Analysis Manual "COBA-9". HMSO-Department of Transport, London.

Foltz, H.L. 1980. User costst versus road roughness in pavement management. FHWA Pavement Design Branch, Washington.

Michels, Th. 1988. Impact assessment and evaluation of road improvement in Dutch land development planning. Proceedings Workshop Minor Rural roads, October 28-30, 1987, Wageningen.

Ministerie van Verkeer en Waterstaat 1984. Het planningsproces van rijkswegen, 's-Gravenhage.

Rademaker, M.P. 1987. Wegenevaluatie Eemland. Government Service for Land and Water Use, Utrecht.

Regterschot, D. and J. Mooy 1981. Capaciteit van plattelandswegen. Government Service for Land and Water Use, Utrecht.

Regterschot, D. and Th. Oenema 1981. Verkeersafwikkeling op plattelandswegen. Government Service for Land and Water Use, Utrecht.

Staatsuitgeverij 1983. De HELP-methode voor de evaluatie van landinrichtingsprojecten; beschrijving en verantwoording, 's-Gravenhage.

Staatsuitgeverij 1983. De HELP-methode voor de evaluatie van landinrichtingsprojecten; toelichting en uitwerking, 's-Gravenhage.

Stichting Wetenschappelijk Onderzoek Verkeersveiligheid (SWOV) 1984. Economische schade ten gevolge van verkeersonveiligheid. Rapport R-84-10, Leidschendam.

Vrolijk, C. 1988. Wegenevaluatie De Hilver (concept). Government Service for Land and Water Use, Tilburg.

Weiland, P. 1988. Wegennota Walcheren (concept). Government Service for Land and Water Use, Utrecht.

Werkgroep Evaluatie Wegen 1986. Handleiding toepassing methode wegenevaluatie, Utrecht/Wageningen.

Land and Water Use, Dodd & Grace (eds), © 1989 Balkema, Rotterdam. ISBN 90 6191 980 0

A proposal for axle load restrictions during spring-thaw period

J.Litzka & F.L.Stiedl
Institute for Transportation Engineering and Road Construction, University of Bodenkultur, Vienna, Austria

ABSTRACT: A new method for axle load restrictions on weak pavements during the spring-thaw period is presented. Based on theoretical stress calculations a value for a restricted axle load is given which reduces the stresses in the pavement during the spring period to the amount caused by the normal non-restricted axle load (10 t) during the autumn period. The assessment of the necessary restriction is made by deflection measurements. Diagrams for axle load restrictions are given for several types of pavements.

RÉSUMÉ: Une nouvelle méthode pour limiter la charge par essieu sur des revêtements faibles pendant la période du degèl au printemps sera presentée. Fondé sur des calcules de contraintes théoretiques on déclare ces charges par essieux admissibles, qui ne provoquent pas des efforts plus grandes du revêtement que la charge par essieu non réduite (10 t) pendant la période de l'automne. La détermination de la réduction de la charge par essieu résulte par des mésures de déflection. Des diagrammes correspondants pour des différentes types de construction ont été indiqués.

KURZFASSUNG: Es wird eine neue Methode zur Achslastbeschränkung auf schwachen Befestigungen während der Frühjahrs-Tauperiode vorgestellt. Aufbauend auf theoretischen Spannungsberechnungen werden jene zulässigen Achslasten angegeben, die während der geringeren Tragfähigkeit im Frühjahr keine größeren Beanspruchungen in der Befestigung hervorrufen als die nicht reduzierte Achslast (10 t) während der Herbstperiode. Die Ermittlung der erforderlichen Achslastbeschränkung erfolgt mittels Einsenkungsmessungen. Entsprechende Diagramme werden für verschiedene Bautypen angegeben.

1 INTRODUCTION AND SCOPE

The dimensioning of the pavements of rural low traffic roads should aim in finding the optimum compromise between a sufficiently solid construction on the one hand, and a construction that meets the economic demands on the other hand. The solving of this problem is made even more difficult by the fact, that the traffic loading occurs irregularly, which means that a generally low traffic load changes with a periodically high loading through construction vehicles, the transportation of wood, harvest vehicles and the like.

Additionally, in thin constructions, variations of the bearing capacity of the subgrade due to seasonal impacts have a strong influence on the wear and thus the servicelife of the pavement. During the spring-thaw period, the proportion of wear in such pavements may amount to 60 and more percent of the wear occuring during a whole year - provided an equal traffic load distribution over the year.

If extreme cases, that is peaks of loading during the time the subgrade has the lowest bearing capacity - form the basis for dimensioning, this will result in high values for the necessary thickness of the pavement which is an uneconomic overdimensioning considering the loading during the rest of the year. For this reason, this technically quite correct solution cannot be financed in the secondary road network.

Because of budgetary constraints, there is usually a compromise resulting in a weaker pavement, which also includes the risk, that high loadings during the critical period will substantially reduce the life expectancy of the pavement.

The risk of a shortened life time can only be eliminated by imposing axle load restrictions during the spring-thaw period and thus avoid too heavy loadings.

The problems in connection with seasonally conditioned axle load restrictions are hardly ever dealt with in the literature on road construction. The few detailled works on this topic were delivered by the Scandinavian countries (NVF, 1986; Reinslett, 1981; Fyksen a. Eirum, 1976) and from North America (Connor, 1982; Gough and Haas, 1985). There is only one proposal delivered by a member of the German-speaking community (v. Becker, 1970) This work, however, does not discuss these methods for the determination of axle load restrictions in any detailled way.

A detailled description and discussion of the methods is contained in the work by Stiedl (1988). The above mentioned methods can either only be applied for rough assessments or are too complicated and impracticable for the application in rural roads. For this reason, it was tried to develop a new method.

2 NEW METHOD FOR THE DETERMINATION OF AXLE LOAD RESTRICTIONS

2.1 Principles

The method described below for the determination of axle load restrictions is based on the calculation of stress and deflection.

The determination of necessary reductions in axle loads for the spring period is based on the assumption, that the wear of the pavement due to this reduced loading will not be greater than the one caused under normal conditions during the autumn period. But it must be taken into account, that there are different values for the stiffness of the unbound as well as the bound layers of the pavement during these two periods. For this reason, the comparison is made on the basis of the cumulative number of standard axle loads admitted in the investigated period (cumulative number of reference axle load of 10 tons).

For the calculations of stresses and deflections the computer programm BISAR was used (De Jong, Peutz, Korswagen, 1979).

Stresses that are of interest are:

- Flexural tensile stress σ_r at the bottom of bituminous layers and cement-treated bases.
- Vertical compressive stress on the subgrade and on unbound bases.

These stresses form the basis for the assessment of the wear of the pavement under differently great axle loads. The values of the stresses are displayed. graphically in a diagram for different axle loads and different bearing capacities of the respective subgrades (i.e. different degrees of the reduction of the bearing capacity). Corresponding to the relevant stress (autumn period) it is now possible to deduct the reduction of axle load that is necessary for the various cases of subgrade situation. As stresses occuring in the pavement can practically not be determined, the corresponding elastic deflections of the road surface under a 5t - wheel load are calculated and used for the assessment. Figure 1 shows, in wich way a diagram for the determination of the permissable axle loads as a function of the calculated deflection can be produced from the stress calculations.

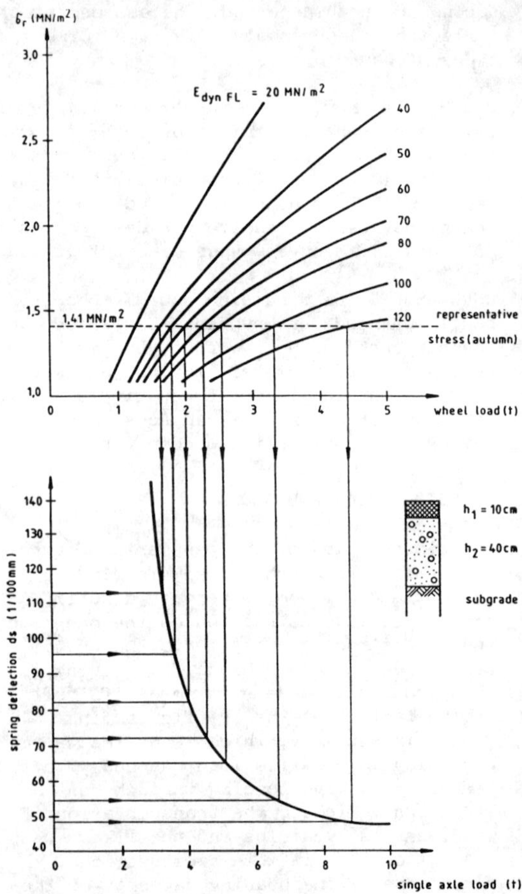

Fig.1 Deduction of a diagram for the determination of the permissable axle load as a function of the elastic deflection (example)

The calculation of stresses and the corresponding deflections are based on assumptions which are not necessarily identical with the actual parameters of the pavement itself. For this reason, it is not the calculated deflection itself, but the ratio (in percent) between spring deflection and autumn deflection that is chosen as the relevant value for the diagrams. It may be assumed that the ratio between the calculated spring and autumn values - within certain limits independently of actual material parameters - is approximately equal to the ratio of the corresponding measured values.

On the basis of this assumption the final diagram for the determination of axle load restrictions can be produced by converting the calculatory spring deflection d_s into a percentage value of the autumn deflection d_a (see figures 2 to 4).

The above description of the procedure for the determination of a diagram of axle load restrictions is in fact only valid for construction types with bound layers, in which the flexural tensile stress is relevant for the assessment. In unbound pavements, the vertical compressive stress operating on the subgrade is the important factor. The vertical compressive stress remains constant for a certain type of construction, independent of the bearing capacity of the subgrade. Thus, there is one single graph for the ratio between axle load and vertical compressive strain. The relevant strain in spring, however, has to be newly calculated for any of the investigated bearing capacities of the subgrades.

2.2 Diagrams for the determination of axle load restrictions

According to the considerations described in 2.1 corresponding diagrams have been elaborated for the determination of axle load restrictions during the spring-thaw period which are applicable for different construction types of rural roads. Altogether, the following three construction types have been considered, for which, in dependence on the structural thickness, several graphs are drawn:

Construction type I:
unbound pavement consisting of a gravel or crushed stone base course, possibly with a surface dressing.

Construction type II:
unbound base course and bituminous surface layer.

Construction type III:
cement-treated base course, possibly with bituminous wearing course.

The corresponding diagrams are shown in figures 2 to 4.

Additionally, investigations have also been carried out for concrete pavements. In these, the relevant stress is the flexural tensile stress σ_r at the bottom of the slab. The calculations were based on the formulas by Westergaard (1926) for the load condition of the slab edge (undowelled transverse joint), considering several bearing capacities of different subgrades and thicknesses of surfacings between 16 cm and 20 cm on a thin unbound levelling course, which is the usual way of constructing rural roads. The calculations confirmed the well known fact, that the wear of a rigid pavement depends to a considerably less degree on the bearing capacity of the subgrade than the one in flexible pavements.

Due to the calculations it could thus be recommended - to determine a general axle load restriction of 7 tons for the spring-thaw period.

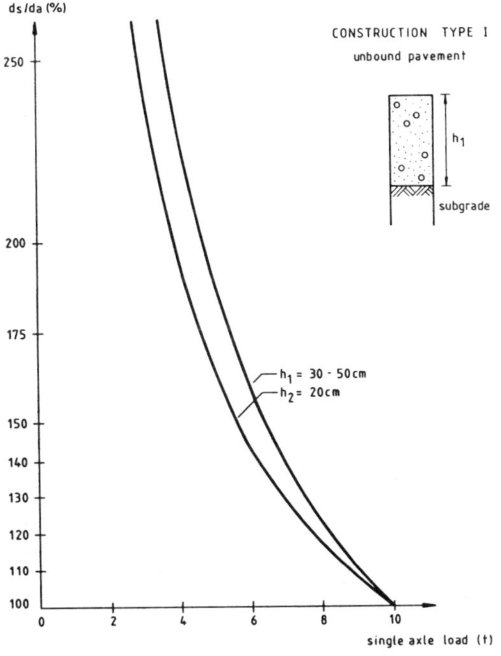

Fig.2 Admittable axle loads for unbound pavements during the spring-thaw period (d_s = spring deflection, d_a = autumn deflection)

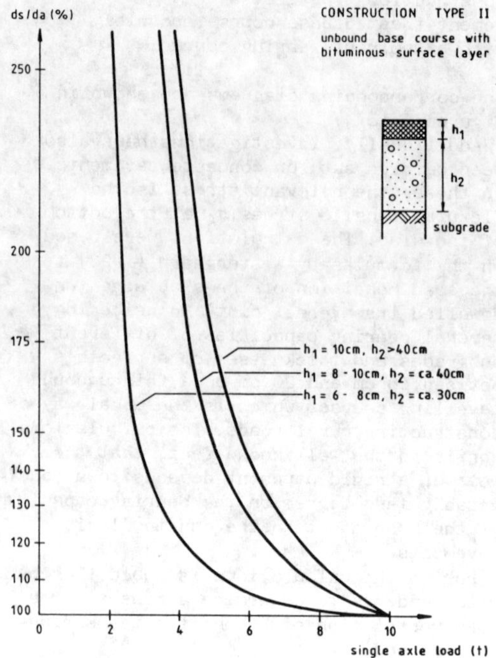

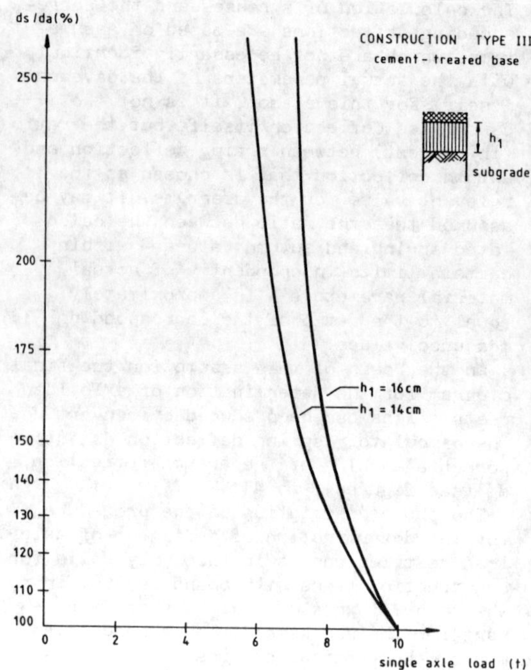

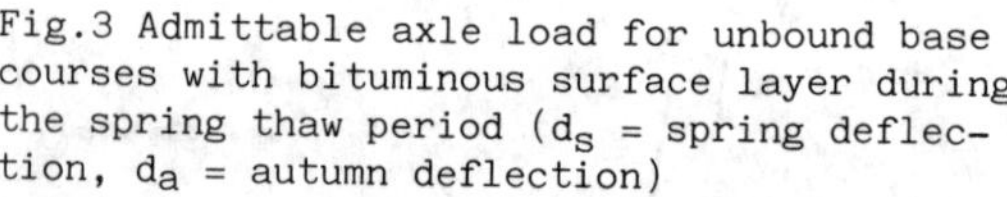

Fig.3 Admittable axle load for unbound base courses with bituminous surface layer during the spring thaw period (d_s = spring deflection, d_a = autumn deflection)

Fig.4 Admittable axle loads for cement-treated base courses during the spring-thaw period (d_s = spring deflection, d_a = autumn deflection)

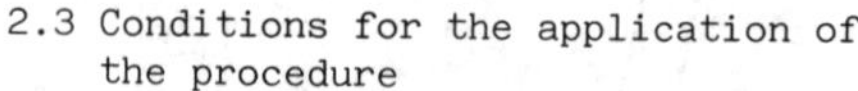

2.3 Conditions for the application of the procedure

From the above explanations it can be seen, that the proposed method is solely based on the wear of the existing pavement under a certain axle load. Thus, it is principally not considered, whether the dimensioning of the existing pavement is sufficiently strong for the actual traffic load.

The procedure deliberately excludes any actual traffic load. It is exclusively based on the principle of reducing load peaks during the spring period, that result from the lower bearing capacity of the subgrade, by means of corresponding axle load restrictions and thus secure a relative lengthening of the pavement's life time. The lengthening of life time that can be obtained in this way for pavements which have already originally been planned for higher loadings is usually less than for pavements planned for medium or weak loadings. In the latter, the lengthening of their life times may naturally be quite considerable.

Before the described procedure for determining axle load restrictions can sensibly be applied, it must in any case be investigated, whether the existing pavement is dimensioned sufficiently strong for the actual traffic loads. Axle load restrictions on roads with pavements that are underdimensioned from the beginning, can admittedly result in a relative lengthening of life time, but they cannot prevent that heavy structural damage of the pavement occurs already prematurely.

3 CONSIDERATIONS CONCERNING THE PRACTICAL APPLICATION

For the application of the diagrams it is necessary to have the results of deflection measurements carried out during the spring-thaw period and the autumn period on the one hand, and to have an approximate knowledge of the structure, respectively the thicknesses of the various layers, to be able to determine the necessary axle load restrictions.

Usually, deflection measurements for this purpose will be carried out with the Benkelman beam.

The practical application of the procedure is possible in different ways, which have been described in detail in a paper by Litzka and Stiedl (1989).

The determination of axle load restrictions can be based on the following considerations:

- Deflection measurements in spring and autumn on every road concerned and application of the diagrams. This is an ideal case which can only be realized when treating special problems.
- Deflection measurements only on representative road sections (reference sections) and application of the diagrams. The results are considered to be valid for the whole road net to which these reference sections belong.
- Approximate determination without deflection measurements. Based on the assumption, that the bearing capacity of the subgrade during the spring period is diminished to about 50 percent of the autumn value, table 1 can be used for the determination of axle load restrictions. This method should be used only exceptionally.

Table 1. Permissable axle loads during the spring-thaw period (simplified method)

Construction type	Pavement	Axle load (t)
I	30-50 cm unbound base	5
	20 cm unbound base	4
II	10 cm bituminous surface layer, >40 cm unbound base	6
	8-10 cm bit. surface layer, appr. 40 cm unbound base	5
	6-8 cm bit. surface layer, appr. 30 cm unbound base	3
III	≥14 cm cement treated base	7
IV	16-20 cm plain concrete slabs	7

The prescription of the axle load restriction should already be made at the time when the soil starts thawing at the surface. To be on the safe side, it is recommended to suspend the restriction in general not earlier than one month after the minimum bearing capacity of the subgrade has been reached, unless measurements show a more rapid improvement of the bearing capacity.

In practice, certain problems may arise with the control of the determined axle load restrictions. Normally, mobile axle load scales are not available. The person authorized for controlling may possibly conclude from the weigh bills which show the gross weight of the vehicle to the weight of the axle loads. Approximately corresponding values for conversion are contained in the tables 2 and 3. In vehicles with rear tandem axles the equivalent values for the single rear axle are the relevant ones.

Table 2. Relation between gross weight and rear axle load for 2 axle trucks

Gross weigth (t)	Single rear axle (t)
8	4,2
9	4,8
10	5,5
11	6,5
12	7,0
13	7,8
14	8,5
15	9,3
16	10,0

Table 3. Relation between gross weight and rear axle load for trucks with tandem axle

Gross weight (t)	Rear tandem axle (t)	Equiv. single axle load (t)
13	7,8	4,3
14	9,4	5,1
15	10,1	5,5
16	10,8	5,9
17	11,6	6,3
18	12,5	6,8
19	13,5	7,4
20	14,3	7,8
21	15,2	8,3
22	16,0	8,7
23	17,0	9,3
24	17,8	9,7
25	18,7	10,2

4 CONCLUSION

The problem of axle load restrictions during the spring-thaw period is of special importance in roads with weak pavements. The described procedure renders it possible to determine the necessary axle load restrictions in a relatively simple way. Thus, too great wear of the existing pavement during the spring period can be avoided. Whenever possible, determinations should be based on previous measurements of the bearing capacity (deflection measurements). The simplified procedure that has also been described should only be applied when it is not possible to carry out measurements.

The application of the method results in a relative lengthening of the pavement's life time, provided the structural dimensioning is sufficient for the actual traffic load.

REFERENCES

Becker, P. v. 1970. Erfahrungen mit Einsenkungsmessungen auf Straßenbefestigungen und bei der Erdbaukontrolle. Straße und Verkehr, 56:161-166.

Connor, B. 1982. Rational Seasonal Load Restrictions and Overload Permits. Transportation Research Records Nr. 846, p. 6-15.

De Jong, D.L., M.G.F. Peutz and A.R. Korswagen, 1979. External Report Computer Program BISAR. Layered Systems under Normal and Tangential Surface Loads. Koninklijke/Shell-Laboratorium, Shell Research B.V., Amsterdam.

F.G. Köln, 1968. Der AASHO-Road-Test. Forschungsarbeiten aus dem Straßenwesen. Neue Folge, Heft 73. Kirschbaum-Verlag, Bad Godesberg.

Fyksen, T. and R. Eirum, 1976. Baereevneregistrering i teleløsningsperioden. Erfaringer ved bruk av Lacroix deflectograf og telegrensemålere. Statens Vegvesen, Veglaboratoriet, Intern rapport Nr. 713, Oslo.

Gough, J.W. and R.C.G. Haas, 1985. Development of a Rational System for Seasonal Load Restrictions. Transportation forum, Vol. 1-4 p. 57-65.

Litzka, J. und F.L. Stiedl, 1989. Achslastbeschränkungen auf schwach befestigten Straßen während der Frühjahrstauperiode. Zeitschrift für Kulturtechnik und Landentwicklung, 30:48-57.

NVF, 1986. Trafiklastbegränsningar under Tjälllosningen. Nordiska Vägtekniska Förbundet, Utskott 31-Vägens uppyggnad, Rapport Nr. 1/1986.

Reinslett, E. 1981. Bestemmelse av tidspunkt for innføring/oppheving av aksellastrestriksjoner i teleløsningsperioden. Vegdirektoratet, Veglaboratoriet, Intern rapport Nr. 992, Oslo.

Stiedl, F.L. 1988. Überlegungen zur Achslastbeschränkung auf schwach befestigten Straßen während der Frühjahrstauperiode. Diplomarbeit am Institut für Geotechnik und Verkehrswesen, Universität für Bodenkultur, Wien.

Westergaard, H.M. 1926. Stresses in Concrete Pavements Computed by Theoretical Analysis. Public Roads 7, Vol. 2.

Land and Water Use, Dodd & Grace (eds), © 1989 Balkema, Rotterdam. ISBN 90 6191 980 0

The minimum transport-work method on the drainage sites

D.Dejas & E.Bondar-Nowakowska
Institute for Land Reclamation and Improvement, Agricultural University, Wroclaw, Poland

ABSTRACT : The transport of ceramic drainage pipes on a building site is very expensive so its cost should be reduced. In this paper the method of finding the best position of the storage place of pipes and calculating the transportwork on any store place position, irrespective of the area and surface, is presented. This method allows to calculate the transportwork in two ways: analitic by computer programme and graphic by nomograms. The graphic way of presentation is not complicated and the simulation verification proves that deviations are to 10%. The method applied in Poland reduced the cost of transport and improved the liquid fuels management.

RÉSUMÉ : La part du transport des drains céramiques dans les coûts de construction d'un investissement d'assainissement agricole est trés grande. Jl existe alors la nécéssité de leur réduction. Dans cette étude nous présentons une méthode de la determination d'une telle localisation du dépôt des drains, pour laquelle le paramétre masse x distance du transport est minimum, independamment de la grandeur de la superficie et de la forme de la surface drainée. Les calculs peuvent étre réalisés par deux voies: voie analytique pour lequelle nous avons construit un programme et voie graphique, á l'aide des nomogrammes élaborés par nous. L'utilisation de la voie graphique est simple. Des calculs de simulation ont prouvé que l'erreur par rapport á la voie analytique ne dépasse pas les 10%. La méthode a été appliquée en Pologne et apporte des economies considérables.

ZUSAMMENFASSUNG : Der Anteil des Tondränrohrtransportes in den Produktionskosten beim Bau von Dränanlagen ist sehr groß. Es besteht also das Bedürfnis diesen Anteil zu vermindern. Dazu wurde eine Methode entwickelt, nach der die günstigste Lage eines Lagerplatzes und die optimale Transportarbeit für jede Dränflächengröße und Dränflächenform bestimmt werden kann. Die Transportarbeit kann nach zwei Verfahren ermittelt werden:analytisch mit Hilfe eines speziellen Rechenprogramm und graphisch mit Hilfe von Nomogrammen. Das graphische Verfahren ist sehr einfach im Gebrauch und die Simulationsberechnungen haben gezeigt,daß die Abweichnung von den Werten nach dem analytischen Verfahren bis zu 10% beträgt. Die Methode ist in Polen in die Praxis eingeführt worden und bringt sichtbare Einsparungen an Transportarbeit.

1 INTRODUCTION

Realization of land melioration projects depends to a much greater extent on the transportation than other branches of industry. Because of constant translocation of work objects, tools and workstands, storage places etc. the transportation conditions the efficiency of the realization process.

During the transportation process there is no change of the work objects except for the defects. According to the definition of the transportation process, the change of the state of the work object consists in the change of its position, for example, it is subjected to the change of the storage place that, in turn, enables its actual change through its installation on the desired site.

Thanks to the continuing progress in the production processes, mainly by introducing modern, multifunctional machines and equipment, productivity icreases and thus also the amount of material installed per unit time. It requires constant improvement in the organization of the transportation and, in particular, the rational exploitation of the transportation means at the disposal of the land melioration companies.

The transportation as an integral part of the production process has a large influence on the costs of production.Its contribution in the drainage projects amounts to 25% in average and in especially difficult operating conditions even 40% . Thus, it becomes indispensable to look for ways to decrease these costs.

2 OBJECTIVE

In the recent years in Poland unnecessary transport amounted to about 30% of all transported loads from which about 18% resulted from enappropriate distribution of productive forces and about 18% from inappropriate decisions.

In the civil engineering and also in the land melioration the share of the transportation costs in the final product is more than 3.5 times higher than in the industry.

It is estimated that the transport-work level is in Poland several times higher than in other countries. The analysis of the foundations of the transport policy of the economically developed countries, the trends in the consumption of crude oil includet, showed that in most of those countries this consumption will´ be in 1980 smaller than or equal to the one in 1973. But in Poland it is going to be 4 times higher, which is the evidence of high energy consumption in our industry.

Transporting the drain pipes from the manufacturer to the place their installation is carried out by the indirect transport. Each time it requires the localization of the storage place from which the pipes are delivered by the building-site-transport to the installation sites. This kind of transport is in Poland neglected and underestimated, but it has deep reserves.

In almost all drainage sites the storage places are localited according to arbitrary criteria which couse unjustified increase of the transport-work.

By analizing a dozen or so installation sites already completed it has been determined that this increase amounts to 100%. Taking into account that on 1000 ha of the drainage field the average mass of the transported materials is 5760 tons, and that in Poland there still remains about 2.7mln ha to drain, a decrease of the transport-work and resulting from that economies, mainly the fuel economy, can have a decisive influence on the realization of the projects.

In literature there is a lack of works in the area of the rationalization of the transportation on the land melioration sites. That very subject, undertaken by the authors, has as its aim working out some methods allowing the determination of the transport-work of delivering the drain pipes from an arbitrary storage place on the drainage field to all sites of their installation.

Applying the results of this work to practice is supposed to enable the management to determine under existing natural conditions such a localization for storing the drain pipes so that the transport work is minimal.

3 METHODS OF CALCULATING THE TRANSPORT-WORK

In the years 1982-87 we carried out research investigations concerning the influence of the location of the storage place on the transport work on drainage fields. They have led to the development of an analytical and a graphical methods of determining the transport-work of distributing the drain pipes on drainage fields of various shapes and various drainage systems.

Those methods are based on the following assumptions :
- the mass of the drain pipes to be installed is centered at the centre of mass of the elementary field
- the movement of the transport means is carried out radially on a flat surface along straight lines from the storage place to the centers of mass of the elementary fields.

3.1 Analytical method

This method should be used mainly to surfaces of very irregular shapes and asymmetric mass distribution of the drain pipes. To calculate the transport-work by the analytical method one has to divide the drainage field into elementary fields on the coordinate system and the division into squares or rectangles is recommendet. The number of such fields cannot be smaller than 100. In each elementary field the mass of installed drainage pipe lines is determined as well as its center of mass coordinates (x_i, y_i). For such data two programs have been written. The WLPS program is designet to calculate the minimum value of the function

$$\tau = \min_{x,y} T(x,y) \qquad (1)$$

where

$$T(x,y) = \sum_{i=1}^{n} \sqrt{(x-x_i)^2+(y-y_i)^2} \circ q_i, \quad (2)$$

where x,y the coordinates of the storage place, x_i, y_i the center of mass coordinates of the i-th elementary field and q_i the mass of the drain pipes installed in the elementary field.

The calculated coordinates x,y denote the localization of the storage place for which the transport-work is minimal. The WLPS program is written in BASIC. It requires the following input data : x_i, y_i the center of mass coordinates of the elementary fields, q_i the mass of drain pipes installed in a given elementary field (i=1....n), dx the average distance of the discretization of the drainage field along the x-axis and dy the average distance of the discretization of the drainage field along the y-axis.

The WLPS program prints the following data : x_i, y_i, q_i the input data, x,y the coordinates of the optimal location of the storage place and T the transport-work for the storage place localized at (x,y).

The calculation of the transport-work is done by the Monte Carlo method and its accuracy depends on the area of the elementary fields assumed.

The OTDP program enables us to calculate the transport-work for the storage places localized at an arbitrary point at the installation site as follows:

$$T_j = T(x_j, y_j) \quad \text{for } j=1,\dots,n \qquad (3)$$

(according to the formula (2))

where x_j, y_j are the coordinates of the arbitrary localization of the storage place.

The TDP program is also written in BASIC. It requires the same input data as the WLPS program whereas it prints the following data : x_i, y_i, q_i the input data (i=1,..n), and T_j the transport-work for the storage place localized at the i-th elementary field (i=1,..n).

The transport-work is expressed in units that depend on the input data; for example for x_i, y_i given in km and q_i in tons(t), T_j is given in tkm.

3.2 Graphical method

Taking into account that the decision about the localization of the storage place is made directly at the installation site, nomograms based on research and calculations were constructed. They enable us to calculate the transport-work of distributing the drain pipes from an arbitrary storage place to all installation sites.

The nomograms are made for the areas that have the shape of a rectangle and equilateral triangle or a right friangle. Those figures can approximate the actual shapes of the drainage fields.

They consist of 2 parts. The first one (Figs 1,2,3) serves to determine the transport-work of distributing the drain pipes from the storage place located at the geometrical center of the drainage field. The second part (Figs 3,4) represents the coefficient n as a dependence between the transport-work from an arbitrary point on the installation site and the transport-work from the geometrical center of the drained surface.

Figure 1 refers to rectangular areas, Figures 2 and 3 refers to triangular areas. Their construction is as follows: on the x-axes are the areas of the drained surface in ha. and on each of the 9 y-axes which correspond to the spacing between the drain pipe lines ranging from 10 to 18 m the values of the transport-work are indicated (tkm).

In Fig. 1 the oblique lines correspond to the drained surface whose shape is an approximation of a square or a rectangle. The bottom line represents a square A/B=1, the top line represents a rectangle with the sides ratio A/B=5. In Figs 2 and 3 oblique lines refer to the triangles with the base to height ratios A/H=1.0, A/H=0.5 and A/H=0.33. To obtain the transport-work from the geometrical center of the drainage field which in the analytical method corresponds to the formula (1) one has to replace the drained surface by rectangles, triangles or combinations of those figures.

For a fixed ratio A/B or A/H on an oblique line one has to determine the point corresponding to the drained surface, transfer it horizontally to the vertical line which corresponds to the average drainage spacing and read the value of the transport-work.

The transport-work from an arbitrary storage place is determined by Fig 4 for rectangular surfaces and Figs 5, 6 for triangular ones. From those Figures one can read the value of the coefficient "n" whose value depends only on the location of the storage place.
The product $\tau \circ n$ is the transport-work from an arbitrary point on the installation site.

The way to use those nomograms is as follows. One has to introduce a coordinata system centered at the geometrical center of the drained surface, fix the coordinates of the storage place as X/B and Y/A and read from the nomogram the value of "n" corresponding to them.

The transport-work of distributing the drain pipes calculated from the nomograms has been subjected to a verification on a large number of real and simulated drainage fields. It has been determined that those results when compared with the analytical method show an error not exceeding 10%. This error depends on the assymmetry of the mass distribution of the drain pipes and the irregularities of the drainage field shape.

REFERENCES

Bondar-Nowakowska, E., Dejas,D. and Witt,F.K. 1983. Rationalisierung des Dränrohrtransportes auf Dränbaustellen durch technologisch begründete Transportorganisation. Rostock: Wissenschaftliche Zeitschrift d.W.-P.Universität Jahrg. XXXII.Naturwissensch. Reihe, H.4.

Bondar-Nowakowska,E. and Dejas, D. 1985.Metoda uproszczonego wyznaczania racjonalnego miejsca składowania rurek na budowach drenarskich. Wrocław: Zeszyty Naukowe Akademii Rolniczej, Melioracja XXVIII Nr.155.

Bondar-Nowakowska, E. 1986. Wpływ położenia placów składowych rurek drenarskich na transportochłoność na budowach melioracyjnych. Wrocław: Akademia Rolnicza, Wydz. Melioracji Wodnych. Doctoral thesis, in press.

Bondar-Nowakowska, E .Dejas, D. and Witt F.-K.1988. Rechnergestützte Senkung des Aufwandes für den Tondränrohrtransport auf Dränbaustellen. Magdeburg: Proceedings

of the International Conference: Anwendung der Schlüsseltechnologien im Meliorationswesen.
Chomicz,J.and Radczuk,L.1985.Zastosowanie złożonego łańcucha Markowa do modelowania hydrologicznych szeregów czasowych. Wrocław: Zeszyty Naukowe Akademii Rolniczej. Melioracja XXVII, Nr. 154.
Hopfer,A.and Urban,M.1977. Wrocław: Geodezyjne urządzenia terenów rolnych. Państwowe Wydawnictwo Naukowe.
Piskozub,A.1982. Gospodarowanie w transporcie. Warszawa: Wydawnictwo Komunikacji i Łączności.

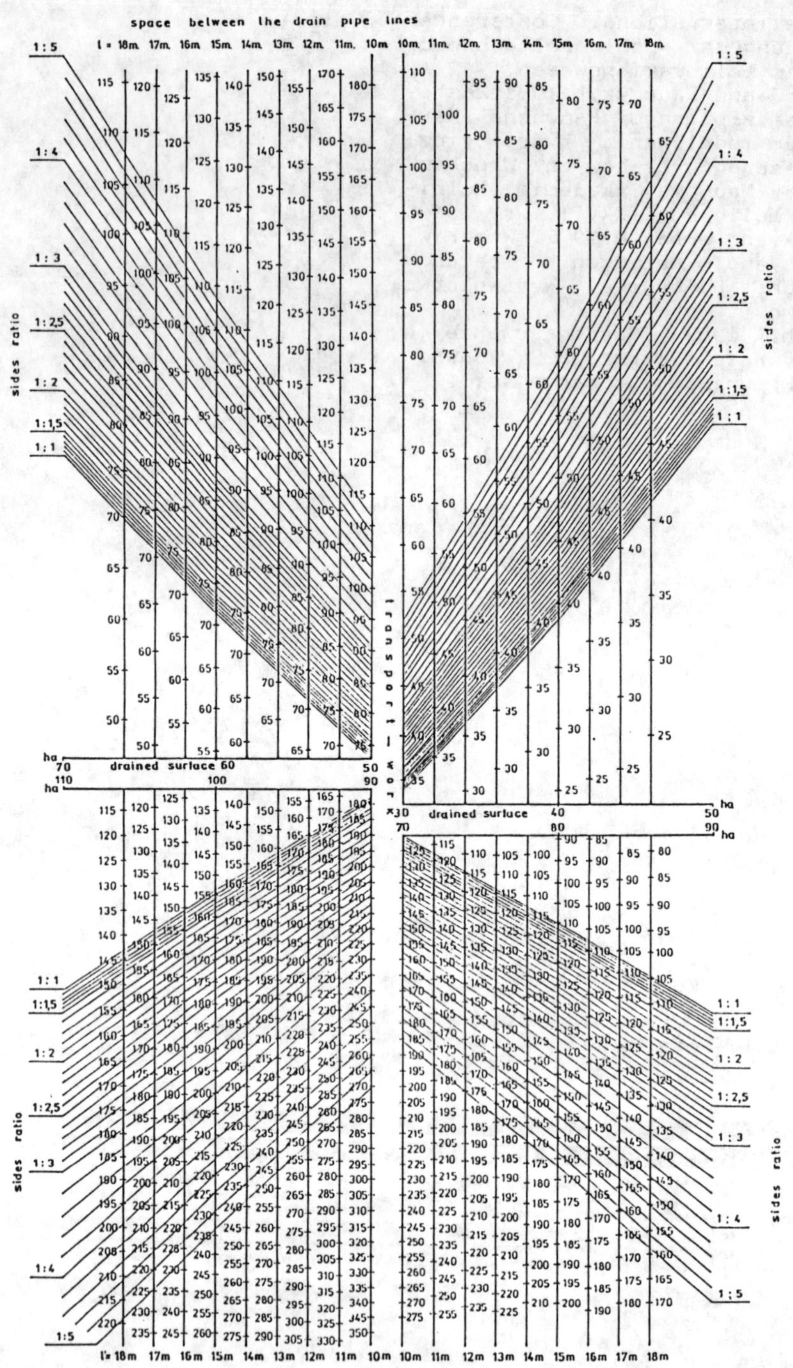

Fig.1 Nomogram for determining the transport-work from the geometrical center of the drained surfaces of approximately rectangular shapes

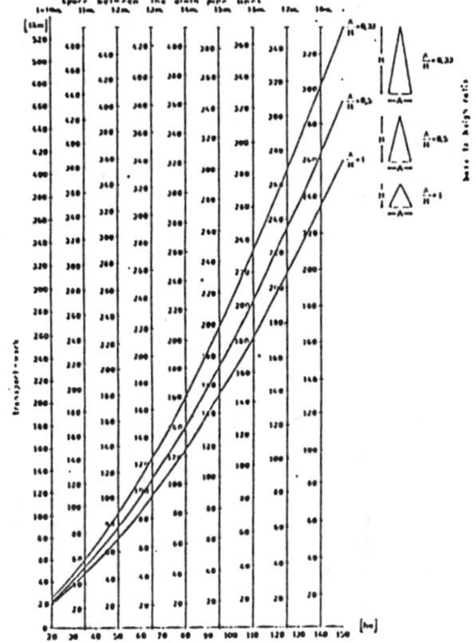

Fig.2 Nomogram for determining the transport-work from the geometrical center of the drained surfaces whose shapes approximate the equilateral triangles

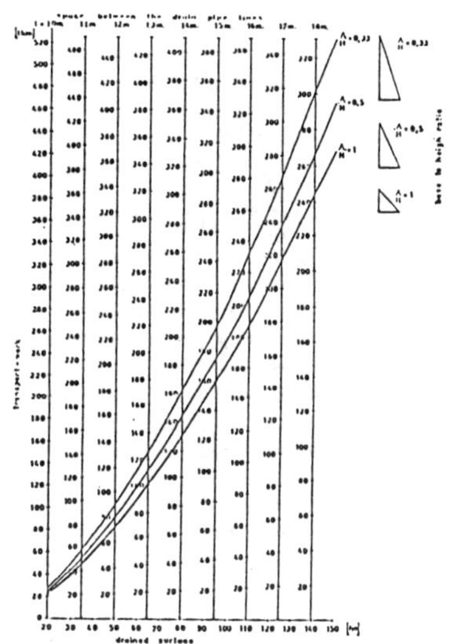

Fig.3 Nomogram for determining the transport-work from the geometrical center of the drained surfaces whose shapes approximate the right triangles

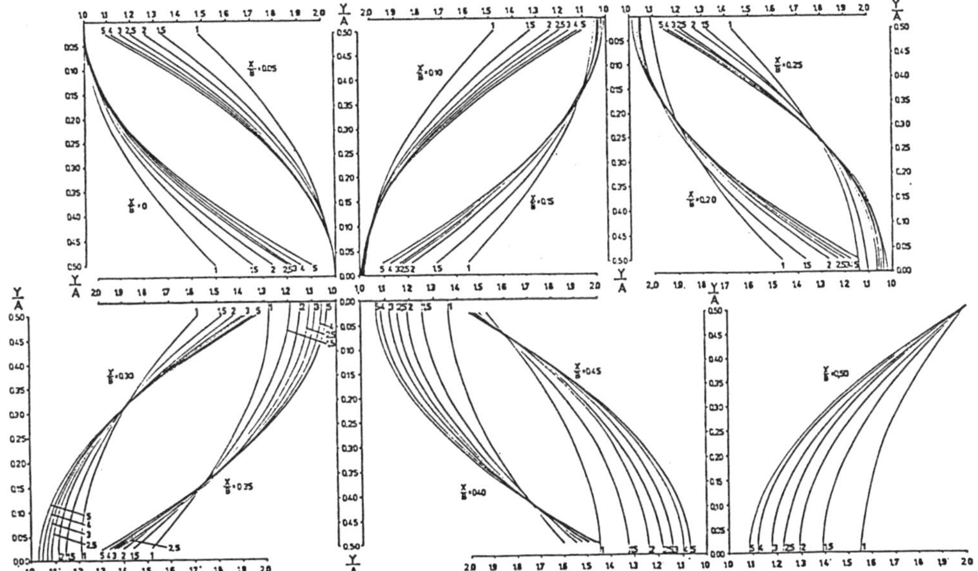

Fig.4 The coefficients n for determining the transport-work from an arbitrary storage place on the drained surfaces whose shapes approximate the rectangles from $\frac{A}{B} = 1$ to $\frac{A}{B} = 5$

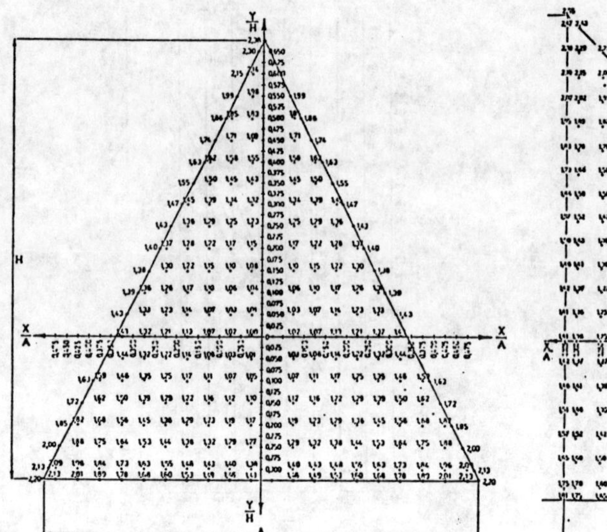

Fig.5 The coefficients for determining the transport-work from an arbitrary storage place on the drained surfaces whose shapes approximate the equilateral triangle $\frac{A}{H} = 1$ (as an example)

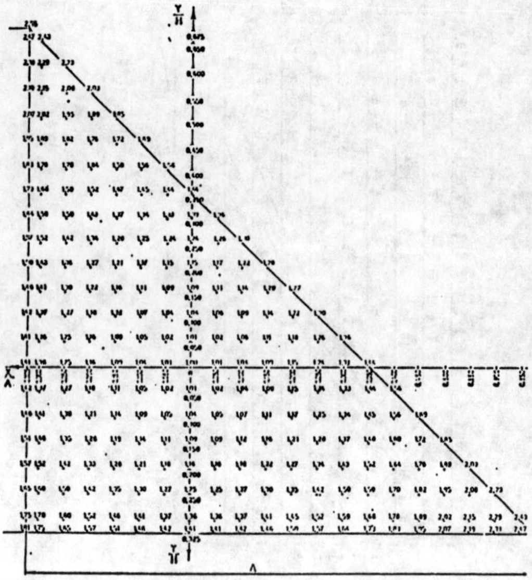

Fig.6 The coefficients for determining the transport-work from an arbitrary storage place on the drained surfaces whose shapes approximate the right triangle $\frac{A}{H} = 1$ (as an example)

Land and Water Use, Dodd & Grace (eds), © 1989 Balkema, Rotterdam. ISBN 90 6191 980 0

A computerized optimal forest road network location procedure

M.A.Nieuwenhuis
University College, Dublin, Ireland

T.J.Corcoran
University of Maine, Orono, USA

ABSTRACT: The high costs of forest road construction, maintenance, and timber transportation warrant extensive planning of forest road network layout and transportation scheduling. The integration of these efforts in the overall management decision making process will result in more efficient timber harvesting operations.

During the last decade, the use of computerized geographic information systems (GIS) for forest management has increased rapidly. The capability of these systems to facilitate the accessibility of diverse data allows for the inclusion of all essential information in management planning procedures.

A forest road network location procedure was developed using both the spatial and descriptive data bases of an existing geographic information system. The objective is the optimal location of a road network in areas not yet serviced by existing roads, constrained by a user-defined maximum service zone width. A local search algorithm was designed, producing optimal or near-optimal solutions in all cases examined.

RÉSUMÉ: Les coûts élevés de construction et d'entretien de chemins et de transport de bois justifient un effort important de planification du réseau routier et d'organisation des cédules de transport. L'intégration de ces activités de planification dans le processus décisionnel d'aménagement forestier résultera en des opérations forestières plus efficaces.

Durant la dernière décénnie, l'utilisation des systèmes d'information géographiques à référence spatiale a connu une croissance rapide. Ces systèmes rendent plus façile l'accès à diverses données et permettent donc la considération d'informations nécessaires pour la confection de plans d'aménagement forestier.

Une procédure de mise en place du réseau routier qui utilise les bases de données spatiales et d'attribus d'un système opérationnel d'information géographique à référence spatiale a été développé. Cette procédure avait pour but de trouver le traçé optimal du réseau routier dans des régions qui n'étaient pas encore servies par des chemins, optimisations sujètte à des contraintes établies par l'utilisateur sur l'espacement des chemins. Un algorythme de recherche locale a été développé et a produit des solutions optimales ou quasi-optimales dans tout les cas étudiés.

ZUSAMMENFASSUNG: Hohe Kosten für Waldwegebau, Unterhalt und Holztransport erfordern detaillierte Planung forstlicher Wegenetze und des Transports. Die Integration dieser Planungsmöglichkeiten in Entscheidungssystemen für Forst Management resultiert in ökonomischerem Holzabbau.

Während des letzten Jahrzehnts hat die Anwendung Computer-gesteuerter geographischer Informationssysteme für Forst Management stark zugenommen. Die Fähigkeit dieser Systeme eine Vielzahl von Daten zugänglich zu machen, ermöglicht es alle wesentlichen Informationen in Planungsverfahren anzuwenden.

Ein Verfahren für die Bestimmung von Waldwegenetzen wurde entwickelt. Es benutzt die Data Banken eines bestehende Informationssystems. Die Absicht war die optimale Placierung eines Wegenetzes in Gebieten ohne bestehende Wege. Der Abstand zwischen den Wegen ist beschränkt von einer maximale Erschliessungsweite. Ein Algorithmus wurde entwickelt, der optimale oder fast optimale Lösungen in allen untersuchten Fällen erstellt.

1 INTRODUCTION

The use of computerized geographic information systems is increasing rapidly in the area of forest land management. These systems provide the decision maker with complete and up-to-date information, required for the effective and efficient management of large timberlands.

Introduction of these information systems can provide the ability to have all data readily accessible in the form of maps and reports. The capability of the systems to combine the different data sets, such as cover type, timber volumes, soil types, cut areas, topographical features, and other management information, enables management to include as many factors as appropriate in the decision-making process.

One aspect of the forest operations which is often not included as an integral part of the present information systems is the harvesting and transportation of timber. Special harvesting decision-making support systems exist, but are uniquely designed for this purpose and do not allow for complete integration into a map-based information system. Making the transportation and road network analysis procedures a part of the overall system enables all data in the data-bases to be shared by these subsystems,enhancing their utility and capabilities.

A particularly interesting part of road network analysis is the location of new roads within the existing network. Because of the growing need to control logging and transportation costs, road networks should be optimized in consideration of the costs of road construction, skidding and trucking. In many cases this will mean network expansion. New roads are also required in areas not previously subjected to intensive operations, and in areas of salvage operations (Corcoran and Nieuwenhuis 1983). Other reasons for new road construction include changes in transportation systems (i.e. river drives are replaced by road haul), upgrading of existing network segments (i.e. straightening of important routes), environmental policies (i.e. minimum distances required between streams and roads), fire control, and recreation. In addition, the possibility exists that, because of vehicle size and weight restrictions for public road systems,it is economically attractive to expand the private road network all the way from the forest to the processing location. Road construction and maintenance costs for this expansion are offset by a reduction in transportation costs. It has been claimed that the most valuable real estate in many operational forests, on a per unit area basis, are the road networks. Because of this high value, combined with the high transportation and road maintenance costs, it is essential to ascertain the right combination of road quality and quantity for successful forest management, regardless of the size of ownership.

2 THE ROAD NETWORK ANALYSIS MODULE

Part of the developed Geographic Information System (GIS) is a road network inventory module (Nieuwenhuis 1983). The GIS is used by a large paper company for the management of almost 1,000,000 hectares of forest land and 3,500 miles of forest roads. Because of the fast-growing complexity and size of the in-place road network, it became necessary to automate the inventory procedures used for transportation and harvesting management. Making the inventory module part of the overall GIS has the further advantage that the information can be used to enhance the usefulness of the system for other procedures, such as accessibility studies or environmental impact analyses. At the same time, the road inventory module can use all capabilities and data of the overall system, which increases its flexibility and accuracy.

After determining the parameters of interest for the existing roads, road segments, culverts, bridges, and other important obstacles, each with their range of possible values, a scheme has to be developed to handle storage and retrieval of the data in the data base. This scheme is commonly known as a data base model. Points of special interest in the development of the data base model are the hierarchical relationships between the entities. Entities are the objects in the data base and are completely defined in terms of attributes and their associated values. In the road and structure data base the entities are roads, road segments, culverts, bridges, gates and obstacles. Hierarchical relationships occur when one entity is part of a higher order entity. For instance, road segments (members) are a part of a road (owner). By defining the owner-member relationships the user has the capability of addressing the member entities directly or through the owner entity.

Two important factors, asides from storage space limitations, have a significant impact on the number of attributes included in the data base.

First, it always must be kept in mind that the primary problem in data base management is the maintenance and updating of the stored information, not the initial loading of the data base. Characteristics which are difficult to quantify and/or qualify, or are not used regularly, are not included in the data base model. Attributes with frequently changing values are also excluded, because of upgrading problems. Second, it is not necessary to include information required for possible future use. The way in which the data base is organized and the data is stored allows for expansion and modification of the data base model without disturbing data already stored.

The GIS is a combination of the Database Management and Retrieval System (DMRS) and the Interactive Graphics Design System (IGDS). A special feature of the DMRS is the possibility to use coded information to represent attribute values. The use of the code lists serves three purposes: storage space reduction, ease of handling, and error detection. The code lists are also an efficient way to assign values to graphic parameters, such as colour, line weight and line code. For the mapping and reporting of elements with discrepancies between structural specifications and functional requirements, extensive use is made of code lists to assign colour and line code values to the graphic elements, enabling the plotting of road segments and structures according to their structural-functional relationship.

The road inventory module was developed with several possible users in mind. The first and most basic purpose was to have a complete and up-to-date inventory of the total road network available. Second, for certain users, it was important to have an inventory capability on township and district levels. Third, the capability to graphically depict the road and road segments (and their associated structures) which show a discrepancy between functional and structural classification is important.

The permanent road network is classified into three functional categories, Class I, II, and III. For each of these categories minimal and/or maximal structural requirements are defined (by company policy). These requirements include maximum up-hill and down-hill grades, minimum widths, maximum curvature, and required surface materials for roads; minimum diameters, materials, and slopes for culverts; and minimum load capacities, minimum widths and required materials for bridges. The ability to identify roads, road segments, and structures which do not satisfy their required structural specifications, is necessary to efficiently bring all roads up to their functional specifications. The capability to do this in the form of maps and/or reports facilitates the upgrading and maintenance scheduling process.

Other specific uses which are taken into consideration, but for which no special accommodations were made, are network analysis procedures. In order to use the road inventory module for these purposes, a number of changes must be made in the way the roads and road segments are stored in the data base. The required changes do not require starting from scratch, but necessitate expansion of the data base model, to include nodes and directed links. These modifications make it possible to use the inventory module for network analysis purposes,such as travel time estimations (taking into account vehicle type, weight, road conditions, etc.) and route selections (accounting for vehicle size and weight, road class limitations, etc.), including both simple shortest route determinations and more complex tour optimization procedures.

Shortest route determination, both for travel time and for travel distance, involves finding the shortest path in a directed network, where the road segments form the links, and the points between segments and the intersections form the nodes. The network must be directed because of the differences in up-hill and down-hill grades for the direction of travel. This can be accomplished by defining the road segments by their beginning and ending nodes. Data handling and programming ease can be facilitated by adoption of the rule that the data base information is correct for travel from a lower to a higher numbered node. This eliminates the need for inspection of the link data base identification number each time the link is examined. If the road network analysis procedures are used frequently, it can become feasible to store the network information in the form of shortest distance and path matrices, rather than developing these matrices each time a route or travel time determination is required.

The road inventory module in its present form is sufficient to be used as the basis for the new road location procedure. If the serviced area by roads of a given class has to be found, the data base and the graphics can be used to find the roads of that class in the area of interest. With this information the road location procedure starts by constructing service zones along these roads.

3 ROAD LOCATION PROBLEM DEFINITION

The problem of road location consists of the optimal location of roads in forest areas not yet served by an existing road network. Each existing road serves a zone on both sides of the road. The service width used is established by company policy. The goal of this policy is to create an equally-spaced, minimum-cost network throughout the company's forest lands.

The problem can be stated in the following manner: locate a minimum-cost road network in areas which are not served by existing roads, using a constant service zone width to determine the served areas. In order to allow the use of combinatorial optimization techniques in the algorithm, a grid representation of the areas involved must be established. The roads to be located consist of a selection of the links in the grid network. The problem can now be expressed in mathematical terms as follows: determine a minimum cost network consisting of the links and nodes of a grid, so that any link and node in the grid is within a specified distance from at least one of the selected network elements. The selected network consists of a so-called 'tree' or a 'forest' (multiple trees), depending on the number of selected network roots. A mathematical notation of the defined problem is given below.

$G = (V,E)$ is the network formed by the grid cell vertices V and all possible road location edges E.

$F = (W,E')$ is the forest consisting of the road network trees. F contains as many trees as there are entry points. $F \subset G$, W = set of vertices in F, E' = set of edges in F ($W \subset V$, $E' \subset E$). Entry cells are selected boundary cells from which the road network trees extend

$i \subset V$, $j \subset V$, $(i,j) \subset E$: c_{ij} = road construction cost for edge (i,j).

WID = the service width expressed in the number of grid cells.

For $v \subset V$: $d(v,w)$ = shortest distance between v and w, $w \subset W$.

$d'(v,W) = \min [\, d(v,w): w \subset W]$.

The road location problem can now be expressed as follows:

$$\text{Min } Z = \sum_{i,j} c_{ij} \quad \text{for } i \subset W,\ j \subset W, (i,j) \subset E'$$

$$\text{s.t. } d'(v,W) < \text{WID} \quad \text{for all } v \subset V.$$

4 THE ROAD NETWORK LOCATION MODULE

The developed road location procedure integrates road network analysis techniques with overall forest management decision making processes, by utilizing the data base and graphic design files of an existing GIS.

The procedure is designed to locate non-served land areas by constructing service zones along existing roads. In the non-served areas new roads are located to serve the area (using similar service zones) at a minimum road construction cost. The selection of the service zone width used for a particular problem depends on several factors. First, it is possible to apply certain rules, as defined by company policy. In this case, service zone widths are predetermined for each of the possible road classes. Second, for small areas, it will be possible to use a service zone width directly related to the optimal yarding distance associated with the local situation and the selected harvesting system.

The developed algorithm is based on a local search principle, using shortest distance and minimum cost matrices. The algorithm does not guarantee an absolute optimal solution in all cases. Because of the complexity of this type of network location problem, a proven optimal algorithm would be extremely time consuming (Tansel, Francis and Lowe 1983). From extensive testing of the developed procedure, it can be concluded that the solutions are optimal or very close to optimal. For small scale problems optimality can be verified. For larger problems it rapidly becomes impossible, and surely impractical, to determine if solutions are optimal. In these cases, only local network cost minimization can be evaluated easily. In the presence of dividing rivers and streams, special attention must be paid to the entry cell configuration and possible bridge locations, to ensure optimality of the solution.

The developed algorithm uses several large data structures to store information pertaining to the minimum cost and distance routes between grid cells. The size of these structures increases exponentially with an increase in grid size. A special data format was used to reduce storage space requirements thus enabling the use of grids of this size. Several steps can be taken to improve algorithm performance, both in the areas of allowable grid size and execution time reduction. Decomposition algorithms exist to break the total area (and the related data structures) down into small sub-units. Direct addressing for data input and output will reduce I/O time considerably. In addition, special data structure format techniques will allow for an increase in grid size.

A 30 by 30 grid allows for the examination of an area of 3 by 3 km, if a grid cell size of 100 by 100 meters is used. By increasing the grid cell size, larger areas can be examined, but the degree of precision of the solution will decrease. In addition, the user has two options in defining the area to be included in the grid. First, only the non-served area is selected, excluding the existing roads and their associated service zones. In this case the user must define the entry cell configuration, and the linkage of the existing road system with the new network is not part of the solution. Second, the neighbouring existing roads and service zones are included in the selected area (figure 1).

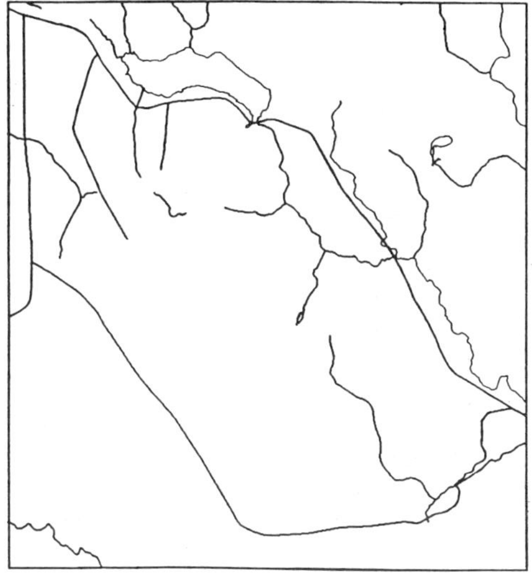

Figure 1. The existing road network and water bodies in the study area (scale: 1 cm = 0.7 km).

In this case, existing roads cells serve as starting points of the network to be located. The first option allows for the examination of larger non-served areas, as the existing roads and service zones do not take up part of the grid space.

The integration of the developed location procedure with the existing GIS is a rather complicated process. The reason is that the GIS is not designed to be used for network analysis purposes. The data transformation from polygon format to grid format is necessary to perform spatial analysis, and requires a complex series of scanning and overlay procedures (Boss 1985). The representation of roads and streams as linear elements further complicates the process, because only complex shapes are recognized by the scanning software (figure 2). If, in the

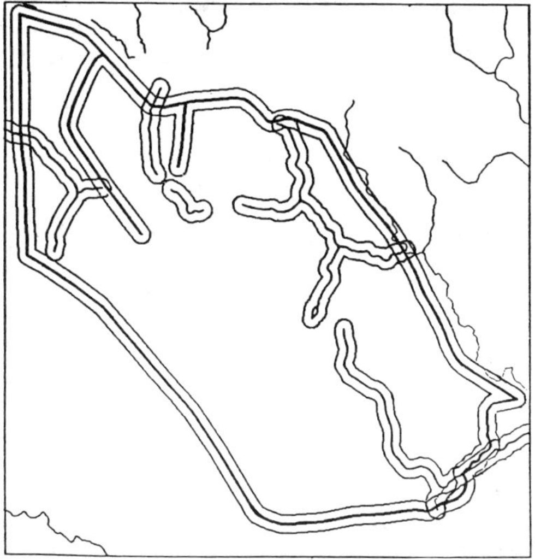

Figure 2. Both roads and streams have been converted into complex shapes.

future, the location procedure is to be used extensively, it may be beneficial to construct layers of graphic information which contain a representation of all linear elements as complex shapes. In addition, scan-line files for frequently used design file and data base information, such as cover types (figures 3 and 4), may be kept readily available. This will reduce the data retrieval and transformation processes needed to establish data files for the location procedure.

The integration of the location procedure output with existing design files is relatively simple. The output consists of a report file and a road segment coordinates file. This is in accordance with the principles of the GIS, which allows for output in both report and map form. The coordinates file has only to be transformed into a design file to display the solution in graphic form, on the screen or as a map (figure 5).

The problem of cost determination, based on a series of color codes relating to different types of information, has been explored, but not fully resolved. Given that information on soil types, elevation,

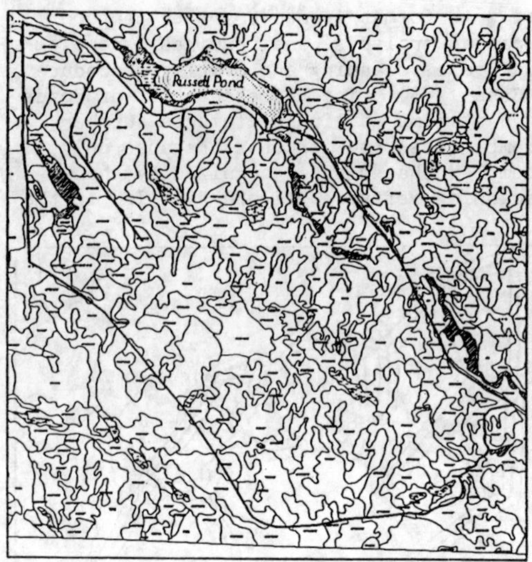

Figure 3. The existing roads overlaid with the cover type polygons.

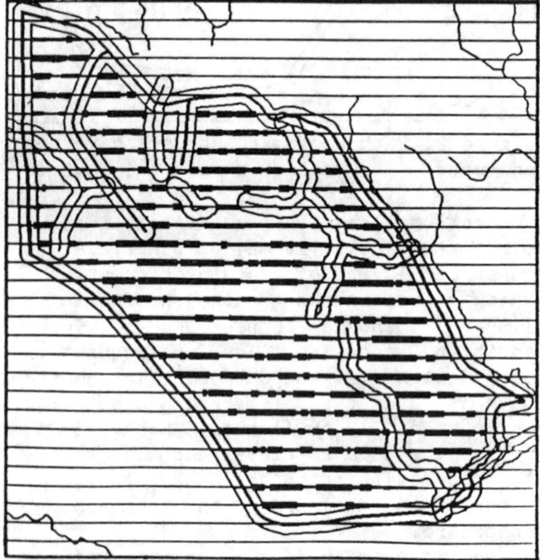

Figure 4. Scan-line file overlaid on road and stream layers. Scan-line thickness represents road construction suitability. Only the area inside of the existing roads is considered.

terrain classification, and other factors is present in the data base, the inclusion of this information would require an extensive data analysis process. A weighting procedure, which evaluates new combinations of, and interactions between the individual data layers has been conceptualized.

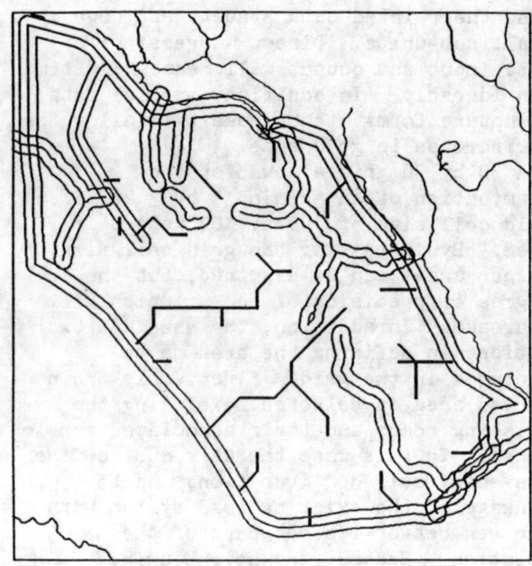

Figure 5. The road segment design file combined with existing road, stream, and zone data layers.

5 CONCLUSIONS

The road network location procedure is developed to be used an a tool, not as a decision maker. In most cases, several runs of the programme, using various entry cell configurations, should be made. The results then serve as an aid in the actual road network lay-out planning. Especially in cases where dividing rivers are present, possible bridge location and access combinations should be carefully examined. The options for area access selection provided in the programme, give it the necessary flexibility to deal with special conditions. Selection of only the non-served area gives the user full control over possible access. In the case that existing roads are included in the data set, the user still has a choice between access from all existing roads, or selection of only part of this network for access. In this case, the total existing network is still used for coverage of the associated service zones. In all of the above cases, the user has the additional option to include required but non-existing roads in the solution network. This allows for the access of selected grid cells, such as future gravel pits, by the resulting road network.

As discussed above, several aspects of the total process of algorithm development and integration in the GIS require additional research. But, both the ideas of computerized road location and integration of the procedure in a GIS have been shown to be feasible. The flexibility of area and entry cell selection in the developed program allows for the optimal use of the procedure under various conditions. Algorithm improvement and data base expansion will further enhance the capabilities of the total system and improve the efficiency and effectiveness of this management decision making tool.

In closing it should be noted that, while the figures in the text are in black and white, terminal displays or hard copies thereof are typically multicolored.

REFERENCES

Boss, D.E. 1985. A cartographic modeling aid to strategic forest planning. M.S.F. thesis, University of Maine, Orono.

Corcoran, T.J. and M.A. Nieuwenhuis. 1985. Modeling wood flow under forest salvage conditions. Council on Forest Engineering Proceedings. Tahoe City, California.

Nieuwenhuis, M.A. 1983. Development of a map-based information system for logging road network analysis. Master of Science Thesis, University of Maine, Orono.

Tansel, B.C., R.L. Francis, and T.J. Lowe 1983. Location on Networks: a survey. Part I and II. In Management Science, volume 29, no.4.

Land and Water Use, Dodd & Grace (eds), © 1989 Balkema, Rotterdam. ISBN 90 6191 980 0

Technical and economical evalutation of a forage transportation system in mountain areas

R.Cavalli & L.Sartori
Department of Territory and Agro-forestal systems, Padua University, Italy

ABSTRACT: A forage transportation system based on containers and a low number of trailer chassis is proposed. Utilizing a supply of containers in harvesting areas, the system can assure the best performance of harvesting machinery, while the trailer chassis travels between field and silo, loading the filled containers and replacing them with empty ones. Technical, energetic and economic performances compared with the side loaded trailers system are pointed out; field capacity of the containers system is similar to side loaded trailers system, but energy and labour requirements are lower. The container system, despite the higher list price, will be more economical than the trailer system if its annual use is increased. In fact the proposed system could be employed for logging operations, slurry distribution, manure and fertilizer spreading and for routine maintenance of the mountain roads.

RESUME: On propose un système de transport basé sur containers et un nombre peu élevé de transporteur. Avec ce système en utilisant une réserve adéquate des containers dans la zone de récolte il est possible de garantir le meilleur rendement opérationnel de la machine de récolte, tandis que les transporteurs peuvent effectuer sans discontinuer la navette entre la zone de récolte et celle d'ensilage, en prélevant les containers pleins et en remplaçant par des containers vides. L'analyse du type organisationnel, énergétique et économique faite sur ce système en comparaison à celui à remorque côte à côte a mis en évidence la validité de la solution adoptée en consentant d'opérer avec une capacité de travail tout à fait semblable à celle du système des remorques côte à côte, par contre la consommation d'énergie et les besioins de main-d'oeuvre résultent inférieurs. Le système à containers pour son prix d'achat élevé peut être plus économique si l'utilisation annuelle est augmentée: il peut être utilisé pour la distribution de lisier et fumier, déboisement et de la manutention hinvernale des ruotes enneigées.

ZUSAMMENFASSUNG: Es wird ein Transportsystem basierend auf Containern und einer niedrigen Anzahl von Transporteuren vorgeschlagen. Mit diesem System, daβ einen angemessenen Bestand an Containern in der Sammelzone benutzt, wird die bestmögliche Ausnutzung der Sammelmaschine garantiert, während die Transporteure kontinuierlich zwischen del Sammelzone und der Lagerstelle pendeln können, wobei die vollen Container entnommen und mit leeren ausgewechselt werden. Organisatorische, Energie-und ökonomische Analysen des so geführten System im Gegensatz zu dem mit verbundenen Anhängern haben den Wert dieser Lösung hervorgehoben, wobei zugegebenermaβen die Arbeitskapazität gleich ist, es sich jedoch herausstellte, daβ der Energiekonsum und die Notwendigheit zur Handarbeit vermindert wird. Der hohe Preis für die Anschaffung des Container System Kann wirtschaftlich interessanter werden, wenn die jährliche Nutzung erhöhtwird; in der Tat könnte sowohl bei der Verteilung von organischen Düngemitteln und gesammelten Holzrückständen als auch bei der winterlichen Instandhaltung der verschneiten Straβen eine Nutzung möglich sein.

1 INTRODUCTION

Conservation of forage in mountain areas by means of silage technique could be limited in forage hauling operations. In fact field location and road conditions could affect timeliness of operations and conservation. One thus needs to have as many trailers at disposal as to fit harvesting capacity to the time available.

However, in some situations a trailer system could be uneconomical because of the large number of transportation means required.

Starting from some equipment adopted for industrial transportation, the Dept. Territory and Agro-forestal Systems of Padova University has developed and tested a system based on containers and a low number of traile chassis, evaluating technical and economical performances.

2 EQUIPMENT DESIGN

The main unit of proposed transportation system, developed by Moro Company (Pordenone - Italy) in co-operation with the Dept. Territory and Agro-forestal Systems of Padova University, consist of a two-wheeled trailer chassis on which is fixed a device with an articulate arm for loading and unloading containers. Trailer chassis sizes are 4900 mm long, 1600 mm wide and 900 mm high.

The loading and unloading device is characterized by a frame, 1600 mm long and 900 mm wide, with an articulate arm and two rolls at the rear; the device is completed by a slide on which the container is mounted. The articulate arm consist of two levers, each one powered by a double-acting hydraulic ram. The main arm is articulate on the frame and it can swing on a 150° arc; the secondary lever, articulated on the main one, can swing on a 30° arc. Combined movement of two levers allows loading and unloading operations of the slide with the container that can be fitted and guided on the frame through the two rolls. For emptying container a 50° tip angle is provided operating only the main lever of articulate arm (figure 1).

The slide could be equipped with several devices for carting and loading agricultural and forestal products; sizes and weights must fit maximum lift capacity of the articulate arm (3000 kg).

During forage harvesting the following operations are performed: a trailer with a container is towed by forage harvester; once the container is filled up, it is uloaded and then loaded by a trailer towed by a tractor and carted to the storage area; meanwhile the forage harvester loads an empty container and continues harvesting. So the forage harvester could operate without any delay and the transportation trailer could act as a shuttle between the harvesting and storage areas, loading the filled container unloaded by the forage harvester and replacing them with empty ones.

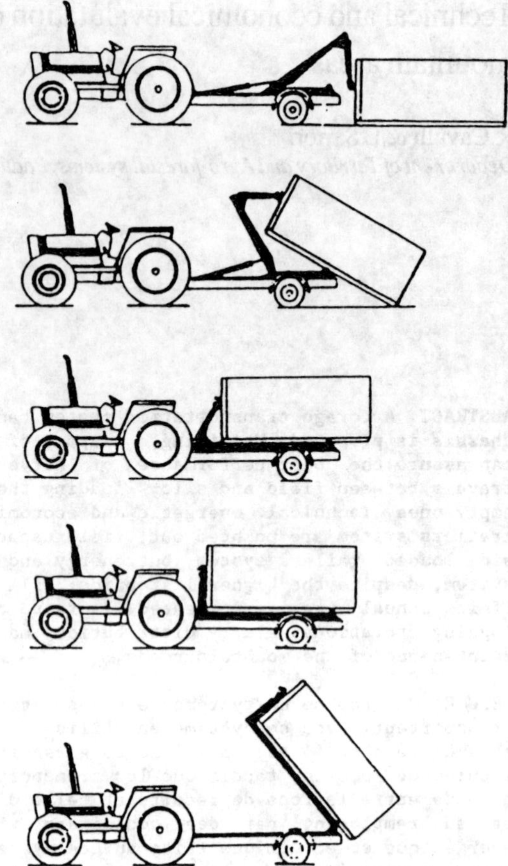

Figure 1. Container loading and unloading operations and container emptying.

3 MATERIALS AND METHODS

The aim of the experiment was the comparison, from the technical and economic points of view, between the new forage transportation system and a conventional system with side loaded trailers. Tests were carried on Phleum pratense crops in sloping fields (10-15°) located in Pian Cansiglio (Belluno - Italy). Forage was mowed and swathed; fine chop forage harvester semimounted and powered by a 51 kW 4WD tractor was used for harvesting; all the transportation trailers were coupled to 38 kW 4WD tractors.

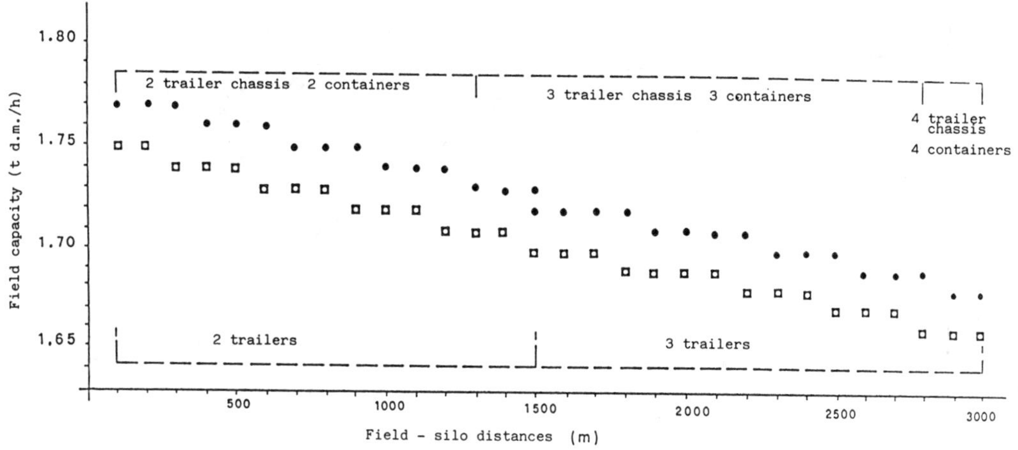

Figure 2. Field capacity of the two harvesting and hauling systems in optimal situation.

Loading capacity was 12.2 m^3 for containers and 13.2 m^3 for conventional trailers, two-wheeled fitted with hydraulic rear tipping.

To analyse the two transportation systems a computer model was developed in order to simulate performance of systems in term of field capacity, energy and labour requirements, total cost of harvesting and hauling.

4 RESULTS

For both technical and economic analysis two operative situations were considered: one with an optimal number of trailers to match maximum field capacity and one with a sub-optimal number of trailers. The optimal number of trailers allows completion of the round-trip from the field to the storage area and back without any delay and it is calculated as follows:

$$NT = \frac{RT}{TF}$$

where:

NT = number of trailers
RT = round-trip time
TF = trailer filling time (included turning time and loading and unloading time for containers)

Without any delay and time lost for waiting for trailers the forage harvester´s field capacity could be improved.

To meet this requirement, the containers system should consist of 2 trailer chassis (one towed by forage harvester) and 2 containers for field-silo distances up to 1300 m, 3 trailer chassis (one towed by foarge harvester) and 3 containers up to 2800 m and 4 containers for longer distances. For side loaded trailers system maximum field capacity is performed using 2 trailers for field-silo distance up to 1500 m, while 3 trailers are necessrary up to 3000 m (figure 2). The containers system shows a fairly higher field capacity than the trailers system (+1.43%).

In the sub-optimal situation, time lost for waiting for trailers affects the forage harvester´s field capacity and hence that of the harvesting rig.

When 2 trailer chassis and 2 containers are provided, field capacity remains high up to 1300 m (1.77 t d.m./h); for longer distances field capacity decreases up to 0.82 t d.m./h for distance of 3000 m. The possibility of increasing the number of containers depending on field-silo distance without increasing the number of trailer chassis allows to lower the reduction of field capacity. Using an optimal number of containers, calculated in relation to the number of trailer chassis and the field-silo distance, it is possible to operate at the maximum field capacity for each combination: 1.69 t d.m./h with 3 containers at 1400 m, 1.68 t d.m./h with 4 containers at 2000 m, 1.30 t d.m./h with 11 containers at 3000 m (figure 3).

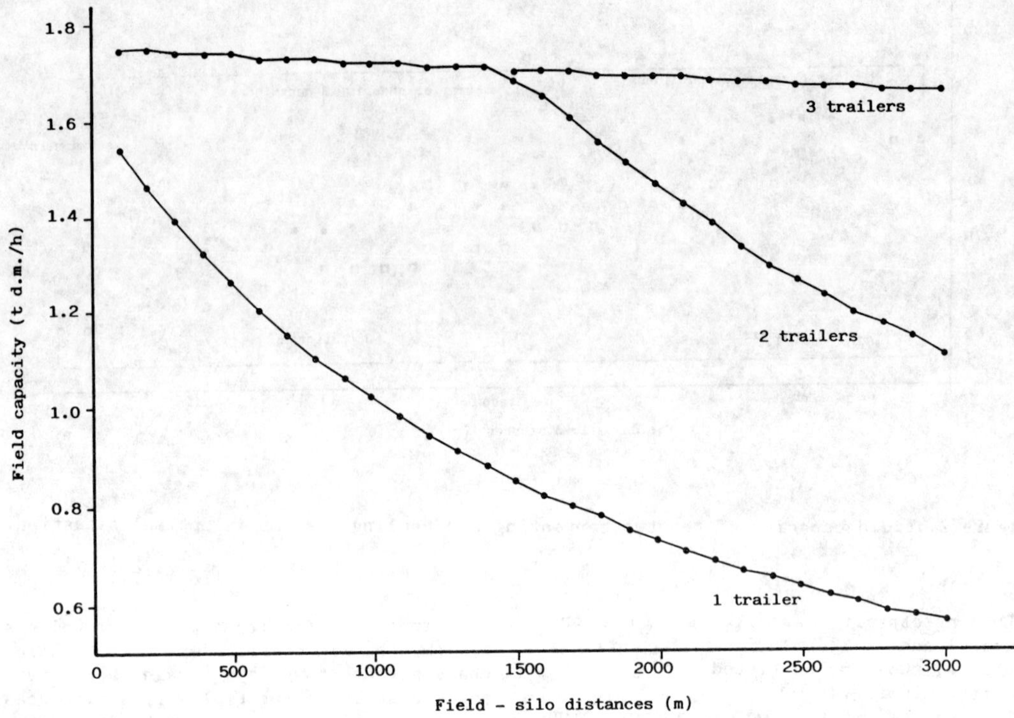

Figure 3. Field capacity of the containers system in sub-optimal situation.

Nevertheless these values of field capacity are lower than those recorded with optimal sizing of transportation rig. In fact, using 2 trailer chassis with a suitable supply of containers, it is possible to reduce waiting time, but time for carting the containers gathered in the field at the end of harvesting day is increased. As regards this fact one can point out that trailer chassis towed by forage harvester can be also used for carting filled containers after harvesting operations. Nevertheless it must be emphasized that the reduction in field capacity due to time for carting the containers at the end of harvesting is lower than that one due to waiting time occurring in sub-optimal situations when the number of containers used is the same as the number of trailer chassis. The containers system shows a greater operative flexibility, fitting different field conditions; deciding when to reduce the number of trailer chassis - the most expensive equipment of the system - must be carefully evaluated because of field capacity reduction, partially recovered through increasing the number of containers.

In the side loaded trailers system, which shows a greater operative rigidity, sub-optimal situations greatly affect field capacity. It is a matter of course that when using one trailer the field capacity rapidly decreases when the field-silo distance increases due to time for waiting for the trailer, thus affecting forage harvester performance.

Using 2 trailers allows one to operate in the best conditions for distances up to 1500 m, beyond which the field capacity decreases up to 1.10 t d.m./h at 3000 m. Also in this case the decrease of field capacity is due to time for waiting for trailers, while the field-silo distance is gradually increased (figure 4).

As far as energy and labour requirements are concerned it could be emphasized that, in optimal situations the containers system is more convenient than the side loaded trailers system which requires, in the same conditions, one more operator and anther tractor. Moreover, in sub-optimal situations, using 2 trailer chassis and from 2 to 11 containers, in relation to field-silo distance, gives the least labour requirements for distances up to 3000 m (figure 5), while requiring the least

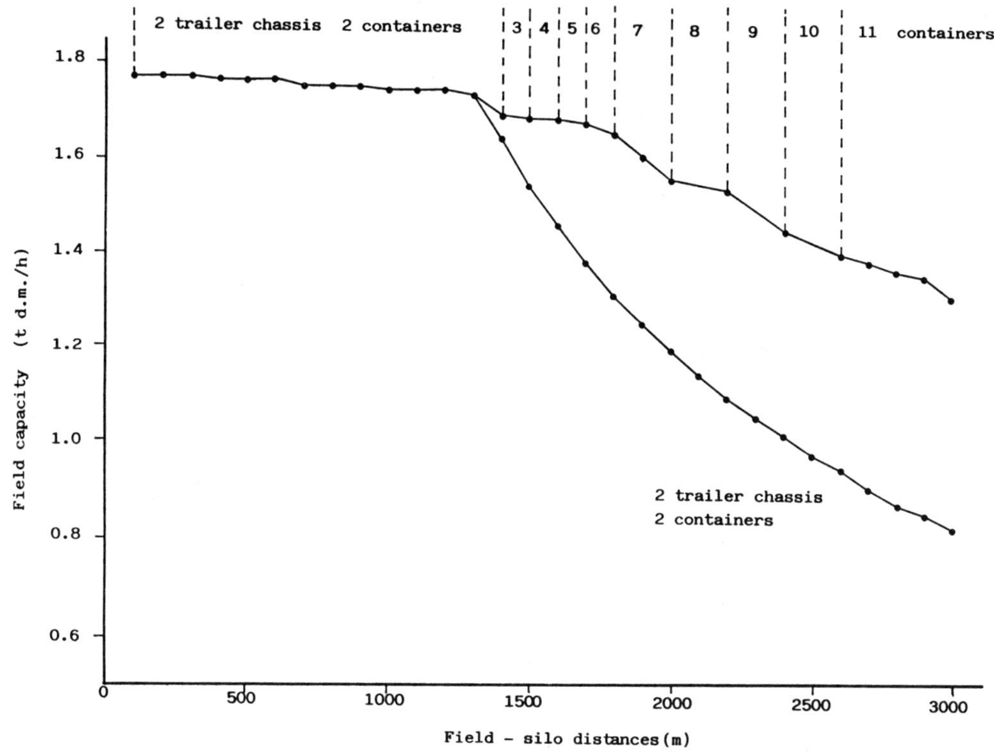

Figure 4. Field capacity of the side loaded trailers system in sub-optimal situation.

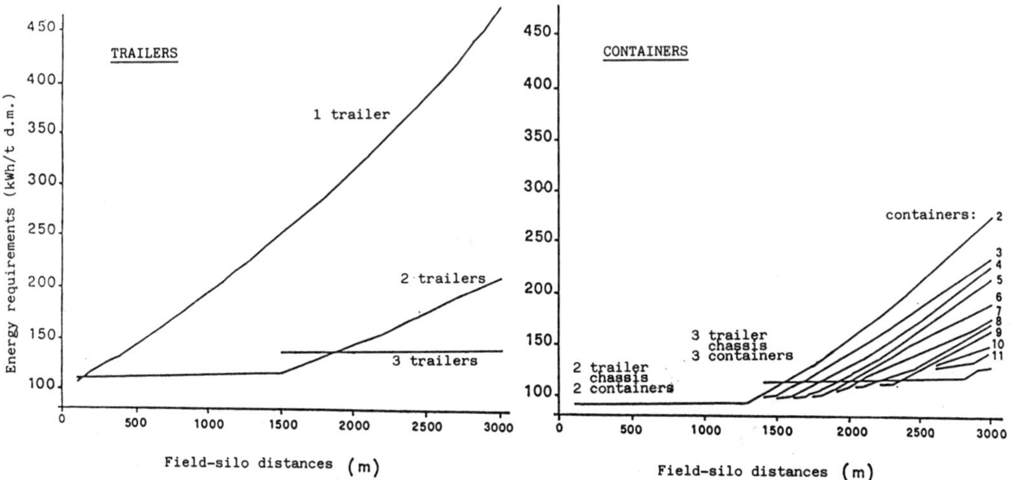

Figure 5. Energy requirements of the two harvesting and hauling systems.

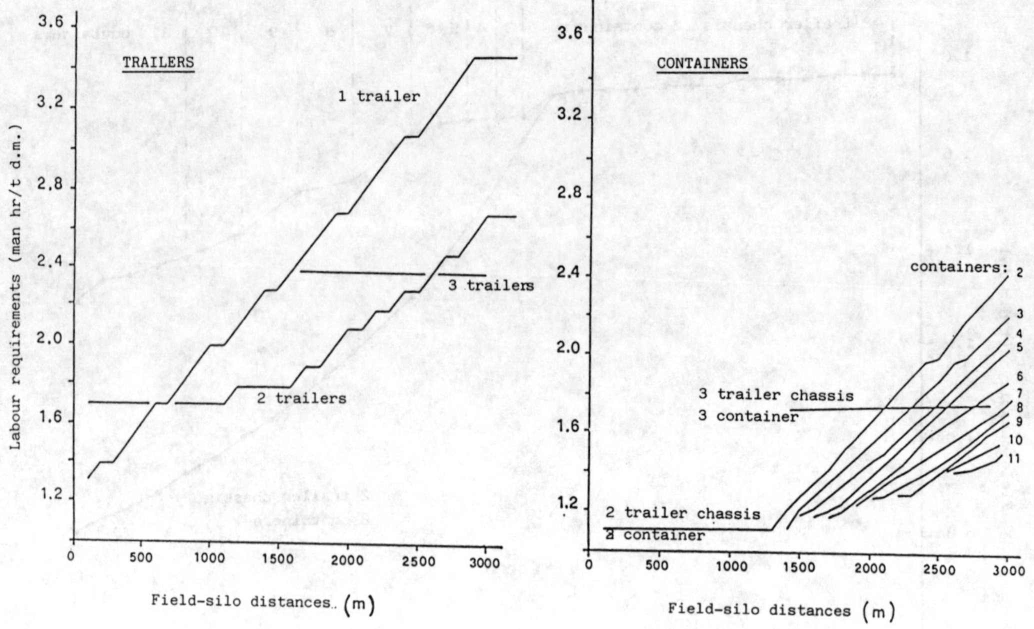

Figure 6. Labour requirements of the two harvesting and hauling systems.

energy for distances up to 2400 m, beyond which 3 trailer chassis and 3 containers must be used (figure 6).

As far as the total cost for harvesting and hauling are concerned it should be emphasized that for the containers systems the least cost always occurs using 2 trailer chassis and an increasing number of containers depending on the field-silo distance, while for side loaded trailers system the least cost occures using 1 trailer up to 400 m, 2 trailers up to 2200 m and 3 trailers for greater distances. For the containers system the total cost for harvesting and hauling is 10% lower than the side loaded trailer system for distances ranging from 500 m to 3000 m; only for short distances (less than 200 m), the least cost occurs with the side loaded trailer system (figure 7).

It should be noted that the technical and economic analyses do not teach the same conclusions, because the system size allowing the maximum field capacity does not always give the least total cost for harvesting and hauling. So the choice of

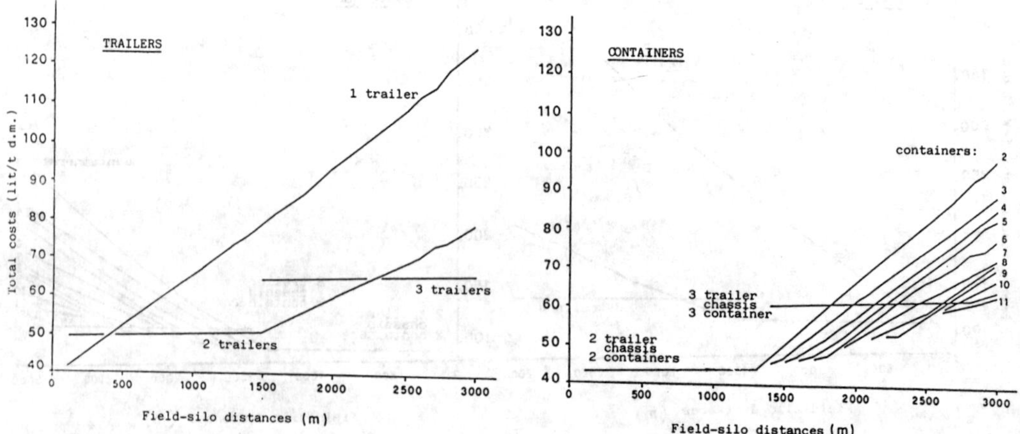

Figure 7. Total cost of the two harvesting and hauling systems.

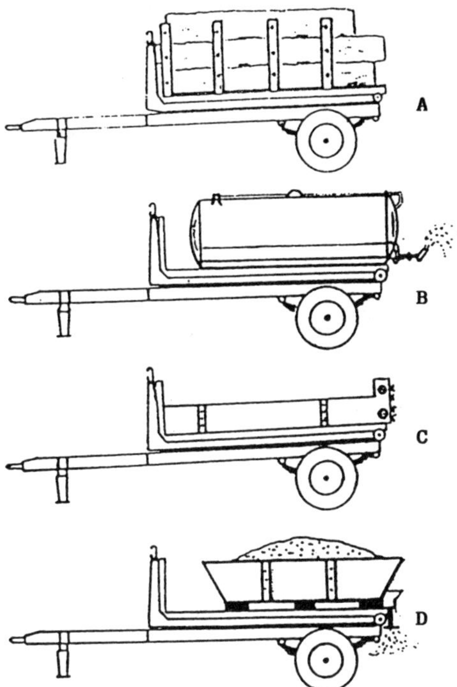

Figure 8. Multi-purpose operations proposed for the containers system: (A) logging operations; (B) slurry distribution; (C) manure spreading; (D) sand and salt spreading.

system size must be related to farm requirements; maximum field capacity will be considered if timeliness in harvesting and hauling is a critical factor. Otherwise one should consider system sizes which allow one to minimize total cost.

5 CONCLUSION

The results of the analyses carried out show better performance of the containers system not only with respect to field capacity and energy and labour requirements, but also operational costs, despite the higher list price of the equipment.

To reduce total system costs it is necessary to increase annual use through operation versatility offered by the containers systems. In fact the trailer chassis with articulated arm and slide must be considered as an unit which, integrated with different equipments, could be used not only for agro-forestal operations, such as manure or slurry spreading and logging operations but also for other typical operations in mountain areas, such as spreading sand or salt on snowy roads (figure 8).

A multi-purpose utilization of the containers system could increase annual usage and reduce total system costs, making it worthwhile in mountain farming.

Land and Water Use, Dodd & Grace (eds), © 1989 Balkema, Rotterdam. ISBN 90 6191 980 0

A low ground pressure silage harvesting system for small farms on low trafficability soils

D.Forristal
Agricultural Engineering Department, Teagasc, Oak Park Research Centre, Carlow, Ireland

ABSTRACT: A light load, low ground pressure harvesting system was developed to overcome the problems of forage harvesting on low bearing strength soils. A 4 m^3 capacity tractor-mounted tipping transporter was designed and constructed to replace a conventional silage trailer in the field. The hopper capacity enables the system to operate in small fields, tipping when full into a trailer parked at the field edge or roadway. On test, in wet conditions the power required to propel the unit was 37% less than a conventional tractor/trailer combination. Rut depth and wheelslip values were also lower. Where both systems could operate, the harvesting rate of the transporter system was 10% to 30% lower than the conventional system, depending on grass yield. The new transporter system was able to operate where conventional systems failed.

RESUME: Un système de récolte à charge légère et faible pression au sol a été mis au point pour surmonter les problèmes de la récolte du fourrage sur des sols peu résistants à un poids élevé. Une trémie basculante sur tracteur d'une capacité de 4 m^3 a été mise au point et construite pour remplacer une remorque classique d'ensilage dans les prés. La capacité de la trémie permet au système de fonctionner dans des prés de surface limitée puisqu'il est possible de la vider dès remplissage dans une remorque garée au bord du pré ou sur la route. Lors d'essais par temps pluvieux la puissance nécessaire au fonctionnement de la machine est de 37% moindre que celle d'un tracteur/remorque conventionnel. La profondeur des ornières et les valeurs de glissement des roues sont également moindres. Lorsque les deux systèmes peuvent fonctionner le taux de récolte du système à trémie était de 30% moindre que celui du système conventionnel selon le rendement en herbe. Le nouveau système à trémie pouvait fonctionner là où les autre systèmes avaient échoué.

AUSZUG: Es wurde ein Kleinlast-Erntesystem mit geringem Bodenaufdruck entwickelt, um Probleme, die beim Futterernten auf niedrigtragendem Kraftboden entstehen zu überkommen. Eine Trecker-Kippmulde mit 4 m^3 Kapazität wurde entwickelt und gebaut, um den herkömmlichen Silageanhänger auf dem Feld zu ersetzen. Mit der o. a. Kapazität der Kippmulde kann man auf kleinen Feldern arbeiten. Wenn die Mulde voll ist, wird der Inhalt in einen am Feldrand abgestellten Lastwagen gekippt. Bei Testversuchen unter nassen Bedingungen lag die erforderliche Antriebskraft für den kleinen Wagen bei weniger als 37% der herkömmlichen Trecker/Anhängerkombination. Die Fahrrinnentiefe und die Radrutschwerte waren ebenfalls geringer. Wo beide Systeme eingesetzt werden konnten war, je nach Grasertrag, die Ernte des Muldensysstems 10% bis 30% niedriger als bei dem herkömmlichen System. Der neue Muldenwagen konnte auch dort eingesetzt werden, wo andere Methoden versagten.

1 INTRODUCTION

One-third of Ireland's agricultural land could be described as wet, because of soil type, glacial and bedrock geology and rainfall (Gardiner and Radford 1980). Much of this land is capable of producing acceptable yields of grass (Lee and Diamond 1972), but utilisation of this grass is a problem because of its frequent inability to support animal or machinery traffic, caused by low bearing strengths. Impermeable soils, in particular, have trafficability problems, even after drainage. All machinery operations are affected by poor trafficability, but heavy loads, high traffic density and the need to harvest at specific times combine to make forage harvesting the most critical.

Trafficability and machine operation on weak soils have been studied at Ballinamore, Co. Leitrim and Kilmaley, Co. Clare. The soils at both sites are poorly drained surface water gleys. Slopes are a feature of the topography. Rainfall at Ballinamore is 1140 mm/annum and at Kilmaley is 1500 mm/annum.

Previous work on this project has concentrated on modifying conventional forage harvesting systems by fitting alternative wheels/tyres to reduce the ground contact pressure of the individual machines (Forristal 1985, 1986). This work has been successful. Higher tractor traction forces have been developed, fewer and shallower ruts have been caused and the modified systems have been able to operate successfully on poorly trafficable soils.

The round bale silage system (1.2 m x 1.2 m bales) was also evaluated in these trials. The small unit loads used allow low ground pressures to be easily achieved. Ensiling costs and labour requirement can be high where wilting is not possible, and bale yield per hectare is high.

A system which would combine the small load concept and conventional bulk handling of grass was envisaged to be ideal for use on soils of low trafficability. A system was developed in Norway which used a tractor-mounted grass transport box to transport the grass from the field to the silage clamp (Landbruksteknisk Institutt 1982). With this system, small unit loads and high grass yields would result in poor workrates where the silage clamp is a moderate distance from the silage field.

This paper describes the development and evaluation of a new low ground pressure silage harvesting system.

2 MACHINE CONCEPT

The aim of the new system was to achieve lower ground pressure by using smaller grass loads, while retaining bulk handling of the grass.

A new machine was designed to replace the silage trailer as the grass carrying unit in the harvesting field. Low weight and small load capacity allow low ground pressures, low rolling resistance and low tractor draught requirement to be achieved. Typical loaded small trailer weights of four to six tonnes are replaced by a loaded transporter weight of less than two tonnes. The silage transporter has sufficient capacity to allow the tractor/harvester/transporter to cut one swath around a small field. The grass load is then transferred to a conventional silage trailer located on an adjacent farm roadway or on the field headland. When full, this trailer transports the grass to the silage pit. The system eliminates or severely restricts conventional trailer traffic in the field.

2.1 Design parameters

A number of different design options utilising the small load concept were considered. Maintaining mobility when very low bearing strength areas are encountered prompted the use of large diameter driven wheels as the primary support and driving mechanism. A side-tipping transporter, mounted on the rear of a conventional tractor fitted with dual wheels or wide section tyres, was the design selected for the first prototype. The following design requirements were established:

1. Capacity of the hopper to be 4 m^3 (1 tonne grass).
2. To be rear mounted on a tractor fitted with a side-mounted forage harvester and capable of being filled by the forage harvester.
3. Machine weight to be kept to a minimum, and the centre of gravity of the loaded transporter to be kept close to the tractor, to minimize the overturning moment.
4. Hopper to be side tipping with a minimum discharge height of 2.4 m, at a distance of 1.8 m from the centre-line of the tractor, enabling a trailer to be filled using a harvesting tractor fitted with dual wheels or wide-section 'Terra-Tires'.
5. Simple construction techniques to be employed to facilitate manufacture by small engineering firms.

3 CONSTRUCTION AND DEVELOPMENT OF PROTOTYPE

The prototype was completed in mid-1986. The transporter was attached to a modified linkage frame of a side-mounted harvester as shown in Fig. 1. Basic specification of the unit is given in Table 1. The original transporter capacity was 3.5 m^3. The sides of the hopper were steeply sloped to assist grass filling and emptying. A 100 mm taper was incorporated in the depth dimension to assist emptying. Two 1.8 m stroke rams were used to effect tipping. As the transporter is carried indirectly on the tractor 3-point linkage, an extra 0.5 m in tipping height can be obtained by raising the lift arms. A hydraulically operated stabilising leg was fitted to provide horizontal lower frame stability when tipping.

Preliminary testing revealed that the discharge angle of the left hopper side at 47° was not sufficient to consistently eject all the grass. This angle was increased to 60°. The hopper also proved difficult to fill without grass loss because of the proximity of the harvester chute. A light-weight canopy with automatic door opening eased the filling operation, increased hopper capacity and gave more controlled grass discharge (Figs. 1-3). At this stage, the transporter operated satisfactorily in the field.

3.1 Weight distribution

As the hopper is mounted on the rear of the tractor, weight distribution between the tractor axles is influenced by the loaded transporter's weight. Weight distribution of the hopper with three different loads is given in Table 2.

The design imposes heavy loads on the tractor rear axle and linkage, in addition to causing weight transfer from the front axle. These factors limit the load capacity to 1,000 kg using a tractor in the 50-60 kW power range, particularly when working on slopes. Sideways stability is good, as the tractor is always fitted with duals or wide single tyres giving a track width in excess of 3 m.

Table 1. Transporter specification

Hopper:	Width (m)	2.1-2.7
	Height (m)	1.2
	Depth (m)	1.2
	Capacity (m^3)	3.5
Extension:	Capacity (m^3)	1.3
Overall width (m)		3.0
Overall height (m)		3.3
Tipping height (m)		2.55
Tipping angle		60°
Tipping cylinder stroke (m)		1.8
Total weight (kg)		800

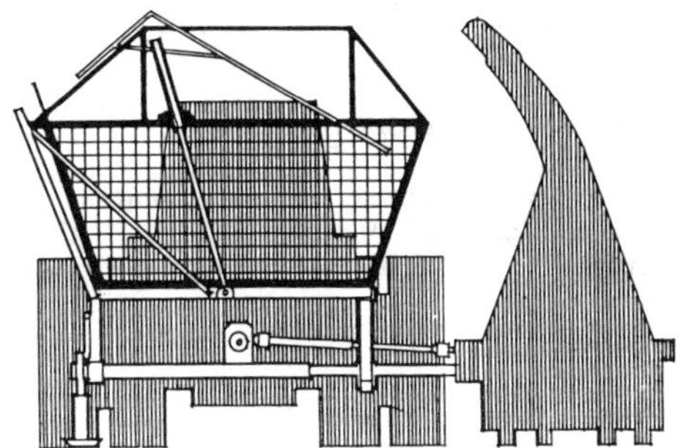

Fig. 1. Prototype transporter: rear view

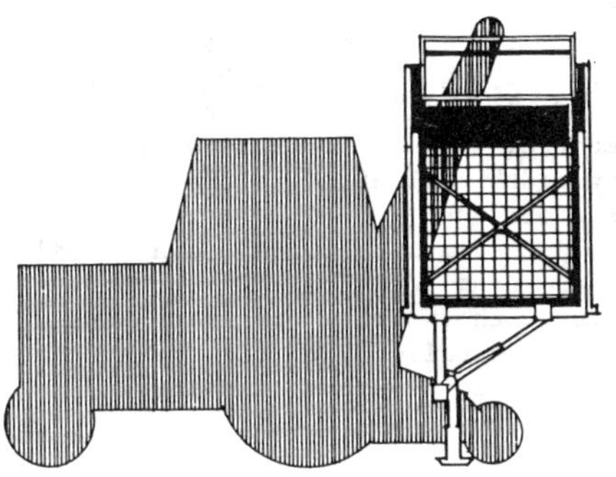

Fig. 2. Prototype transporter: side view

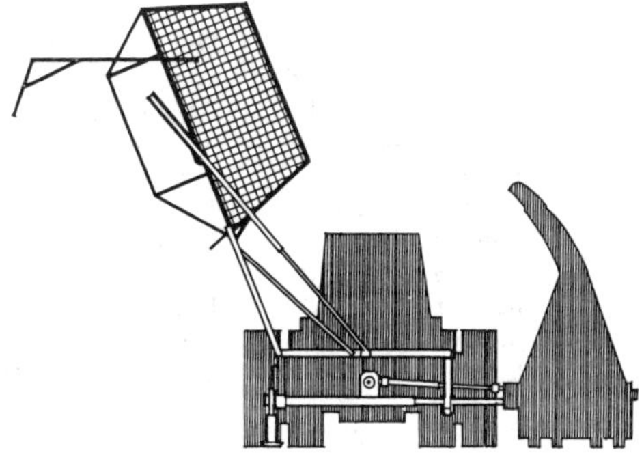

Fig. 3. Prototype transporter: tipped, rear view

Table 2. Weight distribution: Transporter fitted on 56 kW 4WD tractor with 150 kg of front ballast weight

Hopper weight (kg)	Grass load (kg)	Rear axle load (kg)	Front axle load (kg)	Lift requirement at lower links (kg)
800	-	3039	1360	1132
800	500	3882	1017	1887
800	750	4284	865	2264
800	1000	4708	691	2640

4 PERFORMANCE ASSESSMENTS

The transporter system workrate, rut profile effects and draught power requirement were assessed between 1986 and 1989.

4.1 Workrate

The workrate of the silage transporter system was monitored on a number of occasions in a range of field sizes, harvesting grass crops with different yields.

The hopper was fitted to a 56 kW 4 WD tractor fitted with dual 16.9-34 rear tyres and single 13.6-24 front tyres. A side-mounted single-chop harvester with a cutting width of 1.35 m was operated by the tractor.

On two occasions, a conventional trailer system using an identical tractor and harvester with two 12 m^3 trailers was operated in parallel with the new system. Grass yield, load weights and forward speed were recorded in addition to system workrates. Work times for the harvesting and non-harvesting (transporter tipping, trailer changing) parts of the work routine were recorded for each load (Table 3). The overall workrate of the transporter system was between 10% and 30% lower than the conventional system when both systems were operated together. The relatively small load capacity of the hopper necessitated frequent load tipping, particularly in high yielding crops. Field size and layout also influenced the transporter's workrate. Where the harvesting pattern could be arranged to have the hopper filled as it approached the trailer, good workrates were achieved. Forward speed when harvesting was higher with the transporter system because of the lower power required to propel the lighter unit, particularly on slopes. The manoeuvrability of the combined tractor/harvester/transport unit contributed to good harvesting performance, particularly in small fields.

The mean workrate over all the fields harvested was 9.52 t/h and 0.353 ha/h. This workrate is acceptable for a

Table 3. Workrates of the low-load and conventional harvesting systems

Field no.	System	Grass yield (t/ha)	Forward speed (m/s)	Non-filling time/load (min)	Filling time/load (min)	Load weight (kg)	Harvesting rate (t/h)	Harvesting rate (ha/h)
1	Hopper	29.6	1.67	1.90	2.56	920	12.38	0.418
1	Trailer	29.6	1.38	3.08	10.14	3025	13.73	0.464
2	Hopper	34.8	1.55	1.85	1.92	757	10.39	0.299
2	Trailer	34.8	1.40	3.12	8.50	2910	14.97	0.430
3	Hopper	15.0	1.92	2.47	3.85	792	7.59	0.506
4	Hopper	38.5	1.44	2.83	2.18	876	10.50	0.262
5	Hopper	22.2	1.66	2.88	2.66	690	7.47	0.336
6	Hopper	27.5	1.50	2.71	2.42	722	8.42	0.306
7	Hopper	28.4	1.60	2.46	2.47	810	9.85	0.347
Averages	Hopper	28.0	1.62	2.44	2.58	795	9.52	0.353
	Trailer	32.2	1.39	3.10	9.32	2968	14.35	0.447

low-powered single-chop system working in small fields on slopes. Importantly, the hopper system was able to maintain these workrates when ground conditions would have prevented conventional systems from operating. During 1986, the hopper system successfully harvested silage for farmers where conventional equipment had consistently failed to operate because of poor ground conditions.

System workrate could be improved by better design at the transporter's tipping point, good field layout and using an experienced operator.

4.2 Wet land performance

A trial was carried out to assess the transporter system's performance on soils of low bearing strength.

Rut measurements were used to assess the relative effects on the soil of the transporter system compared to conventional systems. Draught power was measured to indicate the system's rolling resistance.

4.2.1 Experimental procedure

The experiment was carried out on a poorly drained surface water gley soil. Soil bearing strength over the experimental area was assessed using a hand-held shear vane instrument fitted with a 19 mm vane at the soil surface and a recording cone penetrometer fitted with a 322 mm^2 base area cone used to a depth of 120 mm. A 50 kW 2WD tractor with a ballast box fitted, or a ballasted trailer attached, was used to simulate harvesting systems using either the transporter or a conventional silage trailer. The ballast box was ballasted to simulate the transporter with a 1,000 kg grass load. The trailer was ballasted to simulate a small capacity trailer with a 2,750 kg grass load. Both conventional and low ground pressure tyres were fitted to the silage trailer. A tractor-only treatment ballasted to give similar axle loads to a tractor drawing a trailer was also included. Axle loads, tyre sizes and inflation pressures used in the experiment are given in Table 4. The tractor was fitted with an electronic data capture unit which accepted data from three tractor sensors: a speed sensor fitted to the pto shaft, monitoring pto speed; a radar unit monitoring forward speed; and a fuel meter monitoring fuel consumption. Values from each sensor were recorded every second. Subsequent to the test the tractor's pto power output at various pto speeds and fuel consumption rates was measured on a dynamometer. This calibration procedure allowed the power required to propel the simulated harvesting combinations in the field to be calculated.

Rut characteristics were recorded using an electronic recording relief meter developed specifically for rut measurements on low bearing strength soils. Fifty 10 mm pins at 30 mm spacing

Table 4. Weights, tyre sizes and inflation pressures for low load and conventional systems

Treatment no.	1	2	3	4
Treatment	Transporter	Trailer	Trailer	Tractor
Tractor front axle:				
Weight (kg)	605	1375	1375	1375
Tyre size	7.50-16	7.50-16	7.50-16	7.50-16
Inflation pressure (kPa)	150	240	240	240
Tractor rear axle:				
Weight (kg)	4765	3065	3065	3065
Tyre size	16.9-34D	16.9-34D	16.9-34D	16.9-34D
Inflation pressure (kPa)	80	80	80	80
Trailer axle:				
Weight (kg)	-	3535	3535	-
Tyre size	-	9.00-20D	500-22.5	-
Inflation pressure (kPa)	-	350	80	-

are used to measure the rut cross-section profile. Subsequent computer analysis yields rut depth and cross-section area values.

In the replicated experiment, each combination was driven over two 50 m plots, at a forward speed of 1.9 m/sec. As the site had a consistent 3% slope, the combinations were driven both up and down the slope.

4.2.2 Rut parameters: results and discussion

Soil bearing strength of the plots used in the trials was very low as indicated by shear vane and cone penetrometer values. Average shear vane strength at 40 locations in the experimental area was 13 kPa. Mean penetration resistance to a depth of 120 mm at the same locations was 303 kPa.

Data from the relief meter was analysed to give mean rut depth and rut area measurements for each treatment (Table 5). Rut characteristics of the different treatments were markedly different. Plot slope had a large effect on the size and shape of the ruts caused. Few ruts were caused when travelling down the slope. Driving uphill produced deeper ruts, larger rut areas and large differences between treatments. As slope affects the draught force on the tractor, it also influences wheelslip, which affects the bearing strength of the soil surface for the following trailer wheel. The differences in rut depth and areas between treatments were significant. The transporter treatment (1) overall created ruts only 25% as deep as the conventionally wheeled trailer system. As expected, the low ground pressure wheeled trailer caused less deep ruts than the conventionally wheeled trailer, but they were almost twice as deep as those on the transporter treatment. The shallow ruts, caused by treatment 4 in which the tractor travelled alone, confirm that the trailer is the principal cause of excessive rutting. The rut depths caused by the conventional wheeled trailer are the maximum likely to be reached on this soil type, as the shallow top soil has been fully penetrated and the vehicles are being supported by the strong compact subsoil. When this occurs, the top soil and its vegetation are totally displaced, exposing bare subsoil.

4.2.3 Draught power requirement: results and discussion

Power needed to propel the various tractor-trailer, and tractor and ballast box combinations, was also markedly different (Table 6). The disparity in power between the two replications of the experiment was accounted for by the difference in slopes of the plots and was similar to that predicted by calculation. The tractor and transporter treatment required less power to propel it than either the conventional wheeled trailer or the trailer fitted with low ground pressure tyres. Differences in power were particularly marked on the uphill plots where the extra weight of the trailer combinations added to the power requirement. Overall, the hopper system required 37% less power than the conventional wheeled trailer and 15% less power than the tractor and trailer fitted with low ground pressure tyres. The higher power requirement of the trailer system is due to the high rolling

Table 5. Rut characteristics of traffic treatments

	Rut depth (mm)			Rut area ($m^2 \times 10^{-3}$)		
Treatment	Slope -3% (downhill)	Slope +3% (uphill)	All	Slope -3% (downhill)	Slope +3% (uphill)	All
1	14.75	20.12	17.44	7.893	9.574	8.733
2	15.0	52.89	33.94	7.363	23.600	15.482
3	40.33	100.67	70.50	14.300	43.867	29.083
4	20.0	25.11	22.55	6.753	11.833	9.293
F-test			***			***
SED			23.17			12.34

*** significant at 0.1% level

Table 6. Power requirement of transporter, trailer and tractor combinations

Treatment	Power (kW)		
	Slope -3% (downhill)	Slope +3% (uphill)	All
1	12.97	21.33	17.15
2	15.01	30.79	22.90
3	19.66	35.00	27.33
4	11.79	16.11	13.95
F-test			***
SED			2.78

*** significant at 0.1% level

resistance caused by the heavier loads and trailer wheel sinkage.

The transporter system, in spite of having a large individual axle load and wide rut cross-section caused by the dual wheels, has a lower power requirement due to: (i) the lower total weight, (ii) elimination of an axle, and (iii) most of the weight being on large diameter driven wheels with low contact pressures. This lower draught power requirement partly explains the transporter's faster spot workrate compared to a trailer system when harvesting silage. The low draught power requirement also contributes to the hopper system's satisfactory performance where trailer systems lose mobility.

Loss of mobility of tractor/trailer combinations on low bearing strength soils is a compound problem which was observed closely during these experiments. Rut-forming trailer tyres create a high rolling resistance, thereby increasing the draught forces on the tractor. This increases wheelslip at the tractor drive tyre which disrupts the bearing surface of the terrain, leading to further increases in trailer rolling resistance and lowering of drive tyre traction. Forward speed and momentum drop as wheelslip increases and engine speed drops. Unless the tractor/trailer combination reaches on an area of markedly higher bearing strength, total loss of mobility occurs quickly. The transporter system performance in these situations is markedly better.

5 FURTHER DEVELOPMENT OF SMALL LOAD SYSTEM

Areas in which the prototype transporter design could be improved were identified. Changes to the discharge plate at the tipping point and the operation of the hydraulic leg would improve the turnaround time at the trailer. Rigid mounting of the hopper to the tractor back axle, independent of the harvester frame, would improve rigidity and obviate the need for a high linkage lifting capacity. However, the tipping height would need to be modified to counteract the loss of linkage lift.

Alternative low-load tipping hopper configurations which could offer improved performance were identified.

1. Load platform mounted hopper

The load capacity of the prototype hopper is restricted on a conventional tractor because of weight transfer caused by its rear mounting position. Mounting the unit on the load platform of a tractor with a mid-mounted or forward-mounted cab (e.g. MB-Trac, Deutz Intrac) would greatly improve the axle load distribution and allow increased hopper capacities. Grass loads of 2 tonnes could be carried, with consequent improvements in workrate.

2. Trailed tipping hopper

The mounting of both the harvester and hopper on the same tractor contribute to the unit's good performance. A trailed, small load tipping hopper, while partly compromising the system's wet land performance, would allow good load capacities to be achieved with reduced axle and linkage loads. The ability to fit up to four large diameter wheels/tyres, and the low total load of the unit should ensure better wet-land performance than conventional trailer systems.

6 CONCLUSIONS

1. The unsatisfactory performance of conventional forage harvesting machinery on soils of low bearing strength prompted the design of a new small load bulk silage harvesting system.

2. Design parameters were established and a 4 m^3 capacity, tractor-mounted side-tipping transporter was constructed. This replaces a conventional silage trailer as the grass transporting unit in the field.

3. Work studies showed that overall workrate, although 10% to 30% lower than conventional trailer systems, at 0.353 ha/h, was acceptable for a single-chop system working in small fields on slopes. The new system was able to maintain these workrates on low bearing strength soils where other systems failed to operate.

4. Trials showed that on low bearing strength soils, rut depths and areas caused by the hopper system were significantly lower and soil structure damage was less than caused by conventional trailer systems. The draught power requirement of the new system was also lower than conventional systems.

5. Alternative low-load tipping hopper configurations, which could improve the performance and adaptability of the system, were identified.

REFERENCES

Forristal, D. 1985. Forage harvesting and trafficability on wet land. Plant Sciences and Crop Husbandry Research Report, p.10-13. Dublin: An Foras Taluntais.

Forristal, D. 1986. Forage harvesting and trafficability on wet land. Plant Sciences and Crop Husbandry Research Report, p.55-56. Dublin: An Foras Taluntais.

Gardiner, M.J. and T. Radford 1980. Soil Associations of Ireland and Their Land Use Potential, p.125-142. Dublin: An Foras Taluntais.

Landbruksteknisk Institutt 1982. Annual Report, p.26-27,57. Norway: LTI.

Lee, J. and S. Diamond 1972. The Potential of Irish Land for Livestock Production, p.6-9. Dublin: An Foras Taluntais.

Land and Water Use, Dodd & Grace (eds), © 1989 Balkema, Rotterdam. ISBN 90 6191 980 0

Rationalisation des systèmes de ramassage et de transport du lait ovin de l'exploitation à la fromagerie

A.Pazzona
Istituto di Meccanica Agraria, Università di Sassari, Italie

RESUME: Le but de cette étude est d'analyser et de comparer du point de vue technico-fonctionnel et économique les deux systèmes communément utilisés pour le ramassage et le transport du lait ovin: le camion avec bidons et le camion-citerne. A travers une enquète biennale auprès de 10 fromageries sardes et des relevés expérimentaux des principaux paramètres opérationnels de 40 différentes tournées de ramassage, on a pu tirer des indications utiles pour la rationnalisation des parcours eux-mêmes et pour un meilleur emploi de la main d'oeuvre. On a réalisé, enfin, un programme de simulation qui a permis d'évaluer les aspects économique et organisatifs des nouveaux systèmes de ramassage réalisé avec une livraison de lait tous les deux jours et/ou avec la création de centres collectifs de ramassage.

Rationalisation of systems for collection and transport of sheep's milk from the farm to chees factory.

SUMMARY: The aim of this study was to analyse and compare the two systems most commonly used for the collection and transport of sheeps milk: the lorry with tanks and the tank-trailer. The analysis concerned technical functional and economic aspects. A biennal investigation at 10 chees factory in Sardinia and an experimental survey of the main operative parameters of 40 different collection routes have led to indications for the rationalisation of the routes themselves, the proper size of means of transport and improved utilization of farm labour. Finally, a simulation program was prepared that makes it possible to evaluate the economic and organizational aspects of innovative collection systems for delivery of the milk on alternate days and/or the establishment of joint collection centers.

Rationalisierung der Einsammlungs- und Transportsysteme der Schafmilch vom landwirtschaftlichen Betrieb zur Käserei

ZUSAMMENFASSUNG: Der Zweck dieser Studien ist, das Analisieren und Vergleichen des technisch-funktionellen und wirtschaftlichen Aspekts der zwei am meisten gebrauchten Systeme fur Einsammlung und Transport der Schfmilch: der Lastwagen mit Behälter und Tankwagen. Durch eine zweijährige Nachforschung bei 10 Käsereien aug Sardinien und durch die experimentelle Feststellung der gültigen Masstabe in 40 verschiedenen Einsammlungsruden, war es möglich, interessante Hinweise für die Rationalisierung der Strecken zu pflücken. Man hat auch ein Simuylationsprogramm vorbereitet, was uns erlaubte, die wirtschaftlichen- und organizatorischen Aspekte der Erneuerungssysteme der Einsammlung zu bewerten. Das Erneuerungssysteme besteht in der milchlieferung alle 2 Tage und oder der Grundung von gemeinsamen Einsammlungszentren.

1 PREFACE

Dans le secteur laitier et fromager, comme dans tout autre secteur agricole, les transports représentent un des facteurs de base de la production, en particulier en ce qui concerne la rapidité d'exécution nécessarie de par la nature même des produits et les caractéristiques du milieu ambiant parfois difficile où on opère.

Il faut rationaliser les systèmes de transport et de ramassage à cause de l'augmentation constante des dépenses soutenues par les centres de transformation pour le ramassage du lait auprès de chaque producteur. C'est un aspect particulièrement important dans le secteur ovin et caprin principalement à cause du morcellement des exploitations d'élevage et des carences du réseau routier rural.

Le but principal de cette étude est d'analyser et de comparer sous l'aspect technico-fonctionnel et économique, les deux systèmes habituellement utiisés pour le ramassage et le transport du lait ovin en Sardaigne: le camion avec bidons et le camion-citerne isolé. Mais, même avec ce dernier système, le lait est souvent pris directement des bidons, car la presque totalité des exploitations est dépourvue de citerne réfrigérant.

Enfin, il a été preparé un programme de simulation qui a permis de définir les caractéristiques opérationnelles et économiques des nouveaux systèmes de ramassage - pour le secteur ovin - tel que l'usage de la réfrigération du lait à l'étable, le ramassage du lait tout les deux jours et la création des centres collectifs de ramassage.

2 L'ENQUETE

2.1 Zone de collecte

Tout projet d'analyse de ramassage du lait nécessite une connaissance approfondie de la zone de collecte considérée et l'analyse de ses caractéristiques.

La première prise de contact avec le problème a été la subdivision du circuit ou de la tournée de ramassage, Pr (km), en parcours homogènes, dans lesquels la vitesse du véhicule peut être presque constante; on a distingué:
- le parcours mort (km) en référence à la distance parcourue entre l'usine et le primier producteur et entre le dernier producteur et l'usine;
- le parcours utile, Pu (km), correnspondant à la distance parcourue entre le premier et le dernier producteur;
- le parcours dans l'exploitation (km), égal à la distance éventuellement parcourue à l'intérieur de chaque exploitation.

En deuxième lieu, afin de synthétiser par une valeur numérique les principales caractéristiques d'une zone de collecte, on a défini les paramètres indice suivants:
- densité de production (t/km): quantité de lait ramassé rapporté au parcours utile;
- densité de ramassage (t/km): quantité de lait recueilli rapporté au circuit de ramassage;
- parcours utile de collecte (km): rapport entre la longueur du parcours utile et le nombre de points de ramassage;
- production par point de collecte (t): rapport entre la quantité de lait ramassé et le nombre de points de collecte.

2.2 Méthodologie de recherche

En ce qui concerne la comparaison technico-économique des systèmes de ramassage utilisés, on a caractérisé les différents parcours au moyen d'un coefficient (*coefficient de ramassage*) qui synthetise les principaux éléments qui ont une influence prépondérante sur le coût de la collecte, c'est-à-dire: la longueur et la nautre des parcours, la quantité de lait livré (Ql) et le nombre des points de collecte (Pcoll).

Ce coefficient (Cr) est donné par la valeur en pourcentage du produit entre la densité de ramassage et le parcours utile de collecte, c'est-à-dire:

$$Cr = (Ql/Pcoll) \times (Pu/Pr) \times 100$$

L'augmentation de ce coefficient, auquel correspond un amélioration des caractéristiques de la tournée, est donc obtenue en réduisant les valeurs du dénominateur (longueur du circuit de ramassge et nombre des points de collecte) et/ou en augmentant les valeurs du numérateur (quantité de lait et longueur du parcours utile).

Quant à l'introduction de nouveaux systèmes de ramassage, on a construit un modèle mathématique - en utilisant exclusivement les résultats d'expérience des principaux paramètres opérationnels de 40 tournées différentes - qui décrit sous forme analytique le cours des phases de ramassage et de transport. La variation de chacun des facteurs a permis de définir et évaluer différentes solutions en fonction des caractéristiques de tournées de ramassage.

3. SYSTEMES TRADITIONNELS DE RAMASSAGE

3.1 Véhicules de ramassage et de transport

C'est un camion qui est le plus couramment utilisé pour le ramassage et le transport du lait en bidon, en général, il est muni d'une caisse de 2,5 t de portée; le plateau peut accueillir 50 bidons de 50 kg chacun. Parfois, au départ de la laiterie une partie des bidons contient le petit-lait qui n'a pas caillé dans la chaudière après avoi fait le fromage ou la brousse, car celui-ci est utilisé par les éleveurs pour l'alimentation des porcs.

La portée du vèhicule de transport n'est exploité qu'en partie car, à cause du grand nombre de producteurs pour chaque tournée, bon nombre de bidons ne sont pas complètement remplis. Le coefficient de charge du camion est en moyenne de 0,88. Dans d'autres cas, en particulier quand une partie du lait est prise à partir de tanks de réfrigération, on utilise un camion-citerne avec réservoir horizontal isolé. Afin d'eviter au lait de mauvais traitements mécaniques, le réservoir est subdivisé en compartiments dont un peut être employé si nécessaire au transport du petit-lait.

La valeur moyenne du coefficient de charge du citerne isolé, de 0,77, est inférieure à la valeur du camion à cause de la présence du petit-lait dans les compartiments du réservoir qui n'est tout-à-fait vidé qu'à la fin du parcours utile de la moitiée environ des tournées considérées.

Le camions-citerne réfrigérés, dont on prévoit l'utilisation dans l'hipothèse d'une généralisation de la réfrigération du lait à l'etable, sont équipés d'un prélévateur automatique pour déterminer les matières grasses, le protéines et la concentration microbienne. Les véhicules les plus modernes sont équipés, entre autre, d'un système électronique pour déterminer instantanéament le pH et la température du lait.

Les opérations de ramassage et transport ont été accopplies par un seul ouvrier dans tous les systèmes considérés.

3.2 Paramètres relevés

Au cours de cette enquète biennale qui a intéressé 10 laiteries sardes, 40 différentes tournées ont été analysées, dans 10% des cas le lait était prélevé de tanks refrigerants par un camion-citerne isolé ou réfrigéré; le lait dans les 90% restants a été ramassé en bidon, 70% utilisant un camion et 30% environ un camion-citerne isolé.

Pour chaque tournée de ramassage on a relevé les principaux éléments suivants:
- longueur des parcours et type de route (départementale, communale etc.);
- durée du parcours selon le type de route et de parcours (utile, mort etc.);
- quantité de lait ramassé à chaque point de collecte;
- durée du ramassage.

L'opération de ramassage a été décomposée de façon analytique en une série d'opérations élémentaires où on été déterminé les deux temps specifiques d'exécution rapporté à une tonne de lait ramassé (fig. 1)..

Pour le camion avec bidon, le temps opérationnel (TO) était subdivisé en:
- temps de préparation, pendant lequel l'ouvrier sort de sa cabine, monte sur le plateau et décharge le petit-lait;
- temps de lavage de bidons ayant contenu le petit-lait;
- temps de ramassage: transvasement du lait des bidons du producteur dans ceux du camion et chargement de ceux-ci sur le plateau;
- temps d'échantillonage: prélèvement des échantillons de lait et détermination du pH;
- temps de mesure: mesure et enregistrement de la quantité de lait apporté;
- temps mort: évitable et inévitable;
- temps de repos.

Avec le camion-citerne prélevant le lait des bidons ou du tank à lait, la routine du travail, comme on le verra par la suite, est différente en ce qui concerne certaines opérations comme la préparation ou le ramassage; d'autres opérations, dont l'échantillonage, sont automatiques fig. 1).

3.3 Analyse des résultats

3.3.1 Ramassage

La productivité du travail de la main d'oeuvre dans les opérations de ramassege varie, non seulement en fonction du système adopté, mais aussi de la quantité de lait ramassé en moyenne à chaque point de collecte. Donc, dans la comparaison des différents systèmes de ramassage, le TO et la productivité de la main d'oeuvre correspondant se rapportent à un point de collecte apportant 400 kg de lait.

Comme le montre le tab. 1 il existe des différences visibles dans la durée des diverses chaines opérationnelles. L'emploi du camion-citerne qui prélève le lait du tank à lait enregistre le meilleur rendement de la main d'oeuvre avec 3,3 t/h suivi par le système utilisant le camion-citerne pour ramasser le lait des bidons (2,4 t/h) puis par l'utilisation du camion avec bidons (1,5 t/h).

L'analyse des temps élémentaires d'exécution montre que ces différences sont dûes, d'un côté à la durée moindre du ramassage en lui-même (3,9 min/t) réalisé en utilisant le tank à lait par rapport aux bidons (4,3 min/t avec le camion et 5,4 min/t avec le camion-citerne) et de l'autre à la simplification de la routine du travail - grâce à l'automatisme de certaines opérations comme l'échantillonage, la mesure et le lavage - ce qu'on obtient par l'emploi du camion-citerne, surtout si le lait est aspiré directement du tank refrigérant.

L'augmentation du niveau de mécanisation permet donc d'enregistrer une réduction du TO de collecte. La seule exception à cette tendance est représentée par le temps de préparation qui est plus élevé quand le camion-citerne doit manoeuvrer pour s'approcher au tank.

Puisque la production par point de collecte est un des éléments qui modifie la productivité du travail, on a construit la graphique de la fig. 2 qui reporte le cours du TO et de la productitité de la main d'oeuvre en fonction de la quantité de lait à ramasser au point de collecte et de la chaine opérationnelle utilisée. Les données expérimentales ont été interpolées avec des courbes de régression qui reproduisent fidélement le cours réel de chaque temps, avec un

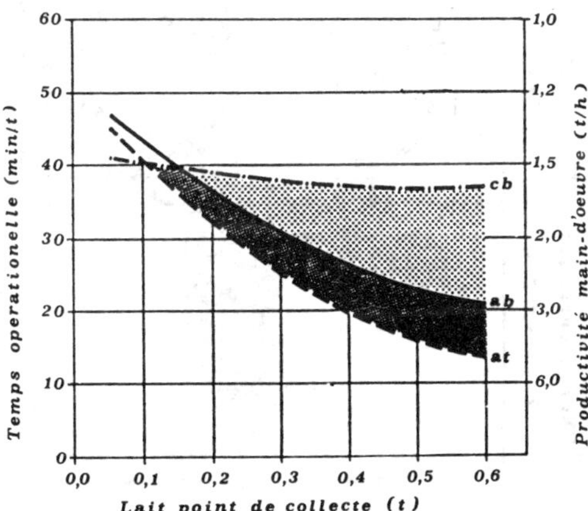

Fig. 2 Influence de la production par point de collecte sur le temps opérationnels et sur la productivité de la main d'oeuvre dans le ramassage du lait en exploitation; -cb: camion avec bidons; -ab: camion citerne à partir de bidon; -at:camion-citerne à partir de tank.

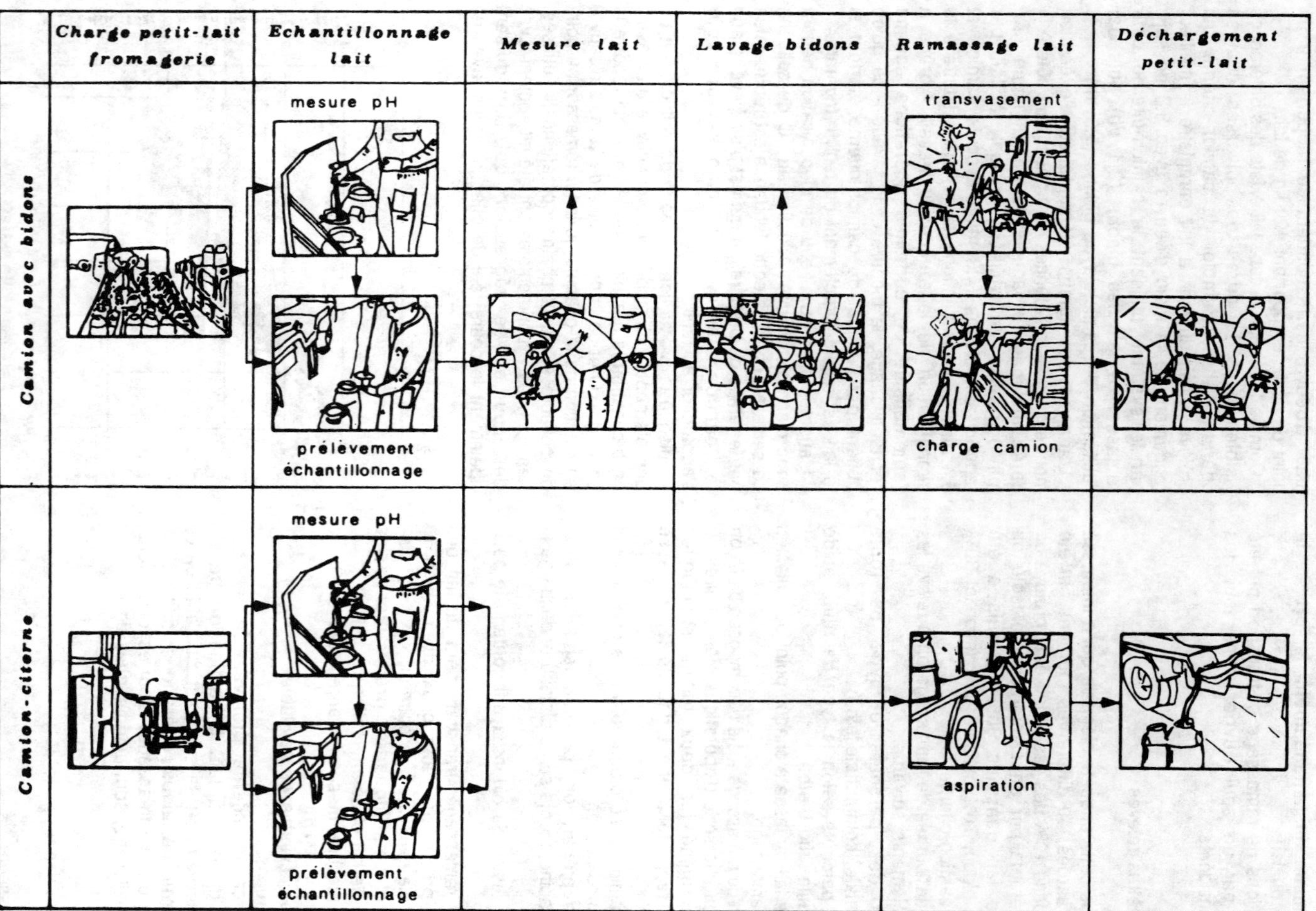

Fig. 1 Systèmes de ramassage et de transport du lait ovin.

Tab.1 - Dureé moyenne de la chaine opérationnelle et productivité de la main d'oeuvre dans le ramassage du lait. Les temps se rapportent à un point de collecte livrant 400 kg.

Système de ramassage	Temps operationnel (min/t)								Productivité main d'oeuvre (t/h*ouvrier)
	préparation	lavage	èchantillonage	ramassage	mesure	mort	repos	total	
camions avec bidons	6,9	4,2	4,7	5,4	8,8	4,6	3,4	38	1,58
camion-citerne à partir de bidons	6,5	-	5,0	4,3	2,2	4,0	3,0	25	2,40
camion-citerne à partir de tank	7,2	-	-	3,9	-	4,0	3,0	18	3,33

coefficient de corrélation entre les variables considérées compris entre 0,90 et 0,94.

Comme prévu, le TO se réduit quand la production par point de collecte augmente. Les variations les plus significatives sont enregistrées avec les camions-citerne qui permittent des augmentations de productivité de la main d'oeuvre de 25-40% par rapport à 4% environ avec le camion, en passant de la production de 300 kg par point de collecte à celle de 500 kg. Par contre, quand la quantité de lait livré ne dépasse pas 100-150 kg, le camion avec bidons est le moyen de ramassage les plus rapide.

3.3.2 Transport

La vitesse du véhicule destiné au transport du lait est influencé tout d'abord par la viabilité et les caractéristique du véhicule (poid et encombrement) mais aussi, en grande partie, par la longueur du parcours et par le nombre des arrêts.

Ces deux aspects ne sont pas assez pris en ligne de compte dans l'élaboration du circuit de ramassage. Par le relevé systématique de la nature des routes et des temps de parcours correspondants, on a essayé de quantifier l'influence de ces paramètres sur la durée de la tournée de ramassage.

Ayant constaté que, pour une même portée, il n'y a pas de différences significatives entre la vitesse du camion et celle du camion-citerne, on a réalisé, par un procédé analogue à celui illustré au § 2.1, un nomogramme pour calculer la vitesse et le temps de parcours du véhicule en fonction du type de parcours (mort, utile et d'exploitation) et du nombre de points de collecte (fig. 3). Les données font référence à un parcours type constitué de routes nationales (10%) départementales (60%) et communales (30%).

A l'aide du nomogramme il est possible d'évaluer, par exemple, l'influence de la longueur du parcours utile et du nombre des points de collecte sur le type de parcours. Sur 20 km prévoyant 4 points de ramassege du lait, le véhicule a une moyenne d'environ 35 km/h avec un temps de parcours d'environ 34 min; en portant le parcours utile à 30 km (+50%) la vitesse moyenne monte à 39 km/h (+11%) et le temps de parcours est de 46 min (+41%).

Si pour ce même parcours utile (20 km) on prévoit 8 points de collecte au lieu de 4, la vitesse du véhicule descend à 30 Km/h (-14%) et le temps de parcours correspondant passe à 40 min (+18%).

De toute évidence, il faut donc contenir la dispersion des points de collecte, phénomène largement dépandu en Sardaigne, pour obtenir un double objectif: maximiser la productivité de la main d'oeuvre et réduire les temps de parcours.

4. NOUVEUAX SYSTEMES DE RAMASSAGE

4.1 Lait en vrac

En Sardaigne, le lait réfrigéré ne représente qu'une minorité sur les 200.000 t de lait ovin destiné à la transformation; en réalité les tanks réfrigerant n'existent que dans les exploitations équipées d'une installation de traite mécanique. On peut estimer qu'il n'y a pas plus de 5% du lait produit qui est livré réfrigéré.

Comme on l'a vu auparavant (fig. 2), quand la densité de collecte est supérieure à 100 kg, le système camion-citerne-tank garantit des temps de ramassage inférieur par rapport à ceux qui prévoit la livraison du lait en bidons.

En outre, il faut considérér une série d'aspects positifs en faveur des producteurs et du centre de transformation, qu'il soit du genre industriel ou coopératif. Avec un tank à lait l'éleveur peut éluder l'obligation des horaires de ramassage et donc rationnaliser le travail à l'interieur de l'élevage. Il suffit de penser que pendant l'été la traite a lieu à 21 heures afin de limiter les dommages causés au lait par la température ambiante.

Avec les tanks les horaires de ramassage sont flexibles et les circuits peuvent être organisés par la fromagerie pour faire arriver le lait avec une plus grande continuité dans la journée et distribuer ainsi, le travail de façon plus rationnelle et économique.

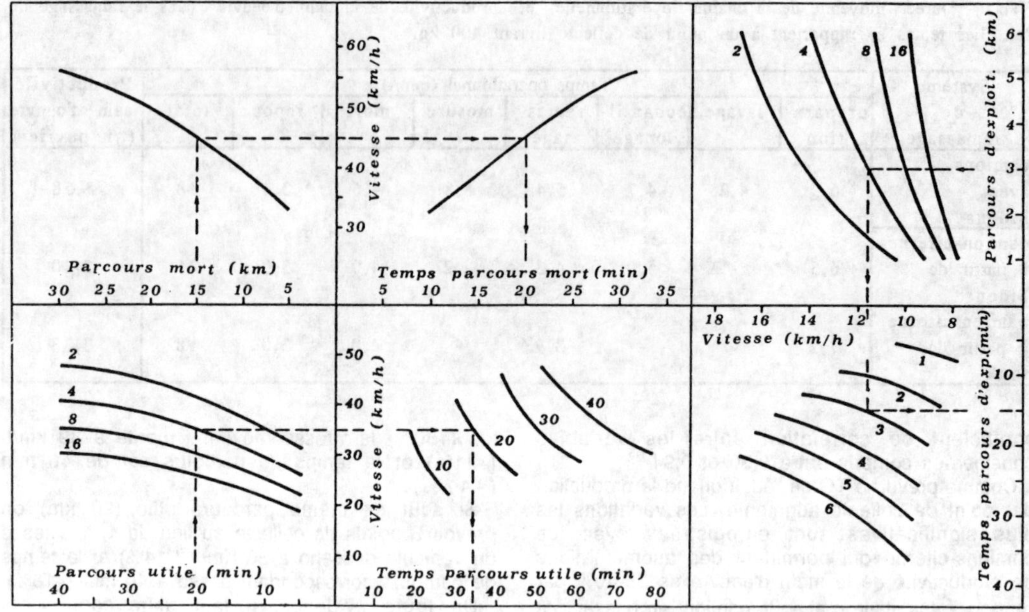

Fig. 3 Nomogramme pour le calcul de la vitesse et du temps de parcours (mort, utile et d'exploitation) et du nombre de points de collecte. les données se rapportent à un parcours type constitué par des routes nationales (10%), des routes départementales (60%) et par des routes communales (30%).

Pour les zones où le lait possède une faible concentration microbienne, il est rentable de ramasser le lait tous les deux jours, toutes les 4 traites. Ceci permettrait de réduire considerablement les coûts de ramassage car le temps de transport seraient réduit de moitié ainsi que ceux de ramassage grâce à l'augmentation de la quantité de produit aux points de collecte. Ce dernier aspect justifirait l'emploi de camion-citerne plus élaborés, c'est-à-dire dotés de rispositifs automatiques pour certaines opérations avec lesquels la productivité de la main d'oeuvre aumenterait ultérieurement.

4.2 Centres collectifs

L'emplacement de tanks à lait en position stratégique sur le parcours de ramassage permet de créer des centres collectifs de ramassage. Cette solution, peu répandue dans le secteur ovin, porte des avantages aussi bien pour les éleveurs que pour la fromagerie provenant de la réfrigération du lait à l'intérieur de l'exploitation; le coût de transport du lait du producteur au centre de ramassage est compensé par une économie de gestion du réservoir de réfrigération. Le lait peut y être porté tous les jours ou tous les deux jours.

Les centres de ramassage se justifient quand les zones de collecte sont loin de la fromagerie, quand elles sont formées de petits élevages ou bien quand l'accès à l'exploitation est difficile à cause de l'état des routes.

Au plus tard deux heures après la traite, chaque producteur doit porter le lait au centre de ramassage où le produit est filtré et dont on contrôle l'acidité; de temps en temps, et sans préavis, il faudrait prélever un échantillon pour une analyse complète du lait. Il serait conseillé, en outre, d'équiper le centre d'une installation pour le lavage des bidons.

5 ASPECTS ECONOMIQUES

5.1 Systèmes traditionnels

Les principaux éléments qui conditionnent le coût de ramassage du lait sont la production par point de collecte et la longueur du parcours. Dans le cas du lait en bidon, comme relevable au diagramme de la fig. 2, pour des quantités supérieures à 150 kg par point de collecte, l'emploi du camion-citerne permet une économie croissante de main d'oeuvre par rapport au camion. Tutefois, il faut se souvenir que les coûts d'utilisation du camion-citerne sont plus élevés que ceux du camion. Pour évaluer la rentabilité économique quant à l'emploi de l'un ou l'autre, il faut donc regarder le coefficient de ramassage (Cr) dont les variations reflètent celle du coût global du système de ramassage de façon inversement proportionnelle.

Pour la Sardaigne, le diagramme de la fig. 4 permet de définir la valeur seuil de Cr, égale à 33,

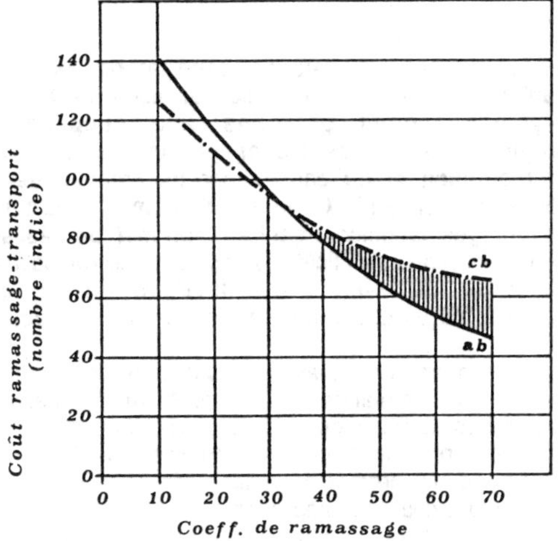

Fig. 4 Influence du coefficient de ramassage sur le coût de ramassage et de transport; -cb: camion avec bidons; -ab: camion-citerne à partir de bidons.

au-dessus de laquelle l'utilisation du camion-citerne est tojours la solution la plus rentable.

Pour les circuits de ramassage avec faible densité de production, comme considéré dans l'histogramme de la fig. 5 qui représente les coûts unitaires du système ramassage-transport, le camion avec bidon, avec une valeur indice de 100 par rapport à la valeur 108 enregistrée par le camion-citerne, est la solution la plus économique.

Dans le choix des véhicules de ramassage, il faut considérer certains aspects positifs, pas toujours facilement quantifiables, en rapport avec l'utilisation d'un camion-citerne isolé, l'économie de main d'oeuvre dans les opérations de décharge à la laiterie et la suppression des risques de déterioration du lait pendant le transport quand la temperatura ambiante est élevée.

5.2 Nouveaux systèmes

5.2.1 Ramassage quotidien

Avant d'introduire un système de ramassage du lait en vrac et donc, avant de modifier les installations des éleveurs et de la fromagerie, il est opportun de faire avec le plus grand soin une estime du résultat financier de l'opération.

Si, ce qui est rare dans le secteur ovin, une zone est caractérisée par une forte densité laitière et par des exploitations attenantes, il semble rentable équiper chaque producteur d'un réservoir de réfrigération. Quand la plus grande partie des producteurs est constituée par des petits élevages et par des exploitations morcelées, on a intérêt à créer un ou plusieurs centres collectifs de ramassage et à effectuer la réfrigération à l'exploitation pour les élevages moyen-grands.

Le rendement économique du ramassage en tank par rapport à celui en bidon est notable même pour une faible densité de production (Cr=28). De l'histogramme de la fig. 5 il ressort que le système camion-citerne-tank permet des économies des 20% environ (coût indice 80), alors que avec la création des centres de ramassage qui réduirait à 2 les 6 points de ramassage actuels, l'économie est de 36% (coût indice 64). Les économies sont encore plus grandes en augmentant la densité de production.

5.2.2 Ramassage tous les deux jours

Avec le ramassage tous les deux jours, la densité de production étant double, on obtient une réduction considérable des coûts de gestion par rapport au ramassage quotidien.

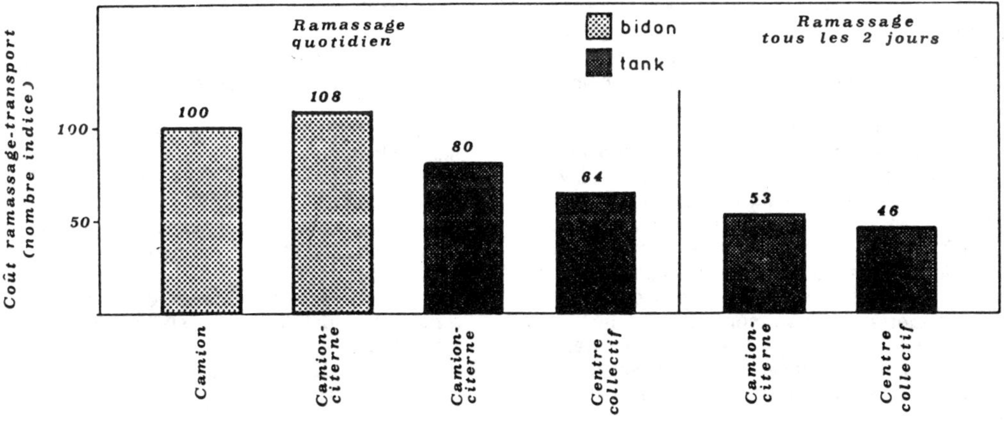

Fig. 5 Comparaison des coûts de ramassage-transport pour les différents systèmes de collecte supposant une tournée avec un coefficient de ramassage de 28.

On estime que le coût d'investissement du moyen de transport avec le redoublement de la capacité du réservoir augmente selon un coefficient d'environ 1,5 alors que les coûts variables (carburant, manutention etc.) se multiplient par un coefficient de 1,4 (tab. 2).

Tab. 2 - Coefficient d'augmentation des couts fixes et variables du camion-citerne isolé en fonction de la capacité du reservoir.

Capacité reservoir		Coefficient augmentation cout camion-citerne	
t	coeff. augment	fixes	variables
2,5	1,0	1,0	1,0
3,5	1,4	1,2	1,2
5,0	2,0	1,5	1,4
7,5	3,0	1,9	1,7
10,0	4,0	2,3	2,0

En contrepartie des économies importantes sont réalisées quant à la main d'oeuvre: l'augmentation de la productivité du travail, en doublant la quantité de lait par point de collecte, est en moyenne de 20% (fig. 2). Puisque le coût de la main d'oeuvre varie en fonction de la zone et de la livraison du lait à la coopérative de transformation ou à la fromagerie industrielle, il vaut mieux établir son incidence selon le cas précis.

En utilisant le modèle mathématique on a procédé à l'estimation des coûts particuliers de ramassage en supposant des conditions défavorables, c'est à dire un Cr égal à 28 (fig. 5). Pour un ramassage chez chaque producteur, en redoublant l'intervalle de temps de la livraison du lait, le coût indice passe de 80 à 53 (-33%). Avec les centres collectifs l'économie est légèrement inférieure (-28) puisque du coût de 64 on passe à 46, ce qui s'explique par le fait que pour les contres collectifs, l'économie sur le transport est moins sensible: la longueur du parcours est inférieure car il y a moins de points de collecte.

On n'a pas tenu compte des coût de gestion dest tanks refrigérants dans les bilans économiques des nouveaux systèmes de ramassage. Pour l'éleveur, ces dépenses sont compensées par une meilleure qualité du lait, qui permet d'augmenter les prix d'un lait de "qualité" et par des avantages provenant de la possibilité de se débarraser des contraintes de l'horaire de la traite et de la livraison du lait. Pour la fromagerie le paiement de primes aux producteurs est compensé par un meilleur rendement dans la transformation du lait et par la possibilité de réaliser un produit fini de qualité.

6. CONCLUSIONS

Au vu des résultats de cette étude, on peut affirmer que la rationnalisation et modernisation du système actuel de ramassage suppose le remplacement des bidons par un réservoir de réfrigération et du camion par un camion-citerne isolé ou réfrigeré. Cette solution, mis à part les avantages technico-organisatif, permet de réaliser des réductions significatives des coûts de ramassage à partir de valeurs de Cr d'environ 20 qui représente un circuit de ramassage type du secteur ovin.

Au moyen terme, avec la diffusion de la traite mécanique l'amélioration des conditions hygiénico-sanitaires de l'élevage et de la qualité des pâturages, on peut prévoir la livraison du lait tous les deux jours, ce qui permet de réaliser des économies au niveau des coûts de ramassage.

En redoublant la quantité de lait à conserver, il faut bien faire attention à la taille de l'installation de réfrigération. En effet, comme on l'a relevé au cours de précédentes recherches Pazzona (1983), la pleine récupération des coûts de gestion de l'installation, y compris la prime payée pour le produit réfrigéré, est obtenue si le coefficient d'utilisation du volume utile du tank est obtenue si le coefficient d'utilisation du volume utile du container est assez élevé. Pour un réservoir de 1,2 t, par exemple, en augmentant le coefficient d'utilisation de 0,7 à 0,9, les coûs de gestion diminuent de 20% environ.

Dans l'immédiat, les centres collectifs de ramassage représentent une alternative valable aux systèmes traditionnels et constituent le premier pas vers la diffusion de l'usage de la réfrigération du lait à l'exploitation. Il ne faut pas sous-estimer, enfin, le rôle éventuel du centre de ramassage dans le domaine de la divulgation et de l'assistance technique.

REFERENCES

Bolli, P. & Scotton, M. (1987). Lineamenti di tecnica della meccanizzazione agricola. Edagricole, Bologna.

Gac, A. (1963). Rapport au sujet de l'intérêt économique du ramassage du lait en citerne et du refroidissement en vrac à la ferme. Bull. Tech. Génie Rural 64.

Ichilczi, K. (1981). Le refroidissement du lait et son comportement en fromagerie. Rev. Lait. Franc. 401.

Pazzona, A. (1983). Scelta ed impiego dei refrigeratori del latte alla stalla. L'Informatore Agrario 26.

Ripault, L. (1982). Rationaliser la collecte. Rev. Lait. Franc. 415.

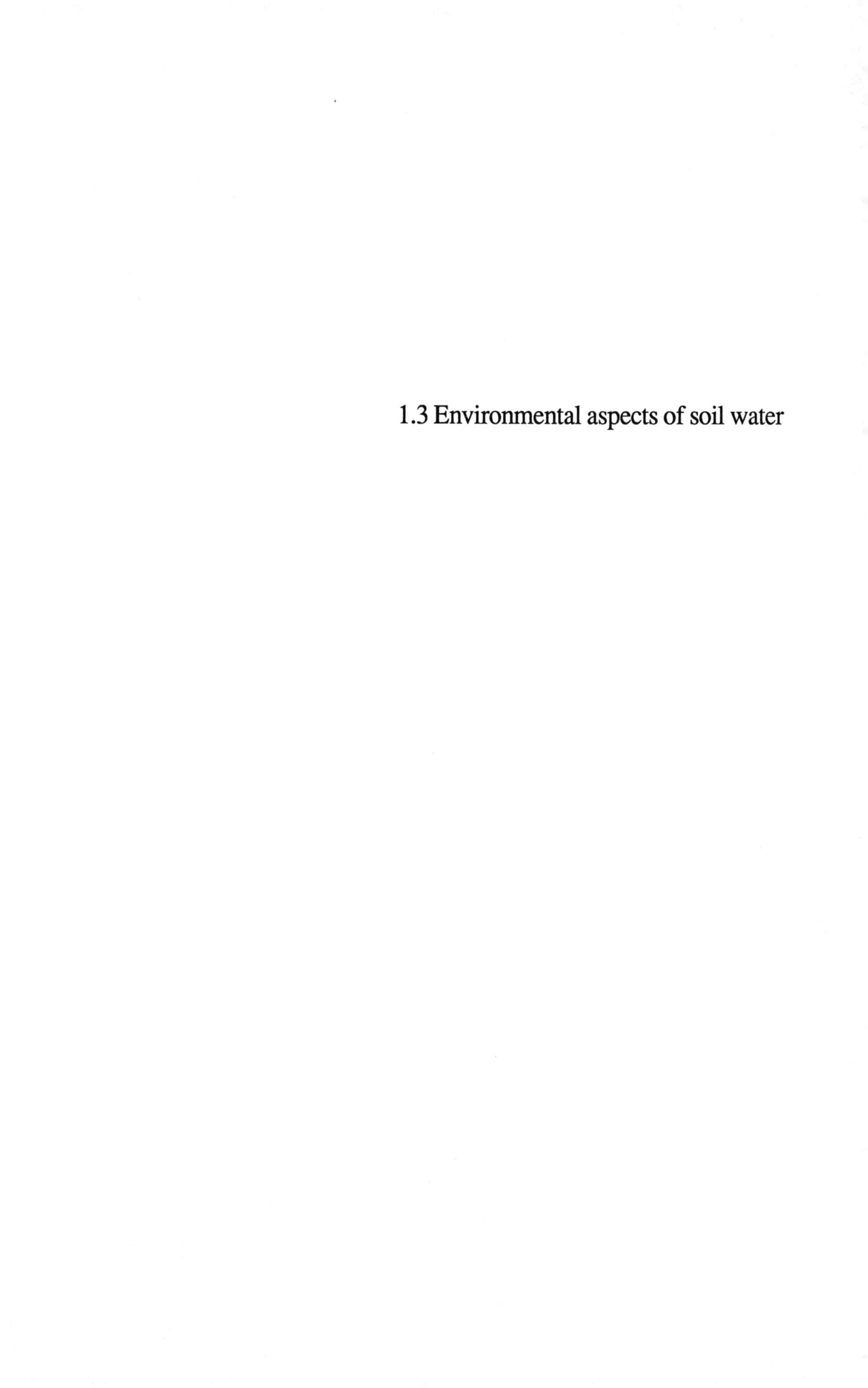

1.3 Environmental aspects of soil water

Land and Water Use, Dodd & Grace (eds), © 1989 Balkema, Rotterdam. ISBN 90 6191 980 0

Keynote paper:
Land and water use: An iterative conservation process

L.W.De Backer
University of Louvain, Louvain-la-Neuve, Belgium

OUTLINE : Conservation is defined in the Webster dictionary as a careful preservation and protection of something; esp : planned management of a natural resource to prevent exploitation, destruction or neglect.

Management is a concept which can be summarized from the dictionary as the act of administering business or affairs with care, skill and success.

The purpose of this paper is to explain how land and water conservation measures should be considered and managed in development projects and/or programmes. References will eventually be made to the state of the art in this matter. A consolidated understanding of sustainable development is a basis for action that we should share with the Stockholm Group (1988).

RESUME : Le mot "conservation" est défini dans le Larousse Sélection comme l'action de maintenir en bon état, de garder avec soin et de aussi ne pas perdre; par exemple : conserver de la viande, sa santé, son calme, un souvenir. Dans le dictionnaire Webster, l'accent est mis surtout sur la gestion planifiée d'une ressource naturelle pour prévenir son exploitation, sa destruction ou sa négligence.

La gestion est un concept qui peut être résumé à partir de ces deux sources de définition comme l'administration des biens, affaires et intérêts de quelqu'un avec soin, talent et succès.

Le but de cette communication est de montrer comment les mesures de conservation de l'eau et du sol devraient être considérées et aménagées dans les projets et programmes de développement. Il sera fait occasionnellement référence à des exemples en cette matière. Le point de départ pour agir est la compréhension globale du concept de développement durable que nous pouvons partager avec le "Stockholm Group"(1988).

ABRISS : Das Wort "Erhaltung" bezeichnet eine Tätigkeit deren Ziel ist etwas zu stützen, zu bewahren und zu schützen; beziehungsweise einer Naturquelle durch geplante Verwaltung ihre Ausbeutung, Zerstärung oder Vernachlässigung zu verhindern.

Verwaltung ist also eine sorgfältige, talentvolle und erfolgreiche Führung von Gütern, Geschäften und Interessen.

Das Ziel dieser Mitteilung ist darzustellen wie Massnahmen für die Erhaltung von Wasser und Boden zu betrachten und in Enwickelungs -projekten und -programmen zu bewirtschaften sind. Diesbezüglick werden einige Beispiele dargestellt. Als Ansatzpunkt zur Aktion sehen wir ein Gesamtverständniss des Begriffes dauerhafte Entwicklung welches wir mit dem "Stockholm Group" (1988) teilen können.

"The Water is as the Land through which it flows"

(Plinius the Elder)

1. MOTIVATION

In 1981, the World Soil Charter stated that "among the principal resources of which man disposes is the earth which consists of soil and water and the animal and vegetable life which is associated with them. The **use** of these resources should not result in their degradation nor their destruction, for the survival of man is tied to the maintenance of their productivity."
As part of our natural environment, soils have then been attributed 5 main functions :

- an agricultural function,
- a supporting function,
- groundwater conservation,
- an ecological function, and
- mining of minerals.

Because water is so intimately associated with soil, one can go one step back and ask "What does water mean in scientific context ?

- to the social scientist, water is an important factor in a number of significant types of group action and is defined at various times as either a means of subsistence, or a method of transportation, or as a technological input or as an important element in religious ritual,
- to the economist, water is a scarce resource to be allocated in some optimal fashion between different users of different economic needs in different places at different times,
- to the biologist, water is of prime importance as a necessity of life and as a shaper of the various forms which living organisms take,
- to the botanist, water is of particular importance as the means by which plant nutrient is taken up and distributed,
- to the zoologist, the K-rich water of the cell is important for the diffusion of chemical substances while the Na-rich extracellular water provides an environment for the cells that is thermostatically controlled."(Dooge, 1986)

Moreover, Brookshire et alii (1986) discussed the implication for valuing water resources in terms of existence values and normative economics.
Existence value is best represented by notions of knowledge for mere existence of natural resources: desires for participation (direct use) and desires for preservation (future use).

Existence values may have two components :
- an economic component that is consistent with utility maximizing behavior (utility means individual satisfaction derived from all sources, economic ((goods and services)) and non economic ((altruism))),
- an ethical component that is consistent with normative benefit-cost assumptions. Benefit-cost analysis relies on an identity between preference (personal ethics), welfare (utility maximization) and choice (commitment). Thus, not all expressions of willingness to pay for preservation of (aquatic) environments are relevant to benefit-cost analysis.

From the above one deduces that motivation is indeed a postulate when one wants to work in a given discipline but that integration of objectives is necessary to fullfil the common goal of conservation.

The ultimate goal of a user of natural resources could be written in short as **the well-being of man on earth.** Figure 1 illustrates the status, relationships and qualifications involved in trying to reach this goal. It ends up that any user should be a manager, eventually he or she is a mismanager, and since everyone of us uses natural resources, all of us must learn how to manage.

Care is achieved through education (culture), skill through training (technology) and success through integration of care and skill in the management (institution) of all projects.

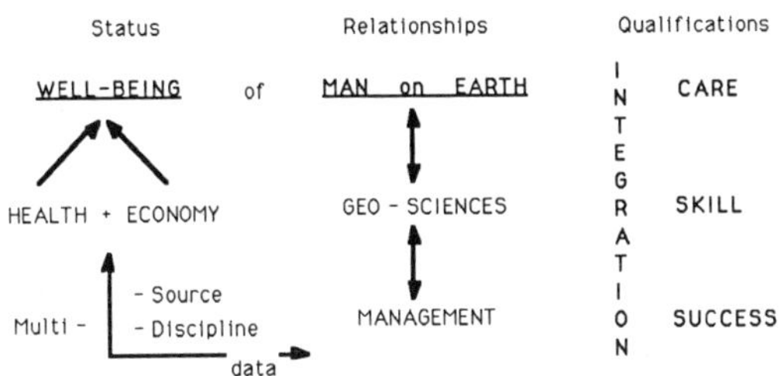

Figure 1.: Scheme of the status, relationships and qualifications required in management to reach the user's ultimate goal : the well-being of man on earth.

2 SPACE AND TIME versus LAND AND WATER

First of all, a space scale is necessary in management. The fact is that under its most **use**ful state water being a fluid, it is unseparable from its container.

In the vapor phase, water is contained in the atmosphere that we breath : a normal human being inhales about 12.000 liters of air and exhales with them some 700 grammes of water a day (De L'île, 1988).

In the liquid phase, water rests on or in and flows on, in or through the ground. Therefore, **land and water conservation measures have to be considered together.**

This has again been emphasized in a recent study on trends and consequences of changes in nonpoint nutrient loading into european freshwaters since 1950 until 2080. Prevention against groundwater contamination seems still possible if the nitrogen and phosphorus fixation in the soils is not exceeded. The capacity of the soil (1 meter deep) to accumulate nitrogen is more limited (10 years) than for phosphorus (100 years) at actual rates of inputs (Behrendt,1988). These estimates pertain to soil and climatic conditions of north-central Europe.

Also, it is rather obvious that land reservoirs must be defined for water retention and that transfer of water from one reservoir to the others will render easier the understanding of the soil-water relationships in the water cycle. Figure 2 shows a water balance scheme with 5 land reservoirs, each containing a given amount of stored water, S.

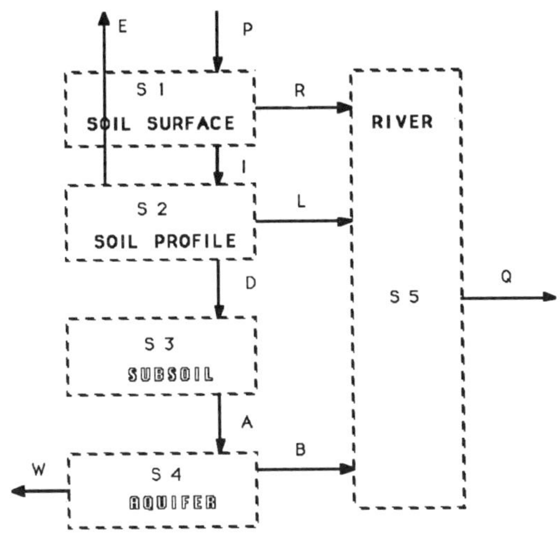

where :

$\Delta S = \Delta S_1 + \Delta S_2 + \Delta S_3 + \Delta S_4 + \Delta S_5$

Δt = f (phenomena, objectives)

P : Precipitation;
E : Evapotranspiration;
I : Infiltration;
D : Drainage;
A : Aquifer recharge;
R : Rapid runoff;
L : Late runoff;
B : Base flow;
Q : River discharge;
W : Inter-Aquifer exchange;

Units : mm/(year, month, day, hour, or...)

Figure 2. : Inputs and outputs of the 5 land reservoirs of water

The inputs and the outputs of each reservoir vary according to the phenomena involved at their boundaries and the quantitative and qualitative accumulations in each of these reservoirs change accordingly.

Within the physical sciences and the earth sciences, there is and can be no universal model for water movement because it would require a unique choice of time and space scales. Looking at Table 1, it is obviously not possible. The significant length and time scales appropriate to the different models of water are approximately as summarized in that Table (Dooge, 1986).

TABLE 1 Significant Length and Time Scales

DISCIPLINE	SUBJECT OF STUDY	LENGTH (m)	TIME (s)
Physical	Water molecule	10^{-10}	10^{-13}
-Chemistry	Water cluster	10^{-8}	10^{-11}
Continuum Mechanics	Continuum point	10^{-5}	10^{-6}
Hydraulics	Turbulent flow	10^{-2}	10^{-2}
Hydrology	Experimental plot	10	10
	Basin module	10^{2}	10^{2}-10^{3}
	Subbasin	10^{3}	10^{3}
	Basin	10^{4}-10^{5}	10^{4}
Climate studies	Gen. atm. circulation	10^{5}	10^{6}

These scale ranges are just a few examples. But Table 1 gives also a comprehensive look at the complexity of the continental water cycle system to which we belong.

Ideally, a water resources manager would like to know the amount and the quality of the available water stored at any given time in each reservoir.

Quantitatively, figure 3 integrates water and land as related to time and space increments. It illustrates the plausible evolution of the water storage variation, ΔS, over a period of time, Δt, in each of the reservoirs of figure 2. Moreover, it shows that the time difference to reach the maximum of filling, Δt_F, of the successive reservoirs is quite different from that necessary to reach the minimum of their emptying, Δt_E, (De Backer,1982). Notice that the water storage variation, ΔS, is a derived space increment because it implies the knowledge of the geometry of the reservoirs and that the time increment, Δt, to be considered in a project depends not only on the phenomena observed (ecology !) but also on the objectives pursued (socioeconomics !).

Qualitatively, considerations about man's activities are necessary especially when one knows that the degradation of our water resources is due only to that fraction of a percent of the water that we bypass from the water cycle for our consumption and return polluted to nature. In that respect, drinking water risks are unvoluntary and unavoidable by individuals, but are more cost effectively avoidable by government or industry action (Jensen,1986).
Regular measurements on a collective basis should therefore be undertaken.

3 INDICES : INDICATORS versus NORMS

Indications about the status of the reservoirs and that of the water they contain are obtained from measurements of appropriate variables,sometimes grouped and called **indicators** . This means that not all the possible variables are measured continuously or at a given frequency. In water resources the frequency of measurements should obviously be smaller than that of the water cycle i.e. the average residence time of water vapor in the atmosphere : 10 days (Taube, 1985).

The interactions between soil, water and man's activities being also unseparable, it is **normal** to examine these interactions in quantitative and qualitative terms. Many **norms** are naturally established on health standards and usable potential of the resources as contained in the acts of the US Congress or in the Directives of the CCE.

Norms can be compared with the evolution of the indicators corresponding to the occuring phenomena. The ratio between the value of the indicator and that of the norm is defined as an index. Hartmann(1987) cites, for example, 2 first rank integrated indices : the Harvard index which is used to describe the nutritional and health status of children, and the average life expectancy. This author presents a.o. the "equal preference" procedure for evaluation of the state of the water related environment. It requires 3 levels of indicators to define the status of the system and the definition of categories of the system performance such as sound, acceptable or poor, especially useful in the case where norms (standards) are not available. According to the same author, from the first level or basic indicators (measurements) emerge only the 2 third-level indicators : ecology and socioeconomics, used in the just mentioned procedure.

While it is very important to have a definition of the **indices**, it is sometimes very difficult to set the number as well as the upper and lower limits of the norms and it can be much time consuming to have them implemented in law terms. Here are a few examples :

- The National Academy of Sciences (NAS) records Recommended Daily Allowances (RDA). Some chemical qualities of water, like a given concentration in major elements contribute to human health in a positive way. Yet doubts remain, decreased CVM (cardio vascular mortality) associated with hardness seems related to ischemic heart disease (Hopps and Feder, 1986).

- In 1986, the Safe Drinking Water Amendements set standards for over 80 specific drinking water contaminants, e.g. nitrates are only one of them. It is expected that 4 times that many will be registered as national standards (over 300) by 1989.
- The CEE directive 86/278 of the 12 june 1986 concerns the protection of the environment and particularly the soils in the case of the use of sludge from waste water treatment plant in agriculture and is applicable by european member states starting June 1989. France is already ahead of time because its19 July 1979 law on the control of fertilizer use ended up 6 years later into a norm (NF U 44 041) enforced since 29 August 1988 (Chalot,1988).

- The International Standards Organisation (ISO) is examining since 1986 the norms relative to the Quality of the Soils(ISO/TC-190). Those relative to air (ISO/TC-146) and water (ISO/TC-147) quality started in 1971. It is interesting to note that the scope of these 3 Technical Committees (TC) is "the standardization in the field of soil, water or air quality including definition of terms, sampling of soils, waters, or airs, measurement and reporting of soil, water or air characteristics, excluded : limits of acceptability for water and air quality and for soil pollution as well as civil engineering aspects dealt by ISO/TC-182."

Those limits of acceptability are left to governmental decisions. "Of the more than 48,000 chemicals listed by EPA (Environmental Protection Agency of the USA), next to nothing is currently known about the toxic effects of almost 38,000. Fewer than 1,000 have been tested for acute effects, and only about 500 for their cancer-causing, reproductive or mutagenic effects. Funding must be increased for such research." (p.32,TIME, 1989)

4 MANAGEMENT : RESEARCH versus TECHNOLOGY

When norms are established projects can be launched within safe limits for the environment i.e. the projects are integrated in the natural system. In other words, they will not impair the earth's heritage any more provided they are (properly) managed. According to the definition given at the beginning of this paper, the act of administrating business cannot be called management if it is not done with care, skill and success. Figure 4 is a memento (De Backer et alii , 1986) which can help the manager to fullfil these conditions.

This scheme is based on the 5 simplest questions any normal human being is supposed to ask himself before he decides to start an action : what, why, how, when and where ? Among these five little words that everybody understands, there is one black sheep which doesn't start with W : **how** is the question on which there is often much controversy and

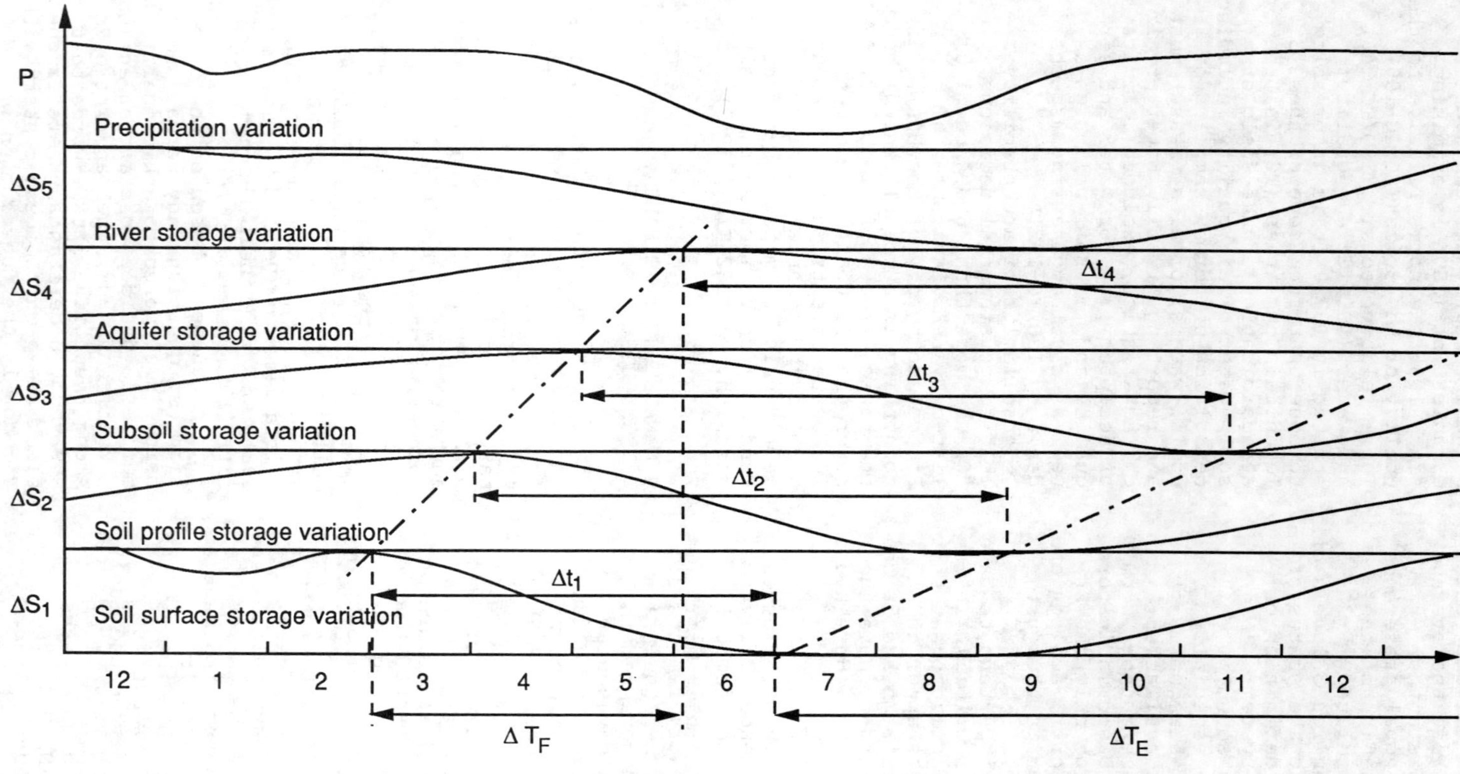

Figure 3 Qualitative evolution of the water storage variations in the 5 land reservoirs with respect to rainfall.

time wasting. But, if the above mentioned well-being is the motivation considered as a postulate behind the action as it is in this paper, then conservation measures can probably be all taken care of automatically during any project.

Before examining briefly these 5 questions it is important to notice also that they have already been answered globally in the previous paragraphs about the conservation of our vital land and water **system**, the following considerations are relevant to the management of a **project** within the system, i.e. its integration. To some extent there is as much difference between a project and the system as there is between a clandestine passenger and the running train. To succeed i.e. reach the train destination, he must integrate the train. When his interference is minimal, he and the train conserve their integrity. But such passenger might have different objectives, and risks are that neither he nor the train ever reach their destinations. Then conservation of the system fails because there is no integration of the project within the system.

The comprehension of all the dimensions of management of a project is therefore of prime importance in minimizing its impact on the natural resources as well as in providing conservation measures. Following the numeration of figure 4:

- 4.1 WHERE is the space dimension. It was already introduced in terms of volume to characterize globally the 5 land reservoirs through which water flows. Although the surface of a watershed is the dimension advisable for water resources development, the scale of a project might be greater or smaller depending on the social needs. With Table 1 in mind for lengths, here the space dimension refers rather to areas within institutional boundaries : site, locality, region, nation, continent and world. Nonetheless they encompass the 5 reservoir dimensions. This renders even more imperative the use of norms (standards) to integrate the project space into that of the natural resources system.

- 4.2 WHY is the aspect dimension. Motivation is the driver of any project. As indicated in section 1, managers and users of a project have an ultimate goal in common : the well-being of man on earth. Unless they are clearly known to everybody in the project, nearer aspects might blur this noblest picture of man's spirit. Up to a recent past, projects were often oriented in the line of a single aspect, particularly an economic or a technological one, or even just for institutional prestige.

Figure 4 lists a few of them, some of which might actually be considered together in what is called now a multi-objective project. The integration of these multi-objective projects into the natural system requires **interdisciplinarity** in all the phases and the stages of management. Data from several sources would be necessary to meet the goal as shown in figure 1. A minimum of 9 crucial aspects to be considered together for the optimization of the impact of the projects on natural resources will be introduced for discussion in Section 5.

- 4.3 WHEN is the time dimension. The initiation of a project supposes long sequences of historical observations such as those illustrated in figure 3 for example, but particularly those relative to the crucial aspects which will be presented later.(see 5)

The time interval during which a project evolves can be divided in 3 stages of management : **planning, organization and operation** which constitute the "time setting cycle" supporting the decision process. In small projects, it still happens that one manager is able to play these 3 roles. In fact, this is what we do perhaps unknowingly each time we perform an action.

In larger projects, these stages have to be well structured and staffed. The overlapping effects induced by the iterative property of the stage sequencing must be properly taken into account, since :

- Planning encompasses also the formulation of sub-systems' designs, plans for the organizational structures, budgeting and operating policies, etc...

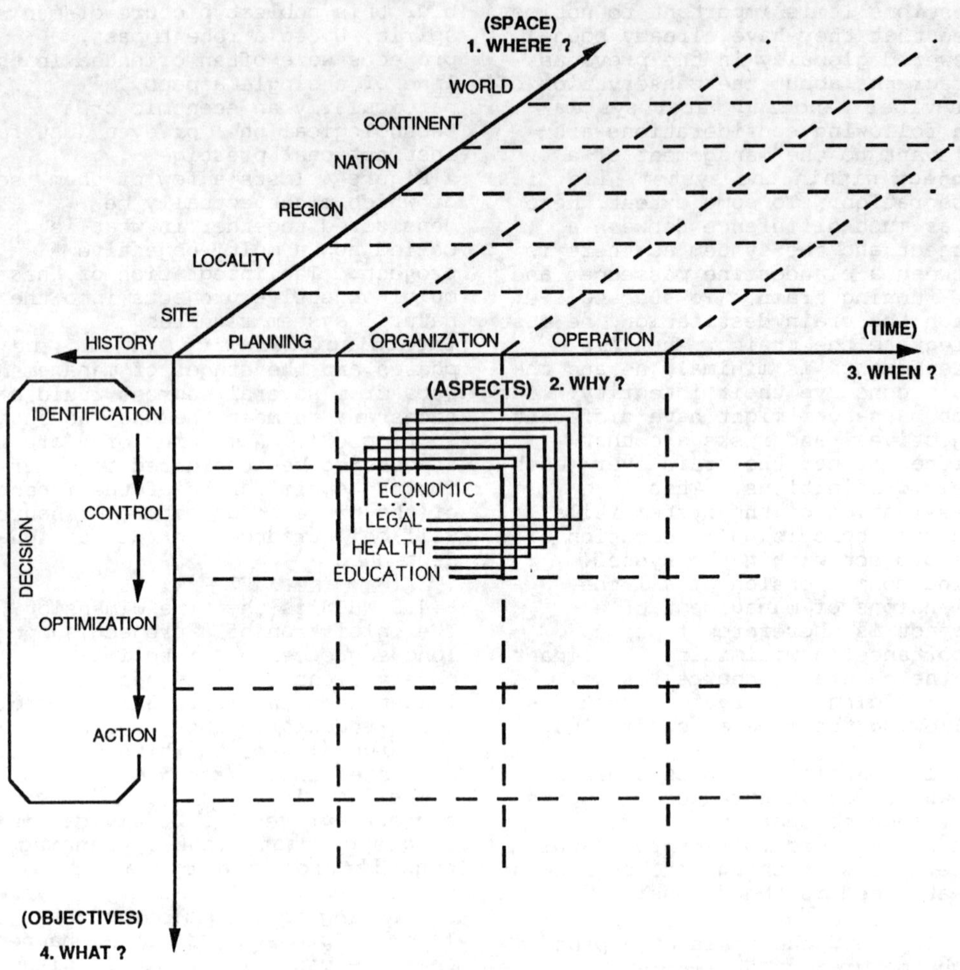

Figure 4. The 5-dimension matrix for the management of a project.

- Organization also embeds the plan's implementation, the programming of operations, etc...
- Informations derived from Operation help monitoring the sub-systems and plans' reformulation. As the management progresses, the staff of the stages must be able to communicate easily.

Feedforward and feedback procedures and networks have to provide the appropriate interfaces between the stages. Actually, desk computers connected to an information system will facilitate greatly these procedures, particularly in the Organization stage that the european Communities want to strengthen during the 90's.

- 4.4 WHAT is the dimension of the management objective : the action. It is therefore the dimension of the decision. Indeed, once the project goals relative to each stage structure have been defined, they must be translated into objectives which are their quantitative expression. Concrete actions can then be derived through an optimization process.The basic successive phases of the decision process must be respected : **identification, control and optimization**, before an action is taken. This process is repeated enough times in each stage before getting to the next up to the moment the final action is done at the end of the operation stage.

The whole process which includes the feedforward and feedback procedures (see 4.3) might involve maintenance as well as rehabilitation and can last several years.

In the case of multi-objective projects, MultiCriterion Decision Making (MCDM) must be adopted (Changkong and Haimes, 1982). In the US, the National Environmental Policy Act of 1979 has directed planners to consider more than the narrow economic aspect used in the past to justify projects. In water resources development benefits cannot always be expressed in monetary terms. In those cases, physical units like number of people protected from floods, amount of sediment or dissolved oxygen, visitors days,etc are used as indicators. Indicators array is available either in tabular form for discrete alternatives or as a generator of alternatives in the case of continuous alternatives like the amount of water treated or stored in a reservoir. The selection among alternatives to find an overall optimum solution is called MCDM. It is possible only by considering tradeoffs among these options (Bogardi et alii, 1984) and ranges from simple engineering judgement to sophisticated multiobjective programming methods (Cohon,1979).

- 4.5 HOW is the means dimension.

Manager decisions are taken according to their skill and care. Usually there are technical solutions for most problems to solve. If the level of technology cannot guarantee proceeding to the next phase, the managers could take the decision with calculated risks or call on research. Because research takes much more time than the mere adaptation of a technology to a project, there is a good reason for researchers to anticipate the needs of the entrepreneurs.

In all cases, research objectives should be considered in the statement of project goals, since they are crucial in the management of the interfaces of the differrent aspects (see 5). Useless to say that in this critical area of land and water management, no specific-matured devices exist. Therefore, valuable effort should be directed towards the elaboration of appropriate tools and techniques. Also considerable advantages could be drawn from an oriented-research.

In the World Strategy for Conservation (UICN-PNUE-WWF, 1980), one can find a list not only of the priority conditions but also of the priority measures which should be taken immediately on a national as well as on an international basis to conserve the environment.

These priorities have been recalled in the Declaration of Policy and other decisions of the United Nations Economic Commission for Europe about the Integrated Water Management in that region (CEE, 1983 and 1985)

5 IMPACT versus INTEGRITY and REMEDY versus PREVENTION

As anyone who has ever been sick or ill knows, **man's well-being** depends not only on his own state of health (physical, emotional and intellectual) but also on the quality state of the natural resources he consumes and returns as refuse in a continuous exchange process with his environment.
Hence, this exchange process seems to determine man's impact on the resources. Land and water conservation on the other hand is (planned) management to prevent exploitation, destruction or neglect of these resources or in other words to **keep their integrity for future use**. Therefore, the conditions which control the exchange process to respect their integrity can be called **conservation measures.**

To take these measures, consumption and refuse have to be identified, con-trolled and optimized according to the decision process described in figure 4.
Although many indicators show that exploitation, destruction and neglect of the natural resources are occuring at a scaring rate, several questions need to be answered before man can be proud of the results of his resources management.
To save time in an attempt to slow the degradation even before research brings up these answers, a framework of man's exchange process with nature synthesizes in figure 5 the management observations of more than 300 papers presented during the fifth world congress of the International Water Resources Association (IWRA).

This framework (De Backer and de Bruyn, 1987) presents the 9 crucial aspects which should always be considered as a minimal iterative sequence in the management of natural resources.

The exchange process, consumption-refuse, takes place during the 3 main activities that involve **man's life, research and development**.

The exchange nature of these activities is conditionned **by the constraints of 3 domains : institutions, culture and technology**, in which man abstracts himself from nature.

The degree of abstraction of these domains depends on the integration that is achieved between the results of the activities and the domains by the 3 behavioral currents : **social, ethical and economic**. The nature of these currents is thus that of integration.
There is not only a time sequence to be respected within these respective 3 groups of 3 aspects but there is also an inter-simultaneity between the groups. These can be called **inter-groups**.

Indeed,

1) man's life (history) is generated through institutions by the social current (basic drive),

2) research (dreams, projects) emanates from culture thanks of ethics (actual mores), and

3) development (which is constrained by institutions) is produced because technology is the seat of the economic current.

FIGURE 5

MAN'S IMPACT ON HIS RESOURCES

WELL-BEING & INTEGRITY

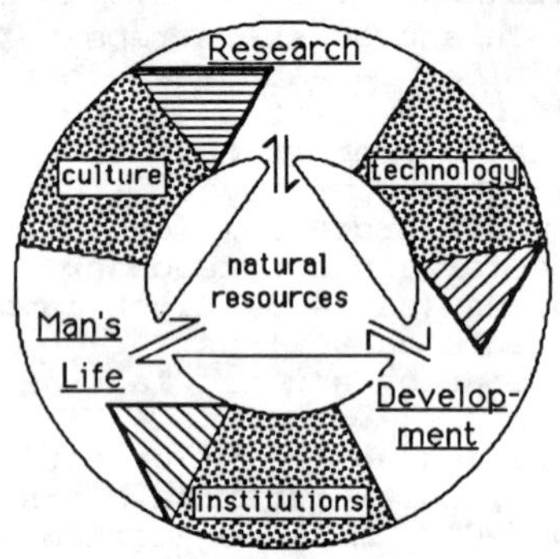

⇆ : Exchange process

3 domains : institutions , culture , technology

3 currents : social , ethical , economic

3 activities : man's life , research , development

Surprisingly, these 3 inter-groups correspond rather well to the 3 famous biorythms which pulsate our individual existence :

1) the intellectual cycle of 33 days,

2) the emotional cycle of 28 days, and

3) the physical cycle of 23 days.

It is interesting to note that some managers already use the biorythms in the everyday life of their enterprises as criteria for their personnel, e.g. airline pilots.

These last observations might be mere coincidence but it seems that a certain equilibrium should be respected between the 3 intergroups if one wants to take effective conservation measures.

Numerous examples of conservation from around the world show that integration of the 9 crucial aspects can be considered as the untapped alternative to our foreseable shortages (Postel, 1985).

Here is how T.A. Sancton (p.17, TIME, 1989) explains the actual mess resulting from not taking care simultaneously of these 9 aspects but from giving too much weight to our cherished technology : "In many pagan societies, the earth was seen as a mother, a fertile giver of life. Nature - the soil, forest, sea - was endowed with divinity, and mortals were subordinate to it. The Judeo-Christian tradition introduced a radically different concept. The earth was the creation of a monotheistic God who... ordered its inhabitants in the words of Genesis : "Be fruitful and multiply, and replenish the earth and subdue it : and have dominion over the fish of the sea and over the fowl of the air and over every living thing that moveth upon the earth." The idea of dominion could be interpreted as an invitation to use nature as a convenience. Thus the spread of Christianity, which is generally considered to have paved the way for the development of technology, may at the same time have carried the seeds of the wanton exploitation of nature that often accompanied technical progress. ...The use of pesticides has increased crop yields but polluted water supplies. ...Yet the advance of technology has never destroyed man's wonder and owe for the beauty of the earth. ... No attempt to protect the environment will be successful in the long run unless ordinary people ...are willing to adjust their life-styles. Our wasteful, careless ways must become things of the past. We must recycle more, procreate less, turn off lights, use mass transit, do a thousand things differently in our everyday lives. We owe this not only to our selves and our children but also to the unborn generations who will one day inherit the earth. ... Now, more than ever, the world needs leaders who can inspire their fellow citizens with a fiery sense of mission, not a nationalistic or military campaign but a universal crusade to save the planet. Unless mankind embraces that cause totally, and **without delay**, it may have no alternative to the bang of nuclear holocaust or the whimper of slow extinction."

This statement illustrated with examples confirms that the conditions for a sound exchange process which satisfies conservation cannot be met by putting the accent on one aspect only, here the technology. A sound exchange process which guarantee conservation implies integrity i.e. a reasonable equilibrium between the 3 domains, the 3 activities and the 3 behaviors described in figure 5 to smooth out the man's impact on his resources. Indeed, this figure shows not only the relationships between them but also the kind of interdisciplinarity necessary to cover all these 9 crucial aspects in multiobjectives projects (4.2), to record, treat and interpret historical observations relative to these aspects (4.3), and to look for the appropriate interfaces in the management process as well as in the feedforward and feedback procedures (5). Unless these requirements are fulfilled it can be said that prevention will not replace definitively the remedies.

6. Conclusions

The ecology-economics indicators necessary to describe the system integrating a project (see 3) can be easily assimilated into the general scheme of figure 6 which summarizes the proposed cyclic methodology for land and water conservation. It shows the iterative process involved in the management of our environment if we care to use our natural resources for sustainable development.

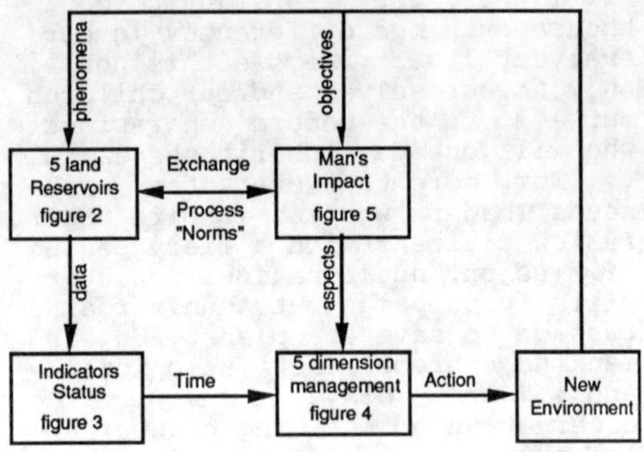

Figure 6. A cyclic methodology for the conservation of natural resources for sustainable development in project management.

Indeed, once a project has been launched and is running at its crusing speed a new environment takes shape. Its development will undoubtedly modify not only the phenomena (ecology) in the 5 land reservoirs but also the objectives (socioeconomics) as a consequence of man's impact on the resources. This double impact of a new environment on our behavior has often been presented as a source of conflicts (Postel, 1984).

However, it is at another level of this iterative process that the conflicts occur.This can happen in the definition of the norms involved in the exchange process as well as in the elaboration of an integrated project.In our country, a better acceptation of the concept of interdisciplinarity is a prerequisite to coming to say **yes** (Priscoli, 1987). It results from what we have observed that the exchange process implies finally 3 types of norms for a sound management of our activities: 1) socio-ecological standards in man's life, 2) ethico-ecological standards in research and 3) economico-ecological standards in development. This is probably the way our institutions, our culture and our technology will remain useful to man in harmony with the limited earth resources.

The well-being of man is at that measure.

Unfortunately up to now, while conflicts have taken most of our efforts, the degradation of our environment went on and on.

We are still living on the safe resources of the past. Therefore the retrieval of a viable environment for sustainable development will require more costly new remedies as well as sound conservation measures.

REFERENCES

Behrendt H., 1988 : Changes in nonpoint nutrient loading into european freshwaters : trends and consequences since 1950 and not-impossible changes until 2080. Intern. Inst. for Appl. Syst. Anal., A-2361 Laxenburg, Austria, WP-88-026, 28 p.

Bogardi I.,L. Duckstein and A. Bardossy, 1984 : Tradeoff between cost and efficiency of pollution control, Proceedings of the VIth Int. Conf. on Multiple-Criteria Decision Making, Cleveland ,Ohio, June 1984.

BROOKSHIRE D.S., L.S. EUBANKS and C.F. SORG, 1986 : Existence values and Normative Economics : implications for valuing Water Resources, Water Resour. Res., Vol.22, 11, pp. 1509-1518

CEE, 1983 : Politiques et Stratégies pour une utilisation rationnelle des ressources en eau de la région de la CEE, ECE/WATER/31, Nations Unies, N.Y., 39 p.

CEE, 1985 : Politiques de Gestion Intégrée des Ressources en Eau, ECE/WATER/38, Nations Unies, N.Y., 48 p.

Chalot F., 1988 : L'Arrêté tant attendu Valorisation Agricole des Boues d'Epuration, Génie Rural n°11, pp. 20-25.

CHANGKONG V. and Y.Y. HAIMES, 1982 : Multiobjective Decision Making Theory and Methodology, 472p., N.Y., Elsevier North Holland

CHANGNON S.A., Jr, 1987 : An assessment of Climate Change, Water Resources, and Policy Research . Water International, Urbana, Illinois, USA, Vol. 12, n°2, pp. 69-76.

COHON J., 1979 : Multiobjective Programming and Planning, N.Y., Academic Press.

DE BACKER L.W., 1982 : Pre-Management water resources study. in P. Laconte and Y.Y. Haimes (ed), Water Resources and Land-Use Planning : A Systems approach, pp.53-67, The Hague, Martinus Nijhoff.

DE BACKER L.W., ANSOULT M. and MALLE K., 1986 : Management concepts and manager roles. Water International, Urbana, Illinois, USA, Vol. 11, n°4, pp. 146-149.

DE BACKER L.W. and W.A. de BRUYN, 1987 : Well-being and Integrity. Water International, Urbana, Illinois, USA, vol. 12, n°3, pp 149-155.

De L'île R., 1988 : Hôtel Solvay - une maison est un concerto. Expression, Bruxelles, 4/88, pp. 94-99.

Dooge J.C.I., 1986 : Looking for Hydrologic Laws, in Trends and Directions in Hydrology, Water Resourc. Res. 22, (9), 46S-58S.

Hartman L., 1987 : Methodological Guidelines for the Integrated Environmental Evaluation of Water Resources Development.,152 p., UNESCO, Paris.

Hopps H.C. and G.L. Feder, 1986 : Chemical qualities of water that contribute to human health in a positive way. The science for the Total Environment, pp. 207-216, Amsterdam, Elsevier.

ISO/TC-190, 1986 : Soil Quality Technical Committee Secretariat, NNI, Delft, the Netherlands. ISO, C.P. 56, CH-1211 Geneva 20 .

JENSEN L.J., 1986 : Risks and Drinking water. Speech to American Water Work Association, June 25, 1986. US EPA, Washington D.C.

POSTEL S., 1984 : Water : Rethinking Management in an age of scarcity, Paper 62, 65 p., Washington D.C, Worldwatch Institute.

POSTEL S., 1985 : Conserving water : the untapped alternative, Paper 67, Washington D.C., Worldwatch Institute, 66 p.

Priscoli J. D., 1987 : Social Science, Engineering and Water Resources Management : a perspective. in The role of social and behavioral sciences in water resources planning and management; Ed. Baumann and Haimes, Amer.Soc.Civ. Eng.,New York, pp.5-39.

Stockholm Group, 1988 : Perspectives of Sustainable Development, Some critical issues related to the Brundtland Report, Stockholm Studies on Natural Resources Management N°1, Univ. of Stockholm,115 p.

Taube M.,1985 : Evolution of Matter and Energy, 289p., N.Y., Springer-Verlag

TIME, 1989 : Planet of the Year - Endangered EARTH, Time Magazine, Vol.133, n° 1, pp.12-47.

UICN-PNUE-WWF, 1980 : Stratégie Mondiale de la Conservation, CH-1196 Gland, Suisse.

Land and Water Use, Dodd & Grace (eds), © 1989 Balkema, Rotterdam. ISBN 90 6191 980 0

Modelling soil-water processes to evaluate environmental pollution

J.W.Biggar & T.X.Lu
University of California, Davis, Calif., USA

ABSTRACT: It is becoming evident that concern for the quality of groundwater is receiving more emphasis in many regions of the world. Agricultural production may be identified as a source of pollution during recharge of groundwater. Management strategies to minimize contamination include models that predict the quality of percolating waters. Such a model is described in this paper including details of the hydrologic processes involved and soil and crop data from a field experiment. A dual line source sprinkler system supplies waters of two qualities and a gradient of salinity. Crop removal of water and chemical processes that occur in the root zone yield different qualities of drainage water. Chemical reactions involved during transport include exchange, precipitation and dissolution. Predicting the quality of drainage water will help decision makers to assess the impact on groundwater quality and improve management of water resources.

Résumé: L'interêt porté à la qualité des eaux de la nappe phréatique grandit dans bien des régions du monde. La production agricole est maintenant reconnue comme une source de pollution durant l'alimentation des eaux souterraines. Les techniques d'aménagement pour réduire la contamination incluent la modèlisation pour prédiré la qualite des eaux d'infiltration. Le modèle presente ici inclut les détails des processus hydrologiques étudiés ainsi que les données sur les sols et les plantes étudiés dans la parcelle expérimentale. Deux systèmes d'arrosage par aspersion fournissent des eaux de deux qualités différentes et deux niveaux de contenu en sels. L'utilisation de l'eau d'arrosage par les cultures et les réactions chimiques dans la zone racinaire affectent différemment la quantite d'eau de percolation. Les réactions chimiques d'interêt incluent le transport et l'échange ionique ainsi que les processus de précipitation et de dissolution. La prédiction de la qualité des eaux de drainage des terrains agricoles va aider les resposables de l'aménagement à évaluer leur impact sur la qualité des eaux souterraines et a` améliorer l'utilisation des ressources en eau

Kurzfassung : Es wird offensichtlich, dab die Sorge um die Grundwasser-beschaffenheit in vielen Gebieten der Welt mehr und mehr wächst. Die land-wirtschaftliche Produktion wurde vielerorts als Verunreinigungsquelle des sich neu bildenden Grundwassers identifiziert. Bewirtschaftungsstrategien zur Kontaminationsminimierung beinhalten Modelle, die die Beschaffenheit des Sickerwassers prognostizieren. Solch ein Modell wird in vorliegendem Beitrag zusammen mit den mabgebenden hydrologischen Prozessen und den Boden- und Produktionskennwerten eines Feldtests beschrieben. Ein Bereg-nugssystem als Doppellinienquelle dient der Wasserzufuhr mit zwei unter-schiedlichen Beschaffenheiten und einem Gradient der Salzkonzentration. Der Pflanzenwasserentzug und chemische Prozesse, die in der Wurzelzone auftreten, verursachen die unterschiedliche Beschaffenheit des Drainwas-sers. Chemische Reaktionen, die beim Transport auftreten, schlieben den Austausch, die Fällung und Lösungsprozesse ein. Die Vorhersage der Drainwasserbeschaffenheit hilft den Entscheidungsträgern den Einflub auf die Grundwasserbeschaffenheit einzuschätzen und die Bewirtschaftung der Wasserressourcen zu verbessern.

1 INTRODUCTION

It is becoming more evident with time that concern for the quality of groundwater is receiving more attention in many regions of the world. While some effort has been directed for a number of years to predict the quantity and direction of flow of groundwater, the quality aspects have received less attention. This increased interest has developed as new sources of surface water are no longer available or are too expensive to develop with population increases. A second factor that has stimulated interest is the recognition that solid and liquid waste disposal practices have lead to localized dangerous water contamination problems that expand with time and are more difficult to rectify than surface water pollution problems. The concept that the land as a living filter is capable of purifying wastes regardless of whatever their constitution has, in some cases, lead to irresponsible waste disposal practices.

The natural recharge of groundwater through the land surface in regions of agricultural production has in recent years been identified with contamination of underlying groundwater. In some areas , the pollutant is nitrogen resulting from excessive use of fertilizer or excessive application of animal wastes. In areas where water is applied for the production of crops, excessive salt left in the soil as a result of water extraction by plants poses a difficult problem. Leaching the salt out to maintain a healthy environment for plant growth leads to salt loading of the groundwater. In order to develop management strategies to minimize contamination of groundwater from many different sources containing various constituents, models that are capable of revealing the quality of the drainage water and the disposition of constituents in the root zone and vadose zone would be most useful. Such models for nitrogen disposition both on a local and regional basis have been reported (Shaffer and Gupta 1981 and Mehranetal 1981). Here a model is described that is useful for predicting localized disposition of constituents in the root zone and beyond. It includes hydrologic processes associated with water flow in soil and extraction by plants and chemical processes that determine the fate of dissolved constituents. A field experiment that includes water of various qualities provides data that serves as input for model predictions of drainage water quality.

2 THE MODEL

The computer model LEACHMS (Leaching Estimation and Chemistry Model, Salt) utilized in this study was developed by R. J. Wagenet and J. L. Hutson (1987) after extensive modification of one by Robbins etal (1980). It is a later generation process based model that includes soil, water, plant and chemistry processes for prediction of transient as well as steady state conditions in field soils. The program is organized on a modular basis consisting of a main program which reads data, calls subroutines and prints outputs. Subroutines for the salt model manage the time step, water flow, solute movement and chemical reactions, evapotranspiration and leaf and root growth.

Inputs that are required, and in the present case come from a field experiment to be discussed later include:

(1) soil properties and initial conditions with depth; (2) hydrologic constants for calculating moisture characteristics and hydraulic properties; (3) water content or water potential.

Surface boundary conditions to be input are sources of water whether by rainfall, irrigation or both, including frequency, duration and rate on a daily basis, and pan evaporation. For those times when crops are present, subroutines that include plant factors can be called. Crop factors include soil water potentials that control water extraction by plants, days from planting to maturity, root growth models and parameters that describe canopy growth.

2.1 Soil profile

The profile is divided into segments that can be of variable thickness although in the present case 12 segments of 10 cm each are considered. That corresponds to the depth of profile sampled in the experimental phase. The flow of water and salt is simulated by finite differencing techniques with variable time intervals possible. This latter fact makes possible better simulation of rapidly changing events corresponding to infiltration as compared with longer periods of drying because of drainage or water extraction by plants. The principle governing flow equation for transient one dimensional vertical flow of water used in the model is

$$\frac{\partial \theta}{\partial \tau} = \frac{\partial}{\partial z}\left(K(\theta) \bullet \frac{\partial H}{\partial z}\right) - z\ u(z,t) \quad (1)$$

where θ is volumetric water content, H is Hydraulic head, K is hydraulic conductivity, t is time, z is depth, positive downward, and u is a sink term representing water lost per unit time by transpiration. With the differential water capacity $C(\theta)$ defined by

$$C(\theta) = \frac{\partial \theta}{\partial h} \quad (2)$$

equation (1) is transformed into

$$\frac{\partial h}{\partial t} C(\theta) = \left(K(\theta) \bullet \frac{\partial H}{\partial z}\right) - u\ (z,t) \quad (3)$$

where h is the soil water pressure head.

The flux controlled boundary condition at the soil surface is changed with time to simulate infiltration of water or evaporation from the soil. In the present case the lower boundary is characterized as that for a freely draining profile with a hydraulic gradient approximating unity.

The conductivity and retentivity functions obviously play an important role in a soil-water-plant system and therefore in the simulation model.

The θ-h relationship utilized in the model is one proposed by Hutson and Cass (1987)

The conductivity function is based on that proposed by (Campbell 1974) and takes on two forms depending on whether the pressure head lies on the parabolic or exponential part of the θ-h curve.

For $h > hi$, the parabolic part of the curve,

$$K = K_s\ (\theta/\theta_s)^{2b+2+p} \quad (4)$$

and when $h < hi$

$$K = K_s\ (a/h)^{2+(2+p)/b} \quad (5)$$

K is hydraulic conductivity, K_s is saturated hydraulic conductivity, and p is a pore interaction parameter.

2.2 Solutes

As previously stated the solutes of interest in this investigation are those derived from irrigation water and agricultural chemicals. The movement of each constituent will be a function of various processes that control that constituent which includes convection and diffusion for all constituents, ion exchange for cations, precipitation and dissolution reactions for some cations and anions and adsorption for individuals such as boron.

The general transport equation for each constituent is

$$\frac{\partial(\theta c)}{\partial t} = \frac{\partial}{\partial t}\left[\theta D\ (\theta,q)\ \frac{\theta c}{\partial z} - qc\right] \quad (6)$$

where c is concentration of constituent, D is apparent diffusion coefficient accounting for diffusion and mixing processes due to variations in pore water velocity, and q is macroscopic water flux.

The concentration (c) in the liquid phase available for transport is calculated in other chemical subroutines that consider competitive ion exchange, for Ca, Mg, Na, K, solid phase $CaCO_3$ and $CaSO_4$, the influence of CO_2 partial pressure, and the effect of ionic strength of the solution on the various reactions. Ion pairs considered significant are $CaCO_3°$, $CaHCO_3^+$, $CaOH^+$, $CaSO_4°$, MgCO3°, $MgHCO_3^+$, $MgOH^+$, $MgSO_4°$, $NaSO_4^-$ and $NaCO_3^-$. For example, if the total calcium in solution is given by

$$Ca_T = Ca^{2+} + CaHCO_3^+ + CaOH^- + CaCO_3^\circ + CaSO_4^\circ \quad (7)$$

then the activity of Ca in solution (Ca) is

$$(Ca) = Ca_T \div \left[\frac{1}{\gamma^2} + \frac{Ka\,KH\,PCO_2}{(H)\,Kd2\,\gamma_1^2} + \frac{Kw}{(H)\,Kd3\,\gamma_1} + \frac{Ka1\,Ka2\,KH\,PCO_2}{(H)\,Kd1} + \frac{(SO4)}{Kd4} \right] \quad (8)$$

The reader will recognize that the Kdi's are stability constants for given ion pairs, the K's are dissociation constants for the carbonic acid system and water, and γ's activity coefficients. The latter are calculated using the Davies equation. Similar expressions are written for the other soluble cations.

The assumption is made that the cation exchange capacity (CEC) is given by

$$CEC = X\,1/2\,Ca + X\,1/2\,Mg + XNa + XK \quad (9)$$

and a selectivity coefficient (K) for each exchangeable cation pair can be expressed as for example by

$$\frac{(Ca)^{1/2}\,X\,1/2\,Mg}{(Mg)^{1/2}\,X\,1/2\,Ca} = K_1,$$

$$\frac{(Na)\,X\,1/2\,Ca}{(Ca)^{1/2}\,X\,Na} = K_2 \quad (10)$$

for divalent pairs and divalent-monovalent pairs respectively. Here the cation activities are used instead of analytical concentrations and they are corrected for ion pairs and ionic strength.

Combining the expressions for CEC (9) and selectivity coefficients (10) an expression for each ion takes the form similar to that for exchangeable Na

$$X\,Na = \frac{CEC}{\left[\frac{Ca^{1/2}K2}{(Na)} + \frac{Mg^{1/2}K5}{(Na)} + \frac{K(K6)}{(Na)} + 1 \right]}$$

where K2, K5, K6 refer to selectivity coefficients for Na-Ca, Na-Mg, and Na-K, respectively.

The initial inputs to the chemical model are obtained from saturation extracts of the soil. In this experiment, both soluble and exchangeable ions were measured so that selectivity coefficients were calculated from data collected in the experiment and not assumed as is often the case for many field investigations. Additional measured input variables are CEC, pH, solid phase $CaSO_4$ and lime, Cl, SO_4, CO_3, HCO_3 and boron. Other inputs include bulk density and volumetric water content.

2.3 Plant growth

Simulation of solute flow from soil profiles through any given period may include times when both fallow and crop conditions exist. Here the simulation begins before the crop is planted and proceeds through to crop harvest. That entails preplant irrigation followed by progressive seeding, root development and canopy growth. Water loss is by evaporation, transpiration, and drainage with emphasis shifting as crop growth progresses.

Here the growth of the roots is described according to the model of (Tillotson etal 1980) and the progression of the crop cover as

$$CRPCOV = \frac{AKI}{1+cxp(6-12TIM/ESTOP-ESTART)}$$

which gives the crop cover at anytime as a fraction. AKI is the crop cover at maturity as a decimal fraction, ESTART is time from planting to seedling emergence (days), ESTOP is time from planting to crop cover maturity (days), and TIM is time from planting to present (days).

3 FIELD EXPERIMENT

In order to examine the spatial variability in the distribution of salinity and solute outflow during the growth of a crop irrigated by waters of different qualities, a dual line source sprinkler system was established on Panoche clay loam soil to irrigate a crop of sorghum (Morkoc etal 1985). The parallel irrigation lines were spaced 15 m apart so that the water from one line just reached the other. The sprinkler heads in each line were closely spaced at 4.3 meters to provide significant overlap in order to apply water as uniformly as possible along any transect parallel to the lines. Simultaneously low salt (LS) water was applied through one line and high salt (HS) water through the other. As a result the depth of water applied between the lines was uniform while a gradient in salt application was simultaneously achieved. In effect the water quality varied across the central zone of irrigation. Of particular interest here is the quality of water leaving the profile as drainage water. After a second summer of irrigation of a sorghum crop, a winter crop of wheat was grown with the winter rain as a source of water. Following removal of the wheat, the soil was sampled with a hydraulic sampler that took cores with a diameter of 0.05 m to a depth of 120 cm. These were sectioned at 0-.15, .15-.30, .30-.60, .60-.90, .90-1.20 m. The samples were air dried, ground, mixed and subsampled. Saturation extracts for cations and anions, electrical conductivity (EC), pH, were obtained. Cation exchange capacity (CEC), gypsum and lime were measured on separate samples.

The sampling scheme provided five samples at each depth in each plot. These data were input for simulations of the solute outflow to be reported here for a subsequent crop of sorghum irrigated with water of quality given in Table 1. In addition to the output expected from the actual irrigations during the growing season, other simulations are presented. Irrigation water applied was measured in each line as was the quality of the water. A weather station nearby provided the necessary data and the hydrologic properties of the soil were obtained from other sources (Nielsen, Biggar and Erh 1973).

4 FINDINGS

The results of the model predictions presented in the following graphs and tables are from inputs to the model of soil sample analyses taken prior to the first irrigation and subsequent application of several irrigations.

Table 1. Typical composition of irrigation waters used in the experiment. (LS) low salt; (HS1) high salt; (HS2) high salt.

	LS	HS1	HS2
		mol/m^3	
Ca	3.6	10.0	9.5
Mg	2.1	6.6	8
Na	7.5	39.2	53
Cl	2.5	13.0	16
SO4	4.9	25.8	35
HCO3	1.7	2.2	2.5
SAR	3.2	9.6	12.6
EC(mS/m)	1.6	5.5	6.9

The plot numbers refer to the location of the plot area between the two outer boundaries. All plots reported here are located between the line sources and would receive some water from each source, but equal total amounts within the uniformity of the sprinkler system. Consequently, Plot 15 is at the low salt line and receives its water from that source with the percentages changing as the numbers increase to Plot 30 at the high salt line. Plots in between receive proportional amounts from each line depending on location.

Table 2 presents the calculated accumulative mass of each constituent drainage water removed during the growing season.

Table 2. Constituents leached from the soil profile from pre-irrigation to harvest using waters LS and HS1.

	Plots					
	18		22		26	
	Constituents (g/m^2 x 10^{-3})					
	N	N+20	N	N+20	N	N+20
Ca	597	2194	582	2086	636	2194
Mg	184	673	366	1381.5	192	685
Na	891	3199	1018	3605	911	3305
Cl	161	548	267	722	286	798
SO_4	1174	4164	1451	5058	1272	4298
Water (mm)	47	171	48	172	51	174

Plots 18, 22 and 26 correspond to low salt to high salt regions respectively. Columns in Table 1 labelled N correspond to actual amounts of water applied, which was 717 mm for the season. Those labelled N+20 are predictions when water applied was increased by 20 percent. For Plot 18 the predicted drainage is 47 mm or 7 percent of that applied. Notice that the removal of each constituent is not directly proportional to the increase in water applied. The reason for this is associated with the combined effects of water extraction by the plants corresponding to wetting and drying periods in the soil profile, and the interaction of competitive ion exchange and dissolution-precipitation reactions associated with lime and gypsum or approximately 7 percent leaching fraction.

The cumulative loss of Ca is slightly increased for N treatments progressing from the lower Ca water applied at Plot 18 to the higher Ca water at Plot 26. However, there is no change for the N+20 cases. By contrast Mg loss proceeds through a maximum at Plot 22 both for N and N+20 water treatments. To a lesser extent Na shows a similar pattern. Sodium can apparently compete quite well with Mg especially when the applied Na concentration is quite high as it is on the high salt side.

The role of the anions should not be overlooked relative to salt storage and outflow. Chloride, the least reactive of the constituents, is very mobile but will be constrained by charge balance considerations. A large increase in outflow occurs between Plot 18 and 22 but with little increase from 22 to 26. Sulfate is more complicated as might be expected because of its solubility product limitations with Ca. The pattern of SO_4 is similar to Mg and Na with a maximum occurring in Plot 22. As the high salt side is approached, increased precipitation of gypsum occurs and the outflow of Ca and SO_4 is restrained.

Table 3 presents the concentration of constituents in drainage water at the 1.05 m depth for three plots, the initial and final concentrations corresponding to quantities of irrigation water of 717 and 860, the latter representing 20 percent additional water. The electrical conductivity (EC) and sodium adsorption ratio (SAR) are widely used quality parameters for drainage and irrigation water (Ayers and Westcot 1985).

Table 3. Concentration of constituents in drainage from the 1.05 m depth for 3 plots irrigated with LS and HS1 waters.

Plot	Ca	Mg	Na	Cl	SO_4	EC	SAR
	mol/m^3					mS/m	
18I[1]	12.6	3.9	18.9	5.6	24.7	253	4.7
18F[2]	12.6	3.9	18.7	2.5	24.7	275	4.6
18F+20[3]	13.6	4.1	18.9	5.7	23.1	276	4.5
22I	12.4	8.6	21.6	9.7	28.9	284	4.7
22F	11.9	8.0	20.8	3.7	30.1	370	4.6
22F+20	12.8	8.1	21.6	6.2	27.8	363	4.7
26I	12.9	3.9	18.1	9.3	23.8	286	4.4
26F	12.4	3.8	18.0	4.1	25.1	287	4.5
26F+20	13.4	4.4	21.6	6.8	24	307	5.1

[1]I = before planting
[2]F = at time of harvest
[3]F+20 = 20 percent more irrigation water.

There is striking similarity in the parameters for all three conditions of plot 18 including EC and SAR. At plot 22 Mg (and to a lesser extent Na) concentrations are elevated above those of 18, but the changes are not large between initial, final and final +20 for the cations. Chloride and sulfate both show the impact of more high salt water in the mixtures added at plot 22. However, at the high salt side represented by plot 26, the concentration of most ions and other parameter are again similar to 18, a condition not expected if one compares the water qualities in Table 1.

In Table 1 a second high salt water (HS2) is included, it being somewhat elevated in sodium and sulfate above HS1. Although not shown, the mass outflow of constituents and concentrations in the drainage water from the LS and HS2 combination were not elevated above the LS and HS1 combination, as one might expect.

The expected changes are considerably moderated by what is occurring in the profile which is demonstrated in figures 1 and 2 with the parameter EC and SAR from Plot 26 as examples.

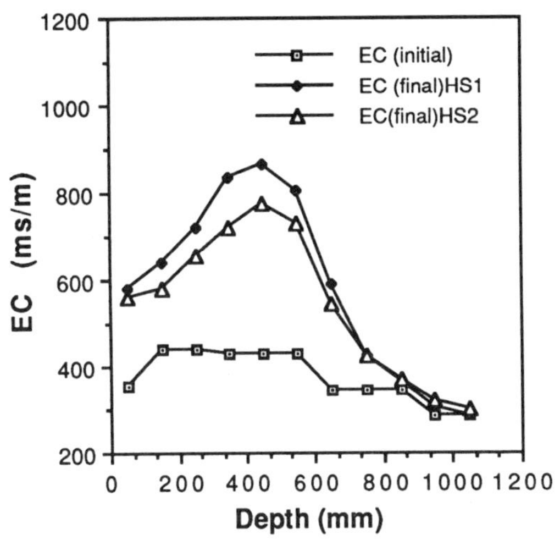

Fig. 1 Calculated electrical conductivity of the soil water in Plot 26 irrigated with two high salt waters.

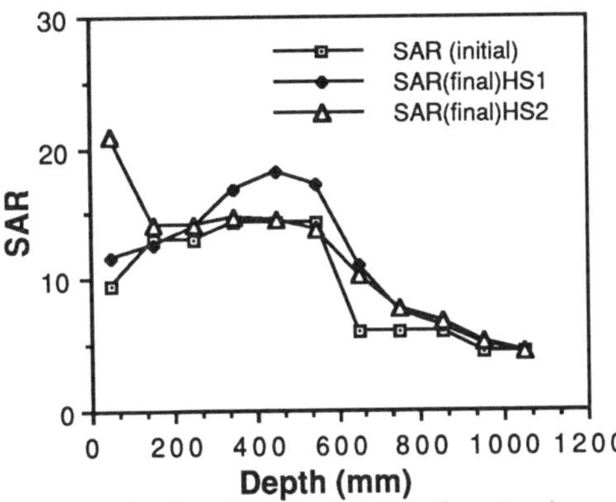

Fig. 2. Calculated SAR of the soil water in Plot 26 irrigated with high salt waters.

In each figure initial represents the profile condition before planting and final refers to the condition at harvest. The moderate changes that have occurred in both salt outflow and concentrations of constituents are the result of significant changes in the profile itself which has buffered the drainage. That is also true for the effect of Na which is quite elevated in water HS2.

Increasing the quantity of irrigation water applied by 20 percent increases the application of salt and expected outflow of water and salt. However the concentration was not greatly affected and again this is primarily due to changes taking place in the profile.

The program LEACHMS demonstrates considerable potential as a tool predicting changes in drainage outflow and corresponding profiles changes. Input data can be quite site specific or more general depending upon the degree of accuracy of the predictions desired. All of the major processes controlling quality of drainage outflow are included. The model has the advantage that it can be executed on a desktop computer with moderate storage and speed.

REFERENCES

Ayers, R. S., Westcot, D. W. 1985. Water quality for agriculture. FAO irrigation and drainage paper 29.

Campbell, G. 1974. A simple method for determining unsaturated conductivity from moisture retention data. Soil Sci. 117:311-314.

Hutson, J. L. and Cass, A. 1987. A retentivity function for use in soil-water simulation models. J. Soil Sci. 38:105-113.

Mehran, M., Tanji, K. K. Ishandar, I. K. 1981. Compartmental modeling for prediction of nitrate leaching losses. In modeling waste water renovation, I. K. Ishander, editor. John Wiley & Sons, Chichester.

Morkoc, F., Biggar, J. W., Miller, R. J., Nielsen, D. R. 1985. Statistical analysis of sorghum yield: A stochastic approach. Soil Sci. Soc. Amer. Journal 49:1342-1348.

Nielsen, D. R., Biggar, J. W., Erh, K. T. 1973. spatial variability of field measured soil water properties. Hilgardia 42:215-259.

Robbins, C. W. Wagenet, R. J. Jurinak, J. J. 1980. A combined salt transport chemical equilibrium model for calcareous and gypsiferous soils. Soil Sci. Soc. Amer. Journal 44:1191-1194.

Shaffer, M. J., Gupta, S. C. 1981. Hydrosalinity models and field validation. In modeling wastewater renovation, I. K. Ishandar, editor. John Wiley and Sons, Chichester.

Tillotson, W. R., Robbins, C. W., Wagenet, R. J., Hanks, R. J. 1980. Soil water, solute, and plant growth simulation. Bulletin 502. Utah State Agr. Exp. Sta., Logan, UT. 53 pp.

Wagenet, R. J. and Hutson, J. L. 1987. LEACHM. Leaching Estimation and Chemistry Model. Water Resources Institute, Center for Environmental Research, Cornell University, Ithaca, NY.

Land and Water Use, Dodd & Grace (eds), © 1989 Balkema, Rotterdam. ISBN 90 6191 980 0

A biotechnological method for elimination of environment polluting hydrocarbons

F.Buday, Z.Gergely, G.Török & S.Szoboszlay
University of Agricultural Sciences, Gödöllő, Hungary

B.Csaplár
Filantrop Service Company for Environment Protection, Kecskemét, Hungary

Z.Szabó
National Institute of Hygiene, Budapest, Hungary

ABSTRACT: The objective of this study was to develop a biotechnological method for the elimination of hydrocarbon contaminations in environment, especially in soil. Therefore hydrocarbon decomposing microorganisms were isolated from oil polluted soil samples which were capable to decompose 80-85 % of oil added to the culture medium in five days. The most effective cultures were mixed cultures with the predominence of Nocardia, Pseudomonas, Flavobacterium and Candida species. After laboratory experiments, inoculation with a starter culture prepared from hydrocarbon decomposing microbe-strains has been found to be effective in small plot experiments and under large-scale application, too. The method was used in case of "Havaria" /an accident due to a pipe-line fracture/. It was also applied for oily waters and sludges by land farming and composting with farmyard manure. According to our investigations there was a 98 % decrease in oil content of surface level in case of the "Havaria" over less than one year. Land farming and composting gave similar results.

Le but de l'étude est la mise au point d'une procédé biotechnologique pour éliminer les salissures d'hydrocarbures dans la nature et surtout dans le sol. A partir des échantillons de sol souillés par huil, nous avons isolé tels microorganismes décomposants les hydrocarbures, qui sont conformes à décomposer le 80-85 % de l'huil additionné à la base nutritive, pendant 5 jours. Les plus éfficaces étaient les propagations mixtes, dans lesquelles ont dominé les espéces Nocardia, Pseudomonas, Flavobacterium et Candida. Après les éssais de laboratoire, l'inoculation par la starterculture préparée des souches décomposantes les hydrocarbures, s'avérait éfficace à l'application sur les petites parcelles et aussi en mésure d'usine. La méthode était utilisée en cas de la salissure à la cassure du tuyau. Nous l'avons appliqué pour la pose des eaux et vases oléagineuses sur le sol, et aussi pour composter avec fumure. Selon nos expériences, en cas de la salissure à la cassure du tuyau, dans les couches près de la surface, le 98 % du teneur en huil s'était décomposé pendant moins d'une année. Le "land farming" et la procédé à composter ont donné les mêmes résultats.

Ziel der Arbeit ist es, ein Verfahren auszuarbeiten, dass zur Eliminierung der in der Umwelt, vor allen dingen im Boden vorkommenden Kohlenwasserstoffverschmutzungen dient. Aus ölverschmutzten Bodenproben isolierten wir solche Kohlenwasserstoff abbauende Mikroorganismen, die fahig sind, zum Nahrboden zugegebenes Öl innerhalb von 5 Tagen zu 80-85 % abzubauen. Am wirksamsten waren die gemischeten Kulturen, in denen die Arten Nocardia, Pseudomonas, Flavobacterium und Candida dominierten. Nach den Laborversuchen wurden mit den aus den Kohlenwasserstoff abbauenden Stammen bereiteten Starterkulturen Beimpfungen vorgenommen, die sich sowohl auf Kleinparzellen, als auch im gorssbetrieblichen Masstab als erfolgreich erwiessen. Die Methode wendeten wir bei "Havariefallen" /Umweltverschmutzung infolge von Rohrbrüchen/ an. Wir nutzten sie auch beim Versprühen von ölhaltigem Wasser oder Schlamm auf Böden, sowie bei der Kompostierung mit Dung. Nach unseren Untersuchungen wurde in "Havariefallen" in den oberflachennahen Bodenschichten in weniger als einem Jahr 98 % des Ölgehaltes abgebaut. "Land farming" und Kompostierung brachten ahnliche Ergebnisse.

1 INTRODUCTION

One of the most important problems in the second part of this century is environmental pollution, especially the hydrocarbon contamination /resulting from oil spill by "havaria", various kind of oily waters and oil sludges/. As it is known, the oil penetrating into the soil contains mutagenic, carcinogenic and growth inhibitory materials which affect adversely the water and terrestrial ecosystems /Chet and Mitchell, 1976./

Since decontamination of oil pollution by the traditional methods is expensive and difficult to realize the use of microorganisms seems to be an adequate approach. Isolation and application of hydrocarbon decomposing starter cultures in case of oil polluted waters have been reported by several authors /Atlas and Bartha, 1972, 1973; Bartha and Atlas, 1977/, but there is little information on the use of microorganisms for decomposing oil contaminations in soil. Biodegradability of hydrocarbon pollution in soil has also been studied by others /Parkinson, 1973; Jobson et al., 1974/. They have reported that addition of oil utilizing bacteria as well as fertilizers to the soil-microflora exerted a favourable effect on the intensity of oil degradation. According to the experiments the NPK fertilizer supplements added to the soil increased the bacterial count and accelerated the degradation of oil and regeneration of vegetation. Some investigators /Dibble and Bartha, 1979/ have reported a case history at New Jersey. An estimated 1,9 million litres of fuel oil spilled out of an underground pipeline when it broke spraying oil over approximately 1,5 ha of a wheat field. After cleaning of surface oil a rehabilitation program was applicated consisting of liming, fertilization and tilling. They have stated that the hydrocarbon content of surface soil decreased to an insignificant level 2 years after the spill.

In our study we have attempted to isolate such hydrocarbon decomposing microorganisms which are suitable for producing a starter culture for the inoculation of oil polluted soil and capable for the elimination of oil pollution from the soil.

2 MATERIALS AND METHODS

2.1 Isolation and identification of hydrocarbon decomposing microorganisms.

The hydrocarbon decomposing microorganisms were isolated from soil and sludge samples collected from oil polluted areas.

The isolation of hydrocarbon decomposing microorganism was carried out on Winogradsky's kieselguhr plates saturated with a liquid medium containing $/NH_4/_2SO_4$ 4,0 g, $CaCl_2$ 2,0 g, $MgSO_4$ 0,5 g, KH_2PO_4 0,5 g $FeSO_4$ 0,1 g, $CaCO_3$ 3,0 g, soil extract 10 ml, tap water 1000 ml, pH=7,0. 1 ml crude oil was added to each petri dish. The inoculation of the kieselguhr plates was carried out with 1 ml of watery suspension prepared from oil polluted soil and sludge samples, the plates were then incubated for 5 days at 28°C. The cultures grown on the surface of the kieselguhr plates, were also inoculated onto kieselguhr plates, then passed on a liquid medium in shake cultures.

The hydrocarbon decomposing activity of the isolated mixed cultures was tested in 500 ml Erlenmeyer flasks, so as to add 150 ml liquid medium into each flask then 1,2, 4 ml crude oil, gasoline or mixed oil per flask. The sterile liquid medium in each flask was inoculated with 5 ml pre-inoculum, then incubated in shake cultures for five days at 28 ^{o}C. After five days shaking the amount of non-decomposed oil in culture medium was determined and the rate of oil degradation calculated.

From the most effective mixed-cultures, by pour-plate method 100 strains were isolated and identified by Bergey's Manual /1975/ and Barnett et al. /1983/.

In addition to the freshly isolated oil degrading strains, 21 cultures were inculded in study, serving as reference cultures. These included: Nocardia opaca CBS 4276, Nocardia erithropolis CBS 4277, Nocardia corallina CBS 14347, Nocardia blackwellii CBS 27253, Nocardia salmonicolor CBS 19657, Nocardia rosea AEMT, Nocardia rubropercincta CBS 14352, Pseudomonas aeruginosa ATCC 10145, Pseudomonas fluorescens ATCC 13525, Pseudomonas maltophilia ATCC 13637, Pseudomonas putida ATCC 12633, Bacillus cereus var mycoides ATCC 6462, Bacillus subtilis ATCC 6051, Micrococcus luteus ATCC 4698, Micrococcus roseus ATCC 418, Enterobacter aerogenes ATCC 13048, Klebsiella aerogenes NCTC 8172, Escherichia coli AEMT, Flavobacterium sp. AEMT, Acinetobacter Lwoffii AEMT, Candida lipolytica CBS 2787.

2.2 Effect of inoculation with oil decomposing starter culture.

The starter culture used for the inoculation of the oil polluted soil was prepared from the most effective mixed cultures by non-sterile fermentation on the spot. The microbe-suspension used for inoculation was sprayed on the oil polluted soil by a dispersing-machine at a rate of 5 l/m^2. The starter culture added to the soil was a mixture of various strains containing 3.5x 10^9 germs/ml. After inoculation, the microbial suspension was mixed with the soil by a mixing disk.

To establish the rate of oil degradation soil samples were taken from the oil polluted area prior to inoculation and subsequently at regular intervals and the oil content was determined. The amounts of oil decomposed by microorganisms were calculated from the quantities of oil recovered from the soil samples. The soil samples taken from different sites by the aid a probe were homogenized and an average sample was dried at 105 ^{0}C to constant weight. To determine the amounts of residual oil content, 50 g aliquots of the average soil sample were extracted with ether for 2 hours in a Soxhlet extractor and subsequently dried to constant weight at 105 ^{0}C.

3 RESULTS AND DISCUSSION

3.1 Isolation and identification of hydrocarbon decomposing microorganisms.

In the first part of our experiments, 14 hydrocarbon decomposing microbe cultures /mixed-cultures/ were isolated. The oil-degrading activity of the isolated cultures was tested and stated that the most effective cultures /5 mixed cultures/ were capable to decompose 80-85 % of oil added to the culture medium, within 5 days under laboratory conditions.

From the mixed cultures having the most effective hydrocarbon decomposing activity, 100 hydrocarbon decomposing strains were isolated. The isolated strains were sorted into 7 groups according to morphology, staining and pigmentation and identified. /Table 1./

Table 1. Identification of hydrocarbon decomposing microorganisms, as indicated by relationship to reference strains and comparison with description given in Bergey's Manual of Determinative Bacteriology.

Taxonomic group	Isolates %
Pseudomonas spp	52
Nocardia spp	26
Flavobacterium spp	8
Micrococcus spp	5
Enterobacteriaceae	4
Yeast /Candida/	3
Unidentified strains	2

As table 1 shows in the microbe community organized in the oil polluted habitat, 7 taxonomic groups have found their essential conditions, with about 80 % contribution of Pseudomonas, Nocardia and Flavobacterium genera.

3.2 Elimination of oil contamination in soil using hydrocarbon decomposing starter culture.

After laboratory experiments a starter culture was prepared from the most effective mixed cultures /in which Pseudomonas, Nocardia and Candida spp. dominated/ for inoculation of oil polluted soil. The effect of inoculation was examined in small plot and large-scale experiments.

Small plot experiments have been carried out in Gödöllő and Kecskemét with a starter culture consisting of the most effective hydrocarbon decomposing microbes. According to the results, 80 and 83 % of oil pollution was decomposed in a year in these areas, compared with a decrease of 36 and 37 %, respect., in the control, uninoculated plots.

The large-scale application of the oil decomposing starter culture has been carried out in the region of Kékcse on an area of several hectares covered by oil due to a pipeline rupture /Buday et.al., 1988/. After applying adequate agrotechnics 2,5 ha were inoculated with the starter culture. According to the results 99 % of the oil contamination was decomposed in the 0-20 cm soil layer in a year. We have stated that the hydrocarbon degradation is an aerobic process and the rate of decomposition shows a proportional decrase with the depth of the soil. The corresponding rates of decomposition obtained for 0-20, 20-40 and 40-60 cm soil layers are 99,71 and 26 %, respectively. After 7 years the oil contamination decrased to ca 0,1 % in the upper soil layers and it decreased to 13 and 27 % respectively, in the deeper layers. Beside measurable changes in oil content, regeneration of the soil has been evidenced by the renewal of the flora and fauna of the inoculated area. Eight months after the inoculation the undergrowth has commenced to grow normally in the microbe--trated area, whereas no signs of renewal have been noted in the uninoculated area.

It is noteworthy, that in Hungary not the casual havarias seem to be the most dangerous problem but rather the oil-polluted waters and sludges due to activities of various factories. We have therefore developed a process for decontamination of oily waters and sludges applying our bio-

Fig.1 Inoculum producing fermentor for starter culture with a working volume of 8 litres.

Fig.2 Non sterile fermentor producing the starter culture with a working volume of 600 litre (developed from a duplicator).

Fig.3 Maize crop on the inoculated area.

Fig.4 Maize crop on the control area.

technological method. To our process at first the oil was separated from the oily waste-water, then the water acceptable for irrigation was transferred into canals. The oily part together with sludges, mixed with peat and supplemented with NPK additives, was applied into well-isolated areas in 20-40 cm layer and inoculated. The starter culture was produced by non-sterile fermentation on the spot. The results of biodegradation is presented in Table 2.

Table 2. Variation in oil content of soil samples taken from plots /polluted with different sort of oil/ after inoculation with oil degrading starter culture.

Date of sampling	A		B		C	
	1	2	1	2	1	2
Nov.	7,56	-	8,42	-	3,45	-
Márc.	5,24	30,7	6,26	25,7	2,64	23,5
Jun.	2,94	61,1	2,31	72,6	0,82	76,2
Sept.	1,25	83,5	0,72	91,5	0,35	89,9

A= plot with gasoline tank oily sediment
B= plot with crude oil tank sediment
C= plot with oily sludge
1= oil content in 100 g sediment/sludge
2= oil degradation in %.

As Table 2 shows this process was successfull and 80-90 % of the oil has been decomposed in a year. It is also clear from the table that rate of oil degradation is less intense during the winter due to lower temperature. We have therefore searched for the possibilities enabling intense oil degradation during the winter months. According to our preliminary experiments combination of this process with composting by farmyard manure seems to be a good approach; the time of oil biodegradation can be reduced to 4-5 months.

Based on our experiments, the biotechnological process presented here can be succesfully applied for the elimination of various oil contamination. As a result of inoculation with hydrocarbon degrading starter culture the environment polluting hydrocarbons were decomposed in short time. The microorganism of starter culture producing growth stimulating materials and having antagonistic effect on the plant pathogenic fungi, influence favourably the regeration of the vegetation on the oil--polluted areas.

4 REFERENCES

Atlas, R.M. and R. Bartha 1972. Degradation and mineralization of petroleum by two bacteria isolated from coastal water. Biotechnol. Bioeng. 14:297-308.

Atlas, R.M. and R. Bartha 1973. Fate and effects of oil pollution in the marine environment. Residue Rev. 49:49-85.

Barnett, J.A., R.W. Payne and D. Yarrow /eds./ 1983. Yeasts: Characteristics and identification. Cambridge: Cambridge Univ. Press

Bartha, R. and R.M. Atlas 1977. The microbiology of aquatic oil spills. Adv. Appl. Microbiol. 22:225-266.

Buchanan, R.E. and N.E. Gibbons /eds./ 1975. Bergey's Manual of Determinative Bacteriology. 8th ed. Baltimore: Williams and Wilkins Comp.

Buday, F., G. Török, Z. Gergely, S. Szoboszlay, Á. Szakács, I. Farkas, B. Csaplár and P. Biliczki 1988. A biotechnological process for the elimination of environment polluting hydrocarbons. In: R. Abbou /ed./, Hazardous Waste: Detection, Control, Treatment, p. 1657-1662. Proceedings of the Worl Conference on Hazardous Waste, Budapest, Hungary, October 25-31, 1987. Amsterdam: Elsevier.

Chet, I. and R. Mitchell 1976. Petroleum hydrocarbons inhibit decomposition of organic matter in seawater. Nature 261:308-309.

Dibble J.T. and R. Bartha 1979. Rehabilitation of oil-inundated agricultural land: a case history. Soil Science 128:56-60.

Jobson, A., M. McLaughlin, F.D. Cook and D.W.S. Westlake 1974. Effect of amendments on the microbial utilization of oil applied to soil. Appl. Microbiol. 27:166-171.

Parkinson, D. 1973. Oil spillage on microorganisms in Northern Canadian soils. Canada Environmental Social Commitee Northern Pipelines. Task Force on Northern Oil Development, Report No. 73-25. Information Canada, Cat. No. R 72-8573.

Land and Water Use, Dodd & Grace (eds), © 1989 Balkema, Rotterdam. ISBN 90 6191 980 0

A criterion for evaluating quality of irrigation water

G.Crescimanno
Dipartimento di Economia, Ingegneria e Tecnologie agrarie, Settore Idraulica, Università di Palermo, Italy

G.Giordano
Istituto di Genio Rurale, Università di Reggio Calabria, Italy

ABSTRACT: In this paper the Authors describe a criterion for evaluating water quality based on the Cass-Sumner and Maas-Hoffman models.These models allow the evaluation of the effects of using saline-sodic waters on soil structural stability and crop yield.
The Authors have applied the two models to some soil-crop combinations existing in Sicilian areas irrigated with reservoirs waters. The applications made so far have shown the importance of considering the hydrological characteristics of soil and the final characteristics of irrigation water in water quality evaluation.

RÉSUMÉ: Dans cette étude les Auteurs considèrent les modelès de Cass-Sumner et Maas-Hoffman pour l'évaluation de la qualité des eaux d'irrigation. Les deux modèles permettent d'évaluer les effets de l'utilisation des eaux salines sur la structure du sol et sur la productivité des cultures.
En particulièr on a effectué des applications en considerant des combinations sol-culture existant en Sicilie dans quelques zones irriguées par les eaux de deux réservoirs.Les résultats obtenus ont confirmé la nécessité de tenir en consideration les caractéristiques hydrologiques du sol et climatiques pour l'évaluation de l'aptitude de l'eau à l'irrigation.

ABSATZ: In der Erwähnung erklären die autoren eine Methode für die bewertung der Qualität von den Bewässerungswasser, bezug auf den Modelle von Cass-Sumner und Maas-Hoffman. Diese Modelle erlauben, die Ergebnise um die Struktur des Boden und um die Leistung des Anbau der Verwendung von Salz-Natrium-Wasser, zu abwägen.
Besonders erklären die Autoren eine Anbringung für einigen boden-anbau Zusammenstellungen in zwei Gagenden des Sizilien, interessiert für die Bewässerung mit Wasser in zwei Tanks. Die Ergebnise bestätigen die Badeutung der Hidroeigenarten der Wasser fur die Bewässerung um die bewertung der Qualität der Bawässerungwasser.

1 INTRODUCTION

As it is known, using saline-sodic waters for irrigation may have a strong influence on the structural and hydrological characteristics of soil and on crop yield. As a consequence the quantitative evaluation of these effects is very important.

Recently Cass and Sumner (1982c) have proposed a model based on the evaluation of the influence of these waters on the hydrological characteristics of soil and in particular on the saturated hydraulic conductivity Ks.

In order to apply this model it is necessary to determine experimentally the relationships between the saturated hydraulic conductivity Ks and the cationic concentration C of the irrigation water, for some fixed SAR values.

By these relationships it is possible to deduce the "threshold concentration curves" which are relationships between C and SAR for some fixed values of the hydraulic conductivity reduction ΔK.

By making these curves to intersect with the initial and final characteristics of irrigation water it is possible to evaluate the influence of using saline-sodic waters on soil structure stability.

A similar criterion has been also developed by Maas and Hoffman (1977) for evaluating the effects of saline-sodic waters on crop yield.

Combining the two models it is possible to obtain a criterion for evaluating quality of irrigation water.

In this paper the Authors, after introducing the theoretical principles of the two models, report the results of some applications carried out on some soil-crop combinations existing in Sicilian areas irrigated with waters from reservoirs. The purpose of the investigation is to evaluate the quality of these waters in relation to soil structure stability and crop yield.

2. THEORETICAL PRINCIPLES OF THE CASS-SUMNER AND MAAS-HOFFMAN MODELS

The Cass-Sumner model is based on the "threshold concentration curves" (Quirk and Schoefield 1955) which are relationships between SAR and C at some fixed ΔK values.

These curves represent, for a certain soil, the points C, SAR discriminating a condition of soil pore stability under the hypothesis that the structure stability is associated to the ΔK value.

These relationships have in general the following form:

$$SAR_t = a_t + b_t\, C_t \qquad (1)$$

The slope b_t is assumed as being representative of the structure stability condition; in fact, for a fixed C_t value, the stability threshold associated to the ΔK value depends on b_t, being generally the intercept a_t very close to zero.

Since b_t depends on ΔK, Cass and Sumner have proposed a model based on the relationship between the slopes b_t and the corresponding ΔK values (Cass and Sumner 1982a). This relationship is linear of the following type: $b_t = a_{Na} + b_{Na}\, \Delta K$

This equation is not associated to any particular ΔK value and so may be used to characterize soil structure stability and to make comparisons among different soils.

Let us consider, for a certain soil, a threshold concentration curve associated to a fixed ΔK value and suppose to represent this straight line in the plane C_t, SAR_t, together with the points C_{iw}, SAR_{iw} and C_{sw}, SAR_{sw} which represent the total cationic concentration C (salinity) and the sodium adsorption ratio (SAR) of the irrigation water (iw) and of the soil solution (sw) respectively.

These values, which can be deduced from chemical analyses of irrigation water and from available models (Cass 1980; Marion & Babcock 1976; Oster & Rhoades 1975), define, for a certain soil, the field of variability of soil solution salinity and sodicity represented by all the points falling into the rectangle having area A_t and vertices SAR_{iw}, SAR_{sw}, C_{iw} and C_{sw}.

For the meaning of the threshold concentration curve, the portion A_t of the rectangle area falling under the curve represents all the possible C, SAR combinations which do not cause hydraulic conductivity reductions greater than ΔK and as a consequence the soil pore structure can be considered stable.

The ratio between the portion A_t and the total area A_T provides a criterion for evaluating quantitatively the structure stability through the stability index:

$$SI = A_t / A_T \qquad (2)$$

SI varies between 0 and 1.

When SI=1 the rectangle defining the water composition lies interely under the threshold concentration line ($A_t = A_T$) and no reductions superior to ΔK take place.

On the contrary, when SI=0, $A_t = 0$ and reductions superior to ΔK take place. SI values ranging from 0 to 1 characterize all the other possible conditions of soil pore structure stability.

In order to avoid any arbitrary choice of the ΔK value discriminating the condition of soil structure stability, it is more correct to define SI by considering all the possible ΔK values ranging from 0 to 100%, that means considering the surface generated by the threshold concentration lines having ΔK as a parameter.

In this case SI is defined as follows:

$$SI = V_t / V_T \qquad (3)$$

where V_t is the volume of the water composition volume element which lies below the threshold concentration plane, and V_T is the total volume of the water composition element.

The water composition volume element is given by:

$$V_T = (SAR_{sw} - SAR_{iw}) \ (C_{sw}-C_{iw}) \ \Delta K_m = \\ = \Delta SAR \ \Delta C \ \Delta K_m \qquad (4)$$

where ΔK_m is the maximum value of ΔK reduction (equal to 100%).

An estimate of V_t may be obtained by integrating over n vertical planes through the volume element in the following manner:

$$V_t = \int A_t(i) \ d[\Delta K(i)] \qquad (5)$$

where i is the sequential number of vertical planes from zero to n along the ΔK axis, each of width $d[\Delta K(i)]$.

By considering equations (4), (5) and (6) we have:

$$SI = \int A_t(i) d[\Delta K(i)] / (\Delta SAR \Delta C \Delta K_m) \qquad (6)$$

and in finite terms:

$$SI = \sum_{i=1}^{n} A_t(i) d[\Delta K(i)] / (\Delta SAR \Delta C \Delta K_m) \qquad (7)$$

$$\text{Since:} \quad \Delta K_m = n \sum_{i=1}^{n} d[\Delta K(i)] \qquad (8)$$

equation (7) reduces to:

$$SI = 1/n \sum_{i=1}^{n} A_t(i) / (\Delta SAR \Delta C) \qquad (9)$$

A similar concept may be applied to evaluate the probability that a given irrigation water will cause a yield reduction ΔY for a specific crop.

Since ΔY is dependent on the water concentration C only, the functions to be considered, which can be represented in the C, ΔY plane, are:

1) the salt sensitivity curve, having the following form:

$$C_c = a_c + b_c \ \Delta Y \qquad (10)$$

2) the rectangle A'_T having height $C_{sw}-C_{iw}$ and base ΔY_m, where ΔY_m is the maximum yield reduction (100%).

Interaction of the salt sensitivity curve with this rectangle supplies the yield index:

$$YI = A_y / A'_T \qquad (11)$$

where A_y is the proportion of the total area A'_T which lies beneath the salt sensitivity curve.

3. MATHERIALS AND METHODS

The Cass-Sumner and the Maas-Hoffman models have been applied by considering two Sicilian reservoirs whose water is used for irrigation of soils for which the relationships $b_t(\Delta K)$ have been determined.

From the pedological point of view, the soils considered are vertisols (Prizzi reservoir) and alfisols (Poma reservoir).

The clay fraction in the two soils is rather elevated and equal to 49.0% and 48.4% respectively.

The relationships $b_t(\Delta K)$ have been deduced by using a battery operated constant head permeameter for measuring the saturated hydraulic conductivity by using solutions at fixed SAR values and decreasing cationic concentration C.

SAR values chosen have been 0,10 and 30 and concentration used have been 0,5 N, 0,05 N, 0,01 N and 0,005 N.

From the thresold concentration curves, having a rather linear form, the relationships b_t (ΔK) have been deduced.

In table 1 the chemical characteristics of the water considered are reported with reference to two different dates of sampling.

Water sampling and chemical analyses have been carried out by the Istituto di

Table 1- Chemical water characteristics and environmental factors

Chemical water characteristics									Environmental factors	
Reservoir	Date of sampling	EC [micromhos/cm]	Ca^{++} [meq/l]	Na^{+} [meq/l]	Mg^{++} [meq/l]	K^{+} [meq/l]	CO_3^{--} [meq/l]	HCO_3^{-} [meq/l]	E-P [mm]	H [mm/year]
Prizzi	27/4/83	580	4,10	1,40	0,00	0,00	2,00	3,40	285	100
Prizzi	14/7/83	470	2,70	1,40	0,90	0,00	1,00	2,80	285	100
Poma	15/2/82	720	3,00	2,43	1,80	1,00	1,20	1,40	252	100
Poma	1/9/82	700	2,50	1,14	2,40	0,40	ass.	2,44	252	100

Agronomia e Coltivazioni erbacee of Palermo University (Scarduzio 1984).

In order to determine C_{iw} [meq/l] and C_{sw} [meq/l] the following expression has been used (Marion and Babcock 1976):

$$\log C = -1{,}023 + 1{,}009 \ (\log EC)$$

by substituting to EC the electrical conductivity EC_{iw} and EC_{sw} of the irrigation water and of soil solution respectively, both expressed in mS/m.

SAR_{sw} and EC_{sw} values have been obtained by the following expressions (Cass 1980):

$$SAR_{sw} = (E/H)^{0,5} \ SAR_{iw} \ [1 + (pH_s - pH_c)$$

$$EC_{sw} = (E-P) \ EC_{iw}/H$$

where P and E are the annual values of rainfall and of maximum evapotranspiration, pH_s is the soil pH (usually assumed to be 8,4, the pH value of a nonsodic soil in equilibrium with calcium carbonate), pH_c is the pH water would adopt if in equilibrium with calcium carbonate; the term (pH_s-pH_c), the saturation index, indicates the tendency of carbonate to precipitate from irrigation water or dissolve in soil; the term E-P is identified with the irrigation water (Cass 1980).

Considering that the irrigation water in the Sicilian environment is generally supplied during the irrigation season, going from may to september, the Authors have thought it convenient, even for cautionary purposes, to consider the difference E-P with reference to this period.

4. RESULTS AND CONCLUSIONS

In table 2 the values of the soil stability index SI obtained for the waters of the Prizzi and Poma reservoirs are reported together with all the elements required for calculating these ones; in the same table the values of the crop yield index YI are also reported by considering some crops existing in the areas irrigated with the considered waters.

As it can be seen, the SI values obtained for the Poma reservoir are very close to 1, this result indicating no danger for structure stability of the considered soils at least in the short term period.

On the contrary the SI values obtained for the Prizzi reservoir are lower than the first ones, being equal to 0,86 (first date of sampling) and 0,68 (second date of sampling).

At this purpose it is important to observe that for the Prizzi reservoir SI values lower than for the Poma one have been obtained although the lower initial and final SAR values of the last one (first date of sampling).

Considering the similarity in the environmental characteristics of the two areas considered, the differences in the SI values obtained for the two reservoirs are due to the differences in the hydrological characteristics of the soils considered from the structure stability point of view (see the b_{NA} values).

As far as is concerned with the crop yield index, the YI values obtained are very close to 1, this result indicating no danger of crop yield reduction with reference to the crops considered.

In conclusion the results obtained seem to confirm the inadequacy of the criteria for evaluating water quality which are based on the water characteristics only (for example the USDA criterion) underlying the importance of considering the hydrological characteristics of soil and the environmental factors too.

Table 2- Inputs and outputs of the Cass-Sumner and Maas-Hoffman models

Reservoir	Date of sampling	C_{iw} [meq/l]	C_{sw} [meq/l]	Sar_{iw} [$\sqrt{mmol/l}$]	Sar_{sw} [$\sqrt{mmol/l}$]	a_{Na}	b_{Na}	SI	Crop	a_c	bc	YI
Prizzi	27/4/83	5,70	16,41	0,51	3,69	0,18	4,08	0,86	tomato	24,92	1,00	1,00
Prizzi	14/7/83	4,61	13,27	0,74	4,51	0,18	4,08	0,68	tomato	24,92	1,00	1,00
Poma	15/2/82	7,10	18,00	1,10	5,98	-0,36	65,40	0,97	grape	14,88	1,03	0,99
Poma	1/9/82	6,89	17,50	0,51	2,65	-0,36	65,40	0,98	grape	14,88	1,03	0,99

REFERENCES

Bower C.A. & M. Maasland 1963. Sodium hazard of Punjab ground waters. Symposium on water-logging and salinity in West Pakistan. p 49-61

Cass A. 1980. Use of environmental data in assessing of irrigation water quality. Soil Sci. 129:45-53

Cass A. & M.E. Sumner 1982a. Soil pore structural stability and irrigation water quality. I Empirical sodium stability data. Soil Sci. Soc. Am. J. 46:503-506

Cass A. & M.E. Sumner 1982b. Soil pore structural stability and irrigation water quality:II Sodium stability data. Soil Sci. Soc.Am. J. 46:507-512

Cass A. and M.E. Sumner 1982c. Soil pore structural stability and irrigation water quality: III Evaluation of soil stability and crop yield in relation to salinity and sodicity. Soil Sci.Soc. Am. J. 46:513-516

Maas E.V. & G.J. Hoffman 1977. Crop salt tolerance current assessment. J. Irrig. Drain. Div. Am.Soc.Civil Eng. 103:115-134

Marion G.M.& Babcock 1976.Predicting specific conductance and salt concentration in dilute aqueous solutions. Soil Sci.122:181-187

Oster J.D. & Rhoades J.D. 1975 Calculating drainage water composition and salt burdens resulting from irrigation river waters in Western United States. J. Environ. Qual.4:73-79

Quirk J.P. and R.K.Schoefield 1955. The effect of electrolyte concentration on soil permeability. J. Soil. Sci. 6:163-178

Scarduzio G. 1984. L'irrigazione in Sicilia.Sviluppo Agricolo N.11.

Land and Water Use, Dodd & Grace (eds), © 1989 Balkema, Rotterdam. ISBN 90 6191 980 0

The regime of salts in the soils near Kisköre reservoir on the basis of investigations carried out since 1981

J.Fekete & Gy.Szalai
University of Agricultural Sciences, Gödöllő, Hungary

ABSTRACT: The Kisköre Reservoir has a considerable effect on the soils of the surrounding area. The examined area runs the risk of secondary salinization. In this paper the salt regime of the soils is reviewed. The investigations of salt contents showed that the drainage system established around the Kisköre Reservoir functions well. Pumping from the drainage canal results in the lateral movement of the water in the soil, so the soluble salts may leach from the profile.

RESUME: Le barrage-réservoir de Kisköre a un effet considérable sur les sols des superficies voisines. Sur les surfaces en question, se manifeste le risque de la salinisation secondaire. Les auteurs du rapport étudient le régime de salinité de ces sols. L'étude de la teneur en sels montre que le système de drainage établi autour du réservoir, fonctionne bien. Il assure le mouvement de la nappe d'eau résultant le lessivage des sels des profils.

ZUSAMMENFASSUNG: Das Reservoir Kisköre nimmt einen bedeutenden Einfluss auf die Böden des umliegenden Gebietes. Das erforschte Gebiet ist durch eine sekundäre Versalzung bedroht. Hier wird die Salzbilanz der Böden dargelegt. Die Untersuchungen bestätigten, dass das um das Reservoir Kisköre angelegte Entwässerungssystem gut funktionierte. Pumpen aus dem Sickergraben ergibt eine seitliche Strömung des Wassers im Boden, so die lösbare Salze aus dem Profil ausgelaugt werden können.

1 INTRODUCTION

The effectiveness of agricultural production is greatly influenced by moisture conditions and by the water supply to soils on the Great Hungarian Plain. The large-scale hydrological regulation works and irrigation projects carried out were to guarantee a favourable moisture regime in the soil. The various effects of hydrological regulation and irrigation systems were studied by among others Szabolcs /1961/, Arany /1956/, Darab /1958/, Fekete /1978/, Herke /1964/ and Várallyai /1967/.
The implementation of the barrage and irrigation system of Kisköre was a significant agricultural investment of recent years. This system provides the Middle-Tisza region, Jászság and Nagykunság with irrigation water.
Along the basin and irrigation canals great changes in the soils' water and salt regimes are expected. In an earlier study we summarised the changes observed between 1968 and 1981 /Fekete and Szalai, 1984/. In a strip 1-1.5 km wide surrounding the reservoir and drain canals the persistently wet state of the soil moderates salt accumulation and reduces the amount of soluble salts. Again in profiles farther from the reservoir salt content was higher and sodium alcalinity increased. This report summarises the results of examinations done since 1981.

The Consulting Engineering Company /VIZITERV/ entrusted us with evaluating the data from the soil examinations of the 1981-85 period. We evaluated the data from ten profiles. The profiles are at different distances from the reservoir. The data received from soil examinations were the following:

Physical examinations: mechanical composition; heaviness index by Arany /K_A, plasticity limit, sticky point/; volume mass; density; pF value; water conductivity.

Chemical examinations: pH; h_y value; organic matter content pc; total salt pc; sodium alcalinity; 1:5 water extract analysis; $CaCO_3$-content; exchangeable cations; S'-value.

Chemical composition of ground water.

In the analysis of the data we used the 1984 report of VIZITERV: 'The effect of 89 mN damming on the surroundings'.

2 EVALUATION

The data were arranged in two groups. One was data showing slowly changing soil properties, the other quickly changing properties. The former group was used for general characterization of the soils, the latter for the detection of the type and extent of soil processes.

2.1 Characterization of the soils involved

The profiles examined over several years well represent the soils of the area. The distribution of the profiles is shown on the sketch map /Fig.1/.

Meadow soil: Profile 137, NW from Tiszaderzs. It is clay loam. Silt phase exceeds 60 pc. Total porosity is above 50 pc in level A, though only 45 pc deeper. The distribution of the volumes of pores of different kinds is unfavourable /Fig.2/. The volume of fine pores containing absorbed water exceeds the volume of capillar and gravity pores. The proportion of medium pores containing disponible water and pores ensuring water conductivity and airing is small. The reaction in upper layers is moderately acidic, and at deeper layers neutral. The total amount of soluble salts is about 0.06-0.07 pc. Salt content is hydrocarbonate, chloride type, and calcic, calcic-magnesium type respectively, in deeper layers calcic-sodic.

Solonietz Meadow Soil: Such are profiles 140 Tiszaszőlős, 2062 Abádszalók and 2353 Tiszabura. These are loams and clay loams. Water conductivity is poor in the upper layers of profiles 140 and 2062 and deeper in profile 2353. This is connected with the smaller proportion of gravity pores /Fig.2/. They contain $CaCO_3$ and Na_2CO_3 in deeper layers. Total salt content is 0.10-0.15 pc.

Alluvial Meadow Soil: Profiles 2070 Tiszaderzs and 5119 Tiszafüred. In heavy clay soils, the amount of silt phase goes up to 70 pc. Total porosity is about 50 pc but the distribution of pores is unfavourable. The proportion of fine pores containing absorbed water is very high, that of capillar and gravity pores is low /Fig.2/. Water conductivity is poor, the value of 'K' is about 1-2 cms/day. Reaction is slightly acidic in upper layers and neutral in lower layers. Salt content is about 0.05 above and 0.10 deeper.

Rough Alluvium Soil: Profile 5010 Tiszafüred is a heavy soil, silt phase is about 60-70 pc. Heavyness index by Arany /K_A/ is 50. Total porosity exceeds 50 pc in the whole profile. The distribution of fine,

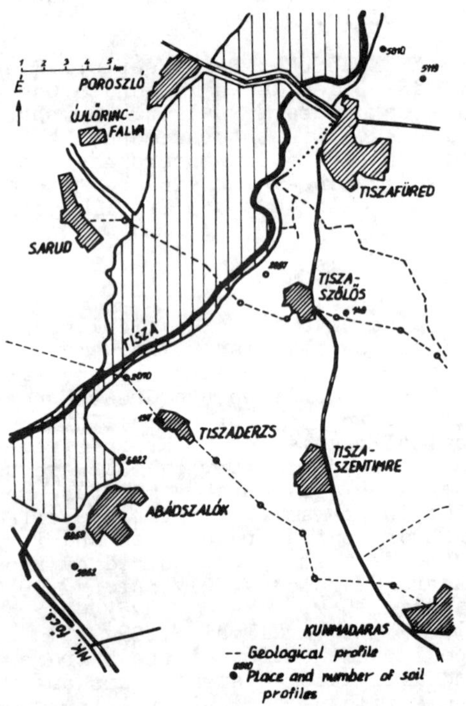

Fig. 1.: DISTRIBUTION OF PROFILES

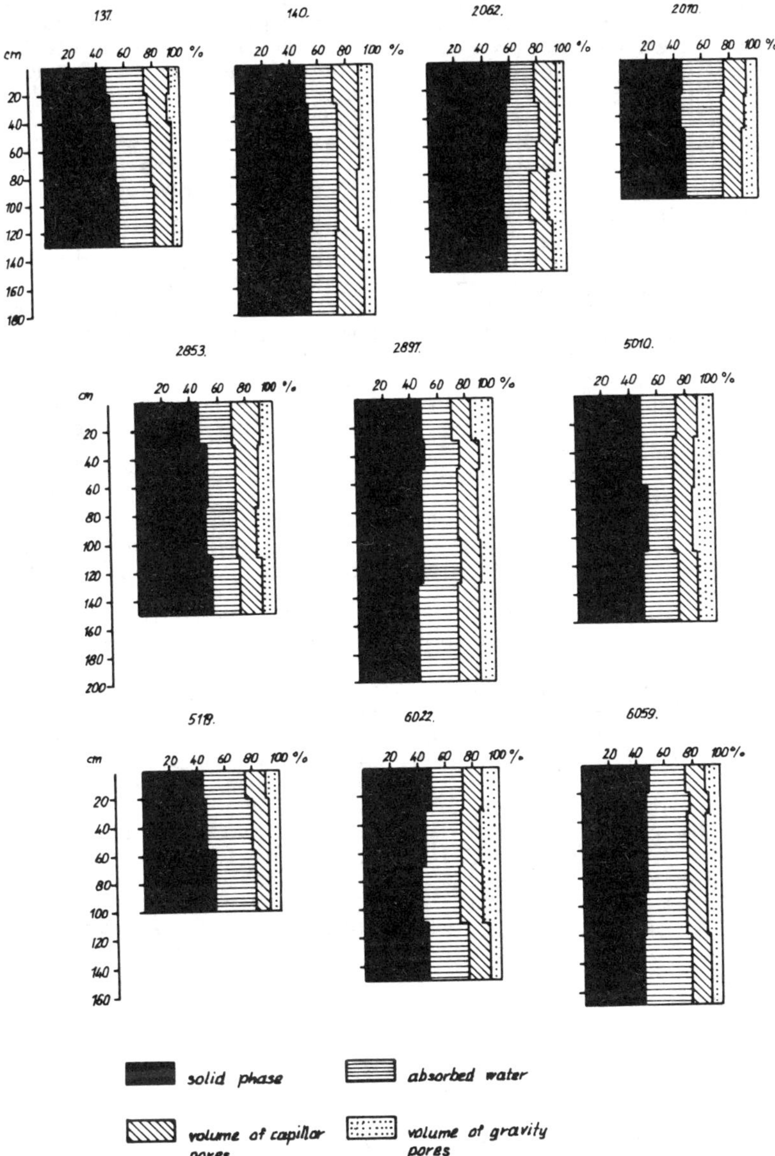

Fig 2.: DIFFERERENTIAL POROSITY (volume pc)

medium and rough pores can be favourable /Fig.2/. Reaction is acidic, the amount of soluble salts is between 0.02-0.03 pc.

Manyfold Humic Allumium Soil: Profiles 2897 of Tiszaszőlős and 6059 of Abádszalók. Heavy clay soils. The silt phase in the upper layers is about 60-70 pc and 70-80 pc deeper. Total porosity is 50-55 pc. The proportion of fine pores containing absorbed water is unfavourably great /Fig.2/. Their water conductivity is often less than 1 cm/day. Reaction is slightly alcalic. Soluble salt content is under 0.05 pc in the upper layers but is more than 0.10 in the deeper layers.

2.2 Changes in salt content

In the region of the Kisköre Reservoir damming produced changes in the water regime of the soil, which also caused changes in the salt regime of soils. The factors of climate, geography, geology and hydrology favour the process of secondary salinization of the Great Plain. The threat of secondary salinization is present in the examined area too. This question has already been considered in detail during the project work /Szabolcs et al.,1968/. The elevated water level of the reservoir and the irrigation canals may dam up groundwater. The higher level of groundwater containing unfavourable salts can cause the accumulation of soluble salts in the soil. The salinization process may damage soil fertility that is why we considered salt examination our main task.

Salt balances give exact information about the transport of salts in the soil. These show salt content changes numerically, measured by t/ha. Salt balances are given for the different layers and are also calculated for the whole profile. They show to what extent salt content increased /+/ or decreased /-/. Besides the data of salt balances the quotients show the proportion of last measured salt content to the initial value /≤1 when decreasing and ≥1 when increasing/.

The annual data of salt balances calculated for the period November 1982 - October 1983. At most profiles significant leaching was detected. In case of profiles 2070 alluvial meadow soil, 2897, 6022 and 6059 manyfold humic alluvium soils the decrease of salt content was about 3.5 t/ha, the quotient of salt regime was 0.6-0.7, which is considered significant over a period of a year. In these soils - except profile 2070 - leaching was a dominant process in the draught year of 1983 and also in the dry year of 1984. It suggests that the drainage system installed by the reservoir fulfilled its function which was to let water filtrate down in the profile; thus causing a leaching process. Between Oct. 1984 and Oct. 1985 the direction of salt transport changed.

Salt balances for two years. In the profile 2062 solonietz meadow soil, 5010 rough alluvium and 5119 meadow soil which were farthest away from the reservoir, accumulation of salts was detected between November 1981 and October 1983. In other profiles leaching was found. The only exception was 6059, where salt content rose slightly. Leaching was most intensive in profile 6022 of Abádszalók on manyfold humic alluvium soil.

Three years salt balances. With the exception of 2062 solonietz meadow, 5010 rough alluvium and 5119 alluvial meadow soil all these salt balances are negative /-/. In profile 2853 solonietz meadow soil significant salt transport was detected. The results of the examination show that leaching was the dominant process between November 1982 and October 1985. The amount of salt leached out of the profiles varied from 1.88 to 2.55 t/ha. Salt regime quotients changed between 0.6-0.8.

Salt balances for 4 years are shown on Fig.3. Salt balances calculated for the period November 1981 - October 1985 already gain positive signs. In profiles 137 meadow, 2062 and 2853 solonietz meadow and 5010 rough alluvium soils - which are relatively far from the reservoir - the accumulation of salts was more intensive. In profiles nearer to the reservoir and to the drain canals that is profiles 2897, 6022 and 6059 manyfold humic alluvium soils salt accumulation was small, about 1-1.3 t/ha. It was characteristic that in the same profile the signs of balances in different layers alternated, detecting significant transport of soluble salts. Salt balances of profiles 140 solonietz meadow soil and near the reservoir 2070 alluvial meadow soil showed leaching.

3 CONCLUSION

Examining the soluble salt contents of the soils the following were found:

- Considerable deleterious salinization processes were not detected in the examined soils by the reservoir in the period 1981-85 compared to 1981. Salt accumulation was observed in profiles farther from the reservoir /these were the profiles 137 meadow, 2853 solonietz meadow and 5010 rough meadow soil/.
- In a number of profiles soluble salt content varies around the same level. Over a longer period neither an increase or decrease could be detected, just seasonal dynamics were observed.

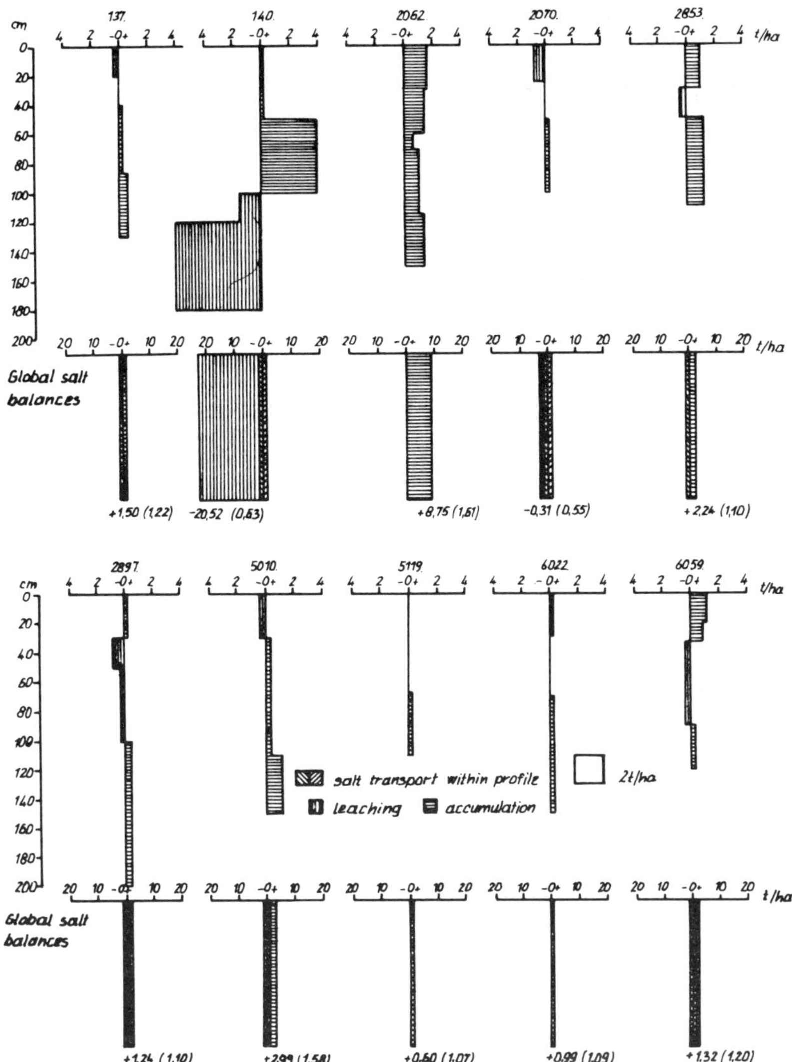

Fig. 3.: SALT BALANCES FOR FOUR YEARS (Nov. 1981 - Oct. 1985)

- At another group of profiles soluble salt contents tended to decrease both at the upper and the lower layers. The reasons for the decrease were the favourable moisture regime or the permanently wet state of the soil and the deep or falling level of groundwater. According to salt balances taken over 3 and 4 years' periods, salt content decreased in profiles 140 solonietz meadow, 2070 alluvial meadow and 6059 manyfold humic alluvium soils.
- It was characteristic of salt regimes in profiles near the reservoir that seasonal changes in salt content showed little fluctuation. This is clear from profiles 2070 alluvial meadow, 5010 rough alluvium and 6022 and 6059 manyfold humic alluvium soils.
- The main factors determining the favourable changes in the soils' salt regime were the falling level of groundwater and the efficiency of the drainage system. The data obtained from observation of groundwater level wells and salt content examinations show that the drainage system fulfilled its task. As a re-

sult of pumping from the drain canal groundwater moved sideways, thus even if groundwater contains unfavourable salts, these can not accumulate in the soil; salinization doesn't occure. Salt balances calculated for both short and long periods conclusively prove that near the drainage system the soluble salt content of the soil decreases.

- Salt accumulation was not detected in the drought year of 1983 or in the dry year of 1984. If the drainage system is kept working correctly, salt accumulation will not occur even during dry periods.

REFERENCES

Arany, S. 1956. Saline soils and their amelioration. Mezőgazdasági Kiadó, Budapest.

Darab, K. 1958. Secondary salinization of irrigated meadow soils east of river Tisza. Agrokémia és Talajtan, 7,1: 53-64.

Fekete, J. 1978. Examination of irrigated meadow chernozem soils' salt regime. Hidrológiai Közlöny, 4. 214-225.

Herke, S. 1964. The role of groundwater regulation in the changes of saline soils' properties between the Danube and the Tisza. Hidrológiai Közlöny, 1. 14-20.

Fekete, J. - Szalai, Gy. 1984. Evaluation of soil examinations' results in the area of Tisza-II reservoir. Hidrológiai Közlöny, 1. 14-20.

Szabolcs, I. 1961. The effects of hydrological regulation and irrigation on soil genetic processes east of the river Tisza. Akadémiai Kiadó, Budapest.

Szabolcs, I. - Darab, K. - Várallyai, Gy. 1968. Irrigation systems of the river Tisza and the fertility of the soils on the Great Plain. Agrokémia és Talajtan, 17. 453--464.

Várallyai, Gy. 1967. Salt balances of soils between the Danube and the Tisza. II. Salt balances of irrigated soils. Agrokémia és Talajtan, 16. 1-2: 7-56.

Land and Water Use, Dodd & Grace (eds), © 1989 Balkema, Rotterdam. *ISBN 90 6191 980 0*

Problems and perspectives of saline soil development in the zone of irrigation

B.B.Shumakov
USSR Academy of Agricultural Sciences, Moscow, USSR

Heavy saline soils are at present increasingly becoming the object of land reclamation, especially in arid zones. According to I.Sabolch, saline soils in need of reclamation amout to 950 million ha worldwide. The world area of strongly saline barren soils occupy 20-25 million ha, while some 50-60 million ha of irrigated lands are resalinized (V.A.Kovda). Irrational irrigation practices cause secondary salinization processes on 200-300 thou ha leaving them outside the use in agriculture. The above data points out the significance of saline soils reclamation in the arid zone of the Globe.

With the development of virgin lands for irrigation in arid and subarid zones the problem of reclamation of strongly saline heavy soils becomes more frequent. Such soils are most labour and investments intensive objects of reclamation. Construction costs of irrigation systems in the above conditions often exceed 4-6 thou Roubles per hectar, while, allowing for agricultural development, this value tops 10 thou Rouble/ha. Reclamation of saline heavy soils and their agricultural development are a time-consuming process, extending to 5-10 years and more. Reclamation and development of soda solonchaks, for instance, in the Ararat Plain of the Armenian SSR takes (in case of light textured saline soils) some 5-8 years, while in case of heavy clays the period extends to 20 years and more. The objective is to considerably enhance the efficiency of reclamation investments and to develop radically new reclamation methods for saline heavy soils, providing drastic reduction of reclamation and development periods and earlies attainment of designed yields of farm crops.

The USSR research centers have developed and widely employed agricultural, engineering and reclamation measures to ensure reliable development of primarily saline soils and effective control of resalinization on irrigated lands. These measures, involving subsurface horizontal drainage, drainage wells, leaching irrigations of farm crops, reclamation tillage, application of reclamation chemicals and cultivation of development crops, allow to reclaime saline soils within sufficiently short time and to control salt balance further on.

Effective implementation of the above measure enabled to develop vast primarily saline areas in Central Asia and Transcaucasus (the Golodnaya, Karshi, Dzhizak and Mugan steppes) formerly not used in acriculture, where as now ever increasing yields of such valuable crops as cotton , alfalfa, maize, grapes, vegetables, etc. are attained.

Further development of irrigation in this country with the progress in science and technology and increasing shortage of fresh water as a background call for radically different and more effective methods for the development of saline soils. This requirement becomes topical in view of the priority formely given to the development and irrigation of slightly saline soils (as less investments-intensive) by general development of irrigation in the country, especially in arid zones. Today the development of irrigation is closely related to the development of strongly saline hard to reclaim heavy soils.

In view of the above research centers are challenged to elaborate, on the basis of obtained experience in the field of saline soils reclamation, more effective methods to develop saline soils. Unlike conventional techniques new methods face the following liabilities:

- to rapidly and effectively remove salts from the active soil layer with the minimum of fresh water assigned;
- to prevent pollution and salinization of fresh water sources with drainage and collector effluents, provide for storage or utilization and processing of salts, removed from the soil profile, in order to exclude pollution of other areas.

The new techniques shall fully comply with environmental protection requirements.

To meet the above basic conditions new techniques shall employ new methods of saline soils drainage, including: combined and vacuum drainage, which considerably enhances drainage rates. This of particular importance by reclamation of heavy saline soils. Application of reclamation chemical agents with irrigation water accelerates exchange reactions as they start in irrigation water proper. Wide employment of electrical reclamation treatments, implying field effects not on soil liqor, but on leaching water instead of to intensify salt leaching from the soil.

Higher permeability of heavy untextured soils, allowing desalinization in the general case, is provided, as is known, by consistent implementation of the following measures: reduction of saline ground water table, mechanical loosening, application of chemical and biological agents, intensifying of aggregation of soil colloids. Revealed by Hedreutz specific role of calcium ions and protons (hydroxonium ions) in aggregation of soil colloids forms the basis for modern methods of chemical reclamation. Hydroxonium ions not only render strong coagulation effect, but also enhance the anolog action of other ions, including those of calcium.

The role of protons in all soil processes is more diversified, than that of other ions. This investigations by M.M.Ovcharenko established their leading role in weathering of primary and secondary minerals, which furnished explanation to the changes of the mineral soil component in course of acidification. Observed changes of water soluble salts' composition and of nutritives by soil acidification account for the buffering capacity of soils and soil liqor (diminishing of carbonate, phosphate and other buffers) and the increment of soluble salts (A.F.Vadyunina).

Positive effect of acidification thus confines to reduction of pH value in the liqor of soil, dissolution of salts, including carbonates and gypsum, higher content of readily hydrolyzed nitrogene and free phosphates, increment of water resistant micro aggregates and to the shift of mechanical composition towards accumulation of the physical sand.

The innumerated changes depend on the amount of acid applied, its initial properties and resistance redox conditions.

Acidification techniques now in practice are generally an irregular measure, and therefore admit of sharp time-bound changes in redox conditions.

A methods of leaching irrigation has been developed for saline soils, which includes in-situ production of acid reclamation agents by means of irrigation water electrolysis, Direct current takes the path via electrolyzer (with cathode, anode plate and poros diaphragm), installed in the flow from the lateral to the reclaimed site. Mineral acids, composition of which depends on the chemical pattern of prime water, material of electrodes and the potential at the plate, are formed at the anode, while mineral alkalis tend to cathode. The diaphragm serves to prevent mixing of catholyte and anolyte.

The anolyte, the acidity of which is determined by current density and contact time between the water and the plate, is used as a reclamation agent. The catholyte is diverted from electrolyzer to drainage collectors, where it is readily neutralized by atmospheric carbon dioxide.

Mineral composition of the anolyte differs from that of prime water in lower content of sodium, hydrocarbonates, chlorine and other elements due to substitution of cation for hydroxonium ions (primarily free sodium ions); chlorine reduces and volatalizes.

By application of iron plates the anolyte is acidified to a lower extent, than by the use of inert anodes (graphite, carbon graphite, etc.), but the solution is enriched by ions, featuring pronounced coagulating capacity.

This method facilitates smooth control of the anolyte acidity by changing the strength of the current in electrolyzer and/or the discharge of the anode section.

This method allows to combine leachings and irrigation in cases, when growing of development crops is possible.

The method retains capability of supplying of acidated water by conventional techniques: surface irrigation, via mole drains, slots and sprinkler units.

Continuous acidification of irrigation water permits to extend admissable salt content limits of ground and drainage water used for irrigation.

Transportation of reclamation chemicals to the site of reclamation is excluded.

Power consumption and lack of high capacity electrolyzers to produce acid are the disadvantages of the suggested method. Power consumption for anolyte production depends on the type of proton yielding reactions.

Though reclamation of strongly saline heavy soils has attained tangible results, much remains to be consistently investigated to promote intensive texturing and desalinization of saline soils.

Land and Water Use, Dodd & Grace (eds), © 1989 Balkema, Rotterdam. ISBN 90 6191 980 0

Drainage efficiency and drain water quality

M.E.Grismer
Department of Land, Air & Water Resources, and Department of Agricultural Engineering, University of California, Davis, USA

ABSTRACT: Problems associated with disposal of potentially toxic, chemically-laden subsurface drain water suggests that engineers consider drain water quality as part of the design of drainage systems. The concept of drainage efficiency, defined as the fraction of deep percolation water actually collected by the drains, can be related to irrigation efficiency and serves as an index of drainage system performance. Herein, drainage efficiency is related to drain water quality at the field and water district scale using data from the San Joaquin Valley of California. Intensive measurements of water movement to drains in a heavy clay soil in the Imperial Valley are used in numerical simulations to demonstrate the effects of drain depth and lower boundary pressure head on drainage efficiency and drain water quality. Continued development of relationships between other such drainage design factors and drainage efficiency should provide insight into alternative drainage system design strategies which, when implemented, yield drain water of acceptable quality.

RÉSUMÉ: Les problè mes creés par l'élimination des eaux souterraines de drainage contenant des élements chimiques potentiellement toxiques montre que les ingénieurs. prennent compte de la qualité des eaux d'ècoulement dans la conception des systèmes de drainage. La notion d'éfficacitédu drainage, définie comme la fraction des eaux de percolation profonde reçue par les drains, se rapproche du concept défficacité d 'irrigation et peut servir d'indice de performance du système de drainage. Ici, l'éfficacité du drainage est mise en relation avec la qualité des eaux d'écoulement à l'echelle de la parcelle et du district en employant des données fournies par différents districts d'alimentation en eau de la San Joaquin Valley en Californie. Des mesures intensives du mouvement de l'eau vers les drains dans un sol très argileux sont employées dans une simulation pour démontrer l'effet de la profondeur des drains et l'effet du potentiel capillaire à la limite inférieure du profil sur l'éfficacité du drainage et la qualité des eaux d'écoulement. La poursuite de l'étude des rapports entre les caractèristiques des systèmes de drainage et leur éfficacité devrait fournir d'autres données permettant d'améliorer la conception et les techniques de construction de ces systèmes et d'obtenir des eaux d'écoulement de meilleure qualité.

KURZFASSUNG: Probleme mit der Beseitigung potentiell toxischer, chemisch belasteter Dränwässer zeigen, dab Ingenieure die Dränwasserbeschaffenheit als Be-standteil des Dränsystementwurfes betrachten sollten. Das Konzept der Leistungsfähigkeit einer Dränung, definiert als Anteil der Tiefenwasserver-sickerung, die tatsächlich von der Dränung gefabt wird, kann auf die Bereg-nungsleistungsfähigkeit bezogen werden und dient dann als Leistungskenn-ziffer eines Dränsystems. Im vorliegenden Beitrag ist die Leistungsfähigkeit einer Dränung auf die Dränwasserbeschaffenheit im Feld- und Gebietsmab-stab bezogen. Hierzu werden Daten aus dem San Joaquin Valley genutzt. In-tensive Messungen der Dränanströmung in schweren Tonböden des Imperical Valley und numerische Simulationen dienen der Demonstration des Einflus-ses der Dräntiefe, der Druckhöhe längs der unteren Berandung und der Drän-wasserbeschaffenheit auf die Leistungsfähigkeit der Dränung. Die weitere Entwicklung dieser und anderer Beziehungen zwischen derartigen Entwurfs-faktoren und der Leistungsfähigkeit einer Dränung sollten den Einblick in al-ternative Strategien des Dränagesystementwurfs ermöglichen, die nach ihrer Realisierung Dränwasser mit akzeptabler Beschaffenheit gewährleisten.

1 INTRODUCTION

Traditionally, subsurface drainage systems have been designed as part of an overall farm water management scheme for control of soil moisture, or aeration, and salinity in the crop root zone. Several books and monographs have been published and models developed which consider the depth and spacing of subsurface drains necessary to achieve the desired control in the root zone (USDA, 1987). Corrective design procedures have also been developed for drainage of heavy clay soils (Hermsmeier 1973; and Smedema 1986) and use of drains for subirrigation (Massey et al. 1983). Problems associated with disposal of drain water having potentially deleterious effects on the environment as part of agricultural development (Deason 1987) suggests that engineers consider water quality aspects of drain discharge in the design of subsurface drainage systems. Though such consideration was not included as a research priority for the 1980's (ASCE Task Committee on Irrigation and Drainage Research 1981), it was alluded to as an area of concern. More recently, an ASCE Task Committee on Management Options for Toxic Trace Elements in Agricultural Drainage Waters has been formed to assess and develop alternative management options for discharges from irrigated agricultural lands having naturally-occurring toxic trace elements.

Though conceptually designed to remove excess deep percolation or leachate not removed by natural drainage, relatively deep subsurface drainage systems in arid and semi-arid regions also collect adjacent ground waters subject to evaporative salination, or displacement by subsurface lateral flows. Drain water quality is determined then by the relative quality and quantity of deep percolation water and other groundwaters collected by the drain. As a means of assessing the performance of subsurface drainage systems with respect to the leaching fraction component of surface irrigation system design, Grismer et al. (1988) defined drainage efficiency (DE) as the ratio of drainflow and deep percolation volumes. Later, Tod and Grismer (1988) defined DE as the ratio of drainflow less other groundwater flow to deep percolation volumes so as to account for contribution of flows other than root zone drainage to total drainflows. Due to this contribution of other groundwater flows to drainflows, changes in irrigation efficiency or uniformity alone may not substantially affect drain water quality.

The purpose of this paper is to examine the relationship between drainage efficiency and drain water quality based on data collect from the Panoche and Broadview Water Districts, and the Imperial Valley of California. Drainage efficiency and drain water quality will then be considered in the context of subsurface drain system design.

2 FIELD STUDIES

Several studies have been initiated, or are in progress related to reduction of drainage flows in the San Joaquin Valley of California. J. M. Lord (1988) reported results of an intensive water balance study on two fields in the Panoche Water District (PWD) and one cotton field in the Broadview Water District (BWD). Deep percolation losses accounted for 16%, 48% and 31% of the applied water in these fields, respectively. In one field, estimated deep percolation and measured drainflow volumes were equivalent, and in the other two, estimated deep percolation was less and greater than measured drainflows. Due to the limited number of fields studied, no relationship between DE and drain water quality could be determined.

Drainflow, salt loads, cropping patterns and irrigation applications have been monitored in both the 14000-ha PWD and 4000-ha BWD since 1986. Hoffman and Schrale (1988) estimated deep percolation losses in cotton fields of the PWD which ranged from 4% to 57% of the applied water with a mean value of approximately 40%. Wichelns and Nelson (1988) made no estimate of deep percolation losses in the BWD, but determined from a spatial linear regression model that subsurface lateral flows, or "baseflows" accounted for 26.9% and 21.2% of total drainflows in the district during 1986 and 1987, respectively. This baseflow occurred in 9 of the 21 drain sumps monitored in the district. Hoffman and Schrale (1988) estimated that "about 11%" of total drainflow in PWD was baseflow in 1987. Unfortunately, estimates of so-called baseflows provide little information about the precise origin of the various drainflow components. Differentiation of these components may be possible using isotopic composition analysis of the water (Deverel and Fujii 1988), but extensive analyses of this type may be limited by the expense of the isotope analysis.

Drainage efficiency for drain sumps in the BWD was determined assuming that deep percolation losses represented 35% of the total applied water during the years 1986 and 1987. This figure is based on the data summarized above for BWD and PWD, and is sufficient to show the relative trends desired here. For drain sumps having no baseflow, DE was calculated as the ratio of annual drainflow volume to 35% of the annual applied water volume. For drain sumps having a baseflow component, the drainflow volume was reduced by the estimated volume of baseflow when calculating DE. Results of DE calculations together with average annual drain sumps salinities for BWD are summarized in Table 1 and shown in Figure 1. Unfortunately, similar calculations were not possible for the PWD due to lack of correspondence between measurements of applied water and locations of subsurface drainage systems and their sumps. In both the BWD and PWD, concentrations of other elements in the drainwater such as boron, molybdenum, and selenium roughly correlate with the salinity of the water. Only salinities are used here because the salinity data set was more complete.

Despite scatter in the data shown in Figure 1, the trend in increasing TDS concentration with decreasing DE is evident. Linear regression equations for sump data without baseflows, with baseflows and overall, respectively, are given below.

$$TDS = 7463 - 32.95\ DE \qquad (r = -0.485)$$
$$TDS = 7563 - 34.11\ DE \qquad (r = -0.709)$$
$$TDS = 7490 - 33.22\ DE \qquad (r = -0.601)$$

Though the correlation coefficient, r, is significantly greater for data from sumps having baseflows, the regression equations are not significantly different from each other, or the overall equation. The decreasing trend in the data is indicative of groundwater hav-

Table 1. Drainage efficiency (DE) and salinity (TDS) for fields in the BWD during 1986 and 1987.

	1986		1987	
Field*	DE (%)	TDS (mg/l)	DE (%)	TDS (mg/l)
1	53.4	6940	31.5	7710
2	93.1	4010	57.7	4740
3	9.6	9280	16.3	8500
4	25.5	5940	25.0	5960
5	10.2	4320	17.5	5730
6	65.4	5160	49.3	5550
7	51.5	4270	63.7	4530
8	23.6	5390	13.6	6200
9	49.1	7180	39.0	8060
10	27.3	6580	25.8	8130
11	40.7	5640	34.4	6500
12	—	5550	60.5	6210
13	42.9	6620	50.4	7410
14	48.7	5660	87.9	6080
15	64.9	5660	64.1	5560
16	14.7	6270	25.6	6340
17	30.7	6860	48.4	7330
18	89.7	4370	78.7	4450
19	27.3	6300	28.1	6460
20	64.4	3520	100.3	3550
21	84.4	4125	73.4	4320

*Field sumps 1-12 have no baseflow component whereas sumps 13-21 have variable baseflows.

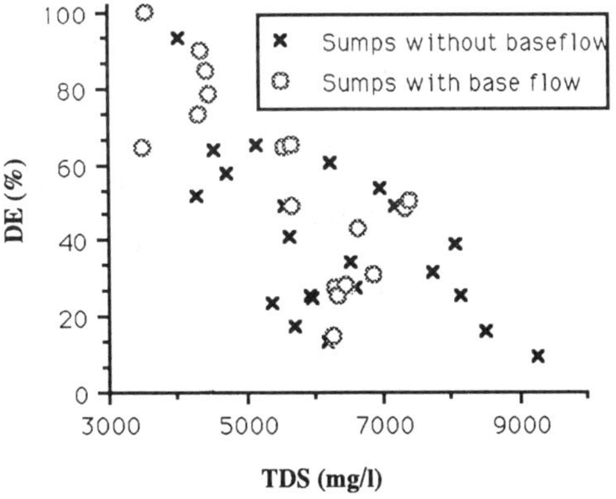

Figure 1. Drainage efficiency and salinity for fields in the BWD.

ing salinity greater than that of the deep percolation water contributing to drainage flows in at least those sumps having baseflows. Though less well defined, it is not clear why this trend also exists for data from sumps not thought to have a baseflow component, unless these sumps too have a baseflow component. In the BWD overall, DE changed little from 1986 to 1987, averaging approximately 38% at a TDS of 5390 mg/l. Such a relatively low DE suggests that drainage systems within the district as a whole are not intercepting all deep percolation losses and that the portion not intercepted continues to move either as lateral flows out of the district, or as deep seepage into deeper aquifers. These results are similar to those obtained by Grismer et al. (1988) for individual drains in the Imperial Valley and Chang et al. (1983) for fields in both the Imperial and San Joaquin Valleys.

Conceptually, the relationship between DE and salinity should be linear based on relative contributions of flow and salt to the drainage system from deep percolation and groundwater. In arid and semi-arid irrigated regions, groundwater is often more saline than deep percolation water leaching the rootzone. Thus, drains performing with low values of DE are typically discharging waters of greater salinity and concentrations of trace elements. From the viewpoint of subsurface drainage system design, reducing, or minimizing salinity and trace element discharge from the drainage system, requires insight into those factors affecting DE such that a system with the greatest efficiency can be constructed. Clearly, the DE of a particular drainage system depends on its hydrogeologic setting. As a means of assessing the influence of some of these factors on DE, numerical simulations of drain system performance were conducted using extensive data collected from a clay field in the Imperial Valley by Tod (1989).

3 NUMERICAL SIMULATIONS OF IMPERIAL VALLEY FIELD DRAINS

Data on water movement and drain performance were collected from three subsurface lateral drains in a heavy clay field in the Imperial Valley having high soil salinity of approximately 12 dS/m. The relatively homogeneous clay soil extends to depths of 1.9 to 2.5 m and is underlain by an artesian aquifer of fine sand. Subsurface drains are spaced 37 m apart and at a depth of 2.1 m. The pressure head in the artesian aquifer remains relatively constant in time at a depth of 1.6 m. Tod and Grismer (1988) found that the primary components of the drainflows in this field were artesian groundwater, water rapidly moving through the excavation trench of the drainline during and immediately following flood irrigation, and deep percolation losses slowly moving through the clay soil. Due to the very small permeability of the clay soil, the effective radius of influence of the drainlines was less than 3 m.

Using output generated by the finite-element model described by Tod (1989), the effect of drain depth and pressure head of the artesian aquifer on DE and drain water quality are examined. These two, of several possible factors were chosen to illustrate the effect of a design factor (drain depth) and hydrogeologic parameter (pressure head) on DE and drain water quality.

Illustrated in Figure 2 is a simulated drainflow hydrograph immediately following an irrigation event for present conditions at the Imperial Valley site (Tod and Grismer 1988). Due to initially large flows through the excavation trench, the DE of this particular system is variable with time. For purposes of analysis, DE is computed on a volumetric basis for a 60-hour period following irrigation, since "trench flow" is insignificant by this time. Computed volumes and DE's are for an 8 m wide region centered on the drainline, since all flows beyond a 3 m wide region around the drain are deep percolation losses to the sand aquifer. Salinity of the drainflow is determined on a mass per volume contribution basis over the same period assuming TDS concentrations for trench flow, deep percolation and groundwater of 3000, 4900 and 9000 mg/l, respectively. Effects of drain depth on DE and TDS of the drainflow for a pressure head of 1.6 m below groundsurface in the sand aquifer is summarized in Table 2.

For artesian aquifer pressure heads near or below the drain depth, the water quality of the drainflow is similar to that of the water leaching the rootzone. DE remains relatively low in this case primarily as a result of the low hydraulic conductivity of the clay soil.

Table 2. Effect of drain depth on DE and TDS of drainflow at Imperial Valley field.

Drain depth (m)	DE*(%)	TDS (mg/l)
1.50	15.0	3920
1.75	27.3	4380
2.00	36.8	8110
2.15	–15.0[a]	--

*DE = (drainflow–groundwater–trenchflow) x 100%/deep percolation.
[a]The negative number results from excessive trenchflow by-passing drain into the sand aquifer.

In fact, as the drain depth decreases, the zone of influence of the drain, in terms of the distance from the drain which flow diverges from vertical toward the drain, also decreases.

The effect of lower boundary pressure head on DE and TDS for two different drain depths commonly used in Imperial is summarized in Table 3. In this case, as the artesian pressure of the sand aquifer increases, the DE also increases due to declining deep percolation losses resulting from reduced hydraulic gradients in the soil. Trends between DE and TDS of the drainflow are not as clear, reflecting the complexities associated with the hydrogeologic setting of the drains. Lack of trends between DE and TDS at the local scale of the single drainline or field, may also be a factor affecting the scatter of data points in Fig. 1.

Table 3. Effect of sand aquifer pressure head on DE and TDS of drainflow.

Drain depth (m)	Pressure head (m)	DE (%)	TDS (mg/l)
1.75	–1.4	41.5	5350
	–1.6	27.3	4380
	–1.8	15.4	3380
2.00	–1.4	47.4	6660
	–1.6	36.8	8110
	–1.8	24.2	5090

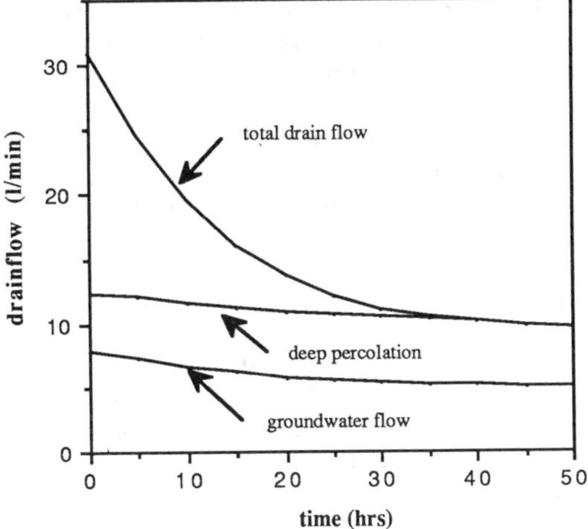

Figure 2. Contribution from different sources to total drain flow following surface irrigation.

Overall, results summarized in Tables 2 and 3 indicate the importance of site-specific characteristics when incorporating drainage efficiency and drain water quality into the design of subsurface drainage systems for irrigated regions having stratified alluvial soils. These results suggest that more intensive field measurements of soil stratification and permeability in conjunction with computer modeling of the system may be necessary for satisfactory drainage system design. For example, modeling of water movement at the Imperial Valley site, showed that most deep percolation losses continue as deep seepage losses and that the drains intercept little, if any, vertical seepage at distances greater than one meter away from the drain. Thus, for this field, lateral subsurface drainlines are of little value in maintaining favorable conditions for plant growth in the rootzone.

4 CLOSURE

Drain water quality can be related to the efficiency at which drainage systems operate in extracting deep percolation losses from the rootzone. At the regional scale, data from water districts in the San Joaquin Valley indicate that water quality improves as the efficiency of the drainage system increases. Improved water quality stems in part from reduced contributions of saline groundwater to drainage flows. Detailed numerical analysis of a specific field site in the Imperial Valley highlighted the importance of the hydrogeologic setting of the drainage system to incorporation of drain water quality aspects into the system design. Results of both analyses suggest that engineers gain a greater understanding of the hydrogeochemical setting of potential subsurface drainage systems when designing drainage systems. Such an understanding should enable alternative designs to be developed which, when implemented, maintain adequate rootzone conditions for plant growth while producing drainwater of acceptable quality.

REFERENCES

ASCE Task Committee on Status of Irrigation and Drainage Research 1981. ASCE J. of Irr. & Drain. 110(1):55-74.

Chang, A. C.,, R. W. Skaggs, L. F. Hermsmeier and W. R. Johnston 1983. Evaluation of a water management model for irrigated agriculture. ASAE Trans. 26:412-418, 422.

Deason, J. P. 1987. Selenium: Its not just in California. In: Irrigation Systems for the 21st Century, Proceedings of ASCE Specialty Conference, Portland, OR. pp. 475-482.

Deverel, S. J. and R. Fujii 1988. Processes affecting the distribution of selenium in shallow groundwater of agricultural areas, western San Joaquin Valley, California. Wat. Res. Res. 24(4):516-524.

Grismer, M. E., I. C. Tod and F. E. Robinson 1988. Drainage system performance after 20 years. Cal. Ag. 42(3):24-25.

Hermsmeier, L. F. 1973. Shallow drain performance in a heavy soil. ASAE Trans. 16:92-94, 96.

Hoffman, G. J. and G. Schrale 1988. Management strategies to reduce drainage from irrigated agriculture. ASAE Paper No. 88-2656.

JM Lord 1988. Innovative techniques to reduce subsurface drainage flows. Phase II Report to the San Joaquin Valley Drainage Program, Sacramento, California. 74p.

Massey, F. C., R. W. Skaggs and R. E. Sneed 1983. Energy and water requirements for subirrigation versus sprinkler irrigation. ASAE Trans. 26:126-133.

Smedema, L. K. 1985. Drainage coefficients for heavy land. ASCE J. of Irr. & Drain. 111(2):101-112.

Tod, I. C. and M. E. Grismer 1988. Drainage efficiency and cracking clay soils. ASAE Paper No. 88-2588.

Tod, I. C. 1989. Irrigation and drainage of cracking clay soils. Unpublished Ph.D. dissertation in Agricultural Engineering, Univ. of Cal., Davis.

Wichelns, D. and D. Nelson 1988. Empirical analysis of spatial variation in subsurface drainage flows. Interim report to the San Joaquin Valley Drainage Program. Sacramento, California. May. 20p.

Land and Water Use, Dodd & Grace (eds), © 1989 Balkema, Rotterdam. ISBN 90 6191 980 0

Die Folgen der Anwendung von Gülle auf Sandböden

R.Kovács
Közalkalmazottak Szakszervezete, Ungarn

AUSZUG: Auf der mit Chemikalien behandelten Gülleausbringungsfläche der 36.000-er Schweinezuchtanlage der CÁT Csemő, suchte ich die Antwort auf die Veränderung in Bezug der Qualität des Grundwassers bei Sandböden mit guter Durchlassungsfähigkeit, als erste wassergebende Schicht. Ich untersuchte die Ausbreitung der Verschmutzungskomponenten bei horizontaler und vertikaler Bewegung des Grundwassers, sowie die daraus folgenden Quantitätsveränderungen des Grundwassers auf der Ausbringungsfläche.

ABSTRACT: Looking for the change of quality of subsoil water in good permeable sandy soil, as the first water giving stratum, in the territory, making use of deluted manure in the CÁT pigfarm in Csemő, keeping 36000 animals. Examination of the spreading of contaminating components with the horizontal and vertical movement of subwater in the area, where deluted manure is used.

EXTRAIT: Je faisais des recherches sur un terrain utilisant le lisier d'une porcherie de 36000 places couchées - CÁT Csemő - pour savoir comment la qualité de la nappe phréatique du sol sablonneux de bonne perméabilité - la premiére couche de sol donnant de l'eau - se change-t-elle. J'examinais l'expansion des composants salissants du lisier, son expansion avec le mouvement horizontal et vertical de la nappe phréatique en étudiant aussi les changements déroulés dans les relations quantitatives de la nappe phréatique de la terre où on utilise le lisier de la porcherie.

1 DIE PROBLEME DER GÜLLEAUFBEREITUNG UND -AUSBRINGUNG DER TIERZUCHTANLAGEN IN GROẞBETRIEBEN

In Tierzuchtanlagen der Großbetriebe war die Haltungstechnologie vorteilhaft, welche die Gülle mittels Wasser entfernte. Die Verarbeitung der entstandenen Gülle entwickelte sich unabhängig von einander in den landwirtschaftlichen Zweigen, die ökologischen Forderungen wurden außer acht gelassen, die Behandlung und Ausbringung der Gülle wird durch die großen Kosten erschwert. Die Überbemessung der Standplätze der Tierzuchtanlagen, sowie die verschwenderische Wasseranwendung erhöht nur die Probleme.
Auf Grund von Berechnungen ist die punktmäßige Belastung durch Verschmutzungsstoffe durch Gülle aus,den 79 Tierhaltungsanlagen im Gebiet der Mittel- Theiß 1.442.360 LEE, die der gesamten komunalen und industriellen Belastung des Gebietes 255.170 LEE jährlich.
Die chemische Zusammensetzung der Gülle kann sich andern, in der Abhängigkeit der Verdünnung, in Abhängigkeit der Lagerungszeit die organischen Komponenten, Sulfid, N-Haushalt, und in der Menge der gesamten Schwebstoffe (Sántha 1981).
Die Mikroorganismus-Komponenten der Oberflächengewässer und des Grundwassers können aus der Sicht des Human- und Tiergesundheitswesen geschädigt werden.
Das technologische Wasser, daß aus Tiefenbrunnen gewonnen wurde, be-

sitzt eine hohe Na^+Konzentration, sowie die natürlichen Verhältnisse der Gewässer der Mitteltheiß begünstigen die schädlichen Auswirkungen der Gülleverregnung bei der sekundären Sodabodenbildung.
Anderseits ist die Gülle der Träger für die wichtigsten Nährstoffe der Pflanzen.
Mit Durchschnittswerten gerechnet, enthalten die 2.095.460 m^3 Gülle/Jahr des Gebietes der Mitteltheiß Nahrstoffe, die bei 10.330 t 34 % Wirkstoffe entsprechen N-Kunstdünger, 8.040 t 18 % Wirkstoffe entsprechen P-Kunstdünger, 8.270 t 40 % Wirkstoffe entsprechen K-Kunstdünger.
Bei den optimal ausgebauten und betriebenen Gülleverwertungsanlagen, ist sowohl wirtschaftlich als auch vom Umweltschutz her mit einer "entlastenden" Wirkung zu rechnen.

2 UNTERSUCHUNG ÜBER DIE WIRKUNG DER GÜLLEVERREGNUNG

2.1 Die Untersuchungsanlage

Für die Untersuchung habe ich die Gülleverwertungsanlage der 36.000 er Schweineanlge des Staatlichen Lehrgutes Cegléd (CÁT) gewählt.
Die Anlage ist seit 1974 in Betrieb.
Die Gülleaufbereitung erfolgt mittels 2 Stück Homogenisierungstürmen von je 25.000 m^3. Die anfallende Güllemenge beträgt 900 m^3/d, 328.000 m^3/Jahr.
Die Beregnungsfläche beträgt 420 ha, davon sind

Schutzfläche	50 ha
Zellulose- Pappel	130 ha
Weinstöcke	98 ha
Ackerfläche	122 ha

Das Belastungsverhältnis der Beregnungsfläche beträgt 120 mm.
Cegléd- Csemő ist ein Teil der Niederung des ungarischen Flachlandes. Die Ur-Donau lagerte in einer Stärke von 100-300 m feinen Sand und tonigen Schlamm in wechselnden Schichten als Ablagerung ab (Weppel 1979). Die Kornzusammensetzung des bodenbildenden karbonaten Löß Sandes ist 0,1 - 0,6 mm. Das Wasser wird gut durch gelassen. Die Wassersteigefähigkeit überschreitet 1m.
Die Ion- Bindefähigkeit ist unbedeutend (Stefanovits 1981). Die Adsorbtionskapazität erreicht nicht die 10 mg ee/ 100 g Bodenwert

Grundwasserverhälnisse in der Zeit 1955- 1981

Werte des Hochwassers	266-117 cm
Werte des Niedrigwassers	311-210 cm
Werte des Mittelwassers	164-289 cm

Wegen dem negativen Druckzustand des Gebietes bieten sich möglichkeiten zur Nachuntensickerung des Grundwassers, in erster Linie durch die strukturellen lücken.
Die klimatische Charakteristik des Gebietes ist eine reichliche Sonnenscheindauer, starke Ausstrahlung, extreme Temperaturverhältnisse. Die Menge der Niederschläge ist klein, im Durchschnitt 573 mm/Jahr, von April bis September 324 mm. Die vorherschende Windrichtung ist N- NW. mit charakteristischer Festlandluftströmung (Csaba-Kiss-Szinay-Vermes 1978).

2.2 Untersuchungsmethode

1. Art der Probeentnahme

Wasserproben wurden mit Hilfe von Pumpen und auch ohne aus dem Beobachtungsbrunnen Nr. 4 genommen, ich untersuchte die Konzentration von NO_3^-, NO_2^-, NH_4^+ und Cl^- und auf alle Bakterien. Das Ergebnis hat bestätigt, daß, die Strömungsverhältnisse des Gebietes, sowie die ständige Verwendung des Wassers der gegrabenen Brunnen - wegen der Tierhaltungso wird bestätigt, daß das Wasser der Beobachtungsbrunnen die Verhältnisse des Grundwassers charakterisiert.

2. Die Bestimmung der charakteristischen Komponenten

Es dienten die Angaben der Wasserqualität der Grundwasser-Beöbachtungsbrunnen als Hintergrundsangaben und die aufgearbeiteten Angaben von Szebell édyné aus dem Jahre 1967.
Das Grundwasser auf Sandböden ist gering salzhaltig. Die Erhöhung der Gesamtsalz- Konzentration ist auf außere Ursachen zurück zu führen.
Die Anwesenheit von NO_3^-, NO_2^- weist auf eine fortlaufende verschmutzung hin.
Das Anwachsen von Cl^- beweist die Veränderung des Ionengleichgewichtes.
Ca^{2+}-Mg^{2+} Komponenten wurden nicht gemessen, obwohl durch die Beregnung eine Wirkung angezeigt werden müßte.
Im Falle eines Na^+ haltigen technologischen Wassers ist das Verhältnis auf Sandboden von Ca: Mg 4:1 vom

Gleichgewicht aus 2:1-1:1, dann wird über das Mg > Ca Verhältnis durch dieWirkung der Beregnung ein Typ mit Na^+ vorherschen. Die Coliform Zahl und die NH_4^+ Komponenten weisen auf frische Fäkalienverschmutzung hin, welche nicht nur durch den Boden ins Grundwasser gelangt sind. Der oxidative Boden bindet die NH_4^+ Teile.

3. Die Art der Auswertung der Grundwasserqualität

Da auf dem Anwendungsgebiet die Bewohner der Gehöfte das Grundwasser, das eine Trinkwasserqualität aufwies, bis 1975 für komunale Zwecke benutzten, führte ich die Überprüfung auf Basis ungarischen Normen für die physische und chemische Beurteilung von Trinkwasser, sowie die bakteriologischen Überprüfungen von Trinkwasser durch "Msz. 450/1-78 und Msz. 22901-71". In den Normen sind die Mengen der charakteristischen Komponenten nach, "entsprechend", "gerade noch zulassig", und "nicht trinkbaren" Kategorien unterteilt.

2.3 Die Qualität des Trinkwassers

Auf dem Anwendungsgebiet habe ich seit dem Jahre 1974 Wasserproben aus 14 Probeentnahmestellen, die Angaben der Qualität ausgewertet. Der Nitratgehalt fällt in die Zone von 120 - 320 mg/l, wenn wir absehen von den hervorspringenden Minimumwert (38 mg/l) und dem Maximumwert (560 mg/l). Vom Gesichtspunkt der Brauchbarkeit überschreiten bei weitem die Komponenten die in der Norm angegebene Kategorie "nicht trinkbar" NO_3^- Grenzwert. Der NO_3^- Gehalt der auszubringenden Gülle erreicht noch nicht den mg/l Wert, wobei 800 mg/l ca. eine NH_4^+ Konzentration zeigt. Der N-Gehalt der Gülle oxidiert im Boden von der Form von NH_4^+ zu der Form NO_3^-.
Im Grundwasser ist auch in geringem Maße NO_2^- anzutreffen. In allen Grundwässern sind Cl^- Ionen im Cleichgewicht mit Na^+ zu finden. Auf äußere Einwirkung hin kann eine Verschiebung von Cl>Na oder Na>Cl erfolgen. Das Cl^- Ion können wir nicht als eine die Verschmutzung anzeigende Komponente betrachten, obwohl in manchen Fällen eine große desinfizierende Wirkung in der Anlage ausführt, der höchste Wert lag bei 196 mg/l. Das leicht bewegbare Anion ist zu einer schnellen Konzentrationsveranderung fahig.
Der erhöhte Grundwasserstand der beregneten Fläche ist vom bakteriologischem Gesichtspunkt aus stark zu beanstanden. Da die infizierung des Grundwassers mit fäkalischem Coliform von der augenblickliche Beregnungssituation abhängt, von der seit der Beregnung vergangenen Zeit, die werte weisen eine starke Wellenform auf. Zur Beseitigung der FC-Verschmutzung ist ein Intervall von einem halben Jahr ausreichend. Eine anhaltendere Wirkung der Coliform-Werte mit der NO_3^- Komponente ist parallel wahrnehmbar.
Im Vergleich zu den Hintergrundwerten (562 mg/l) zeigt die Gesammtsalz-Komponente eine bestimmte Erhöhung auf (max. 1665 mg/l). Die schwach adsorbtiven, gut beweglichen Ionen gelangen schnell ins Grundwasser. Wegen der günstigen Zusammensetzung des technologischen Wassers, ist mit einer Sodabodenbildung durch Gülleberegnung nicht zu rechnen.
Die charakteristischen Grundwasserverhältnisse zeigen einen Hydrogenkarbonat - Chlorid - Sulfat - Amiontyp. Die aus der Schweineanlage ausgebrachte Gülle ist weiterhin unverändert amiontypisch. Die Anionverhältnisse auf der beregneten Fläche gehen im Grundwasser in Sulfat - Hydrogenkarbonat über. Die Hydrogenkarbonat- Menge verringert sich stark während sie durch den Boden hindurch geht (von 25 mgeé/l auf 4,3 mgeé/l) im wesentlichen kommen aber die Sulfationen - Konzentration ohne Veränderung in das Grundwasser.

2.4 Die vertikalische Ausbreitung der Grundwasserverschmutzung

Wegen der Verschlechterung der Wasserqualität der gegrabenen Brunnen im Bereich des Beregnungsgebietes wird das Wasser bei drei Gehöften aus einer Tiefe von 25 - 30 m gewonnen, hier wurden die schlammigen tonigen Sperrschichten mit 8 - 17 m durchbohrt. Der Vergleich der wasserqualität der zwei Wassergewinnungsschichten beweist, daß nach einer Betriebsdauer von 8 Jahren eine Verschmutzungswirkung nicht über die Sperrschicht hindurchging. Bis zur verdichteten Sperrschicht beträgt die hydrologische Durchlaßfähigkeit 10^{-4} cm/sec, die des oberen Sandbodens 10^{-3} cm/sec. Es ist natürlich, daß die Bewgungsrichtung des Wassers horizontal ist. Unsachgerechte Bohrungen können die Sperrschicht zerstören, was zur Folge hätte, daß auf Gebieten mit negativem Druckzustand bei den Tiefenbrunnen die Verschmutzung bis zu einer Tiefe von 150 - 200 m vordringen kann.

2.5 Quantitätswirkung der Gülleberegnung auf das Grundwasser

Die Messung des Grundwasserspiegels nehmen wir an gegrabenen Brunnen vor, das Ergebnis trug ich in baltischer Höhenangabe in die Karte ein. Das Gelände ist gleichmäßig, mit großem Gefälle, auf einer Entfernung von 5,5 km nahezu 9 m. Das ablesbare Geländeniveau 131,7-123,2 m auf baltische Höhe bezogen. Das Grundwassergefälle beträgt hier 130,7 - 121,0 m. Entgegengesetzt den Strömungsverhältnissen, in einer Entfernung von 150 m vom Beregnungsgebiet liegenden Beobachtungsbrunnen, weist die Wasserbewegung keine Wirkung der Beregnung auf. Innerhalb der Beregnungsflache nähert sich der Grundwasserspiegel der Geländeoberfläche. Wir maßen Unterschiede, die sich innerhalb eines halben Meters bewegten. Entfernend von der Beregnungsfläche in Strömungsrichtung wuchs der Unterschied zwischen Gelände und Grundwasserspiegel bis zu 2 m. Also verursacht die Beregnung auf guten wasserleitfähigen sandigen Gebieten einen Grundwasserspiegel-Hügel, welcher nicht den Strömungsverhältnissen entgegengesetzt wirkt, in dessen Richtung aber mit dem groβen Gefälleabschnitt zum originalen Grundwasserspiegel sich anschließt. Wo der Grundwasserspiegel der Oberfläche am besten sich nähert, dort ist auch die Wasserqualität am schlechtesten. Die adsorbtiven Verhältnisse wirken bei der horizontalen Verbreitung der Verschmutzung auf Sandböden nicht so, wie auf bindigen Böden. Die Verringerung der Konzentration wird eher durch die Verdünnungsverhältnisse ermöglicht. Gleiches konnte bei den Überprüfungen von Gebieten um Güllebecken in Erfahrung gebracht /Nádasdy-Budai 1978) werden. In Fällen von tonigen Böden konnten Verschmutzungsmaterialien bis zu einer Tiefe von 200-500 m im Grundwasser nachgewiesen werden, bei Sandböden eine starke Konzentration in Unregelmäßigkeit bis zu einer Entfernung von 3 km.

2.6 Die Veränderung des Bodens durch Einfluß der Beregnung

Der Flugsandboden des Gebietes weist bei Tiefe bis zu 60 cm einen Humusgehalt von 0,3 - 1 % auf. Auf Grund der Beregnung maßen wir in einer Tiefe von 0 - 30 cm 2,7 %, zwischen 30 - 60 cm 0,9 % organische Stoffe. Wo Furchenberegnung angewendet wurde, war auch in tieferen Schichten gleichmäßig verteilt humusbildendende organische Stoffe anzutreffen. Genau die gleiche Verbesserung war bei N, P, K Nährstoffgehalt nachzuweisen.

3 VORSCHLÄGE ZU GÜLLEBEREGNUNG

Die Betriebe müssen gezwungen werden zu einem sparsamen Wasserverbrauch. Die große Menge Gülle erhöht die Probleme der Lagerung und Aufbereitung. Die idealste Lösung wäre natürlich die Streuhaltung. Der große Na^+ Gehalt des technologischen wassers erschwert die Ausbringung. Wo aus hydrogeologischen oder agronomischen Gründen ungeeignete Gelände für die Pflanzenproduktion sind, dort dürfen keine Gülle- technologischen Tieranlagen errichtet werden. An Stelle von "Superkombinaten" sind aus Sicht des Umweltschutzes 8 - 10000-er Schweineställe und 1000-er Rinderställe als optimal zu betrachten.
Die Gülleverarbeitungs - und Ausbringungsanlage soll ein organischer Teil des Betriebes sein.
Bei der komplexen Überprüfung der Belastungsverhältnisse, sind neben dem Nährstoffgehalt auch die hydrogeologischen Bedingungen in Betracht zu ziehen und zu bestimmen.
Die homogenisierte Gülle verursacht selbst auf Sandböden Probleme der Kolmatation. Nach Beregnung kann die Fläche nach 1-2 wochen eingeackert werden, bei Furchebewässerung ist erst nach 3-4 Woche die Bedeckungsarbeit durchzuführen. Bei der

Furchenbewässerung ist das Umweltschutzproblem größer, aber günstiger werden die organischen Stoffe verteilt. Das ist bei Sandböden besonders wichtig.

4 ÜBERPRÜFUNGSERFAHRUNGEN

Es wäre zweckmäßig eine einheitliche Methode zu erarbeiten, zu schaffen für die Auswertung und Aufarbeitung der Grundwasser- Prüfungsergenisse. Die Grundwasserqualitätsprüfung ist nur mit der Nebenordnung der Nebenordnung der Mengenverhältnisse zusammen zu erledigen.

5 SCHLÜßFOLGERUNGEN AUS DEN GEMACHTEN ÜBERPRÜFUNGEN

Bei Prüfungen auf Sandböden sind die Stellen der Probeentnahmepunkte in Richtung der Strömungsverhältnisse bis zu einer Entfernung von 4 - 5 km zu markieren.
Wegen der gute Strömungsverhältnisse sind rechtzeitig dichtere Proben erforderlich.
Die beste veränderliche Komponente ist die frische Vershmutzung anzeigende Fäkalien Coliform.
Die schwach adsorbtiven fähigen, oxidativen Verhältnisse auf Sandböden zeigen neben einer ständigen Verschmutzungsnachschub NO_3^-, eine gleichmäßig verteilte Coliformkomponente.
Die Veränderung der Grundwassermenge ist lokal. Die Verschmutzungskomponente kannleicht mit der Grundwasserströmung befördert werden. Nach einigen km ist mit einer Verdünnung zu rechnen, mit Adsorbtion und mit Eliminierung.
Die Ergebnisse der Untersuchung können bei der Planung und Entwicklung neuer Anlagen von Nutzer sein.

Literaturverzeichnis

Sántha A /1981/1./ A higtrágya kezelő és hasznosító rendszerek rövid összefoglaló értékelése. Pécs.p. 7-23.

Stefanovits P /1977/ Talajvédelem, környezetvédelem. Mezőgazdasági Kiadó p 199-200.

Stefanovits P /1981/ Talajtan Mezőgazdasági Kiadó Bp.p 15-66.

Neppel F /1979/ Talaj és rétegvizek kapcsolatának vizsgálata a Közép-Tisza vidékén. VITUKI

Csaba-Kiss-Szinay-Vermes /1978/ Higtrágya-hasznosítás Mezőgazdasági Kiadó, Bp. p 72-192.

Szebellédy L./1967/ Magyarország első vizadórétegének minőségi térképe és ezek alapján a hasznosítás lehetőségei vizminőségi szempontból. Kandidátusi disszertáció. MTA /Kézirat/

Nádasdy-Budai /1978./ Higtrágyatározók talajvizszennyező hatásának vizsgálata. VMGT-99. Bp. p 137-139.

Kovács R /1980/ Higtrágya kezelési és elhelyezési gondok a Közép-Tisza vidékén. KÖTIVIZIG Kézirat

Kovács R A higtrágya károsító hatása a talajvizekre KÖTIVIZIG Kézirat /1982/

Land and Water Use, Dodd & Grace (eds), © 1989 Balkema, Rotterdam. ISBN 90 6191 980 0

Nitrate in shallow unconfined ground water beneath agricultural fields

W.L.Magette & A.Shirmohammadi
University of Maryland, College Park, USA

ABSTRACT: Several ground water monitoring studies in Maryland and elsewhere indicate that nitrate concentrations in shallow ground water beneath agricultural fields routinely exceed 10 mg/L. This paper discusses the results of these studies and relates ground water quality to agricultural management practices. Possible alternatives for minimizing nitrogen leaching from agricultural fields are described.

Mehrere Grundwasser uberwachungs Forschungen, im Staat Maryland und anderswo, zeigen die Nitrat Konzentration ofters hoher als 10 mg/L im seichtem Grundwasser unter landwirtschaftlichem acker an. Dieser Aufsatz discutiert die Ergebnisse von diese Forschungen und zeigt die Berziehung zwischen grundwasser Qualitat und beste landwirtschaftliche verwaltungs Anwendungen. Die mogliche alternativen zur verkleinerung nitrat Ablaugung von landwirtschaftlichen Feldern sind beschrieben.

Plusieurs études conduites dans l'état de Maryland et ailleurs indiquent que les concentrations de nitrate dans l'eau située près de la surface des champs agricoles sont en excès de 10 mg/L. L'article discute les résultats de ces études et rattache la qualité de l'eau sous-terraine à la gestion des champs agricoles. Des alternatives pour réduire cet l'infiltration dans les champs agricoles sont décrit.

1 INTRODUCTION

Concern about ground water quality is widespread. Research indicates that the quality of ground water in rural areas may be adversely affected by some agricultural activities (Hallberg 1986). Especially at risk are those unconfined aquifers overlain by unconsolidated sediments in coastal plain physiographic provinces or in fluvial deposits near streams and rivers. In these regions, the water table may rise to within a few meters of the soil surface (often < 3 m).

A nitrate standard for public drinking water exists in the United States at 10 mg/L nitrate nitrogen (NO_3-N) to guard against the development of methemoglobinemia in infant humans. Some scientists question the need for a standard at this level (Lehr 1985), citing the rarity at which the disease occurs and noting the simple and efficient cures available.

Research suggests that it may be difficult to maintain NO_3-N concentrations in ground water below 10 mg/L beneath some agricultural production areas. Baker (1981) reviewed a wide variety of scientific literature reporting drainage water from agricultural areas that contained nitrate nitrogen in concentrations above 10 mg/L on a widespread basis. Pacheco and Magette (1987) surveyed the literature and determined that NO_3-N concentrations in subsurface drainage in agricultural areas averaged 21 mg/L. Kanwar, Baker and Baker (1987) reported NO_3-N in subsurface drainage beneath both conventional and notill corn production fields almost always exceeded 10 mg/L, but was influenced by fertilization practices. The quality of water draining from the root zone is not necessarily the same as that which ultimately reaches the water table (Linden, et al. 1984; Staver, Brinsfield and Magette 1988; Leonard, et al. 1988; Shirmohammadi, et al. 1988). Nevertheless, high losses of nitrates below the root zone indicate the potential for high NO_3-N concentrations in ground water, especially when the ground water table is near the root zone (Hallberg, Baker and Randall 1986).

Ritter and Chirnside (1983) reported that nitrate in 32% of wells in Delaware, many of which were in agricultural areas, exceeded 10 mg/L. Robillard (1988) reported that much of the ground water beneath an intensively farmed area in southeastern Pennsylvania contained nitrates in excess of 10 mg/L. Wall and Magner (1988) reported that the 10 mg/L NO_3-N standard was exceeded in 25% and 19% of the wells in surficial sand aquifers beneath intense and moderate agricultural areas, respectively, in Minnesota. Bachman (1984) found that 15% of the wells sampled in a study of wells in the surficial aquifer beneath the DELMARVA penninsula contained more than 10 mg/L nitrate nitrogen. Areas especially affected were those where urban and agricultural land use occurred on well drained soils.

In Maryland, concern about water quality in the Chesapeake Bay and the realization that ground water quality may affect Bay water quality has added to the interest about ground water. This paper summarizes the findings of four studies being conducted in Maryland to trace the fate of nutrients transported from agricultural production systems. The results indicate that shallow, unconfined ground water is being impacted by typical agricultural fertilization practices. Data also indicate that in the coastal plain, it may be difficult to meet a 10 mg/L nitrate standard for shallow ground water beneath corn and soybean production areas.

2 STUDY FEATURES

Three of the four studies are being conducted on Maryland's Eastern Shore, a highly productive agricultural area in the Atlantic coastal plain. One of these is taking place on a University of Maryland Agricultural Experiment Station (MAES) research farm. The fourth study is also located on a MAES farm, but is being conducted in the piedmont physiographic province.

2.1 Study 1

This study is being conducted on a commercial grain farm on Maryland's Eastern Shore. The primary crops grown are corn and soybeans, roughly in 1-year rotations by field. Alfalfa is also grown. The soils at the site are largely Sassafras sandy loams and Elkton silt loams. Twenty-two ground water monitoring wells have been established in the shallow ground water (Magette, et al. 1987). Most of these are approximately 8 m deep with 3-m screens. The wells are sampled on a monthly basis, at which time water level elevations are also recorded.

Split applications of fertilizer are made for corn. In 1986, conventionally tilled corn (chisel plowing and secondary tillage) received 180 kg/ha nitrogen in 4 applications; notill corn received 142 kg/ha nitrogen in 2 applications. In 1987, conventional corn received 171 kg/ha nitrogen and notill corn received 146 kg/ha in 4 and 2 applications, respectively. The first fertilizer applications were made approximately one month prior to planting in May of both years. Corn was harvested for grain in mid-October in 1986 and in late September in 1987.

Conventionally tilled soybeans (chisel or moldboard plowing plus secondary tillage) were planted in late May and harvested in late October and early November both years. Only phosphorus and potash were applied as fertilizer.

2.2 Study 2

This study is being conducted at The University of Maryland Wye Research and Education Center in a long-term effort to evaluate losses of nutrients, sediment and pesticides from notill and conventionally tilled continuous corn systems (Staver, Brinsfield and Magette 1988). Here, conventional tillage consists of chisel plowing only. Soils at the site are Mattapex silt loams. Both systems are fertilized at the same rate of approximately 158 kg/ha nitrogen in two applications. Corn is planted in mid-May and harvested in mid-September.

Ground water monitoring wells have been installed on both watersheds in this study to a depth of 3 to 5 meters below ground surface; screens are 1.5 m long. The water fluctuates between 1 and 3 meters below the ground surface. Wells are sampled monthly, and water level elevations are recorded.

2.3 Study 3

This study was conducted on commercial farms on Maryland's Eastern Shore on

Sassafras sandy loam soils (Weil, Weismiller and Turner 1988). In 1986, all fields were planted to corn and irrigated using surface water supplies. Some fields were fertilized prior to planting with poultry manure (of unknown nitrogen content) and 106 kg/ha nitrogen at planting as a side dressing. One of these fields received additional poultry manure in the fall of 1986 and again in mid-March 1987. Other fields received 225 to 240 kg/ha nitrogen in 1986. Two of the four fields received top dressings in mid-June of 100 and 120 kg/ha nitrogen, respectively. Wheat followed corn on one field, as did barley on another.

Monitoring wells were installed in the shallow ground water beneath all fields and into nearby forested areas at depths ranging from 4 to 8 meters below ground level. Screens extended upward to within 1.5 meters of the ground surface. Wells were monitored monthly at which time samples were collected and water table levels recorded.

2.4 Study 4

This study is underway at The University of Maryland Forage Research Farm on Manor loam soils in the piedmont physiographic province (Angle, Gross and MacIntosh 1988). Since 1978, the study has monitored pollutant movement from watersheds to which conventionally tilled and notill corn has been planted. Conventional tillage pratices have consisted of contour moldboard plowing, disc-harrowing and cultipacking. Both watersheds received 470 kg/ha nitrogen in two separate applications. Corn was planted in mid-April and harvested for grain in October. Stalks were mowed on both watersheds following harvest but only the notill system received a winter cover crop (barley).

Monitoring wells approximately 10 to 11 meters deep were installed with 5 m screens in both watersheds as well as in grassed control areas upgradient of the watersheds. Samples were collected monthly, but water table elevations were not recorded.

3 RESULTS

Precipitation recorded at The University of Maryland Wye Research and Education Center totaled 102.8 cm, 89.4 cm, 78.0 cm and 61.7 cm for the years 1984-87. These amounts were 6.9 cm, 20.3 cm, 31.7 cm and 48.0 cm, respectively, below the 30-year average during the 4 years. These data are especially representative of precipitation that occurred at the sites of studies 1, 2 and 3, and indicative of precipitation at the site of study 4.

Ground water data were collected in Study 1 from June 1986 to December 1987. Frequencies in which NO_3-N concentrations exceeded 10 mg/L beneath selected fields are summarized in Table 1.

Wells were monitored monthly in Study 2 from July 1986 through April 1988. NO_3-N concentrations averaged approximately 15 mg/L throughout the entire period beneath both tillage systems and were not statistically different. Concentrations varied from approximately 8 mg/L to 19 mg/L in the notillage watershed and from 12 mg/L to approximately 18 mg/L in the conventionally tilled watershed.

In Study 3, NO_3-N concentrations for the periord August-November 1986 averaged 18.3 and 15.1 mg/L for manured and nonmanured fields, respectively. In December-May, however, these averages were 41.6 and 14.8 mg/L for the manured and nonmanured areas, respectively, and were statistically different.

Ground water monitoring in Study 4 was conducted from May 1984 to February 1988. NO_3-N concentrations in ground water beneath grassed control areas generally averaged 2 mg/L, whereas in both watersheds nitrate nitrogen averaged approximately 9 mg/L. Of 34 sampling dates for which data are reported, NO_3-N in ground water exceeded 10 mg/L on 2 and 12 events in the notill and conventionally tilled corn systems, respectively.

4 DISCUSSION

Bachman (1984) reported that naturally occurring nitrate nitrogen concentrations in in the surficial aquifer of DELMARVA are 3 mg/L. Based on this measure, data from the four studies reported here indicate that some typical agricultural practices are impacting shallow ground water quality in Maryland's coastal plain. If a similar measure of natural ground water quality holds for unconfined ground water in the piedmont province, the same might be said about some agricultural practices in central Maryland.

In virtually all sampling events for studies conducted in the coastal plain, concentrations of nitrate exceeded the federal limit for nitrate in public drinking water supplies. Sampling was conducted in areas for which a key

Table 1. Percentage of sampling events in which NO_3-N in ground water exceeded 10 mg/L, study 1.

Crop, cover or soil condition	Field	%
Sandy loam soils		
Conventionally tilled corn	B2	100
Corn stalks	B2	100
Plowed	B2	100
Soybeans after corn	B2	100
Soybean stubble	B2	100
Notill corn	H1	0
Corn stalks	H1	0
Notill corn, year 2	H1	0
Corn stalks	H1	0
Corn stalks	D	50
Plowed	D	75
Soybeans after corn	D	100
Soybean stubble	D	100
Silt loam soils		
Soybean stubble	F	0
Plowed	F	100
Conventionally tilled corn	F	88
Corn stalks	F	100

recommended nutrient management practice -- i.e., split applications of nutrients -- was followed. Except for Study 3, fertilization rates also were within realistic and recommended values. Sampling included ground water below both corn and soybeans, crops with dramatically different requirements for supplemental nitrogen. Finally, sampling included areas of both well-drained and moderately well-drained soils.

An important point to remember about the studies reported here is that monitoring wells were designed to withdraw ground water samples from the water table -- i.e., from the most highly impacted, upper boundary of ground water. Numerous studies confirm that NO_3-N concentrations in ground water decrease with increasing well depth. For example, Detroy, Hunt and Holub (1988) found that of the 887 wells tested in Iowa, 50% of those with greater that 10 mg/L NO_3-N were less than 19 m deep. Consequently, nitrate nitrogen concentrations reported here are likely, but not necessarily, higher than might have been found had the wells been screened to extract samples several meters below the water table.

Nevertheless, these data justify concern about the impact of agricultural practices on shallow, unconfined ground water in coastal plain regions. Similarly, although nitrate concentrations found in study 4 were not as high as those found in coastal plain studies, there is reason to examine agricultural practices for their ground water quality impacts in that region as well.

5 CONDITIONS FAVORING NITRATE LEACHING

A logical question to pose is why, when best management practices for nutrient management were followed in these studies, did an appreciable amount of nitrogen leach into ground water. One answer might reflect the generally poor crop yields that were obtained in the three coastal plain studies, even in study 3 which included the use of irrigation. Depressed yields were caused by the abnormally low precipitation that occurred during the study periods. Because of poor crop yields, unused nitrogen remained in the soil profile after harvest, which was subsequently leached in the fall and winter when precipitation exceeded evapotranspiration and soil storage.

Another contributing factor to the findings of the four studies reported here was likely the proximity of the water table to the soil surface. Water tables in studies 1, 2, and 3 were within 1 to 3 meters of the soil surface, whereas those in study 4 were much deeper. Nitrate nitrogen in percolate collected in study 4 at a depth of 1.5 m averaged 15 and 30 mg/L over the study period from the notill and conventionally tilled watersheds, respectively. Had the water table during the study been near 1.5 m, measured nitrate in ground water likely would have been higher.

If low crop yields were the only reason for the high nitrate concentrations in ground water reported in these studies, the same results might not be found under more normal precipitation and crop yield conditions. Data from Roka (1988) were used to examine the potential temporal distribution of high NO_3-N leaching losses as a function of cropping system. Values for total NO_3-N leached and total water drained below the root zone were combined to approximate an annual average concentrations of NO_3-N leached. As shown in Table 2, the distributions of years in which average NO_3-N concentrations in water leaving the root zone vary by cropping system. These simulated results suggest the frequency with which findings from the field studies reported herein might occur.

Results from these simulations also help identify the reasons that leaching of NO_3-N occurred. Leaching data are shown in Tables 3 and 4 for two of the cropping systems that frequently experienced NO_3-N losses in excess of 10 mg/L. Tables 3 and 4 illustrate when during the year NO_3-N leaching and water percolation took place. It is important to recognize in examining Tables 3 and 4 the fertilization patterns for each system. Continuous corn was fertilized on 3 dates -- March 27, May 9 and June 10. Corn in rotation with wheat was fertilized on these 3 dates also; in addition, fertilizer was applied on October 21 to establish the subsequent wheat crop. In the years that wheat was grown, fertilizer was applied on March 27 and October 21.

A comparison of data in Tables 3 and 4 indicate that both cropping systems experienced similar H_2O percolation patterns. Percolation varied by month in response to typical precipitation patterns for the area being simulated. Dramatic differences occurred in the frequency of days when NO_3-N leaching exceeded 10 mg/L. Closer examination of the data indicate that exceedances occurred generally after fertilization dates. The corn-wheat rotation system experienced more frequent losses of NO_3-N in excess of 10 mg/L than the continuous corn production system because of the fall fertilization of wheat. These data illustrate the importance of minimizing the buildup of inorganic nitrogen (i.e. NO_3-N) in the soil during periods when precipitation patterns would suggest that percolation will take place.

Table 2. Number of years in which simulated average annual NO_3-N concentrations in water leaching from the root zone exceeded 10 mg/1[1].

Cropping System	Silt Loam	Sandy Loam
	Conventional Tillage	
Corn	7	4
Soybeans	1	0
Corn-Soybeans	2	1
Corn-Wheat-Soybeans	0	1
Corn-Wheat(Clover)	6	7
Corn with Cover Crop	3	4
Wheat	1	3
Wheat-Soybeans Double Crop	0	2
Grass	0	0
	Notill	
Corn	6	4
Soybeans	0	0
Corn-Soybeans	3	1
Corn-Wheat-Soybeans	0	1
Corn-Wheat(Clover)	6	7
Corn with Cover Crop	5	4
Wheat	2	3
Wheat-Soybeans Double Crop	0	2

[1]Based on data from Roka (1988), 10 years of simulated data using CREAMS simulation model.

Table 3. NO_3-N leaching in excess of 10 mg/L from continuous corn production.

Month	H_2O Percolation, Days	NO_3-N Leaching, Days
Jan	3.7	0.8
Feb	2.7	0.3
Mar	2.7	0.3
Apr	2.7	1.4
May	2.6	2.6
Jun	1.7	1.7
Jul	0.6	0.6
Aug	0.1	0.0
Sep	0.0	0.0
Oct	0.5	0.0
Nov	1.3	0.1
Dec	2.2	1.0

Data are mean values for 10-yr simulation.

Table 4. NO_3-N leaching in excess of 10 mg/L from corn-wheat production in a 2-yr rotation (and clover interseeded with wheat).

Month	H_2O Percolation, Days	NO_3-N Leaching, Days
Jan	3.2	1.7
Feb	3.1	1.8
Mar	1.8	1.0
Apr	2.5	2.2
May	1.4	1.4
Jun	1.1	1.1
Jul	0.5	0.4
Aug	0.3	0.0
Sep	0.0	0.0
Oct	0.4	0.3
Nov	1.2	0.7
Dec	2.8	2.3

Data are mean values for 10-yr simulation.

6 MANAGEMENT OPTIONS

Nitrogen must be managed differently than most other potential agricultural nonpoint source pollutants, especially phosphorus and sediment, because of its high mobility in the soil environment. Whereas sediment and phosphorus control can often be equated, nitrogen control may actually be thwarted by some practices that minimize losses of soil and phosphorus because such techniques often increase the movement of water through the soil profile. In the United States, nutrient management in agricultural systems urges "proper" application rates, placement and timing, as well as the use of practices that keep nutrients in the production system until they are utilized by plants (Weismiller, 1984). Proper nutrient application rates must be based on realistic (not maximum) yield goals, and on an assessment of all nutrient sources in the production system. The latter requires soil analyses for nutrient content in addition to testing of all manures or other nutrient amendments. Proper nutrient placement involves matching the nutrient source and tillage system to the application technology that will promote the highest nutrient use efficiency. Proper timing dictates that the production system be supplied with nutrients at the times that they are needed by growing crops. This generally suggests multiple applications of nutrients during a given growing season. Retention techniques encompass the use of different tillage systems, cover crops and crop rotations.

While these techniques are sound theoretically, and have been proven in practice, three of the studies reported herein confirm that even when such practices are followed, NO_3-N in shallow ground water can still exceed 10 mg/L. Consequently, new management techniques may be necessary to control nitrogen, or existing techniques utilized more intensively.

6.1 Expert systems

Magette, et al. (1989) discussed the problem of managing potential ground water pollutants from agricultural systems. The dilemma facing managers is that whenever nitrate is in the soil environment and water is available to percolate through the profile, movement to ground water can occur. The authors suggested that, because agricultural production occurs in an unconfined environment, a combination of techniques must be used. They recommended that an expert system be developed to help producers use nutrients in ways that reconcile environmental and economic concerns. Borgelt (1988) suggested ways that a geographic information system might help accomplish this objective on individual farms.

Magette, et al. (1989) also recommended the use of targetted education programs for agricultural producers and the general public, as well as public policies that enhanced adoption of some management techniques. Sharing of adoption costs by producers and government was a favored policy.

6.2 Frequent soil and plant tissue testing

Sound nutrient management is predicated on properly assessing the need for nutrients in the production system. Westerman (1987) reported that the best management tool a producer has for making fertilizer use decisions is a soil NO_3-N test. Tissue analysis has also been found to be an effective means of assessing fertilizer needs; however, difficulties are encountered in practice because tissue tests often are made too late in the growing season to correct nutrient deficiencies (Keeney 1982). Because nitrate is so mobile, frequent soil testing would be necessary in humid regions. An economic evaluation would be essential to determine the number of fertilizer applications that could realistically be made for a given crop. Cost-sharing could perhaps be used to favor the shift to a higher frequency of small nutrient applications. Alternatively, incentives could be established to promote the use of irrigation as a means to not only help assure anticipated crop yields, but also to facilitate multiple applications of nitrogen in small doses (i.e. fertigation).

6.3 Use of organic nutrient sources

Both "organic" farming and carbon supplements were discussed by Kenney (1982) as means to 1) immobilize soil nitrogen and 2) reduce needs for commercial fertilizers and recycle animal wastes. Luhr (1986) reported that the use of high rates of organic fertilizer together with inorganic fertilizer resulted in decreased NO_3-N concentrations in ground water the following year. Magette, et al. (1988) indicated in a preliminary study using broiler litter as the primary nitrogen source that ground water quality below the application site was little different than below nearby areas receiving commercial fertilizer. The effect of N immobilization through the use of organic wastes could not occur indefinitely, however (Gilliam and Hoyt 1987). The technique might be useful as a short-term solution, or in combination with other strategies to form a long-term solution. Acceptance of reduced yields or major changes in technology would likely be necessary if organic nutrient sources were relied upon totally to meet all crop needs (Keeney 1982).

6.4 Cover crops

"Cover" crops consisting of grasses or small grains have long been used as an erosion control practice to cover the soil surface after harvesting of a primary crop. The ability of cover crops to take up residual inorganic N in the soil has been documented often, most recently by Brinsfield, Staver and Magette (1988). The ability to plant cover crops is adversely affected by unsuitable weather conditions that may occur following harvesting of the primary crop. In addition, the expense of using cover crops is usually much more than the value of the nitrogen incorporated into plant biomass and prevented from leaching. Cost-sharing may be necessary to promote the use of cover crops for nitrogen control. As discussed previously, if fall fertilization is practiced (perhaps to enhance growth of a winter cover crop), the chances are increased that NO_3-N will leach from the system.

7 CONCLUSIONS

The three coastal plain studies reported here indicate that even when recommended nutrient management practices are utilized, losses of nitrate from agricultural production systems can result in concentrations in shallow ground water that are significantly higher than the 10 mg/L standard for public drinking water. Agricultural activities in the piedmont province may also impact ground water quality to a lesser extent.

These results suggest the need to seriously evaluate the feasibilty of a 10 mg/L nitrate nitrogen standard for shallow ground water beneath some agricultural production areas. The extent to which impacts of the magnitudes reported herein cannot be known with certainty. Computer modelling suggests that corn, whether produced by notill or conventional methods, could result in average nitrate nitrogen concentrations leaving the root zone greater than 10 mg/L in 4 to 6 out of 10 years.

8 REFERENCES

Angle, J. S., C. M Gross and M. S. McIntosh. 1988. Nitrate concentrations in percolate and ground water under conventional and notill corn watersheds. Agric. Ecosyst. Environ. (In Press).

Bachman, L. J. 1984. Nitrate in the columbia aquifer, central Delmarva peninsula, Maryland. USGS WRI Report 84-4322. U. S. Geological Survey, Towson, MD. 51 pp.

Baker, J. L. 1981. Agricultural areas as nonpoint sources of pollution. In: M. R. Overcash and J. M. Davidson (eds.), Environmental impacts of nonpoint source pollution, p. 275-310. Ann Arbor, Anne Arbor Press.

Borgelt, S. C. 1988. Geographic information systems for on-farm use. ASAE Paper 88-1606 Presented at 1988 International Winter Meeting of the American Society of Agricultural Engineers. St. Joseph, Amer. Soc. Ag. Eng.

Brinsfield, R. B., K. Staver and W. L. Magette. 1988. Role of cover crops in reducing nitrate leaching. In Agricultural impacts on ground water, p.127-146. Dublin, National Water Well Association.

Gilliam, J. W. and G. D. Hoyt. 1987. Effect of conservation tillage on fate and transport of nitrogen. In T. M. Logan, et al. (eds.), Effects of conservation tillage on groundwater quality: nitrates and pesticides, p.217-240. Chelsea, Lewis.

Hallberg, G. R. 1986. From hoes to herbicides: agriculture and groundwater quality. J. Soil & Water Cons. 41(6):357-364.

Hallberg, G. R., J. L. Baker and B. W. Randall. 1986. Utility of tile-line studies to evaluate the impact of agricultural practices on ground water. In: Agricultural Impacts on Ground Water -- A Conference, p. 298-326. Dublin, National Water Well

Association.

Kanwar, R. S., J. L. Baker and D. C. Baker. 1987. Tillage and n fertilization management effects on groundwater quality. ASAE Paper 87-2077 Presented at 1987 Summer International Meeting of the American Society of Agricultural Engineers. St. Joseph, Amer. Soc. Ag. Eng.

Keeney, D. R. 1982. Nitrogen management for maximum efficiency and minimum pollution. In F. J. Stevenson (ed), Nitrogen in agricultural soils, p.605-649. Madison, American Society of Agronomy.

Lehr, J. H. 1985. Nitrate: unwarranted fear, unnecessary expense. Ground Water Mon. Rev. 5(2): 4-5.

Leonard, R. A. , A Shirmohammadi, A. W. Johnson and L. R. Marti. 1988. Pesticide transport in shallow groundwater. Trans. of ASAE (In Press).

Linden, D. R., C. E. Clapp and W. E. Larson. 1984. Quality of percolate water after treatment of municipal wastewater effluent by a crop irrigation system. J. Environ. Qual., 13:256-264.

Luhr, H. P. 1986. Water - a natural resource in danger. In F. Bender (ed.), Geo-resources and environment, p.47-58. Stuttgart, E. Schweizerbart'sche Verlagsbuchhandlung.

Magette, W. L., R. A. Weismiller, B. V. Lessley, J. D. Wood and C. F. Miller. 1987. Establishing a demonstration farm for agricultural best management practices. In R. R. Allen, et al. (eds.), Optimum erosion control at least cost, pp. 100-107. ASAE Publication 08-87. St. Joseph, Am. Soc. of Agric. Eng.

Magette, W. L., R. B. Brinsfield and D. A. Hrebenach. 1988. Water quality impacts of land applied broiler litter. ASAE Paper 88-2050 Presented at 1988 Summer International Meeting of the American Society of Agricultural Engineers. St. Joseph, Amer. Soc. Ag. Eng.

Magette, W. L., R. A. Weismiller, and J. D. Wood. 1988. Design and implementation of a farm-scale ground and surface water monitoring network. In: Agricultural impacts on ground water, pp. 529-546. Dublin, National Water Well Association.

Magette, W. L., A. Shirmohammadi, B. V. Lessley and R. A. Weismiller. 1989. Managing ground water in relation to agricultural activities. Am. Water Res. Bull. (in press).

Ritter, W. F. and A. E. M. Chirnside. 1983. Impact of land use on groundwater quality in southern Delaware. Ground Water 22:38-47.

Robillard, P. D. Groundwater quality in an intensively farmed region. ASAE Paper NAR-88-204 Presented at the 1988 North Atlantic Region Annual Meeting of the American Society of Agricultural Engineers. St. Joseph, Amer. Soc. Ag. Eng.

Shirmohammadi, A., W. L. Magette, R. B. Brinsfield and K. R. Staver. Groundwater loading of agrochemicals in the coastal plain region. ASAE Paper 87-2632 Presented at 1987 Winter International Meeting of the American Society of Agricultural Engineers. St. Joseph, Amer. Soc. Ag. Eng.

Staver, K. R., R. B. Brinsfield and W. L. Magette. 1988. Nitrogen export from Atlantic coastal plain soils. ASAE Paper 88-2040 Presented at 1988 Summer International Meeting of the American Society of Agricultural Engineers. St. Joseph, Amer. Soc. Ag. Eng.

Wall, D. B. and J. A. Magner. 1988. A comparison of surficial sand aquifer water quality in intense, moderate and limited agricultural regions of Minnesota. In: Agricultural impacts on round water, p. 297-316. Dublin, National Water Well Association.

Weil, R. R., R. A. Weismiller, and R. S. Turner. 1988. Nitrate contamination of groundwater under irrigated coastal plain soils. Unpublished report to Maryland Water Resources Research Center, The University of Maryland, College Park, Maryland.

Westerman, R. L. 1987. Efficient nitrogen fertilization in agricultural production systems. In D. M Fairchild (ed.), Ground water quality and agricultural practices, p.137-151. Chelsea, Lewis.

Land and Water Use, Dodd & Grace (eds), © 1989 Balkema, Rotterdam. ISBN 90 6191 980 0

Movement of pesticides and nutrients in drainage water

E.J.Monke, E.J.Kladivko, G.E.Van Scoyoc, M.A.Martinez & R.L.Huffman
Purdue University, West Lafayette, Ind., USA

ABSTRACT: Pesticides and nutrients as well as water flow rates are being monitored from six subsurface drains spaced 5, 10 and 20 m apart, in Clermont silt loam, a slowly permeable soil in Southeastern Indiana (USA). Discrete, continuous samples for pesticides and nutrients are being collected. Carbofuran, atrazine, cyanazine, and alachlor have been detected in the outflow water mostly after chemical application in the spring. The relative amounts of pesticides detected in the tile drainage water are consistent with adsorption isotherm data for these pesticides. A finite element model to describe chemical and water movement through the soil to the subsurface drains has been developed. The model compared well to actual flow hydrographs from the drains.

ZUSAMMENFASSUNG: Pestizide und Nährstoffe, sowohl als auch Wasserflussraten, werden von sechs unterirdischen Abflüssen in Abständen von 5, 10 and 20 Metern, in Clermont Lehmerde, einer langsam durchlässigen Erde im südöstlichen Indiana (USA), überwacht. Diskrete, kontinuierliche Stichproben für Pestizide and Nährstoffe werden gesammelt. Carbofuran, Atrazin, Cyanazin, und Alachlor wurden im Abflusswasser, hauptsächlich nach chemischer Applikation im Frühjahr, gefunden. Die relativen Mengen von Pestiziden, die im Plattenabflusswasser entdeckt wurden, stimmen mit den Adsorptionsisothermdaten für diese Pestizide überein. Ein Modell der finiten Elemente wurde entwickelt, um die chemische und die Wasserbeförderung durch die Erde zu den Untergrundabflüssen zu beschreiben. Das Modell war in guter Übereinstimmung mit den wirklich beobachteten Hydrographen der Abflüsse.

ABRÉGÉ: Six drains souterrains éspacés à intervalles de 5, 10, et 20 metres dans un sol de terreau limon contrôlent l'allure de l'écoulement d'eau, de pesticides, et de substances nutritives. Le terreau limon est un sol de perméabilité lente qui se trouve dans la région sud-est de l'état d'Indiana aux Etats-Unis. Nous ramassons des échantillons discrets et continus des substances nutritives et des pesticides. Nous avons distingué du carbofuran, de l'atrazine, du cyanazine, et l'alachlor dans l'eau de l'écoulement le plus souvent après l'application de produits chimiques au printemps. Les quantités relatives de pésticides aperçus dans l'eau de drainage sont compatibles avec les données d'adsorption isothermique pour ces pésticides. Nous avons développé un modèle d'élément fini pour décrire le mouvement de produits chimiques et d'eau à travers le sol au drainage souterrain. Le modèle est comparable aux véritables hydrographiques de l'écoulement dans les drains.

1 INTRODUCTION

Annual pesticide usage on farms in the United States totals around 300 million kilograms. It is not surprising then that detectable levels have been found in the ground water even though most modern pesticides have relatively short half-lives (Holden, 1986). In some areas, elevated nitrate levels have also occurred in ground water due to agricultural operations (Hallberg, 1986). In the Corn Belt Region of the United States, the average rate of applied nitrogen fertilizer has increased from around 70 kg ha^{-1} in 1965 to close to 150 kg ha^{-1} now (Keeney, 1985). Since nitrogen use efficiency rarely exceeds 70% and is more likely to be around 50%, the increased application rates are leaving a large reserve of nitrogen in the soil. Leaching of pesticides and nitrates into ground water is accentuated where the soil is coarse textured, where the organic matter content is low at least for pesticides, and where an abundance of subsidiary pathways (cracks, holes, etc.) can direct the leachate around a finer-textured soil matrix.

Many fields in the humid region of the United States are drained by subsurface collector lines to remove excess soil water mostly in the spring. Pesticides or nutrients that leach below the root zone may move into these lines and be discharged into surface ditches or streams or they may move as deep seepage into the ground water. Chemicals detected in drainage water therefore serve as useful indicators of potential chemical contributors to both surface and ground water contamination (Bottcher et al., 1981; Kanwar et al., 1987).

This paper reports on the results of an experiment in which water samples were collected at the outlets of replicated subsurface drains spaced 5, 10 and 20 m apart. The overall objective was to quantify the sediment, pesticide and nutrient concentrations found in drainage water under typical agricultural management practices for continuous corn grown on Clermont silt loam. A secondary objective was to establish a finite element model from which water and chemical movement to the subsurface drains could be predicted.

2 METHODS OF INVESTIGATION

In 1983, drainage research was initiated on a poorly-drained Clermont silt loam at the Southeast Purdue Agricultural Center (SEPAC) in Indiana. Separate studies included the effects of subsurface drain spacing on watertable drawdown and drain outflow, the influence of crop management factors on drainage response, and a comparison of the effectiveness of subsurface and surface drainage practices. Climatological variables, soil temperature, drain outflow, depths to the water table, piezometric heads, and some water quality parameters are being monitored. Most parameter values are automatically measured in the field and collected with an onsite data logger. The stored information is then transferred on demand by a telephone connection approximately 200 km to a computer facility at Purdue University. Then in 1986, automatic water samplers were installed to detect levels of sediment, nutrients and applied pesticides in the drainage water on a flow-proportional basis. Data on nutrient and pesticide concentrations had already been collected manually in 1985. The samplers have been operating continuously since they were installed (Kladivko et al., 1987).

2.1 Soil description

The Clermont soil (fine silty, mixed, mesic Typic Ochraqualf) is typical of an extensive area of similar soils extending from southwestern Ohio, across southern Indiana and southern Illinois as far west as eastern Kansas. The soil, formed in 50-120 cm of loess over Illinoian glacial till, is characterized by shallow, very slowly permeable layers. The surface soil at the study site is a light grey, low organic matter (1.3%) silt loam. The Ap horizon contains 22% sand, 66% silt and 12% clay. It is often referred to as "crawdad land" or "buttermilk flats" by local residents because of its wetness problem and surface color. King and Franzmeier (1981) identified a borderline fragipan in its silty clay loam B horizon and a distinct paleosol surface at 2-3 m depth. The Clermont soil generally occurs on <1% slopes.

2.2 Experimental set-up

Subsurface plastic drains, 10 cm in diameter, were installed on a 0.4% slope at spacings of 5, 10 and 20 m at an average depth of 75 cm. Three drain lines (225 m long) were installed at each spacing with the outside drain lines on each spacing acting as common drains between treatments. Two replications of each spacing were established in non-randomized blocks. The center drains of the 5, 10 and 20 m plots discharge into observation wells where outflow volumes are monitored continuously with tipping bucket devices connected to a datalogger (Huffman et al., 1986). In 1986, water samples were collected on a flow-proportional basis using commercially-procurred automatic samplers. Generally, composite daily samples were taken, but up to 24 separate samples each day were also obtained for a few weeks after chemical application to assess early movement of chemicals through the soil profile. Other measurements, as a regular part of the drainage experiment, would also be used in analyzing the collected data on pesticides, nutrients and sediment.

2.3 Crop and chemical management

Corn (*Zea mays L.*) was planted over the entire experimental area each spring from 1984 to 1988. Tillage and planting were carried out on each spacing treatment as soon as soil conditions permitted so that potential timeliness benefits of drainage could be assessed. Tillage was comprised of one pass of a chisel plow to a 20-25 cm depth in spring, and 2 passes with a soil finisher (disc or field cultivator). The soil finisher was used on the same day as planting in all cases. The time period between primary tillage and planting varied depending on rainfall amounts in each year. Corn was planted in 0.76 m rows, with rows up-and-down the slope and parallel to the drain lines.

Fertilizer N (285 kg N ha^{-1}) was injected as NH_3 (with the nitrification inhibitor nitrapyrin) before planting in all years. This rate of fertilizer N was considered appropriate for a yield goal of 12.5 Mg ha^{-1}. Pesticides were applied during the planting operation, with herbicides sprayed as a tank mix over the entire surface area and insecticides applied in granular form (15%) in a band over the seed. Herbicides used each year were atrazine, cyanazine and alachlor at rates of 1.1, 2.3 and 2.3 kg active ingredients (A.I.) ha^{-1}.

2.4 Laboratory methodology

Water samples from the drains were frozen until analysis. Portions of the samples were extracted onto a solid-phase (C_{18}), eluted from the column with benzene, and analyzed by gas chromatography for pesticide residue. Nutrients in the water (NH_4^+, NO_3^-, soluble P, soluble K) were analyzed by standard techniques (Keeney and Nelson, 1982; Murphy and Riley, 1962; Greenberg, 1981).

Adsorption isotherms for each pesticide were measured using a 6-hour equilibration time for the herbicides and a 24-hour equilibration time for the insecticides. Column leaching experiments with sieved repacked soil were conducted to determine movement in a uniform soil material (Fuller, 1982). Column soil samples were extracted with a 90% methanol: 10% water solution (Cotterill, 1980). These data helped to interpret the field-scale breakthrough curves for evidence of significant macropore flow.

2.5 Analysis procedure

The total mass of pesticides, nutrients, and sediment removed by each drain was calculated from data on concentrations and water outflow volumes. Because water outflow volumes were recorded continuously and concentrations of sediment or chemicals were not always measured every day in which outflow occurred, we assumed that concentrations remained constant between sampling dates. During the spring, after chemical application, all available samples were analyzed after the select chemicals were applied, so the assumption of constant concentration between sampling dates was quite reasonable. During the fall and winter periods, fewer samples were analyzed. However, pesticide concentrations were extremely low during this time period and so little error was introduced by assuming constant concentrations. On the other hand, relatively high concentrations of nitrate-N continued to be found during the fall and winter periods and so more potential error would be possible in the total nitrate loss calculations. However, rapid changes in nitrate concentrations generally did not occur during these periods.

We began a year-of-record on April 10 because soil preparation for the growing season occurred shortly after this date. Pesticide or nutrient losses for each sampling period were then summed to give the annual yield. Detection limits for the pesticides ranged from 0.5 to 1.6 ppb at the 90% confidence limit. The gas chromatograph which we used could record responses at much lower concentrations but the confidence level would be small. Consequently, we used only those concentrations >1 ppb to calculate the total amount of chemicals discharged from the drains.

3 RESULTS AND DISCUSSION

Weather during the study period (1985-88) was unusually dry especially during the spring months. Despite the lack of rainfall in the early part of the growing season, corn yields have been reasonably good even in 1988, the so-called drouth year in the United States, because of timely rains later on. Average corn yields over the drainage plots were 11.2, 8.2, 11.8, and 7.7 Mg ha^{-1} for 1985 through 1988, respectively. Rainfall amounts for April, May and June for the study period, as a percent of the 30-year normal rainfall, are shown in Table 1.

Table 1. Early growing season rainfall at SEPAC.

Month	Normal Rainfall (mm)	Percent of Normal Rainfall		
		1985	1986	1987
April	100	65	48	61
May	116	90	69	29
June	106	68	34	52

The 1986 and 1987 spring periods were so dry that drain outflow virtually ceased for the growing season at or before planting and chemical applications. Consequently, our analysis of the movement of chemicals during the early part of the growing season are only based on data for the first year. Other consequences of the lack of early growing season rainfall were a corresponding lack of benefits to the closely spaced drains for early planting (the planting date was May 1 ± 2 days for all spacings and for all years) and a more-or-less uniform yield response over all spacings for any one year.

3.1 Water and sediment yields

The amount and distribution of water removed by each drain were easily obtained from the data provided by the tipping buckets. The average discharge for the drains spaced 10 m apart as compared to yearly rainfall (a year begins on April 1) is shown in Table 2.

The average flow depth over the 5 m spacing which was discharged by the subsurface drains was over 40% greater than the flow depth for the 10 m spacing while that for the 20 m spacing was approximately 25% less. As expected the efficiency of a drain for capturing rainfall decreased as the spacing increased. However, the peak discharge from a drain increased as the spacing increased as shown in Figure 1.

Table 2. Average discharge for the drains spaced 10 m apart as compared to yearly rainfall.

Rainfall/Discharge	1985-86	1986-87	1987-88
Rainfall depth (mm)	1189	923	1033
Flow depth (mm)	212	116	152
Percent of rainfall	18	13	15

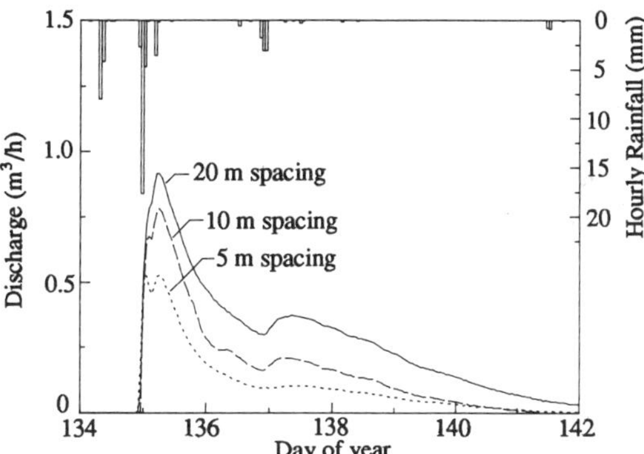

Figure 1. Drain discharge hydrographs and hourly rainfall for May 14 to May 22, 1985.

The drainage water contained very little sediment although at other sites considerable sedimentation of drain lines has occurred in Clermont silt loam because of its high silt content. Total amounts of solids removed in the drainage water varied between 15 and 120 kg ha^{-1} for the study period. Sediment yields were roughly proportional to the discharge volumes. On an area basis, the least sediment loss would be expected from the widest spacing; however, the peak yield would also be expected from drains with the widest spacing. There was no evidence that sediment was accumulating in the drains.

3.2 Nutrient losses to the drains.

The greatest loss of nutrients from the drains occurred as nitrate-N. Within a single cropping season, the most nitrate-N removed on an area basis was from the 5 m spacing while the least was from the 20 m spacing. Nitrate loss ranged from 18 kg ha^{-1} on the 20 m drain spacing in 1986-87 to a high of 74 kg ha^{-1} on the 5 m spacing in 1987-88. The range of total nitrate losses was similar to that found under other fertilized corn fields in the Corn Belt (Logan et al., 1980). In our study, nitrate-N concentrations were seldom less than 10 $\mu g\ ml^{-1}$ in the outflow water at any one time and were far more likely to be in the 20 to 30 $\mu g\ ml^{-1}$ range. Drain outflow usually did not occur from mid-June through September because of the high water demands of the corn crop. Nitrate-N losses then occurred essentially during three periods: after anhydrous ammonia application in early April until the flow from the drains ceased for the summer (Period 1), in the fall when flow resumed until the end of the year (Period 2), and, then maybe after some interruptions due to freezing, in the late winter and early spring (Period 3). Average concentrations of nitrate-N for all the spacings are shown in Table 3 for the three crop years. Many factors could have contributed to the seasonal variation of the concentration of nitrate-N in the drainage water including the residual nitrogen left over from the previous year, crop use and removal in terms of yield, rainfall amounts and distribution, and subsequent differences in drain discharge. At SEPAC, most of the nitrate-N losses to the drains occurred during the fall and winter months. In 1985, when spring rainfall was about 75% of normal, 12% of the nitrate-N that was lost was lost in the spring after nitrogen application. In 1986 and 1987, when spring rainfall was about 50% of normal, only 5% was lost.

Table 3. Seasonal variation of flow-weighted nitrate-N concentrations.

Crop year	Period 1	Period 2	Period 3
		$\mu g\ ml^{-1}$	
1985-86	24.8	20.7	19.0
1986-87	16.9	26.2	30.8
1987-88	31.9	40.2	28.3

In general, the losses of phosphorus and potassium as well as ammonium-N to the drains were very small and as such were quite variable. However, considering all years and all spacings, the average annual loss of NH_4^+–N, P and K were 0.7, 0.04 and 2.6 kg ha^{-1}, respectively, as compared to 41.8 kg ha^{-1} for NO_3^--N.

3.3 Pesticide losses to the drains

No pesticide residuals to the drains >1 ppb were detected after the early part of the growing season for which they were applied. Pesticide losses did not occur then in 1986 and 1987 because flow from the drains was negligible after planting and the application of the pesticides had occurred. Consequently, data on carbofuran, atrazine, cyanazine, and alachlor were only available for 1985 (chloropyrifos was first applied in 1986) when significant late spring outflow took place. Concentration ranges for this period were 5-150 ppb for carbofuran, 1-10 ppb for atrazine, 1-10 ppb for cyanazine, and 1-2 ppb for alachlor.

The pesticide concentrations found in the outflow samples correspond with the rank-order of batch adsorption coefficients and leaching rates measured in the laboratory on the Clermont soil. The sorption values and coefficients for the five pesticides used in the study are given in Table 4.

Table 4. Sorption values and coefficients for Clermont silt loam for five pesticides.

Pesticide	Sorption	K
	(%)	($cm^3 g^{-1}$)
Carbofuran	9	0.29
Atrazine	19	0.64
Cyanazine	20	0.68
Alachlor	35	1.10
Chlorpyrifos	98	198.0

Carbofuran was sorbed the least and it was found in the highest concentrations of all the pesticides in the drain outflow. Although the simple laboratory sorption measurements did a good job predicting the rank order of pesticide concentrations in the outflow, the differences in sorption did not appear to alter the travel time of the leading edge of the solute front for any of the pesticides. This may indicate that both preferential flow and nonequilibrium adsorption are occurring.

The total amount of pesticides lost in the drainage water during the 1985-86 crop year was below 1.0% of the applied carbofuran and below 0.1% of the applied atrazine, cyanazine and alachlor. As with water, sediment and nutrients, most was lost on an area basis from the 5 m spacing with the least from the 20 m spacing. The highest percentages were from one of the 5 m spacings where 0.94% of the carbofuran, 0.06% of the atrazine, 0.04% of the cyanazine, and 0.01% of the alachlor were lost in the drainage water.

The common element in the prevention of pesticides from impacting the quality of ground water is time -- time for more of a given pesticide to be sorbed by the soil or to chemically degrade. Based on our studies, we know that from a water quality standpoint it is best to use a pesticide which is readily sorbed so that only a small amount is temporarily mobile in seepage flow. Also, we know that the pesticides used in the study were not detectable in the outflow from the drains after 4 or 5 months had elapsed. However, we are proposing to further study the dynamics of pesticide movement using tracers both in the field and with mathematical models such as the finite element model which will be briefly described in the following section.

4 FINITE ELEMENT MODEL

The finite element solution of the Richards equation (Richards, 1931) is emerging as a powerful tool in the analysis of transient groundwater flow. With well chosen boundary conditions and grid density, a high degree of accuracy can be obtained for many problems encountered in traditional agriculture, water management, waste disposal, and groundwater quality control. Indeed it offers the opportunity of tracing pollutant movement in soil which may not otherwise be possible with more direct methods.

The finite element model used in this study was one developed by Kaluarachchi and Parker (1987). The model uses finite elements to discretize the spatial domain and finite differences or the fourth-order Runge-Kutta predictor-corrector method to discretize the time domain. The model features the numerical integration and influence coefficient technique of Huyakorn and his associates (1984) to compute element matrices. It also has the capability to accommodate atmospheric and seepage boundary conditions and the option to consider hysteresis in the water retention domain.

4.1 Hydraulic conductivity determination

In order to obtain reasonable accuracy from the finite element method or for, as a matter of fact, any other method, the soil properties, notably hydraulic conductivity, of the flow region must be properly represented. Since soil properties usually have a high spatial variability, two problems must be resolved. One problem is how to collect enough field data, say of hydraulic conductivity, so that we have a good idea of its probabilistic distribution. And the other is how to reduce the gathered information as the number of variables become large so that our analysis methodology does not become completely bogged-down. We therefore chose a relatively new technique for making rapid determinations of hydraulic conductivity in situ (Merva, 1988) and a new analytical method for making probabilistic estimates for multivariate analyses (Harr, 1989).

4.2 Field measurements of hydraulic conductivity

A device called a velocity permeameter was used to make fairly rapid (around 30 minutes) determinations of hydraulic conductivity in the field. For a measurement, a 7.5 cm cup is carefully driven a set distance, either vertically or horizontally, into the soil. The cup is then connected to a small diameter head tube and the entire system is filled with water. The technique is based on monitoring the rate of fall in the water column as the water percolates into the soil enclosed in the cup. A programmable calculator capable of very accurate timing was used to compute hydraulic conductivities as the soil core became more and more saturated. The saturated hydraulic conductivity was noted when a slight increase occurred in the generated curve caused as the wetting front reached the bottom of the cup. We took 6 vertical and 6 horizontal measurements at exact 1 m intervals in 6 discernable soil layers across the 10 m drain spacing in our field study. This device allowed us to obtain the 72 total measurements in a reasonable length of time (about 6 days). Considerable variability occurred in our measurements indicating that some of the measurements were dominated by matrix flow and others by macropore flow.

4.3 Analytical methods

Several probabilistic methods have been developed that yield measures of the distribution function of random variables. The Monte Carlo and the first order, second moment (FOSM) methods are probably two of the best known. The advantages of these methods is that they provide a complete probability distribution of the function of the random variable. However, analyses quickly became difficult, if not impossible, when we have more than only a few uncorrelated random variables. Some of the difficulties associated with the Monte Carlo and FOSM methods have been resolved with the point estimate method (PEM) as proposed by Rosenblueth (1975). Although relatively simple to apply (in general, only the average, the variance, and the coefficient of variation must be known), the PEM requires 2^n realizations (in our case, 2^{12}) to obtain the probabilistic distributions of a function of n random variables. The method can be very cumbersome when the number of random variables becomes large. This difficulty is greatly accentuated when the methodology is combined with a simulation process which takes a large amount of computer time (with us, the use of finite elements for solving a transient flow problem). Looking for a more efficient method, Harr (1989) proposed a modified point estimate method (MPEM) in which the number of realizations needed to obtain the probabilistic distribution of a function is reduced to 2n (for our case, a total of 24). With the MPEM, the variables are transformed to an eigensystem. The eigenvalues of the correlation matrix can be represented by a hypersphere in the standardized coordinate system. Eigenvectors emanating from the origin of expected values in their respective directions intersect the sphere at two points. These points of intersection provide the 2n point estimates for this method.

4.4 Simulation results

The drain flow rate for a saturated, steady condition was simulated by combining the MPEM and the finite element model. (We intend later to use actual rainfall and to generate hydrographs which we will then compare with actual hydrographs). Our objective is to transfer the uncertainties with the hydraulic conductivity measurements to the simulated drain flow rate. This was accomplished by running the finite element model for each of the 24 realizations estimated by the MPEM. In our analysis, the dominate eigenvalue accounts for 35% of the information and the highest five eigenvalues for 99% of the information. The estimated highest and lowest possible drain flow rates were 1.65 and 0.00 m^3/h, respectively. The overall mean drain flow rate was 0.84 m^3/h which compared to a value of around 0.7 m^3/h for the maximum peak flow rate observed so far from a drain with a 10 m spacing.

5 SUMMARY COMMENTS

Our approach to the problem of movement of contaminants in soil water is the modeling/monitoring approach in which a mathematical model is devised and then tested under real-world conditions. Improvements to either the modeling or monitoring effort are then made in a synergetic manner. In one year, around 1% of the applied carbofuran and less than 0.1% of atrazine, cyanazine and alachlor in decreasing order were detected in the outflow from a drain spaced 5 m apart. However, we do not know where the pesticides originated (chances are that a large proportion would have come from the area closest to the drain), how much of the pesticide moved beyond the influence of the drain into the ground water, how different climatic variables would have affected the loss of pesticides to a drain or into the ground water, or what would be the influence of different soil types? These questions can be answered most efficiently with a model that has universal applicability.

The weather during the study period was abnormally dry especially during the spring months. In fact, significant drain outflow only occurred in the late spring and early summer after the pesticides were applied in only one of the three years of record. Our results would indicate that the likelihood of groundwater contamination by pesticides is greatly reduced if, after their application, sufficient rainfall does not occur within a relatively short time to move the pesticides into the drains or presumably also into the ground water. In our study, the pesticides were not detectable in drain outflow in the following fall. Laboratory sorptivity tests are a good indication of the potential of a pesticide to pollute water resources. The loss of any one pesticide to the drains were in rank order with those that were least sorbed in the laboratory experiments.

Although more nitrate nitrogen would have been lost in the early period after nitrogen application with more rainfall, a high percentage of the nitrate-N that was lost to the subsurface drains was lost in the following fall and late winter. Finding management practices to lessen the risk of groundwater pollution by nitrates could then be a much more difficult task than for pesticides where the risk tends to be quickly mitigated with time.

REFERENCES

Bottcher, A.B., E.J. Monke & L.F. Huggins. 1981. Nutrient and sediment loadings from a subsurface drainage system. Trans. Amer. Soc. Agr. Eng. 24: 1221-1226.

Cotterill, E.G. 1980. The efficiency of methanol for the extraction of some herbicide residues from soil. Pest. Sci. 11: 23-28.

Fuller, W.H. 1982. Methods for conducting soil column tests to predict pollutant migration. In D. Schultz (ed.), Land disposal: hazardous wastes, p. 87-105. Proc. Eight Ann. Res. Sym., U.S. Environmental Protection Agency.

Greenberg, A.E. (ed.) 1981. Standard methods for the examination of water and wastewater. Washington: Amer. Public Health Assoc.

Hallberg, G.R. 1986. Overview of agricultural chemicals in ground water. Proc. Conf. on Agr. Impacts on Ground Water, Nat. Water Well Assoc.

Harr, M.E. 1989. Probabilistic estimates for multivariate analyses. Applied Mathematical Modelling, accepted for publication.

Holden, P.W. 1986. Pesticides and ground water quality: issues and problems in four states. Washington: National Academy Press.

Huffman, R.L., E.J. Monke & E.J. Kladivko. 1986. Drainage studies on Clermont silt loam. Paper No. 86-2554, Amer. Soc. Agr. Eng.

Huyakorn, P.S., S.D. Thomas & B.M. Thompson. 1984. Techniques for making finite elements competitive in modeling flow in variable saturated porous media. Water Resources Research 20: 1099-1115.

Kaluarachchi, J.J. & J.C. Parker. 1987. Finite element analysis of water flow in variable saturated soil. Jour. Hydrology 90: 269-291.

Kanwar, R.S., J.L. Baker & D.G. Baker. 1987. Tillage and N-fertilizer management effects on groundwater quality. Paper No. 87-2077, Am. Soc. Agr. Eng.

Keeney, D.R. & D.W. Nelson. 1982. Nitrogen-inorganic forms. Agronomy 9: 643-698.

Keeney, D.R. 1985. Nitrate in ground water: agricultural contribution and control. Proc. Conf. on Agr. Impacts on Ground Water, Nat. Water Well Assoc.

King, J.J. & D.P. Franzmeier. 1981. Morphology, hydrology, and management of Clermont soils. Proc. 1980 Indiana Acad. Sci. 90: 416-422.

Kladivko, E.J., R.L. Huffman, K.M. Oates, G.E. Van Scoyoc, F.L. Larney & E.J. Monke. 1987. Movement of pesticides and nutrients into tile drainage water in Southeastern Indiana. Proc. Conf. on Midwestern Ground Water Issues, Nat. Water Well Assoc.

Logan, T.J., G.W. Randall & D.R. Timmons. 1980. Nutrient content of tile drainage from cropland in the North Central Region. Res. Pub. No. 268, Ohio State Univ.

Merva, G.E. 1988. The velocity permeameter technique for rapid determination of hydraulic conductivity in situ. Unpublished document, Dept. of Agr. Eng., Michigan State Univ.

Murphy, J. & J.P. Riley. 1962. A modified single solution method for the determination of phosphorus in natural waters. Anal. Chim. Acta 27: 31-36.

Richards, L.A. 1931. Capillary conduction of liquids through porous medium. Physics 1: 318-333.

Rosenblueth, E. 1975. Point estimates for probability moments. Proc. Nat. Acad. Sci. 72.

Land and Water Use, Dodd & Grace (eds), © 1989 Balkema, Rotterdam. ISBN 90 6191 980 0

Predicting the fate and movement of bacteria in agricultural runoff

J.A.Moore
Agricultural Engineering Department, Oregon State University, USA

ABSTRACT:
Livestock producers are facing increasing pressure to better manage and dispose of their animal waste. In addition to handling the organic matter and nutrients, today the transportation and fate of microorganisms are a major concern. The impact of livestock operations on a bacterial concentration in runoff and receiving streams has not been widely studied. A model, MWASTE, has been developed to allow the simulation of beef, dairy, poultry, horse, swine, and sheep operations and the monitoring of organisms as the manure moves through the various management options. The organisms' numbers are followed daily as they move from the animals, are collected, stored, treated, and land spread. Die-off as influenced by soil pH, temperature, sunlight, and moisture is monitored until precipitation moves the organisms into the soil or off in runoff. Bacterial concentration is calculated as the runoff leaves the field's edge. The CREAMS model is used to provide the hydrology file which calculates the infiltration and runoff from each storm. The program allows the comparison of the impact of different management practices on runoff quality.

RESUME:
Producteurs de bétail doivent confronter d'enormes exigences en vue d'améliorer la disposition de déchets animal. En plus à le maniement de la matière organique et les eléments nutritifs, audjourd'hui le transport et le sort des microorganismes sont un souci major. Les résultats des opérations d'élevage de bétail sur la concentration de bactérie en ruissellement sont encore en état d'etudes. Un modèle, MWASTE, avait été developé pour la simulation de les opérations de boeuf, laiterie, volaille, cheveaux, cochon et mouton et de servir de moniteur les organismes pendant la mouvement des engrais le long de les options divers de maniement. Le nombre des organismes est suivi quotidiennement durant leur movement ils sont rassemblés, stockés, traités et etalés. La diminution du nombre des organismes influencée par le pH de sol, température, insolation solaire, et himidité serve de moniteur jusq'au moment où organismes sont mu par ruissellement. La concentration de bactérie est calculé quand le ruisselement sort la borde du champ. Le modele CREAMS est utilizé a faire la fichie hydrologique pour le calcul d'infiltration et ruissellement pour chaque orage. La logicel faire la possibilité à comparer la conséquence pour la qualité de ruissellement de pratiques de maniement divers.

KURZFASSUNG:
Viehzuechter stehen unter zunehmendem druck, tierisch abfallprodukte besser zu verarbeiten und zu verteilen. Zusaetzlich zur verarbeitung der organischen substanzen und der nahrungsmittel ist heute die zukunft und der weitertransport der microorganismen von wesentlichem belang. Der einfluss von viehhaltung auf bacterienkonzentration in abflus - und muendungsgewaessern ist bisland weitgehend unerforscht. Ein modell, MWASTE, wurde entwickelt zur simulation von milchwirtschaft und rinder-, gefluegel-, pferde, schwein- und schafzucht sowie zur beobachtung der organismen waehrend der tiermuell die verschiedenen verarbeitungsmoeglichkeiten durchlaueft. Die anzahl der organismen wird taeglich ermittelt, sobald die organismen die tiere verlassen. Sie werden gesammelt, gespeichert, behandelt und ueber das land verteilt. Die absterberate, verursacht durch Boden-pH, temperatur, sonnenlicht, und feuchtigkeit wird festgehalten, bis niederschlag die organismen in die erde waescht oder fortspuelt. Die bacterienkonzentration wird berechnet, sobald der abfluss den feldrand verlaesst. Das CREAMS modell wird zur erzeugung eines hydrologie-files benutzt, das die durchtraenkung und den abfluss nach jedem gewitter berechnet. Das program erlaubt einen vergleich der einwirkung verschiedener behandlungsmethoden auf die abflussqualitet.

1 INTRODUCTION

As water quality became an issue in the United States, the early effort was directed toward regulating and cleaning up the point sources. Today the concern and interest is in evaluating and reducing non-point sources. The agricultural industry is perhaps the greatest contributor to non-point sources of water pollution in the United States.

Recently efforts have been directed to develop an understanding and gather a sound baseline data set of the contributions and movement of bacteria in non-point source pollution. The current interest in protecting water contact activities and improving the bacterial quality of surface water will support additional research investments.

LITERATURE REVIEW

In many studies little difference is seen between areas used as pastures and control areas where manure has not been spread (Doran and Linn, 1979; Kunkle, 1979; McCaskey *et al*., 1971; Robbins *et al*., 1971). This has been attriuted to contamination from wild animals (Schepers and Doran, 1980) and/or to a somewhat stable bacterial background population in the soil. As a rule, agricultural runoff generally exceeds recommended water quality standards for recreational use of the water.

The introduction and transport of enteric bacteria in the natural environment may follow one of several unique pathways. Common to all these cycles is the die-off of bacteria, in storage unis before land application, on and in the soil after application, and in runoff or groundwater after transport from the soil surface. A second pathway for bacterial loss, at least in the sense of being removed from the transport cycle into other segments of the environment, is by filtration and adsorption in the first several inches of the soil profile.

The die-off rate of enteric bacteria in the environment is controlled by many factoss. To summarize the literature on the effect of these parameters is beyond the scope of this paper and would only duplicate the efforts of many previous investigators. Reviews on this topic have been written by Burge and Marsh (1978), Dunlop (1968), Ellis and McCalla (1978), Gerba *et al*. (1975), Krone (1968), Lance (1976), VanDonsel *et al*. (1967), Rudolfs *et el*. (1950), Reddy *et al*. (1981), Mitchell and Starzyk (1975). In general, the factors of temperature, pH, moisture, nutrient supply, and solar radiation seem to have the greatest effect on enteric bacterial survival.

The use of models is helpful in efforts to predict enteric organism populations in soil, water, or waste management systems without the need of expensive and time-intensive field activities.

If models can be developed to accurately predict bacterial die-off after defecation of manure from its source, then the number of bacteria available to contaminate the soil, groundwater, and surface waters can be estimated. Observing data reported from many studies in the literature, it can be seen that several typical patterns are followed during enteric bacterial die-off. A recent review by Crane and Moore (1985) provides a complete summary of data published on die-off rates. Some 15 equations were cited that have been used to describe bacterial die-off. The most common was the proposed model of Chick (1908) known as Chick's Law, which is also the model of a simple first-order reaction in chemical kinetics:

$$\frac{Nt}{No} = 10^{-kt} \qquad (1)$$

where Nt = number of bacteria at time t
No = number of bacteria at time o
t = time in days
k = first order or die-off rate constant

The objective of this project was to develop a computer model that would keep track of bacteria from animal manure as they moved from the animals, through the waste management system. This activity was prompted by a problem of high bacterial counts in the water in a bay in Oregon. This bay has a small oyster industry and harvesting was shut down by the State Department of Health. After the closure several agencies began to look at the probable sources of the bacteria. It was determined that one of the sources was manure from the local dairy operations. This seemed possible because the small coastal county has a dairy population of 18,000 milking cows. The escape in runoff is driven by an annual rainfall of 2.4 m (92 inches).

Our project was then to assist the dairy operations in reducing the loss of bacteria from their systems. Part of the task was to develop a list of best management practices that the operators could look through to select those practices that would be helpful on their dairy. It is no great task to develop the list, but it was difficult to estimate the probable reduction in bacteria loss that each of these practices might cause if implemented.

This led to the development of a computer model which utilized treatment coefficients as they were published in the literature. It examines the environmental factors influencing the die-off and transport of fecal coliform (FC) bacteria present in dairy cattle waste. The model considers the effect of precipitation, season, method of waste storage and application, die-off of the bacteria in storage, die-off on the land surface, infiltration of bacteria into the soil profile, soil characteristics, buffer strips and overland flow characteristics of bacteria. The model was then used to evaluate individual farmsteads to rank them in order to identify which operation should be worked on first. The model was also can on each farmstead to evaluate which management practices would provide the greatest reduction in bacteria loss.

The synthesis of the model in this investigation is based on a "mass balance", or deterministic approach, where the significant events are quantified and summed for a specific time increment. The waste management system determines the number of bacteria that are stored and applied and influences the number that run off. Climatological effects and soil characteristics determine the rate of bacteria die-off on the surface, and the quantity of bacteria finally transported by rainfall water. Assembling these events together in a daily time increment enables a prediction to be made of the concentration of FC in runoff from manure applied lands.

The waste handling practice determines the volume of wastes and the concentration of FC in storage. The model considers three systems. The practice of storing the manure with the animal's bedding (i.e., stacking the wastes) yields a waste that is more concentrated with FC than wastes that have been diluted by milking parlor waste water and/or flush system water. Each of these management practices influences the quantity of FC bacteria applied to the land.

Manure is added to storage on a daily basis, and with each increment in storage volume there is a corresponding increment in the quantity of bacteria found in storage. The bacteria die-off in storage daily is calculated using the simple first order reaction equation.

Surface die-off is the next event in the sequential order of the model. When FC are applied to the land they are subject to adverse environmental conditions and the number of viable bacteria on the land surface declines. The rate constant for surface die-off is a function of several climatic and soil factors as identified and summarized from the literature review. The soil factors do not vary substantially and climatic factors are accounted for by seasonal variations using only two surface die-off rate constants (summer and winter). It is assumed that no growth of fecal coliform bacteria occurs on the land surface. After the bacteria have been land applied they suffer application and die-off losses. The remaining viable bacteria are available to move with infiltration and runoff waters.

Some FC bacteria are lost from overland transport when they enter the soil profile during a precipitation or irrigation event. When the bacteria enter the soil profile they are adsorbed to the soil, and the model assumed they are lost from the system (they may be considered dead).

The ability of the soil to adsorb bacteria in this scheme depends on the soil's infiltration characteristics. Because the vegetative cover in the land receiving manure is predominantly pasture, the ability of different soils to hold bacteria is related only to their profile water capacity and internal drainage rate (at a macroscopic level). The soil profile water holding capacity is the quantity of water that can infiltrate into the soil before saturation and runoff and it depends on water table height. This capacity is increased by reducing the water table level with the addition of subsurface drainlines. When the soil profile is saturated, additional water removes some of the remaining bacteria in overland runoff.

The final process considered in the model is the transport of FC bacteria with runoff. The number of bacteria removed in runoff is a function of the number remaining on the soil surface after die-off and infiltration have occurred, the quantity of runoff, and the rate at which the FC bacteria partition into the runoff. It is necessary to model the partitioning of FC bacteria from land-applied wastes into infiltration or runoff waters to successfully describe the overall process of land disposal of wastes and runoff water quality. The review of the literature indicates that no such modeling has been developed. Consequently, the model used here is based on first-order kinetics and may be represented by

$$Ft = Fo * (1-p)^r \quad (1)$$

where: Ft = Bacteria number remaining on soil
Fo = Original bacteria number on soil
r = Runoff or infiltration water depth
p = Percent reduction factor, or infiltration or runoff

waste type	spreading	infiltration	runoff
solid	--	0.05	0.40
liquid	day 1	0.20	1.00
liquid	>day 1	0.05	0.40

Equation 1. The percent-reduction method equation and coefficients (Moore et al. 1982).

The number of bacteria actually removed by the infiltration or runoff events may be calculated by subtracting the number remaining from the original quantity (i.e., Fo - Ft). The percentage reduction term (P) is a function of the ability of the infiltration, or runoff, waters to partition the fecal coliforms from the wastes, and the environmental factors that influence this process. The (P) term for infiltration is less than that for runoff because the infiltration water must partition the fecal coliforms from the wastes and also distribute them into the soil matrix (i.e., the soil "environment"). The value of (P) for runoff needs only to consider the effects governing the rate at which fecal coliforms become suspended in the runoff water.

The evaluation of this model included a comparison to actual water quality data collected in the basin by the State Department of Environmental Quality and a sensitivity analysis of the die-off, infiltration and runoff parameters. The sensitivity analysis identified that infiltration was the least sensitive to change, and the runoff and storage parameters have the greatest sensitivity.

This model was used successfully to assist dairymen in evaluating management practices. Some time later the Soil Conservation Service (SCS) asked us to expand the model to include all domestic livestock over the range of climates and soils in the whole United States. This resulted in the modification and addition of several components of the model. The second generation model is called MWASTE and builds upon the previous model. The model has been expanded from just dairy to include beef, dairy, swine, sheep, chickens, turkeys and horses. After entering the animal specie and number, the operator inputs the manure type and the types of confinement for the animals from this list of five: roofed/solid, roofed/semi-solid, roofed/liquid, open, and passture. A complete list of the inputs and outputs are listed in Table 1. A flowchart of the new daily bacterial model simulation MWASTE is shown in Figure 1.

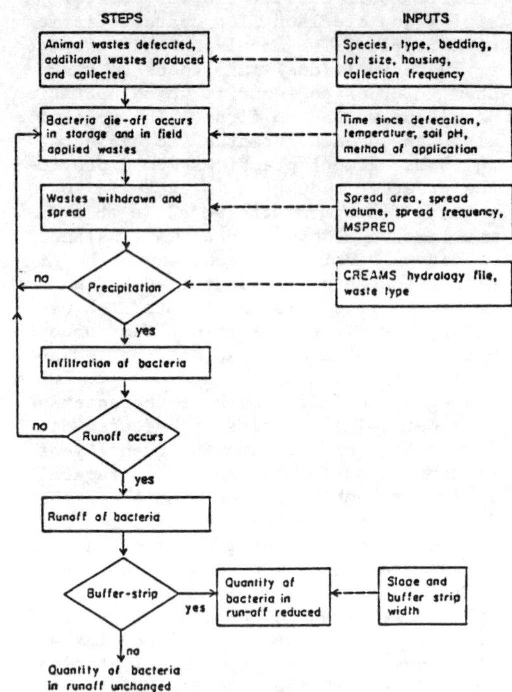

Figure 1. Flowchart of MWASTE Simulation

There are several changes in the model. For waste on the field, the die-off rate of bacteria is now determined using a modified form of the equation developed and reported by Reddy et al. (1981). This modified version considers the influence of temperature, application method, and soil pH. Equation 2 presents the field die-off relationship used in MWASTE and the coefficient values.

$$k = kl * Ft * Fap * FpH \quad (2)$$

where: k = die-off coefficient
kl = base die-off rate (0.50)
Ft = temperature correction factor ($1.0675^{(t-20)}$)
Fap = method of application factor (0.50 for surface)

FpH = soil pH factor

pH	Factor
3-6	1.69 - (0.26 * pH)
6-7	0.25
7-8	(0.21 * pH) - 1.22

The method which calculates the runoff volume is very important in a surface water transport model. The SCS are familiar with and have at their disposal the program CREAMS (Knisel, 1980). This is a field scale model developed to evaluate Chemicals, Runoff, and Erosion from Agricultural Management Systems. The hydrology component of this model is used to generate the runoff values. Local soil characteristics, topography, and weather conditions are input to determine the runoff frequency, magnitude and duration.

The determination of parameter "r", the depth of water infiltrated or in runoff, is determined by CREAMS. All of the environmental parameters which MWASTE uses in the simulation are derived from the CREAMS hydrologic output file. These include temperature and precipitation (including snow).

If a comparison of the resulting runoff water quality between random spreading events (i.e. spreading occurs whenever the storage is full or the operator has time) and spreading on a set schedule (i.e. twice a week) is desirable, the program MSPRED can be utilized. After CREAMS has been used to generate a hydrology pass file, MSPRED can be used to specify spreading dates and volumes. This data is stored in a file called SPRED.DAT. This file can then be used in execution of MWASTE to generate comparison of bacterial run-off values using the same weather data and various spreading dates. Other model results can be generated through interactive program execution when spreading dates and volumes are entered from the keyboard.

Lot runoff is treated as an additional wastewater as it is not generated by CREAMS. The runoff is produced when a lot is used and the runoff from a storm-event is collected. Equation (3) is used to generate lot runoff.

$$RO = (P-F)/*A \quad (3)$$

where: RO = Runoff volume M^3
P = Precipitation M^3
F = Surface factor
A = Area M^2
F = 0.1 for paved lots
F = 0.5 for dirt lots

Equation 3 - Lot Runoff Equation (MWPS-18, 1985)

A buffer strip acts to decrease the number of bacteria in the runoff. The program will also ask the operator if a buffer strip is used, and, if so, its width and slope percentage. An equation was developed based upon research reported by Glenne (1984) in a study of three Utah watersheds. The limits for the minimum value of effective width of the buffer strip is three meters. The slope percentage must be greater than zero and no more than 15 percent. The reduction can be no more than 75 percent of the total bacteria number entering the strip. If the specified parameters do not meet the above criteria, the buffer strip will be ignored and the program will warn the user that the entry does not constitute an effective runoff buffer. The buffer-strip reduction equation used and its parameter limits are shown in Equation 4.

$$PR = 11.77 + 4.26 * S \quad (4)$$

where: PR = Percent removal of bacteria
S = Buffer width x 3.3/slope %

NOTE: $0 < PR < 75\%$
$0\% < slope < 15\%$
$3\ M < \text{buffer strip width}$

Detailed output includes date, temperature, precipitation, runoff, infiltration, animal inventories, duration specifications, and waste allocations. Bacterial information for pasture and field-spread waste, including infiltrated and runoff waste concentrations, are also in the output. This detailed output is printed for each day over the program simulation period (0 to 365 days).

This program will allow the calculation of the concentration of the bacteria in runoff. It can also be used to identify and/or evaluate management practices of a manure system to reduce the movement and loss of bacteria from livestock operations. For a detailed report on the program and its development, the reader is referred to Moore _et al_. (1988) which is available upon request.

Bibliography

Burge, W. D. and P. B. Marsh. 1978. Infectious disease hazards of land-spreading sewage wastes. J. of Environ. Qual. 7(1):1-9.

Chick, H. 1908. Investigation of the laws of disinfection. J. Hyg. 8:655.

Crane, S. R. and J. A. Moore. 1985. Modeling enteric bacterial die-Off: A Review. Water, Air, and Soil Pollution. 27:411-439. 1986.

Doran, J. W. and D. M. Linn. 1979. Bacteriological quality of runoff water from pastureland. Appl. Environ. Microbiol. 37:985.

Dunlop, S. G. 1968. Survival of pathogens and related disease hazards. In: C. W. Wilson and F. E. Beckett (Ed.) Municipal sewage effluent for irrigation. Louisiana Tech. Alumni Foundation, Ruston, LA. pp. 107-122.

Ellis, J. R. and T. M. McCalla. 1978. Fate of pathogens in soils receiving animal wastes--a review. Transactions of American Society of Agricultural Engineers. 21(2): 307-313.

Gerba, D., C. Wallis, and J. Mellnick. 1975. Fate of wastewater bacteria and viruses in soil. J. Irrigation and Drainage Div. ASCE. 101:157-174.

Glenne, B. 1984. Simulation of water pollution generation and abatement on suburban watersheds. Water Resources Bull. 20:2. April.

Knisel, W. G. Editor, 1980. CREAMS, A field scale model for chemicals, runoff, and erosion from agricultural management systems. USDA Conservation Research Report No. 26. 640 pp.

Krone, R. B. 1968. The movement of disease producing organisms through soils. In: C. W. Wilson and F. E. Beckett (eds.) Municipal sewage effluent for irrigation. Louisiana Tech. Alumni Foundation, Ruston, LA. pp. 75-106.

Kunkle, S. H. 1979. Using bacteria to monitor the influences of cattle wastes on water quality. USDA-SEA-ARR. ARR-NE-3, September.

Lance, J. C. 1976. Fate of bacteria and viruses in sewage applied to soil. USDA-ARS Phoenix. ASAE Paper No. 76-2558, ASAE, St. Joseph, MI 49085.

McCaskey, T. A., G. H. Rollins, and J. A. Little. 1971. Water quality of runoff from grassland applied with liquid, semi-liquid and dry dairy waste. In: Livestock Waste Management and Pollution Abatement. Procedures of the Second Int'l. Symp. on Livestock Wastes, ASAE, St. Joseph, MI 49085, pp. 239-242.

Midwest Plan Service. 1985. Livestock Waste Facilities Handbook, MWPS-18, Iowa State University, Ames, IA.

Miner, J. R., L. R. Bernard, L. R. Fina, G. H. Larson and R. I. Lipper. 1966. Cattle feedlot runoff nature and behavior. Proc. 21st Ind. Waste Conf. Purdue Univ. pp. 834-847.

Mitchell, D. O. and M. J. Starzyk. 1975. Survival of salmonella and other indicator microorganisms. Can. Jour. Microbiol. 21:1420.

Moore, J. A., M. E. Grismer, S. R. Crane and J. R. Miner. 1982. Evaluating Dairy Waste Management Systems' Influence on Fecal Coliform Concentration in Runoff. SB 658. Agricultural Experiment Station, Oregon State University, Corvallis, OR.

Moore, J. A., J. Smith, S. Baker, and J. R. Miner. 1988. Evaluating Coliform Concentrations in Runoff from Various Animal Waste Management Systems. Special Report 817. Agric. Exp. Station, Oregon State University, Corvallis, OR.

Reddy, K. R., R. Khaleel, and M. R. Overcash. 1981. Behavior and transport of microbial pathogens and indicator organisms in soils treated with organic wastes. J. Environ. Qual. (10)3:255-266.

Rhodes, R. A. and C. R. Hrubant. 1972. Microbiol. population of feedlot waste and associated sites. Appl. Microbiol. 24(3):269-377.

Robbins, J. W., G. J. Kriz, and D. H. Howells. 1971. Quality of effluent from farm animal production sites. In: Livestock Waste Management and Pollution Abatement. Procedures of the Second International Symposium on Livestock Wastes, April 19-22. American Society of Agricultural Engineers Publication PROC-271. St. Joseph, MI. pp. 166-169.

Rudolfs, W., L. L. Falk, and R. A. Ragotzki. 1950. Literature review on the occurrence and survival of enteric, pathogenic, and related organisms in soil, water, sewage, and sludges and on vegetation. I. Bacterial and virus diseases. Sewage Ind. Wastes 22:1251-1281.

Schepers, J. S. and J. W. Doran. 1980. Effect of grazing management on the chemical aspects of the environment. Procedures of the Nebraska Forage and Grassland Council, Lincoln, NE. January 19. pp. 37-41.

VanDonsel, D. J., E. E. Geldreich, and N. A. Clarke. 1967. Seasonal variations in survival of indicator bacteria in soil and their contribution to storm-water pollution. Appl. Microbiol. 15: 1362-1370.

Young, R. A., T. Huntrods, and W. Anderson. 1980. Effectiveness of vegetated buffer strips in controlling pollution from feedlot runoff. J. Environ. Qual. 9(3):483-487.

Land and Water Use, Dodd & Grace (eds), © 1989 Balkema, Rotterdam. ISBN 90 6191 980 0

Innovative technologies for controlling water pollution by agriculture

T.M.Younos & D.L.Weigmann
Virginia Water Resources Research Center, Virginia Polytechnic Institute & State University, Blacksburg, Va., USA

Abstract: By-products of agricultural activities are major contributors to surface water and groundwater pollution in the United States. In this paper, the different technologies -- watershed monitoring and instrumentation, rainfall simulation, computer models, and database development and screening procedures -- designed for controlling water pollution from agricultural activities are discussed.

Abstrakt: Nebenprodukte in der Landwirtschaft sind die hauptsaechlichen Verursacher der Oberflaechen- und Grundwasserverschmutzung in den USA. In dieser Veroeffentlichung werden die verschiedenen Technologien, die zur Kontrolle der Wasserverschmutzung durch die Landwirtschaft entwickelt wurden, behandelt. Unter anderem werden Ueberwachungsmethoden und Wassereinzugsgebiete, Regensimulatoren, verschiedene Computermodelle und Datenbanken vorgestellt.

Synopsis: Les dérivés des activités agricoles contribuent pour l'essentiel de la pollution des eaux de surface et des nappes d'eau souterraine aux Etats-Unis. Dans cet article les technologies différentes dont le but est de contrôler la pollution de l'eau par ces dérivés sont analysées: surveillance et appareillage dans le bassin hydrographique, simulation de précipitations, simulation par ordinateur, développement de banques de données et procédés de sélection pour des activités agricoles sans risque.

1. INTRODUCTION

In the United States, agricultural chemicals and associated by-products have been identified as major nonpoint sources (NPS) of surface water and groundwater pollution (U.S. EPA 1987). The main contributors to this nonpoint source pollution are sediments, nitrogen, phosphorus, pesticides, and pathogens. Effects of excessive sediments, nutrients, and pesticides on receiving waters include loss of drinking water supplies, epidemics of water-borne diseases, lowered storage capacity in reservoirs, eutrophication, changes in aquatic populations and their food supplies, accumulation of bottom deposits and increased oxygen demands, contamination of economically valuable shellfish populations, and deterioration in aesthetic values.

The Clean Water Act (CWA) of 1972 and subsequent amendments in 1987 authorized the United States Environmental Protection Agency (EPA) to demonstrate water pollution control technologies for agricultural nonpoint source pollution. Two of the earliest programs initiated under the 1972 CWA were the Great Lakes Basin nonpoint source pollution control program and the Model Implementation Program (MIP). The Great Lakes Basin program addressed a wide variety of technologies -- soil conservation, conservation tillage, and animal waste management -- to control nonpoint source pollution (Newell et al. 1986). The MIP implemented similar practices on soil conservation and water quality protection for agricultural watersheds in Indiana, Nebraska, New York, Oklahoma, South Carolina, South Dakota, and Washington (NWQEP 1983). A more recent project, the Rural Clean Water Program (RCWP), funds 21 subprograms that investigate methods of helping agricultural land owners and farmers by developing and testing programs, policies, and procedures for agricultural NPS pollution prevention and control (Humenick et al. 1987). These large scale projects prompted research and the development of technologies and guidelines for NPS pollution control measures known as Best Management Practices (BMPs). A BMP refers to a practice or combination of practices designed for the most effective and practicable control of NPS pollution. BMPs include structural measures, vegetative measures, conservation cropping, animal waste management systems, and management of chemical applications.

Most research projects and technologies developed since 1972 for NPS pollution control evaluate BMP effectiveness. The objective of this paper is to summarize the techniques used for minimizing water pollution by agriculture: advances in watershed monitoring and instrumentation, BMP and NPS modeling techniques, critical area identification, and database development and screening procedures.

2. WATERSHED MONITORING AND INSTRUMENTATION

Watershed monitoring and instrumentation technologies are essential for determining the amount of NPS pollution from agricultural areas and a BMP's effectiveness in controlling water pollution. Watershed monitoring allows the collection of sufficient data for validation of mathematical and computer models. In addition to measurements of precipitation, runoff, and concentrations of sediments, nutrients, and pesticides in runoff, comprehensive watershed monitoring programs should include subsurface monitoring for protecting groundwater.

Advances in instrumentation have led to digital recording gauges which can be interfaced with a computer to greatly facilitate data processing, management, and use. For example, the digital precipitation gauge mechanically converts the depth of accumulated precipitation onto a binary-coded, punched paper tape at selected intervals. The five-minute interval punchout performed by a solid-state timer provides 105,000 readings per gauge during a one-year period. The water level gauge for runoff measurements incorporates a float and counterbalanced pulley mechanism to transfer stream flow depth onto punched paper tape which could be interfaced with a computer.

Sediment, nutrient, and pesticide concentrations in runoff from agricultural watersheds are usually sampled automatically. An automatic sampler is triggered by a magnetic reed switch that responds to magnets mounted on the float wheel of a runoff event recorder. Samples are usually stored in specially designed refrigerators for later laboratory analysis. More samples can be analyzed per unit of time with the recently available multi-channel auto-analyzers.

Sensor technology is being developed for various soil surface measurements. For example, an automatic, non-contact soil profile meter consisting of a laser and a digital camera were coupled for erosion studies of soil surface heights before and after artificial rainfall (Khorashahi et al. 1987). This technology will promote a better quantification of the soil erosion process and the associated nutrient movement from agricultural fields by comparing very small changes in the surface elevation of soils.

Comprehensive watershed monitoring results in an abundance of precipitation, runoff, and water quality data that are essential for a comprehensive analysis of NPS pollution. A number of computer programs -- from simple to complex -- are available for data acquisition, data management, and statistical analysis of nonpoint source pollution studies (Carr et al. 1988).

3. RAINFALL SIMULATION

Appropriate evaluation of NPS pollution effects and the impact of control measures such as BMPs greatly depend on the occurrence and recording of natural rainfall and runoff events. Quite often, researchers are unable to reach statistically valid conclusions using the results of monitoring studies, because of the low or unpredictable frequency of rainfall and runoff events or instrument malfunctions. Furthermore, the characteristics of natural rainfall make it nearly impossible to evaluate its true effects on runoff, sediment, and chemical movements from agricultural watersheds. This has led researchers to simulate rainfall through the development of "rainfall simulators" (Younos & Sewell 1985). Advantages of rainfall simulators include 1) rapid results, 2) maximum control over the rate of rainfall application, 3) efficiency of operation, and 4) adaptability to controlled laboratory and field plot studies (Meyer 1979). Simulated rainfall techniques are being altered and refined by advances in methods of instrumentation and electronics coupled with more knowledge of natural rainfall dynamics. These have introduced new dimensions in the art of simulated rainfall and promoted its use in research on NPS pollution control.

4. COMPUTER MODELS

Determining the sources of NPS pollution from a mixed land use area and evaluating the potential effectiveness of a proposed BMP is difficult. Mathematical or computer-based models are useful tools for predicting the potential NPS pollution load leaving the field and the effectiveness of a proposed BMP. Nonpoint source pollution models offer the easiest and fastest method of selecting the most practical and cost-effective BMP. Without such computer models, a decision maker's options are limited, and these options provide him with little assurance of accomplishing the intended water quality goals. NPS pollution models consisting of the addition of a water quality component to a watershed hydrologic model are natural extensions of hydrologic models. Several NPS pollution models have been developed during the past decade, and they are based on three conceptual types of hydrologic models: regression models, lumped or optimized parameter models, and distributed parameter or spatially responsive models. Detailed reviews of these computer models are given in Haan et al. (1982) and Smolen (1983).

4.1 Regression Models

Simple regression models consist of a single equation derived from experimental data. The USLE, Universal Soil Loss Equation (Wischmeir & Smith 1978), is the best known regression model; it was developed to predict the average annual amount of soil moved by rainfall. The USLE has been modified with loading factors and delivery ratios to estimate yields of sediments, nutrients, and pesticides from agricultural fields (McElroy et al. 1976).

4.2 Lumped or Optimized Parameter Models

In lumped parameter hydrologic models, non-uniform parameters such as rainfall, soils, land use, and topographic characteristics are weighted to yield average values for the entire watershed. Lumped parameter models depend on a truly representative sample of rainfall and runoff records to optimize or calibrate the model. The HSPF, Hydrologic Simulation Program Fortran (U.S. EPA 1980), is the best known lumped parameter model simulating watershed hydrology, sediments, nutrients, and pesticides in runoff and their impact on receiving water quality. HSPF is a comprehensive model, constructed by integrating several less universal NPS pollution models. For example, the fate and transport of pesticides and nutrients from agricultural fields are derived from the Agricultural Runoff Management (ARM) model (U.S. EPA 1978), and the sediment delivery algorithms are derived from the Nonpoint Source (NPS) model (Donigian & Crawford 1977). HSPF has impressive data handling capabilities and a user-oriented interface.

4.3 Distributed Parameter or Spatially Responsive Models

In distributed parameter models, parameters are estimated from measurable watershed characteristics. The most significant advantage of a spatially responsive model is its ability to incorporate many facets of the natural watershed for determining specific responses of the watershed to hydrologic modifications. For example, implementation of many BMPs involve alteration of drainage patterns or changes in soil infiltration properties in sections of a watershed. The effects of these modifications cannot be averaged reliably over the entire watershed and cannot be easily assessed by lumped parameter models. By using a distributed parameter model, the user can assign the parameter estimates to areas of interest within the watershed. The Aerial Nonpoint Source Watershed Environment Response Simulation (ANSWER) model (Beasley et al. 1981) is the best known and widely used distributed parameter model. It has been validated in several regions of the United States.

Other distributed parameter models that also can be viewed as lumped parameter models because of the lumping of parameter values at some level of model resolution are the U.S. Department of Agriculture Hydrograph Laboratory (USDAHL) watershed model (Holtan et al. 1975), the hydrology/sediment transport model of the USDA Sediment Laboratory (SEDLAB) (Borah et al. 1981), and the Finite Element Storm Hydrograph Model (FESHM) (Ross et al. 1976).

Another widely used model for measuring BMP effectiveness is the Chemicals, Runoff, and Erosion from Agricultural Management Systems (CREAMS) model (Knisel 1980). However, unlike the models described above, CREAMS can simulate only individual agricultural fields not entire watersheds. CREAMS can simulate certain user-defined management activities: aerial spraying, soil incorporation of pesticides, animal waste management practices, and alternative agricultural practices such as conservation tillage and terracing. An expanded version of the CREAMS model is the Groundwater Leaching Effects on Agricultural Management Systems (GLEAMS) model (Leonard et al. 1987) that simulates leaching of pesticides and nutrients to groundwater from agricultural fields.

5. DATABASE DEVELOPMENT AND SCREENING TOOLS

Monitoring studies have shown that certain parts of an agricultural watershed may disproportionately affect water quality of the receiving surface water or groundwater. If these critical areas contributing a greater portion of the nonpoint source pollutants can be identified, then BMPs can be targeted for those critical areas to control NPS pollution and achieve receiving water quality goals. Full-scale implementation of voluntary BMPs has not succeeded nationwide, and such programs can be very expensive. More effective methods being developed are computerized databases and procedures to prioritize and target funds for NPS pollution control programs. Procedures involve development, integration, and use of spatial databases to identify and rank areas based on their pollution potential.

Several states have created a Geographic Information System (GIS) database to identify critical areas. Various approaches have been employed for the development of GIS databases; however, the common characteristic of these databases is the computerization of available information to allow the integration and simultaneous mapping and use of several databases. For example, the Virginia Geographic Information System (VirGIS)(Shanholtz et al. 1985) database includes U.S. Geological Survey (USGS) 7.5 minute quadrangle maps, the U.S. Soil Conservation Service (SCS) county soil survey maps, and the USGS Digital Elevation Model (DEM) data. The database is referenced through the Universal Transverse Mercator (UTM) coordinate system. Each dataset, which represents a quadrangle, is stored in map compatible format. This provides flexibility and adaptability of the databases since the files can be merged easily for use in a specific project. Furthermore, overlays of these files can be created using the Map Analyses Package, a cartographic modeling language (Tomlin 1980). VirGIS is being used to identify critical areas within the Chesapeake Bay watershed in the state of Virginia (Hellmand et al. 1986) and as a screening tool to identify and rank sites that have the greatest potential for contributing to off site pollution from animal waste and sludge disposal (Heatwole et al. 1987; Younos & Metz 1988). Although most existing GIS databases are designed to minimize NPS pollution of surface waters, a few have been used for controlling groundwater pollution (USGS 1988).

Recently a screening procedure called DRASTIC was developed to identify regions of the U.S. where groundwater is most susceptible to agrichemical pollution (Aller et al. 1985). DRASTIC is based on seven hydrologic parameters (acronym for **D**epth to groundwater, net groundwater **R**echarge, **A**quifer properties, **S**oil properties, **T**opography, **I**mpact of the vadose zone, and hydraulic **C**onductivity of the aquifer). Each DRASTIC factor receives a rating for the geographic area under consideration. The rating is then multiplied by a weight which reflects the parameter's relative importance to pollution potential. The weighted ratings are totaled, yielding a DRASTIC score. A higher score implies a higher degree of vulnerability to pollution. Researchers at the USDA used the DRASTIC approach (Nielsen & Lee 1987) to identify potential areas of groundwater contamination from pesticide and chemical fertilizer use in U.S. However, their results did not define the pollution potential at the local or county level but provided an analysis of general trends across the country.

To make DRASTIC more useful in designing monitoring programs or taking regulatory actions, some states including Ohio (Hallfrish and Voytek 1987) and Virginia (Wagner 1989) have developed more detailed databases to enter into DRASTIC. The use of GIS and statistical approaches for more reliable data input into DRASTIC is a natural extension of these efforts. The systems will be useful in prioritizing protection, monitoring, and cleanup of groundwater resources.

CONCLUSION

During the past decade, research has greatly advanced and refined knowledge and practices in the area of nonpoint source pollution control. Innovations in instrumentation and computer technology have facilitated the better quantification of nonpoint pollution sources and their impacts on receiving water quality. Several years of data are needed to evaluate NPS pollution and BMP impacts, therefore the results from many research projects are just forthcoming. The applicability of these findings to site specific and local conditions needs to be evaluated before any generalization can be validly adopted. National databases such as National Water Quality Evaluation Project (NWQEP) facilitate the information transfer between research and regulatory agencies and the adaption of technology to local conditions. State and localized databases such as VirGIS are promoting further use and refinement of available information. During the coming decade expanded use of computers in coordinating monitoring, the transfer of data, laboratory analyses, and the statistical analyses of data is expected. Also, the further development of user friendly computer models should promote decision making processes without the need for extensive hydrological and water quality data.

REFERENCES

Aller, L. et al. 1985. DRASTIC: A Standardized System for Evaluating Groundwater Pollution Potential Using Hydrologic Settings. EPA-600/2-85/018. U.S. Environmental Protection Agency, Ada, Oklahoma.

Beasley, D.B. & L.F. Huggins. 1981. ANSWERS User's Manual. EPA-905/9-82-001. Chicago, Illinois.

Borah, D.K., C.V. Alonso & S.N. Parasad. 1981. Single Event Numerical Model for Routing Water and Sediment on Small Catchments. A Report by the U.S. Department of Agriculture, Sedimentation Laboratory, Oxford, Mississippi, to the U.S. Army Corps. of Engineers, Vicksburg District, Vicksburg, Mississippi.

Carr, J.C. et al. 1988. A Computerized Data Management System for Hydrologic Data HAS87. Amer. Soc. Agric. Eng. Paper No. 88-3007. St. Joseph, Michigan.

Donigian, A.S. & N.H. Crawford, 1977. Simulation of Nutrient Loadings in Surface Runoff with the NPS Model. EPA-60013-77-065. U.S. Environmental Protection Agency, Environ. Res. Lab. Athens, Georgia.

Haan, C.T., H.P. Johnson & D.L. Brakensiek. 1982. Hydrologic Modeling of Small Watersheds. Amer. Soc. Agric. Eng. Monograph No. 5. St. Joseph, Michigan.

Hallfrish, M. & J. Voytek. 1987. Groundwater Pollution Potential of Madison County, Ohio. Ohio Dept. of Natural Resources, Division of Water. Columbus, Ohio.

Heatwole, C.D. et al. 1987. Targeting Animal Waste Pollution with Virginia's GIS. Amer. Soc. Agric. Eng. Paper No. 87-4049. St. Joseph, Michigan.

Hellmund, P.A. et al. 1986. A GIS Approach for Identifying Nonpoint Source Pollution. In C.Y. Kuo & T.M. Younos (eds.), Proc. Chesapeake Bay Research Conference: Effects of Upland and Shoreline Activities on the Chesapeake Bay. p.11-20 Virginia Polytechnic Institute & State University. Blacksburg, Virginia.

Holtan, H.N. & N.C. Lopez. 1971. USDAHL-74 Model of Watershed Hydrology. U.S. Department of Agriculture, Agricultural Research Service. Bull. No. 1435.

Humenik, F.J., M.D. Smolen & S.A. Dressing. 1987. Pollution from Nonpoint Sources. Environ. Sci. & Tech. 21:737-742.

Khorashahi, J., R.K. Byler & T.A. Dillaha. 1987. An Opto-Electronic Soil Profile Meter. Computer and Electronics in Agriculture, 2:145-155.

Knisel, W.G. (ed.) 1980. CREAMS: A Field-Scale Model for Chemicals, Runoff, and Erosion from Agricultural Management Systems. U.S. Department of Agriculture, Conserv. Res. Rep. No. 26. Tifton, Georgia.

Leonard, R.A., W.G. Knisel & D.A. Still. 1987. GLEAMS: Groundwater Loading Effects of Agricultural Management Systems. TRANS Amer. Soc. Agric. Eng. 30: 1403-1418.

McElroy, A.P. et al. 1976. Loading Functions for Assessment of Water Pollution Sources. EPA-600/2-76-151. U.S. Environmental Protection Agency. Washington, D.C.

Meyer, L.D. 1979. Methods for Attaining Desired Rainfall Characteristics in Rainfall Simulators. USDA-SEA, Agric. Review & Manuals, Western Series No. 10.

Newell, A.D. et al. 1986. Overview and Evaluation of Section 108a Great Lakes Demonstration Program. EPA-905/9-86-001. U.S. Environmental Protection Agency. Washington, D.C.

Nielsen, E.G. & L.K. Lee. 1987. The Magnitude and Costs of Groundwater Contamination from Agricultural Chemicals: A National Perspective U.S. Department of Agriculture, Economic Res. Service. Agricultural Economic Rep. No. 576. Washington, D.C.

NWQEP 1983. An Evaluation of the Management and Water Quality Aspects of the Model Implementation Programs. National Water Quality Evaluation Project. North Carolina State Univ., Raleigh, North Carolina.

Ross, B.B., V.O. Shanholtz & D.N. Contractor 1976. A Model for Predicting Flood Hazards Due to Specific Land Use Practices. Virginia Water Resources Research Center. Bull. 99. Virginia Polytechnic Institute & State University, Blacksburg, Virginia.

Shanholtz, V.D. et al. 1985. VirGIS Pilot Demonstration Project on MACHODOC Quadrangle of Westmoreland County, Virginia. Depart. of Agric. Eng. Rep. No. AGE-SW85-1. Virginia Polytechnic Institute & State University, Blacksburg, Virginia.

Smolen, M.D. (ed.) 1983. Hydrologic and Water Quality Models for Agriculture and Forestry. Southern Cooperative Ser. Bull. No. 291. Virginia Polytechnic Institute & State University, Blacksburg, Virginia.

Tomlin, C.D. 1980. Map Analysis Package. Unpublished Rep., School of Forestry and Environmental Studies, Yale University. New Haven, Connecticut.

U.S. EPA 1978. User's Manual for Agricultural Runoff Management Model. EPA-600/3-78-080. U.S. Environmental Protection Agency, Environ. Res. Lab. Athens, Georgia.

U.S. EPA 1980. User's Manual for the Hydrological Simulation Program FORTRAN. EPA-600/9-80-015. U.S. Environmental Protection Agency, Environ. Res. Lab. Athens, Georgia.

U.S. EPA 1987. National Water Quality Inventory: 1986 Report to Congress. EPA-440/14-87-008. U.S. Environmental Protection Agency, Washington, D.C.

USGS 1988. Herbicides in Ground and Surface Water: A Mid-Continent Initiative. U.S. Geological Survey, Toxic Waste - Ground Water Contamination Programs, Open File Rep. U.S. Geological Survey, Reston, Virginia.

Wagner 1989. Virginia Water Control Board. Richmond, Virginia, Personal communication.

Wischmeier, W.H. & D.D. Smith 1976. Predicting Rainfall Erosion Losses: A Guide to Conservation Planning. Agric. Handbook 537. U.S. Department of Agriculture. Washington, D.C.

Younos, T.M. & C.R. Sewell 1985. Rainfall Simulators: State-of-the-Art. J. ISEM 7:1-13.

Younos, T.M. & C.D. Metz. 1988. Site Selection for Land Application of Sludge Using a Geographic Information System. Applied Eng. in Agric. 4:56-60.

Land and Water Use, Dodd & Grace (eds), © 1989 Balkema, Rotterdam. ISBN 90 6191 980 0

Small systems of agricultural reuse of municipal wastewater in Sicily*

V.Tamburino
Istituto di Genio Rurale, University of Reggio Calabria, Italy
S.Barbagallo
Istituto di Idraulica Agraria, University of Catania, Italy

SUMMARY: This paper reports the results of experimental investigations and surveys on case-studies of municipal wastewater irrigation. Such results have provided useful indications for planning and operation of small wastewater irrigation systems which regard: contribution of nutrient to soil, content of salts and other dangerous chemical elements, use of localized irrigation to limit sanitary risk. Present wastewater irrigation modalities in Sicily are analyzed together with development opportunities of small systems.

Petits systèmes de utilisation agricole d'eaux usées urbaines en Sicile

RESUME: Cette étude nous indique les resultats des recherches experimentales obtenus dans le cas d'irrigation en utilisant les eaux usées urbaines. Les résultats obtenus ont fait base d'une découverte des moyens de plannification et opération de système à courte échelle d'irrigation de cette sorte. Cette étude consiste d'élements suivants: l'apport d'élements nutritifs au terrain, le contenu du sel et d'autres élements chimiques dangereux, l'emploi d'irrigation localisée en vue de minimiser les dangers de contamination. Actuellement les analyses de ces modalités d'irrigation en SICILE sont prises dans un contexte de development et de possibilités de réalisation de petits systèmes.

Kleine Systeme städtlichen Abwässer für landwirtschaftliche zwecke in Sizilien

ZUSAMMENFASSUNG: In dem Aufsatz wird über die Ergebnisse der experimentellen Forscungen und der verschiedenen Untersuchungen im Bereich der Bewässerung mit städtlichen Abwässer berichtet. Aus diesen Ergebuissen hat man nütaliche Hinweise gezogen für die Planung und die Fuhrung der kleinen mit Abwässer Bewässerungssysteme, die den Nährstoffzusatz für den Boden, den Ihualt der Salze und der anderen gefährlichen chemischen Grundstoffe, den Gebranch der lokaliscerten Bewässerung für die Verminderung des hygienischen Gefahres augehen. Man hat auch die Leutigen mit Abwässer in Sizilien existierenden Bewässerungsbeschaffenheiten und die Entwicklungsmöglichkeiten der kleinen Systemen übergrüft.

(*) This paper has been equally carried out by authors

1. INTRODUCTION

Agricultural wastewater reuse is particularly interesting in arid and semiarid areas, such as Southern Italy, where water resources are limited and represent an important factor for the economic and social development.

Two strategies can be followed to reuse municipal wastewater for irrigation:

1. to carry out advanced wastewater treatment in order to reduce or eliminate harmful elements (for example pathogenous micro-organisms);
2. to bind irrigation modalities (for example irrigation methods, crops, water depths, etc.).

The choice of the most convenient strategy is partly determinated by dimensions of the reuse system.

In Italy big systems of wastewater reuse for agricultural purposes have been planned. They include advanced depuration plants. Such a choice has required previsions of big collecting works of wastewaters in numerous centres in order to get the dimensions suitable for such treatments. Some of these systems have been realized but they have to face many difficulties in their starting, especially because of management problems and high costs of wastewater treatment.

In centres of small and medium dimensions it is possible to follow just the second strategy above mentioned; to this purpose it is necessary to specify modalities of irrigation reuse which require simple and unexpensive pre-treatment plants. Such small systems of wastewater reuse could increase the utilizable water resources avoiding, at the same time, pollution in water bodies. The lack of planning and operation experiences limits, at present, the realization of these systems and encourages, in the small centres, the realization of conventional depuration systems with subsequent discharge of effluent in water bodies (Tamburino & Barbagallo, 1988a).

In order to define characteristics of agricultural wastewater reuse in small systems, farms which have been using wastewater irrigation for many years, have been studied and analyzed. In fact, numerous Sicilian farms spontaneously use untreated or partially treated wastewaters, without any control by Authorities.

This paper reports an analysis of present modalities of agricultural wastewater reuse in Sicily. Moreover, information and useful considerations about planning and operation of small wastewater irrigation wastewater systems are reported. They concern:

1. contribution of nutrients to the soil and the consequent implications for fertilization practice;
2. origin and content of salts and other chemical elements in municipal wastewaters and relative agronomic effects;
3. localized irrigation opportunities aiming at reducing sanitary risk.

The over mentioned considerations derive from results of a survey carried out in 1988 and of preceding investigations on experimental plants and case-studies (Barbagallo etc. 1988, Capra etc. 1985, Indelicato etc. 1981,1984,1988, Tamburino etc. 1988b).

2. AGRICULTURAL USE OF MUNICIPAL WASTEWATERS IN SICILY

In Sicily municipal wastewater reuse has been carried out for many years spontaneously by farmers, without taking into account technical rules and administrative prescriptions. In particular, direct or indirect reuse of wastewaters is widely performed where idoneous conditions exist especially as regards availability of wastewaters and agricultural areas. Some of such case-studies on wastewater irrigation have been studied in order to evidence effects on soils and crops. The results are reassuring also because irrigation lasted many decades without any wastewater treatment.

Surface irrigation methods are generaly used. More rarely localized irrigation methods are used. Spontaneous reuse diffused in Sicily at present can bring both positive and negative consequences, in relation to peculiar modalities which it is carried out with (type of irrigated crops, quantities of delivered wastewaters, presence and vulnerability of underground water, irrigation methods, etc.).

A consistent quantity of wastewater produced in Sicily is directly or indirectly reused by farmers. This is fairly evident from tab. 1 which reports the present final receivers of municipal sewerages (Regione Siciliana 1987). Wastewater reuse, even if partial, of 195 discharges (related to 650,000 inhabitants, 13% of Sicilian population) has been evidenced. Moreover, a great amount of wastewaters from 479 discharges in water courses produced by about 1,7 millions inhabitants, (35% of insular population) is potentially reusable. Such water courses are generally torrent-like courses with minimun flows or no flows almost all year round. A part of wastewaters discharged in water courses is indirectly used by farmers right now; as a matter of fact, in some cases, these waters infiltrate stream beds and are drawn through wells, while in other cases they are drawn by farmers directly from water courses after being partly diluted and self-depurated.

Table 1. Municipal wastewater discharges in Sicily

Receivers	Discharges number	%	Inhabitants n.x 1000	%
Sea	187	22	1.471	33
Water-course	479	55	1.687	37
Soil	195	23	651	14
(*)			728	16
Total	861	100	4.537	100

(*) Population without sewer system whose discharges mainly end in subsoil

In Sicily just a small amount of wastewaters discharged and directly or indirectly used is depurated. In fact, only about the 10% of all discharges (as far as served population is concerned) has a primary or a secondary treatment. Moreover, just a part of plants works efficiently. The analysis of operation modalities and methods has showed that the bad functioning of depuration plants is partly due to organizing deficiencies and partly to the high operation costs (Barbagallo 1987).

3. RESULTS OF INVESTIGATION

3.1 Contribution of nutrients

Municipal wastewater irrigation determines a contribution of fertilizers to soil such as phosphorus, potassium and nitrogen. Both phosphorus and potassium are kept and progressively stored up by soil, while nitrogen is absorbed by crops or denitrificated or lixiviated (Broadbent & Clark 1965). In small systems wastewasters only partially treated are to be used. It is useful to control that nitrogen, phosphorus and potassium concentrations are not so high to damage soil, crops and underground waters; moreover it is necessary to analyze contribution of nutirents to soil in order to take care of it in fertilization practices. Nitrogen, phosphorus and potassium concentrations in wastewaters depend on municipal use and in particular on water supply and on people's habits and life conditions. At the end of 1988 irrigation season a survey has been carried out aiming at getting indications on: - increases of concentrations of such elements in water related to municipal use; - contribution of nutrients to soil with wastewater irrigation.

In every town a composite sample of untreated wastewaters (drawn in 12 day-time hours) and a sample of potable water have been analyzed. Table 2 shows nutrient average content in wastewaters of examined towns; moreover increases found in wastewaters in relation to municipal water supply concentrations are here reported. High variation coefficients show a great variability among various towns.

Phosphorus concentration in wastewaters almost entirely comes from municipal use, being concentrations in supply waters rather low. Higher concentrations in municipal supply waters can be found in potassium and nitrogen; in some cases nitrate concentrations exceed National and European Community standards for potable waters.

Tab. 3 shows quantities of nutrients brought to the soil, calculated with a hypothetic seasonal water depth of 6.000 m^3/ha (usual value in Sicily for citrus orchards). Such data have been calculated according to average concentrations evidenced in 19 untreated municipal effluents in Sicily. In planning systems of wastewater reuse for irrigation prevision of contribution of nutrients has to be carried out by taking into account any reduction in concentration due to wastewater treatment.

Table 2. Content of nutrients in wastewaters of 10 Sicilian towns

Nutrients	Wastewater concentrations (mg/1)	Increases (mg/1)
N	40 (0.31)	21 (0.80)
P	8.2 (0.26)	7.0 (0.27)
K	24 (0.52)	13 (0.74)

Note: Here are reported average values and (into brackets) variation coefficients of nutrient concentrations in wastewaters and concetration increases consequent to municipal use.

Table 3. Contribution of nutrients to soil with untreated wastewater irrigation (Kg/ha year)

Nutrients	min	med	max
N	168	269	450
P	18	55	129
P_2O_5	41	127	297
K	79	143	312
K_2O	95	173	376

Note: Evaluations carried out according to nutrient concentrations in wastewaters and to a seasonal water depth of 6.000 m^3/ha.

Tab. 4 shows increases of P_2O_5 and K_2O content in the first 60cm of soil evidenced in 8 farms irrigated with wastewaters for many years. Entity of accumulation of such elements depends on number of wastewater irrigation years. Yearly average increases of 240 Kg/ha of P_2O_5 and 340 Kg/ha of K_2O have been evidenced. Such increments depend, besides to quality and quantity of wastewaters applied in the years, also on performed fertilization practices and on previous characteristics of soils irrigated with wastewaters and of control soils. As provided no meaningful accumulation of nitrogen in soil has been evidenced.

Table 4. Content of nutrients in soils irrigated with wastewaters

Nutrients	Increases (Kg/ha)	Yearly increases (Kg/ha)
P_2O_5	9,100 (1.45)	240 (1.04)
K_2O	12,100 (1.61)	340 (1.05)

Note: Here are reported average values and (into brackets) variation coefficients of increases (in relation to controls) of nutrient content in soil of 8 farms irrigated with wastewaters.

3.2 Salts and other dangerous chemical elements

Macroelement presence in municipal wastewaters used for irrigation can determine agronomical problems. Such problems are not specific of small systems, but can be found in any case of wastewater reuse for irrigation. They are also typical of conventional waters characterized by a high content of salts; such waters are present in some parts of Sicily.

Main parameters which characterize irrigation fitness in relation to macroelement presence are: electrical conductivity, SAR besides to concentrations of sodium, calcium, magnesium and chlorides. In the case of municipal wastewaters, values taken from such parameters depend on qualities of municipal supply waters and on contribution consequent upon the civil use. Boron is the microelement which more often can create phyto-toxicity problems in wastewater irrigation. Boron concentrations in wastewaters are sometimes high because of the contribution of some kinds of detersives where it is present as sodium perborate. In order to show increases in content of salts and other dangerous chemical elements in municipal wastewaters after munucipal use, during the over-mentioned investigations on 10 Sicilian towns (cfr. 3.1), qualitative analyses both on civil waters and on wastewaters have been carried out. A synthetic picture of results of investigations is reported in tab. 5. Salt concentrations are higher in wastewaters than in civil waters and they differ according to the different parameters and towns. Electrical conductivity, index of the total salt content, goes from values of 0,64 mS/cm in civil waters to values of 1,43 in wastewaters; 9 effluents out of 10 belong to the third class of salinity (high) according to USDA criterium (Wilcox 1955).

Increase in sodium concentration is particularly high and it determines SAR average values in wastewaters of about 15. Seven effluents out of ten show a high or very high alkalinity risk (class 3 and 4 of USDA classification); so the opportunity of irrigation use of such effluents should be further evaluated according to criteria which also take into account the soil characteristics (Crescimanno & Giordano 1988).

Average borum content in wastewaters is about 2mg/l with maximum values of 2,8 mg/l. Such concentrations make irrigation of some sensible cultivations almost risky. For example, under Sicilian weather conditions, fitotoxicity effects on citrus orchards are particularly relevant with borum concentrations in wastewaters over 1,5-2,5 mg/l.

3.3 Use of localized irrigation to reduce sanitary risks

Health risks connected to wastewater agricultural use can be sufficiently reduced by employing localized irrigation methods. Especially in small reuse systems it is not convenient to make use of irrigation methods, such as sprinkling, which involve wastewater contact with crops. In such a case, it would be necessary to depurate wastewaters with advanced treatments which require high operation costs in relation to dimensions of reuse systems. In addition to economic factors, carrying out of advanced depuration plants makes reuse system hardly reliable because of the well-known and relevant operation difficulties particularly evident in small plants. So, in the case of small reuse systems it is convenient to limit wastewater treatment

Table 5. Increases in salt concentration and other chemical elements subsequent to civil use

Salts	Civil water concentrations	Wastewater concentrations	Increases
C.E. (mS/cm)	0.84 (0.43)	1.43 (0.38)	0.59 (0.48)
pH	7.4 (0.02)	8.4 (0.02)	1.0 (0.17)
B (mg/l)	0.54 (0.17)	2.08 (0.21)	1.54 (0.26)
Na (mg/l)	70 (1.16)	174 (0.65)	104 (0.43)
Ca (mg/l)	97 (0.30)	145 (0.51)	48 (1.41)
Mg (mg/l)	31 (0.78)	120 (0.69)	89 (0.77)
SAR	8.3 (1.01)	14.8 (0.39)	6.5 (0.91)
Cl (mg/l)	101.5 (1.20)	191.8 (0.86)	90.3 (0.57)

Note: Data concerning 10 Sicilian towns. Average values and (into brackets) variation coefficients are reported.

(for example just to sedimentation) and to irrigate cultivations by methods which avoid any contact of crops with wastewaters.

Localized irrigation methods, such as capillary sub-irrigation and drip irrigation, are particularly useful to wastewater use. Such methods reduce health risks, especially if they are employed in arboreous cultivations, but they present occlusion of emitters.

Experimental investigations have been carried out in order to study occlusion of emitters. Different emitters have been tested with wastewaters treated at different levels. The best results have been obtained with microtubes with a diameter between 2 and 4 mm.(Capra etc. 1985, Barbagallo etc. 1988). The delivered flows are generally major than those of drippers which are usually employed; this produces some disadvantages such as increase of pipe diameter and cost. Such disadvantages are partly reduced in irrigation networks of arboreous cultivations which have less emitters for each unit area. It is usually possible to reduce delivered flows to acceptable values by using low pressures and enough long microtubes. Good results have also been obtained with elastic fissured emitters (Tournon 1972, 1979). Anyway their use on large scale could be limited by the bigger diameters required by pipeline because of the great emitter flow increase related to small pressure increase.

In order to avoid occlusion risks, it has been enough to perform a filtration by metal cartridges with holes between 0.5 and 1.5 mm. inferior to half of orifices of employed emitters. Cleaning out operations are simplified with these filters (in comparison with filters by nylon fabric with smaller openings). Moreover, cleaning out operation of these filters has not been excessively frequent even if primary effluent has been employed.

Bacteriological investigations carried out on localized irrigation systems have showed a low or inexistent microbic contamination of crops (grapes). Even in those cases where a casual contact of grapes with untreated wastewaters has been simulated, a rapid inactivation of infectious micro-organisms has been observed due to environmental factors (in particular solar radiation) hostile to micro-organism survival (Barbagallo etc. 1988).

4. CONCLUSIONS

The small dimensions of many towns in Southern Italy reveal how interesting it is to carry out small systems of municipal wastewater reuse for irrigation. For example in Sicily about 72% of towns have a number of inhabitants inferior to 10,000 units and a wastewater flow inferior to 30-40 l/s. In many small Sicilian towns a spontaneous use of municipal wastewaters has been evidenced. This shows that there are no psychological resistances to such use.

Reuse systems planned at present generally tend to collect wastewaters from numerous towns in order to reach dimensions suitable to advanced treatment plants. But such a solution is hardly convenient because of high costs, system operation difficulties, and finally because in many cases wastewaters would be withdrawn from present users.

Results of this investigation confirm evidence of the opportunity of developing alternative solutions in order to carry out small systems of wastewater reuse formed of:

1. reliable depuration plant, not expensive and easy to operate (sedimentation tank or biologic pound);
2. farm filter with cartridges by 0,5-1,5 mm diameters opening;
3. localized irrigation network with emitters resistant to occlusion (for example microtubes with diameter of almost 2mm).

In such systems health risk is reduced by localized irrigation especially in presence of arboreous cultivations.

It is not difficult to choose an arboreous cultivation in Sicily where citrus orchards and vineyards absorb the major part of water resources used for irrigation.

Expected contribution of nutrients to soil after wastewater irrigation is generally consistent with such cultivation needs even if it changes according to different cases. This happens also in the case the reduction in nutrient content is low because of simple treatment levels expected in small reuse systems. It is, however, necessary that farmers know the entity of such fertilizing contribution and take it into account during fertilization; in particular, risk of excessive nitrogen supply is to be avoided, as it can pollute natural waters.

During civil use, electric conductivity and SAR values and B, Na and Cl concentrations increase differently according to the town and to the water supply per head. Such parameters constitute indexes of agronomic risks connected to water irrigation use. Some of these parameters (in particular B, Na and SAR) are especially high in effluents of some Sicilian towns; in such cases it is necessary to carry out deep investigations which permit to verify water suitability in relation to soil characteristics.

In order to carry out small systems of wastewater agricultural reuse in Italy, Italian rules should be modified. Present rules do not take into the right account the irrigation methods which are, on the contrary, really important in order to limit health risks. Moreover, there are at present some restrictive constraints on access to reuse spots, the presence of houses and streets near irrigation areas, etc.; the absurdity of such constraints is evidenced by the fact that their application is not expected for wastewater discharges in other receptive bodies.

Finally, it is to be evidenced that a relevant obstacle to the change of spontaneous irrigation reuse into a reuse correctly planned, carried out, operated and controlled in Italy is due to the reluctance by civil officials to get such a responsability. This inertness very often creates health and environmental damages (connected for example to wastewater discharges in water bodies) which are more dangerous than those which civil officials fear. In order to avoid this it is necessary that evaluation of health and environmental risks is always carried out among different solutions of reuse or discharge.

REFERENCES

Barbagallo S. 1987. La gestione degli impianti di depurazione di acque reflue civili della Sicilia. Ingegneria Sanitaria n. 4.

Barbagallo S., Biondi M. & Li Destri Nicosia O. 1988. Prove di irrigazione con acque reflue parzialmente trattate. Secondo contributo. Rivista di Ingegneria Agraria n. 4.

Broadbent F.E., Clark F.E. 1965. Denitrification. American Society of Agronomy, Madison (ed. Bartholomew W.V. Clarke F.E.).

Capra A., Li Destri Nicosia O. & Tamburino V. 1985. Subirrigazione con acque reflue. Ingegneria Agraria n.3.

Crescimanno G. & Giordano G. 1988. Un criterio di valutazione dell'idoneita' dell'acqua all'irrigazione. Atti del IV Convegno Nazionale AIGR "Ingegneria per lo sviluppo dell'Agricoltura".

Indelicato S., Li Destri Nicosia O. & Tamburino V. 1981. A case of raw wastewater irrigation in Sicily. Proccedings "Water reuse Symposium II" vol.1. Washington D.C. August 23-28.

Indelicato S., Li Destri Nicosia O. & Tamburino V. 1984. Effects de l'irrigation avec eaux usees. Etudes en Sicilie. 10eme Congres International du genie rural. Budapest september 3-7.

Indelicato S.,Tamburino V. & Barbagallo S. 1988. Agronomical and health aspects of municipal wastewater irrigation. 2nd International Conference on Environmental Protection. S.Angelo Ischia (Italy).

Regione Siciliana - Assessorato del Territorio e dell'Ambiente 1987. Piano Regionale di Risanamento delle Acque. Palermo.

Tamburino V. & Barbagallo S. 1988a. Indicazioni per la progettazione di piccoli sistemi di riutilizzazione irrigua di acque reflue urbane. Atti del IV Convegno Nazionale AIGR "Ingegneria per lo sviluppo

dell'Agricoltura".
Tamburino V. & Barbagallo S. 1988b. Effetti dell'irrigazione con acque reflue su terreno, colture e acque sotterranee. Atti del Convegno GRUSI "Ricerca e Irrigazione" Bologna.
U.S. Environmental Protection Agency, U.S. Army Corps Of Engineers & U.S. Department of Agriculture 1977. Process design manual for land treatment of municipal wastewater. EPA 625/1-77-008.
Tournon G. 1972. Esperienze di subirrigazione. L'irrigazione, marzo-giugno.
Tournon G. 1979. La subirrigazione capillare. Agricoltura Ricerca II.
Wilcox L.V. 1955. Classification and use of irrigation waters. Circ. n. 969 U.S.D.A. Washington D.C.

Land and Water Use, Dodd & Grace (eds), © 1989 Balkema, Rotterdam. ISBN 90 6191 980 0

Rolle von Pappel- und Weidenbeständen in der Abwässerreinigung

Z.Tihanyi & K.Tompa
Universität für Forst- und Holzwirtschaft, Sopron, Ungarn

ABSTRACT: The autors investigated in a growth pot experiment with poplar and willow clones over two years the effect of irrigation with sewage and with pure water. The applied dosages were between 1200 and 3000 mm/year. Compared to the treatments with the same quantity of pure water, sewage-treated stands produced 16 to 139 percents more wood in function of species and irrigated quantity. Untreated pots yielded only 30 to 36 percents of the pure water irrigated variants. The filter effect of the soil plays an important role in the cleaning process because high dosages cannot be utilized fully by the tree crop. The problem of accumulating heavy metals in leaves and wood, as well as the bacteriological pollution of the groundwater may be avoided if irrigation technology is followed correctly.

ABSTRACT: Le plus ancien récepteur et le metteur en oeuvre de eaux polluées et des autres déchets est le sol. Par la mise en place de l'eau polluée dans le sol, on solutionne non seulement l'éloignement et la réduction a l'impuissance des matériaux contaminés mais on utilise leurs substances alimentaires et aussi l'eau. Le but primaire est l'épuration de l'eau polluée, mais en mettant l'eau polluée de maniére réglée et controlée par voie hydraulique dans le sol du bois, alors onl'utilise à la fois dans la production du bois. En circonstances naturelles il est pour ainsi dire impossible de déterminer la circulation de l'eau polluée dans le sol, dans le bois. Pour mise au point de cela on a fait des recherches en grand pot de culture.

ZUSAMMENFASSUNG: In einem Gefässversuch wurden Pappel- und Weidenklone zwei Jahre hindurch mit verschiedenen grossen Abwasser- bzw. reinen Wassermengen bewässert. Die Dosierung lag zwischen 1200 und 3000 mm/Jahr. Im Vergleich zu gleich hohen Reinwasserbehandlungen stieg der Ertrag an Holzmasse um 16 bis 139 %, in Abhängigkeit von der Behandlungsvariante und von der Holzart. Die unbewässerten Varianten ergaben nur 30 bis 36 % der Holzmasse der Reinwasser-Varianten. Die Filtrierwirkung des Bodens spielt bei der Entsorgung eine bedeutende Rolle, da die hohen Wassergaben nur zum Teil vom Baumbestand aufgenommen werden können. Die Frage der Anreicherung von Schwermetallen in der Blatt- und Holzmasse, sowie die bakteriologische Gefährdung des Grundwassers erwies sich als unbedenklich, falls die vorgeschriebene Technologie eingehalten wird.

1 VERSUCHSMATERIAL UND METHODIK

In 57 cm breite und 85 cm hohe Gefässe wurden in drei Wiederholungen drei Holzarten bzw. Klone ausgepflanzt: die Klone 'I-214' und 'Robusta' der euramerikanischen Pappel /Populus x euramericana /Dode/ Guinier/, sowie die Weissweide /Salix alba L./.Als Pflanzgut dienten einjährige Heister aus Stecklingen. Die Gefässe wurden randomisiert in Blöcken aufgestellt und bis an den Rand in Boden gebettet und starke Temperaturwankungen zu vermeiden.

Die einzelnen Varianten wurden mit den Abwässern bzw. mit dem Trinkwasser der Stadt Sopron zwei Jahre lang zwischen dem 12.April und dem 13.Oktober bewässert.

Die Variantan waren: 1. 1200 mm Abwasser; 2. 2000 mm Abwasser; 3. 3000 mm Abwasser; 4. 1200 mm reines Wasser; 5. unbehandelte Kontrolle. Die gleichen Wassergaben wurden auch Gefässen ohne

Pflanzen verabreicht. Die Variante 1200 mm wurde alle 8 Tage, jene von 2000 und 3000 mm wurden alle 4 Tage verteilt verabreicht in der Vegetationszeit. Der Niederschlag wurde bei den Berechnungen beachtet.

Zur Messung der Flüssigkeitsmenge, welche weder vom Boden zurückgehalten, noch über die Vegetation oder dem Boden verdunstet, oder in das Gewebe der Bäume eingebaut wurde, ist folgene Lösung verwendet worden.

Eine Ablassöffnung wurde 5 cm vom Boden in der Gefässwand eingelassen, welche von innen mit einem Kunststoffsieb und sandigem Kies gegen Verstopfung gesichert war. Ein an die Öffnung angeschlossener Kunststoffschlauch leitete die Flüssigkeit in einen geschlossenem Kanister. Das abfliessende überflüssige Wasser wurde alle 4 Tage entleert und gemessen.

Sowohl das zur Bewässerung gelangede Klärwasser als auch das durch den Boden gefiltertes Wasser wurde monatlich einmal im Labor getestet.

Zur Auffüllung der Gefässe wurde brauner Waldboden mit Karbonatresten im gemischten Zustand verwendet.

Die Kennwerte des Bodens waren: pH /H_2O/ 7,9; pH /KCl/ 7,6; Bindigkeit 35; ges. Salzgehalt 0,04 %, Soda-Laugigkeit 0,027 %, $CaCO_3$ 15,5 %; Humusgehalt 1,72 %, hy 2,08 %, kapillare Wasserhebung /5 Std./ 23 cm. Die mechanische Zusammensetzung war: Sand 64,2 %, Schluff 27,9 %, Lehm 7,9 %. Nährstoffgehalt: loslicher P_2O_5 106 ppm; löslicher K_2O 112 ppm; ges. Stickstoffgehalt 0,11 %.

Nach zwei Jahren wurde der Versuch aufgelöst und die Masse und chemische Zusammensetzung der Bäume wurde ermittelt.

2 ERGEBNISSE

2.1 Reinigung der Abwässer

Der Reinigunsvorgang ist in der Tabelle 1. vorgeführt. Die Tabelle spricht für sich: in Reinigungsprozess spielt nicht der Baum, sondern der Boden die Hauptrolle. Die Bäume werden zum Zwecke der Wasser- und Nährstoffverwertung, zur Transpiration, zur Dämmung der Geruchsauswirkung und des Windes, zw. Belebung des Bodenlebens und aus ästhetischen Gründen benötigt.

Die Bezeichnungen BOI 5 und KOI in der Tabelle 1 stehen für biochemischen bzw. chemischen Sauerstoffmangel. Alle Daten sind Mittelwerte von monatlichen Messwerten.

Bessere Werte wie in Tab. 1 aufgeführt konnten auch durch technische Reinigungsvorrichtungen nicht erreicht werden.

Bei kleinen Abwassergeben konnten nur geringe Mengen überflüssigen Wassers gemessen werden, die Konzentration war allerdings viel höher. Deshalb war auch die im Boden befindliche Menge auch konzentrierter, was sowohl für die Bäume als auch für den Boden schädlich ist.

Tabelle 1. Abwasserklärung bei der Behandlungsvariante 3000 mm Abwasser/Vegetationsperiode

Komponente	Qualität der geklärten Abwässer /ppm/	Reinigungeffekt /%/ Baum und Boden zus.	Boden allein
K+	24	91	87
Ca 2+	122	53	38
Mg 2+	46	60	48
Na+	181	54	38
NH4 +	30	99	97
N-organisch	32	97	95
N-gesamt	56	97	90
PO4 3-	11	99	98
P-organisch	2	97	92
P-gesamt	5	97	92
BOI5	224	97	96
KOI	443	96	90
O2 /KmnO4/	123	96	93
Fenol	0,01	100	70
Detergent	1,10	97	85
Cu 2+	0,55	99	96
Ni 2+	0,09	99	70
Zn 2+	0,76	62	46
Cd 2+	0,06	98	95

Wassergaben, bei denen kein Überschuss in die unteren Schichten durchdringt, sondern nur durch Evapotranspiration aus dem Boden austritt, halten wir für schädlich. Die schädliche Wirkung von zurückbleibenden schädlichen Salzen sowie von der erhöhten Konzentration der Bodenlösung ist unzweifelhaft.

2.2 Wasserhaushalt

Bei der Variante 3 /höchste Dosis/ hat ein Gefäss innerhalb von zwei Vegetationsperioden 1529 l Abwässer und 156 l Niederschlag erhalten. Die Gefässe fassten je 69 l Erde.

Aus dem insgesamt 1685 l Wasser der Variante 3 betrug die Evapotranspiration

1260 l /Transpiration 594 l, Evaporation 666 l/ 425 l Wasser versickerte durch den Boden.

Bei genügender Wasserzufuhr wurde pro kg Holzmasse /atro-Masse/ über dem Boden die folgende Wassermenge verdunstet /Mittel von je 9 Messungen:

Pappelklon I-214	363 l/kg
Pappelklon 'Robusta'	456 l/kg
Weidenklon 'Bédai egyenes'	316 l/kg

Die Transpiration eines jungen, weitgepflanzten Bestandes ist natürlich unbedeutend im Vergleich der Verdunstung des Bodens und der Krautschicht, weil der laufende Zuwachs noch gering ist. In einem älteren Bestand beträgt die Transpiration der Klone 'I-214' und 'Robusta' sowie der Weide 283, 221 bzw. 205 mm /bei einer Trockendichte von 354, 440 bzw. 406 kg/m^3, in der III. E.klasse und bei einem laufenden Zuwachs von 22, 11 bzw. 16 m^3/ha/. Diese Werte erscheinen als verhältnissmässig gering, aber auch die Verdunstung vom Boden und jenes der Krautschicht sowie die direkte Verdunstung vom Laubwerk usw. muss in Betracht gezogen werden.

2.3 Auswirkung der Abwasserbewässerung auf den Holzertrag

Der Vergleich der drei Holzarten ist nur bezogen auf die atro-Masse reell. Diese Werte sind in der Tabelle 2 enthalten.

Tabelle 2. Absolut trockene /atro/ Holzmasse der Bäume in kg

Holzart bzw. Variante		Gesamtmasse über dem Boden	Wurzel	Gesamte Pflanze
Weiss-weide	1	1,06	0,63	1,69
	2	1,50	1,11	2,61
	3	1,98	1,33	3,31
	4	0,79	0,59	1,38
	5	0,28	0,16	0,44
Pappel-klon 'I-214'	1	1,00	0,58	1,58
	2	1,40	0,83	2,23
	3	1,71	0,96	2,67
	4	0,84	0,53	1,37
	5	0,29	0,12	0,41
Pappel-klon 'Robusta'	1	0,76	0,64	1,40
	2	1,12	0,88	2,00
	3	1,38	1,14	2,52
	4	0,65	0,51	1,16
	5	0,26	0,15	0,41

Bei allen drei Arten bzw. Klonen ergab die höchste Dosis /3000 mm/Veg.Zeit/ die besten Resultate, den höchsten Holzertrag. Die Ertragssteigerung betrug bei der Weide 131 %, bei dem Klon 'I-214' 106 %, bei 'Robusta' 100 %.

Verglichen mit der Bewässerung mit klarem Wasser ergab die Abwasserbewässerung des Klons 'I-214' bei der Variante 1200 mm 15,7 %, bei 2000 mm 63,2 %, bei 3000 mm 96,0 % mehr Holzertrag. Der Holzertrag des Klons in der unbewässerten Variante belief auf lediglich 29,7 % der 1200 mm Reinwasser-Kontroll-Variante.

Die gleichen Zahlen beliefen sich bei dem 'Robusta' Klon auf 21,7 % /1200 mm/, 74,3 % /2000 mm/ 118,7 % /3000 mm/ mehr Holzertrag. Der Ertrag der unbewässerten Variante lag bei 35,8 % der Kontrollvariante.

Bei der Weide ergab die Ertragssteigerung 22,7 % /1200 mm/, 88,4 % /2000 mm/, 139,1 % /3000 mm/ im Vergleich zur Variante 1200 mm Reinwasser-Bewässerung. Die unbewässerte Variante lieferte nur 31,8 % der Kontroll-Masse.

2.4 Spezifischer Wasserverbrauch der Bäume

In der Tabelle 3 sind die verbrauchten Wassermengen pro kg Trockenmasse aufgeführt.

Unerwarteter weise verbraucht die wasserliebende Weide weniger Wasser pro kg Trockenmasse als die euramerikanischen Pappeln. Der spezifische Wasserverbrauch ist bei der höchsten Variante am niedrigsten, der Holzertrag ist hier am grössten. Umgekehrt liegt der Fall bei den Bäumen, die nur auf den Niederschlag angewiesen waren /Variante 5/.

Tabelle 3. Wasserverbrauch /l/ zur Produktion von 1 kg absolut trockener Holzmasse

	Varianten				
	1	2	3	4	5
Weissweide	191	200	174	223	446
'I-214'	230	250	232	249	483
'Robusta'	240	275	241	280	475

Bei der Variante 3000 mm fiel auf ein qm Blattoberfläche der I-214-Pappel 184,3 l Verdunstung. Der gleiche Wert lag bei dem Klon 'Robusta' bei 143,9 l und bei der Weide bei 638,1 l.

Die Blattoberfläche des Klons 'I-214'

ist nur 67 % jener der 'Robusta'-Pappel, aber die Intensität der Transpiration ist um 128 % höher. Die Weissweide besitzt nur 20 % der Blattoberfläche der 'Robusta'-Pappel, die Intensität der Transpiration ist aber um 443 % höher. Die Weissweide produzierte die grösste Holzmasse mit der kleinsten Blattoberfläche und Blattmasse und beim geringsten Wasserverbrauch.

2.5 Veränderungen im Boden

Der Vergleich der Bodenanalysen Anfang und Ende des Jahres widerspiegeln die Tendenzen der Abwasserreinigung.

Der Gesamtgehalt der Salze nimmt mit der Zunahme der Abwassergaben zu, aber innerhalb dessen bleibt der Na-Gehalt konstant, und wird bei den baumlosen Varianten sogar ausgelangt. Bei der Variante 3000 mm Abwassergaben verringerte sich der Na-Gelaht sogar in den Gefässen mit Bäumen, da hier die Menge des durchlaufenden Wassers höher lag. Dies deutet auf die Erlangung eines Gleichgewichtszustandes im Boden.

Es ist festzustellen, dass als Folge der Abwasserbehandlung der Boden mit Stoffen aus dem Abwasser angereichert wird. Ein Teil der Nährstoffe wird von den Bäumen verbraucht, deshalb ist die Anreicherung von N,P,K geringer. Es ist auch zu vermerken, dass in Gefässen ohne Bäume das durchlaufende Wasser die leicht lösbaren Salze, besonders NO_3, NO_2, Na, aber auch P_2O_5 und K_2O aus dem Boden auslangt, sogar bei den Varianten mit Abwasserbewässerung. Unter natürlichen Bedingungen verlassen die mobilisierten Salze den Boden und werden im Grundwasser oder in unteren Schichten aufgenommen.

Die bakteriologischen Analysen bezeugten, dass die gesamte Keimzahl des vom Boden filtrierten Wassers nur zehntausendstel der Originalmenge beträgt. Solche geringe Mengen sind auf allen Ackerböden zu messen.

Der bakteriologische Zustand des vom Boden filtrierten Wassers ist sowohl bei Klarwasser- als auch bei Abwasserverwendung gleich geblieben.

2.6 Chemische Untersuchung der Blätter und des Holzes

Die Blatt- und Holzuntersuchungen brachten keinen Beweis für eine Verbindung zwischen den Behandlungsvarianten und der Zusammensetzung, d.h. für die Annahme, dass mit Zunahme der Abwassermengen einzelne Elemente sich anreichern. Im Gegenteil, es ist oft festgestellt worden, dass die Blatt- und Holzmasse der Reinwasser-Varianten oder der unbewässerten Variante die hochsten Werte aufwiesen.

Es muss deshalb nicht damit gerechnet werden, dass die jährlich anfallende Laubmasse die Bodenoberfläche mit Schwermetallen aus tieferen Schichten anreichert. Auch die Anreicherung im Holzkörper konnte nicht nachgewiesen werden. Die Menge der aufgenommenen und eingebauten Stoffe hängt vielmehr von artspezifischen Eigenschaften, als von der Behandlung.

Aus den Ergebnissen ist zu errechnen, wie viel Nährstoff in 1 m^3 Holzmasse enthalten ist, wie viel dem Boden entzogen wird. Gleichermassen kann auch die Nährstoffmenge pro t Blattmasse errechnet werden.

Zum Beispiel enthält ein hiebsreifer Pappelbestand des Klons 'I-214' mit einer Holzmasse von 273 m^3/ha /Ekl. III./ an Nährstoffen 667 kg Stickstoff, 177 kg P_2O_5 und 256 kg K_2O.

Beim Abholzen ist die Nährstoffmenge, die durch den Abtransport des Holzes verlorengeht, unbedeutend, da sie niedriger ist als die Menge, die in einem einzigen Jahr durch Bewässerung mit 3000 mm Abwässer verabreicht wird.

2.7 Ökonomische Berechnungen

Bei den ökonomischen Untersuchungen haben wir mit Hilfe eines komplexen Modells die Frage der Wertproduktion der Pappelbestände analysiert. Weiters wurde berechnet, um wieviel die Entsorgungskosten der Abwässer durch die Ausbringung in Holzplantagen verringert werden können, anstatt der Entsorgung in mechanischen Anlagen. Im ersten Fall kostet die Entsorgung etwa 3 Ft/m^3; in einer mechanischen Anlage liegen die Kosten bei etwa 7 Ft/m^3. Die Einsparung ist also etwa 4 Ft/m^3.

3 SCHLUSSFOLGERUNGEN

Die in den Versuchen gewonnenen Daten wurden mit Produktionsdaten in Kläranlagen verglichen. Die Vergleiche bewiesen die Gültigkeit der Versuchsergebnisse.

Das primäre Ziel der Abwasserausbringung in Wäldern und Holzplantagen ist die Entsorgung und nicht die Nutzbarmachung. Es sollen nicht die vorhandenen Wälder belastet werden, sondern direkt für diesen Zweck angelegte,und mit technischen Vorrichtungen versehene Holzplantagen. Dadurch kann man die Abwässer wesentlich wirtschaftlicher, konzentrierter und bei hozem Wirkungsgrad bei gleichzeitig geringerem Flächenbedarf entsorgen.

Hohe Wassergaben können vom Baumbestand nur zum Teil verdunstet werden. Die Weiterleitung der durch den Boden gedrungenen, filtierten Abwässer muss durch Drainage gesichert sein. Bei der Entsorgung spielt der Boden eine primäre Rolle. Auch die Bäume nehmen eine beträchtliche Menge an Nährstoffen auf. Der Holzertrag erfährt durch die Abwasserbehandlung eine starka Steigerung. In einer für diesen Zweck geplanten Anlage und bei Einhaltung der Technologie wird der Boden nicht geschädigt. Die Blatt- und Holzmasse weist keine Anreicherung an Salzen auf. Die Versuchsdaten haben gleichzeitig wichtige Daten über die Nährstoffaufnahme in Blätter und im Holz, über die Transpirationsmengen und über die notwendige Wassermenge pro Einheit der erzeugten Holzmasse geliefert.

Land and Water Use, Dodd & Grace (eds), © 1989 Balkema, Rotterdam. ISBN 90 6191 980 0

Auswirkungen von flüssigen Klärschlamm auf einen 11 bis 14-jährigen Pappelbestand

Z.Tihanyi & K.Tompa
Universität für Forst- und Holzwirtschaft, Sopron, Ungarn

ABSTRACT: The deposition of the communal sludge quantity of 4.5 million cu.m. per year is still an unsolved environmental problem in Hungary. The authors investigated the effect of liquid communal sludge on an 11 years old Euramerican poplar stand over three years. The applied dosages were 16, 32 and 64 mm/year. The medium intensity around 30 mm proved most seccessful. The yield of the plantation improved by two yield classes /from V to III/ over the three years of application. The most efficient application technology is described.

ABSTRACT: La mise en place de la vase liquide communale environ de 4,5 million m^3 par an en Hongrie est encore un des grands soucis de la protection de la nature. Les auteurs ont fait la recherche de 3 ans pour la mise en place de vase de l'eau polluée communale d'une grande commune dans une culture de peupliers agé de 11-14 ans /Populus x euramericana /Dode/ Guinier cv. I-214/, ils ont fait des essais avec des 16, 32 et 64 mm doses. La dose de 30 mm semblait d'etre juste. Dans la plantation pendant 3 ans il y avait une amélioration de 2 classes de rendement /du V-a III/. Les auteurs publient aussi la plus favorable technologie de la mise en place dans la pratique.

ZUSAMMENFASSUNG: Die Unterbringung des jährlich anfallenden kommunalen Klärschlammes /4,5 Mill.m^3/ ist in Ungarn noch immer ein ungelöstes Umweltschutzproblem. Die Verfasser untersuchten über einen Zeitraum von 3 Jahren die Auswirkungen von flüssigem Klärschlamm einer Grossgemeinde auf einen 11 jährigen Pappelbestand der Zuchtsorte I-214. Die verwendete Dosierung war 16, 32 und 64 mm/Jahr. Die Dosis um 30 mm schien am günstigsten. Der Ertrag der Plantage verbesserte sich um zwei Ertragsklassen /von V. auf III./. Die Verfasser beschreiben auch die günstigste Technologie der Ausbringung.

1 VERSUCHSMATERIAL UND METHODIK

In einem 11 jährigen Pappelbestand der Zuchtsorte I-214 /Gemarkung Bük/ wurde zwischen Herbst 1984 und Ende 1987 ein Versuch mit Klärschlammbewässerung mit 4 Behandlungsvarianten und 3 Widerholungen /mit 576 m^2 grossen Parzellen/ angelegt. Pro Jahr wurden 16, 32 und 64 mm kommunalen Klärschlammes mit einem Trockensubstanzbehalt zwischen 1 und 5 % ausgebracht. Es wurde ein verhältnissmässig ungünstiger Standort gewählt /schwerer Wiesen-Lehmboden, hohes Grundwasser zwischen 1,0 und 2,4 m, 4x4 m Standraum und Anpassungschwierigkeiten der Wurzelsysteme/ um möglichst aus einem Versuch Grenzwerte für anwendbare Dosierungen zu erhalten. Es wurden drei Varianten der Ausbringung angewendet. Im Rahmen der komplexen Untersuchung wurden die Menge der in den Boden ausgebrachten Nährelemente, die Veränderungen der Bodenbeschaffenheit, die Zusammenhänge Niederschlag - Schlammbewässerung - Grundwasserniveau verfolgt. Auch die physische, chemische Verwändung des Grundwassers wurde untersucht, gekoppelt mit ausführlichen bakteriologischen Analysen. Schliesslich konnte auch die Wirkung der Schlammdüngung auf dem Wuchs der Bäume gemessen werden, aber auch die Auswirkungen auf die Holzbeschaffenheit, und auf den Nährstoffhaushalt der Zweige und Blätter wurden analysiert.

Der pH-Wert des Klärschlammes schwankte zwischen 6,4 und 12,6, der Gehalt an organischen Substanzen lag zwischen 50 und 70 % der Trockensubstanz. Der Anteil der Nährelemente /N,P,K/ zeigte eine breite Streuung in Funktion der Trockensubstanzmenge. Schwermetalle konnten auch machwiesen werden /Ni, Cd, Cr/, aber ihr Anteil lag zum Grossteil unter der zugelassenen Grenzwerte.

Gerechnet auf die obersten 20 cm des Bodens wurden mit der Dosis 64 mm/Jahr folgende Wirkstoffmengen pro ha verabreicht: 868 kg N /gesamt/, 527 kg P, 119 kg K, 131 kg Na /die Menge von Ca und Mg blieb unbedeutend/.

2 ERGEBNISSE

2.1 Die Veränderungen im Boden

Die Pappelplantage stockt auf einen nicht karbonathaltigen Schwemmboden. Die Bindigkeit nach Arany lag von der Behandlung zwischen 41 und 58 /zwischen bis 170 cm Tiefe/, der pH-Wert zwischen 6,1 und 7,8. Der Humusgehalt der obersten Schicht lag bei 2,6 bis 3,0 %.

Als Folge der Behandlung stieg der pH-Wert des lehmigen Schluffbodens, hauptsächlich wegen des verabsreichten Kalciums. Die Bindigkeit blieb unverändert. Der Gesamt-Salzgehalt blieb auch nach der Behandlung negativ, aber es ist anzunehmen, dass bei einer Fortführung der Schlammgaben die untere Grenze der Nachweisbarkeit erreicht sein wird. Die Humus-Prozentwerte haben sich im Laufe der 3 Jahre dramatisch, um 26 bis 47 % erhöht. Die Werte der Kontrollparzelle blieben unverändert.

Der Nitratgehalt der Versuchsparzellen ist sehr heterogen, und hängt von der Bodentemperatur und vom Feuchtegehalt ab. Der verwertbare Phosphorgehalt stieg bei den behandelten Parzellen in Funktion der verabreichten Schlammdosen. Die Zunahme des Nährstoffgehaltes kann am besten mit dem Phosphorgehalt charakterisiert werden. Die Veränderung des verwertbaren Kaligehaltes ist nicht eindeutig /der Kaligehalt des Schlammes ist sehr niedrig/. Auch die Zunahme an Mg ist nicht nachweisbar wegen des schon ursprünglich hohen Mg- Gehaltes in Boden. Die übrigen Elemente zeigten keine Veränderungen. Auch der Gehalt an Cd und Cr war im Boden nicht nachweisbar. Die übrigen Schwermetalle zeigten Anteile unter der Zulassungsgrenze, und würden erst in mehreren hundert Jahren diese überschreiten.

Als Folge der Schlammausbringung /Bewässerung/ stieg die Volummasse des unberührten Bodens, d.h. die Gesamtporösität fiel zurück. Diese Abnahme ging auf Kosten des Gravitations-Porenvolumens. Eine Erhöhung des letzteren ist durch Bearbeitung mit Scheibeneggen erforderlich.

2.2 Veränderungen im Grundwasser

Die Veränderungen im Grundwasserniveau zwischen 1985 und 1987 können folgendermassen charakterisiert werden. Maxima des Grundwasserpegels zeigen sich im März /aus dem Winterniederschlag und dem Schmelzwasser/ und als Folge des reichlichen Frühjahrniederschlages im Mai /1985 Ende April-Anfang Mai, 1987 Ende Mai/. 1986 konnten auch die hohen Niederschläge im Mai-Juni den Rückfall des Grundwasserniveaus nicht verhindern; die Schwankung des Grundwassert hängt nicht nur mit den örtlichen Niederschlägen zusammen.

Besonders bei hohem Grundwasserstand folgt das Niveau weitgehend den Konturen der fast ebenen Fläche. Ab Juni fällt der Grundwasserstand im Wald stärker ab, als auf landwirtschaftlichen Flächen, wegen der hohen Transpiration der Bäume. Die verabreichte Wassermenge der Klärschlammbewässerung ist nicht bedeutend, und hat keinen Einfluss auf den Grundwasserstand /berechnet auf die Gesamtfläche, betrug die Menge lediglich 40,5 mm/Jahr/.

Die Richtung der Grundwasserbewegung, die Schneeschmelze, das Aufbauen des gefrorenen Bodens, die Niederschlagsmenge, die Grundwassertiefe und die Schlammbewässerung selbst sind die bestimmenden Faktoren des Nitrit-, Nitrat- und Ammoniumgehaltes im Boden. Gipfel sind zu verzeichnen Ende Januar-Anfang Februar wegen der Schneeschmelze; Ende März wegen der Niederschläge und wegen dem hohen Grundwasserstand; in April oder Mai aus den gleichen Grund, aber zu diesem Zeitpunkt kann auch schon die Auswirkung der Schlammdüngung dazukommen. In den folgenden Monaten nimmt der Nitrit- und Nitratgehalt praktisch nicht mehr zu, die Zunahme von Ammonium kommt allerdings vor. Grund dafür ist in der auslaugenden Wirkung des Niederschlages zu suchen.

Aus der Nährstoff- und Schwermetallanalyse des Grundwassers können folgende Schlüsse gezogen werden. Höchstwerte von N und K wurden in ersten Jahr im Frühjahr, im zweiten und dritten Jahr im August gemessen, obwohl die Möglichkeit der Auslaugung im Sommer auszuschliessen ist. Die geringste Mengen von P und K konnten im Herbst gemessen werden. Der Na-Gehalt zeigte starke Schwankungen, noch stärker an der benachbaren Ackerfläche. Das gleiche gilt für Ca. In den ersten zwei Jahren sind keine Schwermetalle in des Grundwasser gelangt, bzw. nur geringe Mengen von Zn oder Ni. Im dritten Jahr wuchs die Konzentration von Ni, und Cr, Cd erschienen, Zn nahm ab. Allerding kann ungefähr die gleiche Menge an Schwermetallen, oder noch mehr an der benachbarten Ackerfläche nachgewiesen werden. Das Erscheinen der Schwermetalle kann also nicht mit dem Klärschlamm in Zusammenhang gebracht werden.

2.3 Bakteriologische Untersuchungen

Der Klärschlamm wurde - wie erwartet - nach hygienischen und epidemologischen Analysen als schädlich bewertet. Die Gesamtmenge an Bakterien erreichte die Grössenordnung 10^7, die der faecalischen Indikatoren 10^4-10^5. In 10 von 17 Proben konnte Streptococcus faecalis in der Grössenordnung 10-10^3 nachgewiesen werden. Clostridium schwankte zwischen 10 und 10^6. Die Warmeier-Analysen blieben negativ. 2 Monate nach der Ausbringung waren die Ergenisse günstiger. In den verschiedenen stark belasteten Parzellen konnte der Zusammenhang zwischen Schlammdosis und Anzahl von Coliform- und faecalisch Coliform-Erregern nachgewiesen werden. Es waren keine Trychuris sp. vorhanden.

Bezüglich der Grundwasserqualität unter der Ackerfläche konnte man festellen, dass auch dort sowohl die physischen als auch die bakteriellen Merkmale nicht besser sind. Die Grundwasserqualität blieb unterhalb der gesetzlich festgelegten Grenzwerte und wäre dementsprechend in Fliessgewässer zuführbar.

2.4 Chemische Analyse der Pflanzenproben

Die Analyse der Pflanzenteile aus dem Holzkörper lieferte ähnliche Ergebnisse wie bei verhergehenden Untersuchungen.

Gerechnet auf das atro-Masse der Blätter, fiel auf 0,21 kg atro-Blattmasse 1 m^3 Holzmassenzuwachs. Mit dem Herbstlaub sind pro hektar 78 kg N, 17 kg P, 60 kg K, 44 kg Na, 94 kg Ca, 20 kg Mg, 0,7 kg Fe, 0,05 kg Cu, 0,4 kg Zn und 0,3 kg Mn zurück in den Boden gelangt. Diese Mengen sind weitgehend unter jenen die mit den geringsten Klärschlamm-mengen /160 m^3/ha/ ausgebracht wurden, mit Ausnahme von K.

Umgerechnet auf Kunstdünger-Wirkstoffmengen, kommen mit dem Herbstlaub 229 kg/ha 34-prozentiger Stickstoffdünger, 445 kg/ha 17,5-prozentiger Superphosphat und 361 kg/ha 40-prozetinger Kalisalz auf die Bodenoberfläche.

2.5 Auswirkungen auf den Wuchs des Baumbestandes

Es waren auch bei der höchsten Schlammdosis von 64 mm/Jahr keine negativen Auswirkungen nachzuweisen. Der Gesamtzustand des Bestandes verbesserte sich. Dies ist teils mit der Schlammbewässerung, teils mit der Boden- und Baumpflege in Verbindung. In den drei Jahren hat sich der Bestand um zwei Ertragsklassen verbessert, und stieg von der V.Ekl. im Jahre 1984 in die III.Ekl. Ende 1987 /es gibt sechs Ertragsklassen/. Der laufende Zuwachs war 1987 im Durchschnitt 29 m^3/ha. Die positive Auswirkung der höchsten Schlammdosis von 64 mm war im 3. Jahr deutlich zu sehen.

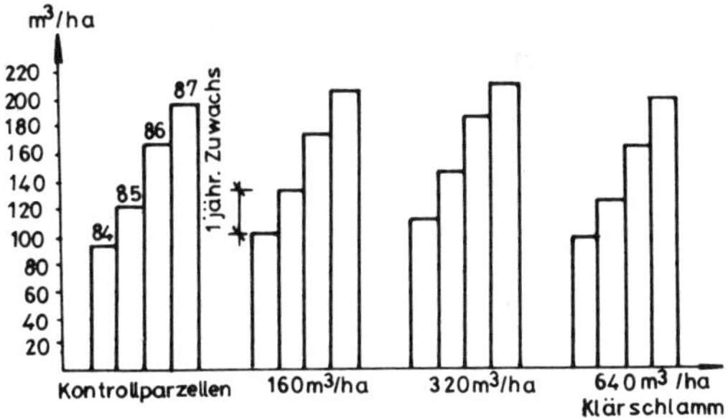

Abbildung 1. Durchschnittliche Hilzvolumendaten

2.6 Technologie der Schlammausbringung

Die drei Varianten der Ausbringung waren:

1. Entleerung des Pumpfahrzeuges zwischen den Reihen;
2. Ausbringung des Klärschlamms in Furchen zwischen den Baumreihen mit Hilfe des 150 mm starken Pumpslauches;
3. Gleichmässige Verteilung der Menge auf der Bodenoberfläche zwischen den Baumreihen mit Hilfe eines 50 mm starken Spritzschlauches.

Vor- und Nachteile der verschiedenen Methoden wurden von den Verfassern analysiert. Die dritte Methode wurde als günstigste beurteilt und als solche empfohlen. Der Spritzschlauch wird auf das verkleinerte Ablassventil des Pumpfahrzeuges montiert. Der Fahrweg soll permanent gekennzeichnet werden, in Abständen von etwa 30 m. Nach Ausbringung des Klärschlammes nach beiden Seiten kann auch in diesem Fall das Eineggen nicht ausbleiben.

3 SCHLUSSFOLGERUNGEN

Der Versuch ist deshalb bedeutsam, weil durch die Vereinheitlichung der biologischen Reinigung der kommunalen Abwässer in Ungarn meistens das Aktivschlamm-Totaloxydations-System angewendet wird, wobei der anfallende Klärschlamm schon durch aerob-Fermentation stabilisiert wird und sehr zweckmässig in Wäldern, die nicht in die menschliche Nahrungskette einbezogen sind, untergebracht werden kann.

Die pflanzliche Nährstoffmente der mit Klärschlamm behandelten Bestandesböden nimmt signifikant zu, das gleiche gilt für den Humusgehalt; die Infektionsgefahr fällt innerhalb weniger Monata allmählich auf das Niveau der nicht verunreinigsten landwirtschaftlichen Böden zurück. Der Boden zeigte auch bei der höchsten Belastung von 64 mm/Jahr keine Anzeichen von Schäden. Im Pappelbestand /Klon I-214/ konnte innerhalb von 3 Jahren eine Erhöhung der Massenleistung um zwei Etragsklassen festgestellt werden, der laufende Zuwachs stieg auf 29 m^3/ha/Jahr.

Die günstigste Ausbringungsmethode ist jene mit dem Spritzschlauch.

Für kleine und mittlere Gemeinden ist die Ausbringung des Klärschlammes in Holzplantagen die einzige wirtschaftliche, unbedenkliche und über das ganze Jahr mögliche Lösung.

Land and Water Use, Dodd & Grace (eds), © 1989 Balkema, Rotterdam. ISBN 90 6191 980 0

Wet detention areas for drainage from citrus groves

D.J.Pitts
Institute of Food and Agricultural Sciences, University of Florida, Fla., USA

ABSTRACT: Extensive citrus development is taking place in southwest Florida adjacent to the environmentally sensitive Big Cypress National Preserve and the Everglades National Park. Developers of these groves are required by state regulation to provide wet detention basins for drainage water. Wet detention basins are water storage areas with bottom elevations at least 30 cm below the elevation of the control discharge structure. The objectives of these wet detention basins and the engineering design criteria employed to meet regulatory guidelines are outlined in this report.

Primary functions of the wet detention basins are: 1) to maintain off-site discharge peak flows at or below pre-development rates, 2) to improve the quality of runoff from the citrus groves, 3) to provide and preserve wetlands habitat, and 4) to provide groundwater recharge.

1 INTRODUCTION

Florida is a major world producer of citrus. Historically, most of this production has been in the upland regions of the central part of the state. However, following a series of devastating freezes in the early 1980's, there has been a major shift in the geographical distribution of citrus occurring as growers seek to reduce the risk of freeze damage by relocating to the Gulf Coast region of southwest Florida. Since 1980, citrus acreage in the region has doubled to the current level of 40,000 hectares. The majority of this development has occurred in the past 3 years (Behr, 1989). This trend is expected to continue through the next decade with a projection of 80,000 hectares in production by the year 2000 (Land, 1988). The on-tree value of this production, given current citrus prices, is estimated to be 380 million dollars annually (Behr, 1989).

There is a concern that the scale of these developments will significantly affect the ecosystems of the region, which borders the environmentally sensitive Everglades National Park and the Big Cypress National Preserve. Much of the current and proposed citrus development is occurring in an area occupied by an endangered species the Florida panther. Consequently, there is concern on the effect of citrus development on the panthers' habitat.

The topography of southwest Florida is very flat with the highest elevation at 12 m. Most of the soils within the region are mineral and very sandy. However, in some areas the sand is overlain with organic muck. These soils were initially formed as marine beds of deep sand, which were deposited by waters of the sea that covered the area several times during the Pleistocene period (Henderson, 1981).

The climate is characterized as sub-humid and tropical. The mean annual precipitation is 135 cm, with approximately 80 percent occurring in the summer months (May through September). Most of the summer precipitation comes from localized thunderstorms. Frequently, tropical cyclones contribute to the summer and fall rainfalls with storms of hurricane intensity occurring about once a decade. The heavy summer rainfall typically produces widespread flooding resulting in significant areas of natural wetlands.

The natural flora of the region consists of pine (south Florida slash) forests and saw palmetto in the higher elevations. Cypress swamps are common in the poorly drained lower areas. Pine growth has been enhanced in some historical wetlands by drainage activities, but it has also been limited in some historic pinelands by increased fires associated with human activity and grazing practices. Cypress grow in strands (broad, shallow, forested rivers) and often are associated with a mixture of hardwoods. Areas where cypress swamps have regenerated after logging are likely to be included in the wet detention basins. The cypress swamp vegetation is distinct from the surrounding pine flatwoods, and species composition is affected primarily by the hydroperiod. Florida law now requires the protection of these wetlands.

An exotic tree, Melaleuca, has become a significant feature of the landscape since its introduction into southern Florida from Australia in the early 1900's. It competes well with cypress and other natural vegetation and grows best on sandy soils that remain water-logged to moderately flooded during the wet season (Myers, 1984). Wetlands invaded by Melaleuca are considered disturbed and are not protected.

Drainage is required for most of the land to be agriculturally productive due to a seasonally high water-table. Currently, most of the area is cattle rangeland and improved pasture. The drainage requirements for this usage are significantly different from those required by citrus groves. Pasture and rangeland are typically drained by shallow ditches placed on fairly wide intervals since the native grasses can survive for significant periods under flooded conditions. On the other hand, citrus is very sensitive to saturated conditions and must be rapidly drained. Therefore, the citrus developments have elaborate and responsive drainage networks.

The citrus developments are causing several repercussions: (1) there is a concern that the drainage of these wetlands will destroy habitat for a number of species that are currently endangered and may result in other species becoming endangered; (2) the pumpage capacity required to have a responsive drainage systems can produce large distortions in the pre-development surface water hydrology of the surrounding area; (3) significant amounts of fertilizer and pesticides are used in the production of citrus which can be transported with the drainage water to contaminate nearby water bodies; and (4) drainage reducing surface storage may affect groundwater recharge rates.

Given these concerns, the governmental body having responsibility for the regulation of storm water, the South Florida Water Management District (SFWMD), has set certain guidelines for the construction of wet detention basins for the impoundment of agricultural drainage water before it can be discharged off-site. Wet detention area is defined as a water storage area with a bottom elevation at least 30 cm below the elevation of the control discharge structure. These wet detention basins for drainage water from the citrus grove, therefore, have four primary functions: 1) to maintain off-site discharge peak flows at or below pre-development levels; 2) to provide detention time for sediment removal and enhancement of the discharged water's quality; 3) to preserve wetland habitat; and (4) to proved groundwater recharge areas.

1.1 Related Studies

For many years studies have been performed that established the importance of drainage for improved crop yield. However, significantly less effort has been expended on examining the consequences of drainage on adjacent lands and water bodies. Skaggs (1980) reported peak runoff rates at the field edge several times higher on agricultural land with improved surface drainage as compared to similar undeveloped land.

Drainage may also increase nutrient outflow resulting in greater transport of phosphorus and other potential pollutants with sediment. Tchobanoglous and Culp (1980) found that the settling of particulate matter is the primary mode of pollutant removal and is largely influenced by flow velocity. Chescheir (1987) reported on a field study that described the hydrology of a wetland buffer area that received pumped agricultural drainage waters. They reported significant nutrient removal from the drainage water by distributing it over a buffer area before discharge into an estuary. Therefore, detention basins may be a practical means of improving the quality of runoff water from drainage of agricultural lands.

Studies have investigated the capacity of

wetlands to treat wastewater. These studies have primarily dealt with municipal sewage and urban stormwater runoff. Their findings should have direct implication on the potential of wet detention areas to improve the quality of agricultural drainage water prior to off-site discharge.

Ewel and Odum (1978) found that cypress swamps have significant capacity for nutrient loading and that improving water quality is one of their natural functions. They concluded that secondarily treated sewage could be applied to cypress swamps at rates up to 2.5 cm per day without causing significant damage to the natural ecosystem. Brown (1981) investigated the influence of inputs of water and nutrients on the structural and functional characteristics of cypress wetlands. She concluded that increased nutrient loading resulted in greater total biomass up to a certain point, beyond which increased nutrients did not result in increased biomass production. This implies other limiting factors, such as available solar energy. Mitsch (1984) found that the annual peak in biomass production in cypress swamps occurred in early spring when maximum solar radiation reaches the water surface. This was thought to be caused by the deciduous nature of cypress that allows maximum sunlight to penetrate when water levels are typically lowest. He also found that due to the low pH of water in the typical cypress swamp, that the environment is not very conducive to aquatic production. Often in the natural state the cypress swamps are hydrologically unconnected to surrounding areas and receive water almost exclusively from rainfall. This condition may contribute to the low levels of dissolved oxygen and low pH limiting aquatic life in many cypress swamps (Dierberg and Brezonik, 1984).

2 CITRUS DRAINAGE MANAGEMENT STRATEGY

2.1 Conceptual Approach

The conceptual approach to grove design, development and management is to maintain, as far as possible, pre-development conditions with respect to the quantity and quality of water discharged off-site. In addition, the development should not adversely affect overland flow or result in the destruction of viable wetlands (Rodgers, 1982).

Since most developments will have relatively low areas (swamps, marshes and bogs) that can be converted to grove only with extensive land forming, the desired approach is to include those areas into the wet detention basin. Since it is considered more viable to have a few larger wet areas than many small unconnected ones, certain modifications are allowed. Small wet areas which still exist after designating the wet detention basin, they can be replaced by a 'tradeoff' with land adjacent land to the wet detention basin.

Floodplain encroachment is to be avoided and should be considered from two aspects: storage reduction and flow interference. A flow channel cross-section must be maintained to adequately carry a 3-day 25-year storm, and the post development hydrograph for that same storm event should conform to the pre-development hydrograph. To maintain the desired water quality, it is required that the wet detention basin be designed to hold the first 2.5 cm of runoff in the detention basin and to release it over a five day period.

2.2 Grove Layout and Drainage System

The size of the typical grove being developed in the region ranges from 500 to 2000 hectares. Obviously, developments of this size can have significant impacts on adjacent lands. Therefore, the grove layout and design of the drainage system must consider numerous factors. The following is a description of procedures employed and factors considered in grove development.

After the proposed site is surveyed and an elevation grid established on 30 m intervals, wet and low areas are identified for possible inclusion in the wet detention area. Natural overland flow paths are located, and a plan is prepared to channel this flow through the grove while avoiding floodplain encroachment. Location and size of the wet detention basin is then determined. Effort is made to include and connect existing wet areas. Exterior levees are located and sized to allow passage of off-site flow while protecting the actual grove area. Interior lateral and collector ditches are located, sized and installed to provide drainage. Typically, lateral ditches 1.5 m deep are placed at 400 m intervals.

Large lift pumps are installed in the main collector ditch adjacent to the wet detention basin. Pump station are typically sized to remove 10 cm of water from the grove area in a 24-hr period.

This pumping capacity along with soil storage is intended to provide protection to the trees from a 10-year 24-hr rainfall event.

The first step in actual grove development is landforming. The area between lateral ditches are precision graded to a slope of 0.05 to 0.10 percent. Landforming is performed after consideration of the natural drainage, and the actual grading design attempts to minimize depths of cuts and fills. After grading, tree beds are formed on 15 m centers. Bed height (distance from top of bed to bottom of furrow) is usually 0.8 m. This bed provides drainage for two tree rows typically planted on 7.5 m centers. Tree spacings usually range from 3 to 7 meters. Finally, 20 cm drainage pipes are installed to convey drainage water from the furrows to the lateral ditches.

Irrigation is required to maintain the trees through the 8 month dry season typical in southwest Florida. Micro irrigation is the accepted practice with micro-sprinklers being most common. This irrigation method has a significant influence on drainage. First, only a portion of the surface area (usually less than 50 percent) is wetted, therefore, significant soil storage capacity is available for rainfall. In addition, fertigation is commonly practiced, thus reducing the opportunity for nutrients to leach during rainfall and subsequent drainage.

2.3 Wet Detention Basin Design Procedure

Accepted engineering procedures are followed in the design of the wet detention basins. Runoff volume from the grove area is estimated by the SCS runoff equation (Eq 1) (USDA-SCS, 1972):

$$Q = (P - 0.02S)^2/(P + 0.8S) \quad \text{Eq [1]}$$

where,

Q = runoff volume,
P = 24-hour rainfall depth,
S = watershed storage parameter depth.

The equation was developed for large, design storms. Methods for determining the value of the storage parameter, S, are given in Capece (1987).

The design storm rainfall is based on a 25-year 3-day event. The area of the wet detention area is determined from estimated runoff volume and the maximum design storage water depth for the impoundment area (usually equal to or less than 1.2 meters). To improve the quality of water discharged off-site, the control structure is designed to allow a only 1.2 cm to discharge in the first 24 hours from storage of the first 2.5 cm of runoff.

In addition, peak discharge through the control structure is limited to the pre-development peak runoff rate that would occur from the design storm. Peak pre-development runoff rates are computed from a graphical technique published by SFWMD (1983). Graphs are presented that were developed from a computer model constructed by Higgins (1976). The model employs the Manning equation combined with an assumed retention depth to estimate peak discharge.

The control structure is set at an elevation of 30 cm above the bottom of the wet detention basin to maintain a flooded condition in a significant portion of the basin during the rainy season. The emergency over-flow structure is sized to discharge the entire pumping inflow capacity and rainfall from a 100-yr 3-day storm on the detention basin, and the overflow is directed <u>on-site</u>. Therefore, during extreme events flood waters is controlled and held on the grove. Finally, the design storm is routed through the wet detention area to insure proper functioning of each component.

Fig. 1 shows an example of the pre-development, post-development, and reservoir hydrographs for a design storm from a 260 ha grove with a 24 ha detention reservoir (Stone, 1988).

3 PRESENT SITUATION

Since the majority of the wet detention basins have been developed in the past few years, there is little reliable data at this point on the prospects of these detention areas actually meeting the described objectives.

The SFWMD has required that citrus growers document, with flora and fauna counts, progressive development of the wet habitats. However, this information has not been systematically collected nor compiled. It is clear, however, by casual observation that there is an abundance of wildlife inhabiting these wet areas, but it is unclear what type of successional development will eventually take place. What will be the diversity of species, and what will be the effect on endangered species?

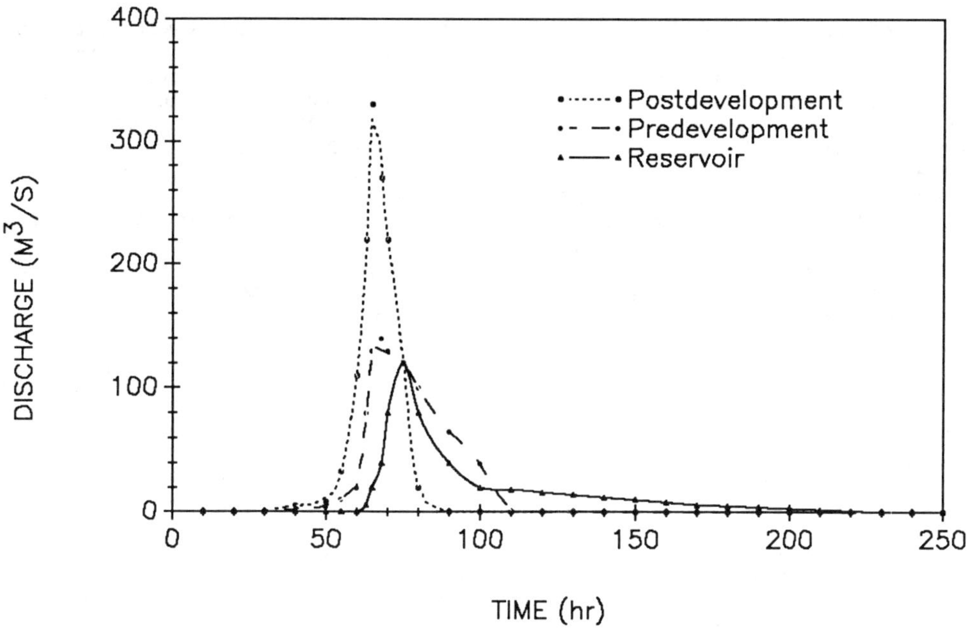

Fig. 2 shows an example of the pre-development, post-development, and reservoir hydrographs for a design storm from a 260 ha grove with a 24 ha detention reservoir.

An initial examination conducted by the Florida State Department of Environmental Regulation (DER) on the quality of water being discharged from wet detention basins indicated significant improvement in quality by passage through the detention basin (Black, 1989).

A model was developed by Stone (1988) that simulated water quality entering and leaving a wet detention area in a citrus grove. The water quality component of the model was tested on a natural wetland site and on a artificial wetland area stocked with water hyacinth and was found to quite closely simulate nitrogen and phosphorus removal rates. Table 1 gives the simulated water quality for a developing citrus grove.

Table 1. Simulated water quality for a 260 ha citrus grove[1].

Nitrogen loss from grove (kg/ha)	3.5
Inlet nitrogen load (kg)	759
Outlet nitrogen load (kg)	33
Percent nitrogen removal	95
Phosphorus loss from grove (kg/ha)	2.5
Inlet phosphorus (kg)	546
Outlet phosphorus load (kg)	198
Percent phosphorus reduction	64

[1](after, Stone, et al, 1988)

However, these preliminary studies require expansion to determine if the desired water quality improvement is occurring. Establishment of game fish in some of the wet detention areas, including Bass which require fairly good water quality, is an indication that water quality problems may not exist.

Peak discharge rates from the groves have been maintained at or below pre-development levels, but low flow rates have in some cases affected downstream wetlands. The size of the detention areas at this point appear sufficient to provide protection to the groves from major storm events. A 3-day 20-cm storm occurred at the end of the 1987 rainy season, and none of the detention basins then in use reached their emergency overflow levels.

Groundwater recharge is likely occurring from the wet areas. Much of the soil in the region has a semi-permeable organic pan at a depth of about 1 m that limits downward percolation. This pan is usually penetrated by the ditch created in providing fill for the dike construction. However, it is likely that over time recharge rates will diminish as sedimentation occurs.

In summary, much scientific investigation is needed to document and help guide future agricultural development. This need is particularly important since the area has a unique ecosystem and has a rapidly increasing population.

REFERENCES

Behr, R.M. 1989. Report presented by Florida Department of Citrus, to the Gulf Coast Growers Association, Labelle, Fl.

Brown, Sandra. 1981. A comparison of the structure, primary productivity, and transpiration of cypress ecosystems in Florida. Ecol. Monogr. 51:403-427.

Black, David. 1989. Personal communication. Water Quality Engineer. South Florida Water Management District. Palm Beach, FL.

Capece, J.C., K.L. Campbell, L.B. Baldwin, K.D. Konyha. 1987. Estimating runoff volumes from flat, high-water-table watersheds. Transactions of the American Society of Agricultural Engineers, Vol 30(5):pp 1397-1402, St. Joseph, MI.

Chescheir, G.M., R.W. Skaggs, J.W. Gilliam and R.G. Broadhead. 1987. Hydrology of wetland buffer area for treatment of pumped agricultural drainage water. Paper presented at the Summer meeting ofthe ASAE, No. 87-2050. 2950 Niles Rd. St. Joseph, MI.

Dierberg, F.E. and P.L. Brezonik. 1984. Water chemistry of a Florida cypress dome, in <u>Cypress Swamps</u>, K.C. Ewel and H.T.Odum, eds., University of Florida, Gainesville, pp 34-50.

Ewel, K. C. and H. T. Odum. 1978. Cypress swamps for nutrient removal and wastewater recycling, in Advances in Water and Wastewater Treatment Biological Nutrient Removal, M. P. Wanielista and W. W. Eckenfelder, Jr., eds., Ann Arbor Sci. Pub., Inc. Ann Arbor, Mich. pp 181-198.

Henderson, W. G. 1981. Soil Survey of Lee County, Florida. United states Department of Agriculture, Soil Conservation Service, Washington, D.C.

Higgins, R.W. 1976. The development of a general watershed-highway mathematical model. Unpublished master's thesis, University of South Florida, Tampa.

Land, David. 1988. Presentation made before the general assembly of the Florida State Horticultural Society, Miami, FL.

Mitsch, W.J. 1984. Seasonal patterns of a cypress dome in Florida, in <u>Cypress Swamps</u>, K.C. Ewel and H.T. Odum, eds., University Presses of Florida, Gainesville, pp 25-33.

Myers, R.L. 1984. Ecological compression of <u>Taxodium distichum var. nutans</u> by <u>Melaleuca quinquenervia</u> in southern Florida, in Cypress Swamps, K.C. Ewel and H. T. Odum, eds., University of Florida, Gainesville, pp 358-364.

Rodgers, D. 1982. Surface water management permitting for new agricultural construction, in <u>SFWMD Permit Information Manual</u>, pp C-VIII2-4, West Palm Beach, FL.

Skaggs, R. W., J.W. Gilliam, T.J. Sheets and J.S. Barnes. 1980. Effects of agricultural land development on drainage waters in the North Carolina Tidewater Region. Report No. 159. Water Resources Research Institute of the University of North Carolina. Raleigh, NC.

SFWMD. 1979. Memorandum report on peak runoff estimates for undeveloped lands. South Florida Water Management District, West Palm Beach, FL.

Stone, K.C., K.L. Campbell, J.C. Capece and L.B. Baldwin. 1988. Southern Region Project S-211 of USDA-CSRS, contribution of IFAS, University of Florida.

Tchobanoglous, G. and G.L. Culp. 1980. Wetland systems for wastewater treatment: An Engineering Assessment. <u>In Aquaculture Systems for Wastewater Treatment - An Engineering Assessment</u>, S.C. Reed and R.K.Basian, eds.,EPA 430/9-80-007.

USDA-SCS. 1972. Soil Conservation Service National engineering handbook, Section 4, Hydrology. U. S. Department of Agriculture, Washington, D.C.

Land and Water Use, Dodd & Grace (eds), © 1989 Balkema, Rotterdam. ISBN 90 6191 980 0

Two alternative ways to deal with silage effluent

P.O'Kiely
Teagasc, Grange Research Centre, Co. Meath, Ireland

ABSTRACT: The results of eleven experiments which studied the acceptibility and feeding value of effluent collected from unwilted acid-treated grass silage are presented. Beef cattle quite readily consumed effluent from well preserved silage but not from badly preserved silage. Effluent was consumed in preference to water and the availability of water did not affect effluent intake. Feeding silage effluent to beef cattle increased total dry matter intake and improved animal performance. Eighteen litres silage effluent (60 g DM/kg and 250 g ash/kg DM) was equivalent in feeding value to approximately 1 kg barley. Ad libitum access to effluent was necessary to permit high intakes.

The ability of the absorbents used to retain effluent within the silo was quite modest. The highest retention rates of effluent were obtained using straw and in general products containing beet pulp based products were superior to barley, distillers grains, corn gluten or soyabean meal. Whereas feeding effluent to cattle, or retaining it within the silo by the use of absorbents, each have distinct advantages, neither are likely to replace landspreading (or wilting) as the main method for dealing with silage effluent.

1. INTRODUCTION

Over 20 million tonnes grass silage are made in Ireland annually. Because of the high relative humidity of the air and the uncertainty of obtaining consecutive rain-free days, silage is normally made without a period of field wilting. Consequently silage effluent production is an integral part of most silage making systems. The production of effluent is undesirable because of the pollution caused if it enters surface or ground water, corrosion to concrete, cost and inconvenience of collection and disposal and the loss of harvested nutrients it represents. It is usually recommended to collect silage effluent and dispose of it by landspreading (normally onto silage aftermath) (Stewart, 1980). If judiciously carried out this avoids pollution problems but the nutrients it contains are lost from the immediate feeding system. If effluent were fed either directly or indirectly (i.e. by use of absorbents) to livestock the nutrients would be used directly. This could significantly reduce conservation losses and increase output of animal product per hectare. This paper reports a series of experiments which studied various aspects of the feeding of silage effluent to beef cattle and the use of absorbents.

2. EXPERIMENTAL

Eleven experiments were carried out studying various aspects of the acceptibility and feed value of grass silage effluent for beef cattle. In a preliminary trial (Experiment 1) 500 kg steers were offered straw (2 kg/day), rolled barley (6 kg/day) and water (ad libitum). One group of five Friesian steers was offered effluent collected from well preserved silage while a second group was offered effluent collected from badly preserved silage. The trial lasted 20 days.

In all subsequent experiments only effluent from well preserved silage was fed. The unwilted silages offered and the silages from which effluent was collected were made using sulphuric acid (450 g/kg) applied at 2.3 l/t (unless otherwise stated). In Experiment 2 finishing Friesian steers (469 kg initial liveweight) were offered silage (170 g DM/kg and pH 4.0) ad libitum plus 4.5 kg rolled barley daily for 99 days. Sixteen animals were offered water and another sixteen offered a choice of water or effluent. In Experiment 3 two groups of Friesian steers (452 kg initial liveweight) were offered silage (194 g DM/kg and pH 3.92) ad libitum together with 3.3 kg corn gluten per head daily for 132 days. One group was offered water (no effluent) and the second group effluent (no water). In a fourth experiment three groups of ten heifers (323 kg initial liveweight) were offered silage (211 g DM/kg, pH 4.22 and 624 g DDM/kg DM) ad libitum and 3 kg concentrates (rolled barley and corn gluten as 50:50 mix) per head daily for 99 days. The three treatments were (a) water, (b) effluent from silage made without acid preservative ("no acid" effluent) or (c) effluent from silage made with acid preservative ("acid" effluent). Experiment 5 had similar treatments to Experiment 4 but was run as two 3 x 3 Latin square designs operating concurrently. Friesian steers (440 kg initial liveweight) were offered well preserved silage ad libitum plus 2 kg barley daily. A 14 day adaptation interval was allowed on each treatment followed by a seven day period during which intake was recorded. In a sixth experiment six heifers were fed silage (216 g DM/kg, pH 3.88, 596 g DDM/kg DM and 74 g ash/kg DM) ad libitum and 2 kg concentrates per head daily (1 kg at 8.00 hrs and 1 kg at 16.00 hrs). Animals were offered water, effluent or a choice of effluent or water, each without or with the addition of salt (20 g/kg concentrates) in a 6 x 6 Latin square design. A ten day adaptation interval and 7 day recording interval occured in each period.

Two groups of six Friesian steers (522 kg initial liveweight) in Experiment 7 were fed chopped hay (860 g DM/kg, 534 g DDM/kg DM and 92 g ash/kg DM) ad libitum, one group being offered water and the second offered effluent. Intakes were recorded over 12 days.

Cattle grazing grass were offered effluent in Experiments 8 and 9. Two paddocks were used in each experiment and the two groups of animals in each case were switched between the paddocks weekly. In Experiment 8 two groups of 10 cattle (331 kg initial liveweight) were offered either water or effluent (no choice). Weather conditions during the experiment were extremely wet and surface water was present on parts of the field for much of the experiment. In Experiment 9 two groups of 11 steers (352 kg initial liveweight) were offered water or a choice of water or effluent over 93 days.

In Experiment 10 (June to August) two groups of seven Friesian steers (434 by initial liveweight) were offered unwilted silage and 2 kg rolled barley per head daily (fed in a.m.). One group had ad libitum access to effluent at all times while the second group had restricted (2 x 30 minutes) access to effluent - from 9.30 to 10.00 hrs and from 15.30 - 16.00 hrs. The effluent used was collected from silage made without preservative treatment. A crossover design was used, with both periods involving 14 days adaptation and 11 days recording. In Experiment 11 (Sept. to Oct.) two groups of 9 Friesian steers (491 kg initial liveweight) grazed two seperate paddocks of grass, alternating between the paddocks weekly. Both groups were offered 0.5 kg barley per head at 9.00 hrs and 15.30 hrs daily. The experiment continued for 45 days. The two effluent feeding treatments and the effluent itself were as in Experiment 10. Water was available ad libitum in Experiments 10 and 11.

Five experiments studied the effects on silage effluent of adding dry materials to grass being ensiled. Unwilted precision-chop harvested grass was used in each case. In Experiment 12 grass was ensiled on five seperate occasions in 200 l plastic barrels. Grass was ensiled either without preservative addition or with the addition of acid additive (85% formic or 45% sulphuric) at 2.5 mls per kg grass or the absorbent Sweet 'n Dry (Nutec, Tallaght, Dublin) at 2.5, 4.0 or 5.5 kg per 100 kg grass. Experiment 13 was similar in design to Experiment 12. Absorbent treatments included molassed sugar beet pulp nuts (7 or 10 mm diameter), loose molassed beet pulp and unmolassed beet pulp nuts (7 mm) each applied at 5.5 kg per 100 kg grass. Molassed beet pulp nuts (10 mm) were also applied at 2.5 and 4.0 kg per 100 kg grass. In Experiment 14 polypropylene sacks each containing 5 kg absorbent were distributed within farm scale silos when the latter were being filled. Three bags of each absorbent were distributed within each of 3 silos. The absorbents used were molassed sugar beet pulp nuts (10 mm), unmolassed beet pulp nuts (7 mm), rolled barley, dried distillers grains, corn gluten and soyabean meal. Grass (70 kg) was ensiled in 2 m high plastic silos in Experiment 15 with either no preservation, 85% formic acid (2.5 ml per kg grass) or rolled barley, Sweet 'n Dry, molassed beet pulp nuts (10 mm), malt culms or a mineral and vitamin mixture, each at 35 g/kg grass. In Experiment 16 one row of straw bales was placed beneath 2 m of grass that was ensiled in a walled silo. Effluent retention was calculated from the change in the dry matter concentration in straw.

3. RESULTS

In Experiment 1 animals consumed 45 litres daily of effluent from a well preserved silage but only 5 litres daily of effluent from a badly preserved silage.

The dry matter, pH, crude protein, ash, lactic acid, total VFA and water soluble carbohydrate values of the effluent in Experiment 2 were 43 g/kg, 4.3, 245 g/kg DM, 315 g/kg DM, 276 g/kg DM, 124 g/kg DM and 24 g/kg DM, respectively. Both groups of animals consumed similar amounts of silage and concentrates (Table 1). Although the animals on the effluent treatment were fed a silage based diet they voluntarily consumed three times more silage effluent than water. Supplementation with effluent increased total dry matter intake by 0.63 kg/day and this extra intake was reflected in an additional 6.2 kg carcass gain.

The intake of silage and concentrates were similar for both treatments in Experiment 3 (silage DM intake was reduced by 4% on the effluent treatment). Animals consumed similar amounts of water and effluent (Table 1) but the effluent fed animals increased total dry matter intake by 0.43 kg/day and had 9.4 kg heavier carcasses at the end of the experiment.

Silage dry matter intake in Experiment 4 (Table 2) was reduced by 14.0% and 18.6% respectively for the "no acid" and "acid" treatments compared to the water treatment. Total dry matter intakes for the no acid and acid treatments were 0.91 kg and 0.41 kg/day higher than the water treatment,

TABLE 1: Intake and performance of steers fed silage and concentrates and offered water or effluent in Experiments 2 and 3

	Experiment 2		Experiment 3	
	Water	Water/ effluent	Water	Effluent
Intake (kg/day)				
Silage DM	6.5	6.4	8.0	7.7
Concentrates	4.5	4.5	3.3	3.3
Effluent	0	16.9	0	16.3
Water	8.2	5.0	15.4	0
Performance				
Carcass wt (kg)	307.1	313.3	374.6	384.0

respectively. The increased total dry matter intake on the no acid effluent treatment and the corresponding improvement in carcass gain agrees with previous experiments. The much smaller increase in total dry matter intake with the acid effluent treatment and the apparent absence of a carcass weight response is difficult to explain, particularly since this was the treatment which in Experiments 2 and 3 gave a positive response. In Experiment 5 silage dry matter intake was similar with the no acid (5.5 kg/day) and acid (5.6 kg/day) effluent treatments and slightly higher (6.0 kg/day) for the animals receiving water instead of effluent. The intake of the no acid effluent, acid effluent and water was 28.4, 30.1 and 14.1 litres per head daily.

TABLE 2: Intake and performance of heifers fed silage and concentrates and offered water or effluent (2 types) in Experiment 4

	Water	Effluent	
		No acid	Acid
Intake (kg/day)			
Silage DM	4.3	3.7	3.5
Concentrates	3.0	3.0	3.0
Effluent	0	30.2	24.1
Water	13.7	0	0
Performance			
DLG (g)	898	981	858
Carcass wt. (kg)	213.6	223.7	214.9

Increasing the intake of salt in Experiment 6 did not lead to an increase in water or effluent intake and did not effect silage or concentrate intake. Animals not fed additional salt and offered water, effluent or a choice of water or effluent consumed 11.2 kg water, 25.9 kg effluent and 6.5 kg water and 26.7 kg effluent, respectively. Animals offered water in Experiment 7 consumed 27.7 kg daily while those offered effluent consumed 53.4 kg daily. The corresponding intakes of hay dry matter were 7.00 and 6.95 kg/day, respectively.

Intakes of water (mean of 0.8 kg/day and CV = 119%) and effluent (mean of 0.9 kg/day and CV = 121%) by grazing cattle in Experiment 8 were low and erratic. The corresponding liveweight gains over 98 days was 1056 and 1092 g/day. In Experiment 9 the intake of water by animals offered water or the water and effluent choice was 11.4 and 8.7 kg per head daily, respectively, while the second group consumed 23.0 kg effluent per head daily. The corresponding liveweight gains were 502 and 705 g/day.

Animals with ad libitum access to effluent in Experiment 10 had mean daily silage and effluent intakes of 4.1 kg DM and 9.6 l respectively. The corresponding values for animals with restricted access to effluent was 4.8 kg DM and 26.4 l. In Experiment 11 grazing cattle with ad libitum or restricted access to effluent consumed 16 l and 7 l per head daily, respectively.

The volume of effluent produced per 100 kg grass ensiled in Experiment 12 was 23.8, 24.0, 20.4, 15.9 and 12.3 l for the no additive, acid and Sweet'n Dry at 2.5, 4.0 and 5.5 kg treatments respectively. This corresponds to effluent retention rates of 1.4, 2.0 and 2.1 l per kg absorbent respectively. Increased levels of Sweet'n Dry addition improved silage digestibility and fermentation characteristics. In Experiment 13 the no additive, acid or molassed beet pulp nuts (10 mm) applied at 2.5, 4.0 and 5.5 kg per 100 kg grass produced 22.6, 20.1, 15.3, 12.9 and 10.0 l effluent respectively (corresponding retention rates of 2.3, 2.1 and 2.0 l/kg). Where no additive, acid (3.2 l/t) molassed pulp nuts (10 mm), molassed loose pulp, molassed pulp nuts (7 mm) and unmolassed pulp nuts (7 mm) were directly compared the corresponding effluent production rates were 20.1, 16.8, 10.5, 11.2, 9.2 and 12.8 l per 100 kg grass respectively. All treatments improved silage digestibility and fermentation characteristics. Effluent retention in Experiment 14 was 0.95, 1.28, 0.39, 0.34, 0.67 and 0.59 l/kg for the molassed pulp, unmolassed pulp, rolled barley, dry distillers grains, corn gluten and soyabean meal treatments respectively. The corresponding figures in Experiment 15 for the first 35 days of ensilage for rolled barley, Sweet 'n Dry, molassed beet pulp nuts (10 mm), malt culms or a mineral and vitamin mixture were 0.34, 1.33, 1.21, 1.71 and 0.53 l/kg respectively. The value for barley straw in Experiment 16 was 3.0 l/kg.

4. CONCLUSIONS

Only effluent from well preserved silage should be fed to cattle. Feeding such effluent is likely to result in increased DM intake and improved animal performance. Eighteen litres silage effluent (60 g DM/kg and 250 g ash/kg DM) was equivalent in feeding value to approximately 1 kg barley. These results are broadly in agreement with Steen (1986). Continuous access to effluent was necessary to permit high intakes. Intakes of effluent were not improved by adding salt (NaCl) to the concentrate component of the diet.

The ability of the absorbents studied to retain effluent within the silo was modest. Quite high rates of addition would be necessary for complete effluent retention. Both Sweet 'n Dry and molassed sugar beet pulp nuts improved silage in vitro digestiblity and fermentation characteristics, the effects being larger at higher rates of addition.

Whereas feeding effluent to cattle or retaining it within the silo (by the use of absorbents) each have distinct advantages, neither, as used in the present experiments, are likely to replace landspreading (or wilting) as the main method for dealing with silage effluent.

REFERENCES

Steen, R. W. J. (1986). An evaluation of effluent from grass silage as a feed for beef cattle offered silage based diets. Grass and Forage Science 41: 39-45.

Stewart, T. A. (1980). Prevention of damage to grass swards from applications of silage effluent by neutralization or dilution of effluent acid. Grass and Forage Science 35: 47-54.

Land and Water Use, Dodd & Grace (eds), © 1989 Balkema, Rotterdam. ISBN 90 6191 980 0

Komplexe Nutzwendung von Schweineflüssigmist und Schlachthausabfälle in Fischteichen

M.Kolláth
Wasserwirtschaftliche Hochschule, Beja, Ungarn

K.Bencze & A.Schmidt
Umweltschutz- und Wasserwirtschaftsdirektion, Beja, Ungarn

ABSTRACT : The use of liquid pig manure and waste meat from the meat industry res- ie ulted resulted in a significant increase in fish production from fish ponds, 4-5 tonnes per hectare compared to 2 tonnes per hectare which is the normal figure for Hungary. The fish species were : carp (Cyprinus carpio), great catfish (Silurnus glanis) and tench (Tinca tinca). The combination of carp and great catfish gave the highest production rate. The success of the method depends on the numbers of a particular species as well as the number of species. Other very important factors include the technical condition in the pooldesign, the water quality and the water temperature (temperate zone). The breeding season was reduced to 2_2.5 years.

ABSTRAIT: L'utilisation komplex des angarais de porc dilliés ét des déchets des charchteries dans les etangs poissonneux ont composer une augmentation importante du debit de poisson (une augmentation nette de 4 á 5 t/ha ala place de 2 t/ha en Hongrie, region tempéré). Les experiences ont été faites avec les especes des poissons suivantes: Carpe (Cyprinus carpio L.), silur (Silurnus glanis L.) et les poissons gitans (Tinca tinca L.).
La policulture carpe-silur a donnée les plus grand debit de poisson. Le succé du procédé depent des proportions des especes de poisson et du poid nette de celles ci. Autre facteurs importants pour le procédé: institutions techniques, qualite de l'eau et la temperature de l'eau (region tempéré). Le temp d'élevation diminue de 2-2,5 áns.

AUSZUG: Durch die komplexe Nutzwendung von Schweineflüssigmist und Schlachthausabfälle in Fischteichen war eine signifikante Zunahme der Fischfleischproduktion erreichbar. (Nettogewinn 4-5 Tonne pro Hektar, doppelt so hoch, wie es sonst in Ungarn - gemässigte Zone! - gewöhnlich wäre). Die Untersuchungen wurden mit den folgenden Fischarten durchgeführt: Karpfen (Cyprinus carpio L.), Schleie (Tinca tinca L.) und Wels (Silurus glanis L.). Die grösste Zunahme und dadurch der höchste Nettogewinn produzierte eine Kombination von Karpfen und Wels in Polykultur. Gute Resultate sind aber nur durch bestimmte Individuenzahl und Individuengewicht beim Wachstumsbeginn erreichbar. Weitere wichtige Faktoren sind noch die technischen Einrichtungen, Wassergüte und Wassertemperatur (gemässigte Zone!). Die Wachstumsdauer wurde auch auf 2-2,5 Jahre abgekürzt.

1 EINLEITUNG

Die umweltfreundliche Lagerung, Behandlung bzw. Verwertung von Abwasser aus Massentierhaltungen und eiweissreichen Abfälle der Fleischindustrie (z.B.Blut, Innereien, Abfallfleisch) ist heutzutage eine sehr wichtige und aktuelle Aufgabe. Die Verwertung des Schweineflüssigmistes in der Pflanzenzucht ist schon gut bekannt. Auch mit gut bekannten Methoden ist es möglich aus Abfallprodukten der Fleischindustrie hochwertige Nahrstoffe (bzw. Tierfutter) zu produzieren. In unserer Arbeit sprechen wir über eine neue Verwendungsmöglichkeit der obengenannten Abfallprodukte in Fischteichen. Die Neuheit unserer Technologie ist die gleichzeitige Ausnutzung der direkten (eiweissreiche Abfälle der Fleischindustrie) und indirekten (Teichdüngung mit Schweinedünger) Faktoren zum Fischfleischgewinn im demselben Fischteich. Durch diese komplexe Verwendungsmöglichkeit der Abfallprodukte aus Massentierhaltungen und Fleischindustrie ist eine sehr intensive Ausnutzung des Fischteiches erreichbar. Durch die Düngung produziert sich im Wasser ein reichliches Phyto- und Zooplankton. Diese Planktonorganismen spielen eine wichtige Rolle als Nahrungsmittel für gewissene Fischarten, die sich gemischt nähren. Durch die natürliche Nahrungskette hat die Dürgung eine indirekte Bedeutung bei der Fischfleischproduktion.

2 MATERIAL UND METHODE

Die optimale Düngermenge für eine maximale natürliche Produktion, die noch ohne Schadeffekte auf die chemischen und biologischen Eigenschaften des Wassers verwendbar ist wurde in kleinen Fischteichen (1oo-1ooo m2) mit Freilanduntersuchungen bestimmt.
Diese optimale Düngermenge wurde in den verschiedenen Fischteichen immer neulich bestimmt, weil es zeigt eine starke Abhängigkeit von der Qualität des Düngers, Bodens und zufliessenden Wassers. Die kalkulierten Portionen wurden ins Teichwasser täglich zugeleitet. Die gekochten und kleingehackten Schlachthausabfälle wurden hauptsächlich von den Raubfischen (in unserer Technologie: Wels) aufgenommen, aber die Karpfen und Schleien, die sich gemischt nähren haben es auch aufgenommen. Die Fische wurden mit diesem Futter täglich gefuttert, aber die täglichen Portionen änderten sich mit der Wassertemperatur und Fischgrösse (bzw.Fischgewicht). Die Untersuchungen wurden seit 1983. in Süd-Ungarn mit der Verwendung von Schweineflüssigmist der Schweinezucht in Nagybaracska (Haladás Mg.Tsz.) und der Abfallprodukten des Schlachthauses in Baja (Bácskai Husipari Vállalat) durchgeführt.

3 BESPRECHUNG DER ERGEBNISSE

Bei den verschiedenen Fischarten wurde 1 kg Gewichtszunahme mit verschiedenen Nahrungsmittelmengen erreicht. Der Wels braucht 7-8 kg Abfallfleisch und Blut zu 1 kg Zunahme in Gewicht, der Karpfen nur 2-2,5 kg Futter (kerniges Futter, Abfallfleisch, Blut). Durch den niedrigen Einkaufspreis der Abfallprodukten war es möglich die Ausgabesumme des Fischfutters niedrig zu halten.
Die Artenzusammensetzung spielt eine sehr wichtige Rolle bei der Futterausnutzung. Es wurden optimale Kombinationen zur gleichzeitigen Zucht von verschiedenen wertvollen Fischarten ausgearbeitet. Die wichtigsten Arten der verwendten Technologie waren: Wels, Karpfen und Schleie. Dazu kamen noch die aus Ostasien stammenden Karpfenfische: Amur-Karpfen (Ctenopharyngodon idella) und Silberkarpfen (Hypothalmichthys molitrix), aber mit deutlich niedrigeren Individuenzahl. Der Wels verwertet das vorbereitete Abfallfleisch und zeigt eine deutliche Gewichtzunahme auch bei einem relativ hochen Individuenzahl. Der Karpfen und die Schleie verwerten hauptsächlich die Fischnährorganismen des Wassers. Es wurde festgestellt, dass diese Fischarten das Abfallfleisch auch aufnehmen und verwerten. Wir haben die Untersuchungen mit verschiedenen Kombinationen durchgeführt. Die erfolgreichsten Kombinationen waren die folgenden: die Wels--Karpfen-Polykultur, die Karpfen--Schleie-Polykultur und die Karpfen-Zentrische-Polykultur. Der optimale prozentuale Anteil der beiden Fischarten in einem Wels--Karpfen-Teich änderte sich zwischen 25-6o % Wels und 4o-75 % Karpfen.
Die gemeinsame, intensive Zucht von Karpfen und Wels mit den obengenannten Prozentzahlen und kombiniert noch mit der küstlichen

Nahrungszufuhr für die Welse zeigt neue Möglichkeiten für die Fischzucht. Fisch polykulturen mit diesen Zusammensetzungsdaten haben sich noch nicht verbreitet, obwohl es mit unseren Untersuchungen festgestellt wurde, daß der Lebensraum und die Entwicklungsmöglichkeit der zusammenlebenden Fischarten ungestört blieb. Es ist möglich mit dieser Technologie auch Zuchtfische und auch Marktfische zu produzieren. Gute Ergebnisse sind aber nur mit zusammengehörenden Individuenzahl und Individuengewicht beim Zuchtbeginn erreichbar. Zeitgewinn zeigt sich dadurch dass Marktfische in einem abgekürzten 2-2,5 Jahren-Zyklus zu produzieren sind (gemässigte Zone!). Nach dem zweiten Wachstumsperiode haben wir die folgenden Ergebnisse erhalten:

Karpfen: o,8o-1,4o kg/stück
Wels : o,6o-1,2o kg/stück

Die Untersuchungen wurden in 27 kleineren Teichen (1oo-1ooo m2) durchgeführt. (Wassertiefe: abhängig von der Arten und Grösse kombination) Wir finden unsere Methode für Teiche bis 5 Hektar noch verwendbar und kontrollierbar. Die kleine Teiche haben der Vorteil: von der Landwirtschaft nicht viel Platz wegzunehmen. Zielorientierte technische Ausrüstung bedeutet für diese neue Technologie folgendes: zum Artenzusammensetzung passenden Teichsystem, gute Zufuhrmöglichkeit von Schweineflüssigmist und Fischfutter, ist adaptierbar und entwickelbar.

Bei unseren Flachland-Untersuchungen haben wir zur Wasserversorgung Druckrohrsystem mit Wasserpumpe verwendet. Schweineflüssig mist wurde mit Wagen oder Druckrohrleitunk in die Teiche transportiert. Künstliche Sauerstoffzufuhr wurde mit Spritzdüsen oder mit anderen speziellen Einrichtungen verwirklicht. Ein geschlossenes Teichsystem ohne Wasserableitung. Eine umweltschonende Technologie, die es uns ermöglicht: umweltproblematische Materien (Produkte) mit Gewinn zu behandeln.

Nettogewin der verschiedenen Untersuchungsteichen
(Tonne / Hektar)

Variation	W	K	S	insgesamt
1.M	1,72	4,oo	-	5,72
2.M	1,68	2,17	o,2o	4,o5
3.M	1,1o	2,27	-	3,37
4.M	o,97	2,49	o,5o	3,96
5.M	o,82	4,23	o,5o	5,55
6.M	o,5o	3,86	-	4,36
7.M	o,o8	3,6o	o,o7	3,75
8.M+Z	1,18	3,33	o,5o	5,o1
9.Z	4,5o	-	-	4,5o
1o.Z	2,41	1,71	-	4,12
11.Z	1,91	o,91	-	2,82
12.Z	1,71	2,8o	-	4,51
13.Z	o,98	1,32	-	3,2o
14.Z	o,55	3,5o	o,67	4,66

Abkürzungen:
W= Wels
K= Karpfen
S= Sonstige Fischarten (Schleie, Amur-Karpfen, Silberkarpfen, Giebel)
M= Marktfisch (2-und 3 jahrig)
Z= Zuchtfisch (1-jahrig)

DANKSAGUNG
Wir danken für die Unterstützung unserer Arbeit und für die aktive Mitwirkung Herrn Pál Varga, Nagybaracska.

Land and Water Use, Dodd & Grace (eds), © 1989 Balkema, Rotterdam. ISBN 90 6191 980 0

Technical development of a low-level boom for spreading of slurry in a growing crop

G.Carlsson, C.Ekfäldt & L.Thyselius
Swedish Institute of Agricultural Engineering, Uppsala, Sweden

ABSTRACT: Since 1985 the Swedish Institute of Agricultural Engineering has been studying methods of spreading slurry in growing crops. The objective has been to develop a low-level spreading technique that places the slurry in strips on the soil surface below the vegetation during the growing period. The spreader that has now been developed is equipped with valves and hoses that take the slurry to the soil. The valves are of the membrane type and pneumatically powered. The new system for spreading has been tested in plant nutrient experiments on grassland and in cereals. Compared with conventional spreading with splash plates, spreading with a boom has given a much better plant nutrient utilization.

ZUSAMMENFASSUNG: Seit 1985 sind beim Schwedisches Institut für Landtechnik (JTI) Methoden für das Spritzen von Flüssigmist in wachsendem Getreide studiert worden. Das Ziel ist die Entwicklung einer niedrigspritzenden Technik gewesen, die den Flüssigmist in Schwaden auf die Bodenoberfläche unter der Vegetation während der Wachstumsperiode anbringt. Det jetzt entwickelte Verteiler ist mit Ventilen und Schlauchen, die den Flüssigmist zum Boden bringen, versehen. Die Ventile sind von Membrantyp und pneumatisch getrieben. Das neue System des Spritzens ist in Pflanzennährstoffsversuchen auf Grasland und in Getreidepflanzen getested worden. Im Vergleich mit konventionellem Spritzen mit Schutzplatten, Spritzen mit Spritzrohr hat eine viel bessere Pflanzennährstoffsausnützung gegen.

RESUME: Depuis 1985 l'Institut Technique Agricole Suèdois (JTI) a étudié des méthodes d'etalage du lisier dans la récolte sur pied. L'objectif a été de developper une technique bas-niveau d'etalage qui dépose le lisier dans des rangées sur le sol sous la vegetation pendant la période de croissance. L'equipment qui a été dévéloppe maintenant est fourni des valves et des tuyaux souples qui placent le lisier sur le sol. Les valves sont d'un espèce de membrane qui marche pneumatiquement. Ce noveau système d'etalage a été éprouvé dans des expériments de nutrition des plantes sur l'herbage et dans des céréales. Comparé à une méthode d'etalage conventionel, l'utilization d'une rampe pour l'effet de la nutrition des plantes a donné un meilleur résultat.

1 INTRODUCTION

The intensive environmental debate in Sweden today also involves agriculture, where the debate has largely focused on the handling of manure.

The greatest positive environmental effect in connection with the spreading of manure would be obtained if manure was spread as slurry during the spring tillage period or soon after crop emergence. If spreading is done in the conventional manner in connection with spring tillage there will, however, be compaction damage and the spring sowing may even be delayed. However, if the spreading is done in a growing crop, e.g. grassland, cereals and row crops, these disadvantages may largely be avoided. As part of the work in correcting the problems involved in manure handling today, the Swedish Institute of Agricultural Engineering (JTI) has been given funds for a number of longterm manure projects. One of these is spreading of slurry during the growing period.

2 DISTRIBUTOR

To reduce the losses at spreading, especially of ammonia, it is important that the transport of the slurry through the air is as short as possible and that the slurry must achieve good contact with the soil. One way to achieve this is to use low-level booms with flexible hoses or injectors. To make it possible for the plant to utilize the nutrition at an optimum, the distance between the outlets must be relatively small which entails that the boom must have a number of outlets. To enable spreading without risking that the outlets become clogged up it is important that they are not to small. As, at the same time, a steady flow is required, it is necessary that the flow is relatively large. Consequently, application rates that are much too large are obtained at realistic speeds .

The principle we have chosen is characterized by large flow areas where the slurry is let out in pulses. This effect is obtained by opening and closing the outlets, which in turn enables a pressure to be built up in the boom ahead of the outlets. Since the late 1970's, JTI has been working on different types of distributors in accordance with this principle.

The first generation of distributors worked together with a slurry injector. The slurry was transported in soft horizontal rubber hoses which could be opened and closed by a spring loaded hose valve. Opening and closing were achieved by a rotating camshaft driven by a hydraulic motor. This distributor worked well but since it only had five outlets the working width was small. With those experiences, a second generation, with 2.6 m working width and 10 vertically placed hose valves, was built.

On both prototypes the flow from the different outlets could be adjusted to give about the same flow by adjusting the spring tension on the hose valves. To obtain a steady flow it was necessary to adjust the spring tension of the hose valves. This was relatively complicated to carry out.

To make a larger working width possible a third generation of distributors with release of the slurry by pulses was developed. We started the work with a study of the functions of nine different valve designs using water and slurry of different qualities. The design finally chosen was a membrane valve built in a 90° tube bend made of plastic. This technique was used both in a 3 m boom (Fig. 1) for plant nutrition trials, and in a 12 m boom.

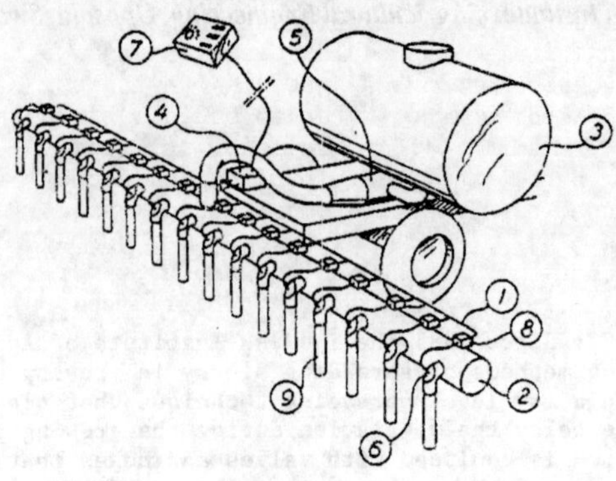

Figure 1. The main parts of a spreading boom with 16 flexible hoses. 1. Pneumatic pipe line, 2. Spreading boom, 3. Slurry tank, 4. Electric power distributor box, 5. Pump, 6. Valve with a rubber membrane, 7. Controller unit, 8. Pneumatic magnetic valve, 9. Flexible hose.

3 DESCRIPTION OF THE CONTROLLER SYSTEM

On the main boom at the rear of the spreading equipment a number of specially developed fast-response knee valves are mounted (see Fig. 2).

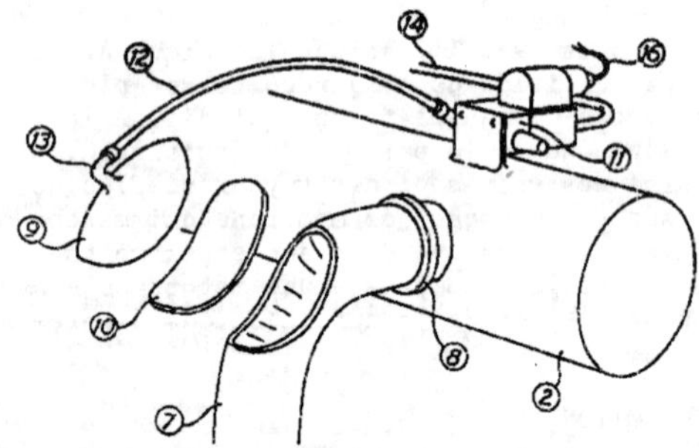

Figure 2. Part of the main boom. 2. Main boom, 7. 1 3/4" PVC-tube, 8. Valve connection ring, 9. Top socket, 10. Rubber membrane, 11. Pneumatic electromagnetic valve, 12. Pressure tube, 13. Air pressure connection to valve, 14. Air pressure inlet, 16. Electric connection.

The knee valves are made of a 1 ¾" PVC-tube with a 90 degrees bend and with an inner membrane of a profiled rubber mixture. The valve is pneumatically controlled by a pneumatic electromagnetic valve. The electromagnetic valve is subordinated an electronically developed controller system in CMOS technology.

Fig. 3 shows the principle layout of the electronic controller system used during the studies reported in this paper.

The number of knee valves controlled by the equipment is interchangeable between 8, 12, 16 or 32.

An attempt to optimize pulse frequency and pulse quote in relation to slurry quality, pump characterstics and valve type indicates a possibility to improve the lateral spreading result.

Each valve can be individually compensated against the pressure losses and dynamic wave patterns which occur in the main boom.

BLOCK LAYOUT OF THE CONTROLLER SYSTEM

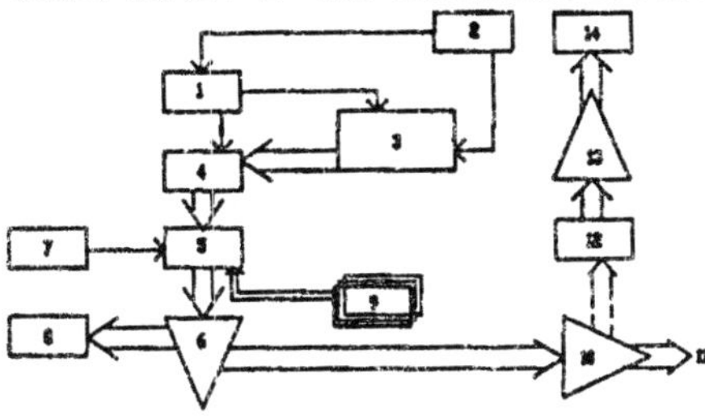

Figure 3. 1. Oscillator, 2. Voltage stabilisation, 3. System initialisation unit, 4. "D"-latches, 5. Monostable multivibrators, 6. Bufferts and LED drivers, 7. Pulse quote master monitoring, 8. LED display, 9. Individual pulse quote monitoring, 10. Power apmplifier, 11. Output to pneumatic electromagnetic valves, 12. Monostable multivibrators, 13. LED drivers, 14. LED display, 1-4 Mothercard, 5-9 Controller card, 10. Powercard, 12-14 Alarmcard.

Fig. 4 shows the time-to-voltage diagram for the controlling of the electromagnetic valve. The knee valve opens during the time the voltage is in a high state (12 V).

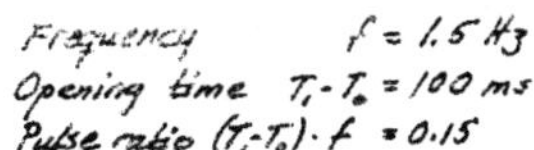

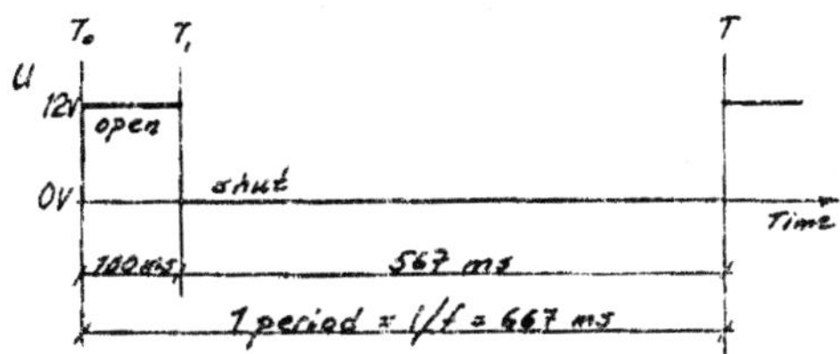

Figure 4. Control counts for pneumatic magnetic valves.

The principle of the time-to-voltage diagram for spreading equipment using 16 valves is shown in Fig. 5.

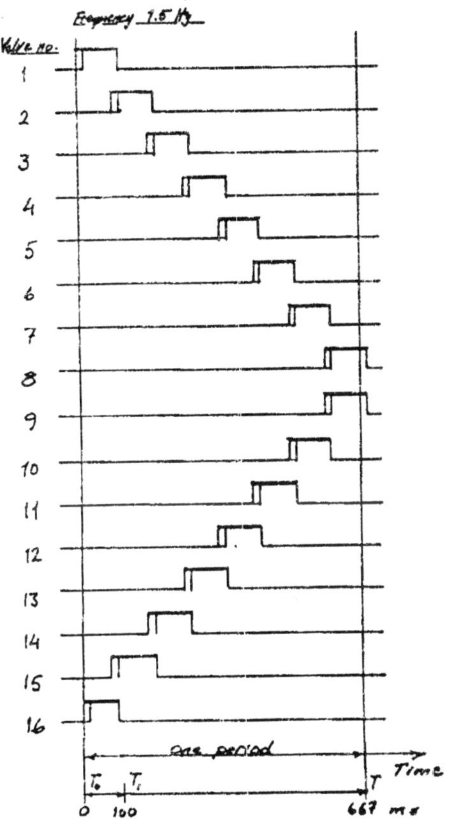

Figure 5. A 16 valve time-to-voltage diagram

The controller pattern in fig.5 is kept in a symmetrical way (the two end valves are opened at the same time) in order to prevent lateral forces contributing to the main boom entering a swinging mode, and also to get a more uniform spreading result.

4 TYPE OF SPREADING DEVICE

In order to study the technique described above together with different spreading methods from aspects of plant nutrition and forage hygiene, slurry was spread on grasslands and growing grain crops.

We have used three different methods of manure spreading.

Broad casting, boom with low-level splash plates about 0.2 - 0.3 m above the vegetation, c/c-spacing between the outlets 0.25 m (Fig. 6 a).

Flexible hoses, boom with hoses dragging on the surface (when spreading on grassland the hoses emerged about 0.2 - 0.3 m above the vegetation), c/c-spacing between the outlets 0.25, 0.375 and on grassland also 0.5 m (Fig. 6 b).

Shoe coulter, boom with shoe coulters which buries the slurry 2 - 3 cm below the surface, c/c-spacing between the shoe coulters 0.25, 0.375 and on grassland also 0.5 m (Fig. 6 c).

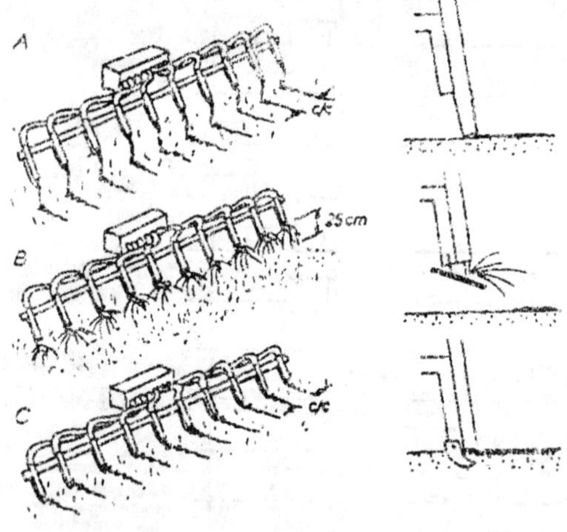

Figure 6. Spreading slurry in growing crops with three different application techniques. A. Flexible hose, B. Splash plate, C. Shoe coulter.

5 RESULTS

5.1 Technical aspects

5.1.1 Liquid pressure in the boom

For the spreader with 12 boom and a pump with an output of 4 000 l/min the pressure in the boom was affected by the opening frequency and the period during which the valves were open. This relationship is shown in Fig. 7 for the ley. The results for the slurry showed the same tendency. A lower frequency and a shorter opening period resulted in a higher pressure than a higher frequency and a longer opening period.

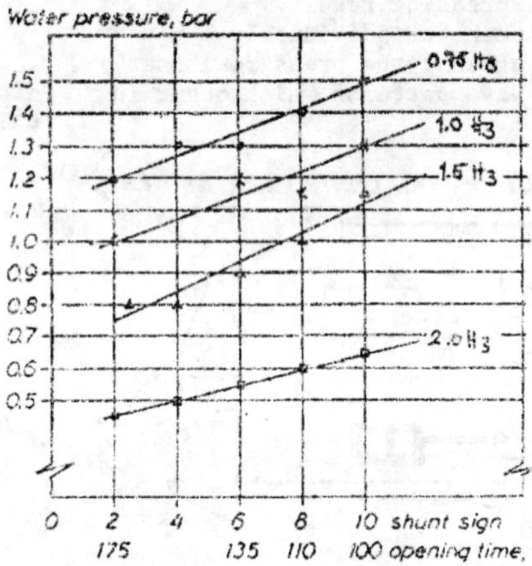

Figure 7. The water pressure in the boom versus the opening times of the valves at four different working frequencies.

5.1.2 Distribution pattern

Using electronics, the flow from each valve can be adjusted individually. These individual adjustments ensure a very good lateral uniformity of spread. When spreading manure with the 3 m boom, the coefficient of variation for lateral uniformity of spreading has been set at 5%. With the 12 m boom the coefficient of variation for the lateral uniformity of spreading of slurry has been set at 10%. This is illustrated in Fig. 8.

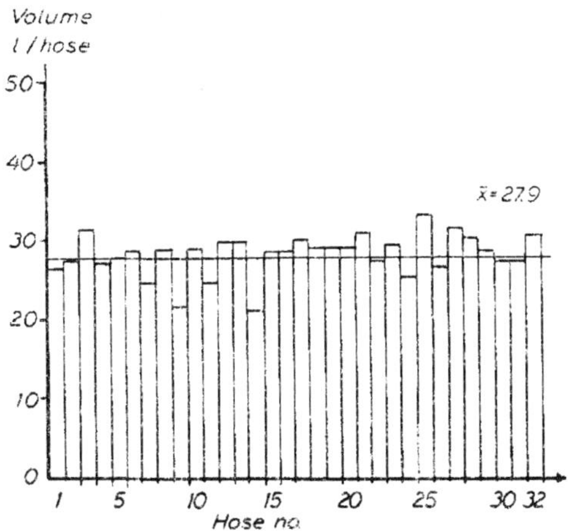

Figure 8. Distribution pattern for the boom with 12 m working width. The test results are calculated 1/hose. The coefficient of variation was calculated to be 10%.

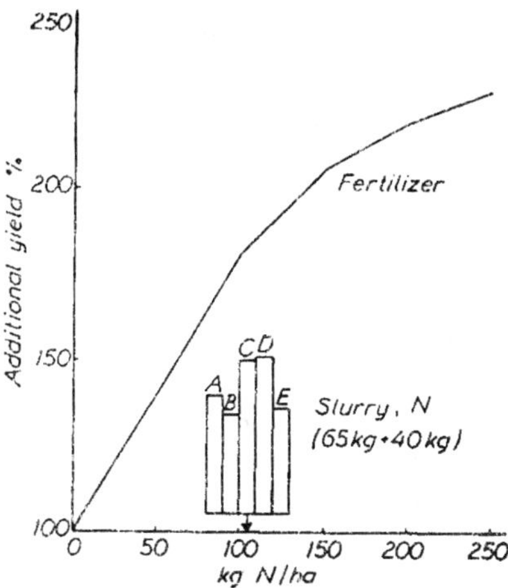

Figure 9. Nitrogen response in grassland after spreading chemical fertiliser and slurry. A. Flexible hoses, c/c 0.25 m, th slurry spread 0.25 m above the vegetation B. Splash plate, c/c 0.25 m, the slurry spread 0.25 m above the vegetation. C-E. Shoe coulters, c/c 0.25, 0.375 and 0.5 m respectively, the slurry spread just below the ground surface.

5.2 Biological aspects

5.2.1 Spreading slurry on grassland

The experiments on grass leys showed that the nitrogen effect of the slurry, regardless of spreading technique, was always poorer than that obtained with the corresponding amount of fertiliser. In most treatments the N-effect of slurry was 40-60% of the N-effect in fertiliser (Fig. 9).

The difference in the yield obtained after various means of spreading was fairly small. The experimental results show that spreading in strips on the soil surface generally gave a better yield than if the manure was spread above the vegetation. In addition, spreading in strips with c/c spacing of 25 and 37.5 cm frequently gave larger yields than spreading with c/c spacing of 50 cm.

In the studies of how different means of spreading slurry can influence the silage-making process, comparisons were made with areas given fertiliser. If the forage was wilted at silage-making the contents of both spores and butyric acid in the manure treatments were much lower. Experimental results also indicated that spreading in strips on the soil surface gave a lower butyric acid content in the silage than if the manure was spread over the entire surface. Addition of formic acid inhibited both the formation of butyric acid and the formation of spores of Clostridium tyrobutyricum. In general, however, silage from directly harvested leys which had been fertilised with slurry over the entire surface and the highest acitivty of harmful microorganisms. This was apparent in the highest ammonia concentration, large amounts of butyric acid and spores of Clostridium tyrobutyricum, small amounts of residual sugars and relatively small amounts of lactic acid.

5.2.2 Spreading slurry on growing grain crops

The experiments in 1987/88 included three treatments: barley on light soil, barley on heavy soil and oats on moderately heavy soil. In each treatment, the slurry was spread at three different times, namely before sowing, when the crop was about 15 cm high and just before heading.

At the two first spreading times all three spreading techniques were used. At the last spreading time the manure was spread only using towed hoses and deflector plates.

In order to obtain comparisons of the

nutrient availability each treatment also included unmanured plots and fertilised plots. Each plot apart from the unmanured plot was given a basic amount of fertiliser amounting to 30 kg N/ha. On each spreading occasion the plots were manured with an amount corresponding to 70 kg N/ha. The manure used was pig slurry with large amounts of straw residues and with a content of ammonium-N which varied between 3 and 3.4 kg/tonne of manure.

A comparison of yields at the same amount of supplied nutritients from fertiliser alone and a combination of fertiliser and slurry, shows that fertiliser alone gives a slightly better gain than the combination of fertiliser and slurry. The best utilization of the plant nutritients in the slurry was obtained when spreading was done before sowing.

A comparison between the different types of spreading equipment suggests that the splash plate gave poorer yields than the other equipment. No particular difference in yield could be discerned between a shallow burial and spreading in strips by flexible hoses.

6 REFERENCES

Rodhe, L. et al. 1988. Spreading of slurry on ley. Swedish Institute of Agricultural Engineering, JTI-Report 93, Uppsala.

Steineck, S. 1988. Slurry applied to grass and miexed ley. Swedish University of Agricultural Sciences, Dept of soil sciences, Div of soil fertility, Report 172, Uppsala.

Land and Water Use, Dodd & Grace (eds), © 1989 Balkema, Rotterdam. ISBN 90 6191 980 0

Rapid field tests for estimating dry matter and fertiliser value of animal slurries

H.Tunney
Teagasc, Johnstown Castle Research Centre, Wexford, Ireland

M.Bertrand
Division Teeepei, Cemagref, Rennes, France

ABSTRACT: Recent developments in rapid field tests for estimating the dry matter and fertiliser value of slurry and other wastes is reviewed. 1) The Slurry-Meter is a hydrometer calibrated in percent dry matter and is based on straight line relationship between the dry matter and specific gravity of animal slurries. The fertiliser value can be estimated from accompanying tables showing the relationship between dry matter and the N, P and K content of the waste. 2) The Nitrogen Meter is a rapid test for the ammonia content of waste, it is based on the release of nitrogen gas after the oxidation of ammonia by calcium hypochlorite the gas pressure is proportional to the ammonia content. 3) The ammonia electrode is used in laboratories for determining ammonia content of waste and waste waters. 4) Water analysis kits can also be used for estimating the ammonia and phosphate content of manures.

1 INTRODUCTION

Legislation is being introduced in some EC countries to limit the quantity of animal slurry that can be spread on land and it is likely that all countries, will have similar laws after 1992. It is therefore important that farmers know the composition of slurry.

Excess application of organic manure has lead to the build up of nutrients in soil, resulting in pollution problems. Often farmers are not aware of the composition of manure and the extent to which nutrients applied are in excess of crop needs. It is now accepted that the input of nutrients should not exceed crop needs. Rapid field tests can be used to control accurately the quantities of nutrients that are being applied.

2 BACKGROUND

The idea of a rapid test for slurry is not new. In Germany over seventy years ago Vogel (1916) showed that there was a significant relationship between density and nitrogen content of slurry. At about the same time Dussare (1915) and Langers (1918) carried out similar studies.

Tovborg Jensen (1929) in Denmark showed that mixing sodium hypochlorite with manure resulted in the release of nitrogen gas due to the oxidation of ammonium. This reaction is the basis of the Nitrogen Meter. This earlier work on rapid tests came to light as a result of recent work on the problems of slurry from the intensive animal rearing.

2.1 Variation in slurry composition

Results of many published papers show a wide variation in slurry composition.

Work at Johnstown Castle Research Centre (Tunney and Molloy, 1975) and at CEMAGREF Rennes (Bertrand and Smagghe, 1985) showed a statistically significant correlation between dry matter and nutrient content of slurry.

Figure 1 gives an example of the relationship between dry matter and the nitrogen content of pig slurry. Similar relationship have been established in other studies (Chescheir and Westerman, 1984; Dube, 1982; Kjellerup, 1985; Villar et al. 1984).

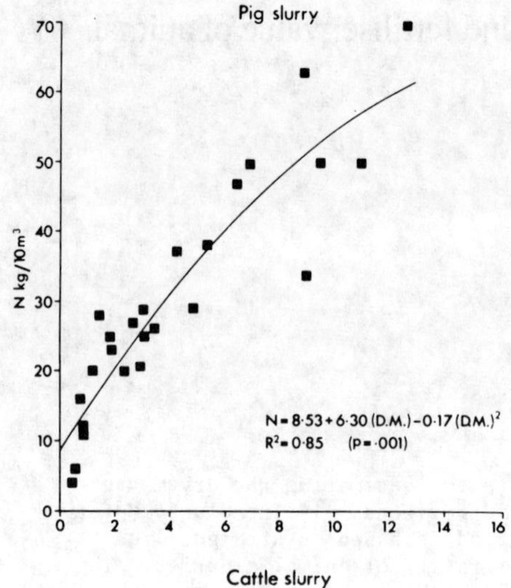

Figiure 1. Relationship between dry matter and total nitrogen content of pig slurry (Tunney, 1979).

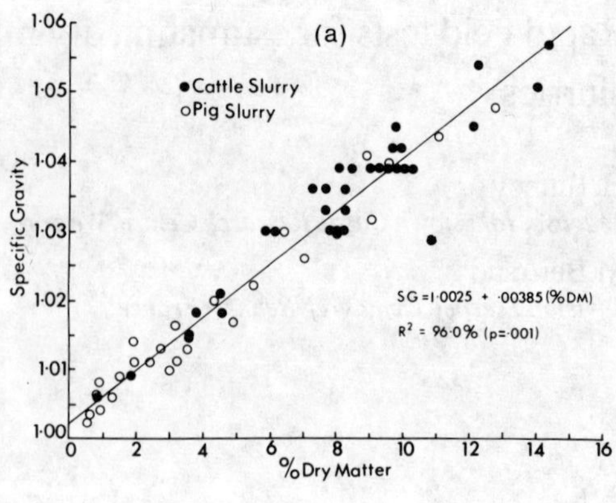

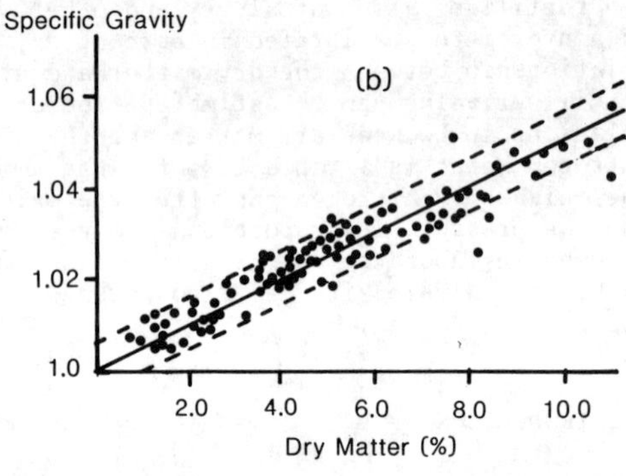

Figure 2. Relationship between dry matter and specific gravity of, (a) pig and cattle slurry (Tunney, 1979), (b) pig slurry (Bertrand & Smagghe, 1985).

3 SLURRY METER

Based on the relationships between dry matter and fertiliser value it was evident that an estimate of slurry dry matter would given a good guide to the nutrient content of slurry.

Therefore a number of possibilities for obtaining a simple test for dry matter were investigated. These included electrical conductivity, colorimetery and density measurements. To our surprise we found a highly significant linear relationship between dry matter and specific gravity of pig slurry. Further studies showed that a similar relationship exists for cattle slurry (Tunney, 1979).

The original relationship between dry matter and specific gravity is shown in figure 2(a). The relationship established by Bertrand an Smagghe (1985) is shown in figure 2(b).

More recently this relationship was also found to hold true for solid poultry manure and the relationship is reproduced in figure 3 (Tunney, 1985). With the poultry manure a four fold dilution with water was necessary before making the specific gravity measurements. Sewage sludge from municipal waste treatment plants also obeyed this relationship as shown in figure 4. However, it is difficult to measure the specific gravity of sewage sludge with a hydrometer because of its viscosity. The relationship sown in figure 4 was established by measuring the density volumetrically.

Based on the relationship shown in figure 2(a) a hydrometer calibrated in percent dry matter was made for direct measurement of the dry matter of slurry. It is known as the Slurry-Meter. The instructions for the Slurry Meter contains a table showing the relationship between dry mater and N, P and K content for cattle and pig slurry (Tunney, 1976).

The relationship between dry matter and nutrient content will depend on the type of slurry. The relationship between dry matter and nutrient content for Ireland is in good agreement with results for Brittany (Bertrand et al. 1988).

In addition to Irish and French work on this method it has also been studied and reported on in Austria (Duetsch and Puchwein, 1981), Canada (Dube, 1982),

Table 1. The equations for relationships between dry matter and specific gravity of wastes adapted from the literature.

Reference	Manure*	Equations	R^2
Tunney 1979	P, C	S.G. = 1 + .0039 (% D.M.)	.96
Dube 1982	P	= 1 + .0041	.72
Chen 1982	C	= 1 + .0036	
Cheschier 1984	P, C	= 1 + .0043	.90
Tunney 1985	H	= 1 + .0036	.92
Kjellerup 1985	P, C	= 1 + .0042	.91
Villar et al. 1985	C	= 1 + .0050	.82
Tunney 1988	S	= 1 + .0038	.94
Bertrand 1988	P	= 1 + .0040	.99
	Mean S.G.	= 1 + .0041 (% D.M.)	

* P = pig, C = cattle, H = hen manure, S = sewage sludge

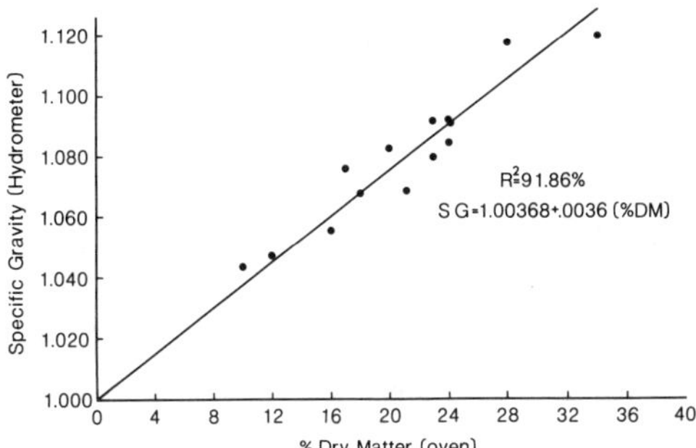

Figure 3. Relationship between dry matter and specific gravity of poultry manure from laying hens (Tunney, 1985).

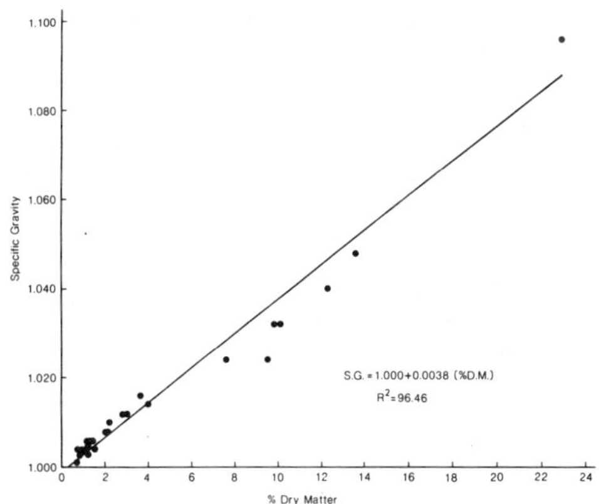

Figure 4. Relationship between dry matter and specific gravity of sewage sludge.

America (Chescheir and Westerman, 1984), Denmark (Kjellerup, 1985), Germany (Schollhorn & Kunz, 1986), Spain (Villar, et al. 1984) and other studies are currently in progress.

A summary of the equations of the relationships between dry matter and specific gravity, adopted from the literature, is shown in table 1. The equations are adapted so a specific gravity of one corresponds to a dry matter of zero. The mean value shown in table 1 can also be expressed using density, D = 1000 + 4.1 (%D.M.).

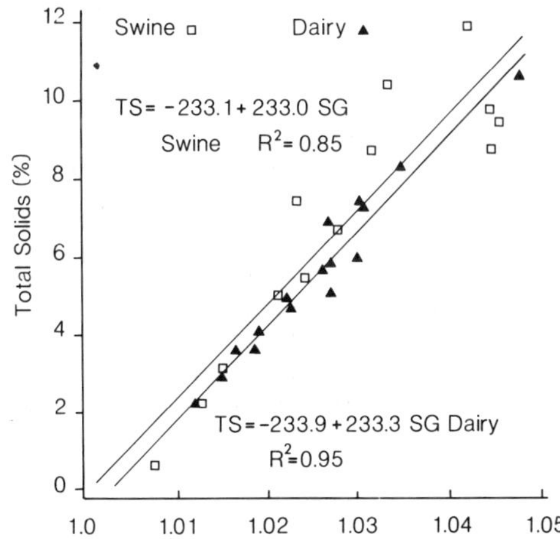

Figure 5. Relationship between total solids (dry matter) and specific gravity from an American study on dairy and swine manure (Chescheir and Westerman, 1984).

Figure 5 shows the relationship between specific gravity and total solids (dry matter) obtained by Chescheir and Westerman (1984) for cattle and pig manure. This work indicated a better correlation between specific gravity and nutrient content than between dry matter and nutrient content. Irish work gave a similar result (Tunney, 1988). The American authors calculated the percentage of samples that gave a result with less than plus or minus 25% error using the direct correlation between specific gravity and nutrient content of slurry, this is summarised in table 2.

Table 2. Percent of estimates for nutrients and total solids within ± 25% error (Chescheir and Westerman, 1984).

	Dairy	Swine
Nitrogen	94	93
Phosphorus	88	72
Potassium	76	82
Total Solids	100	86

Bertrand (1985) has shown how the effect of temperature can be corrected. The Slurry Meter is normally calibrated at a temperature of 15°C. Bertrand's work shows that a slurry of 7.5% dry matter at 15°C will give a reading of 7.0% dry matter if measured at 10°C and 8.0% dry matter at 20°C.

4 OTHER METHODS : Rapid N tests

Ammonium nitrogen is high in slurries (60% in pig slurries, 50% in cattle slurries) and is relatively stable in storage. The quantification of dry matter or specific gravity gives only an approximate evaluation of this important parameter. Three possible rapid methods, to estimate ammoniacal nitrogen are available, 1) nitrogen meters (AGROS and Quantofix), 2) Ammonia electrodes, and 3) water analysis field kit.

4.1 Nitrogen Meters

Tovborg Jensen (1929) showed that mixing sodium hypochlorite with agricultural waste resulted in the release of nitrogen gas, due to the oxidation of ammonia. In an enclosed space the pressure of the released gas is proportional to the ammonia content of the waste and this is the basis of the Nitrogen Meter. The Nitrogen Meter gives an estimate of ammonia nitrogen in the waste.

In recent years a number of companies in Germany and elsewhere have developed Nitrogen Meters based on this principle (Klasse and Werner, 1988). The Nitrogen-Meter comprises an air tight reaction chamber where calcium hypochlorite (Ca (O Cl)2) is mixed with the waste. This is connected to a pressure gauge that measures the pressure of the gas released. The gauge is usually calibrated directly in kg of ammonia nitrogen per tonne of manure.

A comparison of the Nitrogen Meter and standard laboratory analyses for ammonium was made using inorganic salts by Chescheir and Westerman (1984). They found good agreement between the two methods. Work in Denmark (Kjellerup, 1985) on 128 samples of cattle and pig slurry, with a range of 0.5 to 6.0 kg of ammonium nitrogen per tonne of slurry showed good agreement between standard laboratory analyses and the Nitrogen-Meter. The relationship was NH_4-N (laboratory) = 0.5+0.8NH_4-N (Nitrogen Meter) with $R^2 = 0.94$.

In France, works has been in progress since 1983. In particular, we have tested the AGROS apparatus (Klasse and Werner, 1988), more recently, we compared Quantofix with AGROS to see if both give the same result. AGROS uses calcium hypochlorite and Quantofix uses sodium hypchlorite. The results are summarised in table 3.

The R square values for the regression equations, in table 3, varies from 0.94 to 0.98.

AGROS and Quantofix, give broadly similar results. However, sodium hypochlorite also reacts with other nitrogen compounds and releases gaseous nitrogen. Therefore, it is less accurate for the measurement of ammonium nitrogen only. Nevertheless, the measure has a considerable practical value as it quantifies both the easily oxidable (mineral NH_4-N) and easily mineralizable urea and perhaps other simple organic compounds.

4.2 Ammonia Electrode

Orion Research Incorporated have developed an ammonia electrode (Orion, 1979). The ammonia selective electrode has been used successfully in many laboratories to measure the ammonia

Table 3. Comparison of regression equations for measures of NH_4-N by rapid methods and laboratory method (titration after a direct distillation). Dependant variable y=laboratory NH_4-N, independent variable x=rapid test NH_4-N.

Rapid Method g/kg	Regression Equation	Number of samples	Precision Prediction t 0.10
	Pig Slurry		
E*	y=.05+0.89x	43	11%
A	y=.08+0.96x	55	5%
Q	y=.03+0.93x	18	15%
	Cattle Slurry		
A	y=-.11 +1.14x	55	7%
Q	y=-.07 + 1x	17	16%

A = AGROS, E = Electrode, Q = Quantofix

content of organic manures and waste waters.

In one of the earliest studies on cattle and pig slurry (Byrne and Power, 1974) the ammonia electrode was found to give excellent results when compared with conventional analyses for ammonia.

There are several reports of studies with the ammonia electrode for manures and the results indicate that it is an accurate method of ammonia determination (Hills, 1980; Le Duy and Samson, 1982; Sievers et al. 1983).

Chescheir and Westerman (1984) stated that the ammonia electrode was very good for determining the ammonia nitrogen in animal slurries; they suggested, however, that a higher amount of NaOH than recommended in the instructions was necessary. Ion specific electrodes may also be used for rapid tests for potassium and other ions.

4.3 Water Analysis Field Kits

The ammonia content of pig slurry was estimated by a Nesslers field kit (Steward, 1968) with good accuracy. More recently this method has been tested by Chescheir and Westerman (1984) and they found reasonable agreement with conventional analyses. These authors found that large dilutions of slurries were required for analyses with field kits. They also used field kits to estimate the ortho-phosphate content of cattle and pig slurry. In summary they stated that the kit was adequate for measuring ammonia in pig slurry but not in cattle slurry. In contrast the kit was adequate for phosphorus measurement in cattle slurry but not in pig slurry.

The field kits are based on colorimetric measurements after the addition of chemical reagents to the diluted slurry.

5 MEASUREMENT ON SOLID MANURES

The emphasis on rapid tests to date has been on slurry. A high proportion of manure is spread as solid. In general, there is less variation in composition of solid manure and average values will given a more accurate guide to composition than in the case of slurry. However, rapid tests suitable for solid manure could greatly improve the efficient utilisation of nutrients in these manures.

The slurry meter can be used successfully for estimating the fertiliser value of solid poultry manure as illustrated in figure 3. The use of specific gravity and nitrogen meter is more difficult with solid manure than with slurry. The approach of drying in mini ovens and weighing has been studied as a rapid test for the dry matter and fertiliser value of poultry manure without litter by CEMAGREF in France.

5.1 Estimates of dry matter and mineral elements in solid poultry manure

Relatively inexpensive mini-ovens are now available on the market. As a result with this device and accurate weighting we can measure the dry matter content and also fertiliser elements because of their good correlation with dry matter. An example of this approach is summarised in table 4.

5.2 Measurement of ammonium and easily oxidisable nitrogen

The Quantofix apparatus operates with a reagent containing 5 volumes of sodium hypochlorite and 1 volume concentrated sodium hydroixde (Klasse and Werner, 1988). This reagent gives a relatively accurate estimate of ammonium nitrogen when easily oxidisable compounds, such as urea and uric acid are absent.

Table 4. Regression equations used to estimate the elements from dry matter in 60 samples of poultry manure.
Independent variable x = %D.M.

Dependent variable y %	Regression equation	R^2
$y=P_2O_5$	y=-0.082+0.055 x	0.93
$y=K_2O$	y=-0.072+0.036 x	0.89
y=CaO	y=-0.178+0.135 x	0.77
y=MgO	y= 0.054+0.009 x	0.89

Table 5. Relationship between N measured by Quantofix (g/100 g of droppings) and by laboratory analyses poultry manure with dry matter over and under 30%.

Independent variable x = Quantofix measurement

Dependent variable y %	Regression equation	R^2	Number of samples
	Quantofix		
$y = NH_4$-N	y = 0.076 + 0.892 x	0.78	42
$y = NH_4$-N	y = -0.046 + 1.077 x	0.87	26 (<30% D.M.)
$y = NH_4$-N	y = 0.123 + 0.806 x	0.72	16 (>30% D.M.)
	Modified Quantofix		
y = NTK*-N	y = 0.740 + 1.605 x	0.77	20 (>30% D.M.)
y = NTK-N	y = 0.022 + 1.162 x	0.88	26 (<30% D.M.)

*NTK = Total Kjeldal nitrogen

Table 5 shows regression equations for the relationships between ammonium N measured by Quantofix and laboratory methods for poultry slurry separated into samples containing over and under 30% dry matter.

A modification of the Quantofix reagents, using sodium hydroxide and sodium hypochlorite, increased the extraction of easily oxidisable nitrogen in addition to ammonium. All of the urea and 70% uric acid nitrogen was released as nitrogen gas. The relationships between this modified quantofix N and total N is shown in the last 2 lines of table 5.

The results in table 5 indicate that the Quantofix apparatus can give a reasonable estimate of both ammonium and total nitrogen. Droppings high in dry matter are also high in urea and uric acid.

6 CONCLUSIONS

The Slurry-Meter and Nitrogen Meter are now developed to a stage where they can be used at farm level to promote the efficient use of nutrients in organic manures. They are simple to use and inexpensive.

Further work could improve the accuracy of these tests and adoption in farming practice. In particular, there is need for more information on rapid tests for solid manures.

These tests can give the farmer the knowledge required to comply with the regulations controlling the rates of manures that can be spread.

7 REFERENCES

Byrne, E. and T. Power 1974. Determination of ammonia nitrogen in animal slurries by an ammonia electrode. Comm. in Soil Science and Plant Analysis, 5(1): 51-65.

Bertrand, M. 1985. Method d'estimation de la valeur fertilisante des lisiers de porcs. Publ. RNED, MNE 149, rue de Bercy, 75595 Paris.

Bertrand, M. and D. Smagghe 1985. Mise au point methodes rapides d'appreciation de la valeur fertilisante azote et potassique de lisiers de porcs. Journees Rech. Porcine en France 17: 327-338.

Bertrand, M., G. Arroyo, and D. Smagghe, 1988. Methods of fast determination about physio-chemicals animal wastages stocked. Ag. Eng. Int. Confr. Paris, March, 1988.

Chen, Y.R. 1982. Engineering properties of beef cattle manure. Amer. Soc. Agric. Eng. Paper No. 82-4085, presented at ASAE Summer meeting, Univ. Wisconsin, June 1982. Publ. ASAE St. Joseph, MI.

Chescheir, G.M. and P.W. Westerman 1984. Rapid methods for determining fertiliser value of livestock manures. Amer. Soc. Agric. Eng. Paper No. 84-4082, presented at ASAE Summer meeting, Univ. Tennessee, Knoxville, June 1984. Publ. ASAE, St. Joseph, MI.

Deutsch, A. and W. Puchewein, 1981. Praxisgerechte Einschatzung des Nahrstoffgehaltes von Flussigdunger aus der Tierhaltung. Bundesversuchsanstalt fur Alpendlandische Landwirtschaft. Gumpenstein, A-8952 Irdning.

Dube, A. 1982. Methode rapide de determination de la valeur fertilisante des lisiers de porcs a la ferme. Agriculture 39: 1, 32-35.

Dussere, C. 1915. Composition des purins provenant d'exploitations agricoles de la Suisse Romande. Ann. Agric. Suisse, 83-88.

Hills, D.J. 1980. Methane gas production from dairy manure at high solids concentration. Transactions of the ASAE 23, 122-126.

Kjellerup, V. 1985. Agros Nitrogen Meter for estimation of ammonium nitrogen in slurry and liquid manure. In : Efficient Land Use of Sludge and Manure. Ed. A. Dam Kofoed. Elsevier Appl. Sci. Publ. p 216-223.

Klasse, J. and W. Werner, 1988. Method for rapid determination of ammonia nitrogen in animal slurries and sewage sludge. In "Agricultural Waste Management and Environmental Protection". Proc. 4th Int. Symp. of C.I.E.C. Ed. Welte and Szabolcs. Publ. FAL Braunschweig-Volkenrode, D. Vol 1, p 119-123.

Lagers, G.H.G. 1918. Het soortelijk gewicht en het stikstofgehalte van gier. Verslagen Landbouwkundige Onderzoekingen 9-18.

Le Duy, A. and R. Samson 1982. Testing of an ammonia ion selective electrode for ammonia nitrogen measurement in methanogenic sludge. Biotechnology Letters 4: 303-306.

Orion Research 1979. Instruction manual for ammonia electrode model 95-10. Cambridge, MA.

Schollhorn, J. and H.G. Kunz 1986. Schatz und Schnellbestimmungsmethoden fur den Nahrstoffgehalt von Flussigmist. KTBL Arbeitsblatter Landtechnik Nr. 0219. KTBL, Bartningstrasse 49, D-6100 Darstadt 12. p 1-3.

Sievers, D.M., K. Doyle and J.H. Porter, 1983. Ammonia measurement in high organic wastes. ASAE Paper No. 83-4064. ASAE, St. Joseph, MI.

Stewart, T.A. 1968. A rapid method for estimating the soluble nitrogen content of animal slurries. Rec. Minist. Agric. Nth. Ir. 17: 91-96.

Tovborg Jensen, S. 1929. Undersogelser over ammoniakfordampning i forbindelse med kvaelstoftab ved udbringning af naturlig godning. Tidsskrift for Planteavl, 35: 68-71.

Tunney, H. and S. Molloy 1975. Variations between farms in N, P, K, Mg and dry matter composition of cattle pig and poultry manures. Ir. J. Agric. Res. 14, 71-79.

Tunney, H. 1976. A method and device for measuring the dry mater content of slurry. U.K. Patent No. 1,543,223.

Tunney, H. 1979. Dry matter, specific gravity, and nutrient relationships of cattle and pig slurry. In: Engineering Problems with Effluents from Livestock. Ed. J.C. Hawkins. Proc. C.E.C. Seminar, Cambridge, England. Publ. C.E.C. Luxembourg. p 430-447.

Tunney, H. 1985. Manure nutrient composition: rapid methods of assessment. In: Processing and Use of Organic Sludge and Liquid Agricultural Wastes. Ed. P. L'Hermite. Proc. of 4th Int. Symp. Rome, Publ. D. Reidel Publ. Co. p 243-257.

Tunney, H. 1988. Field tests for estimating fertiliser value of agricultural. In "Agricultural Waste Management and Environmental Protection". Proc. 4th Int. Symp. of C.I.E.C. Ed. Wette and Szabolcs. Publ. FAL Braunschweig-Volkenrode, D. Vol. 1, p 119-123.

Villar, M.C. et al. 1984. Methodos rapidos para la characterizacion del purin de vacuno. Anales de Edafologia y Agrobiologia 43: 195-204.

Vogel, F. 1916. Versuche mit Stallmist und Jauche. Jahrb. DLG. 31: 109-116.

RESUME : Les developpements recents des methodes rapides destimation de la matiere seche et de la valeur fertilisante du lisier et des autres dechets sont passes en revue. 1) Le "Slurry-Meter" est un densimetre calibre en pour cent de matiere seche. Il est base sur la regression lineaire entre la matiere seche et la masse volumique des lisiers animaux. La valeur fertisante peut etre estimee a partir des tables montrant la relation entre la matiere seche et les autres elements fertilisants, N, P, et K, de l'effluent etudie. 2) L'analyseur d'azote est une methode rapide d'evaluation du taux d'azote ammoniacal contenu dans des dejections. Il est base sur le degagement d'azote gazeux apres l'oxydation de l'ammonium par l'hypochlorite de calcium ou de sodium. La pression de gaz produite et mesuree est proportionnellee au taux d'azote ammoniacal. 3) L'electrode a diffusion gazeuse de l'ammoniac est utilisee dans les laboratiores pour determiner le taux d'azote ammoniacal des lisiers et des eaux usees. 4) Les trousses d'analyses rapides de l'eau peuvent aussi etre utilisees pour l'evaluation de l'azote ammoniacal et des phosphates contenus dans les dejections animales.

Land and Water Use, Dodd & Grace (eds), © 1989 Balkema, Rotterdam. ISBN 90 6191 980 0

Ferti-irrigation of slurry on feed-lots

R.Jenčič
Biotechnical Faculty, University of Ljubljana, Yugoslavia

ABSTRACT: A rational and ecologically acceptable application of large quantities of slurry on big cattle and pig farms can be carried out by its recycling to the fields. Large quantities require other techniques of capturing, transport and repartition than those used by small farms. Mechanical transport of large quantities by slurry tanks is not rational, it is limited by weather conditions, and does not allow watering. On a pilot farm, a system of fertiirrigation on 156 ha has been established, where 16,000 m^3 of slurry returns to the fields yearly, transported through underground pipe lines. Only 15% of energy needed for mechanical transport is engaged in such a system. Some difficulties, modifications and additional equipment are also discussed.

The development of feed-lots has required and also found a considerable rationalization of zootechnology riched by the use of the up-to date biological knowledge, and the results of research in selection, nutrition, veterinary science, by better organization of work on big farms. However, the problem of rational and ecologically tolerable application of large quantities of slurry has remained unsolved.

Almost without exception, these farm projects were provided with lagoons, primitive storage-earth basins or some depressions for depositing the slurry mass. The temporal capacity of those depressions has not often been determined, assuming that the trickling through soil, evaporation into air, or a similar elimination would occur.

Such depressions represented a considerable ecological burden, and loss of enormous quantity of nutritive material. This caused a big affront dominating today against big farms regardless their contribution to the meat production.

Big farms are reality, and as long as they exist we should find optimal technological solutions for their improvement also in the domaine of rational and ecologically acceptable application of slurry.

Attention has to be payed to the following facts:

- The only rational way to solve this problem is returning the slurry back to the field, disregarding the quantity and the distance. No other way (permanent deposit, draining into waters) can be taken into consideration.
- The mass of these excrements is large and will always represent a certain "overloading" of the same elements (N, P, K).
- Certain agricultural measures can lower the effect of this load to a minimum, or at least to an imperceptible appearance.
- The costs of this procedure are not low but they should always be included in the investment; and in the future, they should mainly be covered through the fertilizing contribution on the implicated acreage.
- The hydraulic recycling, discussed in this paper, is the only variant to the existing feasible methods of fertilization, one solving the existent situation.

The definitive solution represents the returning of the mass to the fields with the lowest consumption of energy, with the smallest dependence on the state of the ground, i.e. the lowest axial pressure of ground and also the lowest "chemical shock". This is reached through stepless diluting of slurry; it would not be possible with the mechanical transport.

PRELIMINARY RESEARCH

We made experiments with tilled lagoons, i.e. field areas taken out of the regular production for 2 - 3 years. Alternate intensive surface cultivation, and intensive fertilization with slurry were made there. We tried to state the influence of normal and abnormal slurry - quantities in regard to the contents, and the accumulation of their individual components, and in different horizons of cultivated land. The soil was brownish, partly gley argil of adequate structure.

The experimental plot was divided into 6 sectors. The following doses: 0, 50, 200, 400, 600, 800 m^3/ha were given during the period lasting from April to July.

The slurry was incorporated into the ground immediately after tank spreading by the standard disc harrow. Samples for chemical analysis were taken twelve days after the last loading.

These preliminary analyses showed us that by a consequent incorporation down to 8 cm, deposits up to 800 m^3/ha are possible without no essential changes in the contents of N and P.

In addition to these analyses, we also followed the contents of these elements in percolated water (collected 80 cm deep in water-tanks). Analytical nitrogen could no be traced, and only minimal traces of phosphorus and potassium were found (N = 0, P_2O_5 = 0.00458%, K_2O = 0.000723%). This means that filter-absorptive ability of soil prevented any contamination of underground water.

Although these results have been achieved through a two-year examination on the same location, they show that even abnormally high doses of slurry have not caused a characteristic modification of the structure of these elements in the stratum down to 60 cm. (The results of these analyses are obtainable by the author)

Corn and kale were sown as experimental crops for further verification on the experiment plots. We have not noticed any stagnation or abnormal behavior of a certain plant or any excessive concentration of certain elements.

We are sure that the results of a similar procedure would not differ very much on any other suitable location. Based on a four years lasting examination, we could conclude: Consequent incorporation on a ploughed and cultivated surface is assured. Doses up to 200 m^3/ha yearly can be used without any risks.

According to our measuring on a beef cattle farm, an average weight of cattle head is 290 kg. The yearly quantity of slurry per head makes:

$$\frac{290 \cdot 365}{500 \cdot 1000} \cdot 50 = 10.6 \text{ m}^3 = 10{,}600 \text{ m}^3$$

yearly on a feed-lot with 1000 heads.

Minimal fertilization acreage of

$$\frac{10600}{200} = 53 \text{ ha has to be provided.}$$

On a pig-breeding farm with an average animal weight of 40 kg, the yearly quantity of slurry (app. of the same structure) per head makes:

$$\frac{40 \cdot 365}{500 \cdot 1000} \cdot 45 = 1.314 \text{ m}^3$$

respectively:

20,000 x 1.314 = 26,280 m^3 for 40,000 pigs yearly in two turns.

Fertilization acreage minimum needed for this farm is:

$$\frac{26280}{200} = 132 \text{ ha.}$$

Absorptive soil and cultivated land have been presumed and built into these parameters.

Ponderous impermeable argil, and low sandy soil with an underground water stream are unsuitable.

If only the criterion of economical fertilization would be decisive, regardless of the investment cost (practically impossible), the recommended specific charge (200 m^3/ha) should be lowered because of high doses of nutritive elements. Through relatively high stock of these elements in soil, they may not bring any special effects.

Our task and aim in this care is innoxious but if possible, a useful application (eliminating) of accumulated substance.

CHOICE OF TRANSPORT

Beside allowed or still tolerable doses, and determination of fertilized areas per head, there still remains a transport problem.

There are two ways:

- mechanical cistern transport,
- pipeline (hydraulic) transport.

The usual way is cistern transport. Its advantages are:

- low investment costs,
- theoretically unlimited length of transport,
- simple organization.

Disadvantages are:

- uselessness for big quantities,
- limited carrying capacity on wet and soaked soils,
- tendency towards the transportation of concentrated matter.

The axial pression per wheel with 5 m^3 useful load and 1600 kg tara amounts to 3400 daN. The touch down surface from the tyre 15 x 12 results app. 1320 cm^2. The specific pressure is 2.57 daN/cm^2.

The rut width per hectare at the spraying width of 7.0 m:

$$\frac{10,000}{7} \times 15 \times 0.0254 \times 2 = 1088 \text{ m}^2/\text{ha}=11\%.$$

The rationalization of mechanical transport requires thickening (concentration of matter), but this is an ecological and agricultural risk.

Pipeline transport has also its:

a) negative sides:

- Investment costs are high.
- It is limited to a short distance.
- It requires good technical knowledge and organization.

b) positive sides:

- It can be applied also for maximal quantity.
- It enables an ecologically and agriculturally satisfactory cultivation in all regards.
- It can be realized regardless to the state (load capacity) of soil.
- It is harmless for soil structure.
- It enables combined irrigation with sprinkling.
- It enables automatization or remote controle for starting, dilution and stopping.

TECHNICAL DATA

Because of these advantages we made the project for a pipeline fertiirrigation system on a new feed-lot. It is in regular exploitation and will be used as pilot object for further research and collecting of experiences.

Main characteristics of this construction are:

- cattle feed-lot = 1050 places in 2 buildings,
- charge = 90%,
- yearly accumulation (with sanitary water) = 16,000 m^3,
- fertiirrigational area =

  27 ha meadow and 129 ha field, that is 156 ha total,
- loading $= \frac{16,000}{156} = 102.56$ m^3/ha,
- water-storage = 6 x 12 x 4.3 = 2916 m^3
- accumulation period = reservoir capacity $= \frac{2916}{16000} \cdot 365 = 67$ days.

Comparison of energy consumption:
Daily capacity of the hydraulic transport (1260 m^3) equals 252 transports of 5 m^3. Each mechanical transport on average distance of 1 km takes 30 minutes of tractor work (30 kW). 125 working hours of 30 kW are needed, together

125 x 30 = 3750 kWh.

Hydraulic transport energy consumption for:

$Q \;\; = 35$ l/s,
$h \;\; = 85$ mVS,
$eta = 0.55$,

$$N = \frac{35 \cdot 85 \cdot 10}{102 \cdot 0.55} = 530 \text{ kWh.}$$

Length of the underground pipeline on the pilot object:

= 2130 m^1 DN 150 AC TC 10
= 10013 m^1 DN 125 AC TC 10
= 77.83 m^1/ha.

On 27 ha of grassland, 74 hydrants are built-in in underground shafts in triangle lay-out on which fixed wide-range sprinklers of 37.5 m range are installed. Each of them covers an active hexagonal area of 3653 m^2.

Intensity of the sprinkling density of sprinklers a 8.7 l/s = 31.32 m^3/h =
= 31320/3653 . 5 = 8.57 mm/h.
200 m^3 are needed for one dose of 20 mm:
20/8.57 = 2.33 h = $2^h20'$.

In 10-hours work each sprinkler makes 4 positions a 1.46 ha = 5.8 ha daily by 4.

Diluting 1 : 2 on this area in one deposit of 66 m^3 slurry with 132 m^3 water is applied.

Generally 3 deposits are given. That gives 200 m^3 slurry yearly on 27 ha of theoretical area using 5400 m^3 = 34% of total quantity. In that system a subsequent flushing with clear water is possible = washing plant blades. This has been applied.

On a ploughed field, the stable system can not be applied, as placing of hydrants is possible only on the borders of the plots - in rectangular lines to the sowing line. On 129 ha field, 54 hydrants are placed for feeding drum irrigators with hydraulic winding, and pulling the sprinkler from the distance up to 350 m. Active width = 70.0 m.

Winding up is automatically kept on the same speed, regardless the coil number on the drum, and operates with the speed up to 50 m/h.

The sprinkled area is cca 0.35 ha/h, and the density of app. 9.0 mm/h is achieved. On ploughed and sown area, the period of possible exploitation is of course shorter (due to the sensitivity of crops, soaked ground for pulling the sledge with the sprinkler). The intensitivity of fertilization is lower here.

Two sprinklers in 10-hour work with maximal speed of 50 m/h cultivate an area of app. 7.0 ha daily. On 129 ha plowed area, the remaining 66% of the total slurry = 10,600 m^3 diluted in the ration 1 : 1 = 21,200 m^3 of "half" slurry is deposited. Therefore:

$$\frac{21{,}200}{8.7 \cdot 3.6 \cdot 2} = 338$$ working hours (minimum) are needed, 35 days, which have to be found in the period before or after sowing. Flushing with clear water is not possible in this case (except repeated procedure), which could be necessary to prevent burning or damage of plants.

Characteristics of the pump station

Pump station must have three functions:
- pumping water to the fertiirrigation system with stepless mixing and diluting with clear water from 1 : 0 to 0 : 1,
- hydraulic mixing and homogenization of decomposed stratum in the basin,
- hydraulic opening of the outlet pipe between the basin and pump with recurrent stream.

Feeding the pump on the entering side with slurry and clear water comes from the same hydrostatic level (Slurry in the basin is lowered but the clear-water level remains the same).

A centrifugal horizontal pump with a suction pipe is built in. Vertical (submersed) pumps are not suitable for such mixing and diluting as they do not have a suction pipe, even if those with submersed rotor ladle the liquid most reliably.

The telecommand and relay switching of pumps, and electromotive valves through UKV transmitter are in a realization phase.

This pilot object, i.e. its first phase has been operating for five seasons by now. In that period, we have come to the conclusions out of which we deducted certain changes in equipment, and the way of work. We use them also for projecting other systems. Our experiences are:

- No disposition of cattle feed-lot can prevent the coming of other matter into slurry.

- Silage parts cause nozzle-stoppage of sprinklers.

- Small parts of corn grain prevent the turbin rotation by coiling the irrigator drum.

- Cattle fur particularly tail fur comes into slurry and then clutches at the sharp edges, particularly on the parts of the automatic hydrant, avoiding the closing.

- Hydraulic parameters being used for clear water to retain the hydraulic loss $J > 10$ m/km have to be revised: While pumping slurry with 8 - 9% dry substance at this low speed, used for clear water (< 1.5 m/s), "lazy" liquid gradually throws the sediment down the pipes.

- We have therefore completed the system with mechanical and biological cultivation through strict dilution at lowest ratio 1 : 1.

ADDITIONAL TREATMENT OF SLURRY

Mechanical slurry cultivation has two supplementing proceedings: maceration = homogenization and separation = removal of solid parts.

We tried to eliminate the critical points of hydraulic transport and ferti-irrigation through introduction of this procedure combined with biologic cultivation:

- prevention of sedimentation in pipes,
- prevention of stoppings in driving turbins of drum irrigators and in additional nozzles of sprinklers,
- approaching of slurry to the hydraulic caracteristics of ideal liquid (water).

The mechanical processing of slurry brings, beside the elimination of the listed critical points, also the following advantages:

- reduction of necessary store-space because of the volume reduction of liquid phase for 10 - 15%,
- simpler execution of biological cultivation = aeration of liquid phase,
- avoiding swimming layers in basins for slurry,
- dry phase being used for composting or as humus manure,
- the liquid phase being used for washing slurry canals in stables (recycling) instead of additional water; the total volume of liquid being lower and the emission reduced.

Two macerators are placed at both collecting basins where slurry from stables flows. After homogenization = maceration slurry goes to the platform, where separator separates dry and liquid phase. Liquid phase flows into basins, where installations for biological cultivating are placed, dry phase is stocked up and transported to the fields (vineyards, orchards).

Aerobic biological treatments, based on entering of fine dispersed air and intensive turbulent mixing should release oxidant exoterm processes, when medium in unisolated reservoirs warms up to 30°C, and in isolated to 40°C. This would represent thermic desinfection with oxidant products: NH_3, CO_2, H_2O, PO_4. Also the main emissions: H_2S, CH_4 are removed.

Ventilators = aerators with added mixing operation are installed for bringing air into the medium and its mixing. Several systems are known up to now, but not yet completely tested and successfully asserted. On the pilot object, submersed aerators "ALDO 300" produced by Alfa-Laval are used in each basin.

The submersed aerator is combined with a radially placed screw blower, sucking the medium from below and forming an intense horizontal turbulent current. This causes the mixing of the whole reservoir content. External air is sucked through the conduit-pipe and blown as tiny bubbles into the moving medium.

The turning is regulated by an automatic timer. Nowadays, the aerators work each 2 hours for 15 minutes.

WATER RESOURCE

Additional water is required for a dilution in the rate from 1 : 0 to 0 : 1.

35 l/s $\approx$ 1260 m^3/10 h of water are needed in our object.

This water quantity can not represent an irrigation, which would require at hydromodul = 0.7 l/s/ha for this area

156 x 0.7 = 110 l/s at minimum.

The water quantity of only 35 l/s enables a dilution, favorable from the ecologic and agricultural point of view.

If sufficient accumulation could be provided, this pump capacity with a hydromodul 0.7 l/s/ha provides an intervening irrigation

$$\frac{35}{0.7} = 50 \text{ ha.}$$

CONCLUSION

Based on the results of a preliminary research concerning the influence of normal and abnormal quantities of slurry applied on experimental plots, the following conclusion can be made Doses up to 200 m^3/ha yearly can be used with consequent incorporation on ploughed cultivated land.

Absorptive soil is supposed, impermeable argil and low sandy soil with underground water stream are unsuitable.

The pipeline transport is not limited by soil wetness, the ecologically friendly dilution of slurry is possible.

On the experimental pilot object, the following experiences were made:

- The energy consumption is lowered to 15% of the quantity needed for tractor powered transport.
- With a pump capacity of 35 l/s for ca 32,000 m^3 of "half slurry" (= diluted 1 : 1) about 50 working days are needed.
- Horizontal pumps with aspiration (suction) pipe allowing dilution are suitable.
- Mechanically prepared (homogenized and separated) slurry prevents stoppages and sedimentation in tubes. The slurry has to be prepared also biologically to release oxidant processes, thermic desinfection and desodoration.
- Additional water ressource equal to the pump capacity has to be assured.

REFERENCES

1. Advice on avoiding pollution from manures & other slurry wastes. Farm waste management. MAFF.ADAS.
2. Profitable use of farm manures. Farm waste management. MAFF.ADAS.
3. Technik der Flüssigmist-Ausbringung Merkblatt DLG.

Land and Water Use, Dodd & Grace (eds), © 1989 Balkema, Rotterdam. ISBN 90 6191 980 0

Intensive reclamation of compacted heavy soils by application of liquid manure

V.S.Kazakov & S.Sh.Babuadze
Moscow Institute of Agricultural Engineering, Moscow, USSR

At present it is not always possible to succeed in getting sufficiently high and stable yields of farm crops on compacted heavy, poorly structured soils. Low permeability of such soils hinders effective drainage, promotes surface runoff of natural precipitation and, in case of sprinkling, precludes soil from saturation to the desired depth. Alkalinized and saline soils even after reclamation operations suffer from reduced permeability and air holding capacity.

The fertility of soils is featured not only by the quality of the topsoil, but also by the texture of the subjacent subsoil, which controls water and air balances.

In the period of rainfalls the subsoil accumulates natural precipitation and subsequently furnishes water to plants during the growing season. If compacted, the subsoil fails to perform this function with resulting disruption of the water balance in general. Overmoistening or water deficit in the topsoil have negative effects on the development of plants.

Mole drainage, slotting, deep ploughing and loosening, fertilizers and reclamation chemicals (lime, gypsum) are employed to upgrade water and physical properties of compacted heavy soils.

High effectiveness of deep loosening in a radical improvement of soil textures has been recently established. Structural improvement of the subsoil may increase soil water-retaining capacity, air permeability and form congeneal habitat for microorganisms.

Large-scale deep loosening operations were started in various parts of the USSR as back as 1977 and 250 thou ha of compacted heavy soils have been reclaimed ever since.

It has been established, that high positive effect of deep reclamation loosening of compacted heavy soils may persist for a long period, provided the stability of newly formed soil texture. There are two ways to attain this: subsoil application of liquid fertilizers and of reclamation agents, including liquid manure, line and gypsum, and soil development by intensive rooting.

Liquid manure renders a telling and durable effect on texture development in compacted heavy soils. Apart from vital nutritives, including microelements, liquid manure comprises active reclamation agents, capable of considerable upgrading of the texture in the subsoil. Manure, if applied regularly, becomes in heavy soils a major source of the organic substance, increases humus content in the soil, improves its physical and chemical properties, enhances buffering and absorbing capacity. Manure-induced effect in heavy soils is traced for a number of years. The experiments have revealed, that application of liquid manure accounted for 50% of the total yield increase for the 1st crop of the rotation and 20-30% for the second. Above all, changes in the texture of heavy soils have a response to calcium and phosphor, which occur in manure in large quantities and in a readily consumed form.

Nitrogene may be treated as an efficient reclamation agent, promoting rooting in deep layers and stabilization of the expanded texture.

Heavy clay soils admit calcium as the best reclamation agent to enhance aggregation and clodding of heavy soils.

According to agrochemical analyses manure, i.e. excrements of animals plus shredded litter, averages 0.5% N; 0.25% P_2O_5; 0.6% K_2O and 0.5% CaO or 5 kg N, 2.5 kg P_2O_5, 6 kg K_2O and 5 kg CaO per ton of manure.

The above data show, that liquid manure possesses high potential effectiveness in reclamation of compacted heavy soils, and in conjunction with deep loosening it may render a significant impact on soil fertility and productivity of arable lands.

A number of USSR research centers have been investigating methods and techniques of deep slot-free loosening, its effect on yields of farm crops, as well as subsoil application of chemical reclamation agents, fertilizers, including liquid manure, to the root zone. Rippers of types PT-0.5: PT - 0.8; PT - 1.2 were developped for deep loosening of soils by means of V-shapes blades to the depth of 0.5; 0.8 and 1.2 m respectively. Unlike other rippers they do not leave vertical slots in expanded soil.

The results of loosening by V-shape rippers of soils with the bulk density of 1.5 kg and above are:

- the bulk density of soils reduces by 20-30% and the air capacity increases by 2-3 times;
- water permeability increases by 10 times;
- time of drainage in the zone of overmoistening reduced due to better conditions; cultivation and sowing on loosened fields in spring starts earlier;
- in areas of insufficient moistening, as for instance in steppes, more water is stored in the soil due to better seepage of natural precipitation.

Soils in the Zavolzhye area, for instance, after deep loosening operations additionally accumulated in spring 1981 1 thou cu.m. of water per ha and 500 cum/ha in spring 1982. Water accumulation in dry steppes totalled 300-400 cum/ha;

- deep reclamation loosening of saline alkali soils provides better desalinization both by means of natural precipitation and at the expense of higher permeability;
- the effective impact of deep reclamation loosening on water and physical properties and fertility in general keeps for 3-4 years and life-time of this impact may be extended by application of reclamation chemicals;
- deep reclamation loosening provides high and stable yields of farm crops on soils with the bulk density of 1.5 g/cu.cm and above. Surplus harvest by irrigation amounts for cereals to 0.3 - 0.6 t/ha, for green forage maize - 10 t/ha and 0.3 - 05. t/ha for cotton.

Deep loosening alone (without irrigation) in the steppe zone accounts for a bit smaller yield increase, totalling 0.2 - 0.3 t/ha for cereals and 5 t/ha for green forage maize. The positive effect in this case is attributed to better aeration of the soil and greater accumulation of natural precipitation.

Pilot tests revealed higher yields of other forage and technical crops. In the Kuibishev Zavolzhye area, for instance, the yields went up by 50% for alfalfa, 70% for sugar beets and 40% for sunflower seeds.

Deep loosening pays back in the first year of crop cultivation and the price of the surplus produce ranges 25-100 Roubles per hectar, depending on the purchasing price of the crop. Labor productivity in deep loosening operations tops 10 hectars per shift by employment of a 220 kw wheel tractor.

Deep loosening of compacted heavy soils has a perspective in the development of solonetz soils, especially of deep ones. Experimental subsoil application of gypsum to the entire 60 cm depth of the loosened layer with subsequent leeching with the total 3 thou cum/ha provided removal of 62% of absorbed sodium, from sodium-rich solonetz soils, while surface spreading of the same amount of gypsum only 44% of sodium were removed.

Wide-span sprinklers proved to be effective in irrigation of loosened solonetz soils by gipsum aplicationinto the subsoil prior ploughing. Better seepage conditions of in 0.8 m deep loosened solonetz soil yields of green forage maize went up by 16.8 t/ha, while the conditions, supplemented by conjunctive subsoil application of gypsum resulted in 28.3 t/ha yield increase.

Feasibility and methods of deep loosening shall by layed down in the design of reclamation works. The above is a proof that deep reclamation loosening shall become part and parcel of industrial-type cultivation of farm crops on reclaimed lands.

It is noteworthy, that deep loosening retards sprouting of weeds for at least 10-12 days and their intensive development comes after the shooting of farm crops. This was verified by field experiments in the Golodnaya steppe, Povolzhye area, Kazakh and Gergian Republics, in course of which no pesticides were used.

Techniques have been worked out for in-soil applications of reclamation chemicals and liquid manure into the loosened subsoil by means of V-shape rippers to the depth of 0.4 - 1.2 m. Equipment for liquid manure application, consisting of apipe system, ganged up to the ripper posts, has been developed. For transportation of liquid manure to the field 8-16 cu. m trailer sprayers for liquid fertilizers were used. Same sprayers may be employed for direct conjunctive application of liquid manure when moving parallel to the route of the ripper. In-field reloading of liquid manure from the trailor sprayer into the ripper's tractor-mounted containers is also possible. In case manure storage is 1-5 km away from the operation site, employment of wheel tractor units with mounted containers and ripper is feasible.

Methods of subsoil application of liquid manure were verified in 1988 on compacted heavy soils in the Lithuanian SSR, Vladimir province of the Russian Federation, the Kolkhida River lowlands in the Georgian SSR.

The seen of investigations in Lithuania was a perennial pasture on compacted heavy turf podzole soils. A ripper of 0.4 - 0.6 m operation depth and an 8 cu. m liquid fertilizer sprayer were employed. The both units were mounted on standard 220 kw wheel tractors. Conventional practice of liquid manure application before the experiment was surface spraying, which led to the pollution of water sources, atmospheric air and ground water.

In course of deep loosening liquid manure was charged by gravity flow from the sprayer container via flexible hoses to the distribution unit, mounted at the hind quarters of the ripper. Loosening occurs by turf layer elevation. Liquid manure is injected into expanded subsoil and the turf settles down, shreaded into transversal strips. The soil surface retains traces of the lateral posts of the ripper and ruptures of the turf and the pasture remains fit for further grazing.

By this mode of application of liquid manure the nutritives do not evaporate, pollute the environment and liquid manure is excluded from rivers and reservoirs. Liquid manure application rates are controlled by changing the operation velocities of the technological units. There are no precise application rate for liquid manure as yet, though existing recomendations, i.e. 100 - 150 cu.m/ha may be employed. Depending on the depth of loosening and nitrogene content in applied liquid manure these rates may be extended to 200 cu.m/ha.

Test-site studies, conducted by the researchers of the Lithuanian Academy for Agricultural Sciences and the author, showed that application of liquid manure into loosened heavy subsoil promoted regeneration of grass in pastures and meadows. In a month after the treatment young bright green grass appeared on the site surface. Soil productivity increased and the service period of pastures and meadow extended for several more years. The time for radical rehabilitation of pastures and meadows, including turf tillage, may be postponed. As a result comes considerable saving of general investments into rehabilitation and regrassing of pastures and meadows.

Under the impact of deep loosening compaction in the solum reduced by 15- 20%. By two mowings, adopted for the test site, the total hay yield (alfalfa and clover) after application of 100 cu.m/ha of liquid manure to the depth of 0.4.m was 72% higher, than that of the reference plot (7.17 t/ha) and amounted to 12.37 t/ha (yield increase 5.2 t/ha). Deep loosening alone without liquid manure applications) under the same conditions (0.4 m deep) increased the hay yield by 15%, totalling 8.28 t/ha (1.11 t/ha increase).

The experiment revealed, that deep loosening to the depth of 0.4 m with conjunctive application of liquid manures application provided hay yields of 8.19 t/ha from the first mowing and 3.46 t/ha from the second with respective increments of 3.85 t/ha and 1.35 t/ha. Application of liquid manure to the depth of 0.5 m under the same conditions resulted in 8.1 t/ha from the first mowing and 3.56 t/ha from the second with respective yield increase of 3.08 t/ha and 1.45 t/ha. The analysis of the obtained data showed, that deeper applications of liquid manure have a retarted effect on the yield of hay.

Subsoil liquid manure application tests were staged in 1988 under the same climate and soil conditions to investigate the impact of the treatment on forage beets growing. Loosening 0.5 m and 0.6 m deep employed conjunctive application of 100 cu.m/ha of liquid manure. By 0.5 m deep loosening the yield of beets reached 142,2 t/ha and of leaves - 76.7 t/ha. Surplus production totalled 46.7 t/ha and 37.4 t/ha respectively. By loosening 0.6 m deep the yields amounted to 162 t/ha for beets and 59.6 t/ha for leaves, respective increment being 66.5 t/ha or 70% and 20.3 t/ha or 52%.

As is seen the changes in the depth of loosening altered the balance of biological masses of the surface and the root part of forage beets.

The above data allow to draw conclusions on the feasibility and effectiveness of subsoil application of liquid manure in course of deep loosening of heavy compacted soils.

Similar results were obtained from heavy turf podzole soils: depth of loosening - 0.6 m, liquid manure application - 40 t/ha. Hay yield after the second mowing reached 25 t/ha, whereas on reference plot - 1.32 t/ha .

In cooperation with S.Sh.Babuadze the author investigated deep loosening operations on heavy soils with ortstein compacted layers in the Kolkhida River lowlands. The compaction is attributed to chemical bonds of composite materials: iron, alluminium, manganese. The bulk of the soil ranges 1.8 - 2.0 g/cu. cm. The compaction occurs at the depth of 20-30 cm. Farm crops and plants do not grow on such soils, as the roots can not develop to the deeper layers and they come out to the surface. Plants perish of overmoistening in spring (annual precipitation rate ranges 1600-2000 mm) and of lack of moisture in summer as the topsoil dries out due to intensive surface evaporation.

Intensive reclamation of such

potentially fertile soils employed loosening to the depth of 1 m and subsoil application of chicken manure into expanded layer.

Characteristic composition of chicken manure averages (in % to wet substance):

water	55- 60
potassium K_2O	0.9 - 1.0
nitrogene	1.3 - 1.5
lime CaO	0.4 - 0.5
phosphor P_2O_5	1.7 - 1.8
magnesium M O	0.7 - 0.8

Chicken manure application amounted to 40-60 t/ha. To provide this a 220 kw wheel tractor was rigged with a V-shape PT - 1.2 ripper, two 2 cu.m containers and an in-soil manure distribution unit. The operation of the ripper was started from an open collector ditch. After the treatment the fields were sown to soya seeds as a transient crop for deep structural development of the expanded soil profile. Soya was kept on the fields for two seasons and then used as a green manure. Subsequently the area was developed for citrus trees in relatively short period of time at low costs and inconsiderable labor expenditures.

The above research data allow to recommend the proposed method of intensive reclamation of compacted heavy soils by means of subsoil application of liquid and chicken manure for places, where large stocks of these valuble fertilizers are available. The yields of hay and root crops increase by 1.5-2 times. Employed technological means - V-shape rippers and pipe systems - are reliable in operation and provide loosening to the depth of 0.5-1.2 m with conjunctive subsoil application of liquid or chicken manure.

Land and Water Use, Dodd & Grace (eds), © 1989 Balkema, Rotterdam. ISBN 90 6191 980 0

Swine slurry disposal through agricultural use and purification

P.Martini
Istituto di Edilizia Zootecnica, University of Bologna, Italy

ABSTRACT: Environmental problems created by intensive pig farming are surveyed in a region in the Province of Modena which is characterized by: a high density of head per unit of area; hydrogeological fragility of the lithological substratum; an increase in the concentration of nitrates in the underground waters. Various projects were developed to give priority to agricultural use of the slurry and aim to minimize its negative impact on the environment. The most important projects were:
- centralized land spreading after effecting treatment to reduce pollutants.
- treatment of excess waste at centralized points for fertilization/irrigation use or disposal into public drains with terminal purifier.
An evaluation of unit and total costs incurred by the various solutions, is then given.

Résumé: Les problèmes d'environnement que pose la production intensive de cochons sont étudiés dans une commune de la Province de Modène, caractérisée par: une densité très élevée de têtes par unité de surface, la fragilité hydrogéologique du substrat lithologique et la concentration croissante des nitrates dans les eaux souterraines. Sur la base d'une prise de contact méthodologique fondée sur la priorité accordée à l'usage agronomique des reflux dans le but de minimiser l'impact environnant, nous avons élaboré plusieurs projets pour l'assainissement du territoire communal dont les aspects principaux sont:
- une distribution agronomique centralisée après avoir effectué des traitements appropriés au purin pour en réduire la capacité de pollution.
- un traitement des reflux en excédence dans des pôles centralisés pour les utiliser comme irrigation fertilisante, et la décharge dans des égouts publics pourvus d'un dépurateur terminal.
Nous fournissons ensuite une évaluation du coût unitaire et global des différentes solutions exposées.

Zusammenfassung: Die Umweltprobleme, die durch intensive Schweineerzeugung geschaffen werden, werden in einer Gemeinde der Provinz Modena erforscht, charakterisiert durch: sehr hohe Dichte der Stückzahl pro Oberflächeneinheit; hydrogeologische Brüchigkeit des lithologischen Substrats; Erhöhung der Nitratkonzentration im Grundwasser. Basierend auf einer methodologischen Annäherung, die auf die Priorität der agronomischen Anlage der Rückflüsse am Ende gegründet ist, um die Umweltbelastung zu minimalisieren, sind verschiedene Projektoptionen zur Wiedersanierung des Gemeindelands ausgearbeitet worden, von denen die wichtigsten folgende sind:
- zentralisierte agronomische Streuung, vorhergehende Ausführung von Schlammbehandlungen, um die Umweltbelastung zu verringern.
- Behandlung der überschüssigen Rückflüsse in zentralisierten Polen für die Düngungsanlage oder den Abfluss in die öffentliche Kanalisation, die mit einer Endkläranlage ausgestattet ist.
Darauf wird eine Schätzung der Einheits- und Gesamtkosten der verschiedenen dargelegten Lösungsmöglichkeiten geliefert.

1 INTRODUCTION

This paper discusses the viability and preliminary planning involved in abating pollution from swine waste in a district of the Province of Modena.

2 THE LAND/WATER RELATIONSHIP IN THE AREA SURVEYED

The district of Formigine is mainly concentrated in the part of the Modenese foothills geologically formed by flooding from the Secchia and Fossa river. The river-beds of these and other minor water courses in the area studied are made up of coarse material (highly permeable gravel and sand). This material is found both on the surface and at a moderate depth in the whole area. It forms a roof for the ample underground water-bearing stratum.

The above-mentioned rivers represent the main source of water for the water-bearing stratum and due to its prevalent stratigraphy the area of Formigine constitutes the preferential water source for the whole province of Modena.

There is a high concentration of pig farms in this hydro-geologically fragile area. Table 1 gives the data relative to live weight on average present and to excess numbers in view of land availability.

Live weight reared was deduced from parameter 1.2 mq/q (c.s.e. = head of stock equivalent to 80 kg).

Land available for spreading is mainly not connected to the farms and situated at fairly long distances from them.

Table 1. Pig farms in the Formigine area (MO), maintainable and excess numbers of livestock.

Farm (n°)	Method of disposal	Live weight reared (q)	Covered area (m^2)	Live weight maintainable (q)	(%)	Live weight in excess (q)	(%)
36	on land	73.240	91.550	39.250	54	33.990	46
1	in surface water	12.720	15.900	13.060	82	2.840 (sludge)	18
37		85.960	107.450	52.310	61	36.830	39

Table 2. Division of territory in the Formigine area for land spreading.

District areas (ha)	"PRG" law (o q/ha) (ha)	Zone A (o q/ha) (ha)	Zone B (20 q/ha) (ha)	Zone C (40 q/ha) (ha)	Live weight gross maintainable
4.698	1.092	666	2.294	646	71.720

Table 2 shows the division of territory into zones for land spreading: the situation is rather precarious and confirms the difficulties inherent in the area's geology.

Along side pig farming in the area surveyed, other types of intensive animal husbandry and viticulture is practised (the population of dairy cows is estimated at about 22.000 q), barley, wheat and fruit-growing as well as industrial cultivation are also carried out.

This environmental situation justifies the worry that the disposal of human and agricultural waste, slurry and chemical fertilizers could, in the long run, compromise the quality of underground water sources: the consequences of which would be dramatic for the whole of the Modenese plain which relies on these for its water supply.

A survey of sample wells in high and middle plain lands started in 1978 by the local health authorities showed that the increase of nitrates, although occurring with seasonal fluctuations, is constant.

For a vast area including the populated area of Formigine, all the results show a constant concentration of nitrates reaching the legal limits for drinking water (50 mg/l EEC directive no.80/778 regarding the quality of water for human consumption).

The remaining area comprises underground water with high concentrations of nitrates of analogous origin, which the river, poor in nitrogen compounds, partially dilutes. The decreased fluvial dispersion of the last 30 years (due to lowering of the river-bed) has caused an increase in the importance of the topographical surface and minor river networks with their higher nitrate content (chemical fertilization, animal and human waste), as a source of imput for the water-bearing stratum.

Whereas human waste is channelled and kept separate from water courses, the problems posed by other forms of pollution are still unsolved.

Solutions designed to clean up the area should take account of urban and industrial waste disposal and utilization of the water.

In the survey zone it is important to limit the use of underground water courses for drinking water, and to consider the river Secchia, both as an alternative source of water for various uses (S.Michele acquaduct) and a source of imput for the underground water-bearing stratum (this can be achieved by eliminating erosive elements and possibly reestablishing various water profiles.

3 METHODOLOGICAL APPROACH

Priority is given to agricultural use of waste (as fertilizer) but damage to the environment must be minimal.

The Technical regional directives for regulating farm animal waste disposal (Art.1 L.R. 42/86) were used as a guide-line for our projects.

Furthermore a district regulation has been proposed regarding the establishment of pig farms in keeping with the projected solutions.

Utilizing the hydro-geological data available, a map of environmental vulnerability to spreading was devised.The subdivision into classes A,B and C (see table 2) was carried out by assigning a score relative to the 4 parameters given below, as dictated by the technical directives.:

- Surface lithology (1.30-1.80 m from the ground)
- Thickness of permeability in the first 50 m of depth
- Depth of the first utilizable water level
- Type of water stratum.

The technical directives allow an increase in livestock number per hectare (from 20 to 40 q/ha in zones B, from 40 to 60 q/ha in zones C) when slurry is treated according to line 2 table 3: that is when concentrations of nitrogen and phosphorous in the waste are reduced to a level of about 50%.

4 SUGGESTED SOLUTIONS

The proposals are advanced in order of increasing complexity, from land spreading (on the farm itself or by a centralized farmers' society) to complete purification

Table 3. Methods of treatment examined for agricultural use.

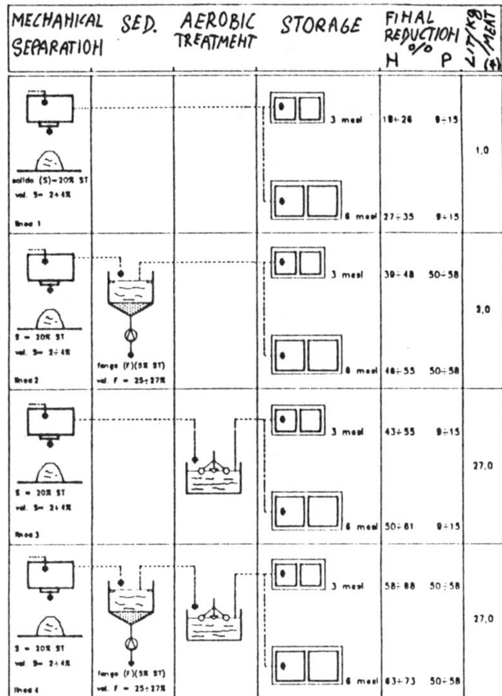

(•)The cost of removing solids from thickened and its amortization not considered.

for disposal into surface waters.

The role of this farmers' society becomes increasingly important with the increasing complexity of the solutions. The most interesting solutions, given the considerable excesses in live weight and the technical, managerial and economic difficulties inherent in complete purification of swine waste would appear to be the intermediate ones.

These are choices which involve tackling the slurry on the farm up to the point where pre-treatment is still economically viable (based on land availability) and then treatment of excesses at centralized stations managed by a cooperative.

The entity of treatment at these stations would vary according to the ultimate

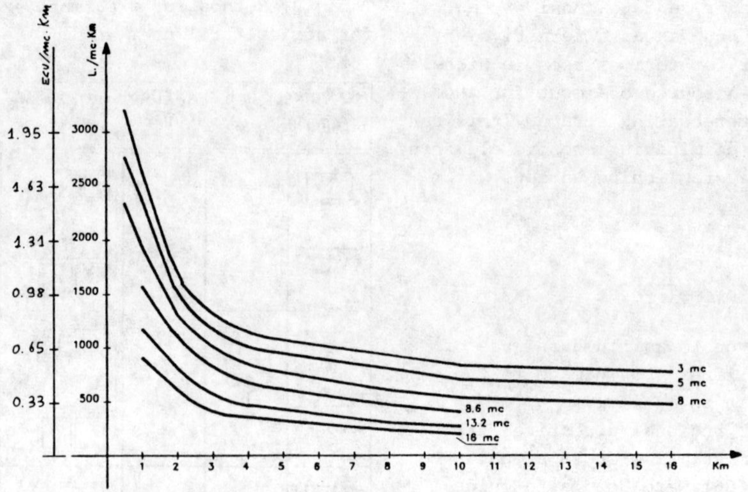

- Lir/mc x km cistern capacity 3,5,8,8.6,13.2,16 mc.
- Inclusive of cost of loading, transportation, spreading and returning under normal conditions.

Fig.1 Transportation and spreading costs using tankers with spray irrigator (1987)

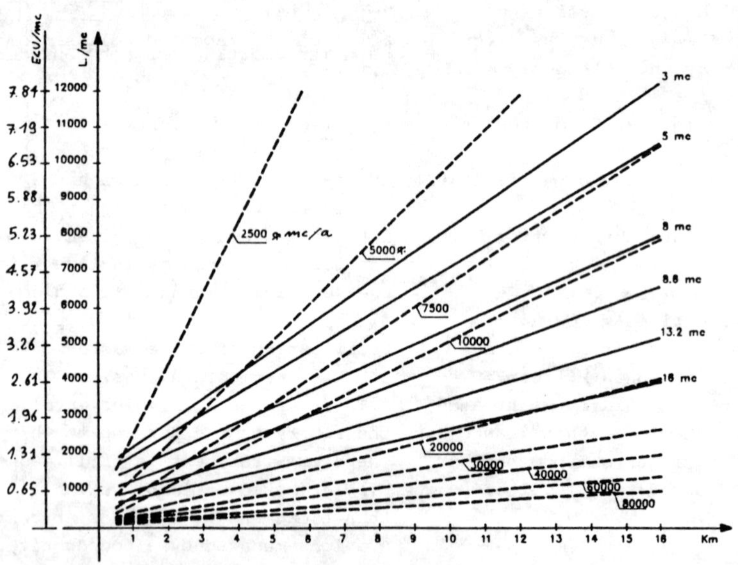

— using cisterns with wheels
-- using underground pipes

Fig.2 Cost of transportation of pigs only (1987)

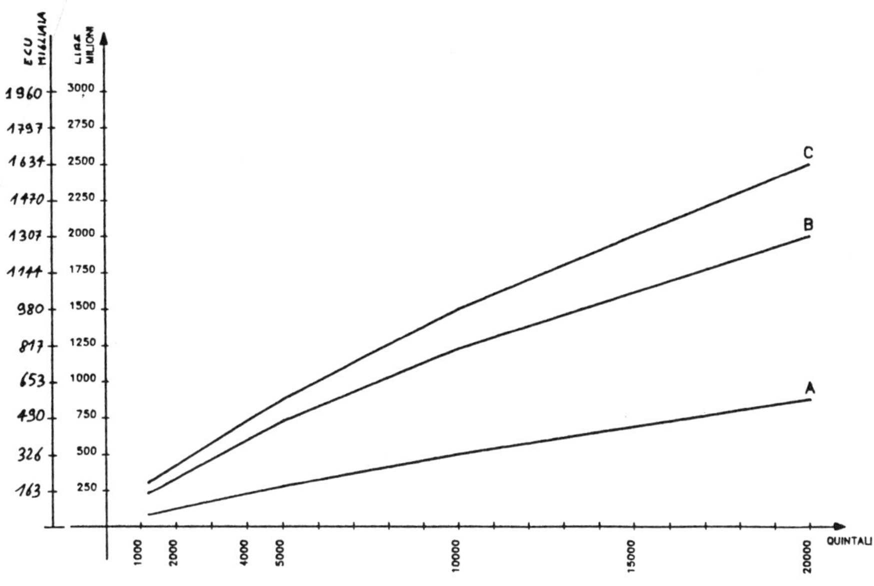

A = pre-treatment
B = partial treatment
C = complete treatment

Fig.3 Options for treatment of pig slurry - Investment costs - (1987)

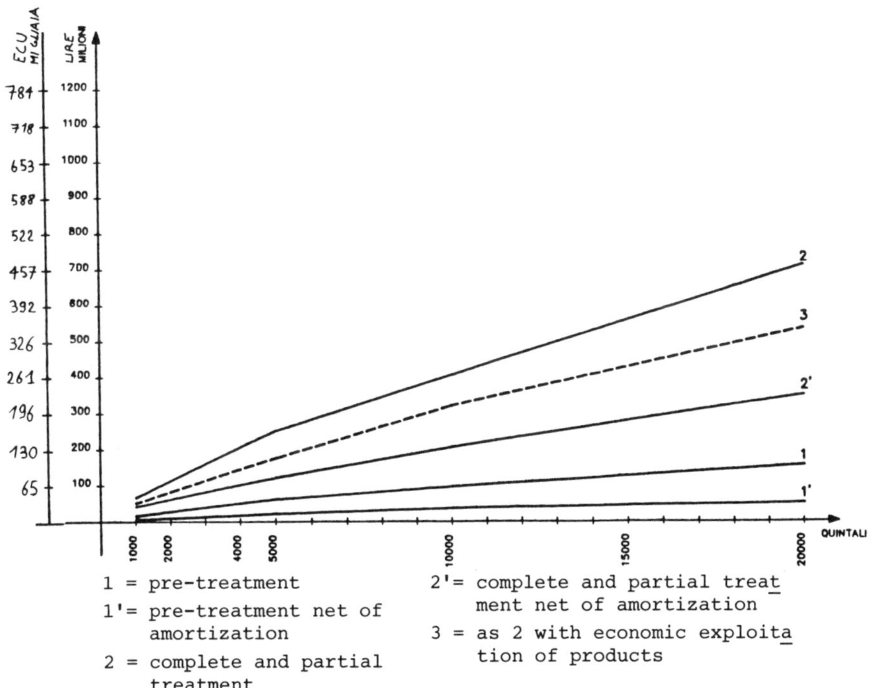

1 = pre-treatment
1'= pre-treatment net of amortization
2 = complete and partial treatment
2'= complete and partial treatment net of amortization
3 = as 2 with economic exploitation of products

Fig.4 Options for treatment of pig slurry - Management costs - (1987)

destination of the waste. There are two suggestions in relation to this:

1. Fertilization/irrigation use

Centralized stations are established in a barycentre position with respect to the area served. The slurry is treated and pipes leading from the station then take it to the areas where it is required. The viability of distribution for fert-irrigation of the light fraction of the slurry is, in turn, dependent on the existence of suitable conditions required for normal irrigation.

However for various reasons the fert/irrigation solution is today still hindered by considerable uncertainties in legal interpretation. A similar problem of legal interpretation arises for agricultural use of residual sludge produced during the purification treatment.

2. Disposal of waste in public drains with terminal purifiers

When the above mentioned requirements can not be satisfied, we should consider the viability of partial treatment for eventual disposal in public drains with a terminal high-powered purifier (treating the urban waste water of Modena).

Degree of purification is much more clearly definable (even from a legal standpoint), since the body which manages the purifier establishes both the acceptable qualitative limits for disposal (Tab.C L. 319/76 or less) and the costs of services rendered.

It is, then, possible to take into account a technical and economic optimization of the solution overall when identifying the level of purification required for admission into the drains.

The large incoming flow in the Modena purifying plant redresses the carbon/nitrogen inbalance in the treated swine slurry.

A preliminary project, based on this latter solution, was formulated and its realization financed by state funds.

The project involves transporting farm wastes to two treatment centres by underground pressure.

The average characteristics of the incoming waste are given in table 4 and the characteristics of the treated outcoming slurry should comply with the national legal limits or those established by the managing body of the purifier as discussed before.

Table 4. Data on pollution parameters on entry of the purifying plant for disposal in the drains planned by the District of Formigine (1988).

Imput data - east plant
(n.c.s.e.) no. of 80 kg head = 50.000
average capacity = 600 mc/day

Parameter	Specific imput (gr/day x c.s.e.)	Total imput (kg/day)	Concentration (mg/l)
Co (BOD)	180	9.000	15.000
Co (COD)	425	21.250	35.500
SS tot	425	25.500	35.500
SS V	75% SS tot	15.950	26.600
TKN	35	1.750	2.900
P	15	750	1.250

Imput data - west plant
no. of 80 kg head = 60.000
average capacity = 720 mc/day

Parameter	Specific imput (gr/day x c.s.e.)	Total imput (kg/day)	Concentration (mg/l)
Co (BOD)	180	10.800	15.000
Co (COD)	425	25.500	35.500
SS tot	425	25.500	35.500
SS V	75% SS tot	19.100	26.600
TKN	35	2.100	2.900
P	15	900	1.250

The plants provide aerobic slurry treatment (including tertiary treatment) and anaerobic sludge treatment with dehydration of the digested sludge to allow maximal agricultural exploitation. Each farm carries out the separation of the coarse, solid fraction, with the combined aim of using an organic product in the fertilizer and reducing pollutants on plant entry. The treated slurry coming out of the plant is also suitable for fert/irrigation use and storage containers already present on the farms may be used for stocking excess.

5 COSTS

The cost of the various solutions regarding treatment of pig slurry are calculated on the basis of standard structural operations.

For specific costs, the following general criteria were considered:

1. Labour costs involved in transportation and spreading are considered part of management costs and refer to average data for normal conditions of agricultural practice (fig.1).

On a territorial scale and where natural slopes exist, it could be economically viable to transport animal waste through underground pressure pipes (Fig.2).

Systems of transporting slurry through pressure pipes have been used in various countries and in Italy, farms have used normal irrigation pipes (PVC, polyethylene, etc) and centrifugal pumps have also been realized.

2. Scale economy has been kept in mind as consequential to the power increase in the plants considered (Fig.3 and 4).

3. Management costs (Fig.4) involved in partial treatment are not significantly less than those involved in complete purification due to the considerable incidence of service cost incurred by the drains and terminal purifier.

4. Investment and management costs (net of amortization) of the centralized treatment plants proposed in the preliminary project mentioned (table 5) are intended to refer to a contract and involve a high investment per head of stock due to the territorial nature of the project (pipeline network, expropriation etc.).

Table 5. Investment and management costs of purification plants for disposal into drains as projected by the District Council of Formigine (1988).

Summary of costs in thousand lire

Item	East plant	Completed plant
A) work offers	11.473.183.000	21.626.840.000
B) money for district use		
- V.A.T.	372.770.000	702.543.000
- Technical costs	632.000.000	1.197.000.000
- Expropriation and servitude	300.000.000	600.000.000
- Non material goods	20.000.000	40.000.000
- Unexpected technical costs	900.000.000	1.680.000.000
- Revision of prices	575.000.000	1.075.000.000
- Rounding off	47.000	1.616.000
C) Total	14.273.000.000 (9.328.000 ECU)	26.923.000.000 (17.596.000 ECU)

The following table shows management costs of the two treatment stations (west and east) corresponding to lots 1 and 2 respectively and expressed in millions/year.

cost item	lot 1	lot 2	Total
Staff	231	198	429
Electrical energy plant	340	290	630
Electrical energy network	15	15	30
Disposal tax	80	66	146
Chemical agents	70	65	135
Maintenance	80	70	150
Analyses	5	5	10
Fuel and lubricants	5	5	10
Total	826	714	1540 (1.006.535 ECU)

It can be seen that the incidence of management costs does not exceed 100 lire/kg (0.065 ECU) saleable meat overall, net of amortization.

6 CONCLUSIONS

The indications arising from the study are essentially concerned with:

1. Maximization of agricultural use after pre-treatment recommended by the technical regional Directives.

2. Adoption of a district regulation with the aim of rationalizing husbandry techniques and water consumption and reducing wastes.

3. Institution of centralized assistance to optimize transport and spreading techniques and computerized management of applications for authorization to dump the waste.

4. Reduction of environmental risks concerning the "light" fraction of treated slurry in view of fert/irrigation utilization or disposal in drains.

5. Agronomical exploitation of the residual fraction through dehydration, stabilization and compounding.

REFERENCES

Bonazzi, G. 1988. Tecnologie innovative per lo smaltimento dei liquami suinicoli. Rivista di Suinicoltura 5.

Bonazzi,G.,Cortellini,L.,Ferrari,A. 1984. Indagine sulla diffusione dei depuratori su scarichi suinicoli in Emilia Romagna. Rivista di Suinicoltura 9.

Bonazzi,G.,Martini,P.P. 1984. Tecniche di trattamento dei liquami suinicoli per lo adeguamento allo scarico. Informatore Zootecnico 13.

Ferrari,A.,Bonazzi,G. 1987. Il trasporto dei liquami zootecnici con condotte in pressione. Informatore Agrario 1.

Martini,P.P. 1988. Smaltimento di reflui suinicoli tra impiego agronomico e depura zione. Convegno Nazionale AIGR, Università degli Studi di Sassari.

Martini,P.P. 1988. Gestione dei reflui suinicoli a fini agronomici e fertirrigui: aspetti tecnici e normativi. Convegno EURECO: Agricoltura e Ambiente, Modena.

Martini,P.P. 1988. Lo smaltimento dei liquami zootecnici nella normativa nazionale e regionale. Genio Rurale 10.

Sangiorgi,F.,Balsari,P.,Bonfanti,P. 1986. Reflui zootecnici: possibili trattamenti in vista dell'impiego agronomico. Bologna: Edagricole.

Tournon,G. 1984. Utilizzazione agricola dei liquami zootecnici mediante impianti consortili di liquamazione-fertirrigazione. Rivista di Suinicoltura 11.

Comprensorio di Modena 1981. Piano per la

tutela e l'Uso delle Risorse Idriche.
Atti del Convegno: "Scarichi suinicoli, problemi tecnici e normativi", CRPA, Reggio Emilia, maggio 1982.
Comune di Modena-Ufficio agricoltura 1986. Studio di utilizzazione di acque superficiali ad uso irriguo nella zona compresa fra le reti del canale S.Pietro e del canale Maestro
Comune di Formigine-Provincia di Modena 1987. Studio di fattibilità per il risanamento del territorio comunale dall'inquinamento da reflui suinicoli.
Provincia di Modena-Agricoltura Ambiente. 1988. Quaderni di tecniche di protezione ambientale. Bologna: Pitagora.
Regione Emilia-Romagna 1988. Progetto preliminare per il trattamento dei reflui zootecnici nel territorio del comune di Formigine ed aree limitrofe.

Land and Water Use, Dodd & Grace (eds), © 1989 Balkema, Rotterdam. ISBN 90 6191 980 0

Utilization of organic wastes for improving grasslands and pastures

O.S.Marchenko & N.N.Harenko
All-Union Research Institute for Mechanization of Agriculture (VIM), Moscow, USSR

ABSTRACT: Pecularities of utilizing liquid organic fertilizer (wastes of cattle production complexes) with the aim of improving soil fertility and productivity of meadows and pastures are considered. Description of tools of the units ABB- Ф - 2,8 (USSR) and M-4 "Alpha-Laval" (Sweden) for applying liquid organic fertilizers into soil of meadows and pastures is given. The usage of these units both at spring and fall extra fertilization (the rate of fertilizer application is up to 120 t/ha) ensures increasing meadow and pasture productivity by 10-30% in comparison with fertilizer application on soil surface.

Utilization of liquid organic wastes (liquid manure) of cattle production complexes is an important link of improving soil fertility, productivity of meadows and pastures and other relative to the farm greenlands.

In practice some variants of utilizing liquid manure as fertilizers for meadows and pastures are used.

1. Separation of liquid manure into fractions. Solid fraction of liquid manure is distributed on the surface of meadows and pastures with the help of spreading trailers and liquid fraction of manure is delivered to an irrigation network or after appropriate treatment it is used for repeated hydrowash-off.
2. Liquid manure distribution on the surface of meadows and pastures with the help of tanker-spreaders which at present are abundant.
3. Application of liquid manure into the soil of meadows and pastures with the help of special tillage implements.

Every varian has its advantages and shorcomings. Application of liquid manure into soil is the least understood form the point of view of its effort on soil fertility and environment.

First of all liquid manure for application into soil must be properly prepared; its moisture content must be not less 92%, it must not contain mechanic inclusions the size of which is more 10 mm, it must contain excitants of especially dangerous diseases and seeds of quarantine weeds.

In comparison with liquid manure application into arable land additional requirements are imposed upon liquid manure application into the soil of grasslands and pastures; they are: minimal damage of sod at operation and penetration of tools into soil, exclusion of liquid manure going out from soil and its distribution on the sod surface.

In comparison with fertilizer application on soil surface fertilizer application into soil demands large labour expenditures and costs and its real efficiency may be defined with due regard for corresponding increasing productivity of grasslands and other positive factors.

With this aim we have studied processes of liquid organic fertilizer application on soil surface and into soil, compared agrotechnicalindices of tools for fertilizer application into soil used on our country unit ABB - Ф - 2,8 and on the Swedish unit M-4 "Alpha-Laval".

Tools of each gang of ABB- Ф - 2,8 (fig. 1a) include a disk coulter, a feeding device in the form of a V-shaped sweep with a branch and a covering roller. Microrelief gauging in vertical plane is carried with the help of a spring-loaded parallelogram mechanism. A gang is hinged and has the possibility to move in horizon plane; due to these factors the unit can by-pass little obstacles and prevent tools from damaging. At operating the disk coulter cuts sod at the depth of 12 cm and the feeding device in the form of a V-shaped sweep which follows the disk coulter undercuts sod at the depth up to 18-20 cm, lifts it, forming a slot for feeding device tine passing. Liquid fertilizers which are distributed into soil in the form of band (width - 30-35 cm) are supplied through the branch into the space formed under a sod layer. After passage of the feeding device the sod layer settles down and closes up, in addition the covering roller presses it, covering the slot in once and for all. The depth of fertilizer application is controlled by moving the device along grooves of arms with the help of adjusting screws as well as moving tines of the feeding device in guides. The unit is designed for work on grasslands and pastures, the hardness of which is not more 3,0 MPa and when moisture content of soil is not more 30%.

After the unit ABB- Ф -2,8 passes, a slot is formed in soil; its cross-section is given on figure 2. The significant width of undercutting a sod layer leads to formation of the space which is sufficient for intersoil and uniform distribution of maximum rates of fertilizer up to 120 t/ha. The width of damaged sod in its top part does not exceed 2-3 cm.

The tools of the Swedish unit M-4 "Alpha-Laval" (fig. 1b) are characterized by the fact that a tine in the form of a chisel with not large undercutting sod layer flaps in its bottom part is installed on hinged spring-loaded gangs after the disc coulter; the covering roller is absent. The depth of fertilizer application is controlled by changing supporting ski and moving tines of ripping teeth along guides. Depending on change of load on tools (change of soil hardness, surface unevenness), each gang has the possibility to move in the vertical direction. However, it upsets stability of tools.

After the passage of the unit M-4 "Alpha-Laval" an open recess in soil is formed; its cross-section is given on figure 2b. The recess width in its top part is about 5 cm, in the bottom part is about 10 cm. The recess volume gives the possibility to distribute in it up to 120 t/ha. However, even at unsignificant level of region slope we can observe going out on soil surface and run-off of fertilizers in low spots.

Field tests on defining agronomic effectiveness of liquid manure application into soil were carried out on grasslands, grass stand of which contains 60% of fescue, 20% of smooth bromgrass and 20% of timothy (humus content - 2,1-2,4%, total nitrogen content - 0,11-0,13%, P_2O_5 - 18-24 mg/100 g of soil, K_2O - 8-10 mg/100 g of soil, pH - 5,4-6,0).

Liquid organic fertilizer application on soil surface and into soil was compared with a control test where fertilizer was not used: test I - in spring; test 2 - in autumn; test 3 - next spring. Minimum rate of fertilizer application was 60 t/ha; maximum rate of fertilizer application was 120 t/ha; as far as 2 and 3 tests concerns - only 120 t/ha.

Prepared cattle manure without litter moisture content of which was 91-96% was used in the tests (table 1).

In test I signigicant gain from fertilizer application into soil has been marked; the first year at applying 120 t/ha the gain was by 10-30% more in comparison with fertilizer application on soil surface and 2-2,5 times higher in comparison with the control test (without fertilizer application).

In tests 2 and 3 both fall and spring fertilizer application was effective. On one hand, it may be

Table 1
Content of Cattle Manure without Litter

N of field test	Dry matter	Total nitrogen content	NH_4	P_2O_5	K_2O	pH
			% in wet mass			
1	4,0	0,25	0,09	0,17	0,28	8,10
2	8,9	0,37	0,17	0,24	0,46	7,80
3	6,0	0,21	0,07	0,15	0,37	7,75

explained by large content of nutrients in manure applied in autum (table 1). However, at fall application of liquid manure (test 2) significant gain was received during the first mowing while at spring application of fertilizer a real gain during the first mowing was not observed. This fact is explained by negative influence of undercutting plant roots in spring, while root damage in autumn does not probably influence grass stand productivity during the first mowing. It is advisable to go on investigations with the aim of choosing the best width of the undercutting part of tools depending on different types of grass stands. It should be mentioned that an essential shortcoming of the tools of M-4 "Alpha -Laval" is liquid manure going out on soil surface at large rate of fertilizer application; this is impermissible in accordance with agrotechnical requirements in relation to contamination by helminthes.

CONCLUSIONS

1. Liquid manure application into soil increases productivity of grassland by 10...30% in comparison with its distribution on soil surface; it is advantageous to apply increased rates of liquid organic fertilizers (up to 120 t/ha).

2. It is established that at fall fertilizer application into soil gain occurs already during the first mowing; at spring fertilizer application gain occurs only from the second mowing.

3. At fall fertilizer application the units ABB- Φ -2,8 have some advantage over the units M-4 "Alpha-Laval", at spring fertilizer application we observe the other phenomenon.

4. The form and size of a slot formed by tools of the unit ABB-Φ -2,8 give the possibility to distribute reliably up to 120 t/ha into soil; while the slot formed by the tools of the unit M-4 "Alpha-Laval" does not exclude contamination of environment due to liquid manure going out on sod surface and its runoff in low spots.

5. It is necessary to go on investigations on defining optimum width of undercutting parts of tools for liquid organic fertilizer application into soil in different types of meadows and pastures.

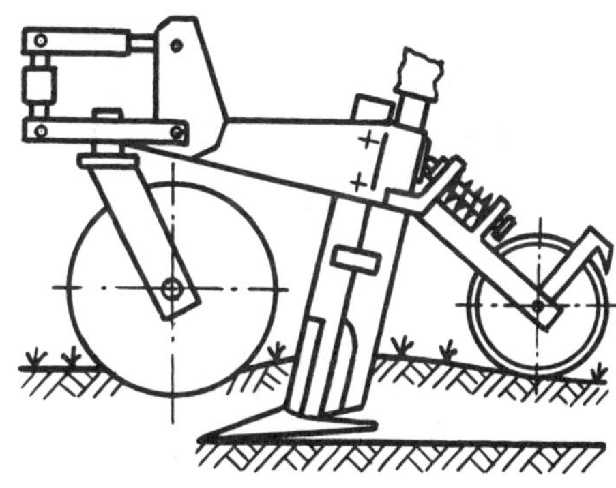

Fig. 1. a) Scheme of installing the tools of the unit ABB- Φ -2,8

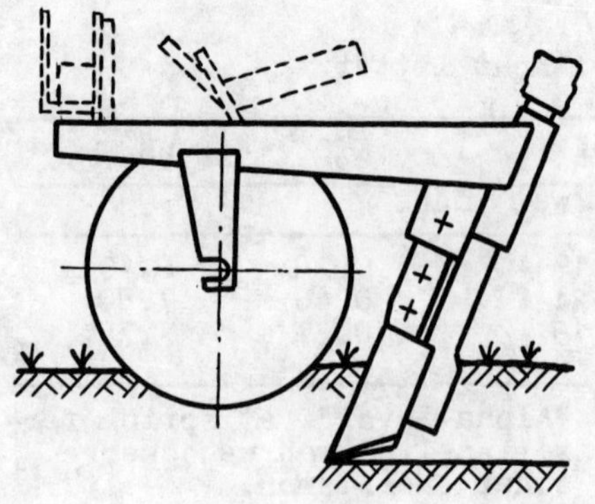

Fig. 1. b) Scheme of installing the tools of the unit M-4 "Alpha-Laval"

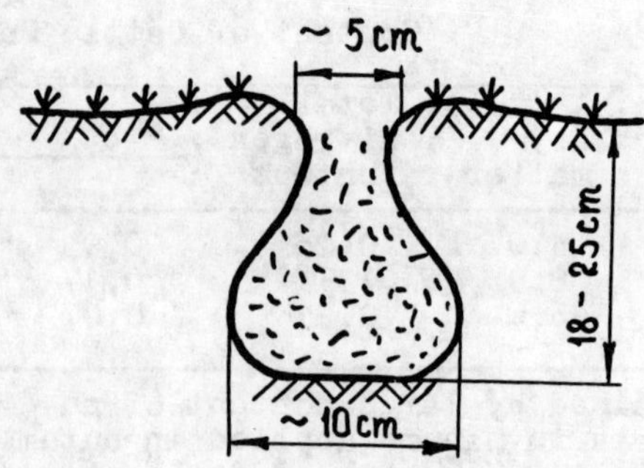

Fig. 2. b) Cross-section of the recess formed by the tools of the unit M-4 "Alpha-Laval"

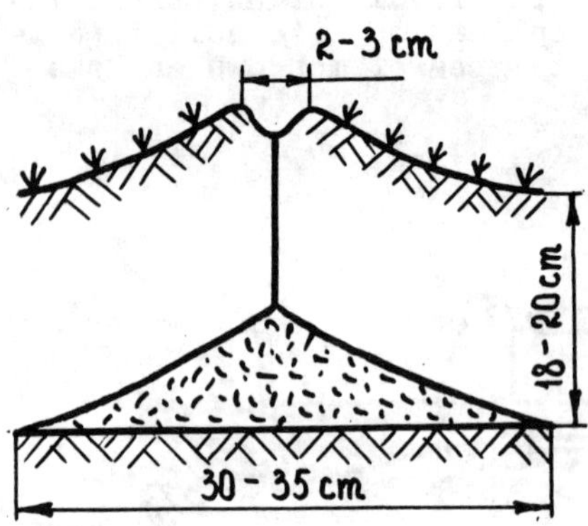

Fig. 2. a) Cross-section of the recess formed by the tools of the unit ABB- Ф -2,8

Land and Water Use, Dodd & Grace (eds), © 1989 Balkema, Rotterdam. ISBN 90 6191 980 0

The problems of utilizing the wastes of livestock production complexes and non-grain part of cereals crops

V.M.Kriazhkov, N.M.Marchenko & E.V.Zhalnin
All-Union Research Institute for Mechanization of Agriculture (VIM), Moscow, USSR

ABSTRACT: The most effective ways of utilizing wastes of livestock production complexes and non-grain part of cereal crops are considered. The technologies and sets of machines for applying solid and liquid fertilizers with the aim of improving soil fertility at tilling, at row plant growing and improving greenlands are suggested. Three variants of utilizing straw and chaff as raw material for preparing feed, for soil mulching with the aim of moisture preservation and ploughing in straw as fertilizer are considered. Technologies and sets of machines for every variant of utilizing non-grain part of yield are considered.

The objective necessity in satisfying population growing demands for agricultural products and demands in industry for raw materials puts a task of systematical improving soil fertility in the forefronts and on this base the task of raising farm crops yielding capacity. A clearly defined system of commercial and organic fertilizer application is a deciding factor in solving the problem of soil fertility improvement.

In the connection with intensive steady raising of the rate of commercial fertilizer production and application thet tendency to depreciate the role of organic fertilizer took place. Meanwhile as early as 1930...1939 academician D.N.Prianishnikov pointed out that no matter how great commercial fertilizer production might be, manure would never loose the importance as a major fertilizer of agriculture. Organic fertilizer is is not only the main source of soil humus reproduction and a main link of recirculating nutrients for plants. It is known that the most part of nutrients for plants consumed by animals with feed is not assimilated. For example, 40...50% of organic matter of feed, 90% of nitrogen, up to 80% of phosphorus, 95...98% of potassium, 70...85% of calcium and the most part of microelements are excreted with cattle manure. Besides, it is established that simultaneous application of commercial and organic fertilizer favours increasing commercial fertilizer pay-back by 20...50%.

In the last few years in connection with rise in oil and mineral raw material prices the interest to organic fertilizer has increased in all countries. Effective application of organic fertilizer gives the possibility to reduce considerably energy expenditure at producing commercial fertilizer and especially nitrogen fertilizer. For example, to produce 1 t of nitrogen fertilizer we are to use energy released at burning 1 t of oil. Meanwhile 100 t of organic fertilizer containes 1 t of nitrogen.

At present in the USSR more than 1 billion tons of organic fertilizer is applied and by 1980 the volume of its application will increase up to 1,5 billion tons.

In connection with the changeover to production of livestock products on the industrial base involving housing system without litter liquid manure accumulation has been sharply increased. Liquid manure accumulation in large quan tities contaminates soil, adjacent water reservoirs, air and gives the possibility for emergence and spreading helminthosis and infection diseases. Methods of disinfection which are worked out at present are power-intensive and that is why they will not be widely used.

Removal and application of solid as well as liquid organic fertilizers is a serious transport-technological problem. 25% of total transport expenditures in agroindustrial complexes is due to the organic fertilizer. In this case expenditures connected only with fertilizer delivery to fields are up to 60% of total labour expenditures and costs for fertilizer production and application.

In perspective soild organic fertilizer will be applied on the surface of soil and subsequent covering the fertilizer with the help of tillage implements according to the straight-run scheme (farm-field) and according to the scheme with intermediate clamping (farm-clamp-field) will follow it. According to the scheme "farm-field" organic fertilizer is prepared and accumulated in relative to the farm storage or on a special relative to the farm platform. With the aim of improving physical-technological properties of organic fertilizer, its desinfection and reduction of germinating power of weed seeds biothermal treatment was included into the technology of organic fertili zer preparations.

The technology of preparing peat-manure and peat-droppings composts finds expanding application. At present the technology of organic-nimeral composts preparation is being mastered.

Application of commercial fertilizers such as phosphorite meal gives the possibility to reduce losses of manure nitrogen. Stationary and mobile units for compost preparation are being worked out; (the output is 80...100 t/h and 50...60 t/h, respectively). The high-productive loader ПНД -250 of continuous action was tested. The distinguishing feature of this machine is a rotary tool. The loader ПНД -250 can be sused both for breaking organic fertilizers with the aim of intensification of biothermal process and for compost preparation. The high productive sets of machines ПРТ-10, ПРТ -16 and МТТ-23 are developed and manufactured with the aim to deliver and apply fertilizers.

According to the scheme "farm-clamp-field" fertilizer is previously accumulated and prepared in a relative to the farm storage, and then it is periodically delivered to a field edge and unloaded into clamps for ageing and storing up to the time of its application. This technology gives the possibility to use machines equipped with low-pressure tyres and auxilliary devices ensuring high-quality fertilizer spreading along all the treated area. The usage of replaceable units on self-propelled machines of chemization for organic fertilizer applying shows promise.

Liquid organic fertilizer is mostly applied on the surface of soil. Fertilizer loaded from a relative to the farm storages is delivered to fields and applied on field surface with the help of tanker-spreaders with subsequent ploughing in. The technology of fertilizer delivery to a field along tubes and its distribution at surface irrigating finds expanding application in large farms and complexes. The technology of fertilizer pumping along tubes to field hydrants and its distribution by tanker-spreaders shows promise.

Development of technologies and means for mechanization of liquid organic fertilizer application directly into soil looks promising. Application of fertilizers into soil gives the possibility to reduce nitrogen losses, to increase the efficiency of fertilizer application at the cost of their distribution in the root zone, to reduce contamination of the environment.

Technologies of preparing, loading and applying liquid organic fertilizer into soil of grasslands, pastures and at row cultivation of row fodder crops were developed and tested, the machines for this purpose were created. The technology of fertilizer preparation involves cutting organic inclusions in liquid manure and homogenization. Delivery of liquid manure to field compensators with the aim of preparing solutions is carried out by mobile tanker-spreaders. Application of the solutions prepared on the base of liquid organic fertilizer, ammonia water and other chemicals is carried out with the help of special machines.

The technological set consisting of a loader-cutter such as ПНЖ -250, 3...4 transport-technological machines such as МЖТ-10, a field compensator and 2...3 special machines such as АВВ-Ф - 2,8; АВМ- Ф - 2,8 ensures liquid fertilizer application into soil at 10...12 ha per day output, The solution dose is 50...150 kg/ha (by nitrogen). The depth of fertilizer ploughing in is 18...15 cm. The tests showed that at fertilizer application into soil fertilizer pay-back is increased by 15...20%.

The technology and a set of machines for applying large doses (100...400 m^3/ha) of liquid organic fertilizer at ploughing were developed. Fertilizers are applied at the depth of an arable layer in 2...3 horizons. It has been established that from technological and energetic view point fertilizer application into soil is most effective at fulfilling the operation simultaneously with tillage.

According to this technology fertilizer delivery to fields is carried out by mobile transport vehicles and application of fertilizers under tillage implements is carried out by machines which move simultaneously parallel to each other. With the aim of liquid manure application alternate fields of ploughing in are being made at farms and complexes. The size of alternate fields of ploughing in is defined by volume and dose of liquid manure intended for applying manure according to this technology.

Rational accumulation, preparation and utilization of organic fertilizer is a key link in the system of technologies of production of agricultural products without wastes. Organic fertilizer has come into importance in connection with urgent substantiation of the perspectiveness of biodynamic system of crop farming free from application of commercial fertilizer, pesticides and other means of plant protection from diseases, insects and weeds.

Biodynamic system of crop farming based on new principles and achievements of scientific-technical progress gives the possibility to produce food and forage of high quality free from pesticides and other detrimental substances.

Utilization of non-grain part of cereal crops is another problem which is considered.

Non-grain part of cereal crop yield is 1,5 higher than the grain yield.

Average annual production of straw and chaff in the country is in the range of 250...300 mln t at average level of laying in 90...100 mln t for animals.

At present the main method of harvesting non-grain part of cereal crops at available park of straw harvesting machines is a shock technology (80%). Swath harvesting technology when bales are formed makes up 5% and flow-line harvesting when straw is cut by special devices of harvesters makes up 15%.

With the aim of complex mechanization of harvesting non-grain part of cereal crops necessary sets of suitable machines and attachments to grain harvesters gathering non-grain part of cereal crops according to the mentioned technologies are incorporated into. The main trend of technical progress in harvesting non-grain part of cereal crops by a shock method is increasing the density and mass of shock; at a swath technology with baling the main trend is development of high-productive machines and means for transporting and keeping bales in storage; at a flow-line technology when straw is cut the main trend is development of an automatic hitch and large transport facilities, improvement of harvester cutting devices.

One of trends of rational utilization of non-grain part of cereal crop yield is its utilization as a fertilizer for improving soil fertility and controlling wind erosion. It is established that in this case climate and zone soil peculiarities are the deciding factors.

In the conditions of temperate climate it is advantageous to carry out straw spreading on the surface of soil. In wet conditions, in the presence of plenty of weeds and in arid zones it is advantageous to carry out shallow straw ploughing in. In the zones of water and wind erosion it is advantageous to carry out straw covering simultaneously with soil and stubble mulching at graded tilling by antierosion implements.

Numerous experiments carried out in our country and abroad showed that good results were achieved at simultaneous applying straw in the mixture with liquid dung, liquid manure and ammonia water.

Another trend of straw utilization is its usage as forage; in this case straw is cut and mixed with other types of feed, enriched by vitamins and mineral substances. With the aim of preparing balanced forage of full value we also carry out thermal-chemical and chemical treatment of forage to improve taste properties and to increase straw and chaff digestion. The first trend is preparation of different feed mixtures. For this purpose the cut straw (2...3 cm) is mixed with green concentrated feed, silo, dried and liquid beet cutting, melasses and enriched by mineral and synthetic feed additions.

The second trend is preparation of feed mixtures with thermal-chemical treatment of straw which is mixed with components according to the first variant.

The third trend is preparation of multicomponent and full-ration feed briquettes and granules on the base of straw and chaff at farms where there are high temperature dries and equipment for feed peletting and briquetting.

Straw is treated by caustic natrium and lime, soda ash, ammonia water or liquified ammonia.

In the last few years the following method of improving feed quality of straw is practised on a large scale. It consists in barohydrothermal treatment of straw in an autoclave at the pressure up to 6 GPa that gives the possibility to improve straw nutritive value (from 0,2 up to 0,35 feed unit) and to increase sugar content up to 8-10%.

The development of mechanized feed preparation gives the possibility to receive full-ration balanced feed prepared on the base of agriculture wastes (straw and chaff, cereal and legume crops, sunflower heads, corn and sorgo stems and corn cobs).

The characteristic feature of all the farms is the quest for increasing in animal ration the quantity of feeds from by-ptoducts of agriculture; it concerns also the farms with large forage reserve based on sowing perennial and annual grasses as well as utilization of natural meadows and pastures.

Land and Water Use, Dodd & Grace (eds), © 1989 Balkema, Rotterdam. ISBN 90 6191 980 0

Enhanced biological sealing for earthen manure storages

A.E.Ghaly
Agricultural Engineering Department, Technical University of Nova Scotia, Canada

ABSTRACT: The possibility of developing an enhanced biological sealing mechanism for earthen manure storages was investigated. A bacterial specie was selected and used for the production of insoluble polysacharides which were effective in plugging the soil pores. The developed mechanism was very effective in reducing the seepage rate as well as the concentrations of total solids, COD and nitrogen in the leachates.

ZUSAMMENFASSUNG: Es wurde die Entwicklungsmöglichkeit eines erhöhten biologischen Abdichtungsmechanismus für erdige Düngerlagerungen untersucht. Man wählte eine Bakteriensorte aus und benützte diese für die Produktion von unlöslichen Polysachariden, die sich in der Verfüllung der Bodenporen wirksam zeigten. Der entwickelte Mechananismus zeigte sich als sehr wirksam in der Verminderung der Durchsicherungsrate sowohl als auch der Konzentrationen von allen Feststoffen, von chemischem Sauerstoffbedarf und von Stickstoff in den Laugenflüssigkeiten.

RÉSUMÉ: Afin d'éliminer les pertes de purin dans les réservoirs de lisier faits de terre, on s'est servi d'une espèce de bactérie pouvant produire le polysaccharides insolubles. Ces derniers sont avérés efficaces dans les blocage des pores du sol. Par conséquent, on a ainsi reussi à développer un méchanisme de blocage biologique très efficace, tant au niveau de la réduction du taux de filtrage de purin, qu'à celui des concentrations respectives de solide total, de demande chimique d'oxygène (D.C.O.) et d'azote dans l'eau d'infiltration.

1 INTRODUCTION

The soil sealing by manure is a result of physical, chemical and biological processes occurring simultaneously. The sealing in ponds, lagoons and other earthen manure storage facilities is thought to be accomplished by the physical clogging of soil pores by the fine solids present in liquid manure, (Davis et al., 1973; and De Tar, 1979). These fine solids, under the hydraulic pressure of the liquid manure, penetrate the macropores of the soil causing reduced hydraulic conductivity and in turn reduced fluid movement. There are several factors influencing the physical sealing process including soil type, time, solid content in manure and hydraulic head of manure (Baier et al. 1974). However, De Tar (1979) questioned the stability of the physical sealing process and observed substantial loss in sealing efficiency immediately after emptying of the manure storage system.

Chemical sealing of soils refers to the modification of the pore geometry in soil by chemical reactions or ionic exchange resulting in a reduced fluid movement. The exchange can occur through soil particle interactions with either salts in the soil solutions or with strong oxidizing or reducing agents of biological origin (Barrington, 1985). Although, it is believed that sealing of soil may occur because of modification of pore geometry, it is not known whether the viscosity and the total potential gradient have an important role to play in reducing the hydraulic conductivity of the soil.

Two biological sealing mechanisms have been reported. The first involves the breakdown of soil structure by strong oxidizing or reducing agents which are the products of microbial metabolism, a proçess known as "gelization" (De Shpande,

1968). The second involves the clogging of soil by "mucus like" polysacharides. Such materials are formed for the purpose of trapping of nutrients or for adhesion to different types of surfaces or for protection against desiccation and predators (Avimelech and Nevo, 1964). Frankenberger (1979) reported good correlations between hydraulic conductivity and quantities of polysacharides and phosphatase enzymes produced by bacteria and the number of bacteria per gram of soil.

In selection of suitable soil sites for the construction of earthen manure reservoirs, researchers suggested a hydraulic conductivity varying from 10^{-3} to 10^{-6} cm/s. Environmental authorities in Canada seem to believe that such a site selection criterion will guarantee groundwater protection because of the sealing of soils by manure. However, field studies have indicated that after a fresh loading of manure into these reservoirs there were an "initial flush" behaviour observed and some of the manure constituents percolated through the soil resulting in high levels of pollutants in groundwater (Reese and Lauden, 1983). De Tar (1979) observed substantial seepage, during the time required for the seal to develop, which resulted in groundwater pollution. In addition, physical sealing is a faster process causing clogging at the soil surface, thereby substantially minimizing the effectiveness of the biological sealing. Therefore, an enhanced biological sealing, deep in the soil layers, is of paramount importance. The production of extracellular polysacharides is important in the plugging process. Microbial polysacharides which have potential in the sealing mechanisms include: dextran, xanthan, curdlan, indican, pullulan, seleroglucan and zanflopolysaccharides (Jang et al., 1983; and Jansheka, 1983).

2 OBJECTIVES

The aim of this study was to investigate the possibility of developing an enhanced biological sealing mechanism for earthen manure storage facilities. The specific objectives were:

(a) to culture a biological species whose metabolic end products will cause the sealing of soil pores,

(b) to develop a method of testing the sealing capabilities of these cultures on various types of soils, and

(c) to evaluate the effectiveness of the sealing mechanism in reducing the infiltration rate and the pollutants concentrations in the leachates.

3 EXPERIMENTAL APPARATUS

Twenty one (21) columns were constructed of 20 cm ID PVC pipes. Each column is assembled as a two-part system as shown in Figure 1. The lower section (60 cm in length) is designed to carry the soil whereas the second section (140 cm in length) is filled with manure to provide a hydraulic head on the soil core. Flanges made of 1.2 cm thick PVC plastic are provided on the lower edge of the top section and the upper edge of the bottom section. The two sections are secured in position by bolting the two flanges together. An O-ring (gasket) placed between the flanges of the two sections provided good sealing. A bottom cap made of 1.2 cm PVC material was fastened to the pipe with PVC glue and screws to obtain a water tight fit. A plastic nipple was mounted in a 0.6 cm diameter hole drilled at the centre of the cap. This provided a port for the collection of seepage water. The inside of the lower section of each column was lined with coarse sand to minimize the channeling effect.

On the lower section of each column, four 0.6 cm diameter holes were also drilled and tapped on the side to allow installation of piezometric tubes which were used to indicate the hydraulic pressure throughout the column and serve as an indicator of the location and thickness of the seal. Four piezometric sensors were installed at different levels along the soil core (at depths of 2, 5, 10 and 15 cm below the soil surface). Each sensor consisted of a stainless steel draining tube of 20 cm length and 0.5 cm diameter and acrylic tubing. Several holes were drilled along a stainless steel tube which provides a path for the pore water to flow. The concept is similar to the idea of tile drainage. Each stainless steel tube was soldered to a threaded brass fitting which was screwed into the column to ensure a water tight fit. An acrylic tubing of 0.6 cm diameter was connected to the brass fitting and held in place vertically along the sides of the columns using clamps made of 1.2 cm thick PVC plastic which were glued to the side of the column.

The columns were mounted vertically on the top of a wooden platform. Under the platform, directly below each column, a 3L sample bottle was used for the collection of the leachate from the column.

4 SOIL COLLECTION, PREPARATION AND ANALYSES

Three types of local agricultural soils were used in the study. These were (a) Stewiacke series, where 50 kg/acre of urea, 75 kg/acre of 17-17-17 fertilizer, and approximately 1500 kg of liquid manure are applied annually; (b) Queens series, where 50 kg/acre of urea, 75 kg/acre of 12-24-24 fertilizer and approximately 1500 kg of liquid manure are applied annually; and (c) Kentville series, taken from virgin land with good crop soil characteristics.

At each site, the top vegetative and trash cover of the soil was scraped away and the soil was removed by a shovel from the top 50 cm. The soils were placed in heavy duty (16 mm thick) polyethylene bags. Each bag was filled with approximately 40 kg of soil and was sealed. The bags were transported from the collection site to the waste management laboratory. Each soil was analyzed to determine field bulk density and classification. Grain size analyses were performed on the soils using the mechanical and hydrometer methods and the grain size distribution was determined. These tests were carried out according to the procedure described by Black et al. (1965) in Methods of Soil Analysis. The results of these tests are presented in Table 1. The three soil series were found to be different from one another and were classified as sandy loam, sandy clay loam and loamy sand.

Table 1. Soil Type and Classification

Soil Series	Field Bulk Density (g/cm^3)	Particle Size			Soil Type
		Clay	Silt	Sand	
Stewiacke Series	1.40	16	20	64	Sandy Loam
Queens Series	1.50	20	16	64	Sandy Clay Loam
Kentville Series	1.30	5	15	80	Loamy Sand

The values are the average of three replicates

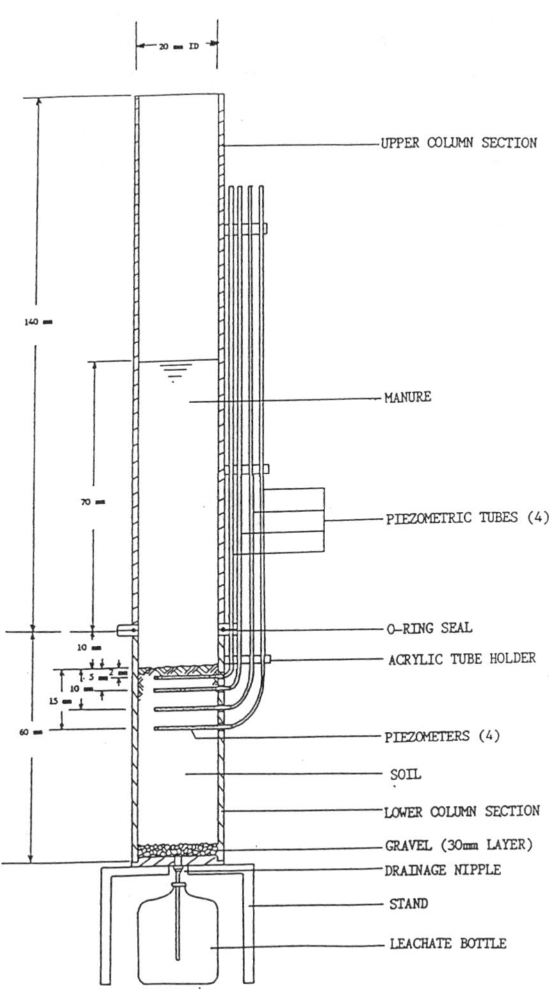

Figure 1. Experimental Apparatus

5 MANURE COLLECTION AND PREPARATION

The manure was collected from a commercial dairy farm (Winding River Farm) located at Stewiacke, approximately 80 km from Halifax. It was screened, diluted (2 manure: 1 water) and placed in 50 L plastic buckets. The buckets were sealed and stored in a cold storage facility at -18 °C. Prior to being applied into the soil columns, the manure was removed from the cold storage facility and allowed to thaw at room temperature for 24 hours. Some characteristics of the manure are presented in Table 2.

Table 2. Some characteristics of the manure used in the study.

Item	Measured Value
Total solids	65 870
Volatile solids	53 960
Ash	11 910
Total chemical oxygen demand	74 880
Soluble chemical oxygen demand	24 100
Total kjeldahl nitrogen	5 460
Ammonium nitrogen	560

6 EXPERIMENTAL PROCEDURE

6.1 Column Preparation

A 3 cm crushed gravel zone was constructed on the bottom of each column. A geotextile screen was then laid over the gravels. This was done to permit freer drainage of the columns and to prevent the soil from clogging the leachate collection outlets.

Soils were then placed in layers (of approximately 2 liters in volume) until the desired height was reached (10 cm below the upper edge of the lower column section). Each layer was compacted until the bulk density equaled that obtained under field condition. The weight of the columns was then recorded for bulk density check. The addition of equal volumes of soil and the consistent soil compaction ensured a uniform solid density through out the column.

The columns were then placed on the support stands and the plastic nipples were installed onto the bottom sections. The piezometric tubes and other components of the apparatus were then assembled as described earlier.

6.2 Selections and Preparation of the Microorganisms

Pure cultures of <u>Bacillus lichenformis</u> NRC-9012 were obtained from the Division of Biological Science of the National Research Council in Ottawa and were grown in the laboratory. The growth medium consisted of 5.0 g sucrose, 2.5 g tryptone, 2.5 g K_2 HPO_4 and 5.9 g yeast extract, all dissolved in 1L distilled water.

Portions of the media were transferred to 5 screw-top test tubes and the remaining were transferred to ten 1L Erlenmeyer flasks. All of the media containers were then sterilized in an autoclave for 15 minutes at 121°C. After the media had been cooled, the cultures were hydrated using sterilized liquid media and the microbes were then transferred from the hydrated viles to 5 covered petri dishes having a solid growth media. The solid growth media consisted of the same constituents as the liquid media with the addition of 15 g agar per liter. The bacterial cultures were then grown in a temperature and humidity controlled growth chamber at 21°C for 48 hours. The cultures were then transferred from the petri dishes to five 500 ml Erlenmeyer flasks (each having 200 ml of sterilized liquid media) and grown in a controlled environment shaker for 3 days at 21°C. The cultures were then transferred from the shake flasks to two stirred tank reactors having 5 L each of sterilized liquid media. Propagation of the bacteria was allowed to continue for five days. A plate count test revealed that there was a count of approximately 8.24×10^{10} microbial cells per ml.

6.3 Experimental Protocol

The soil columns were first brought to saturation by flooding with water, allowing the water to stand for three days and then draining out the excess water. A volume of 250 ml of the prepared culture suspension (media and bacterial culture) was then applied to 18 of the soil columns (6 columns each for each type of soil). This was followed by the application of an equal volume (250 ml each) of sterilized growth medium (without culture). In the 3 control columns (one for each soil type), no bacterial culture was introduced, (rather the columns were each fed with 500 ml of microbe free medium. All the columns including the controls were then left to stand for 2 weeks to enable the growth of microbial culture and the development of the sealing mechanism. At the end of this period, 40 litres of manure were fed to each of the 21 columns using TAT pump model 110-030. The manure hydraulic head was approximately 127 cm.

The infiltration rate of manure through the soil columns was determined by measuring the drop in height (cm) of the manure surface in the column over a period of time, multiplying the cross sectional area of the column by the height and dividing by the time period in hours. The seepage rate was also determined by collecting the total quantity of liquid seeping through the soil over a period of time and dividing this quantity by the time period in hours.

The leachates from each of the columns were collected and stored in BOD bottles under refrigerated conditions until chemical analyses were performed on the samples. The solids, COD, nitrate, nitrite, ammonium and total Kjeldahl nitrogen were performed on the leachate samples according to the procedures described in the Standard Methods (APHA, 1980).

7 RESULTS

7.1 Infiltration and Seepage Rates

The infiltration results are presented in Table 3 whereas the seepage results are presented in Table 4. The values are the average of 6 replicates each. For each type of soil, the infiltration rate in the control column were much greater than those of the culture treated columns. The soil type has significantly affected the infiltration rate in both the control and treated columns. It appears from the results that the infiltration rate was positively correlated to the percentage of sand in the soils, the higher the percentage of the sand the greater was the infiltration rate. However, after 4 days the infiltration rates seemed to be lower than the seepage rates. This was apparently due to the volume expansion of manure in the soil columns. The anaerobic activity in the manure produced gases which resulted in the formation of a layer of suspended solids at the surface of the liquid manure. The formation of this layer, and the continuous increase of its thickness, resulted in continued expansion of the manure volume so that the manure height on day 4 was higher than the initial value measured at the beginning of the experiment.

The seepage rate was also affected by the type of soil. It was correlated to the percentage of sand in the soil, the higher the percentage of sand the greater was the seepage rate. The enhanced biological sealing in the treated columns significantly reduced the seepage rate. The seepage rate continued to decrease with time for both the treated and control columns. However, it would take approximately 20 days for manure to produce the sealing effect that was observed in the treated columns on the first day.

Table 3. Infiltration Rate

Soil Type	Height Reduction (cm)	Infiltration Rate (ml/hr)
Sandy Clay Loam	1.9 (5.3)	5.4 (23.1)
Sandy Loam	2.9 (9.6)	12.8 (33.2)
Loamy Sand	3.5 (11.4)	15.8 (49.7)

The values are the average of 6 replicates
( ) control with manure after 4 hours

7.2 Solids

The results of solid analysis are presented in Table 5. The total solid concentration in manure was 65873 mg/L. Only concentrations of 1130 to 1676 mg/L were found in the leachates collected over the first 3 days. The solid concentrations in the leachates slowly decreased with time. For a given soil type, the leachates collected from the treated columns had significantly lower solid concentrations than those collected from the control column. Most of the soils were in the ash form (87-95% of the total solids in leachates as compared to 18% in the manure). The solid concentrations in leachates collected from loamy sand were the highest whereas those in leachates collected from the sandy clay loam were the lowest. The percentage of sand in the soil appeared to have a positive correlation with the concentration of the solids in the leachates, the higher the percentage of the sand the higher was the solid concentration in leachates.

Generally the sealing mechanism reduced the solid concentration by 99.39-99.64% in the first 3 days as compared to 84.3 - 87.9% in the control columns in 22 days. After 22 days the manure in the control columns did not produce as good sealing as the enhanced biological sealing; the solid concentration in the leachates obtained on day 22 from treated columns were 3 times those obtained on day 1 from the control columns.

Table 4. Seepage Rate

Day	Treatment or Control	Sandy Clay Loam		Sandy Loam		Loamy Sand	
		Leachate Volume (ml)	Seepage Rate (ml/day)	Leachate Volume (ml)	Seepage Rate (ml/day)	Leachate Volume (ml)	Seepage Rate (ml/day)
4	T	416	139	556	185	676	225
	C	(2710)*	(903)*	(3070)*	(1023)*	(3615)*	(1205)*
		[15900]*	[5300]*	[16400]*	[5467]*	[23800]*	[7933]*
7	T	240	80	288	96	363	121
	C	(1980)	(660)	(2430)	(810)	(2480)	(827)
10	T	240	80	282	94	366	122
	C	(715)	(238)	(850)	(283)	(1100)	(367)
13	T	235	78	255	85	371	124
	C	(620)	(206)	(680)	(266)	(960)	(320)
16	T	207	69	251	84	340	113
	C	(590)	196)	(600)	(200)	(860)	(287)
19	T	199	66	226	75	340	113
	C	(270)	(90)	(415)	(138)	(850)	(283)
22	T	183	61	225	75	344	115
	C	(260)	(87)	(400)	(133)	(550)	(183)

The values are the average of 6 replicates
( )* control with manure after 4 hours
[ ]* control with water after 4 hours

Table 5. Solid Concentration in Leachate (mg/L)

Day	Treatment or Control	Sandy Clay Loam			Sandy Loam			Loamy Sand		
		Total	Volatile	Ash	Total	Volatile	Ash	Total	Volatile	Ash
4	T	1130	52	1078	1170	58	1121	1676	209	1467
	C	(7980)*	(5580)*	(2400)*	(8204)*	(5612)*	(2592)*	(10348)*	(7122)*	(3236)*
7	T	302	52	250	317	58	259	722	188	534
	C	(5200)	(2900)	(2300)	(7280)	(4880)	(2400)	(8900)	(6250)	(2650)
10	T	260	60	200	310	60	250	600	188	412
	C	(4880)	(2680)	(2200)	(6950)	(4558)	(2392)	(7500)	(4900)	(2600)
13	T	255	55	200	290	62	228	550	200	350
	C	(4675)	(2575)	(2100)	(6830)	(4610)	(2220)	(7300)	(4745)	(2555)
16	T	270	60	210	288	58	230	500	190	310
	C	(3995)	(1795)	(2200)	(6865)	(4700)	(2165)	(7833)	(5343)	(2490)
19	T	230	50	180	260	60	200	450	180	270
	C	(3060)	(1060)	(2000)	(6100)	(3970)	(2130)	(6950)	(4520)	(2430)
22	T	240	50	190	240	55	185	400	190	210
	C	(3010)	(1000)	(2010)	(5400)	(3260)	(2140)	(6500)	(4030)	(2470)

The values are the averages of 6 replicates
The Total Solid Concentration in manure = 65870 mg/L
( )* control with manure after 4 hours

7.3 COD

The COD results are presented in Table 6. The COD concentration in the manure was 74880 mg/L whereas that in the leachates obtained from the treated columns on day 4 were 56-217 mg/L. After 3 days, the COD was reduced by 99.70 - 99.93% due to the soil sealing process.

The soil type had a significant effect on the COD reduction. The sandy clay loam had the highest COD reduction while the loamy sand had the lowest reduction. At the end of the experiment, for a given type of soil the COD reduction for treated columns was substantially higher (99.70 99.93%) than that of the control columns. However the COD concentrations in the leachate obtained after 22 days from the control columns were 75 times those in the leachates obtained after one day from the treated columns.

Table 6. COD Concentration in Leachates (mg/L)

Day	Treatment or Control	Sandy Loam	Sandy Clay Loam	Loamy Sand
4	T	56	104	217
	C	(9390)*	(9374)*	(9942)*
7	T	34	90	187
	C	(5315)	(5660)	(5924)
10	T	56	95	208
	C	(4841)	(5734)	(6230)
13	T	56	93	193
	C	(4741)	(5897)	(5833)
16	T	50	90	200
	C	(4388)	(5357)	(6460)
19	T	48	79	214
	C	(4476)	(5476)	(5819)
22	T	50	80	92
	C	(4200)	(5480)	(5496)

The values are the aerage of 6 replicates
The COD Concentration in manure=74880 mg/L
( )* control with manure after 4 hours

7.4 Nitrogen

The results for the TKN are presented in Table 7 whereas those for NH_4, NO_2 and NO_3 nitrogen are presented in Tables 8. The concentrations of TKN and NH_4 in the manure were 5463 mg/L and 560 mg/L, respectively. The soil type did not seem to have any significant effect on the reductions in the TKN and NH_4. For a given soil type, the reductions in the TKN and NH_4 after 4 days were significantly higher for the treated columns (99.6 - 99.7% for TKN and 99.1 - 99.6% for NH_4) than those of the control columns (83.3 - 85.7% for TKN and 42.9% for NH_4).

Table 7. Total kjeldahl Nitrogen Concentration in Leachates (mg/L)

Day	Treatment or Control	Sandy Loam	Sandy Clay Loam	Loamy Sand
4	T	15	17	20
	C	(780)*	(790)*	(810)*
7	T	7	8	25
	C	(69)	(132)	(620)
10	T	7	9	20
	C	(58)	(105)	(490)
13	T	8	11	20
	C	(42)	(105)	(358)
16	T	8	10	15
	C	(42)	(85)	(290)
19	T	8	10	20
	C	(39)	(85)	(191)
22	T	8	10	14
	C	(39)	(87)	(132)

The values are the average of 6 replicates
The TKN Concentration in Manure=5463 mg/L
( )* control with manure after 4 days

8 DISCUSSION

The results indicated that special type of microorganism can be used to enhance the soil sealing process by manure. Earthen storage with enhanced biological sealing mechanisms appeared to be a very viable alternative to other expensive manure storage structures.

The sealing process took place from the soil surface throughout the soil column. The piezometric sensors recorded no liquid level in the treated columns whereas liquid levels similar to those in the control columns were noticed in the four piezometric tubes of each column.

Table 8. Inorganic Nitrogen in Leachates (mg/L)

Day	Treatment or Control	Sandy Loam			Sandy Clay Loam			Loamy Sand		
		NH_4	NO_2	NO_3	NH_4	NO_2	NO_3	NH_4	NO_2	NO_3
4	T	2	4	2	3	4	3	5	4	6
	C	(320)*	(27)*	(3)*	(320)*	(27)*	(3)*	(320)*	(36)*	(6)*
7	T	0	2	1	0	3	1	0	3	2
	C	(26)	(2)	(0)	(37)	(3)	(0)	(53)	(3)	(0)
10	T	1	3	0	4	3	0	4	2	0
	C	(31)	(1)	(0)	(35)	(3)	(0)	(66)	(3)	(0)
13	T	1	3	0	3	3	0	2	2	0
	C	(36)	(2)	(0)	(27)	(3)	(0)	(57)	(2)	(0)
16	T	1	4	0	2	3	0	2	3	0
	C	(24)	(3)	(0)	(46)	(3)	(0)	(75)	(2)	(0)
19	T	1	3	0	1	3	0	1	3	0
	C	(37)	(1)	(0)	(46)	(2)	(0)	(63)	(3)	(0)
22	T	2	3	0	1	1	0	1	1	0
	C	(27)	(3)	(0)	(40)	(0)	(0)	(60)	(1)	(0)

The values are the average of 6 replicates
NH_4 Concentration in manure = 560 mg/L
NO_2 Concentration in manure = 0 mg/L
NO_3 Concentration in manure = 0 mg/L
( )* control with manure after 4 hours

The initial flush behaviour observed with the control columns was not observed with the treated ones; as the amount of seepage from the treated columns was very small. At the beginning of the experiment the seepage rate of the treated columns was 168 times lower than that of the control columns. The treated columns also had the ability to reduce the concentrations of the total solids, COD, TKN and NH_4 by up to 99.6%, 99.9%, 99.6% and 99.7%, respectively. These were mostly due to the effectiveness of the biological sealing resulting from the polysacharide produced by the bacteria which plugged the soil pores.

9 CONCLUSIONS

The developed biological sealing mechanism was very effective in reducing the seepage rate and the concentrations of the solids, COD, TKN and NH_4 in the leachates. The sealing process took place from the soil surface throughout the soil core. The technique developed has the potential of substantially reducing the cost of manure storages associated with concrete and lined manure storage facilities.

10 REFERENCES

APHA, 1980. Standard Methods for Examination of Water and Wastewater. American Public Health Association, Washington.

Avimelech, Y. and Z. Nevo. 1964. Biological clogging of sands. Soil Science, 98: 222-226.

Baier, D., J.L. Meyer and D.R. Nielsen. 1974. Manure holding pond sealing study. Proceedings of the Agricultural Waste Management Conference, Cornell University, Ithaca, N.Y.

Barrington, S.T. 1985. The sealing of soils by manure. Unpublished Ph.D. thesis. McGill University, Montreal, Quebec.

Black, H.M., D.D. Evans, L.G. Ensminger, J.L. White and F.E. Clark. Methods of Soil Analysis, Part I: Physical and Mineralogical Properties. American Society of Agronomy, Madison, Wisconsin.

Davis, S., W. Fairbank and H. Weisheit. 1973. Dairy waste ponds effectively self sealing. Transaction of ASAE 16(1):69-71.

De Shdande, T.L. 1968. Changes in soil properties associated with the removal of iron and aluminum oxides. Journal of Soil Science 19(1):108-122.

De Tar, W.R. 1979. Infiltration of liquid dairy manure into soil. Transaction of ASAE 22(2):520-528.

Frankenberger, W.T., F.R. Traeh and L.I.C. Dumenil. 1979. Bacterial effects on hydraulic conductivity of soils. Soil Science Society of America, 43:333-338.

Jang, L.K., M.M. Sharma, J.E. Findlay, P.W. Chang and T.F. Yen. 1983. An investigation of the transport of bacteria through porous media. Proceedings of the International Conference on Microbial Enhancement of Oil Recovery, Bartleville, Oklahoma.

Jansheka, H. 1983. Microbial products for injection. Proceedings of the International Conference on Commercial Applications and Implications of Biotechnology. Online Publications, Northwood, U.K.

Reese, L.E. and T.L. Laudon. 1983. Seepage from earthen manure storages and lagoons. ASAE Paper No. 83-4569, St. Joseph, Michigan.

Land and Water Use, Dodd & Grace (eds), © 1989 Balkema, Rotterdam. ISBN 90 6191 980 0

A decision support system to optimize the application and processing of manure

R.M.de Mol
Institute of Agricultural Engineering, Wageningen, Netherlands

ABSTRACT: In the Netherlands there is a serious manure problem. New legislation restricts the application of manure. It is becoming more difficult to give all manure an acceptable destination (application or processing). To gain a better insight into the logistic problem on a regional level a decision support system (DSS) is being developed. A mathematical model is used to estimate the application on the farm of origin and in the neighbourhood and optimizes the application elsewhere as well as the processing, including storage and processing locations. The DSS can be used on tactical and strategic levels, and is already in use in the practice.

RESUME: Aux Pays-Bas, le fumier constitue un grand problème. La législation prévoit la restriction de l'emploi du fumier. Il devient plus difficule de donner une destination acceptable à tout le fumier. Afin de mieux comprendre le problème logistique au niveau régional, on développe un système de soutien de la décision (DSS). Un modèle mathématique permet d'évaluer l'emploi à l'exploitation agricole et dans les environs et optimalise l'application et ailleurs et le traitement et indique les endroits de stockage et de traitement. On peut utiliser le DSS au niveau tactique et/ou stratégique. Il fonctionne déjà dans la pratique.

ZUSAMMENFASSUNG: In den Niederlanden gibt es ein ernsthaftes Gülleproblem. Die Gesetzgebung beschränkt die Mengen, die ausgefahren werden. Es wird ständig schwieriger, für allen Stalldung eine annehmbare Bestimmung zu finden (Ausbringen oder Aufbereiten). Um auf regionaler Ebene eine bessere Einsicht in die Transport- und Lagerungsproblematik zu gewinnen, wird ein System zur Unterstützung im Entscheidungsprozeß (DSS) entwickelt. Mit Hilfe eines mathematischen Modells wird der Bedarf des eigenen Betriebs und des umliegenden Gebiets geschätzt und werden das Ausbringen in anderen Gebieten und die Lagerung und Verarbeitung, einschl. der benötigten Standorte optimiert. Das DSS kann auf taktischer und auf strategischer Ebene eingesetzt werden und wird bereits in der Praxis angewandt.

1 INTRODUCTION

The strong growth of the livestock in the Netherlands has caused a manure problem. Excessive fertilization with manure has resulted in environmental pollution. Legal measures are imposed to counteract excessive dressings. These measures make it more difficult to find acceptable destinations (by application or processing) for all the manure produced. The need for facilities for the transport, storage and processing of manure will increase, as will the costs related.

Hence, it becomes more important to organize storage, transport and processing in a way which minimizes the long-term cost level. The Institute of Agricultural Engineering (IMAG) and the Agricultural University, both in Wageningen, are developing a Decision Support System (DSS) to be used to optimize the application and processing of manure on a regional level.

This paper deals with the following subjects: the manure problem in the Netherlands, the options of application and processing, the need for a DSS, a mathematical model for the application and processing, and the use in the practice.

2 THE MANURE PROBLEM

The Netherlands has seen a large growth of its livestock in the last decades (see table 1). Feedstuff imports have made animal production independent of the available amount of agricultural land.

Table 1. Cattle, pig and poultry numbers in the Netherlands in 1970, 1980, 1987 (in millions)*

	Cattle	Pigs	Poultry
1970	4.3	5.5	55.4
1980	5.2	12.4	81.2
1987	4.9	14.3	96.8

* Source: CBS 1983,1988

Proportional to this increase in livestock numbers was that in the production of manure. It became no longer possible to find acceptable destinations for all the manure. The farmland in some parts of the Netherlands has become oversaturated with nutrients. The balance of inputs and outputs of minerals (phosphate, nitrate and potash) and heavy metals (copper, cadmium and zinc) has come to be disturbed. The overdressings are found in the regions where the intensive husbandry is concentrated. These 'concentration regions' are the sandy soils in the eastern and southern parts of the country (figure 1).

The consequences of excessive fertilization are:

1. A decline in soil fertility and in quality of crops. Most crops do not stand overdressing, forage maize being an exception to this.
2. A deterioration of the quality of groundwater and surface water. The leaching to groundwater and surface runoff of minerals are consequences of the excessive application of manure. The limit for nitrate has been exceeded already at various groundwater pumping stations. Eutrophication of surface water has come to be a widespread problem.
3. Pollution by acid rain. The emission of ammonia by production, storage and spreading of manure is one of the causes of the acid rain problem.

The Dutch government has enacted legislation in 1987 to counteract these consequences. These laws relate to:

- 'Phosphate limits': The annual application of phosphate (P_2O_5) in manure is limited to a maximum per hectare, dependent on the land use: grassland, forage maize or arable land. The phosphate limits are introduced gradually.

During the first phase (1987-1990) the annual limits per hectare are 250 kg P_2O_5 for grassland, 350 kg P_2O_5 for forage maize and 125 kg P_2O_5 for arable land. With a national manure production of 94 million tonnes, 14 million tonnes cannot be used on the farm of origin. This gross surplus is supposed to be used on other farms.

Figure 1. Locations of concentration regions in the Netherlands (source: Mestwetgeving 1987)

During the fourth and last phase (after 2000) the limits will correspond with the actual phosphate requirements. These limits have not been fixed yet but will be about 110 kg P_2O_5 for grassland, 75 kg P_2O_5 for forage maize and 70 kg P_2O_5 for arable land. The gross surplus is estimated at 36 million tonnes, the net surplus (the national surplus) at 27 million tonnes. These figures are taken from: 'Mestproblematiek' 1985 (manure issue) and 'Afzetmogelijkheden' (markets) 1987.

On top of the general phosphate limits more severe regional limits are possible, for instance in the vicinity of a groundwater pumping station.

- A spreading prohibition: The spreading of manure is restricted in the autumn and winter to curb the leaching and surface runoff.

- A 'work-into-the-ground' obligation: Spreaded manure on open land must be worked into the soil within 24 hours.

- An expanding prohibition: It is prohibited for farms with intensive livestock husbandry to expand.

- An obliged 'manure recording': Every farm has to keep records on the manure production and application.

- 'Manure banks': The manure banks coordinate the application of manure. They are obliged to purchase manure surpluses, with the supplier having to pay for this service.

- A 'surplus levy': A farmer with a manure surplus has to pay a levy dependent on the size of his surplus.

3 APPLICATION AND PROCESSING

In general the following destinations exist for manure:

- Application for manuring on farmland: Most manure will be applied on the farm where it is produced or on other farms. The application is restricted by the phosphate limits. These will not always be reached actually, because of the manurial state of the land or because farmers can prefer fertilizer to manure.

- Processing to harmless effluent or useful products: The manure that cannot be disposed of by land application must receive another destination by being processed. Pig manure, e.g., can be processed to pellets, a fertilizer with additional organic matter. These pellets are intended to be exported. At the moment there are two pilot plants for the processing of pig manure in the Netherlands.

- Export to neighbouring countries: The possibilities for the export of manure are uncertain, so not much attention will be paid to this now.

The following items are of importance by application and processing:

- Transport: Transport is necessary to bridge the distance between the site of origin and that of destination. Dependent on the kind of manure, its destination, and the distance to be covered, several methods of transport are possible: by tractor or by truck (the most common way), by barge (for greater distances, cheaper than by truck), by train (probably too expensive) or by pressure pipe (similar to a pressure sewerage for households, interesting for concentration regions).

- Storage: Because there is a continuous supply and a discontinuous time of application, manure has to be stored. This can be on a site where it is produced or where it is to be applied (e.g. an arable farm); storage can be separately for each farm or central for several farms together. It can be realized in slurry cellars below livestock house floors, in tanks or in lagoons.

- Treatment: It can be useful to treat manure to increase its marketing prospects. Treatments can be: separating, fermenting, settling and drying (especially of poultry droppings). Treated manure is used as an organic fertilizer after all. Most treatment techniques are also a part of processing chains.

The destination of manure (application or processing) is reached by using certain methods of transport, storage and, if required, treatment.

4 THE DECISION SUPPORT SYSTEM

Transport, storage or treatment facilities are necessary to find acceptable destinations for all the manure produced. As there are many options of application and processing, it is not obvious which methods to choose.

To gain a better insight into this logistic problem a Decision Support System (DSS) is being developed by IMAG and Wageningen Agricultural University.

The DSS will determine, in a concrete case, the optimum structure of the logistics, including transport and storage facilities, required and the costs related. This optimization embraces the whole chain from production to spreading or processing.

The optimum structure depends on the options available. By changing the options step by step and re-determining the optimum structure, the consequences of changes become apparent.

This makes it possible to answer questions such as:

- Is it profitable to realize central storage instead of individual storage?

- How much processing capacity is needed in a certain region and what are the best locations for the plants?

The Decision Support System is restricted to a regional level, and two region types are considered: the surplus region and the shortage region. The DSS can support decisions on both tactical and strategic levels. On the tactical level it can help determine the optimum use of the existing infrastructure. On a strategic level it can support choices concerning possible investments.

5 A MATHEMATICAL MODEL

The destination of manure is either application or processing. For the DSS a mathematical model is used to determine the application and processing on an annual basis. The model consists of three steps, which are as follows:

step 1: application on the original farm;

step 2: application on other farms in the same region;

step 3: application in another region, and processing.

The steps will be discussed separately. Step 1 works on a farm level, whereas steps 2 and 3 work on a regional level.

5.1 Application on original farm

Most manure will be used on its farm of origin. On the basis of the livestock numbers of a farm, the annual manure production is calculated. On the basis of the use of the land, the options for application are calculated. The options are first restricted by the statutory phosphate limits. Further, there is an agricultural limit for potassium (K) on grassland (both P_2O_5 and K are expressed in maximum weights per hectare per year). These possibilities form the 'capacity to be received'.

The farmer can choose which kind(s) of manure he will use for a certain land use. The realistic manure use is estimated, and by the DSS the amount to be applied is calculated. There is no optimization in this step, but it is possible to presume an optimum manure use.

On the basis of the manure use the application is calculated for each farm; there are two possibilities:

- There is a manure surplus. The surplus must find a destination outside the farm.
- There is no manure surplus and some capacity to be received is free. Not all of this capacity can be used for manure from other farms. The actual situation will depend on the manure production related to the requirements and on whether the farmers are willing to apply the manure. By estimating these options, the part of the remaining capacity to be used is calculated; this can be called the usable capacity to be received.

For each area the surpluses for each type of manure and the usable capacity to be received for each type of land use will be known by aggregation over the farms in an area.

5.2 Application on other farms in the same region

A part of farm waste surpluses will be absorbed by the usable capacity to be received of other farms in the same region. Also in this step, the manure use is estimated. But not all manure types are suitable to be absorbed by the usable capacity to be received. For instance, the inferior liquid manure of fattening calves will not be readily accepted.

On the basis of the estimated use of the various manure types for the various types of land use, the application on other farms can be calculated. First the usable capacity of an area is filled with manure from the same area, next with manure from neighbouring areas and finally with manure from the remaining areas within the same region.

It is supposed that there will not be any surpluses left in the shortage region at this stage. The manure surpluses in the surplus region are disposed to the shortage region or are processed.

5.3 Application on another region, and processing

It is necessary to optimize the annual manure flows for the application in other regions and for processing. This consists in minimizing the costs of the logistic infrastructure, including transport, storage and processing. The results depend on the options. For each type of manure, the means by which a destination can be reached (transport, storage and the like) have to be given, together with the related costs. Also the structure of the manure flows follows from the options.

When there is only direct transport with no central storage, the structure will be simple. But if central storage and different means of transport are included, both application and processing are possible, the structure can be rather complex. The structure of manure flows is modelled by means of networks, with the flows following arcs between different nodes in the network. An arc represents the transport. A node is an origin and/or a destination of arcs and can represent a storage or a processing location.

EXAMPLE

Suppose that poultry manure surpluses in the surplus region are sent to the shortage region in barges (loading and unloading with trucks). The part of the manure flows network that relates to this example is shown in figure 2.

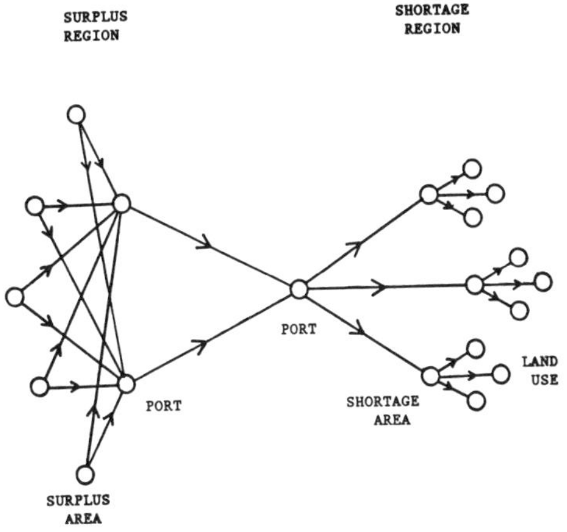

Figure 2. Network structure for poultry manure flows (example)

A node in figure 2 represents a surplus area (production), an inland port either in the surplus region or in the shortage region (trans-shipment), or a shortage area (application). An arc represents transport by barge or truck.

The flows in the network are optimized within the options by using the Linear Programming (LP) technique.

Most of the costs of flows can be taken proportional to the size of a flow. An exception are the processing costs, as the costs of a processing plant will show 'economies of scale' effects and have a 'fixed charge' structure.

The restrictions in the problem are all linear, they are as follows:

- All surpluses in a surplus area must be conveyed.
- The total flow to a storage area or a processing plant may not exceed the plant capacity (if capacity is restricted).
- The input to a storage area equals the output.
- If in the processing organic fertilizer is produced, the output of a processing plant is a certain percentage of the input.
- The application of all kinds of manure to a shortage area is not allowed to exceed the usable capacity to be received.
- The application of a single type of manure in a shortage area is bounded to a maximum.

The choice of optimum locations for storage and processing plants can be added to the linear programming problem. But while the size of the problem tends to be rather large already, this addition would not work. Therefore, it is possible to use a two-step algorithm:

In the first step optimum locations are calculated, using an algorithm for the uncapacitated facility location problem (Erlenkotter, 1978).

In the second step the LP problem as described above is solved using the SCICONIC software package (Scicon Ltd. 1986).

After these two steps have been made feedback and re-calculation can be necessary.

6 PRACTICAL USE

The development of the DSS was started in 1986, and a working version is to be completed in 1989. The part of the DSS referring to steps 1 and 2 in section 5 (application in the same region) was made suited for practical use in 1988. The practical use is on behalf of 'manure groups'. A manure group is an association of farmers who want to tackle the manure problem on the local level. They organize an inquiry among the farmers in their area to gain an insight into the problem. Processing of the inquiry data with the DSS results in figures indicating manure surpluses or capacity to be received and storage balance per farm, per area and for the manure group as a whole. An 'inquiry letter' (2 or 3 pages) to the farmer reports the results for his farm, describing the manure situation of the farm.

The software needed for the inquiry data handling is taken over by the National Manure Bank. A few initial results of step 3 (optimization) will be given in the presentation. These concern an application for the Dutch province of Overijssel.

REFERENCES

Afzetmogelijkheden 1987. De afzetmogelijkheden van dierlijke meststoffen; 19883. No. 2. In: Gedrukte stukken van de Tweede Kamer; Vergaderjaar 1986-1987. The Hague, Staatsuitgeverij.

CBS 1983. Statistisch zakboek 1983. The Hague, Staatsuitgeverij.

CBS 1988. Statistisch zakboek 1988. The Hague, Staatsuitgeverij.

Erlenkotter, D. 1978. A dual-based procedure for uncapacitated facility location. Operations Research 26: 992-1009.

Mestproblematiek 1985. Mestproblematiek; 18897 No. 2. In: Gedrukte stukken van de Tweede Kamer; Vergaderjaar 1984-1985. The Hague, Staatsuitgeverij.

Mestwetgeving 1987. Mestwetgeving; Mestbank; Mestboekhouding; Afleveringsbewijs; Afzetovereenkomst; Uitrijbepalingen. The Hague, Ministerie van Landbouw en Visserij (Ministry of Agriculture and Fisheries).

Scicon Ltd. 1986. SCICONIC/VM; User guide; Version 1.40. Milton Keynes, UK, Scicon Ltd.

Land and Water Use, Dodd & Grace (eds), © 1989 Balkema, Rotterdam. ISBN 90 6191 980 0

Air and ground water pollution from high application of animal manure

A.E.Ghaly
Agricultural Engineering Department, Technical University of Nova Scotia, Canada

ABSTRACT: A controlled experiment was performed to determine the relative amounts of nitrogen losses from land applied manure through the process of leaching and ammonia volatilization. The effects of three manure and three water application rates on nitrogen losses were monitored over a period of one month. The results indicated that hazards of ground water pollution from one season application of manure is minimal. The observed losses of nitrogen through ammonia volatilization was substantial and could constitute a bothersome source of air pollution.

ZUSAMMENFASSUNG: Es wurde ein kontrollierbares Experiment duchgeführt, um die relativen Mengen der Stickstoffverluste des am Boden verteilten Dungers durch den Prozess der Laugung und Ammoniakverflüchtigung zu bestimmen. Man überwachte für einen Monat die Auswirkungen von drei Dünger - und drei Wasserverteilungsraten an Stickstoffverlusten. Die Ergebnisse zeigten eine geringe Gefahr der Grundwasserverschmutzung bei der Verteilung von Dünger während einer Jahreszeit. Die beobachteten Verluste an Nitrogen durch Ammoniakverflüchtigung waren dagegen bedeutend und könnten eine beeinträchtigende Quelle der Luftverunreinigung erzeugen.

RÉSUMÉ: On a dirigé une expérimentation afin d'évaluer les quantites relatives de pertes d'azote, par le processus de filtrage et de volatilisation d'ammoniac, sur une terre préalablement recouverte de fumier. On a mesuré les effets de trois applications respectives de fumier et d'eau sur les pertes d'azote pendant une durée d'un mois. Selon les résultats obtenus, le danger de pollution de l'eau causé par l'application du fumier pendant une durée d'une saison est minimal. Les pertes d'azote obtenues par la volatilisation d'ammoniac sont importantes et pourraient constituer une source de pollution de l'air nuisible.

1 INTRODUCTION

Animal manures have long been recognized for their value as a nutrient source in the production of agronomic crops. Magdoff and Amaddon (1980) and Lund and Doss (1980) reported that certain nutrients, particularly phosphorus, potassium, calcium and magnesium, are more readily available to plants growing in soils receiving animal manure than in soils receiving only inorganic fertilizer. Manure has also been shown to improve the physical properties of the soil. Elson (1943) reported increased aggregate stability while Unger and Stewart (1974) reported increased water-holding capacity 2and reduced evaporation rate. Swader et al. (1972) and Mathers et al. (1980) reported increased soil organic matter contents.

Theoretically, recommended manure application rates are those that will supply nutrients in amounts equal to crop utilization. However, because of the uncertainty in determining actual nutrient availability, crop producers have been hesitant to rely solely on animal manure as a nutrient source for crop production. Farmers have often applied amounts of manure in excess of nutrient requirements, to ensure adequate nutrient availability. Also, in many intensive livestock operations, manure application in excess of crop utilization is often necessitated by the limitation of land available for manure disposal. Excessive manure application rates may constitute certain air and groundwater pollution hazards. Nitrate and nitrite contamination of groundwater is of prime concern to farming communities. The presence of high levels of these inorganic nitrogen compounds in groundwater has resulted in methemoglobinemia in bottle-fed infants (Bosch et al. 1950) and poisoning of farm animals (Campbell et al. 1954).

2 OBJECTIVE

The purpose of this study was to simultaneously monitor the nitrogen-leaching and ammonia volatilization processes occurring in soils under high dairy cattle manure application rates, and to evaluate the resultant air and groundwater pollution potentials of land application of manure.

2 EXPERIMENTAL APPARATUS

Ten 90 cm high columns were constructed from 20 cm I.D. plexiglas pipes. A leachate drain tube with a shut-off valve was connected to the bottom of each column. A piece of fine screen was fixed onto the top of the drain tube inside each column to filter out soil particles from the leachate. The top of each column was closed and sealed with a plexiglas plate. A hole drilled at the center of each plate was plugged with a stopper having two glass tubes. One tube (inlet) was connected to a BOD bottle containing standard acid solution for trapping ammonia released from the manure. Soil was placed in each column to a height of 60 cm. A 30 cm space above the soil surface was left for manure and water applications. The inside of each column was lined with coarse sand to minimize the channeling effect. A labelled diagram of the apparatus is presented in Figure 1.

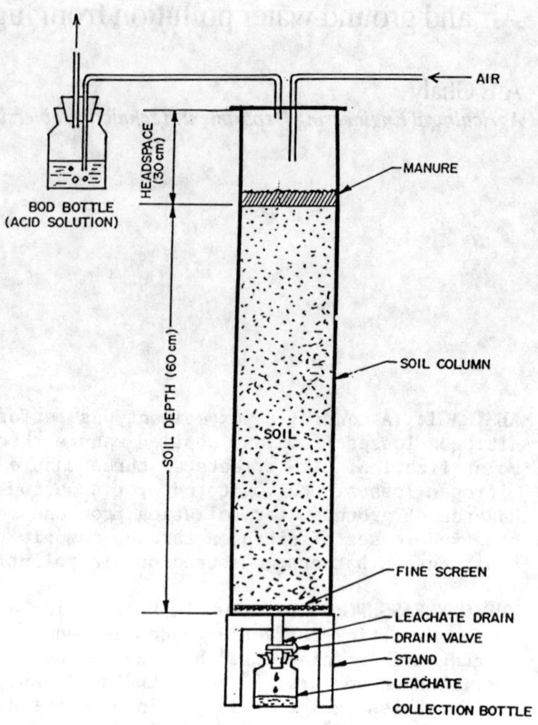

Figure 1. Experimental Apparatus

3 MANURE COLLECTION AND PREPARATION

The manure was obtained from a free stall dairy barn on Winding River Farm, a commercial dairy operation located in Stewiacke East, approximately 80 km from Halifax. It was scraped off the floor in a semi-solid form. It was diluted with water (2 manure:1 water) and mixed thoroughly. The manure was then placed in 18 L plastic bags and transported to the Waste Management Laboratory. The bags were stored in a freezer at -18 °C. Prior to use, the manure was removed from the freezer and allowed to thaw at room temperature for 24 hours. Some characteristics of the manure are presented in Table 1.

Table 1. Characteristics of the Manure Used

Item	Measured Values* (mg/L)
Total Solids	106 000 (4800)
Volatile Solids	86 300 (3700)
Ash	19 500 (950)
Total Chemical Oxygen Demand	64 200 (2100)
Soluble Chemical Oxygen Demand	16 500 (400)
Total Nitrogen	5 530 (120)
Organic Nitrogen	4 960 (80)
Ammonium Nitrogen	570 (10)

* Means of 4 replicates with standard deviation in parenthesis.

4 SOIL COLLECTION AND ANALYSIS

The soil used in this study is classified as Stewiacke series. It was obtained from a good quality pasture in Stewiacke East. The vegetative cover was scraped off and the soil was removed with a shovel to a depth of 50 cm. The soil was then placed in heavy duty polyethylene bags. Each bag was filled with approximately 40 kg of soil. The bags were then transported to the Waste Management Laboratory. The soil was analyzed to determine the field bulk density and grain size distribution. These tests were carried out according to the procedure described by Black et al. (1965). The analyses showed that the soil had a loam texture with a composition of 48% sand, 34% silt and 18% clay. The bulk density was found to be 1320 kg/m^3.

5 EXPERIMENTAL PROCEDURE

5.1 Conditioning the soil columns

The soil was screened to remove large stones and vegetative materials. It was then placed in the columns and compacted in 5 cm layers to achieve a bulk density of 1320 kg/m^3. The drain tube outlet in each column was closed and the soil was saturated with water. After 3 days, the excess water, still left standing at the top of the soil column, was carefully drained off through the drain tube.

5.2 Application of manure

The total nitrogen in the manure was 5530 mg/L, most of which (4960 mg/L or 90%) was in organic form. Application of manure was made based on the assumption that the minimum nitrogen requirement for a corn crop is 250 kg-N/ha. Three manure application rates of 250, 500, and 750 kg-N/ha were used. The amounts of manure required to give the chosen manure application rates were 150, 300, and 450 ml per column, respectively. Three sets of columns (each set having 3 columns) were used for each manure application rate. The 10th column was used as a control and did not receive any manure.

Following the application of the required amount of manure onto the soil surface in each column, the column cover was replaced and the column was sealed. The top outlet glass tube was connected to the BOD bottle containing 100 ml standard acid solution (1N HCL) for trapping ammonia gas. The inlet glass tube was connected to a pressurized air supply through a pressure regulator and manifold system. The air supply was set to give approximately 60 air bubbles per minute which was judged sufficient to ensure adequate removal of the ammonia evolved by volatilization.

5.3 Application of water

Corn farmers usually apply manure to the land in late May or early June which is the beginning of the five-month period (June to October) representing the growing season for corn in Nova Scotia. The average monthly rainfall for this period is 116 mm/month (Dzikoski et al. 1980). This is equivalent to 120 ml of water per column per day. Three application rates of 60, 120, and 180 ml per column per day were used to represent dry, normal and wet years respectively. Each set of soil columns (3 columns receiving the same manure application rate) received the three water application rates (one application rate per column). Water was applied every three days and the amounts applied were therefore 180, 360 and 540 ml per application. The 10th column (control) received the highest water application rate of 540 ml every three days. The experiment was carried out for 30 days.

5.4 Analysis procedure

Samples of the leachate were analyzed for total Kjeldahl, ammonium, nitrite and nitrate nitrogen; total solids; and chemical oxygen demand. The tests were performed on the samples according to the procedures described by APHA (1980). Ammonia volatilization was determined according to the procedure described by Mills et al. (1974).

6 RESULTS

6.1 Total Kjeldahl Nitrogen (TKN) in Leachates

The measured values of TKN concentrations in the leachates are presented in Table 2. The TKN concentrations in the leachate samples varied over time ranging from a high of 135 mg/L on day 3 to a low of 74 mg/L on day 30. The results showed that a large fraction of the TKN in the manure was retained in the soil. For instance, at the highest manure application rate of 750 kg-N/ha and the highest water application rate of 180 ml per day, of the 2488.5 mg TKN added to the soil column, a

Table 2. Totak Kjeldahl Nitrogen Concentrations in Leachate (mg/L)*

Column No.	Manure Application (kg-N/ha)	Water Application (mL/3 days)	Day									
			3	6	9	12	15	18	21	24	27	30
0	0	180	50	52	50	48	43	45	43	48	42	43
1	250	180	135	125	115	105	99	95	93	91	86	78
2	500	180	135	125	115	102	98	95	90	90	86	82
3	750	180	133	127	115	106	104	101	90	92	86	83
4	250	360	130	120	114	105	102	100	93	92	85	79
5	500	360	132	122	115	104	100	100	93	92	86	80
6	750	360	135	121	116	104	106	100	94	90	87	82
7	250	540	132	122	114	102	98	95	91	92	89	77
8	500	540	135	123	118	100	99	90	90	92	89	77
9	750	540	133	120	118	105	95	93	92	90	88	74

* Values are the average of 3 measurements
Totak Kjeldahl nitrogen concentration in manure = 5530 mg/L

Table 3. Ammonium Nitrogen Concentrations in Leachates (mg/L)*

Column No.	Manure Application (kg-N/ha)	Water Application (mL/3 days)	Day									
			3	6	9	12	15	18	21	24	27	30
0	0	180	14	13	14	14	14	11	12	11	10	12
1	250	180	23	21	19	18	19	15	15	14	14	13
2	500	180	24	22	20	19	19	15	15	13	13	14
3	750	180	24	22	20	20	20	15	15	14	14	13
4	250	360	24	22	19	18	17	14	14	13	13	13
5	500	360	26	23	20	19	18	15	15	14	14	14
6	750	360	26	23	20	20	18	14	15	13	13	13
7	250	540	23	21	18	17	14	14	14	13	14	14
8	500	540	25	22	18	16	15	15	14	15	14	13
9	750	540	25	24	19	18	15	15	14	14	13	12

* Values are the average of 3 measurements
Ammonium nitrogen concentration in manure = 570 mg/L

maximum of 560.5 mg TKN was leached out. Approximately 77% of the TKN in the manure was retained by the soil. The results also showed that both the water and the manure application rates had no pronounced effect on the concentration of the total Kjeldahl nitrogen found in the leachates. The average concentration of total kjeldahl nitrogen in the leachates obtained from the control column during the course of the experiment was 46 mg/L. Thus, the concentrations of total kjeldahl nitrogen in the leachates, as a result of manure application, were in the range of 28-89 mg/L.

6.2 Ammonium Nitrogen in Leachates

The measured values of ammonium nitrogen concentrations in leachates are presented in Table 3. The ammonium nitrogen concentration in the manure was 570 mg/L (10.3% of the total nitrogen). The ammonium nitrogen concentrations in the leachates ranged from a high of 26 mg/L on day 3 to a low of 12 mg/L on day 30, indicating that most of the ammonium nitrogen in the manure was either converted to nitrite and nitrate by the nitrification process or absorbed by the soil. The average concentration of ammonium nitrogen in the leachates obtained from the control column was 13 mg/L. This means that the ammonium nitrogen concentration in the leachates, caused by manure application was in the low range of 0-13 mg/L.

6.3 Nitrite Nitrogen in Leachates

The measured values of nitrite nitrogen concentrations in leachates are presented in Table 4. For all treatments, the nitrite nitrogen concentrations in the leachates first increased rapidly to a peak value, then decreased rapidly and finally declined gradually with time. The values ranged from a high of 0.42 mg/L on day 9 to a low of 0.10 mg/L on day 30. The results indicated that the first step of the nitrification process, in which the ammonium is converted to nitrite as described by equation 1, had been established.

$$NH_4^+ + 1\tfrac{1}{2}\,O_2 \xrightarrow{\text{Nitrosomonase}} NO_2 + 2H_2O \quad (1)$$

The average nitrite nitrogen concentration in leachates obtained from the control column was 0.12 mg/L, making the concentrations of nitrite nitrogen in the leachate, caused by manure application, to be in the low range of 0-0.30 mg/L. The low concentration of nitrite was due probably to the rapid conversion of nitrite to nitrate by the soil nitrifying bacteria.

6.4 Nitrate Nitrogen in Leachates

The measured values of nitrate nitrogen concentrations in the leachates are presented in Table 5. The nitrate nitrogen concentrations in the leachates exhibited a similar variation as the nitrite nitrogen. The values were relatively high on day 6 (9.5-16.0 mg/L) and then abruptly decreased with time reaching between 3.0 and 3.5 mg/L at the end of the experiment. The production of nitrate in a similar pattern to that of nitrite indicates that the second step of the nitrification process, in which nitrite is converted to nitrate as described by equation 2, was established.

$$NO_2 + \tfrac{1}{2}\,O_2 \xrightarrow[\text{Bacteria}]{\text{Nitrobacter}} NO_3 \quad (2)$$

The average concentration of nitrate nitrogen in leachates obtained from the control column was 3.8 mg/L, making the concentrations of nitrate nitrogen in the leachates, due to manure application, to be in the range of 0-12.2 mg/L.

6.5 Ammonia Volatilization

The measured values of volatilized ammonium are presented in Table 6. The release of ammonia from the manure was initially high but declined markedly, almost at a uniform rate, with time. Both the manure and the water application rates substantially affected the amount of ammonia released. Generally, increasing the manure application rate or decreasing the amount of water applied,increased the amount of ammonia collected from the columns. A range of between 104 and 268 mg of nitrogen was lost through ammonia volatilization from the soil surface, depending on manure and water application rates. This is equivalent to between 33 and 85 kg-N/ha.

7 DISCUSSION

It has long been established (Jenny, 1946) that the nitrogen contents (N) of undisturbed soils attain a steady state level, which depends on the combined

Table 4. Nitrite Nitrogen Concentrations in Leachates (mg/L)*

Column No.	Manure Application (kg-N/ha)	Water Application (mL/3 days)	Day 3	6	9	12	15	18	21	24	27	30
0	0	180	0.15	0.13	0.15	0.10	0.14	0.10	0.12	0.11	0.10	0.10
1	250	180	0.21	0.33	0.32	0.18	0.16	0.16	0.13	0.13	0.10	0.10
2	500	180	0.19	0.35	0.38	0.23	0.20	0.18	0.15	0.13	0.12	0.10
3	750	180	0.24	0.38	0.42	0.23	0.25	0.24	0.16	0.16	0.13	0.12
4	250	360	0.18	0.23	0.29	0.19	0.16	0.12	0.12	0.13	0.12	0.12
5	500	360	0.19	0.27	0.34	0.22	0.16	0.13	0.13	0.13	0.12	0.10
6	750	360	0.25	0.33	0.36	0.22	0.16	0.14	0.12	0.12	0.13	0.10
7	250	540	0.19	0.32	0.29	0.23	0.14	0.13	0.13	0.14	0.12	0.10
8	500	540	0.23	0.35	0.29	0.23	0.14	0.13	0.13	0.14	0.13	0.12
9	750	540	0.23	0.38	0.30	0.23	0.14	0.13	0.14	0.13	0.12	0.10

* Values are the average of 3 measurements

Table 5. Nitrate Nitrogen Concentrations in Leachates (mg/L)*

Column No.	Manure Application (kg-N/ha)	Water Application (mL/3 days)	Day 3	6	9	12	15	18	21	24	27	30
0	0	180	5.0	5.0	4.0	3.5	4.0	3.5	3.5	3.0	3.5	3.0
1	250	180	5.5	10.0	7.5	5.0	4.0	4.5	4.0	4.0	4.5	3.5
2	500	180	5.5	12.0	8.5	5.0	4.0	5.0	4.5	5.0	4.0	3.0
3	750	180	6.0	15.5	9.5	5.5	4.0	5.0	4.5	5.0	4.0	3.0
4	250	360	5.5	9.5	8.5	5.0	4.5	3.5	3.5	4.0	3.5	3.0
5	500	360	6.0	11.5	10.0	5.5	4.0	3.5	3.5	3.5	4.0	3.0
6	750	360	6.5	16.0	13.5	6.5	4.0	3.5	4.0	4.0	3.5	3.5
7	250	540	6.5	9.5	8.5	5.5	4.0	4.5	3.5	3.5	3.5	3.0
8	500	540	6.5	12.0	8.5	5.5	4.0	4.5	4.0	3.5	3.5	3.0
9	750	540	7.0	16.0	9.5	5.5	4.0	4.5	4.5	3.5	4.0	3.5

* Values are the average of 3 measurements

Table 6. Volatilized Ammonia from Land Applied Manure (mg NH_3/3 days)*

Column No.	Manure Application (kg-N/ha)	Water Application (mL/3 days)	Day 3	6	9	12	15	18	21	24	27	30
0	0	180	0	0	0	0	0	0	0	0	0	0
1	250	180	225	211	165	148	119	85	54	0	0	0
2	500	180	321	278	261	202	154	106	86	54	0	0
3	750	180	378	324	290	231	200	170	122	68	25	0
4	250	360	184	161	117	89	54	0	0	0	0	0
5	500	360	289	219	174	137	99	62	0	0	0	0
6	750	360	348	307	248	198	177	128	55	0	0	0
7	250	540	150	131	99	87	70	21	0	0	0	0
8	500	540	275	201	154	119	92	64	17	0	0	0
9	750	540	315	277	200	171	134	114	64	0	0	0

* Values are the average of 3 measurements

effects and interactions of the soil forming factors of climate (c), topography or relief (r), vegetation and micro-organisms (o), type of soils or parent materials (p) and age or time (t), such that it can be stated that,

$$N = f(c, r, o, p, t, \ldots) \quad (3)$$

Although, a conscious effort was made to make the gross bulk density of the soil columns in this experiment the same as that of the undisturbed soil, it is fully appreciated that the time allowed may have been too short to guarantee that the physical properties of the soil columns, especially with respect to their porosity, absorptivity, permeability, etc., were sufficiently reproduced to resemble the undisturbed soil. But, any deviation from the undisturbed or 'steady state' conditions would only contribute to making the leaching process more severe.

It is known that, other factors being reasonably favourable, crops commonly recover 50-75% of N added to the soil for purposes of fertilization, depending on the soil type and crop type. Even though the manure application rates used in this experiment were purposely in excess of normal levels for fertilization needs alone, it is evident that the leaching process was further facilitated by the absence of crops.

The water application intervals of three days is artificially regular and certainly more frequent than normal rainfall frequency. There is no opportunity for the applied water to be lost by run-off or transpiration. Furthermore, water loss by evaporation, if it occurred at all during the experiment, must have been much less than would occur in the field. So, in these regards, the leaching conditions were made much more severe than would be expected to occur in the field.

The specific arrangement of the soil columns excluded the impeding effects to leaching, of a water table or compacted lower layers of soil. The apparatus thus facilitated drainage which, according to Duxbury and Perverly (1978) and Terry (1980), enhances subsidence as a result of enhanced microbial decomposition of the organic matter, with the accompanying loss of organic N caused by its conversion, first to NH_4^+ through mineralization then to NO_2^- and NO_3^- by nitrification, followed by their subsequent loss through leaching. The rapid initial increase of the concentrations of nitrite and nitrate nitrogen demonstrated in this experiment may be accounted for by the accelerated processes of mineralization and nitrification. It is reasonable to suppose that, in actual field conditions, mineralization and nitrification would not occur quite as fast, so that the higher concentrations of nitrate nitrogen observed at the beginning of the experiment would not cause any real concern of a potential groundwater pollution hazard.

In spite of the above facts to the effect that the conditions of the experiment have been made to favour the leaching process, the inescapable conclusion from the results is that, given the observed low concentrations of the N-compounds in the leachates, the groundwater pollution potential from one season application of manure is very minimal. To provide more complete data on which to base a more general assessment/conclusion, it would be necessary to perform the experiment using different soil types at different soil depths, for longer durations and several consecutive (simulated) seasons.

The observation that the concentrations of the TKN in the leachates were independent of both manure and water application rates may seem curious but it can be explained on the basis of the so-called self-sealing phenomenon of certain types of soils by manure (Barrington, 1985).

On the other hand, the ammonia volatilization process was observed to depend on both water application rate and the amount of manure permitted to decompose on the soil surface, as was also found by Stevenson (1982). Contrary to the findings of Mills et al. (1974) that about 70% of the NH_3 volatilization occurred during the first 3 days after application of commercial fertilizers, the observation in this case was that, although the initial rate was quite high, the ammonia volatilization occurred more gradually and at a fairly uniform rate of decline over a period of three weeks. Nevertheless, the amount of ammonia released by volatilization can be very high, and sufficient to constitute a serious source of air pollution. To reduce this air pollution hazard, as well as the loss in fertilizer value of land applied manure it is recommended that the manure be soil covered or be applied by below-surface placement.

8 CONCLUSION

The concentrations of total kjeldahl, ammonium, nitrite and nitrate nitrogen in the leachates obtained from the soil columns were very low. Neither the manure application rate nor the water application rate had any substantial effect on the concentrations of nitrogen forms in the leachates. Although the concentrations of nitrate nitrogen at the end of the experiment were very small, higher concentration levels at the beginning may appear to have some potentials to cause ground water pollution; but, this is not likely to cause a serious concern under actual field conditions.

Ammonia released by volatilization from the applied manure is considered high enough to constitute a serious source of air pollution. From 33 to 85 kg/ha of nitrogen were lost through ammonia volatilization, depending on the water and manure application rates. Therefore, appropriate measures, such as below surface placement of manure, should be taken to reduce the ammonia volatilized from land applied manure.

9 REFERENCES

APHA. 1980. Standard methods for the examination of water and waste water. American Public Health Association, New York.

Barrington, S.T. 1985. The sealing of soils by manure. Unpublished Ph.D. Thesis. Agricultural Engineering Department. McGill University, Montreal, Quebec, Canada.

Black, H.M., D.D. Evans, L.G. Ensminger, J.L. White and F.E. Clark. 1965. Methods of Soil Analysis, Part I: Physical and Mineralogical Properties. American Society of Agronomy, Madison, Wisconsin.

Bosch, H.M., A.B. Rosenfield, R. Huston, H.R. Shipman and F.L. Woodward. 1950. Methemoglobinemia in Minnesota well supplies. American Water Works Association Journal 42:(1):161-170.

Campbell, A.B., A.N. Davis and P.I. Myhr. 1964. Methemoglobinemia of livestock caused by high nitrate contents of well water. Canadian Journal Comp. Med. Vet. Sci. 18(1):93-101.

Duxbury, J.M. and J.H. Peverly. 1978. Nitrogen and phosphorus loss from organic soils. J. Environ. Qual. 7:566-570.

Dzikoski, P.A., G. Kirlby, G. Read and W.G. Richards. 1984. The climate for agriculture in Atlantic Canada. Publication No. ACA-84-22-500, Advisory Committee on Agrometeorology, Halifax, Nova Scotia.

Elson, J. 1943. A 4-year study of the effects of crop lime, manure and fertilizer on macroaggregation of Dunnmore silt loam. Soil Sci. Soc. Am.Proc. J.,8:87-90.

Jenny, H. 1946. Factors of Soil Formation. Mc-Graw-Hill, New York.

Lund, Z.F. and B.D. Doss. 1980. Residual effects of dairy cattle manure on plant growth. Agron. J., 72:123-130.

Magdoff, F.R. and J.R. Amaddon. 1980. Yield trends and soil chemical changes resulting from continuous N and manure application to corn. Agron. J. 72:161-174.

Mathers, A.C. and B.A. Stewart. 1980. Effects of feedlot manure on soil physical and chemical properties. In Livestock Waste, Proceedings of the 4th International Symposium on Livestock Wastes. ASAE, St. Joseph, Michigan.

Mills, H.A., A.V. Barker and D.N.Maynard.1974. Ammonia volatilization from soils. Agron.J., 66:355-358.

Stevenson, F.J. 1982. Origin and distribution of nitrogen in soil. In Nitrogen in Agricultural Soils (Stevenson, F.J. ed.). Published by Am. Soc. of Agronomy, Inc., Crop Sci. Soc. of Am. In., and Soil Sci. Society of Am. Inc., Madison, Wisconsin, U.S.A.

Swader, F.N. and B.A. Stewart. 1972. The effect of feedlot wastes on water relations of pullman clay loam. ASAE Paper No. 72, 959, St. Joseph, Michigan.

Terry, R.E. 1980. Nitrogen mineralization in Florida Histosols. Soil Sci. Soc. Am. J. 44:747-750.

Unger, P.w. and B.A. Stewart. 1974. Feedlot waste effects on soil conditions and water evaporation. Soil Sci. Soc. Am. Proc., 38:954-957.

Land and Water Use, Dodd & Grace (eds), © 1989 Balkema, Rotterdam. ISBN 90 6191 980 0

Anaerobic processing of stable manure: Situation and development possibilities in Czechoslovakia

J.Wolff
Research Institute of Agricultural Engineering, Praha, Czechoslovakia

ABSTRACT: The classic way of storing excrements, i.e. reservoirs, farmyard heaps, and field storage areas does not guarantee the quality of manure making particularly with regard to the contents of organic matter and nutrients and these have often a negative effect upon the environment due to smell, insects and pollution of ground and surface water. Since the beginning of the 80ies the technology of anaerobic fermentation of excrements developped more intensively. Experience with operating municipal cleaning plants assisted the development of new more efficient digesters. Anaerobic fermentation of cattle manure containing more than 20 % DM makes use of the self-heating capacity of manure in the initial stage of the aerobic process. The technological line consists of a material feeding gantry crane, of fermentation baskets and covering bells. The elimination of exogenous heat supply is energetically convenient but at low ambient temperaturesthe process is less stable and the biogas production lower. Actual research is aimed at improving the process by extending the aerobic phase as far as to reach the thermophilic region and by mixing cattle manure with pig dung or with manure from calf sheds.

ABSTRAIT: Les méthodes traditionnelles de stockage, p.ex. fosses, dépôts de fumier à la ferme et au champs ne garantissent pasla maturation adéquate du fumier, surtout du point de vue des matières organiques et nutritives. Elles dégradent aussi l'environement. En Tchécoslovaquie, on a dévéloppé surtout la technique de la fermentation anaérobique des excréments depuis les années '80, en utilisant les éxpériences de traitement des déchets municipaux. Comme technologie nouvelle et plus éfficace, on a introduit la fermentation anaérobique du fumier des bovins à MS de plus que 20 %. On utilise l'autoréchauffement du fumier sans phase initiale aérobique. Le chantier se compose en principe d'une grue portale à remplir des corbeilles digestrices et des cloches à recouvrir les corbeilles. L'élimination des sources de chaleur exogènes étant favorable du point de vue énérgetique, il y a donc un désavantage en cas des températures basses ambientes qui déstabilisent le processus et abaissent la production du biogaz. La recherche est actuellementconcentrée à la prolongation de la phase aérobique jusq'au domaine thermophile et aux avantages du mélange du fumier bovin avec le lisier des porcs ou avec le fumier des veaux.

ABSTRAKT: Klassische Methoden der Lagerung des Dunges, d.h. Becken, Farm- und Feld-dunghaufen befriedigen öfters nicht im Bezug auf gute Ausreifung, Inhalt organischer Stoffe und Nährwert. Darüberhinaus gefährden sie die Umwelt. Ab Anfang 80.Jahre griff man in der ČSSR zu zu der Technologie der anaeroben Fermentierung der Exkremente unter Ausnützung der Erfahrungen mit der Behandlung städtischer Abfälle. Als neue Methode der Bearbeitung des Rinderdunges mit über 20 % liegendem TS-Gehalt wird anaerobe Fermentierung unter Ausnützung der Eigenerwärmung betrieben. Das Verfahren ist auf Fülleinrichtung (Portalkran), Fermentierungkörbern und Deckglocken gegründet. Energetisch ist es vorteilhaft, jedoch bei niedrigen Aussentemperaturen leidet die Stabilität des Prozesses sowie die Biogasproduktion. Nun richtet sich die Forschung auf die Regelung des Fermentierungszyklus durch Verlängerung der aeroben Phase bis in den thermofilen Bereich und auf die Mischung des Rinderdunges mit Schweinegülle oder mit Kälberdung.

Problems of the storage and application of livestock excrements are important in farming under any production conditions.

The traditional ways of storing straw manure and slurry, practiced in Czechoslovakia, are unsuitable in many cases or are improperly used. Czechoslovak farms work on large-scale conditions (there are, for example, farms with 600-800 dairy cows or 30 000 pigs) and the systems of manure handling on such farms still fail to meet the strict requirements for environment conversation. Improper storage and application of dung affect ground and surface water quality, cause smell, deteriorate the soil and so on.

Anaerobic fermenting is one of the ways how to improve manure management on farms. Fermenting processes are advantageous because of the possibility of controlling the storage conditions and producing biogas as a source of energy. It should be pointed out in this context that problems of adequate storage of livestock dung with its environmental implications are investigated with priority in Czechoslovakia; biogas production is regarded as an additional, though significant, factor of the quality and efficiency of fermenting processes.

These reasons and considerations have motivated the development of anaerobic fermenting methods since the early nineteen-eighties in Czechoslovakia.

Environment, especially in the near proximity of farms, is most significantly affected by pig and cattle slurry. Anaerobic fermentation of these materials is based on methods used in traditional sewage treatment plants. Its future development in Czechoslovakia will be focused on improving the effectiveness of the process and on reducing the detention time.

However, anaerobic fermentation of straw manure is more advantageous in agricultural practice than the processing of slurry, because the equipment is cheaper, costs of operation lower, operation simpler, and manuring with straw dung has a long-established tradition.

The method of anaerobic fermentation of straw manure, used in Czechoslovakia, is based on the direct use of dung from the stable, without crushing or diluting with liquids. No pumps are needed; the material is handled with conveyers, containers and cranes (truck cranes, portal cranes) with grabs. The principle of the technology is the use of the self-heating of straw dung under aerobic conditions. It is not necessary to use any external heat sources to heat the substrate. The technological process is fairly well known (Fig. 1) and has already been shown in literature.

Several biogas stations of this type are now in operation at small stables (up to 100 dairy cows). There are also some stations for straw manure processing on farms with 300, 600 and 1 000 cows.

Operation at these stations is promising and gives prerequisites for further development of this method of processing under the conditions of agriculture in Czechoslovakia and probably also some other countries.

Until now, the most significant experience with the operation of a biogas station has been gathered on the Hustopeče co-operative farm. The station of the Hustopeče farm processes dung from 1 000 cows and has been in operation since 1986. Practical experiments have shown clearly the advantages and drawbacks of the system in different periods.

The main advantage of the technology is the possibility to do without an external heat source. This promotes the effectiveness of the process and allows effective use of the biogas produced. Biogas is used to heat water and generate electric power. Another advantage is the simplicity of the project and low costs of operation.

However, the use of self-heating as the main advantage of the technology brings about some drawbacks. The fermenting processes are affected by ambient temperature and by the variability of straw dung quality. Good quality of the dung, especially dry matter content, appers to be the basic requirement for the process. Thermal insulation and suitable dosage of material in the fermentation baskets can just partially offset the effect of outdoor temperatures.

Full-scale trials were evaluated and fermentation curves were drawn. The following conclusions were drawn :

- at a gigh dry matter content in straw dung (above 24%), the maximum of the fermentation curve is constituted by biogas production between the 10th and 15th day of fermentation,
- at lower dry matter contents (below 24%), the maximum in the fermentation curve is constituted by production from the 1st to 8th of fermentation or the curve declines regularly without any marked peak.
- straw manure with less than 20 % cannot be used in this system with advantage. The adverse effect of these factors on

the fermentation process encouraged efforts for further trials with the mixing of the dung of dairy cows with mellow manure and also with the straw dung of calves or pig slurry.

All the trials are performed on farm scale (fermentor volume 169 m^3) to provide the possibility of working out accurate instructions concerning the operation and use of the equipment in farm practice.

The first trials were conducted with the use of fresh straw manure from the cattle barns and mellow dung, which is in fact fresh manure stored in the dung pit for 5-8 days and used for fermentation afterwards. The dung was put into the fermentation basked only after reaching 45°C at the minimum, which is a temperature at which the process starts reliably. During fermentation (28 to 32 days) the temperature falls maximally to 12°C. The results of biogas production in these trials are shown in Fig. 2. Variant "mixture 1 . 1" denotes the weight ratio of fresh and mellow dung put into the fermentation basket. The differences in the patterns of the curves are significant and indicate clearly that there are differences in biogas output from one case to another. If biogas production from fresh manure is taken as 100 %, then the output of biogas from the mixture is 30 % higher and from mellow dung up to 57 % higher.

The effect of mellow dung is observed until the 17th-20th day of fermentation; afterwards the output of biogas returns to a balanced state. The effect of a prolongation of the anaerobic stage on the manure is also documented by the starting temperatures of the process. When only mellow dung was used the temperature reached up to the thermophilous range (63°C). In the mixture it was 59°C and in fresh manure just 50.3 %.

The moment of gas inflammability is also an important parameter that can be influenced. As the process is discontinuous, has begins to burn only after some time. In fresh manure it burns on the 3rd - 5th day of fermenting, in the mixture already the second day, and in mellow manure the burning begins within several hours after the closing of the fermentation basket.

The subsequent operation trials were focused on further investigation of the effect of mellow manure on the process and on biogas production. The results of the measurements are recorded as fermentation curves (Fig. 3). The curves are denoted by the percentual proportions of mellow manure of the total weight of the batch.

The results confirm the positive influence of mellow manure on the process, especially during the first half of fermentation. Biogas balancing was observed, unlike in the trial described above, above, already on the 16th to 17th day of fermentation.

Generally, biogass production becomes balanced between the different mixtures on the 16th to 22nd day of fermentation; the individual differences are due to dry matter content in the material and outdoor temperatures.

From the practical point of view it should be noted that the use of mixtures is particularly advantageous in the winter season when there is the highest requirement for biogas. The use of mellow manure can also reduce fermentation time from the original 32 days to about 20 days. This increases the throughput of the equipment and accelerates the turnover of fermentation units (eight on the whole).

The experimental efforts also included investigation of the quality of fermented manure, with particular respect to losses of organic substances. Manure quality is the basic requirement, so it is not recommended to use exclusively (100 %) mellow manure. In such cases, the losses of organics are as high as 30 % or even more, which is not desirable with respect to manuring requirements. For the optimization of manure quality, biogas production and process stability, it is best to use a mixture of fresh straw manure with 40 to 60 % mellow dung.

The straw manure of calves and pig slurry are not good supplemental materials for fermentation; their proportion may be 10 - 15 % of the batch at the maximum.

There is quite a large potential for the development of anaerobic fermentation of straw dung and liquid manure in Czechoslovakia, especially in areas with higher requirements for environment conservation and with intensive animal production.

Elaboration and testing of various technologies of processing farm animal excrements on farm scale are now the main task of applied research in this field in Czechoslovakia.

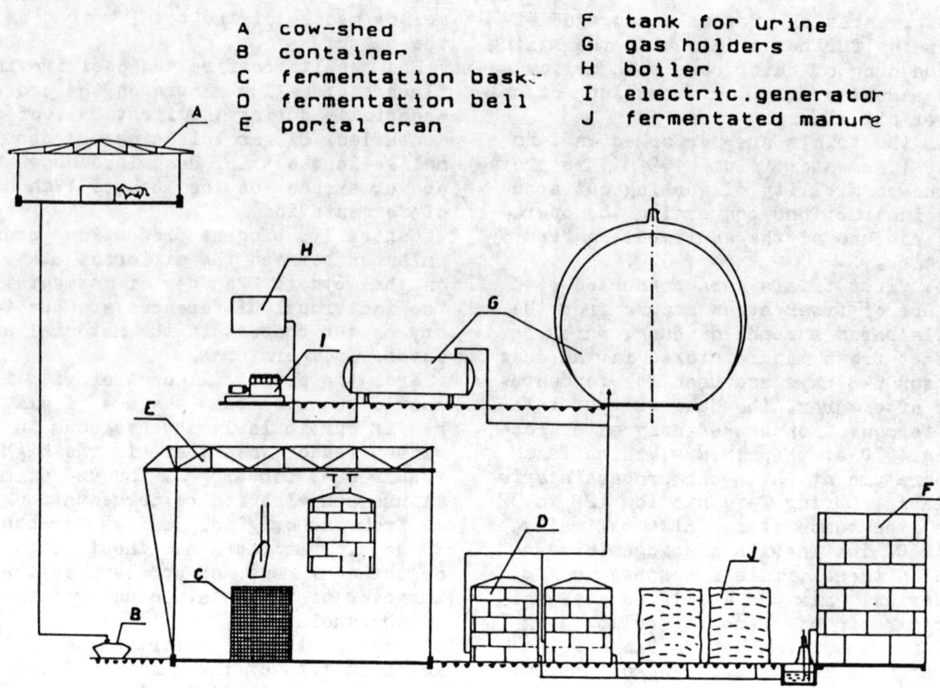

Fig. 1 Diagram of anaerobic treatment of straw manure without using thermic inputs

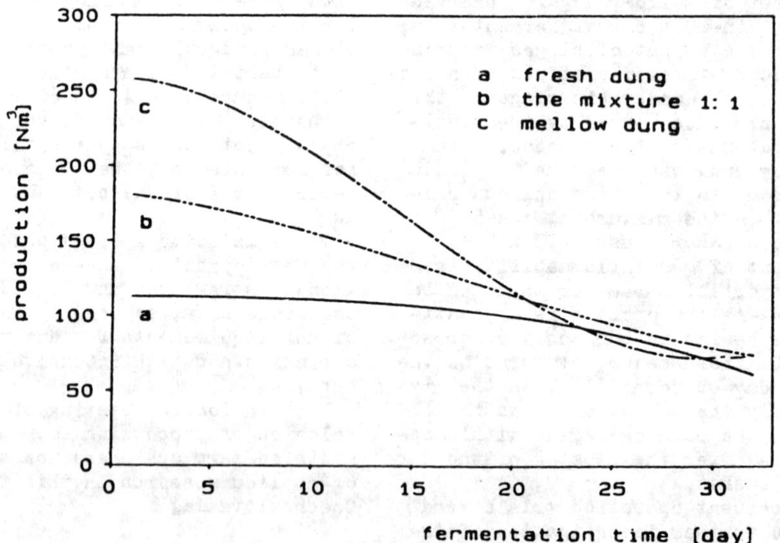

Fig. 2
The course of the fermentation cycles of excrements tested

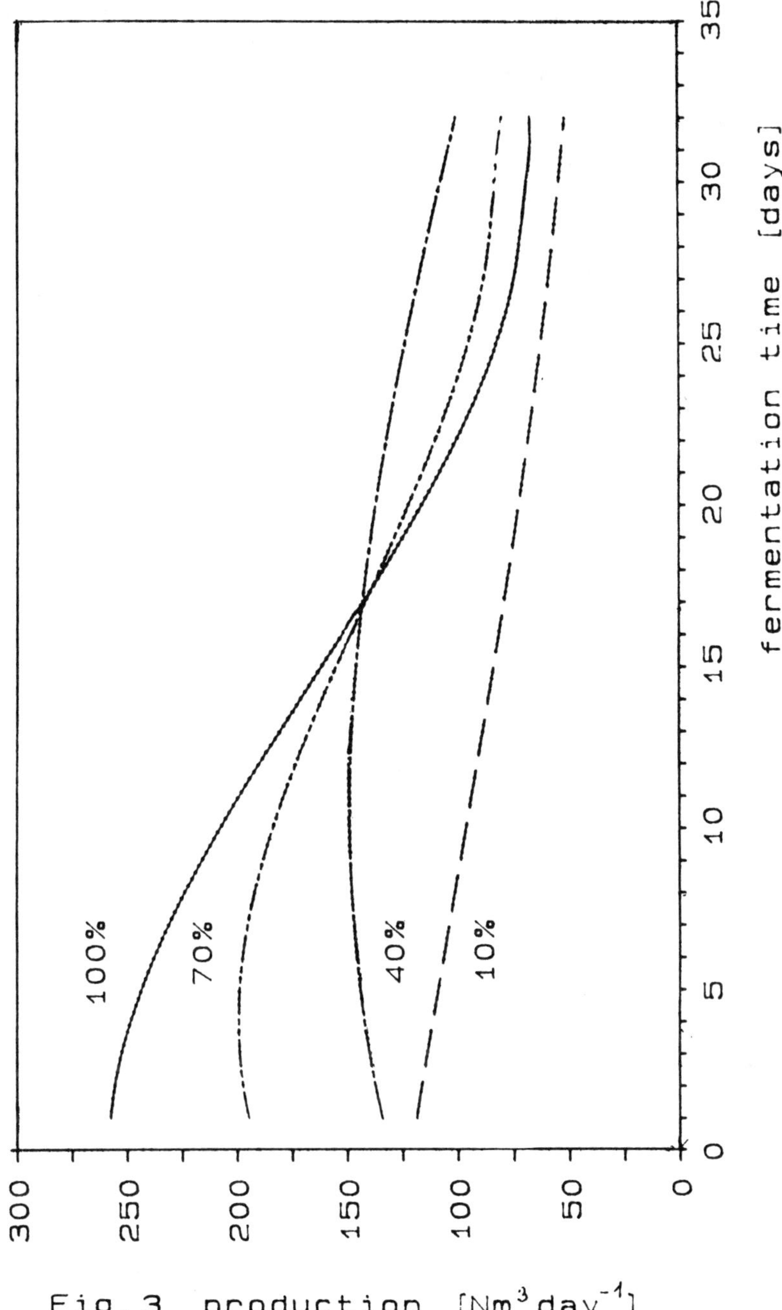

Influence of mellow manure onto biogas production

Land and Water Use, Dodd & Grace (eds), © 1989 Balkema, Rotterdam. ISBN 90 6191 980 0

Poultry manure composting – An evaluation of the process

R.C.Hansen, H.M.Keener & H.A.J.Hoitink
Ohio Agricultural Research and Development Center, Wooster, Ohio, USA

ABSTRACT: Types, proportions, and particle sizes of organic amendments to add to caged egg layer chicken manure were evaluated as they pertain to the kinetics of aerobic composting. Initial fraction of dry solids, mixing methods, and stirring frequency were also evaluated. Experimental equipment included 208 liter reactor vessels designed to recycle air through a plenum with a temperature controlled feedback system. Main effects of various levels of the controllable factors studied in the composting process were determined using an experimental design developed in Japan by Taguchi for industrial process design. The measured responses include compost weight reduction, dry matter reduction, nitrogen retention and quantity of poultry manure processed. The results will assist poultry farmers to design composting systems which lead to economical use or disposal of manure in ways which are environmentally acceptable.

1 INTRODUCTION

Composting continues to be a viable, recommended process for reclaiming or disposing of organic wastes in ways which are environmentally acceptable (M. De Bertoldi 1986). While a number of research projects have evaluated the composting of animal manures (Bell (1970), Galler and Davey (1971) and Willson (1971)), the use of the process is not widely accepted.

The success of any industrial process depends on the monitoring of materials, the setting of process controls, responding to process conditions and changes with sensitive feedback loops, and knowing the desired quality of the end product. The choices that have to be made to set up and operate a process pertain to controllable factors. The choices involve setting or selecting materials or levels for a process based on past experience, guessing, or a combination of both.

A schematic illustration of the composting process for poultry manure is shown in Figure 1. The process requires choices concerning the type of organic amendment to add, the proportion of amendment to add to the mixture, the amount of moisture to add and whether or not to recycle compost product. Table 1 shows a partial listing of controllable factors which pertain to the composting process.

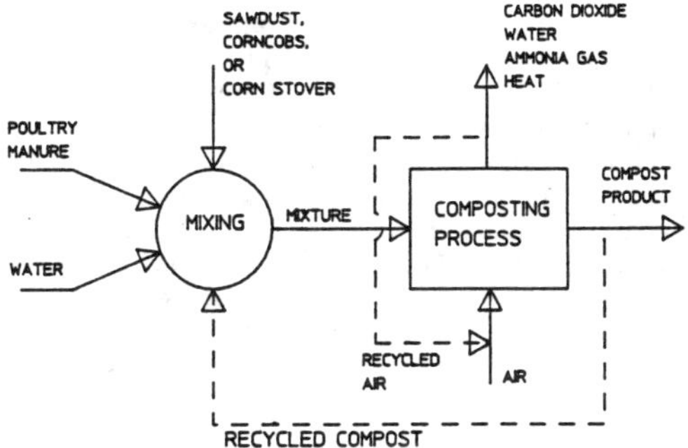

Fig.1 Schematic illustration of materials flow for composting poultry manure.

1.1 Design of experiments

Taguchi (1987) recognized that process engineers had no practical way to optimize settings for many manufacturing processes. However, if experiments could be designed to find those controllable factors which

Table 1. Controllable factors for composting.

Organic amendment	Moisture control
Carbon/Nitrogen ratio	Aeration
Particle size	Ambient temperature
Percent recycled compost	Retention time
Mixing equipment	Depth
Reactor vessel size	Percent recycled air
Stirring frequency	Type of process
Chemical pH moderating agent	Curing time
Initial moisture content	Inoculation
Temperature	Bulking agents

have the greatest impact on the quality of the final product, and if optimum combinations of settings could be identified, improvements could be made in product quality. Thousands of experiments are conducted by Japanese design and process engineers each year by using Taguchi methods of experimental design. The competitive advantage that Japanese products now have established in the world-wide market place is attributed, to a significant degree, to the use of Taguchi experimental design techniques (Box, et al 1987).

If a full factorial experiment were designed for 10 controllable factors set at 2 levels, 1024 tests would have to be run. Composting tests generally require a minimum time period of 2 weeks. A testing program of this magnitude would obviously be impractical in terms of both time and expense. A limited number of factors have to be selected for study, based on the experience of the persons involved. Then these factors are tested at chosen levels.

1.2 Research objectives

The following objectives were identified for this research:

1. To investigate the effects of a few select controllable factors on the composting process by comparing settings at two levels,
2. To evaluate the products of the composting process on the basis of four measured responses: reduction in compost weight, reduction in dry matter weight, reduction in nitrogen content, and quantity of poultry manure processed.

2 PROCEDURE

Seven controllable factors were selected from Table 1 for testing in a pilot study. Two levels were chosen for each factor as shown in Table 2. The choices pertain to decisions a poultry farmer would make if he were designing a composting process. The use of Taguchi's $L_8(2^7)$ orthoganal array permits seven factors to be evaluated at two levels by running eight tests (see Table 3). The letters in the design refer to the controllable factors listed in Table 2 while the numbers refer to either the first or second level to be tested.

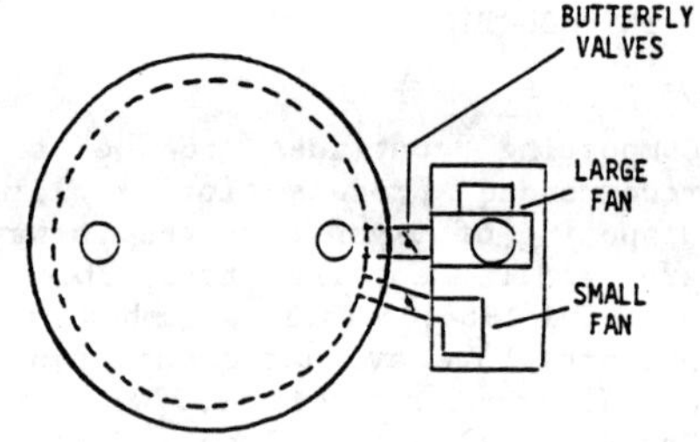

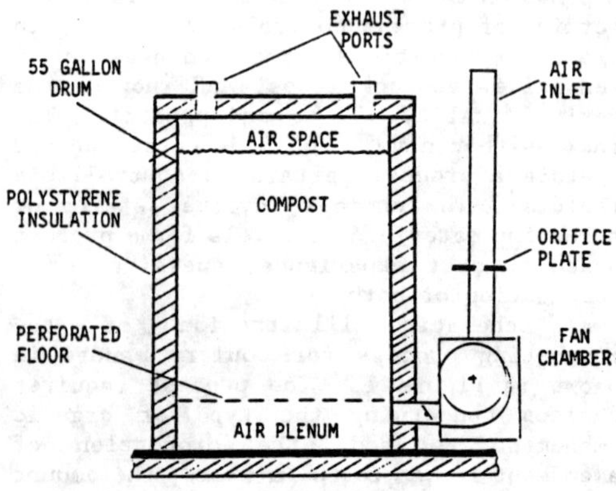

Fig.2 Schematic illustration of a reactor vessel showing air supply system.

Table 2. Controllable factors and levels selected for pilot test.

Controllable Factors	Level 1	Level 2
A. Type of Amendment	Sawdust	Corncobs
B. C/N Ratio	20	25
C. Mixing Method	Tumbler	Kelley-Duplex
D. Stirring Frequency	M-W-F	7th Day
E. Initial Dry Solids	.40	.50
F. Particle Size	Fine	Coarse
G. Temp Setpt	50°C	60°C

2.1 Experimental equipment

Three 208 liter vessels were designed and prepared as experimental reactors so concurrent composting tests could be conducted. Air is supplied to the compost alternately with two fans: low rate at 0.5 kg/hr and high rate at 2.3 kg/hr. The high rate fan is thermostatically controlled to come on at temperature set points of 50°C or 60°C with a differential of 3°C. The vessels hold about 100 kg of compost at a depth of 60 cm (see Figure 2). A more complete description of equipment is included in a previous paper (Hansen, et al 1988).

2.2 Testing method

The combinations of factors and levels designated by the L_8 array (Table 3) were randomly ordered as shown in Table 4. The first three tests (runs) were started on April 22, 1988. Because of mixing problems, Run No. 3 was repeated on May 6. The duration for each series of three runs was fourteen days. All tests were limited to the use of caged, egg-layer, chicken manure obtained from poultry research facilities at The Ohio State University. The chickens were consistently fed a standard, egg-layer ration.

At the completion of the pilot study outlined above, a second series of tests were run. The controllable factors and levels used for Test Series II are shown in Table 5. In these tests the effects of a lower than normal C/N ratio (C/N ratio = 15) are compared as level 1 to a C/N ratio = 20 for level 2.

Mixing methods were re-evaluated in these tests. The superior blending capability of the Kelly-Duplex paddle mixer did not lead to expected advantages over the tumble mixing method in the pilot study. The ribbon blender used for the Series II tests performed similarly to the Kelly-Duplex paddle mixer used in the pilot study.

Products of the composting process are commonly recycled to reduce the quantity of organic amendment required. The Series II tests were designed to compare the addition of 5% recycled compost to the addition of

Table 3. Taguchi experimental design for pilot study of composting process. (Letters refer to factors and numbers refer to levels: See Table 2)

Orthogonal Array $L_8(2^7)$

Test No.	A	B	C	D	E	F	G	Results
1	1	1	1	1	1	1	1	Y_1
2	1	1	1	2	2	2	2	Y_2
3	1	2	2	1	1	2	2	Y_3
4	1	2	2	2	2	1	1	Y_4
5	2	1	2	1	2	1	2	Y_5
6	2	1	2	2	1	2	1	Y_6
7	2	2	1	1	2	2	1	Y_7
8	2	2	1	2	1	1	2	Y_8

Table 4. Combinations of levels that were tested during the pilot study.

	Starting Date April 22, 1988		
	Run No 3	Run No 6	Run No 7
A.	Sawdust	Corncobs	Corncobs
B.	C/N = 25	C/N = 20	C/N = 25
C.	Kelly-Duplex	Kelly-Duplex	Tumbler
D.	M-W-F	7th Day	M-W-F
E.	DS = 0.40	DS = 0.40	DS = 0.50
F.	Coarse	Coarse	Coarse
G.	Temp = 60°C	Temp = 50°C	Temp = 50°C
	Starting Date May 6, 1988		
	Run No 4	Run No 5	Run No 3
A.	Sawdust	Corncobs	(Repeat)
B.	C/N = 25	C/N = 20	
C.	Kelly-Duplex	Kelly-Duplex	
D.	7th Day	M-W-F	
E.	DS = 0.50	DS = 0.50	
F.	Fine	Fine	
G.	Temp = 50°	Temp = 60°C	
	Starting Date May 20, 1988		
	Run No 2	Run No 8	Run No 1
A.	Sawdust	Corncobs	Sawdust
B.	C/N = 20	C/N = 25	C/N = 20
C.	Tumbler	Tumbler	Tumbler
D.	7th Day	7th Day	M-W-F
E.	DS = 0.50	DS = 0.40	DS = 0.40
F.	Coarse	Fine	Fine
G.	Temp = 60°C	Temp = 60°C	Temp = 50°C
	REACTOR 1	REACTOR 2	REACTOR 3

25% recycled compost (by weight).

An additional feature of the Series II tests was an attempt to replicate results. The $L_4(2^3)$ orthoganal array used for the tests is shown in Table 6 while the combinations tested and the testing schedule are shown in Table 7.

3 RESULTS

One purpose for designing a composting process could be to maximize the reduction of compost weight. Successful achievement of this purpose would lead to minimizing the quantity of product that ultimately needs to be handled, stored and hauled away. As much as 85% of compost weight loss results from reduction in moisture content of the compost mixture. The low moisture content of the end product (30 - 40%) along with metabolism of volatile solids leads to cessation of microbial activity and to biological stability of the finished compost.

Table 8 shows the percent reduction in compost weight for each of the eight combinations tested in the pilot study. The Taguchi method requires that the main effects of each of the tested controllable factors be analyzed. The impact of Factor A (the choice of organic amendment) is determined by adding the results for Run Nos. 1, 2, 3 and 4 when Factor A was set at Level 1. Then divide by 4 to find the mean value and enter that value in the response table in the A_1 position (See Table 9)

$\overline{A}_1$ = (43.0 + 31.9 + 33.5 + 27.1) / 4 = 33.9

Table 5. Controllable factors and levels selected for Series II tests.

Controllable Factors	Level 1	Level 2
A. C/N Ratio	15	20
B. Mixing Method	Tumbler	Ribbon Blender
C. Percent Recycled Compost	5	25

Table 6. Taguchi experimental design used for Series II tests

Orthoganal Array $L_4(2^3)$				
Test No.	A	B	C	Results
1	1	1	1	Y_1, Z_1
2	1	2	2	Y_2, Z_2
3	2	1	2	Y_3, Z_3
4	2	2	1	Y_4, Z_4

A similar procedure is followed to determine $\overline{A}_2$.

$\overline{A}_2$ = (43.3 + 51.5 + 42.3 + 56.9 / 4 = 48.5

Note that for $\overline{B}_1$, the results for Run Nos. 1, 2, 5 and 6 are averaged, i.e., these were the tests where Factor B was set at Level 1.

$\overline{B}_1$ = (43.0 + 31.9 + 43.3 + 51.5) / 4 = 42.4

The analysis continues in this fashion until the means for all factors at 2 levels

Table 7. Combinations of levels that were tested during Series II tests.

Starting Date	December 8,1988		
	Run No 1	Run No 2	Run No 3
A.	C/N = 15	C/N = 15	C/N = 20
B.	Tumbler	Ribbon	Tumbler
C.	5% Recycle	25% Recycle	25% Recycle
Starting Date	December 22, 1988		
	Run No 4	Run No 1	Run No 2
A.	C/N = 20	C/N = 15	C/N = 15
B.	Ribbon	Tumbler	Ribbon
C.	5% Recycle	5% Recycle	25% Recycle
Starting Date	January 5, 1989		
	Run No 3	Run No 4	Run No 2
A.	C/N = 20	C/N = 20	(Repeat)
B.	Tumbler	Ribbon	
C.	25% Recycle	5% Recycle	
	REACTOR 1	REACTOR 2	REACTOR 3

Table 8. Results showing percent reduction in compost weight for each of the eight combinations tested.

Orthogonal Array $L_8(2^7)$								
Run No.	A	B	C	D	E	F	G	Results
1	1	1	1	1	1	1	1	43.0
2	1	1	1	2	2	2	2	31.9
3	1	2	2	1	1	2	2	33.5
4	1	2	2	2	2	1	1	27.1
5	2	1	2	1	2	1	2	43.3
6	2	1	2	2	1	2	1	51.5
7	2	2	1	1	2	2	1	42.3
8	2	2	1	2	1	1	2	56.9

are determined and recorded in the response table.

The best combination of controllable factors can now be selected from the response table by noting which level for each factor resulted in the greatest reduction in compost weight. The best combination was:

$$A_2\ B_1\ C_1\ D_2\ E_1\ F_1\ G_2$$

Table 9. Response table summarizing results from pilot study.

	A	B	C	D	E	F	G
Percent reduction in compost weight							
1	33.9	42.4*	43.5*	40.5	46.2*	42.6*	41.0
2	48.5*	40.0	38.9	41.9*	36.2	39.8	41.4*
Δ	14.6	2.4	4.6	1.4	10.0	2.8	0.4
Percent reduction in dry matter							
1	13.5	15.3	18.8*	18.1*	18.5*	16.7	16.8
2	20.1*	18.3*	14.8	15.5	15.1	16.9*	16.8
Δ	6.6	3.0	4.0	2.6	3.4	0.2	0.0
Percent reduction in nitrogen content							
1	28.5*	30.6	27.0*	27.3*	28.7	31.6	29.7
2	29.4	26.6*	30.3	31.1	28.0*	26.9*	27.8*
Δ	0.9	4.0	3.3	3.8	0.7	4.7	1.9
Weight (kg) of poultry manure processed (dry basis)							
1	11.2*	12.6*	-	-	10.7	-	-
2	10.9	9.6	-	-	11.5*	-	-
Δ	0.3	3.0			0.8		

*Best combination for indicated response.

Reference to Table 8 shows that none of the 8 combinations tested were the same as the "best" combination identified with the response table. Confirming tests to evaluate this "best" combination are recommended.

Reduction in compost weight as a measured response was also evaluated in the Series II tests. Table 10 shows two replications of these results for each of four tests. A response table is prepared from the replicated results to determine the main effect of each controllable factor (See Table 11).

By conducting a similar analysis, values for percent reduction in dry matter, percent reduction in nitrogen content, and weight of poultry manure processed were determined and entered into the response tables. Factors that would not be expected to have an effect on quantity of poultry manure processed are left blank (See Tables 9 and 11). The main effects of each controllable factor can be assessed and the factors having the largest impact can be noted.

Table 10. Results from Series II tests showing percent reduction in compost weight for each of the four combinations tested

Orthogonal Array $L_4(2^3)$				
Run No.	A	B	C	Results
1	1	1	1	36.1, 38.6
2	1	2	2	29.1, 32.8
3	2	1	2	26.8, 37.5
4	2	2	1	47.3, 42.9

Table 11. Response table summarizing results from Series II tests

	A	B	C
Percent reduction in compost weight			
1	34.2	34.8	41.2*
2	38.6*	38.0*	31.6
Δ	4.4	3.2	9.6
Percent reduction in dry matter			
1	14.8	14.4	18.6*
2	15.8*	16.2*	12.0
Δ	1.0	1.8	6.6
Percent reduction in nitrogen content			
1	43.7	36.5*	47.7
2	29.7*	36.9	25.7*
Δ	14.0	0.4	22.0
Weight (kg) of poultry manure processed (dry basis)			
1	15.4*	-	16.2*
2	10.4	-	9.7
Δ	5.0	-	6.5

*Best combination for indicated response.

3.1 Response curves

A graphic illustration of measured responses is shown in Figures 3 and 4. The significance of the factors and levels tested can be compared and evaluated with response curves. The greater the slope in the lines, the greater the impact of the controllable factor tested. The best combination of levels can be identified for a given measured response depending on whether larger is better (reduction in compost weight) or whether smaller is better (reduction in nitrogen content).

3.2 Discussion of results

The response curves for reduction in compost weight in Figure 3 show that the choice of organic amendment was more significant than any other factor tested in the pilot study. The dry solids content of the starting mixture was also a significant factor. The results show that a mixture that is initially wetter (dry solids - 0.40) brings about greater weight reduction than a drier starting mixture (dry solids = 0.50). The Series II test results illustrated in Figure 4 show a significant increase in compost weight reduction as C/N ratio increases from 15 to 20. In contrast, results in Figure 3 show a decrease in compost weight reduction as C/N ratio increases from 20 to 25. Therefore, these results point to a possible optimum C/N ratio somewhere close to 20.

Compost products are commonly recycled and used in place of organic amendments as

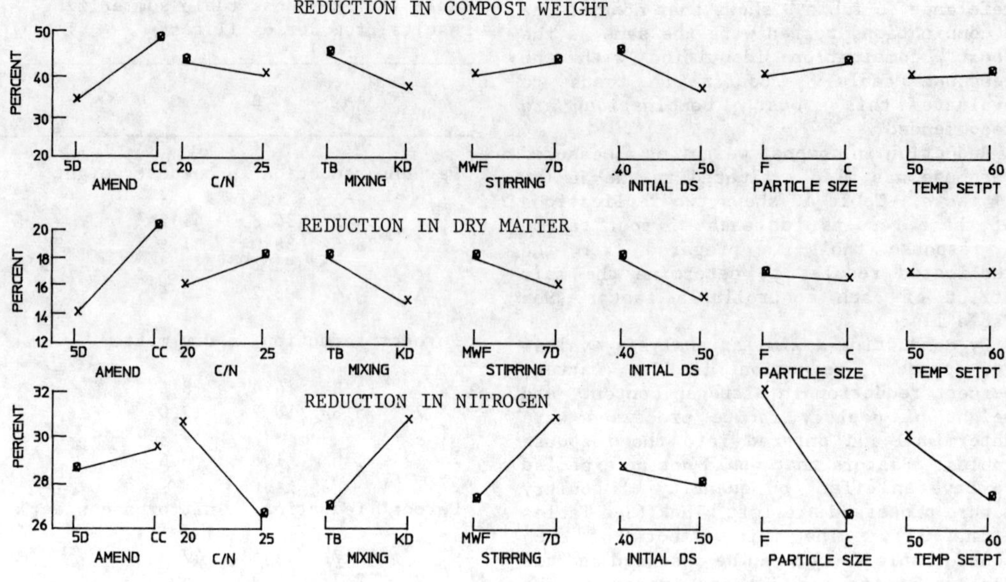

Fig.3 Response curves - composting process/pilot study.
(ⓧ Best combination of levels for factors tested)

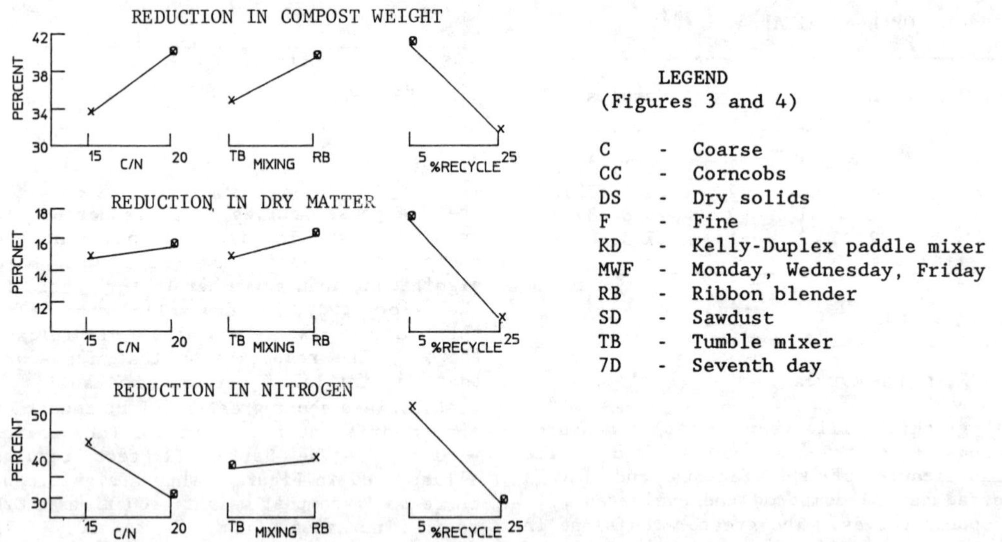

Fig.4 Response curves - composting process/Series II Tests.
(ⓧ Best combination of levels for factors tested)

a means of saving costs. Results illustrated with the response curve in Figure 4 show a significant negative effect on compost weight reduction when the proportion of the initial compost mixture increases from 5 to 25% recycled compost.

The thoroughness of mixing would intuitively appear to have a significant effect on measured responses for the composting process. However, the results did not show consistent, significant advantages for blending versus tumble mixing. During the pilot study, the tumble mixer showed an advantage in dry matter reduction; during Series II tests the tumble mixer showed a disadvantage to the ribbon blender.

Reduction in dry matter weight increased with increasing C/N ratios in both series of tests, while a decrease was noted for increasing percent recycled compost.

Percent reduction in nitrogen content was shown to be significantly affected by four factors during the pilot study: C/N ratio, mixing method, stirring frequency and particle size. Particle size showed the most significant effect on nitrogen retention. Amendments consisting of coarse particles retained nitrogen during composting better than amendments made of fine particles. The next most significant controllable factor was C/N ratio: nitrogen losses increased as C/N ratio decreased for both series of tests. Stirring frequency was also shown to be significant: compost that was remixed three times per week retained more nitrogen than compost that was only remixed once per week. Addition of 25% recycled compost significantly increased nitrogen retention. This result most likely occurred because the nitrogen in the recycled component of the compost mixture was fixed.

Another desired feature for a poultry manure composting process would be to identify the combination of factors that would maximize the quantity of manure that would be composted during given time and space constraints. C/N ratio, initial dry solids content, and percent recycled compost are all controllable factors which have a significant effect on quantity of manure processed. Unfortunately, while a low C/N ratio and a low percent of recycled compost lead to a maximum quantity of manure processed, these levels also lead to maximum nitrogen losses.

4 CONCLUSIONS

Poultry manure composting is a process which consists of controllable factors set at selected levels. The combination of settings which will optimize a process can be effectively evaluated by following experimental design techniques developed by Taguchi for industrial manufacturing processes.

This paper reported the results for four measured responses: (1) reduction in compost weight, (2) reduction in dry matter weight, (3) reduction in nitrogen content, and (4) quantity of poultry manure processed. A best combination of controllable factors and levels was identified for each desired process outcome.

The results of the experiments reported here may not pertain to poultry farmers in other localities. For example, rice straw or peanut hulls may be an economical organic amendment. The manure may already be amended with organic material used as bedding. Therefore, customized tests need to be conducted that pertain to the unique conditions represented by each poultry operation.

This research has demonstrated that the Taguchi method of experimental design can be used to optimize the poultry manure composting process for any set of conditions when the desired outcome for the process is known.

REFERENCES

Bell, R.G. 1970. The influence of aeration on the composting of poultry manure-ground corncob mixtures. J. Agr. Engr. Research 15(1):11-16.

Box, George E.P., R. Kackar, V. Nair, M. Phadke, A. Shoemaker, and C.F. Jeff Wu. 1987. On quality practice in Japan. Center for Quality and Productivity Improvement, Report No. 27, University of Wisconsin, Madison. 16 pp.

De Bertoldi, M., M.P.Ferranti, P.L'Hermite, and F. Zucconi (eds.) 1986. Compost: Production, quality and use. London and New York: Elsevier Applied Science.

Galler, W.S. and C.B. Davey. 1971. High rate poultry manure composting with sawdust. Proceedings of the International Symposium on Livestock Wastes, ASAE. St. Joseph, MI. pp 159-162.

Hansen, R.C., H.M. Keener, and H.A.J. Hoitink. 1988. Poultry manure composting: system design. Presented at the International Summer Meeting of the ASAE, Rapid City, SD. (Pub. Pend.)

Taguchi, G. 1987. The system of experimental design (2 vols.). UNIPUB/Kraus International Publications, Lanham, MD.

Willson, G.B. 1971. Composting dairy cow wastes. Proc. International Symposium on Livestock Wastes, ASAE. St. Joseph, MI. pp. 163-165.

Land and Water Use, Dodd & Grace (eds), © 1989 Balkema, Rotterdam. ISBN 90 6191 980 0

Electrical processes for the treatment of animal slurry

L.Campbell, G.Rahilly & C.Maguire
Electricity Supply Board, Dublin, Ireland

ABSTRACT: This paper describes pilot studies undertaken by the Electricity Supply Board, Ireland, for the treatment of animal slurry and other effluents, using electrical processes.
The processes involved are:
- Digestion by Electro-mechanical Aeration
- Evaporation by Mechanical Vapour Recompression
- Sterilisation or Pasteurisation

Whereas the individual processes are well tried and tested the combination in the pilot study has made the project unique for the disposal of animal slurry. Public concern on the unresolved issues of water and land pollution, was the driving force behind the ESB involvement. It was felt that Aerobic Digestion with its attendant higher reaction temperatures was a route which should be investigated since it had already proved effective with domestic sewage. Mechanical Vapour Recompression already established as a useful tool in the battle against pollution from industrial wastes allows concentration of effluents with high energy efficiency, while producing a clean discharge. These two processes combined with sterilisation or pasteurisation would make it possible to separate out many of the valuable constituents of waste materials at low cost.

RÉSUMÉ: Cet exposé decrit les études pilotes entreprises par l'Electricity Supply Board d'Irlande concernant le traitement des boues animales et d'autres effluents grâce à l'utilisation de procédés électriques.

Ces procédés sont les suivants:

- Digestion par aération électromécanique
- Evaporation par recompression mecanique de vapeur
- Stérilisation ou pasteurisation

Les procédés individuels ont été amplement testés et éprouvés mais leur association dans le cadre de l'étude pilote fait de ce procédé d'élimination des boues animales un projet unique en son genre. L'intérêt public en ce qui concerne les questions non encore résolues de la pollution de la terre et des eaux a été la motivation principale de l'ESB dans cette étude. On a pensé que la digestion aérobique et les températures de réaction plus élevées qui l'accompagnent était une méthode à examiner de façon approfondie puisqu'elle s'etait déjà révélée efficace avec les effluents domestiques. La recompression mécanique de vapeur, déjà bien établie en tant qu'outil utile dans la lutte contre la pollution produite par les déchets industriels permet une concentration d'effluents de haute efficacité énergétique tout en produisant un écoulement propre. Ces deux procédés, associés à la stérilisation ou la pasteurisation permettraient de séparer à moindre coût, un grand nombre des constituants de valeur des résidus.

AUSZUG: In diesem Beitrag wird eine Versuchsstudie über die Behandlung von Tier-Abwässern und anderen Abfallstoffen unter Einsatz von elektrischen Verfahren beschrieben, die von der irischen Stromversorgungsbehörde (Electricity Supply Board) durchgeführt wurde.

Bie den Verfahren handelt es sich um:
- Verfaulung durch elektromechanische Belüftung
- Verdunstung durch mechanische Dampfverdichtung
- Sterilisierung oder Pasteurisierung

Während sich die individuellen Prozesse seit langem bewährt haben, stellt ihre Kombination in der Versuchsstudie ein einmaliges Projekt in der Beseitigung von Tier-Abwässern dar. Die öffentliche Unruhe in Bezug auf die ungelösten Probleme in der Wasser-und Erdverschmutzung waren das Hauptmotiv für die Beteiligung der ESB an diesem Projekt. Man war der Meinung, daß die Luftverfaulung mit den dabei entstehenden höheren Reaktionstemperaturen einen Ausgangspunkt darstellte, der näher untersucht werden sollte, da sich dieses Verfahren bereits bei Haushaltsabwässern bewährt hat. Die mechanische Dampfverdichtung, bereits als wirksames Mittel gegen die Verschmutzung durch Industrieabfälle etabliert, ermöglicht die Konzentration von Abwässern mit hoher Energieeffizienz, wobei jedoch ein sauberer Abfluß entsteht.

Electrical Processes for the Treatment of Animal Slurry

Modern intensive rearing of animals is giving rise to problems with respect to disposal of wastes since the traditional method of land-spreading is no longer applicable owing to the huge quantities of slurry involved. Pollution of lakes, rivers and groundwater is causing concern and in some cases is putting a limit on the possible expansion of existing enterprises. Animal slurry may contain many organic compounds and much potential energy, the value of which cannot be realised with present disposal methods. The proposals outlined in this paper are intended to indicate the application of electrical processes which could help to resolve the pollution problem and at the same time allow the realisation of some of the inherent value of the slurry.

A number of processes exist whose characteristics are well known but which have not been used in conjunction in the treatment or disposal of farm waste. These include:

- Mechanical Aeration
- Pasteurisation or Sterilisation
- Mechanical Vapour Recompression (MVR)
- Reverse Osmosis (RO)
- Irradiation

Briefly, the characteristics of these processes which make them attractive for this application are as follows:

Mechanical Aeration:
Mechanical aeration by means of submerged jets has been used to treat town sewage and many other types of waste. In a suitably designed treatment tank high rates of oxygen absorption can be achieved which results in operation at relatively high temperatures - up to 70°C - favouring the development of thermophyllic bacteria[1]. At these temperatures the efficiency of absorption of O_2 from the injected air is high - up to 50% - yielding in excess of 3kg of O_2 absorbed per kW. H_2 consumed, for a tank of approximately 4 metres depth.

The reaction results in considerable reduction of B.O.D., modification of the nature of sludge and the generation of large quantities of heat.

Pasteurisation & Sterilisation:
Depending upon the eventual end use of the treated material the presence of certain moulds or spores which can survive the thermophyllic temperatures can be detrimental and an electrical pasteurisation or sterilisation process would be proposed for their elimination. With the use of suitable heat exchangers the operating cost can be quite low, with a power consumption of about 6kWhrs per ton of liquid treated.

Mechanical Vapour Recompression:
MVR has been applied to the concentration of foodstuffs[2] and many industrial processes[3] for many years and it is proposed that a suitably modified process could be applied to the removal of water from animal slurries with its attendant low energy consumption of 10 or 12kWhrs per tonne of water evaporated. By this means the slurry can be concentrated and the water content returned as a relatively pure condensate.

Reverse Osmosis:
This process, also well known, can remove water from many product streams for as little as 6kWhrs per tonne. In some cases it may find application where

concentration or extraction of salts is desired.

Irradiation:
Irradiation by Gamma Rays finds application in the foodstuffs industry and could provide an alternative to pasteurisation or sterilisation by traditional means.

Aeration of Pig Slurry:
As an indication of how an electrical process could be applied to the treatment of animal waste, prolonged aeration of a sample of pig slurry was undertaken.

The sample selected had a relatively low Dry Matter content of 2.03% with the following analysis:

C.O.D.	24,644p.p.m.
B.O.D.	18,000p.p.m.
Nitrogen	2.97g/kg
Phusphorus	.149g/kg
Pottasium	1.94g/kg

Also, it carried the usual malodour associated with pig slurry drawn from a holding tank.

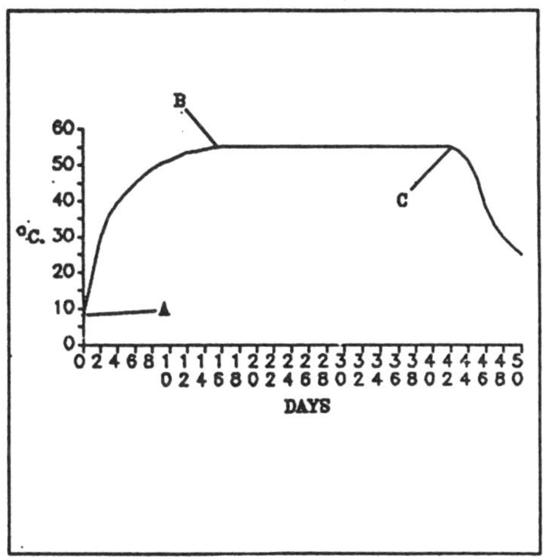

Fig. 1 Heating & Cooling Curve

Aeration:
50 tons of pig slurry was contained in an underground tank with slated covers and fixed with a $7\frac{1}{2}$kW venturi aerator delivering $178m^3/hr$ air.

Aeration commenced at Point A when the bulk temperature was 8°C. At this temperature, O_2 absorption was such that a net output of 28kW was produced due to bacterial action. At Point B the temperature of the slurry ceased rising and the net output was about 52kW.

When the aeration was turned at C the activity within the tank did not cease immediately but continued utilising free or excess oxygen after which the drop of temperature followed the more usual experimental decay.

By insulating the tank walls, a heat output of approx. 50kW could have been obtained and fully utilising this heat would enable a saving of approx £1 per pig to be made at the farm where the experiment was carried out. A further by product of the aeration - thermophyllic bacteria - could be beneficial to the growth of mushrooms and a further experiment has been put in hand to ascertain if the aerated slurry could be of value in making compost.

Mechanical Vapour Recompression:
In many areas land spreading of animal slurry is no longer the answer to the increasing volumes of the material being produced and a combination of aeration and concentration by Mechanical Vapour Recompression is proposed as a means of extracting the valuable fertilisers present to produce a soil conditioner or biological fertiliser when combined with the previously extracted fibres. The condensate from the M.V.R. plant could be reusable in which case it may benefit from sterilisation by U.V. radiation.

These comments and suggestions are presented to indicate where processes, familiar in other sectors of industry, might find valuable application in alleviating the problem associated with the treatment and disposal of animal wastes.

References

1. Morgan, S., Littlewood, M.H., Winstanley, R. March 1984. Electricity Council Research Centre, Morandum, M1808.

2. Rahilly, G. 1987 Mechanical Vapour Recompression in the Food Industry. The Engineers Journal, 40: 20-23.

3. O'Neill, G. Sept. 1987 MVR in the Dairy Industry. Co-Op Ireland: 51-53.

Land and Water Use, Dodd & Grace (eds), © 1989 Balkema, Rotterdam. ISBN 90 6191 980 0

Möglichkeiten des Umwelt- und Resourcenschutzes in Flurbereinigungsverfahren – Erfahrungen aus Deutschland

K.Kollmer
Landesamt für Flurbereinigung, Baden-Württemberg, Kornwestheim, Bundesrepublik Deutschland

KURZFASSUNG: Der Bericht zeigt, wie im Rahmen von Flurbereinigungsverfahren auf intensive Weise Landentwicklung mit dem Ziel einer verbesserten Nutzung bei weitgehendem Interessenausgleich der verschiedenen Nutzungsansprüche erreicht werden kann.
Hierbei kann der Schutz natürlicher Resourcen wie Wasser, Boden, Artenvielfalt durch Planung, Flächenbeschaffung, Bodenordnung, Finanzierung, Maßnahmenausführung wirksam unterstützt und gefördert werden. Beispiele zeigen Ergebnisse dieser Verfahrensweise.

ABSTRACT: The paper deals with methods and procedures of land consolidation aiming to develop rural regions in order to arrange an optimal form of land use. The special goals thereby are the protection of environments and natural resources like water, soil, nature and wildlife. The mainly used tools are careful planing in cooperation with all competent authorities and the groundowners, land acquisition, subvention and financing and finally the realisation by a total reorganisation of the project area. Examples in the field of landscape improvments and the protection of water and soil shall give illustrations by results.

ABRÉGÉ: Le rapport montre comment, dans le cadre des mesures de remembrement, on peut arriver à un développement rural intensif en ayant pour but d'améliorer l'exploitation du sol tout en tenant compte, dans la mesure du possible, les intérêts des différents usagers.
En même temps la protection des ressources naturelles comme l'eau, le sol, la diversité de la faune et de la flore peut être soutenue et favorisée efficacement par la planification, l'acquisition de terrains, l'aménagement du territoire, le financement et la réalisation des mesures en vue. Des exemples montrent les résultats de ce procédé.

1 ZIELE DER FLURBEREINIGUNG

Über 100 Jahre Flurbereinigungsarbeit hatten in Deutschland vorrangig das Ziel, die land- und forstwirtschaftliche Produktion in Quantität und Qualität zu steigern, um

- die Nahrungs- und Rohstoffversorgung zu verbessern,
- die Einkommens- und Lebensverhältnisse der ländlichen Bevölkerung zu verbessern,
- die Kulturlandschaft zu erhalten und auf rentable Weise zu pflegen.

In neuerer Zeit kam dazu die weiter gefaßte Aufgabenstellung, durch Bodenordnung eine verbesserte und sinnvollere, d.h. den potentiellen Möglichkeiten besser angepaßte Flächennutzung herbeizuführen.

Flurbereinigung war und ist somit noch immer eine Maßnahme, die die Landbewirtschaftung erleichtern, verbessern und rentabler gestalten soll. Heute wird jedoch diese Aufgabenstellung umfassender gesehen und ausgeführt. Es gilt die Landnutzung zu optimieren, wobei Nutzung nicht mehr gleich Bewirtschaftung, und Nutzen nicht mehr nur als geldwerter Ertrag gesehen wird.

Flurbereinigung ist damit mehr denn je ein Instrument, das den Gemeinden durch Bodenordnung ermöglicht, die jeweils gewollten und in politischen Entscheidungen festgelegten Entwicklungsziele zu erreichen.

Hier sei auch gleich auf zwei zentrale Probleme der Flurbereinigungsarbeit hingewiesen:

- Nutzen ist nur teilweise monetär oder in technisch-naturwissenschaftlichen Maßsystemen erfaßbar.
- die darin enthaltene Bewertungsfrage ist in hohem Maße abhängig von der politischen Gewichtung und Wertung der verschiedenen Formen des Nutzens.

2 LEISTUNGSANGEBOT DER FLURBEREINIGUNG

Die Möglichkeiten der Flurbereinigung bei dieser Aufgabenstellung, Lösungen und Hilfe anzubieten, liegen in den Bereichen

- der Planung und Planungsabstimmung
- der Flächenbeschaffung, der Umnutzung und Rekultivierung, der Bedarfsreduzierung,
- der Bodenordnung und der Schaffung rechtlich geordneter Verhältnisse (besonders zur Regelung von Unterhaltung und einer evtl. Nutzung)
- der Finanzierung
- der Maßnahmenausführung

Die Flurbereinigungsverwaltung vermag ein gut organisiertes und strukturiertes Angebot mit leistungsfähigem und leistungsbereitem Personal anzubieten.

Unser Problem liegt dabei leider häufig in einer quantitativen Überforderung der verfügbaren Kapazitäten, das dann zu unerträglich langen Bearbeitungszeiten führen muß.

Zur Vervollständigung des Leistungsbildes sei auch aufgezeigt, daß Unterhaltsleistungen auf Dauer, Nutzungs- und Pflegeüberwachung, also die langfristige Erhaltung eines gewünschten Zustandes wegen der verfahrensbedingt befristeten Zuständigkeit von der Flurbereinigungsverwaltung und von der Teilnehmergemeinschaft nicht übernommen und gewährleistet werden kann. Hier müssen Naturschutz- und Liegenschaftsverwaltung, Gemeinde und evtl. auch Verbände eintreten.

3 PLANUNG UND PLANUNGSABSTIMMUNG

Die Planung im Rahmen der Flurbereinigung vollzieht sich in den Stufen Verfahrenseinleitung, Plan nach § 41 FlurbG, Abfindungsentwurf und Flurbereinigungsplan mit seinen Nachträgen.

Jede Phase umfaßt eine gründliche Erfassung und Bewertung des Ist-Zustandes, Erarbeitung der Zielvorgabe mit Alternativen und die erschöpfende Abstimmung mit den Trägern öffentlicher Belange, d.h. insbesondere auch mit den Behörden und Organisationen des Naturschutzes. Dies gilt übrigens keineswegs nur für land- und forstwirtschaftliche Nutzflächen, sondern im Rahmen der Dorferneuerung auch häufig für Ortslagen.

Hervorzuheben ist bei der Planungsarbeit insbesondere

- die gründliche Erfassung und Bewertung der Landschaftselemente,
- die systematische Untersuchung der Erhaltungs- und Verbesserungsmöglichkeiten,
- die Abstimmung mit anderen Planungs und Entwicklungsinteressen im Sinne einer Optimierung,
- die rechtliche Vollzugssicherung.

Die Anforderungen an den Flurbereinigungsingenieur als Planer sind hoch. Deshalb wird auch in Aus- und Fortbildung intensiv für Erhaltung und Steigerung der fachlichen und organisatorischen Qualifikation unseres Personals gearbeitet.

Als Qualitätsmerkmal ausgezeichneter Planungsarbeit gilt, daß jeder wichtige Planungsabschnitt einvernehmlich mit den Trägern öffentlicher Belange abgestimmt ist und nur in ganz wenigen Ausnahmefällen gegen das Votum eines anderen Planungsträgers festgesetzt werden muß. Dies gilt übrigens auch für die betroffenen Grundeigentümer: Nur ein verschwindend geringer Bruchteil von Teilnehmern sucht auf dem Verwaltungsrechtswege eine andere als die getroffene Entscheidung herbeizuführen.

Die wird wegen der Aufbauschung gelegentlicher Einzelfälle in den Medien allzuhäufig übersehen!

4 FLÄCHENBESCHAFFUNG

Eines der Hauptprobleme bei der Wahrung der Interessen des Naturschutzes sind die meist in anderer Richtung laufenden Nutzungsvorstellungen der Grundeigentümer.

Der wirksamste Schutz ist damit oft die Überführung empfindlicher Bereiche in öffentliches Eigentum. Im Rahmen der Flurbereinigung kann hiergeholfen werden, durch Geldabfindung und Kauf, durch Tausch, durch Einsatz öffentlicher Flächen, durch langfristige Pacht, durch Vereinbarung von Nutzungsentschädigungen.

Hierfür sind vor allem Finanzierungsmittel und die Bereitschaft von Land und Gemeinden zur Flächenbereitstellung erforderlich. Für die Zukunft weckt das Flächenstillegungs- und Extensivierungsprogramm der Landwirtschaftsverwaltung auch bei der Flurbereinigungverwaltung Hoffnung auf Arbeitserleichterung.

Auf ein besonderes Problem sei hier aber auch hingewiesen: Jede Umnutzung von Flächen mindert die Wirtschaftsfläche der Landwirtschaftsbetriebe. Um die agrarstrukturellen Verhältnisse nicht zu verschlechtern, sollten in gleichem Maße auch Bewirtschafter ihre Tätigkeit aufgeben. D.h., es sollten ganze Betriebe erworben werden, die dann im Rahmen der Flurbereinigung optimal für Zwecke der Landschaftspflege und der Agrarstrukturverbesserung verwertet werden könnten.

5 BODENORDNUNG UND SCHAFFUNG RECHTLICH GEORDNETER VERHÄLTNISSE

Die zentrale Aufgabe der Flurbereinigung ist die Bodenordnung im Sinne einer optimalen, zukunftsorientierten Bodennutzung. Es sind dies die Maßnahmen, die auf der Basis sorgfältiger Planung alle im Planungsgebiet bestehenden Eigentumsrechte so ordnet, daß kein Eigentümer Wertverluste hinnehmen braucht. Erst in Verbindung mit der Bodenordnung erhält Planung, Flächenbeschaffung, Maßnahmenausführung und -finanzierung ihren eigentlich höchstmöglichen Wert!

6 FINANZIERUNG

Der eigentliche Vorteil einer Flurbereinigung liegt in der Ausschöpfung und Bündelung aller denkbaren Finanzierungs- und Fördermöglichkeiten. Leistungen der Grundeigentümer, der Gemeinde, des Bundes und des Landes, von Unternehmensträgern und Dritten können "maßgeschneidert" kombiniert zur Finanzierung von Grunderwerb und Ausführung der einzelnen Maßnahmen eingesetzt werden. In der Teilnehmergemeinschaft steht ein potenter, kreditwürdiger Ausführungs- und Finanzierungsträger zur Verfügung, der auch gelegentliche Engpässe durch Kreditaufnahme zu überbrücken vermag.

Durch die Aufsicht und Beratung des Flurbereinigungsamtes ist ein Höchstmaß an Effizienz und Sicherheit gewährleistet.

7 MAßNAHMENAUSFÜHRUNG

Ähnlich wie bei der Finanzierung ist auch bei der Maßnahmenausführung die Teilnehmergemeinschaft, oft in Zusammenarbeit mit der Gemeinde, ein zuverlässiger Träger. Die fachliche Beratung und Aufsicht durch das Flurbereinigungsamt bei Kostenermittlung, Finanzierung, Ausschreibung, Vergabe, Ausführung und Abrechnung, gewährleistet eine kostengünstige und effiziente Abwicklung.

Es sei an dieser Stelle betont, daß im Rahmen der Flurbereinigung die Ausführung erst zugelassen ist und beginnt, wenn die planungsrechtliche Grundlage und

Finanzierung gesichert ist und Eigentum, Nutzung, Unterhaltung und Pflege rechtlich befriedigend geregelt werden können. Nur so kann der Einsatz öffentlicher Mittel auf Dauer zweckentsprechend gewährleistet werden.

8 BEISPIELE

Leistungsangebot und Einzelziele der Flurbereinigung in den Bereichen des Natur- und Landschaftsschutzes sowie der Resourcenerhaltung sollen durch 3 Beispiele illustriert und erläutert werden:

1. Beispiel: Natur-und Landschaftsschutz

Biotop-Verbund-System Flurbereinigung Wimsheim (s.Abb. 1 und 2)

Auf Wunsch der Gemeinde und unter Mitwirkung des Landkreises wurde aus der ökologischen Bewertung der Landschaftselemente und dem Entwurf des Planes nach § 41 FlurbG ein Landschaftsplan entwickelt, der systematisch bis ins Detail auch kleine Landschaftsbestandteile erfaßt, nach ökologischen und ökonomischen Gesichtspunkten bewertet und gewichtet, Flächenbedarf, Kosten, Unterhaltungsaufwand und mögliche Nutzung ermittelt, sowie Vorschläge für Eigentumsregelung, Ausführungsträgerschaft und Finanzierung erarbeitet.

Die Untersuchung, ausgeführt von einem Landschaftspfleger, einem Landwirtschaftsfachmann und dem zuständigen Flurbereinigungsingenieur erfolgte nach dem Schema, abgebildet in Abb. 6.

Die Summe dieser Einzeluntersuchungen ergibt eine Bilanz aller planungs- und ausführungsrelevanten Erfordernisse. Dies bildet die Basis für die Planungsabstimmung und die Realisierungsvorbereitung mit dem Ziel, unter Ausschöpfung aller verfügbaren Quellen eine weitgehende Verwirklichung zu ermöglichen.

2. Beispiel: Wasserschutz

Schutz des Trinkwasserbereiches der Wasserversorgung Bruchsal in der Flurbereinigung Bruchsal-Büchenau durch Bodenordnung und Flächenbeschaffung:

In der Unternehmensflurbereinigung Bruchsal-Büchenau ist die Wasserversorgung der Stadt Bruchsal durch intensive Sonderkulturbewirtschaftung (Spargelanbau) und hohe Grundwasserentnahmen für Beregnungszwecke gefährdet. Über Grunderwerb, Bodenordnung unter Einsatz stadteigener Flächen und Ausgleichszahlungen für extensive Bewirtschaftung bzw. Abfindungsnachteile soll die gesamte Wasserschutzzone II der genannten Wassergewinnungsanlage in eine extensive Grünlandbewirtschaftung überführt werden.
Benötigt werden hierfür ca. 12 ha
Verfügbar sind als
1. Vollwertige Tauschflächen 3,5 ha
2. nicht gleichwertige Tauschflächen 4,0 ha
3. Rest 4,5 ha

Der heute noch nicht voll verfügbare Teil nach Pos. 2 und 3 wird über entsprechende Vereinbarung mit Ausgleichszahlung, Kauf- und Dauerpacht abgesichert.

Ferner soll im Rahmen der Neuordnung des Flurbereinigungsgebietes eine Verlegung der Brunnen in andere Bereiche des Flurbereinigungsgebietes erfolgen, so daß im näheren Umfeld der Schutzzone II keine weitere Entnahme mehr erfolgt. Dies wird für dieses Gebiet auch eine weniger intensive Bewirtschaftung zur Folge haben.

Schutzmaßnahmen für Fließgewässer:

Im Rahmen von Flurbereinigungen kann im Zuge der Neuanlage des Wege- und Gewässernetzes ein System von Klein-Rückhaltungen geschaffen werden, das Oberflächenwasser aus intensiv bewirtschafteten Ackerfluren zurückhält. In diesen meist einfach ausgeführten Erdbecken kann sich erodiertes Erdmaterial absetzen. Bis zu einem gewissen Grad wird dadurch wohl auch der Dünge- und Pflanzenschutzmitteleintrag in die Vorfluter verringert. Zum gleichen Zweck können Fließgewässer durch 4-8 m breite extensiv bewirtschaftete Grünlandstreifen als Pufferzone gegen Ackerland abgeschirmt werden; Voraussetzung hierfür ist allerdings, daß ausreichend Land für diesen Zweck beschafft werden kann. (Abb. 2-5).

3. Beispiel: Erosionsschutz

Automatische Erstellung von Erosionsgefährdungskarten aus Orthophotos und den Ergebnissen der Bodenschäzung.

In erosionsgefährdeten Gebieten müssen die Veränderungen der Flurstruktur durch die Schaffung größerer Wirtschaftsflächen besonders sorgfältig geplant werden. Nach heutigem Kenntnisstand liegt zwar das wirksamste Gegenmittel in der Hand des Landwirtes selbst, der mit Bodenbearbeitung und -pflege, Anbau- und Fruchtfolgemethoden am erfolgreichsten Erosionsschutz betreiben kann. Dennoch kann auch der Flurbereinigungsingenieur durch den Aufbau des Wege- und Gewässernetzes und durch die Erhaltung und Neuanschaffung erosionshemmender Anlagen vieles zur erfolgreichen Erosionsbekämpfung beitragen.

Erster Ansatz der Planungsarbeit ist dabei die Ermittlung der Erosionsefährdung und ihrer Ursachen.

Zu diesem Zweck ist unter Mitwirkung der Naturschutz und der Landwirtschaftsverwaltung vom Landesamt für Flurbereinigung der Versuch unternommen worden, auf dem Wege der Datenverarbeitung automatisch erstellte Gefährdungskarten zu fertigen.

Grundlage bildet dabei ein Digitales Geländemodell (das bei der Erstellung der Orthophotos für das Flurbereinigungsgebiet gewonnen wird) und die Bodenartenbeschreibung der Bodenschätzung, wobei die übrigen erosionsbestimmenden Faktoren in mehreren Variationen unter Annahme schlechter, günstiger und mittlerer Werte eingeführt werden.

Die Ergebnisse können nun für die Planung des Wege- und Gewässernetzes, besonderer erosionsverhindernder Anlagen, der Bewirtschaftungsrichtung und dergleichen sowie auch zur Information und Beratung der Landwirte direkt benutzt werden.

9 ZUSAMMENFASSUNG

Der Beitrag des Flurbereinigungsingenieurs zum Schutz unserer Lebensgrundlagen ist vielfältig. Planung, Bodenordnung, Finanzierung und Maßnahmenausführung sind nicht mehr nur als Summe dieser Leistungen zu sehen. In Zusammenwirken und in der Koordination liegt vielmehr auch ein weiterer Steigerungseffekt.

Als Zukunftsaufgabe ist hervorzuheben, daß der Unterhaltung und Pflege des einmal erarbeiteten, oft mit großer Mühe geschaffenen Neuzustandes, noch viel mehr Aufmerksamkeit zu widmen ist. Daraus ergibt sich, daß der Beratung sowie der Überwachung und Erfolgskontrolle, zur Steuerung korrigierender Eingriffe im Rahmen der Erhaltungs- und Pflegemaßnahmen, künftig mehr Bedeutung zukommen muß.

10 BILDNACHWEIS

Landesamt für Flurbereinigung Baden-Württemberg, Stuttgarter Straße 161, 7014 Kornwestheim

Luftbildfreigabe durch Regierung Oberbayern

Abb. 1 Intensiv genutzte, aber dennoch reich strukturierte Agrarlandschaft
Well structured landscape but anyway intensively used

Abb. 2 Renaturierter Wasserlauf begleitet von extensiv genutztem Grünland
Watercourse, reshaped in a natural form accompagnied by extensively used meadows

Abb. 3 Kleine Wasserrückhaltung mit Dauerstau als Feuchtbiotop
Small pool to hold surface water in order to clear it and beeing simultaneously used as natural wet area

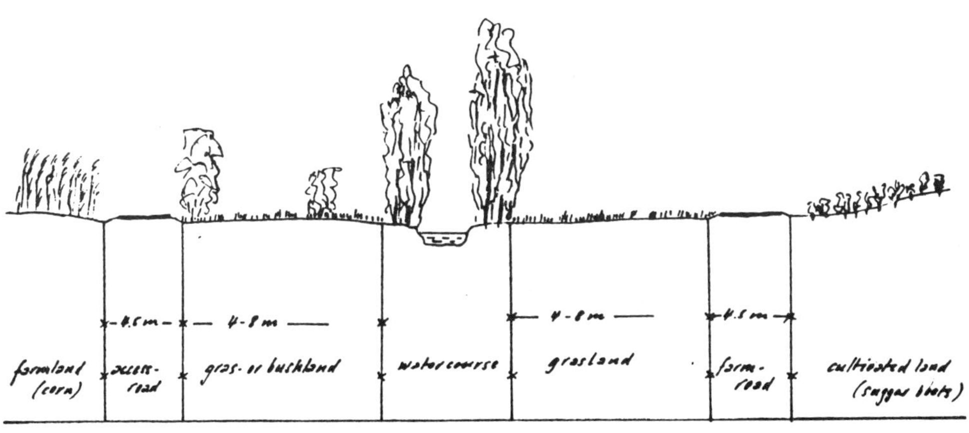

Abb. 4 Regelquerschnitt eines Wasserlaufs mit seitlichem Grünland als Schutzstreifen
Cross-section to show how intensively cultivated land can be separated from watercourse in order to avoid pollution by fertilizer and pesticides

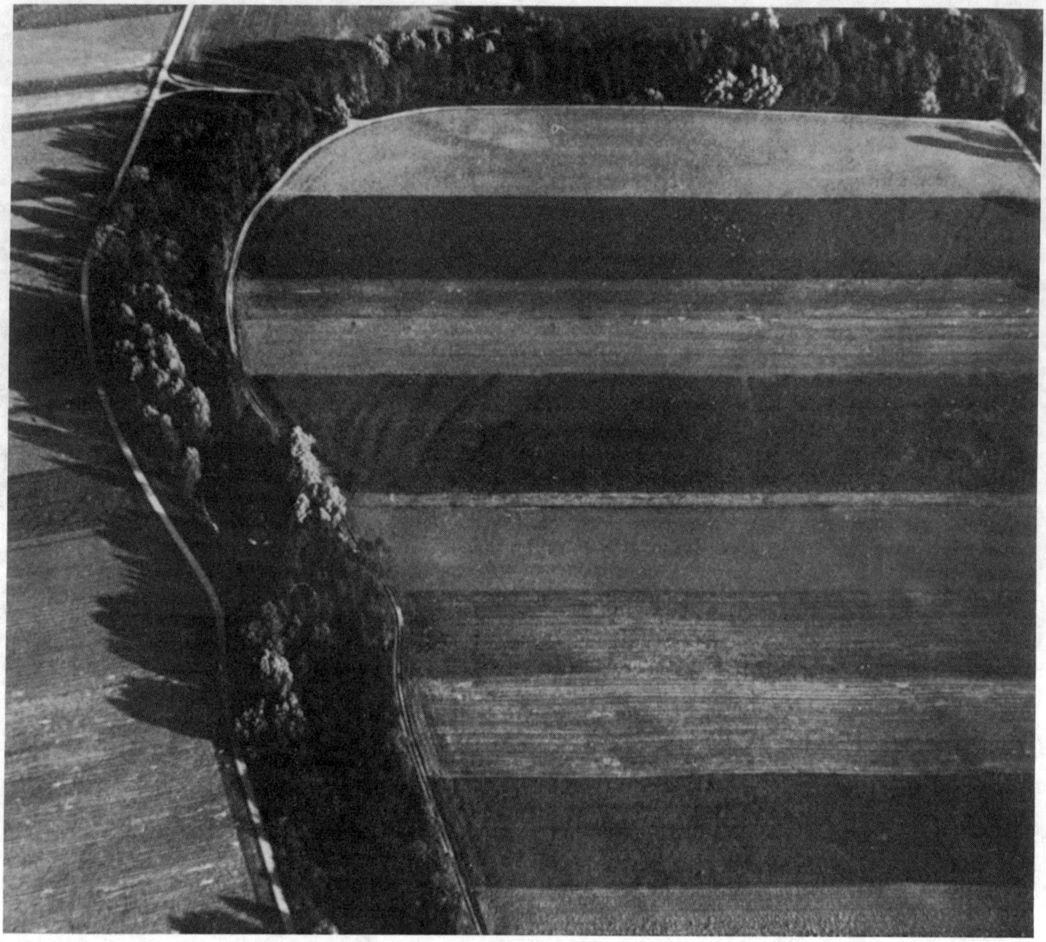

Abb. 5 Beispiel eines Wasserlaufs, der entsprechend Abb. 4 gestaltet ist
Example for watercourse shaped following the principles showed in No. 4

Protokoll über die
- Bewertung der Landschaftselemente
- Ermittlung von Flächen- u. Finanzierungsbedarf
- Eigentums-, Nutzungs- und Unterhaltungsregelung
- sonstige rechtliche Regelung zu Schutz und Erhaltung

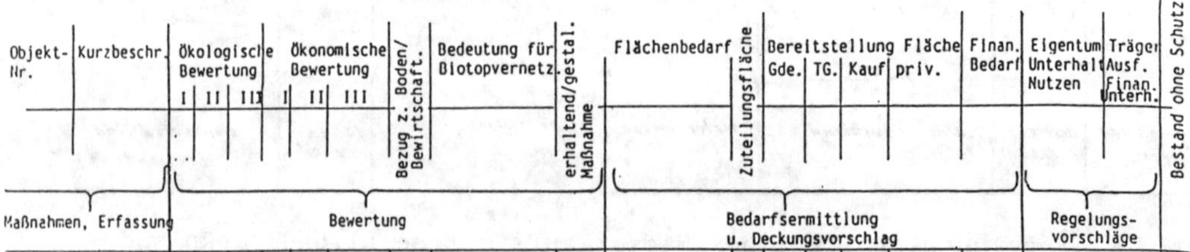

Abb. 6 System ökologischer und ökonomischer Bewertung als Grundlage der Landschaftsplanung
System for ecological and economical evaluation as a base for landscape planning

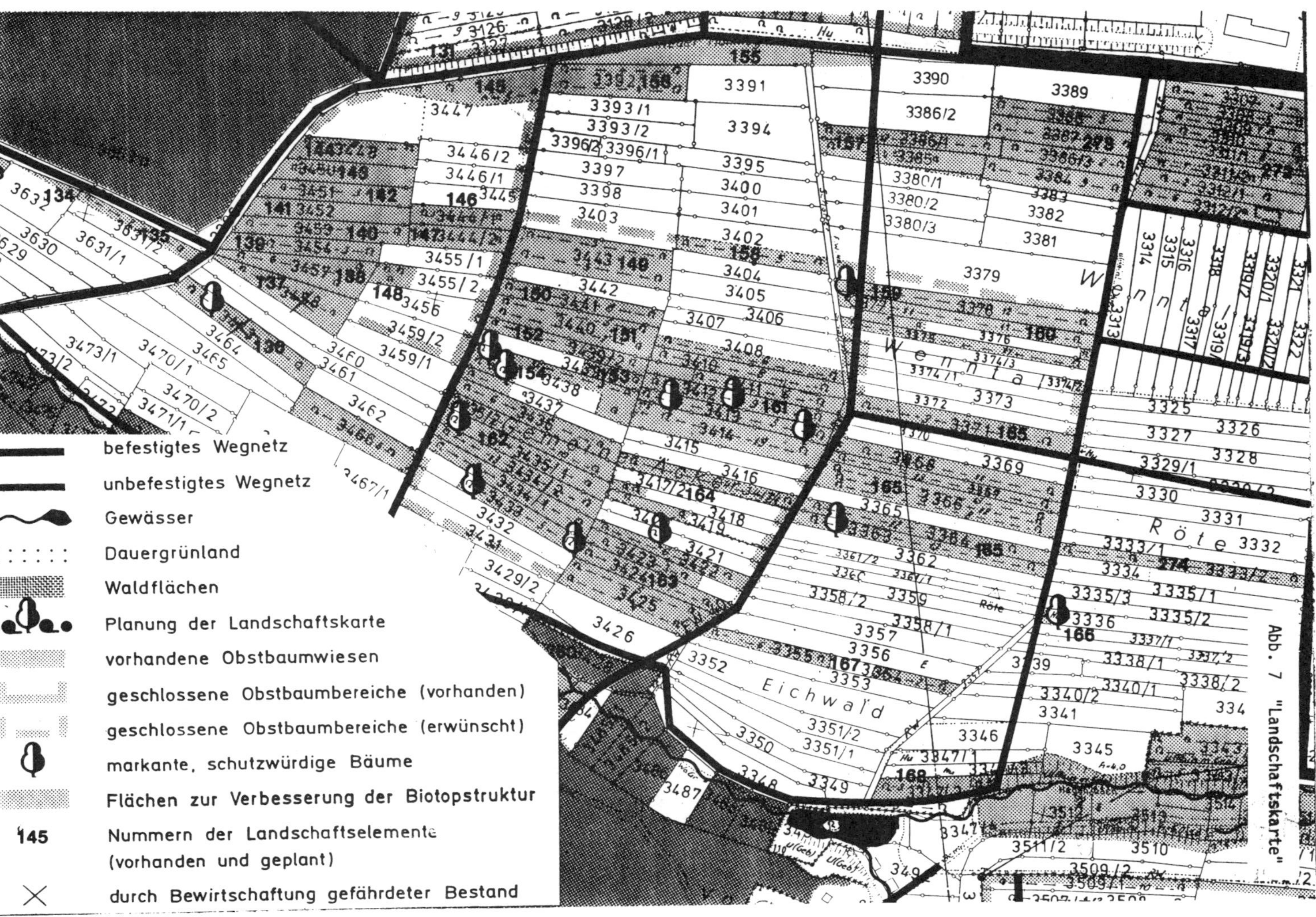
Abb. 7 "Landschaftskarte"
Röte
Eichwald
befestigtes Wegnetz
unbefestigtes Wegnetz
Gewässer
Dauergrünland
Waldflächen
Planung der Landschaftskarte
vorhandene Obstbaumwiesen
geschlossene Obstbaumbereiche (vorhanden)
geschlossene Obstbaumbereiche (erwünscht)
markante, schutzwürdige Bäume
Flächen zur Verbesserung der Biotopstruktur
145 Nummern der Landschaftselemente (vorhanden und geplant)
durch Bewirtschaftung gefährdeter Bestand

Land and Water Use, Dodd & Grace (eds), © *1989 Balkema, Rotterdam. ISBN 90 6191 980 0*

Decision factors in municipal leaf composting method selection

J.J.Kolega
University of Connecticut, USA

R.Rynk
University of Massachusetts, USA

R.L.Leonard & R.F.Jeffrey
University of Connecticut, USA

ABSTRACT: Based upon the experiences of the authors in developing guidelines for leaf composting by municipalities, a number of factors should be evaluated before a method of leaf composting is selected. The decision factors for leaf composting methods are discussed by a multi-disciplinary team from the fields of economics, engineering, and community resource development. Composting methods addressed are passive leaf pile, windrow and turn, aerated static pile, in-vessel processes, and backyard composting by residents. Information was compiled from scientific literature, site visit, a survey of current practices, and meetings with consultants and researchers.

SOMMAIRE: Base sur les expériences des auteurs de cette communication en développant des indications pour le compost de feuilles par les municipalites, un certain nombre de facteurs devraient être évalues avant qu'une méthode de composter soit choisie. Les facteurs décisifs pour les méthodes de composter sont discutées par une équipe multi-disciplinaire choisie dans les disciplines suivantes; l'économie, le engineering, et le développement des resources municipales. Les méthodes de composter adressées ici sont les suivantes: passives, tas de feuilles, file de feuilles et mélange, le tas statique aéré, les processus en vaisseau, et le compost de domicile fait par les habitants. Ces renseignements ont été recueillis dans la littérature scientifique, dans les visites de sites, dans un survol des pratiques courants et dans des réunions avec des experts et des chercheurs.

ZUSAMMENFASSUNG: Nach den Erfahrungen der Authoren zu beurteilen, die mit der Entwicklung der Richtlinien für Blatterkompost in Gemeinden beschäftigt sind, mussen mehrere Fakboren bewertet werden, bevor die endgültige Kompostmethode gewabet werden kann. Die "Entscheidungsfaktoren für Blätterkompostmethoden wurden von einem Auschuss mit Mitgliedern von den vissenschaftlichen Gebieten der Volkswirtschaft, dem Ingenieurwesen und der Gemeindeaktivaentwicklung besprochen. Kompostmethoden, die behandelt werden, sind der passive Blätterhaufen, Schwaden und Wendan, des mit Luft durchsetzte ruhende. Haufen, "in-Behälter"-Vorgange und Hinterhofkompost haufen der Gemeindemitglieder. Information wurde zusammengestellt aus wissenschaftlicher Fachliteratur, Stättebesichtigungen, einer Umfrage nach den gegenwärtigen Verfahrensmethoden, und Sitzungen mit Fachberatern und Forschern.

1 INTRODUCTION

In the Spring of 1988, the University of Connecticut Cooperative Extension Service entered into an agreement with the State of Connecticut Department of Environmental Protection to develop leaf composting guidelines for use by municipalities. This agreement was in response to state legislation that mandates a strategy for achieving a 25 percent reduction in the amount of solid waste going to sanitary landfills by the year 1991. The Cooperative Extension Service is affiliated with colleges of agriculture in the United States, usually at State universities. Historically, extension's role has been one of technology transfer to agricultural clientele, but this role has expanded in recent years to include various areas of community need.

A multi-disciplinary team was formed by the Extension Service to develop the guidelines. The team consisted of an economist, two engineers, and a planner from the community resource development area. This team reviewed the literature, conducted a mail survey, made site visits, and met with consultants and researchers to prepare a leaf composting guidelines document for the state regulatory agency. The document is intended for use by municipalities with primary use by local planning agencies, public works directors, and associated staff.

2 LEAF COMPOSTING SYSTEM OVERVIEW

A municipal leaf composting system involves three components - collection, processing and disposition.

There is no one best way to design and operate a leaf composting system. Each component of the system has a variety of options that a community can choose and match with other selected components. A successful leaf composting program will include those options which are most applicable to the particular situation.

Leaf collection may be in bags or in loose form. Proper management of the leaf composting process is essential to produce a quality compost and to minimize odor problems. Collection and composting can be by individual communities or by a number of communities operating cooperatively on a regional basis. Alternatives for disposition of the finished product (compost) should be considered in the early stages of planning. Options are available for municipal, residential and commercial use.

3 LEAF COLLECTION

The primary requirement of any leaf collection system is that the leaves be collected free of extraneous material, such as glass, metal, paper and household solid waste, so that a high-quality compost can be produced. The choice of a specific collection technique should reflect the volume and quality needs of the end user and the cost of processing the leaves to supply that end user. If the compost is to be used as landfill cover, for instance, some extraneous material may be tolerated and screening of the end product may be unnecessary. Landscapers and nurseries, however, will not accept a product which has extraneous material and screening may be required to achieve product uniformity.

As with other recyclables, the more convenient the collection service the higher the participation rate. Consequently, the volume of leaves arriving at a municipal composting site will vary with the size of the area serviced by municipal collection and the convenience of the collection site to residents and independent haulers.

Population density, estabished practice, accepted levels of public service and municipal costs need to be considered when determining the size of the area to be provided with municipal leaf collection. If a municipality determines that it cannot economically provide collection service throughout the town but wants to provide some convenience to its residents and businesses, a compromise can be achieved by providing curbside leaf collection to the more densely populated areas while providing community collection stations in the rest of the town.

Municipalities which provide leaf collection must make a series of choices about collection techniques and equipment. Since these choices impact collection and processing costs and the quality of the end product, a variety of scenarios should be considered before a decision is made. Leaves can be collected bagged or unbagged. Bagged leaves typically have little extraneous material and can be collected quickly with a standard compactor truck. However, labor is required at the composting site to remove the leaves from nonbiodegradable bags. Biodegradable paper and plastic bags are available, but more information is needed on rates of decomposition. Loose leaves can be collected with a vacuum truck or a front loader. This process is more time-consuming and the choice of equipment is less obvious. The vacuum works well on dry leaves; the front loader is more efficient for wet or frozen leaves. The amount of extraneous material is likely to be higher with loose leaves than when leaves are bagged.

Issues to be considered in the selection of collection techniques include:

1. Effectiveness in excluding extraneous material;
2. Availability and cost of labor;
3. Existing equipment;
4. Capital, operating and maintenance costs of equipment;
5. Cost of bags (plastic, degradable plastic or paper);
6. Convenience for residents and businesses;
7. Susceptibility to adverse weather;
8. Hazards associated with placing leaves at curb or in street;
9. Potential noise and dust from collection equipment.

4 BASICS OF COMPOSTING

Composting is a biological process in which mircoorganisms break down organic materials, like leaves, into a soil-like product called compost. The microorganisms are naturally present among the leaves. If nutrients, oxygen and moisture are present in the proper amounts, the microorganisms will heat the leaves to 38 to 60 degrees C. (100 to 140 degrees F.) and will result in compost with a pH that is neutral to slightly alkaline.

The relative availability of carbon and nitrogen is an important factor in the composting process. The optimum range of the carbon to nitrogen ratio (C:N) is from 20:1 to 30:1. With a ratio greater than 40 to 1, nitrogen is a limiting factor and the metabolic reaction rate slows. With a C:N ratio lower than 15 to 1, excess nitrogen is driven off as ammonia.

The C:N ratio in leaves tends to range between 60:1 to 80:1. Thus, leaf composting is generally slower than most composting applications. By adding nitrogen-rich materials such as, manure, seaweed or grass clippings, the C:N ratio is reduced and process efficiency is improved.

An adequate supply of air is also essential to the maintenance of efficient composting. If the oxygen level in the composting leaves is less than 5%, the aerobic organisms begin to die off and the composting process is taken over by anaerobic organisms which do not require oxygen. Anaerobic compost process conditions can lead to severe odor problems.

Temperature affects biological activity and should be monitored frequently. The metabolism of the microorganisms creates heat and results in a temperature increase. Pile size is a key variable in achieving the optimum temperature range of 38 to 60 degrees C. (100 to 140 degrees F.). In leaf composting, the optimal moisture content is 40% to 60%, by weight. A moisture content below 40% limits bacterial growth. Leaves usually require additional water at the start of the process. If the moisture content exceeds 60%, the flow of oxygen is inhibited and anaerobic conditions begin to develop.

During the composting process, the material will become slightly acidic and then return to near neutral conditions as stability is approached. Decomposition is most efficient with a pH of between 6.0 and 8.0. If the pH is too high, nitrogen is driven off as ammonia. As the pH drops below 6.0, the microorganisms begin to die off and the decomposition slows. If low pH is a problem, the addition of lime may be necessary. To minimize this possibility, keep the pile in an aerobic state. The normal pH range for finished leaf compost is neutral to slightly alkaline (7-7.5).

The microorganisms act on the surface of the composting material. Smaller particles have greater surface area and break down more quickly. However, extremely small particles limit air flow through the material so some compromise is required.

The time required to transform leaves into finished compost varies considerably, depending on the process utilized, from 10 days to 3 years. Frequent aeration, fine particle size and the proper ratio of carbon to nitrogen speed the process. The process is slowed by low temperatures and materials with a high proportion of cellulose and lignin.

5 COMPOSTING METHODS

Selection of the methodology best suited for the community will depend upon a variety of factors, including expected use of the compost, quantity of leaves, site constraints, and opportunities for shared use of equipment. Additionally, the availability and usefulness of other organic wastes may influence the decision-making process within each municipality. The windrow and turn method has been used most often for leaf composting.

5.1 Passive leaf piles

Leaves are deposited in piles ranging in height from 3 to 6 meters (10 to 20 feet) and are left for a minimum of two to three years. Leaf piles that are too small, less than 2 meters (6 feet) high should be combined. An option within this method is to turn and aerate the leaf pile in the early spring or late fall. Odor may be a problem when these piles are disturbed as anerobic conditions may exist in the oxygen starved sections of the pile. Even after two to three years the compost may retain clumps of uncomposted leaves.

5.2 Windrow and turn

Leaves are deposited on a compacted pad to form a windrow measuring 3 to 6 meters (10 to 20 feet) at the base with a height of 2 to 4 meters (6 to 13 feet) or higher. The windrow length is determined by site constraints and the method of material handling. The windrows are turned periodically with a front end bucket loader or a special turning machine, and water is added as needed. The frequency of windrow turning is determined by the temperature and moisture content of the windrow. Windrows are combined as they shrink in size. The periodic turning will result in a relatively uniform compost with less undecomposed leaves as compared to passive leaf piles.

Use of a specialized windrow turning machine improves aeration, resulting in shorter time requirements for composting. Turning machines are either self-propelled or machine driven. If machine driven, it is important that the drive method selected be properly matched to the machine. The machines available limit the windrow height to approximately 2 meters (6 feet). Windrow width varies from 4 to 5 meters (14 to 18 feet) to give a trapezoidal shaped cross section.

5.3 Aerated static pile

An aerated static pile has an air blower, usually temperature controlled, and duct work to aerate the compost pile. This method is being used successfully for leaves composted with sewage sludge. Our research found no current use of the aerated static pile method for composting leaves alone. The system components mentioned represent additional equipment costs which are not needed with either the leaf pile or the windrow and turn methods.

5.4 In-vessel

In-vessel composting is a "factory type" operation. Organic materials are composted in sheltered bins or vessels set on a paved surface. Material turning is accomplished with augers, flails or other devices. Compost temperature is controlled by aeration from an air blower duct system. In-vessel composting of leaves is unlikely unless co-composted with other material.

5.5 Co-composting

Leaves may be composted with other organic materials such as animal or poultry manures, seaweed, or sewage sludge. If co-composting of leaves is being considered, attention should be given to the carbon to nitrogen ratio. When mixed with sewage sludge, the preference is to grind leaves prior to mixing to improve compost pile aeration characteristics.

End use of a co-composted leaves and sewage sludge product may require an evaluation for the presence of organic contaminants and heavy metals. Adequate health and safety provisions are needed for the workers at the co-compost facility.

5.6 Backyard composting

Backyard composting involves the composting of leaves and other yard wastes on a small scale within the confines of one's own property. This method is particularly appropriate for areas where the residences are located on one-quarter hectare (one-half acre) plots or larger. Backyard composting should be encouraged because residents benefit from readily available leaf compost and the municipality benefits by avoiding the cost of handling and processing the leaves.

6 QUESTIONS AND ISSUES

Questions to ask initially are:

1. Is there a site available for leaf

composting? If there is no acceptable site available within the community, will another community or a regional leaf composting site be the solution? Is there a farm that will take leaves for use as a soil amendment?

2. What is going to be done with the compost and are there any time constraints for composting leaves?

3. What are the community resources available for leaf composting in terms of labor, management and equipment?

4. Will community attitude be a factor? Are the residents receptive to leaf composting and the proposed leaf collection scheme?

7 DECISION CRITERIA

If a large site is available at low cost the passive pile method is the least expensive. The windrow and turn method requires less space for a given annual quantity of leaves. With a well managed windrow and turn system the composting process can be completed and the product removed within one year. The compost site size requirement, net of associated buffer area, is approximately 0.4 hectare (1.0 acre) for 2300 cubic meters (3000 cubic yards) of leaves in windrows 2 meters (6 feet) high with sufficient space for turning of windrows. A substantially larger leave volume to site ratio can be achieved with higher and wider windrows.

Two or more composting cycles can be completed in one year through the use of specialized aerating and turning equipment. However, the reduction in the total site size requirement is not proportional to the increase in the annual number of composting cycles. Space is required for stockpiling leaves.

Good drainage is also an important site consideration as are the separation distances from human occupied buildings and surface and groundwater supplies. The separation distances are somewhat arbitrary depending upon the aesthetic, health, nuisance (odor and noise), and environmental concerns.

Leaves will not generally be composted for their market value. A lack of landfill space and the high cost of incineration have been the motivation for leaf composting in Connecticut. The economic value of the leaf compost product is generally not the reason for making a leaf composting decision.

Since leaves are seasonal in nature, the community government may have available manpower and equipment that can be allocated to leaf composting. This in itself can reduce operating and maintenance costs and the initial investment requirement. However, if the community assumes responsibility for leaf composting, the importance of good management and monitoring of the operation can not be over-emphasized. If proper management cannot be provided by the municipality, an alternative is to use a private site management firm, if available, or contract the leaf composting operation to a commercial firm.

The economic feasibility of investing in specialized composting equipment is determined primarily by the volume of leaves to be composted. A general purpose front end loader is usually sufficient for turning and aerating a small volume of leaves. A first step toward specialization at little capital cost can be taken with the purchase of a loader bucket designed for light weight material. At the other extreme, large, specialized aerating and mixing equipment have high annual capital cost and low operating cost per ton of material composted.

A favorable public attitude is essential for a successful leaf composting program. Citizen cooperation is needed in keeping leaves out of other types of solid wastes and in keeping extraneous material out of leaf bags or piles. Public information activities should be carefully planned to both provide information and secure cooperation.

BIBLIOGRAPHY

Division of Solid Waste Management, June, 1988. Leaf Composting Guidance Document. Division of Environmental Quality Engineering, Commonwealth of Massachusetts, Boston, MA, U.S.A.

Jeffrey, R. et. al., 1988. Leaf Composting: A Guide for Municipalities. State of Connecticut, Department of Environmental Protection, Hartford, Connecticut, U.S.A.

Strom, P.F. and M.S. Finstein, 1985. Leaf Composting Manual for New Jersey Municipalities, Department of Environmental Science, Rutgers, The State University, New Brunswick, New Jersey, U.S.A.

1.4 Soil/water engineering

Land and Water Use, Dodd & Grace (eds), © 1989 Balkema, Rotterdam. ISBN 90 6191 980 0

Soil and permeable backfill displacement by the mole plough and winged subsoiler and the effect on water entry into the pipe drain

D.A.Castle
ADAS Field Drainage Experimental Unit, Trumpington, Cambridge, UK

ABSTRACT: The movement of soil into, and permeable backfill from, the pipe drain trench during the installation of mole drainage and deep subsoiling has been investigated. The soil movement caused by two commonly used mole plough expander shapes was examined in detail, and calibrated flows were used to determine the effectiveness of the connection for water entry into the pipe drain from new mole drains. This work has shown that soil displacement into the pipe drain trench is a function of mole plough geometry, and depending upon the effective width of the drain trench can seal the permeable backfill connection, affecting water entry.

The displacement caused by a winged subsoiler working below critical depth was also investigated. Whilst a satisfactory connection with the backfill is achieved, the longer term effect of continued deep subsoiling is still being investigated.

RESUMÉ: Le mouvement de sol dans la tranchée pour drain, et aussi du remblai poreux de celui-ci pendant l'installation de taupage et de sous-solage profond, a été étudié. Le mouvement de sol causé par deux boulets expanseurs employés habituellement, a été examiné en détail, et des débits calibrés ont été utilisés pour determiner l'efficacité du raccord pour l'entrée d'eau dans le drain venant du nouveau taupage. Cette oeuvre a demontré que le déplacement de sol dans la tranchée pour drain est une fonction de la géométrie de la charrue-taupe, et selon la largeur effective de la tranchée pour drain peut boucher le raccord du remblai poreux et ainsi peut influencer l'éntrée d'eau.

Le déplacement produit par une sous-soleuse ailée travaillant au-dessous d'une profondeur critique a été étudié aussi. Tandis qu'un raccord satisfaisant pour le remblai est accompli, on fait l'étude encore du résultat à plus long-terme de sous-solage profond continu.

ZUSAMMENFASSUNG: Wir haben die Bewegung des Bodens und des Kieses im Rohrdrän-Graben während der Installation der Maulwurf-Dranung und des Tieflockerns untersucht. Die Bodenbewegung, die durch zwei weithin benützte Maulwurfpflug-Schleppkegel-Typen verursacht wird, wurde in aller Einzelheit untersucht und kalibrierte Wassermengen wurden verwendet, um die Wirksamkeit des Anschlusses für den Wassereintritt in das Dränrohr von den neuen Maulwurfdränen zu bestimmen. Diese Arbeit hat gezeigt, dass Bodenverdrängung in den Rohrdrän-Graben eine Funkion der Maulwurfpflug-Geometrie ist und sie kann je nach der effectiven Breite des Drängrabens den Kiesanschluss verdichten und somit den Wassereintritt beeinflussen.

Die Verdrängung durch einen beflügelten Tieflockerungspflug, der unterhalb der kritischen Tiefe arbeitet, wurde auch untersucht. Obwohl ein zufriedenstellender Anschluss zum Kies erreicht wird, wird die langfristige Auswirkung einer dauernden Tieflockerung noch untersucht.

1 INTRODUCTION

Mole drainage has been used as a relatively low cost method of draining low permeability soils for over 200 years. Nicholson (1942) advocated the use of pipe drains covered with a permeable backfill material as an outfall, intercepting the mole drain (Castle 1986) rather than have it discharge direct into an open ditch or connect with a moled main. Subsequently the use of mole drainage, connecting with widely spaced pipe drains, has been continually advocated by the Ministry of

Agriculture for the drainage of many clay soils in England and Wales. Where a clay soil is not considered to be suitable for mole drainage, or soil has been reinstated after mineral abstraction, the use of deep subsoiling is recommended in order to increase the permeability by creating fissures in the soil profile to a depth of 400 to 500m. When Godwin and Spoor (1977) showed that a narrow tine working below a critical depth would form a 'square mole channel', the Ministry of Agriculture recommended the use of permeable backfill in the pipe drain trench so that a connection was made with deep subsoiling. Consequently there is considerable capital investment in mole drainage as the dominant drainage treatment in the Midlands, South East and Eastern Counties of England, and for subsoiling in other areas, (Robinson and Armstrong 1987) where high yields are achieved under intensive farming systems.

When promoting the use of permeable backfill as a connector with mole drainage and deep subsoiling, diagrams have shown the backfill material as being undisturbed by the passage of the implement (MAFF 1976, Castle et al 1984). However, Mulqueen and Burke (1967) reported that a plug of soil was drawn tongue-like through the backfill, forming an arch in the roof of the mole channel. The presence of soil smeared across the backfill material to form a base to the mole channel was also noted, but not considered to be important except on very steep slopes. (NB Diagrams indicate that the drain trench was probably at least 225mm wide at mole drain depth having been excavated with a tapered excavator bucket). Similar soil and permeable backfill displacement was reported by Twocock (1971) where the pipe drain trench was nominally 180mm wide when installed. This work included detailed hydrologic monitoring which showed the importance of the backfill connection in that mole drains drawn through the permeable backfill discharged water at 3 times the rate of those drawn through soil backfill in the drain trench.

In the early 1970's the placing of pipe drains became more intensive at 40m spacing with mole drains drawn at 90° across the drain trench at 2-3m intervals for most new drainage. During the same period, machinery development brought the introduction of both narrower open trenches and trenchless machinery to increase efficiency and reduce costs. Consequent to this the cost savings associated with the narrower width of backfill material were considered to be more important than any possible effect on the connection with the mole drain or deep subsoiling.

The effectiveness of the connection with a narrow pipe drain trench was not considered until Whitmore et al (1987) investigating a mole drain on an experimental site found that soil had been drawn into the backfill material and had formed an almost complete seal across a 100mm wide drain trench. Meanwhile other investigations had shown that deep working winged subsoilers severely disturbed the backfill material (Castle 1986). This displacement was typical of the soil failure ahead of a narrow tine described by Godwin and Spoor (1977), but neither this work nor that of Godwin et al (1981) and Spoor and Ford (1987) who had considered the action of the mole plough forming the mole channel, had examined the effect of the implement passing through the drain trench.

2 PRELIMINARY INVESTIGATIONS

With so much agricultural land requiring renewable mole drainage and deep subsoiling to maintain a drainage status essential for continued satisfactory production levels, and where pipe drains had been laid relatively recently by narrow open trench or trenchless machinery, further examination of the apparent problem under commercial conditions was necessary.

2.1 Mole drainage

Existing mole drains were examined on 10 sites with different soils in England and Wales (Castle et al 1987), the criteria for selection being that the soil type was normally considered to be suitable for mole drainage. The objective was to examine the extent of soil and backfill material displacement and assess, subjectively, the effectiveness of the connection with the pipe drain. Detailed information of mole plough geometry was not available so long after installation, other than the farmer/adviser knowing that an expander had, or had not, been used. Pipe drains installed by open trench and trenchless machinery were examined. At least 2, sometimes 4, mole drain/drain trench connections were investigated on each site, so that the effect of installing mole drains in both upslope and downslope direction was also considered.

The topsoil was removed, then the subsoil carefully excavated piecemeal from one side of the leg slot until the top of the backfill material was exposed. The exit side of the pipe drain trench was opened down to the mole drain, and then the entry side exposed. Finally the floor of the mole channel was opened to examine the conditions beneath. Throughout the excavation the horizontal displacement of soil and backfill material, and the soil displacement into the roof and floor of the mole channel, was measured from, or at, the original position of the drain trench entry wall.

the channel, gradually reducing, but forming a substantial sealing of the backfill material, across the drain trench. Excavation below the mole channel also showed that the backfill material had been curved, along with the trench exit wall, into the soil beyond. Soil movement was found to be noticeably less where an expander had not been used.

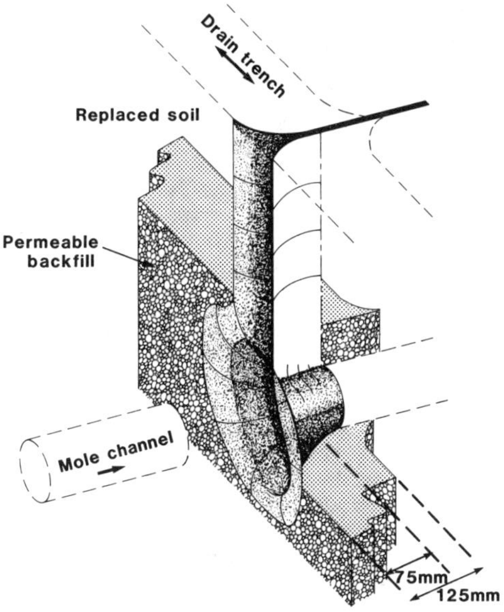

Figure 1. General pattern of soil movement into the drain trench

On every occasion soil displacement was found to be of the form shown at Figure 1, irrespective of the drain trench width. Horizontal displacement in the leg slot was in the range 100 to 160mm, and in the channel roof 150 to 230mm. The channel floor showed two conditions (Figure 2). An almost complete sealing of the backfill material was more common with soil movement in the range 100 to 135mm, and of 40 to 65mm to form a partial seal. Where the soil movement was greatest, a small area of exposed backfill material in the mole channel floor was located just over, or outside, the original position of the trench exit wall. Horizontal displacement of soil to each side at the centre-line of the mole channel was in the range 50 to 90mm from the channel wall, reducing as the soil moved forward into the backfill material to form the channel walls. Measuring vertically, soil movement was 50mm above, and to a depth of 40mm below

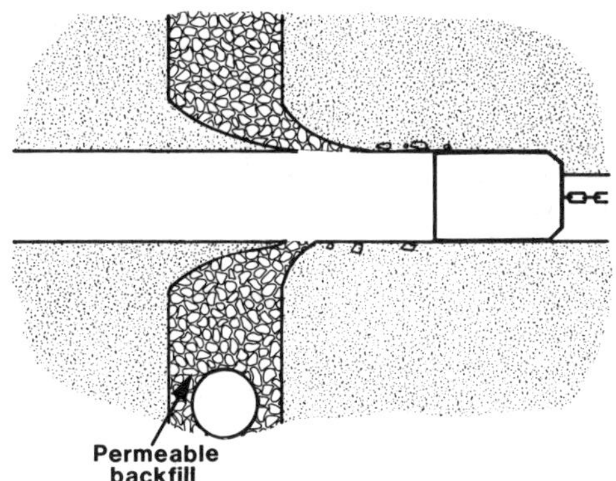

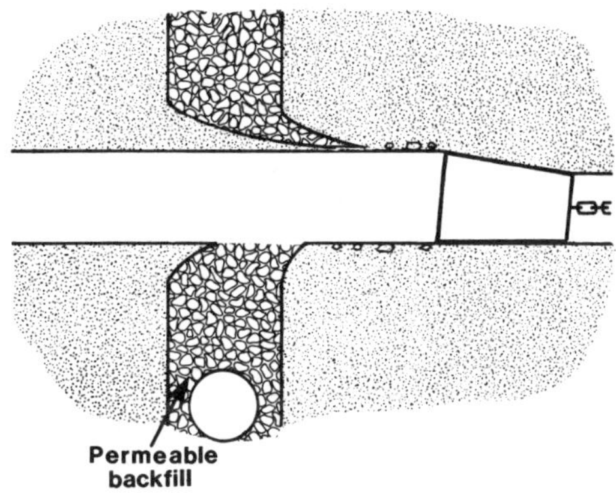

Figure 2. Connection with backfill material created by barrel (above) and taper (below) expanders

Investigation of the narrow 'trench', typically 65 to 75mm wide after settlement, created by trenchless machinery showed the same pattern of soil movement, with soil moved across the backfill material and into the soil on the exit side of the trench for a further 50 to 60mm. This indicated that there was unlikely to be any connection between the mole channel and 'trenchless drainage'.

In each case a distinct hole had been

formed through the drain trench and backfill material moved from the drain trench into the initial 100 to 150mm of the mole channel with individual fragments up to 300mm beyond the trench exit wall. Where the mole channel roof was found to be open to the drain trench above, backfill material in the size range 25 to 40mm was wedged and held firmly by the surrounding soil. Only where pipe drains had been installed by trenchless machinery were large quantities of 5 to 15mm backfill material found to have dropped into the mole channel. This was laying on an already sealed mole channel floor, and so did not provide a connection for the discharge of water.

Whilst it was not possible to draw firm conclusions from sites with different soils, lack of information on mole plough geometry and soil conditions at the time of installation, there were useful indications gained from the investigation. It seemed evident that differences in mole plough geometry, notably expander shape, might affect the magnitude of soil movement. It was also suggested that drain trench width was the important factor, as only the wider trenches indicated the possibility of a satisfactory connection. Of further concern was the effect of direction of installation as the soil movement caused by a mole plough drawn downslope might lead to mole channel flow passing over, or having severely limited entry into, the small area of backfill material visible in the channel floor.

2.2 Deep subsoiling

The effect due to deep subsoiling was also examined (Castle et al 1987) as the displacement caused by a winged subsoiler was expected to be very significant. As better monitoring had to be achieved, fragments of backfill material were painted, and according to colour were placed at different depths below ground level and at different positions across the width to be subsoiled. A winged subsoiler with tines at 600mm spacing was pulled at 360mm effective depth through backfill material installed to 150mm below ground level. When the disturbed profile was excavated it was found that soil had been moved forward through the drain trench zone, and the backfill material moved forward almost 'en bloc' (figure 3). At 150mm below ground level the backfill material was moved 370mm horizontally forward, at 250mm depth by 250mm and at 300 mm depth by 150 to 200mm.

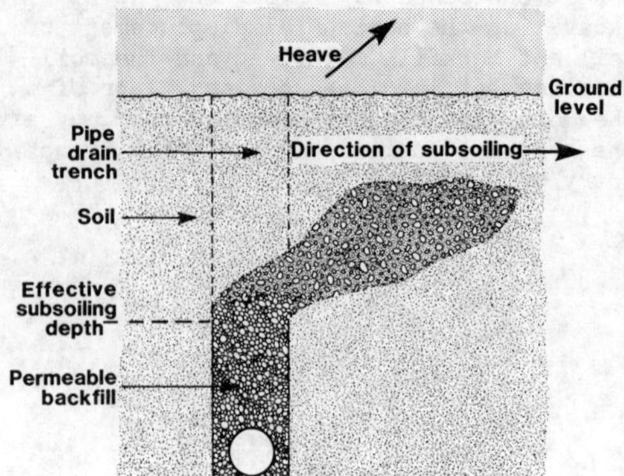

Figure 3. Backfill material moved 'en bloc' by subsoiler

Below subsoiling depth there was no movement. The subsoil leg had also caused backfill material to move laterally up to 170mm on either side. This was, in fact, the typical pattern of soil failure described by Godwin and Spoor (1977). Of significance, however, was the clean connection with the undisturbed backfill material left in the drain trench immediately below subsoiling depth, although there was a 'square mole channel' at the base of the subsoiled zone. This indicated that a sound hydraulic connection could be obtained when subsoiling over a wide drain trench during a single passage of the implement, but the effect with the narrower trenchless connection was still unknown, as was the long term effect of continued subsoiling.

3 FURTHER INVESTIGATIONS

Following the findings of the preliminary investigations a more detailed examination of the soil and permeable backfill displacement caused by known mole plough geometry was necessary. As pipe drains are now installed in narrow open trenches, usually 100mm width, or sometimes by trenchless machinery the possibility of modifying mole plough geometry to create less soil movement had to be considered. There was also a need to investigate whether the direction of installation of mole drains had an effect on the performance of the connection with the drain trench. It was also necessary to investigate the effects of deep subsoiling through the trench formed by trenchless machinery.

3.1 The effect of mole plough geometry

In a collaborative investigation undertaken by the Field Drainage Experimental Unit and Silsoe College (Castle and Spoor 1989) the two most commonly used expander shapes, the barrel and taper, were pulled at 90° and 45° across 125mm and 150mm wide drain trenches, using the same standard mole plough leg and foot, at the same working depth and a constant speed. Excavation was undertaken in the same way as before.

It was found that the differences in soil movement which formed the roof and floor of the mole channel was explained by the different expander shapes (figure 2). A full explanation of the soil mechanics is given in Castle and Spoor (1989). The barrel expander formed an almost total seal where the forward soil movement in the channel floor and roof was similar. The taper expander created a partial seal with less displacement to form the channel floor, and a longer roof arch which extended into the soil profile beyond the drain trench. The respective distances for soil and backfill movement were similar to those previously recorded, confirming the conclusion by Spoor and Ford (1987) that the pattern of soil movement produced by the standard mole plough remains fairly constant over a wide range of soil conditions. Hence, the drain trench width is critical to obtaining a satisfactory connection with the backfill material. As trenches of 100mm width or less would always be suspect, modified mole plough geometry was required.

As soil bin and field trials with alternative shaped mole plough foot and expanders showed no substantial changes of soil movement, a simple small tine was attached 30 to 40mm behind the barrel expander (Castle and Spoor 1989). The tine extends below the bottom face of the expander and cuts a groove 10mm deep and 10mm wide in the mole channel floor. It is deep enough to hook into the backfill material in the drain trench and disturb the soil smeared across by the expander, thereby opening up the connection.

The effectiveness of the connections formed by the barrel, taper and barrel + tine expanders was tested using calibrated flows increased incrementally to determine the threshold level of water entry into the backfill material. It was found that each connection accepted flow rates greater than those recorded on several sites monitored by the Field Drainage Experimental Unit. That formed by the barrel expander on a slope less than 3% crossing the 125mm wide trench at 90° left little margin for deterioration, casting further doubts about narrower trenches. The connection formed by the taper expander accepted a flow rate approximately 2 times greater and that of the barrel + tine some 4 times greater. Subsequent examination of the mole channels after substantial rainfall had created drainflow showed that the short coils of soil raised from the channel floor by the small tine were little different to normal roof fall, and the groove had virtually disappeared due to unconfined recovery. These flow tests also showed that over all treatments there was no significant difference in the effectiveness of the connections due to the direction of installation of the mole drain.

3.2 The effect of deep subsoiling

The Field Drainage Experimental Unit investigated the effect caused by a heavy duty winged subsoiler pulled at depths of 400mm and 450mm across 75mm and 215mm wide drain trenches installed by trenchless and open trench machinery respectively in a restored soil (Hodgkinson 1988). Permeable backfill was placed in the drain trench to depths of 150mm and 300mm below ground level.

Subsequent excavations confirmed the findings of the preliminary investigation within the expected variations due to the differences in subsoiling depth and depth to the backfill material. In each case the square mole channel provided a sound connection with the drain trench. When working below critical depth the subsoiler wings create a void as soil is moved upward and forwards, some of which falls back. Where a block of soil was positioned over the remaining backfill material in the drain trench, a clear connection was found beneath. Due to the inclined subsoiler foot and the different confining forces acting on it, compared with the mole plough, there is very little, or no sliding resistance between the underside of the foot and the soil beneath. Consequently there is no evidence of soil movement which would seal the backfill material in the floor of the 'square mole channel'. These findings were confirmed when the soil and backfill movement created by a standard winged subsoiler in undisturbed soil was examined during work concurrent to that described by Castle and Spoor (1989).

It is evident that with one pass of a winged subsoiler, with tines set at 600mm spacing, approximately 50% of the backfill material placed above subsoiling depth is moved from the drain trench into the soil profile (Hodgkinson 1988). The high cost of this apparent wastage must be considered against the need to achieve a hydrologic connection with the topsoil/soil surface or the provision of a hydraulic connection for deep subsoiling, in which case less backfill material can be placed in the drain trench. Although a satisfactory connection with the drain trench is achieved the long term effect of continued deep subsoiling is still being investigated (Arrowsmith 1988).

4 SUMMARY AND CONCLUSIONS

These investigations have shown that:

for mole drainage
- the movement of soil and backfill material is related to expander shape
- a better connection is likely to be obtained with a taper shaped expander
- a much improved connection can be created by fitting a small tine behind the barrel expander
- narrow drain trenches (<100mm) are unlikely to provide an adequate or lasting connection with the mole drains
- subject to a satisfactory connection there is no significant difference in water entry into the backfill material due to the direction of soil movement related to the direction of installation of mole drains

for deep subsoiling
- backfill material placed above subsoiling depth is totally displaced and subsequently will not provide a hydrologic connection with the topsoil or cultivated layer
- a satisfactory connection will be achieved below the critical depth for which the backfill material need not be higher than 300mm below ground level (NB This level of permeable backfill material allows for 75 to 100mm settlement after installation)

5 ACKNOWLEDGEMENTS

The several investigations reported were variously funded by the Ministry of Agriculture, Fisheries and Food; the Home-Grown Cereals Authority and the British Coal Corporation Opencast Executive. The co-operation of many farmers, the fieldwork undertaken by several colleagues in the Agricultural Development and Advisory Service (ADAS) and the collaboration of Silsoe College was invaluable.

6 REFERENCES

Arrowsmith, R. 1988. The effects of using drainage design standards recommended by MAFF for underdrainage with subsoiling. Unpubl. internal report, ADAS Field Drainage Experimental Unit, Trumpington. Cambridge. UK.

Castle, D.A. 1986. The rationale for permeable backfill placed over pipe drains. Soil Use and Management, Vol 2. No 2 June 1986.

Castle, D.A., G.L. Harris and A.A. Thorburn 1987. The effectiveness of the permeable backfill connection between mole drains or deep subsoiling and the pipe drain. Unpubl. Research Paper, ADAS Field Drainage Experimental Unit, Trumpington, Cambridge, UK.

Castle, D.A., J. McCunnall and I.M. Tring 1984. Field drainage: principles and practices. Batsford Academic.

Castle, D.A. and G. Spoor 1989. Mole plough design factors influencing the hydraulic connection between mole channels and the permeable backfill placed over pipe drains. Mss submitted to Journal of Agricultural Engineering Research.

Godwin, R.J., G. Spoor and P.B. Leeds-Harrison 1981. An experimental investigation into the force mechanics and resulting soil disturbance of mole ploughs. Journal of Agricultural Engineering Research 26:477-497.

Godwin, R.J. and G. Spoor 1977. Soil failure with narrow tines. Journal of Agricultural Engineering Research 22: 213-228.

Hodgkinson, R.A. 1988. The displacement of permeable backfill from the drain trench by winged subsoilers. In 'Ten Years of Research - What Next' Proceedings of 'A Seminar on Land Restoration Investigation and Techniques'. Univ of Newcastle-upon-Tyne, July 1988. British Coal Opencast Executive, Berry Hill, Mansfield, Notts. UK.

MAFF 1976. The use of permeable backfill in field drainage. Getting down to drainage, Leaflet No 20.

Mulqueen, J. and W. Burke 1967. A review of Drumlin soils research 1959-1966. An Foras Talúntais, Dublin.

Nicholson, H.H. 1942. The principles of field drainage. Cambridge University Press.

Robinson, M. and A.C. Armstrong 1988. The extent of agricultural field drainage in England and Wales 1971-80. Trans. Institute of British Geographers NS 13: 19-28.

Spoor, G. and R.A. Ford 1987. Mechanics of mole drainage channel deterioration. Journal of Soil Science 38:369-382.
Twocock, J.G. 1971. Summary of field observations, Scredington, Lincs. 1965-71 Annual Report 1971:87-94. ADAS Field Drainage Experimental Unit, Trumpington, Cambridge, UK.
Whitmore, M.E., G.L. Harris and P. Morgan 1987. The effects of surcharging in the underdrainage system on mole drainage. Unpubl. Research paper, ADAS Field Drainage Experimental Unit, Trumpington, Cambridge, UK.

Land and Water Use, Dodd & Grace (eds), © 1989 Balkema, Rotterdam. ISBN 90 6191 980 0

Methodology for the study of hydraulic properties of soil on a plot scale

G.Ciollaro & V.Comegna
Hydraulic Agriculture Institute, University of Naples, Italy

ABSTRACT: Presented here are the results of the hydraulic characterization on a plot scale of an essentially silty alluvial soil, situated near Policoro (MT). After having pointed up the usefulness and the concrete possibilities of making a survey of the hydraulic properties of soils, taking into account their intrinsic variability, we proceed to a description of measuring instruments and techniques established at the Institute of Agricultural Hydraulics at the University of Naples for the determination of tension curves and water conductivity on undisturbed soil samples. For six profiles of soil sampled from different depths over the course of three years of experiments, we give a complete hydraulic characterization, followed during depletion processes for evaporation, starting from conditions in hydrostatic equilibrium in the liquid phase and measuring the time-course of the sample weight and of the water potential at different depths. The results of the experiments were processed using a technique suggested by the instantaneous profile method, and it was then possible to deduce the values tension h and of water conductivity k for different values of water content in volumes θ. With the aim of being able to use in a more practical way the hydraulic properties of soil in mathematical models, simple analytic expressions were applied to the equations h(θ) and k(θ), the effectiveness of which was verified by simulating the process of imposed motion during the tests and comparing the experimental surveys with the results of the calculations. In short, by employing the concept of geometrical similarity, it was possible to quantify the spatial variability of hydraulic parameters which, within the limits of the particular plot under survey, was seen to be modest.

1 INTRODUCTION

In order to plan the optimum management of irrigation water and to achieve an efficient protection against soil and ground water pollution, it is necessary to understand water and solute transport processes which occur in the soil and in particular in the root zone. Only through studying water dynamics is it possible to characterize the water consumption of plants and to determine water depth application and its frequency in such a way as to make the distribution of soil humidity more effective by reducing the water stress of plants and keeping checks on changes in deep-level water circulation.

Recently, large-scale theoretical applications have been developed to permit a rigorous mathematical modelling of the phenomena of water transport processes, not only within the soil, but also within the complex soil - plant-atmosphere system (Hillel 1980 , Santini 1979). The adoption of modern methods of numerical analysis and automatic calculus has thus made possible the establishment of mathematical models for the study of even more complex flow fields, for the geometry of the boundary conditions as well as for the particular conditions imposed thereon.

Widespread diffusion of these models will arise from the establishment of techniques and test apparatus designed to provide a reasonable and ever less burdensome characterization of the system.

In order to characterize the porous medium, it is necessary to pinpoint the equations connecting the potential h in the liquid phase to the local value of the water content in volume θ and which refer to θ the hydraulic conductivity k. For field applications, an evaluation in statistical terms of the variability of these equations is also necessary.

Numerous researches have proposed testing techniques for laboratory and field which, however, are almost always laborious and time-consuming (Klute 1972 , Hillel et al. 1972 , Ciollaro et al. 1987).

The choice of which methodology to adopt for these determinations depends in each case on the particular soil-type and on the particular conditions on site. For this reason, it is preferable to have estimations carried out pedologically (Bouma 1983 , Ciollaro and Comegna 1981).

The validity of the laboratory measurements will be linked to the number, and dimension of the field samples and to the sampling techniques used thereupon.

Laboratory measurements, compared to those made in the field are seen to be more practicable: they are taken under more controlled conditions, and the employment of more sophisticated equipment leads to more precise results from the samples being analyzed (Ciollaro and Comegna 1979).

Field surveys and laboratory determinations can easily be integrated. Nevertheless laboratory tests in themselves form a tool of undoubted validity in making statistical evaluations of the variability of single parameters, thanks to the high operative capacity offered by automatized modern experimental equipment.

In every case, one is dealing with determinations which are troublesome owing not only to the need for specific equipment and qualified engineers, but also to the need to arrive at a large number of observations, due to the essential variability of porous media (Nielsen 1973 , Russo and Bresler 1980 , Ciollaro and Comegna 1988).

In this paper, we refer to a study directed at the hydraulic characterization of a silty-loam soil of the south of Italy. The study makes use, to determine water conductivity, of a methodology suggested by the instantaneous profile method, which has been shown to be relatively practicable and consequently capable of application also in characterizing spatial variability in the hydraulic parameters of soil.

2 MATERIALS AND METHODS

The area under survey is situated about 10 km from the town of Policoro (MT) and falls under the direction of the experimental unit "E.Pantanelli" of the Institute of Agronomy at the University of Bari.

The unit hosted a collaborative study of water movement in the soil-plant-water system, also when under conditions of water-stress (Santini et al. 1988).

To this end, plants were grown on surfaces of about 3000 m^2, which were subdivided into two plots of equal dimensions and submitted to differing water regimes.

In order to interpret accurately the observations, the establishment of a mathematical model was indispensable. This would take account also of the processes of water-extraction from the soil by roots.

The series of measurements effected by the working unit of the Institute of Hydraulic Agriculture of Naples also considered the determination of the hydraulic characteristics of soil, which are indispensable for the application of the model.

The soil develops on a recent alluvial formation of material of marine and fluvial origin, containing sandy clay lying on gravel beds. With reference to the 7th approximation proposed by the Soil Survey Staff of the U.S. Department of Agriculture, it is an association in which predominate fluvents vertic and vertic soils, with the former represent most strongly.

As to the texture, it is made up of silty-loam soils to a depth of 110 cm approx., changing to loam/sandy-loam at depths over 110 cm. The relevant specifications of the pedological profile are given in the appendix.

Over the years of experiments at the edge of the plots, two sample areas were chosen at random for each year. Therefore six measuring sites were individuated which, taking into account preliminary inquiries into the structure of spatial variability of the hydraulic characteristics, may certainly be considered spatially independent (Gajem et al. 1981).

On each site, by means of the digging of a pit to the depth 10-40 cm, 40-60 cm and 60-80 cm, samples of undisturbed soil of diameter 80 mm and height L, variable between 140 and 160 mm, were taken. For each sample, we determined texture, bulk density, hydraulic conductivity at saturation point, and curves $h(\theta)$ and $k(\theta)$.

The percentage of sand, silt and clay, in keeping with the scheme proposed by the SISS, was determined using the densimetric method (Day 1965). Bulk density was obtained by drying the soil samples in an oven at a temperature of 105° C, basing the calculation on soil weight and on the volume of the cylindrical container.

The determination of the hydraulic properties of the soil samples was made by submitting the samples to a process of evaporation of the upper surface, having taken care to seal off the lower one in order to prevent water-loss from the bottom.

The soil samples, after depth saturation, were dried under thermoregulated conditions employing a small fan with the same diameter as the sample and positioned about 10 cm above its upper surface.

Following the starting of the fan, a constant evaporation flow is rapidly determined, the value of which can be varied within certain limits by regulating the air-flow which strikes the upper surface of the sample.

During the transient state, simultaneous measurements of the weight P of the sample were taken, using strain-gauge load cells with a measuring range from 0 to 6 kg and precision $\pm$ 1 g and of potential h to 4 or 5 depths z of the sample. The measurement of the potential values was effected by boring a fixed number of holes, into which were then inserted miniaturized tensiometers with a porous septum of sinterized glass with a diameter of 10 mm and with air entry to -0,5 bar, joined to pressure transducers by means of small nylon tubes with a diameter of approx. 1,5 mm.

The measuring and recording of the data was effected by the Institute's data control and acquisition system, driven by a numerical calculator (Ciollaro and Comegna 1979), and the depletion process was interrupted at the observation of the entrance of air into two tensiometers.

Following this, with the aim of extending the curve $\theta(h)$ to lower potential values, the sample was removed and, after having been broken down into sections of depth equal to 2 cm, was submitted to further draining on the membrane plate apparatus.

The measurements of evaporation flows and of potentials at different depths permitted the evaluation of both the tension curve $\theta(h)$ and the conductivity curve $k(\theta)$.

In fact, assuming that the sample can be characterized by a single equation:

$$\theta = f(h) \tag{1}$$

and indicating with ΔP_t the weight of evaporated water from the beginning of the transient until time t we have:

$$\Delta P_t = A \int_0^L \gamma \left[\theta(z) - \theta_0(z) \right] dz \tag{2}$$

in which θ_0 (z) indicates the initial distribution of water content, and A the cross-section of the sample.

(2), then, taking into account (1), can be set out in the form:

$$\Delta P_t = A \int_0^L \gamma \left[f\left(h(z)\right) - f\left(h_o(z)\right)\right] dz \quad (3)$$

If finally, the sample is divided into n strata of range Δz_i, and if we assume that the measured values h as taken by each tensiometer correspond to the mean values of the strata, (3) can be approximated by:

$$\Delta P_t \simeq A \gamma \sum_i^n \left[f(h_{t,i}) - f(h_{o,i}) \right] \Delta z_i \quad (4)$$

in which $h_{t,i}$ indicates the potential h measured to time t by the tensiometer of stratum i.

Following the suggestions of Boels et al. (1978) the equation $\theta=f(h)$ was approximated with a poligonal with k vertices for which, when the axes h_j^* of the vertices are fixed, the equation $\theta=f(h)$ will be sectionally linear and can be defined once the values θ_j^* of water content corresponding to potentials h_j^* are known.

Therefore, once the weight variation ΔP_t and the potentials $h_{t,i}$ and $h_{o,i}$ are known, and the values of h_j^* fixed, (4) can be reduced to a linear equation in the unknown quantities θ_j^*. The determination of the values θ_j^* is possible if the potentials and weights are experimentally estimated at different times, and if the system of linear equations, obtained by writing (4) at least for k values of t, is resolved.

The precision of these evaluations can be increased within certain limits by raising the number of equations and therefore of the number of measurements.

The choice of the values h_j^* will be arbitrary, but in each case, in order to obtain a uniform approximation of the tension curves through the entire range being surveyed it is necessary to have the values $h_{t,i}$ when measured, fall uniformly into all the intervals between two values h_j^*.

To make practicable the evaluations of θ_j^*, a calculus programme was prepared on the PC Unisys at this Institute, which is able automatically to fix the most convenient coordinates h_j^* of the vertices of the polygonal $\theta=f(h)$ and to resolve the system of equations (4) by minimizing the sum of the squares of the residuals.

Once the law $\theta(h)$ had been obtained using the foregoing procedure, and using the potential profile h(z,t), it was possible to turn to the water content profiles $\theta(z,t)$ and to calculate the conductivity k in the function of θ, using the instantaneous profile method in the way demonstrated earlier (Ciollaro and Comegna 1981).

3 RESULTS AND DISCUSSION

In Figure 1 is presented, by way of example, the time-flow values of the evaporation of water volume at the top of samples no.34 and no.40, taken from the depths Z=10-40 cm and Z=60-80 cm respectively.

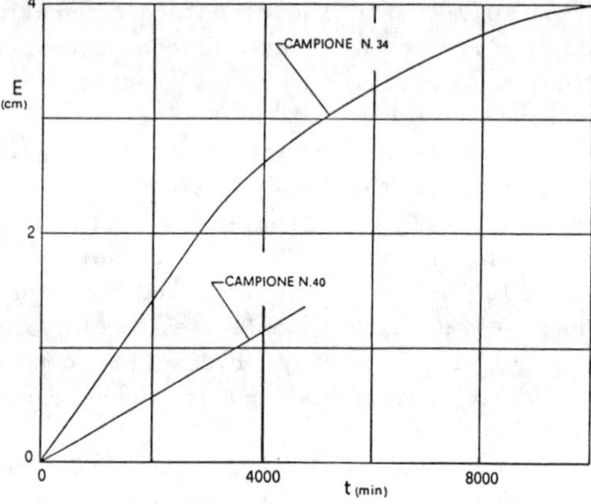

Figure 1 Relation of cumulated evaporation E to time, for samples no.34 and no.40

As has been shown (Ciollaro and Comegna 1983), the process of evaporation is determined by external energy, combined with the capacity of the soil to make available to the surface of evaporation the relative volumes of water.

Therefore, in general the process evolves in two distinct phases, of which the first is characterized by constant flow which depends only on the difference in vapour concentration between the air and the soil surface and on the laminar boundary resistences of the air.

When the soil surface reaches a level of water content such as not to permit the carriage of adequate water from the lower layers, the process becomes characterized by a reduction in time of evaporation flows, which is ever more marked by the reduction in water content.

These two phases are clearly visible in Figure 1, in which the evaporation process in sample no.34 was protracted for longer values of time.

From the graph, it can be seen that the flow of initial evaporation from sample no.40 is reduced. In fact it was necessary, in the case of samples characterized by low conductivity at saturation-point, to reduce the ventilation flow so as to avoid over-rapid depletion of the upper layers, which leads to potential profiles characterized by gradients which are too high at the surface and which cannot therefore be evaluated to an acceptable approximation.

However, in the tests it was necessary, depending on the particular characteristics of the samples, to regulate the flow of air so that water movement within the soil be sufficiently well distributed.

In figures 2 and 3, for the two samples above mentioned, the potential values at different measuring depths are presented From the graphs, we see that the potentials are clearly well differentiated from one another and that they evolve regularly in time, in such a way as to facilitate processing.

The results of the formulations carried out for all the samples according to the methodology illustrated allowed us to arrive at the presentations θ(h) of Figures 4 and 5, in which are shown the points specifying the vertices of the polygonals with which the tension curves were approximated.

Figure 4 in particular refers to samples taken from a depth of 10-40 cm and shows how the values of maximum water content are arranged in a band between 0,46 and 0,52, while for h equal to -250 cm, the values of water content are between 0,32 and 0,38.

Figure 5 refers to all remaining samples taken at a depth of 40-80 cm, in so far that it was not possible to make distinctions between them. From the graph, it can be seen that the values of maximum water content are lower, and fall within a narrower band than those of Figure 4.

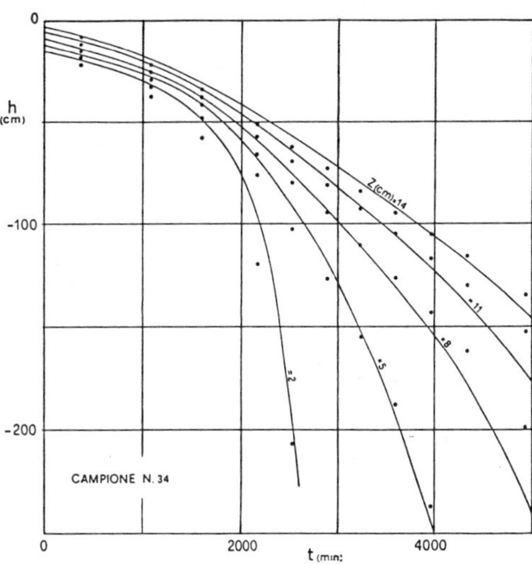

Figure 2 Measured (solid lines) and calculated (circles) h(t) data for different depths of sample no.34

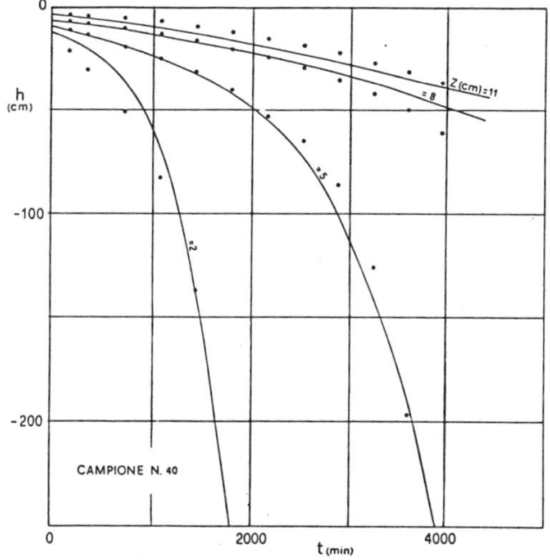

Figure 3 Measured (solid lines) and calculated (circles) h(t) data for different depths of sample no.40

For a more practicable comparison, in Figure 6 are presented the mean tension curves for the two depth bands and the range of values of θ, into which it is expected that 68% of the water content will

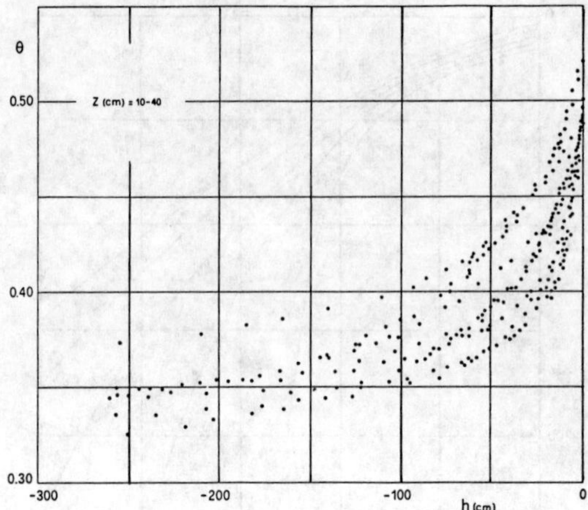

Figure 4 Pooled soil water characteristic data for all samples of selected depths 10-40 cm

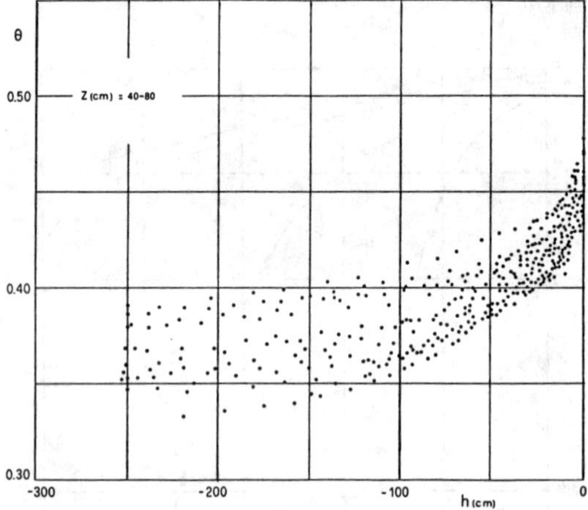

Figure 5 Pooled soil water characteristic data for all samples of selected depths 40-80 cm

fall. The comparison shows how the mean maximum water content is equal to 0,50 for the surface samples and equal to a little over 0,45 for the deeper layer which, with its potential values of -250 cm, is characterized, on average, by higher water content than the superficial layer.

The hydraulic conductivities relative to the two ranges of depth, calculated using the methodology above described, are presented in Figures 7 and 8 in the function of water content. It can be seen that,

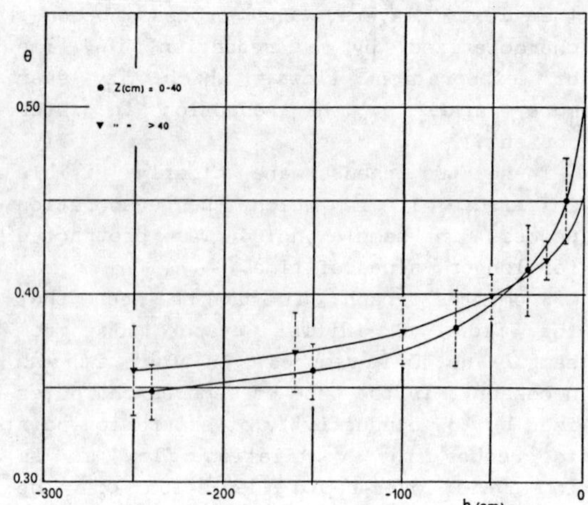

Figure 6 Mean soil water characteristic curves

as for all the samples, water conductivity is considerably reduced with a diminution of water content and that, despite there being a notable dispersion of points, slightly lower conductivities correspond to the greatest depths.

The test methodology was shown to be highly practicable, in that it did not call for the use of sophisticated equipment. The only evaluations necessary were of the weight of the samples and of the potentials at different depths.

To be able to test what the measurements mean and their precision, in view of one of their possible uses in simulation models, calculations were then carried out using the mathematical model, established in order to facilitate the study of evaporation processes in bare soil. The model was used to reconstruct the transient of evaporation effected during the tests, by attributing to each sample the hydraulic properties determined using the methodology previously illustrated.

With the aim of greater practicability, the analytic expression proposed by Van Genuchten et al. (1985) was applied to the tension curve $\theta(h)$; while the conductivity curve was approximated with an exponential function at two parameters. The parameters shown in the above-mentioned equations were then determined by adapting, through the optimization method, the analytic curves to the experimental data.

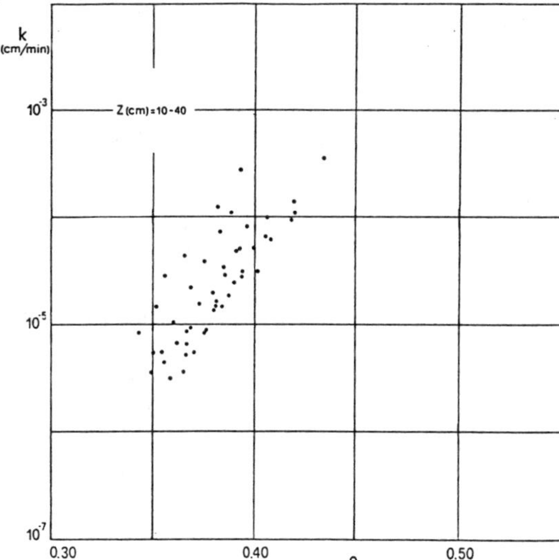

Figure 7 Pooled hydraulic conductivity data for all samples of selected depths 10-40 cm

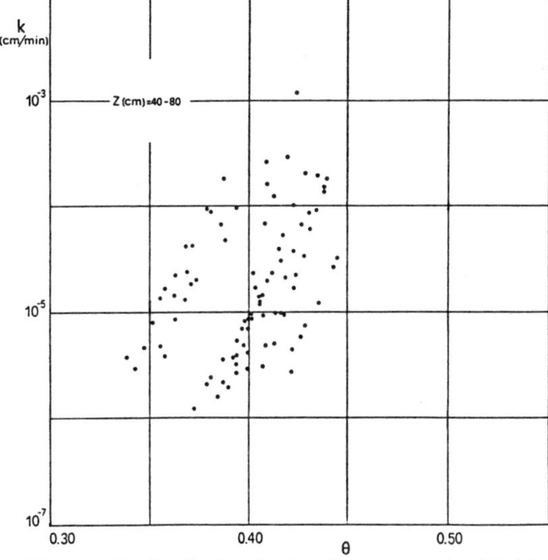

Figure 8 Pooled hydraulic conductivity data for all samples of selected depths 40-80 cm

The results of these calculations were shown to be quite similar to the measurements taken during the tests. In particular, continuing to refer to sample nos. 34 and 40, in Figures 2 and 3, together with the curves measured experimentally, are shown the results of the numerical reconstructions carried out. In these graphs it can be seen that the dispersion of the points is modest: this can, however, be attributed to the imperfect homogeneity of the samples.

Therefore the verification calculations we carried out show how the test method employed is sufficiently precise for application in mathematical models.

The research was extended to include the evaluation of the variability of the hydraulic parameters in the soil under survey.

The data were therefore processed statistically, and shown in Table 1 are the mean values, the standard deviation and the coefficient of variation for the water content relative to prefixed potential values, and for water conductivity, estimated at θ = 0,40. The data shown in the table demonstrate the slight spatial variability of the parameters, depending on the limited extent of the area under survey and on the homogeneity of the soil. Vice versa, a greater spatial variability of water conductivity is seen, which can be approximated to a distribution of the log-normal type (Nielsen et al. 1973; Russo and Bresler 1980; Ciollaro and Comegna 1988)

Table 1 Values of mean, standard deviation and coefficient of variation CV of volumetric water content from saturation (θ_0) to -250 cm (θ_{250}) and of unsaturated hydraulic conductivity ($k_{0,40}$) for the composite

	Z=10-40 cm	Z=40-80 cm
θ_0		
mean	0,502	0,459
standard error	0,0155	0,0133
CV	0,031	0,029
θ_{100}		
mean	0,372	0,390
standard error	0,0187	0,0139
CV	0,050	0,036
θ_{250}		
mean	0,347	0,373
standard error	0,0164	0,0147
CV	0,047	0,039
$k_{\theta,40}$		
mean	$1,28\times10^{-4}$	$8,58\times10^{-6}$
standard error	$1,29\times10^{-4}$	$1,15\times10^{-5}$
CV	1,012	1,344

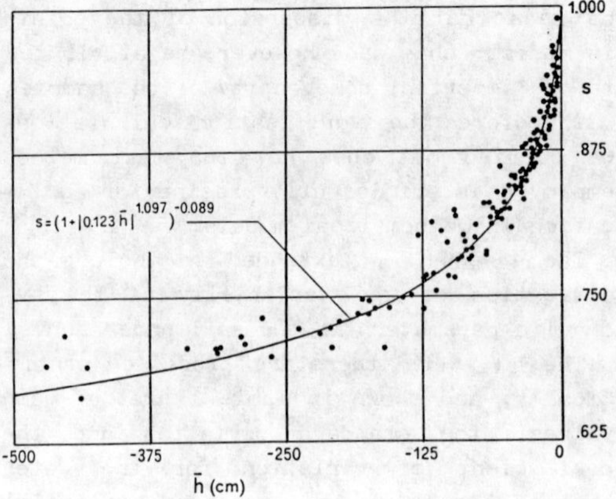

Figure 9 Scaled soil water characteristic data for the composite

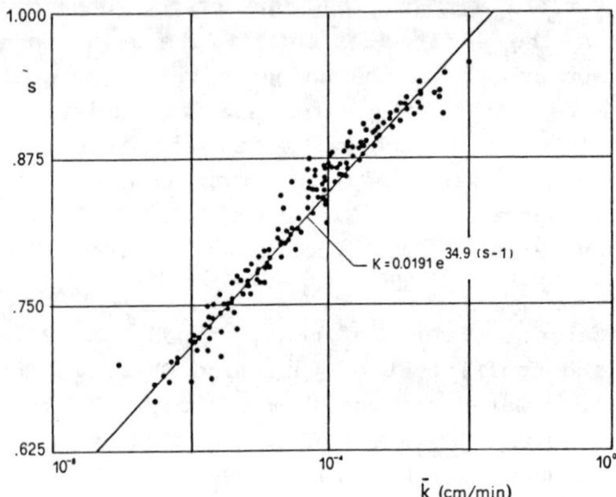

Figure 10 Scaled hydraulic conductivity data for the composite

For a more profitable use of the data in models of the stochastic type, it has been demonstrated (Ciollaro and Comegna 1987) that the heterogeneity of the soils can be defined more simply by taking into consideration the concept of geometrical similarity of porous media. According to the similar media theory of Miller and Miller (1955), the functions $h_r(\theta)$ and $k_r(\theta)$ in each site r can be deduced from the mean equations $\bar{h}(\theta)$ and $\bar{k}(\theta)$ by means of a single scale factor α_r, which is characteristic of the measuring site. As we have already seen (Ciollaro and Comegna 1987), the values α_r can be profitably evaluated from both the tensiometric and conductivity data by means of functional normalization techniques, by referring tensiometric and conductivity data to degree of saturation s rather than to water content and by expressing the equations h(s) and k(s) through applied analytic functions.

For all the samples analyzed in Figure 9, in the function of degree of saturation, the values of potential $\bar{h}=\alpha_{h,r}\, h$, are presented, while in Figure 10 the conductivity values $\bar{k}=\alpha_{k,r}^{-2}k$ are given, obtained by processing the potential and conductivity data respectively. The graphs show how the method we adopted markedly reduces the dispersion of experimental points and allows the evaluation of the variability of the hydraulic properties of soil through the spatial distribution of the sole parameters $\alpha_{h,r}$ and $\alpha_{k,r}$.

The statistical processing thus demonstrated how the log-normal rule is that which best approximates to frequency distribution, not only to the parameters $\alpha_{h,r}$, but also to those $\alpha_{k,r}$. The principal statistical moments of $\alpha_{h,r}$ and $\alpha_{h,r}$ are given for the applications in Table 2.

Table 2 Values of mean, standard deviation and coefficient of variation of $\alpha_{h,r}$ and $\alpha_{k,r}$ scale factors determined respectively from h(s) and k(s) data

	$\alpha_{h,r}$	$\alpha_{k,r}$
mean	0,978	1,098
standard error	0,924	1,968
CV	0,945	1,793

A study of the table reveals that the coefficients of variation obtained by processing the conductivity data are higher than those relative to the processing of the potential data, which is probably due to the fact that conductivity values are more sensitive to degrees of saturation. It must, finally, be underlined that the high degree of correlation between parameters $\alpha_{h,r}$ and $\alpha_{k,r}$ supports the macroscopic similarity hypothesis of Miller and Miller (1955).

4 CONCLUSIONS

A determination of the hydraulic properties of soil is seen to be indispensable if we wish to approach the study of water movement in quantitative terms, using mathematical models. The practical possibilities of extensive application of such models leads to the definition of measuring techniques which render the determination of the laws $\theta(h)$ and $k(\theta)$ less problematic.

This proposed method allows us readily to obtain functions $\theta(h)$ and $k(\theta)$, by submitting the samples, in the laboratory, to a process of evaporation, and requires only the measurement, through weighing, of the mean water content of the sample and of the potentials at different depths of the same.

It was verified that the tension and conductivity curves thus determined and described, using analytic equations, can be used in simulation models with an acceptable degree of approximation. Moreover, it can be observed that the duration of the tests was not excessive: this makes acceptable the labour intensiveness of characterization imposed on statistical evaluations of the variability of individual hydraulic parameters.

Therefore, from this proven association of accuracy and speed in the experimental work of characterization, we are convinced that the adopted methodology can be considered adequate to the demands of modern study at an operative level of problems connected with water movement and of solutes in unsaturated media.

APPENDIX

Description of the modal profile.
Area: Policoro (MT); Locality: Pantanelli.
Substratum: alluvial; unity a_1 on the geological map.
Classification: vertic xerofluvent.
Horizons:

Ap_1 - 0-20cm; colour 10YR4/2; skeleton absent; fine crumb structure and average poliedric evident; friable consistency; not very humid; modest drainage; grassy roots, fine, scarce; notable effervescence; gradual boundary, cracks; silty-loam (sand 25%; silt 51%; clay 24%).

Ap_2 - 20-25 cm; colour 10YR4/2; skeleton absent; subangular poliedric and coarse angular poliedric; not very humid; resistent; moderate drainage; channels and large pores; grassy roots, fine, scarce; notable effervescence; abrupt boundary; cracks; silty-loam (sand 24%, silt 52%, clay 24%).

A_1 - 45-90 cm; colour 10YR4/2 and 4/3; skeleton absent; subangular poliedric and coarse angular poliedric; small pores; slow drainage; cracks; notable effervescence; abrupt boundary; silty-clay-loam (sand 15%, silt 55% clay 30%).

C_1 - 90-100 cm; colour 2,5Y5/2; skeleton absent; humid; notable effervescence; slow drainage; silty-loam (sand 25%, silt 55%, clay 20%).

C_2 - 100-120 cm; colour 2,5Y5/2; skeleton absent; humid; notable effervescence; slow drainage; loam (sand 42%, silt 38%, clay 20%).

C_3 - 120 cm and over; colour 2,5Y5/2; skeleton absent; notable effervescence; sandy-loam (sand 60%, silt 28%, clay 12%).

REFERENCES

Boels, D., J.B.H.M. Van Gils, G.J. Veerman & K.E. Wit. 1978. Theory and system of automatic determination of soil moisture characteristics and unsaturated hydraulic conductivities. Soil Sci. 126:191-199.

Bouma, J. 1983. Use of soil survey data to select measurement techniques for hydraulic conductivity. Agric. Water Managem. 3:235-250.

Ciollaro, G. & V. Comegna. 1979. Definizione delle caratteristiche idruliche del suolo. 3° Convegno Nazionale A.I.G.R. Catania 1:438-459.

Ciollaro, G. & V. Comegna. 1981. La caratterizzazione idraulica del suolo nell'intervento irriguo. Hydraulic Agriculture Institute, University of Naples.

Ciollaro, G. & V. Comegna. 1983. Esaurimento per evaporazione di mezzo poroso di spessore limitato. Hydraulic Agriculture Institute, University of Naples.

Ciollaro, G. & V. Comegna. 1987. Studio della variabilità spaziale delle proprietà idrauliche del suolo mediante il concetto di similitudine. Hydraulic Agriculture Institute, University of Naples.

Ciollaro,G. & V. Comegna. 1988. Spatial variability of soil hydraulic properties of Psammentic Palexeralfs soil of South Italy. 4° International Symposium on water supply and irrigation in the open and protected cultivation. I.S.H.S. Padova. 228:61-71.

Ciollaro,G., V. Comegna & C. Ruggiero. 1987. Confronto tra metodi di campo e di laboratorio per lo studio delle caratteristiche idrauliche del suolo. Hydraulic Agriculture Insitute, University of Naples.

Day, P.R. 1965. Methods of soil analysis. Part 1. Agronomy. C.A. Black Amer. Soc. of Agron. Madison, Wis. 9:545-556.

Gajem, Y., A.W. Warrick & D.E. Myers. 1981. Spatial dependance of physical properties of typic torrifluvent soil. Soil Sci. Soc. Am. J. 45:709-715.

Hillel, D. 1980. Flow of water in unsaturated soil. In Fundamentals of soil physics. D. Hillel (ed.), Academic Press, N.Y.

Hillel, D., W.D. Krentos & Y. Stylianou. 1972. Procedure and test of an internal drainage method for measuring soil hydraulic characteristic in situ. Soil Sci. 114:395-400.

Klute, A. 1972. The determination of hydraulic conductivity and diffusivity of unsaturated soil. Soil Sci. 113:264-277.

Miller, E.E. & R.D. Miller. 1955. Theory of capillary flow. Soil Sci. Soc. Am. Proc. 19:267-275.

Nielsen, D.R., W. Biggar & K.T. Erh. 1973. Spatial variability of field measured soil water properties. Hilgardia. 42:215-259.

Russo, D. & E. Bresler. 1980. Field determinations of soil hydraulic properties for statistical analysis. Soil Sci. Soc. Am. J. 44:697-702.

Santini, A. 1979. Bilancio idrico del sistema suolo-vegetazione: l'impiego di un modello matematico di simulazione. Rivista di Ingegneria Agraria. 10:63-76.

Santini, A. et al. 1988. Impiego di un modello per la simulazione del trasporto idrico nel sistema suolo-pianta-atmosfera. Hydraulic Agriculture Institute, University of Naples.

Van Genuchten, M.Th. & D.R. Nielsen. 1985. On describing and predicting the hydraulic properties of unsaturated soils. Ann. Geophys. 3(5):615-628.

Land and Water Use, Dodd & Grace (eds), © 1989 Balkema, Rotterdam. ISBN 90 6191 980 0

New method to determinate monthly runoff in small basins

R.Dal-Ré Tenreiro & F.Ayuga Téllez
Departamento de Construcción y Vías Rurales, Universidad Politécnica de Madrid, Spain

ABSTRACT: Small basins with less than 50 km^2 of surface provides many times enough runoff water to establish irrigation areas of good productive interest if dammed up. This option is better than others in many situations.
However, a design problem always present in this small basins is the absence of gauging measuring of the stream, necesary to know monthly runoff and hidric plan.
Nevertheless, almost every meteorological observatory has monthly average of temperatures rainfall and raining days data from more than 15 years. So, it is somehow possible to know annual runoff using any of the habitual methods.
Nowdays the best adapted method to spanish conditions is the so-called "curve number" of the United States Soil Conservation Service. (or TR-20). This method obtains surface runoff of a strom from rainfall, soil humidity and soil-vegetation hydrology conditions - (hydrology soil-cover complex).
The proposed new method consist in application of the CN method to monthly runoff instead of storm runoff with a few necesary adaptations.
Annual runoff must be known or obtained by any well adapted method and a list of monthly meteorological data such as expressed before is also necesary.
New method considers that direct runoff is due per a percentage of monthly average rainfall (between 50% and 70%). Using CN method equations monthly runoff is obtained by multiplication of monthly average rainfall with a k factor (0,5 to 0,7) called "Adjusting coefficient". The exact value of this coefficient is obtained by iteration until equaling annual runoff.
It is supposed that complementary rainfall percentage should hold humidity levels of soil. This levels are referred to precedent humidity conditions of CN method (A, B, C). The Adjusting Coefficient method consider. A condition that month of less than 5 rain days (average), B condition that month of 5 to 10 rainy days and C condition that month of more than 10 rainy days.

INTRODUCTION

The short basins up 50 sq. km. long provide in a great number of cases, with enough runoffs, which adequately damed up, allow water available to establish local irrigated land of a remarkable productive interest.
The irrigable zones options with contributions of surface damed up runoffs seems - far from negligible.
However, one of the problems that always come up in those basins is the absence of flow data, in enough series, which would allow to establish the monthly runoffs with the sufficient guarantee so that the hydric balance could be drawp up.
Nevertheless, in the majority of local meteorological observatories there is some - data on the monthly average temperature, monthly average pluviometries and average number of rainy days per month in series of 15 years or more. This allows us in some way to get to determine the annual runoffs with certain accuracy through any of the methods up to date. Nowadays, the most contrasted method is the one called the - "Curve Number" carried out by the "United States Soil Conservation Service" (USSCS). This method estimates the surface direct runoff from an issolated and specific rainfall, according to the value of the total rainfall, the soil and vegetal cover humidity (Hydrologic Complex cover-soil). The influence of the complex cover-soil on - the runoff is measured through the curve

number Nb, which varies according to the type of soil (fallow, pasture, forest, - etc.), the hydrologic soil groups (in reference to their texture and structure). However, it does not take into account the deep runoff.
In the table (nº 1) the Nb values get detailled. The previous conditions of soil humidity are taken into account considering different curve numbers A, B and C, table nº 2 and figure 1.
The A condition is for dry soil and the C for damp one, according to the rain fallen during the 5 days previous to the rain considered and to the season of the year. Curve Nb does not match the one of average - conditions.
One the curve number matching each rainfall is known, the maxium infiltration S, is - determinated:

S = 254 (100/N-1)   S in mm.

Together with this value, where N may adopt the values Na, Nb or Nc, the direct runoff produced by a rainfall P, is obtained:

$$Q = \frac{(P-0,2.S)^2}{P+0,8.S}$$

NUMEROS (N_B) DE LAS CURVAS DE ESCORRENTIA PARA LOS COMPLEJOS HIDROLOGICOS «SUELO-CUBIERTA»
(Según el U.S. Soil Conservation Service)
Condición B

Cubierta del suelo			Números N_B correspondientes a los grupos hidrológicos del suelo			
Clase	Laboreo	Condiciones hidrológicas para la infiltración	I	II	III	IV
Barbecho	—	—	77	86	91	94
Cultivos alineados	R	Pobres	72	81	88	91
	R	Buenas	67	78	85	89
	C	Pobres	70	79	84	88
	C	Buenas	65	75	82	86
	C-T	Pobres	66	74	80	82
	C-T	Buenas	62	71	78	81
Cultivos no alineados, o con surcos pequeños o mal definidos	R	Pobres	65	76	84	88
	R	Buenas	63	75	83	87
	C	Pobres	63	74	82	85
	C	Buenas	61	73	81	84
	C-T	Pobres	61	72	79	82
	C-T	Buenas	59	70	78	81
Cultivos densos de leguminosas o prados en alternativa	R	Pobres	66	77	85	89
	R	Buenas	58	72	81	85
	C	Pobres	64	75	83	85
	C	Buenas	55	69	78	83
	C-T	Pobres	63	73	80	83
	C-T	Buenas	51	67	76	80
Pastizales (pastos naturales)	—	Pobres	68	79	86	89
	—	Regulares	49	69	79	84
	—	Buenas	39	61	74	80
	C	Pobres	47	67	81	88
	C	Regulares	25	59	75	83
	C	Buenas	6	35	70	79
Prados permanentes	—	—	30	58	71	78
Montes con pastos (ganadero-forest.)	—	Pobres	45	66	77	83
		Regulares	36	60	73	79
		Buenas	25	55	70	77
Bosques (forestales)	—	Muy pobres	56	75	86	91
		Pobres	46	68	78	84
		Regulares	36	60	70	76
		Buenas	26	52	63	69
		Muy buenas	15	44	54	61
Caseríos	—	—	59	74	82	86
Caminos en tierra	—	—	72	82	87	89
Caminos con firme	—	—	74	84	90	92

Table 1

CONDICIONES PRECEDENTES DE HUMEDAD DEL SUELO

Condición	Lluvia total caída durante los cinco días anteriores	
	Período húmedo: octubre-marzo	Período seco: abril-septiembre
A	Menos de 12,5 mm.	Menos de 35,5 mm.
B	De 12,5 a 28 mm.	De 35,5 a 53 mm.
C	Más de 28 mm.	Más de 53 mm.

Table 2

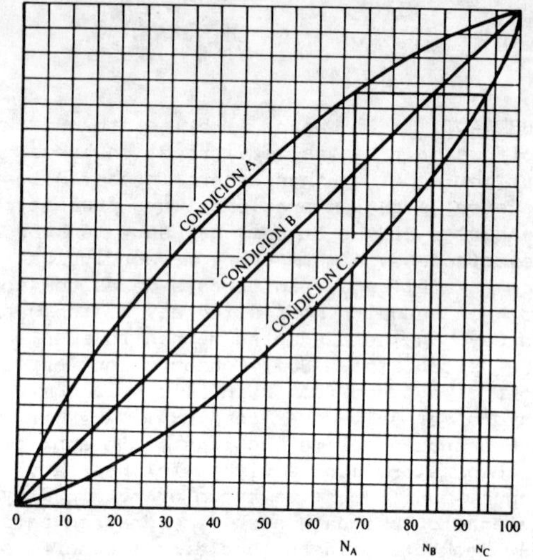

Figure 1

It is estimated that rainfalls of a magnitude inferior to 0.2.S mm. do not produce runoffs. The runoff produced by all and - every rainfall each year is calculated - month by month, and the addition of them, would result in the monthly average runoff for the period under observation (10, 15, 25 years) and as sum of the monthly average figures we would get the annual surface direct runoff. The values adopted for project will be the result of the interval of liability fixed and the statistic method followed.
As it can be noted the stric application of this method requires, among other data, to have available the daily pluviometric series from the observatories in question to be able to determine the previous humidity conditions for each rainfall. The final elaboration turns up extremely boring.
On the other hand the method, in its application on an isolated rainfall, gives place to sharp "jumps" of the runoff produced when passing from condition A to B or

from B to C.

In fact, if we take an isolated rainfall, in less than 24 hours, of 40 mm. with curve number Nb = 70, Na = 50 and Nc = 85, if for the 5 previous days have fallen 27.8 mm. in the damp period, we would be in condition B, but if there had been registered 28.1 mm. it would get to condition C. These 3 thenths will change the calculation of the runoff:

$$QB = \frac{(40-0.2.Sb)^2}{40+0.8.Sb}$$

And as $Sb = 254\ (\frac{100}{70} - 1) =$ then,

$$QB = \frac{(40-21.8)^2}{40+87} = 2.63 \text{ mm.}$$

and in contrast, with 0.3 mm. of rain more it results,

$$QC = \frac{(40-0.2\ Sc)^2}{40+0.8\ Sc}$$

with $Sc = 254\ (\frac{100}{85} - 1) = 44.8$

which gives,

$$QC = \frac{(40-8.9)^2}{40+35.8} = 12.76 \text{ mm.}$$

That is to say, that just an increase of 0.3 mm. multiplies by 4.85 the runoff produced. It does not seem acceptable.

The same way, the limit of conditions A and B could be dealt with if considered 12.4 mm. of rain during the 5 previous days or else 12.7 mm.

$$QA = \frac{(40-50.8)^2}{40+203.2} = 0, \text{ there is no runoff}$$

What looks as an absurd.

Q runoffs grow more than proportionally in relation to the rainfalls that create them; the 2nd degree equations,

$$Q = \frac{(P-0.2.S)^2}{P+0.8.S}$$

have a double root for P = 0.2.S. They are tangent to the axis of abscissas in the point (0.2.S ; 0). Around the downward zone of the axis of abscissas, from P = = 0.2.S , Q has no real meaning although it may have geometrical representation, as rainfalls, of value inferior to 0.2.S do not produce runoff.

Geometrical representation for any N value would be, (figure 2).

Now then, we have:

$$\frac{Q1}{P1} = \frac{BB'}{QB} = \frac{AA'}{OA} \quad \frac{AC}{OA} = \frac{Qo}{Po}$$

therefore,

$$\frac{Q1}{P1} \quad \frac{Qo}{Po}$$

On the other hand, daily pluviometric series from the observatories mentioned before in order to be allowed to determine the previous humidity conditions for each rainfall are hard to get, not only for feasibility studies but for projects too.

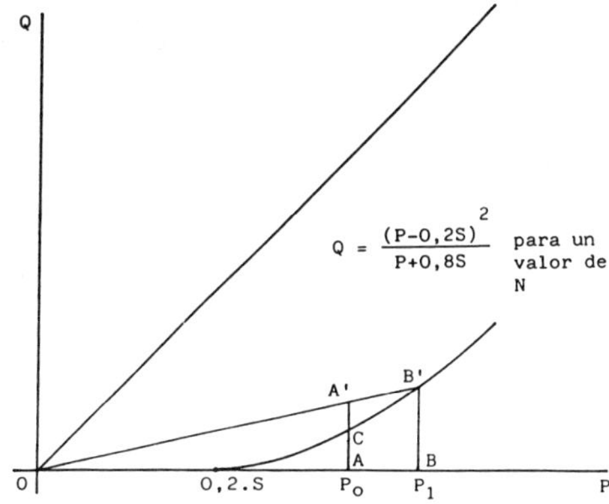

Figure 2

With the simplification of the method next proposed, monthly average numbers of rainy days, which are usual meteorological data. The result, if not as strict as the American method, has the advantage of a greater aplicability, without loosing accuracy as the biggest errors in the calculation of the runoff through the method of the "Curve Number" are precisely due to the settling of N.

Indeed, the changing characteristics of the vegetal cover and the subjectivity in the classification of the complex soil-cover makes the determining of Nb not extremely strict and little variations in this number give place to great variations in the runoff.

Besides, the proposed method can be adapted to calculation systems with personal computers, without needing to resort to any data base, or integrated as a subroutine to programs of more complexity such as computer aided design, as much hydrologic as of irrigable and engineering.

DESCRIPTION OF THE METHOD

Starting data

It is neccessaty to have available the following meteorological data:

Monthly average pluviometry, average number of rainy days per month; an all those that would allow to calculate the <u>annual</u> average specific runoff <u>QF</u> through any procedure in force.

Settling of the damp and dry seasons

The "Curve Number" method sets for damp -

periods will be different ones according to their own characteristics.
As it has been earlier established, to consider previous soil humidity conditions - through the original method it is taken into account the rainfalls fallen during the five days previous to the one on which it is studied the rainfall that fives origin to the runoff.
For the application of the method it will be considered in condition A the months - whose average number of rainy days is inferior to 5; in condition B those months in which the average number of rainy days is within 5 and 10 both included, and in condition C the months in which the averabe number of rainy days is equal to 11 or higher, with the exception later explained in terms of Pm (monthly average precipitation).

Simplification of the "Curve Number" Method

The simplification lies in considering that a percentage of the monthly average pluviometry generally within 50 and 70, produces surface runoff, it will there by be used for calculation as the value of P in the formula for getting Q, a virtual precipitation equal to the monthly Pm multiplied by a rate K within 0.5 and 0.7 (it can exceptionally go downwards to 0.4 or upwards to 0.8). This rate is called "Adjustment rate".
The monthly precipitation left (not producing direct surface runoff) will be used for keeping up the soil humidity level - which corresponds to the condition in which it has been classified that month.
As it has been said according to general - meteorology as long as it could be verified that P 0.2.S there is no runoff produced, hence in our case this circumstance would come expressed by:

K.Pm   0.2.S

In the general runoff formula will therefore show up the expression:

$$Qm,K = \frac{(K.Pm-0.2.S)^2}{K.P+0.8.S}$$

Where Qm,K represent the runoff produced by the virtual rainfall K.Pm on the B, C or A corresponding condition, depending on the number of rainy days we are considering for the month in question.
Using this procedure month after month we will reach a total value

$$QK = \sum Qm,K$$

That us to say, sum of all the monthly runoffs originated by every virtual rainfall K.Pm along the year.
Certeainly, depending on the rate K we - apply, it will give us a different value for QK, but as we have already fixed the value of the total runoff and not only the surface direct one, QF through any of the methods up to date, it is this value the one which will determine the total addition of the different Qm,K that have to be reached in order to check:

$$QF \simeq \sum Qm,K$$

total annual specific runoff calculated by the desired aproximation. This way, adjusting K value, the monthly runoffs get adjusted to the resulting total runoff.
Thus, to the advantage of having available a contrasted value for the total runoff - (through methods less risky than the "Curve Number") it is added a distribution of runoffs along the year which would be very useful for carring out a hydric balance.
For obtaining the adjustment rate through a computer program, it is set out automatically the possible values within the limits 0.5 and 0.7 from hundredth to hundredth (exceptionally 0.4 and 0.8) being selected the rate K from whose application gets derived an annual total runoff bearer to QF than the rest. With this procedure it is obtained the adjusted monthly distribution allowing the drawing up of the monthly hydric balance.

LIMITATION OF THE METHOD

Extension

As it was explained at the beguining this method is valid for basins of no more than 50 km2. of surface. This limit is due to the fact of not many different "Curve Numbers" must been considered.

Pluviometry

It has to be noted the lprinciple that this simplified method can not be applied to - extremely rainy areas, with very uniform rainfalls along the year. There has to be a highly clear difference between damp and dry periods or otherwise damp and less damp. The methods also saves areas with a very - low degree of precipitations, under 250 mm. per year or those in which pluviometric distribution gets concentrated within three - months or less.

Correcting N Values according to Pm and Dm ones

To improve the calculation of monthly runoffs in the damp period through the "adjustment rate" method it has to be taken into account, fot the method considered, - not only the average number of monthly rainy days Dm but the amount of rain, Pm/4 too.

The monthly average rain is thence divided into four parts as supossed probable way - for its distribution along the Dm days. Four "blocks" of rain of Pm/4 mm. are considered in the manner which will make us note, by means of its value, the average - probable humidity of the land in each -- "block" whenever such amount of rain Pm/4 falls.
The following relations get establishec:

$$\alpha = \frac{Nb}{Na} > 1 \quad \text{and} \quad \beta = \frac{Nc}{Nb} > 1$$

Values N'a and N'b are therefrom introduced,
N'a = a.Na and N'b = b.Nb
Thus,

$$1 \leq a \leq \frac{Nb}{Na} \quad \text{and} \quad 1 \leq b \leq \frac{Nc}{Nb}$$

It will hence be reached the N'a and N'b values from the table;

DAMP PERIOD

N'a		N'b	
Pm/4	a	Pm/4	b
<12.5	1	<12.5	1
12.5-28	$0.5+\frac{\alpha}{2}$	12.5-28	$0.5+\frac{\beta}{2}$
>28	α	>28	β

This way, when there is Dm < 5 in the month object of study, that would mean being in condition A. Na value gets modified depending on the monthly average rainfall giving us Pm/4 values either within 12.5 and 28 mm or higher than this value, as it is specified in the enclosed table.
Certainly, if monthly average rainfall is of 30 mm., then Pm/4 = 7.5 mm., we are in strict A condition, but if it had been of 80 mm., with Pm/4 = 20 we would get a value N'a = a.Na where,

$$a = 0.5 + \frac{\alpha}{2}$$

Thereof the runoff gets increased, not only for having increased P when going from 30 mm., to 80 mm., but for the N'a value being higher than Na, which generates an - even superior increase, due to previous humidity conditions being higher.
The same could be said in the case of N'b= =b.Nb as it can be noted in the table.
However, dry period is not considered due to the average amount of rain Pm having to be very high to make Pm change the classification which could hardly be taken into account, with such values, a real dry period.
Note: For the cases in whic $0.5+\alpha/2$ and $0.5+\beta/2$ values are applicable, it is of easy verifying the values reached for N'a and N'b being respectively, (Na+Nb)/2 and (Nb+Nc)/2.
Let's see some esamples:
Considering a short basin in which:
Na=50; Nb=70; Nc= 85
1st Case.- Dm= 10 days - Class B (5<Dm<11) if Pm=100 mm.
Pm/4=24 mm. within 12.5 and 28

$$\text{then,} \quad b = 0.5+\frac{\beta}{2} = 0.5+\frac{Nc}{2.Nb} = 0.5+\frac{85}{2.70} = 1.11$$

There from
N'b=77.7; S'b=72.9
The runoff will be:

$$Q' = \frac{(100 - 14.6)^2}{100 + 58.3} = 46.1 \text{ mm.}$$

If the correction would not have been made, we would have,
Nb = 70; Sb = 108.9

$$Q = \frac{(100 - 21.8)^2}{100 + 87.1} = 32.6 \text{ mm.}$$

Which does not seem to correspond to 100 mm. of monthly average rain in the damp period.

2nd. Case.- Dm= 4 - Class A (Dm< 5); if Pm = 80 mm.
Pm/4 = 20 it is ranged within 12.5 and 28 mm. hence,

$$a = 0.5 + \frac{\alpha}{2} = 0.5 + \frac{70}{2.50} = 1.2; \text{ therefore}$$

N'a = 1.2; Na = 60; S'a = 169.3;
The runoff will be:

$$Q' = \frac{(80 - 33.9)^2}{80 + 135.5} = 9.9 \text{ mm.}$$

If the correction would not have been made:

$$Q = \frac{(80 - 50.8)^2}{283.2} = 3 \text{ mm.}$$

Which is not logical for 80 mm. of monthly average rain for 4 days in the damp period.

3rd Case.- For Dm= 11 - Class C (Dm>11) but with the same amount of rain as in the 1st. case.
Pm = 100 mm. of monthly average

$$Nc = 85 \quad Sc = 254\left(\frac{100}{85} - 1\right) = 44.8;$$

$$Q = \frac{(100 - 8.9)^2}{100 + 35.8} = 61.1 \text{ mm.;}$$

If in the forst case the correction would not have been made, the difference of just a rainy day more would have given place to runoff difference being of:

Q = 61.1 - 32.4 = 28.5 mm.
With the established correction it gets,
Q = 61.1 - 46.1 = 15 mm.
which looks less far from reality

EXEMPLES OF APPLICATION OF THE METHOD

Let's take a basin with the following characteristics:

Area . . . . 525 Ha.
Na Curve Number 67

Nb Curve Number 83
Nc Curve Number 93
Annual average temperature 16.5ºC
Months in the damp period 6

Pluviometric data: the ones which are shown in the corresponding tables:

Calculation of the monthly runoffs through the simplified method without the corrections of N values according to Pm and Dm ones.

MONTH	PRECIPITACION	DAYS	N	RUNOFF
1	48.00	5	83	3.00
2	120.00	10	83	25.09
3	64.00	8	83	6.66
4	135.00	12	93	50.25
5	115.00	11	93	40.65
6	75.00	9	83	9.71
7	72.00	7	83	8.84
8	43.00	6	83	2.09
9	34.00	5	83	0.82
10	6.00	2	67	0.00
11	5.00	1	67	0.00
12	35.00	3	67	0.00
	752.00 mm.			147.11 mm.

Direct runoff rate K = Qa/pa . . . 19.56%
Adjustment rate 0.51

Calculation of the monthly runoffs with the correction of N values according to Pm and Dm ones.

MONTH	PRECIPITATION	DAYS	N	RUNOFF
1	48.00	5	83	2.14
2	120.00	10	93	37.44
3	84.00	8	88	8.87
4	135.00	12	93	43.88
5	115.00	11	93	35.32
6	75.00	9	88	12.22
7	72.00	7	83	6.90
8	43.00	6	83	1.43
9	34.00	5	83	0.48
10	6.00	2	67	0.00
11	5.00	1	67	0.00
12	35.00	3	67	0.00
	752.00 mm.			148.68 mm.

Direct runoff rate K = Qa/Pa . . . 19.77%
Adjustment rate 0.46

It can be remarked that runoffs rearranging is more real than in the simplified method.

It is evident that when Pm/4 values do not imply any correction, the simplified coincides with the corrected one.

BIBLIOGRAPHY

Alvarez Martínez, A. 1981. Apuntes de proyecto y construcción de presas. Tomo I. Estudios del terreno e hidrológicos. Madrid: Serv. Publ. Revista de Obras Públicas.

Ayuga Téllez, F. & García y García, A. 1987. Nuevo método para el cálculo de escorrentías en pequeñas cuencas no aforadas. Tecnología del agua nº 41: 47.52.

Dal-Ré Tenreiro, R.; Segura, R. et al. 1985. Diseño y construcción de pequeños embalses. Manual técnico nº 2. IRYDA. Madrid. Mº de Agricultura, Pesca y Alimentación.

García y García, A. 1986. Diseño óptimo de presas de materiales sueltos para uso agrícola. Tesis Doctoral. Madrid. E.T.S.Ingenieros Agrónomos.

Heras, R. et al. 1981. Manual de ingeniería de regadíos. Heras, R. (ed.)

USDA - Soil Conservation Service. 1964. Hydrology guide for use in watershed planning. National Engineering Handbook. Secc. 4. Suplement A. Beltsville, Md.

USDA - Soil Conservation Service. 1984. Engineering Field Manual. Washington D.C.

USDI - Bureau of Reclamation. 1966. Diseño de presas pequeñas. Ed. Compañía Editorial Continental. México.

Land and Water Use, Dodd & Grace (eds), © 1989 Balkema, Rotterdam. ISBN 90 6191 980 0

The rough set method in water flow models

D.Dejas & A.Reinhard
Institute for Land Reclamation and Improvement, Agricultural University, Wrocław, Poland

ABSTRACT: The rough set theory is applied to the analysis of the field investigation results in order to determine the movement of the underground water as well as the water remaining in the soil layer in the drained or meliorated zone. It attempts to describe how the water-air ratio is attained in the soil in a drained zone; it takes into account such parameters as the precipitation and the water levels in the ditches in the vinicity of the place where the soil humidity reading is taken.

RÉSUMÉ: Dans cette étude on a presenté l'application d'une méthode des ensembles approximatifs à l'analyse des résultats des essais faits sur le terrain. Ces essais déterminent les possibilités de définir le mouvement de l'eau souterraine et de l'eau du sol pour une installation d'assainissement choisie. On a constaté que pour décrire la formation des relations entre l'air et l'eau dans le sol dans un milieu d'assainissement défini, il suffit d'avoir - comme paramètres - des précipitations et des niveaux d'eau dans les fossés proches du lieu où on a mesuré l'humidité du sol.

ZUSAMMENFASSUNG: Zur Ermittlung der Grund u. Bodenwasserbewegung auf einer bestimmten Meliorationsanlage wurde, zur Analyse der im Laufe der Feldversuche erhaltenen Messwerte, die Methode der angenäherten Mengen angewendet. Es konnte nachgewiesen werden, daß es für die Bestimmung der Gestaltung der Wasser-Luftverhältnisse im Boden genügend ist, die Parameter Niederschlag und Wasserstände in den Meliorationsgräben, die sich in der Nähe der Bodenfeuchtigkeitsmesspunkte befinden, zu berücksichtigen.

1 INTRODUCTION

The purpose of the study was to work out a method for the control of water-air relations under the conditions of shallow ground water levels. Water is a major factor affecting plant growth. Both continuous and temporary excess of moisture in the root system layer as well as its deficiency exert and adverse influence on water-air relations and do not favour high yield. These relations are most frequently controlled by digging the ditches. Such a system affects water-air relations in the soil by decreasing the levels of ground water through drainage and by raising the ground water levels during irrigation. It is therefore interesting to know to what extent the climatic (rainfalls and temperature) and soil conditions, levels of ground water and distribution of ditch and drain network can affect moisture in the root system layer. The autors of this paper are trying to answer these questions using a method of rough sets.
The investigations were carried out for the melioration object located in Piwonia Górna. The measurements were taken in the vicinity of the Wieprz-Krzna Canal (KWK). In order to determine the air-water relations for a given point of this object, the following factors were taken into consideration : water levels in ditch distance 30 m from

the point where moisture content was measured, water levels in ditch located on the opposite side in the distance of 74 m from the point where moisture content was measured, water levels in canal (KWK) located 752 m from the point where moisture content was measured, precipitation and temperature of air.

The dynamics of ground water depths and the levels of water in a drainage network were measured from April to October each year. The measurements were made in control wells and on water gauges. The distance of ground water level from the surface of the ground was measured in the control wells, while the water level in ditches was measured on water gauges in conversion to the distance of water from the surface of the ground. The levels of water in KWK were measured from the bottom of the canal. In 1976 the measurements were made at one week intervals and in later years three times a month.

The moisture content of the soil was measured from April to September each year, three times a month, on average. The soil samples were collected from the following depths (from the surface to the ground water level) : 5-10 cm, 15-20 cm, 25-30 cm, 35-40 cm, 55-60 cm, 75-80 cm, 95-100 cm. The moisture content of the soil was determined by drying-weight method and the data were obtained in % of the soil volume. In further parts of this study the authors will take into consideration the mean moisture content noted in the root system layer (from 0 to 30 cm) .The temperatures and distribution of rainfalls for the examined object were received from a meteorogical station in Sosnowica. This comlex was characterized as wet; the soil type is organic. The field trials indicate that moisture content of the soil type mentioned above depends on the amount and distribution of rainfalls.

2 BASIC CONCEPT OF ROUGH SETS THEORY

By an information system we understand the 4-tuple $S=(U,Q,V,\rho)$, where U is a finite set of objects, Q is a finite set of attributes, $V=\bigcup_{q\in Q} V_q$ and V_q is a domain of the atttribute q, and $\rho : U \times Q \Rightarrow V$ is a total function such that $\rho(x,q)\in V_q$ for every $q\in Q$, $x\in U$, called information function.

Thus, information system may be considered as a finite table in which columns are labeled by attributes, rows are labeled by objects, and the entry in column q and row x has the value $\rho(x,q)$. Each row in the table represents an information about an object in S.

Let $S=(U,Q,V,\rho)$ be an information system and let $P\subseteq Q$, $x,y\in U$. We say that x and y are indiscernible by set of attributes P in S (denotation $x \; \mathbb{P} \; y$) if $\rho(x,q)=\rho(y,q)$ for every $q\in P$. One can easily check that $\mathbb{P}$ is an eqivalence ratio on U for every $P\in Q$. Equivalence classes of ratio $\mathbb{P}$ are called P-elementary sets in S. Q-elementary sets are called atoms of S.

Any finite union of P-elementary sets will be called P-definable set in S.

By P-lower (P-upper) approximation of set X in S we understand the sets $\underline{P}X$ ($\overline{P}X$) defined as follows:

$$\underline{P}X = \{ x\in U \mid x\in \underline{x}_j \text{ and } \underline{x}_j \subseteq X, \text{ for } j=1,2,\ldots,k\}$$

$$\overline{P}X = \{x\in U \mid x\in \underline{x}_j \text{ and } \underline{x}_j \cap X \neq \emptyset, \text{ for } j=1,2,\ldots,k\}$$

With every subset $X\subseteq U$ we associate a number $\alpha_p(X)$ called an accuracy of approximation of set X by P in S, or in short accuracy of X, defined as

$$\alpha_p(X) = \frac{\text{card}(\underline{P}X)}{\text{card}(\overline{P}X)}$$

Let S be an information system, $P\leq Q$, and let $X_* = \{X_1, X_2, \ldots, X_n\}$, $X_i \subseteq U$, be a family of subsets of U. If X_* is classification of U then

$$\beta_p(X^*) = \frac{\sum_{i=1}^{n} card(\underline{P}X_i)}{\sum_{i=1}^{n} card(\overline{P}X_i)}$$

is called the accuracy of approximation of X by P in S, or simply accuracy of classification. The coefficient

$$\gamma_p(X^*) = \frac{\sum_{i=1}^{n} card(\underline{P}X)}{card(U)}$$

is called the quality of approximation of set P of attributes.

3 RESULTS

The programme in Turbo Pascal language enables the determination of: lower approximation $\underline{P}X_i$, upper approximation $\overline{P}X_i$, accuracy of approximation $\alpha_p(X_i)$ for each X_i, accuracy of approximation $\beta_p(X^*)$ for X^* and quality of approximation $\gamma_p(X^*)$ for X^*. Moreover, the use of this programme gives a possibility of checking whether the set of attributes is reciprocally dependent.

The data obtained for Piwonia Górna are presented in a table containing each measurement. The rows of the table begin from the date of measurement and subsequent columns contain the following values : 1. air temperatures, 2. rainfalls, 3. water levels in KWK, 4. water levels in ditch, 5. water levels in ditch, 6. ground water levels at the place of sample collection 7. moisture content of the soil in a given point.

Moisture content of the soil 7 is a decision attribute, whereas condition attributes are included in 1-6. The table of measurements has been the basis for a decision table formed on the basis of a division of the values (obtained in the measurements by a user of the object) into sub-ranges (the so called classes). For condition attributes they are as follows :

1. 0-5,5-10,10-15,15-20 (°C)
2. 0-15,15-30,30-45,45-60,60-75, 75-90 (mm)
3. 0-30,30-60,60-90,90-120, 120-150,150-180,180-210 (cm)

4,5,6. 0-15,15-30,30-45,45-60, 60-75,75-90,90-105,105-120 (cm).

On the basis of the field and laboratory measurements it was found that the best conditions for plant growth are when moisture content of the soil ranges from 42 to 58%. Thus, three ranges have been accepted for a decision attribute: excess moisture content of the soil above 58%, proper moisture content from 42% to 58% and too low moisture content below 42%. Then the value of the coefficients γ, β and α for different conditions attributes were calculated. Since no moisture content below 42% was found of measurements, therefore the decision attribute, in this case moisture was divided into two ranges only.

Each condition attribute (1.-6.) was taken into consideration in order to find the effect of a given attribute or group of attributes on the behaviour of moisture content of the soil, and at the next stage, one, two, three, four and five attributes and also their combinations were successively left out.

Analysing the measurement results it can be seen that a single condition attribute is not enough to descibe the moisture content of the soil adequately. This has been confirmed by low values of γ, β and α coefficients. The highest values of the coefficients for two condition attributes were: γ=0.7143, β=0.5556, $\alpha(X_1)$=0.6000 and $\alpha(X_2)$=0.5000.

These values were obtained for the levels observed in KWK and ditch. This situation is demonstrated as a diagram (Fig.1) whose X-axis shows subranges of water levels in ditch and Y-axis shows subranges of water levels in KWK. Each rectangle of the network thus formed contains only those measurements whose values are included in the subranges and form the rectangle.

The subsets totally included in X_1 and X_2 have been the basis for lower approximation of X_1 and X_2 sets and thus reveal those measurements which indicate that moisture content of the soil is too high or proper. The thick line in the Fig. 1 shows the areas belonging to lower approximation of X_1 and X_2 sets. The letter B depicts the area of the soil too high in moisture, the letter G shows the area of proper moisture content of the soil, and the letter N is for the area where moisture content of the soil cannot be determined. The area N (called "boundary") contains the measurements exhibiting both proper moisture content and excess as well. Fig. 1 shows that one subset included in the rectangle of the following coordinates: 105 to 90 and 120 to 150, is located in the wrong place, but it contains only one measurement and it may be loaded with an error. For this reason, it will be excluded from further considerations. The highest values of the coefficients for three condition attributes are as follows: γ=0.9524, β=0.9091, $\alpha(X_1)$=0.9167 and $\alpha(X_2)$=0.9000. They occur when the condition attributes include: water levels in KWK, water levels in ditch and rainfalls. The results of calculations in this particular case are presented in Fig. 2 which shows that addition of another condition attribute, i. e rainfalls generally reduces ambiguity or even eliminates it.(Fig.2)

The curve plotted in Fig.3 shows rainfalls at the level of 15 mm and divides the set of measurements into two subsets: B, where moisture content of the soil is too high, and G, where moisture content is proper (42% to 58%). The results obtained in the study confirm our assumption that it is enough to measure water levels in KWK, water levels in ditches and rainfalls in order to obtain an explicit description of water-air relations in the examined soil.

4 CONCLUSIONS

The method of rough sets used for the analysis of the data obtained during field trials confirms earlier opinions that water-air relations depend on the depth of ground water and rainfall.

Explicit description of moisture content of the soil on the examined melioration object can be done when the following factors will be taken into account: water levels in the Canal Wieprz- Krzna, (KWK) water levels in ditches and rainfall.

The water levels in the Canal Wieprz- Krzna and in ditches for which the moisture content of the soil will belong to the good range (G) for all ranges of rainfalls can be found from Fig.3. Hence, these results can be used for the determination of adequate water levels in the ditches in order to control water-air relations.

The method used in this study is stabile because a slight change in the division of attributes into sub-ranges does not cause any significant changes in subsequent results.

REFERENCES

Pawlak,Z.1984.Rough classification. London: International Journal of Man-Machine Studies.

Pawlak,Z. 1986.Rough Sets and Decision Tables. Lecture Notes. New York: Springer Verlag.

Reinhard,A.,Stawski,B. and Weber,T. Application of rough sets to study the water outflow from the river basin. Warszawa: Bulletin of the Polish Academy of Sciences, in print.

Szajda,J.1980. Opracowanie metody prognozowania nawodnień w warunkach płytkiego zalegania poziomu wody gruntowej. Lublin: Instytut Melioracji i Użytków Zielonych, Zakład Przyrodniczych Podstaw Melioracji, manuscript.

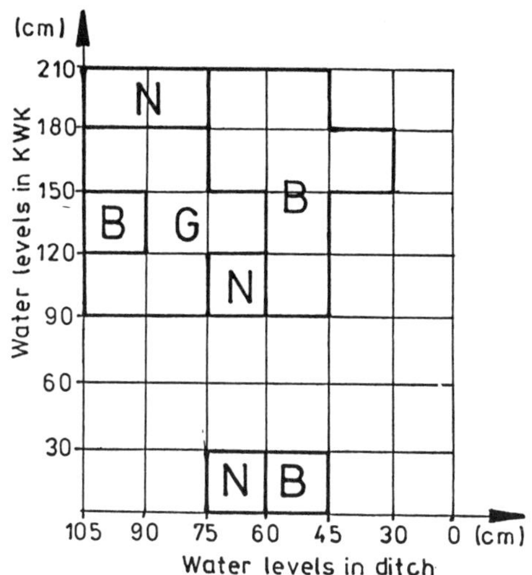

Fig.1 Relationship between moisture content of the soil and water level in KWK and ditch

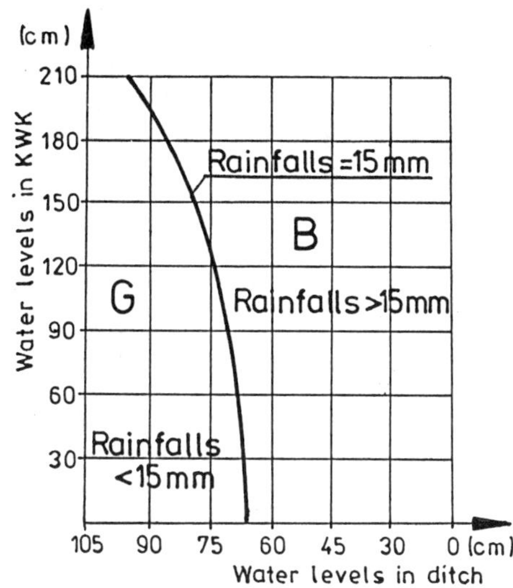

Fig.3 Location of areas whose moisture content is : B - to high, G - proper

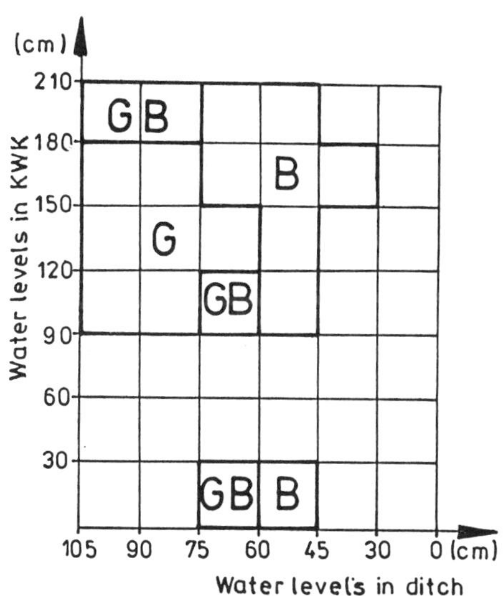

Fig.2 Relationship between moisture content of the soil and water level in ditch and rainfalls

Land and Water Use, Dodd & Grace (eds), © 1989 Balkema, Rotterdam. ISBN 90 6191 980 0

Overland flow under simulated wind-driven rain

J.L.M.P. de Lima
Department of Land and Water Use, and Department of Hydraulics and Catchment Hydrology, Agricultural University Wageningen, Wageningen, Netherlands

ABSTRACT: The effects of inclined rainfall and the wind induced shear-stress on overland flow have been investigated in an experimental study. These processes have been simulated in a wind tunnel having variable slope. Upslope blowing wind was found to delay the occurrence of overland flow and increase the depth of water over an impermeable plane. Even upstream discharge has been observed. The implications of these results to overland flow modelling are discussed.

RESUME: Cet article traite l'effet de la pluie inclinée et de l'effort de cisaillement du vent sur l'écoulement superficiel. Des expériences ont été effectuées au moyen d'une soufflerie aérodynamique à pente variable. L'effet du vent soufflant vers l'amont se manifeste par un retard de l'écoulement superficiel et par une plus grande épaisseur de la couche d'eau sur une surface imperméable. Une décharge à l'amont a aussi été observée. Les conséquences de ces résultats pour la modélisation de l'écoulement superficiel sont discutées.

ZUSAMMENFASSUNG: Einflüsse von schrägfallendem Regen und vom Wind verursachten Schubspannung auf den Oberflächenabfluss wurden in einer experimentelle Studie untersucht. Diese physikalische Vorgänge wurden in einem Windkanal mit variablen Gefälle simuliert. Der bergauf wehende Wind hat nicht nur den Beginn des Abflusses verzögert und die Wassertiefe über die undurchlässige Oberfläche vergrössert, er sorgte sogar für Abfluss gegen Gefälle. Die Folgen obengenannter Resultate auf die Modellierung von Oberflächenabflussvorgänge werden in diesem Aufsatz erörtert.

NOTATION

The following symbols are used in this paper:
a = discharge variation due to wind (l/s)
b = discharge variation due to wind (l/s)
c = discharge variation due to wind (l/s)
d = discharge variation due to wind (l/s)
D = distance at which drops hit the water sheet (m)
e = discharge variation due to wind (l/s)
f = Darcy-Weisbach friction factor
k = von Karman's constant (≈0.4)
L = lenght of working section of the overland flow surface (m)
m = exponent of flow depth (kinematic flow approximation) (approximately 5/3 for turbulent flow)
n = number of observations
Qdown= discharge at x=L (l/s)
Qup= discharge at x=0 (l/s)
r = regression coefficient
R = effective rainfall rate (m/s)
S = slope gradient
t = time (s)
t_r = time from beginning of rain to start of outflow at x=L (s)
Vw = wind speed at the height z (m/s)
Vwater = overland flow velocity (m/s)
x = distance from top of plane (m)
WROF = numerical model (Wind, Rain, and Overland Flow)
z = measuring height of Vw (in this experiments half distance between water sheet and drop formers) (m)
zo= roughness length (m)
α = angle of the surface with the horizontal (degrees)
θ = angle of incidence of the rainfall with the vertical (degrees)
ρ air = density of the air (Kg/m3)
τ = wind frictional shear stress (Pa)

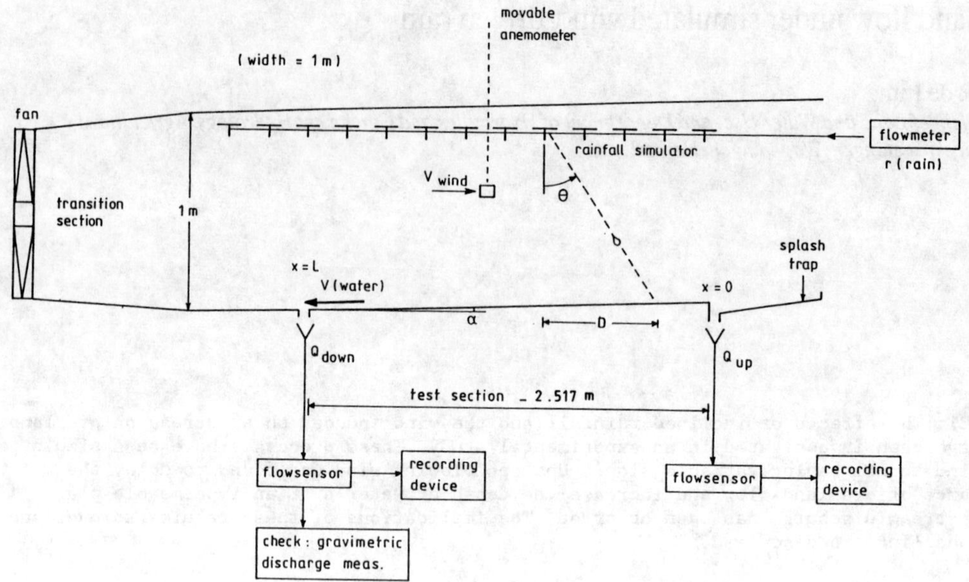

Fig. 1 Schematic representation of the experimental set-up and variables measured.

1 INTRODUCTION

Wind action on the surface of bodies of water produces such phenomena as wind waves, wind driven and wave currents, seiches, wind tides, etc (Baines & Knapp, 1965, Wu, 1975, Kranenburg, 1987, etc). Also wind may cause rain to fall at considerable inclination (Umback & Lembke, 1966, Struzer, 1972, Sharon, 1980). Falling through a logarithmic wind profile (expected profile under rainstorm conditions, Rosenberg et al., 1983), raindrops arrive at the surface retaining most of the horizontal momentum they possessed at 10 m height (Caldwell & Elliott, 1972). All these factors affect the overland flow process, which is further complicated by the effects of impinging raindrops (Shen & Li, 1973). Although many rainfall events are accompanied by strong winds (Lyles et al., 1974, Lyles, 1977), only limited knowledge is available on the influence of wind on overland flow.

This paper deals with the effects of inclined rainfall and wind induced shear-stress on the overland flow process. Wind affects overland flow in the following ways: effective rainfall pattern, shape, size, angle and velocity of impact of raindrops, splash shape, and tangential shear stress in the water-air boundary (Lima, 1989).

Evaluating these phenomena, in the field, is very difficult due to the factors of time and space, and lack of control over the variables involved. In order to provide a preliminary insight into the problem and to test the numerical kinematic model WROF (Lima, 1989) a laboratory experiment was undertaken.

Wind blowing over the overland flow sheet can be divided into two components: (1) longitudinal wind blowing along the steepest line of the surface, and (2) the cross wind blowing perpendicularly to the same line. The longitudinal wind causes a change of the driving force of the overland flow. Moreover, these effects become even more pronounced as an upslope and a downslope wind induce opposite effects. The cross wind causes a change of flow path as the overland flow sheet is pushed sideways. In this experiment wind direction was always upslope and parallel to the surface.

The results of the experimental runs are analysed. In these experiments the effect of the wind shear stress was very pronounced due to the small impact velocity of the raindrops and because of the smoothness of the surface. The implications of these results to overland flow modelling are discussed.

2 EXPERIMENTAL SET-UP

The laboratory experiment was carried out in the Hydraulic Laboratory of the Department of Hydraulics & Catchment Hydrology of the Wageningen Agricultural University. The laboratory set-up was mainly composed of three units:

(a) channel, with an uniform rectangular cross section 1 m wide, in which a wind tunnel was installed. The surface was a smooth impermeable plane (painted plywood). The slope of the channel was adjustable;

(b) rainfall simulator consisting of circular plates with conveyors to the edges of the plates where the drops are formed. The rainfall simulator gives a uniformly distributed and time invariant geometrical rainfall pattern. Drop shape differed significantly from the equilibrium shape at terminal velocity due to oscillations after release from the drop formers and small fall height (Epema & Riezebos, 1984). The equivalent diameter of the drops (defined as the diameter of a spherical drop with the same mass) was rather large at the drop formers (up to 8 mm). However, drop breakup existed, increasing with wind speed;

(c) low-speed, open-circuit wind tunnel made of wood and plastic sheets, having a working section 1 m wide, 1 m high and 2.517 m long. The rainfall simulator was installed inside the wind tunnel.

Fig. 1 shows a schematic representation of the laboratory set-up and the variables measured.

The rainfall intensity was measured by a flowmeter. The outflow hydrographs (of Qup at x=0, and of Qdown at x=L) were registered by a magnetic flow sensor connected to an automatic continuous recorder. At steady state Qdown was checked against a gravimetric discharge measurement (i.e. a weighing drum and a stopwatch). For no-wind conditions, velocity measurements of overland flow (Vwater) were made using dye tracing.

The wind profile obtained was an effect of the tunnel construction and of the presence of the rainfall simulator inside it, and does not agree with natural wind velocity profiles. A decrease of the wind profile at higher level was due to friction with the rainfall simulator. Wind speed was adjusted roughly by means of the velocity of rotation of the fan and checked with a movable anemometer. The anemometer permits measurements at a distance of 25 mm of any border. Note that the wind direction in this experiments was always upslope and parallel to the surface.

The fan (axial fan) had a 1.64 kW electric motor at 1200 r.p.m., which was connected to a variable transformer. This fan type is commonly used for the ventilation of stables and cereal storage buildings. The air sucked to the wind tunnel was dry air from the centrally heated hydraulic laboratory room.

The angle of incidence of the rain was estimated by injection of dye in the drop formers, and measuring the distance at which the coloured drops hit the water sheet.

The angle of incidence of the simulated rain was found to be independent of the rainfall intensity. However this angle was slighly changed by the slope gradient because the slope affected the position of the rainfall simulator (kept horizontal inside the wind tunnel) (Fig. 1 and 2).

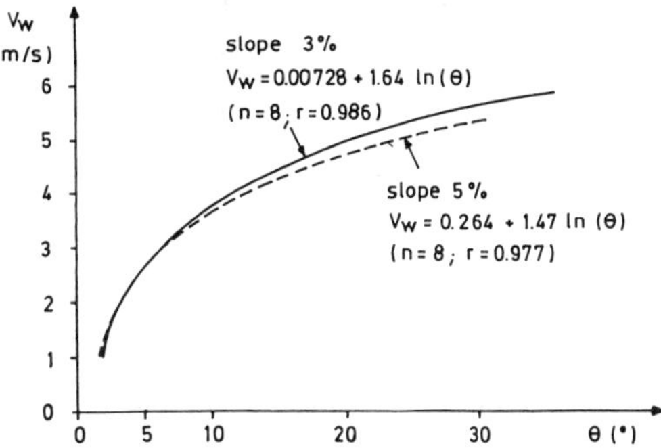

Fig. 2 Wind speed versus angle of incidence of rainfall, for slopes of 3 and 5 %.

Raindrops falling through the wind profile do not adjust their speeds immediately to the wind. The trajectory of the drops was approximately linear due to nearly constant acceleration components in the vertical and wind directions, an initial zero velocity at the drop formers, and a small fall height. The velocity component of the drops in the wind direction at impact point was much smaller than the average wind velocity (Vw): for the slope of 5%, around 0.4 Vw for θ = 30 degrees, and 0.15 Vw for θ = 5 degrees. Thus, stresses developed which deccelerate the air. However, the resulting decrease of wind speed was very slight and was neglected.

Experiments (54 runs) were repeated for each slope (3 and 5%) with different average wind speed (ranging from 0 to 6 m/s), and each wind speed with different rainfall intensities (ranging from 0.04 to 0.18 mm/s). In some runs it was the rainfall intensity that was kept constant for varying wind speeds.

3 RESULTS

Less discharge at x=L (Qdown) for higher wind velocities is due to an increase of discharge at the origin, x=0, of the impermeable plane (Qup)(Fig. 1). In these experiments the effect of the wind shear stress was very pronounced due to the small impact velocity of the raindrops and because of the smoothness of the surface.

The frictional wind shear stress τ, assuming that the mean wind speed increases linearly with the logarithm of the height (Rosenberg et al., 1983), can be expressed by:

$$\tau = \rho_{air} \left[Vw\ k / \ln(z/zo) \right]^2 \quad (1)$$

where z was taken as half the distance between water sheet and drop formers, because of the decrease in wind speed near the rainfall simulator. (See the notation list for symbols and units.)

The value of the hydrodynamical roughness length (zo) was for these experiments 0.0002 m, and was found through calibration of various runs with the model WROF (see section 3.3). In these experiments the water sheet covered the entire surface (a smooth impermeable plane). Higher values of zo should be expected for rougher surfaces.

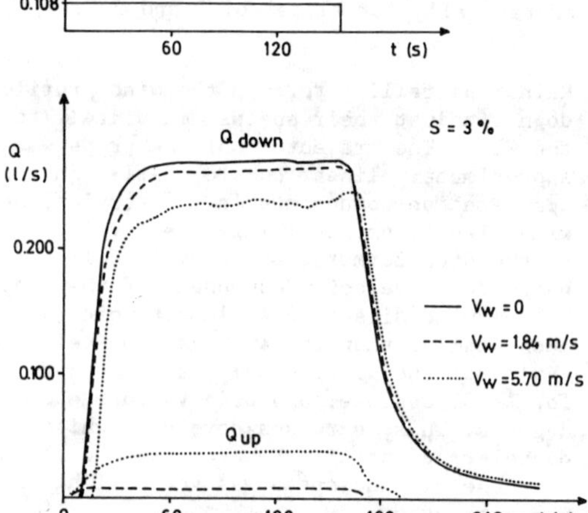

Fig. 3 Influence of wind on observed hydrographs of overland flow (both for Qdown and Qup).

3.1 Experiments with constant wind speed

In Fig. 3, the effect of the wind speed is assessed for approximately equal rainfall patterns. Wind speed delayed the start of overland flow at x=L (t_r), and increased the amount of water pushed slope upwards (Qup). Fig. 4 presents the effect of slope for approximately the same wind and rectangular rainfall patterns. At a steeper slope we obtain more discharge at x=L (Qdown). For the same rainfall intensity the time of initiation of overland flow at x=L decreases for increasing slope steepness. This can be seen in Fig. 5 comparing the solid lines (S = 3%) with the dashed lines (S = 5 %) for approximately the same effective rainfall rate, R. Also the form of τ-t_r (assumed linear due to high regression coefficients: 0.977 to 0.999) is affected. The steeper the slope the smaller is the effect of τ on t_r (Fig. 5).

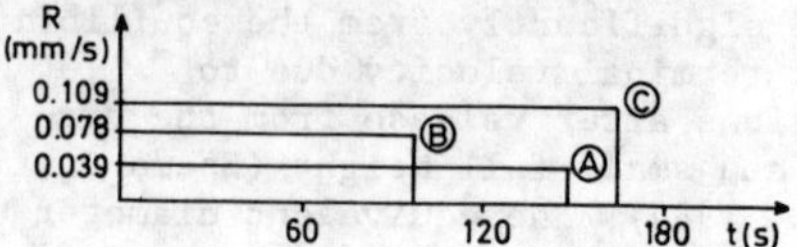

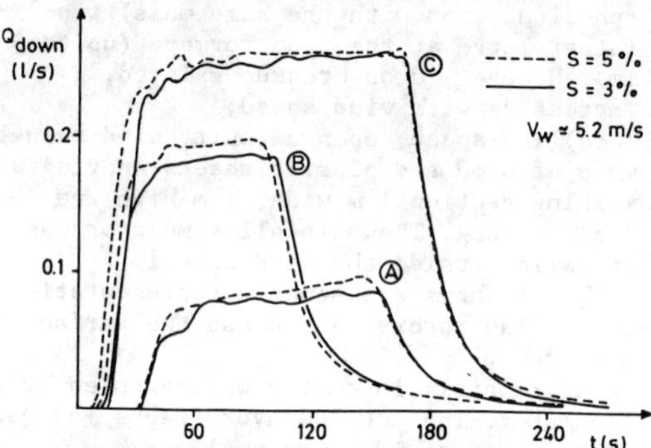

Fig. 4 Effect of slope on observed hydrographs of overland flow (Qdown), for three rainfall rates (R).

3.2 Experiments with variable wind speed

Experiments with variable wind speed consisted mainly in starting and stopping wind at fixed intervals and observing the discharge behaviour at the extremes of the surface (Qdown and Qup). Fig. 6 shows the general behaviour of Qdown in time and for a certain fixed rainfall when a wind speed of 5.9 m/s is suddenly started and stopped after one minute. The decrease of discharge due to sudden start of wind was found to be approximately equal to the increase of discharge due to the sudden cessation of the same wind (b=-d in Figures 6 and 8, and also b=a+c). However <u>b</u> became larger than -<u>d</u> for higher rainfall intensities (Fig. 8). Fig. 11 shows a complete sequence of discharge variations due to a sudden stop and start of a series of different wind speeds.

Discharge variations considered as a function of the effective rainfall rate (R), can be correlated with the wind speed, i.e. with the wind frictional shear stress

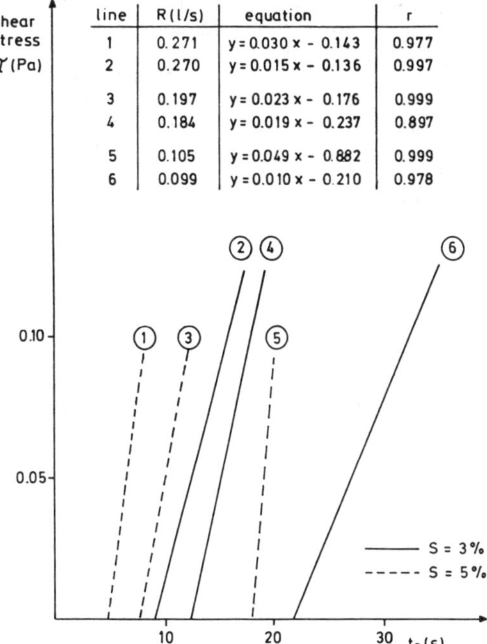

line	R(l/s)	equation	r
1	0.271	y = 0.030 x - 0.143	0.977
2	0.270	y = 0.015 x - 0.136	0.997
3	0.197	y = 0.023 x - 0.176	0.999
4	0.184	y = 0.019 x - 0.237	0.897
5	0.105	y = 0.049 x - 0.882	0.999
6	0.099	y = 0.010 x - 0.210	0.978

Fig. 5 Shear stress as a function of time to start overland flow (t_r), for various rainfall rates (R) and slopes of 3 and 5 %.

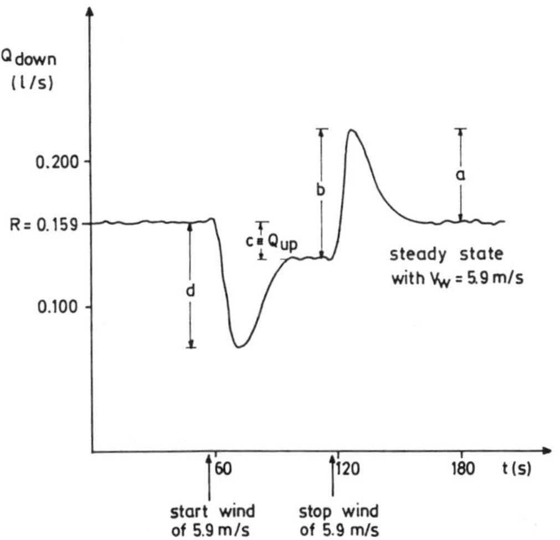

Fig. 6 Discharge variations in time, at x=L, for a sudden start and stop of a wind of 5.9 m/s.

(τ). When plotting (b/R) and (-d/R) against shear stress τ a linear relation is found (Fig. 9), but (Qup/R) against τ yields a curve to which an exponential relation has been fitted (In Fig. 10, simple exponential curve was fitted to the data). Fig. 7 gives discharge variations, i.e., a, b, c and d (see Fig. 6) as a function of wind speed for the 5% slope gradient. Remarkably the shape of a tends to a constant value for higher wind velocities (within the same measured range of Vw).

The different effects which induce the temporary increase of downstream discharge with respect to steady state conditions were (ranking by importance):

(i) sudden stop of wind;
(ii) sudden stop of wind and rain;
(iii) sudden stop of rain.

These effects are shown for two rainfall rates in Fig. 12.

The increase of discharge due to a sudden stop of rain and wind (case ii) is lower than the increase due to only the sudden stop of the same wind (case i), mainly because in the latter case a new higher steady state discharge under rainfall will be reached. Case (iii) created the smallest increase of the three cases.

When rainfall stops (case iii) outflow discharge (Qdown) increases temporarily before recession. This phenomenon can be

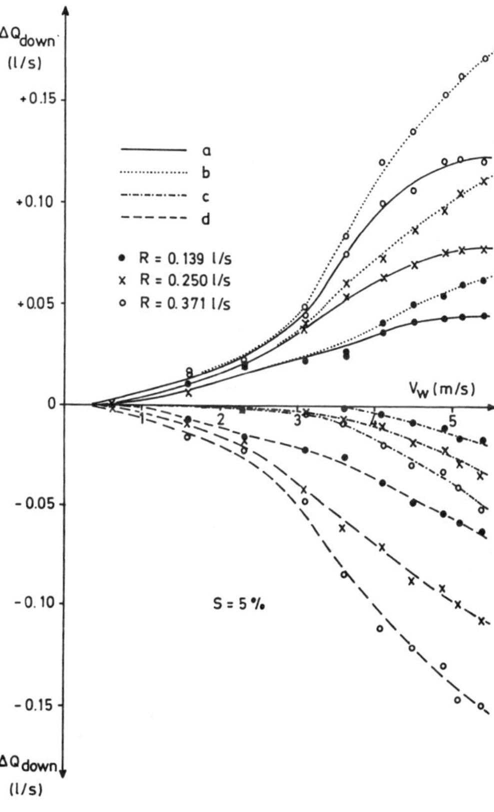

Fig. 7 Discharge variations as a function of wind speed (Vw), for various rainfall rates (R). For a, b, c and d see Fig. 6.

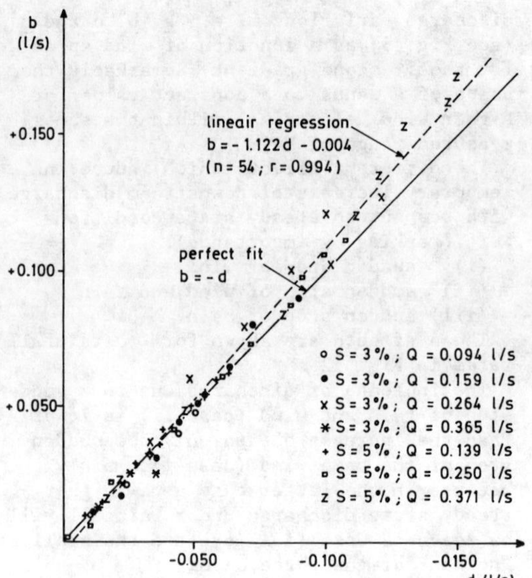

Fig. 8 Plot of b versus -d (see also Fig. 6)

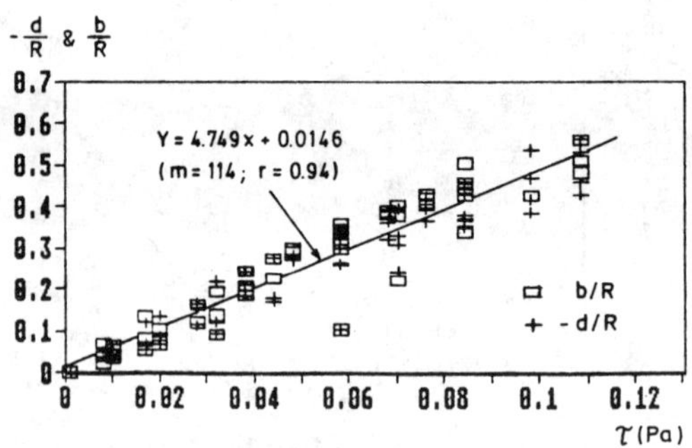

Fig. 9 Plot of b/R and -d/R versus τ.

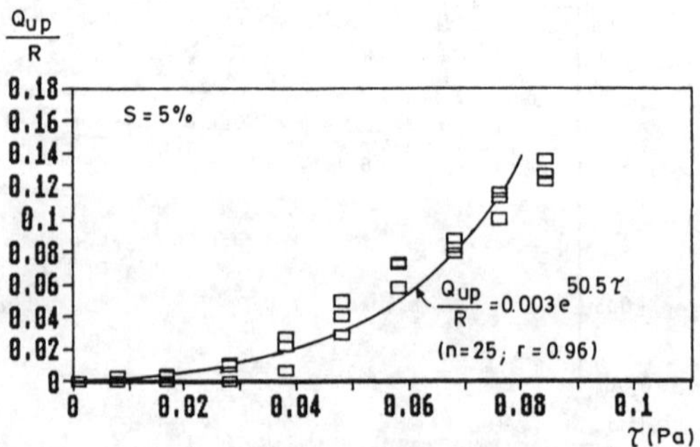

Fig. 10 Plot of Qup/R versus τ for the 5% slope.

explained by the fact that the stopping of raindrop impact causes a change in the friction slope and the energy gradient line, due to reduction of the resistance coefficient to a lower value (different dissipation of energy). As a consequence, the outflow has a temporary increase. This peak was not always observed in the outflow hydrograph.

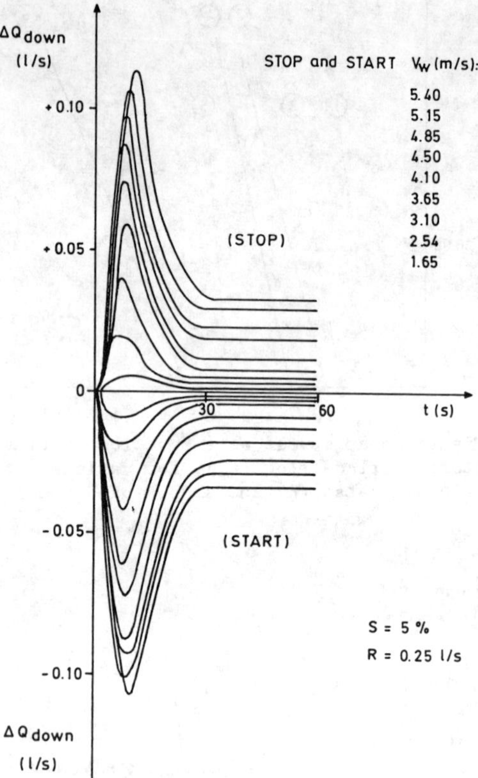

Fig. 11 Observed discharge variations as a function of time, for sudden start and stop of various wind speeds (Vw).

3.3 Testing of the model WROF (Lima, 1989)

Simulation with the numerical kinematic model WROF (Wind, Rain, and Overland Flow) gave good results both in the estimation of the initiation of the overland flow process at x=L, and the shape of the hydrographs, although the stronger the wind speed the more fluctuations were observed (Fig. 13).

The model was run with hydraulic data (i.e. f and m) obtained from vertical rain experiments. Sensitivity analysis showed that the roughness length (zo) is an important parameter in the assessement of wind effects on sheet flow.

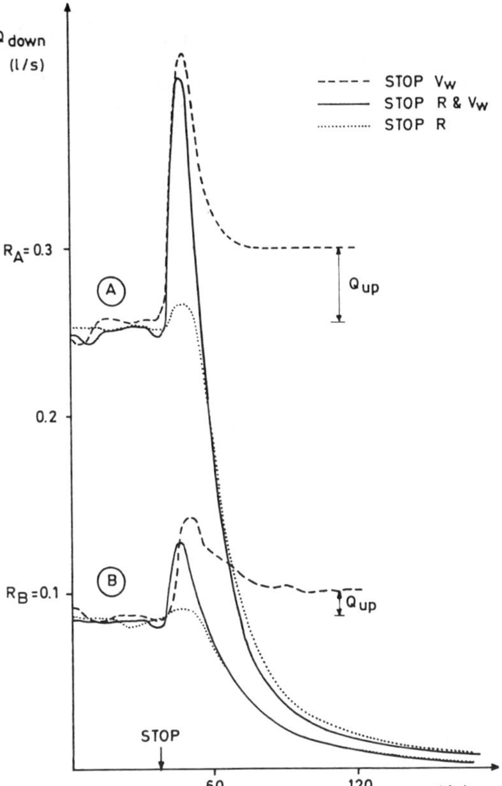

Fig. 12 Effects inducing temporary increase of discharge, at x=L (Qdown), for two rainfall rates (R).

4 CONCLUSIONS

The role of wind on overland flow was discussed. Wind intensity and direction influence both the effective rainfall pattern and the mechanics of the overland flow process.

Wind blowing upslope was found to delay the start of overland flow and increase water depth along the surface, leading to the existence of a finite water depth at x=0 accompanied by upslope discharge (Qup). The decrease of overland flow at x=L due to the start of a certain wind speed was highly correlated with the subsequent increase of flow due to stopping the same wind.

The effect of the wind shear stress and splash were small in comparison with the effect of the inclined impinging raindrops for high intensity rainfalls. However the effect of wind shear stress may be significant when a thin water sheet covers the entire surface and for low kinetic energy of the rain (low intensity rainfalls or, as in these experiments, small drop fall height).

The results of the laboratory experiments describe the importance of wind in the overland flow process under rainfall. Neglecting wind effects in rainfall simulation studies might be a cause of errors. Transferability of these results to the field is not obvious. More research is needed on effects of wind-rain forces on overland flow and soil surface, and on the magnitude and frequency of winds associated with rains in areas susceptible to water erosion.

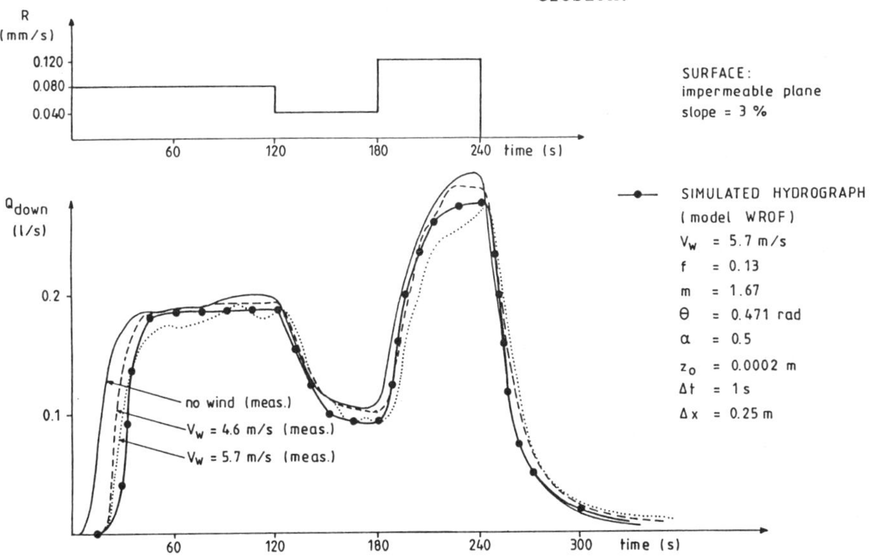

Fig. 13 Simulation with the model WROF (Lima, 1989).

ACKNOWLEDGEMENT

The author wishes to thank Prof. W.H. van der Molen, Dr. R.W.R. Koopmans and Dr. A.F.G. Jacobs for their helpful suggestions, the assistance of Mrs. G. Kranjac-Berisavljevic in the laboratory work, and the staff of the hydraulics laboratory for the installation of the equipment. The author is indebted to "Comissão Permanente INVOTAN" for financial support and is also very grateful to the University of Coimbra (Portugal) for the permission and partial support of his stay at the Wageningen Agricultural University.

REFERENCES

Baines, W.D. & D.J. Knapp 1965. Wind driven water currents. J. of Hydraulics Division, ASCE, 95: 205-221.

Caldwell, D.R. & W.P. Elliott 1972. The effect of rainfall on the wind in the surface layer. Boundary-Layer Meteorology, 3: 146-151.

Epema, G.F. & H. TH. Riezebos 1984. Drop shape and erosivity Part I: experimental set up, theory, and measurements of drop shape. Earth Surfaces Processes and Landforms, 9: 567-572.

Kranenburg, C. 1987. Turbulent surface boundary-layer induced by an off-shore wind. J. of Hydraulic Research, 25: 53-65.

Lima, J.L.M.P. de 1989. The influence of the angle of incidence of the rainfall on the overland flow process. Proc. IAHS Third Scientific Assembly, Baltimor, USA, May 10-19 (accepted).

Lyles, L. 1977. Soil detachment and aggregate disintegration by wind-driven rain. SCSA Special Publ. no. 21: 152-159.

Lyles, L., J.D. Dickerson & N.F. Schmeidler 1974. Soil detachment from clods by rainfall: effects of wind, mulch cover, and initial soil moisture. Transactions of the ASAE, 17: 697-700.

Rosenberg, N.J., B.L. Blad & S.B. Verma 1983. Microclimate: the biological environment. John Wiley & Sons, New York.

Sharon, D. 1980. The distribution of hydrologically effective rainfall incident on sloping ground. J. Hydrol., 46: 165-188.

Shen, H.W. & R.M. Li 1973. Rainfall effect on sheet flow over smooth surfaces. Journal of the Hydraulics Division, ASCE, 99: 771-791.

Struzer, L.R. 1972. Problem of determining precipitation falling on mountain slopes. Sov. Hydrol. Selected Papers, 2: 129-142.

Umback, C.R. & W.D. Lembke 1966. Effect of wind on falling water drops. Transactions of the ASAE, 9: 805-808

Wu, J. 1975. Wind effects on stream flows. J. of Hydraulic Research 13: 405-423.

Land and Water Use, Dodd & Grace (eds), © 1989 Balkema, Rotterdam. ISBN 90 6191 980 0

The evaluation of river maintenance in agricultural areas

T.M.Hess, P.B.Leeds-Harrison & J.Morris
Silsoe College, Silsoe, Bedford, UK

ABSTRACT: Organisations responsible for river maintenance are under increasing pressure to justify and prioritise expenditure. This paper describes an approach to the evaluation of river maintenance strategies for lowland agricultural areas in the U.K. The method involves estimating field drainage status for a given local condition from river water levels, in order to classify floodplain land into one of three productivity categories. Land use scenarios are developed which describe typical farming practices and output for different land use types operating under each productivity category. These scenarios are expressed in financial terms using standardised enterprise budgets. Channel geometry and catchment hydrology are used to estimate average annual flood damage. The impact of changes in maintenance strategies can be assessed in terms of field drainage conditons and the financial consequences for agricultural land use.

ZUSAMMENFASSUNG: Die für Flussinstandhaltung zuständigen Behörden befinden sich zunehmend unter Druck, die Ausgaben zu rechtfertigen und Prioritäten zu setzen. In diesem Beitrag beschreiben wir einen Beurteilungsansatz für Flussinstandhaltungsstrategien für landwirtschaftliche Böden in Tieflandsgebieten des Vereinigten Königreichs. Der Feldentwässerungsstatus für gegebene örtliche Verhältnisse wird aus dem Flusswasserpegelstand geschätzt, um Schwemmebeneböden in eine von drei Produktivitätskategorien einteilen zu können. Bodenbenutzungsszenarios werden entwickelt, um typische landwirtschaftliche Verfahren und Erträge für verschiedene Bodenbenutzungstypen aus jeder Produktivitätskategorie zu Beschreiben. Diese Szenarios werden finanziell mit vereinheitlichten Unternehmensbudgets ausgedrückt. Das durchschnittliche jährliche Hochwasserschadensniveau wird auf der Basis von Flussbettgeometrie und Wassereinzugsgebietshydrologie geschätzt. Der Einfluss von Änderungen der Instandhaltungsstrategien kann auf der Basis von Feldenwa"sserungsverhältnissen und deren finanziellen Auswirkungen auf die landwirtschaftliche Bodenbenutzung beurteilt werden.

RESUME: Les organisations responsables de l'entretien des fleuves se trouvent de plus en plus contraintes de justifier et de prioritiser les dépenses. Dans cet article, on décrit une approche à l'evaluation des stratégies d'entretien des fleuves pour les régions agricoles de plaine du Royaume Uni. Par cette méthode, on prépare une estimation du statut du drainage des champs pour des conditions locales données à partir du niveau de l'eau fluviale, avec l'objectif de classer les terrains des plaines inondables dans une de trois categories de productivité. Des scenarios pour l'utilisation des terres sont développés pour décrire les méthodes culturelles et les rendements agricoles typiques pour plusieurs utilisations des terres dans chaque catégorie de productivité. Ces scénarios sont exprimés en termes financiers avec l'application de budgets d'entreprise normalisés. La géometrie du lit et l'hydrologie du bassin fluviatile servent comme base pour l'évaluation des dommages moyens annuels occasionés par les inondations. L'influence de changements dans les stratégies d'entretien peut être évaluée en termes des conditions de drainage des champs et des conséquences financières pour l'utilisation des terres agricoles.

1. INTRODUCTION

Organisations charged with the responsibility of maintaining rivers are increasingly being asked to justify their activities in financial terms. In addition, limited maintenance budgets require that maintenance work must be ranked in order of priority. For some years, methods have been available for evaluation of land drainage improvement schemes (eg. Penning-Rowsell & Chatterton, 1977; Hess & Morris, 1987) which involve estimating the increase in productivity of agricultural land as a result of the alleviation of land drainage constraints. The purpose of river maintenance, however should be to maintain a given standard of land drainage, in terms of both waterlogging and flood risk. There are, therefore no 'extra benefits' to be set against maintenance costs, but rather the 'avoidance of losses' which would otherwise be associated with the anticipated reduction in agricultural productivity resulting from a decline in drainage standards.

This paper describes an approach to the evaluation of river maintenance strategies for agricultural areas that has been developed for Severn Trent Water Authority (STWA) and is currently in use as part of their River Information Management System (RIMS). The requirement was to produce a method by which the financial implications (at farm level) of changes in the standard maintenance of 'main' rivers could be evaluated. A major constraint was that the method had to be based on very limited site specific information that could be derived from published sources or from a reconnaissance level field survey initiated by STWA.

The method involves estimating the field drainage status from river water levels in order to categorise the floodplain land into one of three productivity classes. Land utilisation scenarios are developed that describe typical farming practices and levels of output for different types of land use in each productivity class on each soil class. Using standardised enterprise budgets, net returns are calculated for each land utilisation scenario.

Consideration of the channel geometry and catchment hydrology allow the flood risk to be quantified and used to calculate an average annual flood loss. This is deducted from the net return to give an estimated net productivity for the flood plain under a particular river regime.

A deterioration in the standard of river maintenance will lead to a reduction of the channel capacity, a rise in the river water level, an increase in the risk of flooding and ultimately a fall in the productivity of the land. The expected reduction in productivity due to no maintenance can be compared with the cost of maintenance to produce a benefit-cost ratio. This can be used to compare and justify alternative maintenance strategies.

2. NET RETURNS

The typical agricultural output from the land can be estimated from the land use, drainage condition and soil type.

2.1 Productivity Classes

Three drainage related productivity classes have been defined on the basis of field water table position. Although the water table will vary throughout the year, the method considers the position at the end of the field capacity season (ie. spring). This is a crucial period for agriculture, being the time when plant growth is most rapid and the requirements for trafficability are at a maximum.

Where the spring water table is more than 50cm below the surface, waterlogging is not considered to be a constraint on agricultural production and the land is allocated to a 'normal' productivity class. If the water table is 50-30cm below the surface, poor drainage may restrict access to the land, limit fertiliser application and so reduce the yield potential of the land. The land is therefore classified as 'low' productivity. With water tables less than 30cm below the surface, drainage forms a major constraint and will dictate the types of agriculture that may be practiced. The land is then classified as 'breakdown'.

These boundaries of 30cm and 50cm may be criticised as being somewhat arbitrary, however they do represent critical values for root growth and trafficability (see Morris et al 1984). Given the generality of the land use data, these figures represent acceptable averages for a variety of edaphic conditions.

2.2 Estimating drainage status

Details are rarely available on water table position in the field. The approach used in this work has been to model the effect of river freeboard on spring water table position for a variety of soil textures and topographies. The freeboard is defined as the distance between normal water level in the river (usually taken to be that observed at the end of the field capacity season) and the ground level in the field adjacent to the river (not bank level).

Two situations are recognised, firstly where underdrainage has been installed and water tables are controlled by the subsurface drains, and secondly where the field water table is controlled by the level of the river.

Where underdrainage is installed, drains are usually placed at depths varying from 0.75m to 1.3m depending on soil type, drainage problem and topography. A fall on the laterals and the collector drain must be provided to the outfall point on the river, therefore if the natural slope is less than 0.2%, this must be added to the drain depth. In addition, some allowance (0.15m to 0.2m) for fluctuating river levels is also required at the river.

Based on these considerations, a minimum acceptable freeboard for the particular soil type and topography can be calculated. Where field drains are designed to keep the water table below 50cm, provided the drain outfalls are above normal river water level the site will fall into the 'normal' productivity class. For clay soils and other soils of low hydraulic conductivity (0.5m d^{-1}), if the freeboard is reduced and the outfalls are submerged, the site automatically falls into the 'breakdown' class as natural drainage alone cannot provide adequate water table control. For soils of higher conductivity there may be an intermediate phase in which the site may be treated as undrained.

With no underdrainage, the field drainage status is controlled by the ability of the site to drain to the river. In this case an hydraulic head will be built up in the soil adjacent to the river and water will drain towards the river. The relationship between hydraulic head and seepage rate is well understood and a relationship between the slope of the land, hydraulic conductivity of the soil, rainfall rate, and water table depth can be found. Here it is assumed that the landward boundary is defined by a ditch at the edge of the floodplain.

Figure 1. Steady state water table model.

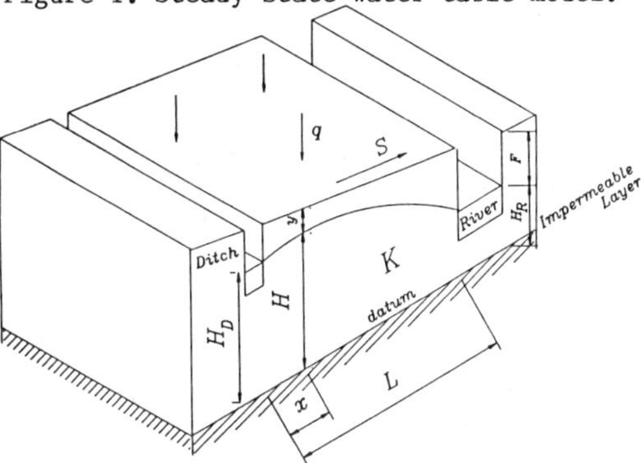

For the assumed steady state situation shown in figure 1, and making the Dupuit-Forchheimer assumptions, we may state;

$$\frac{d^2 H^2}{dx^2} = \frac{-2q}{K} \qquad (1)$$

where K is the saturated hydraulic conductivity of the soil and q, the average daily rainfall rate which can be estimated from regional meteorological data (MAFF 1975). Here the value of the five day average rainfall excess has been used to estimate the daily rainfall rate.

Solving the above equation for the given boundary conditions, we find that;

$$H^2 = \frac{-qx^2}{K} + \frac{qLx}{K} + H_D^2 - (H_D^2 - H_R^2)\frac{x}{L} \qquad (2)$$

which may be rewritten as;

$$(H^2 - H_D^2) = \frac{q}{K}(Lx - x^2) - (H_D^2 - H_R^2)\frac{x}{L} \qquad (3)$$

For deep soils, we may make the approximation that;

$$H + H_D = H_D + H_R = 2D_0 \qquad (4)$$

where D_0 is the average depth of the impermeable layer below the water table.

Equation 3 then reduces to;

$$h = \frac{q}{2KD_0}(Lx - x^2) - \frac{ix}{L} \qquad (5)$$

where h is the height of the water table above ditch water level and i, the hydraulic gradient between the ditch and the river. On land sloping towards the river, i may be taken as being equivalent to the land slope. On a flat floodplain a gradient of 0.5% can be assumed.

In practice, the ditch and the river will not be fully penetrating (ie they do not extend to the impermeable layer). In deep soils (where $D_0 > L/4$) the concept of equivalent depth (Hooghoudt, 1940) is used to account for convergence of flow towards the ditch and the river. An approximate value for the equivalent depth, d, can be found for this situation from;

$$d = \frac{L}{(2.55 \log_e L/4)+4} \qquad (6)$$

and d may be substituted for D_0 in equation 5 above.

The depth to the water table, y, at any distance, x, from the ditch may now be derived from the geometry of the situation as;

$$y = (L-x)\,S + (F - iL) - h \qquad (7)$$

where S is the land slope and F the freeboard at the river.

From equations 5 and 7 the position of the water table at any distance from the river can be estimated and the proportion of the floodplain falling into 'normal', 'low' and 'breakdown' classes can be determined.

2.3 Land use Scenarios

Agricultural land may be classified into permanent pasture, temporary ley or arable. For each land use class, scenarios have been drawn up describing typical utilisation systems and levels of inputs and outputs at three levels of field drainage. For grassland these include estimates of stock type, grass conservation system, nitrogen fertiliser application rate and length of grazing season. A grassland productivity model (Hess & Morris, 1987) was then used to estimate likely stocking rates. For arable land uses the scenarios include typical crop rotations and yields.

The development of these land use scenarios drew heavily upon the observations of a major survey of farming on riparian land in the Midlands region (Morris et al, 1984). For each, the physical information was combined with standardised enterprise budgets using the Silsoe College Agricultural Drainage Evaluation (SCADE) model (Hess & Morris, 1987) to produce typical gross margins, fixed costs and net returns. An example of the net returns for each land use class on clay soils is given in table 1.

Table 1 Net returns (£/ha/yr) for land uses on clay soils

Prod. Class	Permanent Grass	Temporary Grass	Arable
Normal	350	390	390
Low	160	175	260
Breakdown	70	80	140

Where farmers practice arable / ley rotations, these are likely to be reflected in the range of land use in the floodplain at any one time. The average productivity of the floodplain therefore reflects the nature of the rotations.

These scenarios do not attempt to predict the actual land utilisation, nor any form of potential, but are intended to be typical of the type of enterprises and levels of output found in riparian areas under varying conditions of drainage. Movement from one scenario to another, therefore does not imply that farmers will cope with deteriorating drainage standards by changing enterprise in the fashion indicated.

3. FLOOD DAMAGE

The average annual flood damage will depend upon the frequency with which flood events are expected to occur, the seasonal distribution of flood events and the sensitivity of the land use to inundation.

3.1 Flood risk

The likelihood of flooding will depend on the relationship between the maximum capacity of the river channel and the likely runoff from the catchment. The bankfull channel capacity, Qbf may be calculated from Manning's formula;

$$Qbf = \frac{A\,R^{2/3}\,S^{1/2}}{n} \qquad (8)$$

where R is the hydraulic radius, A, the cross sectional area of the channel, S, the energy gradient (taken to be equal to channel slope) and n the rugosity.

The hydrology of the catchment may be summarised by the magnitude of the mean annual flood, $\bar{Q}$. If local gauge data are available, $\bar{Q}$ may be determined from the mean of the observed annual flow maxima. Where site specific data are not available, it may be estimated from the physical characteristics of the catchment using the Flood Studies Report method. This uses a multiplicative equation relating $\bar{Q}$ to catchment area, stream frequency, channel slope, soil type, rainfall and area of lakes within the catchment (NERC,1975 p341).

The return period for flows in excess of channel capacity can be determined from the ratio of channel capacity, Qbf to the mean annual flood, $\bar{Q}$ using the Region Growth Curve (NERC, 1975). This return period (Ta) reflects the interval between years containing floods. For rare events, expected less often than one year in five, this may be assumed to be equal to the mean interval between floods (Tp). For more frequent events, Tp must be estimated using modified growth factors (NERC, 1977).

3.2 Flood seasonality

The average damage from flooding will depend upon the seasonal distribution of floods. An analysis of river flow records by Severn Trent Water Authority suggests that three seasonal distribution curves are applicable;
(i) large catchments ($\bar{Q} > 4\ m^3\ s^{-1}$) dominated by winter events,
(ii) small catchments ($\bar{Q} < 4\ m^3\ s^{-1}$) still dominated by winter events but with more summer events, and
(iii) urban catchments ($>$ 25% urban) with more or less equal flood risk all year.

However it should be noted that the pattern of weed growth is seasonal and field studies have shown a 6 to 10 fold change in rugosity through the year (Dawson, 1978, Brooker, 1978). An abandonment of weed cutting therefore may lead to a seasonal change in the capacity of the channel, in particular increasing the flood risk and reducing the freeboard in the spring and summer when the land is at its most sensitive to inundation and waterlogging.

3.3 Flood loss

A method has been developed (Hess and Morris, 1988) for the estimation of the average annual flood loss for grass and a variety of crop types, depending upon the frequency and seasonality of flooding. The method considers the value of lost production and additional costs incurred (or saved) as a result of inundation at various times of the year. The seasonally weighted flood loss, L can be calculated from;

$$L = \sum_{m=12}^{m=1} L(m)\ P(m) \qquad (9)$$

where L(m) is the loss from a single flood in month m and P(m) the probability of a flood occurring in month m.

Where flooding is expected less frequently than one year in five, the average annual flood loss can be calculated as the loss for a single flood in a year divided by the mean interval between floods (Tp) and the effect of multiple floods can be ignored. Where the risk of flooding is high, some account must be taken of exclusive and compounding effects of multiple flooding within the same year.

Using this method, the average annual flood loss can be calculated for each land utilisation scenario for various flood frequencies and seasonalities.

4. FINANCIAL APPRAISAL

For each block of land on the floodplain, the productivity class is determined from the freeboard level. The output of the land, in terms of net returns per hectare per year, is then determined for the appropriate land use and soil type. This represents a benchmark productivity for that land with no flooding. Given the channel geometry and hydrology, the flood frequency is estimated and the average annual flood loss calculated. The financial productivity (£/ha/yr) is then defined as the net returns minus the average annual flood loss.

If the level of maintenance is reduced there is likely to be an increase in the rugosity and reduction in cross sectional area due to silting and weed growth. These will result in a reduced channel capacity and reduced freeboard, leading to a deterioration in field drainage status (eg moving from 'normal' to 'low' productivity) and an increased flood loss.

The difference in financial productivity between the present (maintained) situation and the future (non-maintained) situation represents the annual 'benefit' due to maintenance.

5. EXAMPLE

Assume a site under arable land use, with no underdrainage, having a silty clay soil ($K = 1.2\ m\ d^{-1}$). The catchment is 'large' and the floodplain is 100m wide with a 1% slope toward the river. Under the present (maintained) condition the freeboard at the river is 1.3m and the return period for flooding is one year in five.

The depth to the field water table at the middle of the floodplain is calculated as 0.89m, placing the site in a 'normal' productivity class, thus the net return is taken to be £390/ha/yr. The average annual flood loss is £8/ha/yr giving a net productivity of £382/ha/yr.

If deteriorating drainage standards lead to a reduction of freeboard to 0.4m and a flood return period of one in two years (Tp), the site would fall into a 'low' productivity class. The net return would fall to £260/ha/yr and the average annual flood loss would be £25/ha/yr, giving a net productivity of £235/ha/yr.

In these circumstances, maintenance work avoids the losses associated with poor drainage conditions and has an estimated value of £147/ha/yr. On a reach of, say, 5km, this represents £14,700/yr (5km x 200m x £138/ha/yr). This can be discounted at the relevant discount rate and set against the present value of maintenance costs to derive a cost / benefit ratio or internal rate of return. The investment period chosen should reflect the number of years to complete a maintenance cycle.

6. DISCUSSION

The preceding methodology has been developed in response to pressures on public sector drainage authorities to justify and prioritise river maintenance expenditure. It draws on methods used for the design and evaluation of drainage capital improvement works, but regards maintenance expenditure as a means of avoiding losses rather than deriving additional benefits.

The approach taken in developing the model has been deliberately simple due to the paucity of soil, climatic and agricultural data likely to be available for a given site. The water table equations are not predictive and cannot be used to make a definite statement concerning the water table position at a particular site in a particular year. Similarly, the land utilisation scenarios are unlikely to correspond to the situation in individual fields and the movement from one scenario to another may be unrealistic. In practice, deteriorating drainage standards are likely at first to lead to diminishing returns within the existing land use system, eg reduced grazing seasons, lower crop yields, until a point is reached where a change in land utilisation becomes inevitable. The nature and timing of this change will depend on a number of farm level factors, such as farm type and the proportion of floodplain land on the farm, about which little, if any information is likely to be available.

Although there may be pressures to the contrary, the method is not designed to be used to determine freeboard requirements or to justify additional maintenance expenditure as a means of achieving extra agricultural benefits (this would imply some form of uptake of improvement potential which is inconsistent with the concept of maintenance). Where major differences exist between the current maintenance standards and those suggested by the model, site investigation may reveal that levels of service may be reduced without prejudicing agricultural standards. This was the case in an example in North Wales (Sutherland, 1988).

Work is now underway to refine and validate the method by the monitoring and evaluation of maintenance strategies on selected rivers in central England. A major concern is with predicting the hydraulic effects of a particular maintenance strategy and the time taken for a given channel to deteriorate. Special attention will be given to monitoring the marginal effects of changes in maintenance practice and the rates of post-maintenance deterioration. Given the environmental value of many river corridors in lowland Britain and the incentives being offered to enhance environmental qualities (such as land set aside payments) river maintenance planning must attempt to reconcile agricultural, environmental and recreational needs through balanced management strategies.

7. REFERENCES

Brooker M P (1978) Plant - flow relationship in the Wye catchment. Proc. European Weed Research Society, 5th symposium on aquatic weeds. pp63-70.

Dawson F H (1978) The seasonal effects of aquatic plant growth on the flow of water in a stream. Proc. European Weed Research Society, 5th symposium on aquatic weeds. pp71-77.

Hess T M & Morris J (1987) A computer model for agricultural land drainage scheme appraisal. Proc. 3rd Int. Workshop on Land Drainage. Dec 7-11, 1987. Columbus, Ohio. ppA77-A86.

Hess T M & Morris J (1988) A model for the economic evaluation of fluvial flood alleviation in agricultural areas. In Economic aspects of flood control and non structural measures. Special technical session of International Commission on Irrigation and Drainage. Dubrovnik, Yugoslavia. pp69-78.

Hess T M & Morris J (1988) Estimating the value of flood alleviation on agricultural grassland. Agricultural Water Management, 15, pp141-153.

Hooghoudt (1940) Bijdragen tot de kemis van enige natuurkundige groot-heden van der grood. No.7 Versl. Landbruwktechnic Onderz, 46, pp515-507.

Ministry of Agriculture, Fisheries and Food (1975) Climate and Drainage. Ref book 434. HMSO London.

Morris J, Hess T M, Leeds-Harrison P B & Ryan A M (1984) Drainage Benefits and Farmer Uptake. Report to Severn Trent Water Authority, Birmingham, UK. July 1984.

Natural Environmental Research Council (1975) Flood Studies Report. Vol I, Hydrological studies. NERC, London.

Natural Environmental Research Council (1977) The estimation of low return period floods. Flood Studies Report, supplementary paper No 2. NERC, London.

Penning-Rowsell E C and Chatterton J B (1977). The benefits of flood alleviation - a manual of assessment techniques. Saxon House Teakfield Ltd. Farnbouough.

Sutherland D C (1988) Land drainage maintenance evaluation, Maltraeth Marsh, Anglesey. Unpublished MSc thesis, Silsoe College.

Land and Water Use, Dodd & Grace (eds), © 1989 Balkema, Rotterdam. ISBN 90 6191 980 0

Ein Verfahren zur Modifikation der Bohrlochmethode für gespanntes Grundwassers

F.Kastanek
Universität für Bodenkultur, Wien, Österreich

ZUSAMMENFASSUNG: Durch die Analyse der Strömungsverhältnisse bei der Zuströmung in ein Bohrloch im gespannten Grundwasser konnte gezeigt werden, daß diese Zuströmung jenem Fall des freien Grundwassers entspricht, bei dem die Oberfläche des freien Grundwassers im und um das Bohrloch vor dessen Absenkung an der Grenzfläche zur undurchlässigen Deckschichte angenommen wird. Mit dieser Modifikation kann die bekannte Bohrlochmethode auch bei gespanntem Grundwasser angewendet werden.

ABSTRACT: The flow pattern of seepage of water into an auger hole in a confined aquifer equals the flow pattern of seepage into an auger hole in an unconfined aquifer, if the water table at rest in and surrounding the auger hole is taken at the boundary between the impervious top layer and the aquifer. With this modification the well known auger hole method may be used for confined aquifers.

RESUME: On a pu montrer par l'analyse de la courantologie dans le cas d'un suintement dans un trou de forage que le cas d'une aquifère captive correspond à celui des eaux libre à condition qu'avant le rabattement, la surface phréatique aussi bien dans le trou de forage que tout autour soit au niveau du surfaçage impermèable. Cette modification rend possible l'application de la mèthode de trou de forage que nous connaissons dans le cas d'une aquifère captive.

1 EINLEITUNG

Zur Beschreibung und Lösung vieler landwirtschaftlicher Probleme ist die Kenntnis der Bodenwasserbewegung Voraussetzung. Insbesondere Meliorationsmaßnahmen mit Hilfe der Dränung benötigen die Bestimmung der Durchlässigkeit der Böden für Wasser an zahlreichen Stellen. Alle Formeln zur Berechnung der Dränabstände grundwasservernäßter Böden enthalten einen Beiwert kf, den Durchlässigkeitsbeiwert, ein für gegebene Verhältnisse konstanter Bodenparameter, welcher die Intensität der Wasserbewegung in einem Boden zur treibenden Kraft, dem hydraulischen Gradienten, beschreibt. Dieser kf-Wert ist in dem von Darcy vor mehr als hundert Jahren experimentell gefundenen Widerstandsgesetz der Grundwasserbewegung enthalten:

$$v = -k \nabla H \qquad (1),$$

wobei v die Durchflußrate, das ist jenes Wasservolumen, welches pro Zeiteinheit und Flächeneinheit normal zur Fließrichtung durchfließt ($m^3/s.m^2$), k der Durchlässigkeitsbeiwert des Bodens für Wasser (m/s) und ∇H der hydraulische Gradient (m/m) bedeuten. Die hydraulische Höhe H setzt sich zusammen aus der Lagenhöhe und der Druckhöhe h

$$H = z + h \qquad (2).$$

Obwohl dieser lineare Zusammenhang zwischen Durchflußrate und treibender Kraft auch für die ungesättigte Wasserbewegung Gültigkeit hat, soll im folgenden nur auf die Wasserbewegung im gesättigten Bereich Bezug genommen werden.

2 DIE METHODE DER MESSUNG DER AUFSTIEGS-GESCHWINDIGKEIT IN BOHRLÖCHERN

Prinzipiell ist es möglich, im Labor Versuchsanordnungen zu bewerkstelligen, welche die Messung von v und ∇H gestatten und daraus die einzige Unbekannte, den Durchlässigkeitsbeiwert kf, zu berechnen gestatten. Abgesehen von den Schwierigkeiten, ungestörte Bodenproben im Feld zu entnehmen, handelt es sich bei diesen Proben um nur sehr kleine Ausschnitte des gesamten Bodenprofiles. Zielführender ist die Messung des kf-Wertes direkt im Feld. Ähnlich den Labormethoden muß eine Versuchsanordnung im Feld gefunden werden, welche aus den Messungen von v und ∇H die Berechnung des Durchlässigkeitsbeiwertes kf ermöglicht. Gesucht ist also eine Versuchsanordnung im Felde, deren Strömungsform eindeutig definiert, mathematisch erfaßbar ist und im Feld mit vertretbarem Aufwand realisiert werden kann.

Dies ist am leichtesten bei der Zuströmung zu einem Brunnen möglich. Es gibt eine Reihe von Brunnengleichungen (z.B. Huisman 1972), welche zumeist aus der Kenntnis der Entnahmemenge Q und den in mehreren zusätzlich abgeteuften Standrohren gemessenen Absenkungen des Wasserspiegels die Berechnung des kf-Wertes mit Hilfe eines Pumpversuches gestatten. Pumpversuche erfassen einen großen Bodenbereich um das Bohrloch, der so ermittelte kf-Wert ist gleich dem mittleren Wert über einem größeren Bereich. Sie haben jedoch den Nachteil, sehr geräte-, zeit- und personalaufwendig zu sein. Pumpversuche sind für landeskulturelle Untersuchungen wegen des hohen Aufwandes daher ungeeignet.

Am einfachsten dürfte es sein, ein lotrechtes Loch - das Bohrloch - bis unter die Grundwasseroberfläche abzuteufen und dieses ganz oder teilweise zu entleeren und dann zu beobachten, wie schnell der Wasserspiegel im Bohrloch wieder ansteigt: die Bohrlochmethode.

Der Meßvorgang bei der Bohrlochmethode ist sehr einfach. Mit einem Handbohrgerät wird ein oder werden mehrere Bohrlöcher mit zumeist acht bis zehn Zentimetern Durchmesser lotrecht bis zu einer willkürlich gewählten aber bekannten Tiefe unter der Grundwasseroberfläche gebohrt. Danach schöpft man einen Teil oder das gesamte Wasser aus dem Bohrloch heraus. Zusickerndes Grundwasser aus dem umgebenden Boden verursacht ein Ansteigen des Wasserstandes im Bohrloch, wobei die Geschwindigkeit des Wasseranstieges gemessen wird, und zwar in der Art, daß zu verschiedenen Zeitpunkten das Abstichmaß des Wasserspiegels im Bohrloch von einem Referenzpunkt eingemessen wird. Die Geschwindigkeit des Wasseranstieges im Bohrloch ist nun eine Funktion der Durchlässigkeit des umgebenden Bodens; sie ist aber auch unter anderem von den Abmessungen des Bohrloches, dem Maß der Absenkung des Wasserspiegels im Bohrloch etc. abhängig. Läßt sich die Zuströmung im Bohrloch durch ein geeignetes Modell mathematisch beschreiben, kann aus der Geschwindigkeit des Wasseranstieges im Bohrloch die Durchlässigkeit berechnet werden.

Durch das in das Bohrloch einsickernde Wasser wird den Bodenhohlräumen an der Bodenoberfläche Wasser entzogen, und die Grundwasseroberfläche im Bereich um das Bohrloch wird abgesenkt. Im weiteren Verlauf wird ein immer größer werdender Bereich erfaßt und mit steigendem Wasserspiegel im Bohrloch die Grundwasseroberfläche im Bohrloch allmählich wieder angehoben. Der gesamte Vorgang ist somit instationär. Die dem Strömungsvorgang zugrundeliegenden Randbedingungen sind somit zeitabhängig und sie können nicht für jeden Zeitpunkt des Strömungsvorganges im vorhinein angegeben werden.

2.1 Die Randbedingungen nach Hooghoudt für freies Grundwasser

Hooghoudt (1936) war der erste, der genau definierte Randbedingungen seinen Überlegungen zugrundelegte. Er schlug vor, nach dem Bohren den Ruhewasserspiegel im Bohrloch abzuwarten und dann erst den Wasserspiegel im Bohrloch abzusenken, denn unmittelbar nach dem Absinken sind die Randbedingungen, welche zugleich auch die Anfangsbedingungen für den weiteren instationären Verlauf des Strömungsproblemes sind, bekannt. Damit legte er den Grundstein für alle später folgenden und auch heute noch angewendeten Methoden. Es stellt sich dann eine Strömung entsprechend den in Fig.1a und Fig.1b dargestellten Randbedingungen ein.

Für diese Randbedingungen fanden Kirkham und van Bavel (1948) und Kirkham (1958) Lösungen mit Hilfe von Fourierreihenentwicklungen.

Ernst (1950) geht von folgender Beziehung zur Berechnung des k-Wertes aus:

$$k = c \frac{\Delta y}{\Delta t} \qquad (3),$$

wobei $\Delta y/\Delta t$ die Aufstiegsgeschwindigkeit des Grundwassers im Bohrloch und c einen von der Strömungsgeometrie abhängigen dimensionslosen Beiwert bedeuten.

Ernst (1950) berechnete mit Hilfe eines numerischen Verfahrens (Relaxationsmethode) c-Werte für verschiedene Strömungsfälle. Diese c-Werte sind vielfach publiziert (z.B. van Beers 1962).

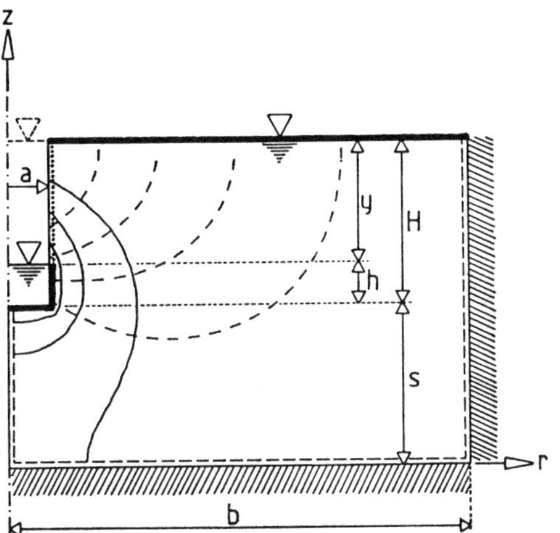

Fig.1a Randbedingungen für die Zuströmung zu einem Bohrloch nach Hooghoudt (1936), rechter Rand: Stromlinie

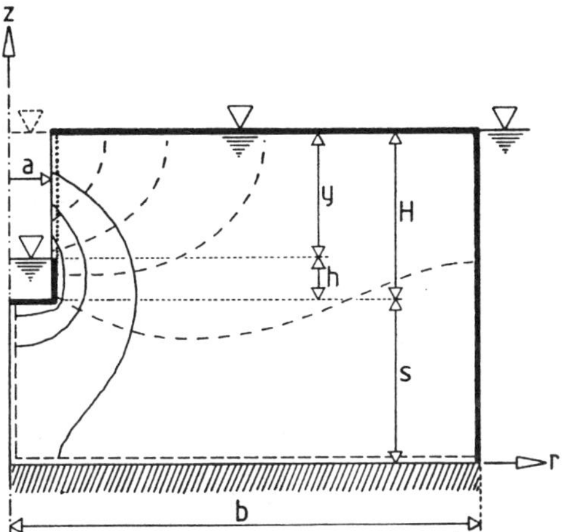

Fig.1b Randbedingungen für die Zuströmung zu einem Bohrloch nach Hooghoudt (1936), rechter Rand: Potentiallinie.
Randbedingungen:
Potentiallinie:
Stromlinie :
Sickerlinie :

Mit Hilfe der Arbeiten von Ernst bzw. Kirkham und van Bavel kann gezeigt werden, daß es belanglos ist, ob die äußerste Randbedingung eine Stromlinie oder eine Potentiallinie ist, wenn B >> C.

2.2 Die Randbedingungen für gespanntes Grundwasser

Die genannten Verfahren sind nur für freies Grundwasser anwendbar. In alpinen Tallagen, aber auch in Beckenlagen findet man vielfach gespanntes Grundwasser. Es entstand daher der dringende Bedarf, die Bohrlochmethode auch für gespanntes Grundwasser zu erweitern.

Bei oberflächlicher Betrachtung könnte man der Meinung sein, daß die obere Randbedingung weiterhin eine Potentiallinie ist, wobei hier auch nach der Wasserspiegelabsenkung im Bohrloch Überdruck wie im Ruhezustand herrscht. Eine genaue Analyse dieses Strömungsproblemes zeigt jedoch, daß diese Vorstellung nicht richtig ist.

Vorerst sei jener Fall betrachtet, bei dem die äußerste Randbedingung eine Stromlinie (Grenze zu einer undurchlässigen Berandung) ist. Bei diesem Problem erfolgt die Wassernachlieferung des in das Bohrloch einsickernden Wassers von der Grundwasseroberfläche aus und bildet einen Absenkungstrichter unmittelbar um das Bohrloch. Diese Grundwasseroberfläche befindet sich an der Grenzfläche zwischen durchlässiger und undurchlässiger Deckschichte und steht im Ruhezustand unter Überdruck. Bevor sich die Bodenhohlräume entleeren können, muß allerdings der vorhandene Überdruck auf jenes Maß abgebaut werden, welches der entsprechenden Lagenhöhe entspricht, oder mit anderen Worten, durch das Absenken des Wasserspiegels im Bohrloch unterhalb der Schichtgrenze zur Deckschichte, wird auch der Druck an der Grundwasseroberfläche reduziert und das Problem vereinfacht sich zu dem Fall, den wir bereits durch Hooghoudt vorgestellt bekamen, allerdings mit dem Unterschied, daß als Bezugsebene für das Maß der Absenkung y nicht der Ruhewasserspiegel im Bohrloch vor dessen Absenkung, sondern die Grenzfläche durchlässige - undurchlässige Schichte (die Grundwasseroberfläche) anzunehmen ist, so als wäre eine freie Grundwasseroberfläche an der Schichtgrenze vorhanden. Da bei diesem Modell idealisierend angenommen wird, daß die Deckschichte vollkommen undurchlässig ist, kann bei einer geringeren Absenkung des Wasserspiegels im Bohrloch bis oberhalb der Schichtgrenze kein Wasser in das Bohrloch nachgeliefert werden, da sich in diesem Fall die Bodenhohlräume des durchlässigen Bodens an der Grenze zur undurchlässigen Deckschichte nicht entleeren können. Um das vorgestellte Verfahren anwenden zu können, muß demnach der Wasserspiegel im Bohrloch stets bis unterhalb der Begrenzung zur undurchlässigen Deckschichte abgesenkt

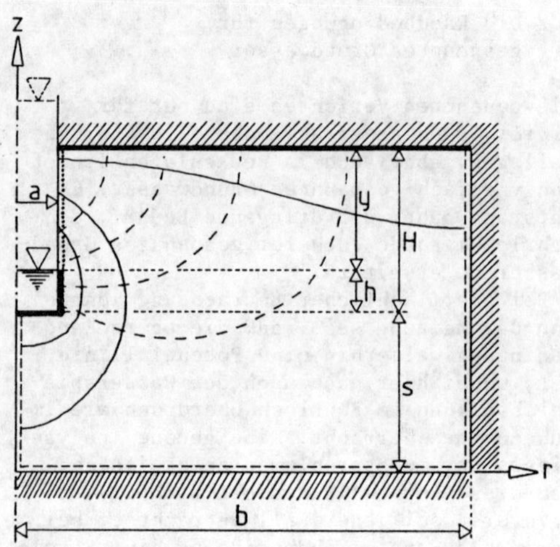

Fig.2 Randbedingungen für die Zuströmung zu einem Bohrloch im gespannten Grundwasser, rechter Rand: Stromlinie (Symbole für Randbedingungen siehe Fig.1b), (Mikulits 1988)

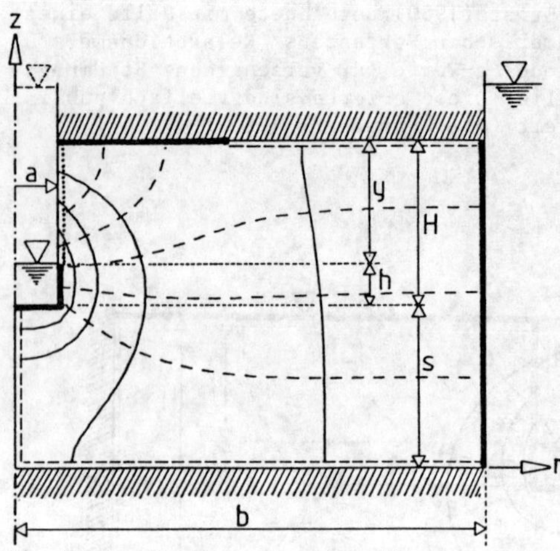

Fig.3 Randbedingungen für die Zuströmung zu einem Bohrloch im gespannten Grundwasser, rechter Rand: Potentiallinie (Symbole für Randbedingungen siehe Fig.1b), (Mikulits 1988)

werden. In Fig.2 sind die Strömungsverhältnisse für den Fall des Druckwassers bei Vorhandensein einer äußeren undurchlässigen Berandung dargestellt.

Wenn die äußerste Randbedingung eine Potentiallinie ist - dies ist der Fall, wenn an das betrachtete Strömungsgebiet ein See anschließt - kann zusätzlich Wasser von diesem See in das Bohrloch zusickern. Dieser Fall ist in Fig.3 dargestellt. In diesem Fall ist die obere Randbedingung geteilt: im Bereich unmittelbar um das Bohrloch wird sich ein Absenkungstrichter einstellen; hier ist die obere Randbedingung eine Potentiallinie. Im äußeren Bereich wird Wasser vom umgebenden See in den Bodenkörper einsickern; hier ist die obere Randbedingung eine Stromlinie. Der Punkt, bei dem an der oberen Begrenzung die Strom-in die Potentiallinie übergeht, kann mit Hilfe eines numerischen Iterationsverfahrens gefunden werden, wenn man voraussetzt, daß das Potential an dieser Stelle nicht kleiner sein kann als das der Lagenhöhe entsprechende Potential (Kastanek 1984).

Die von außen zusickernde Wassermenge wird mit zunehmendem Überdruck des Druckwassers größer, doch wird sie mit zunehmendem Abstand des außen liegenden Sees zum Bohrloch vernachlässigbar, so daß sich auch dieses Problem auf jenes von Hooghoudt definierte reduziert.

Da die dargestellten theoretischen Überlegungen für die axialsymmetrische Zuströmung zu einem Brunnen wie für die zweidimensionale ebene Zuströmung zu einem Graben die gleichen sind, wurden entsprechende Versuche an einem Sandkastenmodell (Loiskandl 1988) beziehungsweise an einem elektrischen Analogmodell (Regelsberger 1986) durchgeführt, welche die dargestellten Überlegungen bestätigten.

3 NACHWORT

Das vorliegende Projekt wird unter Projekt Nr. P6704B vom Fonds zur Förderung der wissenschaftlichen Forschung, Wien, in dankenswerter Weise gefördert.

LITERATURVERZEICHNIS

Ernst, L.F. 1950. A new formula for the calculation of the permeability factor with the auger hole method. TNO Groningen.

Hooghoudt, S.B. 1936. Bijdragen tot de kennis van eenige natuurkundige grootheden van den grond. 4. Bepaling van den doorlaatfactor van den grond met behulp van pompproeven (z.g.boorgatenmethode). Verslagen van Landbouwkundige Onderzoekingen. No.42:449-541. Gravenhage, Algemeene Landsdrukkerij.

Huisman, L. 1972. Groundwater recovery. London, Macmillan.

Kastanek, F. 1984. Numerisches Rechenmodell DURBO. Unveröffentlichtes Programm, Universität für Bodenkultur, Wien.
Kirkham, D. & van Bavel, C.H.M. 1948. Theory of seepage into auger holes. Proc.Soil Science Society Am. 13:75-82.
Kirkham, D. 1958. Theory of seepage into an auger hole above an impermeable Layer. Proc.Soil Science Society Am. 22:204-208.
Loiskandl, W. 1988. Verwendung Polynome höherer Ordnung für ein numerisches Modell der Zuströmung zu einem Brunnen oder Graben. Dissertation, Universität für Bodenkultur, Wien.
Mikulits, R. 1988. Unveröffentlichtes Programm zur Berechnung der Stromfunktionswerte bei der Zuströmung zu einem Brunnen oder Graben. Universität für Bodenkultur, Wien.
Regelsberger, M. 1986. Simulation der Zuströmung von gespanntem Grundwasser zu einem Graben mit elektrischem Analogverfahren. Diplomarbeit, Universität für Bodenkultur, Wien.
van Beers, W.F.J. 1962. Die Bohrlochmethode. Bulletin 1/D, Internationales Institut für Landgewinnung und Kulturtechnik,Wageningen.

Land and Water Use, Dodd & Grace (eds), © 1989 Balkema, Rotterdam. ISBN 90 6191 980 0

Results of installation pipe drainage system by continuous trencher and trenchless crawler machines on soil structure

J.Medyk
Institute of Forestry and Land Reclamation, Agricultural University, Wrocław, Poland

ABSTRACT: Studies have been carried out on I3 drainage facilities South - Western Poland. Factors studied include installation of drain lines, machine performance and the impact of different types of machines on soil structure near drains. Although these factors are inter-related in terms of final drain system performance it was found that trencher(ETC-202) and trenchless (KD-20) machines had different effects on the soil structure near the drain. The purpose of this particular study was to determine the effect of continuous trenching and trenchless crawler machines on the shear strength and permeability of soil near the drain.
This investigation is a part of a study of operations research methods and developing a network model for application of these methods to regional drainage projects. It is planned to develop an index of the quality of drainage construction based on factors related to design parameters, material quality and worker qualifications.

RESUME: Cet article présente l'étude des résultats de 13 chantiers de construction d'amélioration dans la partie Sud-Ouest de la Pologne, sur le facteur des changements de profil du sol après avoir instalé des drains. Ce factewr determine la qualité de la réalisation du systèmes des drains avec des genres particuliers de machines. Ces recherches constituent une partie des études dans le réseau de modèle sur les solutions des projets régionaux et sur leur mise en pratique comme le facteur de la qualité du drainage, de la qualité des matériaux et des qualificatious des travailleurs.

ZUSAMMENVASSUNG: In der Arbeit wurde die Ergebnisse der Forschungen des Koeffizienten der Veranderungen im der Bodenstruktur nach der Dranagelegung mit verschiedenen Maschinen anhand von 13 Meliorationsbaustellen im Sudwesten Polens dargestellt. Dieser Koeffizient bestimmt die Qualitat der Dranagesystemlegung mit verschiedenartigen Maschinen.
Diese Forschungen bilden einen Teil des Studiums, im Modellnetz, der regionaler Projekte and ihrer Anwendung als Koeffizienten der Dranagequalitat, der auf die funktionelle Relation technischer Dranageparameter, der Materialqualitat und der Qualifikationen der Arbeiter zuruckzufuhren ist.

1 INTRODUCTION

Drainage problems differ widely because of the varied nature of physical coditions and crops to be grown. The topography, soils, and quality and source of water of any given area vary greatly both indivitually and in their total effect. A complete and thorought evaluation of these factors is needed before the correct system can be prescribed (Donnad & Bradshaw, 1952).For exemple, a study of topografy might reveal that the problem area is in a closed basin, thus preventing construction of any type of gravity drain. Some soil are almost impossible to drain internally. Thus it would be futile to prescribe a tube drain for effective drainage. Water tables can be lowered, thus minimizing or eliminating the total cost of the drainage, if ground waters that are not too highly minieralized are pumped and used for irrigation (Schwab, 1974).

Technical improvements of installation drainage systems in USA, also in Western Europe too, aim to increase the capacity, to reduce wear and tear,or to save manpower. The trenchless crawler (drainage plow) is designed to increase tear.

About 1941, polyethylene plastic, a British development, was made available for manufacture in the USA. According to Schwab, in thesis cited above investigated as early as 1946 the use of "perforated plastic tubing" installed with "cable-laying machines" for airport drainage.However Schwab's research from 1947 to 1954 (Schwab, 1955) is considered the begining of the development and use of plastic drain tubes in the USA, if not the world.

Its development started new technology draining installation, with trenchless crawler in the Europe from 1960 by Janert and Willner (Voss and Zimmerman, 1974). From 1965 on those machines have been commercially available, and in Western Europe several types of machine have been developed, and in Eastern Europe from 1970 too. The enginepower varies from 57 to 300 kW, the maximum working depth from 1.20 to 3.00 m, and the weight from 2.5 to 20 tons (Van Someren, 1971; Eggelsman, 1973; Diemke, 1973; Schwab, 1974; Medyk, 1976; Naarding, 1977).

Presently, in the USA, Western Europe also in Poland too, trenchless machines are used only on a small scale, because it is exepted that due to deformation and compaction of the soil profile, the drainage system will not function as as well as pipe drainage indigger. However, one has to take into account that, particularly in unstable soil, the losse soil in traditional permebove a certain depth, cracks formed inability (Boels and Havinga, 1974; Havinga, 1975; Medyk, 1982; Medyk and Grismer, 1987; Medyk, 1988).

To compare the performance of drains installed with a trenchless drainage machine (drainage plow) with that of trencher laid drainage, the change in permeability in the soil profile surrounding the drains must be determined. In this paper two a methods draining are presented to predict such changes.

2 METHODS AND PROCEDURES

Due to recurrent failure of recently installated drainage systems by trenchless crawler Poldren KD-20 in the South-western Poland an intensive study was initiated to examine various aspects affecting performance of those systems. Factors studies included installation of drailines, machine performance, and the impact of different types of machines trencher (ETC-202) and trenchless (KD-20) on soil structure near drains. Althought these factors are interrelated in terms of final drain system performance, it was found that trencher and trenchless machines had different effects on soil structure near drain. The purpouse of this particular study was determine the effect of continuous trenching and trenchless crawler machines on the shear strength and permeability of soil near the drain.

Soil shear strength was measured at the different depths,0.05; 0.25; 0.45; 0.65; 0.85 m;showns in the drain depth of 1 m, and at five different distances on either side of the drainline trench: 0.15; 0.30; 0.60; 1 and 2 m.

Mesurements were made using a vane shear similar to that typical apparatus on investigations. Results of this investigations from the shear strength were statistically related by empirical equestions to an index of soil bulk density (g).

Measurments of soil permeability were made at depth of 30, 60,and 90 cm using 1 cm thick cores. Ten such cores were taken with increasing distance from each side of the trench, and used in laboratory filtrations.

3 SOIL DEFORMATION CAUSED BY TRENCHER AND TRENCHLESS

When a machine's ETC body is forced in soil by drainline, and horizontally above a certain depth, cracks formed in two side direction. Figure 1 summirizes typical results obtained from shear strength measurments as reflecan index of soil density.

Aggregatted results shown in figure 1 are from trencher created by the trenchless crawler, and analogous results correspond to continuous trencher up to the depth 0.45 m (fig.2).

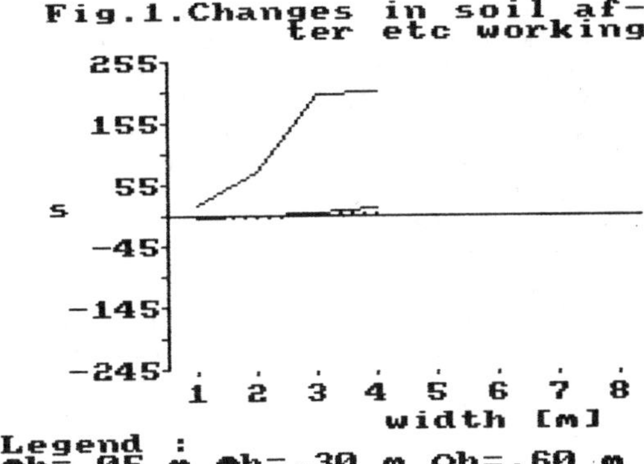

Fig.2.Changes in soil after kd working.

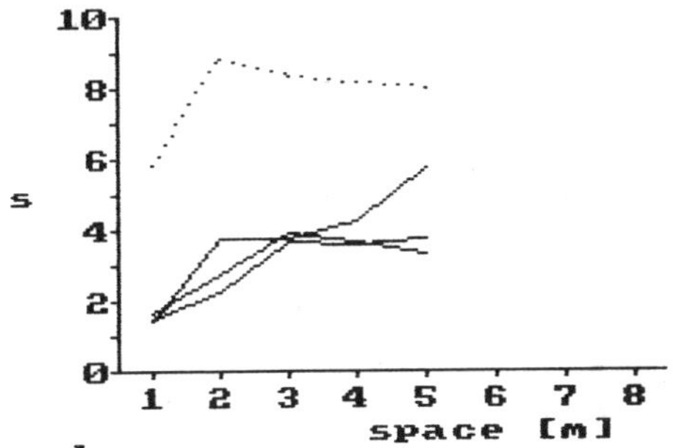

10
8
6
s
4
2
0
1 2 3 4 5 6 7 8
space [m]
Legend :
●h=.25 m ●h=.45 m ●h=.85 m
Oh=1.0 m

In either case, bulk density increases with depth and distance away from the drain. The trenchless crawler appears to uniformly disturb the soil profile at depths greater than about 0.45 m, whereas the continuous trencher has limited impact on soil bulk density at depth between 0.05 and 0.90 m. At a depth of 0.90 m, the continuous trencher compact the soil at distances up to 2 m from drain. Such large compaction in soil at this depth will tend to limit movement of water to the drain in this layer.

4 THE RELATIONSHIP BETWEEN DEFORMATION AND SOIL'S PERMEABILITY

The relation deformation/permeability conductivity was determinated by cores, placed in edometer apparatus. The core was surrounded by a rubber sleeve.In the core a constant pressure head, $h-h_1$ is maintained. To prevent the water from flowing out between sleeve and core, a higher pressure is maintained in the cell. A constant head gradient over the score allows the determination of the hydraulic conductivity, as filtration's index.

As might be expected, both the trenchless crawler and contiuous trencher change soil permeability near the trench. Figure 3 illustrates dependence of the filtration V of the ratio of permeability (before, Vo, and after V, construction) with distance from the trench. Analogous results from continuous trencher installation are shown in figure 4. Permeability after construction incrased at all depths when using the continuous trencher,and only at the two shallower depths when using the trenchless crawler. Large permeabilities for the continuous trencher are associated with backfill materials and the width of the trench created.

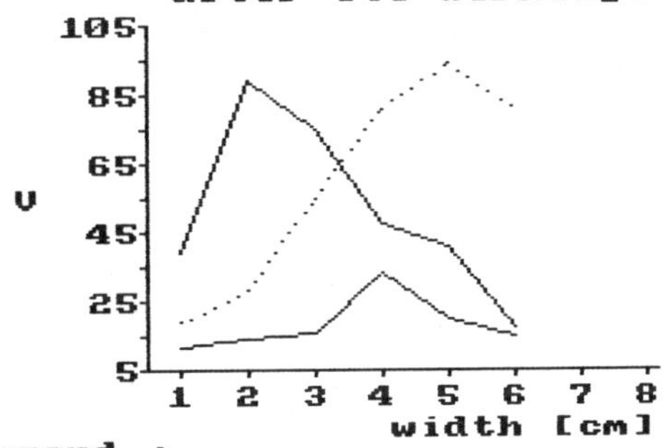

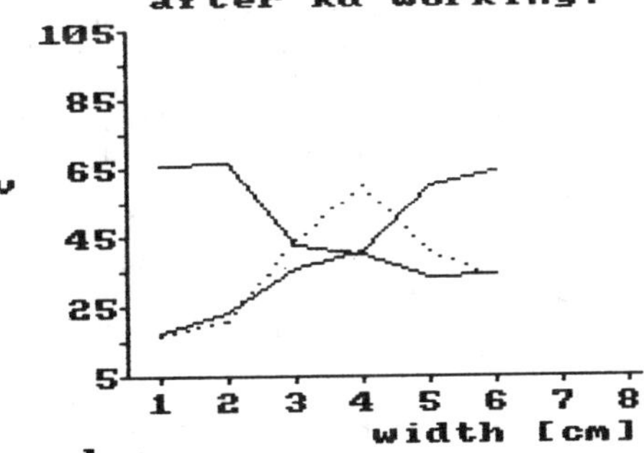

Urfontunately, use of the continuous trencher resulted in substantial compression of the soil at the trench walls, which was simultaneously associated with a decrease in permeability.

5 SUMMARY AND CONCLUSIONS

As far as currently available data show, the hydraulic conductivity of soil, mainly dependent on a small number of large pores, rapidly decreases due to small deformations in the soil. When the hydraulic conductivity is mainly dependent on the texture of the soil, it hardly influenced by soil deformation. The dry bulk density of the unstable backfill of a trench can reach higher values than the original undisturbed soil.

Drainline installation machines have different effects on soil structure depending on the type of machine employed. Generally, continuous trenching machines results in greater disturbance of the soil profile adjacent to the drain as compared to trenchless crawlers. Depending on the soil water content and cohesiveness at the time of installation, trenchless crawlers may be prefered over continuous trenchers while installing drains at depth of approximately 1 m.

REFERENCES

Boels D. 1979. A method to predict in hydraulic conductivity caused by drainage plows and backfilling of trenchless. Institute for Land and Water Management Research at Wageningen, Technical Bulletin, No.111, Holland.

Boels D. and Havinga L., 1974. Effect van woelen op verdichte zavelground. Cultuurtechn. Tijdsch., No.14/1, ICW, Holland.

Dejas D. and Medyk J., 1983. Changes in the filtration property caused by the realization of mechanical draining systems. Inern. Symp.on Comput.Hydr. in Agr.Eng., Agrul.Univ.at Wroclaw, Poland.

Dejas D. and Medyk J., 1987. Einflusch der Qualitat der Dranarbeiten auf Instandhaltungsmaschnahmen. Rostocker Agrarwis.Betrage. W.Pieck's Univ. at Rostock, GDR.

Diemke K., 1973. Die Dranmachine Melioma.Wissenschaftlich-techniche Informationen fur das Meliorationswessen. Veb If.M Bad Freinwalde, GDR.

Eggelsmann R., 1973. Drananleitung. Verlag Wasser u.Boden Alex Lindow & Hamburg, Germany.

Godwin R., 1974. An Investigation into the Mechanics of Narrow Tines in Frictional Soils. Thesis, Univ. of Reading.

Havinga L., 1975. Blijvend effect van diepe grondbewerking. Cultuurtechn. Tijdschr. No.15, Holland.

Medyk J., 1984. Quality assessment of drainage working machinery Poldren and ETC-202 on Soth-west Polish Region. Zesz.Nauk.No.26, Agr.Univ. at Wroclaw, Poland.

Medyk J., 1988. The influence of using continuous trencher and trenchless crawler drainline machines on soil structure. Studying Agr.Univ. Debrecen, Hungary.

Medyk J. and Grismer M.E., 1987. Effects of Continuous Trencher and Trenchless Crawler Drainline Machines on Soil Structure. Soil & Water, Winter No.69, Univ. of California at Davis, USA.

Naarding W.H.,1977. A Reviev on International Experience w with Drainage Machines. Med.Cultuurt. No.121, Utrecht, Holland.

Schwab G.O. and Donnan W.W.,1974. Current Drainage Methods in the USA, Drainage for Agriculture. American Society of Agronomy, Madison, Wisconsin, USA.

Smedana L.K. and Rycroft D.W., 1983. Land drainege - planing and design of agricultural drainage systems. Cornell Univ.Press Ithaca, New York, USA.

Sokolowski J. and etheirs, 1988. Trenchless technology (Technologia drenowania - The Polish translation), Pan.Wyd.Rol.i Les., Warsaw, Poland.

Voss B. and Zimmerman F., 1974. Zugkraft, Bedarf u.Verelegeschwindigkeit grabenlos Drainmaschinen. Waser u.Boden, No.26, Germany.

Land and Water Use, Dodd & Grace (eds), © 1989 Balkema, Rotterdam. ISBN 90 6191 980 0

Un outil pour la maîtrise de l'azote en agriculture: La méthode du bilan

J.C.Muller
INRA Station d'Agronomie, Chalons sur Marne, France

RESUME : Partant de l'écriture de l'équation du bilan de l'azote , un bref rappel du principe de la méthode est fait, et les différents postes de l'équation sont présentés afin d'établir: soit un bilan "prévisionnel", soit un bilan "réel. Puis sont analysés les postes qui correspondent aux besoins de la plante et aux fournitures d'azote par les différentes source du sol, les engrais azotés . A partir d'exemples provenant des expérimentations récentes conduites en France dans le domaine de la fertilisation azotée, les point suivants sont examinés avec plus de détails - Le postulat de l'additivité des différents postes et ses conséquences sur la mise en oeuvre de la méthode et sur la possibilité d'utiliser le reliquat en sortie d'hiver comme indicateur des fournitures d'azote par le sol. - Les notions de minéralisation de l'humus et de potentiel de minéralisation de l'azote du sol, et notamment les exigences à satisfaire pour valider au champ les données de laboratoire. - L'estimation de la fourniture globale d'azote par le sol à l'aide des techniques de marquage isotopique, l'utilisation des témoins "zéro" non fertilisés et la mise en évidence de l'indépendance des quantités prélevées par la plante en provenance du sol et de l'engrais. Enfin pour la rendre opérationnel par l'agriculteur, la possibilité d'une méthode simplifiée du bilan, fondée sur une dose de référence, est envisagée dans un contexte régionalisé du conseil de fumure.MOTS CLES : Azote 15. Bilan azote. Engrais Azoté. Fertilisation.

1 INTRODUCTION

L'enjeu d'une meilleure gestion de l'azote est devenu une nécessité impérieuse pour valoriser au mieux un facteur de production de plus en plus coûteux, et par voie de conséquence limiter au maximum les risques de pollution de l'environnement.

L'apport d'engrais a d'abord été conçu comme une restitution au sol des éléments prélevés par les cultures. C'était une attitude bien adaptée à une époque où les éléments fertilisants étaient peu coûteux. La pratique actuelle consiste à faire l'avance, au plus juste , des éléments qui seront prélevés par une culture, ce qui implique que l'on connaisse, d'une part les besoins unitaires de chaque plante cultivée, et d'autre part la quantité que l'on espère récolter, c'est à dire le rendement "objectif".

Pour satisfaire à ces exigences, dans le cas de l'azote, les modèles proposés de fertilisation peuvent se classer en deux groupes.

-Les modèles de prévision empiriques. L'approche est faite en établissant des corrélations directes entre la fertilisation optimale et le rendement. Les variables les plus influentes sont le climat, le système de culture, le sol, . .

Parmi ces modèles on peut citer ceux proposés par VAN DER PAAUW (1962), DUTIL et BALLIF (1971), KOLENBRANDER(1980), WEHRMAN(1980), de SCORRAILLE(1987). Cette approche consiste à rechercher une fonction de production $y = f(x)$, reliant le rendement et l'azote disponible. Deux types d'équation peuvent être proposées : la parabole $y = y_0+bx-cx^2$, et la courbe exponentielle de MISCHERLICH : $y = A\ [1-e^{-c(x+x_0)}]$. Ces deux relations permettent une analyse "a posteriori" de la situation, mais elles ne peuvent pas s'appliquer "a priori", c'est-à-dire être prédictives, que si l'on connaît, pour une situation expérimentale donnée, les coefficients y_0, b et c dans le premier cas et A, c et x_0 dans le second, et admettre qu'ils sont suffisamment stables dans le temps.

-Les modèles de type "budget prévisionnel". Ils essayent d'estimer les différents postes de fourniture et de consommation de l'azote dans la zone exploitée par les racines. La méthode du "bilan de l'azote", proposée par REMY et HEBERT en 1977, se rattache à cette démarche. Son grand mérite est une prise en compte plus réelle des fournitures d'azote, c'est à dire de la quantité x_0 provenant des diverses sources possibles; celles-ci étant définies par une typologie qui, pour être approximative, n'en est pas moins très pragmatique.

2 RAPPEL DE LA METHODE DU BILAN

Le principe de la méthode du bilan est simple. Sur un cycle cultural (en général de la sortie de l'hiver à la récolte, pour le blé par exemple) il fait équilibrer les entrées et les sorties. Tous les postes étant appréhendés, les deux termes de l'équation du bilan doivent être égaux. Il suffit donc de connaître "n-1" de ces postes pour obtenir le "nième" par différence.

- *Deux types d'équations* sont possibles avec ou sans coefficient d'utilisation de l'azote disponible (tableau 1). Dans le cas d'un *bilan prévisionnel* la quantité "y" correspond à l'objectif de rendement et "b" est la quantité d'azote présente par unité de rendement quand la fumure est optimale, dans le cas d'un *bilan réel,* "y" correspond au rendement constaté et "b" est la quantité d'azote présente par unité de rendement.

L'écriture sans coefficient oblige à inclure dans les sorties différents postes de perte: volatilisation, lessivage, réorganisation difficiles à apprécier. Certains postes du bilan peuvent être mesurés ou évalués à partir de situations de référence avant la période d'application de l'engrais, c'est le cas du reliquat d'azote minéral en sortie d'hiver. D'autres ne peuvent qu'être estimés à partir de situations de référence pour connaître leurs valeurs moyennes et leurs facteurs de variation. Enfin il y a lieu de rappeler que l'équation du bilan donne la dose totale d'azote de la fumure, mais lorsqu'elle doit être fractionnée, d'autres critères, basés sur les stades de végétation, doivent être pris en compte pour ajuster les fournitures aux besoins.Si l'on veut que cette méthode ne reste pas uniquement préconisée par les techniciens et les chercheurs mais soit effectivement utilisée par les agriculteurs il faut s'interroger sur la pertinence de sa formulation.

TABLEAU 1 . EQUATION DU BILAN EN AZOTE D'UNE CULTURE

Sans coefficient d'utilisation de l'azote disponible :

bY = (Rh-Rf) + Mh + Mr + Ma + X

b = azote absorbé par unité de production
Y = objectif de rendement
Rh = azote mineral à l'ouverture du bilan. C'est la quantité d'azote minéral (nitrique et ammoniacal) présente sur une profondeur donnée de sol. Elle est le plus souvent déterminée à la foin de l'hiver au moment de la reprise de la végétation.
Rf = azote minéral à la fermeture du bilan. Elle est déterminée sur la même profondeur que pour la mesure de Rh, immédatement après la récolte.
Mh = minéralisation de l'humus (ou matière organique stable).
Mr = minéralisation des résidus de récolte.
Ma = minéralisation des amendements organiques.
X = azote de l'engrais apporté par l'agriculteur.

Avec coefficient d'utilisation de l'azote disponible :

bY = K(Fs-X)

Fs = fourniture d'azote par le sol; elle est égale à :
(Rh-Rf) + Mh + Mr + Ma
K = coefficient d'utilisation de l'azote disponible; c'est le rapport de la quantité d'azote mobilisé par la culture à la quantité d'azote disponible :

$$K = \frac{\text{N mobilisé}}{\text{N disponible}} \times 100$$

L'écriture de l'équation du bilan suit les recommandations de normalisation proposées par le groupe "AZOTE" du COMIFER, dans le "glossaire agronomiquede l'azote, termes et symboles utilisés(1987).

Initialement le bilan prévisionnel prenait en compte l'efficacité du transfert de l'azote du sol vers les racines. On accordait au paramètre "Rf" une valeur minimale correspondant à la fraction de l'azote non récupérable. La fourniture par le sol "Fs", de même que la fumure "X", était considérée comme utilisée à 100 p. cent. Un certain nombre d'échecs dans la prévision d'une part , et l'utilisation de l'azote 15 d'autre part, ont montré que l'azote n'était pas utilisé à 100 p. cent. Pour rendre le bilan opérationnel, un coefficient "k" a donc été introduit pour exprimer la proportion d'azote valorisée. Ce coefficient fort commode - généralement compris entre 0,9 et 0,5 - traduit de nombreux phénomènes et son interprétation est loin d'être évidente pour apprécier l'efficacité d'un apport d'engrais azoté. C'est pourquoi on a cherché à lui substituer, dans un premier temps, un Coefficient Apparent d'Utilisation (CAU) qui représente le supplément d'azote absorbé par la culture fertilisée comparé à la quantité d'azote provenant de l'engrais appliqué. Or seule l'utilisation de l'azote 15, comme traceur, permet d'accéder au Coefficient Réel d'Utilisation (CRU) de l'azote absorbé par la culture. Il réprésente la fraction de l'azote de l'engrais absorbé par la culture.

Il y aurait identité entre le CRU et le coefficient "k" si l'azote du sol et de l'engrais étaient affectés d'un même coefficient d'utilisation; or les expériences récentes (MACHET 1986) ont montré que tel n'était pas le cas.

Ainsi dans une situation donnée, et pour la gamme des doses agronomiques courantes, le CRU semble indépendant de la quantité d'azote apportée (fig. 1). On observe par contre une relative constance des quantités prélevées provenant du sol. Cela signifie que :

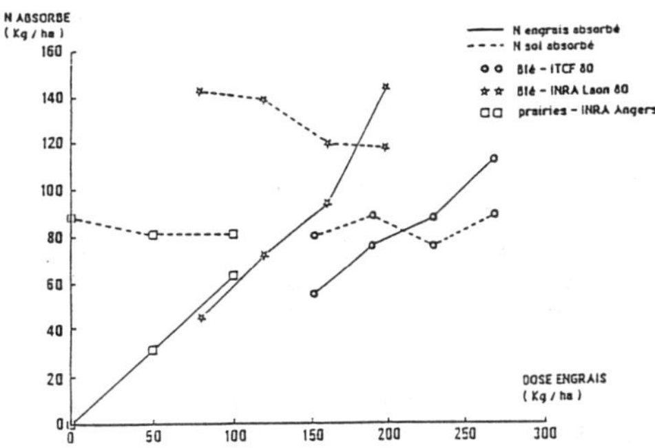

Figure 1 - Effet de la dose d'engrais sur l'utilisation de l'azote du sol et de l'azote de l'engrais.
(d'après J. M. MACHET et col.1987)

- les fournitures par le sol dépendent des conditions pédoclimatiques et de l'historique cultural de la parcelle et ne sont pas affectées par l'apport d'engrais;
- l'azote du sol ne constitue pas une "source relais" en cas de mauvaise utilisation de l'azote de l'engrais.

3 ANALYSE DES TERMES BESOINS ET FOURNITURES DE LA METHODE DU BILAN "AZOTE"

Dans cette analyse les deux termes de l'équation du bilan, besoins et fournitures, doivent être examinés séparément. Le premier se rattache au concept de potentialité parcellaire. Le second est lié essentiellement à la dynamique de l'azote dans le sol.

Le poste des besoins, apprécié à partir de la potentialité parcellaire. Dans un milieu donné défini par le sol et le climat, et caractérisé à l'aide de critères pertinents et facilement accessibles, la potentialité parcellaire peut être définie comme le rendement qu'il est raisonnable d'espérer compte tenu des contraintes auxquelles est soumise la parcelle. Il est important de dissocier cette notion de potentialité parcellaire de celle d'objectif de rendement, car en plus des contraintes de milieu , elle prend en compte les contraintes liées au sytème de culture et aux problèmes de trésorerie (fig.2).

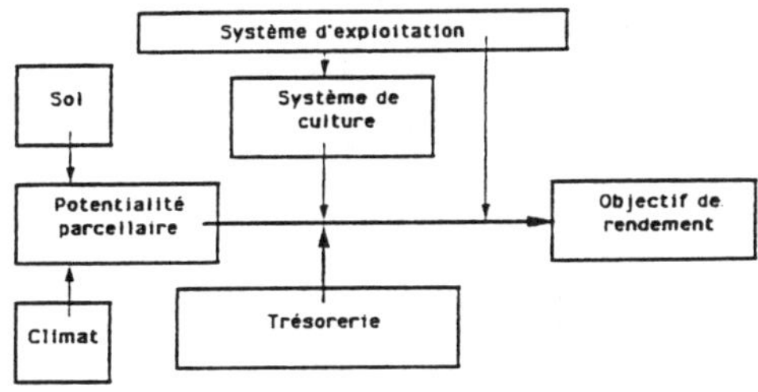

Figure 2 - Schéma des relations entre potentialité parcellaire et objectif de rendement en fonction des contraintes de milieu. (d'après F. LIMAUX 1988)

Pour raisonner l'ajustement de la fertilisation azotée, et notamment satisfaire aux besoins de la culture au cours de sa croissance, il est nécessaire d'introduire une notion supplémentaire qui est celle du potentiel de la culture.

Pour en rendre compte, dans le cas du blé par exemple, on mesure l'état de la croissance du peuplement au stade "épis 1 cm" au travers de la matière séche produite par les parties aériennes, bon indicateur du nombre d'épis potentiel. En effet cette composante du rendement permet de prévoir le nombre de grains potentiel par mètre carré, donc de calculer les besoins en azote

Les problèmes posés pour apprécier les fournitures en azote. Nous allons en examiner successivement quelques uns.

Postulat de l'additivité des différents postes du bilan. Le modèle additif suppose que les différents compartiments fournissant de l'azote minéral sont sans interférence les uns avec les autres. Or dès 1983 nous avions signalé (MULLER et MARTIN 1983) qu'en période hivernale lorsque les conditions climatiques sont favorables à la minéralisation, la matière organique stable, les résidus de cultures et les amendements organiques peuvent réalimenter le poste azote minéral à l'ouverture du bilan.

Cette constatation a deux conséquences. D'abord, elle explique la meilleure réussite de la méthode du bilan dans les régions où les conditions hivernales de température entrainent une faible minéralisation, alors qu'en climat atlantique, où la minéralisation hivernale de l'azote peut être importante, son application est beaucoup plus délicate. Ensuite, si la minéralisation de l'azote organique est suffisamment intense pendant cette période, il devient possible d'en faire une expression des fournitures d'azote minéral (DECROUX et DUGAST 1987).

Pour se rapprocher de la réalité, il faut considérer le sol comme une série de compartiments caractérisés par la nature des composés azotés qu'ils renferment et entre les quels s'effectuent des échanges incessants. C'est le résultat net de ces processus qui est pris en compte dans la méthode du bilan. Il est possible de les améliorer par l'étude des cinétiques à court terme. Or, s'il est évidemment difficile de suivre à chaque instant l'état de l'azote dans le sol, il est encore possible d'améliorer les prévisions en utilisant des techniques de laboratoire et de terrain faisant mieux ressortir la sensibilité des diverses réserves azotées à la minéralisation.

Minéralisation de l'humus et potentiel de minéralisation des sols. La détermination de l'azote potentiellement minéralisable est un important problème que les agronomes tentent de résoudre depuis longtemps. Or dans la pratique l'azote minéralisable est mal pris en compte. Si l'énoncé du problème est simple: "comment arriver à partir d'un échantillon de terre à prévoir la quantité d'azote qui sera minéralisée au champ pendant un cycle cultural?", sa solution est loin d'être évidente.

Le calcul de la fourniture d'azote minéral à partir du coefficient de destruction de l'humus (coefficient K2 du modèle HENIN-DUPUIS) généralement calculé pour le carbone et appliqué à l'azote total du sol ne donne qu'un ordre de grandeur.

Sans sous-estimer les problèmes posés par l'échantillonnage, les méthodes d'analyse, les techniques d'incubation et la modélisation, le problème fondamental est celui de la validation et de l'utilisation pratique des valeurs obtenues au laboratoire. Cette validation ne peut se faire que si l'on assure simultanément sur les mêmes dispositifs de terrain, l'analyse des composantes du rendement, la dynamique de l'azote dans le sol et l'étude au laboratoire de sa fourniture potentielle. De telles études n'ont pas encore été faites.

Minéralisation des résidus de récoltes. Ainsi que le fait remarquer B. MARY (1987), l'estimation de la minéralisation de l'azote des résidus de récolte a été faite empiriquement en fonction du type de précédent cultural pour aboutir aux grilles habituellement utilisées (tableau 2). Récemment de nombreux essais réalisés au champ avec des résidus marqués à N15 ont permis de quantifier la contribution du précédent cultural à la fourniture d'azote minéral. La disponibilité de l'azote du résidu végétal pour la culture suivante est toujours faible, inférieure à 20 pour cent. Le faible coefficient d'utilisation ne résulte pas de pertes par voie gazeuse ou par lessivage, mais essentiellement par incorporation dans des compartiments organiques du sol. Il en résulte que la décomposition des résidus peut affecter fortement les termes Rh et Rf de l'équation du bilan (GUIRAUD et BONIFACE 1897).

TABLEAU 2 . FOURNITURE D'AZOTE MINERAL PAR QUELQUES PRECEDENTS CULTURAUX. VALEURS COURAMMENT UTILISEES DANS LE CALCUL DU BILAN PREVISIONNEL.

Précédent cultural	N kg/ha	Coefficient Réel d'Utilisation (CRU) %
Maïs irrigué	-30	
Maïs grain non irrigué	-25	1 à 5
Blé paille enfouie	-20	
Maïs fourrage	0	
Lin	0	
Betteraves verts enfouis	+20	15
Pomme-de-terre	+20	
Colza	+20	
Tournesol	+20	
Pois protéagineux	+20	
Luzerne	+30	11 à 17

Il faut également rappeler les arrière-effets dus au retournement de prairies se traduisant en première année par des flux extrêmement importants d'azote minéral.

Les chiffres habituellement retenus résultent d'une approche expérimentale et ne sont pas toujours concordants. Une modélisation précise de cette fourniture est nécessaire et pour cette étude les dispositifs de lysimètrie s'avéreraient particulièrement utiles.

Minéralisation des amendements organiques. Jusqu'à présent comme pour les prairies, la fourniture d'azote est estimée à partir de tables provenant de synthèses expérimentales obtenues dans un contexte donné et souvent transposées dans d'autres conditions. Des tentatives de modélisation sont en cours (ZIEGLER1988).

Détermination du reliquat d'azote minéral d'hiver. Rappelons qu'il représente la quantité d'azote minéral (nitrique et ammoniacal) présente sur une profondeur donnée de sol et prise en compte à l'ouverture du bilan. Le plus souvent cette mesure est faite au moment de la reprise de la végétation entre les mois de janvier et février suivant les régions. La méthodologie de cette détermination doit partir de la mesure à la parcelle pour aboutir à l'établissement de modèles régionaux d'estimation du reliquat d'azote minéral d'hiver. En effet la mesure sur une parcelle, non incluse dans un plan régional d'enquête annuelle, est entachée d'une variabilité trop importante si le nombre de prélèvements est insuffisant. Il faut entre 10 et 12 sondages sur un mètre de profondeur pour avoir un prélèvement fiable, à condition de maîtriser les problèmes d'échantillonage sur terre fraîche, ce qui n'est réalisé que sur les essais .

L'enquête régionale par la multiplication des points de prélèvements (au minimum 10 parcelles par précédent) permet, pour un précédent donné, de réduire considérablement cette variabilité. Très développées entre les années 1975 et 1985, ces opérations ont tendance à se réduire, ce qui est dommage car elles fournissaient les données de base pour établir et valider les modèles régionaux. Dans ces modèles les variables explicatives les plus importantes à prendre en compte sont, en dehors du type de sol et du précédent : la pluie lessivante, la somme des températures journalières d'octobre à la date de prélèvement, la date de récolte du précédent et le solde de son bilan azote. Il faut toutefois reconnaître que les tentatives de modélisation réalisées jusqu'à présent sont assez décevantes car elles ne permettent une approche qu'à plus ou moins 40 p. cent de la valeur du reliquat(TAUREAU 1987).

4 EVALUATION DE LA FOURNITURE GLOBALE D'AZOTE PAR LE SOL

Dans ces conditions on peut estimer que l'approche globale de l'offre en azote par type de sol et par précédent serait une démarche plus opérationnelle. Cette démarche est valable dans la mesure où l'on est capable de mesurer les fournitures du sol et où l'on est assuré de l'indépendance des fournitures par le sol et par l'engrais

Mesure des fournitures globales d'azote par le milieu La méthode isotopique utilisant le marquage par l'azote 15 de l'engrais apporté (G. GUIRAUD,1984; J.M MACHET et al. 1987) a montré que la fourniture du sol "Fs" pouvait être calculée à partir de la relation:

Fs = N absorbé par la culture [1-(Excès isotopique plantes/ Excès isotopique engrais)]C'est la seule méthode fiable mais lourde à mettre en oeuvre . Sur le plan pratique il est souhaitable de disposer d'une méthode plus facile à utiliser: les témoins "zéro" non fertilisés en azote. Les résultats obtenus par l'INRA de LAON et l'ITCF (fig. 3) montrent qu'avec certaines précautions les témoins non fertilisés peuvent être utilisés pour apprécier l'offre du sol. Dans les traitements "sans azote" le développement racinaire est toujours plus faible, ce qui peut expliquer que la fourniture d'azote mesurée par cette méthode soit toujours inférieure à celle obtenue par l'utilisation de N 15. Il faudrait essayer de corriger ce biais par une modification des densité de semis ou par le choix de plantes occupant mieux le sol.

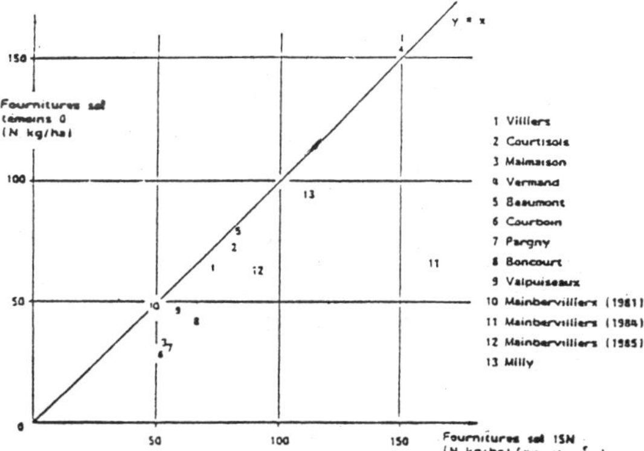

Figure 3 - Estimation comparative de la fourniture du sol (Fs) avec **le traceur 15N et** les témoins non fertilisés (essais 1 à 8 = INRA Laon ; essais 9 à 13 = ITCF),(d'après J. C. TAUREAU 1987)

Indépendance des quantités prélevées provenant du sol et de l'engrais. Il est important de vérifier cette indépendance car elle détermine la validité de la démarche suivie. Les résultats obtenus par J. M. MACHET et col. (1987) semblent bien en apporter la preuve (Fig. 3). En effet dans des essais réalisés sur blé d'hiver et sur prairies l'effet de la dose d'azote sur le coefficient réel d'utilisation (CRU) montre l'indépendance entre la quantité d'azote apporté et celle provenant du sol. Ces auteurs observent une relative constance des quantités prélevées provenant du sol quelle que soit la dose d'azote apporté.

Relations entre fourniture d'azote par le sol et reliquat d'azote d'hiver. Dans une série d'essais bisannuels J. DECROUX et P. DUGAST (1987) ont montré qu'il était possible d'établir une relation entre fourniture du sol (estimée à partir du témoin sans azote : N0) et le reliquat d'azote d'hiver . A partir d'un réseau de parcelles de blé NO bien définies en terme de bilan, ces auteurs montrent que le reliquat est une détermination permettant d'estimer la capacité de fourniture du sol.

5 CONCLUSIONS

Les résultats présentés dans cette communication montrent, au travers des difficultés de mise en oeuvre de la méthode du bilan, la richesse de cette démarche.

En premier lieu elle a permis l'élaboration d'une méthode fixant des rendement objectifs.Pour le poste des fournitures d'azote les résultats obtenus indiquent quels sont les paramètres principaux à prendre en compte :

fourniture globale et *mesure du reliquat d'hiver*. Ils indiquent également les voies dans lesquelles la recherche doit orienter ses travaux : analyse des compartiments de l'azote et du carbone organique du sol, car au delà de la méthode du bilan il faudra aborder une démarche plus mécaniste permettant d'envisager les variations de composition du végétal en fonction de la disponibilité en azote du milieu. Mais il ne faut pas oublier que l'*objectif final* est un *conseil de fumure* qui doit à la fois préserver les *intérêts de l'agriculteur* et le *respect de l'environnement*. Or la formulation du conseil, qui doit être fondée sur les connaissances scientifiques, est déterminante pour la perception du message. Il ne suffit pas qu'une méthode soit connue pour qu'elle soit utilisée. Ainsi que le constate J. M. MEYNARD (1988) il y a inadaptation de la formulation des références.

Par ailleurs on constate que peu d'agriculteurs adaptent l'azote en fonction des parcelles, ce qui conduit à supposer que la méthode serait mieux utilisée si l'on partait d'une dose de référence adaptée aux sols et aux précédents et modulés par un effet "année". On aboutirait ainsi à une régionalisation de l'équation du bilan dans laquelle certains termes pourraient être négligés car peu importants ou pris comme constants (par exemple reliquat azote minéral dans certaines régions). On pourrait alors proposer des tables simplifiées(tableau 3) comme ceux qui ont été proposé par Marianne CERF et J.M. MEYNARD (1988).

TABLEAU 3 . AJUSTEMENT DE LA FERTILISATION AZOTEE. Exemple de présentation de références pour l'aide à la décision.
- La dose "D" de rëférence est établie pour un blé de betterave, de colza ou de tournesol, en sol profond, limoneux ou argileux, une année à hiver de pluviosité moyenne.
- Quantités d'engrais (kg/ha) à ajouter (+) ou à enlever à la dose "D" de référence :

TYPE DE SOL	Limon, Argile	0
	Argile à silex	+10
	Cranette, sable	+30
PRECEDENT	Maïs grain	+50
	Blé paille enfouie	+40
	Maïs fourrage	+20
	Endives, carottes	+10
	Betterave, colza	0
	Pois, haricots	-10
	Luzerne	-20
CONDITIONS DE RECOLTE DU PRECEDENT DE SEMIS DE BLE	Saine	0
	Humide	+30

(d'après Marianne CERF et J. M. MEYNARD, 1988)

Les équations du bilan subsistent, elles servent à calculer la dose de référence. Le message centré sur l'équation et ses limites de validité serait destiné aux conseillers. La méthode simplifiée servirait à la majorité des agriculteurs.

Les travaux menés en France depuis plus de dix ans sur la fertilisation azotée portent sur des analyses soit conceptuelles, soit expérimentales de la trame des relations "azote-plante" et sur l'estimation de la répartition de l'azote dans les différents compartiments du sol dans le but d'optimiser la fertilisation azotée. Par voie de conséquence ils doivent réduire les excédents d'azote post-culturaux, donc réduire les risques de perte par lessivage. Mais par principe la méthode du bilan ne porte que sur un cycle cultural,il faudrait pouvoir l'étendre sur une période plus longue. Or plus que la culture, c'est le système cultural, déterminant des périodes d'interculture sol nu en concordance avec les périodes de lessivage, qui est prépondérant. D'ou la nécessité d'une part d'un approfondissement des connaissances scientifiques sur la dynamique de l'azote, et d'autre part la mise au point d'indicateurs concrets des risques d'excès d'azote qui permettent d'améliorer les pratiques agricoles actuelles.

BIBLIOGRAPHIE

Cerf, Marianne & Meynard J.M. 1988. Enquête sur la mise en oeuvre des méthodes de fertilisation raisonnée. 3ième Forum National de la fertilisation Raisonnée, Nancy 22-23 juin, COMIFER : 5-12.

Decroux, J. & Dugast, P. 1987. Utilisation des essais bisannuels pour l'estimation des fournitures d'azote par le sol. Signification du reliquat d'hiver Rh. C.R. Acad. Agric. Fr., 73, 3 : 91-103.

Dutil, P. & Ballif, J.L. 1971. Prévision de la fumure azotée du blé d'hiver en Champagne crayeuse. C.R. Acad. Agric. Fr., 57, 1 : 88-95.

Guiraud, G. & Boniface Raymonde; 1987; Utilisation de l'azote 15 pour les études de transfert d'azote dans le système sol-plante. C.R. Acad. Agric. Fr., 73, 3 : 23-38.

Kolenbrander, G.J. 1980. The effect of weather pattern on the quantity of mineral nitrogen in the soil profile and the relationship to leaching and fertilizer requirement. C. R. Congrés I.I.R.B., 43 : 381-392.

Machet, J.M. 1986. Contribution à l'étude du coefficient d'utilisation des engrais azotés sur le blé d'hiver. Colloque Franco-Roumain, Clermont-Ferrand, 17-19 mars.

Machet, J.M., Pierre, D, Recous Sylvie, & Remy J.C. Signification du coefficient réel d'utilisation et conséquence pour la fertilisation azotée des cultures. C.R. Acad. Agric. Fr., 73, 3 : 39-55.

Mary, B. 1987. Effet du précédent cultural sur la disponibilité du sol en azote minéral. C.R. Acad. Agric. Fr., 73, 3 : 57-69.

Muller, J.M. & Martin, 1983. Possibilité d'estimer le reliquat d'azote nitrique à la sortie de l'hiver en Champagne crayeuse. Bulletin Fédér. Départ. Groupes d'Etudes et de Développement agricole de la Marne. janvier: 1-7.

de Scorraille, G. 1987. Courbes de réponse à la fumure azotée et méthode du bilan : approches complémentaires. C.R. Acad. Agric. Fr., 73, 3 : 21.

Taureau, G. 1987. Evaluation de l'offre du milieu en azote. Perspectives Agricoles, 115, juin : 43-57.

Van Der Paauw, F. 1962; Effect of winter rainfall on the amount of nitrogen available of crops. Plants and soils.16,3.

Wehrmann, J. & Scharpe, H.C. 1980. Der mineral Stickstoffgehalt des Bodens als Grundlage der Sticktoff dungung bei Zuckerrüben. C. R. Congrès I.I.R.B. 43 : 327-341.

Land and Water Use, Dodd & Grace (eds), © *1989 Balkema, Rotterdam. ISBN 90 6191 980 0*

Investigations of a drainage problem in glacial drift in relation to design of a drainage system

J.Mulqueen
Agriculture and Food Development Authority, Ballinrobe, Co. Mayo., Ireland
A.Otten
Department of Land and Water Use, Agricultural University, Wageningen, Netherlands
D.Daly
Geological Survey of Ireland, Beggars Bush, Haddington Road, Dublin, Ireland
T.Killeen
Department of Agriculture, Hynes Building, Galway, Ireland

ABSTRACT: Enclaves of wet soils, common in otherwise dry land developed on glacial deposits, present problems in drainage design. In a study of one such enclave, developed on limestone glacial deposits, several techniques were used to diagnose the cause of the wetness. Soil physical properties were also measured. It was concluded that test pits excavated by hydraulic digger, which may have to be deep, were necessary and sufficient to classify the drainage problem and provide boundary conditions for design. The results are discussed in relation to drainage design and environmental issues

RESUMÉ: Des enclaves de sols humides, communes d'une autre manière en terre seche formées sur des dépôts glaciaires, presentent des problèmes dans un plan de drainage. À l'étude d'une telle enclave , formée sur des dépôts glaciaires calcaires plusieurs techniques ont été utilisées pour diagnostiquer la cause de cette humidité. Les propriétés physiques du sol ont été aussi mesurées. Il a été conclu que des trous creusés par une pelleteuse hydraulique, qui ont du être profonds, étaient nécessaires et suffisants pour classer le problème du drainage et fournir les conditions pour le plan. Les résultats sont discutés en relation du plan de drainage et des issues de l'environement.

AUSZUG: Enklaven nasser Böden, gewöhnlich in ansonsten trockenem Land, entwickelt aus Eisablagerungen, bringen Probleme in Drainagevorhaben. In einer Enklave, entwickelt aus Kalkeisablagerungen, wurden mehrere Techniken benutzt, um den Grund der Feuchtigkeit zu erkennen. Es wurden auch physikalische Bodeneigenschaften gemessen. Daraus wurde gefolgert, daB Testgruben, die mit einem hydraulischen Grabegerät ausgehoben wurden und tief sein können, ausreichend und notwendig waren, um das Drainageproblem klassifizieren zu können und Grenzkonditionen für das Vorhaben zu liefern. Die Resultate wurden in Beziehung zum Drainagevorhaben und unter Berücksichtigung von Umweltfragen diskutiert.

1 INTRODUCTION

Soils developed on glacial deposits are often variable in Ireland. The deposits themselves may be of drift or meltwater origin and often are stratified. The soils may be wet or dry depending on their thickness, structure, chemical composition, degree of consolidation, slope and elevation in the landscape. Soils developed on glacial deposits of limestone origin are the most widespread and are often stony. They are generally dry but there may be 10-45% wet, often occurring in enclaves and bands (Gardiner and Radford 1980).

These wet soils present special problems in drainage design. Drainage is commonly undertaken in small (4 ha) blocks and this imposes economic limits on the detail of site investigation. One solution is to investigate a few typical sites in detail to establish the range of conditions. This paper describes one such investigation. The objectives were to investigate as fully as possible the soil and hydrological conditions causing the wetness, to identify the necessary and cost-effective investigation methods and to design a drainage solution.

2 LITERATURE REVIEW

Although an extensive literature on land drainage now exists, less attention has been given to site investigations to identify drainage problems. However, the occrrence and investigation of drainage problems in flood plains and deltas are well described in some books (de Ridder 1972, Volker et al. 1973, de Ridder et al. 1974). The hydrogeology of glacial plains is briefly discussed by de Ridder (1972). Causes of excess water in soil are outlined by Eggelsmann (1978), Concaret (1981) and Castle, McCunnall and Tring (1984). The latter describe the use of soil augers and hand spades but clearly these tools have limitations in stony soils of glacial origin. The stony and gravelly nature of many glacial deposits and their sloping and undulating nature has restricted sampling and measurement and therefore limited investigations. Recently, the pit bailing method was developed by Bouwer and Rice (1983) to measure hydraulic conductivity in stony soils with high watertables. When watertables are low as in summer the inversed auger hole and infiltrometer methods can be used (van Hoorn 1979).

3 SITE DETAILS

The site is a sloping 2 ha rectangular block of wet land downslope of the 60 m contour and surrounded by dry land (Figure 1). It forms part of an area used for irrigation of soiled waters from adjoining dairy yards, equivalent to a recharge of 2.3 mm/d when the rainfall is 12 mm/d. The average rainfall is 1110 mm/y and potential evapotranspiration is 350 mm/y leaving an annual surplus of 760 mm water.

The wet soils are on a glacial moraine. Upslope near the 90 m and above the 120 m contours (Figure I) the soil is very thin and there is much bare karstified Burren limestone (Daly 1985) with features such as enlarged joints and bedding planes and swallow holes. An offset borehole record indicated that the glacial deposits are of limestone origin locally. An offset well dug into the hillside (Figure 1) strongly overflows in wet weather and often dries out in summer.

The soils on the wet site are gleyed loams. On the dry land the soils are loamy brown earths. Typical textural analyses are 45% sand, 35% silt and 20% clay with cation exchange capacities of 20 m eq/100 g in the top 0.25 m and 12-17 m eq/100 g in the 1-3 m deep layers. The pH is mostly alkaline with free carbonates in most horizons. The topsoil on the wet block is frequently poached by cows and rutted by tractors. Measured static cone index values are frequently less than 3 bar in the top 0.2 m in the grazing season compared with values of 8-9 bar on the surrounding free draining topsoils.

4 INVESTIGATIONS AND MEASUREMENTS

These were carried out to determine the nature and extent of the glacial deposits and bedrocks and of the groundwater fluctuations and movements. The investigations included: test pits up to 3.3 m deep, well recovery tests in the farm borehole, a geophysical survey, watertable and piezometric elevations, hydraulic conductivity and moisture characteristic curves.

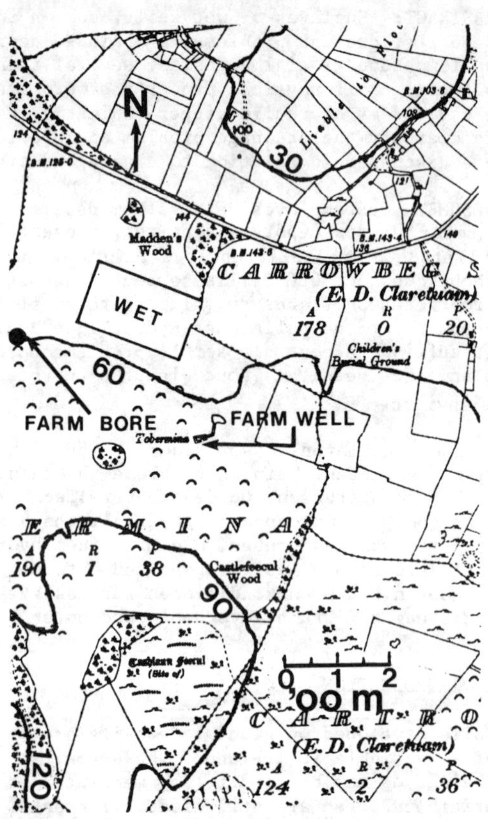

Figure 1. Location of the wet area in relation to contours (m), farm well and farm borehole. Taken from Galway sheet 43.

After a survey, five deep test pits were excavated by hydraulic digger. These were located at the upslope dry boundary and within the wet area. These test pits were about 3 m long x 2 m wide and impervious layers and soil variations could easily be seen.

The farm borehole is offset and upslope of the wet area. As it was drilled into the bedrock (Table 1) well recovery tests were used to determine the elevation of the ground watertable and the transmissivity of the limestone. After prolonged pumping at 92.7 m^3/d, the pump was stopped and the recovery measured over an 80 minute period after which it was minimal.

In the geophysical survey, electrical resistivity and seismic refraction methods were used to indicate the depth to bedrock and layering in the glacial deposits. Electrical resistivity soundings were carried out at 9 different locations in the wet and dry areas. Using the Wenner array, voltage readings were taken at spacings varying from 0.5 m to 128 m. Results were plotted to derive geoelectrical sections from which the geology was interpreted. A north-south traverse across the wet area was made to show profile variations in the upper layers of the glacial deposits. Along this traverse line a seismic survey was also carried out and the travel time of a shock wave against the distance from source to geophone was plotted to indicate the nature of the strata.

Steel tubes of 12 mm bore were used to measure the watertable and piezometric elevations. Tubes to measure watertable were perforated throughout their length and piezometric tubes were open only at the bottom and top. One watertable and 5 piezometer tubes were installed in a cluster at each of 12 locations, 8 of which were in the wet area and 4 in the upslope dry area. Piezometer tubes were installed at depths of 0.6, 1.2, 1.8, 2.4 and 2.7 m. Watertable tubes were installed at 0.9 m depth in the wet area and 3.0 m in the upslope dry area. All tubes were tested for leakage and response time.

Hydraulic conductivity was determined at 16 places by the inversed auger hole and infiltrometer methods (van Hoorn 1979) and by the piezometer method of Kirkham (Luthin and Kirkham 1949). Measurements were made in the summer when the watertable was deep at 5 different locations and at different depths. The piezometer method was used to measure the deeper depths between 1.2 and 2.7 m. Moisture characteristic (pF) curves were determined on undisturbed samples after methods outlined by Stakman (1974).

5 RESULTS

The test pits showed that the soils and subsoils were for the most part stony and gravelly loams derived from glacial deposits of limestone origin. There were also lenses, thin layers and wedges of sand at various depths; these did not appear to be continuous over significant distances. A thick impervious layer was found at depths of 2, 2.5 and 3 m in the wet area and this was deeper than 3 m in the upslope dry ground. The impervious subsoil was rigid and difficult to excavate. This gives rise to a perched watertable in the overlying soil in wet weather.

Table 1. Elevations (E) and depths (D) at the farm borehole.

	E (m OD)	D (m)
Ground surface	60.5	-
Top of bedrock	35.5	25
Bottom of borehole	-32.5	93
Water level May 1987	16.9	43.6
Water level Nov. 1975	14.2	46.3

Tests on the farm borehole showed that the watertable was 43.6 m below ground level and in the bedrock (Table 1). The upper 18.6 m of limestone bedrock was unsaturated. The aquifer is unconfined. This indicates that the high watertable in the wet soil is caused by an impervious layer. It also indicates that the offset farm well and the flooding indicated in the north-east of Figure 1 are due to impervious layers. There is a 1/2000 gradient in the groundwater westwards towards Lough Corrib. The transmissivity of the bedrock was calculated by Logan's approximation method for a yield of 92.7 m^3/d and a drawdown of 5.5 m and by the Theis recovery method for unsteady flow in a confined aquifer (Kruseman and de Ridder 1970). The transmissivity (KD) values were 25 and 27 m^2/d respectively. These values are low for karstic limestones.

Table 2 shows electrical resistivities at wet and dry sites. Values higher than 500 ohm-m were assumed to indicate bedrock (White 1985). Figure 2 indicates a thinning out of the glacial deposits both upslope and downslope. It shows a valley in the bedrock to a maximum depth of 51 m. This could indicate a collapsed karst feature infilled with glacial deposits underlying the wet area. Along the traverse, the apparent resistivities were analysed for variation. The close electrode spacings of 0.5, 1.0 and 2.0 m had co-efficients of variation of 25, 16 and 14 respectively. The distant spacings of 4, 8 and 16 m had co-efficients of variation of 10, 8 and 6 respectively. This indicates variable toplayers and more uniform deeper layers. Seismic tests suggested a 3-layer condition with 3-4 m of loose soil resting on a stiff dense layer with bedrock at a depth of 24 m.

Table 2. Typical resistivity in ohm-m (R) and thickness in m (T) of layers with depth in wet and dry sites.

	Wet		Dry	
	R	T	R	T
	295	0.8	260	0.9
	360	0.3	483	0.2
	254	1.9	245	1.1
	342	16.0	324	1.5
	510	-	472	20.0
			990	-
Total	-	19.0	-	24.0

Watertable contours were more or less parallel to the ground contours and indicated a seepage downslope. At each location the piezometers indicated a more or less static watertable, indicating negligible vertical percolation.

The geometric mean of hydraulic conductivity by the infiltrometer method was 0.2 m/d with a range of 0.05-0.6 m/d. By the inversed auger hole method it was also 0.2 m/d with a range of 0.11-0.35 m/d. Values for the deeper layers by the piezometer method ranged from 0.04 to 0.24 m/d with a geometric mean 0.1 m/d. A value of 0.15 m/d was taken for design purposes.

A flow net was used to compute the amount of seepage into the wet area (de Ridder 1974). The transmissivity of the soil was derived by multiplying the hydraulic conductivity (0.15 m/d) by the saturated thickness of the soil when the watertable is at ground level (3 m) to arrive at 0.45 m^2/d. The seepage amounted to 0.2 mm/d.

Moisture characteristic curves suggested a design crest of watertable at least 0.5 m deep. This is 0.2 m deeper than the design value often used for the drainage of similar areas.

6 DISCUSSION AND CONCLUSIONS

In this study the site survey, test pits excavated by hydraulic digger to sufficient depth to define the hydrological boundary conditions and measurements of the hydraulic conductivity were necessary and sufficient to design a drainage solution. This would appear to hold for similar drainage problems throughout Ireland. In this site the capacity to excavate deep test pits in a cost effective manner was essential to find the impervious layer and moreover only mechanical diggers can excavate fast enough in stony ground so that layering can be seen without being obscured by rising muddy water. Test pits can often be excavated when a machine arrives on site to clean, deepen or provide outfall ditches or drains.

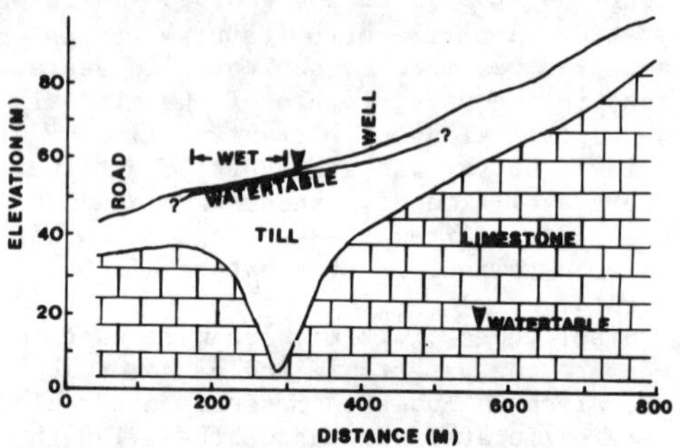

Figure 2. Hydrogeology of the wet site. The farm well is offset from the section but is shown to illustrate the elevation relationships.

The additional measurements on the farm borehole, the geophysical survey and the watertable and piezometric elevation survey provided additional information on the geohydrology and confirmed the findings from the test pits. However, farm boreholes are not often available at wet sites or are not useful but where accessible they should be dipped to find the depth to watertable. The geophysical and watertable and piezometric surveys require a high level of expertise and are time consuming and expensive. They can only be justified for scientific investigations of typesites. The additional measurements carried out on this site were unnecessary for design purposes and would not be cost effective in everyday practice.

The measurement of hydraulic conductivity can be carried out by the inversed auger hole or infiltrometer methods when watertables are low if investigations have to be carried out in summer (van Hoorn 1979) or by the pit bailing method of Bouwer and Rice (1983) when watertables are high. Van Hoorn (1979) in a comparison of the inversed auger hole and infiltrometer methods with the auger hole below a watertable and drain discharge methods states that the former two agree well with the latter two indicating that the former are satisfactory.

A value for hydraulic conductivity is necessary to compute the quantity of seepage. In this site the gradient of the watertable was similar to the ground surface gradient. It appears to be justified to make this assumption in many cases, after which the seepage can be estimated.

The rainfall and irrigation of soiled water are the major recharges. Using a design recharge value of 14.5 mm/d (12 mm/d rainfall + 2.3 mm/d soiled water + 0.2 mm/d seepage) and drain depths of 1.2 and 1.5 m the drain spacings are 6 and 8 m respectively for a design crest of watertable 0.5 m below ground level using the Toksöz and Kirkham (1961) drain spacing nomogram, justifiable since the slope on the wet site is only 6%. With the 1st drain on the boundary with the dry ground, 12 drains 1.5 m deep and spaced at 8 m across the slope are required.

In irrigating soiled water from dairy and stock yards, it is important that direct access into field drains be prevented. Soiled water is preferably spray irrigated on the soil surface. Areas within the catchments of farm wells and areas with thin soils or with very sandy or gravelly soils should not be spray irrigated to minimise the risk of groundwater pollution. Most of the soiled water (1.8 mm/d) in this study arose from uncovered yards and roofing of these areas will reduce this to a minimum.

REFERENCES

Bouwer, H. & R.C. Rice 1983. The pit bailing method for hydraulic conductivity measurement of isotropic or anisotropic soil. Trans. ASAE 26 (5): 1435-1439.

Concaret, J. 1981. Drainage agricole. Bourgogne: Chambre Regionale d'Agriculture de.

Castle, D.A., J. McCunnall & I.M. Tring (eds.) 1984. Field drainage, p.30-41. London: Batsford Academic & Educational.

Daly, E.P. 1985. Geological survey of Ireland, pers. comm.

de Ridder, N.A. 1972. Hydrogeology of different types of plains. In J. Kessler et al. (eds.), Drainage principles and applications 1, p.1-31. Wageningen: Inter. Inst. Land Reclam. & Improvement.

de Ridder, N.A. et al. (eds.) 1974. Drainage principles and applications 111. Wageningen: Inter. Inst. Land Reclam. & Improvement.

de Ridder, N.A. 1974. Groundwater survey. In N.A. de Ridder et al. (eds.), Drainage principles and applications 111, p.187-190. Wageningen: Inter. Inst. Land Reclam. & Improvement.

Eggelsmann, R. 1978. Subsurface drainage instructions. Hamburg and Berlin: Paul Parey.

Gardiner, M.J. & T. Radford 1980. Soil associations of Ireland and their land use potential. Soil Survey Bull. 36. Dublin: An Foras Taluntais.

Kruseman, G.P. & N.A. de Ridder 1970. Analysis and evaluation of pumping test data. Wageningen: Inter. Inst. Land Reclam. & Improvement.

Luthin, J.N. & Don Kirkham 1949. A piezometer method for measuring permeability of soil in situ below a watertable. Soil Science 68: 349-358.

Stakman, W.P. 1974. Measuring soil moisture. In N.A. de Ridder et al. (eds.), Drainage principles and applications 111, p.221-239. Wageningen: Inter. Inst. Land Reclam. & Improvement.

Toksöz, S & D. Kirkham 1961. Graphical solution and interpretation of a new drain spacing formula. J. Geophys. Res. 66 (2): 509-516.

Volker, A. et al. (eds.) 1973. Fieldbook for land and water management experts. Wageningen: Inter. Inst. Land Reclam. & Improvement.

van Hoorn, J.W. 1979. Determining hydraulic conductivity with the inversed auger hole and infiltrometer methods. In J. Wesseling (ed.), Proc. inter. drainage workshop, publication 25, p.150-154. Wageningen: Inter. Inst. Land Reclam. & Improvement.

White, P.A. 1985. Handbook on hydrolgeological applications of earth resistivity measurements. Wellington: New Zealand Water & Soil Conservation Authority.

Land and Water Use, Dodd & Grace (eds), © 1989 Balkema, Rotterdam. ISBN 90 6191 980 0

Design of a vertical drainage system in a shallow aquifer with an underlying semiconfined aquifer

S. van der Schaaf
Agricultural University, Department of Land and Water Use, Wageningen, Netherlands

ABSTRACT: The paper describes the design of a system with a row of vertical drains and an auxiliary open drain in a shallow phreatic aquifer on top of a very permeable semiconfined aquifer. The design is based on field work (drillings and pumping tests) and ground water flow modeling with applicition of a spreadsheet programme. A formula is derived to calculate the distance between the vertical drains and the dept of the collector drain.

RÉSUMÉ: La contribution décrit la mise au point d'un système à drainage verticale avec un drain auxiliaire dans une aquifère sableuse et peu profonde à perméabilité modéré sur une aquifère à haute perméabilité, séparées par une couche d'argile mince. La mise au point se base sur des examens hydrogéologiques (des sondages et des essais de pompage) et des études à modèles numériqes qui utilisent un programme "spreadsheet". Une formule pour le calcul de l'intervalle des puits et la profondeur du tuyeau collecteur est dérivée.

ZUSAMMENFASSUNG: Es wird der Entwurf eines Dränsystems mit vertikalen Filtern und einem Hilfsdrän in einem untiefen mäßig durchlässigen Sandschicht mit unterliegendem sehr durchlässigen Schicht beschrieben. Die beiden Schichten sind von einem dünnen Tonschicht getrennt. Der Entwurf beruht auf Felduntersuchungen (Bohrungen und Pumpversuche) und Grundwassermodellierung mit Verwendung eines Spreadsheetprogramms. Zur Bestimmung des Abstandes zwischen den vertikalen Brunnen und der Tiefe des Verbindungsrohres wurde eine Formel entwickelt.

1. INTRODUCTION

A rectangular area of 120 by 500 metres near the town of Lunteren, The Netherlands, is to be developed into an industrial estate with small, partly agriculture related industries. It is situated as shown in fig. 1.

Soil and geohydrological conditions in the area are relatively unfavourable. Groundwater levels generally are less than 0.40 m below the surface in winter and 0.50-0.80 m in summer. The soil consists of fine sand with a humous topsoil of 0.20 - 0.30 m. Land use is predominantly grassland. The productivity of the grassland partly depends on capillary rise of groundwater during the summer season.

To avoid instability of paved surfaces and moisture problems in buildings, an increased distance between surface and groundwater level is necessary. There are two options: either raise the surface or lower the groundwater table.

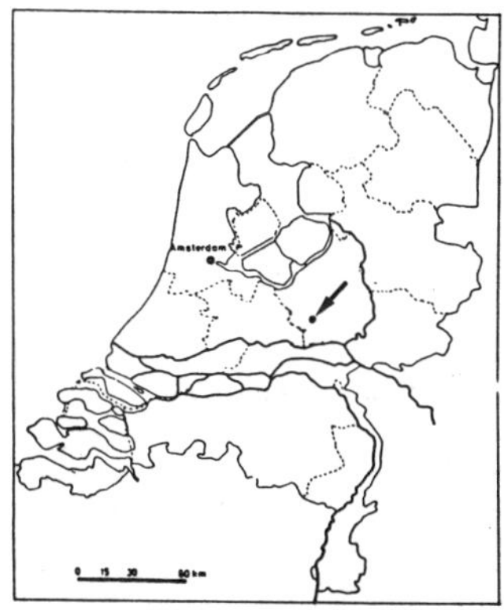

Fig 1. Situation of the area (arrow and dot)

The first option has the advantage that it hardly affects groundwater levels in the surrounding area. However, it was discarded for financial reasons because no suitable sand is available within a reasonably short distance. The second option was therefore worked out. There were, however, some problems, requirements and limitations to be dealt with. They were:

1. The presence of an aquifer with high transmissivity below the upper (phreatic) aquifer. The presence of a separating layer was uncertain and no detailed information was available as to the transmissivity and the composition of the upper part of the aquifer.

2. Groundwater levels in the surrounding grassland should be kept at a minimum.

3. No drains should be placed under buildings, which were planned on both sides of a road at the central long axis of the area (fig. 2).

4. The minimum allowed depth to the groundwater table in the drained situation was 0.50 m. The maximum drain depth allowed was 2 m. Since in the undrained situation the groundwater may reach the surface during wet periods in the winter, the surface level is reference level for the computation of the drawdown caused by drainage. Therefore, the drawdown should be at least 0.50 m at all places under design conditions.

2. HYDROGEOLOGY

2.1 Available information

The area lies in a filled-up glacier tongue basin that was formed in the Saalian. The bottom of the basin can be expected at a depth of 30-35 m below the surface. Below this level predominantly coarse pre-Saalian fluvial deposits occur. They are about 100 m thick and form an aquifer with a transmissivity of about 4000 m^2/day (Meinardi, 1978). The lower 10-15 m of the fill consists of fine to medium coarse fluvioglacial sediments. The upper part comprises relatively fine sands of different origin with intercalating peat and local clay layers. The transmissivity of the fill is generally less than 250 m^2/day and most of it is concentrated in the lower 10 m.

In the southern part of the basin, the fill and the underlying aquifer are separated by a clay layer of several metres thickness, probably of lacustroglacial origin (Verbraeck, 1984). It has a vertical resistance of up to at least 3000 days (van der Hoek & van der Schaaf, 1988). However, further north the layer is reported to be hydrologically unimportant and locally non-existing (Meinardi, 1978). A possible absence of the clay layer at the location of the planned industrial estate could have serious consequences for the design of the drainage system.

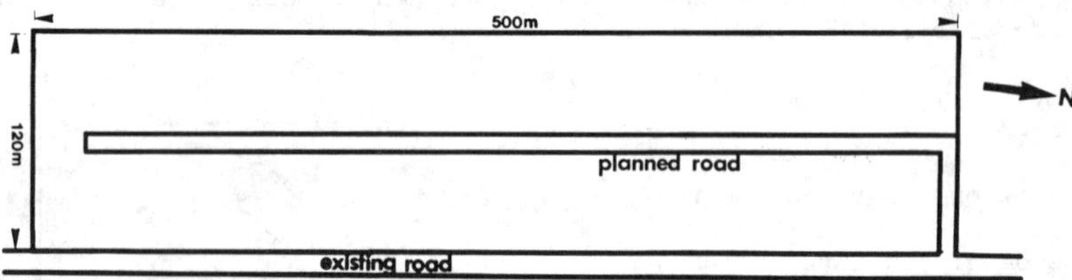

Fig. 2. Schematic outline of the area.

The only place for a drainage system was near or under the planned road. A system at or near two long sides was likely to have an unwanted effect on groundwater levels in the surrounding agricultural land.

The above considerations and the lack of quantitative information on the local hydrogeology made it necessary to investigate the hydrogeological conditions before designing a drainage system.

2.2 Field investigations

The field work included drillings and two pumping tests.

Two drillings were made near the opposite short sides of the field, about 450 m apart. The first one showed a layer of sandy loam between 33.50 and 34.50 m below the surface. In the second drilling no such layer was found, even though it was continued to a depth of 50 m. It could be concluded that there was no confining clay layer.

However, in both drillings, two layers of 0.50-1.50 m of a silty loam occurred. The deeper loam layers were found at different depths (17 and 23 m), so it was unlikely that they belonged to the same layer.
However, the upper layers began at depths of 10-11 m in both drillings and were accompanied by underlying peat in both cases. Both occurrences might therefore well belong to the same - hopefully continuous - layer. In available descriptions of other drillings made within a distance of a few kms, similar layers at similar depths were reported. This meant that the local situation possibly could be described as a phreatic aquifer of 11 m thick on top of a semiconfined aquifer of some 120 m thick, the semiconfining layer being the silty loam and peat at a depth of 11 m.

To verify this, two pumping tests were done; one with a pumping screen in the upper 11 m and another with a screen between the loam layers at 11 and 17 m depth. Apart from the pumping well, 5 observation wells were made. In one of the two exploration drillings observation well screens were installed also. All drillings made for the pumping test confirmed the existence of the silty loam with underlying peat at a depth of 11 metres. The layer at 17 metres was insignificant or not found at all in 3 drillings.

The data from the drillings also showed the occurrence of a layer of fine slightly loamy sand and some peat with an average thickness of about 1 m, beginning at a depth of about 5 m. Because of its granular composition, it is unlikely to act as a semiconfining layer.

The pumping tests were carried out by the Municipal Office of Public Works of Ede, which Lunteren is part of. The analysis was done by the author.

Analysis of the pumping test data showed that the vertical resistance at 17 m depth was insignificant. For the layer at 11 m depth a value of about 50 days was found. Because the observations showed delayed yield phenomena during the first 9 hours of pumping (the analysis was based on observed values after 45 hours of pumping) and the drawdown near the pumping well was not small compared to the thicknes of the phreatic aquifer, it is likely that the vertical resistance found is below the the real value. The value of 50 days therefore is a rather safe one.

For the transmissivity of the phreatic aquifer at 11 m a value of 26 m^2/day was found. The tests and their analysis have earlier been described in more detail by the author (van der Schaaf, 1988).

3. MODEL STUDY

Because of the relatively complex geohydrological situation, a simple flow model was made using a spreadsheet. The technique has been described by Olsthoorn (1985). It uses an explicit finite difference scheme. Two situations were simulated:

1. A horizontal drain along the long axis of the field
2. A single row of vertical drains to a depth of 10 m, horizontally connected by a collector drain, also along the long axis of the field.

In both cases a vertical cross section starting at and perpendicular to the drain system at the lang axis was simulated. The model had 13 vertical by 18 horizontal nodes. Vertical node distance was 1 m, the horizontal distance was 5 m up to a distance of 65 m from the drain. The remaining nodes were at 75, 90, 120 and 200 m. The model assumptions as indicated by the results of the pumping test and boundary conditions were:

- Horizontal permeability: 2.5 m/day from surface to 11 m dept, but 0.2 m/day between 4 and 5 m
- Vertical permeability: 1/5 of the horizontal permeability in order to account for horizontal layering, which is normal in the geological formation (Twente Formation), the upper 11 m is part of
- A vertical resistance of 50 days at a depth of 11 m
- A precipitation of 10 mm/day. Because precipitation on buildings and paved surfaces may be expected not to reach the groundwater, the precipitation on the area itself was reduced by 50%. Outside the area no reduction was applied.
- A constant head below the semiconfining layer at 11 m depth. Because observed heads above and below the layer were not significantly different, this head can be assumed to be equal to surface level, since under design conditions the groundwater table probably reaches the surface.
- At the drain position a discharge boundary condition was applied instead of a potential, because this would allow the drain potential and thus the minimum drain depth to be part of the solution.
- Groundwater levels could not rise above the surface.

The model for situation 1 showed that the drawdown of the phreatic surface at the position of the central horizontal drain is 1.80 m with a discharge of 1.8 m^3/day per metre of drain length (15 mm/day for the area). The groundwater table reaches the surface at 55 m from the central drain. The result is shown in fig. 3.

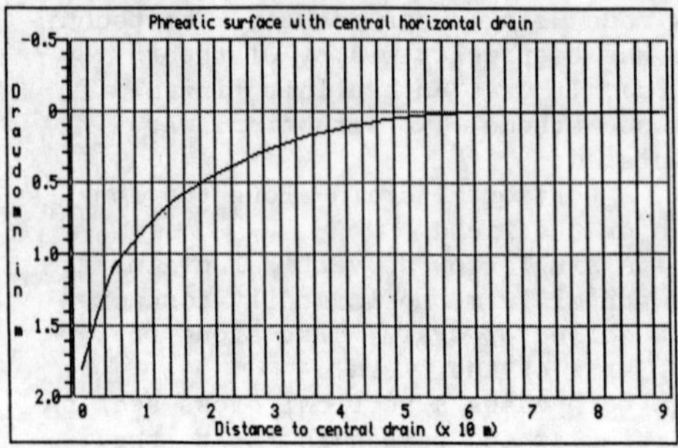

Fig 3. Phreatic surface with central horizontal drain and a discharge of 1.8 m^3/day per metre of horizontal drain

The model for situation 2 showed a slightly more favourable - though still unsatisfactory - result: a drawdown of slightly less than 1.30 m at the drains, but the groundwater table reaches the surface at about 65 m from the central drain. The discharge to obtain this result was higher: 2.4 m^3/day per metre of collector drain, equal to 20 mm/day for the area. The result is shown in fig. 4.

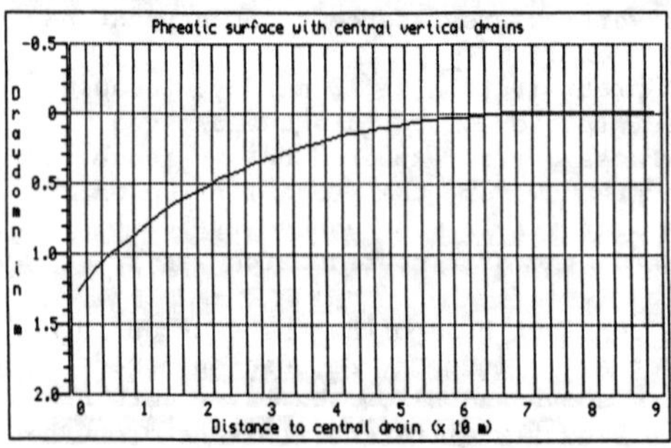

Fig. 4. Phreatic surface with central vertical drain system and a discharge of 2.4 m^3/day per metre of collector drain.

The phreatic surface in fig 4 is less steep than the one in fig. 3. The main reason why the difference with fig. 3 is not larger is seepage through the semi--confining clay layer in the vertical drainage case (the leakage factor is only about 35 m).

The model results show that the drainage requirements are not met in either case.

Application of additional drainage is inevitable. The only place for it are the two long sides of the field. A second deep system cannot be applied, because it would have an unacceptable effect on the groundwater levels in the surrounding agricultural land. Shallow ditches with a water level of approximately 1 m below the surface would, however, be acceptable because they would not considerably influence summer groundwater levels, which normally are already about 0.75 m below the surface.

Both models were therefore rerun with the same discharges, but with a ditch level of 1 m below the surface at 60 m from the central drain system. The result of the model with the horizontal deep drain is shown in fig. 5 and that of the vertical drainage model in fig. 6.

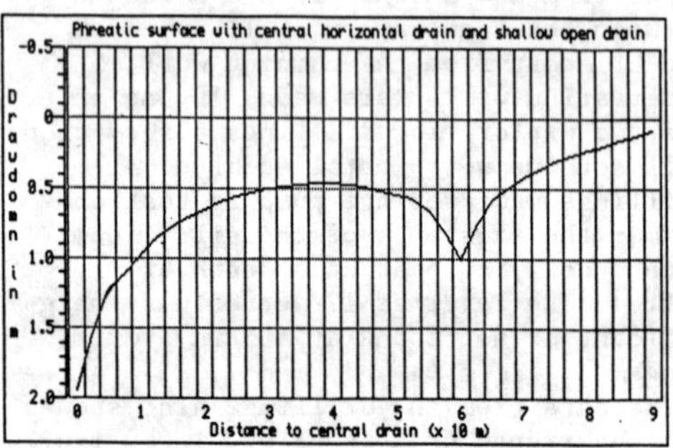

Fig. 5. Phreatic surface with central horizontal drain with a discharge of 1.8 m^3/day per metre of drain and a ditch level of -1 m at a distance of 60 m.

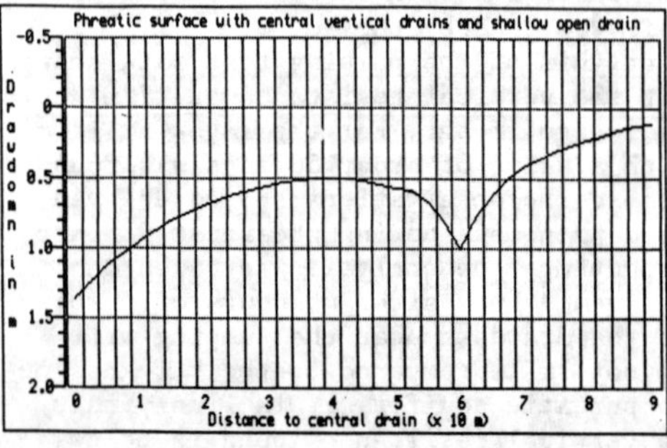

Fig. 6. Phreatic surface with central vertical drain system and a discharge of 2.4 m^3/day per metre of collector drain and a ditch level of -1 m at a distance of 60 m.

The result of the system with horizontal drains dit not meet the requirements. Even with a water table at almost the allowed maximum depth of 2 m at the drains, the highest point of the water table did not completely reach the required depth of 0.50 m.

The vertical drain system did, though rather narrowly. The phreatic level at the drains is only -1.35 m, but in 4, it will be shown that nevertheless the collector drain should be kept at a considerably larger depth.

The vertical system was eventually selected, although the choice was partly based on maintenance considerations: in a horizontal drain, the functions of transport and drainage are integrated in one and the same pipe; in a vertical system with a horizontal collector drain they are separated. If clogging occurs in a horizontal system, the drainage of a considerable area may be impeded; in a system of vertical drains usually only a filter screen does not function properly anymore but the transport function remains unaffected.

The model does not give an answer to the question as to the depth of the collector drain and the distance between the vertical wells. Solving the problem using another model was possible, but it could be done more elegantly by applying the technique of conformal mapping, which would give a generally applicable formula.

4. THE DISTANCE BETWEEN VERTICAL WELLS AND THE DEPTH OF THE COLLECTOR DRAIN

The result of the model includes a head in the plane of the vertical drains. This potential is an average. The maximum occurs halfway between two vertical drains and the minimum in the drains. The difference depends a.o. on the distance between adjacent drains.

The collector drain must be installed at such a level that water from the vertical drains can flow into it. This level should be at or preferably below the head in the vertical drains, which is lower than the head found in the model. It is therefore necessary to find the head in the drains.

Assuming that the value obtained from the model is approximately equal to the arithmetic mean of the maximum and the minimum head, it is sufficient to find the difference between these two.

Fig. 7a shows a horizontal cross section of the flow pattern with some streamlines and equipotential lines near one well. L is the distance between adjacent wells, r is the well radius. The X-direction is the direction of the plane of the vertical drains. The pattern applies to horizontal flow.

In the situation as described the flow contains vertical components from both precipitation and seepage. However, at short distances from the well, the vertical component is usually small compared to the horizontal flow rate.

The pattern of fig. 7a is a half-strip of width in the complex z-plane ($z=x+iy$) with $z=0$ in the centre of the vertical drain. It can be transformed into a half-plane with radial flow in the w-plane by the transform

$$w=\sin(\pi z/L) \qquad (1)$$

(Kober, 1957)

Substitution of $z=x+iy$ in (1) yields:

$$w=\sin(\pi x/L)\cosh(\pi y/L) + i\cos(\pi x/L)\sinh(\pi y/L)$$

or

$$w=u+iv \qquad (2)$$

which is the representation of the flow system in the w-plane.

Some important elements in the z-plane are represented in the w-plane as follows:

The curved stream lines in the z-plane have become straight radial lines in the w-plane; the equipotential lines in the z-plane are represented by concentric half circles in the w-plane; the one from the point $(-L/2,0)$ to $(L/2,0)$ (the points half-way between the drains) in the z-plane is represented in the w-plane by a half circle from $(-1,0)$ to $(1,0)$; the equipotential line through the point $(0,L/2)$ in the z-plane becomes a half circle from $[-\cosh(\pi/2),0]$ to $[\cosh(\pi/2),0]$.

The intersection point $(r,0)$ of the drain perimeter and x-axis in the z-plane is represented in the w-plane by the point $[\sin(\pi r/L)\cosh(\pi r/L),0)$ and $(-r,0)$ by $[-\sin(\pi r/L)\cosh(\pi r/L,0]$. The point $(0,r)$ in the z-plane is represented by $[0,\cos(\pi x/L)\sin(\pi x/L)]$, which shows that the representation of the well in the w-plane is not circular. However, the representation is almost circular in practical situations where $\pi r/L<0.1$, because for small argument values the value of sin is almost equal to that of sinh and cos is almost equal to cosh.

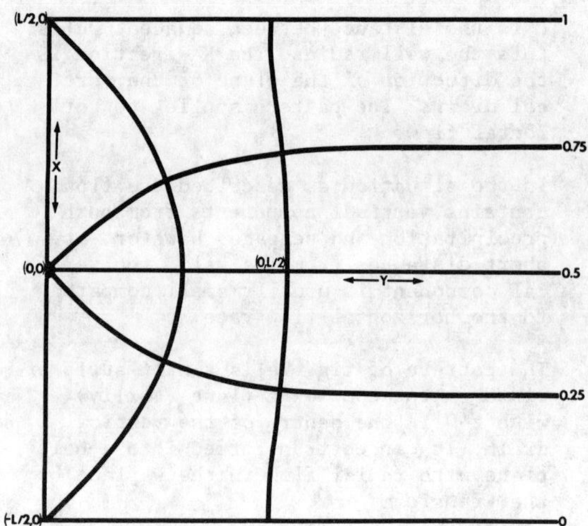

Fig. 7a. Flow pattern near one vertical drain

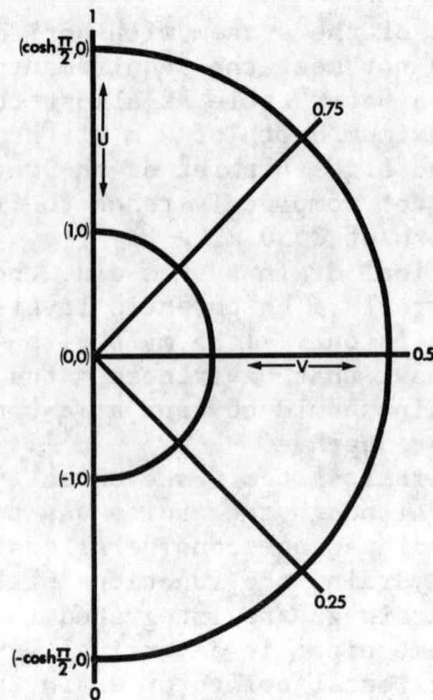

Fig 7b. Representation of the flow pattern of fig. 7a in the w-plane

Now let r_w be the "radius" of the transformed well. An expression for the flow rate Q is

$$Q = \frac{\pi T(H-h)}{\ln(R/r_w)} \tag{3}$$

where T is transmissivity, h is the head in the drain and H the head at a distance R. If R=1, (R,0) is the representation in the w-plane of the point (L/2;0) in the z-plane and H is the head at (L/2,0). Then H-h is the difference in head between the drain and the point halfway between the drains. Substitution of $r_w=\sin(\pi r/L)\cosh(\pi r/L)$ in (3) yields:

$$H-h = \frac{-Q[\ln \sin(\pi r/L)+\ln \cosh(\pi r/L)]}{\pi T} \tag{4}$$

For $\pi r/L<0.1$, the right hand term in the numerator approaches 0 and may be neglected; the left hand term reduces to $\ln(\pi r/L)$. This yields

$$H-h = \frac{Q}{\pi T} \ln(L/\pi r) \tag{5}$$

If the contribution to Q of the part enclosed by the equipotential line through the points (L/2,0) and (-L/2,0) is sufficiently small compared to the flow rate at the drain perimeter, the latter may be considered equal to Q in (5).

Fig. 7b shows the transformed pattern of fig. 7a.

In the flow model, the flow rate was 2.4L m^3/day and T was 25 m^2/day. Because the average phreatic surface in the plane of the drains was 1.35 m and the maximum allowed depth of the collector drain 2 m below the surface, the maximum allowed value of H-h would be about 2x(2-1.35m)=1.3m. If r=0.1 m and L=10 m, H-h will be about 1.05 m and the collector drain should be installed at a depth of about 1.90 m, which is only 0.1 m less than the limit of 2 m.

REFERENCES

Hoek, D. van der & S. van der Schaaf 1988. The influence of water level management and ground water quality on vegetation development in a small nature reserve in the southern Gelderse Vallei (The Netherlands). Agric. Water Management 14:423-437.

Kober, H. 1957. Dictionary of conformal representations. New York: Dover Publications Inc.

Meinardi, C. 1978. Geohydrologische gegevens van Zuidelijk Flevoland en de Gelderse Vallei, RID-meded. 78-4. Leidschendam.

Olsthoorn, T.N. 1985. The power of the electronic worksheet. Modelling without special programs. Ground Water 23(3):381-390.

Schaaf, S. van der 1988. Onderzoek naar de ontwateringsmogelijkheden van het bedrijventerrein "De Stroet" bij Lunteren. Uitgevoerd ten behoeve van de gemeente Ede. Int. Rapp. 38, vakgroep Cultuurtechniek. Wageningen: Landbouwuniversiteit.

Verbraeck, A. 1984. Toelichtingen bij de Geologische Kaart van Nederland 1:50000. Blad Tiel West (39W) en Tiel Oost (390). Haarlem: Rijks Geologische Dienst.

Land and Water Use, Dodd & Grace (eds), © 1989 Balkema, Rotterdam. ISBN 90 6191 980 0

Simulation of the reservoir influence on the groundwater levels in surrounding areas

E. Wiencław & T. Brandyk
Warsaw Agricultural University, Warsaw, Poland

ABSTRACT: Mathematical model which simulates waterlogged conditions arise based on the well-known Boussinesq equation describing groundwater movement is presented in this paper. In the presented model the steady state groundwater movement equation with determined parameters of the flow region and determined boundary conditions with two additional elements is solved. The first of these two elements is internal non-linear condition in the form of $H \leqslant H_T - Z_N$, where H_T represents the soil surface and Z_N is the groundwater table position below the soil surface required for agricultural crops. The second elements is the solution of the equation utilising under-relaxation and the assumption of the initial condition in the form of $H_o < H$. Presented model can be very useful in the design of drainage systems for the groundwater level regulation according to the crops water requirements in the areas surrounding retentional water reservoirs. The model was practically applied for simulation of the influence of the constructed water reservoir Jeziorsko (located on Warta river in Poland) in the surrounding areas. The results of the computations for five different values of $H_T - Z_N$ are shown in this paper.

RESUME: Pour chercher une solution de l'équation permanente avec des paramètres déterminés de zône et des conditions frontières, ainsi qu'avec les élements supplémentaires, on présente un modèle mathématique base sur l'équation de l'écoulement de l'eau souterraine. $H \leq H_T - Z_N$ ou H_T est la surface du terrain et Z_N est la profondeur de l'eau souterraine le plus économique pour une végétation. le deuxième élement est une solution de l'équation basée sur la sous-relaxation de l'état initial H H. Le modèle proposé a été utilisé pour la simulation d'éffect du réservoir JEZIORSKO sur la rivière WARTA et les terrains avoisinants. Pour l'optimalisation de l'épaisseur de la zone d'aération on aréalise des calculs pour 5 surfaces différentes H_T- Z_N.

ZUSAMMENFASSUNG: Die hydrologischen Bedingungen in Fluss-Tälern auf Ebenen haben beträchtlich grösseren Einfluss auf die Groundwasservernässung wie in anderen Geländelagen. Ausser natürlichen Gründen, sehr oft auf Ebenen, die Stauwerke an Flüssen sind die Ursache Grundwasservernässung. Die Notwendigkeit der Beseitigung dieser Ursachen beugt zur eine Simulation - Methoden - Analise. Es wurde ein mathematischer Modell zur Simulation der Entstehung solcher Staunässe beschrieben, welcher auf den bekanten Modell der Strömung des Grundwassers nach der Formel von Boussinesq´ beruht. In diesem Modell wird die stationäre für bestimmte Parametern der Filtrations-Fläche und bestimmter Grenzbedingungen mit zwei zusätzlichen Elementen augelöst. Erster dieser Elemente ist nichtlinearer Innenbedingung $H \leqslant H_T - Z_N$, wo: H_T - ist die Gelendefläche, Z_N - optimale für gegebene Pflanzenart Grundwassertiefe. Zweiter zusätzliches Elemen ist die Methode der Auflösung der Gleichung durch Einsatz der Unterrelaxation und die Einnahme als Ausgangsstand $H_o < H$. Das Modell kann als Basis für Projektierung und Regulierung des Wasserstände für landwirtschaftliche Zwecke ausgewertet werden. Er wurde zu solchen Zwecken für die Modellierung des Einflusses des Wasserbeckens auf dem Fluss Warta bei Jeziorsko Auf die Umgebung eingesätzt. In diesem Fall die Berechnung wurde für finf angenommenen Flächen $H_T - Z_N$ durchgeführt. Dadurch hate man die Möglichkeit Wariante-Optimaliezirung der Aeration-Zone erreicht.

1 INTRODUCTION

The important aim of the constructed system of the retentional water reservoirs on rivers is to increase the available water resources for agriculture. However, a continuous backwater in the river leads usually to the negative changes in soil and water conditions on adjoining areas due to the excessive soil moisture content. This phenomenon is particularly harmful in the case of the water reservoirs located in the lowland river valleys (Poźniak and Wiencław, 1981).

The waterlogging conditions occurring in connection with the construction of the retentional water reservoirs are more troublesome because of their permanent character. If the natural environment is tolerant to the periodical waterlogging conditions caused by the such natural factors, as high water levels in rivers, rainfalls or snowmelts, so the area with the permanent waterlogging conditions is not suitable for agricultural purposes. Ecologic changes occurring in the permanently waterlogged regions lead to the destruction of the plant cover on a large scale and its improvement requires the high costs (Nazaruk and Pawłat 1981).

The best solution would be such a designing of the water reservoirs,which would prevent waterlogging conditions. This is not possible to perform in practice. To prevent the excessive soil moisture content in the areas surrounding retentional reservoirs, the suitable drainage system is designed. This requires the performance of the geo-hydrologic forecast concerning the influence of the water reservoir in the adjoining areas. The areas being under the threat of waterlogging conditions and the amounts of subterranean waters to be removed from these areas for maintenance of their suitability for the agricultural production should be determined in the forecast. Such information is possible to obtain using the numerical model simulating formation of the areas with the waterlogging conditions. This model was described in details by Michalak and Wienclaw (1985).

2 ASSUMPTIONS OF THE NUMERICAL MODEL

The model describes two-dimensional steady-state flow of the groundwater in the x-y plane and is based on the Boussinesq's equation in the following from:

$$\frac{\partial}{\partial x}\left(T_x \frac{\partial H}{\partial x}\right) + \frac{\partial}{\partial y}\left(T_y \frac{\partial H}{\partial y}\right) + q_z = 0 \qquad (1)$$

with the possibility of use the following boundary conditions:

- the first kind $H = F_1(x,y)$ (2)
- the second kind $\frac{\partial H}{\partial n} = F_2(x,y)$ (3)
- the third kind $H + F_3(x,y)\frac{\partial H}{\partial n} = F_n(x,y)$ (4)

where:

H - hydraulic head [L]

T_x, T_y - transmissibility in the x and y direction, respectively $[L^2T^{-1}]$

q_z - sink term $[LT^{-1}]$

F_1, F_2, F_3, F_4 - functions describing boundary conditions [L,-,L,L]

x,y - coordinates [L]

n - directions orthogonal to the boundary [L]

The solution of flow problems by finite difference methods involves modeling the flow region as a grid of points, each separated by finite distances such as shown in Fig.1. The numerical approximation of equation(1) leads to the following finite difference expression which is valid for nodal points except the points at the boundaries:

$$T_{EC}\frac{H_C - H_N}{(\Delta x)^2} + T_{WC}\frac{H_C - H_N}{(\Delta x)^2} + T_{SC}\frac{H_C - H_S}{(\Delta y)^2} + T_{WC}\frac{H_C - H_W}{(\Delta y)^2} = q_{ZC} \qquad (5)$$

The set of the linear equations for determined parameters of the flow region and specified boundary conditions with two additional elements is solved in the model.

The first of them is an internal non-linear condition in the form of:

$$H \leqslant H_T - Z_N \qquad (6)$$

where:

H_T - soil surface position [L]

Z_N - assumed ground-water table position below the soil surface, which is required for the given plant type [L].

If $H \leqslant H_T - Z_N$ the flow in the given block would occur according to equation (5). When $H > H_T - Z_N$, then the value of H is substituted by $(H_T - Z_N)$. This causes in the given block, treated from this moment as fulfiling 1-st kind boundary condition, that the flow would not fulfil the continuity resulting from equation (5). For fulfilment of the flow continuity condition and the condition represented by equation (6) the amount of groundwater equal to the difference between the left and the right side of the equation (5) should be removed from the given block.

The second additional element is the solution of the set of the linear equations (5) by application of under-relaxation and assumption of the initial condition in the form of $H_o < H$, where H_o represents the initial value of the iteration process, which should be lower than the lowest value of

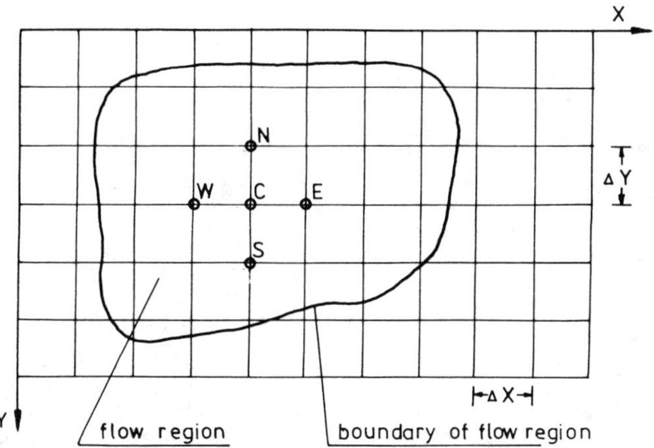

Fig.1 Rectangular mesh superimposed on flow region

the expected solution of H and in subsequent iterations should grow gradually except the blocks, in which the condition (6) becomes to be fulfiled.

As teh result of the solution the values of the hydraulic head for blocks belonging to the inside area are obtained. Some of the blocks, which fulfil the condition $H<H_T-Z_N$ representing the areas not requiring reduction of the groundwater level. The remaining blocks which fulfil the condition $H = H_T - Z_N$ representing the areas requiring drainage. The described algorithm was used for the development of the computer program described by Michalak (1980), who used it for the geo-hydrologic forecast concerning the influence of the presently constructed Jeziorsko water reservoir in surrounding areas.

3 THE MODEL APPLICATION EXAMPLE

To illustrate the practical capabilities of the presented numerical model, the results of the model investigations for one of the six reclamation complexes, adjoining the Jeziorsko water reservoir in the Teleszyna river valley are presented. The investigations were carried out for the following five different cases of the limitation of the groundwater tabel position: 1/ $H\leqslant H_T$, 2/ $H\leqslant H_T-0,5$ m, 3/ $H\leqslant H_T-0,8$ m, 4/ $H\leqslant H_T-1,0$ m, 5/ $H\leqslant H_T-1,2$ m. The assumed cases of calculations were aimed to show the possibility of the location of the areas to be drained and to show the effect of the valley drainage intensity on the groundwater table position.

Introduction of the $H\leqslant H_T$ condition (case 1) in the model causes that the shape and the position of the groundwater table in the valley are limited by the effect of both the water reservoir and the valley area. In order to ensure the occurence of the ground and water conditions in the analyzed area as presented in Fig.2, the groundwater volume of 5000 $[m^3d^{-1}]$ should be removed from the valley area, irrespective of the drainage by embankment ditch. This amount of groundwater is related to the total area of 240 ha requiring drainage, consisting of the nine smaller areas ranging from 4 to 150 ha.

The computation results obtained for the case 1 could be a basis for determining the location of the drained areas in the Teleszyna river valley, but only in the case when the drainage would prevent the appearance of groundwater above the soil surface. Assuming groundwater table position at certain depth below the soil surface, the amounts of the groundwater to be removed must be higher than those resulting from the computed in the case 1 and should be equal to 9000 $[m^3d^{-1}]$ in the case 2, 16000 $[m^3d^{-1}]$ in the case 3, 20500 $[m^3d^{-1}]$ in the case 4 or 23500 $[m^3d^{-1}]$ in the case 5.

4 CONCLUSIONS

Presented numerical model can be applied for the forecasting groundwater levels which are required for designing of the waterlogging threat, on which an intensive agricultural production is planned. Assimption of the different Z_N values gives a possibility of the optimization of the unsaturated zone thickness, the area of the drainage system operation and the amount of water removed by this system.

The model in its present form determines a steady state groundwater flow at the end of the waterlogging process without analysis of the phenomena occuring on the soil surface, such as water accumulation in the

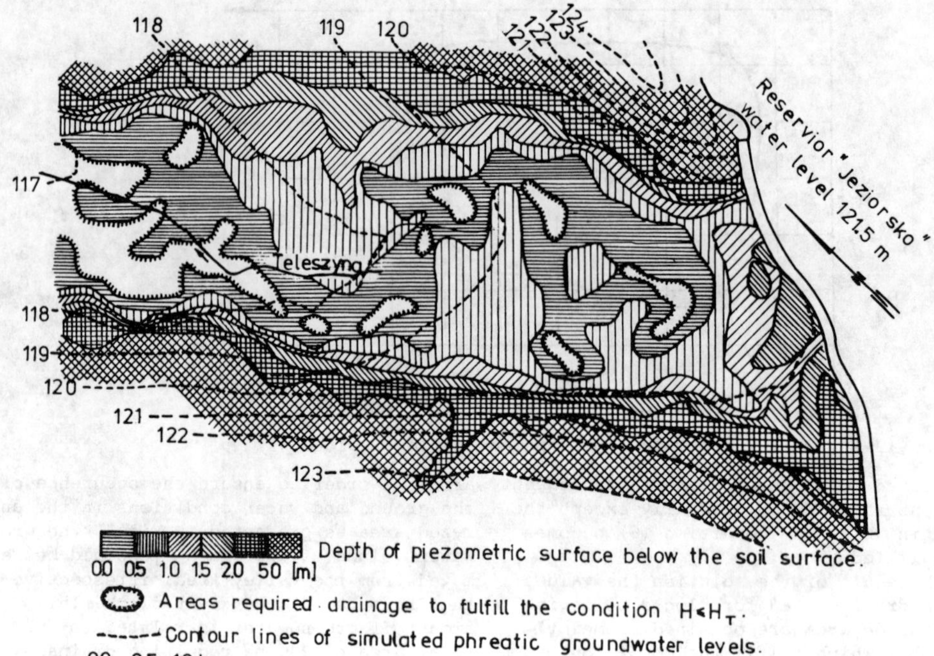

Fig.2 Prognostic geo-hydrological map

surface depressions or water flow on the soil surface from one block into another. Taking into account the above processes in the stationary model only one condition must be fulfiled, which is the change of the interpretation of the internal nonlinear condition ranging from $H \leq H_T - Z_N$ to $H \leq H_G$, where H_G is the position of the water accumulation at the given depression of the soil surface.

REFERENCES

Michalak, J. 1980. ANPLA - a package of the programs for mathematical models in geohydrology (in Polish). Institute of Geohydrology and Geology, University of Warsaw, Warsaw, Poland.

Michalak, J. and Wiencław, E. 1985. Modelling the area waterlogging process by subterranean waters on the areas of lowland river valleys (in Polish). Proceedings of the Symposium on Actual problems in Geo-hydrology . Cracow, Poland.

Nazaruk, M. and Pawłat, H. 1981. Forecasting the changes of the ecologic conditions in the areas surrounding water reservoirs (in Polish). Proceedings of the Conference organized by Polish Soil Science Society, Warsaw, Poland.

Poźniak, R. and Wiencław, E. 1981. Forecasting the changes of the groundwater table in the areas surrounding water reservoirs (in Polish). Proceedings of the Conference organized by Polish Soil Science Society, Warsaw, Poland.

Land and Water Use, Dodd & Grace (eds), © *1989 Balkema, Rotterdam. ISBN 90 6191 980 0*

A computer simulation of water heat salt movement in frozen soils*

D.Zeng & G.Lian
Beijing Agricultural Engineering University, Beijing, People's Republic of China

ABSTRACT: A 2-dimension mathematical model was developed to predict salt accumulation and moisture content of the soil seed-bed layer in Spring germination time before thorough thawing and possible irrigation so as to optimize the Autumn irrigation and drainage scheduling and tillage practices in Hetao Irrigation Scheme, Inner Mongolia, P.R. China. Calculation results of FEM solution with Galerkin approximation while compiled in computer program for typical locations from before freezing through complete thawing were compared to data obtained by field tests, and it showed a good agreement.

ZUSAMMENFASSUNG: Ein 2-dimension mathematisches model wird entwickelt. Die akkumulation des salz und der wassergehalt in der Bodenschicht des Bodensaatbett im voraus ausgerechnet wird, bevor Eis geschmolzen wird und Boden bewassert wird, Wahrend der Germination des Saat im Fruhling, damit die MaBnahme der Bewasserung, Entwasserung und des Ackerbau in der Region Hetao, Inen Mongolia, P.R. China optimiert wird. Fur den typischen Ort, von der gefrorenen bis gescholzenen Zeit. nach Galerkin Approximation Method mit der FEM Losung im Computer der Programm gerechnet wird. Durch die Vergleichung zwischen der Rechnungsergebnis und Daten des Test auf dem Feld wird gut Ubereinstimmung gezeigt.

RESUME: Un modèle mathématique à 2 dimension est développé pour prédire le contenu de l'humidité et l'accumulation du sel dans le lit des semences pendant le temps de gérmination en printemps; avant la congélation complète et une possibilité d'irrigation afin d'optimaliser l'etablissement du project HETAO. Les résultats des calculs de solution FEM avec approximation Galerkin par usage d'ordinateur por des zônes typique avant la congélation jusqu'à la complétion de décongélation sont comparés avec les résultats obtenus sur les éssais faits sur-champs. Ces deux résultats se sont montrés en très bon accord.

1. INTRODUCTION

In most arid and semi-arid areas of Northern and North Western China, the transport process of soil water-heat-salt in response to Winter freezing is of great interest for irrigated agriculture due to the fact that soil freezing has much effect on the soil water-heat-salt movement and hence the water stress and salt accumulation in the seed-bed layer of soils in early Spring sowing and germination period. Although a certain number of laboratory studies have been conducted on water-heat movement in frozen soils (Loch and Miller, 1975; Fuch, Campbell and Papendick, 1978), and some on salt and water flow in frozen soils (Cary and Mayland 1972; Cary, Papendick and Campbell, 1979), there are very few such studies concerned with field applications. The objectives of this paper are (i), to analyse the effects of freezing and thawing processes on the soil water-heat-salt movement and (ii), to investigate how much the soil freezing and thawing processes will effect the salt accumulation and water stress in the seed-bed layer of soils in the early Spring sowing and germination period through simulating the field soil water-heat-salt movement from the late Autumn to next early Spring, so as to lay a scientific ground for the optimization of irrigation and drainage scheduling and tillage practice. The time sequence of ground freezing and thawing processes at Hetao Irrigation Scheme is shown in figure 1.

* The Project supported by National Natural Science Foundation of China.

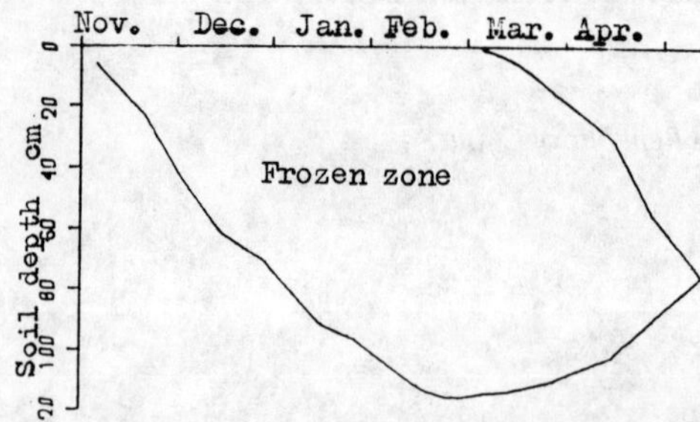

Fig. 1 Time sequence of ground freezing and thawing processes, Hetao Irrigation Scheme.

In handling the quantitative relations for water-heat-salt movement in frozen soils, the following assumptions are followed in this paper for clearness:

1) Discusions are restricted to the relatively wet soils which, in a steady state, involve primarily the liquid and ice phases while the vapour quantity neglected.

2) The movement of ice is negligible relative to the movement of liquid.

3) Only the diffusion and dispersion of salt in liquid phase are considered while that in ice phase neglected.

4) Influence of the ice lense formation and frost heaving processes are not considered.

2. TESTS TO ESTABLISH THE SOIL WATER FREEZING CHARACTERISTIC CURVES

Soil water freezing characteristic curves—liquid water content at different freezing temperature, for four typical soils from Hetao Irrigation Scheme, Inner Mongolia were tested with a devised freezing apparatus as shown in figure 2. The soil previously cooled was put into a compartment of the freezing cell which was kept at different stable freezing temperature by adjusting the concentration of the sodium-nitrate solution mixed with ice. In an elapse of 24 hours to ensure a stable and uniform temperature and phase transform throughout the soil sample, it was then transfered into the heat content measuring chamber and kept there until an equilibrium state established. With the equilibrium temperature and the initial temperatures of the water and the frozen soil sample known (measured by thermocouples), the amount of ice content in the sample was then calculated by the following equation in accord with the law of energy conservation:

$$W_i = \frac{(C_{cw}W_{cw}+K)(T_w-T_e) - (C_sW_s+C_lW_w)(T_e-T_f)}{L_{li} + (C_{lw}-C_i)T_f} \quad (1)$$

where C_{cw} is the specific heat capacity of the water in the heat content measurement chamber, W_{cw} is the weight of the water in the chamber, K is a constant of the chamber which represents its heat capacity per centigrade and can be calibrated by melting an ice sample in it, T_w is the initial temperature of the water, T_e is the equilibrium temperature, C_s is the specific heat capacity of dry soil, C_{lw} is the specific heat capacity of liquid water in the soil sample, W_w is the weight of tatal water in the soil sample (including ice and unfrozen water), T_f is the freezing temperature of the soil sample in the freezing cell, L_{li} is the latent heat of ice fusion and C_i is the specific heat capacity of ice.

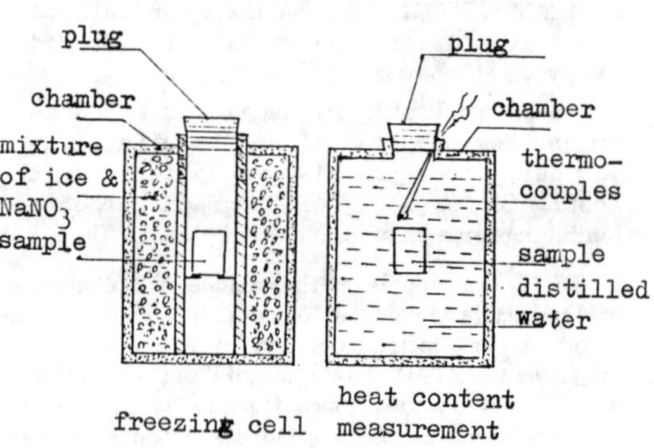

Fig. 2 Schematic representation of the apparatus for measuring soil water freezing characteristic curves.

As the total weight of water (including ice and unfrozen water) and the weight of ice in the soil sample are known, the ice content and the unfrozen water content of the soil can be calculated. Table 1 shows the measured data of the soil water freezing characteristic curves for four types of soil. It was found that for all four soils, freezing phenomenon occurs when the temperature drops below $-0.2^{\circ}C$. The measured data can be reasonablly regressed in the following equation,

$$\theta_l = \theta_o(-T)^{-b} \quad (2)$$

where θ_l represents unfrozen water content, T represents the freezing temperature while θ_o and b are regression coefficients respectivly. Figure 3 gives a comparison between the measured data (denoted by circles in the figure) and regressed curves (denoted by solid lines). The regression coefficients for the four types of soil tested are also given in table 1.

Table 1 Measured data and the regression coefficients for the soil water freezing characteristic curves for the four types of soil.

	T ℃	-0.2	-0.3	-0.8	-1.1	-1.8	-4.1	-10.
A	θ_l %	20.89	19.07	16.69	14.04	13.01	9.91	6.34
		θ_o = 14.15741			b = 0.29090			
	T ℃	-0.3	-0.3	-0.7	-1.0	-1.7	-6.1	-10.1
B	θ_l %	20.49	18.62	15.82	13.11	10.91	7.97	5.03
		θ_o = 12.82003			b = 0.33581			
	T ℃	-0.2	-0.3	-0.7	-1.1	-1.7	-6.0	-10.0
C	θ_l %	17.55	15.16	15.57	9.39	6.18	5.28	4.45
		θ_o = 10.00961			b = 0.37283			
	T ℃	-0.2	-0.3	-0.7	-1.1	-1.7	-4.0	-10.0
D	θ_l %	19.17	20.18	16.07	13.59	12.67	10.62	7.41
		θ_o = 14.12278			b = 0.24895			

A-Clay, B-Sandy loam, C-Sandy soil, D-Loam, the salt contents for the four types of soil are 0.19%, 0.15%, 0.14%, 0.17%.

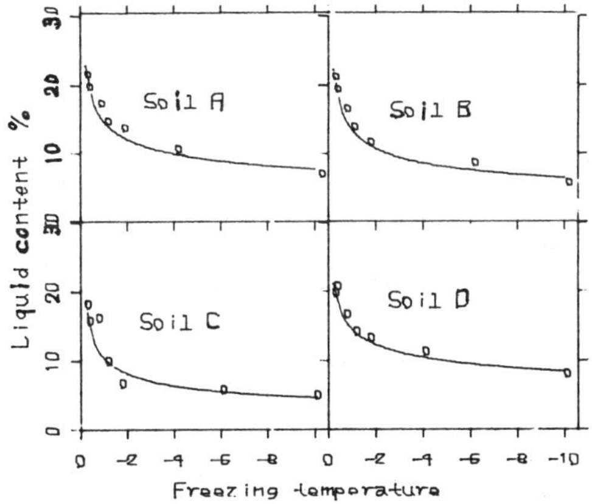

Fig 3. Measured data against regressed soil water freezing characteristic curves for the four types of soil.

3. THE GOVERNING EQUATIONS

The one-dimensional heat balance equation for frozen soil differs from that for the conventional unfrozen soil. In frozen soil, both latent and sensible heat come into play while in unfrozen soil, latent heat of fusion does not. Thus, for frozen soil, according to the law of energy conservation, latent heat of fusion should be incorporated into the conventional soil heat flow equation as,

$$C_h\frac{\partial T}{\partial t} - L_{li}\frac{\partial \theta_i}{\partial t} = \frac{\partial}{\partial z}\left(K_h\frac{\partial T}{\partial z}\right) - C_l\frac{\partial (q_l T)}{\partial z} \tag{3}$$

where C_h is the heat capacity of the soil, K_h is the heat conductivity of soil, T is the temperature, L_{li} is the latent heat of ice fusion, θ_i is the ice content of the frozen soil, C_l is the heat capacity of water, q_l is the conductive water flow, $q_l T$ refers to convective heat flow caused by liquid flow, t is the time, and z is the soil depth.

The conductive water flow in the frozen soil, which is assumed to follow also Darcy's law because the ice is supposed to be immobile and only liquid transport is to be expected, can be written as,

$$q_l = -D_w(\partial\theta/\partial z) \tag{4}$$

where θ_l is the liquid (unfrozen water) content of soil, and D_w is the hydraulic diffusion coefficient of soil which is defined as,

$$D_w = K_w/C_w \tag{5}$$

where K_w is the hydraulic conductivity of soil, and C_w is the specific water capacity of soil.

In a freezing soil, the water balance accounts for the liquid flow and liquid-ice transformation can be best described by incorporating the amount of water from phase transformation into the conventional soil water flow equation which observes the law of continuity, ie.,

$$\frac{\partial \theta_l}{\partial t} + \frac{\partial \theta_i}{\partial t} = \frac{\partial}{\partial z}\left(D_w\frac{\partial \theta_l}{\partial z}\right) + \frac{\partial K_w}{\partial z} \tag{6}$$

Salt movement equation in afrozen soil under the conditions suggested above should be written as,

$$\frac{\partial(\theta_l c)}{\partial t} = \frac{\partial}{\partial z}\left(D_{sh}\frac{\partial c}{\partial z}\right) - \frac{\partial(q_l c)}{\partial z} \tag{7}$$

where c is the salt concentration in soil solution, terms $\theta_l c$ and $q_l c$ refer to salt concentration associated repectivly with liquid content and conductive water flow, and D_{sh} is the salt diffusion-dispersion coefficient of the frozen soil.

Equations (3), (6) and (7i) are the proposed governing equations for water-heat-salt movement in frozen soils. Combining the three equations with the soil water freezing characteristic curves (equation 2), we can eliminate the variable θ_l from the four unknowns in the three equations. Evidently, the proposed mathematical model for water-heat-salt movement in frozen soils applys also to unfrozen soils because in that case, the three equations become the same forms as the conventional equations for water, heat and salt movement in unfrozen soils with the ice content being zero.

4. THE FINITE ELEMENT APPROACH

In field cases, where water-heat-salt movement may occur both in frozen and/or unfrozen soils with complicated boundary conditions, analytical solutions of the governing equations of the model are difficult and consequently resort to the Galerkin-type finite element method for approximation in this paper.

In the case of two-dimension, water-heat-salt movement in frozen and/or unfrozen soils can be describ-ed as following,

$$C_h\frac{\partial T}{\partial t} - L_{li}\frac{\partial \theta_i}{\partial t} = \frac{\partial}{\partial x}\left(K_{hx}\frac{\partial T}{\partial x}\right) + \frac{\partial}{\partial z}\left(K_{hz}\frac{\partial T}{\partial z}\right) - C_l\frac{\partial (q_{lx}T)}{\partial x} - C_l\frac{\partial (q_{lz}T)}{\partial z},$$

$$\frac{\partial \theta_l}{\partial t} + \frac{\partial \theta_i}{\partial t} = \frac{\partial}{\partial x}\left(D_w\frac{\partial \theta_l}{\partial x}\right) + \frac{\partial}{\partial z}\left(D_w\frac{\partial \theta_l}{\partial z}\right) + \frac{\partial K_w}{\partial z},$$

$$\frac{\partial (\theta_l c)}{\partial t} = \frac{\partial}{\partial x}\left(D_{sh}\frac{\partial c}{\partial x}\right) + \frac{\partial}{\partial z}\left(D_{sh}\frac{\partial c}{\partial z}\right) - \frac{\partial (q_{lx}c)}{\partial x} - \frac{\partial (q_{lz}c)}{\partial z},$$

$$\theta_l = \theta_o(-T)^{-b} \text{ (for frozen soil)}; \quad (7)$$

initial conditions,

$$T = T^o(x,z,0)$$
$$\theta_l = \theta_l^o(x,z,0) \quad t=0,\ x,z\in R \quad (8)$$
$$c = c^o(x,z,0)$$

first boundary conditions,

$$T = T^1(x,z,t)$$
$$\theta_l = \theta^1{}_l(x,z,t) \quad t>0,\ x,z\in\Gamma_1 \quad (9)$$
$$c = c^1(x,z,t)$$

second boundary conditions,

$$K_{hx}\frac{\partial T}{\partial x}n_x + K_{hz}\frac{\partial T}{\partial z}n_z - C_l q_l^o T + q_h^o = 0$$

$$D_w\frac{\partial \theta_l}{\partial x}n_x + D_w\frac{\partial \theta_l}{\partial z}n_z + K_w n_z + q_l^o = 0 \quad t>0,\ x,z\in\Gamma_2 \quad (10)$$

$$D_{sh}\frac{\partial c}{\partial x}n_x + D_{sh}\frac{\partial c}{\partial z}n_z + q_l^o c^o - q_l c = 0$$

where n_x, n_z are unit vectors, q_h^o, q_l^o are known flux of heat and water at the second boundary, c^o is the known salt concentration of the known water flux.

For an approximate solution to the problem at any given time, the entire solution domain was devided into a number of subdomains or elements, and by following a Galerkin-type finite element scheme, a set of local approximation analogue for the elements were obtained as,

$$T = \sum_{j=1}^{m} N_j(x,z)T_j^e(t) = \lfloor N\rfloor\{T\}^e$$
$$\theta_l = \sum_{j=1}^{m} N_j(x,z)\theta_{l,j}^e(t) = \lfloor N\rfloor\{\theta_l\}^e$$
$$\theta_i = \sum_{j=1}^{m} N_j(x,z)\theta_{i,j}^e(t) = \lfloor N\rfloor\{\theta_i\}^e \quad (11)$$
$$c = \sum_{j=1}^{m} N_j(x,z)c_j^e(t) = \lfloor N\rfloor\{c\}^e$$

where N_j are linearly independant weighting functions or shape functions, and T_j^e, $\theta_{l,j}^e$, $\theta_{i,j}^e$, c_j^e are unknown independant variables of temperature, liquid content, ice content and salt concentration repectively, at the element nodals.

When equations (11) are substituted into equations (7) and Galerkin's procedure further followed, we can derive a set of first order differential equations as,

$$[H_h]\frac{d\{T\}}{dt} - L_{li}[H]\frac{d\{\theta_i\}}{dt} + [P_h]\{T\} + \{F_h\} = 0$$

$$\frac{d\{\theta_l\}}{dt} + \frac{d\{\theta_i\}}{dt} + [P_w]\{\theta_l\} + \{F_w\} = 0 \quad (12)$$

$$[H_c]\frac{d\{c\}}{dt} + [P_c]\{c\} + \{F_c\} = 0$$

$$\theta_l = \theta_o(-T)^{-b} \quad \text{(for frozen soil)}$$

in which,

$$\{H_h\}=\sum_{e=1}^{ne}\iint_e C_h[N]^T[N]\,dxdz \qquad (13)$$

$$[H]=\sum_{e=1}^{ne}\iint_e [N]^T[N]\,dxdz \qquad (14)$$

$$[P_h]=\sum_{e=1}^{ne}\iint_e K_h\left(\frac{\partial[N]^T}{\partial x}\frac{\partial[N]}{\partial x}+\frac{\partial[N]^T}{\partial z}\frac{\partial[N]}{\partial z}\right)dxdz$$
$$+\sum_{e=1}^{ne}\iint_e C_l\left(\frac{\partial[N]^T}{\partial x}q_{1x}+\frac{\partial[N]^T}{\partial z}q_{1z}\right)[N]\,dxdz \qquad (15)$$

$$\{F_h\}=\sum_{e=1}^{ne}\int_{\Gamma_2^e} q_h^0[N]^T ds \qquad (16)$$

$$[P_w]=\sum_{e=1}^{ne}\iint_e K_w\left(\frac{\partial[N]^T}{\partial x}\frac{\partial[N]}{\partial x}+\frac{\partial[N]^T}{\partial z}\frac{\partial[N]}{\partial z}\right)dxdz \qquad (17)$$

$$\{F_w\}=\sum_{e=1}^{ne}\int_{\Gamma_2^e} q_1^0[N]^T ds+\iint_e K_w\frac{\partial[N]^T}{\partial t}dxdz \qquad (18)$$

$$[H_c]=\sum_{e=1}^{ne}\iint_e \theta_1[N]^T[N]\,dxdz \qquad (19)$$

$$[P_c]=\sum_{e=1}^{ne}\iint_e D_{sh}\left(\frac{\partial[N]^T}{\partial x}\frac{\partial[N]}{\partial x}+\frac{\partial[N]^T}{\partial z}\frac{\partial[N]}{\partial z}\right)dxdz$$
$$+\sum_{e=1}^{ne}\iint_e \frac{\partial\theta_1}{\partial t}[N]^T[N]\,dxdz$$
$$+\sum_{e=1}^{ne}\iint_e\left(\frac{\partial[N]^T}{\partial x}q_{1x}+\frac{\partial[N]^T}{\partial z}q_{1z}\right)[N]\,dxdz \qquad (20)$$

$$\{F_c\}=\sum_{e=1}^{ne}\int_{\Gamma_2^e} q_1^0 c^0[N]^T ds \qquad (21)$$

The set of first order differential equations (12), after discretizing the space and the time domains, were solved by the method of weighted residuals and a computering program in Fortran language for micro computer was compiled.

5. FIELD APPLICATION

The Galerkin-finite element models obtained and the computer program compiled were employed to determine the salt accumulation behaviour during freezing and thawing in a typical field of Hetao, Inner Mongolia. The initial conditions of soil water content and salt concentration in the field were measured on October 30th, 1986 and were shown in figure 4. Functional relationships between pressure head, hydraulic conductivity, apparent salt diffusion coefficient (the sum of the molecular diffusion and the hydrodynamic dispersion) and moisture content of the soil were measured in laboratory. Meteorological data were obtained from a weather station near by.

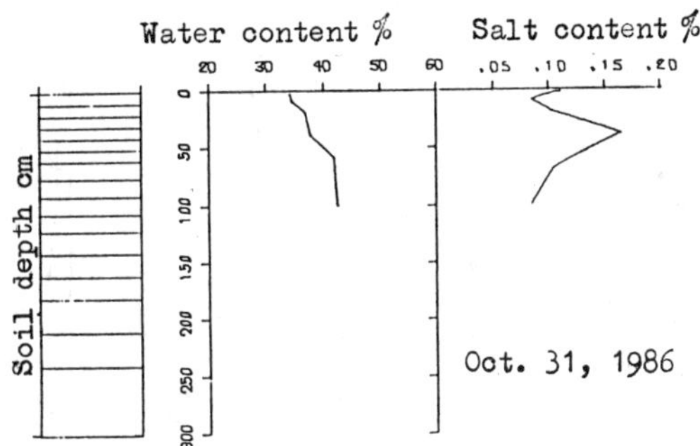

Fig 4 Initial soil water content and salt concentration tested in the field.

The one-dimensional model was used, as under flooded irrigation in that area, horizontal flow was taken as negligible. The numerical model was run from November 1st, 1986 to May 11th, 1987 when the first irrigation was available. Some of the results, indicating the soil water content and salt concentration profiles in the frozen layer obtained by computation, are presented in figure 5. It shows that, under the effects of soil freezing and thawing as well as evaporation, the average salt concentration has increased from 0.0966% to 0.2726% in the top 20-cm layer of soil (see table 2). In figure 5, the computed data (in solid line) are also compared with field measurements (in cross) carried out in the site and it shows a reasonablly good agreement.

Table 2 Computed average water content and salt concentration in the top 20-cm layer of soils at different times during freezing.

	Oct. 30th	Dec. 29th	Feb. 28th	Mar, 27th	May 11th
water content %	35.73	52.74	51.87	43.82	32.09
salt concentration %	.0966	.1014	.1543	.1877	.2726

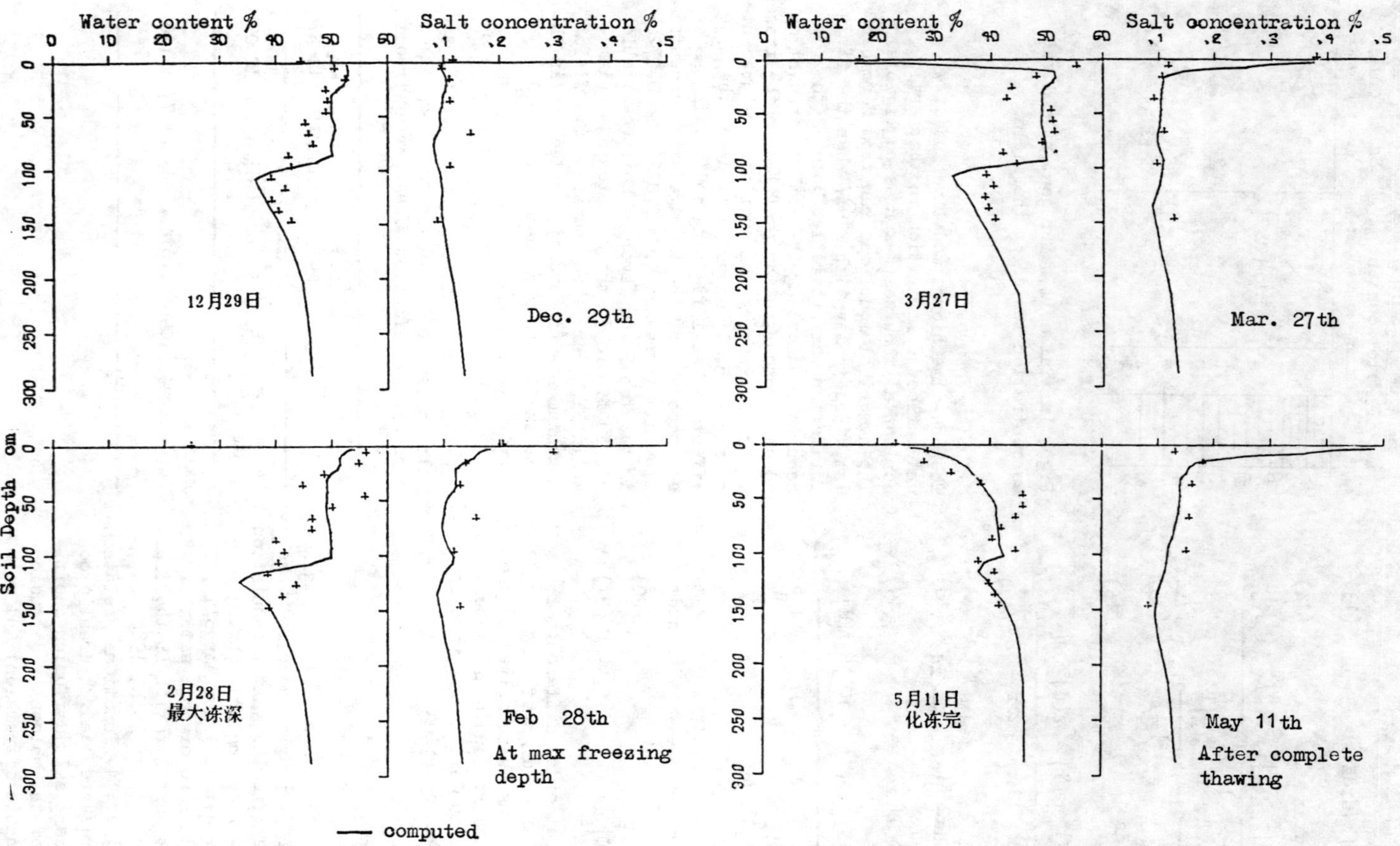

Fig 5 Computed soil water-salt distributions against tested data at different times during freezing

6. CONCLUSIONS

The mathematical model developed in this paper for water-heat-salt movement is applicable to both frozen and unfrozen soils. The Galerkin-type finite element method is a useful tool in analysing water-heat-salt flow problems in field conditions. Prediction with the model developed and the computing program compiled shows, as validated by field test data, that the soil freezing and thawing have much effects on water movement and salt accumulation in the top layer of soil. Application of the model and program to field prediction on soil water-heat-salt movement during Winter and then the salt accumulation in the early Spring should facilitate the optimization of irrigation and drainage scheduling as well as tillage practice for such areas.

REFERENCES

Miller, R.D., J.P.G. Loch & E. Bresler 1975. Transport of water and heat in a frozen permeameter. Soil Sci. Soc. Am. Proc. 39:1029-1036.

Fuchs, M., G.S. Campbell & R.I. Papendick 1978. An analysis of sensible and latent heat flow in a partially frozen unsaturated soil. Soil Sci. Soc. Am. J. 42:379-400.

Cary, J.W., R.I. Papendick & G.S. Campbell 1979. Water and salt movement in unsaturated frozen soil: principle and field observation. Soil Sci. Soc. Am. J. 43:3-7.

Land and Water Use, Dodd & Grace (eds), © 1989 Balkema, Rotterdam. ISBN 90 6191 980 0

Analyse fréquentielle des pluies et estimation des déficits hydriques pour l'irrigation en Tunisie

A.Ben Alaya
Ecole Supérieure d'Agriculture, Mograne, Tunisie

J.Tarhouni, F.Lebdi & N.Ennabli
Institut National Agronomique de Tunis, Tunisie

RESUME: Les pluies sont irrégulièrement réparties dans le temps et dans l'espace. Leur distribution fréquentielle suit dans 80 % des cas la loi de Pearson III. Au Nord-Ouest de la Tunisie, environ 40 %, 32 % et 25 % du total pluviométrique annuel sont les parts respectives de l'hiver, de l'automne et du printemps. L'été est sec. Moins irrégulière que la pluie, l'ETP est faible en hiver (environ 11 % du total annuel) et élevée en été (43% du total annuel). La répartition spatiotemporelle des déficits hydriques est aussi irrégulière. Au Nord-Ouest, les déficits moyens sont quasi nuls pendant cinq mois environ par an et élevés en été (65 % du total annuel). Ils sont plus faibles qu'au Centre.

ABSTRACT : Rainfalls showed irregular trends in time and in space. Their distribution followed in 80 % of the cases Pearson's law III. In the North-West of Tunisia about 40%, 32% and 25 % of the annual rainfall occured during winter, autumn and spring respectively. Summer is a dry period. The Potentiel Evapotranspiration (ETP) is low in winter (11 % of the annuel ETP) and high in summer (43 % of the annuel ETP). In North-West, the average water deficit was around zero during 5 months/year and high in summer (65 % of the annual deficit). However it was less than water déficit averages in Center of Tunisia.

ZUSAMMENFASSUNG : Die regenfälle sind unregulär im zeit und raum verteilt. Ihre fréquentielle distribution folgt das gesetz von Pearson III. Im Nord-Westen Tunesiens fallen 40% 32 % bzw. 25 % der gesamten niederschläge respectiv im winter, im herbst und im frühling. Der sommer ist hingegen trocken. Weniger regulär als die regenfälle, ist aber die potentielle verdunstungstranspiration (ETP). Sie ist gering im winter (11 %) und hoch im sommer (43 %). Die zeiträumliche verteilung der hydrischen defizite ist auch unregulär. Im durschnitt sind die defizite im Nord-Westen Wärend 5 monate des jahres verschwindend klein (65 % vom jahrengesamten). Sie sind viel geringer als im Zentrun des landes.

1 INTRODUCTION

Il est bien entendu que pour combler les déficits en eau d'une culture occupant un type de sol donné, l'élaboration d'un programme d'irrigation doit tenir compte des facteurs pluie et ETP. Si ces derniers sont irréguliers, le programme d'arrosage sera en conséquence caractérisé par des doses et fréquences non constantes. Si le meilleur pilotage de l'irrigation est assuré par le suivi du bilan hydrique, cette méthode qui nécessite des mesures quotidiennes de la pluie, de l'ETP et de l'humidité du sol, ne peut pas être pratiquée facilement par les agriculteurs. La méthode programmée à doses et fréquences d'arrosage systématiques est facilement utilisable. L'élaboration de son programme d'arrosage nécessite une étude fréquentielle des déficits hydriques. C'est cette étude qui constituera l'objet de cet article.

Dans ce qui suit, nous commençons par l'analyse fréquentielle des pluies enregistrées dans différentes régions qui sont réparties dans le Nord et le Centre de la Tunisie et de l'ETP mesurée à Tunis. Par la suite, nous essayons d'estimer la répartition spatiotemporelle des déficits hydriques. Dans les deux cas, nous utiliserons les données décadaires, mensuelles et trimestrielles. Les intervalles d'irrigations utilisés étant généralement de l'ordre dix jours, les valeurs décadaires permettent alors de mieux dégager les variations des pluies, des ETP et des déficits au niveau de chaque mois de chaque saison.

2 ANALYSE FREQUENTIELLE DES PLUIES

2.1 Description

Nous avons étudié les series les plus longues et les plus homogènes possibles (période de mesure de 29 à 107 ans) enregistrées dans 9 stations du Nord et du Centre de la Tunisie (Fig.11).

L'étude fréquentielle a pour objet de calculer les probabilités d'apparition des pluies. Le résultat de cette analyse statistique est d'autant plus significatif que la taille de l'échantillon des données est grande. Celle-ci est déterminée par l'importance de la période de mesure. Des périodes d'au moins égales à 30 ans peuvent faire, en principe, l'objet d'une étude statistique significative. En statistique, il y a de nombreuses lois de probabilité qui peuvent être utilisées pour ajuster la distribution d'une telle variable aléatoire. Ces lois sont différentes et ne peuvent pas ajuster de la même qualité la distribution de la variable. A la recherche de la loi qui assure le meilleur ajustement, nous avons utilisé un programme à dix lois de probabilité qui sont les suivantes :

- Loi de Laplace-Gauss dite Normale,
- Loi de Gumbel, Doublement exponentielle,
- Loi de Galton, Gausso-Logaritmique,
- Loi de Pearson III, Gamma incomplète en X
- Loi de Pearson V, Gamma incomplète en 1/X
- Loi de Goodrich, Exponentielle généralisé en X^a
- Loi de Fréchet, Exponentielle généralisé en $(1/X)^a$
- Loi de WRC-USA, Log-Gamma de première espèce,
- Loi des Fuites,
- Loi de Polaya.

Toutes ces lois ont été tronquées à un seuil unique qui est zéro. Rappelons que la loi tronquée &(x) s'écrit :

$$\&(x) = \frac{F(x) - F(a)}{1 - F(a)} \quad (1)$$

où &(x) peut être la loi de Gauss, de Gubel ..., ou de Polaya qui peut ajuster la distribution des valeurs x supérieures au seuil a (dans notre cas, a=0) et F(x) la fonction de répartition pour toutes les valeurs possibles de la variable x. On a :

$$F(x) = F(a) + [1 - f(a)]\ \&(x) \quad (2)$$

Pour chaque décade, pour chaque mois et pour chaque trimestre, le programme a permis de calculer tout d'abord les paramètres statistiques (moyenne, médiane, mode, variance, écart type, coefficient de variation, coefficient d'asymétrie, coefficient d'aplatissement) de la distribution observée en utilisant uniquement le fichier des données, puis ceux des dix lois utilisées dans l'ajustement en procédant par la méthode du maximum de vraissemblance.

Le traitement numérique est complété par une représentation graphique des valeurs observées en fonction de leurs fréquences d'apparition empiriques et des fonctions d'ajustement. Les graphiques ont servi pour justifier le choix de la fonction qui a fait le meilleur ajustement et qui doit avoir essentiellement des coefficients d'asymétrie et d'aplatissement les plus proches de ceux des valeurs observées. Environ 80 % des échantillons suivent la distribution de Pearson III. Le reste suit celle de Goodrich.

Les lois tronquées retenues ont été utilisées pour déterminer les fonctions de répartition définitives. Celles-ci ont servies pour estimer les pluies décadaires, mensuelles ou trimestrielles en fonction de leurs probabilités de non dépassement.

A signaler que certains décades et mois de l'été (Juin, Juillet, Aout) ont beaucoup de valeurs pluviométriques nulles. Ce qui a réduit la taille de l'échantillons à valeurs non nulles qui ont pour certains cas des nombres inférieurs à 10. Dans ces cas, nous n'avons pas tronqué les lois.

2.2 Répartition de la pluie

Le total pluviométrique annuel est variable et peut passer du simple au double et même au quadriple. En effet, à Tunis, la pluie annuelle a été de 220 mm en 1948 et a atteint 810 mm en 1959. Puis, elle a chuté à 400 mm en 1960. Les années sèches et humides sont irrégulièrement intercalées

L'irrégularité des pluies est observée aussi au niveau de toutes les saisons de l'année. Dans la plus part des régions du Nord, les mois les plus pluvieux sont Novembre, Décembre, Janvier et Février. L'hiver (Déc,Jan,Fév) reçoit la plus grande part (40 %) du total pluviométriques. Ensuite vient l'automne (32 %) puis le printemps (25 %). En été, les pluies sont presque nulles (5 %, fig. 1).

La variabilité des pluies existe au niveau de toutes les stations et elle est d'autant plus importante que les pluies sont rares. Pendant la période pluvieuse (Novembre à Avril), le coefficient de variation (CV) des pluies décadaires est de l'ordre 1 alors que pour la saison sèche, il varie de 1 à 5 (ex: tabl.1,fig 2). Celui des pluies mensuelles est de 0,6 environ pendant la saison pluvieuse et varie de 0,7 à 3 pendant la saison sèche. La variation des totaux pluviométriquestrimestriels est caractérisée par des CV variant de 0,3 à 1 (tabl. 1).

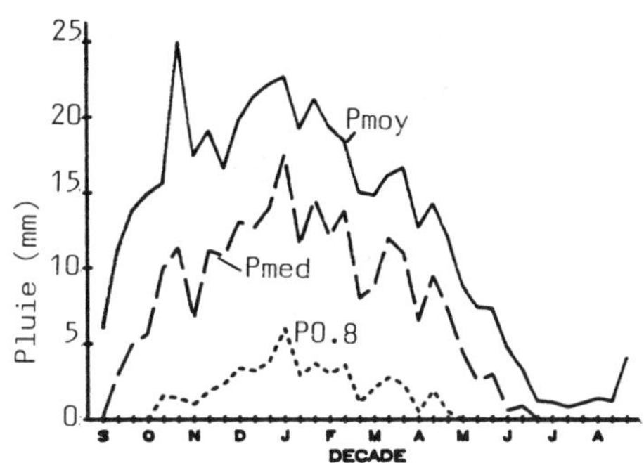

Fig.1 Pluies décadaires à Tunis (Pmoy : pluie décadaire moyenne, Pmed: Pluie décadaire médiane, P0.8: Pluie décadaire de probabilité de dépassement 0.8)

Tableau 1. Coefficients de variation des pluies décadaires, mensuelles et trimestrielles.

Sa.	Moi.	CVdm	CVm	CVm/CVdm	CVmt	CVt	CVt/CVmt
	Sep	1,5	0,9	0,6			
A	Oct	1,4	0,9	0,6	0,8	0,5	0,6
	Nov	1,2	0,7	0,6			
	Dec	1,0	0,6	0,6			
H	Jan	1,0	0,6	0,6	0,6	0,3	0,6
	Fév	1,0	0,6	0,6			
	Mar	1,0	0,6	0,6			
P	Avr	1,1	0,7	0,6	0,7	0,5	0,7
	Mai	1,2	0,7	0,6			
	Jui	1,8	1,1	0,6			
E	Jul	3,1	1,7	0,6	1,4	0,9	0,6
	Aou	2,6	1,5	0,6			

Sa. : Saison, CVdm: Moyenne mensuelle des coefficients de variation (CV) décadaires, CVm: CV mensuel, CVmt: Moyenne trimestrielle des CV mensuels, CVt :CV triemstriel, A: Automne, H: Hiver, P : Printemps, E: Eté.

La variabilité de la pluie diminue si on passe des données décadaires aux données mensuelles et de ces dernières aux données trimestrielles. En effet, pour la même période, les CV décadaires sontplus grands que les CV mensuels qui sont aussi plus grands que les CV trimestriels. Le rapport du CV mensuel sur la moyenne des CV décadaires du même mois est à priori égal au rapport du CV trimestriel sur la moyenne des CV mensuels du même trimestre et a pour valeur 0,6 environ (ex: tabl.1).

$$\frac{CVm}{CVdm} \simeq \frac{CVt}{CVmt} \simeq 0,6$$

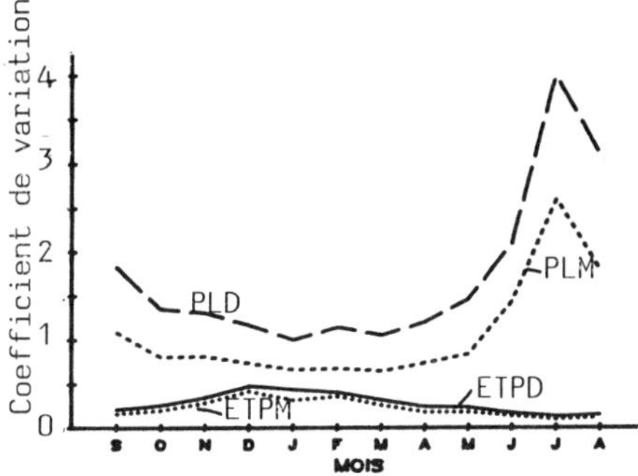

fig.2 Coefficients de variation des pluies et des ETP de Tunis (PLD: Coéfficient de variation (CV) des pluies décadaires, PLM: CV des pluies mensuelles, ETPD: CV des ETP décadaires, ETPM: CV des ETP mensuelles).

Concernant la répartition spatiale, on constate que les pluies sont plus importantes dans le Nord que dans le Centre de la Tunisie(Tabl.2).

Tableau 2. Répartition des pluies saisonnières moyennes.

Station	Autom.	Hiver	Print.	Eté	Total
Bizerte	225	304	140	16	684
Mateur	172	213	130	22	537
Béja	161	232	136	24	553
Jendouba	124	175	127	34	460
Kef	135	181	156	51	525
Tunis	141	180	110	19	450
Mograne	172	175	136	25	508
Fahs	141	155	111	33	440
Kasserine	109	58	94	64	325

3 L'EVAPOTRANSPIRATION POTENTIELLE (ETP)

Seules les ETP quotidienne de la station de Tunis ont été mises à notre disposition et ont fait l'objet d'une étude fréquentielle. Elles ont été mesurées sur cases de kikuyu à l'Institut National Agronomique de Tunis pendant une période de 22 ans (1962-1984). Pour cette station nous avons constaté ce qui suit :

-Dans 65 % des cas, la distribution de l'ETP suit les lois de Pearson III et de Goodrich.

-Contrairement à la pluie, l'ETP est faible en hiver et élevée en été. Son maximum est atteint pendant le mois de juillet (Fig.3)

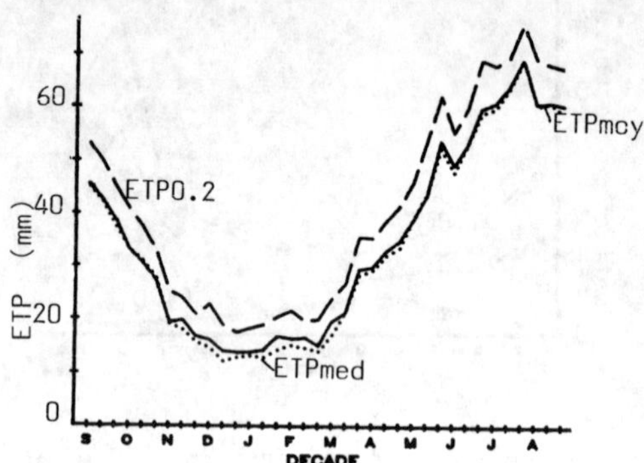

Fig3 ETP décadaires à Tunis (ETPmoy: ETP décadaires moyennes, ETPmed: ETP décadaires médianes, ETP0.2: ETP décadaires de probabilité de dépassement 0.2)

- Comme la pluie, l'ETP est aussi irrégulière. Mais sa variabilité est faible [CV (etp) < CV(pluie), fig.2] et elle est plus faible en saison sèche (CV=0.1 à 0.2) qu'en saison pluvieuse (CV=0.3 à 0.57).

Pour les autres stations, ne sont disponibles que les ETP mensuelles moyennes calculées par la formule de Turc (moyenne sur 13 ans, Cormary Y.1964). Selon Cormary la méthode de Turc conduit à des valeurs trés proches des ETP mesurées. Mougou R. (1978) a noté la même remarque pour Tunis et Gabès. Mais non pour Béja à laquelle elle a effecté, dans le calcul de l'ETP, le rayonnement global de Jendouba.

A Tunis,les ETP mensuelles mesurées et celles calculées par la formule de Turc sont peu différents (Tabl.3).

La variabilité spatiale de l'ETP est plus faible que celle de la pluie. Du Nord au Sud de la Tunisie, le total annuel de l'ETP varie de 1140 mm à 1460 mm environ.

Tableau 3. ETP mensuelles à Tunis.

Mois	ETP (mm) mes.	ETP (mm) calc.	Mois	ETP(mm) mes.	ETP(mm) calc.
Sep	126,9	136,5	Mars	70,5	87
Oct	90,6	95	Avril	97,8	108,5
Nov	55,9	63	Mai	136,8	142
Dec	43,9	42,5	Juin	162,7	176
Janv	45,1	44	Juil	194,6	181
Fev	48,2	62	Aout	182,7	170

4 ESTIMATION DU DEFICIT HYDRIQUE

4.1 Expression générale

Le déficit hydrique (D) d'une culture peut être estimé par l'expression de base suivante :

$$D = ETR - ETM \quad (5)$$

où l'ETM est l'évapotranspiration maximale de la culture. Elle est égale au produit de son coefficient cultural Kc par l'ETP. L'ETR est l'évapotranspiration réelle de la culture produite dans les conditions naturelles du milieu (sans apport d'eau d'irrigation). L'expression (5) est traduite par la suivante :

$$D = Ce\ P + p(RU) - Kc\ ETP \quad (6)$$

avec :

P = la pluie totale tombée sur le sol,

Ce : Le coefficient d'éfficacité de la pluie

p(RU) : part de la réserve utile RU dans le sol.

Le produit (Ce P) donne la valeur de la pluie qui s'infiltre dans le sol et qui sera mise à la disposition de la plante. Des pluies à fortes intensités ou de longues durées ne peuvent être absorbées en totalité par le sol, elles engendrent des ruissellements. Le coefficient Ce utilisé pour éliminer la part ruisselée de la pluie totale enregistrée est difficile à estimer. Mais sachant qu'il dépend essentiellement de l'intensité et la durée de la pluie, de la perméabilité, de l'humidité, de la topo graphie et du couvert végétal du sol. De même, ce coefficient varie d'une saison à l'autre et d'une année à l'autre.

Remarquons que l'expression (6) peut s'écrire sous d'autres formes qui dépendent des cas suivants :

Cas 1 : $p(RU) = 0$ et $Ce\ P < ETM$
$\Rightarrow D = Ce\ P - ETM$

Cas 2 : $p(RU) = 0$ et $Ce\ P \geq ETM$
$\Rightarrow D = 0$

Cas 3 : $p(RU) = 0$ et $CeP + p(RU) < ETM$
$\Rightarrow D = Ce\ P + p(RU) - ETM$

Cas 4 : $p(RU) = 0$ et $CeP+p(RU) \geq ETM \Rightarrow D=0$

4.2 Expression utilisée

Dans cet article, nous avons utilisé l'ETM de référence qui est l'ETP. Le coefficient d'éfficacité Ce est inconnu puisqu'on traite des déficits de régions où il y a des différents types de sols inconnus. Pour des pluies inférieures ou égales aux valeurs décadaires moyennes, nous avons supposé que les intensités sont relativement faibles. Ce qui peut rendre les pluies ruisselantes négligeables. Pour cela nous avons supposé que Ce est à priori égal à 1. L'expression simplifiée du déficit sera la suivante :

$$D = P + p(RU) - ETP \qquad (7)$$

C'est cette formule qui a été utilisée dans le calcul des déficits hydriques qui sont exprimés en valeurs absolues.

4.3 Répartition des déficits

Pour Tunis, nous avons calculé le déficit hydrique décade par décade sur une période de 22 ans pendant laquelle ont été enregistrées la pluie et l'ETP. La distribution aléatoire des déficits obtenus est surtout ajustée pour 13 décades par la loi de Goudrich, pour 9 décades par celle de Gauss et pour 8 par la loi de Pearson III.

Les déficits décadaires moyens correspondant aux différentes capacités d'emmagasinement en eau du sol (RU=50,100,150 et 200 mm) sont quasi nuls pendant 4 à 5 mois environ de l'année (fig.4). Ils sont importants en été la pointe est atteinte pendant la troisième décade de juillet.

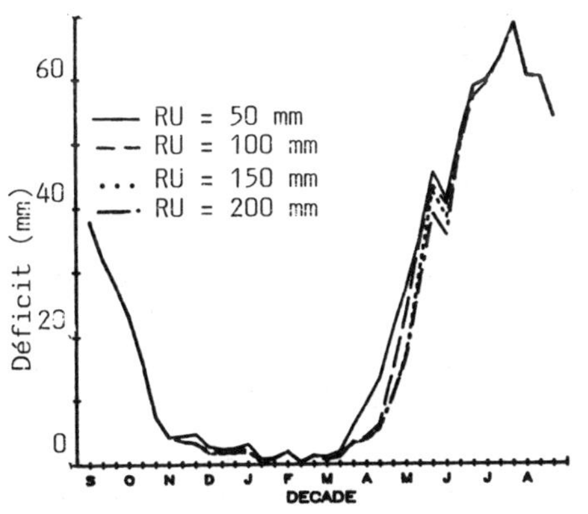

Fig.4 Déficits moyens à Tunis

Les déficits moyens (DEFmoy) et médians (DEFmed) sont presque égaux (fig.5) en saison sèche (avril à octobre). En saison pluvieuse (novembre à mars), ils présentent des écarts ne dépassant guère 5mm.

Si les écarts entre le déficit décadaire de fréquence de dépassement égale à 2/10 (DEF0.2) et le déficit décadaire moyen (DEFmoy) sont faibles (de 0 à 5 mm) en saison pluvieuse, ils ne le sont pas pendant l'autre saison où ils varient de 10 à 20 mm (fig.5)

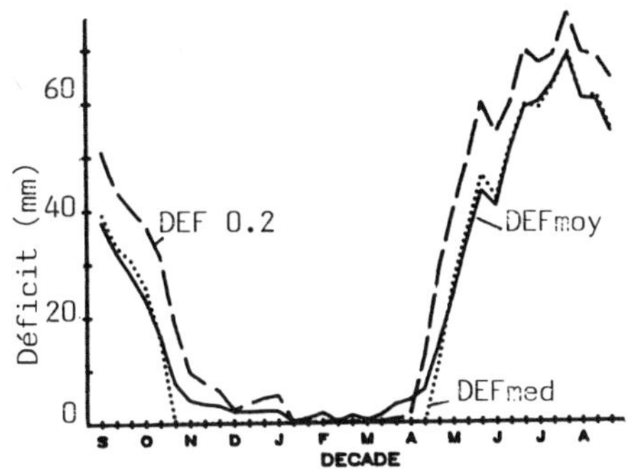

Fig.5 Déficits décadaires à Tunis (RU=100mm)

La figure 6 suivante montre que le déficit saisonnier d'hiver (décembre,janvier et février) s'annulle à partir de la fréquence de dépassement 0.3. Ceux de l'automne et du printemps sont à peu prés égaux et ne s'annullent que rarement. L'été est caractérisé par un déficit restant toujours important. Remarquons en plus que les écarts entre le déficit d'été et ceux d'automne et du printemps restent presque constants depuis les fréquences faibles aux fréquences élevées (courbes presque parallèles).

Le déficit décadaire moyen (Ddm), obtenu en moyennant les déficits calculés décade par décade sur 22 ans, représente en principe le déficit réel moyen. Le déficit décadaire moyen (Ddm), calculé à partir de pluie et d'ETP moyenne, est trés proche de Ddm en saison sèche.

En saison pluvieuse. Ddmm est nul et peu différent de Ddm (écarts variant de 0 à 6 mm, Fig.7).

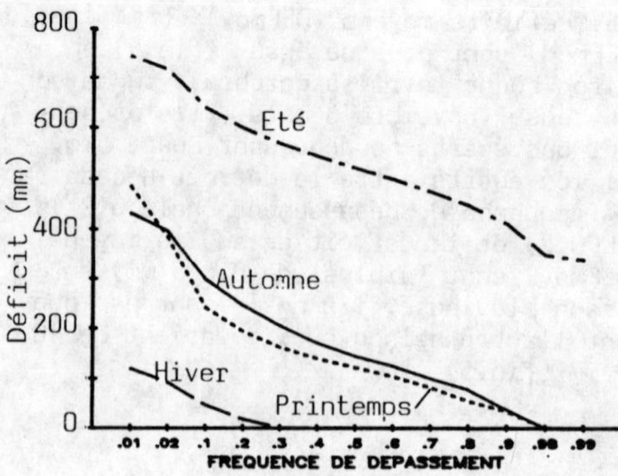

Fig.6 Déficits saisonniers à Tunis (RU= 100mm)

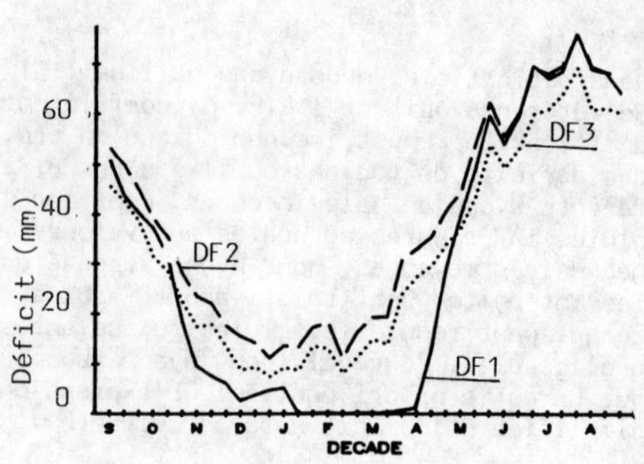

Fig.8 déficits de fréquence de dépassement 0.2 (RU=100mm, station de Tunis)

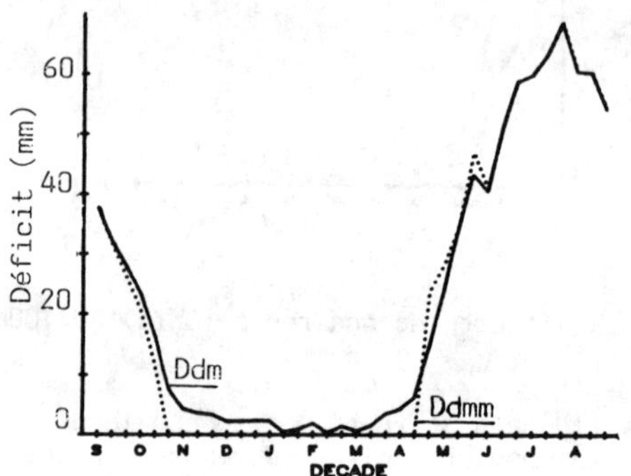

Fig.7 Déficits moyens Ddm et Ddmm (RU= 100mm, Station Tunis)

Si le déficit décadaire de fréquence de dépassement 0.2 (DF1) obtenu à partir de l'analyse fréquentielle des déficits est trés proche du déficit réel à même fréquence, celui (DF2) calculé à partir de la pluie décadaire dépassée 8 fois sur 10 [(P)0.8] et de l'ETP décadaire dépassée 2 fois sur 10 [(ETP)0.2] ne l'est qu'en saison sèche. Le déficit (DF3) calculé à partir de (P)0.8 et de l'ETP moyenne est toujours différent de DF1. En saison pluvieuse DF2 et DF3 surestiment largement le déficit réel (en avril la surestimation dépasse 30mm,Fig.8)

Vu que lesdéficits moyens et médians sont équivalants (fig.5), cela veut dire que le déficit moyen apparaitra pendant une année sur deux environ. Un programme d'arrosage déterminé à partir de ce déficit n'assurera la satisfaction en eau des cultures que pendant une année sur deux environ. Il est assez courant d'adopter en irrigation des programmes d'arrosage permettant de couvrir les déficits de fréquence de dépassement 0.2. Cependant, la question qui se pose c'est de quel déficit à fréquence 0.2 doit-on tenir en compte pour déterminer ces programmes? Est ce que le déficit décadaire, mensuel ou trimestriel?

En réalité, la réponse dépend essentiellement du type, du cycle et de la saison végétatifs de la culture. On peut adopter, par exemple, les déficits trimestriels quand il s'agit des céréales pendant leurs stades les moins sensibles ou les déficits décadaires pour les cultures maraichères lors de leurs stades les plus sensibles. Mais remarquons (fig.9) qu'à la même fréquence (F=0.2), le total trimestriel des déficits décadaires (Dtd2), le total trimestriel des déficits mensuels (Dtm2) et le déficit trimestriel (Dt2) sont peu différents pendant les saisons d'hiver, du printemps et de l'été. En automne, les écarts Dtd2-Dtm2 et Dtd2-Dt2 sont respectivement de l'ordre 30 et 40mm.

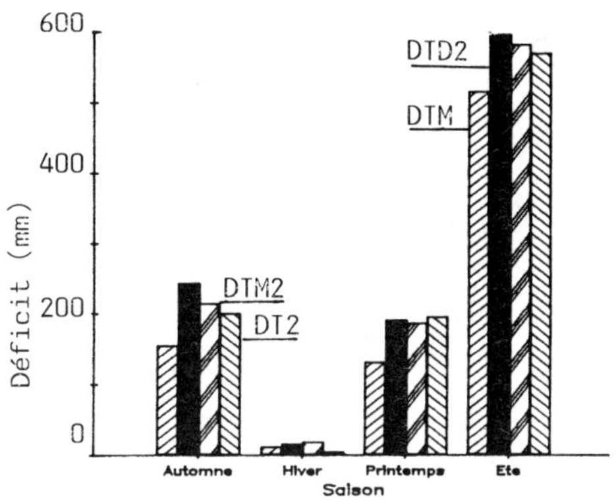

Fig.9 Déficits saisonniers à Tunis (RU=100mm)

Pour les autres stations,où nous ne disposons pas de séries longues d'ETP, nous n'avons pas calculé l'historique des déficits hydriques qui fera l'objet d'une analyse fréquentielle. Pour cela, nous avons calculé les déficits moyens à partir de pluies et d'ETP mensuelles moyennes. Sans oublier que cette méthode de calcul sous-estime légèrement le déficit réel moyen en saison pluvieuse (Fig.7). Les calculs ont commencé à partir du mois de septembre au début duquel le réservoir sol est considéré comme vide (p(RU)=0). Dans ces stations les déficits moyens varient de la même façon que celui de Tunis. Ils sont faibles en hiver et importants en été. Le mois de pointe est toujours juillet. Ces déficits sont plus importants dans le centre que dans le Nord de la Tunisie (Tabl.4,Fig.10). Dans le Nord-Ouest, ils sont quasi nuls pendant 5 mois environ (novembre à avril).

Les courbes d'isodéficits annuels moyens (Fig.11) sont extrapolées à partir des déficits annuels moyens des stations étudiées.

Tableau 4. Déficits saisonniers dans certaines stations (RU=100mm).

Station	Auto	Hiv	Print	Eté	Total
Bizerte	89	0	56	516	661
Mateur	108	0	73	506	687
Béja	103	0	61	517	681
Jendouba	139	0	132	506	777
Kef	118	0	52	461	631
Mograne	110	0	98	503	711
Fahs	115	0	165	495	775
Sousse	195	64	278	505	1042
Kairouan	210	92	254	508	1064
Kasserine	145	63	209	446	862

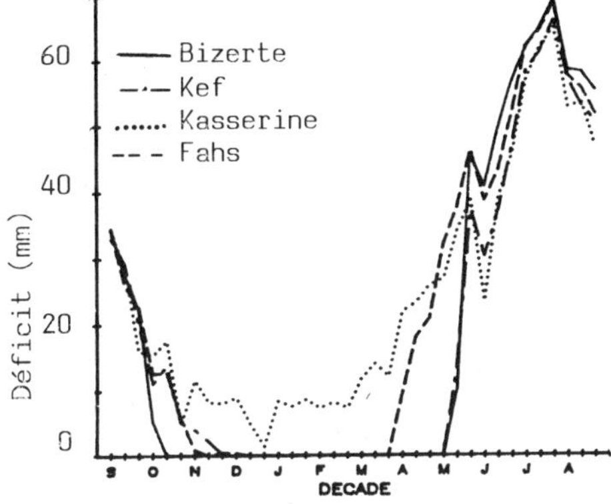

Fig.10 Déficits décadaires moyens dans certaines stations (RU=100mm)

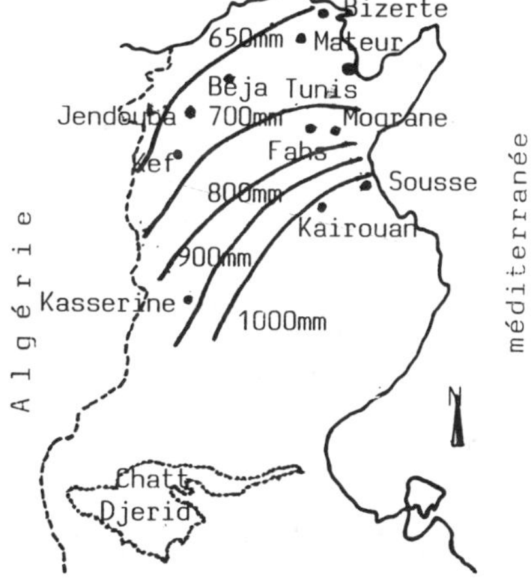

Fig.11 Courbes d'isodéficits annuels moyens dans le Nord et le Centre de la Tunisie (RU=100mm).

5 CONCLUSION

En Tunisie, les pluies sont irrégulièrement réparties dans l'espace et dans le temps. La répartition temporelle est ajustée dans 80 % des cas par la loi de Pearson III et dans le reste des cas par celle de Goodrich. Dans le Nord-Ouest, environ 40%, 32% et 25% de la pluie annuelle moyenne tombent respective ment en hiver, en automne et en printemps. Les pluies d'été sont presque nulles.

L'ETP est aussi irrégulière dans le temps est dans l'espace. Cette irrégularité qui est non négligeable est plus faible que celle de la pluie. Environ 43%, 24%, 22% et 11% de l'ETP annuelle sont les parts respectives de l'été, du printemps, de l'automne et de l'hiver.

De même, l'irrégularité spatiotemporelle des déficits est non négligeable. Dans le Nord-Ouest de la Tunisie, le déficit moyen est plus faible qu'au centre et il est quasi nul pendant 5 mois environ par an. En été, il est important et représente 65% environ du total annuel.

Par ailleurs, en saison sèche ou en saison pluvieuse, un programme d'irrigation établi à partir de données climatiques moyennes n'assurera la satisfaction en eau des cultures que pendant une année sur deux. On recommande généralement de couvrir les déficits hydriques dépassés pendant 2 années sur 10.

REFERENCES

Cormary, Y. & Ayadi,E. 1964. Variabilité du déficit en eau (C.R.G.R étude n°88), Tunisie.

Franquin,R. 1973. Les analyses fréquentielles des déficits hydriques et des pluies pour l'irrigation de complément (Agronomie Tropicale n°9, Spet.1973).

FAO n°33,1980. Réponse des rendements à l'eau.

Gigou,J. 1973. Etude de la pluviosité en Cote d'Ivoire- Application à la riziculture pluviale (Agronomie Tropicale n°9 Sept. 1973).

Mougou,R. 1978. Estimation de l'évapotranspiration potentielle - Cas de la Tunisie.

Land and Water Use, Dodd & Grace (eds), © 1989 Balkema, Rotterdam. ISBN 90 6191 980 0

Computerized irrigation scheduling using spectral feedback

W.C.Bausch
USDA-ARS Irrigation and Drainage Research, Fort Collins, Colo., USA

ABSTRACT: An algorithm to shift the basal crop coefficient (K_{cb}) curve for corn (Zea mays L.) with respect to its time axis was successfully integrated into the USDA-ARS Irrigation Scheduling Program. Spectral feedback to the irrigation scheduling program in the form of a real time reflectance-based crop coefficient periodically shifted the K_{cb} curve to keep it tracking actual corn growth throughout the 1988 growing season. Adjustments made to the K_{cb} curve prior to effective cover for the field using feedback resulted in a one day delay in irrigation dates when compared to irrigations for the field scheduled with the traditional operation of the irrigation scheduling program. This one day delay advantageously used a precipitation event during the latter part of the growing season such that one less irrigation was required for the field utilizing spectral feedback. The adjustments to the K_{cb} curve during vegetative development also made the assumed effective cover date converge on its actual occurrence.

RESUME: Un algorithm pour changer la courbe du coefficient culturel (K_{cb}) du maïs (Zea mays L.) en function du temps a été intégré avec succès dans un programme des besoins en eau au niveau de l'USDA-ARS (United States Department of Agriculture - Agricultural Research Service). La réaction spéctrale au programme des besoins en eau sous forme de temps effectif bas sur la réflectance du K_{cb} change périodiquement la forme de la courbe du K_{cb} en relation du développement végétatif du maïs durant la période d'observation, saison 1988. Des corrections de la courbe du K_{cb} ont été faites avant la période de maximum croissance de la culture utilisant la réaction résultante en retard d'un jour pour les dates d'irrigation comparés au culture utilisant des méthodes traditionelles. Cette méthode retard d'un jour, utilise avantageusement les précipitations qui peuvent résulter en fin de cycle de croissance de la culture induisant de fait une irrigation de moins. Les corrections temporelles de la courbe du K_{cb} pendant la developpement végétatif ont aussi résultées sur la convergence des dates de maximum croissance, supposées et l'actual date de l'évenement.

ZUSAMMENFASSUNG: Ein Algorithmus, der die Kurve des fundamentalen Ertragskoeffizienten (K_{cb}) für Mais (Zea mays L.) mit Respekt zu der Zeit-Achse vershiebt, wurde erfolgreich in das USDA-ARS Bewässerungsplanungsprogramm intergriert. Die spektrale Rückkopplung in das Bewässerungsplanungsprogramm, in der Form eines echt Zeit Ertragskoeffizienten, der auf der Reflektivität beruht, hat die K_{cb} Kurve periodisch verschoben. Damit wurde sichergestellt, daß die K_{cb} Kurve das wirkliche Mais Wachstum während Wachstumszeit in 1988 verfolgt. Die Anpassung der K_{cb} Kurve vor der effektiven Feld Bedeckung resultierte in einer eintägigen Verzögerung der Bewässerungszeiten, verglichen mit der traditionellen Handhabung des Bewässerungsplanungsprogramm. Diese eintägige Verzögerung hat vorteilhafterweise einen Niederschlag nahe dem Ende der Wachstumszeit benutzt um eine Bewässerung einzusparen. Die Anpassungen der K_{cb} Kurve während der vegetativen Entwicklung hat außerdem die Konvergenz des vorhergesagten und tatsächlichen effektiven Bedeckungsdatum sichergestellt.

1 INTRODUCTION

Many computer models have been developed to expedite irrigation scheduling computations. One such model is the USDA-ARS Irrigation Scheduling Program initiated by Jensen (1969) and further developed by Jensen, et al. (1970,1971); Kincaid and Heermann (1974); Harrington and Heermann (1981) and Buchleiter, et al. (1986). In this model and a similar model (Stegman and Coe, 1984), calculated reference crop evapotranspiration and crop coefficients are used to estimate actual crop evapotranspiration (ET). This technique provides a practical method for estimating actual crop ET throughout the growing season to satisfactorily estimate soil water depletions from which irrigations are scheduled (Heermann, et al.,1976).

Crop coefficients are empirically determined from field studies where ET of the crop is measured and ratioed to reference crop ET. Crop coefficients exist for numerous crops either as mean crop coefficients (Jensen, 1973) or basal crop coefficients. The basal crop coefficient (K_{cb}) defined by Wright (1982) represents conditions when the soil surface is dry so that evaporation from the soil is minimal but soil water availability is adequate such that plant growth or transpiration is not limited.

The basal crop coefficient curve can be construed as an ideal growth curve. Due to the empirical nature of this curve, crop coefficients can introduce errors into ET estimates. Various anomalies (weather, nutrient, insect, disease, etc.) cause plant growth to deviate from the ideal represented by the basal crop coefficient curve.

Monitoring crop growth in the field can provide essential feedback to irrigation scheduling models to keep the basal crop coefficient curve tracking actual plant growth. Bausch and Neale (1987) and Neale, et al. (1989) developed a real time crop coefficient for corn (Zea mays L.) from canopy reflectance measurements and the basal crop coefficient concept. This crop coefficient was referred to as a reflectance-based crop coefficient (K_{cr}) which has the capability to respond to periods of slow and/or fast plant growth. Bausch and Neale (1989) developed an algorithm to utilize the K_{cr} to calculate an adjusted basal crop coefficient which in turn could be used in irrigation scheduling models. This paper describes integration of the algorithm into the USDA-ARS Irrigation Scheduling Program SCHED and results of a field test during the 1988 growing season.

2 USDA-ARS IRRIGATION SCHEDULING PROGRAM

The USDA-ARS Irrigation Scheduling Program SCHED maintains a soil water budget for each field specified in the field file. Soil water depletions at the end of each day are calculated by

$$D_i = D_{i-1} + ET_{ai} + E_{wi} - I_i - R_i \quad (1)$$

where D is depletion in the root zone, ET_a is actual crop ET, E_w is evaporation from a wet soil surface, I is irrigation amount, R is rainfall amount, i is the current day and i-1 is the previous day. Actual crop ET is estimated by

$$ET_a = K_{cb} \times K_s \times ET_r \quad (2)$$

where K_{cb} is the basal crop coefficient, K_s is a soil moisture stress coefficient which accounts for the reduction in crop water use resulting from increased soil water depletion and ET_r is reference ET computed from measured climatic data. SCHED utilizes the modified Penman combination equation to compute reference ET. Crop water use calculated using Equation (2) assumes a dry soil surface. Additional evaporation from the soil surface occurs when it is wet following a rain or irrigation. This additional use is estimated by

$$E_w = K_r \times (0.9 - K_c) \times ET_r, \text{ for } K_c \leq 0.9 \quad (3)$$

where $K_c = K_{cb} \times K_s$ and K_r is a coefficient depending on the number of days since the rain or irrigation. K_r is set to 0.8 for the first day, 0.5 for the second day and 0.3 for the third day provided that total E_w does not exceed the rainfall or irrigation event. When K_c becomes greater than 0.9, wet soil evaporation is minimal and ignored.

Basal crop coefficient curves are represented by third order polynomial equations fit to Wright's (1982) tabular data. Regression coefficients used in these equations are those presented by Hill, et al.(1983). Crop coefficients used in SCHED and similar models are time dependent. In other words, planting date and an assumed effective cover date (some future date) are required as independent variables to determine drivers for the curve. The driver used between planting and effective cover is calculated by

$$r = (DOY - PD)/(EC - PD) \times 100 \quad (4)$$

where r is percent of time from planting to effective cover, DOY is the day-of-year in question, PD is planting date and EC is the assumed effective cover date. Elapsed days after effective cover, i.e., r = DOY - EC, is the driver when the assumed effective cover date is reached. Thus, planting date marks the beginning of the time line, consequently, a crop coefficient can be calculated for each calendar day thereafter for the crop species of interest.

Predicted depletions and, hence, the next irrigation date are calculated based on forecast reference ET. Forecast reference ET is determined from an empirical equation fit through ET_r values from several previous years of data (Heermann and Jensen, 1970).

Irrigations are scheduled so that the soil water depletion in the field does not exceed the maximum allowable depletion. This may mean that irrigation must commence before the start position has reached maximum allowable depletion in order that the stop position gets irrigated before it reaches maximum allowable depletion. An overview of program operation is presented by Harrington and Heermann (1981).

2.1 Modifications to SCHED

Crop coefficient curves do not adequately represent actual crop growth during some growing seasons. Therefore, Bausch and Neale (1989) examined the variability in occurrence of various phenological events during a four year period of experimentation with corn in northern Colorado to develop a procedure to shift the basal crop coefficient (K_{cb}) curve with respect to its time axis in order to match the K_{cb} to the reflectance-based crop coefficient (K_{cr}). Based on variations that occurred, a range of days were determined which were used as possible limits for shifting the K_{cb} curve. The end points selected to shift the curve were emergence date ± 11 days and effective cover date ± 7 days. Sixty ± 5 days lapsed between emergence and effective cover which was used as a constraint in the algorithm prior to effective cover. The algorithm to adjust emergence and effective cover dates to match the K_{cb} to actual corn growth is presented in Figure 1. In essence, this algorithm is an iterative procedure to test possible dates from which the K_{cb} is calculated in order to make the K_{cb} converge on the K_{cr} within its limits of error. Candidate dates tested are based on the variation that occurred for the northern Colorado site.

Input data required for the algorithm are day-of-year (DOY), the K_{cr}, and upper and lower bounds on the K_{cr}. DOY directs branching in the algorithm. If the DOY in question is less than the assumed effective cover date, emergence and assumed effective cover dates are varied. Otherwise, only the effective cover date is allowed to vary. Nomenclature used in the algorithm and this paper to represent the independent variables required for the drivers of the K_{cb} curve are emergence date, effective cover date, apparent emergence date, assumed effective cover date and apparent effective cover date. Emergence and effective cover dates are associated with the best determination of their actual occurrence. Apparent emergence date represents a date when emergence should have occurred to match the K_{cb} to actual field conditions. Assumed effective cover is a best guess as to the occurrence of effective cover. Apparent effective cover refers to the date on which effective cover should have occurred to make the K_{cb} track actual crop growth after the occurrence of effective cover. Details are presented by Bausch and Neale (1989) on development of the algorithm and its operation.

To use emergence date as the independent variable to calculate the normalized time driver, i.e., percent of time from emergence to effective cover, the basal crop coefficient curve was redefined starting at emergence. A K_{cb} of 0.15 was used at emergence since soil evaporation is dominant at that time. The daily K_{cb} used in the algorithm is calculated from tabular data and is considered to vary linearly between tabulated values. Tabulated values occurred at 10% intervals for the period between emergence (zero percent) and effective cover (100%) and at intervals of 10 days from effective cover (zero days) to maturation.

Integration of the K_{cb} curve shift algorithm into SCHED required an additional data entry subroutine and the creation of an update file as well as saving certain parameters in the field file that change with each update. Each time the program is executed, a question appears on the computer monitor asking if K_{cr} data are available to shift the K_{cb} curve. For a yes response, the <Enter> key is pressed to input the data. If no K_{cr} data exists, the <Esc> key is pressed to run the program in its normal mode. If <Enter> was pressed, a prompt appears requesting the K_{cr}, then the upper bound on K_{cr}, then the lower bound on K_{cr}, and then the above entered values are displayed along with the DOY with a chance to change them if entered incorrectly.

With each run of SCHED, an update file is used to save the DOY of update, the K_{cb} for that day, the emergence or apparent emergence date (whichever is appropriate), and

```
IF day-of-year < AEC THEN
   FOR CEMG = (EMG - 11) TO (EMG + 11)
      FOR CEC = (AEC - 7) TO (AEC + 7)
            IF 55 ≤ (CEC - CEMG) ≤ 65 THEN
               calculate Kcb based on CEMG and CEC
               IF lower Kcr bound ≤ Kcb ≤ upper Kcr bound THEN
                  set APEMG = CEMG
                  set AEC = CEC
                  quit
               set Kcb error = absolute value of (Kcr - Kcb)
               IF Kcb error < minimum error THEN
                  set APEMG = CEMG
                  set AEC = CEC
                  set minimum error = Kcb error
      NEXT CEC
   NEXT CEMG
   quit

ELSE IF day-of-year is within 20 days after EC THEN
   set Kcb = 0.93
   quit

ELSE IF day-of-year > 20 days after EC THEN
   FOR CEC = (EC - 7) TO (EC + 7)
      calculate Kcb based on CEC
      IF lower Kcr bound ≤ Kcb ≤ upper Kcr bound THEN
         set APEC = CEC
         quit
      set Kcb error = absolute value of (Kcr - Kcb)
      IF Kcb error < minimum error THEN
         set APEC = CEC
         set minimum error = Kcb error
   NEXT CEC
   quit
```

*******************definitions of acronyms used in the algorithm*******************

EMG	emergence date (the actual value)
CEMG	candidate emergence date (a value being tested for emergence)
EC	effective cover date (the actual value)
AEC	assumed effective cover date (the best guessed value for effective cover)
CEC	candidate effective cover date (a value being tested for effective cover)
APEMG	apparent emergence date (a best fit value after emergence occurs)
APEC	apparent effective cover date (a best fit value after effective cover occurs)

Figure 1. Outline of algorithm to shift the basal crop coefficient (K_{cb}) curve for corn with respect to its time axis to make the K_{cb} conform to actual crop conditions.

the assumed effective cover, effective cover, or apparent effective cover date (again whichever is appropriate) for each field. Following each run of the program, the parameters that change in the field file, i.e., last update, soil water depletion, accounting of rainfall and/or irrigation for the previous three days, seasonal ET, and accumulated net application depth, are saved for each field being scheduled.

After the K_{cb} is calculated for an update using K_{cr} data, the K_{cb} value for the day of the current update is compared to the K_{cb} on the day of the last update. Assuming there is no physical damage to the plant during the crop development period prior to effective cover, the K_{cb} for the day of the current update must be greater than or equal to the K_{cb} on the day of the previous update. After effective cover, the K_{cb} on the day of the current update must be less than or equal to the K_{cb} on the day of the last update. If this condition is not met in either situation, then comparisons are made between the K_{cb} for the current update and the K_{cb} from previous updates in the update file until the comparison is true. At that point a message flashes on the computer monitor to replace information in the field file with that saved on the update day corresponding to the DOY that made the K_{cb} comparison true. It is then necessary to delete records in the update file back to that particular DOY and rerun the program. Editing of records in the update and field files are accomplished by selecting the appropriate options in the program's main menu.

3 METHODS

Corn (Zea mays L.) was planted at the Northern Colorado Research and Demonstration Center near Greeley, Colorado on April 27, 1988 in rows spaced 0.76 m. Row orientation was north/south. Two field plots designated as field A and field B were established that were 18 rows wide and 150 m in length. Field A irrigations were scheduled using the traditional mode of operating the irrigation scheduling model, i.e., no feedback required. Field B irrigation scheduling procedures utilized the reflectance-based crop coefficient as feedback to the model.

Radiometer voltage response to incoming and reflected light was measured simultaneously near solar noon eleven times in field B during the growing season with a mobile boom mounted radiometer system. Data of interest to this study were obtained for Thematic Mapper (TM) band 3 (0.63 - 0.69 μm) and TM band 4 (0.76 - 0.90 μm). These data were taken from a 10 m height at eight random locations on the east side of the plot. Bidirectional reflectance of the scene in TM 3 and TM 4 was calculated using a procedure similar to that presented by Duggin (1980). The $BaSO_4$ reference panel used for intercalibration of the radiometers was calibrated using procedures described by Jackson, et al. (1987). Bidirectional reflectance factors were used to calculate the normalized difference (ND) vegetation index as

$$ND = (TM4-TM3)/(TM4+TM3) \quad (5)$$

The ND was then used to compute the reflectance-based crop coefficient for corn. It is expressed as

$$K_{cr} = 1.181 \times ND - 0.026 \quad (6)$$

where K_{cr} is the reflectance-based crop coefficient. This equation was obtained via a linear transformation that equated the ND for dry bare soil and the ND at effective cover to the basal crop coefficient (K_{cb}) for bare soil and at effective cover, respectively (Neale and Bausch, 1983). Several years of data acquired at the northern Colorado experimental site were used to develop the equation. Upper and lower bounds on the K_{cr} were computed as the 95% confidence interval about the mean K_{cr} on the day of question.

Climatic data were measured in an adjacent alfalfa field with an automated weather station and transferred via modem to an office computer. Daily weather summaries were computed at midnight which consisted of maximun and minimum air temperature, average vapor pressure, solar radiation, wind run and precipitation. These data were stored in a file which was read into the irrigation scheduling program whenever the program was executed.

Emergence (which was designated as a detectable green line) occurred on May 14 (DOY 135) 17 days after planting. Assumed effective cover was set at DOY 190 (July 8) 72 days after planting as suggested by Wright (1982) and verified by Bausch and Neale (1989). SCHED, the USDA-ARS Irrigation Scheduling Program, was initiated for each field beginning at emergence. The program was updated thereafter at least thrice weekly and occasionally more frequently. Water holding capacity of the soil in the effective root zone (0.76 m) was 0.12 mm/mm. The field plots were surface irrigated using gated pipe; all furrows were irrigated. Runoff was not measured, consequently, net application depth for each irrigation was not calculated.

4 RESULTS AND DISCUSSION

Irrigation dates for fields A and B during the 1988 corn growing season are listed in the following table with DOY in parenthesis following the calendar date.

Field A		Field B	
June 5	(157)	June 5	(157)
June 16	(168)	June 17	(169)
June 28	(180)	June 29	(181)
July 9	(191)	July 10	(192)
July 17	(199)	July 18	(200)
July 26	(208)	July 27	(209)
August 3	(216)	August 12	(225)
August 14	(227)	August 21	(234)
August 24	(237)	August 31	(244)
September 3	(247)		

One less irrigation was required for field B. This was due in part to 27 mm of rainfall that occurred during the evening of August 3 (DOY 216) which delayed the irrigation scheduled for August 4. Tracking actual crop growth for feedback to the irrigation scheduling model created the one day delay in irrigation events following the first irrigation which advantageously utilized the August 3 precipitation event.

Figure 2 shows the basal crop coefficient (K_{cb}) curves used in the irrigation scheduling program for fields A and B. Both K_{cb} curves are based on Wright's (1982) tabulated data redefined to start at emergence. In essence, field A's K_{cb} curve could have been derived at the onset of emergence since it is based strictly on the emergence and assumed effective cover dates, however, field B's K_{cb} curve was generated as the growing season progressed based on feedback to the irrigation scheduling model. Both curves started on the emergence date (DOY 135) and used DOY 190 as the assumed effective cover date.

The first adjustment to field B's K_{cb} curve occurred on DOY 159 (June 7) following canopy radiance data acquired on that day. As shown in Figure 2, the two curves deviated prior to DOY 159. The reflectance-based crop coefficient (K_{cr}) calculated for DOY 159 was 0.174 with upper and lower bounds of 0.181 and 0.168, respectively. When these data were input into SCHED for the update on DOY 159, the adjusted K_{cb} calculated by the curve shift algorithm was 0.185. This value (K_{cb} = 0.185) was then compared to the K_{cb} on the day of the previous update. The previous update of the model on DOY 157 ended with a K_{cb} = 0.228 which was greater than the adjusted K_{cb} on DOY 159. Thus, a message on the computer monitor indicated that the model must be restarted from DOY 151 (May 30) when the K_{cb} equaled 0.180. Therefore, the field and update files were edited. Pertinent information in the field file for field B from the update on DOY 157 (the previous update) was replaced with data saved from the update on DOY 151. Update records for DOY 154 and 157 were deleted in the update file such that the record for DOY 151 was the last data in that file for field B. The program was then rerun for the June 7 (DOY 159) update. Apparent emergence and new assumed effective cover dates associated with the adjusted K_{cb} were DOY 142 (May 21) and DOY 197 (July 15), respectively. These adjusted dates shifted the basal crop coefficient curve for field B to the right by seven days to account for slow plant growth caused by cool weather following emergence.

The adjusted K_{cb} of 0.185 calculated by the curve shift algorithm for DOY 159 was outside the 95% confidence limits on the K_{cr}. Unfortunately, one day changes in candidate apparent emergence and assumed effective cover dates could not produce a K_{cb} that fell within those limits. Consequently, the minimum difference criteria between the K_{cr} and the possible K_{cb} was used to select the apparent emergence date and the new assumed effective cover date. These dates (DOY 142 and 197) were used henceforth in the irrigation scheduling program for field B until the next update with feedback changed them.

The second update with feedback occurred on June 14 (DOY 166). Candidate apparent emergence dates tested were actual emergence date ± 11 days. Candidate assumed effective cover dates for this update and succesive updates prior to effective cover are based on the most recent assumed effective cover date ± 7 days. Apparent emergence and new assumed effective cover dates calculated on this update were DOY 136 (May 15) and DOY 201 (July 19), respectively. The K_{cb} corresponding to these dates was 0.274 which fell within the limits on the K_{cr}. Two weeks lapsed before reflectance data were acquired for the third update using feedback. Equipment malfunctions were the main cause for the delay, however, various field operations were also occuring during this time. Regardless of all that, a major change in plant growth occurred during this period. On June 28 (DOY 180), the apparent emergence date was shifted to DOY 129 (May 8) and the new assumed effective cover date was considered to occur on DOY 194 (July 12) to account for rapid growth since the last update with feedback. The next update using feedback to the irrigation scheduling model for field B was on June 28 (DOY 187). Apparent emergence and assumed effective cover dates were each shifted back two days such that they were considered to occur on DOY 127 (May 6) and DOY 192 (July 10), respectively. Canopy reflectance data acquired on DOY 193 produced a K_{cr} = 0.937. Leaf area data and intercepted photosynthetically active radiation data acquired on that day indicated a leaf area index (LAI) of 3.1 and canopy ground cover of 78%. Effective cover for agricultural crops has been considered to occur around a LAI of 3 and/or 75% ground cover (Stegman, et al.,1980). Hinkle, et al. (1984) reported LAI values ranging from 2.2 to 3.2 for various varities of corn at effective cover. For the variety used in this study (Pioneer 3732), Bausch and Neale (1987,1989) indicated that the LAI was 3.2 at effective cover from previous field experiments. Thus, effective cover does not occur at a specific LAI, consequently, its

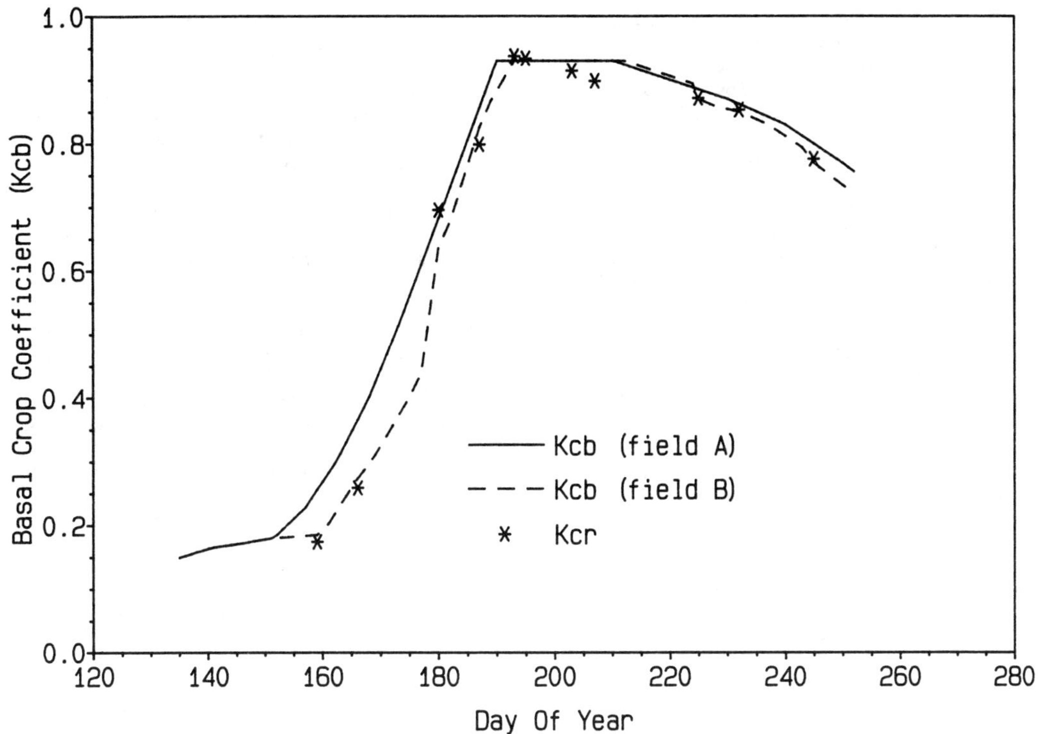

Figure 2. Basal crop coefficient curves for corn used in the irrigation scheduling program for fields A and B during the 1988 growing season. Reflectance-based crop coefficient (K_{cr}) for field B also shown.

actual occurrence is difficult to detect. Nevertheless, DOY 192 was tagged as the effective cover date.

Updates using feedback to the irrigation scheduling program during the 20 day period after effective cover were not used to adjust the basal crop coefficient. This assured that the K_{cb} remained at its maximum of 0.93 during this time in accordance with Wright's (1982) data for corn regardless of the calculated K_{cr} value. The update using feedback on August 12 (DOY 225) changed the apparent effective cover date to DOY 185 in an attempt to make the K_{cb} follow actual crop growth in the field. This update and successive updates with feedback consider candidate apparent effective cover dates as the effective cover date ± 7 days. On DOY 232 (August 19), apparent effective cover was changed to DOY 187; it was again set at DOY 185 with the update on September 1 (DOY 245).

The last irrigation for either field A or field B was a judgement call as to its usefulness. Corn development was near hard dough and the next scheduled irrigation was September 11 for field B and September 15 for field A. Black layer formation at the base of the kernel which indicates maturation occurred on September 16 (DOY 260). Thus, the last irrigation was probably beneficial. Seasonal ET computed for field A was 508 mm; for field B, it was 490 mm.

Hand sampled grain yields were obtained from ten locations in each field. Mean yield for field A was 12,497 kg/ha; field B produced 12,054 kg/ha. Kernel weight per 1000 kernels was 299.5 g and 301.3 g for field A and B, respectively. Yield as well as kernel weight was not significantly different between fields.

5 SUMMARY AND CONCLUSIONS

The USDA-ARS Irrigation Scheduling Program, SCHED, was modified to accept real time data to keep the basal crop coefficient (K_{cb}) for corn tracking actual crop growth in the field. Modifications were: (1) integration of an algorithm to shift the K_{cb}

curve with respect to its time axis, (2) addition of a data entry subroutine to enter the reflectance-based crop coefficient and its upper and lower bounds when these data were available, and (3) creation of a history file to save pertinent information for each field to restart the model if necessary.

Irrigations were scheduled in two fields during the 1988 growing season to test modifications made to the model. Field A irrigations were scheduled with the model operated in its traditional mode, whereas, field B irrigations were scheduled utilizing the real time reflectance-based crop coefficient as feedback to the model. Adjustments to the K_{cb} curve prior to effective cover delayed the irrigations in field B one day as compared to field A. This one day delay advantageously utilized rainfall that occurred on August 3 such that one less irrigation was required for field B.

The four adjustments to the K_{cb} curve during vegetative growth shifted the curve about its time axis in attempts to make the curve track periods of slow and rapid growth. These adjustments also made the assumed effective cover date converge on its actual occurrence. Computed seasonal ET was 508 mm for field A and 490 mm for field B.

ACKNOWLEDGEMENTS

The author gratefully acknowledges Mr. Daniel M. Lund for his expertise in computer programming as well as for his assistance in data collection.

REFERENCES

Bausch, W.C. & C.M.U. Neale. 1987. Crop coefficients derived from reflected canopy radiation: a concept. Trans. of the ASAE 30:703-709.

Bausch, W.C. & C.M.U. Neale. 1989. Spectral inputs improve corn crop coefficients and irrigation scheduling (to appear in Trans. of the ASAE).

Buchleiter, G.W., H.R. Duke & D.F. Heermann. 1986. User's guide for USDA-ARS irrigation scheduling program SCHED. USDA ARS

Duggin, M.J. 1980. The field measurement of reflectance factors. Photogram. Eng. and Remote Sens. 46:643-647.

Harrington, G.J. & D.F. Heermann. 1981. State of the art irrigation scheduling computer program. Irrigation Scheduling for Water and Energy Conservation in the 80's. Proc. ASAE Irrig. Sched. Conf., p. 171-178. Chicago, IL.

Heermann, D.F. & M.E. Jensen. 1970. Meteorological approaches in irrigation scheduling. National Symposium Proc. ASAE. NN1-NN11. St. Joseph, MI.

Heermann, D.F., H.R. Haise & R.H. Mickelson. 1976. Scheduling center pivot sprinkler irrigation systems for corn production in Eastern Colorado. Trans. of the ASAE 19:284287,293.

Hinkle, S.E., J.R. Gilley & D.G. Watts. 1984. Improved crop coefficients for irrigation scheduling. Final Report Project 58-9AH2-9-454. Agric. Engr. Dept. Univ. of Nebr., Lincoln,NE.

Hill, R.W., E.L. Johns & D.K. Frevert. 1983. Comparisons of equations used for estimating agricultural crop evapotranspiration with field research. U.S. BOR, Technical Report.

Jackson, R.D., M.S. Moran, P.N. Slater & S.F. Biggar. 1987. Field calibration of reference reflectance panels. Remote Sens. of Environ. 22:145-158.

Jensen, M.E. 1969. Scheduling irrigations with computers. J. Soil and Water Conserv 24: 193-195.

Jensen, M.E. 1973. Consumptive use of water and irrigation water requirements. ASCE. New York, NY.

Jensen, M.E., D.C.N. Robb & C.E. Franzoy. 1970. Scheduling irrigations using climate-crop-soil data. J. Irrig. and Drain. Div. ASCE. 96:25-38.

Jensen, M.E., J.L. Wright & B.J. Pratt. 1971. Estimating soil moisture depletion from climate, crop, and soil data. Trans. of the ASAE. 14: 954-959.

Kincaid, D.C. & D.F. Heermann. 1974. Scheduling irrigations using a programmable calculator. USDA Bulletin ARS-NC-12

Neale, C.M.U., W.C. Bausch & D.F. Heermann. 1989. Development of reflectance-based crop coefficients for corn. (to appear in Trans. of the ASAE).

Neale, C.M.U. & W.C. Bausch. 1983. Crop coefficients derived from reflected canopy radiation. ASAE Paper No. 83-2527. ASAE Winter Meeting. Chicago, IL.

Stegman, E.C. & D.A. Coe. 1984. Water balance irrigation scheduling based on Jensen-Haise equation: software for Apple II, II+ and IIe computers. Agric. Exper. Sta. North Dakota State Univ. Research Report 100.

Stegman, E.C., J.T. Musick & J.I. Stewart. 1980. Irrigation water management. In: Jensen, M.E. (ed.), Design and operation of farm irrigation systems. p. 763-816. ASAE Monograph No. 3. St. Joseph, MI.

Wright, J.L. 1982. New evapotranspiration crop coefficients. J. Irrig. and Drain. Div. ASCE 108: 57-74.

Land and Water Use, Dodd & Grace (eds), © 1989 Balkema, Rotterdam. ISBN 90 6191 980 0

PC computer model for matching and evaluating an irrigation application system, pump and power plant

R.G.Curley, G.D.Knutson & B.R.Hanson
University of California, Davis, Calif., USA

ABSTRACT: This paper describes a personal computer model which predicts how a pump will perform when coupled with an irrigation application system. The model will determine if the pump is worn by comparing "as is" performance with the performance of an identical pump operating "like new". It will predict how changes in an irrigation system will affect pressures and flows throughout the system. The model will also select an internal combustion engine from an engine data base and predict performance characteristics of both the engine and the pump/application system over a range of engine speeds.

RESUME: Ce travail décrit un modèle d'ordinateur personnel capable d'évaluer la performance d'une pompe en combinaison avec un système appliqué d'irrigation. Le modèle determinera si la pompe est déteriorée en en comparant le rendement "tel que" avec celui d'une pompe identique opérant "comme neuve". Il décidera comment des changements dans un système d'irrigation affecteront pressions et débits a travers le système. Le modèle selectionnera aussi un moteur à combustion interne à partir d'une base de données de moteur et décidera aux caractéristiques operationnelles du moteur ainsi que du groupe pompe/système d'application pour une variété de vitesses de moteur.

ABRIß: Diese Arbeit stellt ein Mikrocomputer Model dar, welches die Leistung einer Pumpe ermittelt, die an ein Bewasserungssystem angeschlossen ist. Das Model bestimmt ob die Pumpe abgenutzt ist, indem es die "bestehende" Leistung mit den Leistungswerten einer identischen Pumpe, die "wie neu" funktioniert, vergleicht. Es ermittelt inwiefern Veranderungen in einem Bewasserungssystem Druck and Fließung in der Anlage beeinflußen. Zusatzlich wahlt das Model einen Verbrennungsmotor von einer Motor-Datenbasis und bestimmt die Lestungsfaktoren des Motors sowie des Pumpen/ Anwendungssystems uber eine Spanne von Motorgeschwindigkeiten.

1 INTRODUCTION

The energy for operating irrigation pumps is a significant and extremely vital part of total agricultural energy use by California's irrigated agriculture. Annual energy use for irrigation pumping in California is approximately 12.5 million barrels of oil equivalent, representing 26.3 percent of the total energy used by agriculture (Cervinka, et. al 1981). Electric motors supply approximately 93 percent of the total irrigation pumping energy. The remaining seven percent is supplied by internal combustion engines, primarily diesel, natural gas and and propane.

Figure 1 shows how the cost of energy for irrigation pumping has risen in California during the past 35 years. It also shows the relative changes that have taken place between electricity costs for the electric motor and petroleum costs for diesel, natural gas and propane engines. Diesel fuel costs are currently enough lower than electricity costs to make the diesel engine competitive with the electric motor from a total cost standpoint. Natural gas and propane fuels aren't as cost competitive as diesel, however, these fuels have also decreased in price with respect to electricity.

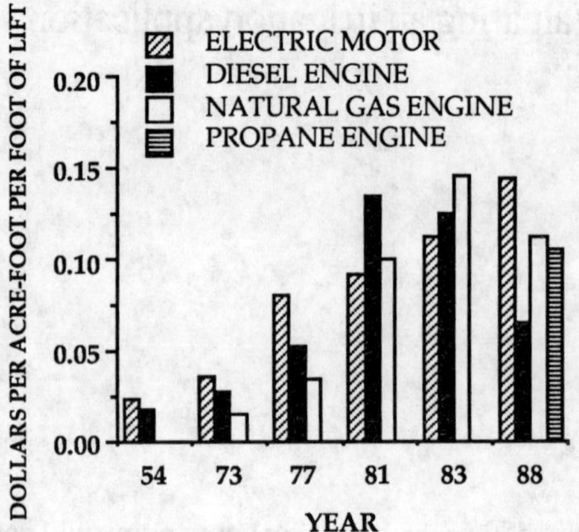

Figure 1. Relative energy costs for pumping by electric motor or engine power units.

High energy costs have combined with high irrigation energy requirements to substantially increase the cost of producing irrigated crops. This has brought about increased efforts to improve the efficiency of pumping plants and irrigation systems. The relative changes in costs between various energy sources and power units have caused farmers to carefully consider alternative power units and some are replacing their electric motors with internal combustion engines, primarily diesel.

The computer model described in this paper was developed as part of a University of California research and extension program to help farmers improve irrigation system efficiency and reduce costs. It can be used to evaluate the performance of an existing system or as an aid in designing a new system.

2 DESCRIPTION OF THE MODEL

The computer model will match the major components of an irrigation system and evaluate system performance under various operating conditions. The model develops mathematical equations which represent the performance characteristics of the pump, the irrigation application system and the engine (in cases where an engine is used to power the pump). The pump performance equations represent head vs. capacity and efficiency curves. Application system performance equations represent head vs. capacity curves. Typical curves for the pump and application system are shown in Figure 2. Engine performance equations represent

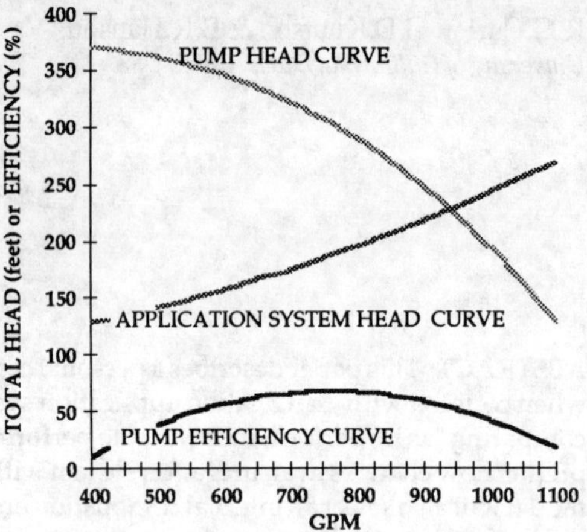

Figure 2. Typical performance curves for an irrigation pump and a pressurized application system.

brake horsepower and fuel consumption vs. engine rpm (Figure 3). From these curves, the model can predict how changes in the system will affect various performance factors.

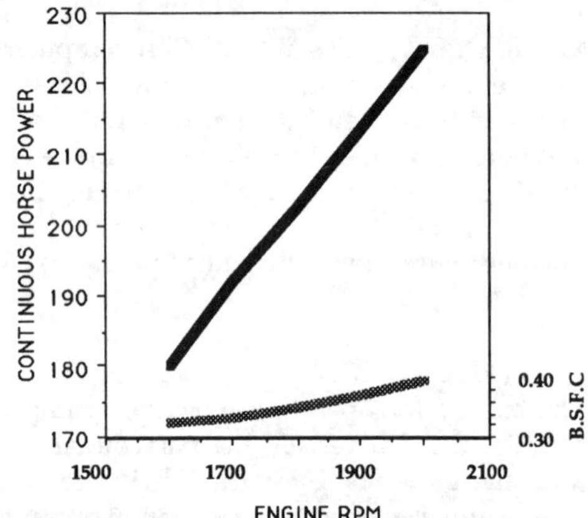

Figure 3. A manufacturer's engine performance curves for continuous horsepower and brake specific fuel consumption.

2.1 Input data
The model requires the following input data

1. Pump - At least four data points from the manufacturer's pump curves plus the results of a recent pump test. The pump test data is used to adjust manufacturer's performance data for wear, improper adjustment, etc.
2. Irrigation application systems - Lengths of the main line and each lateral; inside diameters of all pipes; pipe materials; number and location of sprinklers on each lateral and on the main line; make model and inside diameter of sprinkler nozzles.
3. Engine Matching - Manufacturer's performance curves for continuous brake horsepower and fuel consumption vs. engine rpm.

3 TYPES OF USES FOR THE MODEL
The model can be used to make the following analyses:

3.1 Pump Performance
The "as is" performance of a pump can be compared with predicted performance of an identical pump operating "like new" or with a different pump. This type of comparison predicts what will happen to various performance factors including flow rate, system pressures and energy costs in dollars per acre foot and dollars per year.

Table 1 is an abridged printout of the results of a pump performance comparison. In this example the pump is operating a sprinkler system 2000 hours per year. Columns in the table refer to pump condition or pump selected (Pump), pumping rate in gallons per minute (GPM), application system pressure (PSI), overall plant efficiency which is the combined efficiency of pump and electric motor based on an electric motor efficiency of 90 percent (OPE), acre feet of water pumped in 2000 hours of operation per year (AF) and the cost of electricity in dollars per acre foot of water pumped at an electric rate of $.08/kWh ($/AF).

Table 1. Pump performance comparison.

Pump	GPM	PSI	OPE	AF	$/AF
As is	940	50	45	346	41.45
Like new	1002	56	45	369	43.98
Different	940	50	65	346	28.70

The three lines of the table represent the following pump comparisons: 1) a pump operating in a worn condition, "as is", 2) the same pump operating in a "like new" condition and 3) a different pump that is better matched to the application system. In this comparison, an overhaul of the pump to bring it to a "like new" condition would increase the output and the pressure, however, the pump efficiency would remain the same. These changes would cause the energy cost in dollars per acre foot to increase from $41.45 to $43.98. Since pump output would increase with the "like new" pump the total amount of water applied in 2000 hours would also increase from 346 to 369 acre feet. This means that the "like new" pump could theoretically be operated fewer hours per season to deliver the same amount of water or could be used to irrigate more acres per season. A different pump, better matched to the system, would provide the same flow rate and pressure as the "as is" pump but would increase the overall plant efficiency to 65 percent. This would result in an energy cost of $28.70 per acre foot of water applied which is significantly lower than the original pump operating either "as is" or "like new".

3.2 Total system performance
This feature of the model will provide an analysis of pressures, flow rates, friction losses in the main lines and laterals of a sprinkler system and energy costs in dollars per acre foot and dollars per year. It is possible to use this analysis on a "what if" basis to predict the effect of various changes on these performance factors. Types of "what if" changes include the use of low pressure nozzles, increased or decreased pipe sizes and the addition of laterals to the system.

The unique feature of the performance analysis program is the ability to predict how a change will affect both the pump and the application system. For example, a change to low pressure nozzles in an attempt to reduce energy consumption will affect pump performance. This may result in increased energy consumption if the change results in a lower efficiency.

Table 2 is an abridged printout from the analysis of a single lateral in a sprinkler system. Pipe diameter is four inches and total flow rate is 911 GPM. Each line in the table represents a 300 ft section of lateral. The table shows the following data for each lateral: 1) pressure and flow rate for the last nozzle in the section, 2) friction loss in feet of head per 100 feet of length for each section and 3) the cost of the energy that would be required to overcome friction loss based on 2000 hours of operation per year. The overall plant efficiency is 65 percent and the price of electricity is $0.08 per kWh.

Table 2. Analysis of a single lateral in a sprinkler irrigation system

Section	Pressure*	Flow*	Friction loss	
			per 100 ft	per sect.
	(PSI)	(GPM)	(ft)	($)
1 (last)	35	6.73	.1	19
2	35	6.75	.6	120
3	36	6.83	1.5	369
4 (first)	38	7.03	2.8	845

* At last nozzle in the section.

This analysis makes it possible to quickly compare friction losses and nozzle flow rates in various sections of the lateral. In this example, friction loss would be 2.8 ft per 100 ft of lateral in the first section versus 0.1 ft per 100 ft of lateral in the last section. If the system were operated 2000 hours per year, friction loss in the first section would represent an electrical energy cost of $845 versus only $19 for the last section. It should be emphasized however, that an analysis of the application system by itself is inadequate because it doesn't indicate how the changes will affect the pump.

Table 3 is an abridged, three-part printout which shows how the model can be used to predict how changes in the application system will affect both the application system and pump. The pump in Table 3 is the same pump used for "as is" and "like new" comparisons in Table 1. The table shows performance predictions for three conditions, as fol lows: 1) sprinkler system unchanged, 2) standard pressure nozzles replaced with low pressure nozzles and 3) lateral section number 4 changed from four inch to six inch inside diameter pipe. The model makes it easy to change various design parameters such as pipe diameter and compare the results.

Table 3. The effect of changing to low pressure nozzles or to a larger diamter lateral.

Irrigation system unchanged					
Pump	**GPM**	**PSI**	**OPE**	**AF**	**$/AF**
As is	940	50	45	346	41.45
Like new	1002	56	45	369	43.98
Nozzles changed to low pressure					
Pump	**GPM**	**PSI**	**OPE**	**AF**	**$/AF**
As is	983	39	38	362	43.45
Like new	1044	44	37	384	47.48
Lateral section 4 changed from 4 to 6 inch inside diameter					
Pump	**GPM**	**PSI**	**OPE**	**AF**	**$/AF**
As is	953	47	43	351	41.93
Like new	1015	53	43	374	44.84

In recent years, some proponents of energy conservation have made blanket recommendations for low pressure nozzles as a means of reducing the energy requirements of sprinkler systems. The comparison between standard pressure nozzles and low pressure nozzles in Table 3 shows that low pressure nozzles would decrease overall plant efficiency by seven or eight percent and significantly increase the energy cost per acre foot for this pump and application system. The increase in energy cost would be even greater if the pump were operating in a "like new" condition. Table 3 also shows the effect of increasing the inside pipe diameter of lateral section number 4 from 4 inches to 6 inches. This change would increase flow rate and decrease both pressure and overall plant efficiency. The result would be a slight increase in the energy cost per acre foot of water applied. Both of those changes could have had positive benefits in other pump and application systems. This illustrates the importance of evaluating each system on a case by case basis.

3.3 Engine Matching

This feature of the model will match an internal combustion engine (diesel, natural gas or propane) to a pump and application system. It uses the procedure under "performance analysis", along with engine manufacturers' performance data to predict engine load, pump horsepower, pump efficiency, pressure at the pump and energy costs for the full range of engine rpm. It also flags rpm levels which will overload the engine. The model can utilize engine data supplied by a user or can draw on data for one or more of the 110 diesel and 40 gas engines stored in the data base. If necessary, the model will derate manufacturer's performance data for specific operating conditions of ambient temperature and attitude.

Table 4 is an abridged printout from the engine matching feature of the model. It shows the effect of engine RPM on pump horsepower, engine horsepower, engine load as a percent of maximum continuous load, pressure at the pump, pump efficiency and energy cost per acre foot of water applied.

Table 4. Effect of RPM on engine, pump and application system performance

RPM	Pump HP	Eng. HP	Eng. Load (%)	PSI	Pump Eff. (%)	$/AF
1600	100	145	75	50	50	11
1650	121	150	81	54	49	20
1700	134	154	87	59	47	22
1750	148	159	99	62	46	23
1800	160	163	100	67	45	25
1850	188	168	107	71	44	
1900	197	172	114	76	43	
1950	215	177	122	80	42	
2000	235	182	129	85	42	

4 LIMITATIONS ON USE OF THE MODEL

There are some limitations or restrictions on using the model that should be pointed out. The primary ones are as follows:

1. Pump input data requires at least four points from the manufacturer's head-capacity and efficiency curves. In some cases, farmers no longer have manufacturer's data for pumps that have been in use for several years.
2. Pump input data requires the results of a recent pump test for determining "as is" performance curves. Pump tests can usually be made without difficulty, however, there are exceptions.
3. The irrigation application system can have only a single non-branching main line.
4. The model is programmed to run on an IBM type computer using MBASIC programming language.

5 SUMMARY

Test results with the model to date have been satisfactory, however, further testing will be conducted prior to its release. It is our intent to make the model available for use by farmers, equipment suppliers, Extension staff and others.

REFERENCES

Cervinka, V., W. J. Chancellor, P. K. Avlani, N. J. Rupp, N. C. Thai, E.Q. Yee, 1981. Energy Requirements for Agriculture in California, 1978 Data Base. Report of joint study by California Department of Food and Agriculture and University of California, Davis.

Curley, R. G., E. B. Roberts, G. D. Knutson. Power Unit Costs for Irrigation Pumping, 1988. American Society of Agricultural Engineers Paper #88-2047.

Land and Water Use, Dodd & Grace (eds), © 1989 Balkema, Rotterdam. ISBN 90 6191 980 0

Design of check basin size for the irrigation of a wheat crop in clayey soils

K.P.Gore
Marathawada Agricultural University, Parbhani, India

ABSTRACT: An experiment was conducted during the years 1985, 1986 and 1987 at New Agricultural College farm, Marathwada Agricultural University in levelled medium black cotton soil (vertisol). The experiment was laid out in randomized block design with four sizes of check basin namely 10, 25, 50 and 100 sq. metre and inflow rate of 7.2 litres/second. Wheat crop was grown. The data on average time of water spread from second to sixth irrigations for various sizes of check basins was recorded. Values of emperical constants α, β and correlation coefficient were 10.58×10^{-2}, 1.10532 and 0.8577 respectively. Check basin sizes 50 sq. metre and 100 sq. metre gave significantly higher grain yield over 10 sq. metre and 25 sq. metre check basin sizes. From management point of view, 50 sq. metre check basin size is suitable for efficient use of irrigation water.

1 INTRODUCTION

Although check basin irrigation method is adopted by large number of farmers in India, pattern of check basin differs considerably in various parts of the country, depending upon the conventional practices followed. The input data required for design of irrigation systems can be classified into seven groups (Karmeli, D. et al. 1985), climatic data, soil characteristics, crop characterists, field, area and agrotechnology, equipment data, economic and sociological information and water information. For a successful irrigation system design, all these various parameters are to be considered to the maximum possible extent, and therefore, the problem of design of irrigation system or determining size of individual plot becomes complex. The farmers in general do not have correct ideas regarding proper size of check basin, inflow rates and characteristics of soils. With the suitable combinations of plot size, inflow rate, soil surface conditions and characteristics of soil, it is possible to apply adequate amount of water uniformly throughout the check basin. The study was, therefore, undertaken to find out proper size of check basin for irrigation wheat crop in clayey soil.

2 REVIEW OF LITERATURE

Information to determine the optimum size of check basin for efficient use of water is rather scanty. Following references are from Annual Report of I.C.A.R. scheme on Water Management and Soil Salinity from 1971 to 1975. It has been reported that on sandy loam soil of Roorkee check basin size of 7.5 m x 7.5 m gave higher wheat yield over 100 sq. m., 225 sq. m. and 400 sq. m. sizes. While at Hissar on similar type of soil with non-erosive stream check basin sizes of 10 m x 10 m, 15 m x 15 m, and 20 m x 20 m gave significantly higher grain yield over 5 m x 5 m size. It has been reported that 5 m x 5 m. check basin size gave highest wheat grain yield at Jobner on sandy soil higher wheat grain yields were reported in check basin size of 7.5 m x 7.5 m in black cotton soil at Sirguppa and in check basin size of 50 sq. m. with 10 lit/sec. inflow rate on sandy loam soil of Chackuli. Some varying results were also obtained at few other locations. Maximum wheat yield (25.46 q/ha) and water use efficiency (100 kg/ha/cm) has been reported in check basin size of 250 sq.m. and non-erosive inflow rates on sandy loam soil at Hissar. It was further reported that square shape was superior to other shapes. It has been indicated from study of four years data that 22 sq. m. check basin was found to be suitable for similar type of soil at Jobner as it has given higher water use efficiency with 2.77 lit/sec. discharge. Also irrigation with discharge rate 5.54 lit/sec. during 1974-75 resulted in increased yields (Yadav, 1971, 72, 73, 75). At Mahatma Phule Agricultural University, Rahuri on medium soil two check basin sizes i.e. 50 sq. m. and 100 sq. m. with inflow rates from 4.8 to 8.3 lit/sec. were tried, 50 sq. m. size was found to be optimum, higher efficiency of water application, water storage and water spread was observed as was indicated by higher grain and straw yields Wayse, (1980) from the review of literature, it can be clearly concluded that soil type, soil characteristics and inflow rate are the major factors affecting check basin size.

3 MATERIALS AND METHODS

The experiment was conducted at New College Farm, Marathwada Agricultural University, Parbhani during 1985, 1986 and 1987 on levelled medium black cotton soil (vertisol) having field capacity 35.03 per cent and permanent wilting point 19 per cent, porosity 52 per cent, and apparanent specific gravity 1.19. The experiment was laid out in randomized block design with four sizes of check basin.

Table 1. Details of check basin size

Length (m)	Width (m)	Area (m^2)
5	2	10
6.25	4	25
8.33	6	50
12.50	8	100

Land grading was carried out with tractor drawn leveller, final border bunds were constructed and finishing was done with mannual labour. Existing discharge of 7.2 lit/sec. was used and five replications were taken. Infiltration rate was measured with concentric cylinder infiltrometer. Wheat variety HDM 2189 was used for the experiment.

4 RESULTS AND DISCUSSION

Studies on infiltration rate and cumulative infiltration were carried out. The results are tabulated.

Table 2. Cumulative infiltration and infiltration rate (cm/hr)

Particulars	Time in minutes				
	0-15	30	60	90	120
Cumulative infiltration	0.70	1.35	1.91	2.20	2.49
Difference (R_2-R_1)	0.70	0.65	0.56	0.29	0.29
Infiltration rate	2.80	2.60	1.12	5.8	5.8

Data on average time of water spread from second to sixth irrigation for various sizes of check basin with inflow rated 7.2 lit/sec. is given in Table No. 3.

Table 3. Average time of water spread (minutes) for each plot size for inflow rate of 7.2 lit/sec.

Check basin are (m^2)	Irrigation number				
	2	3	4	5	6
10	1.33	1.36	1.35	1.34	1.36
25	3.74	3.72	3.69	3.68	3.71
50	7.81	7.94	7.98	7.96	8.24
100	17.20	17.24	17.16	17.16	17.25

Analysis of data indicated that average time of water spread for the stream size was parabolic function of plot size and can be presented by equation $TS = \alpha A^{\beta}$ where, TS = time of water spread in minutes, A = area of check basin in quare metre and α and β are empirical constants. The values of emperical constant α and β and correlation coefficient are given in

Table 4. Values of empirical constants and correlation coefficient with inflow rate 7.2 lit/sec.

Inflow rate lit/sec.	empirical constants α	β	Correlation coefficients
7.2	10.58×10^{-2}	1.10532	0.8577

From equation $TS = \alpha A^{\beta}$ time of water spread can be predicted. The yield of wheat obtained in various check basin size treatments is presented in Table 5.

Table 5. Wheat grain yields obtained in various check basin sizes

Check basin sizes (m^2)	Grain yield (q/ha)
10	20.69
25	25.12
50	29.10
100	29.43
SE ± 1.22	C.D. + 3.55

The yields obtained in 50 sq. m. and 100 sq. m. size check basin were statistically significant over 10 sq. m. and 25 sq. m. check basin size. Within the treatments of 50 and 100 sq. m. check basin sizes, there was no statistically significant differences in the yield.

5 CONCLUSIONS

Keeping in view practical preposition of preparing check basins, it may be advisable to use 50 sq. m. check basin size for efficient irrigation and for uniform application of water to wheat crop in clayey soil (vertisol) with discharge rate 7.2 lit/sec.

ACKNOWLEDGEMENT

The authors are grateful to Dr. S.V. Raikhelkar, Head, Department of Agronomy and Dr. V.K. Patil, Director of Instruction and Dean, Faculty of Agriculture, Marathwada Agricultural University, Parbhani for providing facilities and for constant encouragement during the course of study.

REFERENCES

Anonymous (1970). A Guide for estimating on irrigation water requirements, Water Management Division, Ministry of Agriculture, New Delhi, 17-20.

Gandhi, R.T. et at. (1971) Hand Book on irrigation management, Water Management Division, Ministry of Agriculture, New Delhi, III-2, IV-8.

Gore, K.P. and Vaishnava, V.G. (1987) Design of check basin size for efficient use of irrigation water, paper presented at 23rd Annual Convention of Indian Soc. of Agril. Engineers.

Israelson, O.W. and Hensen, V.E. (1962) Irrigation principles and practices, 3rd edition. John Wiley and Sons, New York.

Karmeli, D., Peri, G., Todes, M. (1985) Irrigation systems design and operation. Oxford University Press, New York, 7-19.

Khushalni, K.B. (1971) Irrigation practice and design. Vol. I to VII, Oxford and IBH, New Delhi.

Mazumdar, S.K. (1983) Irrigation engineering. Tata McGrawhill Co. Ltd., New Delhi, 13-19.

Michael, A.M. Mohan and Swaminathan, K.P. (1972). Design and evaluation of irrigation methods, Water Technology Centre, Indian Agril. Res. Institute, New Delhi.

Michael, A.M. (1982) Irrigation theory and practice, Vikas publishing House, New Delhi, 603-610.

Sharma, S.K. (1984) Principles and Practices of irrigation engineering, S. Chand and Co. Ltd., New Delhi, 109-134.

Yadav, J.S.P. (1971). Annual Progress Report of ICAR Scheme on Water Management and Salinity, ICAR, New Delhi.
- do - (1971-72)
- do - (1972-73)
- do - (1973-75)

Wayse, V.S. Gore, K.P. and Singh J.K. (1980) Paper published in journal of Institute of Engineers (India). Vol. 61.

Land and Water Use, Dodd & Grace (eds), © 1989 Balkema, Rotterdam. ISBN 90 6191 980 0

Comparison of model applicability for wheat crops

K.P.Gore
Marathawada Agricultural University, Parbhani, India

ABSTRACT: Historically and still in very large measure, irrigation planning is predicted on complete avoidance of water deficits. But, Agronomists, Economists and Engineers have long understood that not all water deficits are undesirable. It therefore, appears certain that irrigation planning should be based on purposeful imposition of water deficits. Though several theories and mathematical models have been developed between yield and water, the yield varies according to IW/CPE rations at different stages of crop growth. The present study is taken upto to develop and compare the yield model for wheat crop. The yield models (water production functions) were developed by using CSF for wheat for Parbhani (Maharashtra) for the year 1975-76 and 1976-77 and also it was developed for Kharagpur (West Bengal) region for the year 1977-78. The model developed at Parbhani can be applicable to Kharagpur data and vice-versa for giving the best yield predictions.

1 EXPERIMENTAL DETAILS

Experimental data from Parbhani for the year 1975-76 and 1976-77 is used to develop and test the mathematical model.

Table 1. Experimental data for the year 1975-76 at Parbhani (grain yield in q/ha)

Replication/ Treatment		R_1	R_{II}	R_{III}	Mean (q/ha)
0.45	S_1	.27.90	24.99	26.54	26.48
0.45	S_2	25.61	25.61	27.16	26.12
0.45	S_3	30.61	27.77	32.71	30.38
0.60	S_1	33.95	29.19	30.80	31.31
0.60	S_2	34.56	31.11	31.79	32.48
0.60	S_3	33.64	31.04	34.87	33.16
0.75	S_1	35.80	34.25	33.33	34.46
0.75	S_2	33.33	36.91	35.24	35.16
0.75	S_3	36.79	36.33	39.87	37.71
0.90		41.04	38.20	39.01	39.41

Table 2. Experimental data for the year 1976-77 at Parbhani (grain yield in q/ha).

Replications/ Treatment		R_I	R_{II}	R_{III}	Mean (q/ha)
0.45	S_1	27.18	27.03	20.24	24.81
0.45	S_2	24.30	29.62	22.83	25.92
0.45	S_3	27.22	29.62	22.83	28.20
0.60	S_1	26.54	26.79	30.24	27.83
0.60	S_2	28.39	28.82	30.55	29.25
0.60	S_3	39.25	32.28	32.71	34.75
0.75	S_1	31.66	29.32	33.33	31.43
0.75	S_2	35.74	38.27	31.72	35.24
0.75	S_3	33.33	33.95	38.58	35.28
0.90		40.53	47.53	40.49	42.81

2 MATHEMATICAL MODELS DEVELOPED BY USING CSF

The moisture of soil at field capacity is taken as a measure of standard value because, above it will be the percolation loss and below it plant will be subjected to deficiency in soil moisture resulting in reduction of both evapotranspiration and yield of crop. If the irrigation is frequent the soil moisture will not be in deficiency. If the irrigation interval will be more the soil will be subjected to soil moisture stress resulting in lack of available soil moisture and reduction in yield. If the irrigation will be made on the basis of irrigation water applied and pan evaporation ratio, there can be a definite trend in which the yield of a crop increase or decrease considering the different growth phases of crop if the moisture deficiency is made in one phase and in other phases. The soil is not in deficiency of moisture, the reduction in yield due to soil moisture stress at this phase of growth can be a function of IW/CPE ratio in which the irrigation has been made.

The relationship with the crop yield and IW/CPE to which the crop was subjected to some predetermined IW/CPE ratio for particular stage can be assumed as follows :

$$(CSF)_i = a + b\, s_i + c\, s^2_i + d\, s^3_i \quad (1)$$

When $(CSF)_i = 1 - \frac{y_i}{y_m} = 1 - a_i$ (crop susceptibility factor)

s_i = IW/CPE ratio in stage i

a_i = production function for stage i

y_i = yield when particular IW/CPE ratio is maintained at stage i

y_m = maximum yield

a, b, c and d are constants

The equation (i) can be expressed in the form of production functions of different stages :

$$a_i = 1 - (a+b\, s_i + c\, s^2_i + d\, s^3 i) \quad (2)$$

$$i = 1, 2, 3, \ldots\ldots n$$

considering the yield is multiplicative in nature yield model can be expressed as :

$$\frac{y_o}{y_m} = \sum_{i=1}^{n} \frac{y_i}{y_m} = \sum_{i=1}^{n} a_i \quad (3)$$

when y_0 = yield, when different or same IW/CPE ratios are maintained in all stages, the yield model in the form of production function can be expressed as follows :

$$y_0 = a_1, a_2, a_3, y_m \quad (4)$$

3 COMPUTATION AND RESULTS

The yield data Table 1 and 2 at Parbhani reveals that yield was affected by IW/CPE ratios at different growth stages of the crop. As the IW/CPE ratio decreased the yield gets reduced in every stage of growth in the year 1975-76, the average yield varies from 26.48 to 39.41 q/ha while in the year 1976-77 it ranged from 24.81 to 42.81 q/ha. From this, it is clear that the total irrigation water being same, yield is affected more due to stage one when compared to 2nd and 3rd stages. In other words, stage one is more susceptible for stress than any other stage. The values of the coefficients a, b, c and d as described in the equation (1) are evaluated in terms of IW/CPE ratio. The solution of equation (1) for different stages are given below for the year 1975-76 and 1976-77.

$$CSF = 2.57 - 9.18\, s_1 + 12.57\, s^2_1 - 6.04\, s^3_1 \quad (5)$$

$$CSF = 3.85 - 15.34\, s_2 + 21.96\, s^2_2 - 0.65\, s^3_2 \quad (6)$$

$$CSF = 0.368 + 1.14\, s_3 - 3.45\, s_3 2 + 2.06\, s^3_3 \quad (7)$$

and for the year 1976-77

$$CSF = 2.08 - 7.93\, s_1 + 12.99\, s_1^2 - 7.37\, s_1^3 \quad (8)$$

$$CSF = 0.104 + 2.01\, s_2 - 3.24\, s_2 2 - 1.11\, s_2^3 \quad (9)$$

$$CSF = 4.98 - 20.90\, s_3 + 30.88\, s_3^2 - 15.15\, s_3^3 \quad (10)$$

The values of crop susceptibility factors for the different stages based on experimental results are presented in Table 3 and Table 4 for the year 1975-76 and 1976-77 respectively.

Table 3. Crop susceptibility factor using data collected for the year 1975-76.

IW/CPE x	Stage I CSF y	Stage II CSF y	Stage III CSF y
0.45	0.4272	0.4676	0.4277
0.45	0.4314	0.3768	0.3768
0.45	0.5741	0.5196	0.5196
0.60	0.4417	0.4027	0.1743
0.60	0.4364	0.3937	0.3209
0.60	0.3637	0.3572	0.3119
0.75	0.3305	0.2481	0.2976
0.75	0.3832	0.1949	0.2856
0.75	0.2987	0.3326	0.1883

Table 4. Crop susceptibility factors using data collected for the year 1976-77.

IW/CPE x	Stage I y	Stage II y	Stage III y
0.45	0.3202	0.3760	0.2527
0.45	0.3911	0.3760	0.3234
0.45	0.5741	0.5196	0.5196
0.60	0.1728	0.1579	0.1804
0.60	0.2888	0.2420	0.2437
0.60	0.3637	0.3572	0.3119
0.75	0.1277	0.1877	0.0992
0.75	0.1654	0.1007	0.1148
0.75	0.2987	0.3326	0.1182

Where S_1, S_2, and S_3 are IW/CPE ratios for stages one, two and three respectively. These equations for the years 1975-76 and 1976-77 serve the basic functional relationship for the solution of the model. The yield function for the years 1975-76 and 1976-77 for different stages can be expressed as :

$$a_1 = \frac{y_o}{y_m} = -1.57 + 9.18\, s_1 - 12.57\, s_1^2 + 6.05\, s_1^3 \qquad (11)$$

$$a_2 = \frac{y_o}{y_m} = -2.85 + 15.34\, s_2 - 21.96\, s_2^2 + 10.65\, s_2^3 \qquad (12)$$

$$a_3 = \frac{y_o}{y_m} = 0.632 - 1.14\, s_3 + 3.45\, s_2^2 - 2.02\, s_3^3 \qquad (13)$$

and

$$a_1 = \frac{y_o}{y_m} = -1.08 + 7.93\, s_1 - 12.99\, S_1^2 + 7.37\, s_1^3 \qquad (14)$$

$$a_2 = \frac{y_o}{y_m} = 0.895 - 2.01\, s_2 + 3.24\, s_2^2 - 1.11\, s_2^3 \qquad (15)$$

$$a_3 = \frac{y_o}{y_m} = -3.98 + 20.90\, s_3 - 30.83\, s_3^2 + 15.15\, s_3^3 \qquad (16)$$

Therefore, yield models will be in the form for the year 1975-76 and 1976-77 for Parbhani are :

$$yo = (-1.57 + 9.18\, s_1 - 12.57\, s_1^2 + 6.04\, s_1^3)\,(-2.85 + 15.34\, s_2 - 21.96\, s_2^2 - 2.02\, s_3^3)\,(0.632 - 1.14\, s_3 + 3.45\, s_3^2 - 2.02\, s_3^3)\ y\ max \qquad (17)$$

$$y_o = (-1.08 + 7.93\, s_1 - 12.99\, s_1^2 + 7.37\, s_1^3)\,(0.895 - 2.01\, s_2 + 3.24\, s_2^2 - 1.11\, s_2^3)\,(-3.98 + 20.90\, S_3 - 30.83\, S_3^2 + 15.15\, S_3^3)\ y\ max \qquad (18)$$

subjected to

$$0.45 \leq s_1 \leq 0.9$$

$$0.45 \leq s_2 \leq 0.9$$

$$0.45 \leq s_3 \leq 0.9$$

Calculated (computed) values of yield using these models are presented in Tables 5 and 6 for both the years.

Tables 5. Calculated and observed values of yield using CSF model for the year 1975-76.

IW/CPE s_1	s_2	s_3	Calculated yield	Observed yield
0.45	0.90	0.90	23.37	26.98
0.90	0.45	0.90	25.95	26.12
0.90	0.90	0.45	29.95	30.38
0.60	0.90	0.90	29.65	31.31
0.90	0.60	0.90	30.65	32.48

0.90	0.90	0.60	30.86	33.16
0.75	0.90	0.90	32.86	34.46
0.90	0.75	0.90	32.47	35.16
0.90	0.90	0.75	35.44	37.71
0.90	0.90	0.90	40.95	39.41

(regression coefficient : 0.91)

Table 6. Calculated and observed values of yield using CSF model for the year 1976-77

IW/CPE s_1	s_2	s_3	Calculated yield	Observed yield
0.45	0.90	0.90	24.81	24.81
0.90	0.45	0.90	26.31	25.92
0.90	0.90	0.45	25.81	28.20
0.60	0.90	0.90	27.95	27.83
0.90	0.60	0.90	29.66	29.25
0.90	0.90	0.60	35.12	34.75
0.75	0.90	0.90	32.17	31.43
0.90	0.75	0.90	35.69	35.24
0.90	0.90	0.75	35.71	35.28
0.90	0.90	0.90	48.10	42.81

(regression coefficient : 0.93)

From these tables, it could be seen that observed and calculated values of yield are closer to each other, so that the models developed have given good results for both the years for the same location. The model developed in the year 1975-76 can be applied to the next year 1976-77 at the same location to give the best yield prediction. However, the values of coefficients and constants differed to a considerable extent. The difference may be due to climatic variations in these two years. Also these mathematical models were compared to I.I.T. Kharagpur data to test their validity at other location and the results are presented in Table 7.

Table 7. Comparison of the model applicability to Kharagpur data (1977-78).

IW/CPE S_1	S_2	S_3	Calculated Yield by CSF model Parbhani (1975-76)	Calculated yield by CSF model Parbhani (1976-77)	Observed yield by CSF model Kharagpur (1977-78)
0.9	0.9	0.9	32.38	31.92	31.90
0.9	0.9	0.8	25.29	28.93	26.85
0.9	0.9	0.7	23.67	26.40	26.65
0.9	0.8	0.9	26.21	26.48	28.80
0.9	0.7	0.9	22.32	24.56	24.65
0.8	0.9	0.9	23.72	26.99	29.20
0.7	0.9	0.9	20.99	24.67	28.40
0.8	0.7	0.9	18.35	24.64	25.45
0.7	0.8	0.9	18.19	20.46	25.75
0.7	0.7	0.9	19.78	19.10	24.13
0.8	0.8	0.9	19.20	22.39	26.00

(regression coefficient : 0.93 and 0.92)

It is seen that the model with the coefficient developed from the experimental data of Parbhani station can be applied to Kharagpur to give the best yield prediction for both the years. The coefficients constant of mathematical model were also calculated by taking the data of Kharagpur to test the variation in the model constant. The yield equation at Kharagpur is given as follows :

$$Y_o = (0.554 + 0.401\,S_1 + 0.218\,S_1^2 - 0.153\,S_1^3)\,(0.881 - 0.912\,S_2 + 1.07\,S_2^2 - 0.139\,S_2^3)\,(0.526 + 0.684\,S_3 - 0.337\,S_3^2 + 0.126\,S_3^3)\;y_{max} \quad (19)$$

Subjected to

$$0.45 \leqslant s_1 \leqslant 0.9$$

$$0.45 \leqslant s_2 \leqslant 0.9$$

$$0.45 \leqslant s_3 \leqslant 0.9$$

Equation (19) has been applicable to Parbhani data for both the years 1975-76 and 1976-77 and results are given in Tables 8 and 9.

Table 8. Comparison of the model applicability to parbhani data (1975-76)

IW/CPE S_1	S_2	S_3	Calculated yield by CSF model Kharagpur q/ha (1977-78)	Observed yield 1/ha at Parbhani (1975-76)
0.45	0.90	0.90	24.96	26.48
0.90	0.45	0.90	26.07	26.12
0.90	0.90	0.45	25.85	30.38
0.60	0.90	0.90	27.41	31.31
0.90	0.60	0.90	26.65	32.48
0.90	0.90	0.60	28.02	33.16
0.75	0.90	0.90	29.78	34.46
0.90	0.75	0.90	28.89	35.16
0.90	0.90	0.75	30.03	37.71

0.90	0.90	0.90	31.96	39.41

(regression coefficient : 0.93)

Table 9. Comparison of the model applicability to Parbhani data (1976-77)

IW/CPE S_1	S_2	S_3	Calculated yield by CSF model Kharagpur q/ha (1977-78)	Observed yield q/ha at Parbhan (1976-77)
0.45	0.90	0.90	28.89	24.81
0.90	0.45	0.90	30.21	25.92
0.90	0.90	0.45	29.94	28.20
0.60	0.90	0.90	31.74	27.83
0.90	0.60	0.90	30.90	29.25
0.90	0.90	0.60	32.45	34.75
0.75	0.90	0.90	34.50	31.43
0.90	0.75	0.90	33.50	35.24
0.90	0.90	0.75	34.80	35.28
0.90	0.90	0.90	37.10	42.81

(regression coefficient : 0.91)

4 CONCLUSIONS

Based on the results obtained it can be concluded that the observed and calculated values are closer to each other. It is seen from these tables that with the model constants calculated from Kharagpur (West Bengal) data yield function can effectively be used to Parbhani for best yield prediction and vice-versa. Equations (17), (18) and (19) present a lot of variations, in the model constants though every one of them could predict yield either of the location at a statistically significant level, structurally the model is suitable and can effectively predict yield under different stress constraints. But the variation in the model constants perhaps is due to the polinominal nature of the model. In fact a slight change in either a constant or coefficient of any power brings change in all the other constant and coefficient values.

ACKNOWLEDGEMENT

The author expresses his heart felt thanks to Dr. K.A. Rao and Prof. Jaswant Singh from I.I.T. Kharagpur for their valuable guidance, constant encouragement, timely suggestions in developing and testing the mathematical models (water production functions)

REFERENCES

Aher, A.T. 1976. Optimal irrigation planning for wheat crop of Maharashtra State. Unpublished M.Tech. Thesis. Agril. Engg. Dept. I.I.T. Kharagpur.

Arya, Y.C. and Hagan, R.M. 1969. Programming in water use. Journal of Agricultural Engineering, ISAE 6(1), 64-68.

Hagan, R.M. and Stewart, J.I. 1972. Water deficits. irrigation design and programming. Journal of the irrigation and drainage division. ASCE 98 (IR-2) 215-237.

Hall and Buras, N. 1961. Optimum irrigated practice under conditions of deficit water supply. Trans. ASAE 41(2) 131-134.

Hall, W.A. and Butcher, W.S. 1968. Optimal timing for irrigation. Journal of irrigation and drainage division. Prof. ASCE, 94 (IR-2) 267-276.

Hall, W.A. and Dracup, J.A. 1970. Water resources system engineering. McGraw Hill Book Company, New York.

Hanks, R.J. 1973 Model for Predicting Plant yield enfluenced by water use. Utah. Agr. Expt. station journal paper No. 1804-20 P.

Hiler, E.A. and Clark R.N. 1971. Stessday index to characterised effect of water stress on crop field. Trans. ASAF 14(4), 757-759.

Howell, T.A. and Hiler, E.A. 1976. Optimization of water use efficiency under high frequency irrigation. Evapotranspiration and yield relationship trans. ASAE 873-878.

Israelson, O.W. and Hanson, V.E. 1972. Irrigation principles and practices. Johnwilley and Sons, Inc., New York.

McCulloch, J.S.G. 1965. Tables for rapid computation of the penman estimate of evaporation 286-295.

Minhas, B.S., Parikh, K.S. and Shrinivasan, T.N. 1974. Towards the structure of production function of wheat yield with dated input of irrigation water. Water Resources Research 10(3), 383-386.

Patel, K.S. 1978. Development of production functins for wheat. Unpublished M.Tech. Thesis. Agril. Engg. Dept. I.I.T. Kharagpur.

Stewart, J.I. and Hagan, R.M. 1973. Functions to predict effects of crop water deficits, journal of irrigation and drainage division ASCE 99 (IR. 4) 421-439.

Stewart et al. 1975. Irrigating corn and grain sorghum with deficit water supply. Trans ASAE, 18(2) 270-275.
Wu, I.P. and Tung Liong 1972. Optimal irrigation policy and frequency. J. Irri. and Drainage. Div. Proc. ASCE 98 (1), 117-121.

Land and Water Use, Dodd & Grace (eds), © 1989 Balkema, Rotterdam. ISBN 90 6191 980 0

Applicability of selected methods of determining hydraulic properties of a tropical soil

F.I.Idike
Department of Agricultural Engineering, University of Nigeria, Nsukka, Nigeria

C.J.Ejieji
Department of Agricultural Engineering, University of Ilorin, Nigeria

ABSTRACT: The Gupta and Larson (1979) regression equation for determining soil moisture characteristics, the Campbell (1974) equation for evaluating conductivity - moisture content relationship, and the Mein and Larson (1971) method for determining average capillary suction at the wetting front were all applied to a largely predominant local soil in Nigeria and the results compared favourably with laboratory determined values. Kutilek's (1980) simple field method for determining soil water diffusivity was also experimentally evaluated on the same soil and the result compared to that obtained from the widely acknowledged Arya et al (1975) laboratory method. The comparison showed that Kutilek's field method did not yield satisfactory results due to the unreliability of the techniques for obtaining the soil parameters required in practical application of the method.

Die Regressionsgleichung von Gupta und Larson (1979) zur Bestimmung von Charakeristika der Bodenfeuchtigkeit; die Gleichung von Campbell (1974) zur Bewertung des Verhältnisses von Durchlässigkeit und Feuchktigkeitsgehalt; die Methode von Mein und Larson (1971) zur Bestimmung der durchschhittlichen Kapillaraufnahme an der Befeuchtigungsfront wurden alle auf einen überwiegend örtlichen Boden angewandt, und die Ergebnisse waren gut vergleichbar mit den Laborwerten. Kutileks einfache Feldmethode (1980) zu Bestimmung der Verteilung von Wasser im Boden wurde auch experimentell erprobt auf demselben Boden mit vergleichbaren Ergebnissen zu denen der allgemein anerkannten Labormethode von Arya et al (1975). Des Vergleich zeigte, dass Kutileks Feldmethode keine befriedigenden Ergebnisse erzielte, wegen der Unzuverlässigkeit der Techniken zur Bestimmung der Bodenparameter die zu der praktischen Anwendung dieser Methode benötigt werden.

L'équation de regression Gupta et Larson (1979) pour déterminer les caractéristiques de l'humidité du sol; l'équation Campel (1974) pour l'évaluation du rapport conductivité - humudité; et la méthode Meine et Larson (1971) pour déterminer l'aspiration capillaire moyenne au front de mouillage, autant de théories ont été appliquées à l'analyse du sol largement prédominant au Nigéria, et les résultats sont bien comparables a ceux obtenus au laboratoire. A titre expérimental, la méthode Kutilek (1980) pour déterminer la qualité diffusive de l'eau du sol, a été appliquée dans l'evaluation du même sol, et le résultat beaucoup similaire à celui obtenu de la méthode de laboratoire bien connu par Arya et al (1975). Une étude comparée des résultats a révélé que la Méthode Kutilek n'a pas donné de résultats satisfaisants, ce qui est dû à l'inexactitude des téchniques relatives aux paramètres du sol exigées dans l'application pratique de la méthode.

1 INTRODUCTION AND LITERATURE REVIEW

Soil water balance is affected by such processes as infiltration, evaporation, redistribution and drainage, among others. Hydraulic properties such as hydraulic conductivity, soil water retention characteristics, capillary suction at the wetting front, soil water diffusivity, etc, are usually required for the practical application of theories describing these processes.

Simple methods for determining some of these hydraulic properties have been deviced. Arya et al (1975) proposed a relatively inexpensive laboratory method for determining soil water diffusivity and unsaturated hydraulic conductivity of undisturbed soil samples. Ehlers (1976) found that this method gave rapid results comparable to those obtained in-situ. Kutilek (1980), presented a simple field method for determining soil water diffusivity using sorptivity measured with ponded ring infiltrometer. An equation for predicting hydraulic conductivity - moisture content relationship from soil water retention characteristics was proposed by Campbell (1974) and modified by Gosh (1977). Gupta and Larson (1979) have presented a regression equation for predicting soil water retention characteristics from particle size distribution, organic matter percent and bulk density while Mein and Larson (1971) have suggested a simple method of obtaining average capillary suction at the wetting front. In his infiltration studies, Idike (1977, 1981) applied the above last three methods with satisfactory results.

These simple and inexpensive methods which have given good results elsewhere appear attractive in situations where facilities and resources for sophisticated instrumentation are lacking. The objective of this study is to evaluate the applicability of such simple methods on a Nigerian soil. Specifically the following experiments and investigations were carried out:

1. Soil water diffusivity - moisture content relationship determination employing the laboratory method of Arya et al (1975) and the field method of Kutilek (1980) and comparing the results obtained.
2. Comparison of the laboratory determined soil hydraulic conductivity - moisture content relationship with that predicted by the Campbell (1974) equation.
3. Verification of the applicability of the equation of Gupta and Larson (1979) to the local soil by comparing the predicted moisture characteristics with that determined in the laboratory.
4. Determination of the wetting front capillary suction value for the soil using the Mein and Larson (1971) method.

2 EXPERIMENTAL PROCEDURE, MATERIALS AND METHODS

The experimental site is a deep, porous and ferrallitic soil that belongs to the group broadly classified as oxisol and predominates in the Nsukka zone of Anambra State of Nigeria.

For the laboratory tests and determinations, disturbed and undisturbed soil samples were randomly taken, upto the depth of 60 cm at 15 cm depth intervals, from two 2x2 m blocks, 6.0 m apart as reported by Ejieji (1983). The disturbed samples were obtained for the determination of the mechanical and chemical properties of the experimental soil while undisturbed samples were taken for the determination of saturated moisture content, bulk density, saturated hydraulic conductivity using the falling head permeameter, moisture retention characteristics using the pressure plate apparatus and soil water diffusivity. A minimum of four samples were taken from each depth interval of each block. Two sets of double ring infiltrometers, 15 cm thick and 20 cm high and made of steel sheet were used in the Kutilek's (1980) approximate field method for determining soil water diffusivity. The internal diameters of the inner and outer rings of the infiltrometer sets were 10/30 cm and 20/50 cm respectively. They were eight infiltrometers per set and they were randomly installed around the sampling block.

Arya et al (1975) method of determining diffusivity and unsaturated hydraulic conductivity uses the drying technique based on Equ.1 and derived from the Boltzmann transform method of determining soil water diffusivity.

$$D(\theta_x) = \frac{1}{2t} \left(\frac{dx}{d\theta}\right) \theta_x \int_{\theta_x}^{\theta_i} x d\theta \quad \ldots\ldots(1)$$

with boundary conditions

$$\theta = \theta_i : x > 0, t = 0$$
$$\theta = \theta_o : x = 0, t > 0$$

where $D(\theta_x)$ is diffusivity (cm^2/min) as a function of moisture content, θ; x is the distance, θ_i and θ_o are the initial and air dry moisture contents respectively. Details of the experimental procedure adopted in this investigation has been described by Ejieji (1983). Briefly stated, each undisturbed soil core, after being thoroughly wetted and allowed to equilibrate, was exposed at one end to hot air drying in such a manner as to satisfy the requirement that the cumulative evaporation be proportional to the square root of time and the initial moisture content of the unexposed end of the soil core remain unchanged at the end of the drying process. Thereafter, the moisture distribution of each successfully dried sample was plotted and diffusivity as a function of moisture content, θ calculated using Equ.1.

Calculated diffusivities were used to determine the best fitting $D(\theta)$ curve from which hydraulic conductivities $K(\theta)$ (cm/min) were calculated according to the equation

$$K(\theta) = D(\theta) \frac{d\theta}{d\Psi} \qquad \ldots\ldots(2)$$

where $\frac{d\theta}{d\Psi}$ is the slope of the soil water retention characteristics curve and Ψ is the matric potential (cm of water).

Kutilek's field method for determining diffusivity using sorptivity measured with ponded ring infiltrometer utilizes the flux concentration theory of Philip and Knight (1974) to obtain the expression

$$S = \left\{2(\theta s-\theta i) \frac{a}{B} [\exp(B\theta s) - \exp(B\theta i)]\right\}^{1/2} \qquad \ldots\ldots(3)$$

where S is the sorptivity; θi and θs are the initial and saturated volumetric water contents respectively while a and B are the parameters of the soil water diffusivity model which Kutilek assumed to be of the form

$$D(\theta) = a \exp(B\theta) \qquad \ldots\ldots(4)$$

where $D(\theta)$ is diffusivity as a function of moisture content. The field procedure suggested by Kutilek has been described by Ejieji (1983) and involves:

1. Determination of sorptivity, S by fitting the result of a short-time ponded ring infiltrometer test to the Philip (1957b) equation

$$I = St^{1/2} + At \qquad \ldots\ldots(5)$$

where I is the cumulative infiltration, t is the time and A is a constant.

2. Estimation of minimum soil water content, θmin at minimum soil water diffusivity, Dmin where Dmin is defined as the point on the $D(\theta)$ plot where liquid flow starts to be substantially lower than the vapour flux. Kutilek suggests

$$Dmin = 1 \times 10^{-3} \text{ cm}^2/\text{min and}$$
$$\theta min = 0.003 \text{ (\% clay)}.$$

3. Evaluation of a and B using an interative procedure outlined by Kutilek (1980) and involving Equs. 3 and 4 and the values of θmin and Dmin.

The Gupta and Larson (1979) regression equation for soil water retention characteristics is of the form

$$\theta_p = a \times (\% \text{ sand}) + b \times (\% \text{ Silt}) + c \times (\% \text{ clay}) + d \times (\% \text{ organic matter}) + e \times \text{bulk density } (g/cm^3) \qquad \ldots\ldots(6)$$

where θ_p is the predicted water content (cm^3/cm^3) for a given matric potential and a, b, c, d, and e are regression coefficients for a given matric potential established by the authors (Gupta and Larson). Laboratory determined particle size distribution, organic matter percent and bulk density of the soil were used in Equ.6.

The Campbell (1974) method for determining conductivity - moisture content relationship assumes that the moisture retention function can be represented by the equation

$$\Psi = \Psi_e(\theta/\theta s)^{-b} \qquad \ldots\ldots(7)$$

where Ψ_e is the air entry water potential, θs is saturated moisture content and b is an empirically determined constant. The hydraulic conductivity, K, as modified by Gosh (1977), is given by

$$K = Ks\left(\frac{\theta}{\theta s}\right)^{1.5b+3} \qquad \ldots\ldots(8)$$

where Ks is saturated hydraulic conductivity. The data points of the soil water retention characteristics plotted on a log-log scale should produce a straight line with a negative slope numerically equal to 'b'. The equation of this straight line can be expressed as

$$Su = x\theta^{-b} \qquad \ldots\ldots(9)$$

where x and b are the intercept and slope respectively and Su is the capillary suction. Using the laboratory determine value of Ks and Equs.8 and 9, values of capillary suction, Su and conductivity, K were calculated for a series of moisture contents.

In applying the Mein and Larson (1971) suggestion on appropriate wetting front capillary suction value, the area under the capillary suction-relative conductivity (Su - Kr) curve for Kr range of

0.01 - 1.0 where $Kr = K(\theta)/Ks$ was computed.

3 RESULTS AND DISCUSSION

The mechanical and chemical properties of the experimental soil revealed that the 0 - 15 cm soil depth may be designated as sandy clay loam while below this depth the soil is sandy clay. The soil has low organic matter and is considerably acidic.

3.1 Soil water diffusivity

The condition imposed on the drying process of the laboratory method of Arya et al (1975) was achieved by maintaining a hot air (60°C) hair dryer at 15 cm from the exposed end of the soil sample. Linearity between cumulative evaporation and the square root of time was generally achieved within 2 minutes while the drying could be continued for about 12 minutes on the average, before the water content at the unexposed end of the soil sample dropped significantly. Diffusivity values were calculated only for the 0 - 15 cm soil depth as the differences in moisture contents between the exposed and unexposed ends of the soil samples from the other soil depths did not, on the average, exceed 0.05 cm^3/cm^3. The higher clay contents of these depths would appear to confirm the doubt raised by Ehlers (1976) about the applicability of the method on high clay soils.

The calculated diffusivity values showed large variability at a given moisture content over the replicate samples probably due to stochastic heterogeneity of the samples resulting from differences in mechanical/physical properties. Within the moisture content range of 0.16 to 0.32 cm^3/cm^3 where the replicate samples gave comparable results, the geometric means of the replicate diffusivities at each moisture content gave a reasonable approximation of the arithmetic mean that would result from discarding extremely large and small diffusivity values. A plot of the geometric means of the replicate diffusivities at the various moisture contents is shown in Fig.1 and the best fitting curve developed using the logarithm (base 10) of these means is of the form

$$D(\theta) = 10^{-0.081-20.742\theta+61.084\theta^2} \qquad \text{......(10)}$$

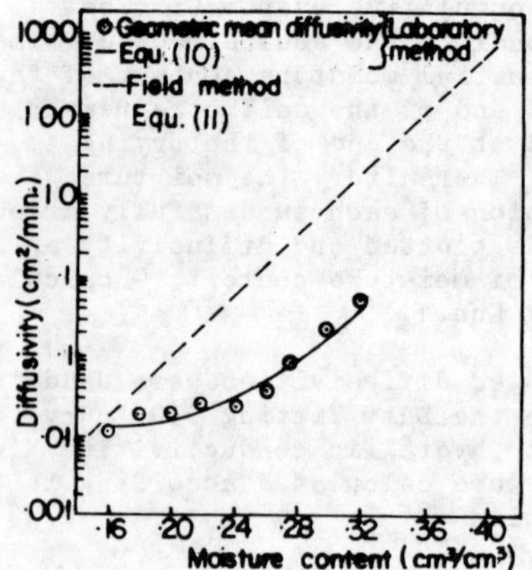

Fig.1 Soil water diffusivity - moisture content relationship for the top soil (0-15 cm) determined in the laboratory (Arya et al, 1975) and in the field (Kutilek, 1980).

with an R^2 of 0.98.

A major short-coming of the Arya et al (1975) laboratory method is the effect of high temperature of the drying source which can cause temperature gradients and their associated vapour fluxes. In this study, a relatively low drying temperature (60°C) was achieved and the undisturbed soil cores (samples) were contained in copper cylinders which have high thermal conductivity. It is therefore, likely that the effect of thermal gradients would be less than in those reported by Arya et al (1975) and Ehlers (1976). Thus, apart from the observation already made to the effect that the method may not be suitable for high clay soils, the soil water diffusivity values obtained in this study are considered reliable.

The Kutilek's field method was applied to the top soil (0 - 15 cm) depth where laboratory determined diffusivity values were available for comparison. The average sorptivity value for the top soil using the infiltrometer set with an internal inner ring diameter of 10 cm was estimated as 2.10 cm $min^{-½}$ while the other set of infiltrometers gave an average sorptivity value of 2.40 cm $min^{-½}$. The analysis of variance of individual determinations of sorptivities which gave the above averages showed that at 95% probability the average sorptivities were not affected by infiltrometer sizes. The measured initial moisture contents before the infiltrometer runs did not vary significantly and averaged 0.017 cm^3/cm^3. By using the interative procedures proposed by Kutilek (1980) and the appropriate values of Θmin and Dmin, Equ.4 was established as

$$D(\Theta) = 1.973 \times 10^{-5} \exp(42.67\Theta) \quad \text{.....(11)}$$

Fig.1 shows that, except at moisture content of 0.16 cm^3/cm^3, there is no agreement between the plot of diffusivity values generated with Equ.11 and the laboratory determined values. In establishing Equ.3, Kutilek did not consider the effect of hydrostatic head at the absorption interface and assumed short time infiltration where gravity effect is negligible. Consequently an appropriate sorptivity ought to be determined with a technique that permits the soil matrix to absorb water to saturation without surface ponding. The ponded infiltrometer method will yield sorptivity under ponding condition when sorption takes place preferentially in the vughs, planar voids and channels constituting the soil macropores (Clothier and White, 1981). Since diffusivity is proportional to the square of sorptivity, Kutilek's method will be very sensitive to errors in sorptivity estimation.

To ascertain whether an erronous field determination of sorptivity was the source of the discrepancy in the plots on Fig.1, the matrix sorptivity of the top soil was alternatively estimated with the expression for predicting sorptivity; S of soils with infinitely steep wetting front given by Philip (1958).

$$S = \left\{2\, Ks(C_f + d)(\Theta_s - \Theta_i)\right\}^{1/2} \quad \text{......(12)}$$

where Cf is the wetting front capillary suction, d is the depth of water above the soil surface, Θ_i is the initial moisture content and Ks is the hydraulic conductivity at saturated moisture content, Θ_s. The approximation involved in the use of Equ.12 is consistent with the simplifying assumptions made by Kutilek (1980) in deriving Equ.3. By applying Equ.12, the soil matrix sorptivity was calculated as 0.18 cm $min^{-1/2}$ while Equ.4 was established as

$$D(\Theta) = 1.172 \times 10^{-4} \exp(23.3\Theta) \quad \text{......(13)}$$

The closer agreement between the new plot of diffusivity values generated with Equ.13 and the laboratory determined values (Fig.2) suggests that the latter estimate of sorptivity is more

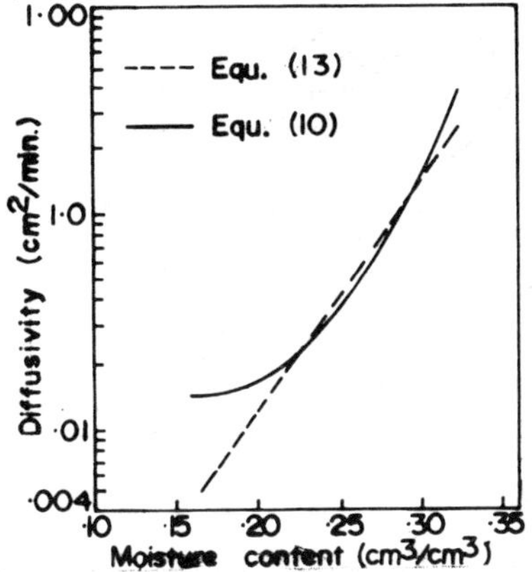

Fig.2 Laboratory determined soil water diffusivity - moisture content relationship for the top soil (Equ.10) compared with that calculated by the method of Kutilek (1980) using an improved estimate of sorptivity (Equ.13).

realistic than that determined with ponded infiltrometer.

3.2 Soil water characteristics

In all the sampled depths, the regression equation of Gupta and Larson (1979) overpredicted moisture contents at all matric suctions, with the overall best agreement shown in the top (10 - 15 cm) soil depth (Figs.3 & 4). Generally, better agreement between measured and predicted values was reflected in the 0.05 - 1.0 bar matric suction range. Similar over-predictions noted in the original work of Gupta and Larson (1979) were attributed to the presence of coarse (greater than 2 mm) fragments in the experimental soil sample. In the present study, the undisturbed samples not only contained fractions coarser than 2 mm but also visible ant and root holes. Moreover the pressure plate apparatus approach employed in this study involved rewetting a sample and subsequently desorping at a higher pressure; a process that introduces scanning hysteresis loops normally enveloped by the primary loops. Hence, except at the initial 0.05 bar suction, the moisture content recorded at each pressure could have been lower than would have been obtained by the equilibration procedure of Gupta and Larson (1979).

A major limiation of the regression equation of Gupta and Larson is that the

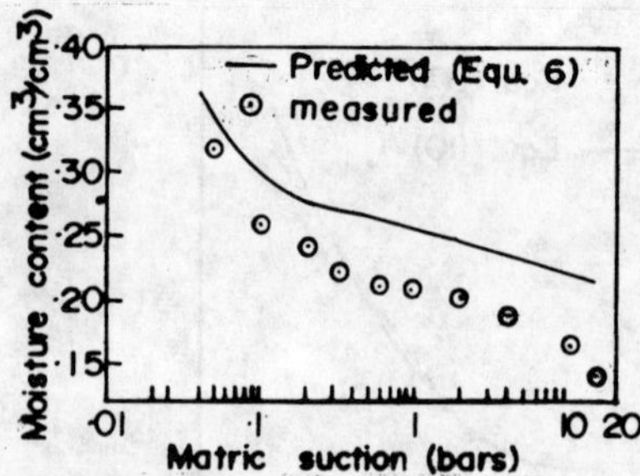

Fig.3 Measured and predicted soil moisture characteristics for the 0-15 cm depth of Block I.

soil samples used in generating the regression coefficients were spatially uniform in bulk density and texture while the particle sizes did not exceed 2 mm. This does not represent actual field condition where particle sizes greater than 2 mm occur and the natural composition of the soil as well as activities of soil flora and fauna can cause random point variations of texture and bulk density even in a small body of soil. Perhaps the performance of the equation can be improved by using undisturbed soil cores to generate the regression coefficients. Despite the limitation and procedural problems discussed above, the regression equation of Gupta and Larson (1979) represents very accurately, the trend of the measured soil water retention characteristics. The predictions are not only considered suitable for modeling soil water flow processes on the experimental soil but are on the safe side for use in irrigation scheduling based on the available water concept.

3.3 Hydraulic conductivity - moisture content relationship

The relationship between unsaturated hydraulic conductivity and soil moisture content (Equ.8) was established as

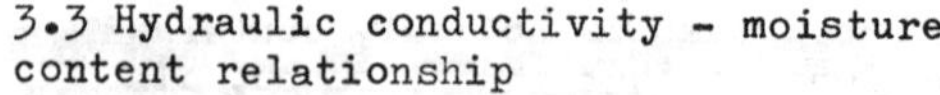

$$K(\theta) = 0.149(\theta/0.431)^{20.342} \quad ......(14)$$

Fig.5 compares the hydraulic conductivity - moisture content relationship predicted by Equ.14 with that computed from the Arya et al. (1975) laboratory determined diffusivities (Equ.2).

The figure shows that over the range of recorded moisture content, the

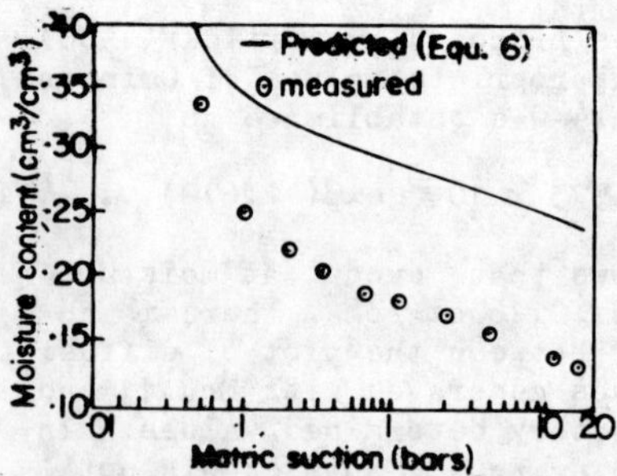

Fig.4 Measured and predicted soil moisture characteristics for the 30-45 cm depth of Block II.

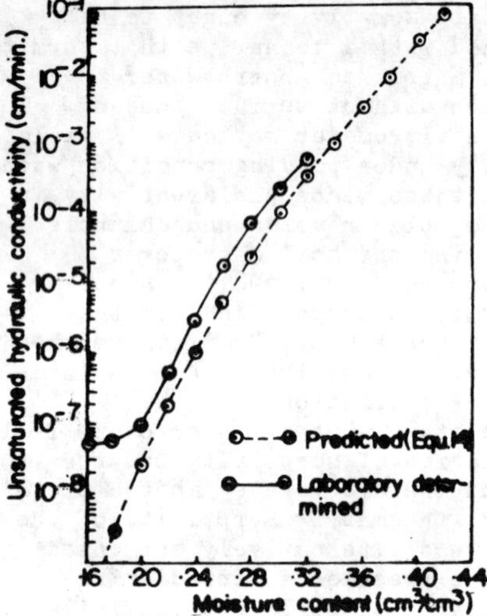

Fig.5 Predicted and laboratory determined unsaturated hydraulic conductivity - moisture content relationship for the top soil (0-15 cm) depth.

conductivity values computed from laboratory measurements were less than those predicted by the Gosh-modified Campbell equation. The above discrepancy, however, appear to be within the range of variability that can reasonably be expected from already discussed variability in the measured diffusivity values from which the conductivity values were calculated (Equ.2). The Campbell equation has therefore performed well in predicting the hydraulic conductivity - moisture content relationship for the top (0 - 15 cm) soil depth of the experimental soil.

3.4 Wetting front capillary suction

Table 1 shows the values of average capillary suctions at the wetting front for the various depths of the two blocks of the experimental soil. Based on the already discussed performance of Equ.2 in reasonably predicting conductivity values of the experimental soil and the close agreement (except for the 45 - 60 cm depth) of the values presented in table 1, the computed wetting front average capillary suction values are considered reliable for modeling infiltration on the experimental soil.

Table 1. Average capillary suction at the wetting front for the various depths of the two blocks of the experimental soil.

Soil depth (cm)	Average capillary suction at the wetting front (cm of water)	
	Block 1	Block II
0 - 15	2.50	2.50
15 - 30	4.75	3.75
30 - 45	3.25	3.25
45 - 60	6.25	0.88

4 CONCLUSION

The results of this study show that the Gupta and Larson (1979) regression equation for determining soil moisture characteristics; the Campbell (1974) equation for evaluating conductivity - moisture content relationship; and the Mein and Larson (1971) suggestion for evaluating wetting front average capillary suction, all performed reasonably well when they were applied to a largely predominant soil in Nigeria.

While the laboratory method of Arya et al (1975) for determining soil water diffusivity gave reasonable result for only the top soil depth, the field method proposed by Kutilek (1980) did not yield satisfactory result due to the unreliability of the techniques for obtaining the soil parameters required in practical application of the method.

REFERENCES

Arya, L.M.; G.R. Blake and D.A. Farrel 1975. A field study of Soil Water depletion patterns in presence of growing soya bean roots: I. Determination of hydraulic properties of the soil. Soil Sc. Soc. Amer. Proc. 39: 424 - 438.

Campbell, G.S. 1974. A simple method of determining unsaturated conductivity from moisture retention data. Soil Sci. 117(6): 311 - 314.

Clothier, B.E. and I. White 1981. Measurement of sorptivity and soil water diffusivity in the field. Soil Sci. Soc. Am. J. 45: 241 - 245.

Ehlers, W. 1976. Rapid determination of unsaturated conductivity in tilled and untilled loss soil. Soil Sci. Soc. Am. J. 40: 837 - 840.

Ejieji, C.J. 1983. Test of the applicability of selected methods for determining soil hydraulic properties in Nsukka area. Unpub. M.Eng. Project Report, Dept. of Agric. Engineering, University of Nigeria, Nsukka.

Gosh, R.K. 1977. Determination of unsaturated hydraulic conductivity from moisture retention function - A Note. Soil Sci. 124: 122 - 124.

Gupta, S. C. and W. E. Larson 1979. Estimating soil water retention characteristics from particle size distribution, organic matter percent and bulk density. Water Resour. Res. 15(6): 1633 - 1635.

Idike, F.I. 1977. Experimental evaluation of infiltration equations. Unpub. M.S. thesis, Dept. of Agric. Engineering, Univ. of Minnesota, St Paul, Minnesota.

Idike, F. I. 1981. Modeling the effects of conservation practices on soil moisture. Unpub. Ph.D thesis, Dept. of Agric. Engineering, Univ. of Minnesota, St Paul, Minnesota.

Kutilek, M. 1980. Approximate field determination of diffusivity and unsaturated hydraulic conductivity. Proc. 2nd International Conf. on Agro Physics, Goddollo, Hungary, pag. 121.

Mein, R. G. and C. I. Larson 1971. Modeling the infiltration component of the rainfall-runoff process. W.R.R.C., Res. Univ. of Minnesota, Graduate School Bull, 43.

Philip, J.R. 1957b. Theory of Infiltration. 4 Sorptivity and algebraic infiltration equations. Soil Sci. 84: 257 - 264.

Philip, J.R. 1958. Theory of Infiltration. Soil Sci. 85: 333 - 337.

Philip, J. R. and J. H. Knight 1974. On solving the unsaturated flow equation 3. A new quasi-analytical technique. Soil Sci. 117: 1 - 13.

Land and Water Use, Dodd & Grace (eds), © 1989 Balkema, Rotterdam. ISBN 90 6191 980 0

Technique de démaillage et équilibrage des réseaux hydrauliques maillés

Y. Labye
GREF, Cemagref, France

F. Lebdi
Institut National Agronomique, Tunis, Tunisie

E. Parent
GREF, Engref, Paris, France

RESUME: On propose une méthode générale d'équilibrage des réseaux maillés, fondée sur la minimisation d'une fonction convexe et une interprétation physique d'un paramétrage du domaine des solutions admissibles.

ABSTRACT: We study a general method for computation of looped networks, based on properties of convex functions.

I. NOTATIONS GENERALES

Soit :
- R un réseau maillé connexe composé de tronçons
- A un arbre ramifié recouvrant R quelconque mais fixé
- M = R - A est dit "ensemble des tronçons maillants"
- i une arête (tronçon) de R ; i un arc de R,
- m une arete de M ; (m) un arc de M,
- a un chemin de R (voir FAURE 1
- N une extrémité d'arc (noeud) de R donc aussi de A

E désigne l'ensemble des noeuds de R

O (pour origine) désigne un noeud quelconque mais fixé : on peut ainsi définir, pour tout i de A, son extrémité amont (la plus proche de O) qu'on notera désormais N_i et son extrémité aval notée N_i''. Les tronçons de M seront orientes de façon quelconque.

On écrira donc, par la suite : N'_i, N''_i extrémités de (i) : $(i) = \overrightarrow{N'_i\ N''_i}$

E_i pour i A désigne l'ensemble des noeuds situés à l'aval de i.

Appelons C le cycle engendré par le tronçon maillant orienté (m) et complété par des tronçons de A (il est unique car d'après les propriétés de l'arbre recouvrant A il existe un chemin et un seul de A qui relie les extrémités N'_m et N''_m du tronçon (m).

$q = (q, q_2, \ldots, q_N, \ldots)$, tel que $\sum q_N = 0$, N E et $-\infty < q_N < +\infty$

sera appelée distribution d'appels sur les noeuds du réseau et restera fixée pour le reste du problème.

$Q = (Q, Q_2, \ldots Q_i, \ldots)$ i $\in$ R, sera appelée distribution de flux sur les tronçons du réseau, si, en tout noeud N, on vérifie la conservation des flux :

(1) $q_N + \sum Q_i = \sum Q_j$, pour i tel que $N'_j = N$ et j tel que $N''_j = N$

Pour tout tronçon i, on donne une fonction "différence de potentiel" $X_i = X_i(Q_i)$ définie pour $-\infty < Q_i < +\infty$, non décroissante et intégrable :

$F^X_i = F^X_i(Q_i)$ désigne une primitive de $X_i(Q_i)$ On définit $F^X(Q)$ par :

$$F^X(Q) = \sum F^X_i(Q_i) \ ; \ I \in R.$$

On notera encore F (Q) lorsqu'il n'y aura pas d'ambiguité sur les pertes de charges adoptées

On pose : $H_c(Q) = \sum_i \varepsilon^c_i . X_i(Q_i)$, $i \in C$ avec ε^c_i égal à $\pm$ 1 suivant que c et (i) sont de même sens ou de sens opposé. On prendra $\varepsilon^c_i = \emptyset$ lorsque i n'appartient pas à c. Pour un cycle c_m, on écrira $H_m(Q)$ au lieu de Hc_m (Q) et ε^m_i au lieu de ε^c_im.

II. HYPOTHSE.H1

On suppose que pour tout cycle du réseau, il existe au moins un tronçon i où $X_i(Q_i)$ est une fonction strictement croissante de Q_i. Ceci exprime que plus la perte de charge appliquée sur ce tronçon est élevée, plus le flux qui y transite est important en réponse à cet appel. Cette hypothèse est largement vérifiée en hydraulique, électricité, thermodynamique...

III. PROBLEME D'EQUILIBRAGE

3.1 Définition du problème P1

On dit que le problème P1 est résolu si on peut trouver un Card (R) - uplet $Q^* : (Q^*, Q_i^*,)$ tel que

- Q^* soit une distribution de flux sur le réseau, donc vérifie
- pour tout m de M, Q^* vérifie l'équation

$$H_m(Q^*) = 0 \tag{2}$$

3.2 Enoncé du problème P2

Le problème P2 consiste à trouver une distribution Q^* de flux qui minimise F (Q^*). Les équations (1) sont les équations de contraintes (linéaires) de l'optimisation.

3.3 Equivalence des problèmes P1 et P2

L'existence et l'unicité de la solution du problème P1 sous l'hypothèse H1 a été démontrée directement (voir LABYE [2]). Pour le problème P2, on va exprimer les flux qui transitent sur les tronçons de l'arbre A en fonction des flux des tronçons de M.

Pour cela, imaginons que l'on "démaille" le réseau en remplaçant les tronçons m par un appel + Q_m au noeud N'_m et un appel - Q_m au noeud N'_m et un appel - Q_m au noeud N" (les équations (1) sont respectées). Pour les tronçons i de A, on a, de la même façon qu'on calcule les flux sur un réseau ramifié :

(3) $Q_i = \sum q_n + \sum \varepsilon^m_i . Q_m$ pou i dans A, N dans E_i, m dans M

L'équation (3) permet de considérer le problème P2 comme un problème d'optimisation "libre" en fonction des variables indépendantes non bornées Q_m, qui permettent de déterminer complètement tous les flux sur le réseau. F (Q), est une fonction convexe différentiable des Q_i qui varient sur un domaine convexe fermé non vide. (En effet si $Q_m = \emptyset$ pour tout m nous nous ramenons au cas ramifié). Elle présente donc un minimum (voir LIONS 3), unique dans le cas de stricte convexité. Lorsque l'on utilise les variables "libre" Q_m, paramètres naturels du problème les conditions d'EULER se calculent par :

$$\text{(4)} \quad \frac{DF(Q)}{dQ_m} = \sum_{i \in R} \frac{\partial F}{\partial Q_i} . \frac{dQ_i}{dQ_m} = \sum_{i \in R} X_i(Q_i) . \varepsilon^m_i = H_m(Q)$$

F sera minimu si, et seulement si, pour tout $m \in M$, $H_m(Q) = \emptyset$. Les problèmes P1 et P2 sont donc identiques. Sous l'hypothèse H1, il est facile de montrer que la solution de P2 est unique (donc aussi celle de P1) : en effet F ne peut pas être à la fois non strictement convexe et présenter la propriété H1 (voir 5.1)

Le minimum de F, alors strictement convexe est unique.

IV. DES ALGORITHMES DE CALCUL POUR L'EQUILIBRAGE DES RESEAUX MAILLES

4.1 Notations et hypothèses

L'indice supérieur n désigne la valeur des quantités, fonctions et variables à l'issue de la nième itération; avant toute itération n = $\emptyset$.

Les seules variables seront les Q_m qui paramètrent le sous-espace affine des flux admissibles. On pose $\Delta Q^n_m + Q^n_m = Q^{n+1}_m$, pour tout m de M.

Les Q_i se calculent par (3), qui s'écrit sous la forme :

$\Delta Q^n_i = \sum \varepsilon^m_i . \Delta Q^n_m$, la somme porte sur tout m de R, pour i dans A; grad F (Q) est un vecteur de dimension card M dont les coordonnées sont les H_m (Q), m élément de M.

Pour assurer la convergence des algorithmes de minimisation, il est commode de "durcir" l'hypothèse H1 sous la forme H2 :

"Les fonctions pertes de potentiel sont dérivables, de dérivée continue positive et sur tout cycle, il en existe au moins une dont la dérivée admet une borne inférieure strictement positive". Sous ces conditions, grad (F) est lipschitzienne sur tout borné et il existe une valeur α telle qu'en tout point le graphe de F(Q) en fonction des Q_m ait une courbure au moins égale à celle du paraboloïde défini par

$$\alpha \sum (Q_m)^2, \text{ m dans M.}$$

4.2 Méthodes par descente le long du gradient

On explore :

$$\Delta Q_m^n) = -\lambda a_m^n \; H_m(Q^n) \text{ avec } a_m^n \geqslant \emptyset;$$

$$\lambda \geqslant \emptyset, \text{ pour } \lambda = \emptyset$$

$$\frac{d\,F}{d\,\lambda} = -\sum a_m^n \; H_m \; (Q)^{n\,2} \leqslant \emptyset,$$ la somme portante sur m dans M.

On cherche λ tel qu'il provoque une diminution maximale de F dans la direction explorée (on sait que F décroit au voisinage de $\lambda = \emptyset$).

Pour $a_m^n = 1$, pour tout m on retrouve la méthode du gradient à descente maxima.

Pour $a_m^n = \emptyset$, pour tout m, sauf celui qui varie au cours de l'itération, c'est la technique de relaxation. La convergence de ces méthodes dans les conditions H2 est bien connue.

4.3 Méthode de descente optimisée

Supposons que l'on soit à l'itération n et que l'on ait un nouvel ensemble de pertes de charge $Y_i \; (Q_i)$ tel que :

$$(5) \quad Y_i(Q_i) = X_i(\,Q_i^n\,) + a_i^n(Q_i - Q_i^n),$$

pour tout i de R avec $a_i^n > \emptyset$ et non nul sur au moins un tronçon de tout cycle de R. La distribution $\bar{Q}$ des flux d'équilibrage du réseau avec les pertes de charge données par (5) s'obtient facilement : il s'agit en effet de résoudre le système, maintenant linéaire, donné par (1) et (2). On supposera en outre que la somme des a_i^n est uniformément bornée en n par une constante D pour des problèmes de convergence.

Si on pose $\Delta Q_m^n = \lambda(\bar{Q}_m - Q_m^n)$; pour $\lambda = 1$ le réseau avec les pertes de charges (5) est équilibré. Pour $\lambda = \emptyset$ il n'est pas à l'équilibre.

F^Y étant une fonction strictement convexe et dérivable de λ : on sait

$$F^Y\;(\lambda = 1) < F^Y\;(\lambda = 0) \text{ d'où } \frac{d\,F^Y}{d\,\lambda}\;(\lambda = 0) < 0$$

Notons aussi que $Y_i \;(Q_i^n) = X_i(Q_i^n)$ et

donc $$H_m\;(Q^n) = \sum \varepsilon_i^m \; Y_i\;(Q_i^n) \quad I \in R$$

Avec des notations évidentes on a :

$$\frac{d\,F^X}{d\,Q_m}\;(Q^n) = \frac{d\,F^Y}{d\,Q_m}\;(Q^n) \text{ pour } m \in M$$

F^X aussi commence par décroître, à partir de Q^n, dans la direction $(\bar{Q}_m - Q_m^n)$, m appartenant à M. On détermine alors λ pour aller au minimum de F (fonction convexe) dans cette direction comme dans les méthodes de gradient. La démonstration de convergence est bâtie sur le même principe (voir annexe).

V. CONCLUSION

Sur le plan pratique, les calculs par méthode de gradient 4.2. ne nécessitent plus d'inversion itérative de matrice. La méthode décrite au 4.2., bien que nécessitant une résolution de système linéaire, est d'un emploi fort commode (notamment en posant $a_i^n = \emptyset$ quel que soit $i \in A$). La méthode HARDY-CROSS [5] s'apparente à cette dernière technique mais elle n'équilibre qu'une seule maille c_m au cours de chaque itération, elle prend pour a_i^n :

$$a_i^n = \frac{d\;X_i}{d\;Q_i}\;(Q_i) \text{ en } Q_i = Q_i^n \text{ pour tout}$$

i de c_m et $\emptyset$ ailleurs (donc $a_i^n \geqslant \emptyset$)

Elle pose $\lambda = 1$. ce qui est regrettable, car λ doit être ajusté ~~sinon~~ il risque de dépasser le minimum local que l'on cherche à cette itération : c'est pour cela que seule, la "convergence locale" de la méthode classique de HARDY-CROSS a été établie. En effet avec $\lambda = 1$, on trouvera

$$Q^{n+1} = Q^n + (\bar{Q} - Q^n) = \bar{Q}\;;$$

$\bar{Q}$ est tel que sur la maille c_m qui varie :

$$H_m^Y(\,\bar{Q}\,) = \emptyset$$

d'après (5) + (2) : $\emptyset = \sum \varepsilon_i^m \, y_i(\bar{Q}_i) = \sum \varepsilon_i^m \, X_i (Q_i^n) + a_i^n \varepsilon_i^m \, (Q_i - Q_i^n)$

l'équation (3) nous donne :

$$\bar{Q}_m - Q_m^n = \varepsilon_i^m (Q_i - Q_i^n) = \Delta Q_m^n \text{ pour}$$

le m qui varie et i sur c_m.

On a bien finalement :

$$(\bar{Q}_m - Q_m^n) = - \frac{H_m(Q^n)}{\sum a_i^n} \text{ avec } a_i^n = \frac{dX_i(Q^n)}{dQ_i} > 0 \quad i \in c_m$$

Un contre-exemple pour lequel cette méthode ne converge pas peut d'ailleurs être facilement construit (voir 5.1)

L'interprétation de cette fonction F(Q) est évidemment liée au principe de minimisation de l'énergie d'un système. Pour le cas de l'hydraulique, la signification de cette fonction est étudiée dans la thèse de F. LEBDI [6] . La méthode s'applique également aux problèmes de réseaux en électricité et thermodynamique en régime permanent où des flux (grandeurs extensives) circulent sous l'action de différences de potentiels (grandeux intensives). Enfin, signalons que l'extension au cas où les Q_i sont bornés (problème de transport à coût marginal fixe en recherche opérationnelle) ne présente pas de difficulté majeure.

5.1 Illustration des notations générales utilisées en 1 de la note principale

Soit le réseau R de la figure 1. Il comprend 13 tronçons numérotés de 1 à 13, et 11 noeuds désignés par les 11 premières lettres de l'alphabet.

$$E = \{A,B,C,D,E,F,G,H,I,J,K\}$$

On choisit arbitrairement un arbre recouvrant A, par exemple

$$A = \{2, 3, 4, 5, 6, 7, 9, 10, 12, 13\}$$

M ensemble des tronçons maillants comprend alors les éléments 1,8 et 11

On choisit alors une origine O, par exemple O = D : sur l'arbre A figure 2. Ainsi par exemple :

(3) = $\overrightarrow{DA}$; (4) = $\overrightarrow{DB}$,; (2) = $\overrightarrow{BC}$; ...
(13 = $\overrightarrow{GH}$ .. ; (12) = $\overrightarrow{JK}$

On oriente enfin arbitrairement les éléments de M; on peut par exemple prendre :

(1) = $\overrightarrow{AB}$; (8) = $\overrightarrow{EG}$; (11) = $\overrightarrow{CJ}$. Sur la figure 2 ces tronçons sont représentés en pointillés.

On en déduit alors $C_{(1)} = \{\overrightarrow{DA}, \overrightarrow{AB}, \overrightarrow{DB}\}$ et

$\varepsilon_1^1 = 1$, $\varepsilon_3^1 = 1$, $\varepsilon_4^1 = -1$ et $\varepsilon_i^1 = 0$

pour $i \neq 1,3$ ou 4

de même $C_8 = \{(5), (6), (7), (8)\}$ et

$\varepsilon_8^8 = \varepsilon_5^8 = 1$

$\varepsilon_7^8 = \varepsilon_6^8 = -1$ tandis que $\varepsilon_i^8 = 0$ pour $\neq$ 5, 6, 7 ou 8.

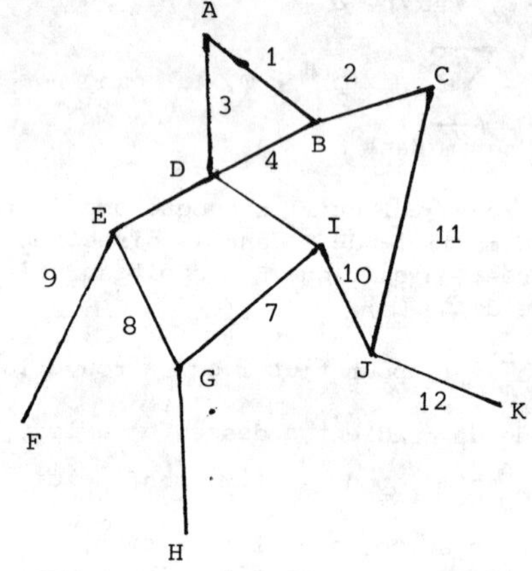

Fig.1: Réseau Initial

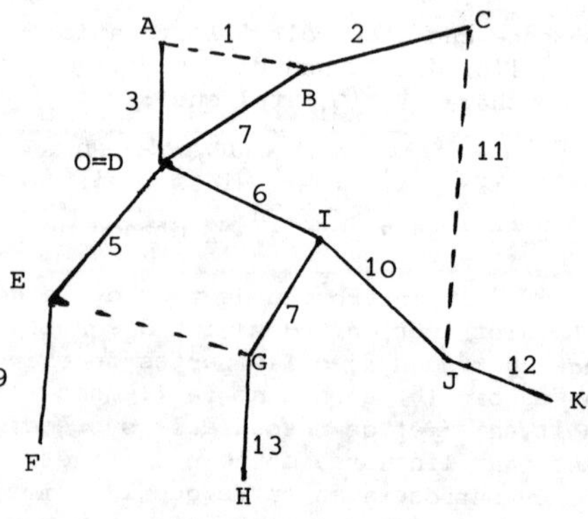

Fig.2: Réseau Grienté après choix de A, O et orientation de M

Enfin $C_{11} = \{(2), (4), (6), (10), (11)\}$ et

$\mathcal{E}^{11}_{4} = \mathcal{E}^{11}_{2} = \mathcal{E}^{11}_{11} = 1$

$\mathcal{E}^{11}_{6} = \mathcal{E}^{11}_{10} = -1$ tandis que $\mathcal{E}^{11}_{i} = 0$ pour $(i) \notin C_{11}$

Le choix de l'arbre A et de l'origine O nous permettent aussi de définir :

$E_5 = \{E, F\}$; $E_6 = \{I, G, H, J, K\}$

$E_7 = \{G, H\}$ $E_{13} = \{H\}$

$E_9 = \{F\}$; $E_{10} = \{J, K\}$; $E_{12} = \{K\}$

$E_4 = \{B, C\}$ $E_3 = \{A\}$; $E_2 = \{C\}$

H_m (Q) représente la perte de charge calculée selon l'orientation définie ci-dessus le long de la maille m.

Notons enfin que la fonction F ne change pas de valeur lorsqu'on choisit une orientation quelconque pour le réseau : en effet, si on oriente un tronçon i dans un sens opposé, à la fois Q_i et X_i (Q_i) changent de signe.

6. Stricte Convexité de F, conséquence de H1

6.1. Propriété H'1

La proposition H1 peut encore se formuler ainsi : "Du réseau maillé R, on peut dégager un arbre A, tel que les tronçons de R-A aient une perte de charge strictement croissante du flux qui y transite" (forme H'1)
En effet H'1 implique H1, car toute maille de R comprend nécessairement un élément de R-A (sinon A n'est pas un arbre ramifié).

De même H1 implique H'1; on sélectionne un ensemble M de tronçons dits "stricts", c'est à dire vérifiant une stricte croissan e de la perte de charge avec le flux, tels que R-M soit un arbre, de la façon suivante :

Si R n'est pas un arbre (cas pour lequel l'équilibrage est sans objet), on peut trouver une maille et, donc d'après H_1, un tronçon "strict",notre premier élément de M. Si R-M n'est pas un arbre, on peut trouver une maille de R-M (donc de R) avec un tronçon "strict" d'après H1. On affecte cet élément à M. R-M reste connexe. On réitère cette opération jusqu'à ce que R-M soit un arbre ramifié.

6.2 Convexité Stricte

Si F n'est pas strictement convexe, il existe deux distributions de débits Q^1 et Q^2 distinctes.

$Q^1 = (., Q^1_i, \ldots)$ i dans R ; $Q^2 = (\ldots, Q^2_i \ldots)$, i dans R avec au moins un i de R pour lequel $Q^1_i \neq Q^2_i$ telles que :

$$F(Q^1) = F(Q^2) + \sum_{i \in R} X_i(Q^1_i) \cdot (Q^2_i - Q^1_i)$$

$$\text{Soit} \sum_{i \in R} \int_{Q^1_i}^{Q^2_i} X_i(t) - X_i(Q^1_i)\, dt = 0$$

(6)

On sait que, d'après la propriété de non-décroissance des X_i :

$$\int_{Q^1_i}^{Q^2_i} \left(X_i(t) - X_i(Q^1_i) \right) dt \geq 0$$

(c'est d'ailleurs ce qui garantit la convexité de F). On doit donc avoir, en cas de non stricte convexité, pour tout i :

$$\int_{Q^1_i}^{Q^2_i} \left(X_i(t) - X_i(Q^1_i) \right) dt = 0$$ c'est-à-dire :

soit $Q^1_i = Q^2_i$, soit $X_i(Q^1_i) = X_i(Q^2_i)$.

Examinons alors la propriété H'1 : Il existe un tronçon m de M (donc "strict") pour lequel $Q^1_m \neq Q^2_m$. (Sinon on aurait d'après l'équation (3), pour tout i de R, $Q^1_i = Q^2_i$ donc $Q^1 = Q^2$). D'après le paragraphe ci-dessus on a donc :

$Q^1_m \neq Q^2_m$ et $X_m.(Q^1_m) = X_m(Q^2_m)$ pour un tronçon m "strict", ce qui est absurde.

F(Q) est donc une fonction strictement convexe puisque en développant les intégrales on peut écrire :

$$(7) \quad F(Q^2) > F(Q^1) + \sum_{i \in R} X_i(Q^1_i)\ (Q^2_i - Q^1_i)$$

pour $Q^2 \neq Q^1$.

VII. CONVERGENCE DE LA METHODE PROPOSEE DU 4.2. DE LA NOTE PRINCIPALE

7.1 Hypothèse H'2

De la même façon que pour la transformation H1 en H'1, l'hypothèse H2 peut s'énoncer sous la forme H'2 : "Du réseau R où toutes les pertes de charges sont dérivables, on peut extraire un arbre A tel que la dérivée des pertes de charges sur tout tronçon de M = R-A admette une borne inférieure strictement positive". Pour la suite de la démonstration de convergence, on utilisera cet arbre A et cet ensemble de tronçons maillants M.

7.2 α Convexité sous l'hypothèse H2

Sous l'hypothèse H2, pour tout i de R, on peut écrire :

$$\int_1^{Q^2_i} \left(X_i(t) - X_i(Q^1_i) \right) dt \geqslant \left\{ \underset{i \in R}{\text{Min}}\ X'_i \right\} \int_{Q^1_i}^{Q^2_i} (t - Q^1_i)\ dt$$

d'après le théorème de Rolle.

En appelant $\alpha > 0$, la borne inférieure strictement positive des

$$\frac{DX_r}{dQ_m}$$

$$\sum_{i \in R} \int_{1\,Q_i}^{Q^2_i} \left(X_i(t) - X_i(Q^1_i) \right) dt \geqslant \frac{\alpha}{2} \cdot \sum_{m \in M} (Q^2_m - Q^1_m)^2$$

cette inégalité est stricte lorsque $Q^1 \neq Q^2$ (car alors il faut au moins un des Q^1_m différent de Q^2_m).

Soit en développant :

$$F(Q^2) > F(Q^1) + \sum_{i \in R} X_i(Q^1_i).(Q^2_i - Q^1_i) + \frac{\alpha}{2} \sum_{m \in M} (Q^2_m = Q^1_m)^2$$

ou encore :

$$(8) \quad F(Q^2) \quad F(Q^1) + \sum_{m \in M} H_m(Q^1).(Q^2 - Q^1_m) + \frac{\alpha}{2} \sum_{m \in M} (Q^2_m - Q^1_m)^2$$

7.3 Montrons que $\| Q^{n+1} - Q^n \|$ a une limite nulle quand n tend vers l'infini

On travaillera par la suite avec les notations du 4.2 de la notre principale. On utilisera la norme sup ainsi :

$$\| \Delta_Q \| = \sup_{m \in M} | \Delta Q_m |$$

$$\| \text{grad}\ F \| = \sup_{m \in M} | Hm(Q) | \quad ;$$

Les a^n_i sont tels que $\sum_{i \in R} a^n_i < D$ pour tout n. $\bar{Q}$ est tel que pour tout m de M, (on sait qu'il existe car pour tout cycle il existe un a_i strictement positif) :

$$(9) \quad H_m(Q^n) + \sum_{i \in C_m} \varepsilon^m_i\ a^n_i\ (\bar{Q}_i - Q^n_i) = 0$$

$$(10) \quad \bar{Q}_i - Q^n_i = \sum_{m \in M} \varepsilon^m_i . (\bar{Q}_m - Q^n_m).$$

Q^{n+1} est défini par λ tel que $Q^{n+1} = Q^n + \lambda(\bar{Q} - Q^n)$ avec λ rendant $F(Q^n + \lambda(\bar{Q} - Q^n))$ minimum soit :

$$(11) \quad \text{grad}\ F(Q^{n+1}).(\bar{Q} - Q^n) = 0 = \text{grad}\ F(Q^{n+1} - Q^n)$$

Q^{n+1} est distinct de Q^n car $\bar{Q} \neq Q^n$ sinon Q^n est la solution cherchée.

L'équation (8) d'α- convexité appliquée en Q^n et Q^{n+1} permet d'écrire :

$$F(Q^n) - F(Q^{n+1}) \geq \text{grad } F(Q^{n+1}) * (Q^{n+1} - Q^n) + \frac{\alpha}{2} \sum_{m \in M} (Q_m^{n+1} - Q_m^n)^2$$

Soit (12) $F(Q^n) - F(Q^{n+1}) \geq \frac{\alpha}{2} * \| Q^n - Q^{n+1} \|^2$

La suite $n \rightarrow F(Q^n)$ est donc une suite décroissante, bornée (car F possède un minimum) ; on en déduit donc :

$$\lim_{n \to \infty} F(Q^n) - F(Q^{n+1}) = 0 = \lim_{n \to \infty} \| Q^n - Q^{n+1} \|^2$$

Remarquons que F étant continue, à partir d'un certain moment d'itération, les Q^n restent tous ensemble dans une boule fermée.

7.4 Montrons que $\lim_{n \to \infty} \text{grad } F(Q^n) = 0$

En appliquant la formule d'α- convexité (8) à Q^n et Q^{n+1} puis à Q^{n+1} et Q^n et en additionnant les 2 inégalités, il vient

(13) $\text{grad } F(Q^{n+1}) - \text{grad } F(Q^n) . (Q^{n+1} - Q^n) \geq \alpha \| Q^{n+1} - Q^n \|^2$

c'est à dire combinée avec (11).

$(-\text{grad } F(Q^n) . (Q^{n+1} - Q^n) \geq \alpha \| Q^{n+1} - Q^n \|^2$

et

$$-\sum_{m \in M} H_m (Q_m^n) . (Q_m^{n+1} - Q_m^n) \geq \alpha \| Q^{n+1} - Q^n \|^2$$

Soit encore d'après l'équation (9) définissant $\bar{Q}$ tel que $Q^n = (\bar{Q} - Q^n)$

$$\sum_{m \in M} \Delta Q_m^n . (\sum_{i \in R}^{m} a_i^n (\bar{Q}_i - Q_i^n)) \geq \alpha \| \Delta Q^n \|^2$$

L'équation (10) permet d'écrire :

$$|\bar{Q}_i - Q_i^n| \leq \text{Card } M \; \| \bar{Q} - Q^n \|$$

On a donc en définitive :

$$(\sum_{i \in R} a_i^n) (\text{Card } M)^2 \| \Delta Q^n \|^2 . \frac{1}{\lambda} \geq \alpha \| \Delta Q^n \|^2$$

d'où

(14) $$\lambda \leq (\sum_{i \in R} a_i^n) \frac{(\text{Card } M)^2}{\alpha}$$

De la même façon l'équation (9) permet d'écrire :

$$| H_m (Q^n) | \leq \sum_{i \in R} a_i^n \; \| \bar{Q} - Q^n \|$$

Soit $\| \text{grad } F(Q^n) \| \leq \sum_{i \in R} a_i^n) . \lambda \| \Delta Q^n \|$

$$\leq (\sum_{i \in R} a_i^{m\,2}) \frac{(\text{Card } M)^2}{\alpha} \| \Delta Q^n \|$$

(15) $$\| \text{grad } F(Q^n) \| \leq \frac{D^2 . (\text{Card} M)^2}{\alpha} \| \Delta Q^n \|$$

Comme au paragraphe précédent, on a montré que ΔQ^n tendait vers 0 quan n tendait ver l'infini, on en déduit :

$$\lim_{n \to \infty} \text{grad } F(Q^n) = 0$$

7.5 Montrons que Q^n converge vers la distribution Q d'équilibre qui donne le minimum de F

Q est telle que $\text{grad } F(Q) = 0$

Dans la formule (13) en remplaçant Q^{n+1} par Q il vient

$(\text{Card } M) . \| \text{grad } F(Q^n) \| . \| Q^n - Q \| \geq$

$\text{grad } F(Q^n) . (Q^n - Q) \geq \alpha \| Q^n - Q \|^2$

D'où $$\| Q^n - Q \| \leq \frac{\text{Card } M}{\alpha} \| \text{grad } F(Q^n) \|$$

Q^n tend donc vers Q lorsque n croit : l'algorithme proposé converge donc.

VIII. UN CONTRE-EXEMPLE POUR LA METHODE HARDY-CROSS

Considérons pour cela une maille (Cm)

d'un réseau pour lequel les pertes sont de la forme courante en hydraulique.

$f_i > 0 \qquad X_i \left((Q)_i \right) = f_i \left| Q_i \right| - Q_i$

avec

pour solution de départ au pas d'itération n . $Q^n_m = 0$ pour la maille ΔC_m que l'on doit faire varier.

On se réserve le droit de faire varier le coefficient f_m.

La variation de débit adoptée par Hardy-Cross pour la maille, ΔQ^n_m est telle que : (linéarisation au 1er ordre sur la maille).

$$\sum_{i \in R} {}^m_i \left\{ X_i (Q^n_i) + (\xi^m_i . \Delta Q^n_m) . \frac{dX_i}{dQ_i} (Q^n_i) \right\} = 0$$

Soit $\Delta Q^n_m = - \dfrac{H_m (Q^n)}{\sum_{i \in R} (\xi^m_i)^2 \dfrac{dX_i}{dQ_i} (Q^n_i)}$

Comme $\dfrac{dX_i}{dQ_m} (Q^n_m) = f_m \left| Q^n_m \right| = 0$

et $X_m (Q^n_m) = 0$,

ΔQ^n_m ne dépend de f_m

$F (Q^n) = \sum_{i \in R} f_i \dfrac{\left| Q^n_i \right|}{3}$. On peut donc écrire :

$F (Q^{n+1}) - F (Q^n)$ sous la forme

$$\sum_{i \in R} f_i \left\{ \frac{\left| Q^n_i + {}^m_i Q^n_m \right|}{3} - \frac{\left| Q^n_i \right|}{3} \right. + f_m \frac{\left| \Delta Q^n_i \right|^3}{3} \quad i \neq m$$

pour f_m assez grand cette expression peut être rendue arbitrairement positive puisque

fi, Q^n_i, ΔQ^n_m ne dépendent pas de de f_m.

Autrement au cours d'une intération, avec la méthode Hardy-Cross, il est possible d'augmenter arbitrairement F et donc de s'éloigner beaucoup de la distribution d'équilibrage.

REFERENCES

Faure,R. Précis de Recherche Opérationnelle, Dunod.

Labye,Y. Cours de calcul d'Eaus Sous Pression, Institut National Agronomique de Tunisie.

Lions,J.L. Cours d'analyse Numérique, Ecole Polytechnique.

Collins,M. et Co. Solving The Pipe Network Analysis Problem Using Optimisation Technics, Management Science, V24 N 7 Mars 1978.

Cross,H. Analysis of Flow in Networks of Conduits or Conductors, Engineering Experiment Station, Bulletin n° 286 University of Illinois 1936.

Lebdi,F. Recherche d'une Méthode d'Optimisation des Réseaux Maillés, These de Doctorat ENSEIT Toulouse, 1985.

CEMAGREF, Parc de de Tourvoie 92160 - ANTONY.

- Exemple pratique

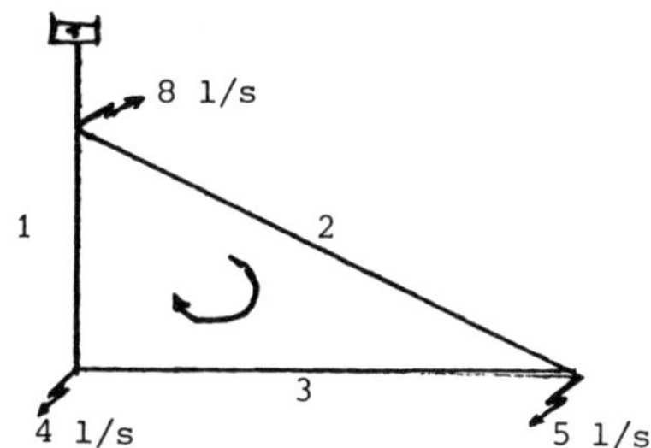

-Caractéristiques du réseau

Tronçon	1	2	3
Diamètre (mm)	200	150	125
Longueur (m)	310	610	100

- Algorithme

a - Choisir une solution initiale $\underset{\sim}{Q}_o$

b - Construire $\underset{\sim}{Q}_1$

c - Si $\|grad\ F(\underset{\sim}{Q}_1)\| - \|grad\ F(\underset{\sim}{Q}_o)\| < 0$ allera (d) si non $H = H/2$ allera b

d - Si $\|grad\ F(\underset{\sim}{Q}_1)\| - \|grad\ F(\underset{\sim}{Q}_o)\| < A$ fin si non $\underset{\sim}{Q}_o = \underset{\sim}{Q}_1$ allera (b)

H: pas d'itération
A: variation maximale du gradient entre 2 itérations successives.

- Résultats de l'équilibrage

- Solution du départ

Tronçon	1	2	3
Débit l/s	6	3	2

Précision = $\underset{\sim}{Q} = 0.01$ l/s

Tronçon	1	2	3
Débit l/s Itération 1	6.12	2.88	2.12
Itération 2	6.23	2.76	2.23

- Résultats de l'équilibrage par la méthode Hardy-Cross (même solution de départ)

Tronçon	1	2	3
Itération 3	6.2	2.8	2.2

Land and Water Use, Dodd & Grace (eds), © 1989 Balkema, Rotterdam. ISBN 90 6191 980 0

A low cost irrigation package for micro-scale irrigation in Africa

R.A.Lambert & R.D.Faulkner
Civil Engineering Department, Loughborough University of Technology, UK

ABSTRACT: If one of the most important goals of our time, that of self-sufficient and environmentally sustainable food security in Africa, is to be realised then intensification of agriculture in the region is necessary. Many attempts to bring about this intensification through the use of modern irrigation have been costly and discouraging failures. However there are many examples of the successful use of indigenous technical knowledge in locally based irrigated farming ranging from Nigeria's fadamas through Tanzanian irrigation on the slopes of Kilimanjaro to dambo cultivation in Zimbabwe. Recent research work in Zimbabwe has clearly indicated the great potential for increasing the scale and intensity of what is already an African success story. A low-cost irrigation package involving manually operated pumps, has been developed in Zimbabwe which offers small farmers the capacity to improve their family food security and cash income generation capacity through the use of micro-scale irrigation.

1 INTRODUCTION

In the last few years serious concerns have been expressed about the role of irrigation in improving food supplies in Africa (e.g. World Bank, 1981; Moris and Thom, 1987). These concerns arise principally because of the poor performance of many large irrigation projects. A notable example is the Bura scheme in Kenya where project costs in 1986 were reported to be between US$40,000 and US$50,000 per hectare (African Business,1986). Much of this cost relates to unproductive but necessary infrastructural development associated with the remoteness of the project and the need to relocate almost all of the people who were supposed to be the beneficiaries.In some cases, such unproductive facilities may cost more than the actual irrigation development. In an extreme example, Biswas (1986) notes that for the Tshovane project in Zimbabwe out of the total development cost of US$10,000/ha only US$750 was for the engineering works directly associated with the irrigation development, the rest being for agricultural buildings, machinery and road networks.

However the cost of such schemes is only part of the problem. Large scale schemes require sophisticated management and highly skilled staff. The FAO (1986) states that "It has been estimated that a single project, the Kano river project, will absorb all the agricultural graduates from Kano State for the next twenty five years...this could jeopardise the management of the whole of the state's agricultural sector for the foreseeable future." The sudden transition from traditional agriculture,which was essentially a household activity, to a modern, alien form of agriculture requiring the adoption of a completely new form of social organisation has all too often presented insuperable problems. Even when the scheme succeeds in producing crops, the problems of marketing, both within the country and to outside markets, remain. In Tanzania, food surpluses in some parts of the country very commonly go hand in hand with near famine in other parts (Park, 1985).

The problems associated with large-scale schemes have prompted an increasing degree of interest in small-scale schemes. However in many such schemes, the only small-scale element is the size of the plots allocated to individual farmers. A scheme extending over as little as 80 hectares may require effective co-ordination among 200 farmers (Lambert et al, 1987).Such small-scale irrigation is often "the usual thoughtless bureaucratic large scale development in little chunks" (Adams, 1988). With the failure of much of Africa's planned, formal schemes it is interesting to look at what has been happening in the unplanned informal sector. Underhill (1984) discusses the main features of this type of irrigation and outlines why it has been successful in so many cases. Bell et al(1987) report that there are 20,000 hectares of dambo cultivation in Zimbabwe, an area far greater than the country's 3,000 - 5,000 hectares of government small-scale shemes. In Nigeria, fadama cultivation rose seven-fold in the period 1958 -1978 (Turner, 1984) and according to Moris and Thom (1987) produced over five times as much rice as the modern irrigation sector in 1982.

It is clear that there are proven alternatives to conventional irrigation. Yet many of these alternatives are still being ignored by irrigation planners, partly because some of the techniques used do not conform to the planners view of irrigation as a system requiring dams, canals, furrows or sprinklers. The watering can or land husbandry practices which make use of shallow groundwater reserves are often not considered worthy of recognition by "proper" engineers. If we use the definition of irrigation proposed by Stern (1979) as "any process, other than natural precipitation, which supplies water to crops, orchards, grass or any other cultivated crops" then these traditional methods are clearly in the realm of irrigation.

Another problem is that of scale. Much of the successful irrigation in Africa takes place on plots covering as little as 0.1 hectares - hence they are overlooked although taken together they may aggregate to a very significant area. In order to clarify this problem of definition and scale the term "micro-scale irrigation" has been proposed (Lambert et al, 1987). Central to the idea of micro-scale irrigation is the concept of farmer control, which is exercised by the individual, family or small group that does the work. Because of the lack of attention paid to this important sector, little has been done to address its many needs, of which the technological needs are but one part. For example the problem of providing small amounts of financial credit to peasant farmers with little collateral remains largely unsolved. From the technical point of view, the need to develop appropriate water lifting and distribution systems has been identified (FAO,1986; Bell et al 1987).

In response to this need, the Overseas Development Administration in Britain is funding a project aimed at developing a low-cost irrigation package. The project is based at Loughborugh University in the UK and the University of Zimbabwe with field-work in Zimbabwe's communal areas.

2 WATER REQUIREMENTS

Crop water requirements were calculated for two sites in Zimbabwe: Grasslands research station with 19 years of data representing the cooler and wetter highveld areas and Buffalo Range with 21 years of data representing the hotter and drier lowveld areas (MWRD,1980). Figure 1 shows the rainfall and class A pan evaporation data from the two stations.It will be noted that Zimbabwe is fortunate in that the dry season is the cooler, winter season with relatively low evaporation losses.

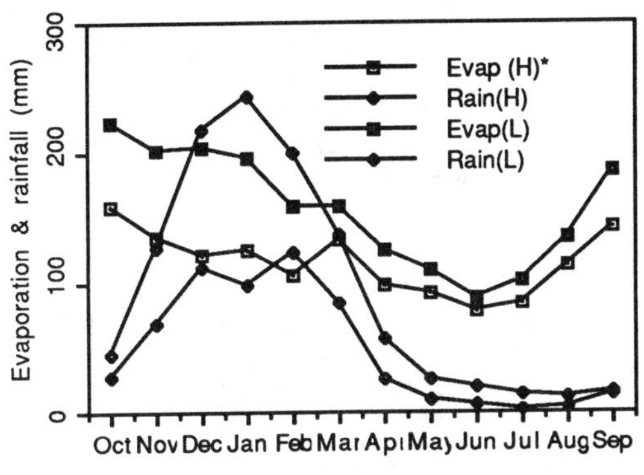

Figure 1 Evaporation and rainfall data Zimbabwe

Peak crop irrigation requirements normally occur in October when there is little rain and high evapotranspiration. Crop evapotranspiration requirements may be calculated according to the formula:

$$ET_c = k_p \times k_c \times ET_{pan}$$

where ET_c is the crop evapotranspiration requirement; k_p is the pan coefficient, in this example taken as 0.75; k_c is the crop coefficient,taken as 0.80 (a mean value for a mixed plot with crops at different stages of growth) and ET_{pan} is the class A pan evaporation loss recorded at the nearest meteorological station (see FAO, 1977). Using the data illustrated in Figure 1 the crop evapotranspiration requirements were calculated. They are shown in Figure 2 for the dry season months, May to October.

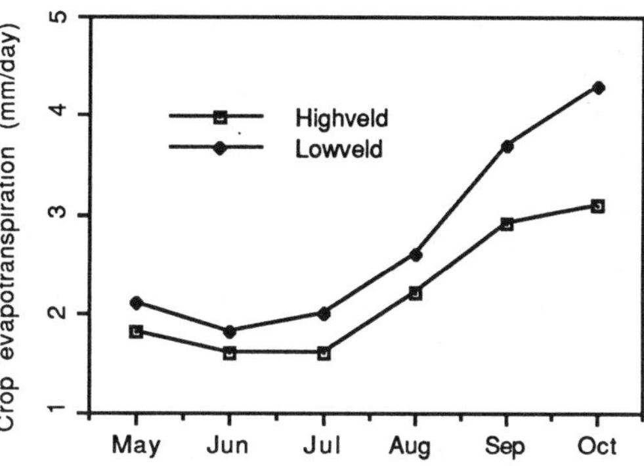

Figure 2 Crop evapotranspiration requirements - Zimbabwe

3 SELECTION OF PUMPS

In order to supply these crop requirements the most commonly available source of water is often shallow groundwater. The case of Zimbabwe's dambos, seasonally inundated valley wetlands, has been documented by Bell et al (1987) and current research at the University of Zimbabwe indicates that the groundwater potential of alluvium and sand rivers is very promising (Owen, 1988).

Much of this groundwater is within 10 metres of the arable surface, generally less than 5 metres in the case of dambos. This water must be lifted to the surface and distributed to the crops. At present the only real alternative to the cheap and effective but backbreaking watering can is the expensive diesel powered pump - an alternative which is out of reach of all but a tiny minority of farmers. In the immediate future the solution seems to lie with manual or animal powered pumps. While it is often argued that using human beings as power sources is undesirable in the long term (Fraenkel, 1986) on the grounds that human energy is a very expensive recurrent cost, it is often overlooked that the initial capital cost of any agricultural machine is critical to its uptake particularly in village economies where capital is very scarce. As a consequence, in Zimbabwe watering cans are widely used whereas diesel pumps are rare. Animal powered pumps require significantly larger amounts of capital than manually powered machines and present very real difficulties in terms of cultural acceptability and accessiblity to the poorer members of the community. Cattle are an indicator of wealth, with animal power being largely available to the relatively wealthy village elite. During the dry season, when irrigation is required, fodder is scarce and the cattle may well be in very poor condition and unsuitable for heavy work.

Human energy is a scarce resource in rural Africa and very often is the limiting resource in agricultural activities. Therefore an anlaysis was undertaken in the initial stage of this project to determine the amount of time spent on irrigation as a percentage of the total crop labour time, including irrigation and other crop husbandry activities such as sowing, weeding, fertilising and harvesting. The analysis was carried out for a common horticultural crop, cabbage, with data from the Agritex handbook (1982). The analysis was based on a daily energy input of 200 Watt-hours (W-hr) which is reported as reasonable for human power(Fraenkel, 1986).

A summary of the results of this analysis is presented graphically in Figure 3. It shows that the proportion of time spent on irrigation can be significantly reduced from almost 90% of total crop time when using a watering can to just over 50% when using a manual pump and distribution system.As the amount of "other" time may be increased from 10% to 50% this means that the plot area can be increased at least five-fold when converting from a watering can to a manually pumped system. With an upper limit of 20 hours per week spent on irrigation, which is reasonable considering the effort involved, the potential increase in area is almost ten-fold. It would appear therefore, that manual pumps offer significant improvements over the present watering can system and it was decided to concentrate on developing manual pumps.

Much research has gone into the development of handpumps for domestic water supply in developing countries. However the requirements for domestic supply are quite different to the requirements for irrigation, particularly in terms of cost and output. An irrigation pump must be affordable to the farmer and give a high output of water, of the order of one litre/second at lifts in the range of 2-10 metres. Domestic pumps are usually installed as a community pump with funding provided by an outside institution and often have to cope with lifts of 40 metres or more, although flowrates as low as 0.1 litres/second may be acceptable. With an irrigation pump, the VLOM (village level operation and maintenance) criteria now accepted for all domestic pumps, are vital.

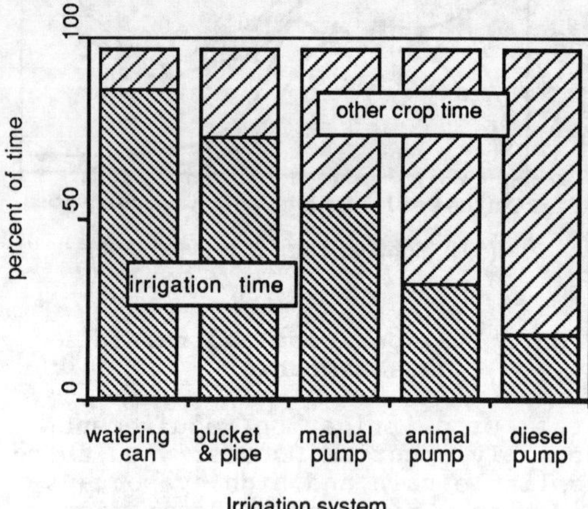

Figure 3 Irrigation time as a percentage of total crop time

Following a preliminary analysis, four pumps were selected for testing:

- the Rower pump which is a single cylinder, reciprocating, hand operated suction pump
- the Treadle pump which is a twin cylinder, reciprocating, foot operated pump
- the Bumi pump which is a hand operated diaphragm pump
- the Rope-Washer pump which is a hand operated, continuous motion, lift pump

At present the only pump that is being commercially manufactured in Zimbabwe is the Bumi pump although the Rower pump has been made to order. The Treadle pump, although widely used in Bangladesh, has not yet been commercially produced in Zimbabwe. The Rope-Washer pump is suited to the informal sector and is unlikely to be manufactured by the formal commercial sector although a number of pumps have been made in selected villages around the country.

Laboratory tests were carried out

in the University of Zimbabwe to determine the performance of these pumps. These tests included an assessment of the actual human energy going into the pumps as well as the hydraulic output. The actual human energy was assessed by calibrating heartrate against a known work ouput on a Treadmill or bicycle ergometer. The output performance for one operator, which involved continuous pumping for 15 minutes in each test, is illustrated graphically in Figure 4. The hydraulic ouptut is shown in units of m^3m/hr, a convenient unit in this case which is essentially a power unit being equivalent to 2.73 Watts for water.An output of $10m^3m/hr$ indicates that $10m^3$of water can be raised 1m in one hour or $5m^3$ raised 2m and so forth.

In terms of hydraulic output the Treadle (110mm cylinder diameter) and the Rope and Washer pumps give comparable results of approximately 15 m^3m/hr (equivalent to $3m^3/hr$ at a lift of 5m) and functioned well up to the maximum test height of 6.5 metres. The Bumi performs well at about 12 m^3m/hr until the lift exceeded 4 metres which was its effective maximum operating suction lift. A number of Rower pumps were tested and their output, at about 4 m^3m/hr, was found to be significantly lower than the other three pumps while the maximum effective lift was also limited to about 4 metres. It was notable that the Treadle pump achieved its high output through a greater input of energy, possible since it utilises the more powerful leg muscles rather than arm muscles, and not through greater efficiency. The most efficient pump, in terms of translating human energy, as measued through heartrate, into hydraulic output, proved to be the Rope-Washer. With a simple modification the Rope-Washer could be modified to use leg muscles.thereby increasing significantly the amount of energy available for pumping.

In addition to the pump performance a number of other considerations are important.The Bumi is robust and durable but heavy and quite expensive and has a limited suction lift; the Rower is light, portable and relatively inexpensive but gives a much lower output than the others and is the only one which does not allow operation by more than one person. The Rope-Washer can be manufactured at village level with the most basic tools and the lift is not limited to the maximum effective suction head of about 7 metres in Zimbabwe (altitude 1000 - 1500m). The maximum lift for the current design remains to be determined although historically, lifts of 80 metres have been reported. The Treadle pump is best manufactured by an engineering firm and while it is limited to a suction lift on the intake, it is capable of pumping water above its own level making it suitable for use in extracting irrigation water from sand rivers. Recent laboratory results indicate that the Treadle is capable of pumping at a total head in excess of 15 metres with a sustainable hydraulic output of 20 m^3m/hr with one person operation.

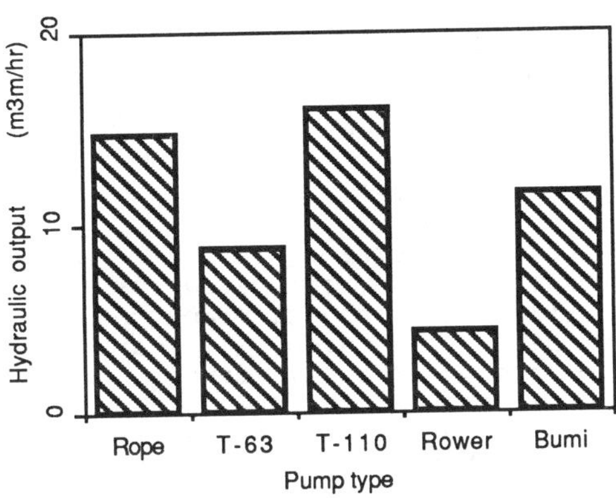

Figure 4 Manual pump output

These considerations pointed to the Treadle and the Rope-Washer pumps as the two most suitable for further development.

4 DISTRIBUTION SYSTEM

Raising water to the surface is the first step in getting it to the crops. In micro-scale irrigation the water source is located close to the irrigated area and so there is little need for major canals or furrows. In sandy soils, which are common in Zimbabwe's dambos for example, infiltration rates in short field furrows may be high even over short distances. Such furrows need to be accurately laid out and may conflict with wet season practices such as the construction of high beds, necessary to avoid waterlogging during heavy rain. Lining the furrows is expensive and leaves a permanent, inflexible structure. Nevertheless simple furrows may often be the best answer to in-field distribution.

In some situations an alternative must be sought. Zimbabwe is fortunate in that it has a pipe manufacturing industry which produces lightweight black poly pipe suitable for field irrigation. This piping is a flexible, simple and economical answer to the problem of water distribution over short distances. The water is pumped into a small header tank from which it may flow, under very low heads (of the order of one metre), in sufficient quantities to irrigate small plots of up to 0.25ha. For adequate flow the minimum pipe diameter at these low heads is 40mm.

5 IRRIGABLE AREA

Using the laboratory results described above, the area that can be irrigated with one pump can be calculated.This area will vary with the crop water requirements and the head against which the water must be lifted. Figure 5 shows the variation in irrigable area with pumping lift and crop water requirements. In the analysis a hydraulic output of 12 m^3m/hr is used, which corresponds to an power output of just 33 watts. This implies a power input of 50 watts at 60% efficiency and a daily energy input, at 4 hours pumping per day, of 200 Watt-hours. This corresponds well with data reported elsewhere in the literature (eg Fraenkel, 1986) for sustainable human effort and is less than the figures of 15m^3m/hr and 20m^3m/hr reported above. The amount of time spent pumping is taken as 20 hours per week. This time, and hence the irrigable area could be increased if more than one person were to use the pump. The distribution efficiency using pipes in small plots can be high and a figure of 80% is considered reasonable. Lifts can be well under 3 metres in dambo areas, increasing to over 10 metres in areas with sand rivers. Where the groundwater must be lifted more than 10 metres then the irrigable area becomes very small although irrigation may be even more important in such areas for dry season vegetable production.

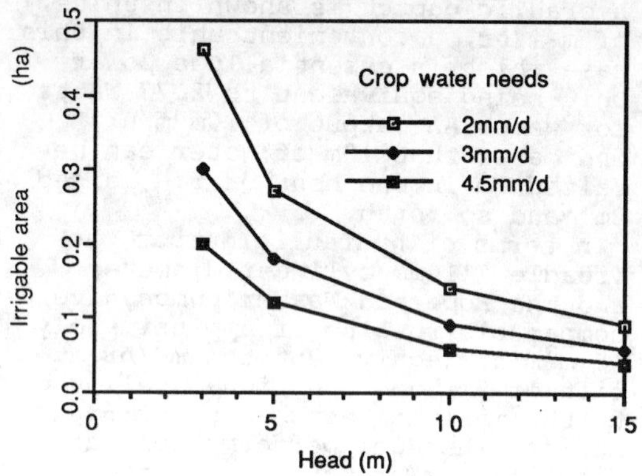

Figure 5 Irrigable areas for manual pumps at different pumping heads and crop water needs

6 COSTS AND BENEFITS

It is beyond the scope of this paper to present a detailed cost-benefit analysis of the irrigation package in comparison with other systems. However it is worth noting that in terms of intitial capital costs a manual pump package compares well on a unit and a total cost basis. The initial total investment required for a Rope-Washer pump is about Z$100 (1988 prices) and with piping the total cost for a plot of 0.25ha would be about Z$250 (Z$1.00 = UK£0.30 = US$0.55 approx). The total cost for a Treadle pump system would be about Z$750.00 for a similar plot. With a plot of 0.25ha the initial unit costs range from Z$1,000/ha for the Rope-Washer to Z$3,000 for the Treadle. As the plot size decreases, either because of a greater pumping lift or a greater crop water requirement, so the unit cost will increase. Conventional small-scale irrigation in Zimbabwe is considered low-cost if the unit cost drops below Z$10,000/ha (Froude, 1988) indicating that the cost of the manual pump package is very competitive.

The benefits of having access to an irrigation pumping system are difficult to quantify. Near larger urban areas, the potential to generate returns of Z$10,000/ha and more through the cultivation of high value crops such as vegetables is already being realised by a significant number of farmers. In such cases a manual pump package is very attractive as it generates very quick cash returns. In areas further from urban centres, the market for vegetables will be more limited. However the ability to guarantee staple food production is very important. When a farmer is sure of the water supply, investment in the correct fertilisers and pesticides is easy to justify. Many small farmers in Zimbabwe produce high yields of maize, sometimes up to 10 tonnes/ha, from small well watered plots. During the maize growing season a break in the rains of 2-3 weeks is not uncommon and frequently disastrous. An irrigation pump could guarantee that at least 0.2ha could be irrigated (or more depending on the degree of desperation). At a yield of 5 tonnes per hectare this means a total of 1 tonne could be harvested, which is enough for the yearly food requirements of an average Zimbabwe household. This is in addition to any benefits accruing through other irrigation activities during the dry season.

7 CONCLUSION

With the poor performance and high cost of many of Africa's conventional irrigation schemes it is clear that new thinking is needed to alleviate the problems of food production. The successes of Africa's informal sector and household based agricultural production has long been overlooked. Micro-scale irrigation, as a farmer controlled system, deserves serious attention to extend its success. This fact must be recognised by the irrigation planners and mechanisms found to provide appropriate technical, financial, marketing and extension support services.

The problem of water lifting is a major technical constraint but in areas with relatively shallow groundwater the use of human powered pumps, together with a low-cost distribution system, such as lightweight poly piping, is a viable solution. Laboratory and field tests indicate that such pumps can provide enough water at an affordable cost to sustain micro-scale irrigation. They can help to improve a family's food security in times of erratic rainfall and can make a significant contribution to rural incomes.

ACKNOWLEDGEMENTS

The authors would like to acknowledge the funding provided by the Overseas Development Administration (ODA) of the UK, the facilities provided by the University of Zimbabwe and the laboratory tests carried out by Mr Manuel Chidziva, final year student in Civil Engineering, University of Zimbabwe.

REFERENCES

Adams, W.M. & Anderson, D.M. 1988. Irrigation before development: Indigenous and induced change in agricultural water management in East Africa. African Affairs,87. 349:519-535

AGRITEX. 1982. Farm Management Handbook. Editor Jacobus de Jong. Harare: Government Publications.

Bell, M et al. 1987. The use of Dambos in rural development with reference to Zimbabwe. Final Project report to ODA. Loughborough University.

Biswas, A.K. 1986. Irrigation in Africa. Land Use Policy. October 1986: p.269-285.

Doorenbos,J. and W.O. Pruitt 1977. Crop water requirements. FAO Paper 24. Rome 1977.

FAO. 1986. Irrigation in Africa south of the Sahara. FAO Investment Centre Technical Paper 5. Rome.

Fraenkel, P. 1986. Water Pumping Devices. London: Intermediate Technology Publications and FAO.

Froude, 1988. Personal communication, Masvingo, Zimbabwe.

Gunnell, B. 1986. The great Bura irrigation scheme disaster. African Business, April 1986.

Lambert, R.A. et al. 1987. Micro-Scale irrigation in Africa. 13th WEDC conference, Malawi 1987.

Moris, J.R. and D.J. Thom. 1987. African Irrigation Overview: Main Report. Water Management Synthesis Report 37. Logan, Utah: Utah State University.

MWRD. 1980. Hydrological Summaries 1980. Harare: Ministry of Water Resources and Development.

Owen, R. Shallow alluvial groundwater for micro-scale irrigation. Paper presented at Ministry of Water conference, Harare 1988.

Park, P. & Jackson, T. 1985. Lands of plenty: lands of scarcity. Agriculture policy and peasant farmers in Zimbabwe and Tanzania. Oxfam report.

Stern, P.H. 1979. Small-Scale Irrigation. London: Intermediate Technology Publications and International Irrigation Information Center.

Turner, B. 1984. Changing land-use patterns in fadamas of northern Nigeria. In: Scott, E.(ed) Life before the drought pp 149-170. London: Allen and Unwin.

Underhill, H.W. et al. 1984. Small-scale irrigation in Africa in the context of rural development. Rome: FAO

The World Bank. 1981. Accelerated Development in Sub-Saharan Africa: An Agenda for Action. Washington DC: The World Bank.

ABREGE: Si l'un des objectifs les plus importants de notre temps, celui d'une Afrique qui se suffit a elle-même et dont le milieu maintient et assure la production de sa propre nourriture, devait être atteint, il est donc essentiel d'y intensifier l'agriculture. On a essayé maintes fois de ce faire en utilisant des méthodes d'irrigation modernes qui se sont soldées en échecs onéreux et décourageants. Cependant, il y a beaucoup de cas où l'utilisation des connaissances techniques indigènes dans une agriculture irriguée à la façon du pays a été couronnée de succès, depuis les fadamas du Nigeria jusqu'aux méthodes tanzaniennes d'irrigation sur les versants du Kilimanjaro en passant par la culture dambo au Zimbabwe. Un travail de recherche récent au Zimbabwe a clairement montré que d'accroître en échelle et en intensité ce qui est déjà une réussite africaine serait vraiment prommetteur. Un système d'irrigation a coût réduit qui utilise des pompes à main, y a été mis sur pied. Il propose aux petits cultivateurs d'améliorer leurs chances de produire suffisamment de nourriture pour leurs familles et celles d'accroître leurs revenus, en utilisant une irrigation à échelle très réduite (micro-scale irrigation).

AUSZUG: Sofern eines der bedeutendsten Ziele unserer Zeit, nämlich die sichere Versorgung mit Lebensmitteln - basierend auf Selbstversorgung und Umweltvertraglichkeit - in Afrika realisiert werden kann, so ist notwendigerweise die landwirtschaftliche Produktion in der Region zu intensivieren. Viele Versuche, diese Steigerung durch moderne Bewässerungsanlagen zu erreichen waren teuere und entmutigende Mißerfolge. Es gibt jedoch viele Beispiele von dem erfolgreichen Gebrauch einheimischer Technik in local gelegenen bewässerten Landwirtschaften, die sich von Nigerias fadamas über Bewässerungsanlagen in Tanzania am Hange des Kilimanjaro zu dambo Ackerbau in Zimbabwe erstrecken. Forschungsarbeiten über die letzten Jahre in Zimbabwe haben deutlich gezeigt mehr und größere Anlagen zu bauen, die schon eine erfolgreiche Geschichte in Afrika haben. In Zimbabwe ist ein Bewässerungssystem zu einem niedrigen Preis entwickelt worden, das Handbetriebene Pumpen enthält, welche Kleinbauern die Fähigkeit gibt, die Lebensmittelversorgung und das Einkommen für ihre Familien sicherzustellen; durch den Gebrauch dieser Bewässerungsanlagen in Kleinstformat (micro-scale irrigation).

Land and Water Use, Dodd & Grace (eds), © 1989 Balkema, Rotterdam. ISBN 90 6191 980 0

A real time automated control system for surface irrigation

E.A.Latimer & D.L.Reddell
Texas A&M University, College Station, Tex., USA

ABSTRACT: Spatial and temporal variation in infiltration create serious obstacles in the design, operation, and management of surface irrigation systems. An intelligent surface irrigation system (ARFIS) that can automatically sense the advance of water at two known distances down the field is described. These advance-rate data are used to determine field-representative infiltration characteristics. The predicted infiltration function is then used to design the irrigation event to achieve a desired net infiltration depth.

RESUME: Les variations spatiales et temporelles de l'infiltration créent de sérieux obstacles pour le design et la conduite des systèmes d'irrigation de surface. Un système informatique d'irrigation de surface (ARFIS) qui enregistre automatiquement l'avancée de l'eau à deux points de pente est décrit. Ces données de vitesse d'avancée sont utilisées pour déterminer les paramêtres d'infiltration caractéristiques d'un terrain. La fonction infiltration est utilisée pour planifier l'irrigation et obtenir la profondeur moyenne d'infiltration désirée.

ZUSAMMENFASSUNG: Räumliche und zeitliche Variationen in der Infiltration verursachen ernsthafte Schwierigkeiten in Bezug auf den Entwurf, den Betrieb und dem Management eines Oberflächen - Bewässerungssystems. Ein Oberflächen - Bewässerungssystem (ARFIS), das automatisch das Vordringen des Wassers an zwei bestimmten Punkten am anderen Ende des Feldes wahrnimmt, ist hier beschrieben. Die Entfernung und die Zeit, die das Wasser zum Vordringen benötigt, werden benutzt um die Infiltrationseigenschaften des Feldes zu bestimmon. Die errechnete Infiltrationsformel wird dann angewandt, um den restlichen Bewässerungsvorgang zu entwerfen, um somit die erwünschte Infiltrationstiefe zu erhalten.

1 INTRODUCTION

Surface irrigation has traditionally resulted in excessive labor and nonuniformity in water application across a field. Large water losses through percolation beyond the root zone are due to large differences in infiltration opportunity time between the upper end and the lower end of the field. In furrow irrigation, water is often allowed to advance past the end of the furrow to insure adequate infiltration time at the lower end of the field, thus causing significant tailwater losses. Compounding these problems in managing a furrow irrigation system are the spatial and temporal variations in infiltration that naturally exist across a field.

A considerable amount of research has been done to develop management techniques to improve irrigation efficiencies of surface irrigation. The most effective practices have resulted in the regulation of the field inlet discharge (cutback irrigation, surge flow irrigation, etc.). Yet, infiltration is the most crucial and often the most difficult parameter to asses accurately in the field. The importance of the infiltration function for describing the hydraulics of a surface irrigation event makes the assessment of infiltration vital for the design of surface irrigation systems. For years, researchers have studied methods to obtain field representative infiltration functions based on the response of the field to an actual irrigation. Some have proposed incorporating real time measurement of advance data into an automated surface irrigation system.

An experimental advance rate feedback irrigation system (ARFIS) incorporates some of these innovative techniques to improve application and distribution efficiencies of furrow irrigation. ARFIS is based on sensing the advance of water to two known distances along the furrow. These advance times are telemetered to an intelligent system which uses the information to define the furrow's infiltration characteristics. The predicted infiltration function becomes the building block for the regulation of inflow discharge to the furrow. This system seems to be a feasible way of increasing surface irrigation efficiencies without increasing labor inputs.

The objectives of this paper are threefold. First, to describe all system components which characterize ARFIS. Second, to define two working volume balance models used with ARFIS. Lastly, to discuss field evaluations of ARFIS using these models. Trade names that appear in this paper are used to supply information only and their use does not infer endorsement or preferential treatment of the product listed by the authors or the Texas A&M University System.

2 ARFIS COMPONENTS

ARFIS is composed of four main systems. These are: (1) an intelligent system, (2) a flow control system, (3) a sensing system, and (4) a telemetry system. Figure 1 presents the interrelation between these systems.

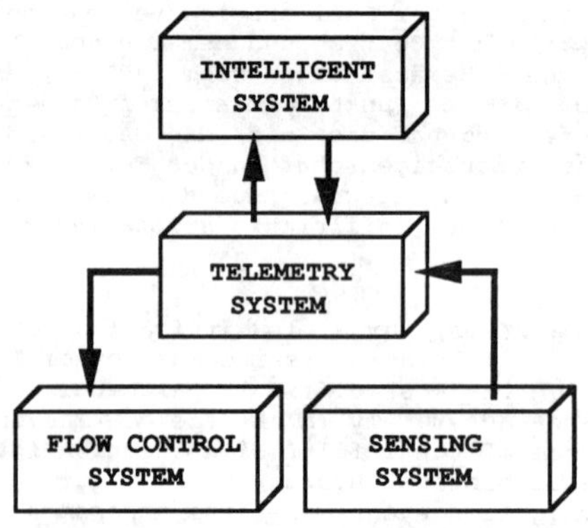

Fig.1 Schematic of the operating interrelation between systems in ARFIS. Arrows depict direction of communication links.

2.1 Intelligent system

This is the main control unit of the ARFIS system. It runs the software which enables communication with the telemetry system and which also controls the irrigation event (volume balance model). ARFIS is presently controlled by a 40 MB Zenith 248 workstation computer which has MS-DOS. A compatible monitor and printer are used to assist with the display and collection of data when running ARFIS.

The communication between the telemetry system and the intelligent system, serial digital data communication, is done using the I/O card available with the computer. The computer and accessories are housed in a 40 cubic meter mobile research trailer to provide the electronic equipment with the proper elements for operation. When in the field, a mobile 20KW diesel generator is used to power the trailer.

2.2 Flow control system

Inflow to individual furrows is regulated by a Rain Bird® 24VAC solenoid valve. Each valve is actuated by the telemetry system using a special relay switch. All valve settings are governed at all times by the intelligent system. Immediately following each of the solenoid valves is a Dole® valve to maintain a constant flow rate over a range of operating pressures. With ARFIS, it is imperative that a steady flow rate is maintained throughout the irrigation event. Because of the inherent high turbulence levels at which water leaves a Dole® valve, energy dissipaters are placed immediately after each valve to dissipate the erosive forces in the water before it enters the furrow. A special PVC plastic header pipe was constructed to hold and supply water to all valves.

2.3 Sensing system

The present sensing system consists of two water advance sensors in each furrow. Sensors are located at distances of 50 and 90 percent of the total furrow length. Research has determined that locating the sensors at these distances optimizes the estimated values of the infiltration parameters.

The water sensors are simple in design and inexpensive. A sensor consists of a steel rod support shaped in the form of a "T". The top of the T-shaped structure provides the user with handles to push the sensor into the ground. Attached to the lower end of the vertical steel rod are two small, rectangular aluminum plates. Each aluminum plate is soldered to an insulated single, small gage strand of electrical wire which runs from the sensor to a telemetry field station a distance away (Figure 2). The plates are positioned so that they are near each other but do not make contact. Immediately below where the aluminum plates are located is a small, square steel plate welded across the T-shaped structure to prevent the aluminum plates from being damaged when driving the water sensor into the ground.

When ARFIS is in operation, the two aluminum plates in a sensor act as a

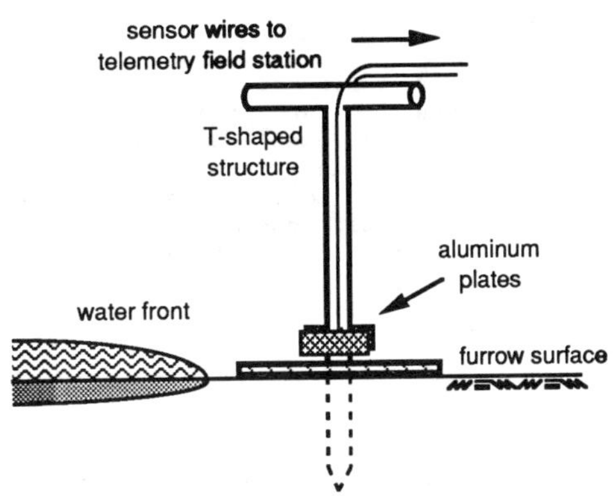

Fig.2 Sketch of the water sensor. The water front shorts the two aluminum plates sending a signal to the telemetry system.

circuit switch. In the absence of water between the plates, the corresponding internal circuit in the telemetry station remains open. When water advances to a sensor, thus creating a "short" between the two aluminum plates, the internal circuit in the telemetry station is closed. This condition informs the intelligent system that water has reached that particular location in the field.

2.4 Telemetry system

This system is the communication link between the intelligent system and the other systems. ARFIS can use two types of telemetry: radio and infrared. The present radio telemetry system uses low power radio (166 MHZ), requiring no permits or licenses for operation. Its transmission range is limited however, and possible noise from any surrounding electrical equipment and other radio signals can interrupt communication. These problems are avoided with a high power signal. The infrared telemetry system is cheaper, rugged and dependable, but it requires a clear line-of-site to operate which results in added labor.

Both the radio (RF) and infrared (IR) systems are Automata, Inc. Data Lynx telemetry systems and are compatible with the intelligent system. The software that runs the RF system will also run the IR system. The telemetry system in ARFIS consists of a base station and three field stations. The base station transceiver (receiver/transmitter) is primarily used to receive the RF or IR signals and convert them to serial data for computer entry, as well as receiving data from the computer and transmitting it via RF or IR signals. Field stations have a number of dedicated input/output terminals that are capable of remotely controlling equipment as well as reading sensors. Of the three field stations, two are used to monitor the two sets of water sensors (50% and 90% of total furrow length) and the third is used to control the flow control valves at the head of the field. Each telemetry station is powered by a four amp-hour battery. An optional solar charger is available to keep the battery charged while being used. A 5cm diameter pipe tripod mount is required for proper outdoor installation of a field station.

A special identification number (ID#) is assigned to each field station for both the IR and RF systems. The ID# is the key to a command recognition by a field station. Thus, an ID# is appended to each command made by the intelligent system. Command functions include interrogation and control. Any particular sensor value or control status can be interrogated at any time. Each input channel in a field station has a unique ID# which is computed by adding the channel number to the field station ID#. Present input channels are assigned to be normally high (5 volts) and become grounded (0 volts) when shorted.

Each RF field station, including the base station, is composed of a transceiver, a Central Terminal Unit (CTU), and an antenna. The transceiver can receive signals from a transmitter and convert the RF signals to encoded audio signals, and vice versa. The encoding and decoding scheme is called phase shift modulation.

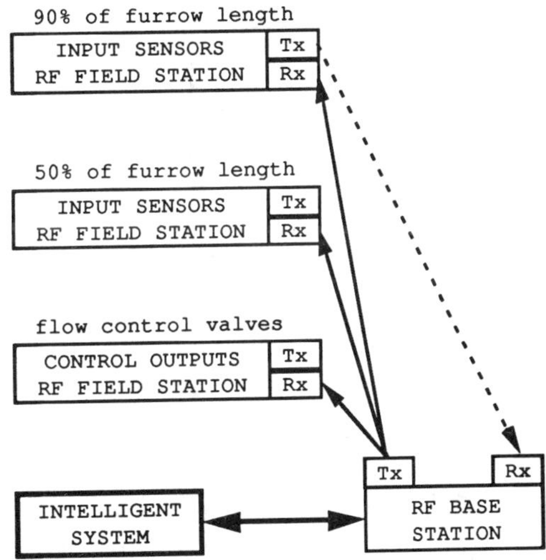

Fig.3 Command functions sent out from the RF base station are simultaneously read by all field stations (solid line). Only the field station with an ID# matching the ID# appended to the message will execute the command and respond (dashed line).

This means that the phase of the audio signal contains the information or message content of the data. The CTU is housed in a separate, plastic weather proof chassis and it contains all the intelligence and data storage for the telemetry unit. The CTU converts audio signals to digital data, and vice versa. Figure 3 shows the overall configuration of RF communication. These particular RF field stations have up to eleven sensor inputs and six control outputs (on/off) that can be monitored.

The components of an IR station include an integral receiver and transmitter, and a processor unit. Communication between stations is by means of an invisible infrared light beam which is modulated by the transmitter according to input signals received from the processor unit. At the other end, the receiver converts the modulated light beam into a serial electrical signal. Both, transmitter and receiver, come with sun shields and are mounted on gimbals so that they can be aimed in any direction. In addition, the receiver operates with a baffle to restrict the angle from which light may enter the receiver. IR field stations have up to eleven sensor inputs and eight control outputs for system use. All IR communication is daisy chained through every operating field station and is returned to the computer in a closed loop (Figure 4). In this configuration the computer can confirm completion of each operation (echo) before awaiting for field station response.

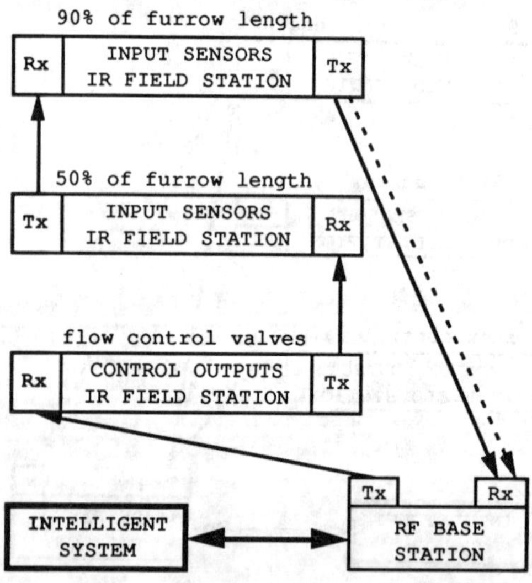

Fig.4 A message sent from the IR base station to an IR field station is daisy chained through the system echoing back to the computer (solid line). Only the appropriate field station executes the identified command (dashed line).

3 VOLUME BALANCE MODELS

Volume balance models utilize only the continuity equation, and thus require simpler solution techniques than other types of models (e.g., Kinematic-Wave) used to evaluate surface irrigation systems. The volume balance relationship for the advance phase of surface irrigation was given by Lewis & Milne (1938) as:

$$qt_x = Cx + \int_0^{t_x} z(t_x - t_s)\, \frac{\partial s}{\partial t_s}\, dt_s \;, \tag{1}$$

where q is the inflow rate to the furrow, C is the average cross-sectional area of flow in the furrow, x is the distance water has advanced during time t_x, s is the distance water has advanced at a prior time t_s, t_x and t_s are the time of advance to distances x and s, respectively, and $z(t_x-t_s)$ is the cumulative infiltration per unit length of furrow at distance x during the infiltration opportunity time (t_x-t_s) (refer to Figure 5). The term on the left side of the equality is the volume of water applied to the furrow. The first term on the right of the equality is the volume of water stored on the surface of the furrow and the second term represents the volume of water infiltrated.

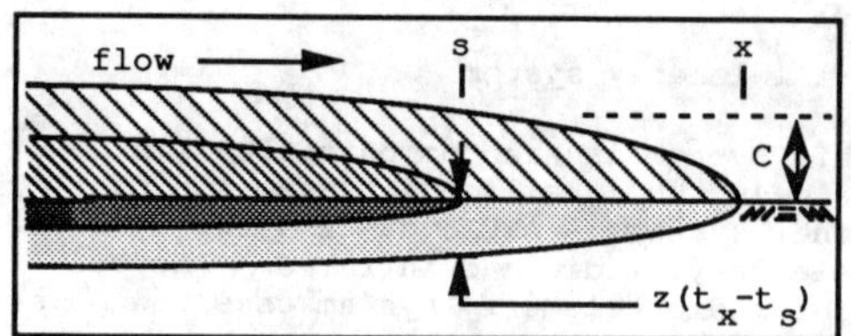

Fig.5 Sketch showing the advance of the water to arbitrary locations s and x in the furrow. The cumulative infiltrated volume at point s and the average cross-sectional area are shown.

The soil's infiltration behavior, the integral term in equation (1), is the most critical component in the relationship and, perhaps, the most difficult to determine. Of the many empirical infiltration equations available, the one most often used in the design and evaluation of furrow irrigation systems is the Kostiakov infiltration equation:

$$z = Kt_o^a \;, \tag{2}$$

where z is the cumulative volume of water infiltrated per unit furrow length, t_o is the infiltration opportunity time at a point along the furrow, and K and a are empirical infiltration parameters. Using work from Philip & Farrell (1964), Wilke & Smerdon (1965), and Reddell (1981), equation (1) can be rewritten in a simpler form as:

$$qt_x = Cx + BxKt_x^a \, , \qquad (3)$$

where B is a dimensionless profile shape factor (0<B<1), and all other parameters have been previously defined. B is a function of the parameter a and is calculated from:

$$B = \Gamma(a+1)\Gamma(2-a) \, , \qquad (4)$$

where Γ( ) is the gamma function of the parameter in parentheses. Given the temporal and spatial variability of the empirical infiltration parameters K and a, they are best approximated when estimated during an actual field irrigation. ARFIS uses the advance data to two points along the furrow (50% & 90% of the furrow length) to obtain field-representative infiltration parameters.

Surface storage, the volume of water stored on the soil surface of the furrow, is dependent on furrow geometry, hydraulic roughness, field slope, soil infiltration characteristics, and furrow inflow rate. Each of these conditions affects flow depth and therefore wetted perimeter; and these field conditions can change rapidly throughout an irrigation event or season, causing substantial variations in surface storage.

3.1 Zero surface storage model

Assuming that the volume of water stored on the furrow surface of soils with a moderate to high infiltration rate was small when the time of irrigation was long (the total volume of water stored is a small fraction of the total volume applied), the surface storage term becomes negligible in the volume balance relationship. Based on this assumption, Reddell (1981) proposed a modified version of equation (3) to describe the advance phase:

$$\frac{q}{x} = BKt_x^{a-1} \, . \qquad (5)$$

ARFIS uses advance distances of 50% and 90% of the irrigated furrow length and their corresponding advance times, and the constant inflow rate to the furrow to determine all the unknown parameters in equation (5). A relationship of log(q/x) versus log(t_x) will plot as a straight line (Figure 6). The slope of the line will equal a-1. Thus, the parameter a is calculated using equation (6), where L is the total furrow length, 0.5L and 0.9L are the fractions of the furrow length to the first and second sensors, respectively, and $t_{0.5L}$ and $t_{0.9L}$ are the corresponding advance times. The parameter B is calculated using equation (4). The parameter K is obtained using equation (5), provided the distance and corresponding water advance time to one of the two sensors is used.

$$a = \frac{\log_{10}\left[\frac{0.5L}{0.9L}\right]}{\log_{10}\left[\frac{t_{0.9L}}{t_{0.5L}}\right]} + 1 \, , \qquad (6)$$

When all of the infiltration parameters in equation (5) are known, the advance time for any given distance can be determined using the following relationship:

$$t_x = \left[\frac{q}{xBK}\right]^{\frac{1}{a-1}} \, , \qquad (7)$$

where t_x is the time for the wetting front to reach distance x in the furrow and all other terms have been previously defined. The time for water to advance to the end of the field (a distance L) is calculated using equation (7).

Instrumental in the management of the irrigation event is the time it takes for the average net application depth, Z_a, to infiltrate. Converting Z_a, a design input, to the units of cumulative infiltration (Y_a) and using equation (2), the time to infiltrate (t_a) the desired irrigation application is calculated as follows:

$$t_a = \left(\frac{Y_a}{K}\right)^{\frac{1}{a}} \, , \qquad (8)$$

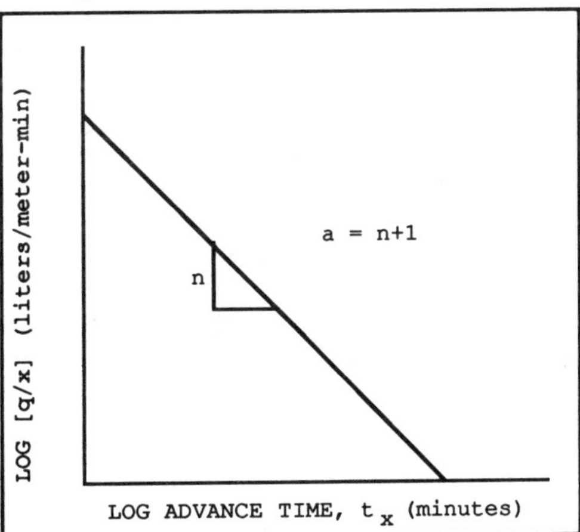

Fig.6 Log-log plot of q/x versus t_x. The **slope is negative and bounded by -1<n<0, while a is positive and bounded by 0<a<1.**

where the term Y_a is the average cumulative net irrigation application. The depth Za is assumed to occur at the point x = 0.6L. The total time water is applied to the furrow (t_w) is calculated as follows:

$$t_w = t_a + t_{0.6L} - t_r \ . \qquad (9)$$

It is assumed that the time of recession, t_r, is small compared to t_a and $t_{0.6L}$, thus negligible. The term $t_{0.6L}$ is the time of advance to the point x = 0.6L, and is calculated using equation (7).

A constant flow rate is applied until water reaches the end of the furrow. At this time, ARFIS begins to regulate the inflow to the furrow to match the existing infiltration rate. This is accomplished by repeatedly closing and opening the discharge valve using a set cycle time and varying the duty cycle (open time for each cycle = duty cycle x cycle time). Because infiltration along the furrow decreases with time, the required inflow to the furrow, q_{req}, decreases accordingly. The desired flow rate q_{req} is obtained by numerically integrating equation (2) along the field length (L) for two application times (irrigation elapsed time, t_e, and $t_e+\Delta t$), and taking the difference to obtain the infiltrated volume over the time increment Δt. The computer then reduces the inflow rate to the furrow by decreasing the duty cycle of the valve to q_{req}/q_{valve}. This cutback procedure is continued until the irrigation event is terminated. Valve cycle times between 10 and 20 minutes have commonly been used with ARFIS.

After the inflow valve is permanently shut off, a final numerical integration of equation (2) over the entire field for the elapsed irrigation time is used to estimate the final volume of infiltration, the application efficiency, and the distribution efficiency. American Society of Agricultural Engineers (ASAE) standards are used to evaluate the furrow irrigation event (ASAE, 1987).

3.2 Complete volume balance model

Studies by Reddell & Latimer (1986) showed that neglecting surface storage in the volume balance model caused an overestimation of the volume of infiltrated water. The effects of the surface storage term on the performance of ARFIS was presented by Latimer & Reddell (1987).

Using equation (3), the volume balance model introduces a third unknown parameter. This parameter, C, is the average cross-sectional area of flow in the furrow. An approach for approximating C was given by ASAE (1987). A surface storage shape factor of 0.77 is assumed, and C was determined as 0.77A, where A is the cross-sectional area of flow at the furrow inlet of the furrow. The value of A is measured and entered into the computer shortly after the water level at the furrow head reaches a steady height.

The parameter a is determined from the log-log relationship of $(qt_x/x - C)$ versus time of advance, t_x. This relationship also plots as a straight line. The slope of this line equals the parameter a (Figure 7). To numerically calculate the value of this parameter, the following equation is used:

$$a = \frac{\log_{10}\left[\dfrac{q\dfrac{t_{0.9L}}{0.9L} - C}{q\dfrac{t_{0.5L}}{0.5L} - C}\right]}{\log_{10}\left[\dfrac{t_{0.9L}}{t_{0.5L}}\right]} \ . \qquad (10)$$

The value of the coefficient B is determined using equation (4). Subsequently, the parameter K is calculated using equation (3), and the advance data to one of the two known points in the furrow, either 0.5L or 0.9L. The time for water to advance to any point x in the furrow is calculated using the following relationship:

$$t_x = \frac{BxKt_x^a + Cx}{q} \ . \qquad (11)$$

The above equation has a time term on both sides of the expression, thus it must be solved iteratively. Hereafter, this model is basically the same as the zero surface storage model previously described.

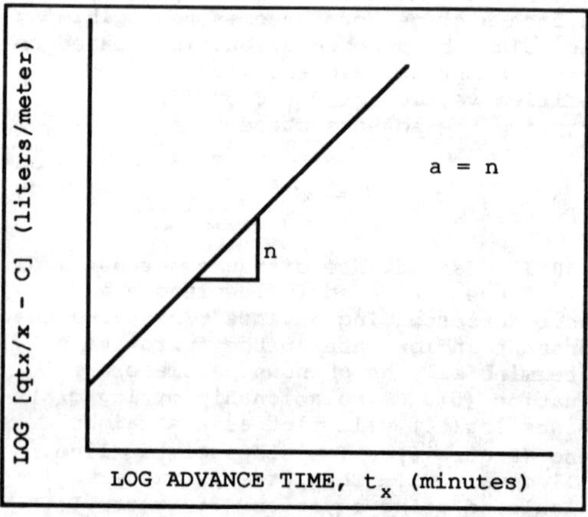

Fig.7 Log-log plot of $(qt_x/x - C)$ versus (t_x). The slope of the relationship is positive and bounded by 0<n<1.

4 FIELD EVALUATIONS

Field evaluations of ARFIS using the Zero Surface Storage (ZSS) model and the Complete Volume Balance (CVB) model were conducted near College Station, TX during the summers of 1986 and 1987, respectively. The experimental plot contained furrows constructed with standard heavy-tillage equipment, and were 245m long and spaced 1m apart. No distinctions were made between track (hard) and non-track (soft) furrows.

Advance data were recorded manually each 7.6m along the length of the furrow. These advance data were used to produce plots of $\log(q/x)$ versus $\log(t_x)$ when using the ZSS model and $\log(qt_x/x - C)$ versus $\log(t_x)$ when using the CVB model, and compare the rate of advance predicted with the two-point method used in ARFIS with the actual advance data. The points x=0.5L and x=0.9L consistently appeared to be good choices for estimating the infiltration parameters. On the average, all plots of advance data exhibited a linear relationship starting at about x=0.5L and continuing throughout the remainder of the advance data. This seems to support the use of these points to evaluate K and a.

Table 1 contains a brief summary of the results obtained for the field evaluations of ARFIS using the ZSS and CVB models. The furrow inflow rate and furrow length are given for each evaluation. Each irrigation event consisted of a single furrow. Irrigation events A thru D, E1, and F1 correspond to the ZSS model whereas E2 and F2 correspond to the CVB model. Events E1 and E2, as well as F1 and F2, were performed at the same time on consecutive furrows for purposes of comparison. For a more thorough discussion, the reader should refer to Reddell & Latimer (1986) and (1987) for the 1986 events and Latimer & Reddell (1987) for the 1987 events.

The predicted advance time to the furrow end made by ARFIS and the measured advance time for each event are shown in Table 1. The predicted advance times compared favorably with the measured times for all irrigation events. The wide range of advance rates among the irrigation events was due to the different field conditions existing for each case. These including variations in soil moisture content and channel roughness.

A constant flow rate was applied to each furrow and the time each valve was left open was measured by the computer. This provided an accurate determination of the total volume of water applied to each furrow. The volume of tailwater was measured volumetrically. Thus, the infiltration volume was determined by subtracting the measured tailwater volume from the measured inflow volume. The desired volume of infiltration was that used in the model to design the irrigation event, mainly a function of the desired net infiltration depth. The ARFIS-estimated infiltration volume, which was determined using a simple numerical integration scheme, is also presented in Table 1. In each of the events where the ZSS model was used this volume was overestimated whereas for event E2 (CVB model) the estimated volume was less than that measured. Event F was performed under wet field condition which produced excessive amounts of tailwater, explaining the low volumes of infiltration. Also, poor measurement of the cross-sectional area of flow at the furrow inlet caused the system to overestimate the volume of infiltration for event F2.

Application efficiency was defined as the volume applied to the field minus the volumes of percolation and tailwater, all divided by the volume applied. The distribution efficiency was determined using the infiltration parameters calculated from the advance data. The infiltrated depth was then calculated over a large number of points along the furrow. The summation of the mean absolute deviation in infiltrated depth from the mean infiltrated depth for each point was used to obtain the distribution efficiency.

Table 1. Results of field evaluations of ARFIS using the ZSS and CVB models during the summers of 1986 and 1987, respectively.

YEAR/ EVENT	INFLOW RATE q (l/min)	FURROW LENGTH L (meters)	TIME TO FURROW END: FIELD MEASURED (minutes)	TIME TO FURROW END: ARFIS ESTIMATED (minutes)	VOLUMES OF INFILTRATION: FIELD MEASURED (liters)	VOLUMES OF INFILTRATION: DESIRED (liters)	VOLUMES OF INFILTRATION: ARFIS ESTIMATED (liters)	IRRIGATION EFFICIENCIES: Ea (%)	IRRIGATION EFFICIENCIES: Ed (%)
86/A	68	244	180	145	12288	12585	12875	97	92
86/B	52	244	216	229	11210	12585	13303	90	85
86/C	76	183	196	202	17735	18878	19023	95	86
86/D	95	183	91	92	8823	9439	10161	94	86
87/E1	93	244	73	70	11524	12585	12552	92	99
87/E2	93	244	106	101	13388	12585	12612	92	90
87/F1	89	244	28	27	3798	12585	4435	85	98
87/F2	89	244	32	31	4806	12585	6981	60	95

5 COMMENTS

The ARFIS system will undergo additional modifications to the volume balance model and then be given more field tests. Automation has clearly shown that improved irrigation efficiencies occurred when ARFIS was used. Expansion of the system to irrigate large land areas is under development.

REFERENCES

ASAE Standards 1987. American Society of Agricultural Engineers, St. Joseph, MI.

Latimer, E.A. & D.L.Reddell 1987. Effect of neglecting surface storage on performance of an automated furrow irrigation system. ASAE/St. Joseph, MI. Paper No.87-2639.

Lewis, M.R. & W.E.Milne 1938. Analysis of border irrigation. Agricultural Engineering. 19(5):267-272.

Philip, J.R. & D.A.Farrell 1964. General solution of the infiltration advance problem in irrigation hydraulics. J. of Geophysical Research, 69:621-631.

Reddell, D.L. 1981. Modified rate-of-advance method for an automatic furrow irrigation system. ASAE/St. Joseph, MI. Paper No.81-2552.

Reddell, D.L. & E.A.Latimer 1986. Advance rate feedback irrigation system (ARFIS). ASAE/St. Joseph, MI. Paper No.86-2578.

Reddell, D.L. & E.A.Latimer 1987. Field evaluation of an advance rate feedback irrigation system. Irrigation Systems for the 21st Century Proceedings, Irrigation & Drainage Division/ASCE, Portlad, Oregon U.S.A., July 28-30.

Wilke, O.C. & E.T.Smerdon 1965. A solution of the Irrigation Advance Problem. J. of the Irrigation & Drainage Division/ASCE. 91(IR1):23-34.

Land and Water Use, Dodd & Grace (eds), © 1989 Balkema, Rotterdam. ISBN 90 6191 980 0

Utilisation de l'équation différentielle générale de Jacob pour la conception d'un réseau de drainage dans le cas d'une nappe libre à substratum imperméable horizontal

M.Mechergui & M.H.Snane
Institut National Agronomique, Tunis, Tunisie

RESUME : La première partie de cet article comporte un rappel des deux solutions de l'équation de Boussinesq présentées par Glover dans le cas d'une nappe libre et des drains placés ou non sur un substratum horizontal imperméable. Ces solutions sont trouvées après avoir négligé le terme $\partial h/\partial x$ dans l'équation différentielle générale de Jacob.
La seconde partie présente une nouvelle solution de cette dernière équation en tenant compte du terme $\partial h/\partial x$.
En utilisant la méthode de Glover - Dumm nous avons développé deux solutions qui permettent de déterminer l'espacement des drains. Une comparaison des deux résultats a montré que l'écartement calculé à partir des nouvelles solutions est plus faible que celui déterminé par les formules de Glover.

ABSTRACT : The first part of this study presents the Glover's solution to Boussinesq's equation. This solution is based on the assumption that the term $\partial h/\partial x$ can be neglected in Jacob's differential equation.
In the second part, a solution to this same equation was developed taking this time into account the gradient $\partial h/\partial x$.
Compar ison of the two results shows that drain spacing computed with the second solution is lower than the one given by Glover's solution when drains are placed on a horizontal impervious layer with an unconfined aquifer.

ZUSAMMENFASSUNG : Der erst teil der arbeit erwähnt beide lösungen der Gleichung von Boussinesq, dargestellt von Glover, für den Fall einer Wasserlamelle mit oder ohne Drainage auf waagerechtem undurchlässigem Substratum. Diese Lösungen ergeben sich indem man das Glied $\partial h/\partial x$ der generellen differentialgleichung von Jacob unberücksichtigt lässt.
Im zweiten teil wird eine neue Lösung letzterer Gleichung dargestellt, wobei das Glied $\partial h/\partial x$ mit einbezogen wird. An hand der Glover-Dumm-Methode wurden zwei Lösungen entwickelt, die erlauben den Abstand zwischen den Drainagen zu bestimmen. Beim Vergleich der beiden Ergebnisse wurde festgestellt, dass der nach den nueuen Lösungen bestimmte Abstand geringer ist als der an Hand der Glover Formeln errechnete.

1. INTRODUCTION

La conception d'un réseau de drainage se base sur les modèles d'écoulement dans les nappes aquifères caractérisées par leurs profondeurs piézométriques leurs conductivitées hydrauliques et leur coefficient d'emmagasinement. Ces modèles d'écoulement et de retention d'eau dans les aquifères que nous allons passer en revue utilisent des hypothèses qui introduisent des limites à leur validité et leur fiabilité.

Jules Dupuit (1963) a été parmi les premiers à trouver une solution approximative au problème de l'écoulement de l'eau dans le sol en supposant que le gradient hydraulique en tout point est égal à la pente de la surface piézométrique au dessus de ce point tout en reconnaissant que cette condition n'est respectée que pour une ligne piézométrique à faible pente. Les solutions à certains problèmes de drainage et d'infiltration trouvées par P.H. Forcheimer (1930) sont basées sur l'hypothèse de Dupuit.

Boussinesq (1904) a utilisé la même hypothèse avec la loi de Darcy pour développer l'équation de l'écoulement de l'eau en régime transitoire.

Il a considéré une tranche de la nappe d'eau telle que présentée dans la fig. (1)

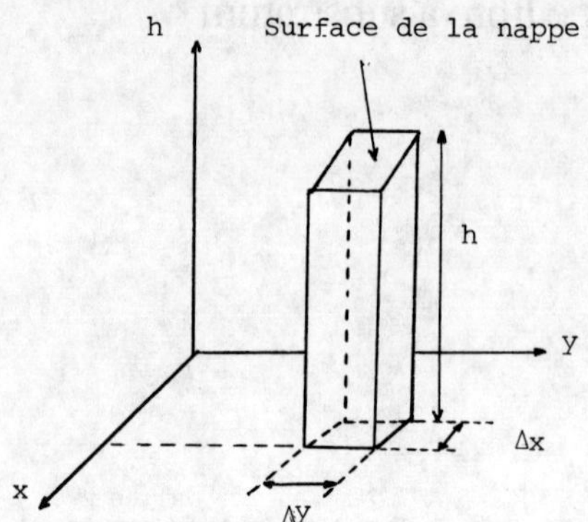

Fig. 1 : Volume élémentaire représentatif

dans laquelle l'eau circule dans la direction O X et O Y.

Les différences entre les débits d'eau entrants et sortants de cette tranche dans ces deux directions sont données respectivement par les deux expressions suivantes:

$$\Delta q_x = -h\frac{\partial J_x}{\partial x}\Delta x\Delta y - J_x\frac{\partial h}{\partial x}\Delta x\Delta y - \frac{\partial J_x}{\partial x}\frac{\partial h}{\partial x}(\Delta x)^2\Delta y \quad (1)$$

$$\Delta q_y = -h\frac{\partial J_y}{\partial y}\Delta x\Delta y - J_y\frac{\partial h}{\partial y}\Delta x\Delta y - \frac{\partial J_y}{\partial y}\frac{\partial h}{\partial y}(\Delta y)^2\Delta x \quad (2)$$

Nous avons en application de la loi de Darcy :

$$J_x = -K_x\frac{\partial h}{\partial x} \qquad J_y = -K_y\frac{\partial h}{\partial y} \quad (3)$$

h = hauteur piézométrique au dessus des drains

K_x et K_y sont les conductivités hydrauliques respectivement dans les directions OX et OY.

La somme des débits donnés par ces deux dernières expressions donne le taux de variation du volume d'eau dans cette tranche de sol ;

$$\Delta q_x + \Delta q_y = S\frac{\partial h}{\partial t}\Delta x\Delta y \quad (4)$$

S étant la porosité de drainage ou coefficient de stockage.

Ainsi, en négligeant les termes de second degré, en simplifiant par $\Delta x\Delta y$ et en remplaçant J_x et J_y par leur valeur nous aurons l'equation suivante :

$$S\frac{\partial h}{\partial t} = K_x\left(\frac{\partial h}{\partial x}\right)^2 + h\frac{\partial}{\partial x}\left(K_x\frac{\partial h}{\partial x}\right) + K_y\left(\frac{\partial h}{\partial y}\right)^2 + h\frac{\partial}{\partial y}\left(K_y\frac{\partial h}{\partial y}\right) \quad (5)$$

La première équation de Boussinesq a été établie à partir de l'équation (5) en considérant les conditions suivantes :

-l'hypothèse de Dupuit - Forcheimer à savoir que $\partial h/\partial x$ et $\partial h/\partial y$ sont très petits permettant de négliger $(\partial h/\partial y)^2$ et $(\partial h/\partial y)^2$

-les conductivités hydrauliques K_x et K_y sont constantes et toutes les deux égales à K(milieu isotrope et homogène);

-la profondeur moyenne de la nappe est égale à D valeur obtenue à partir de la linéarisation de l'équation différentielle partielle.

L'équation (5) donne alors la première équation de Boussinesq qui s'énonce comme suit :

$$S\frac{\partial h}{\partial t} = K D\left(\frac{\partial^2 h}{\partial x^2} + \frac{\partial^2 h}{\partial y^2}\right) \quad (6)$$

La deuxième équation de Boussinesq a été établie toujours à partir de la même équation (5) mais en remplacant $(\partial h/\partial x)^2$ et $(\partial h/\partial y)^2$ par les expressions développées ci-dessous :

$$\frac{\partial(h^2)}{\partial x} = 2h\frac{\partial h}{\partial x}$$

$$\frac{\partial^2 h^2}{\partial x^2} = 2h\frac{\partial^2 h}{\partial x^2} + 2\left(\frac{\partial h}{\partial x}\right)^2$$

d'où $\left(\frac{\partial h}{\partial x}\right)^2 = \frac{1}{2}\frac{\partial^2 h^2}{\partial x^2} - h\frac{\partial^2 h}{\partial x^2}$

de même $\frac{\partial h}{\partial y}^2 = \frac{1}{2}\frac{\partial^2 h^2}{\partial y^2} - h\frac{\partial^2 h}{\partial y^2}$

En substituant ces valeurs dans l'equation (5) et en considérant que :

$K_x = K_y$

$H = h^2$ et

$$S\frac{\partial h}{\partial t} = \frac{S}{2h}\frac{\partial h^2}{\partial t}$$

Nous obtenons la deuxième équation de Boussinesq qui s'énonce comme suit :

$$S\frac{\partial H}{\partial t} = K D\left(\frac{\partial^2 H}{\partial x^2} + \frac{\partial^2 H}{\partial y^2}\right) \quad (7)$$

Cette équation diffère de l'équation (6) seulement par le remplacement de h par $h^2 \equiv H$

Elle découle de l'équation différentielle générale de Jacob (1950) qui a été simplifiée par M. Mariro et L. Luthin (1980).

Ces équations (6) et (7) sont obtenues en considérant $\partial K_x/\partial x = \partial K_y/\partial y = 0$

(sol homogène et isotrope). Elles diffèrent par le fait que la première a été obtenue en considérant $\partial h/\partial x = \partial h/\partial y = 0$

Elles perdent de leur précision dans la conception d'un réseau de drainage lorsqu'on se trouve dans les conditions d'un écoulement préférentiel tel que schématisé dans la figure 2

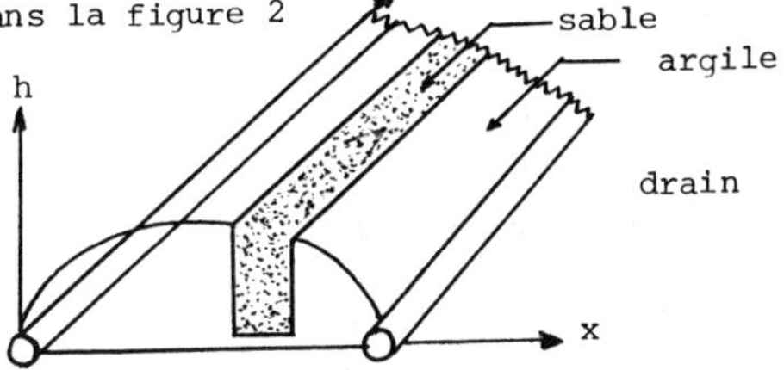

(a) : présence d'un écoulement préférentiel

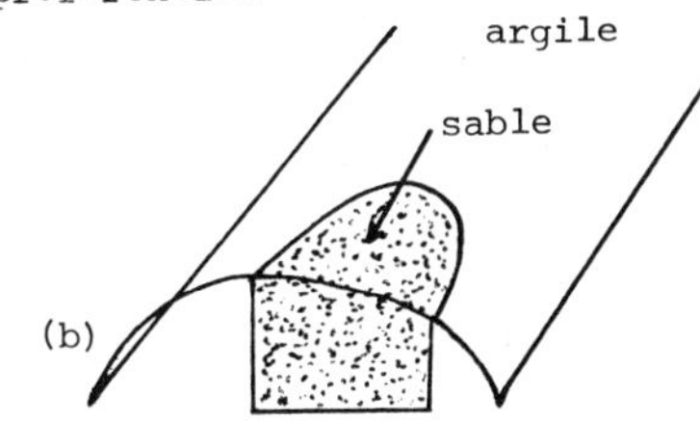

Fig. 2 : Ecoulement préférentiel et variabilité spatiale

Toujours est-il que l'équation (7) reste plus fiable que l'équation (6) parce qu'elle ne néglige pas les gradients $\partial h/\partial x$ et $\partial h/\partial y$ même s'ils sont petits. Pourtant elle a été peu utilisée dans la conception d'un réseau de drainage. Néanmoins elle a été utilisée par Marino (1974 a,b) dans des problèmes de modélisation des écoulements dans des nappes phréatiques pour des conditions limites infinies, et de remontée de la surface de la nappe phréatique suite à une recharge verticale. Nous allons utiliser cette équation (7) pour l'étude théorique de l'écoulement d'une nappe libre en présence d'un réseau de drainage et procéder à une analyse comparative de ces solutions avec celles de l'équation (6).

2. SOLUTIONS DES EQUATIONS DE BOUSSINESQ EN REGIME TRANSITOIRE

2.1. Les solutions de Glover

Les hypothèses de Dupuit-Forecheimer utilisées tout d'abord aux écoulements permanents ont été ensuite appliquées au régime transitoire. Des solutions ont été trouvées à l'équation (6) de Boussinesq. Les plus importantes sont celles de Van Schilfgaarde (1963) et de Glover (1964).

Celles de Glover présentées ci-dessous sont les plus pratiques à utiliser dans la conception des réseaux de drainage.

La première solution est basée sur les conditions aux limites suivantes :

$$h = h_{max}, \quad t = 0, \quad 0<x<L$$
$$h = 0 \quad , t > 0, \quad x = 0$$
$$h = 0 \quad , t > 0, \quad x = L$$

La solution trouvée est la suivante :

$$h = \frac{4\, h_{max}}{\pi} \sum_{n=1}^{\infty} \frac{1}{(2n-1)} \sin(2n-1)\frac{\pi}{L} x \; \exp\left(\frac{-(2n-1)^2 \pi^2 \alpha t}{L^2}\right) \qquad (8)$$

où nous avons

h_{max} = hauteur piézométrique initiale maximale

L = Distance entre les deux drains

α = Diffusivité = K D/ S

h = hauteur piézométrique

K = perméabilité

S = coefficient d'emmagasinement

La deuxième solution est basée sur les conditions aux limites suivantes :

$$h = \frac{8 y_{co}}{4} (L^3 x - 3L^2 x^2 + 4Lx^3 - 2x^4),$$
$$t = 0 \ , \ 0<x<L$$
$$h = 0 \ , \ t>0 \ , \ x = 0$$
$$h = 0 \ , \ t>0 \ , \ x = L$$

La solution est la suivante :

$$h = \frac{192 y_{co}}{\pi^5} \sum_{n=0}^{\infty} \frac{(2n+1)^2 \pi^2 - 8}{(2n+1)^5} \sin (2n+1) \frac{\pi x}{L} \; \exp\left(\frac{-(2n+1)^2 \pi^2 \alpha t}{L^2}\right) \qquad (9)$$

où nous avons :

y_{co} = hauteur piézométrique a mi-distance entre les deux drains (x = L/2)

h = niveau de la nappe au dessus d'un plan référence se trouvant sur les drains.

A notre connaissance aucune solution de l'équation (7) de Boussinesq en régime transitoire n'a été donnée jusqu'à ce jour. Aussi allons-nous la solutionner dans ce qui suit :

2.2. Solutions de l'équation (7)

Pour la résolution de l'équation (7) de Boussinesq nous avons suivi la même procédure que Glover. La première solution que nous avons trouvée et qui s'énonce comme suit a été basée sur les conditions aux limites suivantes :

$$H = h^2_{max} \ , \ t = 0 \ , \ 0< x< L$$
$$H = 0 \quad , \ t > 0 \ , \quad x = 0$$
$$H = 0 \quad , \ t > 0 \ , \quad x = L$$

La solution est :

$$H = \frac{4h^2_{max}}{\pi} \sum_{n=1}^{\infty} \frac{1}{(2n-1)} \sin(2n-1) \frac{\pi x}{L}$$

$$\exp\left(\frac{-(2n-1)^2 \pi^2 \alpha t}{L^2}\right) \qquad (10)$$

où nous avons $H = h^2$

La deuxième solution qui s'énonce comme suit a été basée sur les conditions limites suivantes :

$$H = \frac{64y^2_{co}}{L^8} (L^3 x - 3L^2 x^2 + 4Lx^3 - 2x^4)^2$$

$t = 0$, $0 < x < L$

$H = 0$, $t > 0$, $x = 0$

$H = 0$, $t > 0$, $x = L$

$$H = \frac{256y^2_{co}}{\pi^9} \sum_{n=0}^{\infty} \frac{(161280 - 20160(2n+1)^2\pi^2 +}{(2n+1)^9}$$

$$\frac{408(2n+1)^4\pi^4 - 2(2n+1)^6\pi^6)}{} \sin\left(\frac{2n+1}{L}\right)\pi x$$

$$\exp\left(\frac{-(2n+1)^2 \pi^2 \alpha t}{L^2}\right) \qquad (11)$$

Dans la diffusivité il y a la charge de la nappe D. Pour entrer dans le détail de modélisation pour les cas des drains reposant ou non sur le substratum imperméable il faut se référer au développement donné par Dumm et Glover.

Nous avons reporté sur la figure 3 les

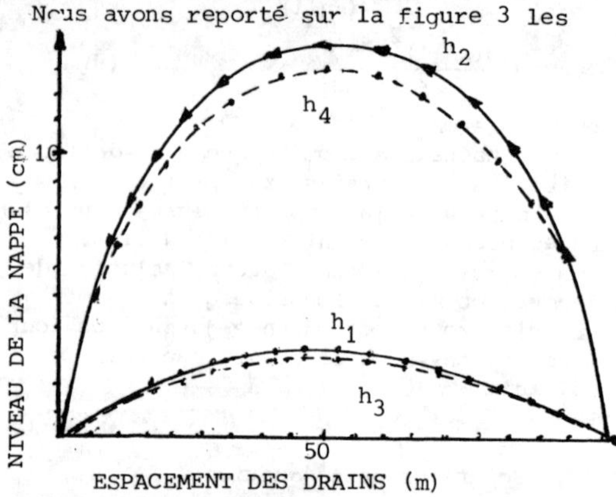

Fig. 3 : Solutions des équations 6 et 7

résultats du calcul donné à partir des solutions de Glover de l'équation (6) et nos solutions de l'équation (7) pour l'exemple suivant :

$t = 35$ jours

$\alpha = 85{,}7 m^2/s$

$L = 95$ m

$y_{co} = 0{,}60$ m

La comparaison entre les résultats donnés par les solutions de Glover (h_1 et h_3) et nos solutions (h_2 et h_4) présentés dans la figure 3 montre que :

$h_2 = \sqrt{y_{co} h_1}$

$h_1 = C_1 h_3$, $h_2 = C_2 h_4$

$h_4 = \frac{1}{C_2} \sqrt{y_{co} C_1 h_3}$

avec $C_1 > C_2$

3. ECARTEMENT DES DRAINS

Les formules suivantes donnant l'écartement des drains sont tirées respectivement des équations (8), (9), (10) et (11), en ne tenant compte que du premier terme de l'expression à sommer (hypothèse de Glover Dumm). Le niveau piézométrique à l'instant $t = 0$ est pris égal à celui se trouvant à mi-distance entre les drains.

$$L_1 = \pi \sqrt{\frac{\alpha t_i}{\ln\left(\frac{4h_{max}}{\pi y_c}\right)}} \qquad (12)$$

$$L_3 = \pi \sqrt{\frac{\alpha t_i}{\ln\left(\frac{192 y_{co}(\pi^2 - 8)}{\pi^5 y_c}\right)}} \qquad (13)$$

$$L_2 = \pi \sqrt{\frac{\alpha t_i}{\ln\left(\frac{4h^2_{max}}{\pi y^2_c}\right)}} \qquad (14)$$

$$L_4 = \pi \sqrt{\frac{\alpha t_i}{\ln\left(\frac{z}{\pi^9 y^2_c}\right)}} \qquad (15)$$

avec $z = 256 y^2_{co} (161280 - 20160\pi^2 - 408\pi^4 - 2\pi^6)$

où nous avons :

t_i = temps nécessaire pour abaisser le niveau piézométrique à une hauteur y_c donnée (la période d'irrigation) .

En utilisant les données suivantes :

$y_c = 1$m, $t_i = 10$ jours, $\alpha = 85$ m^2/s, $h_{max} = y_{co} = 2$m.

Nous avons obtenu les écartements suivants $L_1 = 94{,}66$ m , $L_3 = 99{,}3$ m ; $L_2 = 71{,}74$ m et $L_4 = 62{,}03$ m.

Pour les mêmes données d'autres écartements sont trouvés en évitant de négliger les 20 premiers termes de la sommation des équations (8), (9), (10) et (11) soit

L_1 = 95,09m , L_3 = 99m , L_2 = 72,17m , L_4 = 75,7m.

A l'exception de l'écartement donné par l'equation (11), L_4 tous les autres écartements justifient l'hypothèse de Glover-Dumm qui n'utilise que le premier terme de la sommation. Cependant les écartements (L_2 et L_4) donnés par nos solutions (10) et (11) aux équations de Boussinesq sont inférieurs aux écartements donnés par les solutions de Glover (8) et (9).

Pour déterminer L_4 l'utilisation de plus d'un terme de la sommation de l'équation (11) est conseillée.

4. CONCLUSION

La deuxième équation de Boussinesq n'a jamais été utilisée pour la conception d'un réseau de drainage alors que la première équation de Boussinesq a été résolue par Glover pour deux types de conditions limites .

Nous avons utilisé ces mêmes conditions limites et en se basant toujours sur les hypothèses de Dupuit Forecheimer pour trouver des solutions à la deuxième équation de Boussinesq.

La confrontation de ces quatres solutions a permis de montrer, que l'écartement des drains donné par la première équation de Boussinesq est plus grand que celui de la deuxième équation pour un même rabattement de la nappe dans une période donnée.

Par ailleurs la deuxième équation semble mieux respecter les hypothèses de Dupuit Forcheimer puisque l'écoulement horizontal dû à la pente faible de la nappe doit être le plus lent possible. Ceci a été constaté avec la deuxième equation qui a donné un écoulement plus lent que celui de la première.

Néanmoins les solutions trouvées restent à expérimenter sur le terrain.

BIBLIOGRAPHIE

Boussinesq,M.J. 1904 Recherches théoriques sur l'écoulement des nappes d'eau infiltrées dans le sol et sur débit des sources. Jour. de Math. Pures.et Appl.10:5-78

Dupuit,J.1863. Etudes théoriques et pratiques sur le mouvement des eaux dans les canaux découverts et à travers les terrains perméables. 2nd.ed.Dunod,Paris, 304 pp.

Forcheimer,P.H.1930. Hydraulik. 3rd ed., B.G. Teubner, Leipzig and Bulin, 595 pp.

Glover, R.E.1964. Groundwater movement. Eng.Monogr 31,chief Engineer, U.S.Bureau of reclamation,Denver, Colo. 81 pp.

Marino,M.A. 1974. a.Growth and decay of groundwater mounds induced by percolation J.Hydrol., 22 : 295-301

Marino,M.A. 1974 b. Rise and decline of the water table induced by vertical recharge. J. Hydrol., 23 : 289-298

Marino,M.A. Luthin,J.N. 1980. Seepage and Ground water Developments in water Science El sevier 13 p 144-169.

Land and Water Use, Dodd & Grace (eds), © 1989 Balkema, Rotterdam. ISBN 90 6191 980 0

Diffusion de l'eau à travers une banquette à retention totale: Impact sur le bilan d'eau et le rendement d'une culture

M.Mechergui, N.Ennabli & K.Fadhel
Institut National Agronomique, Tunis, Tunisie

RESUME: La zone d'influence due à la diffusion de l'eau à travers une banquette à retention totale en amont du barrage de Siliana situé dans une région semi aride de la Tunisie a été étudiée durant l'année agricole 1986/87. L'impact sur le rendement de blé dur dû à l'amélioration du bilan hydrique dans cette zone d'influence a été déterminé. L'analyse du bilan hydrique dans une parcelle de 780m^2 en aval de la banquette et à la suite de 10 évènements pluvieux ayant provoqué un ruissellement a montré que la localisation de la ligne saturée dépend de la hauteur d'eau dans le fossé de la banquette, des paramètres pluviométriques et du stock d'eau initial. Une augmentation de 50 mm du stock d'eau entre Janvier et Avril a été constatée dans cette zone d'influence qui a intéressée 70% de la superficie de la parcelle et enregistrée une augmentation de 700kg/ha du rendement en blé dur.

ABSTRACT: The impact of bench terraces on the soil water budget and durum wheat yield was studied in a semi arid area located up the Siliana dam in Tunisie. Soil water diffusion was measured on a 780m^2 parcel down the bench terrace following 10 runoff events. It was found that the position of the wetting front is dependant on the volume of runoff water, rainfall parameters and initial water content. An increase of 50mm in the stock of water was recorded between January and April. In the wetted area which covered 70% of the total area the increase in durum wheat yield was as high as 700 kg/ha.

ZUSAMMENFASSUNG: Oberhalb der Talsperre Siliana, in einer semi-ariden Zone Tunesiens, beoinflusst das Wasser das durch eine Rückhaltebankette dringt, ein Gebiet wo wir während des landwirtschaftlichen Jahres 86/87 Untersuchungen durchflührten. Die Auswirkung der verbesserten Wasserbilanz auf den Hartweizenertrag wurde bestimmt. Die Analyse der Wasserbilanz für eine 780m^2 grosse Parzelle unterhalb der Bankette nach 10 Niederschlagsereighnissen die einen Wasserablauf verursachten, zeigte, dass der Verlauf der Sättigungslinie von
- der Wasserhöhe im Bankettgraben,
- den Niederschlagsparametern und von,
- der ursprünglichen Wassermenge abhängig ist.

Zwischen Januar and April wurde eine Erhöhung des Wasservorrats von 50mm auf 70% der Parzellenfläche festgestellt und einen Ertragszuwachs von 700 kg/ha Hartweizen.

1. INTRODUCTION

Les ressources en eau potentielles du haut tell sont suffisantes pour subvenir aux besoins actuels de cette région. Mais leur distribution dans l'espace reste non satisfaisante. Cette situation est due particulièrement aux conditions climatiques et physiques du milieu et à l'action de l'homme. On assiste à la dégradation du sol, à la chute de la production agricole et à la diminution des revenus.

Pour rémedier à cette situation, plusieurs actions ont été entreprises par le gouvernement. Parmis ces actions nous citons l'intervention par des techniques qui conservent l'eau et le sol sur place.C'est dans ce contexte que nous nous sommes intéressés à l'évaluation des solutions techniques appropriées à l'échelle de l'exploitation, et plus précisemment au système des banquettes construites dans la zone de Mazreg es-chams du Gouvernorat de Siliana, par la Direction de Conservation des Eaux et des Sols.

Le système des terrasses et des banquettes n'est pas nouveau. Il est utilisé dans plusieurs régions du monde et depuis des siècles. Plusieurs études ont été faites sur le système des banquettes. C'est en 1955 que Zinq et Hauser modifient et adaptent ce système pour les terres drainées des grandes plaines et l'appelent "Zinq conservation Bench terrace". Zinq et Hauser(1959) décrivèrent le système des banquettes réduisant l'érosion hydrique et la manière de conserver et de distribuer l'eau pour les cultures céréalières. D'autres scientifiques : Black (1968), Cox (1968), Hauser (1968), Haas et al (1968), Jones et al (1975), Luebs et al (1971) observèrent des résultats importants avec la technique dans des régions où la pluviométrie et inférieure à 500 mm/an. Leurs recherches revèlent la possibilité de stabiliser et d'augmenter la production avec le système des banquettes. Haas et al (1966) étudièrent l'effet des différents types d'impluvium du sol ou sans couverture végétale, sur l'humidité du sol et la production du blé dans la retenue de la banquette dans une région où la pluviométrie est de 377mm/an. Ils trouvèrent que l'humidité du sol et la production du blé dans la banquette étaient supérieures dans le cas d'un impluvium sans couverture végétale ; Bhushan (1979) a étudié l'effet des banquettes sur la production du riz dans le subhumide. Il a trouvé que l'humidité du sol est plus importante dans le cas où l'impluvium est trois fois plus grand que la retenue.

Tous ces chercheurs ont étudié l'impact de l'impluvium sur le rendement et le bilan d'eau à l'amont de la terrasse. Nous allons nous intéresser à l'aval de la terrasse. Nous allons étudier la diffusion de l'eau souterraine à travers la terrasse et son impact sur le bilan hydrique et la production de la zone aval à cette terrasse.

2. PROTOCOLE EXPERIMENTAL ET METHODES D'ANALYSE

La zone d'étude est aménagée par des banquettes de rétention totale dont les objectifs sont au nombre de deux : arret de l'érosion et amélioration du stock hydrique.

Une terrasse a été choisie pour cette étude dont la zone avale a été divisée en mailles fig. 1

Les lignes sont distantes de 5 mètres et les colonnes sont distantes de 6 mètres sauf pour les colonnes 5 et 6 espacées de 15 mètres.

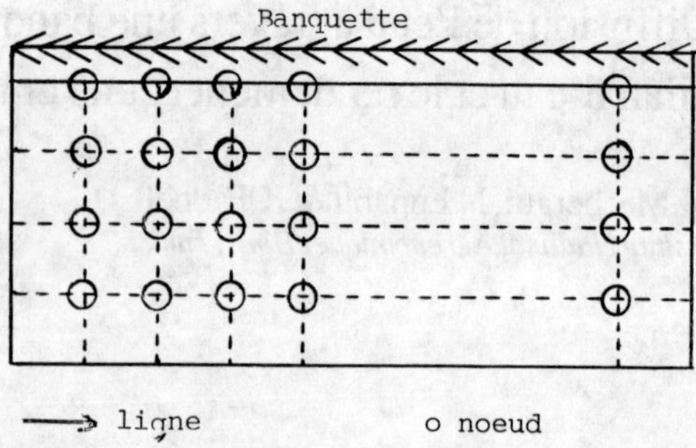

Fig.1 Maillage de la parcelle

Les profils des sols représentés en Fig. 2 montrent que les sols ne sont pas très profonds, ils ont une pente générale vers l'aval et une grande variabilité dans l'épaisseur des couches.

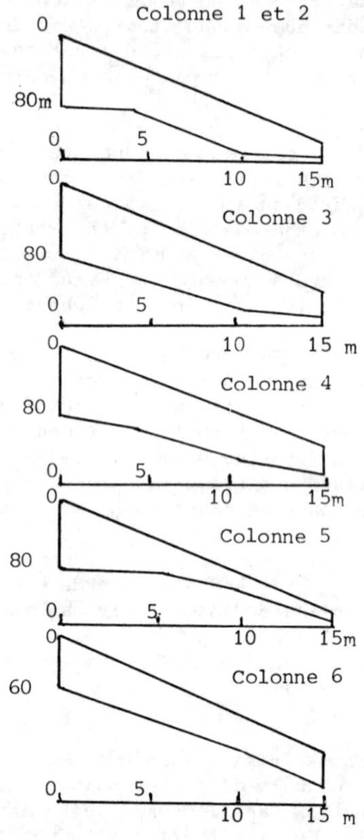

Fig.2 Profondeur du sol dans la parcelle d'étude

Les colonnes 1 à 5 sont représentées par un sol de nature pédologique peu évolué (zone 1) alors que la colonne 6 est représentée par un sol calcimagnésique (zone 2). L'analyse granulométrique de ces deux classes est représentée sur les tableaux 1 et 2.

Tableau 1. Analyse granulométrique de la zone 1

Distance par rapport à la banquette (m)	Zcm	Gravier %	Sable %	A+L %
0	0	7	16	77
5	0	6	20	74
5	20	2	13	85
5	50	2	11	87
10	0	7	17	76
10	20	0	12	88
15	0	8	12	74
15	20	2	12,5	85,5

Tableau 2. Analyse granulométrique de la zone 2

Distance par rapport à la banquette (m)	Zcm	Gravier %	Sable %	A+L %
0	10	6	17	77
0	30	7	15,5	77,5
0	50	3	11,5	85,5
5	10	7	14	79
5	40	6	45,5	48,5
5	50	1	11,5	87,5
10	30	5	18,5	76,5
10	40	0	13,5	86,5
15	10	5	19,5	75,5
15	30	0	17	83

La capacité au champ moyenne est de l'ordre de 23% alors que le point de flétrissement est de 15,5%.

Pour accomplir les objectifs mentionnés plus haut nous avons procédé à la mesure des paramètres suivants :

-sur chaque maille, l'humidité du sol a été mesurée en fonction du temps et de la profondeur.

-le niveau d'eau en amont de la banquette a été mesuré après chaque pluie à l'aide de 3 échelles limnimétriques implantées dans la retenue.

-le front d'humectation horizontal a été suivi en fonction du temps et de l'éloignement de la banquette vers l'aval.

-le rendement en graine et en matière verte en fonction de la distance ont été suivis respectivement pour le blé dur et pour le paturage.

-l'évaporation à partir d'un bac colorado a été suivie durant toute la période de mesure.

-des essais d'infiltration par simple et double anneaux ont été effectués pour déterminer les paramètres de l'infiltration de Philip.

Dans chaque noeud le stock d'eau a été calculé et le bilan d'eau dans la période de mesure a été déterminé par la formule suivante :

$$S = (P_n + I) - E$$

avec S : variation du stock, E : évaporation, P_n : pluie nette et I : infiltration à travers la banquette.

L'évaporation d'un sol nu est calculée par la méthode de Riou (1977).

L'ETR du paturage est calculée en utilisant la méthode de Cruiziat (1980) qui s'énonce comme suit :

M S = a.ETR+b avec MS : matière sèche a et b sont des coefficients déterminés en supposant que sur la banquette l'ETR = ETP en avril et à x = 15m, l'ETR = évaporation d'un sol nu en avril. Ceci nous permettra d'avoir la relation suivante :

M S = 0,7 ETR - 16,47

La courbe expérimentale utilisée, représentée en fig. 3, correspond à l'ensemble de la matière sèche trouvée sur le terrain.

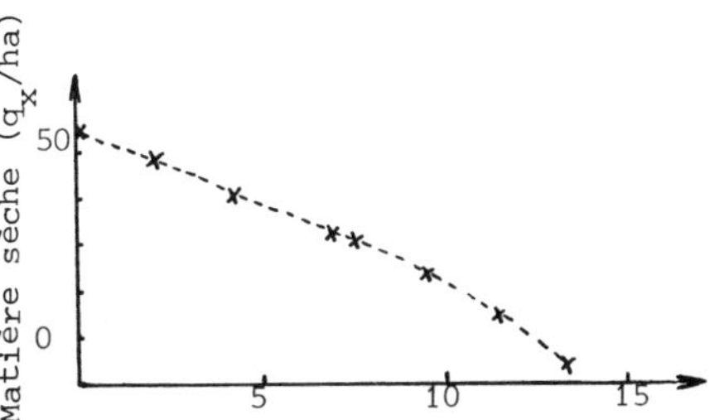

Fig.3 Matière sèche du couvert végétal en fonction de la distance

3. RESULTATS ET INTERPRETATION

3.1 Les essais d'infiltration

Les paramètres de la loi de Philip ($I = St^{1/2} + At$), ajustés sur les valeurs expérimentales pour les essais "simple anneau" sont reportés sur le tableau 3.

Tableau 3. Paramètres de la loi de Philip

Noeud	13	23	33	43	52	53	54	63
S	1,26	1,57	1,16	1,96	0,62	1,66	0,53	2,6
A	1,3	3,0	1,2	4,9	4,2	2,0	3,2	3,0
R	0,98	0,99	0,97	0,97	0,99	0,56	0,99	0,9
N Couples	45	38	43	34	30	37	37	42

Nous remarquons une grande variabilité spatiale dans les paramètres A et S.

La différence entre la lame d'eau infiltrée

en surface, et la différence entre le stock d'eau initial et final nous permettent de déterminer la lame d'eau infiltrée latéralement : Cet écoulement latéral est variable dans l'espace comme le montre le tableau 4.

Tableau 4. Ecoulement latéral

Noeuds	13	23	33	43	52	53	54	63
1	29	46	95	63	42	49	5	62
2	21	28	50	26	16	16	45	29
3	8	18	45	37	26	33	5	33

(1) : lame d'eau infiltrée à la surface(mm)
(2) : stock d'eau sous l'anneau (mm)
(3) : écoulement latéral (mm)

Ces résultats nous permettent d'avancer les hypothèses suivantes :

-en amont d'une banquette l'eau accumulée s'infiltre verticalement et diffuse latéralement dans toutes les directions.

-vu la pente du terrain naturel et l'existence d'une assise imperméable présentant aussi une pente, l'eau s'infiltrera préférentiellement vers l'aval de la banquette.

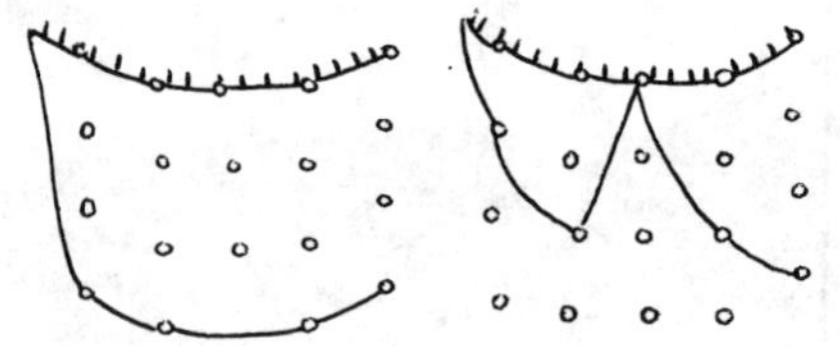

au niveau de l'assise   à 10cm de l'assise

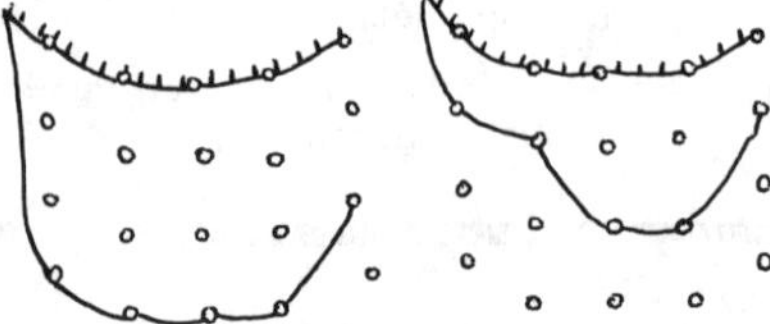

à 20cm de l'assise   à 30cm de l'assise

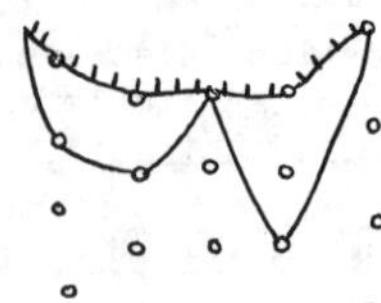

Fig.4a Δt=1 jour

à 40cm de l'assise

Fig. 4 Surfaces imbibées à Δt=1 jour, 3 jours et 7 jours

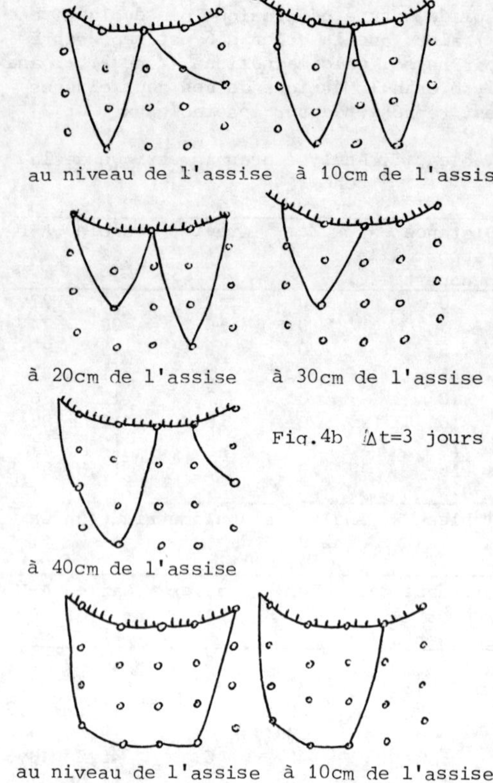

au niveau de l'assise   à 10cm de l'assise

à 20cm de l'assise   à 30cm de l'assise

Fig.4b Δt=3 jours

à 40cm de l'assise

au niveau de l'assise   à 10cm de l'assise

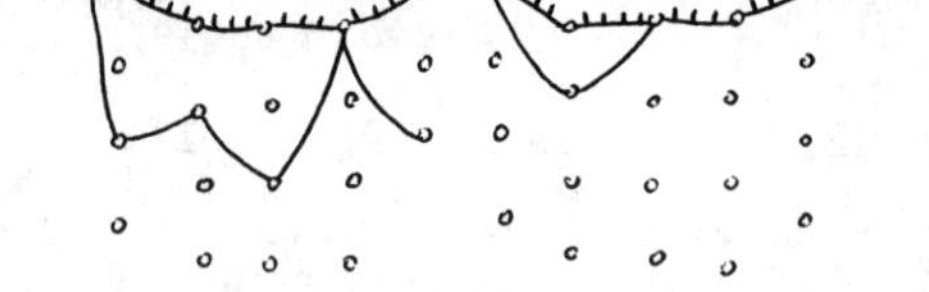

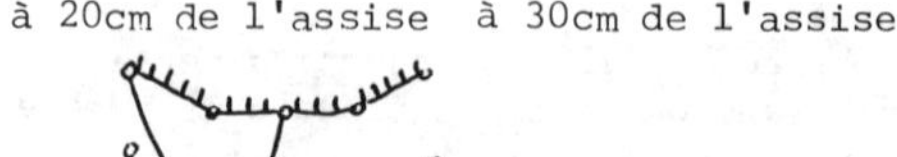

à 20cm de l'assise   à 30cm de l'assise

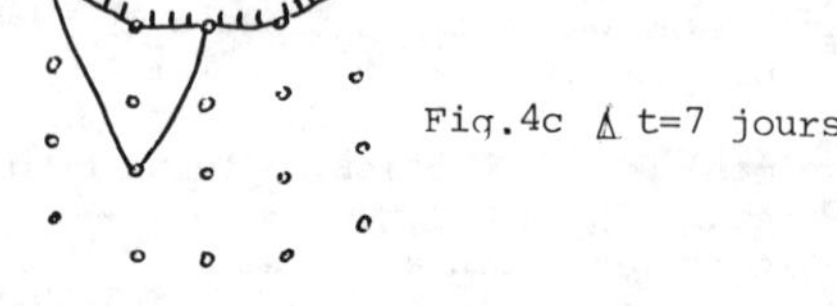

Fig.4c Δt=7 jours

à 50cm de l'assise

3.2. Caractérisation de la zone d'influence

3.2.1. Le front d'humectation

Après une crue provoquant le remplissage de la retenue et à des intervalles de temps variant de 1 à 7 jours, nous avons mesuré la teneur en eau volumique et nous avons déterminé la surface imbibée en aval de la banquette provoquée par la diffusion de l'eau de la retenue. Le front d'humectation pour chaque profondeur est présenté sur la fig. 4a. Nous avons observé le retrait de ce front pour t = 3 jours (fig. 4b) puis au 7ème jour nous avons constaté le retour du front comme le montre la fig. 4c. Ce front n'est pas dûà une autre pluie mais plutôt à un écoulement hypodermique qui vient de la zone supérieure à cette terrasse. Les valeurs de la surface imbibée présentées dans le tableau 5 montrent bien l'arrivée du front hypodermique au 7ème jour.

Tableau 5. Surface imbibées en m^2

T en jour	1	2	3	7
au niveau de l'assise	360	330	180	180
à 10cm de l'assise	345	240	210	240
à 20cm de l'assise	180	240	225	180
à 30cm de l'assise	165	60	30	90
à 40cm de l'assise	150	90	15	150

3.2.2. Le stock hydrique

La zone d'influence est donc caractérisée par des avancements du front en fonction de la variation temporelle de l'humidité. La méthode du bilan hydrique basée sur la méthode gravimétrique permet d'évaluer l'humidité au niveau de ce front. Les valeurs du stock d'eau à chaque noeud présentées pour différents temps dans le tableau 6, diminuent au fur et à mesure qu'on s'éloigne de la banquette vers l'aval. Ceci illustre l'alimentation souterraine à partir de la retenue derrière la banquette.

3.3. Détermination de l'apport d'eau par infiltration à travers la banquette

Deux techniques peuvent être utilisées pour déterminer la quantité d'eau apportée par diffusion à travers la banquette.

-le calcul du bilan d'eau

-la mesure directe de la lame d'eau ruisselée vers la retenue

3.3.1. Le bilan hydrique

Connaissant l'évaporation, la pluie nette après soustraction de la perte par ruissellement, la variation du stock, on peut calculer le bilan aux différentes distances. Les valeurs du tableau 7 correspondent aux lames d'eau ilfiltrées pour la période du 6-1 au 24-1.

3.3.2. Mesure directe

La lame d'eau (H_T) mesurée dans la retenue de la banquette à laquelle on soustrait l'évaporation bac nous donne la lame d'eau infiltrée en amont de la banquette (tableau 8). La quantité d'eau totale infiltrée mesurée par cette méthode nous donne une valeur de 30mm dans la période considérée.

Tableau 6. Stocks hydriques (mm) dans chaque noeud et en fonction du temps

Noeuds	Temps					
	6-1-87	20	10-2	17	17-3	24-4
11	292	331	311	300	242	174
12	138	173	162	140	132	108
13	117	145	130	156	174	116
14	73	69	59	66	63	36
21	353	400	268	399	256	165
22	218	231	169	187	169	124
23	163	206	192	185	142	104
24	76	80	59	79	80	57
31	282	315	276	318	230	152
32	251	294	243	231	244	167
33	197	221	188	208	194	108
34	96	67	50	72	70	52
41	225	290	277	264	211	.
42	206	297	231	224	218	.
43	184	201	141	175	165	132
44	101	122	94	100	98	61
51	271	309	309	274	227	142
52	135	181	116	125	121	84
53	72	90	88	72	68	35
54	34	33	21	25	24	16
61	220	246	206	186	177	201
62	262	213	208	189	185	113
63	166	187	136	165	147	97
64	159	177	157	146	157	127

Tableau 7. Lames d'eau infiltrées calculées

	Périodes				
	1	2	3	4	5
I mm	21	3,5	14	7	5
	Total = 50 mm				

1 : 6/1 au 20/1 ; 2 : 20/1 au 10/2 ; 3 : 10/2 au 17/2 ; 4 : 17/2 au 17/3 ; 5 : 17/3 au 24/4

Tableau 8. Lames d'eau mesurées

Date	18/1	19	14/2	16	26	10/3	21	31	1/4	14
Htmm	4	2	2	1,5	6	2,5	3	9,5	2	4
E mm	0,5	0,2	0,4	0,5	0,5	0,5	0,5	1	1	1,5
I mes	3,5	1,8	1,6	0,5	5,5	2	2,5			

3.3.3. Comparaison entre les deux méthodes

Le tableau 9 illustre cette comparaison. Nous observons un apport supplémentaire venant de l'amont de la banquette dans les trois premières périodes. Pour les deux dernières périodes on observe que Imes est supérieure à Ical. Ceci s'explique par le fait que à partir du début du mois de mars, la végétation apparue autour de la retenue diminue la diffusion de l'eau à travers la banquette.

Tableau 9. Comparaison entre l'infiltration calculée et celle mesurée

	Périodes				
	1	2	3	4	5
Ical (mm)	21	3,5	14	7	5
Imes (mm)	5	0	2,5	7,5	15,5
Ical-Imes	16	3,5	11,5	0,5	10,5

1,2,3,4, 5 : même périodes que tableau 7

3.4. Rendement de blé dur dans l'espace inter-banquettes

Lors de la moisson de blé, nous avons effectué des récoltes de blé dur dans un champ non traité et dans l'espace inter-banquettes voisin de la parcelle. Les résultats sont regroupés dans la fig. 5.

Nous remarquons une chute des rendements de l'espace inter-banquettes en fonction de l'éloignement de cette dernière jusqu'à la valeur du rendement témoin. Ceci illustre encore la diffusion de l'eau à travers la banquette.

3.5. Conclusion

La dégradation des ressources naturelles dans la zone d'étude est un phénomène visible surtout au niveau de la perte en eau et en sol arable. Devant cette situation dangereuse, il a paru nécessaire de concevoir des aménagements telles que les banquettes. L'étude de l'impact du système des banquettes sur les potentiels eau et production nous a intéressé de plus près, et nous a révelé les points suivants :

-l'existence de l'assise imperméable à faible profondeur et à pente forte à crée une zone d'influence qui augmente en fonction de la profondeur du sol. L'écoulement hypodermique venant de l'amont de la banquette a engendré une augmentation de la surface de la zone d'influence.

-l'apport supplémentaire accumulé dans la retenue de la banquette a amélioré le stock d'eau dans la parcelle située en aval.

-Cette augmentation du stock a engendré une amélioration du rendement. De ce fait, le système des banquettes a entrainé les cultures du mode extensif vers un mode qu'on peut qualifier"semi-intensif"

-l'intéret des banquettes dans l'apport d'eau supplémentaire est convaincant, mais le degré d'efficacité reste à analyser si on considère des sols plus profonds que ceux étudiés dans notre zone. Il serait souhaitable aussi de voir comment réagir ce système pour une période plus longue

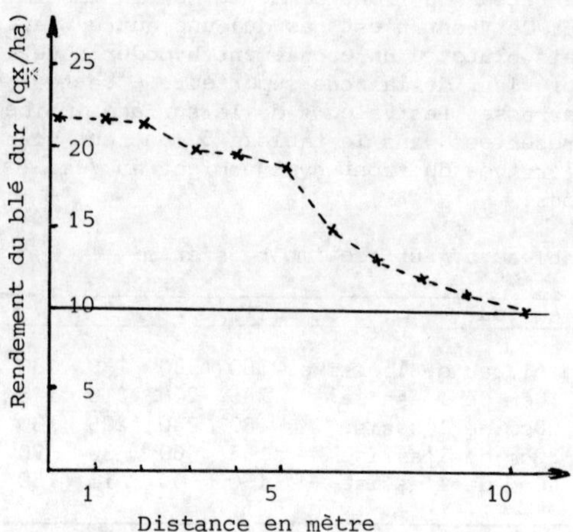

Fig. 5 Rendement du blé bur dans la parcelle d'étude

que celle étudiée.

BIBLIOGRAPHIE

Bhushan.L.S.1979. Conservation Bench Terraces for rice in a subhumid climate.S.S. S. AP. Vol. 43

Criuziat.P.1980. L'eau et les cultures. Techniques Agricoles

Haas, H.T., and Willis, W.O. 1962. Moisture storage and use by dryland spring wheat cropping systems. Soi. Sci. Soc. Am. Proc. 26 : 506) 509

Haas, H.J., Willis, W.O. and Bowtright, G. O. 1966. Moisture storage and spring wheat yields on level - bench terraces as influenced by contributing area cover and evaporation control. Agr. J. Vol. 58: 297-299

Hauser, V.L. and Cox, M.B. 1962. Evaluation of zingg conservation bench terraces. Agr. Eng. 43 : 462-464

Riou, C. 1977. Evaporation du sol nu et répartition des pluies : relations établies en Tunisie à partir des cases lysimétriques. R. INRAT Vol. 50. F4

Zing, A.W. , and Hauser, V.L. 1959. Terrace benching to save potential runoff for semiarid land. Agron. J. 51 : 289-292

Land and Water Use, Dodd & Grace (eds), © 1989 Balkema, Rotterdam. ISBN 90 6191 980 0

Aménagement du sous bassin versant de l'Oued Slimane en vue d'un contrôle total des eaux de surface*

F.Lebdi, J.Tarhouni, J.Braham & N.Ennabli
Institut National Agronomique, Tunis, Tunisie
A.Ben Alaya
Ecole Supérieure d'Agriculture, Mograne, Tunisie

RESUME: Le captage des eaux de crue de période de retour deux ans afin n'avoir qu'un débit faible ou nul à l'aval du bassin versant de l'Oued Slimane à necessité les aménagements suivants :

- un ouvrage de stockage en terre ayant une retenue de 43 550 m^3 environ. La superficie irrigable par le volume d'eau stocké est de 11 hectares environ.
- deux ouvrages de dérivation des eaux de crue assurant l'irrigation par épandage de 51 hectares environ.

L'étude de l'impact de ces aménagements sur le bilan hydrique a montré que les besoins en eau des cultures d'hiver, sont couverts à plus de 75% en moyenne et à plus de 50 % pour les années sèches.

ABSTRACT: Catching the total amount of water relative to floods occuring every two years at the catchment area of the Slimane River requires the following management:
- small dams of 43,550 cubic metres capacity: this permits us to irrigate circa II ha.
- water spreading dams permits the irrigation of circa 5I ha.

The impact of such management of the water balancehas showed that the water consumption of winter crops is provided for up to 75% on average and up to 50% in dry years.

ZUSAMMENFASSUNG: Die Einfassung des Uberschwemmungswassers unterhalb des Beckens, das sich in den Oued Slimane ergiesst, verlangte folgende Massnahmen :

- Ausbau einer Speicherung von ungefähr 43 550 m^3, was ungefähr einer Beregnungsfläche von 11 Hektaren entspricht.
- Zwei Ableitungen des Uberschemmungswassers, was die Beregnung weiterer 51 Hektaren garantiert.

Die wasserbaulich Studie ergab, dass der Wasserbedarf der Winterfrüchte im Mittel zu 75 % und in Jahren der Trockenheit zu 50 % gedekct waren.

*Ce travail a été fait dans le cadre du Projet de Recherches Agricoles sur les systèmes de productions (Coopération USAID)

I INTRODUCTION

L'irrigation par épandage est une forme particulière d'irrigation de surface, elle consiste à dériver les eaux de crue des canaux ou de cours d'eau naturels et à les épandre sur des zones de plaines. La dérivation est assurée par la construction d'une digue sur le cours d'eau et un ouvrage partiteur. Les systèmes d'irrigation par épandage sont conçus pour tirer le maximum de profit des eaux de ruissellement traversant la zone.

En Tunisie, cette technique de mobilisation de l'eau est très ancienne, elle été pratiquée essentiellement dans la région de Sidi Bouzid pour l'irrigation de complément des grandes cultures (les Mougouds), et dans la région du sahel pour l'irrigation de l'olivier (Miskat). Les plans d'aménagements hydro-agricoles ne se sont intéressés à la mobilisation de l'eau à petite échelle que très récemment après la réalisation de grands barrages et la création de plusieurs périmètres irriqués. Cette technique a l'avantage d'être à la portée des agriculteurs, elle peut être réalisée même à l'échelle de la petite exploitation.

L'étude proposée porte sur l'aménagement de l'Oued Slimane dans le Gouvernorat de Siliana dont l'objectif est de stocker le volume de la crue bi-annuelle et d'évaluer son impact sur le bilan hydrique des cultures d'Hiver.

II PRESENTATION DE LA ZONE ETUDIEE

La zone concernée par l'étude de mobilisation à petite échelle est le bassin versant de l'Oued Slimane (cf Figure 1) situé dans le Gouvernorat de Siliana.

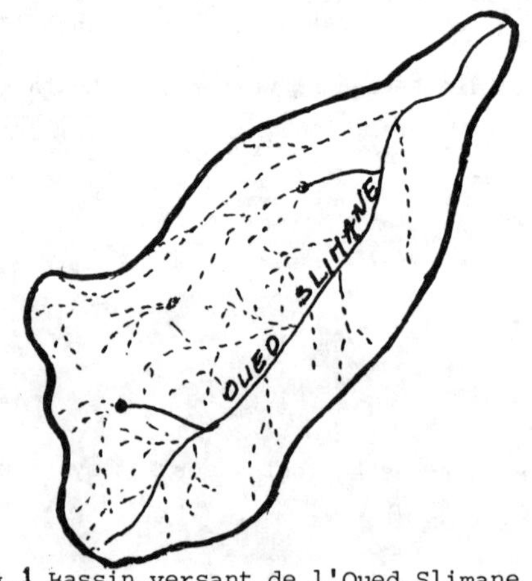

Fig. 1 Bassin versant de l'Oued Slimane

Le diagramme de Bertoli(1969) établi pour différentes stations de la Tunisie a permis de situer la zone dans l'étage semi-aride moyen du bioclimat méditerraneen à tendance continentale.

2.1 Estimation de l'évapotranspiration potentielle (ETP)

L'estimation de l'ETP est faite à partir de la formule de Riou (1980) établie pour les zones côtières et rectifiée pour les zones intérieures du pays :

$$ETP_n = 0.31 \left(\frac{T_n + 2\,T_{n+1}}{3} \right) - B \quad (1)$$

avec :

ETP_n : ETP journalière du mois du rang n exprimée en mm/j

T_n et T_{n+1} : température maximale interannuelle du mois du rang n et n+1 respectivement (°C)

B : 7.1 - 0.1 L où L est la latitude de la station exprimée en degré.

2.2 La pluviométrie

La répartition temporelle de la pluie est irrégulière (coefficient de variance CV= 44%). Le total pluviométrique annuel moyen est de l'ordre de 410 mm réparti entre l'Automne (31 %), l'Hiver (28.1 %) et le Printemps (31 %). L'Eté est sec.

2.3 Bilan hydrique

La pluviométrie mensuelle moyenne est inférieure à l'évapotranspiration potentielle mensuelle moyenne sur toute l'année; il n'y a donc pas de reconstitution de la réserve hydrique du sol (cf.Figure 2) et le déficit hydrique moyen s'identifie au déficit climatique (ETP - P) moyen.

L'évapotranspiration maximale (ETM) du blé, dont le cycle végétatif s'étend du mois de Novembre (semis) jusqu'au mois de Juin(récolte)n'est satisfaite qu'en Hiver (cf.Figure 2). Dans l'objectif de régulariser la production agricole, un apport d'une quantité supplémentaire d'eau de 250 mm à partir du mois de Mars s'avère donc indispensable.

2.4 Hydrologie

Le bassin versant de l'Oued Slimane ne présente pas à son aval une grande retenue pour stocker la crue annuelle, il a été découpé en trois sous bassin versants (cf.Fig.3).

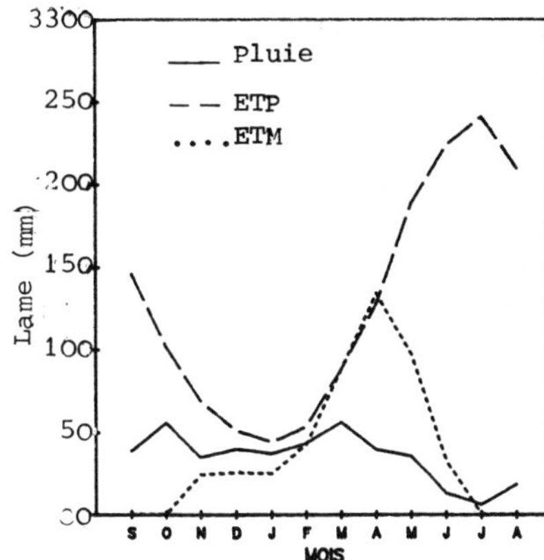

Fig.2 Bilan hydrique

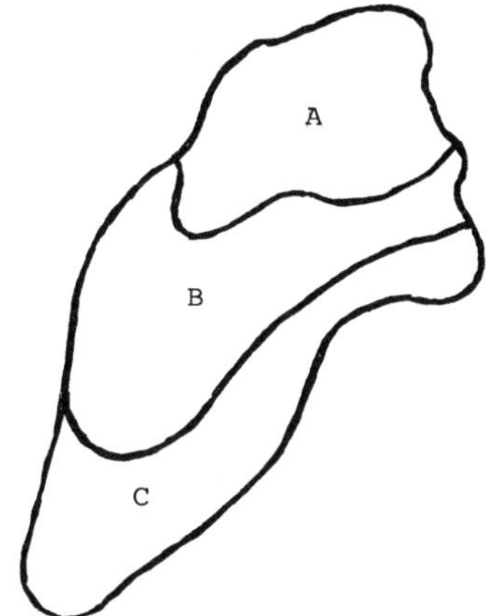

Fig.3 Sous bassins versants de l'Oued Slimane

L'estimation de l'apport en eau moyen annuel (AM) a été faite à partir de la formule de Tixeront (1960) :

$$AM = \frac{P^3}{3E^2} \times S \qquad (2)$$

avec :
E : évaporation moyenne (m)
P : pluviométrie moyenne annuelle (m)
S : superficie du bassin versant (m^2)
Pour les trois sous-bassins versants l'apport moyen annuel total est de 466 000 m^3 environ. (cf.Tableau 1).

2.4.1 Estimation du débit de crue

Aucune étude statistique directe ne peut être envisagée, étant donné qu'on ne dispose pas de série de mesures dans la zone. L'approche suivie est fondée sur l'hydrogramme triangulaire
Cette méthode simplifiée est suffisamment précise pour les petits bassins versants; elle repose sur les deux hypothèses suivantes :

- Le temps de base (t_b) de l'hydrogramme est égal à deux fois le temps de concentration (t_c).
- Le rapport du temps de montée (t_m) au temps de décrue (t_d) est de 0.5.

Le débit spécifique de pointe est donné par l'expression suivante :

$$q_{max} = \frac{2P_n}{t_b} \qquad (2)$$

où P_n : la pluie nette (en m) est estimée comme suit :

$$P_n = P_{tot} \times C_r \qquad (3)$$

avec P_{tot} : pluie totale de durée égale à t_c pour une période de retour T.
C_r : coefficient de ruissellement
- t_b en seconde
- q_{max} en $m^3/s/m^2$.

Le débit de pointe Q_{max} vaut alors :

$$Q_{max} = A \times q_{max} \text{ (en } m^3/s) \qquad (4)$$

A : Superficie du sous-bassin versant

Quant au volume ruisselé V_r, il est déterminé par considération géométrique de l'hydrogramme triangulaire.

$$V_r = Q_{max} \times \frac{t_b}{2} \quad (m^3) \qquad (5)$$

2.4.2 Estimation du temps de concentration

Il est estimé à partir de la formule de Ventura

$$t_c = 76 \sqrt{\frac{A}{I}} \qquad (6)$$

I : pente moyenne du sous bassin versant en %
t_c : temps de concentration en minute
A : superficie du sous bassin en km^2.

Le réseau d'irrigation et les ouvrages (seuils de dérivation, partiteurs...) sont dimensionnés pour une période de retour de deux ans. Quant aux ouvrages de protection (évacuateurs de crues) ils ont été conçus pour resister à la crue de période de retour 20 ans.

Table.1 Caractéristiques des sous bassins et débit ma et volumé ruisselé max. pour T = 2ans et T = 10 ans

Sous bassin versant	A	B	C
Superficie (km^2)	4.7	6.65	5.10
AM ($10^3 m^3$)	133.3	168.61	144.65
t_c (mn)	52.0	70.0	81.0
Q_{max2} (m^3/s)	6.8	8.1	5.6
V_{r2} ($10^3 m^3$)	21.2	34.8	27.20
Q_{max20} (m^3/s)	12.6	15.0	10.3
V_{r20} ($10^3 m^3$)	39.30	63.00	50.10

Q_{max2} et Q_{max} : débit max de T = 2 et T=20 ans respectivement

V_{r2} et V_{r20}: volume ruisselé pour T= 2 et T = 20 respectivement

III LES AMENAGEMENTS PROPOSES

Le choix des sites et des types d'aménagement a été fait en tenant compte :

- de la localisation et de l'étendu des terres à irriguer
- des volumes d'eau à stocker
- de la nature et de la forme du fond et des berges de l'Oued

Les aménagements retenus sont les suivants :

- Une digue en terre A permettant le stockage des eaux ruisselées du sous bassin versant A. L'eau stockée sera utilisée pour l'irrigation par aspersion d'un périmètre de 11 hectares environ.
- Deux ouvrages en terre B et C permettant la dérivation des eaux ruisselées des sous bassins B et C. Les eaux dérivées seront par la suite réparties sur des terres agricoles (51 hectares) situées sur les rives de l'Oued.

3.1. Digue de stockage A

3.1.1 Schéma type

Le schéma principal du barrage proposé est celui d'une digue en terre homogène.Les principales caractéristiques de la digue A sont données sur la figure 3.

3.1.2 Utilisation de l'eau de la retenue A

La culture des céréales necessite un apport supplémentaire d'eau de 250 mm à partir du mois de Mars.

L'étude hydrologique de la zone fait ressortir deux crues : une en Automne et une au Printemps.

Etant donné que le stock hydrique dans le sol est nul pour tous les mois de l'année (Fig.2) on a envisagé une première

digue en terre

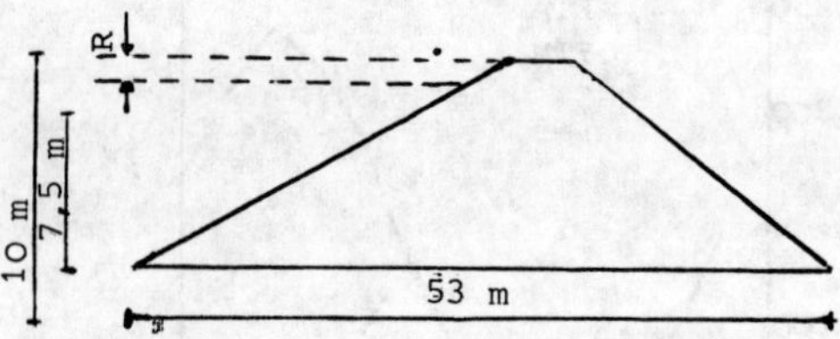

Fig.4 Profil de la digue A

irrigation au mois de Février pour reconstituer la réserve utile du sol (dose = 140mm); ce qui permet de diminuer le volume d'eau dans la retenue avant l'arrivée de la deuxième crue. Cette réserve hydrique emmagasinée dans le sol couvrira largement les besoins hydriques du mois de Mars et de la première quinzaine du mois d'Avril. Le volume stocké pour une période de retour T = 2 ans (V_r = 21 200 m^3) doit permettre, par conséquent, deux irrigations d'une dose égale à 95mm au début de la troisième semaine de chacun des mois d'Avril et de Mai. Ainsi la superficie irriguable à partir de l'ouvrage A est de 11 hectares environ.

3.1.3 Impact sur le bilan hydrique

Pour une année de pluviométrie égale à la moyenne, le volume d'eau qui sera retenu est de 18 000 m^3. Si on considère les pertes par infiltration et par évaporation (20% du volume disponible), la quantité qui reste à la disposition de l'irrigation correspond à un taux de satisfaction en eau des cultures de 80%, pour la période allant du 15 Avril à la fin Juin.

Pour une période de retour d'un an, le volume d'eau apporté par la crue du Printemps est de 16 000 m^3. Pour des pertes par infiltration et par évaporation de 20%, du volume disponible, le taux de satisfaction des besoins en eau des cultures est d'environ 70%, pour la période 15 Avril - fin Juin.

3.2 Epandage des eaux de crues

Les eaux de ruisellement des sous bassins versants B et C sont dérivées vers des terres agricoles en vue de l'irrigation de complément de céréales.

3.2.1 Ouvrages de dérivation B et C

L'ouvrage de dérivation est constitué des

ouvrages suivants :

- Une digue en terre de profondeur identique à celle du cours d'eau qui est protégée contre la submersion et les contournements
- Un évacuateur de crue assurant la protection de la digue
- Un seuil déversoir équipé de vannes latérales permettant de dériver un débit constant vers le réseau d'épandage. Ceci a été conçu dans l'objectif d'éviter le surdimensionnement du réseau et l'inondation excessive des sols à perméabilité faible ($K_s = 8\ 10^{-5}$m/s).

digue en terre

ouvrage de vannage

déversoir

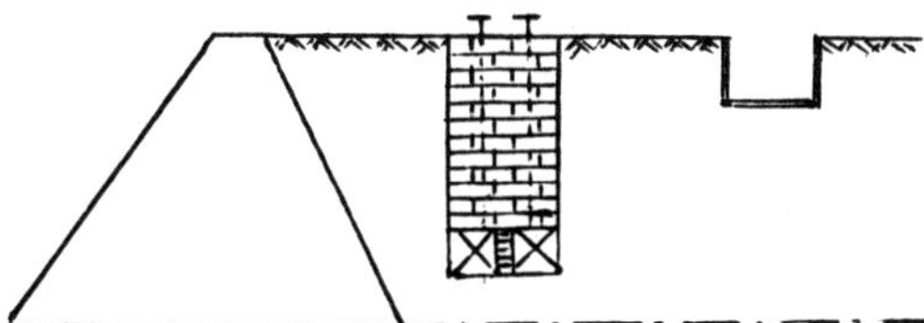

Fig. 5 Ouvrages d'épandage B et C

3.2.2 Périmètres d'épandage

A. Superficie à aménager :

Sur la base d'une réserve facilement utilisable (RFU) de 100 mm et une perte par infiltration et par évaporation estimée à 20%, les superficies à aménager ont les valeurs suivantes :

- 27 hectares environ à partir de l'ouvrage B
- 24 hectares environ à partir de l'ouvrage C

B. Temps de submersion

Le temps de submersion maximale (T_s) calculé à partir de la perméabilité à saturation K_s est de 19 heures, valeur qui est largement inférieure au temps de submersion tolérable par les grandes cultures (> 3 jours).

C. Aménagement au niveau de la parcelle :

Les périmètres d'épandage sont découpés en bandes délimitées par des diguettes en terre de 0.45m de hauteur. La largeur de la bande est obtenue pour une charge nulle au niveau de la digue amont et une charge de 0.25m au niveau de la diguette aval.

3.2.3 Impact sur le bilan hydrique

Les volumes ruisselés sur chacun des sous bassins B et C pour la crue d'Automne et du Printemps ont été estimés pour deux périodes de retour (T= 1an et T = 2 ans) :

	T	C.Aut.	C.Print	LR
B	1	24.730	20.000	132
	2	34.02	30.110	190
	1	21.420	19.130	135
C	2	27.200	19.300	155

LR: Lame d'eau dérivée en mm sur le périmètre

Pour le périmètre d'épandage à partir de l'ouvrage B, la lame moyenne apportée par ruissellement des eaux des deux crues est de 132 et de 190 mm pour une période de retour de 1 et 2 ans respectivement; soit une couverture des besoins de 53% et de 76% respectivement.

Pour le périmètre d'épandage à partir de l'ouvrage C, la couverture des besoins hydriques est de 54% et de 62% pour les mêmes périodes de retour.

CONCLUSION

L'une des caractéristiques principales des zones semi-aride est le déséquilibre entre les apports pluviométriques et les besoins en eau des cultures; pour contribuer à la régularisation de la production des cultures de l'hiver, dans la région de Siliana les aménagements suivants ont été projetés sur l'oued Slimane :

- une digue de stockage à l'aval du sous-bassin versant A, la quantité d'eau emmagasinée permet l'apport de deux irrigations de complément pour 11 hectares environ.
- deux ouvrages de dérivation des eaux de ruissellement des sous-bassins versants B et C; ils permettent l'irrigation par épandage de 51 hectares environ.

En satisfaisant à plus de 50% le déficit hydrique pour une année sèche de fréquence décennale, ces ouvrages assureront une production suffisante pour l'agriculteur.

La généralisation de ces micro-aménagements sur des petits cours d'eau en zone semi-aride ou aride, assure un apport de complément, régularise la production et maintien les hommes et le sol en place.

REFERENCES

Institut National de Recherche Agronomique de Tunisie. Climatologie et Bioclimatologie de la Tunisie septentrionale 1969

Riou,C. 1982. Formule empirique pour estimer l'E.T.P. moyenne en Tunisie, Tunis, INAT - ORSTOM.

Riou,C.1977 . Evapotranspiration et bilan hydrique en Tunisie. Tunis, INAT- DRE.

Tixeront,J. 1960. Petits barrages d'irrigation permanente en Tunisie, Paris, 31p.

Berkaloff,E. et Tixeront,J. Carte du ruissellement annuel moyen en Tunisie, 1956 11p.

Land and Water Use, Dodd & Grace (eds), © 1989 Balkema, Rotterdam. ISBN 90 6191 980 0

Bilan hydrique d'un système de micro-lacs collinaires et impact sur le bassin versant de Siliana*

F.Lebdi, J.Tarhouni, S.Touzi & N.Ennabli
Institut National Agronomique, Tunis, Tunisie

A.Ben Alaya
Ecole Supérieure d'Agriculture, Mograne, Tunisie

RESUME: L'étude de l'impact de l'aménagement hydraulique pour la mobilisation des eaux à l'échelle d'un petit bassin versant dans le Gouvernorat de Siliana est fondée sur la détermination des termes du bilan hydrique du système de micro-lacs. Il ressort que :

- les pertes mensuelles par évaporation et par infiltration sont très importantes (> 70% des apports mensuels par ruissellement). Ceci constitue une contrainte majeure pour le stockage de l'eau en vue de l'irrigation de complément des céréales à partir du mois d'Avril.
- le poids total du dépôt annuel intercepté est de 188 tonnes. Ce qui correspond à une érosion spécifique de 2500 tonnes/km^2/an. La généralisation de ce type d'ouvrage à l'ensemble du bassin versant de l'Oued Siliana constitue un sautien au barrage réalisé sur celui-ci par la retention des matériaux solides et le développement du couvert végétal qu'il permet.

ABSTRACT: The water usage in a small catchment area in the Siliana governate was based on the determination of the water balance of micro-lakes system. It has been concluded that:

- monthly losses by evaporation and infiltration are very important (70% of monthly rainfall). This was a major constraint for water storage to be utilised for complementary irrigation of cereals from April to June.
- the annual total weight of solids was measured at I88 tonnes; this was equivalent to a specific erosion of 2500 tonnes/km /year.

The generalisation of this type of management over the catchment area of the Siliana river protects the barrage against solids and permits agricultural development in the area.

ZUSAMMENFASSUNG: Die wasserbauliche Studie eines Abflussbeckens im Gouvernorat Siliana ergab - unter der Praemisse eines kleinstauseen-Systems-- folgendes :

- die monatlichen Verluste durch Evaporation und Versickerung sind hoch (>70 % der monatlichen Zuflüsse). Besonders in Hinblick auf die Zusatzberegnung von Getreide ab April erscheint dies als wenig sinnvoll;
- die Gesamtablagerungen liegen bei 188 Tonnen, was einer Erosion von 2500 Tonnen pro km^2 und Jahr entspricht.

*Ce travail a été fait dans le cadre du projet de recherches agricoles sur les systèmes de production (Coopération USAID)

I. INTRODUCTION

Après la réalisation des grands barrages, pour la plupart concentrés dans le Nord du pays relativement bien arrosé et la généralisation des ouvrages de dimensions plus modestes tels que les lacs collinaires visant à maitriser l'eau à une échelle plus réduite, on s'oriente vers la maitrise à une échelle encore plus réduite du potentiel restant, en mettant en oeuvre des ouvrages de micro-hydraulique à l'échelle de la ferme.

Dans cette étude, l'on se propose d'étudier à partir d'un système de micro-lacs collinaires construit à Siliana (zone de Mazreg Echems), l'efficacité de ces micro-aménagement et surtout leur éventuel impact sur les ressources prévues par le barrage de Siliana.

Ce travail constitue une modeste contribution à l'éclaircissement du domaine d'interférence (s'il existe), entre la "grande hydraulique" et la "petite hydraulique".

II. PRESENTATION DE LA ZONE D'ETUDE ET DES AMENAGEMENTS PROJETES

La zone d'étude "Mazreg Echems" se trouve à huit kilomètre de la ville de Siliana au Nord -Ouest de la Tunisie. Elle se situe dans l'étage semi-aride moyen à hiver tempéré (coefficient d'emberger $Q = 44.6$). La pluviométrie présente une irrégularité à l'échelle annuelle, saisonnière et mensuelle. La saison la plus humide est l'automne suivie du printemps et de l'hiver (cf tableau 1).

L'étude statistique de la pluviométrie annuelle sur une période de 18 ans fait ressortir une moyenne d'environ 410 mm (cf Tableau 2).

Table.1 Répartition saisonnière de la précipitation (station Siliana Agricole 1968-1987).

Saison	PMS(mm)	(% du T V)
Automne	132	32.0
Hiver	120	29.0
Printemps	131	31.0
Eté	34	8.0

T.V =Total annuel
PMS. Pluies moyennes saisonnières

Table.2.Pluviométrie annuelle (mm)

Probabilité de dépassement de la pluviométrie annuelle		
1/10	5/10	9/10
607	379	282

Quant à l'évapotranspiration potentielle (ETP), elle varie en moyenne de 2mm/jour en Janvier à 7.9mm/j en Juillet.

Le déficit hydrique moyen annuel calculé à partir des moyennes mensuelles de la pluie et de l'ETP est de 1130 mm environ. Pour la période Novembre - Mai couvrant tous les cycles végétatifs de céréales et des cultures fourragères le déficit hydrique est seulement de 337 mm.

Dans le but de sécuriser et d'augmenter la production de ces cultures et de réduire l'érosion menaçant le barrage Siliana situé à l'aval de la zone, la Direction Régional de Conservation des Eaux et du Sol a réalisé les ouvrages suivants :

- Une série de banquettes à rétention totale destinée à :
  . intercepter le ruissellement de surface avant qu'il n'atteigne une vitesse érosive.
  . favoriser l'infiltration de l'ean dans le sol en vue d'améliorer sa réserve hydrique.
- Un système de micro-lacs collinaires constitué de cinq réservoirs de ptites dimensions, réalisés avec des matériaux meubles disponibles dans les alentours immédits de l'ouvrage.

Tableau.3 Caractéristiques des micro-lacs

Micro-lac n°	1	2	3	4	5
Surface du bassin versant (ha)	1.0	2.92	1.12	1.30	1.16
Longueur en crête de la digue (m)	37.5	50.0	34.0	36.0	45.0
Largeur en crête de la digue (m)	5.5	3.5	2.5	3.5	2.5
Hauteur de la digue(m)	3.0	3.0	3.0	3.0	3.0
Capacité des retenues(m^3)	866.0	1015.0	500.0	500	670.0

En vue d'étudier l'impact des microlacs sur la réserve hydrique de la zone de "Mazreg-Echems" et sur l'envasement du barrage Siliana l'on se propose d'effectuer le bilan hydrique et d'estimer la quantité de matériau solide interceptée par ce type d'aménagement.

III. BILAN HYDRIQUE

L'équation du bilan hydrique, pour un réservoir donné, peut s'écrire comme suit :

$$DV = P + R - E - I. \quad (1)$$

DV : Variation du volume d'eau dans la retenue pendant un jour.

P,E: Pluie et évaporation, pour l'intervalle de temps considéré.
R : Ruissellement superficiel
I: Pertes par infiltration et par prélèvements par les plantes.

3.1 Détermination des paramètres du bilan hydrique

Ces paramètres sont déterminés pour un intervalle de temps égal à une journée en raison de leur faible variation.

3.1.1 La pluviométrie

Etant donné l'absence d'hydrogramme pour la station météorologique de la zone d'étude, chaque averse journalière est considérée comme un seul évènement pluvieux. Cette hypothèse est acceptable puisque pour un climat semi-aride, l'intervalle d'une journée n'est pas assez long pour qu'il y ait des averses à traiter comme des évènements indépendants.

3.1.2 L'évaporation

Si l'estimation de l'évaporation directe dans les grandes retenues se heurte à des difficultés, il n'en est pas de même pour les micro-lacs pour les raisons suivantes:

- La surface du plan d'eau n'est pas très grande, elle ne dépasse guère quelques dizaines de m^2.
- La profondeur maximale de l'eau est faible ($<4m$).

Nous pouvons donc considéré l'évaporation mesurée sur bac sans risque des erreurs importantes.

3.1.3 Les volumes d'eau retenus

Pour chaque micro-lac, le volume d'eau stocké a été estimé à partir des mesures journalières de la profondeur de l'eau et de la courbe caractéristique volume - profondeur du micro-lac.

3.1.4 Le ruissellement

Si on définit par :

V : l'apport par pluviométrie directe
V_t: la variation du volume d'eau dans la retenue après l'averse
V_r: le volume ruisselé pour une averse p
P: la hauteur de l'averse.

On aura $V_r = V_t - V = V_t - P.S$ (2)

L'apport journalier par ruissellement sera la somme des apports par ruissellement (V_r) suite aux averses enregistrées à l'échelle d'une journée.

3.1.5 Les pertes par infiltration et par prélèvements (I)

Ces pertes ne peuvent être quantifiées directement, elles sont déterminées à partir de l'équation du bilan hydrique, appliquée à un intervalle de temps égal à une journée (i) :

$$I(i) = P(i).S(i)+R(i).A-E(i)+V(i)-V(i+1) \quad (3)$$

où :

V(i+1),V(i):sont respectivement, les volumes d'eau retenus le jour (i+1) et (i)
P(i), E(i), R(i): sont respectivement, la pluviométrie, l'évaporation et le ruissellement du jour (i).
S(i):surface du plan d'eau retenue au jour (i)
A:surface du bassin versant alimentant les lacs.

3.2 Résultats et interprétations

Les pertes mensuelles par infiltration sous les micro-lacs et par évaporation sont très importantes (cf tableau) ce qui constitue une contrainte majeure pour le stockage des eaux en vue de leur utilisation ultérieure dans l'irrigation.

Au cours de la période Décembre 1986 - Mai 1987, les pertes totales par infiltration et par évaporation ont été estimées à 637 870m^3 dont 73% par infiltration.

Les pertes par infiltration sont dues essentiellement au choix des sites : les digues reposent sur un substratum formé par une alternance de couches calcaires et marno-calcaires dont le sens de pendage est le même que celui du terrain naturel ce qui favorise les pertes à travers les bancs calcaires (drains naturels).

IV. LES APPORTS HYDRIQUES

Faute d'une longue série de données expérimentales, l'estimation des apports hydriques aux micro-lacs a été faite à partir des pluies enregistrées à la station Siliana Agricole et par application de la méthode S.C.S. Celle-ci a été retenue parce qu'elle intègre la plupart des paramètres morphologiques du bassin versant :

$$Q = \frac{(P-I_a)^2}{P-I_a+S} \quad (4)$$

avec
Q = lame ruisselée(mm)
P = précipitation totale (mm)
I_a= capacité initiale de stockage de la pluie par le sol : seuil de déclenchement du ruissellement
S=capacité maximale d'emmagasinement du sol

Nous avons vérifié que la méthode S.C.S donne une bonne estimation des apports annuels d'eau aux micro-lacs en comparant les valeurs prédites par cette méthode pour la période Décembre - Juin 87 aux valeurs mesurées pour la même période :

Table.4 Pertes totale par rapport au volume stocké (%)

Mois	M.L.1	M.L.2	M.L.3	M.L.4	Moyenne
Déc	79.0	72.0	69.0	80.00	75.0
Jan	88.0	83.0	93.0	82	87.0
Fév	65.0	75.0	86.0	67	73.0
Mars	57.0	85.0	66.0	74	71.0
Avril	90.0	90.0	100.0	100	95.0

M.L: Micro-lac

Table.5 Apports mesurés (AM) et Apports caculés (AC) pour la période Des-Juin 87.

Micro-lac n°1	1	2	3	4	5
AM (m^3)	157	193	75	84	129
AC (m^3)	160	198	73	95	116

L'apport total correspondant à une période de retour de deux ans est d'environ 1200 m^3 dont la répartition par micro-lac est donnée dans le tableau 6.

Table.6 Répartition saisonnière des apports bi-annuels

Saison	Aut	Hiv	Print	Eté
AS (m^3)	575.0	260.0	285.0	60.0
$\frac{AS}{AA}$ %	48.5	22.0	24.2	5.1

AS : Apport saisonnier
AA: Apport bi-annuel

L'apport bi-annuel présente un maximum au cours de la saison de l'Automne et un minimum au cours de la saison estivale.

Par suite des limitations suivantes :

- Le bassin versant alimentant les réservoirs est de dimension trop réduite (7,5 hectares).
- Les banquettes situées à l'amont des réservoirs interceptent une partie des eaux ruisselées acheminées vers les lacs collinaires (d'où une réduction de 54% du volume d'eau annuel retenu par chacun des réservoirs (si la surface du bassin était plus grande 7,5hectares).
- La non concordance, entre la période des apports (Automne), et celle des besoins de la céréaléculture essentiellement à partir du mois d'Avril (cf.Figure 2).
- Capacité limitée des réservoirs (volume bi-annuel =1200 m^3), alors que les besoins des cultures sont de loin supérieurs (ETM du blé =5000 m^3/ha).

L'utilisation des eaux retenues en irrigation, ne peut contribuer fortement à l'amélioration des rendements des cultures d'où une incompatibilité entre l'aménagement conçu et l'objectif d'irrigation.

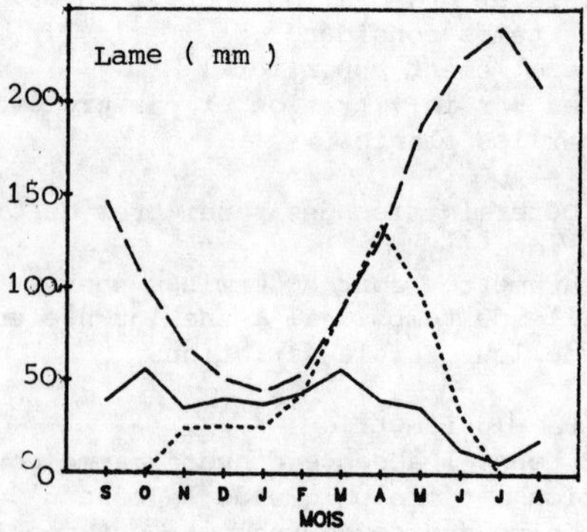

Fig.2 Bilan hydrique

4. Les apports solides

L'envasement des micro-lacs est calculé par analogie avec le barrage de Lakmes situé dans les mêmes conditions à partir de l'expression suivante :

$$A = E \frac{S}{S_o} \qquad (5)$$

avec A = envasement à estimer (en T/cm)
E = envasement mesuré sur le barrage de Lakmes égal à 0.375 10^6T/an
S = surface du bassin versant dominant le micro-lac (en km^2)
S_o= surface du bassin versant de Lakmes égaleà 131km^2

Le poids du dépôt annuel, intercepté par les cinq réservoirs et déterminé sous cette hypothèse est de:188 T/an, soit une érosion spécifique de 2500 T/km^2/an.

Les cinq réservoirs perdent chaque année, un volume d'eau égal à 117m^3, d'où un rythme de comblement annuel de 3.3%. La durée de vie moyenne de ce système de micro-lacs collinaires est d'environ 30 ans.

V. IMPACT SUR LE BARRAGE DE SILIANA

Situé à l'amont du barrage de Siliana, le système de micro-lacs collinaires conçu dans la zone du projet, n'a pas un effet considérable sur les apports hydriques et solides du barrage. En effet, le rapport entre les apports solides aux micro-lacs et les apports solides au barrage est d'environ 0.063%, valeur qui est négligeable. Il en est de même pour le rapport des apports hydriques ($\simeq$0.024%).

L'étude géologique et géomorphologique du bassin versant du barrage Siliana nous a permis d'identifier une vingtaine de sites pour la réalisation de petits lacs collinaires correspondant à une surface totale de bassin versant de 330 km^2 environ. Le volume total d'eau annuel qui sera stocké dans l'ensemble de ces lacs est d'environ $10^5 m^3$ ce qui correspond à 20% des apports hydriques annuels du barrage. Quant à la quantité des apports solides interceptée, elle est estimée à 950 000 tonnes/an soit 32% des apports solides annuels du barrage; d'où une prolongation de vie du barrage de 7 ans.

Les quantités d'eau interceptées par ce type d'ouvrages permettront l'aménagement du bassin versant par la plantation d'arbres et le développement des cultures annuelles; ceci réduira au minimum les pertes en terre et par conséquent l'envasement du barrage. L'infiltration de l'eau dans le sol sera également favorisée d'où une meilleure alimentation des nappes. La généralisation des micro-aménagements à l'amont du barrage peuvent constituer donc, un soutien au grand ouvrage.

CONCLUSION

Le caractère orageux des pluies, la faible perméabilité du sol et la couverture végétale réduite font que la zone d'étude ne profite que très peu des pluies de l'automne et qu'elle est soumise à une érosion intense.

Dans le but de régulariser la production agricole et de protéger le barrage Siliana situé à l'aval de la zone, contre l'envasement, les ouvrages suivants ont été réalisés :

- une série de banquette à rétention totale.
- un système de micro-lacs constitué de cinq digues en terre.

Les micro-lacs réalisés constituent un soutien au barrage Siliana par la rétention des matériaux solides et par le développement d'un couvert végétal qu'ils favorisent. Néaumoins, certains problèmes apparaissent avec la réalisation de ces micro-aménagements :

- les pertes par évaporation sont d'autant plus grandes que la surface d'eau intéressée est faible.

Ce phénomène se trouvant encore renforcé dans les régions arides et semi-arides.

- les pertes d'eau par infiltration ne sont plus négligeables surtout dans le cas où il existe des problèmes de perméabilité du fond du réservoir.

Sur le plan socio-économique, ces micro-aménagements permettent d'aider aux paysans des zones éloignées dont l'intégration dans un projet d'aménagement hydro-agricole classique est non justifiée.

REFERENCES

Bediot,G. et Cormary, Y: Calcul de la lame d'eau écoulée d'après les caractéristiques physiques d'un bassin versant; SCET, 1964, 84p. Vol.1

Berkaloff,E. et Tixeront,J. Carte du ruissellement annuel moyen en Tunisie, 1956, 11p.

Ghalleb,Kh. Guide d'étude et de réalisation des retenues collinaires en Tunisie, 1986, 84 p.

Ghorbel,A. Méthodes d'estimation du ruissellement, deuxième séminaire de l'hydraulique, Tunis, 1981, 9p.

Gloor,R. Journées d'étude en hydrologie de surface, compte rendu, deuxième partie, Ecole polytechnique fédérale de l'Ausanne, Suisse, 1982.

Riou,C. Evapotranspiration et bilan hydrique en Tunisie, Tunis, DRE, 1977, 10p.

Tixeront,J. Note sur la mise en valeur hydraulique des territoires du Sud Tunisien, 1961.

Tixeront,J. Petits barrages d'irrigation permanente en Tunisie, Paris, 1960, 31p.

Land and Water Use, Dodd & Grace (eds), © 1989 Balkema, Rotterdam. ISBN 90 6191 980 0

Increasing vegetable production with trickle irrigation

C.A.Madramootoo & M.Rigby
Agricultural Engineering Department, McGill University, Canada

ABSTRACT: There is a need for supplemental irrigation in eastern Canada, during the growing season. Some sprinkler irrigation of fruit and vegetable crops is practiced. However, the majority of farms is not irrigated, hence, crop yields decline due to moisture stress. Trickle irrigation offers tremendous potential for increasing crop yields. However, this technology is relatively new to the region. There are no criteria for the optimum wetted fractional soil volume (FSV) and nitrogen fertigation rate for maximum plant yield. A field study was therefore conducted in Quebec to determine the response of bell pepper plants to four FSV treatments and three nitrogen fertilizer rates. The parameter, FSV, is governed by the number of trickle emitters along a lateral. It was found that marketable yield increased with FSV, regardless of fertilizer rate.

RESUME: L'est du Canada a besoin d'irrigation d'appoint durant la période de croissance. Des arroseurs à tête tournante sont parfois utilisés pour irriguer fruits et légumes. La plupart des fermes ne sont cependant pas irriguées et les rendements y déclinent à cause du stress hydrique. L'irrigation en goutte à goutte offre un potentiel énorme pour augmenter le rendement des cultures. A l'heure actuelle, cette technique est relativement neuve dans la région. Il n'éxiste pas de critères pour la détermination de la fraction de volume de sol mouillé (FSV) ni pour celle du taux de fertigation à l'azote qui permettent un rendement maximal. Pour ces raisons, une étude pratique a été conduite au Québec. L'influence de quatre traitements FSV et de trois taux de fertilisation à l'azote sur le rendement des poivrons verts fut étudié. Le paramètre FSV est régi par le nombre d'émetteurs de chaque ligne de distribution latérale. On a observé que le nombre de poivrons commercialisables augmente avec la FSV indépendemment du taux de fertilisation.

ZUSAMMENFASSUNG: Während der Pflauzenwachstumszeit besteht in Ost-Kanada ein Bedarf an zusätzlicher Bewässerung. Einige verschiedene Arten der Wasserberieselung des Bodens werden für Obst und Gemüsebau angewandt. Da jedoch die Mehrzahl der Aubanflächen nicht bewässert werden, sind die Ernteergebnisse durch Mangel an Feuchtigkeit vermindert. Ausserordentliche Möglichkeiten zur Erhöhung der Ernten werden durch Aulage von Tropfbewässerung geboten. Diese Technik ist allerdiugs verhältnismässig neu für das Ost-Kanadische Gebiet. Eine Bestimmung des dem Wachstum am günstigsten Feuchtbodenvolumens (FSV), sowie eine Festsetzung des optimalen Nitrogenzusatzes zun Wasser für Maximelernten hat bisher noch nicht stattzefunden. Aus diesem Grunde wurde in Quebek eine untersuchung durchgeführt, inder man die wirkung von vier verschiedenen FSV, als auch von drei verschiedenen Raten von Nitrogen-Wasserzusätzen auf Paprikaschoten studiert hat. Die Zahl der Wassertropföffnungen eutlang einer Horizontalen ist der bestimmende Faktor für den Feuchtbodenvolumen. Es wurde festgestellt, dass die marktbaren Ernteergebnisse entsprechend des Feuchtbodenvolumens zunahmen, ungeachted der Nitrogen-Düngungstrate.

1. INTRODUCTION

In the humid, temperate region of eastern Canada, potential evapotranspiration normally exceeds rainfall during the growing season, from May to September. Supplemental irrigation is therefore required to assist with crop germination and reduce moisture stress. The majority of vegetable growers utilize sprinkler irrigation. Fruit growers have recently installed drip/trickle irrigation systems in orchards. Now, some vegetable producers are becoming interested in trickle irrigation of tomatoes, celery, lettuce, sweet peppers and sweet corn, amongst other crops.

However, there are currently no design criteria for trickle irrigation in eastern Canada. For example, the appropriate design fractional soil volume (FSV) must be determined. The parameter, FSV, may be defined as the percentage of the soil volume within the crop root zone which is wetted during an irrigation cycle. When emitter discharge, irrigation duration and soil type are kept constant, FSV can be altered by varying the number of emitters along a length of lateral tubing (Howell 1983). Keller and Karmeli (1975) recommended wetting 90% of the soil volume in closely spaced vegetable crops. The American Society of Agricultural Engineering (1987) has arbitrarily proposed wetting from 25 to 50% FSV in humid regions. However, these recommendations have not been adequately tested in the field where complex soil-water-plant-atmosphere exchanges prevail.

The situation is further confounded by the introduction of fertilizers through the trickle system (fertigation). High nutrient-use efficiency is associated with trickle fertilization, in that soluble nutrients are transported directly to plant roots. In this manner, FSV and nitrogen fertilization are interrelated. Therefore, it is also important to determine the optimum nitrogen fertigation rate which produces maximum plant yield.

In order to address these questions and arrive at design criteria, a research project was conducted at the Horticulture Research Station, Macdonald College of McGill University in Southern Quebec during the 1987 and 1988 growing seasons. Four FSV treatments and three nitrogen fertilizer levels were examined on a sandy loam soil, using bell pepper (Capsicum annum L.) as a test crop. Several soil and plant parameters, including nitrate distribution and uptake and yield, were measured.

2. MATERIALS AND METHODS

2.1 Study area and field layout

A field experiment was established, consisting of four FSV levels completely cross-classified and factorially combined with three nitrogen fertilizer levels in a randomized complete block design. The twelve treatment combinations were replicated in four blocks, each separated by a 2 m wide machinery path. The treatment combinations were randomly assigned to 12 beds within each block. The beds were 8.1 m long and 1.1 m wide, raised to a height of 0.2 to 0.25 m. Two staggered rows of 18 plants per row with an inter-plant spacing of 0.45 m and an inter-row spacing of 0.55 m were placed in each bed. Seven-week-old seedlings were transplanted from the greenhouse during the first week of June, each year. The bell pepper plants were grown on a St. Bernard sandy loam. The bulk density was 1.2 g/cm³ and the moisture contents at field capapcity (0.1 bar) and permanent wilting point (15 bar) were 23 and 12%, respectively, on a volumetric basis. Nutrient status, as determined in the autumns of 1986 and 1987, revealed the field to be rich in phosphorus and potassium (776 kg P/ha and 489 kg K/ha). In both years, nutrient levels were maintained by working 90 kg/ha each of K_2O and P_2O_5 into the soil. Calcium carbonate (2.86 t/ha) was added in the autumns of 1987 and 1988, in order to raise the soil pH from 6.1 to 6.3.

A single 12.5 mm diameter polyethylene irrigation lateral was placed along the centre length of each bed, between plant rows. Laterals from each bed within a block were connected to a 12.5 mm diameter submain. The submains from all blocks were joined to a 25 mm diameter mainline. The mainline was connected directly to the Campus treated water system. Therefore no field filtration equipment was required. A flow meter, pressure gauge and pressure reducer were inserted in the mainline. The operating pressure of the system was maintained at 170 kPa.

2.2 Experimental treatments

The four FSV treatments examined were: 25, 45, 70 and 90%. These were governed by the number of emitters per lateral, corresponding to 5, 9, 14 and 18, respectively. The emitters were pressure compensating with a discharge of 2 L/h.

The three nitrogen fertilizer levels applied on July 1, 1987 and July 4, 1988 were 1.5, 1.0 and 0.5 times the current recommended rates, resulting in dosages of 67.5 (A), 45.0 (B) and 22.5 (C) kg N/ha. One half of the above rates were also applied on August 3, 1987 and August 12, 1988, when the majority of the plants began to show signs of nitrogen deficiency by the paling of their leaves. The total amount of nitrogen supplied is given in Table 1. Ammonium nitrate (34% N) was the source of nitrogen in all cases.

Table 1. Nitrogen treatments in 1987 and 1988.

Treatment	Month		
	*May	**July	**August
A	80	67.5	33.8
B	80	45.0	22.5
C	80	22.5	11.3

*Broadcast, preplant
**In solution, midseason

The 12 treatments were labelled according to their irrigation/nitrogen combination. For example, 90A represents the 90% FSV treatment (18 emitters/lateral) in combination with the highest fertilizer rate.

Rainfall and Class 'A' pan evaporation were measured daily at the study site, and these data were used in conjunction with a water balance equation, to provide a daily irrigation schedule. A monthly summary of the meteorological data and irrigation water applied is shown in Table 2.

Table 2. Annual rainfall, pan evaporation and irrigation (mm).

	1987			
	June	July	Aug.	Sept.
Rainfall	122	106	52	125
Pan Evaporation	151	173	151	88
Irrigation				
25% FSV	13	21	19	0
45% FSV	24	37	34	0
70% FSV	37	58	53	0
90% FSV	48	75	68	0

	1988			
	June	July	Aug.	Sept.
Rainfall	75	43	110	73
Pan Evaporation	192	156	139	97
Irrigation				
25% FSV	29	19	6	10
45% FSV	52	33	11	19
70% FSV	81	52	17	29
90% FSV	104	67	22	37

2.3 Evaluation of plant growth parameters

Bell peppers were harvested every 10 days from mid-July until the first chilling frost in October, for a total of 8 harvests in 1987 and 7 in 1988. Only marketable fruit, i.e. peppers of a uniform shape and a mass of 100 g or more, was utilized in evaluating yield response. The incidence of sunscald on fruit was also noted.

Plant growth was assessed at two to three week intervals from June until September. Measurement of height and canopy spread were obtained from 18 plants in each bed, nine from every row. Height was evaluated from the cotyledonary node to the base of the petiole at the tallest growing point. Canopy spread was quantified as the diameter projected by each plant's surface area.

Total leaf production was estimated by sacrificing whole plants, minus their fruit, at the end of the experiment and drying them at 165°C for 24 hours. Four plants from each bed were selected for this purpose. Leaves and stems were then separated and weighed to obtain final leaf weight.

2.4 Nitrogen uptake

Plant nitrogen was monitored through tissue analysis of leaves obtained on June 24, July 3 and August 5, 1987 and June 28, July 6 and August 1, 1988. Nitrogen was estimated by removing the youngest fully mature leaf (blade and petiole) from each experimental plant. Samples were taken in the late morning, to allow for some translocation of leaf nitrogen accumulated during the night. A total of 28 leaves were combined from each plot, dried at 80°C for 24 hours and ground to a fine powder. Plant material was then digested in sulphuric acid, distilled and titrated for percent nitrogen as described by the Kjeldahl method of tissue analysis (Bremner 1965).

Extensive measurements of soil nitrate distribution in each treatment combination were conducted during the two growing seasons. Nitrate was sampled at six locations between emitters, and at four depths. These samples were taken at regular intervals during the growing seasons. Details on the sampling procedures, and results, can be found in Rigby (1989).

3. RESULTS AND DISCUSSION

Statistical analyses of total marketable pepper yield indicated no significant interaction between FSV and nitrogen application level for either 1987 or 1988 ($P < 0.05$). The lack of influence of fertilizer level on pepper production, under all FSV treatments, may indicate higher nitrogen availability, when applied via the trickle system. Only a fraction of the recommended rate may be required when nitrogen is transported directly to the plant roots. Alternately, the preplant nitrogen might have been sufficient to mask the effect of mid-season fertilization. This phenomenon was observed with peppers cultivated in Ontario, Canada (O'Sullivan 1979) and Puerto Rico (King et al. 1981).

An F-test for the homogeneity of variance disclosed no significant difference in harvest results between the two sampling years. Both annual and pooled treatment means of marketable pepper yields are presented in Table 3.

Table 3. Total marketable bell pepper yield.

Irrigation % FSV	Marketable Pepper Yield (kg/bed) 1987	1988	Average
25	13.2 a*	19.8 a	16.5 a
45	22.4 b	23.6 b	23.0 b
70	26.0 c	24.3 bc	25.2 b
90	29.3 d	27.7 c	28.5 c

*Treatments with the same letter within the same column are not significantly different, $p < 0.05$.

Bell pepper yields ranged from a low of 13.2 kg/bed for 25% FSV in 1987 to a high of 29.3 kg/bed for 90% FSV in the same year. In all cases, marketable pepper yield increased when a proportionally larger FSV was trickle irrigated. When years were combined, the average total harvest of the 45% and 70% FSV treatments were not significantly different from one another. Irrigating only 25% FSV did, however, curtail productivity while 90% FSV resulted in significantly superior yields. As a point of reference, non-irrigated pepper plants yielded considerably lower than any of the irrigated treatments (5.4 and 8.1 kg/bed in 1987 and 1988, respectively).

Maximum variation in yield was less than 7% between years for the 45, 70 and 90% FSV treatments. In contrast, the 1988 yield was 1.5 times that of 1987 for 25% FSV. This is likely due to the supplemental water given during the critical dry period in June, 1988. It can be observed from Table 2 that June, July and September, 1988, were extremely dry. There the plants in the 25% FSV treatments received additional water to ensure their survival.

Production figures from the Bureau des Statistiques du Québec (1988) indicated an average 1987 pepper yield for Quebec growers equal to 10,456 kg/ha. The data in Table 3 were projected to reflect expected yields on a kg/ha basis. This information is given in Table 4.

Table 4. Projected marketable pepper yields, kg/ha.

Irrigation % FSV	Marketable pepper yield (kg/ha) 1987	1988
25	8,148	12,222
45	13,827	14,568
70	16,049	15,000
90	18,086	17,099

It is evident that trickle irrigating 90% of the plant root volume resulted in a 73 percent increase in yield over the 1987 Quebec average. Unfortunately, statistics for 1988, which was a drier year, were unavailable.

3.1 Production function of yield versus irrigation

Production functions were developed from a polynomial regression analysis of total marketable yield per bed versus % FSV, for both 1987 and 1988 (Equations 1 and 2). Results indicate a significant linear and quadratic effect which follows the classic Mithscherlich diminishing yield response pattern. From the 1987 yield data:

$$Y = 0.47 + 0.608(FSV) - 0.0032(FSV)^2, \quad r^2 = 0.84 \tag{1}$$

and from the 1988 data:

$$Y = 17.52 + 0.108(FSV), \tag{2}$$

$r^2 = 0.70$

where Y is marketable yield, kg/bed and r^2 is the correlation coefficient.

From equation 1, it is possible to predict maximum harvest at 95% FSV. Irrigating beyond this point would result in a gradual decrease in yield. The y-intercept (0.47) represents the expected yield due to rainfall alone. However, non-irrigated yields of 5.4 kg/bed were attained. Whilst the production curve for 1987 was quadratic, a linear relationship was found significant in 1988, indicating that the irrigation-production function was not unique. Furthermore, this demonstrates that the maximum yield for Quebec was not attained in 1988.

As a rule, regression techniques can only be used to predict and compare when all management practices and growing conditions are held constant. Changes in production functions from year to year reflect annual variations in weather or cultural practices (Wierenga and Hendrickx 1984). As previously noted, weather varied significantly from 1987 to 1988 (Table 2).

In general, crop yields increase linearly with water applied until maximum production is attained. This is followed by a plateau phase and eventual drop in production as plant water requirements are surpassed. According to Stewart and Hagan (1973) yield versus evapotranspiration (ET) is a straight line function. Hence, the 1988 linear relationship between yield and irrigation water applied suggests that the amount and frequency of irrigation was sufficient to meet plant ET demands. Maximum yield was, however, difficult to determine without a plateau phase to reveal at what water quantity production could no longer benefit. Conversely, the 1987 curvilinear function may have been indicative of errors in irrigation scheduling or inefficient irrigation methods. It should also be recognized that the plants suffered a physiological set-back in June and July, 1988, due to warm, dry and windy weather conditions.

3.2 Water use efficiency

For humid regions, water use efficiency (WUE) is calculated as the yield increase over unirrigated controls divided by the supplemental water supplied in the form of irrigation. Yields from non-irrigated plants in guard beds surrounding the experimental area were utilized in the computation of WUE. Marketable yield was averaged for each irrigation treatment and the difference between treatment and guard bed yield was divided by millimeters of irrigation applied from June to September (Table 5). Mean water use efficiency was calculated as the average WUE for 1987 and 1988.

Table 5. Water use efficiency (kg/bed/mm irrigation).

FSV (%)	1987	1988	Mean
25	0.15	0.18	0.17
45	0.18	0.14	0.16
70	0.14	0.09	0.12
90	0.13	0.09	0.11

Water use efficiency ranged from 0.13 to 0.18 in 1987 and from 0.09 to 0.18 in 1988. Lowest values coincided with highest irrigation treatments in both years. The drop in water use efficiency from an average of 0.17 to 0.11 with increasing FSV illustrates that yield improvement with incremental increases in trickle irrigation is greatest at low water applications. Irrigating at an FSV below that required to maximize marketable pepper production allowed for a larger proportion of the water applied to be taken up by the plants. This is in agreement with a study by Hiler and Howell (1973) who found that reducing trickle irrigation resulted in a decline in sorghum yield but a gain in water use efficiency.

Unlike arid climates, humid region trickle irrigation favours maximum production over optimum efficiency in water use. Nonetheless, WUE could prove useful at the farm level where periods of drought or low capacity reservoirs necessitate water prioritization.

3.3 Plant size

Monitoring plant height and canopy spread throughout the field seasons produced a characteristically sigmoidal growth curve. When sampling dates were combined and averaged, an analysis of variance indicated no effect due to nitrogen fertilizer and no irrigation/fertilizer interaction in either year. Irrigation did, however, prove significant.

Differences in main effects for plant

height and canopy spread in each year were located. In all cases, the 25% FSV irrigation treatment produced significantly smaller plants. This is in corroboration with a study conducted by Tan et al. (1981) in which various proportions of greenhouse grown tomato roots were irrigated. The treatments in which only one quadrant, or 25%, of the root volume was wetted, resulted in the shortest plants.

Irrigating more than 25% FSV seemed to have less of a significant effect on plant size, although a distinct, positive relationship did exist between FSV and plant height and spread. Horton et al. (1982) reported a similar association linking growth of chili peppers to water status. They concluded that the difference in size is an adaptive response, in that smaller plants require less water to maintain leaf water potential and stomatal resistance.

Smaller plants, produced in 1988, when compared to 1987, emphasized the severe winds in June, 1988, which assaulted the newly transplanted seedlings. By stimulating the proliferation of branching, overall growth was effectively restrained.

The lack of fertilizer effect on plant growth is further indication that nitrogen was not a limiting factor. Evidently, nitrogen levels were not so high as to impair plant development, which is often the case when high nitrogen levels are combined with deficit water (Thomas and Heilman 1967).

3.4 Nitrogen uptake

The average nitrogen content in the youngest, fully mature leaf (blade and petiole) was approximately equal for all fertigation treatments. In keeping with previous analyses, percent nitrogen (by dry leaf weight) was averaged over the three fertilizer levels for each of the FSV treatments. Leaf nitrogen was also averaged over the four FSV levels for each of the nitrogen fertilizer applications and is presented in Table 6. Percent nitrogen is given for each of the leaf sampling dates in addition to an annual mean for 1987 and 1988.

There were no obvious differences in percent nitrogen for any of the irrigation or fertilizer treatments. Nitrogen accumulation in the leaves was considerably above the commonly reported range of 3.0 to 4.5% (Geraldson et al. 1973). It is apparent that even the lowest fertilizer level was sufficient to produce high levels of nitrogen in the leaves. Thomas and Heilman (1964) documented no significant differences in leaf nitrogen of Texas grown pepper plants receiving more than 134 kg N/ha (120 lb N/A). This corresponds to a total nitrogen application midway between the low C total (114 kg N/ha) and recommended B total (147 kg N/ha) rate used in this experiment.

Table 6. Percent nitrogen (dry weight basis) in bell pepper leaves.

Treatment	Percent Nitrogen in Leaves 1987			
	June	July	August	Mean
% FSV				
25	6.1	6.7	5.5	6.1
45	6.4	6.8	5.2	6.1
70	6.1	6.7	5.1	6.0
90	6.1	6.6	5.3	6.0
Fertilizer				
A	6.4	6.9	5.4	6.2
B	6.1	6.6	5.4	6.0
C	6.1	6.7	5.1	6.0

Treatment	Percent Nitrogen in Leaves 1988			
	June	July	August	Mean
% FSV				
25	7.4	8.0	7.1	7.6
45	7.6	7.8	7.1	7.5
70	7.7	7.6	6.8	7.4
90	7.6	7.7	6.9	7.4
Fertilizer				
A	7.6	7.8	7.2	7.5
B	7.6	7.7	6.9	7.4
C	7.6	7.7	6.9	7.4

Elevated nitrogen assimilation may be indicative of superior nitrogen uptake when applied with a trickle system. Kesner et al. (1985) found no significant differences in the high leaf nitrogen content of cherry trees when 50%, 100% or 200% of the recommended rate was trickle applied. The same phenomenon was verified by Aitken and Gorsuch (1985) with no significant effect on nitrogen in tissue of pecan trees receiving 33%, 67% and 100% of the suggested rate.

Substantial leaf nitrogen content before fertigating in July, 1988, corresponds to a relatively moist period prior to June sampling which would have improved the accessibility of preplant nitrogen fertilizer. Percent leaf nitrogen increased slightly following July mid-season fertigation. Nitrate is generally absorbed with irrigation water and is quickly reduced and assimilated. It is therefore not unusual for leaf nitrogen to concentrate at the time of sampling, 48 hours after fertigating in July. A subsequent decrease in leaf nitrogen from July to August reflects the translocation of nitrogen from the leaves to the developing fruit during this period of peak production.

Regardless that percent nitrogen in the leaves was similar for all irrigation treatments, FSV had a notable effect on plant size, which has considerable bearing on the total leaf nitrogen content. Estimates of total leaf nitrogen per plant were calculated by combining data from both years and multiplying mean percent nitrogen content in the leaves by the average final dry leaf weight for each FSV treatment. Total leaf nitrogen content per plant was plotted against FSV (Figure 1).

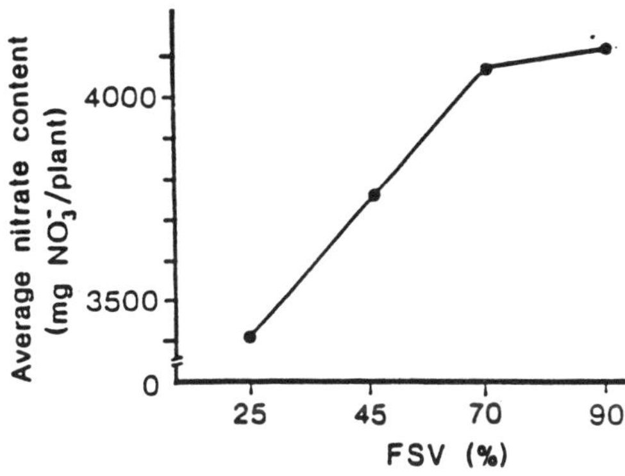

Fig.1 Total leaf nitrogen versus irrigation treatment.

Total leaf nitrogen increased with FSV until a constant was reached at 70% FSV. In this respect, providing a greater root volume with access to available nitrogen ensured higher overall plant uptake. The increase in nitrogen content became less pronounced with irrigation in excess of 70% FSV, which presumably signifies the limit of leaf production and nitrogen accumulation for this region. Even though maximum leaf nitrogen uptake occured at 70% FSV, bell pepper fruit production continued to rise (Table 3).

4. CONCLUSIONS

Bell peppers were grown in southern Quebec, Canada, under four wetted fractional soil volumes (FSV) and three nitrogen fertilizer rates, using a trickle irrigation system. Crop response, nitrate concentration and moisture content of the sandy loam soil, and plant nitrogen uptake, were monitored over two growing seasons.

Varying nitrogen levels 0.5, 1.0 and 1.5 times the recommended rate did not invoke a significant plant response. This could translate into a substantial reduction in fertilizer costs, as no more than one-half the suggested mid-season rate need be applied with the trickle system.

Marketable yield of bell peppers increased significantly with trickle irrigation treatments of 25, 45, 70 and 90% FSV ($p < 0.05$). Maximum productivity was obtained at 90% FSV, though this was not statistically greater than the 70% FSV yield in the drier year, 1988. Production functions developed from the data demonstrated that the curve of marketable yield versus FSV was not unique, as results followed a quadratic relationship in 1987 and a linear relationship in 1988. Annual differences in the shape of the curve were thought to reflect inequalities in the amount and distribution of rainfall during the two years. Total leaf nitrogen content also differed from 25% to 70% FSV, and remained constant to 90% FSV as accumulation of leaf mass became constant. Elevating design FSV beyond 70% was not consistently beneficial in augmenting nitrogen uptake.

It may therefore be concluded that a minimum design FSV of 70% is required for optimizing pepper yields under the conditions encountered in this study. Further increasing FSV may result in significantly elevated yields in some years. However, the effect of high FSV during wet summers might, in the future, prove detrimental to overall plant productivity.

ACKNOWLEDGEMENT

The authors gratefully acknowledge the financial support of the Natural Sciences and Engineering Research Council of Canada towards this research project.

REFERENCES

Aitken, J.B. and C.S. Gorsuch. 1985. Chemigation in pecans. Proc. 3rd International Trickle Congress. 2:495-499.

ASAE. 1987. Design, installation and performance of trickle irrigation systems. In: R.H. Hahn and E.E. Rosentreter (ed.) ASAE Standards. Am. Soc.Agric.Eng., St. Joseph, Michigan. p.522-530.

Bremner, J.M. 1965. Nitrogen-Total. In: C.A. Black (ed.) Methods of Soil Analysis, Part II. Agronomy 9. Madison, WI. p.595-624.

Bureau des Statistiques du Québec. 1988. Cultures Maraîchèrs.

Geraldson, C.M., G.R. Klacan and O.A. Lorenz. 1973. Plant analysis as an aid in fertilizing vegetable crops. In: L.M. Walsh and J.D. Beaton (ed.) Soil Testing and Plant Analysis. Soil Sci.Soc.of Amer.Inc. WI. pp.365-379.

Hiler, E.A. and T.A. Howell. 1973. Grain sorghum response to trickle and subsurface irrigation. Trans. ASAE. 16:700-803.

Horton, R., F. Beese and P.J. Wierenga. 1982. Physiological response of chili pepper to trickle irrigation. Agron. J. 74:551-555.

Howell, T.A. 1983. Design and operation of trickle (drip) systems. In: M.E. Jensen (ed.) Design and Operation of Farm Irrigation Systems. ASAE Monograph No.3. St.Joseph, MI.

Keller, J. and D. Karmeli. 1975. Trickle Irrigation Design. Rain Bird Sprinkler Manufacturing Corporation. CA. 133 p.

Keng, J.C.W., T.W. Scott and M.A. Lugo-Lopez. 1981. Fertilizer for sweet pepper under drip irrigation in an oxisol in Northwestern Puerto Rico. J. of Agr.Univ.of Puerto Rico. 65(1):123-128.

Kesner, D.K., B.R. Hahn, W.M. Klein and V.F. Bralts. 1985. Nitrogen application with trickle irrigation on sour cherry trees. Proc.3rd.International Trickle Irrigation Congress. 1:350-356.

O'Sullivan, J. 1979. Response of peppers to irrigation and nitrogen. Can.J.Plant Sci. 59:1085-1091.

Rigby, M. 1989. Trickle fertigation of bell peppers in southern Quebec. M.Sc. thesis, McGill University, 106 pp.

Stewart, J.I. and R.M. Hagan. 1973. Functions to predict effects of crop deficits. J.Irrig.Drain.Div.Am.Soc. Civ.Eng. 99(IR4):421-429.

Tan, C.S., A. Cornelisse and B.R. Buttery. 1981. Transpiration, stomatal conductance, and photosynthesis of tomato plants with various proportions of root system supplied with water. J.Amer.Soc.Hort.Sci. 106(2):147-151.

Thomas, J.R. and M.D. Heilman. 1964. Nitrogen and phosphorus content of leaf tissue in relation to sweet pepper yields. Am.Soc.Hort.Sci. 85:419-425.

Thomas, J.R. and M.D. Heilman. 1967. Influence of moisture and fertilizer on growth and N and P uptake by sweet peppers. Agronomy J. 59:27-30.

Wierenga, P.J. and J.M.H. Hendrickx. 1985. Yield and quality of trickle-irrigated chile peppers. Agric.Water Management. 9:339-356.

Land and Water Use, Dodd & Grace (eds), © 1989 Balkema, Rotterdam. ISBN 90 6191 980 0

Field application of border irrigation models

B.L.Maheshwari, T.A.McMahon & A.K.Turner
University of Melbourne, Australia

ABSTRACT: Six models of border irrigation, namely Jobling-Turner, Strelkoff, Walker, Jaynes, Schmitz and Ross, were applied to 67 irrigation events, the data being obtained at five field sites in south-east Australia. The models were evaluated for both advance and recession times by simple statistical analyses. In addition, other factors, including volume balance error, computer processing time, model output and user friendliness, were reviewed and it was concluded that overall the Strelkoff model is the most satisfactory for simulating border irrigation in south-east Australia.

ABRÉGÉ: Six modèles d'irrigation par rigoles - Jobling-Turner, Strelkoff, Walker, Jaynes, Schmitz et Ross - ont été appliqués à 67 expériences d'irrigation, les données ayant été obtenues a partir de cinq terrains d'études en Australia du Sud-Est. Les modèles ont été évalués pour leur temps d'avance et de recul par simples analyses statistiques. De plus, d'autres facteurs, comprenant l'erreur de balance de volume, le temps de traitement par ordinateur, le rendement du modèle et la facilité d'utilisation ont été étudiés, et on en a conclu que dans l'ensemble, le modèle Strelkoff s'avère le plus satisfaisant pour la simulation d'irrigation par rigoles en Australie de Sud-Est.

ZUSAMMENFASSUNG: Sechs Grenzbewasserungsmodelle, Jobling-Turner, Strelkoff, Walker, Jaynes, Schmitz, und Ross, wurden zu 67 Bewasserungsereignissen angewandt. Die Daten wurden in funf Feldstudien in sud-ost Australien gesammelt. Die Modelle wurden sowohl fur Vorschritts- als auch Ruckzugs-zeiten durch einfache statistische Methoden getestet. Andere Faktoren, einschliesslich Volumbalanzierungs Abweichungen, Computer Prozessionszeit, Modell Leistung, und Gebraucherfreundlichkeit, wurden auch getestet. Die Schlussforlgerung ist dass das Strelkoff Modell das meist befriedigende Modell zur Simulierung von Grenzbewasserung in sud-ost Australien ist.

INTRODUCTION

Surface irrigation methods are commonly used in Australia, accounting for about 76% of the total irrigated land. The largest irrigation developments are centred in the Murray-Darling basin in northern Victoria and southern New South Wales, where pasture dominates and dairying has been of major importance (Smith et al 1983). Poor design and lack of criteria for surface irrigation systems in, for example, the Goulburn Murray irrigation district in northern Victoria, are generally responsible for uneven irrigation, leading to wastage of water, waterlogging and salinity problems. The overall irrigation efficiency for a project in Australia with surface irrigation is placed around 33% (Watson et al 1983). This means about 67% of the total water released from storage is lost, either on its way from the source to the field or through drainage; it is unavailable for uptake by the crop. There is considerable scope for reducing some of these losses through improvements in irrigation design, particularly the water application aspect.

Border irrigation is one of the surface irrigation methods widely used for irrigating pastures. The method generally has low capital investment and a low energy requirement, as compared to pressure methods such as sprinkler irrigation (Down et al 1989). Therefore this method of irrigation has an important place in present and future irrigated crop production. Precision levelling of the land enclosed by borders or earthen banks is critical for uniform advance and recession of the water front during irrigation. This has attracted many farmers in irrigation areas to do land levelling and field layouts through laser-controlled equipment, which involves a considerable investment of money and labour. It is desirable, before any decision on such an investment is made, to know accurately the design dimensions and operational criteria of borders which will best suit the given field conditions and, at the same time, result in minimum water loss and uniform application.

Because overland flow in border irrigation is an example of unsteady, non-uniform and gradually varied flow over a porous bed, analysis is complicated. Recently, mathematical models have been developed to simulate the border process and this paper evaluates six such models, using field data from south-east Australia.

MODELS EVALUATED

Analysis and design of border irrigation systems have been the subject of investigation for several decades. Hydraulic principles related to border flows are now reasonably understood and known. Development of models for their possible applications in design and selection of management practices has been a goal of researchers. The models so far produced vary in their approach to the problem - different number of parameters, accuracy of prediction and the range of applicability in the field.

The models used in this study vary in their sophistication, approaches, assumptions and details of output. The main input data include border geometry, inflow rate, time to cutoff, hydraulic resistance and infiltration characteristics. The salient features of each model are described briefly.

Jobling-Turner Model

This model is the simplest of the six models examined and is based on a physical model study of border irrigation (Jobling & Turner 1973). The infiltration characteristics in the model are described by the Horton equation:

$$Z = \gamma(1 - e^{r.t}) + \delta.t \qquad (1)$$

where Z = cumulative depth of infiltration,
γ = initial value of infiltration depth,
δ = long-term infiltration rate,
t = time from start of infiltration, and
r = exponent.

The hydraulic resistance in the model is indirectly taken into account by the average depth of flow in the strip. The advance is predicted for both finite and zero long-term infiltration rate, but the recession time can be predicted only for blocked-end borders. The output from the model includes advance and recession times along the strip. In the present study, due to negligible long-term infiltration rates for most of the data sets and irrigation with open-ended strips, the model was applied for the advance phase only. .

Strelkoff Model

Three main approaches to the solution of the hydrodynamic equations are available, viz. zero-inertia (ZI), kinematic wave (KW) and a hybrid, which are solved numerically using the Newton-Raphson technique (Strelkoff 1985). The hydraulic resistance in the model is described by Manning n or Manning n varying as a power function of depth. The infiltration characteristics are described by a general equation of the form:

$$Z = A\,t + K\,t^{B} + C \qquad (2)$$

where A, K, B and C are characteristic parameters.

The output from the model includes advance and recession times, the ultimate distribution of infiltrated water and parameters related to water application, storage and distribution efficiencies and runoff (open-ended borders).

Walker Model

Three main approaches for the solution of the hydrodynamic equations are available, viz. hydrodynamic (HD), zero-inertia and kinematic wave (Walker & Gichuki 1985). The hydraulic resistance is described by Manning n and the infiltration characterized by the modified Kostiakov equation:

$$Z = k\,t^{a} + b\,t \qquad (3)$$

where k, a and b are characteristic parameters. The output from the model is similar to the Strelkoff model.

Jaynes Model

This model is based on the zero-inertia approach using the finite difference procedure for the solution of the hydrodynamic equations (Jaynes 1986). The model is claimed to be simpler than Strelkoff and Walker models in terms of solution technique. The hydraulic resistance is described by Manning n and the infiltration characteristics by the modified Kostiakov equation. The output from the model includes advance and recession times and the ultimate distribution of infiltrated water along the strip.

Schmitz Model

Using an unconditionally stable method of characteristics, this model is based on the solution of the hydrodynamic equations (Schmitz et al 1985). The hydraulic resistance in the model is described by Manning n and the infiltration by using either the modified Kostiakov equation or a quasi-analytical solution of the unsaturated flow equation. The output from the model is similar to that of the Jaynes model.

Ross Model

The Ross model is based on both zero-inertia and kinematic wave approaches (Ross 1986) and its general formulation is based on the work of Strelkoff

& Katapodes (1977), Elliot et al (1982) and Walker & Humphrey (1983). However, the solution technique and assumptions used in this model are somewhat different from those used in the Strelkoff and Walker models. The hydraulic resistance and the infiltration characteristics are similar to those in the Strelkoff model. The output from the model is similar to that of the Walker model.

Hydraulic Resistance Options

Hydraulic resistance is one of the important variables that affect the flow in border irrigation. Most models adopt Manning's equation with constant roughness to represent hydraulic resistance during irrigation. However, the Strelkoff and Ross models allow Manning n to vary as a power function of depth. The implications of this option are not pursued herein, but a detailed discussion of this and a general review of hydraulic resistance can be found in Maheshwari et al (1989).

FIELD EXPERIMENTS

In order to evaluate the accuracy of the models as published, predicted advance and recession times and other parameters were compared with field-derived results. In all, 67 irrigation events were monitored at five sites, namely the Mt. Derrimut Field Station of the University of Melbourne (designated MD), three sites of the Rural Water Commission in northern Victoria - Mitchell (MIT), Kerang Agricultural Research Farm (KARF) and Shepparton (SPT) - and the Whitton site of CSIRO, Division of Water Resources, Griffith (GR).

All the strips at these sites were laid out using laser-controlled machinery. The soils of the sites are characterized by high clay contents (up to 70%). At MD, GR and SPT, there are red duplex soils exhibiting a distinct texture contrast between the A horizon and the dominantly red B horizon. On the other hand, the soils at MIT and KARF are self-mulching grey clays with uniform fine textured profiles. Generally for these soils, except those at SPT, extensive cracks have developed on the surface as well as into the soil profile at the time of irrigation and, as a result, there is an initial fast infiltration, mainly through crack filling and some adsorption at the surface soil, but little long-term infiltration.

The general information for the irrigations carried out at the sites is given in Table 1. All the irrigations have been analysed and are reported here. The experiments carried out at the Whitton site were part of a long-term project on 'Evaluation of gypsum treatments using slotting techniques for soil amelioration'. This aspect is not seen to be an issue for the present study, since the method for estimating infiltration characteristics is based on the volume-balance approach which provides 'representative' infiltration behaviour for the strip.

EVALUATION PROCEDURE

Evaluation of models involves determining their overall accuracy to predict various parameters for given field conditions and to ascertain their relative usefulness as a tool for the design and management of an irrigation system. The parameters which can be measured in the field and used for evaluation include advance and recession times, ultimate distribution of infiltrated water, variation of depth of flow with time at a selected section in the strip, and runoff.

For the purposes of this paper, two major criteria are examined, namely advance and recession times, along with several other factors.

The observed values of measurement (e.g. advance time), X_o, and those predicted by a model, X_p, can be compared and analysed in several ways. However, the procedure used here includes fitting the data to a regression equation of the form:

$$X_p = \lambda X_o \tag{4}$$

To further assist in this evaluation, the standard error of λ (σ), coefficient of determination (R^2) and coefficient of efficiency (C_e), as defined by Nash & Sutcliff (1970), were calculated.

The value of λ generally indicates the overall trend of prediction (under- or over-prediction) and σ and R^2 the scatter of points about the regression equation (4). The value of C_e is a coefficient of determination of the regression equation, $X_p = X_o$. Therefore, if R^2 and C_e are close to each other, then the model is free from any bias for all or part of the data, depending on the value of R^2. In general, a value of λ close to 1.0, accompanied by a low σ and a high R^2 and C_e, would indicate satisfactory prediction by the model.

RESULTS AND DISCUSSION

Preliminary Comments

During preliminary simulation runs of the models, it was found that the Strelkoff model did not converge for the kinematic wave and hybrid approaches, probably because the convergence criterion in the model is not compatible with the conditions at the field sites. Therefore, only the zero-inertia approach of the Strelkoff model was used in the comprehensive evaluation. Also, it was found that there was no difference between the HD and ZI approaches of the Walker model, so the results for these cases have been listed as one and described as "Walker HD and ZI". In all, eight model variations were evaluated - Jobling-Turner, Strelkoff-ZI, Walker-HD and ZI, Walker-KW, Jaynes-ZI, Schmitz-HD, Ross-ZI and Ross-KW.

Table 1. General details of field experiments.

Site	No. of irrigations	Length of strips (m)	Width of strips (m)	Slope (m/m)	Unit flow rate (l/s.m)	Cutoff time (min)
Mt.Derrimut (MD)	23	50	4.5	0.0011 - 0.0021	1.92 - 3.30	25 - 50
Mitchells (MIT)	4	400	40	0.0005 - 0.0013	1.77 - 2.11	202 - 333
KARF (KARF)	18	240	30	0.0006 - 0.0012	1.32 - 3.80	88 - 275
Shepparton (SPT)	4	360	60	0.0019 - 0.0027	1.78 - 1.95	162 - 226
Griffith (GR)	18	85	10	0.0002 - 0.0009	0.61 - 1.11	117 - 268

Advance Phase

Amalgamated results of the application of the eight models are given in Table 2 and regional summaries are tabulated in Table 3. Overall, the models under-predict advance times by 6 to 15%. However, for individual sites, there is a trend of over-prediction for MD for all models except Jaynes and under-prediction for the other sites.

On a site by site basis, the R^2 values for all the models are generally > 90%. Also, the C_e values are close to R^2, which suggests that there is little apparent bias in the predictions.

It should be noted that because of lack of convergence, a solution for the Walker-KW model was not obtained for any event at MIT and SPT, nor for two events at KARF.

In general, considering the values of the various indices for the events, sites and models, it is concluded that the performance of the models with HD and ZI approaches was superior to the Jobling-Turner and KW approach models. Furthermore, the Walker-HD and ZI model performed the best for the advance phase and was closely followed by the Schmitz-HD, Ross-ZI, Strelkoff-ZI and Jaynes-ZI models. The results for the Walker-HD and ZI model are presented graphically in Fig. 1.

Table 2. Average values of λ, σ, R^2 and C_e for the models during the advance phase.

Model	N*	λ	σ (mins)	R^2	C_e
Jobling-Turner	474	0.85	36	0.89	0.84
Strelkoff - ZI	474	0.90	26	0.95	0.93
Walker - HD & ZI	472	0.94	26	0.95	0.95
Walker - KW	377	0.91	20	0.92	0.90
Jaynes - ZI	471	0.88	23	0.95	0.92
Schmitz - HD	458	0.93	23	0.95	0.95
Ross - ZI	455	0.92	23	0.95	0.93
Ross - KW	472	0.90	29	0.93	0.91

* N = number of points (i.e. values of observed and predicted advance times) used in the regression analysis.

Table 3. Average site values of λ, σ, R^2 and C_e for the advance phase.

Model	Site	N	λ	σ (mins)	R^2	C_e
Jobling-Turner	MD	161	1.13	2	0.98	0.94
	MIT	37	0.82	77	0.84	0.69
	KARF	125	0.74	39	0.75	0.44
	SPT	44	1.00	38	0.90	0.90
	GR	107	0.99	16	0.95	0.95
Strelkoff - ZI	MD	161	1.03	1	0.99	0.99
	MIT	37	0.85	60	0.91	0.82
	KARF	125	0.95	27	0.94	0.94
	SPT	44	0.93	28	0.94	0.92
	GR	107	0.94	17	0.93	0.92
Walker - HD & ZI	MD	161	1.01	1	0.99	0.99
	MIT	35	0.93	68	0.90	0.88
	KARF	125	0.96	25	0.95	0.94
	SPT	44	0.97	33	0.92	0.92
	GR	107	0.92	15	0.95	0.93
Walker - KW	MD	161	1.20	2	0.98	0.89
	KARF	109	0.85	30	0.85	0.76
	GR	107	0.99	16	0.94	0.94
Jaynes - ZI	MD	161	0.94	1	0.99	0.98
	MIT	34	0.84	51	0.90	0.78
	KARF	125	0.88	26	0.93	0.89
	SPT	44	0.96	31	0.93	0.92
	GR	107	0.90	14	0.95	0.91
Schmitz - HD	MD	161	1.02	1	0.99	0.99
	MIT	32	0.88	55	0.88	0.82
	KARF	125	0.96	24	0.95	0.95
	SPT	44	0.99	34	0.92	0.92
	GR	96	0.92	14	0.95	0.92
Ross - ZI	MD	161	1.02	1	0.99	0.99
	MIT	31	0.86	54	0.85	0.75
	KARF	122	0.95	25	0.93	0.93
	SPT	44	0.96	33	0.92	0.92
	GR	97	0.93	13	0.95	0.93
Ross - KW	MD	161	1.18	11	0.98	0.91
	MIT	36	0.87	59	0.90	0.81
	KARF	124	0.87	37	0.84	0.79
	SPT	44	0.98	33	0.92	0.92
	GR	107	0.98	16	0.94	0.94

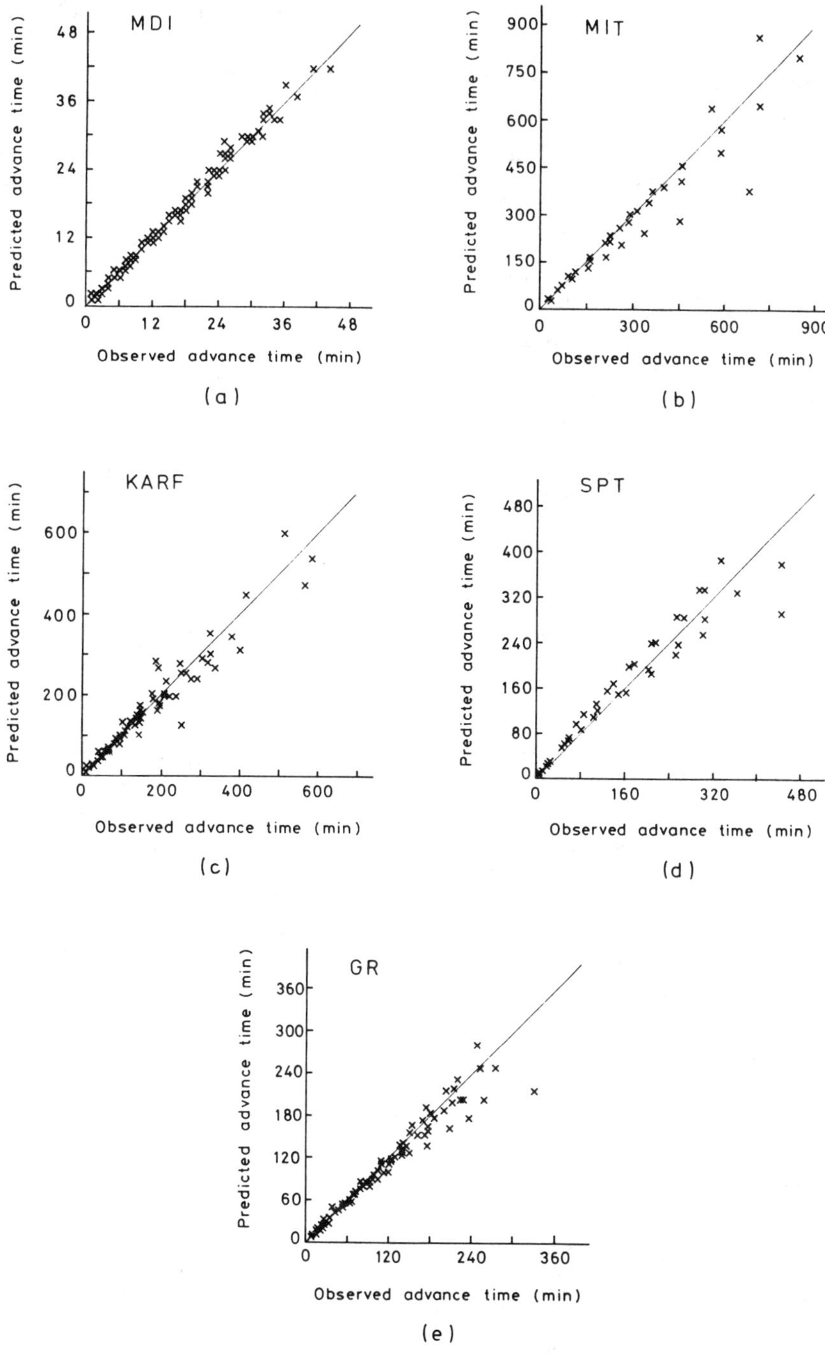

Figure 1. Observed versus predicted advance times for the various sites with the Walker-HD and ZI model.

Recession Phase

In general, the accuracy of prediction for the recession phase is less satisfactory than that for the advance phase. Overall, several models - Walker-HD and ZI, Walker-KW, Ross-ZI and Ross-KW - seriously under-predicted recession times (Table 4). In addition, both Jaynes-ZI and Schmitz-HD models showed relatively low R^2 values. Moreover, these results are confirmed by the individual site averages (Table 5). Generally, it can be stated that the predictions of the Strelkoff-ZI model are superior to the other models, although for the GR site the results were inferior. The performance of the Strelkoff-ZI model by site is shown in Fig. 2. These figures illustrate the level of uncertainty in prediction.

The recession at a point or section results from an interaction of several field variables. In addition to the factors generally accounted for in the models, the recession is also affected by the undulations of the ground surface, the uniformity of vegetative cover over the strip and, above all, the influence of the subjective judgement of an observer in estimating the recession in the field. These factors have important consequences on the performance of the models, although their effects are generally masked by the recession criteria and the method of solution employed.

Infiltration and Runoff

As part of the input data to the models, infiltration characteristics are supplied by either equations (2) or (3). For each simulation run, the models (except Jobling-Turner) compute the water infiltrated. In Fig. 3, these values are compared with measured values for sites SPT and GR, and are summarized in Table 6. (Soil moisture contents before and after irrigation were available only at SPT and GR.)

The results indicate that the models generally under-predict infiltration for most events (Fig. 3). Nevertheless, considering the spatial variability of infiltration and errors associated with measurements of soil-water content and therefore errors in the observed values of infiltration depth, the prediction of infiltration by all models is regarded as satisfactory. The general trend of under-prediction of infiltration by the models may be related to under-prediction of infiltration characteristics by the volume balance method used in the study. (Details are given in Maheshwari et al 1988.) Furthermore, the under-prediction could have been due to underestimation of inflow rate and/or overestimation of surface storage. Further field work is required to elucidate these sources of error.

Runoff values for a total of 13 irrigation events at GR were available from the field measurements. The observed runoff for each event and that predicted by each model except the Schmitz model, which as programmed does not provide such values, are compared in Table 7. The prediction error, if calculated as for the infiltration depths, may be

Table 4. Average values of λ, σ, R^2 and C_e for the models during the recession phase.

Model	N*	λ	σ (mins)	R^2	C_e
Strelkoff - ZI	531	1.06	145	0.80	0.79
Walker - HD & ZI	529	0.68	120	0.65	-0.03
Walker - KW	434	0.52	98	0.60	-1.36
Jaynes - ZI	529	0.86	141	0.65	0.56
Schmitz - HD	514	0.90	150	0.64	0.59
Ross - ZI	513	0.63	110	0.66	-0.33
Ross - KW	529	0.45	105	0.55	-2.65

* N = number of points (i.e. values of observed and predicted advance times) used in the regression analysis.

Table 5. Average site values of λ, σ, R^2 and C_e for the recession phase.

Model	Site	N	λ	σ (mins)	R^2	C_e
Strelkoff - ZI	MD	184	1.02	54	0.56	0.55
	MIT	31	0.90	108	0.85	0.74
	KARF	143	1.07	175	0.76	0.74
	SPT	48	1.37	126	0.73	0.17
	GR	125	1.10	163	-0.25	-0.33
Walker - HD & ZI	MD	184	0.55	27	0.44	-5.8
	MIT	29	0.62	76	0.82	-3.3
	KARF	143	0.60	112	0.62	-1.7
	SPT	48	0.92	72	0.81	0.75
	GR	125	0.95	143	-0.37	-0.40
Walker - KW	MD	184	0.32	21	0.26	-33.
	KARF	125	0.48	108	0.50	-5.2
	GR	125	0.72	106	-0.31	-1.8
Jaynes - ZI	MD	184	1.07	61	0.55	0.53
	MIT	29	0.64	64	0.87	-3.1
	KARF	143	0.83	153	0.70	0.50
	SPT	48	1.06	98	0.72	0.69
	GR	125	1.07	152	-0.08	-0.13
Schmitz - HD	MD	184	0.95	81	-0.46	-0.47
	MIT	27	0.66	53	0.90	-2.8
	KARF	143	0.90	180	0.56	0.48
	SPT	48	1.08	101	0.69	0.64
	GR	112	1.06	166	-0.44	-0.47
Ross - ZI	MD	184	0.49	25	0.26	-12.
	MIT	27	0.60	63	0.85	-4.5
	KARF	140	0.56	109	0.63	-2.6
	SPT	48	0.94	81	0.68	0.65
	GR	114	0.84	118	-0.61	-1.1
Ross - KW	MD	184	0.27	16	-0.02	-87.
	MIT	30	0.56	127	0.62	-3.6
	KARF	142	0.34	82	0.55	-15.
	SPT	48	0.81	65	0.80	0.39
	GR	125	0.60	86	-0.74	-6.9

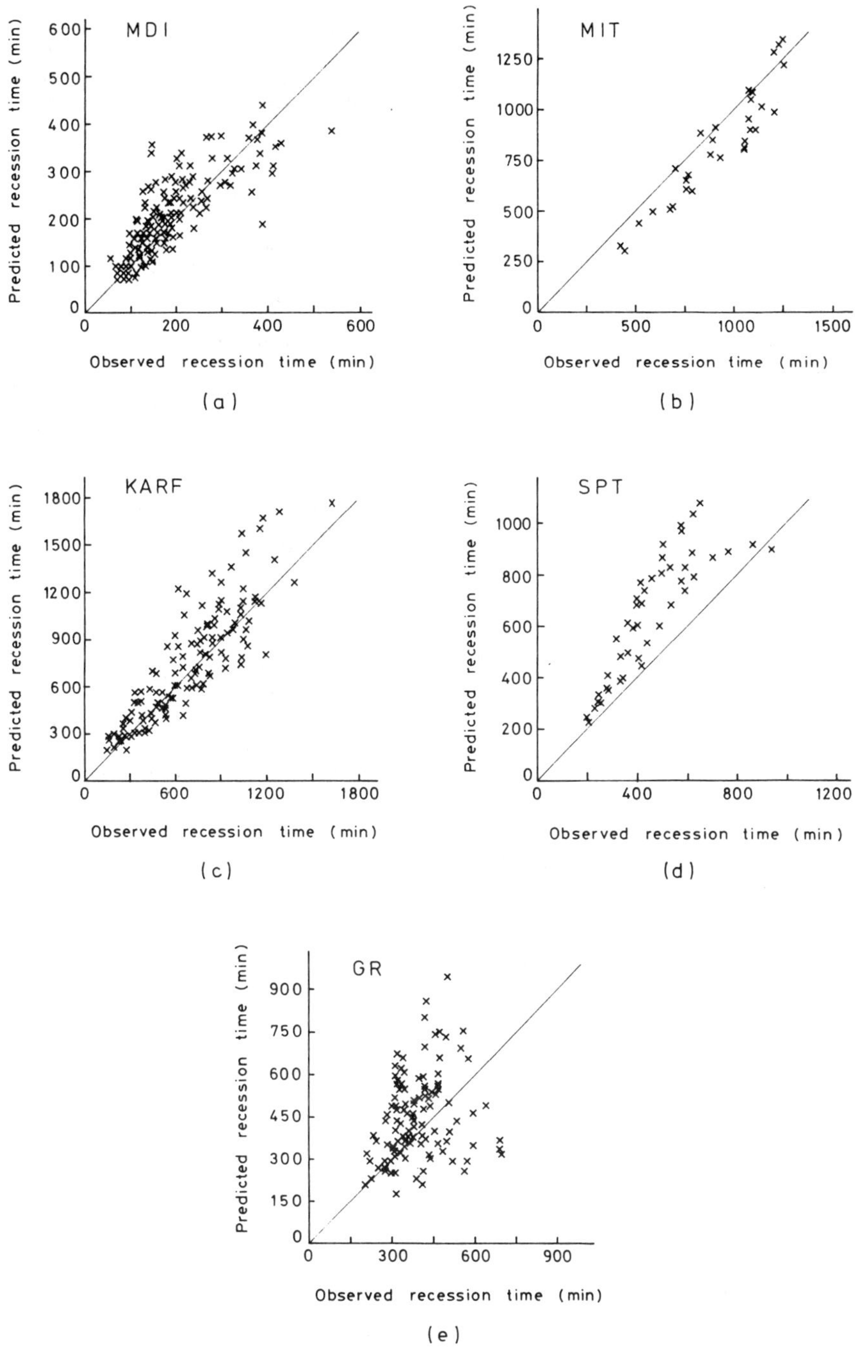

Figure 2. Observed versus predicted recession times for the various sites with the Strelkoff-ZI model.

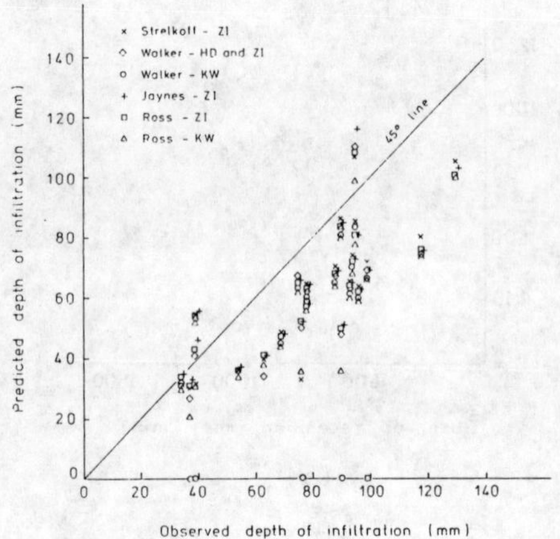

Figure 3. Observed versus predicted depths of infiltration with the various models.

Table 6. Absolute average difference between observed and predicted infiltration (%).

Model	Shepparton (4 events)	Griffith (18 events)
Strelkoff - ZI	35	23
Walker - HD & ZI	28	25
Walker - KW	$	26
Jaynes - ZI	27	25
Ross - ZI	26	24
Ross - KW	44	27

$ = no model solution.

Table 7. Absolute average difference between observed and predicted runoff (m^3).

Model	Griffith (13 events)
koff - ZI	0.8
- HD & ZI	1.0
- KW	1.1
ZI	1.0
	1.2
	1.0

ause the observed and predicted
all compared with the total
ed during an irrigation.
ssessed by examining only the
bserved and predicted values.

The data show that there is no obvious bias in prediction and the absolute difference between observed and predicted values is generally < 2 m^3, which is about 2% of the total water applied during an irrigation.

Other Considerations

Space does not permit a detailed discussion of other factors affecting evaluation of the models. Nevertheless, the following comments need to be included:

1. A detailed comparison was made of the usefulness between a discharge-depth equation (Turner et al 1978) and Manning's n for describing hydraulic roughness of a strip. The Strelkoff and Ross models cater for both methods. No significant advantage of one approach over the other could be established (Maheshwari 1988).

2. A volume balance error indicates whether the continuity equation was satisfied during the simulation of an event. Results suggest that the accuracy of solution in the Walker and Schmitz models is not satisfactory (Maheshwari 1988).

3. The average central processing unit (CPU) time required for a model run is given in Table 8. Excluding the Jobling-Turner model (which can be solved graphically), the differences in computer time range over an order of magnitude. Where multiple runs are required, the relatively large amount of computer time for some models may inhibit their application.

Table 8. Average CPU time for model run on VAX 8650.

Model	CPU time/run (sec)
Jobling-Turner	0.4
Strelkoff - ZI	5.2
Walker - HD & ZI	7.0
Walker - KW	3.7
Jaynes - ZI	11.0
Schmitz - HD	26.6
Ross - ZI	39.4
Ross - KW	13.9

4. With respect to output details and user friendliness, the Strelkoff-ZI model is generally superior. For example, guidance is provided to users when there is no convergence of solution, there is a user manual, and advice is provided on how to deal with unusual occurrences during computation.

In summary then, it can be stated that the Walker-HD and ZI model is the best model for the prediction of advance, but not for the recession. On the other hand, the Strelkoff-ZI model is satisfactory or superior for both the advance and recession phases.

Furthermore, the prediction by the Strelkoff model of infiltration and runoff is generally superior to that by the other models and the average CPU time per simulation and the volume balance error for the model are low. The Strelkoff-ZI model is also superior with respect to other features mentioned earlier and therefore, in general, the Strelkoff program is the most satisfactory model for simulating border irrigation for the field conditions encountered in this study.

CONCLUSIONS

Six mathematical models of border irrigation were applied to data for 67 irrigation events carried out at five locations in south-east Australia. Based on a comprehensive comparison of the predicted advance and recession times with field measurements and evaluation of other features of the models, it was concluded that the Strelkoff-ZI model is the most satisfactory model for simulating border irrigation in south-east Australia.

REFERENCES

Down, M.J., A.K. Turner & T.A. McMahon 1989. On-farm energy use for irrigation in Australia. Bull. ICID (in press).

Elliott, R.L., W.R. Walker & G.V. Skogerboe 1982. Zero-inertia modelling of furrow irrigation advance. ASCE J.Irri.Drain.Div. 108:179-195.

Jaynes, D.B. 1986. A simple model of border irrigation. ASCE J.Irri.Drain.Div. 112:172-184.

Jobling, G.A. & A.K. Turner 1973. Physical model study of border strip irrigation. ASCE J.Irri.Drain.Div. 99:493-510.

Maheshwari, B.L. 1988. Evaluation of border irrigation models for south-east Australia. PhD thesis, University of Melbourne, Australia.

Maheshwari, B.L., T.A. McMahon & A.K. Turner 1989. Hydraulic resistance in surface irrigation. Inter. Symp. on Agricultural Engineering, Beijing, China.

Maheshwari, B.L., A.K. Turner, T.A. McMahon & B.J. Campbell 1988. An optimization technique for estimating infiltration characteristics in border irrigation. Agr. Water Management. 13:13-24.

Nash, J.E. & J.V. Sutcliffe 1970. River flow forecasting through conceptual models. 1. A discussion of principles. J. Hydrol. 10:282-324.

Ross, P. 1986. Zero-inertia and kinematic wave models for furrow and border irrigation. Unpublished notes, CSIRO Division of Soils, Townsville, Australia.

Schmitz, G., R. Haverkamp & O. Palacios-Velez 1985. A coupled surface-subsurface model for shallow water flow over initially dry soil. Proc. 21st IAHR Congress, August 19-23, Melbourne, Australia, p.23-30.

Smith, R.C.G., W.K. Mason, W.S. Meyer & H.D. Barrs 1983. Irrigation in Australia: developments and prospects. In D. Hillel (ed.), Advance in Irrigation, p.99-155. New York: Academic Press.

Strelkoff, T. 1985. BRDRFLW: a mathematical model of border irrigation. USDA-ARS, Washington D.C.

Strelkoff, T. & N.D. Katabodes 1977. Border irrigation hydraulics with zero-inertia. ASCE J.Irri.Drain.Div. 103:325-342.

Turner, A.K., K.J. Langford, Myo Win & T.R. Clift 1978. The discharge-depth equation for shallow flow. ASCE J.Irri.Drain.Div. 104:95-110.

Walker, W.R. & F. Gichuki 1985. Documentation of surface irrigation models. Unpublished report, Utah State University.

Walker, W.R. & A.S. Humphrey 1983. Kinematic-wave furrow irrigation model. ASCE J.Irri.Drain.Div. 109:377-392.

Watson, W.D., R.G. Reynolds, D.J. Collings & R.D. Hunter 1983. Agricultural water demand and issues. Water 2000: Consultants Report No. 5. Dept. of Resources & Energy, Australian Govt. Publishing Service, Canberra.

Land and Water Use, Dodd & Grace (eds), © 1989 Balkema, Rotterdam. ISBN 90 6191 980 0

Influence of drip system on soil heat flux, soil moisture tension and crop growth in a cotton field

O.Padmakumari & R.K.Sivanappan
Water Technology Centre, Tamil Nadu Agricultural University, Coimbatore, India

ABSTRACT: Studies were conducted at Tamil Nadu Agricultural University, Coimbatore to find the crop response under drip irrigation for cotton. The irrigation rate, frequency, crop age and soil depth are so interrelated that evapotranspiration and soil moisture regime with respect to cotton cultivation resulted in complex moisture pattern in the soil profile. The soil heat flux resulted in introducing decisive effect on changing the maturity pattern. Beyond 90th day after sowing cotton the soil heat flux was maintained constantly by maintaining an isothermal temperature of 22°C. Drip irrigation with 1 cm Epan frequency was found to record a soil moisture tension of 20 centibars at the end of one irrigation cycle irrespective of the quantity of water applied. The leaf area index was reduced in frequently irrigated fields while it was found to increase with increased level of water application. Bartlett's index was found highest for low level water application (0.6) and was lowest for the highest level of water application indicating early maturity in the plots of lower water application. Dry matter production, number of fruiting points and number of bolls set were found affected by decrease in the quantity of water applied. Cotton yield was reduced when the level of water application was increased resulting in high vegetative growth.

1 INTRODUCTION

The drip irrigation system provides an entirely different soil environmental regime for plant growth. The water gets distributed in the plants near the root zone through a network of low cost polythene tube. Since watering is at a specified location on the land surface only a small volume of soil gets wetted during each irrigation. Compared to conventional irrigation methods the drip irrigation cycle consists of a relatively long period of infiltration followed by a short period of extraction and redistribution when the frequency of irrigation is increased under drip system. It helps to eliminate low average soil water content caused by high water fluctuations, which situation affects the plant growth and crop yields. Investigation on the effect of this soil regime on the crop response of cotton is presented in this paper.

2 MATERIALS AND METHODS

Since drip system would save water it is important to know the extent to which the savings in water can be obtained without any reduction in yield. For achieving this four levels of water application L_1, L_2, L_3 and L_4 (37.5, 50, 75 and 100 per cent Eo levels) and three intervals of water application I_1, I_2 and I_3 (1 cm Ep, 2 CmEp and 3 cm Ep) were selected where Eo refers to free water evaporation and Ep refers to class A pan evaporation. Gossypium hirsutum L popularly known as MCU 5 cotton was selected as the test crop. The soil was of sandy clay loam with a bulk density of 1.28 g/cc. The field capacity was 24.20 per cent by weight. The irrigation water had an Ec of 4.8 mmhos/cm. The watering was done through drip method with 50 mm diameter polythene pipes for mains and submains, and 12 mm diameter for laterals. 1 mm diameter microtubes were used as emitters. Flow rates were adjusted with gate-valves and water meters. The treatments were laid out as randomised block design with three replications. Soil moisture was observed by using neutron access tubes at different points along and across the row from the emitter. The temperature was measured with a soil thermometer. Five plants at random from each plot was tagged for recording all biometric observations. Leaf area index was noted on 90th and 135th day after sowing. Number of fruiting points, number of bolls per plant, plant height, yield Bartlett's index, root weight and dry matter produced were observed.

3 RESULTS AND DISCUSSION

Soil moisture was observed on different dates at 15, 30, 45 and 60 cm depth in all the treatments. Drip irrigation with 1 cm Ep frequency enabled deeper penetration of wetting front whereas the frequency of 2 and 3 cm Ep resulted in greater sorption and retarded the depth of penetration. The 100% Eo application level alone was found to supply water beyond 60 cm depth. Large variation was noticed in 0-15 cm depth of soil profile with reference to the moisture storage pattern. This is attributed to the root activities. Moisture content in between the row was found to be less than that along the row.

3.1 Soil moisture tension

Moisture tension was measured using tensiometers at 30 cm depth. It was found to record a value of 20 centibars at the end of one irrigation cycle in all I_1 plots irrespective of quantity of water applied.

In I_2 interval it was found to record a value of 30 centibars in L_1 and L_2 levels and 25 centibars in L_3 and L_4 levels at the end of one irrigation cycle. In I_3 interval L_1 and L_2 levels recorded 30 to 35 centibars at 30 cm depth at the end of one irrigation.

From this the following inferences are drawn.

1. Frequent application does not cause wide variation in matric tension at 30 cm depth.

2. Level of water application and moisture extraction influence the tension developed when the interval of irrigation is beyond I_1.

3.2 Influence of drip system as energy balance

The temperature of the soil profile was influenced by drip irrigation as induced by level of application and frequency of irrigation. Microclimatic effect was induced by modifying the thermal regime. The soil temperature at 5, 10, 15 and 30 cm depth (D_1, D_2, D_3 and D_4) was measured for all treatments on 60, 75, 90, 120 and 135th day after sowing. The temperature data observed on 75th and 135th day after sowing is presented below in tables 1 and 2.

Table 1. Soil temperature on 75th day (°C)

Level	L_1	L_2	L_3	L_4
	24.0	23.9	23.8	23.9
SE	0.04			
CD (P=0.05)	0.11			
Interval		I_1	I_2	I_3
		23.58	24.06	24.20
SE	0.03			
CD (P=0.05)	0.10			
Depth	D_1	D_2	D_3	D_4
	23.72	23.73	23.83	24.22
SE	0.04			
CD (P=0.05)	0.11			

where, SE - Standard Error
CD - Critical Difference

Interaction effects

Level-Depth

	D_1	D_2	D_3	D_4
L_1	23.81	23.80	24.00	24.00
L_2	23.70	23.75	23.90	24.50
L_3	23.75	23.75	23.90	24.00
L_4	23.65	23.70	23.50	24.00

SE 0.03
CD(P=0.05) 0.09

Interval-Depth

I_1	23.00	23.05	23.60	23.90
I_2	24.00	24.10	24.30	24.50
I_3	24.10	24.10	24.50	24.55

SE 0.04
CD(P=0.05) 0.11

Interval-Level

	L_1	L_2	L_3	L_4
I_1	23.50	23.60	23.50	23.50
I_2	24.00	24.00	24.00	24.16
I_3	24.45	24.20	24.00	24.12

SE 0.04
CD(P=0.05) 0.11

Table 2. Soil temperature on 135th day (°C)

Level	L_1	L_2	L_3	L_4
	21.63	21.31	21.12	21.11

SE 0.03
CD(P=0.05) 0.01

Interval	I_1	I_2	I_3
	21.17	21.33	21.38

SE 0.03
CD(P=0.05) 0.10

Interaction effects
Depth-Level

	L_1	L_2	L_3	L_4
D_1	21.27	21.05	21.00	21.00
D_2	21.38	21.22	21.05	21.11
D_3	21.72	21.76	21.11	21.16
D_4	22.16	21.83	21.33	21.16

SE 0.07
CD(P=0.05) 0.20

Interval-Level

	L_1	L_2	L_3	L_4
I_1	21.41	21.23	21.08	20.93
I_2	21.87	21.25	21.08	21.12
I_3	21.62	21.45	21.20	21.25

SE 0.06
CD(P=0.05) 0.17

Interval-Depth

	D_1	D_2	D_3	D_4
I_1	20.87	21.16	21.16	21.50
I_2	21.25	21.08	21.33	21.66
I_3	21.12	21.33	21.37	21.70

SE Not significant
CD Not significant

The following inferences are drawn from the results.

1. On 75th day there was no significant difference in soil temperature at different depths. Isothermal regime prevailed at 24.01 ± 0.29°C.

2. Neither the level of irrigation nor the interval of application changed the soil temperature.

3. On 90th day the soil temperature reduced to 22.04 ± 0.64°C. The isothermal nature was however maintained.

4. On 120th day soil temperature differences could be detected at different depths. Temperature gradients were setup by larger irrigation intervals and higher levels of application.

5. On 135th day the temperature gradients got diminished.

6. On 60th day L_1 resulted in a temperature of 22.8°C while in L_4 22.4°C.

The investigation revealed that the temperature of the soil was high when the level of water application was reduced. These results agree with the findings of Grewal and Singh (1978) according to whom the moist soil is responsible for maintenance of isothermal regime.

3.3 Crop response to drip irrigation

Levels of irrigation did not affect germination. Even for low level of water application the germination was better when the frequency of water application was higher. Dry matter production increased proportionately with increase in the level of water application. At low level of water application high dry matter production was observed for 3 cm Ep interval. Plant height was higher in 1 cm Ep interval compared to other intervals for all levels of water application.

3.3.1 Leaf area index

The leaf area index was observed on 90th and 135th days after sowing. The results are summarised below.

Table 3. Leaf area index

Interval	Level			
	L_1	L_2	L_3	L_4
I_1	1.28	1.38	1.45	1.99
I_2	1.29	1.50	1.60	2.11
I_3	1.29	1.45	1.60	2.23

From the results it can be seen that

1. Reduced leaf area is noticed in frequently irrigated fields.

2. Increased level of application of irrigation water has resulted in increasing leaf area index.

3. Beyond a level of irrigation frequency, the influence of level of application of water is not significant.

3.3.2 Root weight

Root weight at different depths indicated that it was maximum in 0-15 cm depth zone. Not much impact could be induced on root growth pattern by varying irrigation application frequencies.

The combined effects of different irrigation levels and intervals resulted in rooting pattern as below.

Table 4. Weight of root (g)

Irrigation at Ep	Levels of irrigation			
	37.5% Eo	50.0% Eo	75.0% Eo	100.0% Eo
1 cm	4.3	4.7	6.4	7.1
2 cm	4.4	5.2	6.3	7.2
3 cm	4.0	5.7	6.5	7.0

As the levels of irrigation increased, root production increased correspondingly. Slight increase in root weight is noticed when the frequency of irrigation is reduced.

The study brings out the fact that more the water available at deeper zone, the greater is the root activities. The lower level of water application at longer interval has resulted in deeper root development. Shallower roots of hairy nature were identified in high level irrigation application and in frequently irrigated fields. Less number of hairy roots at lower levels of application although in greater frequency has been observed. This is in confirmity with the observations of Goldberg, Gornat and Bar (1974). They found that 76% of total root weight was in the top 10 cm of soil under drip system.

3.3.3 Fruiting point and the bolls set

The number of fruiting points were found to be high (33.4) in L_3 treatment followed by L_4 treatment in which the number was 32.8 at the end of 135th day. Minimum number of fruiting points were noticed in low level of application.

The number of bolls obtained through different levels and intervals of irrigation application are summarised below.

Table 5. Number of bolls per plant on 135th day

Intervals	Level			
	L_1	L_2	L_3	L_4
I_1	13.2	31.5	33.8	32.3
I_2	25.3	30.8	33.1	33.0
I_3	24.8	30.0	33.5	32.1

The drastic reduction in boll is observed when the level of irrigation is 37.5% Eo. Maximum boll production is found at 75% Eo level of application irrespective of irrigation intervals.

3.3.4 Cotton yield

The cotton yield in different pickings and the Bartlett's index worked out for the different treatments are presented below

Table 6. Cotton yield (g/plant) in different pickings and Bartlett's index

Treatments	1st picking	3rd picking	5th picking	Bartlett's index
L_1I_1	600	1320	732	0.62
L_2I_1	450	1471	750	0.59
L_3I_1	480	1600	850	0.58
L_4I_1	400	1003	1100	0.53
L_1I_2	430	1500	708	0.61
L_2I_2	430	1500	708	0.61
L_3I_2	449	1500	800	0.60
L_4I_2	380	1066	902	0.55
L_1I_3	450	1300	600	0.56
L_2I_3	449	1247	1077	0.52
L_3I_3	449	1500	800	0.52
L_4I_3	381	1228	750	0.50
Level				
SE	1.24	13.8	3.75	0.03
CD(P=.05)	3.65	40.8	10.99	0.06
Interval				
SE	1.08	11.95	3.24	0.01
CD(P=.05)	3.16	35.05	9.52	0.05
Level x Interval				
SE	2.16	23.9	6.49	0.03
CD(P=.05)	6.33	70.11	19.08	0.11

From the results the following points were observed.

1. For all levels of water application, I_1 interval yielded maximum under first picking (483 g/plot) followed by I_2 (435 g/plot) and I_3 (433 g/plot).

2. For I_1 interval, L_1 gave higher yield (600 g/plot). Maturity in L_1 level application was earlier irrespective of the intervals of water application. Lower water application has hastened up the maturity by inducing moisture stress and modifying ground thermal regime.

3. Increased irrigation level (L_4) delayed the bursting resulting in reduced yield in the first picking. However, in the third picking, higher yield of 1500-1600 kg/plot could be obtained.

4. L_4-I_1 and L_2-I_3 yielded maximum yield in 5th picking. This could be attributed to the delayed maturity resulting from rich soil moisture status at the root zone.

5. Yield reduction has been noticed beyond 4th picking in all the cases.

6. Prolonged irrigation interval in the lowest (L_1) and highest (L_4) levels of application has resulted in poorer yield (1973 ± 20 kg/ha).

7. Bartlett's index had been found to be highest for L_1 level (0.62) and lowest for L_4 level (0.5). This indicated that early maturity was evident in the plots of lower water application.

The water applied and total yield obtained are presented below in Table 7. The different treatments resulted in 48 irrigations in 1 cm Ep, 25 irrigations in 2 cm Ep and 16 irrigations in 3 cm Ep intervals. The bursting of bolls were influenced by the treatments. In the plots of higher water application the cotton yield was affected by delay in the bursting of bolls.

Table 7. Water applied and yield under different treatments

Treat-ments	Total water applied including effective rainfall (cm)	Total yield (kg/ha)
L_1I_1	24.59	2142
L_2I_1	26.60	2298
L_3I_1	31.30	2576
L_4I_1	40.30	2215
L_1I_2	24.40	2090
L_2I_2	26.60	2175
L_3I_2	32.20	2453
L_4I_2	40.32	2161
L_1I_3	24.00	1952
L_2I_3	26.00	2268
L_3I_3	32.20	2398
L_4I_3	40.40	1993

Total yield was found to be highest in L_3 level (2576 kg/ha) with I_1 interval and least in L_1 level (1952 kg/ha) with I_3 interval.

4 CONCLUSION

The study has revealed that 75% Eo level and 1 cm Ep interval has recorded highest cotton yield. Higher application level (100% Eo) reduced cotton yield and increased vegetative growth. I_1 interval (1 cm Ep) has caused increased yield in all levels. Soil temperature reduced from 24°C on 75th day to 22°C on 90th day. But irrigation water influenced the soil heat flux at the early stage of crop growth. Low level and frequent interval of irrigation scheduling resulted in maintenance of optimum temperature.

5 ACKNOWLEDGEMENTS

The studies reported here form a part of the thesis submitted to the Tamil Nadu Agricultural University by the first author. The facilities and finances for completing the research work was provided by the TNAU and the same is gratefully acknowledged.

REFERENCES

Goldberg, D., B.Gornat and Y.Bar. 1974. The distribution of roots, water and minerals as a result of tricble irrigation. American J. Hort.Science 965: 645-648.

Grewal, S.S. and N.T.Singh. 1978. Effect of different moisture regimes on soil temperature and yield and quality of potatoes. Indian J.Agron. 32(2): 130-136.

Padmakumari, O. 1982. Water optimization studies under drip irrigation system for cotton. M.E.(Ag.) thesis Tamil Nadu Agricultural University, India.

Padmakumari, O. and R.K.Sivanappan. 1985. Drip irrigation for cotton. Proceedings of Third International Drip Irrigation Congress, Fresno California, Vol.1, 262-267.

Padmakumari, O. and R.K.Sivanappan. 1988. Studies on the soil moisture regime of cotton under drip irrigation system. Proceedings of Fourth International Microirrigation Congress, Australia, 4C-4.

Land and Water Use, Dodd & Grace (eds), © 1989 Balkema, Rotterdam. ISBN 90 6191 980 0

Drip irrigation for papaya

O.Padmakumari & R.K.Sivanappan
Water Technology Centre, Tamil Nadu Agricultural University, Coimbatore, India

ABSTRACT: Drip irrigation is a method which helps to distribute the water to the plants near its root zone through a network of pipes at frequent intervals. The soil is maintained near field capacity which results in good crop response. Studies were conducted at Tamil Nadu Agricultural University, Coimbatore to compare the crop response of papaya under drip system with the conventional check basin system. It was seen that the germination of papaya seeds were 90% under drip system while it was 67.5% under basin system. The yield per tree was higher in drip compared to basin irrigated papaya trees. The size of fruits were larger in drip irrigated plants compared to the fruits under control method. The water utilized in drip system was only one third of the water consumed in 180 cm check basin and half the water utilized in 90 cm check basin. Considerable water saving was observed by adopting drip method in raising papaya trees

1 INTRODUCTION

In drip method water is let on the ground through drippers. The dripper achieves a three dimensional differential spread of water maintaining low levels of soil water tension. According to Vehimeyer and Handrikson (1929) the utilization of soil moisture by plant would be effectively uniform throughout the whole range between the permanent wilting point and field capacity. Later experiments by Hagan Vadia and Russel (1959) **showed** that the plant respond differentially to variations in soil moisture content within the range and that depletion of soil moisture even above permanent wilting point would reduce growth rate and yield considerably. To find the water requirement of papaya trees and to compare it with conventional irrigation practices two drip treatments and two basin treatments were selected for the study. Papaya is a tropical fruit grown in many parts of India. In ten months after planting harvesting of fruits can be started.

2 MATERIALS AND METHODS

The study was conducted in the orchard of Tamil Nadu Agricultural University campus. Papaya Co.1 variety was selected as the test crop. The drip lines were laid out with 50 mm diameter polythene pipe for mains and submains, 12 mm diameter for laterals and 1 mm diameter microtubes for emitters. The system was connected to a sump with a 3 HP monoblock pumpset. The water flow was regulated by gatevalves and water meters. There were four treatments selected as shown below.

T_1 Drip with single lateral line per row

T_2 Drip with double lateral lines per row

T_3 Check basin of 90 cm x 90 cm size

T_4 Check basin of 180 cm x 180 cm size

The above treatments were laid out as a randomised block in 5 replications. In between the treatments buffer space was provided in the field to eliminate the effect of different treatment getting overlapped. In the drip plots the water to be applied was calculated based on the daily mean class A pan evaporation value of the month. Irrigation was given daily. For both T_1 and T_2 treatments the quantity applied was kept the same by adjusting the time of irrigation. For basin treatments watering was given at 50% depletion of available soil moisture

3 RESULTS AND DISCUSSION

From the experiment data on the pattern of soil wetting, water applied, germination response, weed growth and crop yield were collected. On examining the soil moisture regime of various treatments there was marked difference in the volume and depth of wetting of the soil. In the plots with single lateral/row (T_1) watering was given through 3 microtubes which helped to establish a wetted band of 75 cm on one side of the plant, On the other side the soil was relatively dry on the surface. The total area of wetting was less compared to double lateral wetted plots. Such restricted wetting of the soil surface has additional benefit such as reducing water evaporation by preventing evaporation of water from outside the wetted surface zone. In the plots with double lateral/row watering was given through four microtubes, two in each line. It was seen that alround the papaya plant the soil was wet as watering was given daily. The depth of wetting was found to vary from 60 to 65 cm under the emitter in the double lateral wetted plot (T_2). In T_1 plot the depth of wetting was found to range from 80 to 90 cm below the emitter. In the basin system the soil moisture was found to decrease from field capacity to 21% by weight in one irrigation cycle.

3.1 Crop response

The germination response of papaya and weed growth observed in the different treatments are presented in Table 1 below.

Table 1. Germination and weed growth

Treatments	Germination (per cent)	Weed growth (kg/ha)
Drip with single lateral/row	87.25	35
Drip with double lateral/row	90.00	37
Check basin 90 cm x 90 cm	67.50	70
Check basin 180 cm x 180 cm	65.50	101
CD (P=0.05)	3.6	4.1

It was seen that the percentage of germination of papaya seeds in drip treatments were significantly high compared to the values observed in the basin methods. The germination percentage was on par in both basin treatments.

The weed growth was found reduced by 50% in T_1 and T_2 plots compared to the weed growth in T_4 plots. The partial wetting of the soil restricts the weed growth in drip plots. In addition to reducing the weeding cost the water uptake is also reduced drastically.

The crop response judged by the height and number of leaves also indicated that papaya plant is responding well in drip method. The flowering in drip irrigated plots started fairly one month before the control plots started flowering. The water applied and yield obtained in the different treatments are given below in Table 2.

From the results it was seen that the yield of papaya in T_2 plots were on par with T_1 plots. T_1 and T_2 (drip) yields were significantly superior to T_3 and T_4 (basin) plot yields. The drip irrigation has helped to maintain a suitable moisture regime for better crop growth.

But there was no difference in yield between the two basin sizes. But the water consumed was higher in 180 cm basin compared to the water consumed in the 90 cm basin plot. The water used in both drip treatments were much less compared to the basin plots.

Table 2. Yield and water utilized by papaya trees

Treatments	Water applied cm	Effective rainfall cm	Total water utilized cm	Yield kg per plant
T_1	26	36	62	28.0
T_2	26	36	62	30.5
T_3	75	36	111	24.5
T_4	160	36	196	23.0

From visual observation it was seen that the root growth was less in T_1 plants. But it was not reflected in any significant difference in yield. This may be because there was no difference in water applied to both drip treatments. This confirms the finding of Lunin and Gallatin (1965). They have reported that partial wetting did not affect water uptake.

Black and West (1974) tested the effect on water use of young apple trees when varying proportions of the root systems were supplied with an optimum water regime. They found that when the fraction of wetted root system was decreased from 1.0 to 0.25 the relative transpiration rate reduced from 1.0 to only 0.75.

4 CONCLUSION

The water saving achieved in drip treatments are nearly 50% of the water utilized in 90 cm check basin and 66% of water consumed in 180 cm check basin. The papaya yield was found highest in the plants with double row laterals. As the water applied remaining same single lateral application causes higher downward movement while double lateral helps to spread water uniformly over a larger surface area. But the system cost will increase with double laterals/row. As the yield in single lateral wetted papaya are on par with double lateral yields it can be adopted with a few measures to improve the uniform water application and to reduce any downward loss of water which can occur. The wetting uniformity can be improved by connecting emitters to plastic sleeves of different length and placed at 3 or 4 points around the tree. This will help to avoid the laying of double rows of laterals and help increase in good crop response.

5 ACKNOWLEDGEMENTS

The authors are grateful to the Indian Council of Agricultural Research for providing the financial support for carrying out this piece of research work.

REFERENCES

Black, J.D.F. and D.W.West. 1974. In Proceedings of Inter Drip Irrigation Congress 2nd, pp.432-436.

Hagan, R.M, Y.Vadia and M.B.Russel. 1959. Interrelation of plant response to soil moisture regime, water and its relation to soil and crops. Advances in Agronomy, 11: 77-78.

Lunin, J. and H.M.Gallatin. 1965. Zonal salinization of the root system in relation to plant growth. Soil Science American Proceedings, 26: 608-612.

Padmakumari, O. and R.K.Sivanappan. 1980. Final report of the ICAR Adhoc Scheme on drip irrigation studies for water scarcity areas of Tamil Nadu, 1976-1980, pp.1-78.

Vehimeyer Fy and A.H.Hendrickson. 1929. Soil moisture in relation to plant growth. Annual Review of Plant Physiology 1: 285-304.

Land and Water Use, Dodd & Grace (eds), © 1989 Balkema, Rotterdam. ISBN 90 6191 980 0

Comparison of sprinkler uniformity models using a single lateral line sprinkler

N.A.Sabti
University of Basrah, Iraq

ABSTRACT: Linear, normal and log normal statistical models were fitted to ovelapped sprinkler patterns obtained from a single sprinkler line. The RMS error was used as a measure for comparison. It was concluded that the linear model fitted the water application depths better than did the normal or log normal models for all values of uniformity coefficients.

1 INTRODUCTION

The main objective of sprinkler irrigation systems is to efficiently apply an appropriate depth of water, distributed as uniformly as possible. The distribution of water in an irrigated field under sprinkler irrigation is of primary importance from the point of view of economic and environmental consideration. If the actual distribution of applied water can be discribed, then it is possible to determine the performance parameters of the irrigation system, such as water application efficiency. It is necessary to characterize the water distribution pattern of irrigation systems. Sprinkler irrigation distribution patterns have been discribed using various statistical distribution models, such as linear regression, normal, log normal, beta and gamma distribution models. The objevtives of this study are to evaluate the fittness of the linear regression, normal and log-normal distribution models to the actual water application depths and to recommend an appropriate statistical distribution to be used in the system design.

Several attempts have been made to evaluate the goodness of fit of many statistical distribution models to the precipitation from sprinkler. The suggestion of normal (Gaussian) distribution has been supported by various researchers such as Hart and Reynolds (1965), Walker (1979) and Elliot et.al. (1980), which provide a good fit. Also attempts have been made to measure the water distribution from sprinklers with asymmetric distribution (Chaudhry (1976, 1978), Howell (1964) and Seniwong et.al. (1972)). Also linear regression was used to describe sprinkler distribution patterns (Karmeli (1978)) and the Pearson type III distribution was used to simulate water distribution from rotating sprinklers (Tsakeris et.al. (1985)).

2 BASIC NOTIONS

The individual measurement of sprinkler depth, y_i, was made nondimensional by dividing each depth by the mean depth, $\bar{y}$. Assuming that the area which is irrigated by two sequence lateral lines is divided into N equal squares and each square receives a uniform water depth, y_i, the uniformity coefficient u_c can be defined as (Wilcox and Swailes (1947).

$$u_c = 1-(\frac{s}{\bar{y}}) \quad (1)$$

in which $s/\bar{y}$=the coefficient of variation, s=the standard deviation and $\bar{y}$= mean of the observations.

3 LINEAR REGRESSION MODEL

The linear regression model gives the relation between the dimensionless cumulative frequency of the water depth, $Y_i^*(Y_i^*=y_i/\bar{y})$, and the fraction of area, x_i, which is represented by the linear regression function

$$Y^* = a + b\ x \qquad (2)$$

where a and b are linear regression coefficients (constants).

The least squares method is used to fit a straight line to the frequency curve.

4 NORMAL DISTRIBUTION MODEL

The standarized normal distribution model is given by

$$f(x)= \frac{1}{\sqrt{2\pi}}\, e^{-x^2/2} \qquad (3)$$

The area under the function f(x) between x and infinity, Q(x), which represents the probability, can also be given by a convenient polynomial approximation as (Abranowitz and Stegun (1968))

$$Q(x)=f(x)(b_1t+b_2t^2+b_3t^3+b_4t^4) \qquad (4)$$

in which $t = \frac{1}{1+0.2316419x}$

b_1= 0.31938153 ; b_2=-0.56563782 ; b_3= 1.781477937; b_4=-1.821255978; and b_5= 1.330274429.

The polynomial expression is limited to positive values of x, so that for the negative values, the characteristic symmetry of the distribution can be used.

5 LOG NORMAL DISTRIBUTION MODEL

The log normal distribution is often appropriate for highly variable cases like, intake rates and hydraulic conductivity values. The standarized variable, u, in this distribution is defined as

$$u = \frac{\ln Y^* - \bar{u}}{v_u} \qquad (5)$$

in the function, u is normally distributed with mean, $\bar{u}$, and standard deviation, v_u,

6 EXPERIMENTAL DATA AND COMPARISON

To satisfy the objectives of this study, actual water distributions (i.e water caught in cans placed equidistantly in the sprinkler area) were obtained from side-roll sprinkler irrigation system. The tests were conducted in accordance with Americal Society of Agricultural Engineers Standard.

The lateral of the system was 125 mm in diameter and has 8 sprinkler spaced 12 meters apart. A 3-meters grid was established for a distance of 24 meters parallel to the lateral, which covered three squence sprinklers.The grids were extended laterally from the sprinkler line throughout the wetted area on both side of the lateral. The water caught in cans at the centre of grid was recorded when all the sprinklers on the lateral were in operation.

Two runs were made with this lateral setup. The wind speed varied from test to test. Average wind speeds recorded were 8.0 and 30.0 km/hr.

The sprinkler irrigation distribution patterns obtained through overlapping of observations made from the single lateral of the side-roll sprinkler irrigation system, are employed to evaluate the linear, normal and log normal distribution models. The overlapping was done on rectangular grid spacing of 12m x 21m and 24m x 6m -21m, with a total of 8 different overlapped patterns for each test. In this matter,the range of uniformity coefficients, u_c,was varying from 50.6% to 90.2%. The overlapped water depths were arranged from maximum to minimum and were nondimensionalized by assumed that each depth represented in equal fraction of the total irrigated area. Each overlapped sprinkler pattern thus gave rise to the cumulative function of dimensionless application depth versus the fraction of area receiving at least that depth of water.

Linear regression was used to find the best straight line fit to the cumulative frequency data, however that line represents the linear regression model. As a matter of fact, the normal distribution model fit is defined by the mean and standard deviation of the data. While the log normal model fit is defined by the mean and standard deviation of the transfered data.

For all models, the root mean squares (RMS) deviation the model performance from the actual pattern was calculated and used as a basis for comparison.

Each of the three models yields a continuous distribution, but the deviations were determine only at those fractional areas corresponding to the sprinkler data. The RMS error is given by

$$e = \sqrt{\frac{\sum_{i=1} (Y_i^* - Z_i^*)^2}{N-1}} \qquad (6)$$

in which e=RMS error, Z_i^*=the theoretical dimensionless water application depth, as predicated by the assumed statistical distribution and Y_i^*=the ith actual dimensionless precipitation depth.

The relationship between u_c and e are shown in Figures 1,2 and 3 for linear, normal and log normal models respectively. The data points indicated considerable scatterness. However, certain order in their spread is observed. It is apparent that the error increased as the uniformities decreased, in spite of the fact that one uniformity having more than one value of e. Using the mean of the RMS errors for classes of uniformities then Fig.4 is obtained. This figure shows that the linear model consistently was the best in performances among the three models. The normal model has the lowest degree of fittness among the three models.

Because of the simplicity and good performance, the linear model can be used to described the observed water distribution. Two parameters slope and intercept, of the linear function can be determined by linear regression of the cumulative distribution function. For the log normal model, the calculation of the mean and standard deviation of the data are needed. This can be performed by transfering the data to natural log (y=ln x) so to obtained the parameters of the distribution.

The linear model can be recommended as model distribution to be used for the sprinkler system design. It is more practical and simple for the calculation of water volumes, which are used to determine the irrigation efficiences.

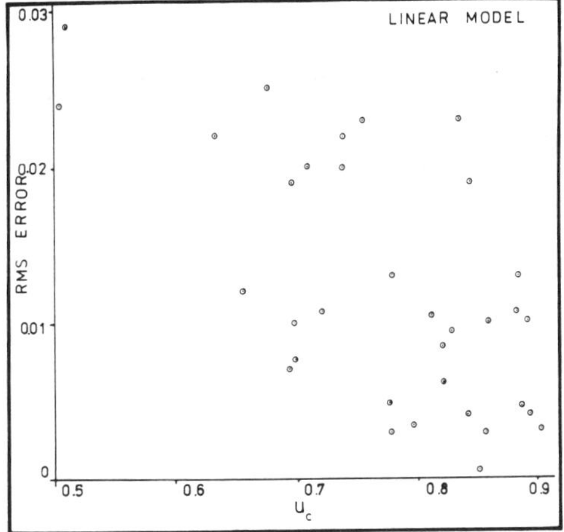

Fig.1 RMS Error of linear model from observed sprinkler distribution versus u_c

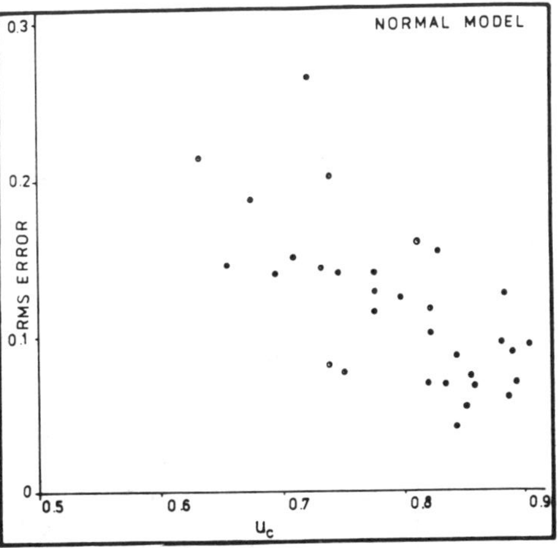

Fig.2 RMS error of normal model from observed sprinkler distribution versus u_c

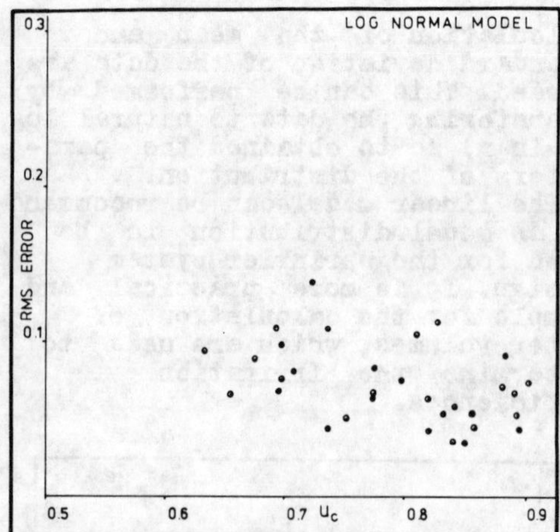

Fig.3 RMS error of log normal model from observed sprinkler distribution versus u_c

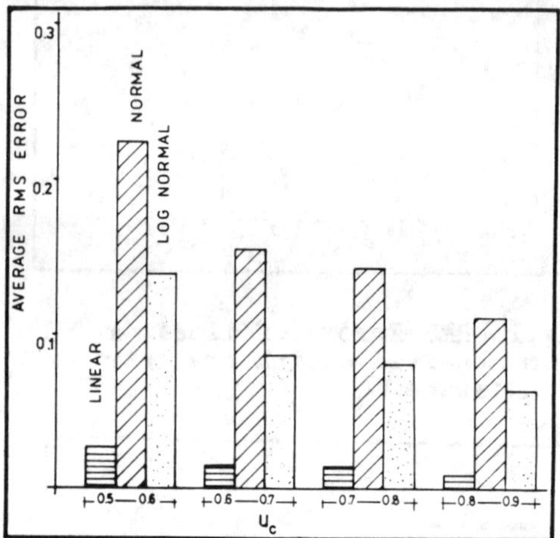

Fig.4 Average RMS error of three statistical models from observed sprinkler distribution versus u_c

7 SUMMARY AND CONCLUSIONS

Linear, normal and log normal statistical model were fitted to overlapped sprinkler patterns obtained from a single sprinkler line. The RMS error of each distribution from the observed water distribution was calculated, and was used for comparison. It was concluded that the linear model fit the water application depths better than did the normal or log normal for all values of uniformity coefficients. For uniformity coefficient values desired in the design and evaluation of sprinkler systems, the linear model is recommended for the calculation of irrigation efficiencies.

ACKNOWLEDGMENTS

The writer expresses his thanks to University of Basrah, Iraq and Hydrographical Studies Center, Spain for making this study possible.

REFERENCES

Abranowitz, M. and Stegun, I. A. (eds.), Handbook of Mathematical Function, National Bureau of United States Goverment Printing Office, Washington, D.C., (1968).

Chaudhry, F.H., Sprinkler Uniformity Measures and Skewness, Journal of the Irrigation and Drainage Division, ASCE, Vol.102, No.IR4, 425-433 (1976).

Chaudhry, F.H.,Nonuniform Sprinkler Irrigation Application Efficiency, Journal of the Irrigation and Drainage Division, ASCE, Vol.104, No.IR2, 165-178 (1978).

Elliot, R.L., Nelson, J.D.,Loftis, J.C. and Hart W.E., Comparison of Sprinkler Uniformity models, Journal of the Irrigation and Drainage Division, ASCE,Vol.106, No.321-330 (1980).

Hart, W.E., and Reynolds, W.N., Analytical Design of Sprinkler Systems, Transactions of the American Society of Agricultural Engineers, Vol.8, No.1, 83-85 (1985).

Howell, D.T., Nonuniformity and Sprinkler Application Efficiency, Journal of the Irrigation and Drainage Division, ASCE, Vol.90 No.IR3, 41-53 (1964).

Karmeli, D., Estimating Sprinkler Distribution Patterns Using Linear Regression, Transactions of the American Society of Agricultural Engineers, Vol.21, No.4, 682-686 (1978).

Seniwonge, C., Wu, I.P., and Reynolds, W.N., Skewness and Kurtosis Influence on Uniformity Coefficient, and Application to Sprinkler Irrigarion Design, Transatctions of the American Society of Agricultural Engineers, Vol.15, No.2, 266-271 (1972).

Tsakiris, G.P., Terzis G.C., Karakos A.S., A Person Distribution Model Describing Sprinkler Irrigation Efficiency, Agric. Water Mangage., 9:325-337 (1985).

Walker, W.R., Explicit Sprinkler Irrigation Uniformity:Efficiency Model, Journal of the Irrigation and Drainage Division, ASCE, Vol.105, No.IR2, 129-135 (1979).

Wilcox, J.C., Swailes, G.E., Uniformity of Water Distribution by Some Undertree Orchard Sprinklers, Scientific Agriculture, Vol.27, No.11 565-583 (1947).

Land and Water Use, Dodd & Grace (eds), © 1989 Balkema, Rotterdam. ISBN 90 6191 980 0

Rehabilitation of disturbed tropical land with trickle irrigation for agricultural production

S.bin Said
Universiti Pertanian Malaysia, Malaysia

ABSTRACT: There are about 200,000 hacteres of tin tailings in Malaysia laid waste by the mining operations. Being mostly sandy, the total available moisture is only 8.3% by volume. A water balance study is conducted to determine the fraction of trickle irrigation water consumed by a crop.

INTRODUCTION

Malaysia is a tropical country situated just north of the Equator. Being in the tropics with temperatures varying from 21°-32°C, we can practise agriculture all year round. Our rainfall is also plentiful (about 2500 mm/year). Nonetheless, we cannot rely solely on rainfall for our agricultural production. During certain periods of the year irrigation is needed to supplement the rain.

In Malaysia, trickle irrigation is gaining widespread acceptance. This is especially true with the drive towards commercialization of our fruit and ornamental crops for export. Government agencies as well as the private sector have set up trials and pilot projects monitoring the performance of trickle irrigation on crop production. Some serious farmers have also installed trickle irrigation system in their orchards.

Trickle irrigation is being practised on all types of soils. However, from the research point of view, trickle irrigation may become one of the useful tools in the rehabilitation of large tracts of tailings laid waste by the tin mining operations.

TIN TAILINGS

Tin mining operations in Malaysia started in a big way with the British intervention at the turn of this century. Today, Malaysia is still the world's number one producer of tin, and the industry is still one of the most important contributors to the nation's economy. Also significant and very visible are the tailings left behind. There were (and still are) no stringent regulations, binding the mining operators to rehabilitate the spoils after a mining operation was completed. At the present, it is estimated that there are about 200,000 hacteres of tin tailings left idling for the nature to reclaim.

PHYSICAL PROPERTIES OF TAILINGS

Depending on the methods of mining operation,tin tailings show a considerable variation in their mechanical composition. Large mines in Malaysia usually employ dredging for the extraction of tin-ores. Under this method distinct separation of the various soil particles take place, since the sand fraction is discharged via a sand chute, and the finer silt and clay particles in suspension are pumped into a slime retention area for settlement. The sand tailings are estimated to occupy an average of 80% of the tailing surfaces. It is devoid of organic matter and characterised mainly by sand or inert quartz particles, leading to excessive drainage, low water and nutrient retention capabilities. Under these circumstances, very high leaching losses of applied nutrients are to be expected. Generally, devoid of natural vegetation, the sand tailings experience high surface soil temperatures (40 - 50°C) and consequently high rates of evaporation.

The slime areas with high content of fine soil perticles, consolidated upon drying into compact masses. Infiltration is restricted and often water-logged conditions prevail. When dry, large cracks appear on the suface.

Smaller mines in Malaysia usually practise gravel pumping. After the tin concentrate is deposited, the tailings are usually discharged on a sloping ground contoured on the lower end by a retaining bank. The coarser materials, are therefore, found near the point of discharge with a continuous gradation to a fine clay at the lower end of the slope. Thus, the tailings have all the gradations between sand and slime. These mixed tailings have less agricultural constraints than the pure sand or slime tailings but the soil heterogeneity poses a problem in designing a suitable irrigation system for the area.

Tables I and II show some of the physical properties of the various tailings.

Table I : Particle size distribution and bulk density of tin tailings

Tailings	% clay <0.002 mm	% silt 0.002-0.02 mm	% fine sand 0.02-0.2 mm	% coarse sand 0.2-2 mm	Bulk density
Sand	9.6	7.5	9.0	73.9	1.48
Slime	47.5	37.6	10.6	4.3	1.20
Mixture (20sa : 80sl)	44.5	35.1	10.4	10.0	1.30

Table II : Soil moisture characteristics (% w/w) of tin tailings

Tailings	Water content at 0.1 kPa	1 kPa	10 kPa	33 kPa	1500 kPa
Sand	24.0	19.2	15.7	13.4	9.1
Slime	50.8	40.6	30.5	27.5	21.4
Mixture (20sa : 80sl)	50.5	41.1	32.0	28.5	24.9

TRICKLE IRRIGATION ON TAILINGS

Obviously, the sand tailings will provide the most problematic conditions in terms of its irrigation requirement. Figure 1 shows the soil moisture characteristic curve of the sand tailings. If we assume the amount of moisture available for plant use as those held between 0.1 bar to 10 bar, then, the total moisture available is only 8.3% by volume.

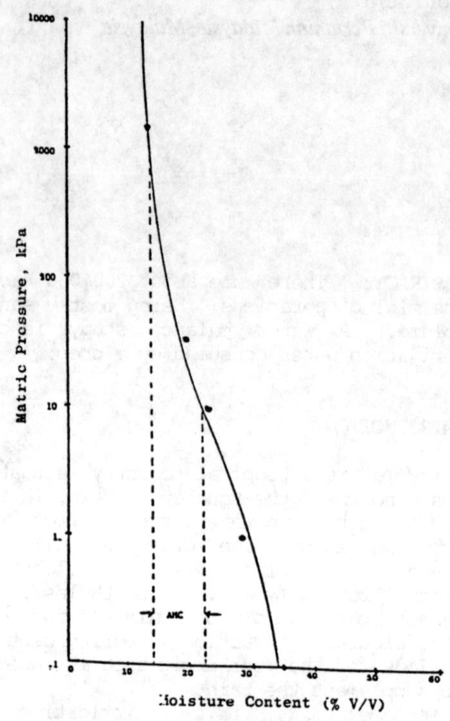

Fig. 1 : Soil moisture characteristic curve of sand tailings

In order to ascertain how water from a trickle irrigation system is being consumed by a certain crop, a water balance study is being carried out in four specially built soil tanks. A schematic diagram of the soil tank is shown in Fig. 2.

The equations for the water balance study are as follows :

$$Qin = Qout$$

$$RF + I = ET + P + \Delta MC$$

where

RF	=	water from rain
I	=	water from trickle irrigation
ET	=	Evapotranspiration loss
P	=	Percolation loss
ΔMC	=	change in moisture content of soil

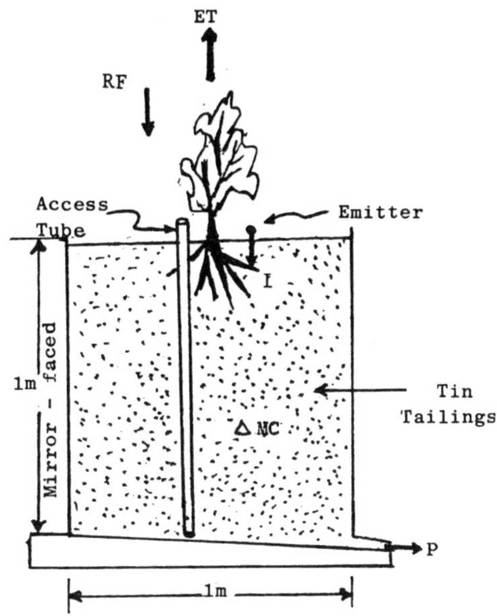

Fig. 2 : Schematic diagram of the water balance model

All parameters, except for the evapotranspiration factor, can be measured directly. Since there are no other factors affecting the discharge, in or out, the evapotranspiration rate can be deduced from the equation.

Three levels of irrigation treatment plus a control are used in the study. Treatment T2 is designed to meet the daily water requirement of the crop. Treatment T1 is for underirrigation effects and treatment T3 is for overirrigation effects.

REMARKS

The soil tanks are useful in the water balance study, since losses due to percolation can be measured directly.

Results from this water balance study can be used to provide guidelines for trickle irrigating crops grown on tin-tailings. With efficient water management practices, plus some other soil conditioning measures, idle tin tailings may be reclaimed for crop production.

REFERENCES

ASAE. 1986. Trickle Irrigation for Crop Production. Ed. Nakayama, F.S & D.A. Bucks. Development in Agricultural Engineering 9. St. Joseph, Michigan.

Barfield, B.J., R.C. Warner & C.T. Haan. 1985. Applied Hydrology and Sedimentology for Disturbed Areas. Oklahoma Technical Press, Oklahoma.

Salim Said. 1988. Trickle Irrigation Design for Malaysian Agriculture. Paper presented at Seminar on Research in Agricultural Engineering. UPM. Malaysia.

UPM. 1981. Reclamation of Tin tailings for Agriculture in Malaysia. Soil Science Dept. Technical Bulletin. Malaysia.

Land and Water Use, Dodd & Grace (eds), © 1989 Balkema, Rotterdam. ISBN 90 6191 980 0

Uniformity of applications from lateral move irrigation machines

R.J.Smith
Darling Downs Institute of Advanced Education, Queensland, Australia

ABSTRACT: Spatial variations in the depths of water applied over fields irrigated by lateral move irrigation machines occur as a result of a large number of factors. Most of these factors have been analysed individually and sometimes in combination but a model incorporating all (or even most) factors has not been available to designers. A model for calculating both the outflows from spray nozzles along a lateral move and the uniformity of the resulting applied depths is described, incorporating all the factors which influence applied depths. Data on the performance of lateral move spraylines and the sampling required to accurately define the uniformity of applied depths are discussed.

Les variations spatiales de profondeur de l'eau appliquée sur des champs irrigués par de machines d'arrosage à mouvement latéral proviennent d'un grand nombre de facteurs. Des analyses de la plupart de ces facteurs ont été faites séparément et quelquefois en combinaison, mais un modèle incorporant tous (ou même la plupart de) ces facteurs n'a pas été disponible aux contracteurs. Un modèle, permettant l'evaluation du d'échargement d'ajutage d'arrosage au cours d'un mouvement latéral et de l'uniformité da profendeur qui provient, est décrit. Tout les facteurs qui influencent le profondeur de l'eau appliquée sont incorporés. Les données sur la preformance des machines d'arrosage à mouvement latéral et l'échantillonage requis pour définir avec précision l'uniformité du profondeur de l'eau appliquée sont discutés.

Räumliche Unterschiede in der Menge des verregneten Wassers auf einem durch eine seitwärts wandernde Anlage bewässerten Feld hängen von einer Anzahl von Faktoren ab! Die meisten dieser Faktoren sind entweder einzeln oder als kombinationen analysiert worden, allerdings hat ein Modell, das alle Faktoren einschließt, bisher nicht zur Verfügung gestanden. Die Berechnung des Ausflusses aus den Düsen über die gesamte Länge der sich seitwärts fortbewegenden Anlage und für die Gleichmäßigkeit der Beregnungsstärke wird in einem Modell beschrieben einschließlich aller Faktoren die Einfluß auf die Beregnungsstärke haben. Daten über die Arbeitsweise des seitwärts wandernden Systems und die benötigten Regenmesser zur genauen Bestimmung der Verteilung der Wassermengen werden außerdem diskutiert.

1 INTRODUCTION

Lateral move irrigation machines have the potential to apply irrigation water to crops both efficiently and uniformly. Here efficiency is used as the measure, in volumetric terms, of the losses incurred in delivering water to the plant and includes losses during conveyance and application, to evaporation, to deep percolation and as runoff. The uniformity of applications, the subject of this paper, refers to how spatially uniformly the water is applied over the entire irrigated field.

Estimates (based on measurement) of the uniformities of applied depths under lateral move and similar (centre pivot) machines would indicate that the potential for uniform applications is not being realized. For example, of the results summarized in Table 1 only those reported by John et al. (1986) and Hills et al. (1988) could be described as uniform. They are also the only reported estimates representative of the uniformity of depths over an entire field. It is also useful to note that no measurements are available for Australian designed and manufactured machines.

The accuracy of the uniformity estimates in Table 1 is also open to question. In most cases the sampling procedure appears to conform to the ASAE standard (ASAE,

1987). However it can be argued, from sampling theory, that more closely spaced samples are required to adequately define the uniformity.

Table 1. Measured uniformities.

Machine	Sampling	Cu	Source
Pivots (two)	Radial 10 ft spacing	0.87 to 0.91*	Heermann & Hein (1968)
Pivots (two)	Radial 6 m	0.90 to 0.94*	Rapp et al. (1979)
Lateral &	⊥'r to machine	0.75 to 0.89	Hanson & Wallender
Pivot	0.3 m	0.67 to 0.77	(1986)
Pivot	Radial 3 m	0.77	"
Lateral	Parallel 3 m	0.73	"
2 spans	0.6 m	0.59 & 0.76	"
Short Lateral	Field grid	0.96	John et al. (1986)
Pivots (six)	Radial 5 m	0.54 to 0.88*	Thooyamani & Norum (1987)
Pivots (eight)	Radial 5 m	0.82 to 0.88	Thooyamani et al.
	⊥'r 0.3 m	0.81 to 0.94	(1987)
Lateral 4 spans	Parallel & ⊥'r	0.94 (mean)	Hills et al. (1988)

Note: Uniformity is expressed here by the Christiansen Uniformity Coefficient (Cu):

$$Cu = 1 - (M/x_m) \quad (1)$$

where M is the mean absolute deviation of the applied depths x_i and equals $(\sum|x_i - x_m|)/n$, x_m is the mean applied depth, and n is the number of depth measurements. The asterisk indicates that the areal weighted coefficient of Heermann and Hein (1968) was used.

2 PREDICTING THE UNIFORMITY OF APPLICATIONS

2.1 Sources of Non-Uniformity

The spatial variability (non-uniformity) of irrigation applications on fields supplied by lateral move machines results from the combined effects of:

- variations in the spray nozzle discharges along the length of the machine, as determined by the pipeline hydraulics,
- variations in spray nozzle or pressure regulator performance due to size or other variations occuring during manufacture,
- variations in applied depth due to the spray pattern, including the effect of spray pattern overlap,
- variations in the land surface and hence pipeline elevation,
- the stop-start pattern of movement of these "continuously" moving systems, and
- other causes such as temperature variations, wear of spray nozzles, wind effects and fluctuations in pump performance.

More important than the uniformity of applications is the uniformity of soil water available to the plant population. The two will differ because measureable pre- and post-infiltration improvements in uniformity can occur, for example during redistribution on the soil surface (John et al., 1986) and the redistribution of soil moisture within the root zone following infiltration (Perrens, 1984). In addition, effective uniformity is thought to be highly dependent upon the horizontal extent of the crop root zone (Seginer, 1979). The uniformity will also increase with time because of varying distribution patterns from one irrigation to the next (Jensen, 1980). In regions with significant rainfall during the irrigation season this rainfall will also serve to increase effective uniformity of the applied water (irrigation plus rainfall).

2.2 Previous Attempts to Predict Uniformities

Over the years since the development of mobile spraylines a number of attempts have been made to predict the pressures and sprinkler discharges along the lines and the resulting depths of applied water. The pipeline hydraulic theory necessary for this purpose is common to other multi-orificed pipelines and has been the subject of many papers. These have been reviewed by Smith et al. (1986) and repetition here is unwarranted. Only that work relating directly to the prediction of the depths applied by mobile systems (lateral moves and centre pivots) will be considered here.

Three significant groups of work emerge. The first commenced with Heermann and Hein (1968) who predicted applied depths along a centre pivot sprayline by using measured sprayline pressures and idealized (triangular and elliptical) spray patterns. This was followed by Kincaid and Heermann (1970) who used a correct but simplified hydraulic theory to calculate the sprayline pressures. James (1982) extended this hydraulic model to include variable topography and combined it with the idealized spray patterns of Heermann and Hein to calculate applied depths over an entire field. James and Blair (1984) further adapted the model to accomodate the closely spaced sprays used on low pressure pivots.

For centre pivots operating on horizontal terrain Chu and Moe (1972) developed an approximate (and incorrect) hydraulic model. Thooyamani et al. (1987) used this model and idealized spray patterns to calculate applied depths and uniformities in the radial and angular directions.

The third and final group (Gu and Hills, 1988 and Hills et al., 1988) modelled the effect of machine motion on the uniformity of applications both parallel and perpendicular to the machine. Here the spray pattern was supplied to the model as a square grid of measured point application rates at 1 m spacings. No mechanism was provided for varying the discharge and spray pattern from one nozzle to the next.

In addition to the over simplified and approximate hydraulic models used in these studies the other significant limitations can be summarized simply in that they:

- considered only a limited number of the sources of variability listed previously,
- usually used idealized spray patterns,
- usually only calculated applied depths and uniformities along a single transect rather than over the entire area covered by the machine, and
- in many cases calculated applied depths at spacings to large to accurately describe the uniformity.

3 THE LATERAL MOVE MODEL

3.1 Objectives of the Model

The proposed model is designed to calculate the uniformity of applied depths over a field irrigated by a lateral move machine. All of the sources of variability listed in the earlier section 2.1 will be included as appropriate algorithms are developed.

The objectives of the model are as follows:

1. To quantify the contributions of each of the sources of variability to the overall field uniformity.
2. To identify those sources which cannot be described adequately and whose contribution cannot be accurately assessed.
3. To identify areas where design modifications may lead to improvements in the uniformity of applications.
4. To provide a possible basis for an improved hydraulic design programme.
5. To investigate the effect of sample frequency and spacing and of root zone scale on estimates of the uniformity of applied depths.

3.2 Model Structure

The overall structure of the model is illustrated in Figure 1. It consists of two sub-models, viz:

- a fully deterministic pipeline hydraulic sub-model for the calculation of the pressures at and discharges from each of the spray nozzles, and
- an applied depth sub-model which predicts and overlaps the spray patterns from the nozzles to determine the applied depths at points along transects crossed by the machine.

While the applied depth sub-model is deterministic in form it is not intended to reproduce exactly the depth of water that would be applied at any point. The intention of the model is a statistical representation of the applied depths over an irrigated field.

A third sub-model also shown in Figure 1, and designed to account for the effects of surface and subsurface redistribution and of rainfall on the uniformity of soil moisture, is a possible future addition.

4. HYDRAULIC SUB-MODEL

4.1 Theory

The pipeline hydraulic theory used in the model is essentially as described in Smith et al. (1986) and Smith (1988) and will only be summarized here.

The six basic equations required are firstly the continuity equation at each sprinkler tee:

$$Q = Q_c + Q_o \qquad (2)$$

where Q is the discharge entering the tee, Q_c is the discharge continuing past the tee, and Q_o is the discharge branching to the sprinkler.

Second the energy (or Bernoulli) equation is applied along the main pipeline assuming no loss of energy across the tee (Smith et al., 1986) giving:

$$\partial h_p/h_v = 1 - (V_c/V)^2 \qquad (3)$$

where ∂h_p is the pressure head recovery across the tee, h_v is the velocity head upstream of the tee, V is the flow velocity upstream of the tee, and V_c is the velocity downstream of the tee.

Third a flow equation (in this case the Hazen-Williams equation) is required to describe the energy loss due to friction in each length of pipe between sprinkler tees.

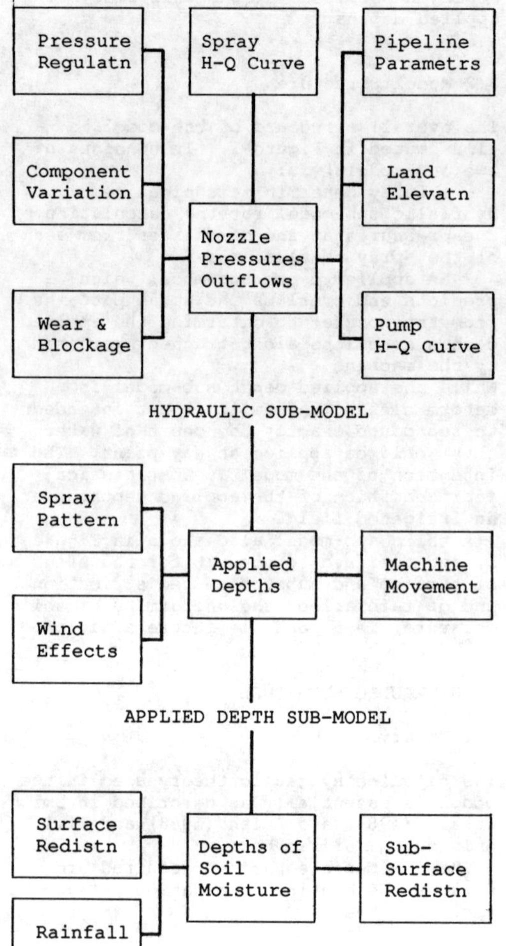

Figure 1. Model Structure.

Fourth an equation, developed from the momentum and energy equations applied at the tee (Smith, 1988), is used to describe the energy loss in the branching flow.

Fifth an outlet or nozzle characteristic is required, relating the discharge Q_o from any spray nozzle to the nozzle area A_o and pressure head H_p by an equation of form similar to the standard orifice equation:

$$Q_o = C_d\ A_o\ (2g\ H_p)^{0.5} \qquad (4)$$

where C_d is an empirical coefficient.

The present generation lateral move machines almost universally use Nelson Spray-I 360° spray nozzles. Performance data (pressure vs discharge) for these nozzles, obtained from the manufacturers catalogues, were used in a least squares regression to obtain the particular relationship:

$$Q_o = 3.6\ A_o^{0.984}\ H_p^{0.489} \qquad (5)$$

where the coefficient 3.6 includes all of the coefficients in equation 4.

Finally to minimize the effects of pressure and elevation variations along the machine it is common practice to fit a pressure regulator on each dropper immediately upstream of the spray nozzle. In the model pressure regulation is optional.

Pressure regulators in current use are also a product of the Nelson Corporation and the data used to characterize their performance were again obtained from the manufacturers catalogue. The performance of the four common pressure regulators, the PRS-10, 15, 20 and 25, can be adequately described (over their normal operating range) by a single relationship:

$$H_{po}=0.71(P-2.5)+0.046(H_{pi}+71)/e^{1645Q_o} \qquad (6)$$

where H_{po} and H_{pi} are the pressures on the outlet and inlet sides of the regulator, P is the model number of the regulator, and Q_o is the nozzle (and regulator) discharge.

Other features of this model not included in the earlier programme of Smith et al. (1986) are; arched pipeline spans, end overhang past the final tower, long droppers, variations in the pipeline diameter along the machine, nozzles of different size along the line and uneven topography.

4.2 Computational Procedure

Data on the machine parameters (span geometry, pipeline diameters, nozzle sizes and spacings, pressure regulation) and land surface elevations are read from file.

Computations commence at the downstream or closed end of the pipeline and proceed upstream as follows:

(i) An initial value for the discharge Q_o at the first (downstream) nozzle is selected and the pressure at that nozzle determined from the nozzle characteristic.

(ii) The total energy in the main line at the first tee is calculated, taking into account the energy losses in the tee and dropper, pressure regulation and the elevation difference between pipeline and nozzle.

(iii) The energy at the next tee in the upstream direction is then found by the addition of the pipe friction loss over the length of pipe between the tees.

(iv) An iterative technique is then

used to solve for Q_o at this nozzle, again requiring estimation of tee and dropper energy losses and accounting for the elevation of the nozzle and for pressure regulation.

(v) Steps (iii) and (iv) are repeated for all of the remaining nozzles.

(vi) The computations are re-initialized (step (i)) and the whole process repeated until the pressure and discharge at the inlet to the machine converge on a nominated pump head-discharge characteristic.

In its present form the hydraulic sub-model is suitable only for the analysis of lateral move machines. However modification of the program for the design of systems, that is, the selection of pipe diameters and nozzle sizes to maximize uniformity of outflows, would take little effort.

5. APPLIED DEPTH SUB-MODEL

5.1 Structure

The applied depth sub-model takes the nozzle data (size, pressure, discharge and height above ground) from the hydraulic sub-model and calculates the depth of water applied at points along a transect parallel to the machine, as the machine passes over that transect. This calculation can be performed for points at any nominated spacing along the transect. However at present the spacing is constrained to give applied depths at 2^n points to allow analysis of the data in the frequency domain via a discrete Fourier transform.

The user nominates the required mean depth of application I_m for the field. The average machine speed required to apply that depth of water is then given by:

$$\text{Speed} = \overline{Q}_o / I_m S \qquad (7)$$

where $\overline{Q}_o$ is the mean nozzle discharge, and S is the nozzle spacing.

Motion of the machine is currently as described by Gu and Hills (1988) and Hills et al. (1988). Travel velocities for each tower and a maximum angular misalignment are specified. The nominated depth of application I_m is obtained by varying the fraction of time (in a 60 second cycle) that the machine is in motion, the guide (end) towers moving according to this calculated on-off cycle. The central towers begin to move when the allowable misalignment is reached and stop when they draw level with the adjacent towers.

For a particular point on the ground a check is made to determine which of the adjacent nozzles contribute to applications at that point. As the machine approaches, passes over and leaves that point the radial distance to, and depth of application from each of the contributing nozzles is determined for each 2 sec time increment of machine movement. The applied depth at a point is the summation of these incremental depths.

5.2 Spray Pattern Representation

A radial spray pattern is required for each nozzle in the form of an equation linking the application rate at a point with the radius from the nozzle.

A weakness of most sprinkler analysis or design programmes which model the effects of overlapping spray patterns is the manner in which the spray pattern is described. Some, for example the SPRINKPAC programme of NZIAE (1987) and the model of Gu and Hills (1988), require a spray pattern measured for a particular pressure and assume incorrectly that the spray patterns for each nozzle are identical. Others such as those reviewed earlier in this paper admit variation but use patterns of idealized shape.

For a lateral move machine the pressures, discharges and nozzle sizes will vary along the length of the machine. This means that the spray pattern will be different for each nozzle. Further, the unusual doughnut shaped pattern from the Nelson Spray-I nozzles do not lend themselves to representation by simple geometric shapes. A method for synthesizing the spray pattern for each individual nozzle is therefore required, utilizing the known nozzle pressure, discharge, size and height above ground (or crop).

A simple approach to this problem has been used in the model, the development of a better method not being possible until measured spray patterns are available for comparison.

The approach uses a simple non-dimensional representation of the radial pattern, where the application rate i at radius r from the nozzle is given by:

$$i/i_m = a_o + a_1(r/R) + a_2(r/R)^2 + .. + a_8(r/R)^8 \qquad (8)$$

The coefficients a_o to a_8 were determined by a least squares fit to an average spray pattern taken from a limited sample of measured patterns. The mean application rate i_m for the particular nozzle is a simple function of the nozzle discharge Q_o and the wetted radius R:

$$i_m = Q_o / \pi R^2 \qquad (9)$$

Solomon et al. (1986) showed that measured drop size distributions from low pressure spray nozzles can be described by an upper limit log normal distribution (ULLN). They also provided expressions to determine the ULLN parameters from the nozzle diameter and pressure, allowing calculation of the drop size distribution for any nozzle.

The trajectory of the largest drop size and hence the wetted radius R (horizontal distance travelled) is calculated using a Runge Kutta 4th order solution of the equations of motion, in the horizontal direction:

$$dV_x/dt = - C\ V_x\ V \qquad (10)$$

and in the vertical direction:

$$dV_z/dt = g - C\ V_z\ V \qquad (11)$$

where V_x and V_z are the horizontal and vertical components of the droplet velocity V. The coefficient C includes the drag coefficient for the droplet and is a function of the droplet size and Reynolds number (Edling, 1985). It is assumed that each drop leaves the spray plate horizontally and with a velocity equal to the velocity of the nozzle jet.

One difficulty with this approach as yet to be resolved is the unavailability of values of the drag coefficient for water droplets travelling at velocities in excess of their terminal velocity.

5.3 Sampling Frequency

The applied depth sub-model calculates depths at any nominated spacing along a transect. At large spacings the distribution of applied depths may not be adequately defined while at small spacings computational time becomes excessive (for example, 2 minutes per spray nozzle if depths are calculated at 0.1 m spacing).

How many applied depth measurements or estimates are required to define the uniformity along a transect? Should they be regularly or randomly spaced? Should we consider the depth applied at a point or the mean depth applied over some small area (approximating the lateral extent of the plant root zone)?

The earlier models, reviewed in section 2.2, calculated applied depths at regular and relatively large spacings, often selected to match the spacing of depth measurements in corresponding field trials. In no case was the spacing justified statistically.

Applied depths under lateral move machines are periodic in the spatial domain. Among the periods present are: one corresponding to the nozzle spacing in a direction along the machine and another related to the on-off cycle of machine motion perpendicular to the machine. According to the data reported by Hanson and Wallender (1986) other longer periods are present, the sources of which were not readily apparent.

Sampling theory (Harris and Ledwidge, 1974) suggests that samples (measured or calculated) should be taken at spacings of less than 1/2 (and preferably 1/4) of the shortest period present in the data. For a machine with spray nozzles at about 2 m spacing the samples would be required at less than 1 m intervals along the machine. If the depths are sampled at some larger interval then all information at periods less than the sampling interval will be lost. The ASAE standard (ASAE, 1987) would recommend a sampling interval of about 3 m for this machine.

To illustrate the sampling problem consider a lateral move machine 420 m long operating on a level field. The machine comprises 7 spans each 58 m long plus a 14 m end overhang. Mean discharge from the 196 spray nozzles is 0.65 Ls^{-1}, with nozzle sizes selected to minimize the variation in discharges along the length of the machine.

The hydraulic sub-model predicts a coefficient of variation of nozzle outflows of 0.055 (equivalent to a Cu of 0.96 or a range of ± 16%). The distribution of the nozzle discharges shows a slight positive skew (skew coefficient of 0.85).

Applied depths were calculated initially at 0.1 m spacings along a single transect parallel to the machine. Depths ranged from 0.011 m to 0.027 m for a nominated mean depth of 0.025 m. The coefficient of variation of the applied depths was 0.052 (Cu of 0.98) with the distribution exhibiting a high negative skew (skew coefficient of -6.34).

The effect of regularly spaced sampling is shown in Figure 2. Applied depths were sampled at spacings ranging from 0.4 to 16.8 m (equivalent to the number of depth samples ranging from approximately 25 to 1000). Clearly, once the spacing exceeds about 2.8 m (or the number of samples is less than 150) the resulting coefficient of variation is no longer a reasonable estimate of the population variability.

6. APPLICATIONS OVER A FIELD OF VARYING TOPOGRAPHY

The model as presently written calculates the uniformity of the nozzle discharges and applied depths along a single transect. The main factor which will cause the whole field uniformity to vary from that for a single transect is the topographical variation in the direction of the machine travel. The result should be a decrease in uniformity.

Modification of the model to include this factor in the prediction of the field uniformity is the next (and final) significant step in the model development.

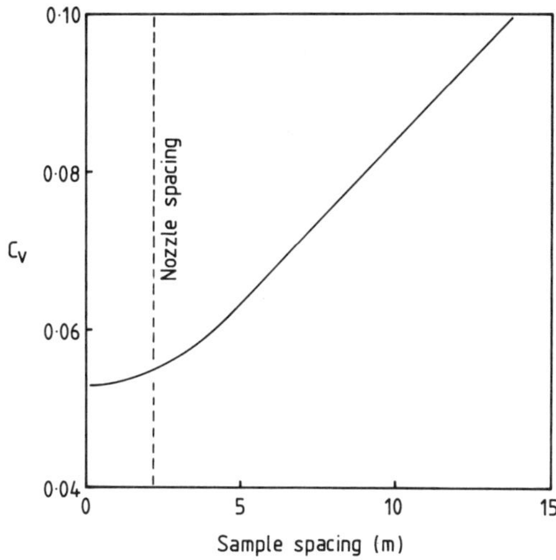

Figure 2. Effect of Sample Spacing

7. REFERENCES

ASAE 1987. Test procedure for determining the uniformity of water distribution of centre pivot, corner pivot, and moving lateral irrigation machines equipped with spray or sprinkler nozzles. ASAE Standard: S436, ASAE, St Joseph, Michigan.

Chu, S.T. and Moe, D.L. 1972. Hydraulics of a centre pivot system. Trans ASAE, 15: 894-896.

Edling, R.J. 1985. Kinetic energy, evaporation and wind drift of droplets from low pressure irrigation nozzles. Trans ASAE, 28: 1543-1550.

Gu, Y. and Hills, D.J. 1988. Irrigation uniformity with a linear move system. Microsoftware for Engineers, 3: 206-212.

Hanson, B.R. and Wallender, W.W. 1986. Bidirectional uniformity of water applied by continuous move sprinkler machines. Trans ASAE, 29: 1047-1053.

Harris, R.W. and Ledwidge, T.J. 1974. Introduction to Noise Analysis. Pion, London, 102pp.

Heermann, D.F. and Hein, P.R. 1968. Performance characteristics of self-propelled centre-pivot sprinkler irrigation systems. Trans ASAE, 11: 11-15.

Hills, D.J., Gu, Y., Rumsey, J.W. and Wallender, W.W. 1988. Lateral move uniformity relative to machine speed. Trans ASAE, 31: 527-530.

James, L.G. 1982. Modeling the performance of centre pivot irrigation systems operating on variable topography. Trans ASAE, 25: 143-149.

James, L.G. and Blair, S.K. 1984. Performance of low pressure centre pivot systems. Trans ASAE, 27:1753-1757,1762.

Jensen, M.E.(ed) 1980. Design and Operation of Farm Irrigation Systems. ASAE Monograph 3, St Joseph, Michigan, 829pp.

John, P.H., Lees, D.M. and English, G.M. 1985. Application performance of travelling irrigators. NZAEI Project Report No 35.

Kincaid, D.C. and Heermann, D.F. 1970. Pressure distribution on a centre-pivot irrigation system. Trans ASAE, 13: 556-558.

NZAEI 1987. SPRINKPAC - A program for overlapping sprinkler distribution patterns - User guide. New Zealand Agricultural Engineering Institute, Publication No CR345.

Perrens, S.J. 1984. Numerical analysis of soil water uniformity under sprinkler irrigation. J Agric Engng Res, 30: 23-27.

Rapp, E., Chanasyk, D.S. and Horvarth, B.C. 1979. Simulated model of centre pivot irrigation systems. Canadian Agric Engng, 21: 141-145.

Seginer, I. 1979. Irrigation uniformity related to horizontal extent of root zone - A computational study. Irrig Sci, 1: 89-96.

Solomon, K.H., Kincaid, D.C. and Bezdek, J.C. 1986. Drop size distributions for irrigation spray nozzles. Trans ASAE, 28: 1966-1974.

Smith, R.J. 1988. Energy loss in branching flow and its application to irrigation pipeline design. J Agric Engng Res, 41: (in press).

Smith, R.J., Watts, P.J. and Mulder, S.J. 1986. Analysis and design of gated irrigation pipelines. Agricultural Water Management, 12: 99-115.

Thooyamani, K.P. and Norum, D.I. 1987. Performance of low-pressure centre-pivot sprinkler systems in Saskatchewan. Canadian Agric Engng, 29: 143-148.

Thooyamani, K.P., Norum, D.I. and Dubetz, S. 1987. Application rates and uniformity under centre-pivot sprinkler irrigation systems using spray nozzles. Canadian Agric Engng, 29: 149-154.

Land and Water Use, Dodd & Grace (eds), © 1989 Balkema, Rotterdam. ISBN 90 6191 980 0

Modelling a farm canal for automated basin irrigation

P.L.Sousa & L.S.Pereira
Department of Agricultural Engineering, Institute of Agronomy, Technical University of Lisbon, Portugal

ABSTRACT: This paper reports on modelling an automated on-farm distribution canal delivering water to ten level basins throughout lateral weirs. This semicircular canal is divided into reaches, each one with a long crest lateral weir and regulated by remote controlled lift check gates. After a brief presentation of the system, the hydraulic conditions are analysed and the consequent mathematical model is described. This aims the simulation of the hydraulic behaviour of this farm canal and consequent design purposes. Results herein presented concern a comparision of field observed and simulated data and they confirm the potentialities of such canal system to supply water to surface irrigated fields.

RÉSUMÉ: Ce rapport concerne la modélisation d'un canal automatisé, avec déversoirs lateraux, pour le fournissement d'eau d'irrigation à dix basins à fond plat. Ce canal semi-circulaire est divisé en dix biefs, chacun avec un deversoir lateral à crète longue et réglé par vannes verticales tele-commandées. Aprés une brève présentation du système, on le caractérise hydrauliquement et on décrit le modèle mathématique respectif, utilisé pour simuler le comportement hydraulique et à des fins de projet. Les resultats présentés concernent la comparaison entre mesures de terrain et valeurs simulées, permettant de confirmer les potencialités d'une telle structure pour le fournissement d'eau à des parcelles en irrigation gravitaire.

ZUSAMMENFASSUNG: Dieser Artikel beschreibt die Modellisierung eines automatiziertes Kanales mit lateralen Stauwehren für die Wasserversorgung von zehn flachen Parzellenweisen. Dieser Halbkreisformkanal wird in zehn Strekken geteilt. Ieder Teil besteht aus Lateralwehren und wird von fern controlierten Hubtören gestenerte. Nachdem dieses System beschrieben ist, stellen wir die hydraulichen Eigenschaften und den mathematischen Modell. Dieser Modell wird für die simulation des hydraulischer Verhalten und der Kanal Bauzeichnung benutzt. Die vorgezeigt Ergebnisse erlauben die Vergleichung der Feld - und der simulierten werte. Der Interesse an diesem System wird mit der Vergleichung bestätigt, dass es geeignet ist für die Wasserversorgung nach Felden mit obererdische Bewässerung.

1. INTRODUCTION

Recent advances in surface irrigation (Pereira, 1987) led to a revised interest in furrow and basin irrigation. Efficiencies and uniformities of water application improve using evolved precise levelling techniques and controlled inflow discharges adequate to length and form of furrows and basins, while manpower efficiencies increase through proper use of farm hydraulic systems and automation.

Research is now being developped in Portugal on these aspects in relation to both the on-farm water management and irrigation systems management in the perspective of making available the technologies needed to modernize the surface irrigation projects (Pereira, 1988).

Among others, one main aspect of this research programme is the development of on-farm automated structures for level basin irrigation (Dedrick et al., 1982; Sousa, 1984). Related solutions concern open channels for on-farm water distribution and several mechanized and auto-

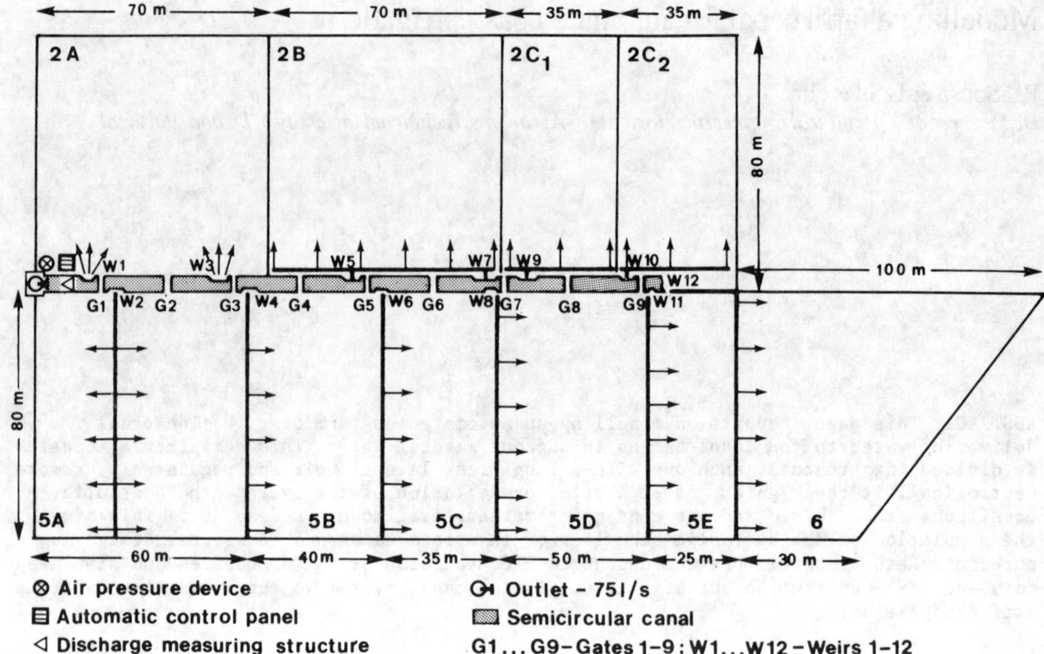

Fig.1 Layout of the experimental level basin plots and the farm distribution canal with lateral weirs, at Coruche Irrigation Experiment Station (2A,2B,....basin numbers)

mated control gates have been developped (Humpherys et al., 1970; Dedrick & Erie, 1978; Haise et al., 1980; Dedrick & Zimbelman, 1981). The combination of automated gates with lateral weirs is very recent (Eftekharzadeh et al., 1987) and can also be applied to furrow irrigation.

This study was initiated in 1986 aiming the study of performances of different sizes and forms of basins and of the respective on-farm distribution channel system. This paper reports on the farm canal, in particular on its design and simulation model.

2. THE ON-FARM CANAL SYSTEM WITH LATERAL WEIRS

To study the performances of level basin irrigation, a large plot of 4.22 ha has been precisely laser levelled and divided into 10 basins of different area, form, surface characteristics (flate or with beds), and distribution ditch systems (Figure 1).

A semicircular farm canal with diameter D=0.60 m was built to distribute water to all the basins. It is connected with an outlet of a low pressure pipe distributor of the irrigation district system of Sorraia Valley (summarized information on this irrigation project is given in Pereira, 1988). Discharges from the outlet are regulated by an hand operated valve associated with a modified broad crested weir (Replogle & Bos, 1982) in the upstream section of the canal, allowing to study the performances of level basin irrigation under selected combinations of discharges and irrigation durations (Sousa, 1987).

The turnouts of the farm distribution canal are 11 lateral weirs of selected lengths and a trapezoidal weir at the terminal section. The canal is divided into 10 reaches separated by lift check gates (Figure 2). They operate under air compression and are remotely controlled by an electronic device. A similar remote control system has been reported by Dedrick & Zimbelman (1981).

The turnouts work sequentially from up to downstream. All gates are opened at begining of operation but the first one, that is closed for a predefined time, regulating the head over the first side weir. When the selected water volume has been delivered to the field, the automatic electric control acts on the pressure system and lifts the corresponding check gate. Then water flows into next reach, where another check gate is simultaneously closed regulating the flow over the lateral weir until the automatic control device commands its opening. The operation repeates until irrigation of all fields is completed. If irrigation is not desired in a given basin this instruction is introduced in the automatic control device and the corresponding check gate is lowered just during sixty seconds, time not enought to fill the reach, assuring that no water flows through the side weir.

In order to make sure that no flow occurs in one lateral weir when the dowstream next is operating, special design care has to be taken relating the elevation of weir crests with the design heads of the sucessive ones. Therefore the elevation and slope of the canal reaches depend on such conditions and small drops of the bed elevation (of about 5 cm) have been adopted at location of check gates.

3. HYDRAULIC CHARACTERISTICS OF THE CANAL WITH LATERAL WEIRS

The flow regime in an open canal with lateral weirs and divided into reaches with small drops on the bed elevation between reaches in gradually varied gravity flow. According to the operation concept described above, the canal is blocked at the end of each reach, where a weir is delivering water to the field.

If the canal bed slope is mild, the flow will be subcritical and a backwater curve of type M1 (Mendonça, 1945, Lencastre, 1987) forms. Under such conditions one may admit that the specific energy is constant along the canal length (De Marchi, 1948), than

$$\frac{dE}{dx} = 0 \qquad [1]$$

in which E = specific energy; and x = the distance dowstream. Considering the significant energy components - canal bed elevation, depth of flow and velocity head - this equation, for a prismatic channel, becomes

$$cos\ \theta \frac{dy}{dx} - \alpha \frac{BQ^2}{gA^3}\frac{dy}{dx} + \alpha \frac{Q}{gA^2}\frac{dQ}{dx} = 0 \qquad [2]$$

in which θ = angle of the bottom of the channel with the horizontal; y = the depth

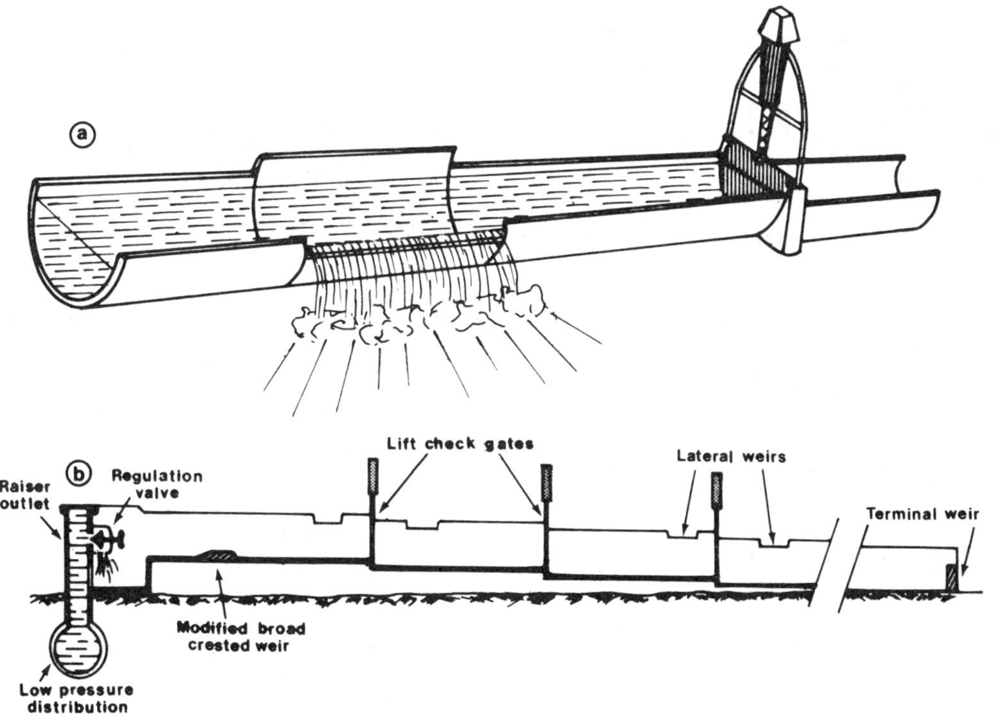

Fig.2 The farm canal with lateral weirs and automatic controlled lift check gates: a) a view of weir and gate; b) the longitudinal profile of the canal

of flow; α = the Coriollis coefficient; B = the width of the water surface; Q = the flow rate in the canal; A = the flow cross section; g = gravitational constant. For gentle slopes one can assume cos θ = 1, and, in general, α = 1. Introducing the Froude number, that allows a more general approach, including cilindrical sections (Mendonça, 1945), the profile slope dy/dx may be calculated by

$$\frac{dy}{dx} = -\frac{\frac{Q}{gA^2}\frac{dQ}{dx}}{1-F^2} \qquad [3]$$

where F = Froude number.

When the canal bed slope is steep the flow will be supercritical, forming a S_1 type backwater curve. Under such conditions the specific energy varies along the distance

$$\frac{dE}{dx} = S_0 - S \qquad [4]$$

in which S_0 = the canal bed slope; and S = the friction slope. Therefore, with the admissible simplifications, the profile slope may be given by

$$\frac{dy}{dx} = \frac{S_0 - S - \frac{Q}{gA^2}\frac{dQ}{dx}}{1 - F^2} \qquad [5]$$

Considering that in the blocked reach the water flows over a rectangular lateral weir of crest length L and long enough to permit discharge with low head above it, than one may calculate the variation of discharge per unit of downstream distance (Manzanares, 1980):

$$\frac{dQ}{dx} = C\sqrt{2g}(y-p)^{3/2} \qquad [6]$$

where C = discharge coefficient; and p = elevation of the weir crest above the canal bottom. The discharge coefficient depends on the crest geometry, on the mean head $(h_0 + h_1)/2$ and on the ratio h_1/h_0 (Lencastre, 1987), according Figure 3.

Introducing equation [6] into the profile slope equations [3] and [5], they become respectively

$$\frac{dy}{dx} = -\frac{\frac{Q}{gA^2}}{1 - F^2}\, c\sqrt{2g}(y-p)^{3/2} \qquad [7]$$

and

$$\frac{dy}{dx} = \frac{S_0 - S - \frac{Q}{gA^2}}{1 - F^2}\, c\sqrt{2g}(y-p)^{3/2} \qquad [8]$$

4. MODELLING

A simple model has been developed to solve

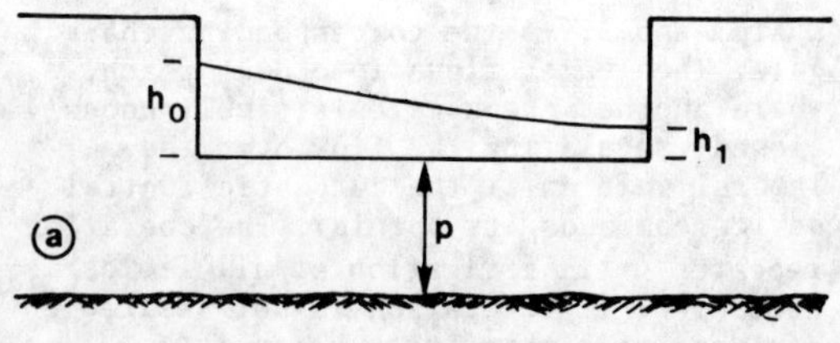

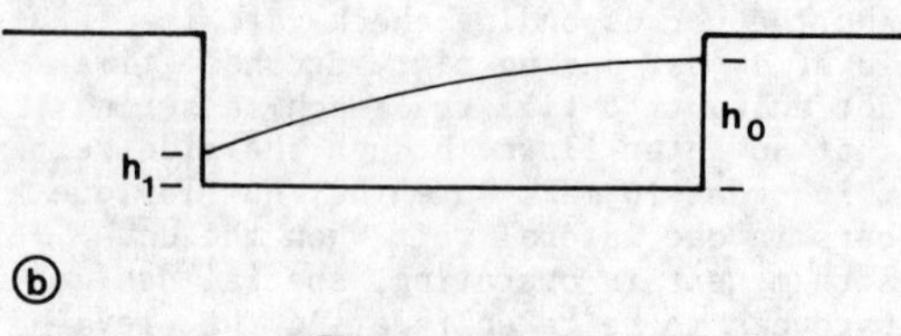

Fig.3 Scheme of discharge conditions for a lateral weir: a) supercritical flow ; b) subcritical flow

the above presented equations [7] and [8] using a finite differences approach combined with an iteractive calculation of the balance

$$Q_{j+1} = Q_j - QL_j \qquad [9]$$

in which Q_{j+1} = flow rate at section j+1; Q_j = flow rate at section j; and Q_{Lj} = discharge over the lateral weir between sections j and j+1. When design discharge Q is imposed convergence is reached for $Q - \Sigma Q_L \leq 1\%$ of Q; if weir crest length is imposed convergence is reached for $L - \Sigma\Delta L \leq 0.01$ m, in which ΔL = the length increments of the weir crest corresponding to the above discharges Q_L.

It is therefore possible to know the law of variation of flow and of the elevation above the crest along the lateral weir, as well as to simulate the water depth as a function of distance, upstream of the blocked section.

The model requires the following input data:

- canal geometry:
  - .bottom width (b) and side slope (s) for rectangular and trapezoidal canals;
  - .diameter (D), for semicircular canals;
- bed slope (S_o);
- roughness coefficients (K_s);
- length of reaches (l_i, i=1,...,n);
- number of reaches (N);
- drop of elevation between reaches (d_i);
- maximum flow depth, downstream (Y_m);
- distance of the downstream edge of the

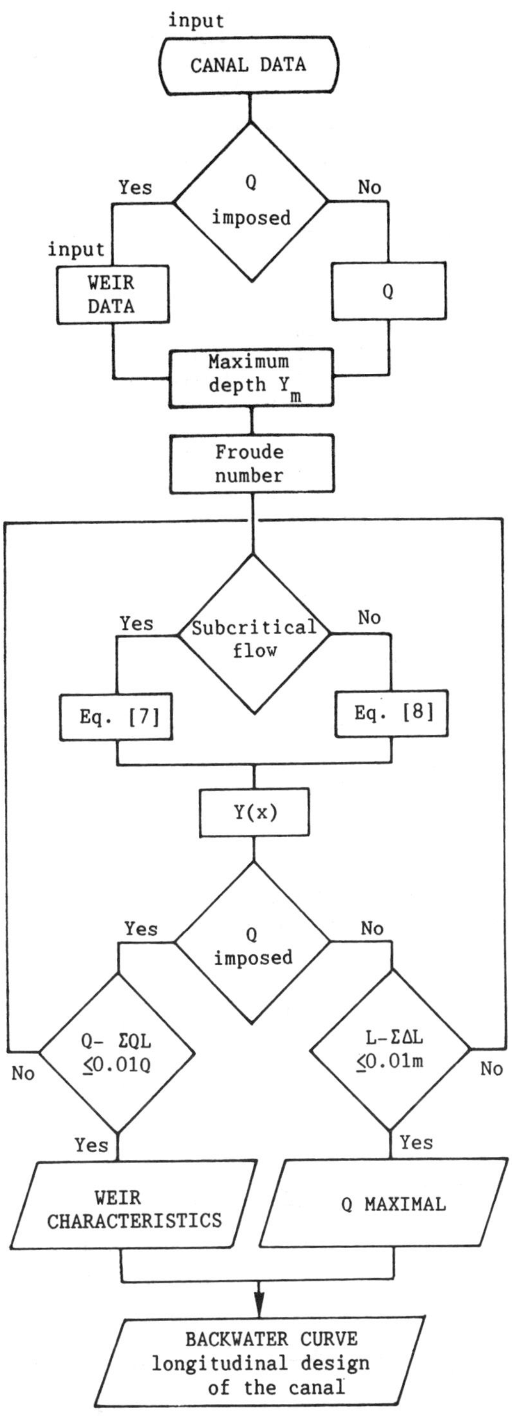

Fig.4 Schematic representation of the model

weir to the check gate (X_i);
- discharge coefficient (c).

Calculation is done using an imposed design discharge or imposed characteristics of the lateral weirs:

- length of weir crests (L_i);
- crest elevation above canal bottom (p_i).

The model (figure 4) starts with the imposed maximum flow depth Y_m and, using the above input data, calculates the Froude number, therefore recognizing if the flow regime is subcritical or supercritical.

The model simulates the discharge law above the weirs and the backwater curves giving the characteristics of the side weir when Q has been imposed, and calculating the maximum discharge when the weir characteristics are known.

Results are presented both in numerical and graphical form and include the y = y(x,Q) curves, the longitudinal profile of the canal bottom, of the side wall crest and of the weir crest.

5. RESULTS OF THE SIMULATION

The model has been tested in the canal described above (section 2).

Discharges were measured through a modified broad crested weir (MBCW) carefully installed in the upstream section. The accuracy of this measurement device is very good for irrigation operation purposes (Replogle & Bos, 1982). According to estimates previously done it corresponds to less then 2.0 %, therefore also acceptable for this purpose. The stabilisation of the required flow rates by handly operating the outlet gate was obtained very easily. But the authors are aware of errors that could be introduced by occasional variations of pression in the irrigation district distributor delivering water to the farm canal, therefore inducing slight variations in discharge during the experiments.

The head above the MCBW, as well as the flow depth along the canal were measured with water stage sensors connected with a pressure transducer and a multichannel data acquisition system. It has been therefore possible to record the backwater curves of the first reaches of the canal for several discharges, ranging from 10 to 60 l/s.

In table 1 and Figure 5 are presented the observed and the simulated heads above the weir crest (W3) for Q = 60 l/s. The results show a good agreement between observed and simulated heads.

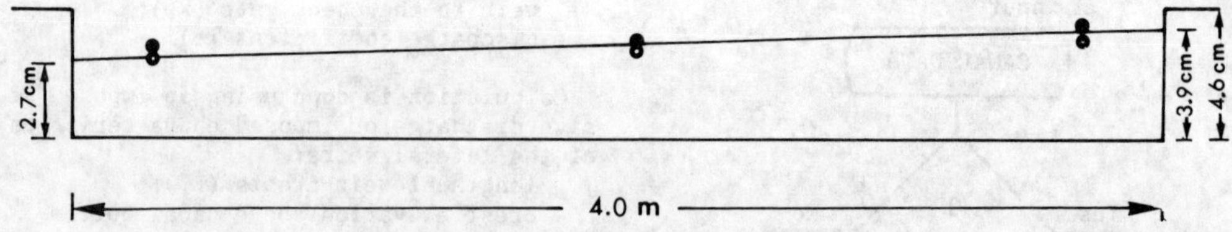

Fig.5 Simulated head profile above crest of the lateral weir W3 for Q=60l/s. Simulated (——) and observed (o - 25/5/88; • - 26/9/88)

Table 1. Observed and simulated heads above the lateral weir crest W3 for Q = 60 l/s.

Distance (m)	Head (cm) Measured 29/04	25/05	26/09	Simulated
-0.30	2.73	2.70	2.60	-
0.00	-	-	-	2.70
0.30	2.76	2.58	3.35	2.85
0.60	-	-	-	2.95
0.90	-	-	-	3.05
1.20	-	-	-	3.20
1.50	-	-	-	3.30
1.80	-	-	-	3.35
2.10	3.19	2.71	3.35	3.45
2.40	-	-	-	3.55
2.70	-	-	-	3.65
3.00	-	-	-	3.70
3.30	-	-	-	3.80
3.70	3.13	3.12	3.99	3.85
4.00	-	-	-	3.90

Table 2. Observed and simulated back water curves for Q = 60 l/s at lateral weir W3 since the begining of the second reach until the end of the third reach.

Distance (m)	Flow depth (cm) Measured	Simulated	
1.2	21.0	21.9	
11.53	21.1	21.3	
19.13	20.65	20.7	
19.395	-	20.6	drop
19.4	-	26.99	drop
19.67	26.6	27.0	
26.12	27.45	27.50	
32.635	28.00	28.10	
33.235	28.75	28.25	W3
35.0	28.75	28.80	W3
36.635	29.39	29.30	W3

In Table 2 and Figure 6 are the observed and simulated backwater curves when flow

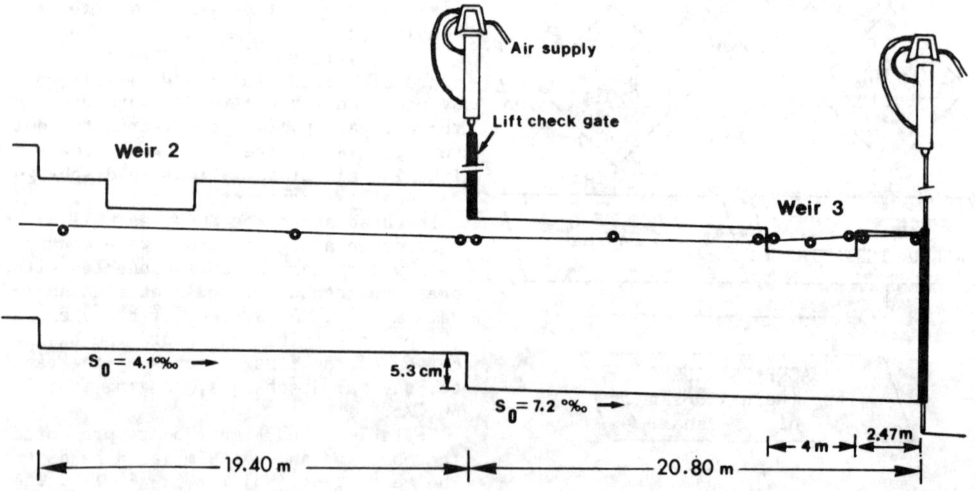

Fig.6 Simulated backwater curve for the canal when W3 is delivering Q=60l/s. Observed values (o)

rate throughout the lateral weir W3 is 60 l/s. Results make evident the applicability of the model as field observed and simulated data agree very satisfactorily.

6. CONCLUSIONS

The simple modelling approach allows an accurate simulation of an on-farm distribution canal with lateral weirs regulated by lift check gates, the canal reaches presenting a drop on the bottom elevation where gates separate these reaches. These positive results led to develop at present a more complete version of the model, allowing other iterative calculations for optimizing several design parameters of the canal. Therefore some of data input will serve only to start calculations and will be changed during the iterative process. A menu type of system will be developed with this design aim.

It can also be concluded that the presented form of side weirs is well adapted for level basin irrigation, where the delivery stream is distributed further inside the basin (see Figure 1). For furrow irrigation it is needed an uniform flow rate along the side weir, then also an uniform head. This only can be achieved with very long weir crests and further tests will be done in the near future.

ACKNOWLEDGMENTS

Research is being carried out in cooperation with the Department of Irrigation of INIA, at Coruche Irrigation Experimental Station and is a part of project NATO-PO-Irrigation (NATO SFS Programme) and JNICT Project 890.86.132.

REFERENCES

Dedrick, A.R. & L.J. Erie 1978. Automation of on-farm irrigation turnouts utilizing jackgates. Trans. ASAE, 21(1):92-96.

Dedrick, A.R., L.J. Erie, A.J. Clemmens 1982. Level basin irrigation. In Advances in irrigation , vol. 1, p. 105-145. New York, Academic Press.

Dedrick, A.R. & D.D. Zimbelman 1981. Automatic control of irrigation water delivery to and on-farm in open channels. In Principles of designing control systems for water resources and irrigation using modern techniques (ICID Symposium, Grenoble), p 113-128. New Delhi: ICID.

De Marchi, G. 1948. Nozioni di Idaraulica. Bologna: Ed. Agricole.

Effekharzadeh, S., A.J. Clemmens & D.D. Fangmeir 1987. Furrow irrigation using canal side weirs. J. Irrig. Drain. Eng.113(2):251-265.

Haise, H.R., E.G. Kruse, M.L. Payne & H.R. Duke 1980. Automation of surface irrigation: 15 years of USDA research and development at Fort Collins. Washington:USDA-SEA Production Res. Rep. 179.

Humpherys, A.S., J.E. Garton & E.G. Kruse 1970. Auto-mechanization of open channel distribution systems. In Proceed. National Irrigation Symposium, p. L1-20. Lincoln, NE.

Lencastre, A. 1987. Handbook of hydraulic engineering. Chichester: Ellis Horwood Limited.

Manzanares, A.A. 1980. Hidráulica geral II - Escoamentos líquidos. Lisboa. Técnica, AEIST.

Mendonça, P.V. 1945. Curvas de regolfo. Canais com leito cilíndrico de pequeno declive e rugosidade uniforme. Lisboa: Liv. Ferin.

Pereira, L.S. 1987. Advances in surface irrigation.Water and Irrigation Review, 7(3):24-29.

Pereira, L.S. 1988. Modernization of irrigation systems: a case of research oriented to improve management. Irrigation and Drainage Systems, 2:63-77.

Replogle, J.A. & M.G. Bos 1982.Flow measurement flumes: applications to irrigation water management.In Advances in Irrigation, vol I, p. 147-217.New York: Academic Press.

Sousa, P.L. 1984. Rega por canteiros de nível. Recursos Hídricos. Vol. 5, nº 2, pp. 73-102.

Sousa, P.L. 1987. Rega por canteiros de nível. In: L.S. Pereira (ed) Gestão da água em regadio, p. 163-191. Lisboa: INIC/Dep. Eng. Rural, ISA.

Land and Water Use, Dodd & Grace (eds), © 1989 Balkema, Rotterdam. ISBN 90 6191 980 0

Création de points d'eau à partir des digues de travaux de conservation de l'eau et du sol*

J.Tarhouni, F.Lebdi, F.Sfar & N.Ennabli
Institut National Agronomique, Tunis, Tunisie

A.Chawachi
Société Nationale d'Exploitation et de Distribution des Eaux, Tunisie

RESUME: L'évaluation quantitative des débits de résurgences apparues à l'aval des digues a montré que l'écrètement des eaux par ce type d'ouvrage ne peut subvenir aux besoins annuels d'un groupement rural d'une dizaine de familles bien que le volume d'eau mobilisé est plus grand. Pour la période humide (Novembre, Janvier) le débit de chaque résurgence dèpasse de loin la consommation journalière de 10 familles, raison pour laquelle le stockage intersaisonnier dans des citernes est proposé. L'analyse chimique a montré que l'eau des micro-lacs est d'une meilleure qualité, sa filtration à travers un filtre sableux permettra de récupérer une eau de qualité acceptable qui sera stockée dans des citernes.

ABSTRACT: Quantitative evaluation of the water flow at the dam level showed that the storage of water by this type of management could not meet the annual needs of a group of a dozen rural families, even if the volume of water stored is more important. For the damp period, Nov. to Jan, the flow of each source was greater than daily consumption of IO families and is thus the reason why storage between seasons in cisterns is proposed. It's filtration through sandy soils resulted in an acceptable quality of water to be stored in cisterns.

ZUSAMMENFASSUNG: Die Auswerküng ergab, dass die Wassermenge unterhabb des Dammes eingefasst, den Bedarf von 10 Familien über das gange fahr hin nicht decken kann. Die ungleiche Niederschlagverteilung ergibt einen uherochüb wahrend der Regenmonate (November his Januar), im sommer ist fedoch im Deficit zü verzichnen. Dem zinfolge wird eine üher saisonate wasserspeicherung im Ziternen vorgeschlagen.

Die chemische Analyse cryab cine reichnde Wasserqualität dieser klein - staüseen, die Reinügung mittels sandfilterung erlaubt die speicherung in Ziternen.

*Ce travail a été fait dans le cadre du projet de recherches agricoles sur les systèmes de production (Coopération USAID)

I. INTRODUCTION

Si pour les centres urbains le taux de desserte en eau potable est à plus de 80% en moyenne actuellement, il n'est pas de même pour les groupements ruraux qui continuent à poser des problèmes sérieux d'alimentation en raison de leur situation éloignée des sources d'eau. En effet, la recherche de l'eau pour subvenir aux besoins quotidiens constitue un souci permanent pour les paysans dans certaines régions de la Tunisie et même dans le Nord où la pluviométrie dépasse les 600 mm. La construction de petits et de micro-barrages permet, en améliorant l'utilisation de l'eau et des sols dans les zones en pente de sécuriser et de créer des points d'eau de réserve. Ces points pourront être utilisés, entre autre, pour desservir en eau potable les habitants de la zone.

L'objectif de cette étude consiste à quantifier les débits de résurgences apparrues suite au micro-aménagements réalisés à "Mazreg Echems" dans la région de Siliana, à étudier leur variabilité temporelle et à caractériser l'eau récupérée d'un point de vue quantitatif et qualitatif.

Ceci nous permettra de se prononcer sur l'exploitation éventuelle des sources.

II. PRESENTATION DE LA ZONE D'ETUDE ET DES AMENAGEMENTS

La zone d'étude "Mazreg Echems" (Gouvernorat de Siliana) est caractérisée par un relief accidenté, une pente forte et un sol nu. Pour le stockage de l'eau et la réduction de l'érosion les ouvrages suivants ont été réalisés :

- des banquettes implantées à contre sens de la pente, distantes de 20 à 30 mètres et d'une longueur de 50 à 80 mètres découpent la zone en plusieurs parcelles.
- quatre diguettes construites au pied du bassin, elles sont destinées à la rétention des eaux de ruissellement. Elles ont respectivement une longueur et une largeur en crête de 45 et de 35 mètres, une hauteur de 4 mètres et un fruit amont et aval de 1,9 et 2,9 respectivement.

Ces ouvrages visent un double objectif :

- rétention du sol en place et ralentissement par conséquent du risque d'envasement du nouveau barrage de Siliana situé à l'aval de la zone.
- amélioration d'utilisation de l'eau et des sols ce qui permet de sécuriser et d'augmenter la production agricole.

III. IMPACT DES OUVRAGES SUR LA ZONE

Tout au tour des banquettes et surtout à l'aval, nous avons noté l'apparition d'une zone humidifiée, sur laquelle il y a eu un développement d'un couvert végétal.

Les diguettes ont permis, outre la formation des micro-lacs, l'apparition des résurgences à l'aval tout le long des cours d'eau barrés par les diguettes.

Ces résurgences qui n'existaient pas avant ces travaux, sont probablement dûes à l'infiltration de l'eau retenue.

Le débit des sources bien que faible, est loin d'être négligeable (cf Figure 1), son exploitation directe ou après filtration permet de répondre même partiellement à certains besoins domestiques de quelques familles rurales.

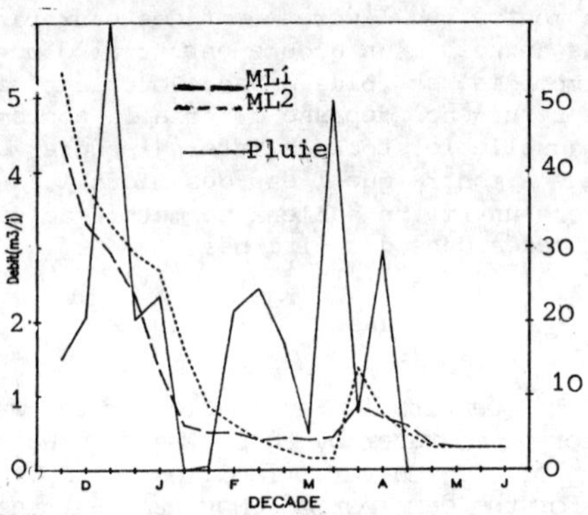

Fig.1 Débits des sources en fonction du temps Micro-lacs 1 et 2 (exemple)

IV. MESURES

Pour chacun des quatres micro-lacs formés suite à la construction des diguettes, nous avons mesuré conjointement le débit de la résurgence la plus éloignée de la diguette et la profondeur de l'eau dans le micro-lac . La variation est très lente, raison pour laquelle nous avons choisi un jour comme intervalle de temps entre deux mesures successives.

4.1. Mesure du tirant d'eau dans les micro-lacs

Deux méthodes de mesures ont été employées :

- Lecture directe de la hauteur d'eau à l'aide des échelles limnitriques installées au préalable dans la retenue.
- Repérages des périmètres mouillés avec des réglettes en bois ancrées dans les berges. Un levé topographique a permis de

déterminer les cotes correspondantes.

4.2. Mesure du débit de sources

L'appareil le plus communément utilisé pour la mesure de faible débit des écoulements de surface est le déversoir triangulaire. Il présente l'avantage d'être adaptable à toute situation de débit en jouant sur l'ouverture de l'angle. Un étalonnage au laboratoire du déversoir utilisé a permis d'exprimer le débit (Q) en fonction de la charge (H) au-dessus de la pointe de l'échancrure comme suit :

$$Q = 0.59H^{2.48} \quad (1)$$
(l/s) (cm)

La compagne de mesures a démarré le 1er Décembre 86 et s'est arrêtée début Juin 87, date de tarissement total des sources.

4.3. Variation du débit des sources avec la pluie

Les ouvrages de CES construits permettent le lamimage et l'écrètement des eaux de ruissellement (cf fig. 1). L'écart T_R entre deux pointes successives respectivement de pluie et de débit constitue le temps de réponse du système aquifère - digue filtrante. Il est d'autant plus grand que la pluie est faible ou que le système est moins perméable.

V. ETABLISSEMENT DES LOIS DE TARISSEMENT DES SOURCES

En régime non influencé (alimentation nulle de la nappe), les courbes de tarissement ont une allure exponentielle (A.Dupont, 81 M.Besbes, 78, G. Shneebeli 78):

$$Q = Q_o e^{-\alpha t} \quad (2)$$

avec :- Q_o = débit initial à l'instant t

- Q = dèbit au temps t_i
- α = coefficient de tarissement qui peut être déduit de l'expression :

$$\alpha = Q_o / V_o \quad (3)$$

où V_o est la capacité d'emmagasinement ou le volume disponible dans la nappe.

L'expression (2) est fréquemment utilisée pour décrire la vidange d'un réservoir souterrain en régime laminaire.

Les paramètres d'ajustement des mesures expérimentales (Q,t) par la loi expotentielle :

$$Q = Ae^{-Bt} \quad (4)$$

pour les quatres micro-lacs sont recapitulés dans le tableau n°1

Table.1 Paramètres d'ajustement des mesures expérimentales (Q,t)

Source du micro-lac	I	II	III	IV
Paramètres				
$A (m^3/j)$	4.24	6.77	4.06	1.77
$B (j^{-1})$	0.30	0.30	0.30	0.30
r^2	0.86	0.92	0.96	0.92
$Q_o (m^3/j)$	5.10	6.70	3.00	1.72
Q_o/A	1.20	0.99	0.74	0.99

Pour les micro-lacs II et IV le paramètre A est pratiquement égal au débit initial Q_o ce qui confirme l'expression (2). Quant au paramètre B, il a la même valeur pour les quatres sources, ce résultat parait logique puisqu'il s'agit de la même formation aquifère.

VI. POSSIBILITE D'EXPLOITATION DES RESURGENCES POUR L'ALIMENTATION EN EAU

6.1. Estimation des besoins en eau

Le groupement rural concerné par cette étude compte une dizaine de familles, il est loin des centres urbains et non équipés. D'après les statistiques faites par la Société Nationale d'Exploitation et d'Alimentation en eau potable (Sonede 87) la norme de consommation spécifique à adopter pour ce genre de groupement est de 13 l/j/hab en moyenne.

Pour une famille de 10 personnes (INS 84) et un coefficient de pointe de 1,5, la consommation de pointe sera de 200 litres/j/famille.

6.2. Mode d'alimentation en eau potable

Le tarissement global des sources a eu lieu au débit du mois de Juin date au delà de laquelle l'alimentation directe à partir des sources n'est plus possible (cf fig.2, et Fig.3). D'autre part à partir de la deuxième décade du mois de Mars leur débit suffit à peine à l'alimentation d'une seule famille, alors que, pour la période du premier Décembre jusqu'au 20 Janvier le débit de chacune des sources est bien supérieur à la consommation d'une dizaine de famille (cf fig.2 et Fig.3). Pour subvenir aux besoins en eau de cette collectivité, il ne faut pas compter sur l'écrètement des eaux de ruissellement par les ouvrages de CES mais il faudrait prévoir des citernes de stockage intersaisonnier. Ces citernes de collecte de l'excès de débit en période humide seront utilisées comme de puisard en période sèche. Cette technique n'est pas nouvelle en Tunisie, elle était pratiquée un peu partout et notamment au Sud-Est où le nombre de citerne est très élevé.

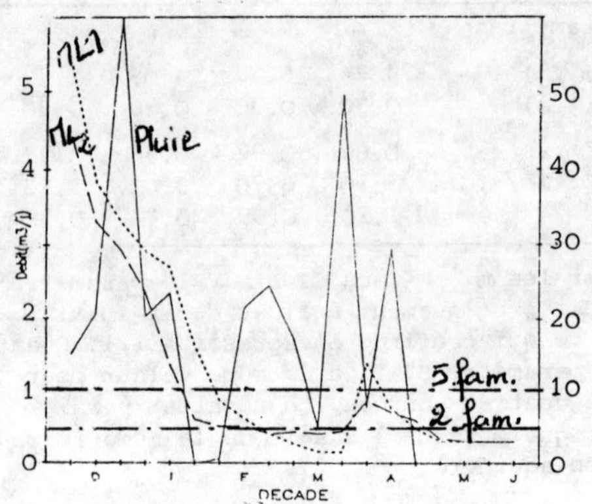

Fig.2 Variation des débits des sources à l'aval des micro-lacs 1 et 2

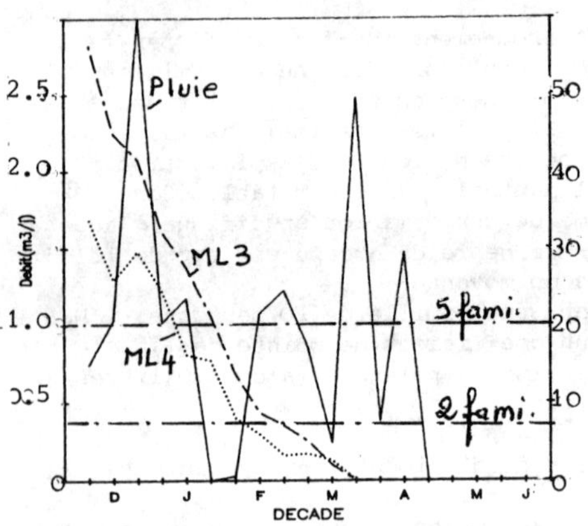

Fig.3 Variation de débit des sources à l'aval des micro-lacs 3 et 4

6.3. Détermination de la capacité de stockage des citernes

La confrontation de la courbe traduisant le tarissement d'une source au besoin en eau d'une famille, de deux, de trois familles (Fig.2 et 3) et en se fixant comme objectif la satisfaction sur toutes l'année on peut déterminer la capacité de stockage des citernes et le nombre de familles qui pourront être desservies.

Le nombre de familles qui pourront être alimentées à partir d'une source est tel que :

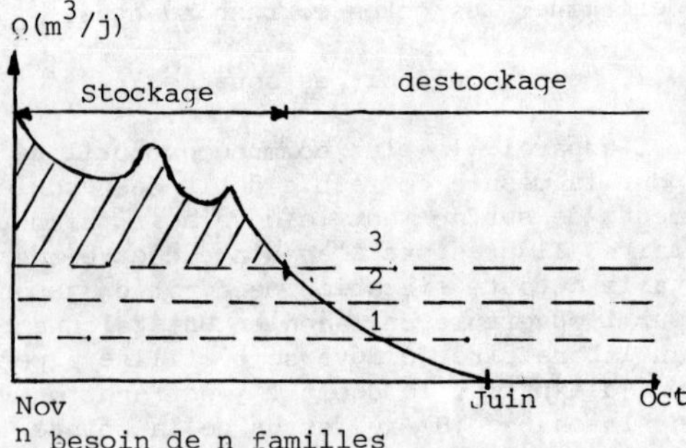

Fig.4 Estimation de la capacité de la citerne de stockage

Volume stocké en période humide = volume destocké en période sèche.

Table.2 Nombre (N) de familles désservies sur toute l'année par chaque micro-lac (année de mesures) et capacités des citernes de stockage

M.CN°	V.T.A.S	N	V.D	C.C
I	280	3.85	70.000	210
II	300	4.11	75.00	225
III	180	2.47	45.00	135
IV	110	1.50	27.00	82
Total =	870	11.92	217.00	652

Soient environ 12 familles au total

M.C:Micro-lac,V.T.A.S :Volume total(m^3) apporté par les sources
V.D:Volume pris directe, C.C :Capacité de la citerne.

La pluviométrie de l'année de mesures (86-87) est d'environ 440mm, valeur qui est supérieure à la moyenne de la zone (410mm).
Une étude statistiques de la pluviométrie annuelle sur 18 années nous a permis d'estimer le nombre de familles qui pourront être désservies pour différentes périodes de retour :

Probabilité	0.10	0.50	0.90
Apport annuels estimés(m^3)	1170	740	550
Nombre de familles satisfaites	16	10	7.5

VII. QUALITE DE L'EAU

L'analyse chimique de l'eau des sources montre qu'il s'agit d'une eau sulfatosodique, trop chargée en résidu sec et ne répondant pas aux normes de l'OMS. Quant à l'eau des microlacs elle est d'une qualité meilleure, ses paramètres physicochimiques sont à la limite des normes fixées par l'OMS (cf Tabl.3). Il sera moins onéreux de procéder à un traitement rustique de cette eau plutôt que celle des sources. Pour la rendre potable il suffit de la filtrer à travers des alluvions sableuses ou graveleuses.

Table.3 Résultats d'analyse chimique

	TB	Ca mg/l	Mg mg/l	Hco mg/l	So mg/l
ES	0.7	256	383	423	2042
ER	24.0	67	26	90	280
NOMS	25	200	150		400

ES:Eau de la source,ER: Eau de la retenue, NOMS: Normes OMS*, TB: Turbidité,

VIII. PROCEDE DE TRAITEMENT PROPOSE

Pour purifier l'eau et la débarasser des matières indésirables, nous avons envisagés des améliorations au niveau de la retenue. Le principe de cette opération consiste à (cf.Fig.4).

- revêtir la surface de la retenue par des alluvions grossières (alternance de couches de sable et de graviers) en vue de filtrer l'eau.
- placer des drains de captage dans la couche de sable et les raccorder à un collecteur commun qui traverse la digue.
- construire une citerne de stockage en vue de la régularisation muni d'un regard de visite.
- canaliser toutes les eaux de ruissellement vers la retenue de la diguette.

Ce procédé sera complété par une chloration. Il nécessite un entretien systématique et notamment de la couche filtrante.

*OMS:Organisation Mondiale de la Santé

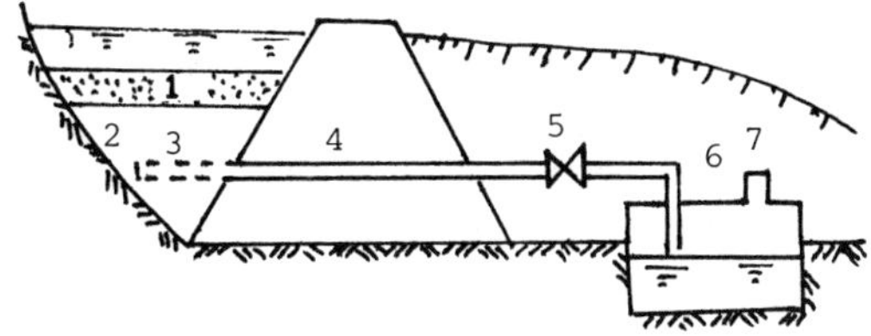

1. Couche de gravier 20cm
2. Couche de sable 40cm
3. Drains de captage
4. Collecteur d'acheminement
5. Robinet vanne
6. Reservoir de stockage
7. Regard de visite (puisard)

CONCLUSION

Les digues réalisées interceptant les eaux de ruissellement d'un bassin versant d'une superficie de 7.5 hectare environ, elles ont permis en plus de l'augmentation du potentiel hydrique du sol, l'apparition de points d'eau.

Le volume total récupéré au niveau des résurgences durant la période Novembre - Mai permet l'alimentation de :

- 7.5 familles pour une pluie annuelle de période de retour 9/10.
- 10 familles pour une pluie annuelle égale à la moyenne.
- 16 familles pour une pluie annuelle de période de retour de 1/10.

Le stockage intersaisonnier des eaux dans des citernes, à l'instar de ce qui fait un peu partout dans le pays, permet alors d'assurer l'alimentation d'un groupement rural sur toute l'année.

REFERENCES

Besbes,M. 1978. L'estimation des apports aux nappes souterraines. Thèse de Doctorat es-sciences naturelles (256P.). Fontainbleau - Paris.

Dupont,A., 1981: Hydrologie - captage et traitement des eaux (262p).

Schneebelli,G. 1978. Hydraulique souterraine (357p).

Société Nationale d'Exploitation et de Distribution des Eaux (SONEDE), 1986; Rapport statistique (interne).

Institut National de Statistiques(INS), 1984 : rapport de ressencement.

Land and Water Use, Dodd & Grace (eds), © 1989 Balkema, Rotterdam. ISBN 90 6191 980 0

Computer aided optimization of rural engineering project-size: Application to irrigation earth dams design

I.Trueba, R.Dal-Ré Tenreiro, A.García & F.Ayuga Téllez
Universidad Politécnica de Madrid, Spain

ABSTRACT: Earth dams projects for agricultural purposes have gone throught a spectacular development during the last few years. In particular, in Spain several fulfilments of small irrigation and water supply projects for rural villages have been accomplished. In general they handle dams which have under 30 m high and reservoirs which obviously vary with the basin topography but usually between 0,5 and 15 Hm3. size.
Considering a creative perspective in the rural engineering projects formulation the generation of alternative solutions is a normal practice. The multi-criterion analysis of each of them allows the designer to choose the most suitable according to some assumptions, information, criteria and design method. In this process, the definition of the optimal dam size, considering the topographycal and economic conditions is one of the design objectives.
This paper has been prepared in two steps: In the first one, the developed method allows the design and the calculation of the dock, spillway and the water outlet and additionaly identifies, quantifies and estimates the corresponding costs and benefits. In the second stage an automatic generation system of alternative solutions has been prepared in terms of the size variable (height or capacity) which sets the media and marginal internal rates of return of every solution and represents graphically the optimization curves of this parameter in terms of the opportunity cost of capital.
All of that contribute as far as possible to the improvement of the techniques and methodology of the rural engineering projects design.
The process has been operated throught a software-package which runs on a PC or compatible computer.

INTRODUCTION

Within the rural engineering projects scope the application of the "Computer aided design" is getting more and more usual. There are several programmes available in the software market that are commercialized as design programmes, they are, however, drawing specific programmes.
Carrying out a remarkable differences profound analysis is not the purpose of this paper. Nevertheless, it can be stated that design is a bit more sophisticated than - drawing and tha latter only cooperate to the better representation of the former.
Projects designing tries the solving of real life specific problems through a complex process of decision making always considering and resources limit constraints. Therefore, it constitutes a creative activity because it has empirically proved that human problems and specifically the rural world ones have not an only solution. It is about an indeterminate problem with difficulties in its raising as the information, the technology, the obstacles and constrains are of difficult establishing with total definition. That is, they are "ill defined problems" (W.J. Mitchell,1977).
The present paper tries to contribute to a certain degree to strenghthen the CAD in the projects dimensioning field (size, dimension). Technology analysis, organization, location, temporality, multiple purpose, etc., which are relevant design aspect are not studied in depth here. However, a methodology with computer treatment that allows to define minimum, optimum and maximum size of rural engineering projects conditions are developed.

2 PROJECTS DESIGN

With project design, real life problems

are tried to be solved. However, when facing a specific rural world problem it - gets empirically proved that different solutions exist. That is, not a mathematic problem with one solution but an indeterminate one with a range of solutions equally valid is tried to be solved.
Thereof, it can be said that design is intimately related with creativity.
Design as creative action, with different solutions to a problem was already known in the Antiquity. Thus, Aristóteles laid out the design for a city through multiple alternatives generated according to morfologic model, that he explained by a biological comparison. Leonardo da Vinci's works on design alternatives for church plans or Leibnitz's on machines designs are also - known. And nowadays there are remarkable contributions in this field, among them it has to be mentioned the ones by the Spanish architect Rafael Leoz, creator of the module HELE and W.J. Mitchell with his isometric shape grammar and its applications to computer aided design.
Alternative solutions must be technically feasible and comply with a series of conditions imposed by the rural world. They cannot be utopian but compatible with reality.
Now then, how can it be designed? Designing is carried out intuitively or through a method? What procedure exist for the development of this activity? These are questions usually set forth, above all when it is not known where to start.
The so-called "design cycle" gets identified with a first answer. Three stages can be considered in this cycle:
These stages are three:
- Generation of alternatives. Synthesis.
- Analysis of options. Criticism.
- Alternatives choice. Selection.
The cycle can be represented the following way:

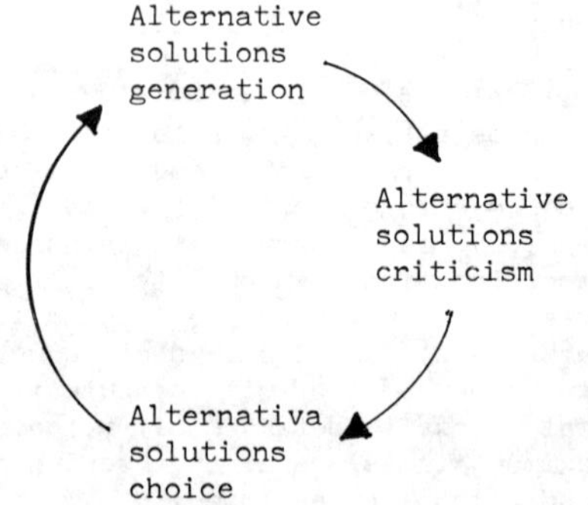

Circle-shaped diagram

Alternatives compatible with reality get generated in the first stage. A review is made of all of them in which between others, aims and constraints are taken into account assessing the alternatives some way in order to make a comparison among them and - choose finally the best one, getting the - cycle closed by the generation of new alternative solutions in the following design level; and like that all the way up to detalilled and definitive design.
Through the spiral diagram of the design - cycle, introduction of the consecutive cycles in the different design levels can be shown:

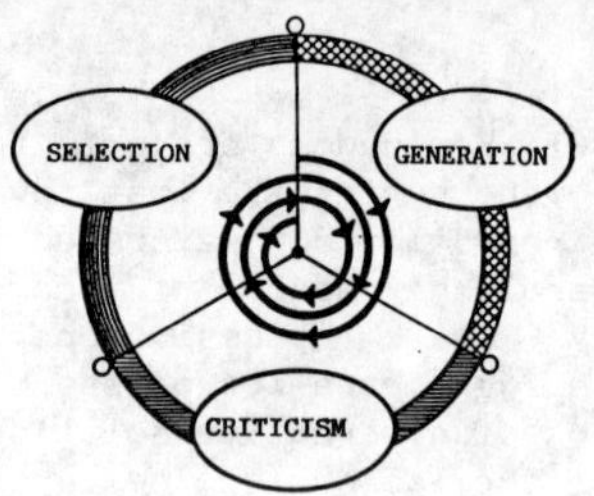

In every one of these generation, criticism and selection stages it has to be available:

1. Information
2. Hypothesis
3. Criteria
4. Method

This information, hypothesis and method - need not being coinciding in every stage - of the cycle.

3. ALTERNATIVE SOLUTIONS GENERATION AND ANALYSIS

This is the stage where it get more clearly shown design creative activity.
How are solutions generated?
Alternative solutions generation can be carried out intuitively or getting based on experience, but a methodology is better to be followed. Some are as old as the morphologic method already employed by Aristóteles, but we are going to follow an alternative solutions generation system based on a series of questions on design variables.
Alternative solutions can be generated by: space (location), dimension, technology, - organization, time and priorities (when - there are multiple aims), in an individual or combined way.
Now then, what are the questions to be asqued according to each variable? These questions and variables are shown in the following
To generate alternative solutions for the variable space the questions where? Where is the dam going to be located? will be ma

de. For the dimension variable, the question would be, How much? How much is it going to be produced? What area is going to become irrigable perimeter?
For technology, How? How is it going to be irrigated? By sprinklers, by dropping, by gravity, etc.?
For time, When? When is the transformation going to be carried out? which is going to be the execution programme?.
For organization Who? Who is going to be - charge of the working of the project? ¿An irrigators community? Who are going to take advantage of the irrigation?
For priority, How much from each aim?. For example, a dam with three goals; supplying water for irrigation, electricity generation, flood control is going to be carried out. One or another alternative will get - generated according to the percentage attributed to every aim.
As it can be observed all variables come - defined by a key question.
Some of this questions may already have answering in terms of reference imposed by the sponsor, in whose case there would not be alternatives generation for that variable.

3.1 ALTERNATIVE ANALYSIS AND SELECTION

As we have already said for analyzing and selecting an information, hypothesis, criteria and some methods are necessary. There are some different systems so to value the advantages and disadvantages of every alternative solution, like:

DESIGN ALTERNATIVES ANALYSIS METHODS

1. Cost-benefit analysis
2. Cost-efficiency analysis
3. Multicriterion analysis

Cost-Beneficт Analysis

It is an economic balance including identifying, attribution, cuantification, valuation and comparison of costs and benefits. That is, the project advantages and disandvantages for an organization, an enterprise, or the economy as a whole. The valuation has an economic-financing meaning and must be measured in real terms (without inflation).
The valuation gets carried out through some profitability indicators, that is TIR, Internal Rate of Return (IRR); VAN, Net - Present Value (NPV); relation B/C, Benefit/ Cost Ratio; and getting based on the capital opportunity cost.
The selected alternative will be the most profitable one, that is the one given by a maxium NPV.

Cost-Efficiency

This analysis is carried out when benefit can be identified, attributed and quantified but it cannot be valued. Through the Net Present Value Cost (NPVC) discounted - from the capital opportunity cost, alternative options can be economically compared if they have the same benefits.
The alternative selected will be the one with the lowest or minimum NPVC.

Multicriterion analysis

Based on an analysis with multiple perspective, criticizing every alternative in terms of different criteria simultaneously. These criteria may be:

- Tecnic criteria
- Economic criteria - financing ones.- Export socioeconomic ones.
- Social criteria: workers, population basic necessities, rent distribution.
- Environmental criteria: vegetation, fauna, climate, earth, water, landscape.
- Esthetic criteria: proportion
- Others: functionality, flexibility, security in work, etc.

Each criterion has a different weight therefrom a pondering coeficient is applied. Alternatives are given a score according to those ponderings, getting a total score after considering those coeficients. The alternative which has the best scoring will be the optimum.
This analysis due to its having a strong subjective component, is convenient to be carried out independently by several design team experts and later on put in common.
These analysis alternative systems are not the only ones but they really are the most usual, we are going to develop in detail the cost-benefit analysis of them which is the one employed for the optimization line.

THE COST-BENEFIT ANALYSIS

With this system, cost and benefits of every alternative solution, produced along - the working life of the project, get identified, quantified, valued and compared.
As cost is identified all those goods and services compromised in the carrying out of the project and as benefit all those - goods and services generated by the project.
COST = project negative aspects
BENEFIT = project positive aspects
Followingly costs and benefits of a project get chronologically enumerated.

Costs identification

C-1 Project fees
C-2 Permits and licences
C-3 Investments (original and deferred ones)
C-4 Works direction

C-5 Operation and Maintenance
C-6 Working capital
C-7 Loans paying-off

Benefits identification

B-1 Production brute value
B-2 By products Brute value
B-3 Waste values (When carried out deferred investiments)
B-4 Residual values (at the end of the project working life)
B-5 Subsidies
B-6 Loans

The valuation of all of them is carried out in constant monetory units pesetas in reference to the project zero year.

Those costs and benefits are the ones corresponding to the so-called situation "without project" which do not coincide with the situation"before" the project, as a future situation of that situation has to be foreseen, in case the project is not carried out.

This can be observed in the following graphic:

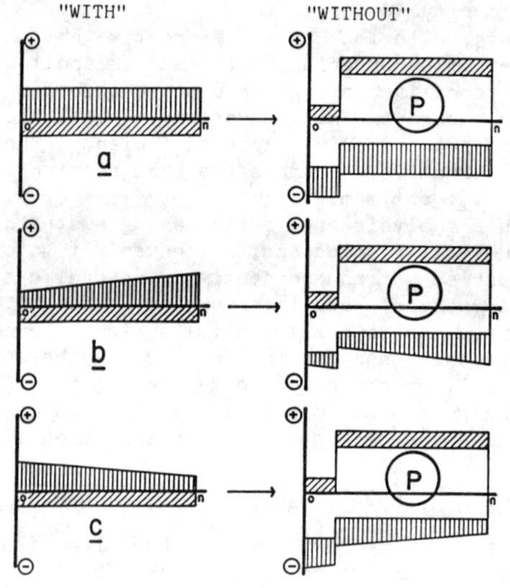

In which can be seen some examples from the ones observed in the situation "with project" and the situation "without project" in case costs do not get developped and benefits remain the same, grow or diminish.

All those costs and benefits, identified, quantified and valued for every alternative solution and for every year of working life have to be compared, taking into account that they are future magnitudes having to be homogneized applying discounting technics.

COMPARISON BETWEEN TWO ALTERNATIVES

Although the IRR (Internal rate of Return) is one of the most employed projects profitability indicators, its employment in the selection of mutually excluding alternatives must be done cautiously. It is not true that the alternative given by a higher IRR is preferable, as it could seem firstly. If we draw the VAN-r curves for two alternatives A1, and A2, it may happen:

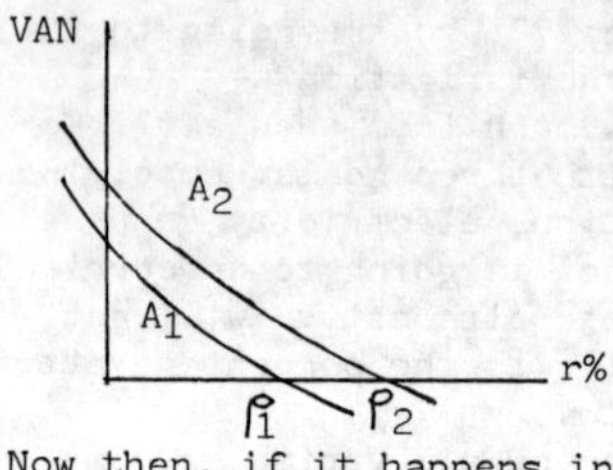

VAN A2>VAN A1
In this case
ρ2 > ρ1
A2 better than A1 from the economic point of view.

Now then, if it happens instead that:

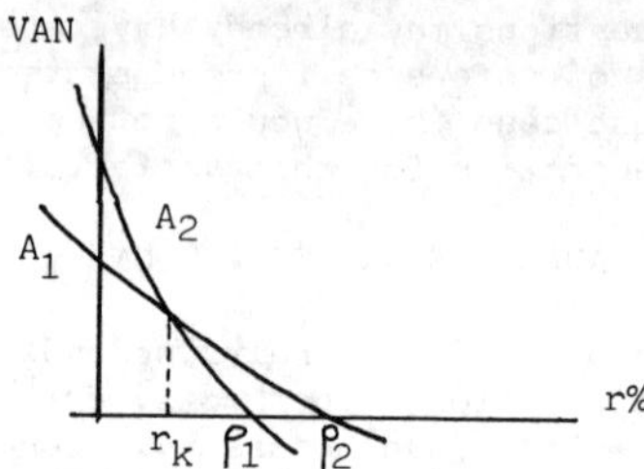

Although ρ1 (of A1) is superior to ρ2 (of A2) it happens that for r values inferior to rk (equilibrium rate) VAN A2<VAN A1. Therefrom it is necessary cost, this way if the c.o.p. is smaller than rk the best alternative would be A2 despite its having a TIR inferior to the A2 one. That is, the best alternative from the economic point of view will be the one that gives a higher Net Updated Value discounted from the developper Opportunity Cost.

4. DIMENSIONING

Within the project preliminary design, one of the most relevant characteristics is its dimension or size.

Due to that when formulating a project, and between the first questions which demand logical answering on the formulating team side are these two:

- What is the most adequate dimension to develop the activities which most rationally and economically accomplish the project aims?
- From all the dimension alternatives mutually excluding which is the most promising one?

To answer these questions the explained above can be applied.

In a very general way dimension can be defined as a quantified function of one or several technic and/or economic variables. A variable to define the dimension or size of rural engineering projects is the production capacity, being menat by that the product physical volume obtained per time unit. Fot other cases the physical capacity as a dam water content or an irrigable area in hectareas.

Other projects examples in which dimension gets defined through another variable are:

. Area variable
  for example. Transformation of 100 hectareas in irrigable land.
. Height variable
  for example. Construction of a reservoir with a 20 m. high dam for irrigation.

5. ECONOMIC METHODOLOGY FOR THE DIMENSION OPTIMIZATION

If we call the project dimension X and r the dicounting rate made to coincide with the opportunity cost of capital and for a given group of parameters, as technology, organization, etc., the X dimension project Net Present Value (NPV).

$$\left. \begin{matrix} x \\ r \end{matrix} \right| \text{VAN}= \sum_{i=0}^{i=n} \frac{b_i(x)}{(1+r)^i} - \sum_{i=0}^{i=n} \frac{c_i(x)}{(1+r)^i}$$

Being $b_i(x)$= year i benefits for dimension x.
$c_i(x)$= year i cost for dimension x

The X dimension IRR yielding inner rate - will be r(x) that complies with the condition:

$$\sum_{i=0}^{i=n} \frac{b_i(x)}{(1+(x))^i} = \sum_{i=0}^{i=n} \frac{c_i(x)}{(1+(x))^i}$$

And the marginal internal rate of return (marginal IRR) of the X dimension project will be r(X), provided that $b_i(X)$ and - $c_i(x)$ are continued and derived:

$$\sum_{i=0}^{i=n} \frac{db_i(x)}{(1+\rho_1(x))^i} = \sum_{i=0}^{i=n} \frac{dc_i(x)}{(1+\rho_1(x))^i}$$

All this indicators that can be represented in some coordinated axis, where abcissas are the dimension and ordinates the discount rates

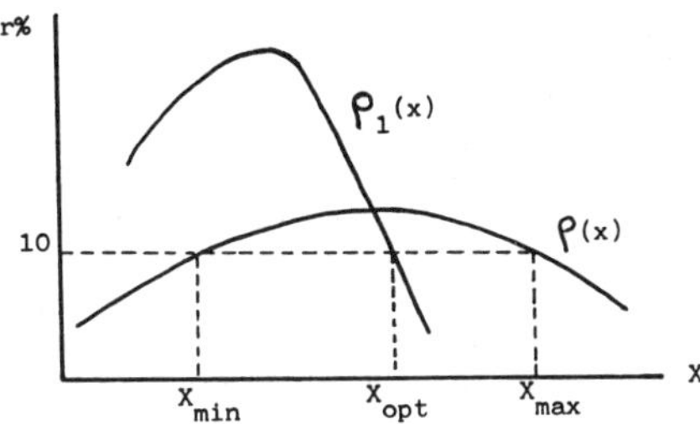

A detailled analysis of the figure allows to assert the following: Let's suppose - that the opportunity cost of capital is of 10%.: the cut point of the straight line r=10 with the upward branche of (x) is the project minimum dimension, why?, because for (x) lower than X_{min} the IRR ( ) is inferior to the opportunity cost of capital.

The cut point of r=10 with the downward - branche of the curve (x) is the maximum dimension due to (x) X_{max} : the IRR OCC. The cut point of r=10 with the curve that represents the marginal IRR is the optimum dimension. To explain it we are going to look at the graphic more deeply:

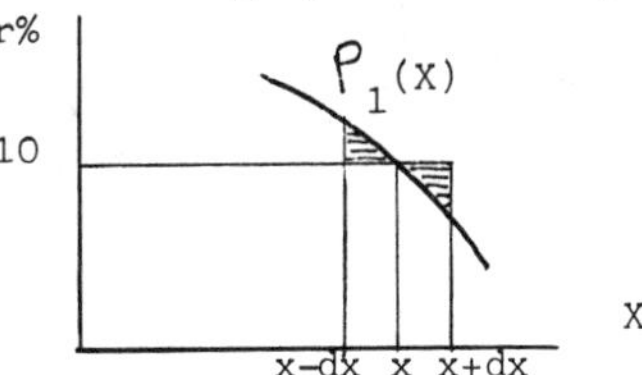

Let's suppone a dimension lower than X but infinitely close to X-dX. The dimensions difference or the dimension increasing - (X-(X-dX)) has a marginal IRR of 10%, that is that the NPV of that increasing is positive, however when X passing to X+dX the NPV < 0. That makes clear that X is the - optimum dimension because it supposes the maximum NPV of the whole project. That is, the project economic yielding improves if dimension from X-dX to increases and worsens if it increases from X to X+dX.

What happens if the OCC is of 20%:
(see figure)

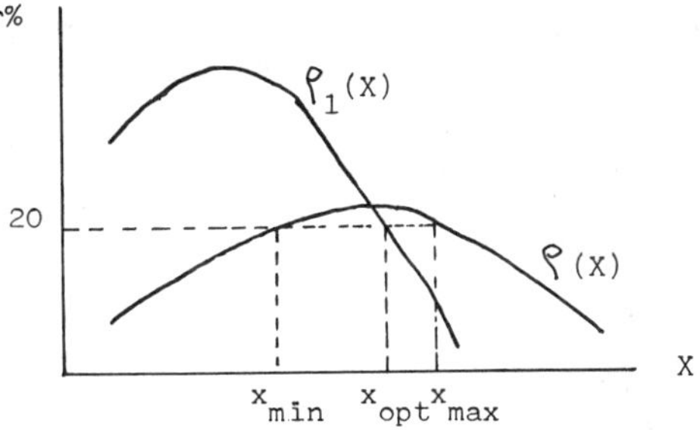

It should be a X_{min} X_{opt} and X_{max}

What happens if the OCC is of 25%:

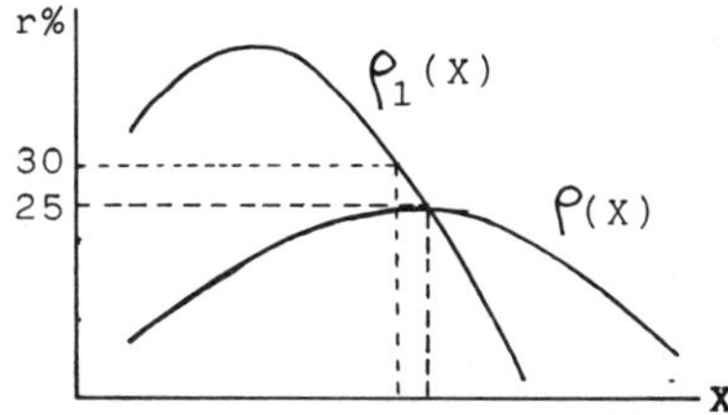

In this case x will be at some times, maximum, optimum and minimum dimension.
If the oportunity cost of capital is 30% (see figure) and the maximum of IRR is 25% it ins't relevant to implement the project in any case. Say, there is not maximum, minimum, optimum dimension.
Apart from the IRR average and marginal graphics we can also represent NPV curves, in which the following can be observed:

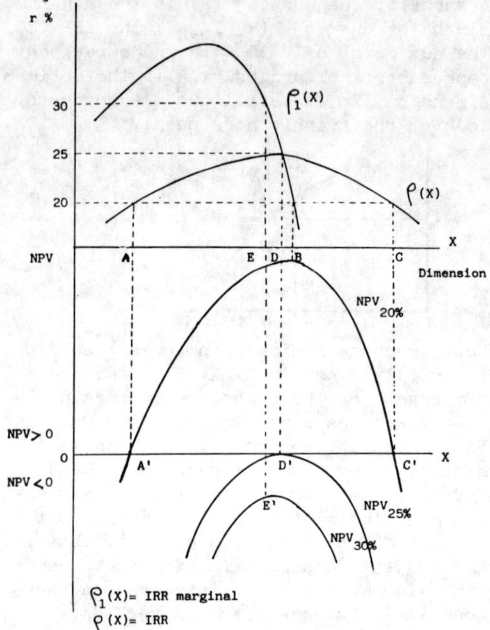

for a OCC of 20% the minimum dimension is A as dimensions lower to this one have a NPV 20% under zero, the optimum dimension is B corresponding to a NPV $_{20\%}$ under zero.
For a OCC of 25% the dimension D is at the same time maximum, optimum and minimum corresponding to a NPV 25%=0 and maximum, for all the rest of dimensions the NPV 25% is negative.
If, as it is usual in projects, we have a top dimension, that we cannot overpass due to technic or other reasons, for example, a dam we cannot go over 50 m. because the valley has no more height.

Optimum dimension will coincide with that maximum as it gives us the highest NPV for every Oppportunity Cost of Capital.
(see figure I)

6. COMPUTER TREATMENT

This methodology, for its being effective, is necessary to give it a computer treatment, simply carrying out the adequate software. Personal computers (PC) can be employed and for simple cases in a few minutes the best alternative can be get.

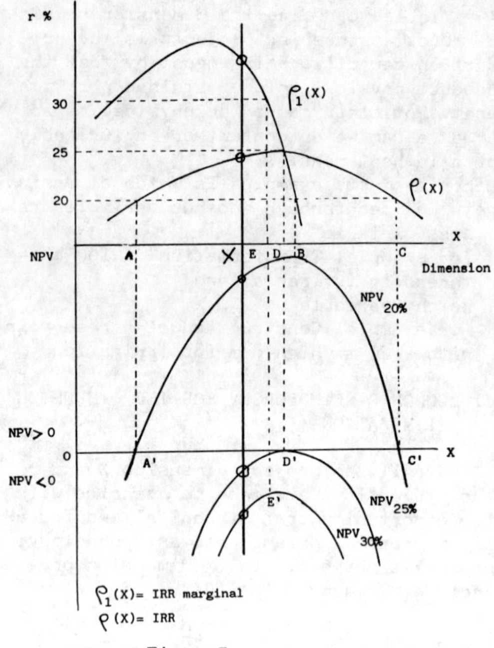

Figure I

7. APPLICATIONTO THE DIMENSIONING OF EARTH DAMS.

In the Escuela Técnica Superior of Ingenieros Agrónomos of Politechnical University of Madrid, the programme DEMAG has been elaborated which allows to get the optimum dimension for the case of earth dams employed only for agricultural use and more specifically for irrigation land.

7.1. HYPOTHESIS

For the development of the aforesaid programme a series of hypothesis have been the starting point, which are:
- The reservoirs and dam places processed through the programme must be of adequate geology for the construction of the dam, - that is, to have impreviousness assured.
- Adequate material for the construction of the dam in quantity enough for any dimension subject to be studied, and at a distance from work not as high as to make it not viable, is available.

It is also possible to optimize the dam - without taking into account economic benefits derived from their construction with investment, operation and upkeeping costs. Investments are the main cost of the dam - project. They have been divided into three parts which are: the material cost of the

dam already done, the cost of the concrete spillway and the cost of the bottom pipes and draunage. Operation and upkeeping costs of these elements are also taken into acco unt, as well as the irrigation implantation and replacement costs and the basin - reafforestation one in case it is necessa ry.
For each one of these groups it has been necessary to design a subroutine in which study necessities (available data of very varied scope, simple enters and in low ti me, etc.) and the strictiness of the latest knowledge on the subject could be taken in to account.
Every one of these subroutines and the hypothesis and models selected will be described.
- Runoffs calculation.
The basic calculation method is the "Adjusting rate" (Dal-Ré and col., 1986) based itself on the "Bend Number" from the USSCS contrasted with some other one.
The method is valid for small basins son gauged, as are the majority of the ones in cluded in the work. The global data can be introduced if it is known through another source.
-Reservoir capacity.
It is derived from the areas closed in by the level curves available (one is enough). An average laminar evaporation according to the annual average precipitation is taken into account.
The strictness of programme results is highly conditioned by this calculation and therefore a good vessel topograpgy availability is fundamental, it does not condition the programme though.
- Dam volume calculation.
It is carried out according to aproximate formulas in terms of the closed form and dam slopes. These slopes can be assumed by the programme or decided by the user. In this second case they can be verified through a stability calculation subroutine based on the Bishop method with some variations.
- Spillway calculation.
The spillway can be known in dimensions or not. In the second case the necessary data get calculated in terms of maximum avenue, As much in a case as in the other the different heights of the dam demand different investment calculations.
Maximum avenue can be considered as data, but there is a subroutine which allows its calculation in terms of the basin characteristics. This subroutine is based on the american hydrologic system HYMO.
- Taking and bottom drainage calculation.
The bottom taking is get through the pipe hydraulic calculation in the most unfavourable situation, acting as taking or as drainage. The same way a pipe protection subsytem which can be concrete beam or visited galery is taken into account.
- Operation and maintenance costs.
They get introduced as an annual cost percentage of every investment great groups.
- Irrigation systems investment costs.
This cost must be added which will show - all de works necessary to carry the water to every countrymen plot. If the dam was not for irrigation zero will be introduced. Every exploitation costs are taken into - account when considering as benefits exploitation cash-flow.
It is believed that all these data are - known and constant be the dam size what - may. If they are unkown or variable an aproximate value can be used, as it will do - comparatively for the optimization. However in this case profitability data indicated by the programme will also be aproximate being the optimum valid though.
- Replacement xost.
The programme assumes automatically replacement and upkeeping costs in terms of irrigation systems investments.
- Reaforestation costs.
In case of carrying out basin reforestation the programme assumes some average costs - and benefits derived from the reforestation that get added to the irrigation ones.
Once get the annual costs and benefits they are processed by means of adequate subroutines in order to get flows for the different prefixed dimensions.

7.2. OPTIMIZATION METHODOLOGY

A programme which carries out the graphics is employed:
1.- Dimension - average IRR marginal IRR
2.- Dimension - NPV (OCC)
3.- Dimension water cubic meters price.

With these graphics and the optimization methodology set forth in detail in the begining dam optimum dimension for every closing is obtained and it is possible to compared closings between themselves.

7.3. CONCLUSIONS

Through the DEMAG programme a great quantity of alternative solutions for agricultural use reservoirs projects with separate materials dams can be analized and valued simultaneously and in a few minutes.
This provides the engineer with a poerful tool and rationalizes the decision making process.
The DEMAG programme is in itself used as - integrating part of a more complex system which will give place to an irrigation engineering expert system.
Several applications of the computer program DEMAG has been processed by the autors

in order to define the optimal dimension of irrigation dams.
This studies has been developped in the Dpto. of Proyectos y Planificación Rural Of Polytechnical University of Madrid. (Spain).

BIBLIOGRAPHY

Evans, M. Garzia,M. Trueba,I. et al.1986. Guide for training in the formulation of agricultural and rural investment projects. ESPT. FAO.

García y García, A. 1986..Diseño óptimo de presas de materiales sueltos para uso agrí cola. Tesis Doctoral. E.T.S.A. Madrid.

Gittinger, J.P. 1982.Análisis económico - de proyectos agrícolas. 2ª ed. Ed. Tecnos. Banco Mundial.

Harberger, A.C. 1976. Project evaluation. Collected Papers. University of Chicago.

Marco Gutiérrez, J.L. 1982. Generación de alternativas en el diseño de sistemas de proyectos tipo de centros de bombeo de - agua para riego. Tesis Doctoral. E.T.S.I.A. Madrid.

Mitchell, W.J. 1977. Computer-Aided Arquitectural design. Petrocelli/Charter. New York.

Murther, R. 1968. Planificación y proyección de la empresa industrial. Ed. Tecnos asociados. Barcelona.

Muther, N. 1981, Distribución en planta. Ordenación racional de los elementos de - producción industrial. Biblioteca de direc ción, organizacion y administración de em- presas. Ed. Hispano. Barcelona.

Trueba Jainaga, J.I. 1975. Evaluación de proyectos agrarios. Instituto de estudios fiscales. Mº de Hacienda. Madrid.

Trueba Jainaga, J.I.; Marco Gutiérrez, J.L. 1985. 1985. Proyectos agrarios y de desarrollo rural. Servicio de publicaciones de la E.T.S.I.A. Madrid.

Trueba Jainaga, J.I.; García y García, A.; Ayuga Téllez, F.; Dal-Ré Tenreiro, R.1986. Aplicación de la informática a los estudios de viabilidad de regadíos. TECNICAE. Polytechnical University of Madrid. Spain.

Land and Water Use, Dodd & Grace (eds), © 1989 Balkema, Rotterdam. ISBN 90 6191 980 0

An initial comparison of three methods of producing short term irrigation schedules for hosereel-raingun irrigation systems

M.J.Upcraft, D.H.Noble & M.K.V.Carr
Silsoe College, Silsoe, Bedford, UK

ABSTRACT: Irrigation scheduling for a hosereel-raingun system involves answering three fundamental questions:

1. When to irrigate each field?
2. How much water to apply?
3. Which field to irrigate first on a particular day?

Three methods have been developed for producing short term irrigation schedules for a hosereel-raingun irrigation system. In each, the objective is to maximise the financial return from irrigation over the season, subject to constraints based on water, equipment and labour availability.

The three methods used for producing irrigation schedules from the modelled hosereel-raingun system are:

1. a mixed linear program, which produces an optimum schedule for 7 days in terms of when to irrigate each field and also how much water to apply, but due to its complexity can only be run on a mini or mainframe computer.
2. a much simpler heuristic, which also produces a schedule for 7 days in terms of when to irrigate each field and how much water to apply, which can be run on a microcomputer.
3. a day to day heuristic which can also be run on a microcomputer, which looks at the irrigation requirements one day at a time.

The three methods are compared using data from a typical hosereel-raingun irrigation system.

RESUME: En développant des programmes d'irrigation pour un système d'arrosage("hosereel-raingun"), il faut tenir compte de trois questions fondamentales:

1. Quand faut-il irriguer chaque champ?
2. Quelle quantité d'eau faut-il utiliser?
3. Quel champ faut-il irriguer le premier tel ou tel jour?

Trois méthodes pour produire des programmes d'irrigation à court terme pour ce système d'arrosage ont été développées. Chacune à pour but de porter au maximum le rapport de capital provenant de l'irrigation au cours de la saison, ce qui dépend pourtant de certain facteurs modifiants, par exemple la disponibilité d'eau, d'équipement et de main d'oeuvre.

Les trois méthodes qui ont été employées pour produire des programmes d'irrigation pour le système modèle sont:

1. un programme multilinéaire qui produit un plan de sept jours, précisant pour chaque champ l'heure d'irrigation et la quantité nécessaire d'eau. A cause de sa complexité ce programme ne peut être exécuté que sur un mini- ou un gros ordinateur.
2. un programme heuristique plus simple qui produit également un plan de sept jours, précisant pour chaque champ l'heure d'irrigation et la quantité nécessaire d'eau. Un mini- ou micro-ordinateur suffit pour exécuter ce programme.
3. un programme heuristique journalier qui peut être exécuté aussi sur un micro-ordinateur, et qui s'occupe des besoins d'irrigation au-jour-au-jour.

Les trois méthodes ont été comparées en employant des données obtenues d'un exemple typique d'un système d'arrosage("hosereel-raingun").

ZUSAMMENFASSUNG: Die Bewässerungsplanung für ein Kreisregnerberegnungssystem wirft drei grundsätzliche Fragen auf, und zwar:
1. Wann jedes Feld bewässert werden muß?
2. Wieviel Wasser gebraucht wird?
3. Welches Feld an einem bestimmten Tag zuerst zu bewässern ist?
Drei Methoden sind entwickelt worden, um kurzzeitige Bewässerungsprogramme für ein Kreisregnerberegnungssystem hervorzubringen. Das Ziel jeder dieser Methoden ist, den Ertrag aus der Bewässerung im Laufe der Jahreszeit zu maximieren, wenn sich auch Beschränkungen durch Wasser-, Ausrüstungs- und Arbeitsverfügbarkeit ergeben.
Die drei Methoden, Bewässerungsprogramme für das modellierte Kreisregnerberegnungssystem zu entwickeln, sind:
1. ein gemischtes Linearprogramm, das einen optimalen Zeitplan für sieben Tage liefert. Dieser schreibt vor, wann jedes Feld bewässert werden muß, und wieviel Wasser zu gebrauchen ist. Wegen seiner Komplexität aber kann dieses Programm nur auf einem Mini- oder Hauptcomputer bearbeitet werden.
2. eine viel einfachere Heuristik, die auch einen Zeitplan für sieben Tage liefert. Dieser schreibt vor, wann jedes Feld bewässert werden muß, und wieviel Wasser zu gebrauchen ist. Diese Heuristik kann auf einem Mikrocomputer bearbeitet werden.
3. eine tägliche Heuristik, die auch für ein Mikrocomputer geeignet ist, und die die Bewässerungsbedürfnisse tageweise behandelt.
Diese drei Methoden werden aufgrund Daten von einem eigentlichen Kreisregnerberegnungssystem verglichen.

1 INTRODUCTION

In recent years, there has been an increase in the use of irrigation in the U.K. This has been due in part to the need for farmers to produce reliable yields of good quality crops for an increasingly discerning market. Generally irrigation is used to limit the risk of low yields and low returns as a result of insufficient rainfall during the growing season. This is known as supplementary irrigation.

In the UK, irrigation is most commonly practised using mobile hosereel-raingun irrigators. Fields are split up into predefined strips, depending on the size and shape of the field, the capacity of the irrigator, and the preferred irrigation depth (the depth of water to be applied). A raingun sprinkler, at the end of a 50-100mm diameter, 200-400m long hosepipe, is pulled out from the hosereel drum to the far end of the strip using a tractor. The drum rotates to pull in the raingun sprinkler at a fixed rate (on older machines the rate may vary), and irrigation is applied along the length and width of the strip. The irrigators are moved from one strip of the field to another, once or twice a day, until the whole field is irrigated. Labour is required to move the irrigators between strips, and from field to field. The farmer may have labour which is dedicated to irrigation activities. There may also be labour primarily employed on the farm for other work, which can be used for irrigation activities. These will be referred to as dedicated and supplemental labour respectively. The main crops irrigated in the U.K., using hosereel-raingun irrigators, are potatoes, sugar beet and field grown vegetables.

If high yields of good quality are to be obtained, the soil water deficit (SWD) of a particular field should not exceed a pre-determined critical value. If the SWD exceeds this critical SWD, then the crop is stressed, due to lack of water, with resultant losses in marketable yield. The critical SWD varies with the stage of growth of the crop, the soil type, the depth of rooting and potential evaporation rates.

SWD is the main decision variable which farmers monitor in order to decide when to irrigate a particular field and how much water to apply. These decisions are usually based on the SWD in the first strip to be irrigated within each field. Models to calculate SWD for irrigation scheduling purposes have been developed, originally by Jensen (1969).

For short term irrigation schedules (of the order of 3 to 14 days) forecast SWD's are required for the scheduling period, and this requires accurate predictions of potential evapotranspiration and rainfall. In the UK, forecasting rainfall is difficult. It is usually assumed to be zero during the scheduling period, and the resultant irrigation schedule is then delayed by one day for each daily rainfall total equivalent to the average potential evapotranspiration. Potential evapotranspiration is thus the only variable left to be forecast. Common practice is to assume that daily rates of potential evapotranspiration for the scheduling period will be equal to the average of the last four days for which evapotranspiration rates are available. This is usually a reasonable assumption.

The problem facing most farmers is not necessarily how to schedule one field, but how to schedule several fields, if the forecast requires them to be irrigated at similar times using the classical SWD models. The problem is possibly further constrained by a limited number of irrigators, insufficient labour to move the irrigators and limited water availability.

The short term irrigation scheduling problem for a hosereel-raingun irrigation system has been modelled. The philosophy is that by attempting to optimise the benefit from irrigation over each short term period of the season, the financial return from irrigation over the whole season may approach its optimum value.

Three complementary solution methods have been developed:

1. A mixed linear programme, which will produce optimum schedules for the modelled situation, but due to its complexity can only be implemented on a mini or mainframe computer. However, solution times may still be excessive when a large irrigation system is modelled.

2. An heuristic method, which uses a much simpler algorithm than that of linear programming, to produce near optimal solutions in a reasonable time. In addition, the heuristic method can be implemented on a microcomputer, which would be of greater benefit to farmers.

3. A day to day heuristic method, which uses a simple algorithm to decide one day at a time which field, if any, should be irrigated. Again, this can be implemented on a microcomputer, but it requires running nearly every day as opposed to only once a week for the other two methods. This may be a deterrent to use by farmers, particularly if they are unable to obtain meteorological data on a daily basis.

An initial analysis has been carried out to compare the three solution methods using data from a typical U.K. hosereel-raingun irrigation system.

2 DESCRIPTION OF MODELLED HOSEREEL-RAINGUN IRRIGATION SYSTEM

A model of the short term irrigation scheduling problem for a hosereel-raingun irrigation system, typical of that used by many farmers in the U.K., has been constructed.

2.1 Parameters

The model considers I field units, each of which consists of an area of land with the same crop and cultivar, planted at approximately the same time, on the same soil type (all of these variables can affect the calculation of actual rates of evapotranspiration) and with the same final market (which affects the selling price of the crop). The scheduling period covers J shifts (with P shifts per day). The farmer has the use of K irrigators. Supplemental irrigation labour is available in two categories (standard and overtime). The farmer has the use of W supplemental labour types (i.e. W/2 workers).

2.2 Decision variables

The major decision variable is the binary irrigation timing variable X_{ijk} with the interpretation:

$$X_{ijk} = \begin{cases} 1 & \text{if irrigation of field unit } i \text{ starts in shift } j \text{ using irrigator } k \\ 0 & \text{otherwise} \end{cases}$$

In conjunction with these are the binary no irrigation variables with the interpretation:

$$Z_i = \begin{cases} 1 & \text{if no irrigation of field unit } i \text{ occurs during the scheduling period} \\ 0 & \text{otherwise} \end{cases}$$

Integer variables Y_{ijk} (mm) are used to indicate one-fifth of the irrigation depth to be applied to field unit i starting irrigation in shift j using irrigator k. The application depths chosen by the model are thus 0, 5, 10, 15, 20, 25 mm, etc., and take into account the poor accuracy, often due to the influence of wind, with which water can be applied using hosereel-raingun irrigators.

Continuous variables E_{jw} indicate the hours of supplemental labour w required during shift j.

2.3 Constraints

The constraints of the modelled problem are split into five components:

(a) Irrigators

1. For each field unit i a maximum of one irrigation can take place during the scheduling period. The option of no irrigation taking place during this period is also included.
2. For each shift j of irrigator k, a maximum of one irrigation can occur.
3. Once an irrigation starts using irrigator k, no other irrigation can begin using irrigator k until the first irrigation has been completed. It may take a number of shifts to irrigate a particular field unit.

(b) Irrigation depth

1. When an irrigation takes place the irrigation depth applied should not exceed the forecast SWD for the field unit, or the preferred irrigation depth, which is specified for each growth stage of the crop. In addition, as rainfall may occur immediately after an irrigation, most farmers prefer to apply less than the forecast SWD, so that no water is 'lost' through drainage.
2. If an irrigation takes place during the scheduling period the irrigation depth applied to field unit i must exceed the minimum irrigation depth, which is specified for each growth stage of the crop.

(c) Labour

1. For each shift j of the scheduling period the labour required to move the irrigators must not exceed the labour available.

(d) Water

1. The volume of water required during the scheduling period must not exceed the volume of water that can be pumped through the irrigation system.
2. For each shift j of the scheduling period the volume of water required must not exceed the volume of water that can be pumped through irrigator k.

(e) Decision variables

1. All irrigations must be started and finished during the scheduling period.
2. If the forecast SWD is below the minimum irrigation depth, for field unit i during shift j, then irrigation cannot commence.
3. Irrigator k may only be used to irrigate certain field units.
4. Field to field moves of the irrigators may only occur in the first shift of each day.
5. The irrigation depth applied should not exceed the forecast SWD, or the preferred irrigation depth.
6. Labour required from supplemental irrigation worker w in shift j must not exceed that available.

2.4 Objective function

The objective function of the modelled problem is split into four components each corresponding to one of the four decision variables:

(a) The cost of yield lost before irrigation of field unit i starting in shift j, COST(i,j).

(b) The return from irrigating field unit i starting in shift j to a depth of 1mm minus the cost of applying 1mm of water (includes cost of water and cost of pumping), RETNY(i,j).

(c) The cost of no irrigation for field unit i during the scheduling period, COSTZ(i).

(d) The cost of labour for supplemental irrigation worker w used in shift j, COSTL(j,w).

3 SOLUTION METHODS

The model has been developed so that three solutions methods are available:

3.1 Mixed linear programme(MLP)

A mixed linear programme has been built which maximises the objective function subject to the constraints (Upcraft et al 1989).

3.2 Heuristic method

An heuristic method has been built with an identical objective function and identical constraints to that of the mixed linear programme. However, the objective function is not maximised by linear programming, but maximisation is approached by the use of a much simpler algorithm. The steps of the algorithm are:

1. Find field unit i with the largest cost of not irrigating during the scheduling period (COSTZ(i)).
2. For field unit i find the shift j and irrigator k combination which is both:
   (a) feasible and
   (b) produces the greatest benefit (RETNY(i,j) * Y_{ijk} - COST(i,j) - Cost of labour).
3. If this benefit exceeds the cost of not irrigating during the scheduling period, then schedule irrigation of field unit i to commence in shift j using irrigator k. Otherwise, field unit i is not irrigated during the scheduling period.
4. Repeat steps 1 to 3, with all the remaining field units which have not been scheduled.

3.3 Day to day heuristic method

A day to day heuristic method has been built with identical constraints to that of the mixed linear programme. The heuristic method has no objective function in the conventional sense. The steps of the algorithm are:

1. For each field unit i calculate the benefit (cost of lost yield - cost of irrigating) from commencing irrigation in shift 1 (COST(i,1) - Cost of applying water - Cost of labour). If the cost of irrigating field unit i is less than the cost of lost yield, then field unit i is a candidate for irrigation.
2. Schedule that field unit i which has the greatest benefit, as long as this exceeds zero, and irrigator k is not already busy.
3. Repeat step 2 until all K irrigators have been scheduled.

4 EXAMPLE DATA

In this initial examination of the three methods data from a typical hosereel-raingun irrigation system have been used. The example considers a farmer with one hosereel-raingun irrigator (K=1), and seven fields (I=7) which may require irrigation during the season. The farmer is prepared to move the irrigator twice a day (P=2, and thus there are two 12-hour shifts per day), but field to field moves of the irrigator only occur in the first shift of each day. Each 12-hour shift includes time for moving the irrigator to the strip to be irrigated, and time for applying the irrigation. The farmer has no dedicated irrigation labour, but one supplemental irrigation worker is available, who has different wage rates for standard and overtime working (W=2).

Field unit data are shown in Table 1, which includes the crop, area and the number of 12 hour shifts required to irrigate the field when the maximum irrigation depth is applied.

Table 1. Field unit data.

Field	Crop	Area(ha)	Shifts
1	Potatoes	7.5	3
2	Potatoes	8.3	5
3	Potatoes	9.4	4
4	Sugar Beet	2.0	1
5	Sugar Beet	5.4	3
6	Spring Wheat	4.0	2
7	Spring Wheat	11.3	4

All fields are assumed to have been planted on the 1st April. The potatoes are assumed to be harvested in mid September, the spring wheat in mid August and the sugar beet during the autumn. The irrigation season is from the 1st April to the 31st August.

The irrigation regime for each field unit depends on the crop and its irrigation dependent growth stages. The potatoes are assumed to have two growth stages. The first is for scab control, during and immediately after tuber initiation, and the second is for (a) before tuber initiation and (b) when the tubers are bulking. The sugar beet is assumed to have three growth stages, the critical SWD increasing as the roots extend in depth. The spring wheat is assumed to have only one growth stage during the irrigation season. The irrigation regime data are shown in Table 2. At all other times during the irrigation season, no irrigation is required.

Table 2. Irrigation regime.

Fields	Growth Stage	Dates	Critical SWD (mm)	Preferred/ Minimum Application (mm)
1-3	1	25/4-31/5	18	15/15
	2	1/4-24/4 and 1/6-30/8	30	25/20
4-5	1	1/6-30/6	30	25/20
	2	1/7-31/7	40	25/20
	3	1/8-31/8	60	25/25
6-7	1	1/4-31/7	60	25/25

Actual and forecast SWD's are calculated by a model developed at Silsoe College (Hess and Mathieson, 1988).

The cost of yield lost is calculated by an inverted use of the ADAS (1977) values of average yield response to irrigation. Instead of increasing the yield (tonnes/ha), whenever 1 mm of irrigation is applied, by the yield response value, a similar loss is recorded for each mm that the SWD exceeds the critical value. This is known as crop stress. The selling price, the costs of the yield lost, and of applying 1mm of irrigation water (includes water and pumping) are shown in Table 3. All values refer to the whole field.

Table 3. Field unit cost data.

Field	Selling Price (£/tonne)	Cost (£) of Yield Lost Per mm of Stress	Water Cost (£/mm)
1	60	36.00	1.33
2	60	39.84	1.41
3	60	45.12	1.69
4	30	0.78	0.36
5	30	2.11	0.92
6	100	7.20	0.70
7	100	20.34	1.98

Typical crop growth data, soil data, irrigation strip sizes, time required to move irrigators, labour availability and costs and irrigation water availability have been used for the 1988 season in the Silsoe area. For the example situation under consideration the availability of irrigation labour and water are non-limiting but their costs are included in the analysis.

The mixed linear programme and the heuristic method are used to produce short term irrigation schedules for 7 day periods. Thus the model considers the situation where I=7, J=14 , K=1 and W=2. The mixed linear programme and the heuristic method are executed on Tuesday mornings, when meteorological data are available up to and including Monday, to produce schedules for Tuesday through to the following Monday.

The day to day heuristic method is used to decide one day at a time which field, if any, should be irrigated.

In order to evaluate the three solution methods under different conditions, three weather scenarios have been considered:

(a) the wettest April to September in the last 25 years - 1968

(b) an average April to September from the last 25 years - 1980

(c) the driest April to September in the last 25 years - 1976

5 RESULTS

The three solution methods have been used to produce irrigation schedules for the three weather scenarios:

1. Assuming no forecast information is available for evapotranspiration and rainfall.

2. With perfect knowledge of evapotranspiration and rainfall for the following week.

Tables 4, 5 and 6 show the cost of lost yield, the cost of irrigating (cost of water and pumping, plus cost of labour) and the total cost of the operation, for wettest (1968), average (1980), and driest (1976) weather patterns respectively. For completeness the cost of rainfed production only is also included. All costs have been rounded to the nearest £10.

Table 4. Irrigation costs wettest (1968).

Method	Lost Yield Cost (£)	Irrigation Cost (£)	Total Cost (£)
No irrigation	5070		5070
Without forecast information:			
MLP	830	820	1650
Heuristic	770	850	1620
Day Heuristic	1220	500	1720
With perfect forecast information:			
MLP	530	330	860
Heuristic	890	360	1250
Day Heuristic	710	410	1120

Table 5. Irrigation costs average (1980).

Method	Lost Yield Cost (£)	Irrigation Cost (£)	Total Cost (£)
No irrigation	5220		5220
Without forecast information:			
MLP	1300	1090	2390
Heuristic	1320	1050	2370
Day Heuristic	1200	950	2150
With perfect forecast information:			
MLP	1300	690	1990
Heuristic	1340	710	2050
Day Heuristic	1310	750	2060

Table 6. Irrigation costs driest (1976).

Method	Lost Yield Cost (£)	Irrigation Cost (£)	Total Cost (£)
No irrigation	11800		11800
Without forecast information:			
MLP	7110	2290	9400
Heuristic	8270	2110	10380
Day Heuristic	7140	2190	9330
With perfect forecast information:			
MLP	7970	2110	10080
Heuristic	7330	2160	9490
Day Heuristic	7010	2200	9210

6 DISCUSSION AND CONCLUSIONS

A number of the results are interesting and some, at first glance, surprising. At this stage in the study, we have chosen to present them and to offer our explanation as to what has caused them. Further work is underway to see if these explanations can be substantiated.

Before looking at the results in detail, it is important to note that:

1. The cost of lost yield is totally dependent on the crop response to irrigation function used and on the selling price. The ADAS (1977) functions can be criticised, but they are the only U.K. specific functions currently available, which have been developed for a number of crops, using the same methodology. Research is now underway at Silsoe College to produce improved functions (Wright et al 1989). The selling price of most crops, varies between seasons, depending on the relative supply and demand. Potatoes are very sensitive to such changes, sugar beet less so and cereals hardly at all. In the driest (1976) year, the selling price of potatoes should be substantially increased, but this would only have become apparent as the season progressed. However, in this initial analysis we have kept the same selling price in each year, and have compared the solution methods within each season, rather than between seasons.

2. The cost of irrigating is dependent on the costs of water, pumping and labour. These are likely to remain constant between years, except for inflationary increases.

To put the cost of lost yield values into perspective the maximum possible value for any year is £13400. This value is directly related to our choice of response functions and selling prices, and assumes the SWD in each field approaches its maximum value.

6.1 Without forecast information

In all three seasons irrigation is beneficial. In both the wettest (1968) and average (1980) years, the costs of lost yield with no irrigation are similar. This is due to our choice of response function, which is dependent on the maximum SWD, and not the length of time for which it occurs.

In both years, the total cost of all three solution methods are very similar, and give substantial cost savings over no irrigation. In both years, the heuristic method has a slightly lower total cost over the whole irrigation season, than the mixed linear programme, which would be expected to be the best. We believe that this is because the actual rainfall patterns that occurred in these years were different from the forecasts used by the models, which assumed zero rainfall over the forecast period.

In the driest (1976) year, the total costs of all three methods are again similar. However, there are no substantial cost savings over no irrigation. This could be because:

(a) the farmer only has the use of one irrigator. With this limited system capacity it is not possible to supply enough water to meet crop demands, and the crops remain under stress for most of the season.

(b) there is a fixed scheduling period of seven days for the mixed linear programme and heuristic methods. With a shorter scheduling period (i.e. five days), fields with a large demand for water could be irrigated more often. In particular, with the high levels of evapotranspiration in mid-summer 1976 (i.e. 5mm/day), a field would build up a 35mm SWD over seven days, but could only receive 25mm of irrigation during the same period. The SWD would thus slowly build up, even with the maximum number of irrigations allowed by the model.

(c) of the assumed fixed selling price.

6.2 Perfect forecast information

For both the wettest (1968) and average (1980) years, the total costs of all three solution methods are again similar. All register cost savings over the situation with no forecast information. Greater savings were obtained in the wettest season, because there is advance knowledge of high rainfall. The mixed linear programme produces the lowest total cost in both these years.

For the driest (1976) year, the total costs of all three methods are again similar. There are no substantial cost savings compared with when there is no forecast information, because there is only a small amount of rain. Surprisingly the mixed linear programme produces a greater total cost with forecast information than without.

6.3 Comparison of solution methods

It is not possible, with the limited evaluation reported here, to specify which of the three short term irrigation scheduling methods is likely to be the best in practice. Further work is needed to evaluate the usefulness of each of these approaches under a wide range of conditions, and to operate the models in real time situations.

ACKNOWLEDGEMENTS

The authors wish to acknowledge the support of the Agriculture and Food Research Council who sponsored this research with grant number PG63/229.

REFERENCES

ADAS 1977. Economics of irrigation. Report of ADAS working party. London, MAFF.

Hess, T.M. & I.K. Mathieson 1988. Irrigation management services. Irrig. News 13:25-32.

Jensen, M.E. 1969. Scheduling irrigation using computers. J. Soil Water Conserv. 24:193-195.

Upcraft, M.J., D.H. Noble & M.K.V. Carr 1989. A mixed linear programme for short term irrigation scheduling. Submitted to J. Opl Res. Soc.

Wright, E., M.K.V. Carr & P.J.C. Hamer 1989. Yield/water-use response functions for potatoes and sugar beet. Final report to MAFF. Department of Agricultural Water Management, Silsoe College.

Land and Water Use, Dodd & Grace (eds), © 1989 Balkema, Rotterdam. ISBN 90 6191 980 0

Effect of magnification factor, supply conditions and valve clearance on performance of an hydraulic ram

K.C.Das
Department of Irrigation, Government of Orissa, Bhubaneswar, India

K.N.Sharma, B.Panigrahi & S.D.Sharma
Department of SWCE, CAET, Bhubaneswar, India

ABSTRACT: Studies were conducted to study the effect of magnification factor, supply conditions and valve clearance on performance of hydraulic ram. Measure of performance of the ram was discharge delivered and various efficiencies viz. Volumetric, Rankin's and D' Aubuisson. It was revealed that at 7.25° supply pipe inclination, ram gave maximum discharge and efficiency. Valve clearance of 2.5 cm gave the maximum discharge whereas valve clearance of 2.0 cm resulted into maximum efficiency. Magnification factor affected the discharge inversely.

1 INTRODUCTION

Hydraulic ram is a pump which utilizes the kinetic energy of moving column of water to pump part of the same water to a greater height. Source of energy for operation of this pump is the potential energy of water which, due to gravitational effect, gets converted into kinetic energy inside the hydraulic ram.

The pump does not need any external source of energy and it can be used effectively in the area where sufficient fall or grade exists in the bed of the stream. Studies carried out by Planning Research and Action Division, State Planning Institute, Lucknow, U.P. indicated that annual cost of pumping water by hydraulic ram was much less than pumping water by diesel run pump or electrically operated pump (Ghate 1977). The reason for low operational cost can be attributed to the following:

1 the hydraulic ram does not require any additional source of energy for its operation;
2 maintenance cost is very low as there are not many moving parts in it.

The hilly terrains of Orissa have number of rivers and rivulets flowing at steep bed slopes. The areas where such rivers and rivulets are situated are in interior places and it is very difficult to arrange external source of energy in these areas to operate pumps. However, in such areas hydraulic ram can be successfully utilised to lift water from these rivulets for irrigation purposes. Performance of hydraulic ram is affected by supply line conditions, magnification factor, valve clearance etc. Hence, it is necessary to carry out indepth studies regarding effect of these factors on performance of hydraulic ram.

Keeping the above facts in view, laboratory studies were conducted to find the effect of magnification factor, supply line conditions and valve clearance on the performance of hydraulic ram. The work was carried out at the College of Agricultural Engineering and Technology, O.U.A.T., Bhubaneswar, India.

2 MATERIALS AND METHODS

Experimental set up consisted of a hydraulic ram assembly, a sump, a 5 H.P.(variable speed motor) pump set, 34 mm G.I. pipe assembly (to supply water from pump to reservoir tank), a waste water measuring channel, a delivery water measuring system as well as water recirculatory system (to avoid wastage of water). Details of various components are discussed in following paragraphs:

2.1 Hydraulic ram assembly

The hydraulic ram used in the study was of 32 x 12 mm size. Various parts of the hydraulic ram with their materials of construction and size which were used are given in Table 1.

Table 1. Various parts of hydraulic ram with their materials of construction and size.

Name of the parts	Materials	Size
1. Supply pipe line	G.I.pipe	32 mm diameter
2. Delivery pipe	G.I.pipe	15 mm diameter
3. Ram chamber	Cast iron	
a. Length		385 mm
b. Outside diameter		85 mm
c. Inner diameter		60 mm
4. Waste valve	Steel	
a. Diameter		48 mm
b. Thickness		10 mm
c. Angle of valve sheet		45°
5. Delivery valve	Steel	
a. Diameter		33 mm
b. Thickness		9 mm
c. Angle of valve sheet		45°
6. Air chamber	Cast iron	
a. Capacity		8 litres
b. Wall thickness		10 mm
7. Supply tank (reservoir)		50 x 50 x 70cm

Details of experimental set-up are shown in Figure 1. In the supply line of hydraulic ram, pairs of bends of 5°, 6°, 7°, 8° and 10.5° were used to provide desired angle of inclination. These bends were made

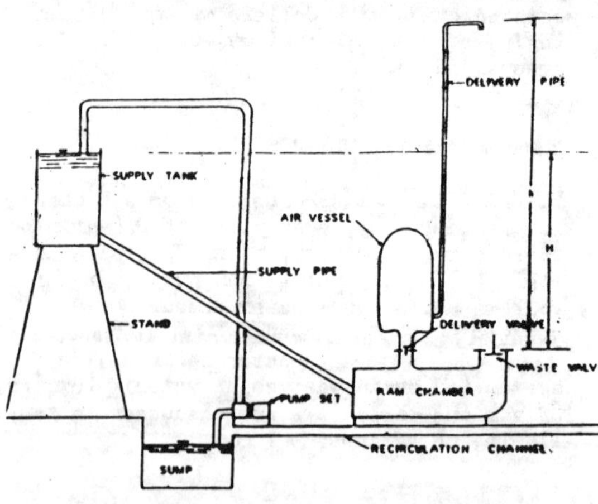

Fig. 1. Schematic diagram of experimental set-up

by cutting medium duty G.I.pipe of 32 mm diameter at an angle half the desired angle of inclination and then welding the two pieces together.

2.2 Waste water measuring channel

A channel of 1700mm x 420mm x 300mm (length x depth x width) was made of G.I. sheet. A 60° V notch of 240 mm depth was fitted at one end of the channel and on other end a reservoir of 800 mm x 800 mm x 420 mm was provided as an integral part of the channel. The crest of V notch was kept at 160 mm above the channel bed. In order to reduce wave action developed by the splash of waste valve water, two baffle walls (perforated) were fitted just after the reservoir. To accurately measure the head over V notch crest, a point gauge was installed at a distance of 570 mm upstream of V notch. The system was calibrated and head discharge curve was plotted in order to determine the discharge through the waste valve accurately.

2.3 Delivery discharge measuring system

This consisted of a tank of 300 mm x 300 mm x 600 mm (length x width x height) size made of M.S.sheet. The discharge rate was measured by observing the time taken to fill up the tank by a stop watch.

The supply tank (reservoir) was kept at an elevated position and a constant supply head of 1.63 m was maintained. In order to vary magnification factor, delivery head was changed.

Experiments were conducted on five supply pipe inclinations, three valve clearances and three magnification factors. Thus different treatments were as follows:

1 supply pipe inclination: 4.90°, 6.02°, 7.03°, 7.75° and 10.64°;
2 valve clearance: 2.00, 2.50 and 3.00 cm;
3 magnification factor: 2.055, 3.180 and 3.960.

Each treatment was replicated minimum three times .

3 RESULT AND DISCUSSIONS

Following paragraphs discuss the effects of various variables such as supply pipe inclination, valve clearance and magnification factor on the performance of the hydraulic ram.

3.1 Effect of variables on discharge

Figure 2 shows the effect of valve

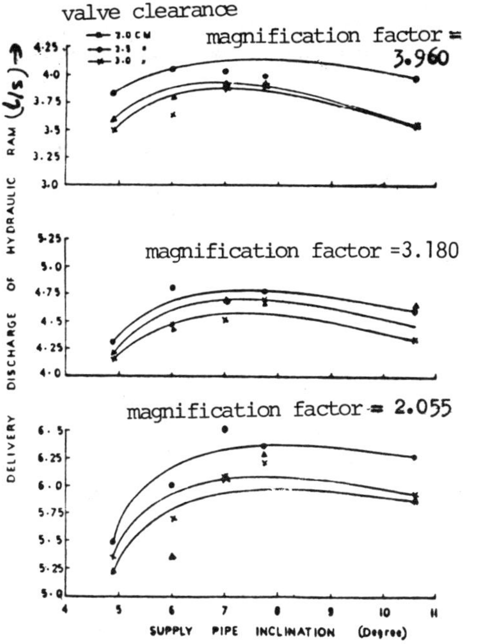

Fig. 2. Effect of supply pipe inclination on delivery discharge of hydraulic ram with different magnification factors

clearance, magnification factor and supply pipe inclination on discharge of the ram. It can be seen from various curves that discharge increases in all the cases when inclination increases from 4.90° to 7.00°, remains constant between 7.00° to 8.00° and starts decreasing when inclination increases beyond 8.00°. Maximum discharge was found to be 6.300, 4.775 and 4.025 lit. per minute at magnification factors of 2.055, 3.180 and 3.986 respectively at supply pipe inclination of 7.25° and valve clearance of 2.50 cm. The above findings are in confirmity with the works reported by Singh (1977), Anonymous (1981) and Sharma and Sharma (1984).

3.2 Effect of variables on efficiencies

Hydraulic ram efficiency is expressed in terms of volumetric efficiency, i.e. ratio of volume of water delivered to the total volume of water supplied through supply pipe line, Rankin's efficiency, i.e. ratio of folume of water delivered per unit time multiplied by head difference between delivery and supply point to the water passed through waste valve multiplied by supply head and D'Aubuisson's efficiency, i.e. ratio of water delivered multiplied by delivery head to the total water supplied through supply line multiplied by supply head (Lal 1982).

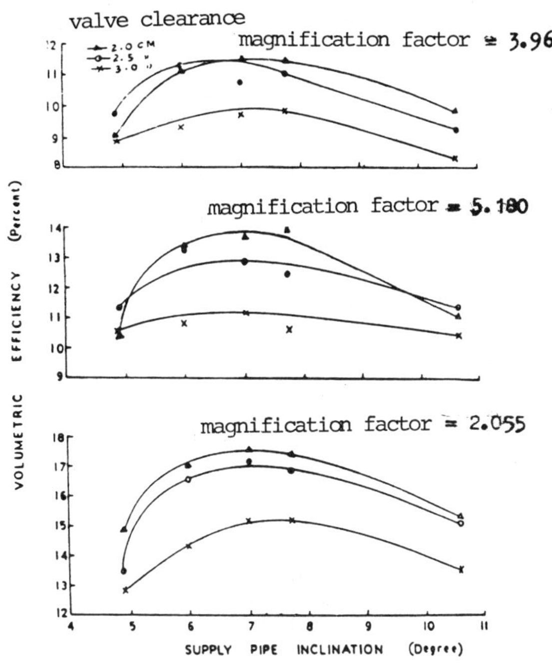

Fig. 3. Effectof supply pipe inclination on volumetric efficiency for different magnification factors

Figures 3, 4 and 5 show the effect of supply pipe inclination, magnification factor and valve clearance on different efficiencies of hydraulic ram.

It can be observed from Figure 3 that in general volumetric efficiency increases with increase in supply pipe inclination from 4.90° to 6.50°, it remains constant between 6.50° to 7.50° and decreases beyond 7.50° inclination. The highest volumetric efficiency (Figure 3) is generally found between 7.00° to 7.25° inclination at a valve clearance of 2.0 cm. Maximum volumetric efficiency of 17.25, 13.85 and 11.50 per cent was observed at supply pipe inclination of 7.25° and valve clearance of 2.00 cm at magnification factor of 2.055, 3.180 and 3.960 respectively.

Rankine's efficiency (Figure 4) and D'Aubuisson's efficiency (Figure 5) were

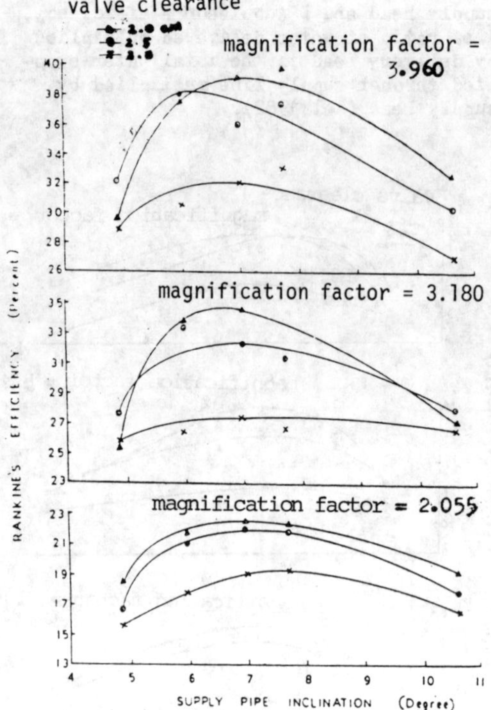

Fig.4. Effect of supply pipe inclination on Rankine's efficiency at different magnification factors

also found to have similar trend with supply pipe inclination and valve clearance as volumetric efficiency. But in case of magnification factor, reverse trend is observed. Maximum Rankine's efficiency was found to be 22.50, 35.00 and 39.35 per cent at magnification factor of 2.055, 3.180 and 3.960 respectively when supply pipe inclination was 7.25° and valve clearance was 2.00 cm. Similarly D'Aubuisson's efficiency was found to be 36.00, 44.00 and 45.80 per cent at magnification factor of 2.055, 3.180 and 3.960 respectively where supply pipe inclination was 7.25° and valve clearance was 2.00 cm.

The maximum efficiency was obtained at 7.25° supply pipe inclination. The inclinations at which other workers have reported maximum efficiency are 7.00° (Singh 1977) and 6.75° (Anonymous 1982) . These values are in close proximity with the values obtained in the present study.

The increase in efficiency with decrease in valve clearance is a usual phenomenon and has also been reported by College of Agricultural Engineering, Jabalpur (Anonymous 1982) . The results in the present study also show that efficiency increases with decrease in valve clearance.

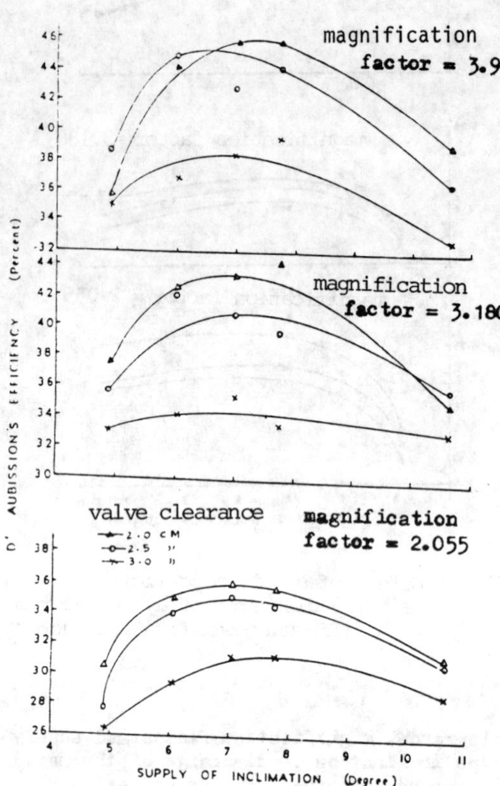

Fig. 5. Effect of supply pipe inclination on D'Aubisseyion's efficieny for different magnification factors

Makhija (1977) has reported that efficiency decreases with increase in magnification factor. But in the present case it was found that Rankine's efficiency and D'Aubuisson's efficiency increase with increase in magnification factor from 2.055 to 3.960 . According to Clark (1973), maximum efficiency can be obtained at magnification factor of 4.00. In the present study of hydraulic ram the maximum magnification was 3.960. Hence, it is likely that the trend upto magnification factor of 4.0 may be that the efficiency increases with increase in magnification factor. However, it may follow trend as reported by Makhija (1977) if observations at higher magnification factor are taken.

CONCLUSIONS

Supply pipe inclination of 7.25^{o} gives the maximum discharge and maximum efficiency for the given hydraulic ram study, The waste valve clearance of 2.5 cm and 2.0 cm gives maximum discharge and maximum efficiency respectively. Efficiency of hydraulic ram is found to be increasing with decrease in waste valve clearance from 3.00 cm to 2.00 cm. With increase in magnification factor delivery discharge was found to decrease. For low magnification factor, efficiency was found to increase with increase in the magnification factor.

REFERENCES

Anonymous. 1981. Optimization of Ground Water Resources through Well and Pump. Research digest, ICAR Coordinated Research Scheme, Department of Soil and Water Conservation Engineering, College of Agricultural Engineering, J.N.K.V.V., Jabalpur, India.

Anonymous, 1982. Research digest, ICAR Coordinated Project on Wells and Pumps, Department of Agricultural Engineering, Pant College of Technology, G.B.P.U.A.T., Pantnagar, India.

Das, K.C. . 1985. Effect of magnification factor and supply conditions performance of hydraulic ram. An unpublished M.Sc. (Agril. Engg. and Tech.) thesis OUAT., Bhubaneswar, India.

Ghate, P. 1977. The scope of irrigation t through hydraulic ram in the hills of India. Agricultural Engineering Today, I (12).

Lal, J. 1981. Hydraulic Machine. Metropolitan Book Co., Pvt. Ltd., New Delhi.

Sharma, A. and Sharma, P.K. 1984. Assess nt of performance of hydraulic ram. An unpublished B.Tech. (Agril. Engg.) Project report College of Agricultural Engineering, J.N.K.V.V., Jabalpur, India.

Singh, D. 1977. Hydraulic ram: their design, operation and installation. Agricultural Engineering Today, I(12).

Land and Water Use, Dodd & Grace (eds), © 1989 Balkema, Rotterdam. ISBN 90 6191 980 0

Performance of a water driven wheel for low lift irrigation

N.A.Egharevba
Department of Agricultural Engineering, Federal University of Technology, Minna, Nigeria

ABSTRACT: A water driven wheel was designed, locally fabricated and tested in the distributory canal at the Kadwa station of the Kano river basin, Nigeria. The device is operated by the stream current with no additional power source. At load condition (water - can attached to each wheel-vane) and threshold stream velocity of 0.55 m/s, it moves at a speed of 2.15 rpm. With increase in the stream velocity up to 0.8 m/s, the speed of rotation was 3 rpm, and the corresponding average discharge was 9 1/min at 1.5m head. The device generates power at the rate of 5.25W when rotating at 3 rpm. Though the power output and the discharge capacity were low, it is without operational cost and also have a two - fold advantage of functional reliability and simplicity of design and fabrication at the village level.

1 INTRODUCTION

Irrigation is an artifical application of water to land for the purpose of growing and producing food and fibre. The objective of irrigation can be one or a combination of the following:

1. to serve as an insurance against crop failure due to unforeseeable droughts.
2. to make crop production possible, that is the case in desert climates where crop production depends mainly on irrigation.
3. to supplement natural precipitation when necessary.

The major determinant in design and suitability of system for irrigation is the soil - plant - water relationship. Soil physical properties which are important are: location (topography), depth, texture, and structures. These properties determine the infiltration rate, capacity, soil moisture relention and erodibility of the Soil.

Numerous irrigation devices have been developed and reported in literatures. More and more smaller Projects such as these that pump water from rivers and ground water to supply individual farm of ½ to 3ha have been established (Godo fredo and moss, 1979). Energy is a prime requirement for progress and development. The close relationship between the proximity of energy resources to the potential user coupled with increasing costs of conventional energy source have led to a considerable interest and activities among developed and developing countries in the development and application of renewable energy resources. Man and animal power may be considered for irrigation of small subsistence productions where lifts are low and there is a low alternative demand for labour. Where large amount of water are to be pumped to higher elevations (for either irrigation or rural water supply), mechanical pumps are necessary. A comparison of performance of some manual and animal - operated water lifting devices is given in table 1.

Table 1. Comparison of performance of some manual and animal operated water lifting devices

Equipment	Height lifted m	Range of lift m	Capacity m^3/ hour	Men for 8-hr operation	Animals for 8-hour operation
A. Manual					
1. Scoop (swing type)	1.0	-	8.0	1	
2. Swing basket	0.6	-	5.0	4	
3. Counter poise lift	2.5	-	3.0	2	
4. Archmedian Screw	0.5	.25-.75	22.5	2	
5. Improved Paddle wheel	0.3	.15-.45	100.0	2	
6. Water ladder	0.9	-	8.0	2	
7.a) chain pump	1.5	-	30.0	4	
b) chain pump	3.0	-	20.0	4	
B. Animal-operated					
1. Self emptying bucket	8.0		8.0	1	2
2.a) Two-bucket lift	3.0		22.0	1	2
b) Two-bucket lift	10.0		5.5	1	2
3. Persian wheel	2.0	1.5-9.	9.0	1	2
4. Sakia	0.3	.3-1.8	110	1	2

Source: Molenar (1956)

2. TEST PROCEDURE AND METHODOLOGY

The theoretical consideration and the design and fabricating procedures of the water driven wheel had earlier being given in a separate report (Egharevba, 1988). Fig. 1 show the prototype of the fabricated water wheel.

The water lifting device was tested in the distributory canal of the Kano River Basin Kadawa, about 12km south of Kano, Nigeria. The test done were preliminary test runs since the device was not permanently installed. The Parameters tested were the stream velocity, Blade (vane) submergence, rotational speed (RPM) of the water-wheel and the discharge capacity. The sequential steps are as follows:

2.1 Site Selection

A reach of distributoral canal of uniform depth, free of weed and near a drop structure was selected. The drop structure with an adjustable metal gate was used to vary the flow regime in the canal. In addition, the canal pavement up stream at the drop structure provided a firm ground for an easy installation in the canal bed.

2.2 Stream Velocity

The velocity of water flow in the stream was measured by float method. Three difference readings were recorded for each flow regimes created in the canal section. The ratio of the distance travelled by the float material to the time taken was recorded as the stream surface velocity.

2.3 RPM of water wheel

The blade (vanes) were first given initial force manually and the rotation started. The number of revolutions of the wheel was counted and cumulative time taken was recorded. The revolution of the water wheel was noted for no load condition (without attaching water cans to the blades) and also for load condition (with water cans fixed to the vanes).

2.4 Blade (vane) Submergence

The width of blade which receive water thrust at each setting of wheel rotation was measured with a scale. This submerged width is referred to as blade or vane submergence. The blade submergence setting during field test was 41 cm, corresponding to a total head of 1.5m for the device.

2.5 Discharge Capacity

Discharge from the water cans was collected in a calibrated bucket. The time taken to fill a 15 litre bucket was noted as well as the corresponding number revolutions made by the device. Average volume of discharge per second was then computed from these values.

2.6 Power input - Power output

The device generates power of 1.77 w at a threshold velocity of 0.55 m/s, at no load condition (see Table 3). At higher velocity of 0.8 m/s, the power increased to 5.25 W. The system efficiency however, dropped from 57%

to 29%. At load conditions, the power output was 1.5w. The power generated by the device is computed by the formula (design equation)

$$WQh = 2.735\ NV^2e,$$

where e = system efficiency in %, V = stream velocity in m/s, N = number of revolutions per minute (RPM), h = lift (total head) in m, Q = discharge in m/s and W = unit weight of irrigation water, 1000 kg/m^3.

3. RESULTS, DISCUSSIONS AND CONCLUSIONS

The performance data of the tested device are presented in Table 2 and 3. As can be observed in Table 2, the RPM of the water-wheel are higher at no load conditions when compared with load condition. This is due to additional weights (watercans and its content) which the device has to sustain as the wheel rotates. The minimum stream velocity to maintain rotation at load condition is 0.55 m/s. The system efficiency decreases with increase wheel rotational speed and discharge (Table 3).

This being a preliminary test, only one blade (vane) Submergence setting was exhaustively investigated. The stream velocity were in the range of 0.3 to 0.8 m/s. At a higher velocity, the device became unstable - as test was conducted without permanent installation. As indicated in Table 3, discharge increases with increase in velocity. However, an optimum blade submergence and stream velocity combination for maximum system efficiency can be obtained only after further research and development of the water-wheel.

Based on the study conducted on the water-wheel, it can be concluded that:

Table 2. Stream Velocity and RPM of water-wheel

Stream	Revolution Per minute (RPM	
vel.m/s	No Load condition	load condition
0.30	2.0	-
0.35	2.5	-
0.40	3.0	-
0.50	4.0	-
0.55	4.6	2.15
0.60	5.2	2.25
0.65	5.8	2.45
0.70	6.3	2.65
0.75	6.9	2.85
0.80	7.4	3.00

Table 3. Relationship between Discharge (Q), Power input (PIN) Power output (pout) and system efficiency

Stream Vel (M/S)	N (RPM)	Q (I/S)	PIN (W)	POUT (W)	EFF (%)
0.55	2.15	0.10	1.77	1.00	57
0.60	2.25	0.11	2.22	1.10	50
0.65	2.45	0.12	2.84	1.20	42
0.75	2.85	0.14	4.38	1.40	32
0.80	3.00	0.15	5.25	1.50	29

1. The device can lift water without additional operating cost as long as the stream velocity is greater or equal to 0.55 m/s.

2. Use of this water lifting device at the remote village level will reduce the use of human energy and ensure higher crop yield.

3. The device can easily be locally fabricated and maintained wherever it's adopted. It also have the advantage of low capital investment.

4. With further research and development, an optimum stream velocity and blade submergence combination can

Figure I Water driven water wheel

be obtained for maximum operating efficiency.

5. For greater reliability especially at crop peak demand-period and when stream velocity may reduce during dry season provision of a back up pumping capacity may be necessary.

6. Besides water lifting the power generated can be used for lighting huts or running transistors when a suitable alternator is developed.

REFERENCES

1. Melenar, A. 1956 Water Lifting Devices for Irrigation. FAO, Agric. Development Paper No 60.

2. Egharevba, N. A. 1988 Design of a water-wheel for Rural use in Nigeria, paper presented at the 12th annual Conference of the Nigerian Society of Agric. Engineers, National Centre for Agric. Mechanization, Ilorin, Nigeria.

Land and Water Use, Dodd & Grace (eds), © 1989 Balkema, Rotterdam. ISBN 90 6191 980 0

Simulation de pluies en laboratoire sur des sols choisis du plateau vaudois, Suisse

P.Baril
Institut de Génie Rural, Ecole Polytechnique Fédérale, Lausanne, Suisse

RESUME: L'érodibilité de sols agricoles du plateau vaudois, en Suisse, est étudiée à l'aide d'un banc d'essai d'érosion mesurant 6 m X 2 m. Chaque sol subit des séquences identique de simulation de pluie comprenant 2 intensités pluviométriques, 2 pentes et sous 3 états hydriques avant essai. Les résultats d'érosion et de ruissellement sont expliqués par les propriétés des sols testés.

ZUSAMMENFASSUNG: Die Erodibilität landwirtschaftlicher Böden aus der Waadtländischen Ebene wird anhand einer 2 X 6 Meter grossen Erosionsverschuchsanlage studiert. Jeder Bodentyp wird anhand identischer Simulationssequenzen auf seine Erodibilität analysirt. Dabei werden für die Simulationen pro Bodentyp 2 veschiedene Regenintensitäten, 2 veschiedene Neigungen des Beckens und 3 veschiedene hydrische Bodenzustände verwendet. Die Resultate bezüglich Erosion und Oberflächenabfluss werden anhand der Eigenschaften des getesteten Böden erklärt.

ABSTRACT: Erodibility of cultivated soils of the "plateau vaudois", in Switzerland, is studied using an erosion bench of 6 m X 2m. Each soil received identical simulated rainfall including 2 rainfall intensity, 2 slope angle and 3 moisture content preceding a run. Erosion and runoff results are explained by the soil properties of the tested soils.

1 INTRODUCTION

Les études sur l'érodibilité des sols peuvent être réparties en deux catégories selon les approches et les moyens utilisés : en parcelles expérimentales soumises au condition naturelle ou en laboratoire avec tests d'érodibilité effectués sur banc d'essai. La première approche, effectuée par Wischmeier et al.(1969,1971) a permis d'exprimer empiriquement l'érodibilité de sols en fonction de 15 propriétés distinctes et de leurs interactions; les plus importantes étant la texture, la structure, la perméabilité et le taux de matière organique. Des indices ont été développés, statistiquement représentatifs, surtout pour les sols à texture moyenne. D'autres expérimentations de ce type ont été effectuées à plus petite échelle et de manière localisée, pour en citer quelques-unes, Yamamoto and Anderson (1973), Bolline (1982), Kenneth, Trott and Singer (1983) et Lelong et al. (1984). Ces auteurs signalent que d'autres propriétés, telles que la minéralogie des argiles, la densité du sol, le taux de sesquioxydes, le type de matière organique peuvent aussi être corrélées avec l'érosion.

En second lieu, les méthodes par simulation de pluie permettent de suivre le phénomène lorsqu'il se produit et de mesurer l'effet de propriétés spécifiques en contrôlant les autres facteurs (pluie, pente et végétation). Des comparaisons directes avec les processus sous pluie naturelle ne sont pas judicieuses, cependant des relations entre propriétés des sols et pertes en terre peuvent être révélées. Ces expérimentations servent à une compréhension préalable de l'érodibilité. Elles sont effectuées, soit en laboratoire sur des bancs d'essai, soit au champ avec des simulateurs de pluie mobiles sur des petites surfaces d'érosion. Il faut signaler les travaux de Bryant (1968,1974), de De Ploey (1981), de Luk (1979) et les publications de Asseline et Valentin (1978), Casenave (1982) et Chevalier (1982) entre autres. Ceux-ci ont considéré des indices d'érodibilité fixes, c'est-à-dire sans variation temporelle.

En outre, les actions de transport et de détachement du sol par l'érosion dépendent de ses propriétés physique chimique et biologique, considérées comme intrinsèques. Elles évoluent dans le temps et affectent le degré de sensibilité des sols. Plusieurs auteurs considèrent que l'érodibilité est une notion à variation temporelle. Poesen (1981) a montré que le détachement varie au cours d'une averse. La durée de l'averse est importante car une fois l'état d'équilibre atteint, le détachement varie peu. L'érodibilité, telle que définie dans l'USLE (facteur K), est aussi dépendante du temps selon Lal (1981) et de l'évolution des propriétés comme la stabilité structurale, l'humidité superficielle, la résistance mécanique au cisaillement ou du degré de battance d'un sol.
A notre connaissance, peu d'études sur l'érodibilité ont été realisées en Suisse. L'érosion est étudiée avec différentes approches dans quelques instituts suisse; les plus importantes étant le Geographisches Intitut (Université de Bâle), le Laboratorium für Bodenkunde ETH de Zürich, la Station Fédérale Agronomique de Changins et l'Institut de Génie Rural de l'Ecole Polytechnique Fédérale de Lausanne comme le souligne Marti (1986).
L'érodibilité est un facteur complexe. Le fait qu'elle est maintenant reconnue comme une notion dynamique a des implications importantes dans la gestion des terres agricoles. Des informations complémentaires sur l'évolution de l'érodibilité en fonction des différents systèmes de culture, devraient être obtenues pour les grands groupe de sols, Hadley et al. (1985). Les objectifs de cette publication sont à considérer autant dans la démarche et l'approche utilisée que dans les résultats obtenus.

2 MATERIEL ET METHODE

L'érodibilité des sols est étudiée en laboratoire à l'aide d'un banc d'essai d'érosion (fig.1). Ce banc consiste en une cuve de 6m. X 2m. X 0.6m. surmonté d'un simulateur de pluie (type à buses). Cette cuve, drainée, peut être inclinée jusqu'à 24 % et l'intensité pluviométrique simulée peut atteindre 150 mm/h. Les sols testés sont choisis selon les critères suivant : horizon Ap de parcelles soumises à une agriculture intensive (grande culture), position géomorphologique en milieu de coteau, taux de matière organique compris entre 2% et 4% et à texture variable (gamme de sol léger à lourd). Les volumes de sol sont prélevés à la même époque de l'année, tamisés à 20mm et chargés dans la cuve. Une période de deux semaines est maintenue avant le premier et entre les essais. Chaque sol subit le même protocole de simulation de pluie comportant deux intensités pluviométriques (40 mm/h pendant 60 min. et 100 mm/h pendant 24 min.), deux pentes (5% et 10%) et trois états d'humidité (sec, humide et très humide). Un total de 12 essais est effectué par sol et un travail superficiel est fait avec un rateau avant chaque essai à sec. L'essai humide est provoqué 24 heures plus tard et l'essai très humide 1 heure après ce dernier. Le ruissellement et l'érosion sont mesurés en continu au cours des essais et l'infiltration à la fin de chacun d'eux.
Huits sols ont été testés avec ce protocole. Les principales analyses sont décrites dans le tableau suivant (tab. 1).

Tableau 1. Propriétés des sols testés

Nom	% sab	% lim	% arg	% m.o	pH
Bussigny	42.0	37.9	20.1	2.8	7.5
Echandens	33.0	45.9	21.1	3.8	6.7
Pampigny	48.5	36.3	15.2	3.6	6.7
Dizy	44.6	50.8	4.6	3.5	7.6
Sullens	32.5	38.9	28.6	2.1	6.9
Cuarnens	36.2	51.8	12.0	2.9	7.6
Montheron	46.6	35.6	17.8	1.7	5.8
Allens	34.1	41.0	24.9	2.9	7.5

3 RESULTATS ET DISCUSSION

3.1 L'"effet de sol"

La figure 1 représente les quantités d'érosion pour tous les essais confondus. Il est possible de définir trois classes de sol : les moins érodibles (Bussigny et Echandens), les moyennement érodibles (Dizy, Sullens et Allens) et les plus érodibles (Pampigny, Cuarnens et Montheron). La même remarque peut être faite en ce qui concerne le ruissellement : les moins ruissellant étant toujours Bussigny et Echandens, les moyennement ruissellant proviennent de Dizy, Sullens, Allens avec en plus Montheron et Pampigny, Cuarnens demeurant dans une classe à part de sols les plus ruissellant. Ces résultats brutes doivent être expliqués par une description détaillée de ces sols.

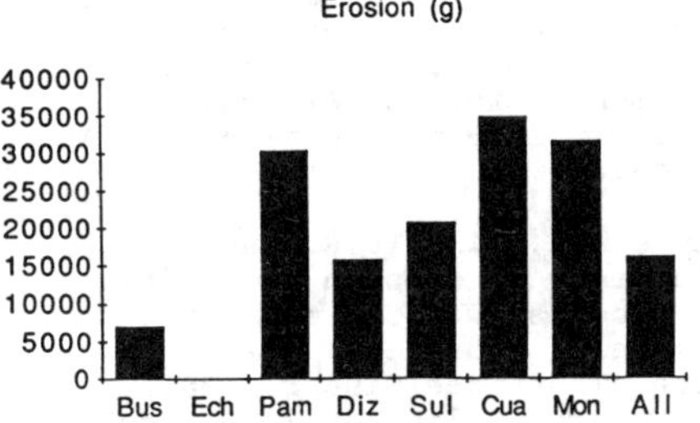

Figure 1. Erosion totale (g) des 12 simulations cumulées par sol.

3.2 L'"effet de pente"

L'influence de la pente est représentée sur la figure 2 suivante. Les sols de Cuarnens et de Montheron ont fortement réagi à ce paramètre alors que tous les autres ont donné des quantités d'érosion plus grande pour la plus forte pente (10%) dans des proportions relativement moindres.

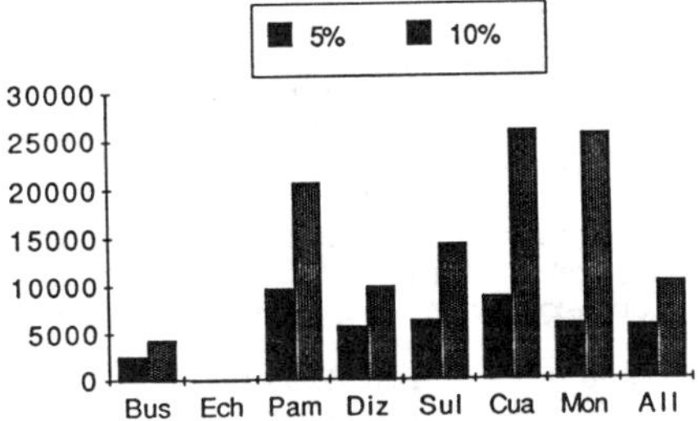

Figure 2. Erosion totale (g) pour les essais à 5% et à 10% de pente par sol.

3.3 L'"effet de pluie"

Les sols se sont comportés différemment lorsque l'intensité des pluies simulées a été augmenté (40 mm/h et 100 mm/h). Les sols d'Echandens, de Bussigny et d'Allens ont sensiblement été érodé dans les même proportions pour les 2 pluies, figure 3. Tous les autres ont substantiellement eu plus d'érosion avec une intensité pluviométrique plus grande et ce dans des proportions diverses.

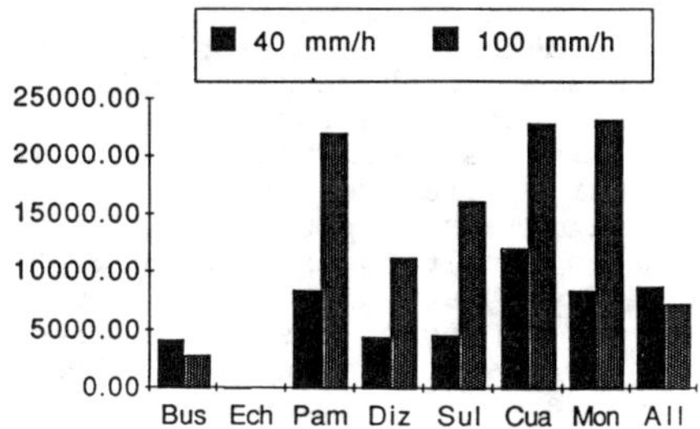

Figure 3. Erosion totale (g) pour les pluies d'intensité de 40 mm/h et de 100 mm/h par sol.

3.4 L'"effet d'humidité"

La figure 4 représente l'influence de l'état hydrique du sol avant chaque essai de simulation. Tous les sols se sont érodés en plus grande quantité à l'état humide et très humide, les sols de Bussigny, Sullens, Montheron et d'Allens ayant réagit le plus fortement.

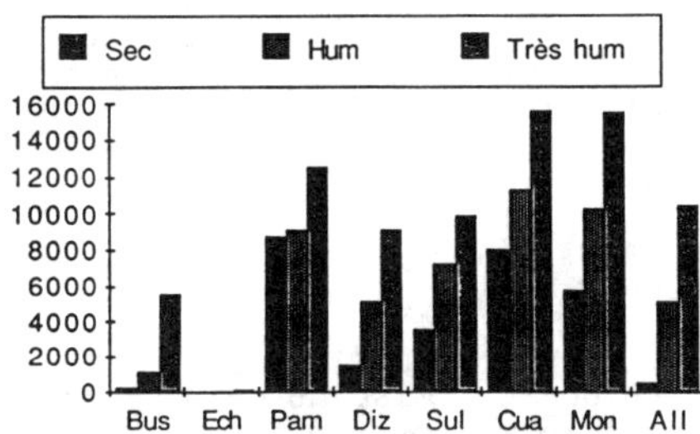

Figure 4. Erosion totale (g) pour les trois états hydriques avant essais, sec, humide et très humide par sol.

4 CONCLUSION

Cette première description des résultats nous permet de classer les sols en fonction de leur comportement et des paramètres pluie, pente et humidité. La présentation de ces résultats bruts sert de base de départ à une interprétation future plus détaillée de tous ces essais.Il est

possible de bien distinguer les différences de quantité d'érosion pour chaque essai. L'effet de pente est marqué dans toutes les conditions expérimentales de simulation. Il en est de même pour l'effet de l'humidité du sol et de l'intensité pluviométrique. A la lueur des résultats expérimentaux obtenus avec ce banc d'essai, il est possible de distinguer des érodibilités propres pour ces sols même si au départ il y a peu de différences entre les propriétés physiques et chimiques de ceux-ci. L'utilisation répétée de ce test amènera les explications nécessaires du comportement des sols du plateau face aux processus érosifs.
En conclusion, nous pouvons dire que les essais en cuve d'érosion permettent de classer les sols selon leur degré d'érodibilité et de distinguer des comportements propres en fonction des paramètres préalables de chaque simulation. Les essais futurs permettront de mettre en évidence les relations entre les propriétés des sols du Plateau Vaudois et les processus d'érosion hydrique. Ces tests de sensibilité à l'érosion se sont avérés concluant et atteignent les objectifs fixés au départ de l'étude. Cependant, comme il est défini auparavant, la dynamique de l'érodibilité des sols implique que des suivis de parcelles en condition naturelle doivent compléter ces essais "in vitro" afin de bien intégrer tous les paramètres régissant le phénomène.

BIBLIOGRAPHIE

Asseline et Valentin (1978). Construction et mise au point d'un infiltromètre à aspersion. Cah. ORSTOM, série Hydrologie, XV, no. 4 p. 321-324.
Bolline (1982). Etude et prévision de l'érosion des sols limoneux cultivés en moyenne Belgique. Thèse de doctorat, Université de Liège, pp356.
Bryant (1968). Development of laboratory instrumentation for the study of soil erodibilty. Earth Sci.J., vol. 2, no. 1, p 38-50.
Bryant (1974). A simulated rainfall test for the prediction of soil erodibility. Z. Geomor. N.F., supp band 21, p138-150.
Casenave (1982). Le mini-simulateur de pluie : conditions d'utilisation et principes d'interprétation des mesures. Cah. ORSTOM, série Hydrologie, XIX, no. 4, p 207-227.

Chevalier (1982). Simulation de pluie sur deux bassins versants sahéliens (Mare d'Oursi-Haute Volta). Cah. ORSTOM, série Hydrlogie, XIX, no. 4, p 253-297.

De Ploey (1981). Some laboratory techniques for invertigating land erosion. In Erosion and sediment transport measurement (Proc. of Florence Symposium), IASH publication, no. 133, p 423-431.

Hadley et al. (1985). Recent developments in erosion and sediment yield studies. UNESCO technical documents in hydrology, p 21-29.
Kenneth, Trott and Singer (1983). Relative erodibility of 20 Californian range and forest soil. Soil Sci. Soc. Am. J., no. 47, p 753-759.
Lal (1981). Analysis of different processes governing soil erosion by water in the tropics. IASH publication no. 13, p 351-366.
Lelong et al. (1984). Géodynamique actuelle de différents sols à végétation naturelle ou cultivée de l'afrique de l'ouest. Catena, vol. 11, p 343-376.
Luk (1979). Effect of soil properties on erosion by wash and splash. Earth Surface Processes, vol. 4, p 241-255.
Marti (1986). Erosion in der Schwiez : ein beitrag zur standort-bestimmung. IUSA, Zürich, Schweiz, pp 53.
Poesen (1981). Rainwash experiment on the erodibility of loose sediments. Earth Surface Processes and Landforms, vol. 6, p 285-307.
Wischmeier and Mannering (1969). Relation of soil properties to its erodibolity. Soil Sci. Soc. Am. J.,vol. 33, no. 1, p 131-137.
Wischmeier et al. (1971). A soil erodibility nomograph for farmland and construction sites. J. of Soil and Water Conservation, sept-oct, p 189-192.
Yamamoto and Anderson (1973). Splah erosion related to soil erodibility indices and other soil properties in Hawaii. Water Resources Research, vol. 9, p 336-345.

Land and Water Use, Dodd & Grace (eds), © 1989 Balkema, Rotterdam. ISBN 90 6191 980 0

Design and development of a rainfall simulator and studies on runoff and soil loss

S.Santhana Bosu & R.K.Sivanappan
Tamil Nadu Agricultural University, Coimbatore, India

ABSTRACT: A tubing tip type rainfall simulator was designed and fabricated to study the runoff and soil loss characteristics of different soil series of Coimbatore district, India. The intensities for simulation and for the design of drop former were selected based on the analysis of rainfall records of Tamil Nadu Agricultural University from 1951 to 1980. Hypodermic needles of 20 gauge size used as drop formers. The four legged frame of angle iron to support entire assembly was designed on the basis of maximum allowable bending stress. The rainfall simulator was taken to the experimental sites at Pollachi, Veerappanur and Pappampatti of Coimbatore district, India for tests. Three intensities viz. 5 cm/hr, 12 cm/hr and 30 cm/hr were selected for simulation studies. Experiments were conducted to study the runoff and soil loss initially for dry state for the intensities for a duration of half an hour. After a break of ten minutes, wet run was given for another 30 minutes and sample collected. Generally, runoff and soil loss bear direct relationship with intensity.

1 INTRODUCTION

In India the amount of soil that is lost as a result of water erosion alone is in the order of 6000 million tonnes annually and it contains nutrients equivalent to 5.37 million tonnes of NPK. Soil erosion on a large scale is not only detrimental to current agricultural production but is a threat to the survival of mankind. To determine land use practices, which will ensure long term productivity, it is quite essential to have the soil loss estimation from individual fields and slopes. Collecting adequate research data involving natural rainfall is time consuming because natural weather is highly variable. Through simulation, the hydrologic processes can be accelerated and the full range of rainfall amounts and hydrologic occurrences can be studied. It is generally more rapid, efficiently controlled and easily adoptable. The rainfall simulators are divided according to the drop forming method as (i) hanging yarn, (ii) tubing tip and (iii) nozzle type (Mutchler & Hermsmeier, 1965). Rainfall simulators have been used for the past half century to accelerate research in soil erosion. This simulation study was taken up to findout the runoff and soil loss of the three major soil series of Coimbatore district in India under three different intensities of rainfall.

2 MATERIALS AND METHODS

To find out the runoff and soil loss under simulated rainfall condition, a tubing tip type rainfall simulator was designed and fabricated. Tubing tip type was selected since it meets all the necessary requirements within acceptable limits (Munn & Huntington, 1976). For the selection of intensity for simulation, thirty years of automatic raingauge charts from 1951 to

1980 available at Meteorological Observatory of Tamil Nadu Agricultural University, Coimbatore were analysed. From the frequency distribution formed for the above data, it was found that the rainfall intensity occurring mostly in Coimbatore is 12 cm/hr, while minimum and maximum of the years are 5 cm/hr and 30 cm/hr respectively. The storms greater than 12.5 mm of rain were considered for computation. Likewise, kinetic energy of the rainfall records 1951 to 1980 were analysed and a frequency distribution was formed. It was found that the kinetic energy of rainfall occurs mostly in the range of 200 to 400 metre tonnes/ha cm.

2.1 Design of diameter of drop former

2.1.1 Determination of drop diameter

Rainfall drop size distribution has been parameterized with the D_{50} drop size where 50 per cent is larger. Relationships have been derived for D_{50} as a function of rainfall intensity. The equation recommended by McGregor and Mutchler (1976) has been made use of to find the median drop size.

$$D_{50}=2.76+11.4\ e^{-1.04I}-13.16\ e^{-1.17I}$$

where D_{50} is median drop size in mm and I is intensity in inches/hr.

2.1.2 Determination of hypodermic needle diameter

From the median drop diameter, the weight of single drop, assuming spherical shape worked out. Mutchler (1965) showed that the diameter of tube, water surface tension and kinematic viscosity could be used as power functions in predicting drop weight. The following equation derived by Mutchler (1965) is made use of to determine the diameter of the drop former.

$$W = 4.924\ \frac{\sigma^{.943}\ d^{.832}\ Q^{.057}\ r^{.093}}{g^{1.018}}$$

where W is drop weight in gm, σ is surface tension of water in gm/sec^2, d is tube diameter in cm, Q is flow

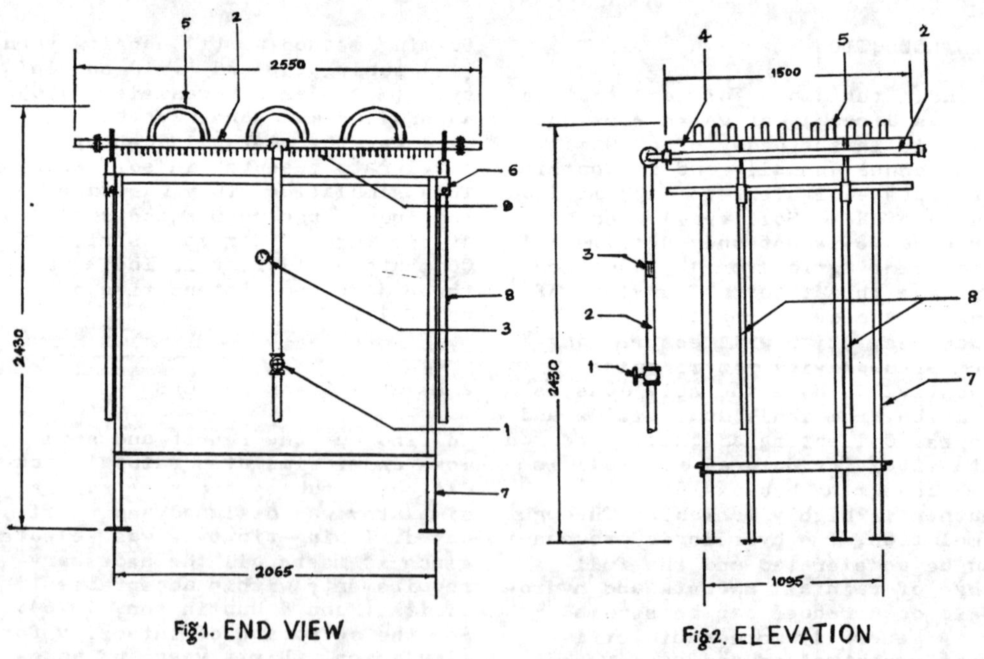

Fig.1. END VIEW

Fig.2. ELEVATION

rate in gm/sec, r is kinematic viscosity in cm^2/sec and g gravitational constant. From this, for the three intensities selected, the D_{50} drop size, volume and weight of drops and the drop former sizes were calculated.

Table 1. Size of drop former

Intensity cm/hr	D_{50} mm	Drop		Drop former size mm
		Volume mm^3	Weight gm	
5	2.916	12.98	0.0130	0.58
12	2.792	11.39	0.0114	0.47
30	2.760	11.01	0.0110	0.42

LEGEND FOR THE FIGURE

No.	Legend description
1	Regulator 25.4 mm size
2	G.I. pipe 25.4 mm diameter
3	Pressure gauge
4	Holder plate
5	Plastic tube
6	Height adjusting screw
7	Angle iron frame
8	Height adjusting rods
9	Hypodermic needle 20 gauge

(All dimensions in mm)

2.2 Design of supporting stand

The 20 gauge hypodermic needles used as drop formers, were soldered to the horizontal G.I.pipes of 12 numbers. They were fixed on two end plates, i.e., holder plates. The entire assembly was supported by a four legged frame of angle iron of 40 x 40 x 3 mm. The design of the frame was made on maximum allowable bending stress.

A portable centrifugal pump with a diesel engine was used for pumping the water to the rainfall simulator. A barrel was used for taking water to the site for testing. A gate valve was provided at the delivery end of the pump to adjust the supply of water.

The rainfall simulator was taken to the farmers' fields at Pollachi, Veerappanur and Pappampatti experimental sites in Coimbatore district to study the runoff and soil loss characteristics of Irugur, Pichanur soil series (both are non calcareous red soils - Typic ustorthents) and Dasarapatti soil series (black soil - Typic chromusterts) respectively. Metallic boundary was made around the test plot (2 x 1 m). A collector unit was fixed at the lower end of the test plot and a small pit was taken for keeping the runoff and soil loss collecting tank at the downstream. The runoff was measured in the collecting tank. Bottle samples were collected for the estimation of soil loss. With all tests, the load was measured by determining the concentrations in grams per litre by pouring off and by evaporating the captured runoff. Then the total soil loss was estimated. The soil loss was estimated for the three intensities 5 cm/hr, 12 cm/hr and 30 cm/hr under two initial conditions, viz. dry and wet conditions. The dry run was given for 30 minutes and after 10 minutes interval wet run was applied for another 30 minutes. Runoff and soil loss samples were collected.

3 TESTING OF RAINFALL SIMULATOR

The drop size was measured using an ocular microscope. Glass plate smeared with silica gel was suddenly inserted and taken back, while raining. The drops formed on the plates were placed under an ocular microscope. The size of drops was measured for various intensities.

Testing was carried out for maximum, medium and minimum intensities (30, 12, 5 cm/hr). The desired intensities were maintained by suitably adjusting the pressure of water supply to the rainfall simulator. The intensity of rainfall was measure by a standard weather bureau raingauge placed inside the simulated area. The intensity has been closely related to total kinetic energy by Wischmeier and Smith (1958) as shown in equation

$$K.E. = 210.3 + 89 \log_{10} I$$

where, K.E. is the kinetic energy in metre tonnes per hectare-centimetre and I is the rainfall intensity in cm per hour. This relationship was used to work out the kinetic energy of storms.

4 RESULTS AND DISCUSSION

The pressure of water supplied to the rainfall simulator was varied and the intensity and drop size at various pressures were measured. As the pressure increases the intensity is also increasing. The pressure vs intensity is having a quadratic relationship given by the equation

$$I = -606.67\ p^2 + 366.51\ p - 10.4$$

where, I is the intensity in cm/hr and p pressure in kg/cm^2. The rainfall simulator was designed to produce rain of any intensity.

Since the most severe erosion problems were usually associated with high intensity storms, this simulator was also designed to apply water at relatively high intensities. The increase in intensity was obtained by increasing the pressure which inturn increased the flow rate supplied to the drop formers (without changing the number or size of drop formers). Meyer (1958) also reported that where the pressure on water source to the drop former could be varied, flow control of resultant intensities could often be achieved over rather wide ranges without varying the number or size of the formers.

The drop size decreases with increase in intensity. However, the intensity and drop size are having a linear relationship.

$$D = 2.387 - 0.033\ I \qquad (r = -0.99)$$

where D is drop size in mm and I is the intensity in cm/hr. At maximum intensity, because of high pressure and the same drop former size, the drop diameter decreases. Hudson (1963) has demonstrated that the median drop diameter decreases when the intensity exceeds 10 cm/hr.

4.1 Runoff

The mean performance of runoff of the three soil series viz., Irugur, Pichanur and Dasarapatti is presented in Table 2.

Table 2. Mean performance of runoff (m^3/ha)

Sl. No.	Intensity cm/hr	Igr	Pch	Dpt
Under dry condition				
1	5	51.55	68.45	86.79
2	12	70.09	266.70	459.09
3	30	86.80	573.52	676.07
Under wet condition				
1	5	84.10	138.41	207.85
2	12	94.95	420.11	476.90
3	30	124.25	686.45	719.47

Igr = Irugur; Pch = Pichanur; Dpt = Dasarapatti

In all the three soil series, the runoff is found to be high in both dry and wet conditions for the intensity 30 cm/hr. During any rain, if the rainfall intensity is high and duration is long, runoff would be produced after meeting the requirements of infiltration to fill up the empty pore spaces. As the intensity increases the runoff rate also increases. For any of the three soil series, the runoff is more in wet condition than in dry condition for the three intensities tested. This behaviour, as expected, is because of the fact that a soil which is already wet will attain its final infiltration rate more quickly when rain falls upon it. Therefore runoff contribution at wet condition is greater than in dry condition. Among the three soil series, the Dasarapatti soil series has more runoff when compared to other two soil series in all the three intensities. This may be due to higher clay content and low permeability of the Dasarapatti series. The percentage runoff from the field plots on different soil types were in the inverse order of the permeability (Peele, 1937). By analysis of variance for runoff and interaction effects for the soil series, intensities and moisture conditions, the runoff showed significance at 1% level for all parameters, and the runoff is highly influenced by intensity than the conditions and interactions.

4.2 Soil loss

The mean performance of soil loss of the three soil series is presented in Table 3.

Table 3. Mean performance of soil loss in kg/ha

Sl. No.	Intensity cm/hr	Igr	Pch	Dpt
Under dry condition				
1	5	335	551	2223
2	12	357	1750	17618
3	30	423	2734	34147
Under wet condition				
1	5	256	585	5400
2	12	527	3013	10739
3	30	1050	6102	43098

Igr = Irugur; Pch = Pichanur; Dpt = Dasarapatti

High soil loss is exhibited in 30 cm/hr intensity for the three soil series in both wet and dry conditions. In the Irugur soil series, except for 5 cm/hr intensity, the rest have more soil loss in wet condition than in dry condition. The soil loss increases with increase in intensity in both the conditions, which is in confirmation with the findings of Free (1960), Bubenzer and Jones (1971) and Elwell and Stocking (1973). For the Pichanur series, the soil loss is more in wet than in dry condition for all the intensities. The soil loss is found to be increasing with intensity. Dasarapatti series shows increasing trend of soil loss with intensity in both conditions. Except for intensity 12 cm/hr wet condition contributes more soil loss than dry condition for all other intensities. Among the series Dasarapatti series, i.e. black soil is subjected to severe erosion than red soils. The loss may be due to high clay per cent which results in poor infiltration rate. Bubenzer and Jones (1971) also expressed the importance of per cent clay in the soil for the prediction of soil splash.

5 CONCLUSION

From this simulation study conducted on fields using rainfall simulator, the following conclusions were made: The highest intensity of rainfall 30 cm/hr yielded the maximum runoff in all the soil series under both dry and wet conditions. In all the three intensities tested, for all the soil series, the runoff was more under wet condition than in dry condition. Black soil produced more runoff than the red soils due to low permeability of the former. Runoff contribution was primarily affected by the variation in intensities than the other parameters.

Soil loss was severe under 30 cm/hr intensity in both conditions. Black soil series subjected to more erosion than red soil series for any intensity under given moisture condition. Soil loss was more pronounced in wet condition than in dry condition. As the intensity increases, both runoff and soil loss increases in all the three soil series.

ACKNOWLEDGEMENT

The authors are thankful to the United States Department of Agriculture, Office of the International Co-operation and Development authorized by Public Law 480 for financing this research study.

REFERENCES

Bubenzer, G.D. & B.A.Jones Jr. 1971. Drop size and impact velocity effects on the detachment of soils under simulated rainfall. Trans. ASAE. 14(4): 625-628.

Elwell, H.A. & M.A.Stocking. 1973. Rainfall parameters for soil loss estimation in a subtrópical climate. J.agric.Engg.Res. 18: 169-177.

Free, G.R. 1960. Erosion characteristics of rainfall. Ag.Engg. 41: 447-449, 455.

Hudson, N.W. 1963. Raindrop size distribution in high intensity storms. Rhod.J.agric.Res. 1(1): 6-11.

McGregor, K.C. & C.K.Mutchler. 1976. Status of the R factor in northern Mississippi in soil erosion: Prediction and control. Proceedings of National Soil Erosion Conference at Purdue University, SCSA, Ankeny, Iowa, 135-142.

Meyer, L.D. 1958. An investigation of methods for simulating rainfall on standard runoff plots and a study of the drop size, velocity and KE of selected spray nozzle. USDA-ARS Special Report No.81, 43.

Munn, J.R. & G.L.Huntington. 1976. A portable rainfall simulator for erodibility and infiltration measurements on rugged terrain. Soil Sci.Soc.Am.J. 40: 622-624.

Mutchler, C.K. 1965. Water drop formation from capillary tubes. USDA-ARS Report No.41, 107: 12.

Mutchler, C.K. & L.F.Hermsmeier. 1965. A review of rainfall simulators. Trans.ASAE. 8(1): 67-68.

Peele, T.C. 1937. The relation of certain physical characteristics to the erodibility of soils. Soil Sci.Soc.Am.Proc. 2: 97-100.

Wischmeier, W.H. & D.D.Smith. 1958. Rainfall energy and its relationship to soil loss. Trans.Am. Geophys.Union, 39: 285-291.

Land and Water Use, Dodd & Grace (eds), © 1989 Balkema, Rotterdam. ISBN 90 6191 980 0

Measurement and modeling of cool season soil erosion

J.R.Burney
Technical University of Nova Scotia, Halifax, Canada
L.M.Edwards
Agriculture Canada Research Station, Charlottetown, Canada

ABSTRACT: A series of soil erosion measurements have been conducted ranging from laboratory to field scale, with emphasis on the cool season of fall through spring. Equipment and data are presented for aggregate stability tests and interrill erosion under freeze-thaw cycling, for unit erosion plots and for agricultural watersheds.

RESUME: On a éffectué une série de me sures de l'érosion du sol au laboratoire et au champ en mettant l'accent sur les saisons froides soit de l'automne au printemps. Le matériel et les données sont presenté.s à la fin des testes de stabilité des aggrégats du sol et pour l'érosion en nappe en cycle de gel-dégel dans des parcelles de measure de l'érosion et des basins versants.

ABSTRAKT: Eine Reihe von Boden-Erosions-Messungen - in Labor und Feld - sind durchgeführt worden, wobei wir besonders die kalte Jahreszeit von Herbst bis Frühjahr berücksichtigten. Geräte und Daten werden vorgestellt: für Aggregat-Stabilitätstests und Mikro-Erosion, bei wechselnden Klimaverhältnissen (Gefrieren/Tauen), für Erosions-Teststücke und für landwirtschaftliche Einzugsgebiete.

1 INTRODUCTION

The Province of Prince Edward Island (P.E.I.) in the Atlantic region of Canada experiences a cool, temperate climate in which extremes in temperature are moderated by the surrounding ocean. Rainfall tends to be of low intensity with few convection storms. Soils, in the rolling topography, are predominantly sandy loam in texture and the dominant annual crops are potatoes followed by cereals, tobacco and soybeans.

However, in spite of the above, soil erosion is widespread in P.E.I.; particularly on potato lands. During summer the soils vary from moist to dryish and are protected by the growing crops. Following potato harvesting in autumn, through until a reasonable cover has again been established by the succeeding crop, planted in late April to mid-May, soils are often left bare. Rains during autumn, and fluctuating temperatures during late winter and early spring, in which snowfall alternates with snowmelt often accompanied by light rainfall, results in a frozen, saturated soil in which the top layer is subjected to several freeze-thaw cycles.

Potatoes are cultivated up-and-down the slope based on historical layout and for operational ease. Fields are mostly in long, narrow strips and can have slopes approaching 12 per cent. In recent years the trend also has been towards heavier field equipment, thus compounding the deleterious effects of naturally occurring compaction at depths of 300 to 600 mm. Natural drainage is therefore further impeded (Edwards 1988), leading to increased runoff. Potato fields left bare following harvesting commonly show marked rilling in the spring.

2 PREVIOUS STUDIES

Zuzel et al (1982) noted that in the

Pacific north-west of the U.S.A., light intensity winter storms and overland flow dominates erosivity. Intermittent snow accumulations and/or frozen soils are common during winter. They found that snowmelt and/or frozen soils were responsible for 86 per cent of measured soil loss on field plots over winter, and noted that rainfall energy was not a major factor in soil erosion. Analogously, Morgan (1979) presented European data which showed substantial soil loss for low intensity rainfall of long duration on saturated soils.

McCool et al (1982) and Onstad and Young (1982) in an examination of the applicability of the USLE to winter conditions, found that, as suggested by Wischmeier (1976), the erosivity factor (R) requires substantial upward modification under conditions in which snowmelt runoff is a significant erosive factor.

Spomer and Hjelmfelt (1983) found that snowmelt and resultant runoff produced less than one per cent of the average annual soil loss from loess soils in Iowa. Kirby and Mehuys (1987) reported that soil texture had a marked effect on relative seasonal soil losses from fallow soils in Quebec. For coarse textured soils approximately 80 per cent of the annual soil loss occurred in summer, while for a clay soil approximately 90 per cent of the annual soil loss occurred in winter.

Data by Pol (1988) indicate that for the Wilmot River in central P.E.I. over half of the average annual suspended sediment load for the upper 45 sq. km. reach occurs in late winter to early spring (March and April), while only 8 per cent was recorded during the summer/fall months of June through October.

3 RESEARCH PROGRAM

A program aimed at evaluation of alternate land use practices, with emphasis on viable methods to protect soils, primarily on potato lands, during the 'cool season', was initiated in 1984. Microscale tests have involved evaluation of aggregate stability under freeze-thaw cycling (Edwards and Burney 1988) and the use of a laboratory cold room facility to separately evaluate interrill (Burney and Edwards 1987) and rill erosion under varying temperature, winter cover crop, residue and compaction effects. At the field scale (Frame et al 1988), three watersheds were selected for detailed measurement and modeling within the upper Wilmot River basin, a major potato production area in central P.E.I., which has been monitored for flow and suspended sediment since 1972 by Environment Canada (Pol 1988). Specifically, the three selected watersheds are 140 ha of mainly potato monoculture lands, 203 ha of mainly potato lands in rotation and 412 ha of mixed farming predominantly under permanent pasture. Within each of these watersheds a set of three unit erosion plots was installed in late 1986. At each site, a hut located next to the erosion plots, houses the data acquisition recording equipment for the watershed site. Electric power and data acquisition lines run from the central hut to each of the erosion plot bunkers and the stream monitoring facility.

Meteorological instrumentation includes rain/snow gauges, wind anemometer, evaporation pan, soil and air temperature sensors, and soil mosture blocks. Snow depth and water content is monitored manually.

3.1 Microscale Tests

Microscale tests have been conducted on three soil textural classes (loam, fine sandy loam and sandy loam) which comprise the range for soils on P.E.I. However, due to logistical problems the same soil series was not used for the specified textural classes in each test. These tests comprised measurements of aggregate stability under freeze-thaw cycling, two sets of factorial experiments using a rainfall simulator in a temperature controlled enclosure to evaluate interrill erosion, and a factorial experiment using the same cold room facility to evaluate rill erosion.

A modified procedure (Kemper 1965) was used to test the effects of daily freeze-thaw cycling on the stability of soil aggregates having a high initial moisture content as occurs naturally in the late winter and early spring in P.E.I. (Edwards 1988). As shown in Figure 1, a rapid decline in stability and therefore increased susceptibility to erosion occurred during the initial period for all three soils, following which little further breakdown occurred.

In the first of two sets of factorial experiments to evaluate cover effects on interrill erosion, 72 of 0.3 m wide by 0.9 m long and 0.3 m deep soil boxes were tested under a rainfall simulator in a temperature controlled enclosure (Edwards and Burney 1987). The experimental design comprised three soils (loam, fine sandy loam and sandy loam), two temperature effects (frozen for 5 days at -5 C and room temperature, 20 C), three cover effects (bare, early and late planted winter rye) and two erosion agents (rainfall, with and without added upslope flow). Tests were run (at 1 C for the frozen soils) for a period of 30 minutes under a rainfall application rate of 50 mm/h, and upslope flow rate of 150 mm/h where applicable. Both splash and runoff were measured in terms of volume and sediment content.

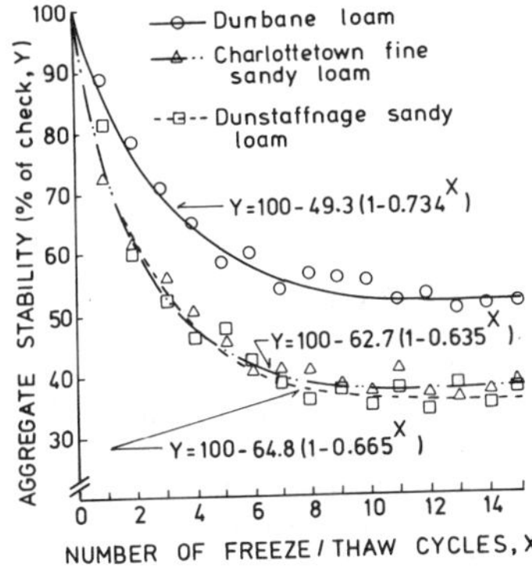

Fig. 1 Effect of number of freeze/thaw cycles on aggregate stability

Sediment in splash was affected only by cover. Sediment in runoff was affected by all the main treatments, and clearly indicated the benefit of a cover crop in protecting soil from interrill erosion during the more vulnerable winter period. Soil loss from the bare soil boxes averaged about six times that from the early seeded winter rye boxes, whereas runoff was not significantly different. Soil loss from the frozen boxes averaged nearly double that from the unfrozen boxes, while runoff was about 25 per cent higher.

In a follow-up factorial experiment, 144 boxes were used to test treatment effects on the same three textural classes of soils. The treatments were temperature (continuously frozen at -5 ℃ and daily freeze-thaw cycling, each for 10 days), cover (bare, early seeded winter rye and incorporated cereal residue) and layering (homogeneous profile and an induced compacted layer at 150 mm depth). Incorporated cereal straw decreased both runoff and soil loss to below that of the winter rye, thus proving to be a most effective protection practice having limited viability in potato lands. The continuously frozen soils produced nearly three times the volume of both runoff and sediment compared to the freeze-thaw treatment. This is attributed to an observation during the freezing pre-treatment that the boxes subjected to freeze-thaw cycling generally developed marked cracking. Although the compacted layer did result in slightly increased runoff and sediment loss, the fact that it did not have a more significant effect was attributed to the relative impermeability of ice layers which still existed at a depth of about 20 mm after the rainfall simulator runs. This was most marked in the continuously frozen boxes which showed little effect of sub-surface compaction on runoff or sediment.

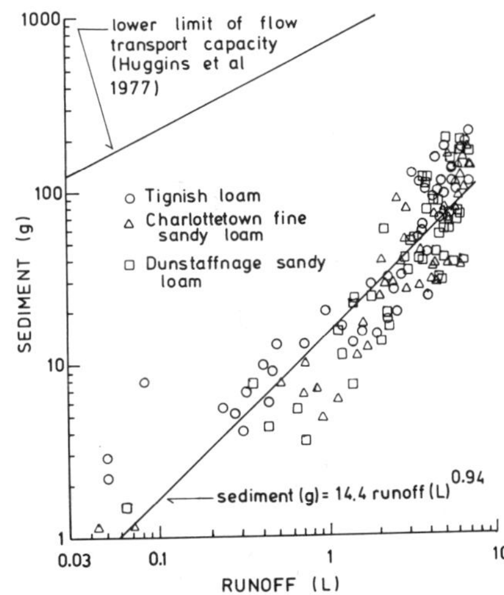

Fig. 2 Runoff versus sediment for all 144 soil boxes

A linear equation, shown in Figure 2, fitted by regression to the logarithms of sediment mass (dependent variable) and runoff volume for all 144 soil boxes, had a correlation coefficient of 0.82 with reasonable uniformity of residual dispersion. The power value was 0.94, which is very close to unity and indicates that sediment concentration was unaffected by runoff (no flow detachment) and that rate of detachment of sediment was constant with variable flow division between runoff and infiltration. Comparison with a flow transport equation (Huggins et al 1977) transformed to mass form, indicated that transport capacity of the flow was unsatisfied.

Overall, the presence of cover had little effect on runoff but markedly affected sediment mass loss. However, runoff from interrill areas under good cover protection has an increased unsatisfied transport capacity and may therefore tend to create wider and deeper rills as it erodes soil in concentrated flow where surface residue is sparce.

Particle size distributions of the original soils and the sediment indicated marked enrichment of the sediment, as shown in Figure 3. To better quantify treatment effects two parameters, fines ratio (FR) and relative enrichment (RE) were defined as below:

$$FR = \frac{SD75}{SD} \quad (1)$$

and,

$$RE = \frac{FR - \frac{SL75}{SL}}{\frac{SL75}{SL}} \quad (2)$$

where, SD75 = mass less than 75 microns in sediment sample,
SD = mass of sediment sample,
SL75 = mass less than 75 microns in original soil sample,
SL = mass of original soil sample.

The fines ratio represents a measure of the particle size distribution in flow, while the relative enrichment represents a measure of the enrichment of sediment relative to the original soil. The effect of runoff on each of these parameters was removed by covariate analysis prior to analysis of

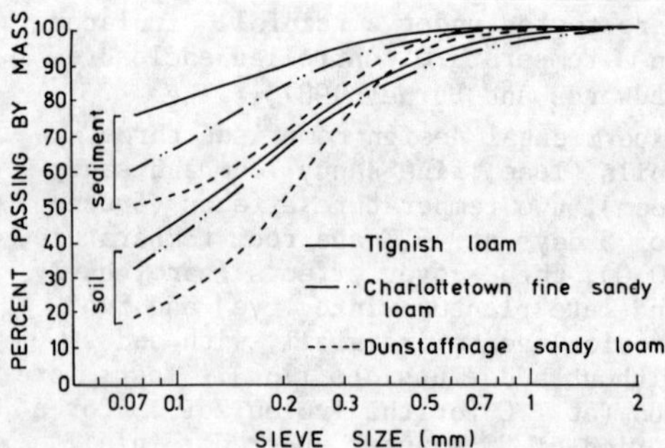

Fig. 3 Particle size distributions for original soils and sediment

variance. The results indicated that for each parameter there was no cover or temperature effect; however, layering and soil type were highly significant.

For the homogeneous (no compact layer) treatment both the FR and RE values were higher than for the layered treatment, and were reflective of the lower total sediment loss. For the soils, FR was highest for the loam soil followed by the fine sandy loam and sandy loam: however, for RE this order was reversed, indicating that relative loss of fine material increased from loam to sandy loam. In terms of productivity therefore, equivalent mass loss from each of the soils is much more severe as texture becomes coarser.

Using the same cold room facility, a factorial experiment was run to evaluate treatment effects on rill erosion on a saturated fine sandy loam soil. Eight troughs, each 3.5 m in length by 0.2 m in width and 0.15 m deep were constructed and insulated on the sides and bottom. Treatments comprised four replications of: residue (bare and incorporated straw), compaction (surface compacted and homogeneous), slope (3.5 % and 7 %) and temperature (room temperature and daily freeze-thaw cycling for 10 days). Water was applied at a rate of 4 L/min to each box. Runoff samples were taken during each run of 20 minutes duration. As anticipated (Frame, personal communication), mass of sediment loss increased markedly with slope and decreased under compaction and residue incorporation. For the normal temperature treatments rate of sediment loss tended to be low initially, increase to a peak and then decline to an equilibrium. For the freeze-thaw pre-treatments, sediment concentration was initially high and decreased to an equilibrium. Mass sediment loss was higher overall for the freeze-thaw treatments.

3.2 Erosion Plots

As previously indicated, three sets of three erosion plots were set out on 3 to 5 per cent slopes, one set adjacent to, each of three monitored watersheds, one of which is shown in Figure 4. Each is 22.1 m downslope by 4 m wide. For each set of plots the soil is a fine sandy loam but varies in terms of previous cropping history. One set is in potato monoculture land, the second in a field where rotation was practiced and the third in an old pasture. Runoff at the lower end of each plot is concentrated into a short channel. This channel passes through the wall of a heated bunker and terminates in a HS flume. Stage in the flume is recorded on the site datalogger and flow through the flume triggers a discrete sediment sampler. Flow from the end of the flume enters the first of a sequence of three barrels. One-seventh of the overflow from barrel 1 enters barrel 2 from which, in turn, one-ninth of the overflow passes to barrel 3. Volume of runoff and mass sediment are measured manually by depth measurement and sampling, respectively, on an event (or daily) basis throughout the year.

Plot treatments at each site comprise a three year rotation of early maturing potatoes (summer) followed by winter rye (winter), late maturing potatoes (summer) followed by bare (winter) and barley (summer) followed by stubble (winter). During the period of May, 1987 to May, 1988 (the first year of recording) average soil loss from all plots was 9.8 Mg/ha ranging from 15.4 Mg/ha for potatoes-winter rye on previously potato monoculture land to a low of 5.6 Mg/ha for barley-stubble on previous permanent pasture land. On average potatoes-winter rye resulted in 11.7 Mg/ha, potatoes-bare 10.9 Mg/ha and barley-stubble 6.7 Mg/ha. In terms of previous cropping history potato monoculture land yielded an average of 12.1 Mg/ha, rotation land 9.8 Mg/ha and permanent pasture 6.7 Mg/ha.

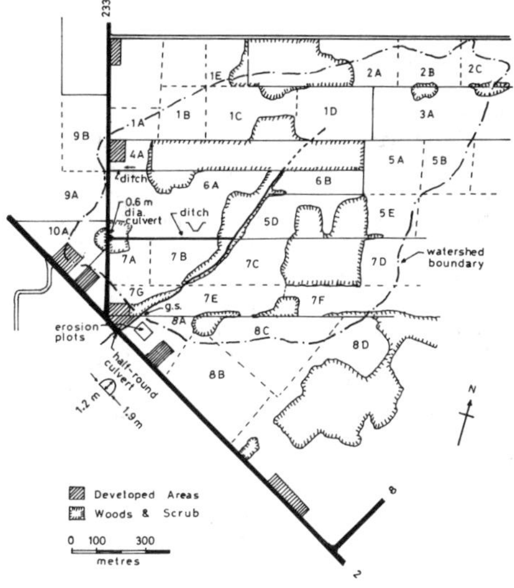

Fig. 4 Property and field boundaries in 140 ha watershed monitored, and associated erosion plots location

Approximately 70 per cent of the soil loss occurred during the cool season.

For the year of recording, the fact that potatoes followed by winter rye resulted in more soil loss than potatoes followed by bare soil on two of the three sets of plots was surprising. Winter cover cropping to protect against soil loss over the cool season is often advocated, and the interrill erosion tests indicated its effectiveness. However, at the larger scale unit plot level it would appear that cultivation which accompanied planting of the winter rye following harvesting created a loose soil surface more easily eroded by rilling in particular. This showed in the late fall period, whereas in the usually more crucial late winter/spring period, soil losses from the bare plots were consistently higher followed by winter rye and barley stubble. Interseeding of winter rye between the potato rows just prior to harvesting would avoid the loose seedbed and soil loss associated with seeding after harvesting.

The data generated from the erosion plots are being utilized in applying the CREAMS model (Knisel 1980) and also provide sample data in applying the

ANSWERS model (Huggins et al 1977) to the associated watersheds.

3.3 Watersheds

Flow records for the upper 45 sq. km of the Wilmot River, dating back to 1972 (Pol 1988), indicated a strong annual periodicity in discharge and suspended sediment load, with peaks occurring in late winter/early spring.

For monitoring the outflow from the three selected sub-watersheds of 140, 203 and 412 ha it was essential that the facilities be insensitive to icing. Additional constraints were no impedance of natural flow, ability to evaluate discharge under conditions of submergence caused by downstream blockage (due to snowbanks, debreis, etc.) and that total sediment (bed load plus suspended sediment) be monitored. Accuracy of low baseflow measurement was not a major concern as such flow carries little sediment.

Based on the above criteria, a Parshall flume section was selected for each site. These sections were each constructed to a 1.22 m depth to allow for underestimation of peak discharges (above the 0.9 m rated depth). A heated hut to house the stage sensor and sediment sampling instrumentation is located alongside each flume.

Water stage at each of the two Parshall flume monitoring locations is recorded using a float and pulley connected to a potentiometer. The stilling well for each stage recorder comprises a 150 mm section of aluminum pipe around which is wound a heating cable, and the whole section then encased in styrofoam blocks. The stage signal is converted to a current proportional signal and relayed over buried lines to the site datalogger.

Sediment sampling is accomplished by means of an in-stream sampling section and a sediment sampler unit similar to that used for the erosion plots. A section, referred to as a 'flow splitter', is mounted vertically on the exit to the Parshall flume. Flow enters a 3 mm wide vertical slit, which faces directly upstream. This widens to a section of 25 mm in width and 300 mm in the downstream direction in which considerable mixing occurs. The sampling port is mounted at the lower end of this mixing section, which is followed by a 50 mm pipe leading to an in-stream sump pump. Tubing, attached to a heating cable and the whole then encased in PVC pipe, runs from the sampling port to a peristaltic pump and then to the sediment sampler unit. At the time of sampling the sump pump starts up and runs for a period of 20 seconds to ensure full drawdown in the mixing section, and to clear deposited sediment. With the sump pump continuing to run, the peristaltic pump then runs for a period of 20 seconds to flush out the lines before switching to sampling. The run time for each pump and the sampling time may be independently set by DIP switches. Additionally, the time between samples is set to an integer number of datalogger scans. Sampling is triggered by the upstream stage recorder indicating a stage above a set threshold. Time of sampling is recorded on the datalogger.

A regular inventory is kept of cropping and cultural practices on each field within each of the watersheds. Overlay maps of field boundaries, topography and soils were used to generate the data base for adaptation of the ANSWERS model (Huggins et al 1977) to cool season soil erosion.

Installation and testing of the sediment samplers has only recently been completed. Coupled with a lack of significant events, stream sampling data collected to date has been sparce.

REFERENCES

Burney, J.R. & L.M. Edwards 1987. Soil erosion under freeze/thaw conditions. Paper No. 87-2601, Amer. Soc. of Agr. Eng., St. Joseph, MI, U.S.A.

Edwards, L.M. 1988. The effects of slope position and cropping sequence on soil physical properties in Prince Edward Island. Can. J. Soil Sci. 68:763-774.

Edwards, L.M. & J.R. Burney 1988. Erosion of soil under frozen and freeze-thaw conditions. Proc. of Conf. XIX, Inter. Erosion Control Assoc., New Orleans, LA, U.S.A.. 355-366.

Edwards, L.M. & J.R. Burney 1987. Determination of soil erosion losses under freeze/thaw and winter ground cover using soil cassettes and a laboratory rainfall simulator. Can. Agr. Eng. 29:109-115.

Frame, P.A., L.M. Edwards & J.R. Burney 1988. Soil erosion measurement facilities for cool season recording. Paper No. NAR-88-405, Amer. Soc. of

Agr. Eng., St. Joseph, MI, U.S.A.
Huggins, L.F., D.B. Beasley & J.R. Burney 1977. ANSWERS model concepts. In: Final report on the Black Creek project. USEPA Rep. No. EPA-905/9-77-007-B.
Kemper, W.D. 1965. Aggregate stability. In: C.A. Black (ed.), Methods of soil analysis. Part 1. Agron. 9, Amer. Soc. Agron., Madison, WI, U.S.A.
Kirby, P.C. & G.R. Mehuys 1987. The seasonal variation of soil erosion by water in southwestern Quebec. Can. J. Soil Sci. 67(1):55-63.
Knisel, W.G. 1980. CREAMS: A field-scale model for chemicals, runoff and erosion from agricultural management systems. USDA Conservation Research Rep. No. 26, USDA-ARS, Washington, D.C., U.S.A.
McCool, D.K., W.H. Wischmeier & L.C. Johnson 1982. Adapting the universal soil loss equation to the Pacific northwest. Trans. Amer. Soc. Agr. Eng. 25:928-934.
Morgan, R.P.C. 1979. Soil erosion. Academic Press, Don Mills, Ont., Canada.
Onstad, C.A. & R.A. Young 1982. Erosion characteristics of three northwestern soils. Trans. Amer. Soc. Agr. Eng. 25:367-371.
Pol, R.A. 1988. Interpretation of sediment data, 1972-1985, Wilmot River, Prince Edward Island. Rep. No. IW/L-AR-WRB-88-137, Environment Canada, Ottawa, Canada.
Spomer, R.G. & A.T. Hjelmfelt 1983. Snowmelt runoff and erosion on Iowa loess soils. Trans. Amer. Soc. Agr. Eng. 26:1109-1111,1116.
Wischmeier, W.H. 1976. Ch.3. Cropland erosion and sedimentation. In: Contol of water pollution from cropland. Vol II. An overview. Rep. No. ARS-H-5-2, USDA-ARS, Washington, D.C., U.S.A.
Zuzel, J.F., R.R. Allmaras & R. Greenwalt 1982. Runoff and soil erosion on frozen soils in northeastern Oregon. J. of Soil and Water Cons. 37:351-354.

Land and Water Use, Dodd & Grace (eds), © 1989 Balkema, Rotterdam. ISBN 90 6191 980 0

Erosionsmindernde Bestellverfahren für Reihenfrüchte

M.C.Estler
Institut für Landtechnik, Freising-Weihenstephan, Bundesrepublik Deutschland

ABSTRACT: Soil erosion not only means a loss of valuable arable soil, but also serious secondary problems. Therefore it is necessary to prohibit soil erosion effectively. The use of new systems with combination of plant production and agri-technical measures guarantees a sucessful control of erosion.

ZUSAMMENFASSUNG: Bodenerosion bewirkt nicht nur den Verlust an wertvollem Ackerboden, es können auch schwerwiegende Folgeprobleme entstehen. Deshalb ist es erforderlich, das Entstehen von Bodenabtrag wirksam zu verhindern. Die Anwendung neuer Verfahren mit einer Kombination von pflanzenbaulichen und landtechnischen Maßnahmen bietet die Chance, einen erfolgreichen Erosionsschutz zu erreichen.

Die Bodenerosion hat sich weltweit zu einem ernst zu nehmenden Problem entwickelt. Unter den mitteleuropäischen Klimabedingungen ist vor allem mit Wassererosion zu rechnen, währenddessen Winderosionen nur auf extrem leichten Sandböden auftreten. Mit dem Auftreten von Bodenabtrag ist vor allem dann zu rechnen, wenn folgende Faktoren vorliegen: hohe Niederschlagsmenge und Intensität, Hangneigungen über ca. 5 %, Hanglängen über ca. 100 m, Bodenarten mit hohem Anteil an leicht abschlämmbaren Teilchen, das Fehlen von Pflanzenbewuchs oder Pflanzenresten an der Bodenoberfläche, schlechter Bodenstrukturzustand, zu intensive und feinkrümelige Bodenbearbeitung.

In Ackerbaubetrieben ist das Ausmaß der Schäden und auch die Geschwindigkeit, mit welcher die Erosionsvorgänge ablaufen, vor allem abhängig von den im Rahmen der Bodenbewirtschaftung vorgenommenen mechanischen Eingriffen in das natürliche Bodengefüge. Konzepte zur Verminderung der Bodenerosion können jedoch nur dann langfristig eine Problemlösung gewährleisten, wenn sie die gesamte Organisation eines Betriebes erfassen und nicht auf einzelne Arbeitsgänge oder Fruchtarten beschränkt bleiben. Im Einzelfall ist zu entscheiden, wo und wann die höchste Erosionsanfälligkeit besteht, wo also die Anwendung geeigneter Maßnahmen die größten Erfolgsaussichten bieten.

Die Erfahrungen der letzten Jahre zeigen, daß vor allem in Jahren mit hohen und intensiven Niederschlägen auch in Getreidebeständen teilweise erheblicher Bodenabtrag auftrat. Wesentlich schwerwiegender sind jedoch Erosionsprobleme bei Reihenfrüchten (vor allem Mais und Zuckerrüben, in abgeschwächter Form auch bei Raps, Ackerbohnen etc.), da infolge der vergleichsweise großen Reihenabstände ein geringerer Schutz der Bodenoberfläche durch oberirdische Pflanzenteile vorliegt. Die nachfolgenden Ausführungen befassen sich deshalb schwerpunktmäßig mit landtechnischen Erosionsschutzmaßnahmen für Mais, Zuckerrüben und ähnliche Reihenfrüchte.

Aus verfahrenstechnischer Sicht ist zu unterscheiden zwischen Erosionsschutzmaßnahmen, die im Rahmen der konventionellen Bodenbewirtschaftung durchgeführt werden, sowie Spezialverfahren mit einer Kombination von pflanzenbaulichen und landtechnischen Maßnahmen.

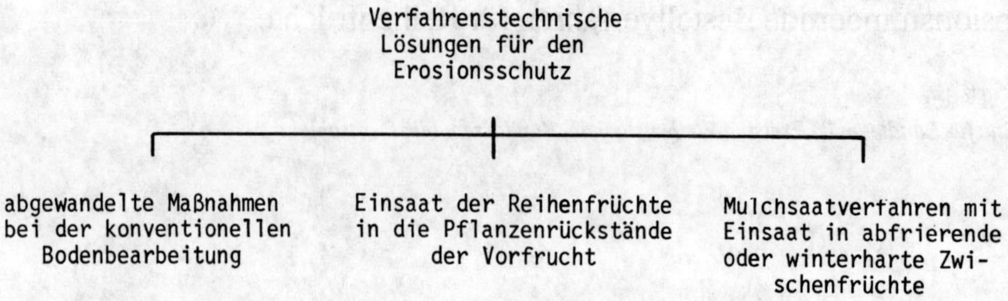

Verfahrenstechnische Lösungen für den Erosionsschutz

- abgewandelte Maßnahmen bei der konventionellen Bodenbearbeitung
- Einsaat der Reihenfrüchte in die Pflanzenrückstände der Vorfrucht
- Mulchsaatverfahren mit Einsaat in abfrierende oder winterharte Zwischenfrüchte

Abbildung 1: Verfahrenstechnische Lösungen für den Erosionsschutz

EROSIONSSCHUTZ BEI KONVENTIONELLER BODENBEWIRTSCHAFTUNG

Die konventionelle Bodenbewirtschaftung ist gekennzeichnet durch den regelmäßigen Einsatz des Pfluges für die Primär-Bodenbearbeitung auf volle Krumentiefe. Die für die gesamte Bodenbearbeitung und Feldbestellung erforderlichen Maßnahmen erfolgen in einzelnen, getrennt und nacheinander ablaufenden Arbeitsgängen.

Erosionsmindernde Maßnahmen beginnen bereits beim Pflugeinsatz. Aus der Sicht der Bodenerosion sind beim Pflügen vor allem zwei Effekte besonders problematisch: das tiefe Einbringen von organischen Substanzen und die starke Überlockerung der Krume. Durch das Anlegen einer rauhen Herbstpflugfurche läßt sich vor allem während der Wintermonate der Wasserabfluß an der Bodenoberfläche bremsen und der Bodenabtrag reduzieren. Auf besonders erosionsgefährdeten Standorten sollte darüber hinaus geprüft werden, ob nicht zumindest zeitweise auf den Pflugeinsatz verzichtet und die tiefgreifende Bodenbearbeitung mit dem Grubber durchgeführt werden sollte. Damit ließen sich zwei Vorteile, die ausreichend intensive und tiefe Lockerung des Bodens sowie der Verzicht auf die Bodenwendung vorteilhaft verbinden.

Auf vielen Standorten kann eine rauhe Saatbettbereitung ebenfalls einen wirksamen Erosionsschutz, insbesondere in der ersten Vegetationsphase nach der Saat, gewährleisten. Diese Maßnahme bleibt jedoch weitgehend auf Fruchtarten beschränkt, die trotz einer grobkrümeligen Saatbettbereitung gute Feldaufgänge erwarten läßt. Bei der Aussaat von Zuckerrüben, bei welcher üblicherweise eine sehr feinkrümelige Saatbettvorbereitung erfolgt, lassen sich schmale Krümelwalzen für eine Streifen-Feinbearbeitung einsetzen. Der eigentliche Bereich der Saatgutablage wird dann ausreichend fein gekrümelt, die Zwischenstreifen bleiben in einer erosionsmindernden, rauhen Struktur liegen.

Besonderes Augenmerk ist auf das Reduzieren der Fahrspuren zu legen, da die vorhandenen Vertiefungen und die Verdichtungshorizonte unter den Fahrspuren eine besonders hohe Erosionsgefahr bewirken. Hierfür bieten sich mehrere Maßnahmen an: Einsatz von Geräten oder Gerätekombinationen mit möglichst großer Arbeitsbreite, Verwendung einer ausreichend dimensionierten Bereifung, Geräteeinsatz zu Zeitpunkten, in denen der Boden weniger druckempfindlich ist.

Nach der Saat lassen sich verschiedene Maßnahmen anwenden, mit denen durch eine spezielle Gestaltung der Bodenoberfläche der Abfluß des Niederschlagswassers gebremst und ein rasches Versickern in den Boden gewährleistet werden soll.

Für die Oberflächen-Profilierung hat sich der Einsatz von Lochstern-Aggregaten oder flach arbeitenden Spatenmaschinen bewährt. Diese schaffen zwischen den Saatreihen flache Vertiefungen im Boden, in denen sich der Wasserabfluß beruhigen kann. Eine Lockerung der Fahrspuren ist überall dort zweckmäßig, wo bei den Bestellarbeiten relativ tiefe Fahrspuren verursacht wurden. Hierfür werden an der Einzelkorn-Sämaschine senkrechte Zinken mit einem schmalen Lockerungsschar angebracht. Der Boden wird in einer Tiefe von etwa 20 bis 25 cm grob aufgebrochen und gewährleistet ein Abbremsen des Wasserabflusses und eine rasche Wasserabführung in tiefere Bodenschichten. Diese Maßnahme sollte jedoch nur bei ausreichend trockenen Bodenverhältnissen durchgeführt werden, um ggf. zusätzliche Verdichtungshorizonte in tieferen Zonen zu vermeiden.

Die Verwendung von mechanischen Pflegemaßnahmen zur Erosionsminderung wird sehr unterschiedlich beurteilt. Nach vorliegenden Erfahrungen lassen sich positive Effekte nur beim Einsatz von Hackmaschinen mit Federzinken-Hackwerkzeugen erwarten. Sie hinterlassen den Boden in einer rauhen Oberflächenstruktur und bewirken durch die federnden Zinken einen unebeneren Bearbeitungshorizont.

Um beim Einsatz konventioneller Bodenbewirtschaftungssysteme auf erosionsgefährdeten Standorten langfristig eine Verringerung des Bodenabtrages zu erreichen, ist es erforderlich, eine Anreicherung organischer Substanzen in den oberen Bodenschichten zu erreichen. Hierfür eignet sich der Grubber eindeutig besser als der Streichblechpflug. Wenn aufgrund spezieller Anforderungen des Standortes oder bestimmter Fruchtarten der Boden dennoch tiefgreifend gelockert werden soll, bietet sich die Kombination von Kurzgrubber und nachgeschaltetem Zapfwellengerät an. Der mit speziellen Lockerungsscharen ausgerüstete Grubber übernimmt das tiefe und grobe Aufbrechen des Bodens, das nachgeschaltete Zapfwellengerät gewährleistet ein intensives Einarbeiten der Pflanzenreste in den oberen, umsetzungsaktiven Bodenhorizont.

KOMBINIERTE PFLANZENBAULICH-LANDTECHNISCHE EROSIONSSCHUTZMASSNAHMEN (MULCHSAATVERFAHREN)

Neuere Untersuchungen zeigen, daß vor allem mit einer Kombination von pflanzenbaulichen und landtechnischen Maßnahmen ein wirksamer und nachhaltiger Erosionsschutz zu erreichen ist. Diese "konservierende Bodenbearbeitung" ist dadurch gekennzeichnet, daß die Einsaat der Reihenfrüchte in einen Pflanzenmulch erfolgt. Diesem kommt die Aufgabe zu, durch die oberirdische Stengel- und Blattmasse einen wirksamen Schutz der Bodenoberfläche zu gewährleisten und durch die intensive Durchwurzelung der gesamten Krume das Bodengefüge zu stabilisieren.

Im einfachsten Fall kann der Mulch aus Pflanzenresten der vorhergehenden Hauptfrucht oder einem aufgelaufenen Bestand von Ausfallgetreide, Unkräutern etc. bestehen. Da der Aufwuchs des Pflanzenbestandes jedoch nicht zu steuern und eine chemische Abtötung unerläßlich ist, wird dieses Verfahren nur in Sonderfällen angewandt.

Dagegen bietet der gezielte Anbau von Zwischenfrüchten eine größere Sicherheit, eine gut entwickelte Pflanzenmulchdecke und gute Durchwurzelung der Krume zu erzielen. Der Arbeitsablauf für die Mulchsaatverfahren bei Zuckerrüben und bei Mais ist in Abbildung 2 und 3 schematisch dargestellt.

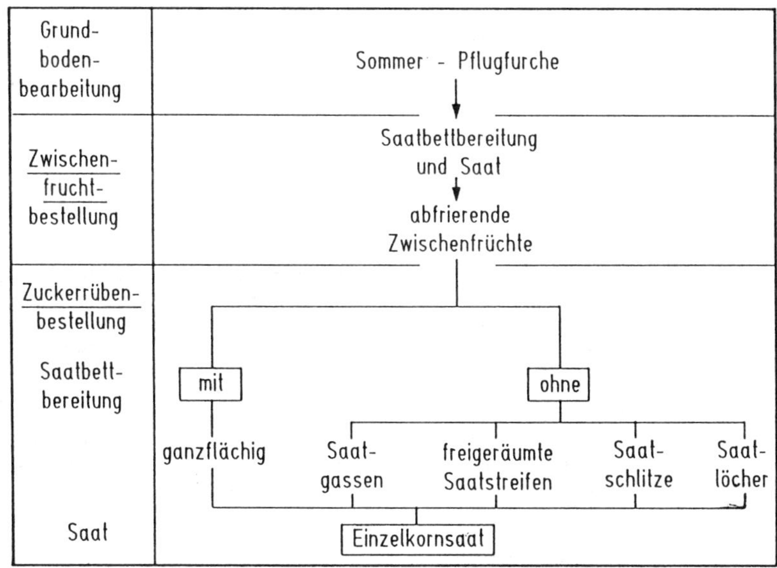

Abbildung 2: Ablauf-Schema für die Mulchsaat bei Zuckerrüben

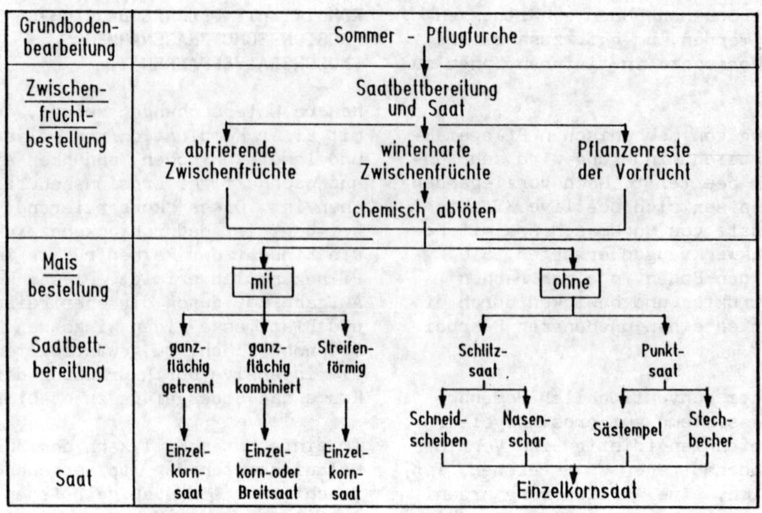

Abbildung 3: Ablauf-Schema für die Mulchsaat bei Mais

Nach der Vorfrucht (im allgemeinen Wintergerste oder früh abreifender Weizen) wird der Boden gepflügt und anschließend die Zwischenfruchtbestellung durchgeführt.

Als Zwischenfrüchte eignen sich winterharte Pflanzenarten, die jedoch mit geeigneten Total-Herbiziden vor der Saat totgespritzt werden müssen, um eine Konkurrenzwirkung mit den späteren Hauptfruchtarten zu beseitigen. Besonderes Interesse finden nicht winterharte, durch die Frostwirkung abgetötete Zwischenfrüchte, da hier das kostenintensive und umweltbelastende chemische Abtöten entfällt.

Für die Mulchsaat bei Zuckerrüben und bei Mais stehen heute bereits praxiserprobte Gerätelösungen zur Verfügung. Sie unterscheiden sich sowohl durch die technische Ausstattung, als auch durch eine unterschiedliche Bearbeitung des Saathorizontes.

MULCHSAAT BEI ZUCKERRÜBEN

Für die Aussaat von Zuckerrüben in einen Pflanzenmulch lassen sich derzeit im wesentlichen fünf Techniken anwenden:

- Mulchsaat mit ganzflächtiger, flacher Saatbettbereitung. Hierbei erfolgt die Vorbereitung des Saatbettes unter Verwendung von rotierenden, zapfwellengetriebenen Bodenbearbeitungsgeräten. Für die in einem zweiten Arbeitagang durchgeführte Zuckerrübensaat lassen sich Standard-Einzelkorn-Sämaschinen verwenden.
- Aussaat in Saatgassen. Hierbei werden bereits bei der Aussaat der Zwischenfrüchte im Abstand der späteren Zuckerrüben-Saatreihen ca. 20 cm breite Streifen frei gelassen. In diese wird im Frühjahr ohne weitere Bodenbearbeitung das Zuckerrübensaatgut abgelegt.
- Freiräumen eines Saatstreifens unter Verwendung von Doppelscheiben-Räumscharen und Saatgutablage in diesen Saatstreifen. Die Räumstreifenbreite kann den Standortbedingungen entsprechend eingestellt werden.
- Vor dem Einzelkorn-Sägerät angeordnete Schneidscheiben (teilweise mit Stützrad) durchschneiden das oberflächliche Pflanzenmaterial und bereiten die Saatfurche vor.
- Spezial-Nasenschar enstelle des normalen Säschares.

Einen Überblick über die gesamte Verfahrenstechnik der Mulchsaat bei Zuckerrüben gibt Abbildung 4.

Bei der Mulchsaat für Zuckerrüben ist zu berücksichtigen, daß die Bodenbearbeitung und -lockerung auf volle Krumentiefe im Vergleich zum konventionellen Bodenbearbeitungsverfahren um mehrere Monate nach vorne verlegt wird. Um eine ungestörte Entwicklung der Rübenkörper zu erreichen, muß durch begleitende ackerbauliche Maßnahmen dafür gesorgt werden, daß eine ausreichende Aggregatstabilität geschaffen wird.

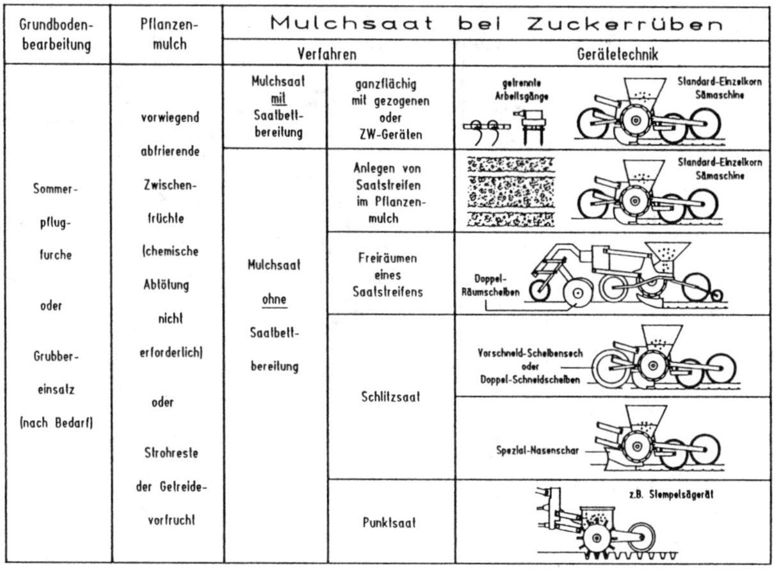

Grundbodenbearbeitung	Pflanzenmulch	Mulchsaat bei Zuckerrüben		
		Verfahren		Gerätetechnik
Sommerpflugfurche oder Grubbereinsatz (nach Bedarf)	vorwiegend abfrierende Zwischenfrüchte (chemische Abtötung nicht erforderlich) oder Strohreste der Getreidevorfrucht	Mulchsaat mit Saatbettbereitung	ganzflächig mit gezogenen oder ZW-Geräten	getrennte Arbeitsgänge; Standard-Einzelkorn Sämaschine
		Mulchsaat ohne Saatbettbereitung	Anlegen von Saatstreifen im Pflanzenmulch	Standard-Einzelkorn Sämaschine
			Freiräumen eines Saatstreifens	Doppel-Räumscheiben
			Schlitzsaat	Vorschneid-Scheibensech oder Doppel-Schneidscheiben
				Spezial-Nasenschar
			Punktsaat	z.B. Stempelsägerät

Abbildung 4: Verfahrenstechnik der Mulchsaat bei Zuckerrüben

MULCHSAATVERFAHREN FÜR MAIS

Auch für die speziellen Einsatzbedingungen der Mulchsaat bei Mais wurden verschiedene Verfahren entwickelt:

- Ganzflächige Saatbettbereitung und Einzelkornsaat in getrennten Arbeitsgängen oder als kombinierte Bestellmaschine (rotierendes Zapfwellengerät mit angebauter Einzelkorn-Sämaschine).
- Streifenförmige Saatbettbereitung unter Verwendung von Streifenfräse oder Streifenzinkenrotor als kombinierte oder Kompakt-Bestellmaschine.
- Anlegen eines Saatschlitzes mit anschließender Einzelkornsaat unter Verwendung von Vorschneidscheiben, kombinierten Schneid-Tast-Rädern oder Spezial-Nasenschar.
- Punktförmige Saatgutablage in den Pflanzenmulch unter Verwendung von Stempel-Sägeräten, Bechersägeräten etc.
- Eine spezielle Variante stellt der Einsatz einer Bodenfräse mit Breitsaat-Säschiene dar. Bei diesem Verfahren erfolgt keinerlei Grundbodenbearbeitung, der Pflanzenmulch besteht aus den Pflanzenresten der Vorfrucht.

Einen Überblick über die Verfahrenstechnik der Mulchsaat bei Mais gibt Abbildung 5.

EINSATZERGEBNISSE UND VERFAHRENSVERGLEICH

Die verfahrenstechnische Entwicklung zur Verminderung der Bodenerosion hat noch keinen endgültigen Abschluß gefunden. Auf der Basis der bisher vorliegenden Erfahrungen läßt sich jedoch bereits eine gewisse Verfahrenseinordnung vornehmen. Im Rahmen der konventionellen Bodenbewirtschaftung können die empfohlenen gerätetechnischen und verfahrenstechnischen Lösungen nur mit Einschränkungen einen nachhaltig wirksamen erosionsmindernden Effekt gewährleisten. Auf Dauer ist davon auszugehen, daß die grob lockernde Bodenbearbeitung mit dem Grubber bei gleichzeitig flacher Einmischung vorhandener Pflanzenreste eine gute Basis für die Verminderung von Bodenerosionen darstellt. Währenddessen die Einsaat von Mais in derart vorbereitete Felder kaum Probleme bereitet, ist bei Zuckerrüben eine Feinkrümelung eines schmalen Saatstreifens und exakte Tiefenführung der Einzelkorn-Sämaschinen eine wesentliche Voraussetzung für die ordnungsgemäße Aussaat. Je nach Standortbedingungen besteht die Möglichkeit, durch begleitende Maßnahmen (z.B. Radspurlockerung, Oberflächenprofilierung), Feldbestellung in der Schichtlinie und Reduzierung des Fahrverkehrs die Erosionsprobleme zu verringern.

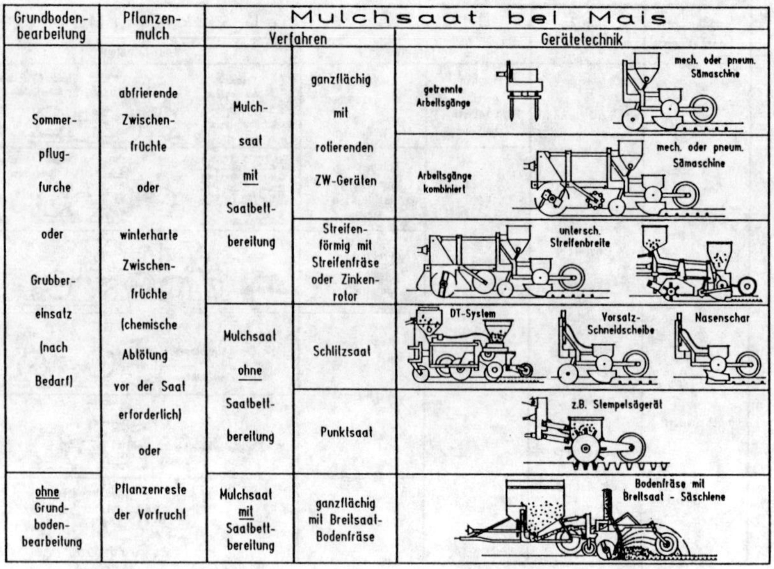

Grundbodenbearbeitung	Pflanzenmulch	Mulchsaat bei Mais		
		Verfahren		Gerätetechnik
Sommerpflugfurche oder Grubbereinsatz (nach Bedarf)	abfrierende Zwischenfrüchte oder winterharte Zwischenfrüchte (chemische Abtötung vor der Saat erforderlich) oder	Mulchsaat mit Saatbettbereitung	ganzflächig mit rotierenden ZW-Geräten	getrennte Arbeitsgänge; mech. oder pneum. Sämaschine
				Arbeitsgänge kombiniert; mech. oder pneum. Sämaschine
			Streifenförmig mit Streifenfräse oder Zinkenrotor	untersch. Streifenbreite
		Mulchsaat ohne Saatbettbereitung	Schlitzsaat	DT-System; Vorsatz-Schneidscheibe; Nasenschar
			Punktsaat	z.B. Stempelsägerät
ohne Grundbodenbearbeitung	Pflanzenreste der Vorfrucht	Mulchsaat mit Saatbettbereitung	ganzflächig mit Breitsaat-Bodenfräse	Bodenfräse mit Breitsaat - Säschiene

Abbildung 5: Verfahrenstechnik der Mulchsaat bei Mais

Beim Mulchsaatverfahren bildet das Herstellen eines möglichst gut entwickelten oberflächlichen Pflanzenmulches mit gleichzeitig intensiver Durchwurzelung der Krume die Basis für die angestrebte Erosionsminderung. Die erosionsverhütende Wirkung des Pflanzenmulches wird umso besser sein, je weniger der Mulch durch die Maßnahmen für Saatbettvorbereitung und Saat verletzt wird. Diese Zielvorstellung kollidiert jedoch mit der Forderung nach rascher Erwärmung sowie ausreichender Abtrocknung des Bodens in Saattiefe. Bei Temperaturmessungen im Bereich der Saattiefe (5 cm) ergaben, daß vor allem an Tagen mit hoher Lufttemperatur die Bestellvarianten mit Saatbettbereitung eine deutlich höhere Bodentemperatur erreichten, als die schlitzförmigen Saatverfahren. An Tagen mit niedriger Lufttemperatur sind diese Unterschiede dagegen nicht erkennbar (Abb. 6).

Bei sehr massig entwickelten Zwischenfruchtbeständen kann die langsamere Bodenerwärmung eine gewisse Verzögerung des Saattermines bewirken. Auf sehr leichten, durchlässigen Böden kann sich allerdings eine gebremste Wasserverdunstung positiv auswirken, da keine unkontrollierte Verdunstung der wertvollen Winterfeuchtigkeit entsteht.

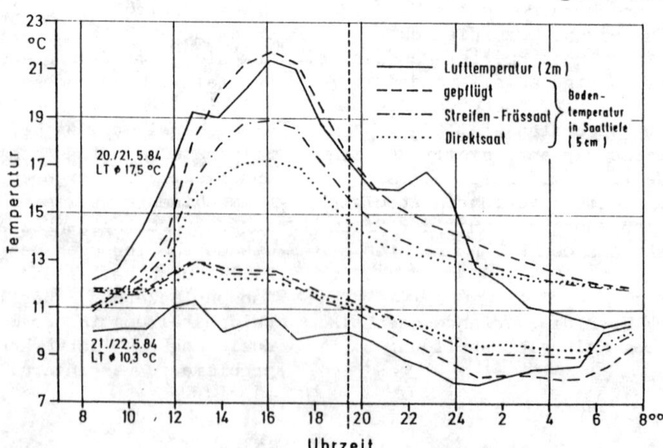

Abbildung 6: Temperaturverlauf in Saattiefe bei verschiedenen bestellverfahren an zwei Tagen mit sehr unterschiedlicher Durchschnitts-Luft-Temperatur (17,5 °C bzw. 10,3 °C)

Zu der Frage, welcher Bodenabtrag bei verschiedenen Bestellvarianten zu erwarten ist, wurden spezielle Untersuchungen unter Verwendung eines Regensimulators durchgeführt (Lehrstuhl für Bodenkunde, Weihenstephan). Dabei ergab sich, daß der höchste Bodenabtrag bei der konventionellen Bestellmethode mit Pflug und normaler Saatbettbereitung entsteht. Der geringste Bodenabtrag wurde auf den Mulchsaatparzellen mit einem weitgehend unverletzten Pflanzenmulch erzielt (Abb. 7).

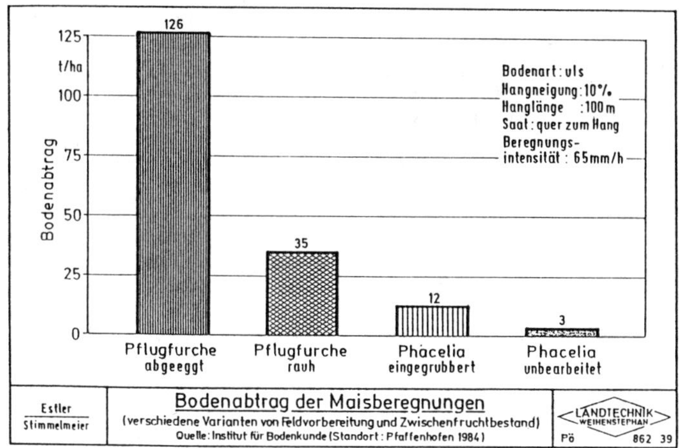

Abbildung 7: Bodenabtrag bei verschiedenen Varianten von Feldvorbereitung und Zwischenfruchtbestand nach dem Regensimulator-Einsatz

Unter normalen Einsatzbedingungen läßt sich mit der geschilderten Gerätetechnik ein weitgehend problemloser Einsatz der Mulchsaattechnik bei den verschiedenen Reihenfrüchten erreichen. Für spezielle Einsatzbedingungen wird in Abbildung 8 anhand von ausgewählten Kriterien eine Beurteilung und Einordnung der Gerätetechnik vorgenommen.

Mulchsaat Mais	Verwendbarkeit bei....Bodenverhältnissen					Einhalten der optimalen Saattiefe bei		Erosionsschutzwirkung	Reduzierung der Wasserverdunstung
	normal	schwer/verhärtet	Fahrspuren	rauhe Bodenoberfläche	bei zähstengeligen Mulchpflanzen	dichtlagernden Böden	massigem Pflanzenmulch		
ganzflächig getrennte Arbeitsgänge	+	+	+	+	+	+	+	O	o bis -
kombiniert	+	+	+	+	+ bis o	+	+	O	o bis -
Streifen-Frässaat/Zinkenrotor	+	+	+	+ bis o	+ bis o	+	+	o bis +	o bis +
Schlitzsaat mit Nasenschar	+	O	–	O	–	O	O	+	+
Schneidrad	+	–	–	O	o bis -	–	O	+	+
Schneid-Stützrad	+	–	–	O	o bis -	–	O	+	+
DT-System	+	O	o bis -	+	O	O	O	+	+
Punktsaat mit Stechbecher	+	O	o bis -	O	+	O	O	++	++
Stechspaten	+	–	–	–	+	–	–	++	++
Säslempel	+	o bis -	–	–	+	0 bis -	–	++	++

+ = gut o = durchschnittlich - = ungenügend

Abbildung 8: Beurteilung der Verwendbarkeit verschiedener Mulchsaatverfahren unter speziellen Bedingungen

FAZIT

Nach wie vor gelten Reihenfrüchte, insbesondere der Zuckerrüben- und Maisanbau als sehr attraktive Betriebszweige. Auf problematischen oder erosionsgefährdeten Standorten sollten deshalb alle Maßnahmen ergriffen werden, um auch unter derartigen Standortbedingungen einen erfolgreichen Anbau aufrecht erhalten zu können.

Aus landtechnischer Sicht sind zur Verminderung der Bodenerosion neben einigen ausgewählten Varianten der konventionellen Bestelltechnik vor allem die Mulchsaatverfahren für Zuckerrüben und Mais besonders erfolgversprechend. Neben den unbestreitbaren Vorteilen bestehen jedoch auch eine Reihe von Problemen, die in Zukunft gelöst werden müssen.

Mit der Anwendung von Mulchsaatverfahren läßt sich nicht nur eine Verminderung der Bodenerosion, sondern generell eine Verbesserung der Bodenstruktur und -fruchtbarkeit erzielen.

ANWENDUNG VON MULCHSAATVERFAHREN

VORTEILE
- längere Bodenruhe
- bessere "Regenverdaulichkeit"
- Schutz der Bodenoberfläche, Stabilisierung der durchwurzelten Krume
- bessere Befahrbarkeit und Tragfähigkeit der Felder
- auf Trockenstandorten langsamere Wasserverdunstung
- Bindung des Gülle-Stickstoffes, geringe Nitratauswaschung
- verschiedene verfahrenstechnische Lösungen vorhanden und erprobt
- Schlagkraft ähnlich wie bei konventioneller Bestelltechnik
- überbetrieblicher Einsatz möglich, bei Spezialmaschinen sinnvoll.

PROBLEME
- langsamere Bodenerwärmung, verzögerte Aussaat
- bei winterharten Zwischenfrüchten Herbizidaufwand für Abtöten
- mechanische Unkrautbekämpfung erschwert
- Probleme bei der Düngung (keine Einarbeitung, erhöhter N-Bedarf, Geruchsentwicklung bei Gülledüngung)
- teilweise Spezialmaschinen erforderlich
- standortspezifische Erfahrungen erforderlich (geeignete Bodenart, Feldaufgang, Ernteertrag)

LITERATUR

Buchner, W. 1986. Maßnahmen zur Eindämmung der Bodenerosion. Landtechnik 41:393-396.

Diez, T. 1982. Erosionsschutzmaßnahmen im Maisanbau. Mais 10:10-11.

Estler, M. 1984. Landtechnische Erosionsschutzmaßnahmen. Berichte aus der Flurbereinigung. Band 52:171-173.

Estler, M. 1986. Landtechnische Maßnahmen zur Verminderung der Bodenerosion. In Österr. Ges. für Land- und Forstwirtschaftspolitik (Hrsg.), Berichte der Wintertagung, S. 138-156. Wien.

Estler, M. 1987. Verfahrenstechnische Lösungen zur Verminderung der Bodenerosion. Landtechnik 42:118-119

Estler, M. 1987. Verfahrenstechnische Lösungen zur Verminderung der Bodenerosion. Landtechnik 42: 166:170

Frede, H.G. 1986. Erosionsgefährdung in der Landwirtschaft. KTBL-Arbeitspapier 104

Köller, K. 1986. Mulchsaat für Mais und Rüben. top agrar 1:116-120

KTBL-Arbeitsgruppe (Hrsg.). Definition und Einordnung von Bodenbearbeitungsverfahren. In Bodenbearbeitung und Bestellung Nr. 0236

Merkes, R. 1987. Mulchsaat bei Zuckerrüben - aber wie? DLG-Mitteilungen 7:368-369

Schwertmann, U. 1980. Bodenerosion durch Wasser - Ursachen, Ausmaß, Vorhersage. Landw. Forschung, Sonderheft 37:117-121

Sommer, C.; M. Zach und M. Dambroth. 1985. Konservierende Bodenbearbeitung. Agrar-Übersicht 36:14-18

Land and Water Use, Dodd & Grace (eds), © 1989 Balkema, Rotterdam. ISBN 90 6191 980 0

Identification of land cover and physiographic parameters in computer simulation of soil erosion

E.R.N.Gunawardena
University of Peradeniya, Sri Lanka

ABSTRACT: A study was carried out to simulate soil erosion from four small catchments. The stanford sediment model was modified and tested. Results showed that the modified sediment model provided reasonably accurate estimates of sediment discharge from upland areas for different vegetative and topographic conditions. The values of the coefficients and the variables of process equations were related to these characteristics and, hence, the applicability of the modified model for the prediction of soil loss from ungauged upland areas is very promising.

RESUME: On a procédé a une étude de simulation d'érosion du sol à partir de quatre petits captage (basin hydrographique). Le modèle Stanford (sédiments) a été modifié et testé. Les resultats ont montré que ce modèle modifié fournissait des éstimations relativement précises des décharge de sédiments provenant de hautes terres de situation différentes sur le plan de la végétation et de la topographie. Les valeurs des coefficients et les variables des pcocessus des mises en équations ont été rapprochées de ces caractéristiques et, de ce fait, l'applicabilité de ce modèle modifié pour la prédiction des pertes de sol des hautes terres non encore mesurées est très prometteuse.

Zusammenfassung: Die vorliegende Studie beschreibt eine Simulation von Bodenerosion in 4 kleinen Wassereinzugsgebieten im zentralen Bergland Sri Lankas. Das Stanford Sediment Modell wurde hierfür modifiziert und auf Anwendbarkeit getestet. Die Resultate zeigen, dass das modifizierte Sedimentmodell hinreichend genaue Schätzungen des Bodenabtrages für unterschiedliche vegetative und topographische parameter liefert. Werte für Koeffizienten und Variablen der Prozessgleichungen wurden in Bezug zu diesen parametern gesetzt. Eine breitere Anwendbarkeit des Modells zur Einschätzung des Bodenabtrages fur das zentrale Hochland Sri Lankas wird vorgeschlagen.

1 INTRODUCTION

Soil erosion has become a serious problem in the hill country of Sri Lanka due to denudation of forest for agricultural development. Proper land use aiming to reduce soil loss and prevent reservoir siltation demands a simple but valid erosion prediction method as a tool for planning. Computer simulation using a sediment transport model is the best way of fulfilling this task. The main draw back of digital simulation model is that processes and process parameters differ in regions of different climatic conditions and, therefore, have to be evaluated locally.

Soil erosion from agricultural catchments depends on catchment characteristics such as land cover, topography, soil etc. and identification of these parameters in the models may assist in developing methods for assessing sediment delivery from ungauged catchments.

The Stanford Sediment Model (SSM) developed by Negev (1967) as a sub-model to the Stanford Watershed Model (Crawford and Linsley, 1966), was selected to simulate the sediment yield. The process parameters in the SSM do not bear any relation to the land cover and topographic characteristics and the identification of these parameters in the model will increase the possibility to use it for ungauged catchments. Soil erosion and sedimentation theories have recently been developed considerably and this also points to the need for improvements in the model. This paper describe the modified sediment model (MSM) and compare the original and the modified sediment models with the data obtained from a field study carried out in Sri Lanka. Detail descriptions of modifications to the Stanford watershed and sediment models and their validation to the Sri Lankan watersheds are given by Gunawardena (1987).

2 MODIFIED SEDIMENT MODEL

The processes described by Meyer and Wischmeier (1969) were used as basic sub processes representing the upland erosion components. They are a) detachment by rainfall, b) detachment by runoff, c) transport by rainfall and d) transport by runoff. The relative importance of each of the four sub processes varies greatly, depending on the condition under which erosion occurs. Therefore, knowledge of the significance of each process for specific situation is essential for accurate prediction of sediment yield (Meyer et al, 1975).

The MSM combines the above processes in to two, namely interrill and rill erosion. Detachment and transport by rainfall and transport by runoff is considered in the interrill erosion process while detachment and transport by runoff is included in the rill erosion process. The flow chart of the MSM, superimposed on the watershed model, is given in figure 1. Only the land surface erosion processes are considered in this study and, hence, described below.

2.1 Interrill erosion

Soil erosion in interrill areas combines the processes of soil detachment by raindrop impact, transport by splash, and transport by very thin flow (Meyer, 1981). Young and Wiersma (1973) concluded that rainfall energy, rather than surface flow energy, was the major force initiating soil detachment on interrill areas. Meyer et al (1975) agreed that thin flow on interrill areas has very little detachment capacity in the absence of raindrop impact. Martinez (1980) stated that detachment by runoff on interrill areas may be minimal and neglected due to shallow depths and low flow rates occurring on these areas. Therefore, interrill detachment due to overland flow was not considered.

2.2 Detachment by rainfall

Bubenzer and Jones (1971) investigated the importance of rainfall mass and impact velocity on the detachment and splash of soils. They presented the relation between soil splash and kinetic energy of simulated rainfall through regression equation of the form:

$$D_1 = B_1 \, KE^n \, I^m \qquad (1)$$

where, D_1 = amount of soil splash, KE = kinetic energy of rain drops, I = rainfall intensity, B_1 = constant, and m,n = exponents having ranges of 0.27 to 0.55 and 0.83 to 1.49 respectively.

The kinetic energy of rain is estimated by Mihara (1951) as:

$$KE = C_0 \, I^{1.2} \qquad (2)$$

where C_0 is a constant.

Combining equation (1) and (2) gives:

$$D_1 = B_1 \, G \, I^{m + 1.2n} \qquad (3)$$

where $G = C_0^{\,n}$

Bubenzer and Jones (1971) tabulated experimental values of m and n for different soils. Their analysis showed mean values of 0.42 and 1.29 for m and n respectively for all soils studied. Substituting these values in equation (3) gives:

$$D_1 = B_1 \, G \, I^h \qquad (4)$$

where h is approximately equal to 2. This is the same relationship Meyer and Wischmeier (1969) obtained after carefully reviewing earlier research findings. This value has been used successfully for most of the erosion modelling applications.

Interrill erosion is also a function of soil characteristics. Primary particle size distribution, amount and type of clay, and clod size after tillage (Moldenhauer and Koswara, 1968) are some of the soil factors that influence interrill detachment. Different soils with different textures and cohesiveness may vary greatly in their susceptibility to rainfall detachment (Meyer, 1981). Relationships have been proposed that relate detachment by rainfall

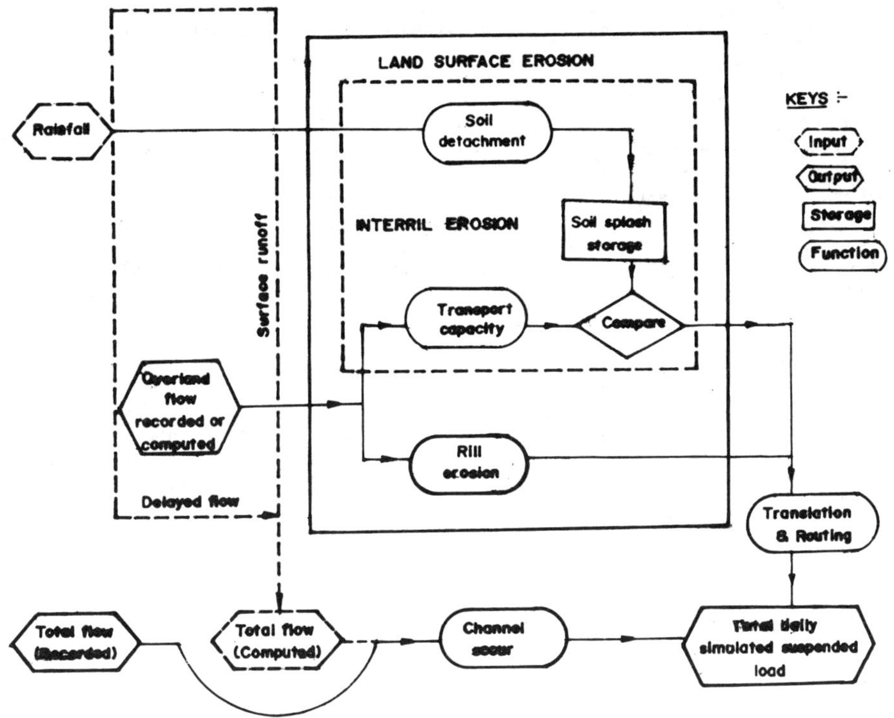

Fig. 1. Modified Sediment model superimposed on Watershed model

to both rainfall and soil characteristics. Based on Free's (1960) observation, Meyer and Wischmeier (1969) suggested the following empirical relationship for detachment by rainfall:

$$D_1 = S_e A_i I^2 \qquad (5)$$

where, D_1 = detachment by rainfall, S_e = soil effect constant, A_i = area of detachment, and I = rainfall intensity.

The detachment capacity of rainfall is influenced by the type of cover . Foster (1982) included cover management factor on interrill detachment as a coefficient. Therefore the following equation is suggested to estimate the splash detachment.

$$D_1 = C_1 A K I^2 \qquad (6)$$

where,
D_1 = interrill detachment rate, C_1 = cover factor, A = area of the catchment, K = soil erodibility factor from the Universal Soil Loss Equation (USLE), and I = rainfall intensity.

The above equation is similar to the one described in the AMSWERS (Beasley, 1983) model.

2.2 Transport by runoff

The detached material from raindrop splash may be redeposited on the ground or may remain in suspension and be transported downslope in case overland flow occurs. The loosened soil particles will be carried into the stream according to the ability of overland flow to carry them. If the transport capacity of overland flow is less than the particles available for transport, the soil loss from this process is equal to the former. The remaining detached soil particles will be left loosely on the ground for sometime as detachment storage. Upon the occurrence of the next overland flow, they may be picked up and transported downslope.

The ability of overland flow to transport the detached soil particles depends on the flow depth, flow velocity, land surface and soil characteristics. David and Beer (1975) suggested the following expression as the transport capacity of the overland flow.

$$T = N\,S^{e}\,Y^{f} \qquad (7)$$

where, T = overland flow transport capacity, N = soil and surface roughness factor, S = slope, Y = overland flow depth, and e,f = exponents both approximately equal to 1.67 (Meyer and Wischmeier, 1969).

Therefore the equation suggested for the transport capacity is:

$$T = M\,A\,S^{1.67}\,Y^{1.67} \qquad (8)$$

where, T = overland flow transport capacity, M = roughness factor, A = area of the catchment, S = slope, and Y = overland flow depth.

In equation (8) roughness factor M indicates the influence of the cover on overland flow. A higher value of M indicated a lower resistance to the overland flow resulting in a higher transport capacity.

2.3 Rill erosion.

The major sub processes controlling rill erosion are detachment by runoff and transport by runoff (Foster and Meyer, 1975). Rill erosion begins when the eroding capacity of the flow at some point exceeds the ability of the soil particles to resist detachment by flow (Meyer et al, 1975). Once rilling begins, the concentrated flow tends to increase the detachment rate, and rilling progress.

In most modelling applications, erosion in many single rills are lumped and described as gross rill erosion. Rill erosion and flow are assumed to be uniformly distributed across the slope, although physically the flow and erosion are concentrated in small channels. With this assumption, the equation assumed for rill erosion is given by Foster and Meyer (1972) as:

$$D_r = a\,(\tau - \tau_{cr})^{b} \qquad (9)$$

where, D_r = rill erosion detachment capacity rate per unit catchment area, τ = shear stress of flow, assuming broad shallow flow, τ_{cr} = critical shear stress, and a,b = constants.

In many modelling applications, can be assumed zero (Foster, 1982). Using data obtained by Wischmeier et al (1971), Foster derived rill erosion equation as:

$$D_r = a_1\,C_r\,K_r\,\tau^{1.5} \qquad (10)$$

where, τ = average shear stress, C_r = cover management factor, K_r = soil erodibility factor for rill erosion, and a_1 = constant.

The relationship for rill erosion and its sub factors are complex and somewhat cumbersome. As a result, a simpler alternative to equation 10 is expressed by Foster (1982) as:

$$D_r = C_b\,C_{slr}\,S_s\,K\,q \qquad (11)$$

where, D_r = rill erosion detachment capacity rate per unit catchment area, C_b = coefficient, C_{slr} = soil loss ratio from the USLE which includes the influence of cover management of runoff, S_s = sine of slope angle, K = soil erodibility factor from the USLE, and q = runoff rate.

Rill erosion usually occurs in significant amounts only in the rilled areas of the catchment where flow converges on steeper overland flow slopes. The combination of these two factors results in greatly increased transport capacity, which is not reflected in the equation (8). Under such conditions, the overland flow transport capacity is not a limiting factor to sediment movements in rills. Hence, it is assumed that the rill detachment capacity rate represents the total soil loss due to rill erosion. Therefore, the following equation is suggested for modelling rill erosion

$$D_2 = C_2\,S\,A\,K\,Y^{r} \qquad (12)$$

where, D_2 = rill erosion, C_2 = cover management factor, S = slope, A = area of the catchment, K = soil erodibility factor from the USLE' Y = depth of overland flow, and r = exponent.

The sum of interrill and rill erosion is considered as the total inflow of sediment from the land surface into the stream.

3 STUDY SITE

A vast majority of the sites available for reforestation in Sri Lankan hill country occupies what is called "patana vegetation". It is a collective term and the type of vegetation dominated in these sites can vary from short, sparsely grown ferns to dense Mana grass (Imperata cylindrica). Four catchments were selected for the study (table 1). The catchment no. Gl is dominated by ferns (Gleicheniacea linearis) and was planted with pinus (Pinus carribea) during the monsoon season in November 1984. The no. G3, which is fully covered by Mana grass, was planted with pinus during June 1983. The catchment no.G2 is dominated by mana which developed on an abandoned tea land. The catchment no.G4 was planted with pinus in 1977 and has crown closure and developed a dense understory of native herbs and woody plants.

Table 1. Physiographic characteristics of catchments.

Catchment no.	Slope %	Overland flow length(m)	Area (ha)
G1	0.54	100	0.459
G2	0.45	94	0.765
G3	0.37	130	0.876
G4	0.34	100	0.541

4 MEASUREMENTS AND ANALYSIS

The rainfall was measured with a weighing type recording rain gauge with weekly charts. Hourly rainfall values were abstracted from rainfall charts and used as hourly inputs to the models for all the catchments. All four catchments were identically instrumented. Runoff was measured by a 2 feet (610 mm) H- flume with an automatic water level recorder installed at the out let of the catchment. Hourly overland flow values were abstracted from these recorder charts. Suspended samples were collected with a two feet (610 mm) diameter Coshocton wheel sampler which is designed to collect a 0.5 percent of runoff volume. This sample then passed through a 10:1 proportional splitter. The resultant sample was stored in a 1 m x 1 m x 1 m tank. Samples were collected daily and analyzed for sediment concentration.

5 SIMULATION RESULTS

The MSM and SSM were tested with the data observed from the four catchments. Hourly recorded rainfall and overland flow and daily observed sediment load were used as inputs to the models. The testing of the sediment models was by the trial and error methods. Parameters were assigned different values until a satisfactory reproduction of the recorded sediment load was achieved.

The relative monthly performances are shown in figure 2. MSM performed well for most of the months. The annual summary of observed and simulated sediment loads, given in table 2, shows that catchments Gl

Table 2. Annual observed and simulated sediment yields and daily correlation coefficients.

catchment no.	Sediment yield (kg)		Daily Correlation coefficient
G1	Obs.	1711.3	
	MSM	1648.7	0.73
	SSM	1624.3	0.58
G2	Obs.	2535.5	
	MSM	2593.5	0.86
	SSM	2584.5	0.73
G3	Obs.	594.5	
	MSM	591.5	0.86
	SSM	625.8	0.87
G4	Obs.	263.0	
	MSM	277.4	0.85
	SSM	291.6	0.71

and G2 gave the highest amount of suspended sediment load. It was expected from catchment G2 as it has a land cover of abandoned tea and mana. There were exposed surfaces and resistent to the falling drops of rainfall was minimal in this catchment when compared to others. Catchment Gl has a good cover when compared to catchment G2, but the steeper slope compensated the protective effect of the cover. Therefore, there was not much difference between the total sediment loads. As expected, catchment G3 and G4 gave a much lower sediment load compared to catchment G2: a reduction of 80 and 85 percent respectively, due to their improved land cover and lesser slope. Both catchment provided good cover and higher flow resistance due to it's under growth. Though expected, catchment G4 did not give much lower sediment load compared to catchment G3. This was because of the unexpected leakage of the catchment G3. Substantial amounts of runoff which could have gone through H-flume was seeped underneath and, as a result, sediment load was under estimated.

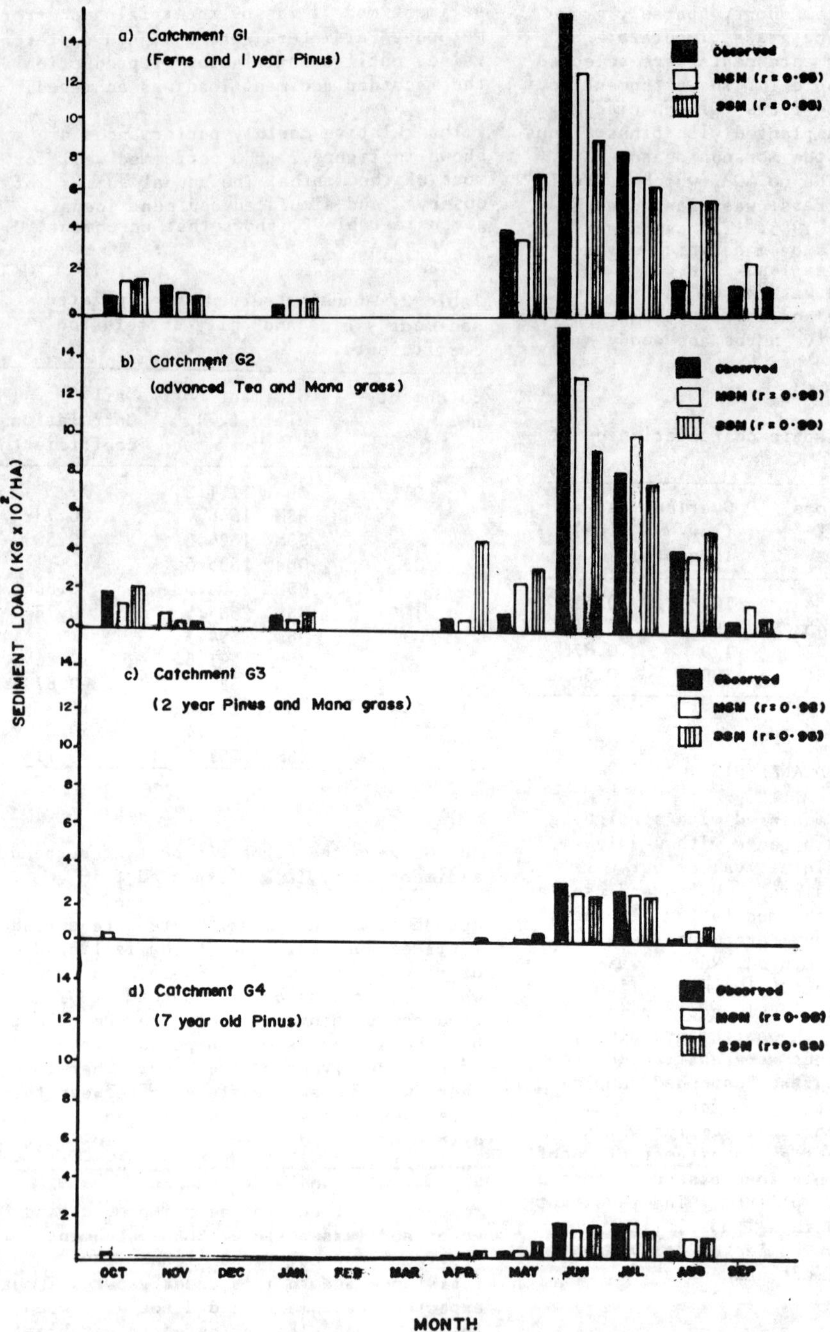

Fig. 2. Comparison of observed and simulated (MSM & SSM) monthly sediment load (1984/85)

The improvements to the MSM can be shown by comparing the plots of simulated and observed daily sediment loads for the MSM and SSM as shown in figure 4. The daily sediment load was seriously over estimated by a factor of 22.8 by SSM, on 8/4/1984 which was simulated much better by MSM. This improvement is achieved by the MSM by introducing the "Detachment and transport

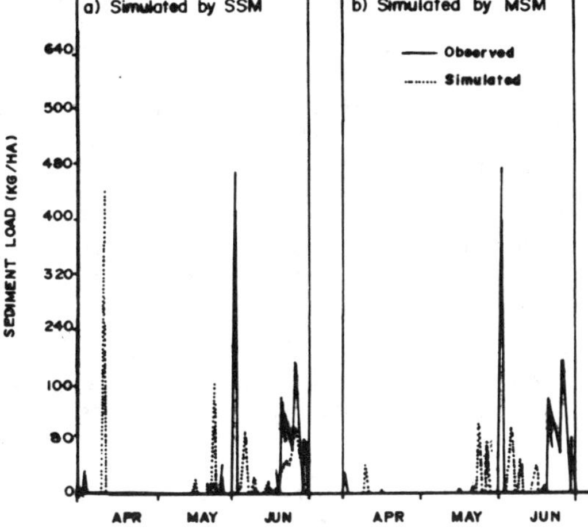

Fig. 3. Daily observed and recorded Sediment loads for catchment G2

limited" concept (Wischmeier and Smith, 1969). The rainfall in this day was high and, as a result a substantial amount of soil particles were detached. According to the MSM, only part of them had left the catchment by overland flow, as it was limited by the carrying capacity. In contrast, SSM assumes that total soil that was to be detached should be carried away by the overland flow when it occurs regardless of the amount of soil detached. This is the reason of highly over synthesized values on that day for the catchment G2 by the SSM.

5.1 Estimation of land cover parameters.

The measured physiographic parameters and the estimated land cover parameters are given in table 3. The estimated values agreed well with the observed land cover characteristics in the catchments. The cover factor represents the resistance of crop cover to the falling rain. A higher resistance to the falling rain is shown by the lower value of cover factor, resulting a lesser amount of soil detachment. As expected, the highest resistance was shown by a lower value for catchment G4. The catchment G2 offers the lowest resistance and, hence, has the highest value.

The roughness factor indicates the flow resistance offered due to land cover. A higher value of roughness factor in the interrill transport equation indicates a lower resistance to the overland flow resulting in a higher transport capacity. Sediment delivery from an interrill area is the lesser between the transport capacity of the interrill flow and the amount of detached particles available for transport. Transport of particles by interrill flow should be directly related to flow characteristics like velocity and depth. Therefore, if flow is ponded by depressions of obstacles by plant stems, transport capacity of interrill flow is greatly reduced. As expected, the highest resistance was shown by the catchment G4 while G2 was the lowest. The value is more or less the same for both G1 and G2.

Table 3. Measured physiographic parameters and estimated land cover parameter values.

Parameter	Catchment no.			
	G1	G2	G3	G4
Cover factor (C_1)	0.165	0.0357	0.182	0.101
Roughness factor (M)	0.813	1.240	0.846	0.611
Cover management factor (C_2)	0.485	0.729	0.490	0.297
Slope	0.54	0.45	0.37	0.34
Area	0.469	0.765	0.876	0.541
Erodibility	0.22	0.22	0.22	0.22

The cover management factor is included as a parameter in the rill erosion equation. This factor affects the rill erosion by affecting the flow's shear stress. Good cover in direct contact with the soil surface reduces the critical shear stress for rill erosion and hence results in the reduced extent of rill formation. At the same time litter on the surface tends to spread the flow across the field rather than producing irregular flows creating shear stress concentration. If stress at these concentration is greater than the soil's critical shear stress, erosion occurs. As expected, the lower value is observed for catchment G4 while the highest is the G2. Again, a very little difference is observed for G1 and G2.

These values were estimated by the trial and error method in this study using the MSM. With further research, guide lines could be prepared to estimate these parameters according to the land cover and, hence, would be able to predict the soil loss without requirement of any calibration.

REFERENCES

Beasley, D.B. 1983. The ANSWERS model. In M.D.Smolen (ed.) Hydrologic water quality models for agriculture and forestry. Virginia polytechnic institute and state university, Blacksburg, Virginia. bull. no. 291, p.46-59.

Bubenzer, G.D. & B.A.Jones 1971. Dropsize and impact velocity effect on the detachment of soils under simulated rainfall. Trans. ASAE, 14(4):625-628.

Crawford, N.H. & R.K.Linsley 1966. Digital simulation in hydrology: Stanford watershed model IV. Dept. of civil engineering, Stanford Univ., Tech. report 39.

David, W.P. & C.E.Beer 1975. Simulation of soil erosion - part 1: Development of a mathematical erosion model. Trans. ASAE 18(1): 126-129,133.

Foster, G.R. & L.D.Meyer 1972. A closed form erosion equation for upland areas. In H.W.Shen (ed.) Sedimentation. Fort collins, Colorado. Chap.12.

Foster, G.R. & L.D.Meyer 1975. Mathematical simulation of upland erosion using fundamental erosion mechanics. In present and prospective technology for prediction of sediment yields and sources. USDA-ARS, ARS-S-40. p.190-207.

Foster, G.R. 1982. Modelling the erosion process. In C.T.Hann,H.P.Johnson, & D.L.Drankensiek (eds.) Hydrological modelling of small watersheds, ASAE. p.297-330.

Free, G.R. 1960. Erosion characteristics of rainfall. Agricultural engineering, 41:447-449,455.

Gunawardena, E.R.N. 1987. Computer simulation of runoff and soil erosion from small agricultural catchments in Sri Lanka. Unpublished Ph.D. Thesis, Silsoe College, Bedford, UK.

Martinez-Merz, M.R. 1980. Modelling for upland areas. Ph.D. thesis, Faculty of renewable resources, University of Arizona.

Meyer, L.D. & W.H.Wischmeier 1969. Mathematical simulation of the processes of soil erosion by water. Trans. ASAE 12(6):754-758.

Meyer, L.D., G.R.Foster & M.J.Romkins 1975. Source of soil eroded from upland slopes. In present and prospective technology for predicting sediment yields and sources. USDA-ARS, ARS-S-40. p.177-189.

Meyer, L.D. 1981. How rain intensity affects interrill erosion. Trans. ASAE 24(1):1472-1475.

Mihara, H. 1951. Raindrop and soil erosion. Nat. inst. agric. sci. bull. series A, no.1.

Moldenhauer, W.C. & J.C.Koswara 1968. Effect of initial clod size on characteristics of splash and wash erosion. Soil sci. soc. of Ame. proc., 32(6):875-879.

Negev, M. 1967. A sediment model on a digital computer. Dept. civil engineering, Stanford univ., Tech. report. 76.

Wischmeier, W.H., C.B.Johnson & B.V.Cross 1971 A soil erodibility nomograph for farm lands and construction sites. J. soil and water cons., 26(5):189-192.

Young, R.A. & J.L.Wiersma 1973. The role of rainfall impact in soil detachment and transport. Water resour. res., 9(6): 1629-1636.

Land and Water Use, Dodd & Grace (eds), © *1989 Balkema, Rotterdam. ISBN 90 6191 980 0*

Modelling vegetative effects on interrill (splash) erosion

A.A.Ogunmokun
Cranfield Institute of Technology, Silsoe, Bedford, UK

ABSTRACT: A process based treatment was used to incorporate vegetative effects directly into a splash erosion model. These effects were accounted for by considering the transformation of drop size distributions and drop impact velocities under crop canopies as different from that under bare soil. This is an improvement in the hitherto practice of accounting for these effects by using a single crop factor that lumps all the effects together. The model is expected to increase our understanding of the interaction between vegetation and splash erosion

1 INTRODUCTION

The trend in soil erosion modelling is to describe the erosion processes physically rather than empirically, using fundamental principles.

Vegetation is known to have considerable influence on soil erosion, and in many cases it is the most easily controllable factor of erosion. Vegetation effects on splash erosion is known to be the interception and redistribution of raindrops by canopy and the transformation of the drop velocities.

However, the infinite variety of species and the complex dynamic relationships between soil, the atmosphere and vegetation have made the study of vegetational effect on splash erosion less detailed in physical models. Thus, most splash models still account for these effects by the use of a single crop coefficient such as the USLE C factor (Wischmeier and Smith (1978)) which compares the erosion under a particular vegetation with that under fallow. This coefficient, apart from lumping all the separate effects together, is always less than one thus implying that the presence of vegetation always results in erosion reduction. Some recent studies however, have shown that at times, and under certain conditions, vegetation may enhance rather than reduce erosion. (Morgan (1982), Noble and Morgan (1983), Finney (1984)).

A better approach to represent vegetation effects in soil erosion models was suggested by Morgan (1986) and Rickson and Morgan (1987) as a process based treatment which shows how individual components of vegetation (canopy, stems, roots, etc.) directly or indirectly influence the processes of soil erosion (detachment and transport). This approach may explain why different vegetation types or the same vegetation under different conditions give different erosion losses in the field.

The objective of this study is to use this approach to incorporate vegetation effects into an interrill (splash) erosion model.

2 MODELLING SPLASH EROSION

Soil erosion on interrill areas combines soil detachment by raindrop impact and transport by thin shallow flow (interrill overland flow) only. Vegetation can and does affect both of these processes but only the effect on splash detachment is modelled here.

2.1 Model concept

Splash detachment depends on raindrop characteristics such as intensity, drop sizes and the distribution, and drop impact velocities (Meyer (1965)). It has been correlated with the kinetic energy and/or the intensity of rainfall (Wischmeier and Smith (1978)), as well as the momentum of the rain (Bubenzer (1979)). In this study a relationship presented by Gilley (1982) based on drop impact pressure as described by Wang and Wenzel (1970) was used. This relationship was preferred to others since it incorporates the effect of water layer depths on the erosivity of raindrops of different sizes. As shown by Palmer (1964), a depth of water layer on soil surface acts as a cushion against raindrop impacts.

2.2 The splash model

Gilley (1982) assumed splash detachment to be directly proportional to drop impact pressure thus:

$$DS = Kd * DIP \quad (1)$$

where DS is the splash detachment rate (in $Kg/m^2/s$), Kd is a proportionality factor named splash detachment factor (s/m), and DIP is the drop impact pressure (N/m^2), given as:

$$DIP = 0.2\ \rho \cos^2\theta \left(\sum_{i=1}^{n} a_i V_i^{\,2} (d_i/h)^{1.83}\right) \quad (2)$$

where ρ is the density of water (Kg/m^3), θ the slope angle (in degrees), i the drop size classes, n the total number of drop size classes, a_i the number of drops in size class i, V_i the drop impact velocity for drop size i (m/s), di the equivalent drop diameter for size class i (m), and h the depth of overland flow (m).

Equation (2) as given by Gilley (1982), is valid only for situations where flow depth h is greater than drop diameter d_i (i.e. when d_i/h 1.0). In natural rainfall condition there will be situations where d_i/h 1.0. Palmer (1964) found that maximum splash detachment occurs when d equals h. Below and above this point, he noted a decrease in splash amount. The drop impact pressure equation (2) was therefore modified in this study by replacing the d/h term by a general term called water depth correction factor; DH in order to take care of the range of drop sizes not considered by Gilley. The modified equation is:

$$DIP = 0.2\ \rho \cos^2\theta (\Sigma\, a_i V_i^{\,2} DH_i) \quad (3)$$

$$DH_i = (d_i/h)^{1.83} \quad \text{for } 0.2\ \ d_i/h\ \ 1.0 \quad (4a)$$

$$DH_i = e^{(h/d_i - 1)} \quad \text{for } d_i/h\ \ 1.0 \quad (4b)$$

$$DH_i = 0.0 \quad \text{for } d_i/h\ \ 0.2 \quad (4c)$$

Equation (4b) is adopted in a modified form from the work of Park et al (1982). Palmer (1964) also noted in his study that there is negligible splash detachment when flow depth is more than five times the drop diameter (ie d/h 0.2). This is the basis for the 0.2 in the conditions for equations (4a) and (4c).

The drop impact pressure model above (equation (3)), depends on slope angle which can be obtained, on raindrop characteristics which will be discussed later, and on Kd, the interrill splash detachment factor, a model parameter which should be evaluated for individual sites. The modified Gilley's equation was tested using data from the work of Menez (1979) for bare soil and was found to give adequate prediction of splash detachment.

2.3 Incorporating vegetative effects into the splash model

Vegetation can transform all the rainfall characteristic that the splash detachment model depends on through its canopy, stem and roots. Canopy intercepts rainfall and retains some, leading to a change in the amount and intensity of rainfall reaching the soil surface. This process of interception also leads to a redistribution of drops under the canopy thus:

1. Some drops splash on contact with leaves to smaller drops (splash droplets);
2. Some coalesce on the leaves to form larger and usually more erosive drops (leaf drips);
3. Some run down the stems, contributing to soil moisture but not to splash detachment (stem flow); and
4. Others fall between the leaves unaffected (direct throughfall).

Vegetation being of limited height also transforms drop velocity, as the transformed drops seldom regain their terminal velocity by the time they strike the soil.

The capability of a vegetation to make these transformations (and thus the amount of splash detachment) depends on the extent of canopy cover, the effective height of the canopy above the soil surface, and the flexibility and arrangement of canopy type. Different vegetation types and even the same vegetation under different conditions or at different growth stages, have different structures and morphology thus leading to different rates of splash detachment.

In equation (3), the terms in parenthesis are essentially the ones that will vary with vegetation. If the drop size distribution under a vegetal cover can be separated into direct throughfall and transformed throughfall (or leaf drainage), then the drop impact pressure under vegetation canopy can be expressed as:

$$DIP= 0.2\rho\cos^2\theta\left[\sum_{i=1}^{n} a_i V_i^2 DH_i(1-TC) + \sum_{j=1}^{n} a_j V_j^2 DH_j(1-GC)\right] \quad (5)$$

where j is drop size classes for transformed throughfall, GC is the fraction of soil surface protected from splash by ground cover, and TC the fraction of soil surface protected by both canopy and ground covers (but not necessarily the addition of both as there may be some over lap).

This type of expression was also used by Styzen and Hogh-Schmidt (1986) to represent drop momentum necessary for splash detachment under canopy cover.

3 RAINFALL AND RAINDROP CHARACTERISTICS

Natural rainfall consists of drops of various sizes ranging from less than 0.25mm to about 6mm in diameter. (Laws and Parsons (1943)). Under vegetation, the range of drop sizes can extend beyond this 6mm to about 9mm due to coalesced drops (Chapman (1948)). In order to get a good prediction of splash detachment amount, the raindrop characteristics of drop size distribution and drop velocity must be accurately measured. When this is not available, models representing these characteristics can be used.

Many investigators recognized the significance of raindrop impact as a soil detachment agent and have carried out studies of and presented models for raindrop sizes, distribution, and impact velocities for bare soils. (Laws and Parsons (1943), Marshall and Palmer (1948), Carter et al (1974), Mualem and Assouline (1987)).

Even though some work has been done on drop size distribution under vegetative canopies (Morgan (1985), Mosley (1982), Quinn and Laflen (1981), Armstrong (1986)), not much has been done however on presenting it in model form. In fact only two attempts, one by Armstrong (1986) and the other by Morgan, Brandt and Stuttard (1988) were found in my literature review.

3.1 Modelling drop size distribution under vegetative canopy

Armstrong (1986) presented a scheme for representing the drop size distribution under crop canopy by using a plot of the ratio of the transformed median drop size (dG) to input median drop size (dR) against a dimensionless canopy descriptor, which he called gammal, to model canopy transformation of drop sizes. He concluded however that dG/dR is not sufficient to represent changes in drop number or volume redistribution within the spectrum of drop sizes which are important in modelling splash erosion. This scheme was therefore not pursued further.

Morgan, Brandt and Stuttard (1988) presented another scheme for modelling the drop size distribution under canopies. They noticed in their study, as did others (Chapman (1948), Mosley (1982), Brandt (1986), and Armstrong (1986)), that rainfall intensity does not seem to have a notable effect on transformed raindrop distribution. They also noticed that transformed drop size distribution are independent of canopy type. This was based on remarkable similarities found between the drop size distribution under canopies of geranium plants (3 sizes) a Schleffera and a Fatsia japonica used in their study. So their model is based only on knowing the percentage canopy cover (CC) for a particular crop with any given input rainfall intensity.

Starting with a known volume (VOL) of water reaching the soil surface, the transformed throughfall volume (INTVOL) and direct throughfall volume (CLRVOL) were determined as follows:

$$INTVOL = VOL.CC \quad (8)$$

$$CLRVOL = VOL.(100-CC) \quad (9)$$

Direct throughfall distribution is assumed to follow the Marshall and Palmer (1948) distribution given by:

$$N_i.d_i = N_o e^{-bi} \quad (10)$$

where

$$b_i = 4.1 * I^{-0.21} \quad (10a)$$

I is the rainfall intensity (mm/h), No is a constant given as 0.08 and $N_i.d_i$ is the number of drops (N) of diameter (d) in m^3 of air in a size class i with d_i usually taken as increasing by 0.25 mm. The number of drops in each size class i per m^2 of ground surface is given by:

$$a_i = N_i.d_i.V_i \quad (10b)$$

where V_i is the terminal velocity of the drops in size class i.

To simulate the direct throughfall component of drop size distribution under canopy, by their model, a random selection of drops greater than 0.25 mm is taken using the Marshall and Palmer drop size distribution for a given rainfall intensity, until the total volume of drops equal CLRVOL.

The transformed throughfall component is based on a normal distribution with a mean of 5mm and a standard deviation of 0.8mm found with measurement under the above canopies. This distribution is constrained to fall within realistic upper and lower limits. The model simulates drop size distribution of transformed throughfall by sampling drops at random from the normal distribution until the total volume of selected drops equals INTVOL. The suggestion by Morgan, Brandt and Stuttard (1988) that transformed throughfall drop size distribution is independent of canopy type was only noticed by them and not in other studies involving agricultural crops. Whether the suggestion will still hold for agricultural crops is doubtful, even though this has not yet been determined. Meanwhile the model can still be used in its present form to predict drop size distribution under crop canopy where no measured data is available.

3.2 Modelling drop velocity at different drop heights

For bare soils, raindrops are usually assumed to fall at their terminal velocities. Under vegetative covers, some drops fall at a height at which their velocities do not reach terminal values. An algorithm was therefore developed in this study for predicting drop velocities (V_i) for different drop size classes (d_i) falling at different drop heights (H) by fitting a regression equation into the data presented by Epema and Riezebos (1983) with R^2=93%. The equation is of the form:

$$V_i=1/(0.08+0.138/H+0.014/d_i) \quad (R^2=93\%) \qquad (11)$$

The fall height is estimated by a method presented by Morgan, Brandt and Stuttard (1988) by which the integer part of the leaf area index (LAI) of the crop is used as an indicator of the number of layers in the canopy of the crop. If these layers are assumed distributed uniformly with height throughout the canopy, the transformed drops are then allocated randomly to height from the relationship:

$$H = H_b+j(H_t-H_b/LAI) \qquad (12)$$

where H_b is height of the bottom canopy layer (m), H_t the height of the top canopy layer (m), and j a randomly selected integer between 1 and the value of LAI.

4. MODEL TESTING

In order to demonstrate the potential of this model in predicting vegetative effects on splash erosion, simulation runs were carried out with hypothetical data for series of percentage canopy covers and fall heights. A rainfall intensity of 63.5mm/h and a slope of 9% were assumed, these being values usually used in most erosion experiments. A constant depth of flow of 1.8 mm (being the median value of the depths quoted in several references for shallow flow conditions as found under agricultural crops) was also assumed for all the simulation runs. Stem flow and permanent interception amounts were not taken into consideration. While their neglect will affect the prediction of actual splash amounts, it is believed that their effect on predicting trends may not be too significant.

4.1 Results and discussions

Since splash detachment is assumed to be directly proportional to drop impact pressure, the simulations were run to give predictions of drop impact pressure amounts.

Figure 1 shows the plot of drop impact pressure against percentage canopy cover at a fall height of 0.5m. Total drop impact pressure decrease with increase in percentage cover, while contribution from leaf drainage to total drop impact pressure increased with increasing canopy cover and at the expense of direct throughfall. Similar trends were noticed by Morgan et al (1986) in their experimental study of splash detachment under sugar beet, brussell sprouts and potatoes. It is also evident from Figure 1 that while leaf drainage contribution to total drop impact pressure increased more at 70% canopy cover and above, the total value actually reduced because of the extent of coverage.

Figure 2 shows the plot of drop impact pressure against fall height at 70% canopy cover. With increase in fall height, both total drop impact pressure and the contribution of leaf drainage to the total value increase appreciably. The rate at which drop impact pressure reduced with reduction in fall height below 1.5m is more rapid than the rate at which drop impact pressure increased with increase in fall height above 1.5m. Also, at a fall height of 8m the predicted drop impact pressure of 1.95MN/m^2 is about 90% of the predicted value of 2.16MN/m^2 for bare soil (0% canopy cover in Figure 1) despite the fact that there was a canopy cover of 70%. These suggest that there is a special relationship between fall height and splash erosion, and as noticed by Brandt (1986), it is perhaps the difference in fall height more than anything else which governs the effect of vegetation on changing the erosivity of the rainfall beneath it.

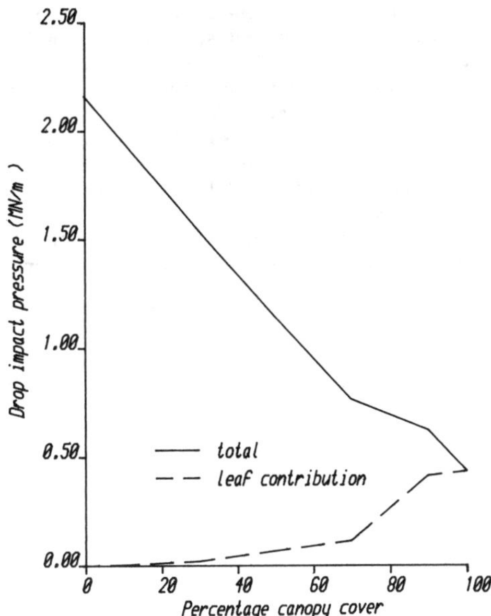

Fig.1 Drop impact pressure versus canopy cover at a fall height of 0.5m

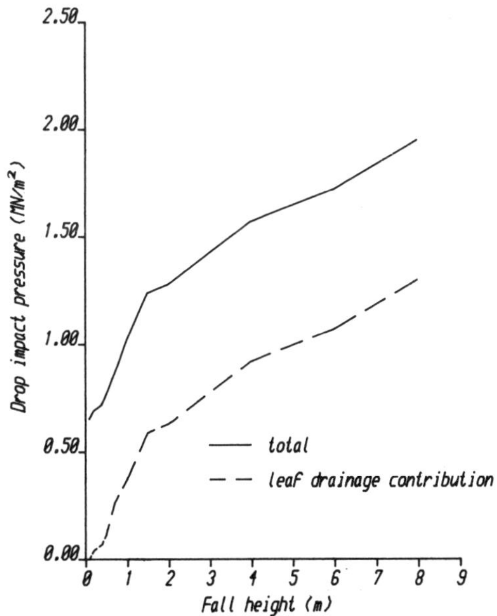

Fig.2 Drop impact pressure versus fall height at 70% canopy cover

The interaction between fall height and percentage canopy cover in affecting splash erosion was not studied analytically here. A cursory look at Figures 1 and 2 together however, suggests that a low growing vegetation with an extensive canopy cover (probably above 70%) is necessary if vegetation is expected to be used for erosion control.

5. SUMMARY AND CONCLUSIONS

Vegetation effects were incorporated into a physically based splash erosion model as presented by Gilley (1982) using relations from literature where available and deriving others by analysing data from relevant works.

Testing the model with hypothetical data shows it has potentials for predicting rational trends of vegetative effects on splash erosion. Fall height (which is a function of crop height and canopy layers) is shown to be an important parameter in describing vegetative effects on splash erosion, especially in situations where greater proportion of the drops under the vegetation falls as leaf drips. The model has the potential for improving our understanding of the interactions between vegetation and erosion from the point of view of erosion control.

Before this model can be used in predicting actual amount of splash detachment under different crops, the drop size distribution model for transformed throughfall of Morgan, Brandt and Stuttard (1988) should be validated for agricultural crops. Also a model of vegetative effects on interrill overland flow should be coupled with it and both studied together. Work on these is currently going on and will be reported in subsequent papers.

REFERENCES

Armstrong, C.L. 1986. Transformation of rainfall by plant canopy. Unpub. Ph.D. thesis. Univ. of Illinois

Brandt, C.J. 1986. Transformation of kinetic energy of rainfall with variable tree canopies. Unpub. Ph.D. thesis. Univ. of London

Bubenzer,G.D. 1979. Rainfall characteristics important for simulation. Proceedings of the rainfall simulation workshop. Tucson Arizona. 22-34

Carter,C.E., J.D.Greer, H.J.Brand & J.M. Floyd. 1974. Raindrop characteristics in south central United States. Trans. Am. Soc. Agric. Engnrs. 17: 1033-1037

Chapman,G. 1948. Size of raindrops and their striking force at the soil surface in a red pine plantation. Trans. Am. Geo. Union 29: 664-670

Epema,G.F. & H.Th. Riezebos. 1983. Fall velocity of water drops at different heights as a factor influencing erosivity of simulated rain. Catena Supp. 4: 1-17

Finney,H.J.1984. The effect of crop covers on rainfall characteristics and splash detachment. J. Agric. Engnr. Re. 29: 337-343

Gilley,J.E. 1982. Soil erosion by sheet flow. Unpub. Ph.D. thesis. Colorado State Univ. Fort Collins.

Laws,J.O. & D.A. Parsons. 1943. The relation of raindrop size to intensity. Trans. Am. Geo. Union 24: 452-459

Marshall,J.S. & W.McK.Palmer. 1948. Relations of raindrop size to intensity. J. of Meteorology 5: 165-166

Menez,R.M. 1979. Erosional modelling for upland areas. Unpub. Ph.D. thesis. Univ. of Arizona. Tucson.

Meyer,L.D. 1965. Mathematical relationships governing soil erosion by water. J. Soil and water cons. 20: 149-150

Morgan,R.P.C. 1982. Splash detachment under plant cover: Results and implications of a field study. Trans. Am. Soc. Agric. Engnrs. 25: 987-991

Morgan,R.P.C. 1986. Soil erosion and conservation. Longmans

Morgan,R.P.C., C.J. Brandt & M.J. Stuttard. 1988. The role of plant covers in controlling soil detachment by rainfall. Final report to the AFRC. U.K.

Morgan,R.P.C., H.J. Finney, H. Lavee, E. Merritt, & C.A. Noble. 1986. Plant cover effects on hillslope runoff and erosion: evidence from two laboratory experiments. In Abrahams, A.D. (ed). Hillslope hydrology. 77-96. Allen and Unwin.

Mosley,M.P. 1982. The effect of a New Zealand beech forest canopy on the kinetic energy of water drops and on surface erosion. Earth and surface processes and landforms 7: 103-107

Noble,C.A. & R.P.C.Morgan. 1983. Rainfall interception and splash detachment with a brussells sprout plant: a laboratory simulation. Earth and surface processes and landforms 8: 569-577

Palmer,R.S. 1964. The influence of a thin water layer on water drop impact forces. IASH Pub. 65: 141-148

Park,S.W., J.K.Mitchell & J.N. Scarborough. 1982. Soil erosion simulation on small watersheds: a modified ANSWERS model. Trans. Am. Soc. Agric. Engnrs. 25: 1581-1588

Quinn,N.W. & J.M. Laflen. 1983. Characteristics of raindrop throughfall under corn canopy. Trans. Am. Soc. Agric. Engnrs. 26: 1445-1450

Rickson,R.J. & R.P.C.Morgan. 1987. Approaches to modelling the effects of vegetation on soil erosion by water. In: Morgan R.P.C. & R.J. Rickson (eds.) Erosion assessment and modelling. Commission of the European Communities: 237-253

Styzen,M. & K. Hogh-Schmidt. 1986. A new description of the relation between drop sizes, vegetation and splash erosion. In: Hasholt, B. (ed.) Particulaert bundet softransport i vand og jorderosion. Nordisk Program. Rapport No 14, 255-271

Want,R.C.T. & H.G. Wenzel. 1970. The mechanics of a drop after striking a stagnant water layer. Univ. of Illinois Water Res. Center. Res. Re. No. 30

Wischmeier,W.H. & D.D. Smith. 1978. Predicting rainfall erosion losses. USDA Agricultural handbook No. 537

Land and Water Use, Dodd & Grace (eds), © 1989 Balkema, Rotterdam. ISBN 90 6191 980 0

Computation of silt production rate for small watershed – A case study of Machkund basin

K.N.Sharma
Department of SWCE, CAET, Bhubaneswar, India

N.C.Pandey
Government of Crissa, Department of Agriculture, India

B.Panigrahi
Department of SWCE, CAET, Bhubaneswar, India

ABSTRACT: Silt load of a basin is an indication of soil erosion and loss of fertility of the catchment. Soil Conservation and watershed management practices are aimed to control silt production rate alongwith peak rate of flow. It is also a guide for priority fixation for watershed development. Rate of silt load depends on rainfall and catchment characteristics like drainage density, stream frequency, stream grade, shape index etc. . Under the present study, catchment characteristics and rainfall of five sub-watersheds of Machkund basin have been correlated with silt loss individually and collectively. Relationship gives high values of correlation coefficient indicating close correlation between observed and computed values of silt loss.

1 INTRODUCTION

Food production demand of the world is increasing at faster rate with growth of population. Every new mouth needs additional food grains. Soil is the basic vital natural source for grain production. With faster rate of deforestation, this natural resource is running out to lower reaches. Thus area in upper reach losses its capacity for grain production. As long as this removal of soil is equal to or less than rate of formation of soil, it may be tolerable, but in most of the basins, it has caused exposer of rock surfaces. Large numbers of dams were constructed for irrigation, flood control, electricity generation etc.. These purposes of dams depend on the life storage of the reservoirs. Life and working capacity of these reservoirs are decreasing at faster rate by silt flow carried by the river systems.

Thus silt production of a basin reduces fertility level of basin, life and capacity of reservoirs and causes flood havoc by clogging river courses in down stream end. Silt production is also a good indication for priority fixation of a watershed for its treatment.

Watershed treatment can be planned well if silt production rate of the basin is known. It may also be useful in designing and planning a reservoir. Practically silt load are being measured on very few sites. In planning reservoirs for the basins which have not been gauged, the only source of silt data is estimation.

Silt production rate of a basin may depends on catchment and hydrological characteristics. Rainfall data can be observed from locally available observatory. But all catchment characteristics cannot be assessed easily without extensive survey of the area. However there are numbers of catchment characteristics which can be measured by topo sheets and remote sensing techniques. Correlation between silt production rate, rainfall and various catchment characteristics can therefore be an extremely useful tool for estimation of silt production rate and in turn planning a reservoir or watershed effectively.

Keeping the objectives in view, regression equations were developed between silt production rate, precipitation and catchment characteristics (individually and collectively) for Machkund basin of South Orissa (India).

2 REVIEW OF LITERATURE

Efforts have been made by several scientists to estimate soil erosion and silt flow on the basis of catchment characteristics and rainfall. Most popular equation for estimation of annual soil loss is Universal soil loss equation (Frevert et al. 1955) which is:

$X_a = f(K,S,L,R,F,E,C,P)$

where X_a is annual soil loss in tons per acre and K, S, L, R, F, E, P and C are factors related to soil, degree of slope, length of slope and rotation, fertility degree of erosion, conservation practice and rainfall respectively. Flaxeman (Pandey 1984) suggested an equation for sediment yield as:

$$\log(100+y) = 6.63792 - \log(100+X_1)^{2.40504} + \log(100+X_2)^{6.066567} - \log(100+X_3)^{6.61820} + \log(100+X_4)^{0.04019}$$

where y is sediment yield in acre.ft/sq.mile, X_1 is $\frac{\text{average annual rainfall in inches}}{\text{average annual temperature in } ^\circ F}$,

X_2 is average slope, X_3 is percentage of soil particles coarser than 1 mm in surface soil profile of 2 inch depth and X_4 is erodibility index of 2 inch depth surface soil profile. Sediment yield of ungauged watershed has been correlated to that of gauged one (Pandey 1984) as:

$$S_c = S_m\ (A_c/A_m)^{0.8}$$

where S_c and S_m are sendiment yield of gauged and ungauged watershed in tons/year and A_c and A_m are drainage area of gauged and ungauged watershed.

Models have also been developed to estimate sediment yield of watershed (Varshney 1974) as:

(1) For catchment of mountain areas

$$y = \frac{0.395}{A^{0.311}}$$

(2) For plain catchment

$$y = \frac{0.392}{A^{0.202}}$$

(3) For catchment having more than 5000 sq.km.

$$y = \frac{1.534}{A^{0.254}}$$

(4) Based on river of South India

$$y = \frac{0.277}{A^{0.194}}$$

where y is annual silting rate in Million Cu. m per 100 sq. km. and A is catchment area in sq. km.

3 MATERIALS AND METHODS

Sediment production rate measured from five sub-watersheds of Machkund basin was considered as a basis for the present study. Silt production from a watershed depends on rainfall and catchment characteristics. Characteristics are many in number and are difficult to evaluate. Some of them which can be easily evaluated from toposheets and land-sat-imaginery are, area (A), drainage density (D_d), stream frequency (S_f), stream grade (S_g), average land slope (S_l), shape index (S_p), circulatory index (C_i) and time of concentration (T_c). These characteristics alongwith rainfall were chosen as variables for development of regression equations for silt production rate estimation. For the present study following five watersheds of Machkund basin of South Orissa (India) were selected.

(1) Kinchore Watershed
(2) Jallagada Watershed
(3) Pakanagada Watershed
(4) Antasula Watershed
(5) Ghodabeda Watershed

River system and contour of above five watersheds are shown in Figures 1 to 5. Data for average annual silt production rate and rainfall for the period of 20 years (1965-84) were obtained, excepting Kinchore Watershed for which data were available from 1972 only.

4 RESULT AND DISCUSSIONS

Silt production rate (SPR) can be explained as a function of rainfall and catchment characteristics as:

$$SPR=f(A,D_d,S_f,S_g,S_l,S_p,T_c,C_i,P) \qquad (1)$$

Regression equation was developed between silt production rate, chatchment characteristics and rainfall of the watersheds individually. Regression equations with correlation coefficients are presented in Table 1. The symbol y in all the regression equations (Table 1) represents silt production rate in HM/100 sq.km.. Correlation coefficients of regression equations indicate that silt production rate is closely related with drainage density, stream frequency, stream grade, shape index and time of concentration. Correlation coefficient of regression with area, average land slope, circulatory index and rainfall shows either fair or poor correlation. These regression equations indicate relation of silt production rate with individual variables.

In practice silt production rate does not depend on any single variable. Hence multiple regression equation was developed considering variables having higher value of correlation coefficients i.e. drainage density, stream frequency, stream grade, shape index and time of concentration.

Table 1. Regression equation between silt production rate and individual variable.

Variables	Regression equation	Coefficient of correlation, r
Area	$y = 1.9611A^{0.2480}$	0.50
Drainage density	$y = -10.4615+4.5518\ D_d$	0.98
Stream frequency	$y = -2.3844+0.8354\ S_f$	0.91
Stream grade	$y = -5.0951+8.3744\ \ln S_g$	0.92
Average land slope	$y = -4.9736+0.5098\ S_1$	0.78
Shape index	$y = 27.3947-48.3262\ S_p$	-0.98
Time of concentration	$y = 57453233.36\ T_c^{-3.8623}$	-0.82
Circulatory index	$y = 0.8664\ C_i^{-3.1450}$	-0.61
Rainfall	$y = 45.8343 - 5.7673\ \ln P$	-0.53

Average annual rainfall was also considered for the multiple regression analysis though it does not indicate high value of correlation coefficient. Area was ignored as silt production rate involves area in itself. Circulatory shape index and average land slope were also ignored as shape index, drainage density, stream frequency which are related factors were indluded in the regression.

Considering above variables, a multiple linear regression equation was developed as:

$$SPR = 9.297117 + 2.018913\ D_d + 0.04561\ S_f - 0.005634\ S_g - 34.31370\ S_p + 0.06244\ T_c - 0.0000003\ P \qquad (2)$$

Above multiple linear regression equation shows that there is maximum variation of 1.42 per cent (Table 2) between observed and computed values of silt production rate.

On χ^2 test the value so obtained was found to be significant at very high probability level. But this equation indicates coefficient with rainfall, P as 0.0000003 which is too low and so the effect of rainfall may be ignored for computation of silt production. This is probable as variation in annual rainfall was very nominal. Similar observation was made in Universal soil loss equation (Frevert 1955).

Table 2. Variation between observed and computed values of silt production rate when rainfall was considered.

Watershed	Observed SPR (HM per 100 sq.km)	Computed SPR (HM per 100 sq.km)	Per cent variation
Kinchore	4.92	4.9205	0.01
Jalagada	8.96	8.9706	0.12
Pakanagada	3.17	3.1753	0.17
Antasula	4.77	4.7831	0.27
Chodabeda	2.07	2.0406	1.42

Considering above fact in view a multiple regression equation was again developed considering all variables (as in equation 2) except rainfall. The equation so obtained is:

$$SPR = 9.206659 + 2.02947\ D_d + 0.045093\ S_f - 0.004113\ S_g - 34.203012\ S_p + 0.062477\ T_c \qquad (3)$$

In both equation 2 and 3, SPR is silt production rate in HM per 100 sq.km., D_d is drainage density in km. per sq.km., S_f is stream frequency asnumber of streams per sq.km., S_g is stream grade in percent, S_p is shape index and T_c is time of concentration in minute. Equation 3 shows maximum variation of 1.448 per cent (Table 3). Multiple correlation coefficient of both equation 2 and 3 is found to be 0.9999 and on χ^2 test they are found to be significant at 99 per cent probability level.

Table 3. Variation between observed and computed values of silt production rate when effect of rainfall was not considered.

Watershed	Observed SPR (HM per 100 sq.km)	Computed SPR (HM per 100 sq.km)	Per cent variation
Kinchore	4.92	4.92337	0.069
Jallagada	8.96	8.96878	0.098
Pakanagada	3.17	3.17681	0.215
Antasula	4.77	4.78101	0.231
Chodabeda	2.07	2.04002	1.448

5 CONCLUSIONS

Silt production of a basin is a complex phenomenon. Any relationship for prediction of silt production rate must depend on different catchment and hydrological characteristics of the basin. Under present study, multiple linear regression equation was developed considering different characteristics. The developed equation shows very high degree of correlation coefficient. The variation between observed and computed values are found to be less than 1.5 per cent . Further on χ^2 test, the equation so developed was found to be significant at 99 per cent probability level. So this equation can be used for prediction of silt production in basins having identical characteristics.

BIBLIOGRAPHY

Frevert, R.K., Schwab, G.O., Edminster, T.W. and Barnes, K.K. 1955. Soil and Water Conservation Engineering, John Wiley and Sons, INC., New York.

Pandey, N.C. 1984. Priority fixation of watershed in river valley project. M.Sc. (Agril. Engg. and Tech.) thesis College of Agricultural Engineering and Tech., O.U.A.T., Bhubaneswar, India.

Raghunath, H.M. 1985. Hydrology-Principle Analysis and Design, Wiley Eastern Limited, New Delhi, India.

Sharma, K.N., Pandey, N.C., Senapati, S.C. and Sharma, S.D. 1988. Estimation of runoff from small watersheds. Proc. U.G.C. National Seminar on Works, M.M.M. Engineering College, Gorakpur, U.P., India.

Varshney, R.S. 1974. Engineering Hydrology, Nemchand and Bros, Roorkee, U.P., India.

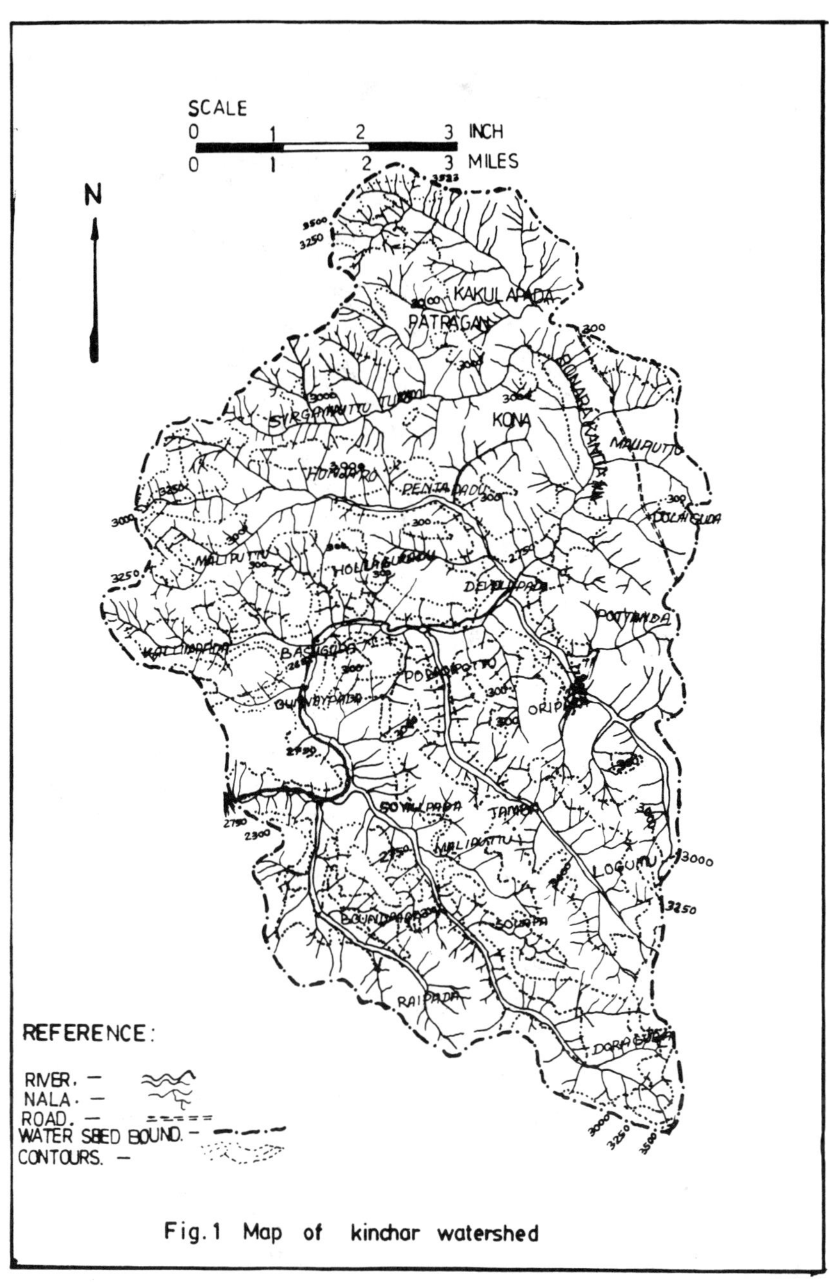

Fig.1 Map of kinchar watershed

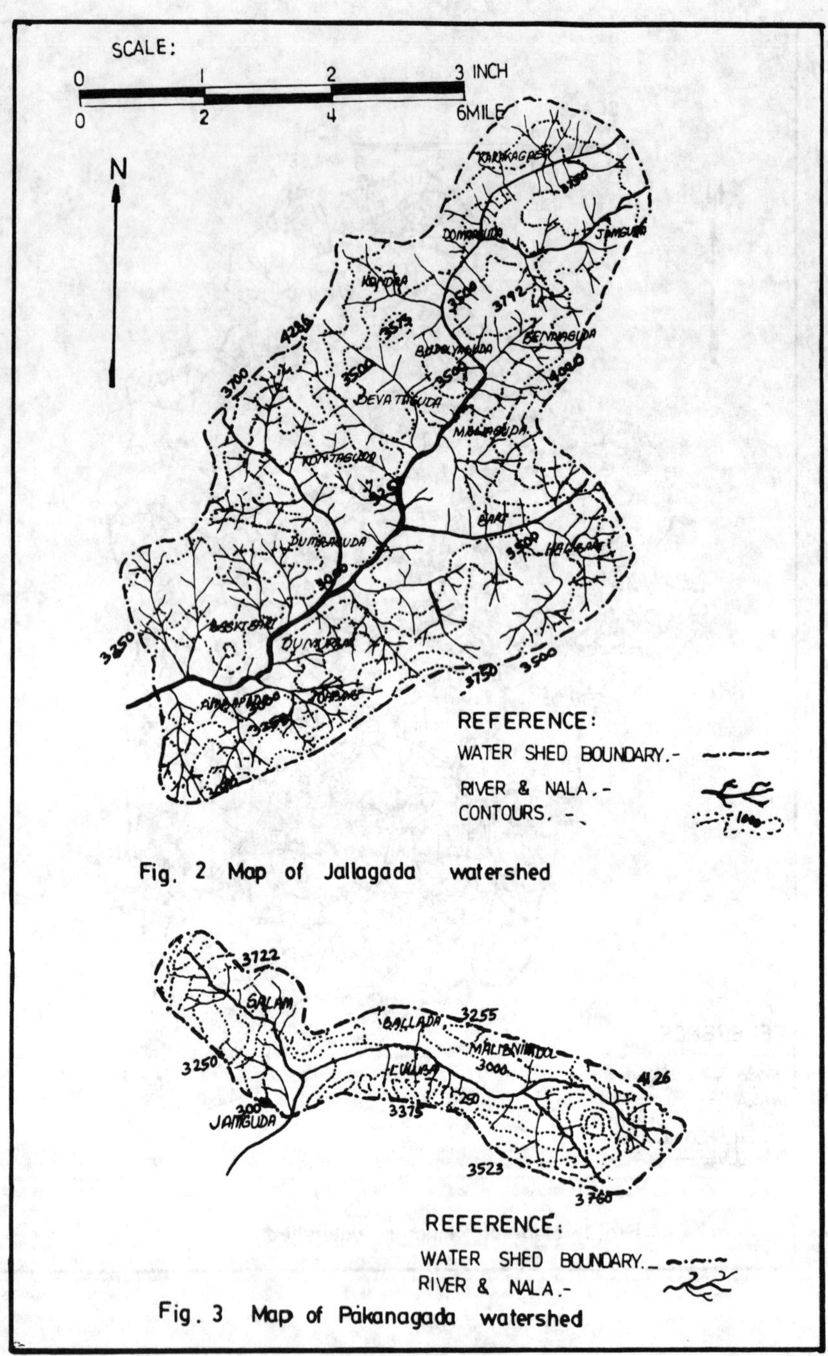

Fig. 2 Map of Jallagada watershed

Fig. 3 Map of Pakanagada watershed

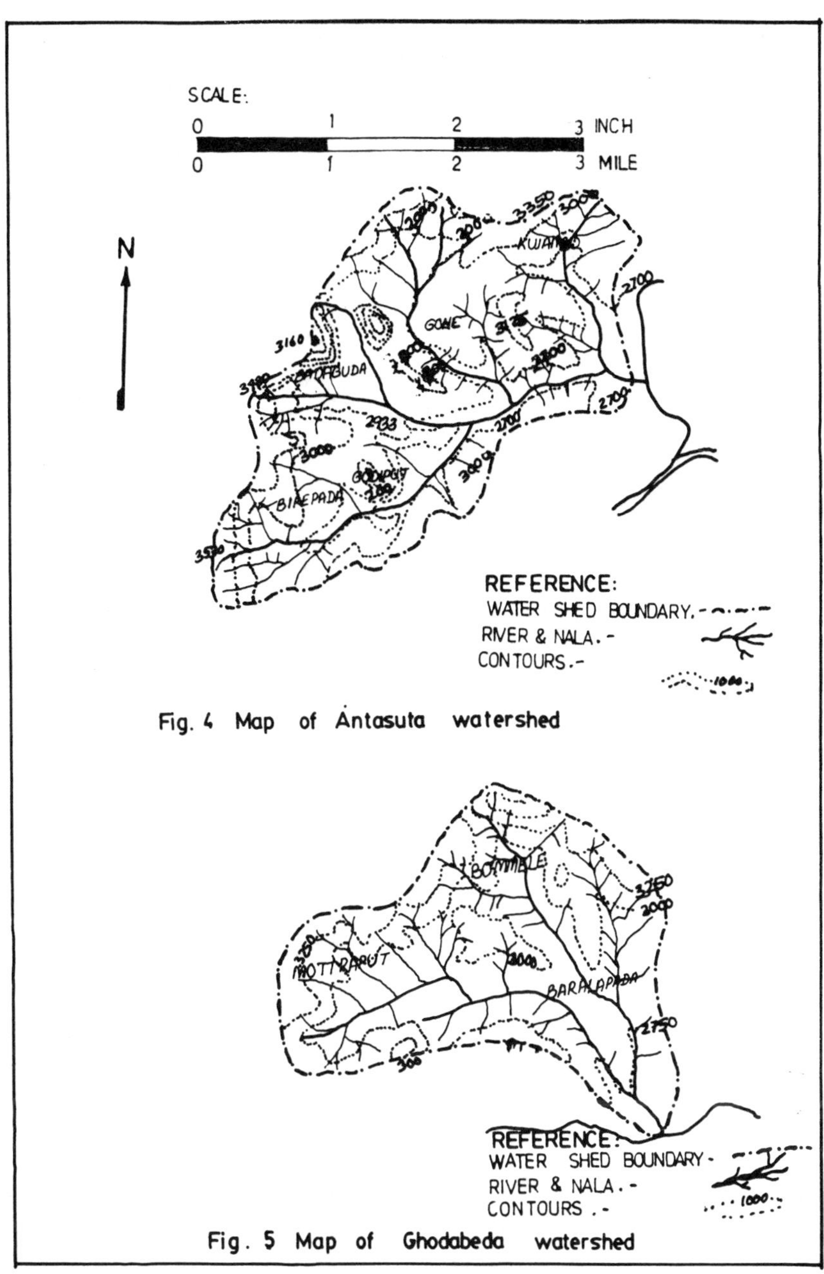

Fig. 4 Map of Antasuta watershed

Fig. 5 Map of Ghodabeda watershed

Land and Water Use, Dodd & Grace (eds), © 1989 Balkema, Rotterdam. ISBN 90 6191 980 0

Impact des banquettes à retention totale sur la fertilité et l'érosion des sols à l'amont du barrage de l'oued Siliana en Tunisie (zone semi-aride)

M.H.Snane, N.Ennabli, M.Mechergui & A.Chaabi
Institut National Agronomique, Tunis, Tunisie

RESUME : La sédimentation du canal des banquettes à retention est l'un des facteurs qui contribue à leur destruction. Pour des évènements pluvieux de l'année 1987 nous avons constaté que ni la formule de Wischmeier et Smith ni celle de Williams ne donnent une bonne estimation du transport solide dans le canal de la banquette. Nous avons alors établi deux formules pour les deux types de milieu les plus fréquemment rencontrés à l'amont du barrage de Siliana. De plus nous avons trouvé que la construction des banquettes donne lieu à une ligne de démarcation de part et d'autre de laquelle les taux de matière organique et d'argile vont varier en sens inverse.

ABSTRACT : Terrace channel sedimendation is contributing to the destruction of level terraces in Tunisia. For the 1987 rainfall data we have found that neither Wischmeier and Smith nor Williams equations give a good estimate of sediment deposition in the channel. Therefore we have established two appropriate formulae for the two most frequent soil conditions upstream of the dam of Siliana. Morever we found that terrace construction has changed the clay and organic matter distribution on the parcel. Beyond a transition line located 6 to 10 m up the channel the clay and organic matter concentration have increased whereas below this line it has decreased.

ZUSAMMENFASSUNG : Einer der Faktoren die in Tunesien zur Zerstörung der Banketts beigetragen haben ist die Auffullung der Gräben durch Sedimente. Für die Niederschläge von 1987 wurde festgestellt, dass weder die Formel von Wischmeier und Smith noch die von Williams eine gute Schätzung der Geschiebeführung in den Bankettkanal ergaben. Daraufhin haben wir für die beiden, oberhalb der Talsperre von Siliana am meisten vorkommenden Bodentypen, zwei Formeln erstellt. Ausserdem wurde festgestellt, dass durch die Errichtung von Banketts eine Abgrenzungslinie entsteht. Auf beiden Seiten dieser. Linie werden die organischen und lehmigen Substanzraten in entgegengesetzter Richtung variieren.

1. INTRODUCTION

Parmi les techniques mécaniques de conservation des eaux et du sol, la banquette reste la plus utilisée en Tunisie bien que l'enquête faite sur ces travaux effectués durant la période 1962-1975 a montré que sur les 758698 ha traités surtout par des banquettes 28 pourcent des travaux seulement sont restés en bon état (FAO-SIDA 1976).

Parmi les facteurs qui ont contribué à la destruction de ces ouvrages nous pouvons retenir les suivants :

1. Non réparation des banquettes cassées à la suite d'évènements pluvieux exceptionnels comme ceux des années 1969 et 1973.

2. Le transport solide dans le fossé de la banquette que ce soit à partir du talus ou de la parcelle interbanquette est souvent sous estimé. Le comblement rapide du fossé par les sédiments le rend vite vulnérable aux pluies les plus fréquentes.

3. Le manque à gagner dû à l'emprise des banquettes et au décapage de la couche du sol au moment de leur construction n'a pas incité les exploitants agricoles à veiller à leur sauvegarde et leur entretien qui s'avèrent d'autant plus nécessaire que ces ouvrages sont construits à la main d'oeuvre.

Ainsi nous nous sommes proposés dans

ce travail de déterminer :

-Des procédés d'estimation des dépôts de sédiment dans les fossés des banquettes à retention totale construites sur les sols érodés des régions du Centre Ouest de la Tunisie et particulièrement à l'amont du barrage de l'Oued Siliana.

-L'impact de la construction mécanique de ces banquettes sur les taux de matière organique et d'argile dans les horizons superficiels (0-20 cm) des sols calcimagnésiques du type rendzine sur encroûtement calcaire.

2. ANALYSE BIBLIOGRAPHIQUE

L'équation universelle de perte de terre (Wischmeier et al 1978, 1959) qui a été trouvée à partir de moyennes établies sur de longues périodes donne des résultats d'autant plus biaisés que la période d'expérimentation est plus courte.

Ainsi Williams (1975) a introduit des modifications à cette équation afin de pouvoir l'appliquer à la perte de terre induite par un évènement pluvieux en modifiant le facteur d'érosivité de la pluie (Rw) par un facteur relié au volume de ruissellement et au débit de crue de l'averse.

Néanmoins les données concernant la pluviométrie étant généralement plus disponibles que les données relatives au ruissellement, il serait utile de trouver des relations reliant la perte en terre au facteur d'érosivité calculé pour des évènements pluvieux particuliers.

Par ailleurs le décapage de la couche superficielle d'un sol peut réduire son potentiel productif, Frye et al (1982) ont trouvé qu'une réduction absolue de cette couche de 4 à 5 % peut conduire à une réduction de 12 à 36 % du rendement selon le type de culture et de sol. Becher et al (1983) travaillant sur un loess ont abouti à une conclusion analogue pour le maïs. Ceci est dû au fait que le décapage de la couche superficielle est accompagné d'une diminution de l'eau disponible pour les plantes. Cette réduction est irréversible parce que la fertilisation ne peut pas compenser la perte de disponibilité en eau surtout dans les zones semi-arides.

3. PROTOCOLE EXPERIMENTAL

Deux petits bassin versants, l'un traité depuis 1983 par des ouvrages de CES et l'autre non traité ont été choisis en 1986 à l'amont du barrage de l'Oued Siliana en Tunisie avec des caractéristiques climatiques, topographiques et pédologiques aussi similaires que possible.

Quatre banquettes ont été choisies dans le bassin versant traité avec l'occupation et le travail du sol suivants :

B_1 et B_2 : Jachère non travaillée

B_3 : blé labouré dans le sens de la pente.

B_4 : blé labouré en courbe de niveau.

Des reglettes graduées disposées au centre de mailles de $4,5 m^2$ chacune ont été placées dans le dossé de chaque banquette. Après chaque pluie engendrant un ruissellement on lit directement sur chaque réglette la hauteur (h_i) de terre déposée dans le fossé. Le dépot total apporté par une pluie donnée est calculé ainsi :

$$E = \sum_{i=1}^{i=N} h_i . s . d_a$$

où nous avons :

E = Poids total de sédiments déposés dans la retenue

N = Nombre de reglettes par banquette

S = Surface de la maille ($4,5 m^2$)

d_a = Densité apparente des sédiments.

une autre parcelle de 2550 m^2 de superficie a été entourée de plaques en éternit, équipée d'un dispositif de collecte des eaux de ruissellement et d'un limnigraphe et installée près des parcelles traitées par les banquettes afin de pouvoir déterminer les volumes de ruissellement et les débits de crue. Ce dispositif n'a commencé à fonctionner que vers le 30 Mars 1987.

Les conditions hydrologiques et la superficie de cette parcelle se rapprochent assez de celles des parcelles interbanquettes pour qu'on puisse supposer que les volumes de ruissellement et les débits de crue vont varier proportionnellement à la surface du bassin versant interbanquette. Ainsi nous aurons :

$$q_B = q_P \frac{S_B}{S_P}$$

q_B = débit de pointe sur la parcelle interbanquette

q_P = débit de pointe sur la parcelle équipée de limnigraphe

S_B = Superficie de la parcelle interbanquette

S_P = Superficie de la parcelle équipée de limnigraphe.

Un pluviographe qui n'a été fonctionnel qu'à partir du 25 février 1987 a été installé près des parcelles .

4. RESULTATS ET INTERPRETATION

4.1. Impact sur le transport solide

Les hauteurs de sédiment déposées dans le fossé de la banquette et lues sur les réglettes enfoncées dans le sol du canal ont montré des apports positifs, nuls et négatifs. Les valeurs négatives se trouvent généralement sur les réglettes installées dans les zones à pente raide tel que le pied des talus de la banquette où l'eau stagne rarement. Le tableau 1 montre les dépots de sédiment dans le canal engendrés par les pluies qui sont tombées entre le 19 janvier et le 14 avril 1987 (A_{obs}). Nous avons aussi présenté dans ce tableau les estimations de perte de terre calculées par la formule de Wischmeier et Smith (A_{wc}) et celle de Williams (A_{wl})

Bien que cette dernière soit considérée comme donner des résultats plus fiables que la première équation à l'échelle d'un évènement pluvieux, nous avons constaté que celle-ci a donné des résultats plus rapprochés des mesures observées. En effet le tableau 1 montre qu'en moyenne :

$$A_{obs} = 0,69\ A_{wc}$$
$$A_{obs} = 3\ A_{wl}$$

Les écarts types de 0,48 pour la première moyenne de 0,69 et de 1,65 pour la deuxième moyenne de 3 sont importants. Ainsi l'utilisation de ces équations pour l'estimation des dépôts de sédiment dans le fossé des banquettes reste aléatoire. L'analyse du tableau 1 montre aussi deux grands groupes d'érosion, celui des interbanquettes B1, B2 et B3 d'une part et celui de l'interbanquette B4 d'autre part. La perte en terre a été toujours la plus faible dans B4 pour tous les épisodes pluvieux car cette parcelle diffère de la parcelle B3 par le labour en courbe de niveau et des parcelles B1 et B2 par la couverture végétale et le travail du sol. En effet la moyenne des pertes en terre est de 8,506 x 10^{-2}t/ha par averse pour la parcelle B4 alors qu'elle atteint 22,192 x 10^{-2}; 20,631 x 10^{-2} et 21,449x10^{-2} t/ha respectivement pour les parcelles B1, B2 et B3. Le labour en courbe de niveau a réduit de plus de 50% les pertes en terre en comparant les parcelles B3 et B4.

La faible différence dans le transport solide entre les parcelles B1, B2 et B3 pourrait être dûe à la faible variation du couvert végétal au cours de l'année d'une part et d'autre part au labour effectué sur l'interbanquette B3 au moment du semis. en effet la végétation naturelle qui s'est installée sur les parcelles en jachère non travaillée B1 et B2 a assuré une protection du sol contre l'érosion similaire à celle assurée par la culture du blé. Le labour dans le sens de la pente de la parcelle B3 a permis de créer des rigoles et des zones préférentielles d'écoulement des eaux de ruissellement dès les premières pluies et dans lesquelles le décapage du sol a été plus intense.

Tableau 1. Relation entre les pertes en terre observées dans les banquettes et celles calculées.

Date de la pluie en 1987	B.V.	A_{obs} (1)	A_{wc} (2)	A_{wl} (3)	A_{obs}/A_{wc}	A_{obs}/A_{wl}
19/1	B1	17,26	-	-	-	-
	B2	10,07	-	-	-	-
	B3	13,05	-	-	-	-
	B4	8,50	-	-	-	-
	B1	4,31	-	-	-	-
16/2	B2	6,71	-	-	-	-
	B3	6,52	-	-	-	-
	B4	-	-	-	-	-
	B1	21,57	39,81	-	0,54	-
25 et 26/2	B2	13,43	37,04	-	0,36	-
	B3	16,31	21,50	-	0,76	-
	B4	4,25	11,72	-	0,36	-
	B1	17,26	33,01	-	0,52	-
9 et 10/3	B2	16,79	30,72	-	0,55	-
	B3	13,05	17,83	-	0,76	-
	B4	8,50	9,72	-	0,87	-
	B1	25,89	28,16	-	0,92	-
21/3	B2	26,86	26,20	-	1,03	-
	B3	22,84	15,21	-	1,50	-
	B4	12,75	8,29	-	1,54	-
	B1	34,52	221,42	9,79	0,16	3,52
30 et 31/3	B2	33,58	206,00	23,33	0,16	1,44
	B3	35,90	119,59	19,79	0,30	1,81
	B4	17,01	65,22	2,86	0,26	5,94
	B1	34,52	69,92	14,34	0,49	2,41
14/4	B2	36,94	65,05	38,93	0,57	0,95
	B3	42,43	22,03	10,62	1,93	3,99
	B4	8,50	12,01	2,18	0,71	3,90
Moyenne					0,69	3,00
Ecart-type					0,48	1,65

(1) Dépôts des sédiments mesurés dans le canal de la banquette (10^2t/ha)
(2) Dépôt de sédiments calculés par la formule de Wischmeier et Smith (10^2t/ha)
(3) Dépôts de sédiments calculés par la formule de Williams (10^2t/ha)

4.2. Variation interpluviale des pertes en terre

Afin de pouvoir diffencier entre les contributions des facteurs de la pluie et de la couverture végétale sur les pertes en terre nous avons comparé le transport solide dû à deux pluies successives mais assez rapprochées entre elles de telle sorte que la couverture végétale pourrait être considérée comme similaire au sein d'un même couple de deux pluies successives Les résultats ont été reportés dans le tableau 2 en considérant deux types de milieu celui des parcelles B1,B2 et B3 et celui de B4, et les deux facteurs d'érosion à savoir la hauteur de la lame ruisselée et l'indice d'érosivité de la pluie déterminé ainsi :

$$R_w = E_g I_{30}/1702$$

R_w = Erosivité de la pluie tel que déterminé par Wischmeier et Smith (pied-tonne. pouce/acre - heure)

E_g = Energie cinétique de la pluie en joules/m²

I_{30} = Intensité maximale de la pluie durant 30 minutes en mm/heure

Tant qu'il n'y a pas de ruissellement, l'effet de l'érosivité de la pluie reste non détectable au niveau du canal comme le montre le couple de pluie n°6 pour lequel la perte en terre a été nulle pour une pluie dont l'indice d'érosivité a été égal à 1,19 et le ruissellement nul.

Les tableaux 1 et 2 et la fig.1 montrent qu'on se trouve en présence de deux sites différents. Le premier est celui des interbanquettes B1, B2 et B3 et pour lequel la relation trouvée entre la perte en terre et l'érosivité de la pluie est la suivante:

$$A = 295{,}1(R_w)^{0{,}34} \qquad (R^2 = 0{,}65)$$

Le deuxième est celui de l'interbanquette B4 avec la relation suivante

$$A = 107{,}8(R_w)^{0{,}32} \qquad (R^2 = 0{,}39)$$

A = la perte en terre en kg/ha.

Ce sont cependant les fortes valeurs de R_w qui permettent de mieux différencier les sites entre eux faisant ainsi ressortir l'importance de ce facteur.

4.3. Impact sur les taux de matière organique et d'argile

Les tableaux 3 et 4 montrent les résultats d'analyse de la matière organique et de l'argile des 20 premiers centimètres de sol le long de trois transects et à des distances de 0,1,5,3,6,10,20 et 30m à partir du fossé de la banquette aval.

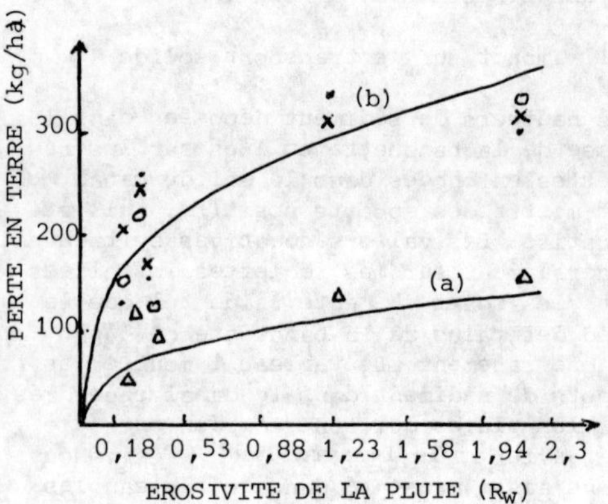

Fig.1 Relation perte en terre-érosivité . pour B1, x pour B2, o pour B3 et Δ pour B4
(a) interbanquette B4
(b) interbanquettes B1, B2 et B3

Tableau 2. Contribution des facteurs de la pluie sur les pertes en terre par couple de pluies successives

N (1)	Date en 1987	pluie (mm)	Ruissellement (mm) (3)	Ruissellement (mm) (4)	R_w (5)	Sediment (kg/ha) (3)	Sediment (kg/ha) (4)
	16/2	3,5	-	-	-	-	-
1	25et						
	26/2	14,4	0,17	0,025	0,41	171	42
2	9 et						
	10/3	13,2	0,11	0,027	0,34	157	85
3							
	21/3	9,5	0,25	0,075	0,29	252	127
4	30et						
	31/3	30,6	3,77	0,780	2,28	346	170
5							
	14/4	16,5	4,51	1,534	1,26	379	85
6							
	15/5	12	0	0	1,19	0	0

(1) Numéro du couple de pluies
(3) Valeur moyenne pour B1, B2 et B3
(4) Valeurs pour l'interbanquette B4
(5) Erosivité de la pluie (Wischmeier,1959)

Ces mêmes analyses ont été faites sur le bourrelet de la banquette et sur un transect choisi dans un sol témoin d'une parcelle limitrophe non aménagée par des banquettes à titre de comparaison. Le tableau 5 montre la moyenne de l'analyse effectuée sur des échantillons pris à 7 endroits différents.

En ce qui concerne la matière organique on a constaté qu'il existe une ligne de transition qui se situe à environ 6 m du fossé. En deçà de cette ligne le taux de matière organique surtout dans la couche superficielle de 10cm de sol est assez variable entre les points de mesure.

La moyenne de cette partie de l'interbanquette est d'environ 0,7% et 0,5% de matière organique respectivement pour les 10 et 20 cm de profondeur de la couche arable. Cette grande variation du taux de matière organique entre les différents point de mesure peut être dûe aussi bien à la variabilité spatiale qu'aux travaux non uniformes de deblai et de remblai effectués par le bulldozer pour la construction des banquettes.

Au delà de la ligne de démarcation de 6m, le taux de matière organique devient plus stable pour les trois transects et les deux profondeurs de sol. Dans cette zone amont de l'interbanquette située entre 10 et 30m du fossé, le taux de matière organique croit de 1,4% à environ 1,9% , la moyenne se situant à 1,65%.

Tableau 3. Pourcentage de matière organique dans l'interbanquette

D (1)	Ligne 1 (2)	Ligne 1 (3)	Ligne 2 (2)	Ligne 2 (3)	Ligne 3 (2)	Ligne 3 (3)	Moyenne (2)	Moyenne (3)
0	0,3	0,3	0,2	0,2	1,2	0,8	0,6	0,4
1,5	0,4	0,3	0,6	0,5	1,0	0,9	0,7	0,6
3	0,5	0,4	1,1	0,4	0,6	0,5	0,7	0,4
6	0,8	0,7	1,7	1,1	1,5	1,1	1,3	1,0
10	1,4	1,2	1,6	1,1	1,6	1,5	1,5	1,3
20	2	1,5	1,9	1,5	1,8	1,7	1,9	1,6
30	1,9	1,6	1,9	1,8	2,1	1,9	2	1,8

(1) distance par rapport au fossé de la banquette en mètre
(2) Horizon du sol de 0-10 cm
(3) Horizon du sol de 10-20 cm

Tableau 4. Pourcentage d'argile dans l'interbanquette.

D (1)	Ligne 1 (2)	Ligne 1 (3)	Ligne 2 (2)	Ligne 2 (3)	Ligne 3 (2)	Ligne 3 (3)	Moyenne (2)	Moyenne (3)
0	6	4	5	4	8	5	6	4
1,5	15	6	13	9	8	7	12	7
3	6	7	55	32	15	11	25	17
6	2	7	37	25	32	9	24	14
10	34	21	35	28	34	31	34	27
20	29	23	31	19	35	28	32	23
30	36	22	34	24	40	31	37	26

(1) distance par rapport au fossé de la banquette en mètre
(2) Horizon du sol de 0-10 cm
(3) Horizon du sol de 10-20 cm

Par conséquent la construction des banquettes a révélé un certain impact sur la richesse en matière organique de la parcelle agricole.

Tableau 5. Pourcentage d'argile et de matière organique dans un sol témoin et sur le bourrelet d'une banquette

	sol témoin		Bourrelet de la banquette	
Profondeur de l'horizon	0-10 cm	10-20 cm	face amont	face aval
% argile (1)	33	28	41	33
% matière organique(2)	1,5	1,1	1,6	2,1

(1), (2) moyenne des résultats d'analyse de 7 échantillons de sol.

Par rapport au témoin qui présente un taux de matière organique relativement constant de 1,3% le tiers du sol arable de l'interbanquette s'est appauvri de0,6% et les deux tiers se sont enrichis de 0,38% Le bilan final est un enrichissement de 0,053% au niveau global de parcelle et qui est une valeur assez faible pour pouvoir tirer des conclusions transposables à d'autres sites.

Les mesures effectuées sur le bourrelet ont donné un taux de matière organique moyen de 1,9% comparable à celui de la zone amont de l'interbanquette et supérieur à celui du sol témoin.

La supériorité du taux de matière organique par rapport au sol témoin peut être dûe à un ou plusieurs des facteurs suivants :

1.l'erreur d'échantillonnage.

2.l'importance et la localisation des déblais et remblais

3.l'augmentation relative de la capacité de transport des éléments fins et légers (matière organique, argile) par les eaux de ruissellement en descendant le long de la pente.

4.le développement plus important de la végétation protectrice du sol dans la zone d'influence de la banquette amont et sur le bourrelet de la banquette.

En ce qui concerne le taux d'argile, on constate qu'il existe à 10m du fossé une ligne de transition nette pour les deux transects gauche et droite de l'interbanquette. Pour le transect du milieu, cette ligne est apparue à 3m du fossé. La moyenne dans la partie aval de l'interbanquette est d'environ 16,75% et 10,5% respectivement pour les 10 et 20cm de profondeur. Au delà de la ligne de transition de 10m, le taux d'argile devient plus stable pour les deux profondeurs et les trois transects avec une moyenne se situant à 31%.

Relativement au témoin qui présente une moyenne de 30,5% d'argile, le tiers aval du sol arable de l'interbanquette s'est appauvri de 16,87% tandis que les deux tiers amont se sont enrichis de 0,5%. Le bilan final sera donc de -5,3% montrant que la construction de la banquette a contribué à une réduction moyenne de 5,3% du taux d'argile de la parcelle cultivée.

Comme pour la matière organique, l'argile perdue dans l'interbanquette s'est retrouvée dans le bourrelet de la banquette puisque le taux d'argile de 37% sur le bourrelet se rapproche de celui du sol non décapé.

Nous pouvons ainsi conclure que la construction des banquettes a un effet néfaste sur la richesse et la distribution de l'argile et de la matière organique qui constituent les deux éléments primordiaux caractérisant la fertilité d'un sol.

5. CONCLUSION

Nous avons trouvé que la formule de Wischmeier et Smith appliquée à quelques évènements pluvieux durant l'année 1987 a donné des résultats qui se rapprochent plus de la réalité que ceux calculés par la formule de Williams. Les conditions du milieu des 4 banquettes étant fréquemment rencontrées dans les régions du Centre Ouest de la Tunisie nous avons cru utile d'établir des équations reliant la sédimentation du fossé des banquettes à l'érosivité des évènements pluvieux qui sont les données les plus susceptibles d'être disponibles.

Nous avons aussi trouvé que le décapage de la couche superficielle du sol pour la construction des banquettes donne naissance à une ligne de démarcation de part et d'autre de laquelle les taux de matière organique et d'argile vont varier en sens inverse.

Néanmoins ces banquettes constituent un moyen efficace de retention des eaux et des sédiments moyennant un dimensionnement qui tiennent compte du transport de sédiment dans leur canal et un entretien périodique.

BIBLIOGRAPHIE

Becher, H.H. Schwertmann, U. and Sturmer, H. 1983. Crop yield reduction due to reduced plant available water caused by water erosion. Soil Erosion and Conservation. Editor El Swaify, S.A. Moldenhauer, W.C. and Andrew, Lo.

Chaabi, A. 1988. Méthodes d'évaluation de l'érosion sur un versant traité en banquettes dans une zone du semi-aride (sous pluies naturelles et simulées).Mémoire de fin d'étude du cycle de spécialisation de l'I.N.A. Tunis

FAO-SIDA - Direction des Forêts. 1976. Inventaire des travaux de conservation des eaux et des Sols en Tunisie.

Frye, W.W. Ebelhar, S.A. Mirdock, L.W. and Blevins, R.L. 1982. Soil erosion effects on properties and productivity of two kentucky soils. Soil Science Society of America Journal. 46 : 1051-1055

Williams, J.R. 1975. Sediment yield prediction with universal equation using runoff energy factor. Present and Prospective Technology for Predicting sediment yield and sources. ARS. 5-40. Agr. Res. Serv, U.S. Dept. Agr., Washington, D.C. : 244-252

Wischmeier, W.H. 1959. A rainfall erosion index for a universal soil loss equation. Soil Science Society of America Proceeding 23 : 246-249

Wischmeier, W.H. and Smith, D.D. 1978. Predicting rainfall erosion losses.U.S. Dept. Agr. Hanbooks. 537

Land and Water Use, Dodd & Grace (eds), © 1989 Balkema, Rotterdam. ISBN 90 6191 980 0

Role of agroforestry in soil conservation

V.G.Vaishnava & P.R.Narwadkar
Marathwada Agricultural University, Parbhani, India

ABSTRACT : Besides other benefits of Agroforestry it plays a pivotal role in soil and Water Conservation. In India many Research workers have tried to study this aspect. Narain (1985) reported that run off and soil losses were greately reduced under Agroforestry systems aimed at nutrient and water budgeting. Pradhan and Srivastava (1986) in Agro Silivi Pastural land use study found that the runoff and soil loss differences between cultivated fallow and grass cover were 20 and 73 times respectively. Sud and Mittal (1987) observed that the average run off was reduced from 25% to 5.6% under Bhabhar Grass cover. Singh et al. (1981) reported that with the introduction of grass species (Bhabhar) in strips the slope was reduced from 4 to 2.7% in 5 years. Chinnamani et al. (1965) observed that the run off accounted for 1.3% and 1.1% of rain fall and soil loss of 0.06 and 0.02 t/ha from Shola forest and Eucalyptus globulus respectively. Similar observations were reported by Sharma et al. (1980). Singh (1986) illustrated that the annual soil loss in T/ha was 24.02 to 62.60 for Pine Apple Cultivation while it was only 0.04 to 0.52 t/ha under natural bamboo forest. Dhruvanarayan et al. (1986) reported that the run off (% rain-fall) and soil loss (T/ha) was 6.45 and 40.95 in case of Agriculture; 3.11 and 2.96 with Agri. Horticulture mixed land uses and bench terraicing, while it was only 0.26 and 0.29 for forest (bamboo).

1. INTRODUCTION

The soil and water conservation problems in India are varied and complex. Major land use in India must be understood before effective soil and water conservation practices can be applied. The seven catagories of land use in India as of 1970-71 were as follows (Anonymous, 1976).

Table 1. Land use in India (1970-71).

Land use	Hectarege (Millon ha)	Percentage of total area
Cropped	165	50.3
Forests	66	20.1
Deserts, mountains	29	8.8
Culturable waste	16	4.9
Non agricultural	16	4.9
Permanent pasture	13	4.0
Other	23	7.0
Total	328	100.0

According to Das and Mukherjee (1980), out of total geographical area of 328 million hectares 175 million hectares is affected by erosion hazards. This is because of wrong land management, heavy pressure of ever increasing demand on limited agricultural lands, faulty agricultural practices, over grazing of grass lands, over exploitation of forest and natural vegetation, faulty irrigation and drainage systems have confined to give rise to high degree of erosion hazards in India. The importance of soil conservation through different methods and significance of forests and grasses in the protection of soils from erosion has been stressed from time to time by various workers (Anonymous, 1964; Kanwar, 1971, Singh et al. 1981; Dhruvanarayana, 1986).

Agroforestry is an age old practice being adopted by Indian farmers in one way or the other. Agroforestry of today is an improvement of Taungya system adopted in some parts (indo burma region) for raising teak, pine and cypress plantations (Jean Combe, 1982). Agroforestry has been defined as sustainable land management system, which increases the yield of land, confines production of crops (including tree crops) and forest plants and/or animals simultaneously or management practices that are compatible with culturable practices of local people (King and Chandler, 1978; Bene et al. 1977).

Besides other benefits of Agroforestry, it plays pivotal role in soil and water conservation. Different Agroforestry systems provide means of covering soil surface through vegetation.

2. VEGETATIVE COVER AND EROSION

The vegetative cover plays very significant role in the conservation of soil surface. The role of forests and grass lands in soil conservation is well understood.

2.1 Influence of forest on rainfall disposition

The data on the research work carried on Acacia species and Shorea robusta indicated that interception of precipitation was high in case of light showers (37.3% when the rainfall was 4.2 mm) and was low in case of heavy showers (4.1% for 140.3 mm rainfall and also 3.7% when rainfall of 64.4 mm was accompanied by strong wind). The average interception loss was 14 percent (Dalal et al. 1967).

2.2 Infiltration studies

Infiltration studies under different vegetative covers carried in the country, indicated that the infiltration rate was the highest under woodland and lowest under Agricultural land (Tejwani et al. 1975). The infiltration rates under different forest areas varied considerably depending upon the condition of forest floor in the Doon Valley and adjoining uplands in the Himalayas and Siwaliks (Patnaik and Virdi, 1962); the data are given in Table 2.

From the following table it is evident that the leaf litter increased whereas the compactness of forest soils decreased the infiltration rates. The infiltration rates for second and third hour are almost uniform both for leaf litter coverd soil and for siwalik range.

Table 2. Infiltration rates in sal forests

Forest type	Infiltration rate (cm)		
	1 hour	2 hour	3 hour
1. Bidhauli(with good leaf litter)	9.00	5.90	5.85
2. Harawala(with little litter and compact surface)	3.71	2.00	2.20
3. Silwalik Range	4.00	2.00	2.10
4. Compact eroded soils of sal	2.80	1.20	1.40

2.3 Run off and soil loss studies

2.3.1 Role of forests

Vegetation plays an important role in the protection of soil. Numerous reports are available on this aspect. The different vegetative covers studied at oatacamand were Shola (natural forest of the region), Eucalyptus globulus (blue gum), cytisus scoparius (broom) and indignous grasses. The run off accounted for 1.3 and 1.1% of rainfall and soil loss 0.06 and 0.02 t/ha from Shola forests and Eucalyptus globulus respectively, whereas run off was only 0.02% of rainfall with virtually no soil loss from protected grass land (chinnamani et al. 1965). Similar results were reported by Sharma et al. (1980). Chakravarti (1971) reported that forest lands showed lower values of erosion ratio than aerable lands. Pathak (1974) found that the sediment load of 0.18 ha m/sqKm was reduced to 0.15 ha m/sqKm in Ramganga watershed when agricultrual land was treated with soil and water conservation measures and 16000 heactares were afforested. Mathur et al. (1976)reported that there was reduction of 28% in run off and 73% in peak rate in watershed due to afforestation resulting in fully stocked plantation and dense growth. Das (1976) reported that due to following an integrated approch for treating land subject to shifting cultivation in North estern India by afforestation, plantations of coffee, cashew, pine apple, banana, sisal, lemon grass etc, the sediment production rate in the reservoir was fallen from 3.80 to 2.36 ha mm/100 sqKm.

Bhatia and Srivastava (1979) observed occurrance of highest amount of clay fraction in forest soil followed by grass cover which they attributed to the fact that these two measures on eroded soils helped in binding soil particles together which minimised loss of this fraction due to run off. Das (1980) reported that Alnus nepalensis trees were grown at hill slopes for fuel wood purpose which subsequently acted as the terraces for paddy fields. The usefulness of populus ciliata as promising tree species for soil conservation in hilly areas was reported by Mathur et al. (1982). Sharma and Agarwal (1980) compared the erodability of soils under different land uses and they found that the erosion ratio ranged from 14 to 83 in cultivated soils 10 to 84 in forest soils considering value of 10 as the significant boundry between erodive and nonerodive soils.

The results obtained by Dhruvanarayana et al. (1986) are given in Table 3.

Table 3. Runoff and soil loss in different land uses.

Type of cultivation	Run off loss (% rainfall)	Soil loss (t/ha)
Agriculture (Jhuming)	6.45	40.95
Agri-horti (mixed) with terracing	3.11	2.96
Forests (Bamboo)	0.26	0.29

It is clear from the data that the run off loss (% rainfall) and soil loss (t/ha) was highest in case of agriculture (Jhuming), medium in Agri-horti (mixed land use) and lowest in case of forest (Bamboo).

2.3.2 Role of grasses in soil conservation

The grasses play very significant role in the protection of soil. In India, due to over grazing, the pasture lands are subjected to soil erosion. The cattle population is increasing and the pastures are to be developed on scientific lines and controlled grazing is required. Goats cause lesser damage to the pastures as compared to cattle and sheep (Shankranarayan et al. 1985; Saini and Sharma,1986). Singh et al. (1981) recommended different grasses and leguminous tree species for various soil types and regions which are given in Table 4.

Table 4. Leguminous tree species and associated grasses recomended for various regions of the country.

Soil type	Plant species	Associated gress(es)
Poor stony soils of Siwaliks and Outer Himalaya.	Acacia catechu	Eulaliop-sis binata
Loam to sandy loam soils of semi-arid regions on field boundaries and roadsides.	A.Cupressiformis	E.binata Cenchrus Ciliars, C.setig-erus
Sandy soils of riverbed and sand dune.	A.Jacquemontii	C.ciliaris, Lasiurus sindicus
Black soils of pennisular India, loamy to sandy loam	A.nilotica spp. indica	Dichanthium annulatum Sehima nervosum
Gravelly soils on pedimount regions.	A.leucophloea	D.annulatum Sehima nervosum
Hills of arid and semi-arid regions and sand dunes.	A.sengeal	D.annulatum, C.ciliaris & Lasiutus sindicus
Sandy soils of arid regions.	A.tortilis	C.ciliaris, C.setigerus
Sandy to sandy loam soils (Calcareous)	Albizzia spp.	D.annulatum
Loamy to loamy sand	B.variegata	C.ciliaris, C.setigerus
Loamy to gravelly soils of hill regions	Butea frondosa	D.annulatum
Pennisular India	Glyricidia maculata	C.ciliaris
Pennisular India	Hardwickia binata	D.annulatum
Loamy to sandy loam	Leucaena leucoephala	C.ciliaris
Sandy loam to loamy sand	Prosopis cineraria	Lasiurus Sindicus

Soil type	Plant species	Associated gress(es)
		C.ciliaris C.setigerus
Saline areas, gravelley loamy to clay loam	P. Juliflora	D.annulatum Sporobolus spp.
Clay loam to loamy soils	Sesbania grandiflora	D.annulatum

The grasses prevent the soil losses in run off (Bhatta et al. 1971; Bhola et al. 1975). Anonymous (1977) reported that even on 11% scope, run off and soil losses were reduced under different grass/legume covers. The results are given Table 5.

Table 5. Effect of different grasses on run off and soil loss.

Grasses	Run off(%)	Soil loss(%)
Cynodon plectostachys	35	56
Cenohrus ciliaris	33	137
Panicum antidotale	36	67
Pennisetum polystachyum	27	67
Urochloa stolonifera	32	78

Grasses prevent the soil loss through soil cover and soil binding capacity. On the basis of study of 11 grasses, it was found that panicum repens had very high soil binding factor of 1287, Dichanthium and Ischaemum having 106 and 103 respctively (Tejwani et al. 1975). Singh et al. (1986) reported that Defregeasia hypoleuca grass grow 98 cm in height within six months with 4167 sqcm crown area, roots grew fast and penetrated deep in the soil and strongly bound the soil, even on steep slopes against the erosive forces. The effect of root system of forest vegetation including harbaceous flora in terms of their capacity to bind the soil particles has also been reported by other workers (Basu, 1953; Chatterjee and Sen, 1964, Sinha and Chatterjee, 1966 and Singh and Chatterjee, 1968.

Sud and Mittal (1987) obsvered that the average run off was reduced from 25% to 5.6% under bhabhar grass cover. Singh et al. (1981) reported that with the introduction of grass species (Bhabhar) in strips the slope was reduced from 4 to 2.7% in 5 years.

3. AGRO FORESTRY SYSTEMS AND SOIL CONSERVATION

Importance of agro forestry in control of soil erosion has been well recognised by different workers (Lundren, 1982; Huxley, 1983; Nair, 1984; Gupta, 1984; Lal, 1986 and Young, 1986). Narain (1985) reported that run off and soil losses were greatly reduced under Agroforestry systems aimed at nutrient and water budegeting. Pradhan and Srivastava (1986) in a Agri-silvipastroal land use study found that the run off and soil loss differences between cultivated fallow and grass cover were 20 and 73 times, respectively. Arora and Mohan (1986) reported that in the young orchards of peach, Flordasun Agri-horti systems with short duration companion crops (cow pea, soybean, sesame) and long duration companion crops (cymbopogon, pine apple, turmeric)can be adopted for soil and moisture conservation purpose. Singh and Reddy (1986) discussed the appropriate agroforestry systems viz; grass-legume-forge system or silvipastoral system for achieving sustained production and proper soil and water conservation in the class 4 soils of rainfed watershed. The use of different agro forestry systems in soil conservation and watershed management was stressed by Narwadkar et al. (1988). Singh (1986) reported the importance of forage-cum-mulch alley cropping system in the conservation of production base. Bavappa (1986) illustrated the benefits of coconut banana and ginger, pineapple multi storied cropping system for economic land use. Itnal (1986) reported that in an agri-silvi study in black soils, the run off loss of rain water was least with Acacia auriculiformis, Albixia lebbeck and Acacia nilotica in that order, while casuarina, Eucalyptus, Acacia catechu and leucanea control plot without tress. Dhruvanarayan et al. (1986) reported that the run off (% rainfall) and soil loss (T/ha) was 6.45% and 40.95% in case of Agriculture, 3.11 and 2.96 with Agri-horticulture mixed land uses and bench terracing while it was only 0.26 and 0.29 for forests (bamboo). Singh (1986) illustrated that annual soilloss in (T/ha) was 24.02 to 62.60 for pineapple cultivation while it was only 0.04 to 0.52 T/ha under natural bamboo forest. Basu et al. (1987) reported that Acacia mangium an Australian exotic was found to be effective for soil conservation in coastal area.

It can be concluded that the type of land use and vegetative cover plays significant role in soil conservation. The Agroforestry, a new concept of multiple land use is very important in the tropical world, not only of economic viability of farming, but also for the conservation of eco-system and conservation of top fertile soil.

REFERENCES

Anonymous 1964. Study of soil conservation programme for agricultural lands. India Planning Commission Programme Evaluation. Govt. of India Publi.

Anonymous 1972. A review of research works Deochanda experimental Station, Damodar Valley Corporation, Vol 3 Mimeo.62.

Anonymous 1976. Report of the National Commission on Agriculture, Part 5, Resource Development, Govt. of India Press, New Delhi, India.

Arora, Y.K. and S.C. Mohan 1986. Agri-Horti systems for water shed management. Indian J. Soil Conserv. 14(3):99-104.

Basu, J.K. 1953. Application of science to the problem of soil and moisture conservation. J.Indian Soc.Soil Sci. J.N. Mukharjee's 60th Birthday commemeration Volume.

Basu, P.K. D.K. Ganguly and G.S. Mandal 1987 Intorduction of exotic in south west Bengal Acacia mangium in coastal area Indian forester 113 10 : .675-680.

Bavappa, KVA. 1986. Mixed multi-storied cropping ICAR-ICRAF Training-cum-workshop on Agroforestry held at CRIDA Hyderabad, Sept. 142

Bene, J.G., H.W.Beall and A.Cote 1977. Trees, Food and People Land Management in the Tropics, Ottawa, IDRC., 52.

Bhatia, K.S. and A.K.Gupta and Tejwani, K.G. 1971. Influence of cropping pattern and land use on plant nutrient losses in Doon Valley. Proc. Int. Symp. Soil Fert. Evaluation 1: 447-451.

Bhatia K.S. and A.K.Shrivastava 1979 Effect of Various soil Conservation measures on fertility status of erodated alluvial soil A note J. Soil Conserev. 7(2) :87-89.

Bhatta P.N., O.P. Gupta and K.G. Tejwani 1971 Influence of crpping pattern and land use on plant nutrient losses in Doon vally Proc. Int. Symp. Soil Fert. evaluation 1:447-451.

Bhola, S.N., M.L.Khybri and R.Dayal 1975. Run off and soil loss studies under different cropping patterns J. Indian Soc. Soil Sci. 23: 298-301.

Chakravarti,D.C. 1971. Investigation on erodibility and water stable aggregates of certain soils of eastern Nepal J. of Indian Soc. Soil scince 19: 441-446.

Chattarjee, B.N. and M.K.Sen 1964. Studies on the effect of grasses on red soil of Ranchi (India) . J. British grassland Soc. 19: 340-342.

Chinnamani,S., S.C.Gupte, N.D.Rege and P.K.Thomas. 1965. Run off and soil studies under different vegetative covers. Indian Forester 91(9):672-682.

Dalal,S.S., Wasiullah and H.N.Mathur 1967. Rainfall interception by sal forest. Indian Forester, 93:

Das,D.C. 1976. An integrated approach for treating the land subject to shifting cultivation in North Eastern India NE Indian Council of Social Sci. Res. Shillong, 34-39.

Das,D.C. 1980 Some aspects of shifting cultivation related to soil and water conservation. Indian J. Soil Conserv. 8(1): 53-59.

Das,D.C. and B.K. Mukherjee 1980. "National prespective on soil and moisture conservation " Agril. Eng. Today, 4(3): 11-17.

Dhruvanarayana,V.V. 1986. Soil and water conservation research in India. Indian J. Soil Conserv. 14(3):22-31.

Dhruvanarayana,V.V., Ram Babu and C. Venkataramanan 1986. Soil erosion under different Agro climatic conditions. Indian J. Soil Conserv. 14(2): 39-50.

Gupta,R.K. 1984. Role of Euclyptus in

soil and water conservation with reference to social/agro forestry. Eucalyptus in India -Past, Present and Future, Kerala Forest Research Institute, Peechi, Kerala, India.

Huxley,P.A. 1983. Plant research and agroforestry ICRAF Nairobi Keniya. Proceedings of a consultative meeting held in Nairobi, 8-15 April, 1981.

Itnal,C.J. 1986. Agroforestry in different climatic/ edaphic zones in India-some examples of indigenous practices in karnataka ICAR_ICRAF Training -Cum-Workshop on Agroforestry, held at CRIDA Hydrabad, Sept. 13.

Jean Combe 1982. Agroforestry techniques in tropical countries. Potential and limitations. Agroforestry System, 1:13-27.

Kanwar,J.S. 1971. Review of soil and water research in India-in retrospect and prospect. ICAR New Delhi.

King, K.F.S. and M.T. Chandler 1978. The wested lands. I.C.R.A.F., Nairobi, Kenya.

Lal, R. 1986. Deforestation and soil erosion. Land cleaning and development in the tropics. Rotterdam, Netherlands.

Lundren,B.O. 1982. Agroforestry approaches to land use in the tropics. ICRAF Nairobi, Kenya, Paper presented at 4th International Congress INTERFORST, Munich, June 29,1982.

Mathur,H.N., J.P.Rambabu and J.B. Singh. 1976. Effect of clear felling and reforestation on run off and peak rates in small watersheds. Indian Forester, 104(4):398-406.

Mathur,H.N., R.P.Singh and K.C. Sharma. 1982. Populus ciliata - A promising tree species for soil conservation in hilly areas. Indian Forester, 108(9):599-604.

Nair,P.K.R. 1984. Soil productivity aspects of Agroforestry Science and Practice of Agro forestry, ICRAF, Nairobi, Kenya.

Narain, Pratap 1985. Agroforestry research in the semi arid tropics. A report on the working group meeting held at ICRISAT Centre, India, 5-6 Aug. 32.

Narwadkar,P.R., L.N.Digrase and V.G. Vaishnava 1988. Role of Agro forestry in the soil conservation and watershed management. Lectures delivered in the Training Programme for Rural Youths and Soil Conservation Officers at Marathwada Agril. Univ. Parbhani, India.

Pathak,S. 1974. Role of forests in soil conservation with special reference to Ramganga watershed. Soil Conserv. digest, 2(1) : 44-47.

Patnaik,N. and S.S.Virdi 1962. Field infiltration studies in Damodar valley. Irrigation and Power, 19(12): 1003-1012.

Pradhan,I.P. and A.K.Srivastava 1986. Silivipasture and Agroforestry to support dry land agriculture in arid and semi arid tropics. A paper presented at the National seminar on Agroforestry Research CRIDA Hyderabad (AP) from 17-18 Feb.

Saini,A.L. and K.Sharma, 1986. The grazing Behaviour of goats vis-a-vis land degradation. Ext. Leaf let No. 18, Makhdoom, Dist. Mathura, Central Institute for research on Goats, India.

Shankranarayan, K.A., H.C. Bohra and P.K. Ghosh. 1985. The goat-an appropriate animal for arid and semi arid regions. Economic and Political Weekly 20: 1965-1972.

Sharma,P.K. and G.C.Agrawal 1980. Soil erodibility under different land uses. Indian J. Soil. Conserv. 8(2):101-105.

Sharma, S., P.D. Gupta and S.K. Das 1980. Nat. Symposium on soil Conservation and water management, Dehradun.

Singh,A. 1986. Farming systems in hilly water sheds. Lecture delivered in the summer Institute on water management at, Dehradun.

Singh, B.P. and Chatterjee 1968. Soil moisture status and root content of grasses at different soil depths. J. Soil and Water Conserv. India. 16: 28-32.

Singh,G., C.Venkatraman and G.Sastry 1981. Manual of Soil and Water Conservation practices in India. Central Soil and Water Conservation Research and Training Institute, Dehradun.

Singh,R.P. 1986. Alley cropping in semi-arid tropics of India. The CRIDA experience. ICAR-ICRAF Training cum-Workshop on Agroforestry held at CRIDA Hyderabad Sept. 13.

Singh,R.P. and G.S. Reddy 1986. Resource management in a rainfed watershed. Indian J. Soil Conserv. 14(3) 52-65.

Singh,R.P., M.K.Gupta and H.N. Mathur 1986. Ecomorphology and soil binding studies on Debregeasia hypoleuca Wedd in temperate region of Himachal Pradesh. Indian J. Soil Conserv. 14(1): 37-42.

Sinha,B.P. and Chatterjee,B.N. 1966. Fertility building under grasses. J. British Grassland Soc. 21: 153-161.

Sud, A.D. and S.P. Mittal 1987.A note on run off and soil loss studies under Bhabhar grass (Eulaliopsis binata Retz.) plantation. Indian J. Soil Conserv. 15(1) :59-60.

Tejwani,K.G., S.K.Gupta and H.N.Mathur 1975. Soil and water conservation research 1956-1971 ICAR, New Delhi, 1-358.

Young,A. 1986. The potential of Agroforestry for soil conservation. Part-I Erosion Control. Working Paper -42 ICRAF Nairobi Kenya.

Land and Water Use, Dodd & Grace (eds), © 1989 Balkema, Rotterdam. ISBN 90 6191 980 0

Optimal design of water management systems

S.O.Prasher & A.Moindarbary
Macdonald College, McGill University, Ste-Anne de Bellevue, Quebec, Canada

ABSTRACT: A PC-based computer software package has been developed for designing water management systems in humid regions. The package employs DRAINMOD, a water management model, to estimate the average crop yields that would result from different water management systems. Next, the package estimates the yearly cost of agricultural production with the help of a comprehensive cost calculation module. The net benefits that would result from different water management systems are thus calculated. The package can be used to optimize the design of water management systems in humid regions by comparing the net benefits that would result from the various water management methods. The operation of the package is described with the help of an example case study.

Resumé: Un logiciel pour micro-ordinateur a été développé pour aider à la conception de systèmes de gestion de l'eau en région humide. Le logiciel emploie DRAINMOD, un modèle de gestion de l'eau, pour évaluer les rendements moyens résultant de différents systèmes de gestion de l'eau. Ensuite, le logiciel évalue le coût annuel de la production agricole avec l'aide d'un programme détaillé de calcul des coûts. Les bénéfices nets résultant des différents systèmes de gestion de l'eau sont ainsi calculés. Le logiciel peut être utilisé pour optimiser la conception de systèmes de gestion de l'eau en région humide en comparant les bénéfices nets résultant de différentes méthodes de gestion de l'eau. L'utilisation du logiciel est illustrée à l'aide d'une étude de cas.

Zusammenfassung: Ein auf PC Gebrauch basiertes Computer Software Program wurde entwickelt, um fuer feuchte Laendereien geeignete Wasserwirtschafts-Systeme zu entwerfen. Zur Veranschlagung von Durchschnittsernten der verschiedenen Wasserwirtschafts-Systeme benutzt das Program ein Wasserregulierungs-Modell genannt DRAINMOD. Durch eine umfassende Kostenberechnungsmethode ermittelt das Program obendrein die Jahreskosten der Ernteproduction. Dadurch werden die Reingewinne ermittelt, die durch Anwendung der einzelnen Wasserwirtschafts-Systeme erreicht werden koennen. Eine Vergleichung dieser durch die Systeme erzielten Reingewinne erlaubt die Bestimmung des jeweilig fuer feuchten Boden geeigneten Optimalsystems. Mittels einer Beispiel-Studie wird die Handhabung des Programs erklaert.

1 INTRODUCTION

Subsurface drainage systems are used extensively in both humid and arid regions of the world to reclaim agricultural soils for economic crop production. In humid regions, the goal of subsurface drainage systems is to ensure trafficable and workable conditions during seed-bed preparations and harvesting, and maintain an aerated root zone in the top soil for good crop production. In arid or semi-arid regions, the goal of subsurface drainage systems is to prevent or alleviate problems related with soil salinization. Many farmers in North America have come to realize the benefits of drainage systems in the last two decades.

In humid regions, attempts are now under way to use subsurface drainage systems as underground irrigation systems. The latter are called subirrigation systems. A dual-purpose system that acts both as a drainage and an irrigation system is sometimes called a water management system. The advantages of a water management system are obvious: the drainage function of the system ensures minimum excess water stresses, on-time planting and harvesting of crops, and prevention of soil salinization; the irrigation function of the system ensures minimum drought conditions in the crop root zone. The cost of this system will be minimal because most of the investment has already occurred during the installation of the subsurface drainage system.

Another very significant benefit of water management systems that has not yet received the attention it deserves is the control of environmental pollution. Since water will be kept within the farm boundaries for extended periods of time, plant roots might uptake some of the leached-out fertilizer along with the water during the upward flow of soil moisture from the water table. Furthermore, some pesticides might become less toxic if they are kept within farm boundaries for extended periods of time (greater than their half-life).

In this paper, we describe a PC-based software package that can be used to design water management systems in an optimal way. The package employs DRAINMOD, a water management model, to estimate crop yields that would result with different water management systems. The crop yield estimates are then combined with the agricultural production costs to estimate the net benefits to a farming enterprise. The combination of water management systems that maximizes the net benefits for the farm can thus be identified.

2 BACKGROUND

The design of water management systems is generally performed by using one of the numerous formulae that relate design parameters to soil properties, hydrologic conditions, and water table fluctuations (Hooghoudt, 1940; van Schilfgaarde, 1965; Young and Legion, 1972; Wiser et al.,1974; Skaggs, 1978; and Belmans et al., 1986). In humid regions of North America, DRAINMOD has become very popular with the designers of water management systems. The model has been verified under field conditions at several locations and its performance has been found to be satisfactory (Skaggs and Nassehzadeh-Tabrizi, 1982; Shirmohammadi et al., 1985; MacKenzie et al., 1988). Since the package described in this paper uses DRAINMOD for designing water management systems, a brief description of this model is given below.

DRAINMOD was developed by Skaggs (1978) to evaluate multi-component water management systems which may include facilities for surface drainage, subsurface drainage, and subirrigation. The model is capable of characterizing the response of the soil water regime to various combinations of water management systems. It can simulate the response of the water table and the soil water above the water table to rainfall and evapotranspiration (ET) for a given degree of surface and subsurface drainage intensity. Climatological data are used in the model to simulate the day-to-day performance of a given water management system over several years of historical record. The model, in its present form, can be used to optimize the design of water management systems with respect to crop yields. The reader is referred to Skaggs (1978; 1985; and 1987) for further details on DRAINMOD.

The major inputs to DRAINMOD include the following :

a) Weather data

- hourly rainfall
- daily maximum and minimum temperature

b) Drainage system parameters

- drain spacing
- drain depth
- effective depth of the impermeable layer
- drainage coefficient
- depth of surface depressional storage

c) Crop parameters

- length of growing season
- depth of root zone as a function of time
- stress susceptibility factors for excess and deficient soil water conditions as a function of the crop growth period

d) Soil parameters

- saturated hydraulic conductivity as a function of depth below the soil surface
- soil water retention data
- drainage volume as a function of water table depth
- steady state upward flux as a function of water table depth
- infiltration parameters of the Green-Ampt equation as a function of water table depth

DRAINMOD is a user-friendly computer software package. Sufficient instructions are given to the user during the execution of the program. The user inputs are checked throughout the program and a chance to re-enter any faulty entries is also provided. It runs on an IBM PC with 640 Kb memory. Both FORTRAN 77 and BASIC versions of the program are available. A hard disk will speed up the I/O steps of the program and an 8087 math co-processor would be an absolute necessity in view of the long term simulations that are often needed to design water management systems.

Though DRAINMOD can be used to design efficient water management systems in humid regions, we believe that there is a great need for software packages that can also take into account the financial aspects of agricultural production. The design methodology should incorporate other costs of agricultural production and the recommended design should be the one that maximizes net benefits for the farm. In this paper, we describe a software package that attempts to design water management systems by maximizing net benefits to the farm.

3 PACKAGE DESCRIPTION

The computer package developed in this study has been written in complied BASIC language. The user is given two options at the beginning of the program: design a new water management system or evaluate an existing subsurface drainage system for water management. In either case, the program chains to DRAINMOD. However, if we are designing a new water management system, we may want to go through the program at least five times, each time with a different drain spacing in order to determine the design drain spacing. In the second case, the program can help in determining the best water management practice for the farm by comparing the expected net benefits that would result from each practice.

For a given field, once the drain spacing is fixed, it

is possible to estimate the length of the drain pipes that will be required to drain the field (e.g. in metres per hectare). Combining the installation cost of the water management system with the other costs of production, it is then possible to estimate the total cost of production for a farming enterprise. Furthermore, as previously mentioned, one of the outputs from DRAINMOD is the expected **long-term** relative crop yield with a given water management system. The actual crop yield can be calculated by multiplying the relative yield with the potential crop yield. The potential crop yield is the yield that would occur if there were no water stresses, whatsoever, during the growing season, good trafficability at the harvest time, and the planting of the crop was not delayed due to poor soil workability in spring.

The program calculates the net benefits that would result with a given water management system by deducting the total yearly cost of production from the yearly revenue through the sale of agricultural produce. The program assumes that the field is located in Quebec, Canada, that it is a corn field, and that the area of the field is 160 ha. However, it can be adopted easily for other locations, crops, or field sizes because the program is menu-driven and it prompts the user for input. If some of the items are not valid for a given location, the user may simply enter zero for that selection. There are default values for each selection in various menus. The latter are given in parentheses. The field size does not enter into calculations as the costs and benefits are calculated on $/ha/year basis.

In the following pages, various menus are described to give the reader an insight of the items considered by the program.

The main menu of the cost module consists of seven items. They are:

1. Investment costs
2. Water management costs
3. Fixed costs
4. Variable costs
5. Income
6. Net benefit/loss
7. Exit to DOS

Each item in the main menu has sub-items. After the execution of each item in a sub-menu, control is transferred back to that sub-menu and then to the main menu. The items in the main menu work independently from each other. The costs associated with a menu were taken from the CREAQ (1985) report based on a 160 ha corn field. However, as stated before, they can be easily modified to reflect the size of the farm, the geographic location, and the current market prices.

3.1 Investment Costs

The first option in the main menu, Investment costs, contains three items. They are :

1. Cost of buildings
2. Cost of machinery and equipment
3. Exit to the main menu

The above options have several associated costs. The latter are given below.

3.1.1 Cost of buildings

The cost of buildings option has two items.

1. Cost of storage room for machinery ($ 25,650)
2. Cost of silos to store corn ($ 43,500)

Once again, these costs assume that the farm is located in Quebec, Canada. For any other location, they will need to be modified to reflect local conditions.

3.1.2 Cost of machinery and equipment

There are twenty selections in this menu:

1. Combine ($ 65,000)
2. Seeder (6 rows; $ 60,000)
3. Tractor (150 hp; $ 32,000)
4. Tractor (80 hp; $ 22,000)
5. Tractor (50 hp; $ 12,500)
6. Sprayer ($ 5,000)
7. Snow blower ($ 1,500)
8. Conveyor (200 mm x 16.5 m; $ 2,500)
9. Conveyor (200 mm x 7.5 m; $ 1,000)
10. Loading conveyor ($ 800)
11. Continuous dryer ($ 8,600)
12. Cultivator ($ 9,200)
13. Wagon ($ 4,700)
14. Plough ($ 2,800)
15. Cutter 3.66 m ($ 7,400)
16. Harrow ($ 600)
17. Silo ($ 9,000)
18. Rotor tiller ($ 35,000)
19. Leveller ($ 3,200)
20. Miscellaneous ($ 2,500)

The sum of the above two menus is later amortized and returned to the user as the investment cost in $/ha/yr. The default prices for each item are shown in parentheses. After the completion of this menu, the user is chained to the main menu display program.

3.2 Water Management Costs

In this section, water management costs are estimated for a given farm. The cost of the subsurface drainage system is always calculated. To this cost, either the cost of a controlled drainage or a subirrigation system is added, but not both. There is a minimal additional cost for adapting a subsurface drainage system to also work as a controlled drainage system. However, the cost of adapting it to work as a subirrigation system is considerably more.

There are four selections in this menu.

1. Subsurface drainage
2. Controlled drainage
3. Subsurface drainage and subirrigation
4. Exit to the main menu

The costs associated with items 1 or 2 in this menu are basically the product of the drainage pipe installation cost and the total length of the pipe. The user is asked to input the installation cost (e.g., $1.98 per metre installed in Quebec). The total length of drainage pipe required is estimated on per hectare basis.

For subirrigation, a farmer may have access to different types of pumps. Therefore, item 3 in the water management menu display has three types of pumps. They are :

1. Diesel engine pump
2. Electric pump
3. Pumping water by some other means

Each item in the pump menu display has fixed and operating costs associated with it. There is also the cost of water table control chambers with each kind of pumping option.

Item 1 in the pump menu, i.e. Diesel engine pump, has a sub-menu consisting of five selections. They are

Fixed costs:

1. Diesel engine pump ($ 50/ha/yr)
2. Control chamber ($ 20/ha/yr)

Operating costs:

3. Maintenance ($ 10/ha/yr)
4. Fuel and lubricants ($ 50/ha/yr)
5. Labour ($ 20/ha/yr)

Item 2 of the pump menu, i.e. Electric pump, is divided into six selections. They are :

Fixed costs:

1. Electric pump ($ 75/ha/yr)
2. Control chamber ($ 20/ha/yr)
3. Electric poles and wiring

Operating costs:

4. Energy ($ 30/ha/yr)
5. Maintenance ($ 10/ha/yr)
6. Labour ($ 20/ha/yr)

The third item in the pump menu, i.e. Pumping water by some other means such as using a tractor, has five options :

Fixed costs:

1. Pump (special case; $ 20/ha/yr)
2. Control chamber ($ 20/ha/yr)

Operating costs:

3. Maintenance ($ 15/ha/yr)
4. Labour ($ 30/ha/yr)
5. Miscellaneous ($ 10/ha/yr)

The fixed costs in the above menus are amortized over the expected life span of the item, supplied by the user. The sum of the fixed and the operating costs in any of the above menus is the cost of the subirrigation system. This cost plus the cost of the subsurface drainage is returned to the user as the cost of the water management system. It may be noted here that only the total cost of one of the pump menu displays would be taken into account. So if a user goes through two different pump menus, the latter one will be used in the final calculations.

3.3 Fixed Costs

Under the fixed costs option in the main menu, there are eight selections. They are:

1. Taxes ($ 6/ha/yr)
2. Fire insurance ($ 16/ha/yr)
3. Maintenance of buildings ($ 4/ha/yr)
4. Maintenance of land ($ 10/ha/yr)
5. Electricity and telephone ($ 4/ha/yr)
6. Automobiles and trucks ($ 10/ha/yr)
7. Professional help ($ 6/ha/yr)
8. Miscellaneous ($ 6/ha/yr)

The total fixed cost is the sum of the above items and the amortized value of the investment costs, i.e. item 1 in the main menu. The Uniform Series Present Value (USPV) is used to calculate the amortized value.

$$AV = PV - SV/((1+R)**(N/USPV))$$
$$USPV = 1 - ((1+R)**(-N/R))$$

Where AV = amortized value ($), PV = present value ($), SV = salvage rate (%), R = interest rate (%), and N = expected life span (yr).

All inputs to the program come from the user and they reflect the prevailing market and local conditions.

3.4 Variable Costs

The variable costs are divided into fifteen items. They are :

1. Seed ($ 60/ha/yr)
2. Fertilizer (0-0-60; $ 35/ha/yr)
3. Fertilizer (18-46-0; $ 70/ha/yr)
4. Fertilizer (41-0-0; $ 115/ha/yr)
5. Lime ($ 20/ha/yr)
6. Pesticides (Atrazine; $ 10/ha/yr)
7. Pesticides (Lasso; $ 40/ha/yr)
8. Maintenance and reparation ($ 60/ha/yr)
9. Fuel and lubricants ($ 50/ha/yr)
10. Drying with Propane gas ($ 70/ha/yr)
11. Drying with Electricity ($ 5/ha/yr)
12. Grain transportation ($ 30/ha/yr)
13. Labour ($ 20/ha/yr)
14. Crop insurance ($ 30/ha/yr)
15. Interest on financing ($ 50/ha/yr)

The sum of the above costs is returned to the user as the variable costs of agricultural production.

3.5 Income

The income menu, option 5 of the main menu, contains three items. They are:

1. Revenue from selling corn ($0.19/kg)
2. Any other revenue
3. Exit to the main menu

The revenue is the product of the actual yield and the marketing price of corn. The price of corn is supplied by the user. The actual yield (kg/ha) is determined by multiplying the potential yield (kg/ha) with the relative yield (%). The latter is estimated with DRAINMOD for a given water management system.

A farmer may use his machinery and equipment to generate some extra income, i.e. renting his equipment. This can be considered by option 2 in the above menu.

3.6 Net Benefits

The net benefits are estimated by subtracting the cost of agricultural production from the revenue from selling the crop. There are six items in this menu. They are

1. Water management costs
2. Fixed costs
3. Variable costs
4. Income
5. Net benefit/loss
6. Exit to the main menu

The cost of each item, determined previously by the program, is shown to the user in a tabular form. The net benefit/loss is calculated by subtracting the costs, items 1 to 3, from the revenues, item 4.

4 APPLICATIONS OF THE PACKAGE

To describe the applications of the package, a hypothetical field site with good surface drainage was assumed near Montreal. It was assumed that the soil is silt loam, with a saturated hydraulic conductivity of 2.7 cm/h. Furthermore, it was assumed that the drains will be laid at a depth of 90 cm below the soil surface and that a restrictive soil layer exists at a depth of 180 cm from the soil surface. The size of the field does not enter into the calculations because the latter are performed on a per hectare basis. It was further assumed that both water and electricity are available at the farm.

The results from DRAINMOD are given in Figure 1 where relative crop yields are plotted against drain spacing. The model was run with 21 years of climatological data for Montreal, from 1963 to 1983. It appears that subirrigation will be the best water management practice for this farm. The yields are consistently higher with subirrigation than with controlled or conventional drainage. The design drain spacing for this farm appears to be 30 m which will, on the average, result in 94% relative crop yield. It is

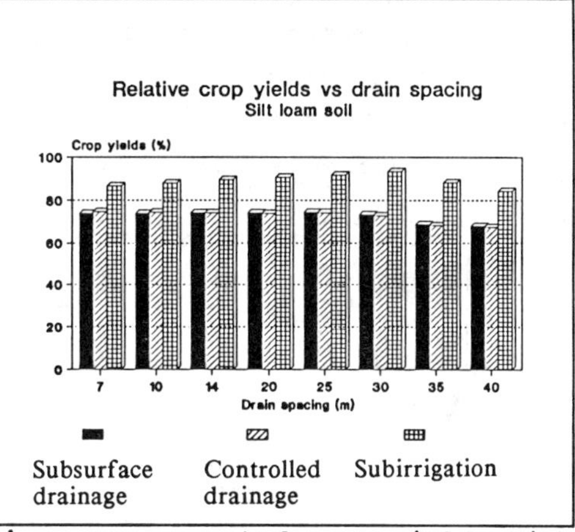

Figure 1. Expected crop yields with different water management systems.

interesting to note that crop yields will be lower if we increase or decrease the drain spacing. For example, we should expect 87% crop yields with a 7 m drain spacing and 84% with a 40 m drain spacing. For the narrower drain spacing, the reasons for the lower crop yield could be due to over-drainage during summer months thus increasing the number of dry days during the growing season. For the wider drain spacing, it could be due to the increased wetness caused by the inefficient drainage of the soil.

Another interesting observation can be made from Figure 1 in that the crop yields from conventional drainage and controlled drainage systems are almost similar. We were expecting higher crop yields with controlled drainage systems as they can conserve excess moisture in the soil profile that could be used by plants during the growing season. However, on checking the DRAINMOD output we found that the reduction in the number of dry days with controlled drainage was somewhat counteracted by the increased wetness in the crop root zone due to the higher weir settings at the outlet. The SEW30 value for controlled drainage was higher than the SEW30 value for conventional drainage. There were lesser number of dry days with controlled drainage but the difference between the two modes of water management was small. With subirrigation, however, there was a greater reduction in the number of dry days and therefore it resulted in higher crop yields in spite of increased wetness in the crop root zone due to occasional rainfall during the growing season.

The net benefits are plotted against drain spacing for the various water management systems in Figure 2. For the most part, the net benefits with subsurface drainage or controlled drainage systems are negative, i.e. the farming enterprise will be losing money every year with these systems. However, with subirrigation it appears that the farm will be making money if the drains are spaced at 30 m interval. Furthermore, the net benefits or losses with subirrigation are consistently more favorable than with the other water management systems.

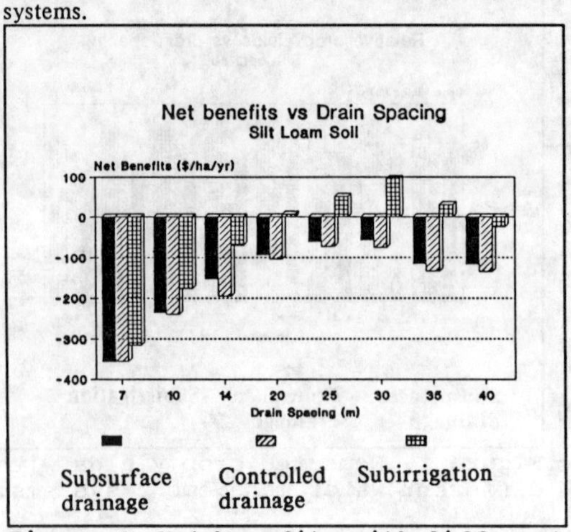

Figure 2. Net benefits with different water management systems.

It appears, however, that the income prospects for this farming enterprise are not very good. An income of about $95 per hectare per year is quite low. Considering an average farm size of 160 ha, the annual income would be $ 15,200 which is below the poverty line in Canada for a family of four. However, it may be noted that in our calculations we have considered a variety of agricultural machinery and other inputs that may or may not be applicable in a given situation. At the same time, however, given the current market prices for agricultural produce in North America, e.g. $0.19 per kg of corn, it appears that farming is not a profitable business these days. The above scenario was not evident from Figure 1 which is useful in comparing crop yields with different water management systems. However, Figure 2 clearly depicts the expected net benefits from different water management systems and also the expected financial health of the farming enterprise. We believe that our approach takes the design of water management systems one step further by providing a way of estimating net benefits or losses to the farming enterprise from a given water management system.

5 CONCLUSIONS

A comprehensive computer package has been developed that can aid in the design of efficient water management systems in humid regions. The package employs DRAINMOD to evaluate different water management practices in terms of their impact on expected crop yields. Once the crop yields are known, the package computes the net benefits that can be expected from the various water management systems by going through a series of cost calculation steps associated with agricultural production. The drain spacing that maximizes the overall net benefits is selected as the design drain spacing and the corresponding water management system is selected as the best water management system for the farm.

REFERENCES

Anonymous, 1985. Le Comite de References Economiques en agriculture du Quebec (CREAQ). Quebec Government. Ministry of Agriculture. Agdex 111/821.

Belmans, C., J.G. Wesseling and R.A. Feddes 1981. Simulation model of the water balance of a cropped soil providing different types of water conditions (SWATRE). Institute for Land and Water Management Research, Wageningen, The Netherlands. 62 pp.

Hooghoudt, S.B. 1940. Bijdragen tot de kinnis van enige natuurkundige grootheden van de grond nr. 7. Versl. Landbouwk. Onderz. 46:515-707.

McCuen, R.H. 1982. A Guide to Hydrologic Analysis using SCS Methods. Prentice-Hall, Inc., Englewood Cliffs, New Jersey. PP 145.

Prasher, S.O., Mackenzie R. and A. Moindarbary 1988. Design of water management systems in eastern Canada using DRAINMOD. Paper No. 88-2107 presented at the Summer meeting of the ASAE in Rapid City, South Dakota.

Skaggs, R.W. 1978. A water management model for shallow water table soils. Report No. 134. Water Resources Research Institute of the University of North Carolina, North Carolina State University, Raleigh, North Carolina.

Skaggs, R.W. 1985. Program design documentation for the water management simulation model in the North Central States. North Central Computer Institute, Madison, Wisconsin.

Skaggs, R.W. 1987. Design and management of the drainage systems. In: Drainage Design and Management. Proceedings of the 5th National Drainage Symposium of the ASAE. Chicago, Ill. P1-10.

Skaggs, R.W. and A. Nassehzadeh-Tabrizi 1982. Optimizing drainage system design for corn. In: Advances in drainage. Proceedings of the 4th National Drainage Symposium of the ASAE. pp 50-61.

Van Schilfgaarde, J. 1965. Transient design of drainage systems. ASCE Proc. 91(IR3):9-22.

Wiser, E.H., R.C. Ward and D.A. Link 1974. Optimizing design of a subsurface drainage system. Transactions of the ASAE, Vol. 17(1):175-178.

Young, T.C. and J.T. Ligon 1974. Water table and soil moisture probabilities with tile drainage. Transactions of the ASAE, Vol. 15(3):448-451.

Land and Water Use, Dodd & Grace (eds), © 1989 Balkema, Rotterdam. ISBN 90 6191 980 0

Etude par simulation du comportement d'une nappe drainée en sols tourbeux

R.Prélaz-Droux & J.F.Jaton
Institut de Génie rural, Ecole Polytechnique Fédérale de Lausanne, Suisse

RESUME: La région des lacs du pied du Jura (Suisse) a fait l'objet d'énormes travaux d'aménagement des eaux au cours de ce dernier siècle. Les lacs de Neuchâtel, Bienne et Morat, d'une surface totale d'environ 300 km2, forment un énorme bassin de compensation en cas de hautes eaux. Un règlement de régulation des lacs permet d'éviter une trop grande variation du niveau des eaux. Ce règlement qui résulte d'un compromis entre de nombreuses contraintes, a fait l'objet d'une étude d'impact dans différents domaines (pêche, ornithologie, flore, entretien des rives, etc). Nous présentons dans ce rapport les questions relatives au domaine hydro-agricole. L'étude a mis en évidence les influences d'une modification du niveau des lacs sur la nappe phréatique voisine, et par conséquent sur la production agricole des plaines bordières des lacs. Les résultats des différentes simulations effectuées ont permis de déterminer le scénario de variation des niveaux d'eau dans les lacs qui soit optimal du point de vue de l'agriculture, tout en intégrant les exigences des autres partenaires.

ZUSAMMENFASSUNG: Im Seeland am Jurafuss (Schweiz) wurden im Laufe des letzten Jahrhunderts wichtige Bodenverbesserungsarbeiten durchgeführt. Der Neuenburger-, Bieler- und Murtensee, die zusammen eine Fläche von ungefähr 300 km^2 ausmachen, dienen im Falle von Hochwasser als grosse Ausgleichsbecken. Ein Regulierungsreglement der Seen erlaubt, allzugrosse Seeschwankungen zu verhindern. Dieses Reglement kam aufgrund mehrerer Umweltverträglichkeitsprüfungen auf den Gebieten Fischerei, Ornithologie, Flora, Uferunterhalt, etc., zustande. In diesem Bericht präsentieren wir die agro-hydrologischen Aspekte. Die Studie hat den Einfluss einer Veränderung des Seewasserstandes auf den benachbarten Grundwasserspiegel hervorgebracht, als auch die daraus entstandenen Folgen auf die landwirtschaftliche Produktion in den angrenzenden Ebenen. Die verschiedenen Simulationen haben erlaubt, die Szenarien für unterschiedliche Wasserspegelstände der Seen optimal für die Landwirtschaft zu bestimmen. Dabei wurden die Ansprüche der anderen Partner mitberücksichtigt.

ABSTRACT: During the last century, the lake areas beneath the Jura (Switzerland) mountains have been subjected to significant changes in land use. The lakes of Neuchâtel, Bienne and Morat, with a surface area of 300 km^2, behave as huge storage reservoirs during flood events. Water levels are controlled by way of a regulation policy which has been implemented to avoid sudden and significant changes of surface elevations. Environmental impacts of this regulation have been assesed (fishing, ornithology, plant life, etc.). This paper deals with impacts with respect to agricultural practice. This study has shown the influence of water surface levels on the nearby water table and consequently on agricultural productivity within the area of concern. A comparison study of various regulation alternatives using simulation models, defined an optimal scenario from an agricultural point of view including the requirements from the other partners.

1. PRESENTATION DE LA ZONE D'ETUDE ET DES PROBLEMES QUI S'Y POSENT

Au cours de ce dernier siècle, d'énormes travaux ont été entrepris en Suisse dans le but d'éviter les débordements périodiques de certains cours d'eau. Parmi les entreprises remarquables, il convient de citer les deux corrections des eaux du Jura. Ces corrections, dont la première a été réalisée il y a environ 100 ans (de 1868 à 1891) et la seconde il y a une vingtaine d'années (de 1962 à 1973), concernent l'aménagement des eaux des lacs de Bienne, Neuchâtel et Morat, qui couvrent respectivememnt 42, 224 et 24 km^2.

La zone étudiée dans ce document se situe dans le Grand Marais bordier des trois lacs (cf. fig. 1). Avant les corrections fluviales, les périmètres des marais étaient périodoquement inondés. Par la suite, de nombreux travaux d'améliorations foncières ont permis la mise en valeur de ces sols tourbeux. La réalisation d'ouvrages de drainage en particulier, est à l'origine de l'assainissement et de la transformation en terres fertiles d'une surface de plus de 6'000 ha. Alors que les premiers drainages étaient gravitaires, la plupart des réseaux d'assainissement actuellement à l'étude ou en cours de réalisation font appel aux stations de pompage en raison du tassement progressif des sols organiques. Il est intéressant de noter que ces ouvrages élévatoires sont souvent conçus de telle sorte que les canaux d'assainissement puissent servir à l'irrigation en saison sèche, et au maintien d'une nappe phréatique élevée en hiver, de manière à diminuer l'aération des horizons tourbeux et à ralentir ainsi leur abaissement. Malgré tout, on observe que la cote des terrains organiques diminue bon an mal an de 0.5 à 2.0 cm en moyenne, ce qui oblige d'effectuer une réfection des drainages tous les 25 à 30 ans environ. Cet abaissement des sols accroît les risques d'inondation et nécessite le rabattement du niveau phréatique afin de maintenir ce dernier à une profondeur optimale pour les plantes.

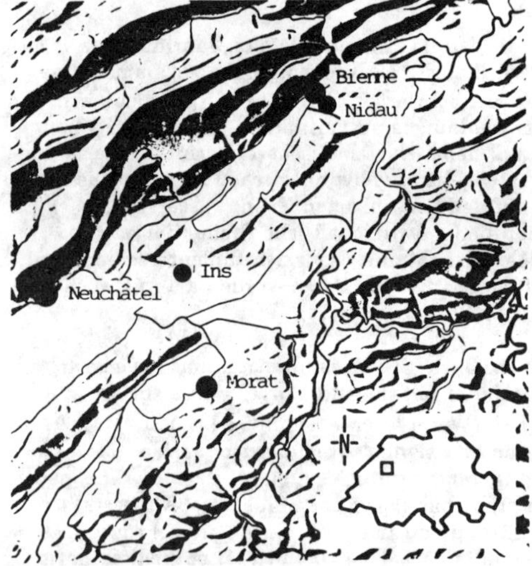

Figure 1 : Situation de la zone d'étude

Le barrage de Nidau-Port, situé sur le canal exutoire des 3 lacs, permet de modifier l'écoulement à la sortie des lacs et d'agir, de ce fait, sur leurs niveaux. Le débit à régler au barrage est fonction du niveau des lacs et de la saison; il est prescrit par un règlement qui tient compte des intérêts des riverains des lacs et de l'agriculture, des souhaits des pêcheurs et des écologistes, ainsi que des débits minimum et moyens à assurer dans la rivière Aar, en aval de Bienne. Différentes courbes des niveaux classés du lac de Neuchâtel sont fournies à la figure 2.

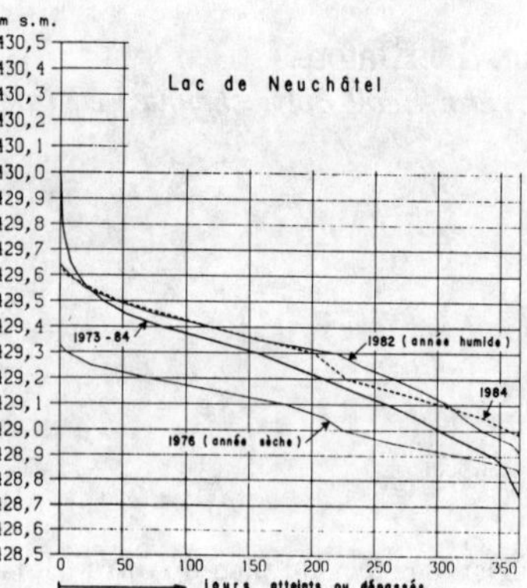

Figure 2: Courbes des niveaux classés du lac de Neuchâtel

Les lacs servent d'exutoire naturel aux réseaux de drainage. Dans les zones drainées par pompage, le niveau de la nappe est indépendant de la cote des lacs. Cette dernière influence par contre les hauteurs de relevage et les volumes de pompage. Dans les zones drainées gravitairement, elle est directement responsable de l'importance des refoulements dans les drainages qui occasionnent actuellement de fréquents dégâts aux cultures.

Dans le cadre d'une vaste étude pluridisciplinaire visant à déterminer l'effet du règlement actuel de régulation du niveau des lacs, ainsi que celui de différents scénarios de variation, sur des domaines aussi variés que l'hydro-électricité, la protection des rives, de la faune et de la flore, l'agriculture, la navigation, la pêche, etc, l'Institut de génie rural de l'EPFL a été mandaté par les Services des améliorations foncières des cantons de Berne, Fribourg, Vaud, Neuchâtel et Soleure, afin d'évaluer l'impact des modifications des variations de la cote des lacs sur l'agriculture des plaines avoisinantes et de proposer un scénario de variation des niveaux d'eau des lacs optimal du point de vue de l'agriculture, tout en intégrant les exigences des autres partenaires.

L'ampleur et la diversité des questions qui caractérisent cette étude (hydrologie, pédologie, agronomie, étude du réseau hydrographique, eaux souterraines, etc) nous ont amené à nous concentrer sur quelques points essentiels relatifs à l'impact de la modification du niveau des lacs sur les systèmes d'assainissement, et les drainages gravitaires en particuliers. En effet, les zones drainées par pompage sont rendues indépendantes des lacs et les résultats des études montrent que les volumes pompés sont principalement fonction des précipitations. Par contre, dans les zones drainées gravitairement et sises à une altitude proche de celle des lacs, les problèmes hydriques rencontrés par les agriculteurs sont en grande partie imputables au phénomène du refoulement dans les drainages, phénomène dont l'intensité est directement liée à la cote des lacs.

2. PRESENTATION DU MODELE DE SIMULATION

Pour décrire le comportement de la nappe en fonction du niveau des lacs dans les zones drainées gravitairement, nous avons choisi le *modèle de simulation de drainage DRAINMOD*, qui a été développé par le Prof. R.W. Skaggs, à l'Université de Caroline du Nord (USA). Ce modèle, de type déterministe, a pour but la conception et l'évaluation des systèmes de drainage. A partir de données climatiques (pluies horaires et températures), pédologiques (conductivité hydraulique, courbe tension-humidité), géométriques (écartement et profondeur des drains, profondeur de la couche imperméable, débit spécifique), agronomiques (coefficient cultural, longueur des racines en fonction du temps), il calcule pour chaque pas de temps Δt (généralement l'heure), le bilan hydrique du sol et en déduit le niveau de la nappe et les volumes drainés. Les termes du bilan sont montrés à la figure suivante.

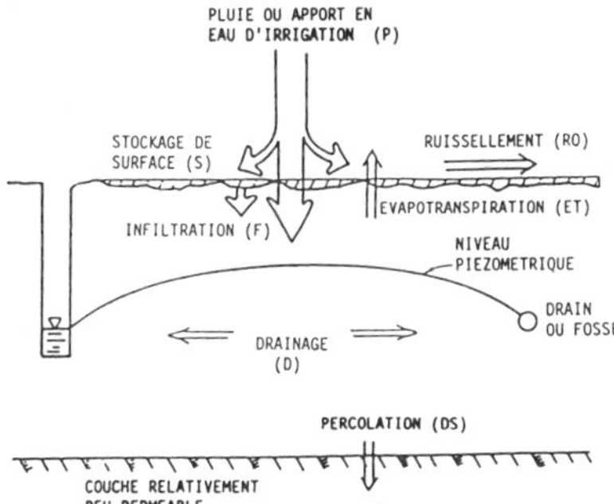

Figure 3 : Schématisation des différentes composantes du bilan hydrique

En modifiant les conditions à l'exutoire du système de drainage, le modèle permet de simuler les cas de figure suivants :

a) *le drainage conventionnel* : l'eau s'écoule librement du système de drainage dans le canal en supposant que le niveau d'eau dans ce dernier reste toujours inférieur au niveau de l'arrivée du système de drainage;

b) *le drainage contrôlé* : la mise en place, dans le canal exutoire, d'un déversoir permet le stockage d'une partie des eaux drainées. Ainsi, ces dernières ne sont pas perdues, mais permettent, en choisissant adroitement le niveau du déversoir, de maintenir la nappe haute entre les épisodes pluvieux par refoulement dans les drainages;

c) *la gestion de la nappe* : la mise en place d'un ouvrage de pompage dans le canal exutoire permet de maintenir le niveau d'eau dans ce dernier à une cote fixée. Le choix judicieux de cette dernière permet de garantir l'efficacité maximale des drainages au printemps et en automne et d'employer les drains pour effectuer de l'irrigation souterraine durant l'été.

Le modèle DRAINMOD a été choisi parce que, pour chaque type de drainage décrit ci-dessus, il fournit non seulement des *résultats quantitatifs* (niveau piézométrique, volume d'eau drainé, etc), mais aussi des *résultats qualitatifs*, tels que:

a) *les jours et les périodes de submersion*, qui permettent d'estimer les dégâts aux cultures. Il est généralement admis qu'une submersion de trois jours provoque déjà des dégâts irrémédiables aux cultures;

b) *les jours de travail*, qui sont définis en fonction de la quantité de pluie qui arrête les travaux agricoles, du volume d'eau à drainer avant de pouvoir recommencer le travail et du nombre de jours de ressuyage;

c) *les jours secs*, qui sont définis comme les jours où le système sol-nappe n'arrive pas à fournir suffisamment d'eau pour couvrir la demande évaporative (égale à l'évapotranspiration potentielle, calculée par la formule de Turc, corrigée par le coefficient cultural);

d) *le paramètre* SEW_{30} qui comptabilise, pour chaque jour où la profondeur de la nappe est inférieure à 30 cm, les différences entre cette profondeur de 30 cm et celle de la nappe.

Le calage et le test du modèle ont été effectués sur une parcelle d'essai de 94 ha, dont les caractéristiques nous sont bien connues et dont les composantes sont facilement mesurables.

Les simulations ont porté sur six ans de données climatiques (1981 à 1986) relevées à la station météorologique de Neuchâtel. Les caractéristiques des systèmes de drainage ont été obtenues à partir des plans d'exécution. Les campagnes de terrain ont fourni les informations nécessaires à la caractérisation des sols, au niveau de la conductivité hydraulique notamment. Cependant, vu sa grande variabilité spatiale, la courbe tension-humidité a été choisie parmi les courbes types données dans la littérature pour des sols identiques. Plusieurs nivellements effectués dans ces zones ont permis de déterminer les altitudes moyennes et minimales du sol (entre 430 et 431 msm).

Les mesures suivies du niveau piézométrique durant l'automne 1987 et l'hiver 1987-1988 ont servi de base à la vérification des résultats de DRAINMOD. De plus, l'enquête agronomique a permis d'évaluer, à partir des discussions avec les exploitants, l'exactitude des paramètres de qualité qui font partie des résultats de DRAINMOD.

Dans le cadre de l'étude générale citée plus haut, les résultats des simulations effectuées avec le modèle DRAINMOD devaient intervenir dans la définition d'un *scénario optimal de régulation du niveau des lacs.* Aussi, dans un premier temps, nous avons étudié les conséquences sur la nappe phréatique et sur les conditions d'exploitations (périodes de submersion, nombre de jours de travail, etc) d'une modification du niveau moyen des 3 lacs aussi bien dans le temps (mois par mois) que dans l'espace (variation du niveau des lacs de 20 cm en 20 cm sur une amplitude de 1.4 m autour de la cote moyenne actuelle qui est de 429.4 msm).

3. EVOLUTION DES PARAMETRES DE QUALITE EN FONCTION DU NIVEAU DES LACS

L'étude de l'évolution des paramètres de qualité en fonction du niveau des lacs a porté sur l'ensemble des zones drainées gravitairement au sein du périmètre défini à la figure 1. Dans le cadre de cette publication, nous nous limiterons à la présentation des résultats obtenus dans une seule de ces zones. Celle-ci se caractérise par une profondeur de drainage d'environ 1.1 m et un écartement entre files de drains de l'ordre de 25 m. Le niveau d'eau à l'exutoire du système de drainage est égal au niveau des lacs. La cote du terrain varie entre 430.0 msm et 431.0 msm.

L'influence de la cote des 3 lacs sur le niveau piézométrique par l'intermédiaire du refoulement dans les drainages est bien mis en évidence à la figure 4. Ces modifications du comportement de la nappe ont des conséquences importantes sur les valeurs des paramètres de qualité, soit le nombre de jours de submersion, le nombre de jours de travail et le paramètre SEW_{30}. La figure 5 montre l'évolution de ces trois paramètres pour un sol sis à une cote moyenne de 430.5 msm et pour les mois les plus critiques au niveau agricole, soit avril, mai, septembre et octobre.

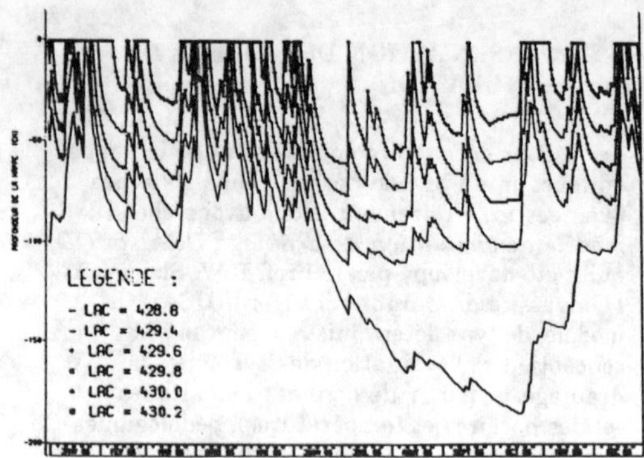

Figure 4 : Variations temporelles du niveau piézométrique en fonction du niveau des lacs.

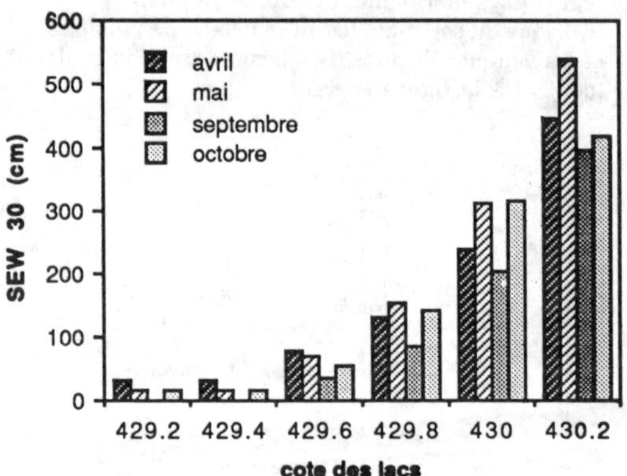

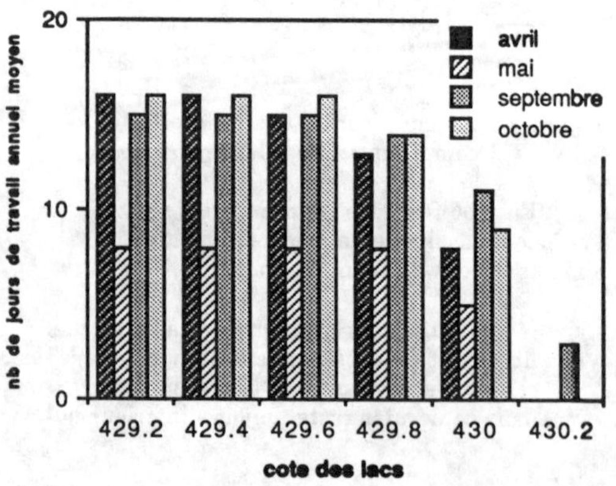

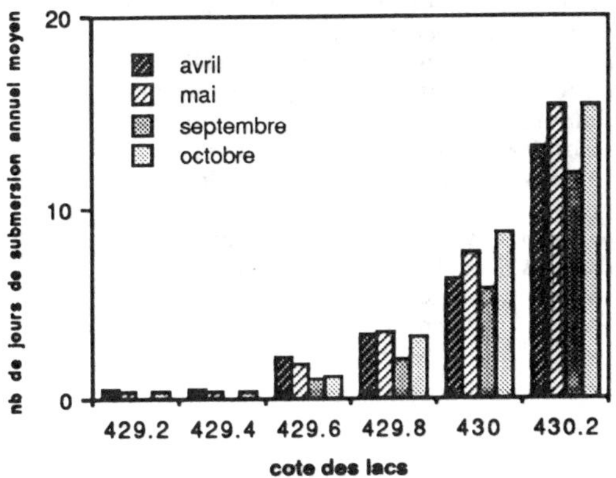

Figure 5 : Evolution des paramètres de qualité

On remarque que dès que le niveau d'eau dans les lacs dépasse 429.4 msm, l'effet du refoulement dans les drainage est manifeste. Ainsi, une cote de 429.6 msm provoque déjà des dégâts irréversibles aux cultures par submersion aussi bien au printemps qu'en automne. Pour de tels niveaux d'eau dans les lacs, SEW_{30} montre bien que la nappe se situe le plus souvent à moins de 30 cm de profondeur.

4. DEFINITION D'UN SCENARIO OPTIMAL POUR L'AGRICULTURE

L'ensemble des résultats obtenus nous a permis de définir des cotes critiques des lacs au-dessus desquelles une élévation du niveau d'eau d'une dizaine de centimètres seulement peut déjà se traduire par des dégâts importants aux cultures, pouvant aller jusqu'à la perte totale de la récolte. Ainsi, le niveau des lacs ne doit pas dépasser le niveau des drains si l'on veut éviter des dégâts importants aux cultures dans les zones composées de sols à dominante tourbeuse ou silteuse. Par contre, la cote des lacs doit rester inférieure d'au moins 50 à 60 cm au niveau du sol pour éviter des impacts néfastes aux cultures sises dans des zones à texture plus grossière, caractérisée par une teneur appréciable en sable. Ces infor–mations, obtenues pour chaque mois de l'année et pour chaque type de sol concerné, dont les sols tourbeux, en fonction des caractéristiques des réseaux de drainage, nous ont permis de déterminer un scénario optimal de variation des niveaux des lacs du point de vue de l'agriculture.

Dans un premier temps, on a recherché à définir un scénario qui protège l'ensemble des terres du périmètre d'étude, ou plus précisément qui permette une bonne exploitation agricole des terrains sis à une cote de 430.0 msm déjà (scénario A). Ensuite, nous avons recherché un scénario permettant une bonne exploitation agricole des terrains sis à 430.5 msm et au-dessus (scénario B), terrains dont l'étendue couvre la majeure partie du périmètre. La figure 6 présente les courbes relatives à ces deux scénarios ainsi que la courbe des niveaux moyens observés entre 1980 et 1987 (scénario C).

On remarque que le scénario A s'écarte nettement des pratiques actuelles de régulation des lacs. Le scénario B s'en différencie par des valeurs sensiblement inférieures durant les mois d'avril, mai et juin. La figure 7 présente l'évolution du nombre de jours de submersion annuel moyen au cours de l'année, en fonction des trois scénarios.

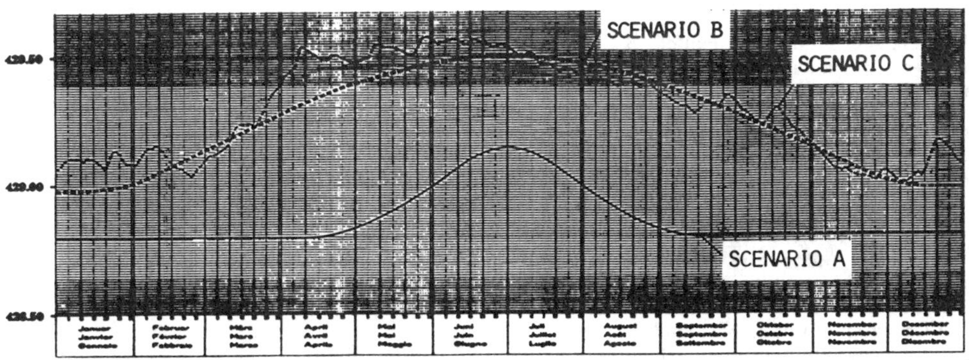

Figure 6 : Scénarios de variation des niveaux des lacs

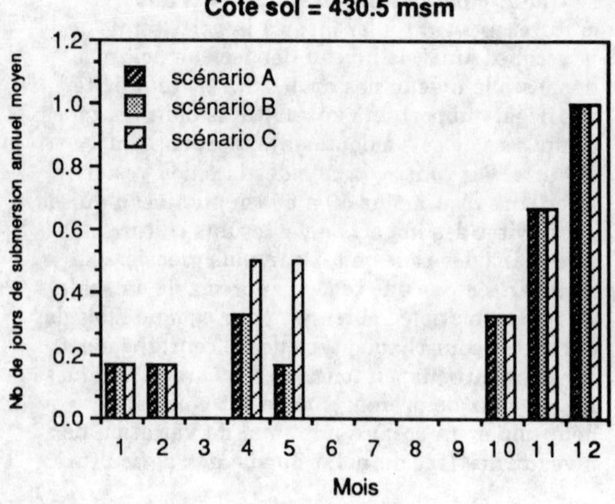

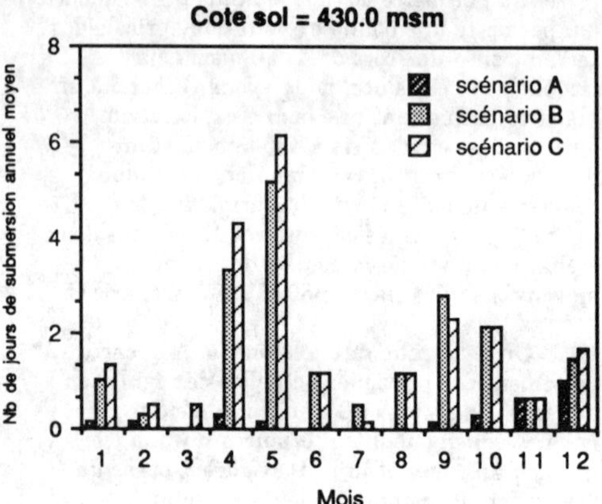

Figure 7 : Nombre de jours de submersion annuel moyen en fonction des différents scénarios

Le choix d'un scénario optimal de variation des niveaux d'eau dans les lacs doit non seulement tenir compte des critères agricoles, mais aussi des exigences de sécurité (volume de stockage suffisant pour assurer la rétention des crues de printemps), des conditions hydro-électriques, de l'écologie, de la pêche, etc. De ce point de vue-là, le scénario A n'est manifestement pas acceptable. D'ailleurs, même au niveau agricole, il suscite certaines réserves parmis lesquelles on peut citer l'augmentation du tassement des tourbes et la nécessité d'irriguer les terrains situés à plus haute altitude, sans pour autant offrir de meilleures conditions de travail que le scénario B aux terrains sis au-dessus de la cote 430.5 msm. De ce fait, le respect des exigences des autres partenaires a pour conséquence que les problèmes rencontrés dans l'exploitation agricole des zones sises aux environs de 430.0 msm ne peuvent être résolus qu'en les équipant d'un système de drainage par pompage. Par contre, le scénario B permet de réduire fortement les problèmes rencontrés actuellement au printemps même pour des terrains sis à 430.5 msm, tout en satisfaisant aux exigences des autres partenaires. De ce fait, tout autre scénario ne présentant pas les mêmes garanties, en particulier au cours des mois critiques que sont mars, avril et mai, devrait être écarté, sous réserve d'une généralisation de l'assainissement par pompage à l'ensemble du périmètre.

5. CONCLUSIONS

A la demande des cantons de Berne, Fribourg, Vaud, Neuchâtel et Soleure, l'Institut de génie rural de l'EPFL avait pour tâche de déterminer le scénario optimal pour l'agriculture de variation des niveaux d'eau dans les lacs de Neuchâtel, Bienne et Morat.L'utilisation du modèle de simulation DRAINMOD nous a permis, dans un premier temps, de définir les cotes des lacs critiques pour l'agriculture pour chaque mois de l'année, pour chaque type de sol rencontré (dont les sols tourbeux), et ceci en fonction des caractéristiques des réseaux de drainage. Dans un deuxième temps, nous avons intégré les exigences des autres partenaires (sécurité des riverains, hydro-électricité, pédologie, agronomie, écologie, pêche, tourisme, etc) et nous avons analysé les conséquences de l'application de divers scénarios. Sur la base de ces simulations, nous avons pu déterminer le scénario capable de garantir une bonne exploitation de la majorité des terres agricoles du périmètre tout en remplissant les conditions émanant des autres partenaires. Les résultats fournis par DRAINMOD, notamment ceux relatifs aux paramètres de qualité, ont donc servi de base aux réflections qui ont conduit au choix du scénario idéal. Ils ont également permis de déterminer l'étendue des zones où une exploitation rationnelle des terres agricoles n'est pas compatible avec les autres exigences (hydro-électricité, protection des tourbes et écologie notamment) sans la mise en place de stations de relevage. DRAINMOD s'est donc révélé être un outil puissant dans l'étude pluridisciplinaire qui a conduit à la détermination de la régulation optimale des niveaux des lacs de Neuchâtel, Bienne et Morat.